ENCYCLOPEDIA OF PHYSICS

EDITED BY

S. FLÜGGE

VOLUME XXXVIII/2

NEUTRONS AND RELATED GAMMA RAY PROBLEMS

WITH 338 FIGURES

SPRINGER-VERLAG BERLIN HEIDELBERG GMBH

1959

HANDBUCH DER PHYSIK

HERAUSGEGEBEN VON

S. FLÜGGE

BAND XXXVIII/2

NEUTRONEN UND VERWANDTE GAMMASTRAHLPROBLEME

MIT 338 FIGUREN

SPRINGER-VERLAG BERLIN HEIDELBERG GMBH

1959

ISBN 978-3-642-45922-1 ISBN 978-3-642-45920-7 (eBook)
DOI 10.1007/978-3-642-45920-7

Ursprünglich erschienen bei Springer-Verlag OHG.Berlin
Göttingen . Heidelberg 1959

Contents.

The Production and Slowing Down of Neutrons.

By

Edoardo Amaldi.

With 281 Figures.

A. Introduction.

In a note sent to Nature the 17th of February 1932 and entitled "Possible existence of a neutron", Chadwick announced the discovery of a new fundamental particle, electrically neutral whose mass was very close to the mass of the proton.

This particle, *the neutron*, is a component of the penetrating radiations, first observed by Bothe and Becker and further investigated by I. Curie and F. Joliot, emitted by beryllium when bombarded by polonium α-particles. Some unexpected results of the latter authors led Chadwick to suspect the existence of the neutron. He submitted this assumption to experimental check: the neutron was discovered.

This discovery is a milestone in the history of nuclear physics and in its applications.

The experiments of Chadwick show that the neutron is a particle of subatomic dimensions which is emitted when beryllium is bombarded with α-particles. We know today that, together with the proton, it is one of the constituents of atomic nuclei and that it is emitted in many nuclear reactions with an energy of the order of a few million electron-volts.

We know that neutrons once emitted, can collide with nuclei and undergo scattering or produce various types of reactions. The neutron's efficiency for producing these last processes, is, generally speaking, very large, as a consequence of its electrical neutrality.

As was shown by Fermi in 1934, the high initial energies of the neutrons emitted in a nuclear reaction can be reduced to energies of the order of those corresponding to thermal agitation, by means of successive elastic collisions with nuclei of hydrogen or of other light elements.

These low energy neutrons, usually known as *slow neutrons*, are much more efficient than fast neutrons in producing certain types of reactions. The investigation of their properties has provided new means of studying nuclear structure and the structure of matter in general.

Today neutron physics is a very large part of fundamental nuclear physics. It is also the starting point of applied nuclear physics, which was initiated by the discovery of Hahn and Strassmann in 1939, of a new type of reaction produced by neutrons, the *fission of heavy elements*.

This article, of a marked experimental character, can be considered as a general introduction to the more specialized ones dealing with particularly important branches of neutron physics.

Resonance processes of neutrons are treated by J. Rainwater in Vol. XL, Neutron diffraction and interference by G. R. Ringo in Vol. XXXII. Fission by J. A. Wheeler in Vol. XLI and Pile technique by D. J. Hughes in Vol.XLIV of this Encyclopedia. Other information about neutrons can be found in Vols. XXXIX, XL, XLI and XLII which refer to nuclear reactions in general and their general theory.

In spite of the fact that this chapter is devoted to a rather restricted part of fundamental neutron physics, this part is already sufficiently large and complicated to make it desirable to have an introduction in which all the basic phenomena are briefly presented. These will then be treated in detail in the succeeding parts, which have been drawn up under the assumption that the reader is familiar with the phenomena presented in the introduction. This aims at the same time to give an outline of the historical development of the part of nuclear physics which we consider. The historical approach, however, has been abandoned whenever it was necessary for reasons of clarity.

The author owes a special debt to Prof. G. C. WICK for permission to use an unpublished work on neutron physics prepared, a few years ago, in collaboration with the author, and to Prof. H. H. BARSCHALL for reading, revising and completing part B· Reactions with emission of neutrons and neutron sources. To him are due the introductory "Comparison between neutron sources" and the following sections: 37, 38, 44, 59, 60 and parts of 39, 41, 42, 53, 54.

I. The discovery of the neutron.

1. In 1930 W. BOTHE and H. BECKER[1] observed the emission of a penetrating secondary radiation by various light elements (Li, Be, B, F, Mg, Al) bombarded with polonium α-particles and interpreted it in terms of hard γ-rays. That this secondary radiation really includes a γ-ray component was proved later by the same authors[2] and by RASETTI[3] by means of experiments in which two Geiger counters in coincidence were arranged to detect fast electrons produced by Compton effect. By interposing between the two counters aluminium absorbers, BOTHE and BECKER were able to establish the presence of electrons of about 5 Mev energy.

The investigation of the penetrating radiation emitted by Be and B, the two elements in which the effect was more intense, was continued by I. CURIE and F. JOLIOT[4] who detected the secondary radiation by means of a thin-walled ionization chamber connected with an electrometer. They confirmed the results of BOTHE and BECKER about the existence of a secondary penetrating radiation and its interpretation and estimated, on the basis of absorption measurements, that the energy of the corresponding photons was still higher.

Thinking that γ-rays of such high energy could perhaps produce some kind of transmutation, they placed close to the thin wall of the ionization chamber thin layers of various substances. While no change in the ionization current was observed in all other cases studied (C, Al Cu, Ag, Pb), in the case of paraffin (or of other hydrogeneous substance, such as water or cellophane) an increase by a factor of about two was observed[5].

By means of a series of classical experiments these authors were able to show that the secondary penetrating radiation emitted from Be when bombarded with polonium α-particles was capable of transferring to the protons present in the hydrogeneous substances kinetic energies of about 5 Mev.

---

[1] W. BOTHE and H. BECKER. Z. Physik **66**, 289 (1930).

[2] H. BECKER and W. BOTHE. Congresso di Fisica Nucleare, Roma 1931, p. 153. — Naturwiss. **20**, 349 (1932). — Z. Physik **76**, 421 (1932).

[3] F. RASETTI: Naturwiss. **20**, 252 (1932).

[4] I. CURIE. C. R. Acad. Sci. Paris **193**, 1412 (1932). — F. JOLIOT: C. R. Acad. Sci. Paris **193**, 1415 (1932).

[5] I. CURIE and F. JOLIOT. C. R Acad. Sci. Paris **194**, 273 (1932) — J. Phys. Radium **4**, 21 (1933).

In order to explain this production of recoil protons by γ-rays, CURIE and JOLIOT thought first that the observed effect was perhaps similar to the Compton effect and concluded, by applying the laws of conservation of energy and momentum, that the photons emitted from Be must have an energy of at least 50 Mev. Later, however, they realized the serious difficulties involved in such an interpretation, due to the fact that according to the Klein-Nishina formula, the cross section for Compton effect is inversely proportional to the square of the mass of the recoiling particle. In the case of protons the evaluated intensity was much too small to explain the experimental results.

They then put forward the assumption that the observed effect was due to a type of interaction between γ-rays and protons different from that of the Compton effect[1].

But in the meantime, the first results of CURIE and JOLIOT[2], presented at the Academie des Sciences of Paris, on the 11th of January 1932, had induced CHADWICK to make a series of fundamental observations.

These were performed with an ionization chamber connected to a linear amplifier, so that the ionization produced by a single ionizing particle could be measured[3].

With this apparatus CHADWICK proved that the secondary penetrating radiation emitted from Be, could produce recoil atoms not only from layers of hydrogeneous substances, but also from layers of Li, Be, B, C, N. In the case of the last element, he found that the kinetic energy of the recoiling atoms was between 1 and 1.4 Mev, that is, between two and three times the value expected according to the conservation laws, under the assumption that the incident radiation was composed of photons of 50 Mev energy. In other words the kinetic energy transferred to the atoms of hydrogen and nitrogen did not correspond to the same energy of the primary radiation if this was composed of photons. Besides this argument based on the conservation laws applied to the production of recoil atoms, CHADWICK developed a second argument, based again on the conservation laws, but applied this time to the collision process between an α-particle and a Be nucleus and the consequent emission of a photon of 50 Mev. The conclusion of both these arguments was that the interpretation of the corresponding effects in terms of photons was inacceptable. But all difficulties disappeared and the observed effects could be fitted into a satisfactory picture under the assumption that the beryllium nucleus, bombarded with α-particles, emitted a neutral particle of mass very close to that of the proton.

These experiments and arguments not only proved definitely the existence of the neutron but showed also that this new particle, indicated usually with the symbol ${}_0n^1$, could be produced in light elements through what we call (α, n) processes. In particular the production reaction in the case of beryllium—which is monoisotopic—was necessarily

$$ {}_2\mathrm{He}^4 + {}_4\mathrm{Be}^9 \rightarrow {}_6\mathrm{C}^{12} + {}_0n^1 . \tag{1.1} $$

The high energy photons originally observed by BOTHE and BECKER, are due to the fact that the residual nuclei of the (α, n) processes are frequently left in excited states.

---

[1] I. CURIE and F. JOLIOT: C. R. Acad. Sci. Paris **194**, 708 (1932).

[2] See footnote 5, p. 2.

[3] J. CHADWICK: Nature, Lond. **129**, 312 (1932). — Proc. Roy. Soc. Lond., Ser. A **136**, 692 (1932).

II. The properties of the neutron as a fundamental particle and its role in nuclear structure.

2. The mass. Besides the first rough evaluation of the mass of the neutron, based on the energy of the recoiling atoms of hydrogen and nitrogen, CHADWICK[1,2] made other determinations of this fundamental quantity by applying the conservation laws to the production processes ${}_5B^{11}(\alpha, n)\,{}_7N^{14}$, ${}_3Li^7(\alpha, n)\,{}_5B^{10}$. These, as well as other determinations of the mass made, on the same principle, shortly afterwards by various authors[3], gave values very close to one unit of atomic weight, but they were all soon superseded by the determination based on the photo-effect of the deuteron.

Not long after the discovery of the neutron, CHADWICK and GOLDHABER[4] discovered this important effect, which allows a precise determination of the binding energy E_3 of the deuteron (in its triplet ground state) to be made. Once E_3 is known, one can deduce the neutron-hydrogen atom mass difference, and the mass of the neutron, making use of the obvious relation

$$\Delta = m_n - m_H = E_3 - a \tag{2.1}$$

where

$$a = H_2^+ - D^+ \tag{2.2}$$

is the separation of the mass spectroscopic doublet of the molecular ions of hydrogen and the atomic ions of deuterium [*17e*].

For a long time the photodisintegration of the deuteron (Sect. 46) was the only method which allowed a determination of E_3 of sufficient accuracy[5]. Later it became possible to obtain an accurate value of E_3 also by measuring the energy of the photons emitted in the radiative capture of slow neutrons by protons (Sect. 29). Table 1 shows a few recent determinations of E_3 made by various methods. The results of older measurements are collected in the review article by STEPHENS[5].

The neutron-hydrogen atom mass difference deduced by combining the value of E_3 given, for instance, by MOBLEY and LAUBENSTEIN with the value of $a = (1.442 \pm 0.002)$ Mev given by ROBERTS and NIER[6], is

$$\Delta = m_n - m_H = (784 \pm 4)\ \text{kev}.$$

This can be compared with other determinations of the same quantity obtained from various reactions assuming that the rest mass of the neutrino is zero (Table 2).

The value of the mass of the neutron adopted today is[7]

$$\left.\begin{aligned} m_n &= (1.008982 \pm 0.000003)\ \text{units of mass} = (939.505 \pm 0.010)\ \text{Mev} \\ &= (1.67470 \pm 0.00004) \times 10^{-24}\ \text{g} \end{aligned}\right\} \tag{2.3}$$

---

[1] J. CHADWICK: Proc. Roy. Soc. Lond., Ser. A **136**, 692 (1932).
[2] J. CHADWICK: Proc. Roy. Soc. Lond., Ser. A **142**, 1 (1933).
[3] I. CURIE and F. JOLIOT. C. R. Acad. Sci. Paris **197**, 237 (1933). — J. Phys. Radium **4**, 494 (1933). — W. D. HARKINS and D. M. GANS: Nature, Lond. **134**, 968 (1934). — G. N. LEWIS, M. S. LIVINGSTON, M. C. HENDERSON and E. O. LAWRENCE' Phys. Rev. **45**, 242, 497 (1934).
[4] J. CHADWICK and M. GOLDHABER: Nature, Lond. **134**, 237 (1934). — Proc. Roy. Soc. Lond., Ser. A **151**, 479 (1935).
[5] W. E. STEPHENS: Rev. Mod Phys. **19**, 19 (1947).
[6] T. R. ROBERTS and A. O. NIER: Phys. Rev. **77**, 746 (1950).
[7] E. R. COHEN, J. W. M. DUMOND, T. W. LAYTON and J. R. ROLLETT. Rev. Mod. Phys **27**, 363 (1955).

to be compared with the values of the masses of hydrogen atom, proton and electron

$$m_{\mathrm{H}} = (1.008142 \pm 0.000003) \text{ units of mass} = (938.722 \pm 0.010) \text{ Mev}, \quad (2.4)$$

$$m_p = (1.007593 \pm 0.000003) \text{ units of mass} = (938.211 \pm 0.010) \text{ Mev}, \quad (2.5)$$

$$\left.\begin{aligned} m_e &= [1836.12 \pm 0.02]^{-1} \times m_p = (0.510976 \pm 0.000007) \text{ Mev}, \\ 1 &\text{ unit of mass} = (931.141 \pm 0.010) \text{ Mev}. \end{aligned}\right\} \quad (2.6)$$

3. The nuclear model based on protons and neutrons. Immediately after the discovery that neutrons are emitted in certain nuclear reactions, several authors advanced the natural hypothesis that they were among the constituents of the nucleus[1] and proposed various nuclear models in which α-particles, protons,

Table 1. *Determinations of the binding energy of the deuteron:* E_3.

Reference	$-E_3$ (Mev)	Method
(a)	2.181 ± 0.005	From the energy of photoprotons produced by γ-rays of $h\nu = 2.62$ Mev
(b)	2 229 ± 0.020	From energy of photoneutrons produced by γ-rays of $h\nu = 2.62$ Mev
(c)	2.227 ± 0.010	From the Q value of $^2D + ^1H \rightarrow n + ^1H + ^1H$
(d)	2.230 ± 0.007	From the energy of the photons emitted in capture of slow neutrons by protons (Sect. 29)
(e)	2 226 ± 0 003	From threshold of photodisintegration of the deuteron (Sect 46)
(f)	2.227 ± 0.003	From threshold of photodisintegration of the deuteron

(a) P. MEYER: Z. Physik **126**, 336 (1949).
(b) A. O. HANSON: Phys. Rev. **75**, 1794 (1949).
(c) R. V. SMITH and D. H. MARTIN: Phys. Rev. **77**, 752 (1950).
(d) R. E. BELL and L. G. ELLIOT: Phys. Rev. **79**, 282 (1950).
(e) R. C. MOBLEY and R. A. LAUBENSTEIN: Phys. Rev. **80**, 309 (1950).
(f) J.C. NOYES, J.E. VAN HOOMISSEN, W.C. MILLER and B. WALDMAN: Phys. Rev. **95**, 396 (1954).

Table 2. *A few determinations of the neutron-hydrogen atom mass difference from various reactions under the assumption that the rest mass of the neutrino is zero.*

Reference	$m_n - m_{\mathrm{H}}$ (kev)	Used reactions
A. V. TOLLESTRUP, F. A. JANKINS, W. A. FOWLER and C. C. LAURITSEN: Phys. Rev. **75**, 1947 (1949)	786 ± 6	$D + D \rightarrow {}^3He + n$ $D + D \rightarrow {}^3H + n$ ${}^3H \rightarrow {}^3He + e^- + \nu$
W. FRANZEN, J. HALPERN and W. E. STEPHENS: Phys. Rev. **76**, 317 (1949)	783 ± 4	${}^3He + n \rightarrow {}^3H + p$ ${}^3H \rightarrow {}^3He + e^- + \nu$ and ${}^{14}N + n \rightarrow {}^{14}C + p$ ${}^{14}C \rightarrow {}^{14}N + e^- + \nu$
R. F. TASCHEK, H. V. ARGO, A. HEMMENDINGER and G. A. JARVIS: Phys. Rev. **76**, 325 (1949)	782 ± 2	${}^3H + p \rightarrow {}^3He + n$ ${}^3H \rightarrow {}^3He + e^- + \nu$

electrons and also neutrons appeared. IWANENKO[2] was probably the first to publish the theory that neutrons and protons were the *only* fundamental constituents of the nucleus, while electrons were definitively excluded. This point was of considerable importance because many of the serious difficulties of all

[1] F. PERRIN: C. R. Acad. Sci. Paris **195**, 236 (1932); see also W. HEISENBERG, Rapports et discussions du 7ème Congrès de l'Institut International de Physique Solvay, p. 289, 1934.
[2] D. IWANENKO: Nature, Lond. **129**, 798 (1932).

preceding nuclear models were due to the assumption that there existed "nuclear electrons" [*5*], [*3*].

HEISENBERG[1] and MAJORANA[2] were the first to appreciate fully the importance of this new model and they attempted to explain qualitatively, and in part quantitatively, nuclear properties in terms of exchange forces between pairs of nucleons (i.e. protons and neutrons).

These attempts, as well as the recognition by WIGNER[3] that nuclear forces must have a very short range of action, constitute the first steps in the very wide chapter of nuclear physics dealing with the investigation of the nucleon-nucleon interaction [*4*], [*5*], [*3*], [*16*] and of its relations with the mesonic field[4]. These problems of paramount importance are completely outside the scope of the present chapter. The only point that has to be stressed here is that all successive developments of our knowledge on nuclear structures have confirmed the basic conclusion that protons and neutrons are the only constituents of all nuclei.

4. The spin and statistics. In this model two very important properties were attributed to the neutrons, to make them as similar as possible to protons: namely, that their spin is $\frac{1}{2}\hbar$ and that they obey Fermi-Dirac statistics.

These two assumptions are not independent of each other, at least from the point of view of the relativistic theory of wave fields, where it can be shown that particles with half integral spin must obey Fermi-Dirac statistics. But independently of this theorem, there were already many arguments in favour of these two assumptions: from the investigation of the Raman effect of the rotational spectra of homonuclear diatomic molecules[5] it had been concluded that *all nuclei with an odd mass number A obey the Fermi-Dirac statistics while nuclei with even A obey Bose-Einstein statistics*[6]. While this experimental result was in many special cases in open contradiction with the predictions of the theory in which the nucleus is built up of protons and electrons, it was completely in agreement with the new model where A represents the total number of nucleons, provided neutrons, as well as protons, obey Fermi-Dirac statistics. Moreover all empirical rules found for obtaining nuclear spins were clearly explained if the neutron had half integral spin. This argument, as well as the above mentioned theorem, requires that the neutron has half integral spin but not necessarily spin $\frac{1}{2}\hbar$, although this value is that which explains all experimental results in the simplest way.

A few other arguments also indicate that the neutron spin must be half integral and favour definitely the value $\frac{1}{2}$ rather than the value $\frac{3}{2}$. Such are for instance the energy-dependence of neutron-proton-scattering taken in connection with the properties of the deuteron (see HULTHÉN and SUGAWARA'S contribution to Vol. XXXIX), the coherent scattering of neutrons by parahydrogen and by crystals containing hydrogen, and finally the total reflection of neutrons by mirrors containing hydrogen (RINGO'S article in Vol. XXXII).

[1] W. HEISENBERG: Z. Physik **77**, 1 (1932); **78**, 156 (1932); **80**, 587 (1933).

[2] E. MAJORANA: Z. Physik **82**, 137 (1933). — We owe it to the memory of MAJORANA, who died so prematurely, to recall how he communicated to FERMI some of his considerations about the structure of the nucleus, immediately after the first note of CHADWICK. It was only his sense of self criticism that prevented him from publishing earlier these results.

[3] E. P. WIGNER: Phys. Rev. **43**, 252 (1933).

[4] See for instance E. FERMI: Elementary particles. New Haven. Yale University Press 1951. — H. H. BETHE and F. DE HOFFMANN: Mesons and fields Row, Peterson & Co. Evanston, Ill. 1955.

[5] W. HEITLER and G. HERZBERG: Naturwiss. **17**, 673 (1929). — F. RASETTI: Z. Physik **61**, 598 (1930).

[6] For a discussion of these effects see for instance J. MATTAUCH and S. FLÜGGE: Kernphysikalische Tabellen. Berlin: Springer 1942.

5. The magnetic moment. Another fundamental property of the neutron is its intrinsic dipole magnetic moment which contributes, together with that of the proton to the magnetic moments of the various nuclei and in particular to that of the deuteron. This last one being smaller than the magnetic moment of the proton, gives an indication of the fact that the magnetic moment of the neutron must be negative. It is true that the magnetic moments of the nucleons do not simply add, as the corresponding mechanical moments do, to give the magnetic moment of the nucleus: the currents associated with the exchange of mesons among the various nucleons also give a contribution which, however, is not very large, especially in the case of the deuteron. Therefore the argument is still valid and this indication of the sign is confirmed by direct measurement of the magnetic moment of the free neutron which gives[1]

$$\mu_n = (-1.913148 \pm 0.000066)\,\mu_N \tag{5.1}$$

where μ_N indicates a nuclear magneton[2]

$$\mu_N = \frac{e}{2m_p c} = (0.505038 \pm 0.000018) \times 10^{-23}\ \mathrm{erg} \times \mathrm{gauss}^{-1}. \tag{5.2}$$

A discussion of the experimental determinations of this quantity as well as of its interpretation in terms of meson field theories, would take us too far away from the scope of the present chapter. Therefore we limit ourselves to giving the value (5.1) of μ_n and to mentioning very briefly and qualitatively the type of arguments usually presented in connection with it.

The value (5.1) is considered anomalous in the following sense. After the recognition that the behaviour of electrons was described, if not exactly at least with good accuracy, by means of the Dirac equation, it was natural to assume that all other particles, different from electrons, but with spin $\frac{1}{2}$, could also be described by the same equation. But if one applies the Dirac equation to the nucleons one finds that the proton must have a magnetic moment equal to one nuclear magneton, and the neutron a magnetic moment equal to zero. We know that these two conclusions do not correspond to the experimental results: the proton has the magnetic moment[2]

$$\mu_p = (2.79275 \pm 0.00003)\,\mu_N \tag{5.3}$$

and the neutron the magnetic moment given by (5.1).

This is expressed by saying that both nucleons have anomalous magnetic moments (with respect to the expectation based on the Dirac equation) and the following type of argument has been invoked in order to explain the experimental results.

The neutron existing in nature, indicated as physical neutron, is a mixture of states, for part of the time it is really a neutron (a bare neutron $\boldsymbol{n}$) with zero magnetic moment, but for a fraction of the time it is (virtually) dissociated into a bare proton $\boldsymbol{p}$ and a negative π-meson according to the equation

$$\boldsymbol{n} \rightleftharpoons \boldsymbol{p} + \pi^- . \tag{5.4}$$

Here $\boldsymbol{p}$ is a proton with $\mu_{\boldsymbol{p}} = \mu_N$ and π^- is a negative meson moving around $\boldsymbol{p}$ so that its orbital magnetic moment, greater and antiparallel to that of the bare proton, prevails and gives rise to the observed magnetic moment (5.1).

[1] V. W. Cohen, N. R. Corngold and N. F. Ramsey: Phys. Rev. **104**, 283 (1956).
[2] See footnote 7, p. 4.

The process (5.4) is only one of an infinite number of possible processes of dissociation, but is generally believed to be one of greatest importance.

Similarly, the physical proton p is a mixture of two states, the first that of a bare proton $\boldsymbol{p}$, with $\mu_{\boldsymbol{p}} = \mu_N$, and the second a state in which it is (virtually) dissociated into a bare neutron (with zero magnetic moment) and a positive π-meson. The presence of the π^+-meson would explain the anomalous magnetic moment (5.3).

6. The beta radioactivity. The fact that the mass of the neutron is greater than that of the hydrogen atom (Table 2) was recognized a long time ago and makes energetically possible the spontaneous disintegration of a neutron into a proton and an electron.

Such a possible instability of the neutron, first suggested by CHADWICK and GOLDHABER[1] was proved experimentally only many years later[2-4].

We know today that free neutrons undergo a spontaneous beta decay process with emission of an electron and a neutrino

$$_0n^1 \rightarrow {}_1p^1 + e^- + \nu. \tag{6.1}$$

Beside this process of transformation of a neutron into a proton, one has to consider a process of transformation of a proton into a neutron with emission of a positive electron and a neutrino. This is energetically impossible in the case of a free proton, but can take place in the presence of other nucleons, whenever it is allowed by the various conservation laws. These two processes of transformation of one nucleon into another with emission of negative or positive electrons not only are the basis of FERMI's theory of beta decay [*5*], [*9*], [3], but also represent important examples of the so-called Fermi universal interaction in which four particles, obeying the Pauli principle, are involved[5].

This fundamental property of the neutron has no part in the various phenomena treated in this chapter, and so we refer for details of process (5.4) and of the experimental determinations of the corresponding β-ray spectrum and half-life, to Vol. XLI; here we limit ourselves to giving in Table 3 the experimental values of the half-life and of the maximum energy of the β-ray spectrum as they result from direct measurements[3,4].

Table 3. *Properties of the neutron.*

Charge	$< 10^{-18}\, e = 4.8 \times 10^{-28}$ e s.u.
Mass	(1.008982 ± 0.000003) units of mass $\left(\frac{1}{16}\,O^{16}\right)$ $= (939.505 \pm 0.010)$ Mev $= (1.67470 \pm 0.00004) \times 10^{-24}$ g
Spin	$\frac{1}{2}\hbar$
Statistics	Fermi-Dirac
Magnetic moment	(-1.913148 ± 0.000066) nuclear magnetons
Half life	(12.2 ± 1.3) minutes
Maximum energy of β-ray spectrum	$E_{\max}^{(-)} = (782 \pm 1)$ kev

[1] See footnote 4, p. 4.
[2] A. H. SNELL and L. C. MILLER: Phys. Rev. **74**, 1217 A (1948). — A. H. SNELL, F. PLEASANTON and R. V. MCCORD: Phys. Rev. **78**, 310 (1950).
[3] J. M. ROBSON: Phys. Rev. **78**, 311 (1950); **83**, 349 (1951).
[4] P. E. SPIVAK, A. N. SOSNOVSKY, A. Y. PROKOFIEV and V. S. SOKOLOV, p. 33 of Ref. [*15a*]
[5] See for instance L. MICHELL: Progress in cosmic ray Physics. Windfork 1952.

III. General remarks on the interaction of neutrons with matter. The first neutron reactions.

7. The neutron-electron interaction. When neutrons move through matter, they produce ions indirectly, by the action of the recoil atoms. This is just the effect that made the discovery of the neutron possible, and it obviously gives a clear proof of its interaction with nuclei.

One can, however, ask, and the answer is of a considerable interest, whether some kind of *direct* interaction exists between neutrons and electrons which might be responsible for some observable effect, in particular for the direct production of ions by neutrons.

This problem was originally investigated by CHADWICK[1] and by DEE[2]. Both these authors gave an upper limit for the number of pairs of ions produced by a neutron per unit length of path. The value given by DEE is not more than one pair of ions per 3 meters of path in air. By comparing this upper limit for the energy loss through ionization, with that of protons, FLÜGGE established an upper limit for the possible electric charge of the neutron at about $^1/_{700}$ of the electronic charge [*18b*].

Much lower limits for the value of the possible charge of the neutron ($\lesssim 10^{-12} e$) can be established through various indirect arguments given by FELD [*17b*] which are based on the neutrality of atoms, or on the neutron decay process (6.1), or finally on the observed small deviation ($\sim 1\%$) from spherical symmetry observed by FERMI and MARSHALL[3] in the scattering of slow neutrons by xenon atoms ($\lesssim 10^{-18} e$).

This experiment was the first one of a series devoted to establishing the existence of a neutron-electron interaction which have been described in detail by G. R. RINGO in Vol. XXXII because they necessarily make use of the methods of neutron optics. These experiments are also discussed by HOFSTADTER and FUBINI in Vol. XLIII. Here we limit ourself to mentioning the main considerations and conclusions which are of general interest from the point of view of the interaction of neutrons with matter.

When in this connection we speak of the interaction between neutrons and electrons we do *not* refer to the interaction between the magnetic moment of the neutron (5.1) and the magnetic field produced by the atom as a whole. This interaction, usually called the magnetic interaction, depends on the orientation of the spin of the neutron with respect to the atomic magnetic field and is used, for instance, to produce polarized beams of slow neutrons by scattering them in magnetized ferromagnetics (see RINGO's article in Vol. XXXII). The magnetic interaction should not exist for diamagnetic atoms such as those of the noble gases in which the electrons are bound in closed shells.

Beside the magnetic interaction one might expect also the existence of a potential energy between neutron and electron which gives rise, even in the case of diamagnetic atoms, to a very weak scattering of the incident neutrons superimposed on the uncomparably bigger scattering due to the nuclei.

As we shall discuss later, this interaction is expected to exist as a consequence of two effects: the anomalous magnetic moment (5.1), and other structure terms of the neutron arising from the mechanism of Eq. (5.4) which were in fact, introduced in order to explain its anomalous magnetic moment.

---

[1] J. CHADWICK: Proc. Roy. Soc. Lond., Ser. A **136**, 692 (1932).
[2] P. I. DEE: Proc. Roy. Soc. Lond. **136**, 727 (1932).
[3] E. FERMI and L. MARSHALL: Phys. Rev. **72**, 1139 (1947).

Such an interaction is certainly very weak and has a short range of action; therefore, its effect on the neutron can be calculated by means of the Born approximation. The corresponding cross section increases with increasing wavelength $\lambda\!\!\!^{-}$ of the neutron, and for neutrons of very low energy ($\lambda\!\!\!^{-} \to \infty$) it does not depend on λ nor on the shape of the potential well but only on the volume integral of the interaction potential. The situation is similar to that considered later in Sects. 18 and 21.

This remark justifies the use of the rather arbitrary convention of representing the neutron-electron interaction in terms of the depth V_0 of a fictitious spherical potential well of radius equal to the classical electron radius

$$r_e = \frac{e^2}{m_e c^2} = 2.82 \times 10^{-13}\,\mathrm{cm}. \tag{7.1}$$

As already mentioned above the effect is so small that in all the experiments made to reveal its existence it was necessary to take advantage of the coherent scattering of slow neutrons by the peripheral electrons surrounding heavy nuclei in order to obtain an enhanced effect.

The method of FERMI and MARSHALL is based on the experimental determination of the deviation from spherical symmetry of the slow neutrons scattered by xenon. For neutrons of wavelength comparable with atomic dimensions and therefore much larger than nuclear dimensions, the scattering by the nucleus is spherically symmetrical (see Sect. 21) while the scattering by the electrons shows an asymmetry with respect to the equatorial plane, due to the form factor $\mathcal{F}(\vartheta)$ of the electron distribution which is known from X-ray experiments or from the corresponding well-known calculations.

The differential scattering cross section of a neutral atom of atomic number Z and nuclear cross section σ_s, may be written as follows

$$\left(\frac{d\sigma}{d\omega}\right)_{\mathrm{atom}} = \frac{\sigma_s}{4\pi}\{1 + cZ\,\mathcal{F}(\vartheta)\}^2 \tag{7.2}$$

where c represents the relative contribution of a single electron

$$c = \frac{\frac{4\pi}{3} r_e^3 V_0}{\left[\frac{\pi \hbar^4}{m^2}\sigma_s\right]^{\frac{1}{2}}} = 1.28 \times 10^{-7} \frac{V_0\,(\mathrm{ev})}{\sqrt{\sigma_s\,(\mathrm{barns})}}. \tag{7.3}$$

Its value is so small that only the interference term between the nuclear and the electronic scattering is appreciable. The sign of this term can be used in order to establish the sign of the neutron-electron interaction, that is its repulsive or attractive nature. The method of HAVENS, RAINWATER and RABI[1] is based on the measurement of the variation with the wavelength λ of the incident neutrons of the total cross section of liquid bismuth. In these experiments advantage is taken of the fact that while the nuclear scattering is constant in the interval of variation of λ considered, the form factor of the electron distribution is a known function of the wavelength.

Finally, the method of HUGHES, HARVEY, GOLDBERG and STAFNE[2] makes use of the so-called balanced mirror technique, in which the total reflection of a beam of slow neutrons by the surface of separation between liquid oxygen and bismuth is studied [*12*], [*13*].

[1] W. W. HAVENS, L. J. RAINWATER and I. I. RABI: Phys. Rev. **82**, 345 (1951).

[2] D. J. HUGHES, J. A. HARVEY, M. D. GOLDBERG and M. J. STAFNE: Phys. Rev. **90**, 497 (1953).

These substances are chosen because their corresponding nuclear scatterings are almost equal while the electron scattering is much higher in bismuth than in oxygen because of the higher Z.

Table 4 contains the results of these authors expressed by giving the corresponding values of V_0. The negative sign means that this short range interaction between neutron and electron is attractive.

Table 4. *Neutron-electron interaction.*

Depth of a potential well of radius $r_e = \frac{e^2}{m_e c^2}$.

	Used material	Method	Result V_0 in ev
FERMI and MARSHALL: Phys. Rev. **72**, 1139 (1947)	Xe, gas	angular asymmetry	-300 ± 5000
HAVENS, RAINWATER and RABI: Phys. Rev. **82**, 345 (1951)	Bi, liquid	total cross section for λ varing between $0.3-3$ Å	-5300 ± 1000
HAMMERMESH, RINGO and WATTENBERG: Phys. Rev. **85**, 483 (1952)	Xe and Kr gases	angular asymmetry	-4100 ± 1000
HUGHES, HARVEY, GOLDBERG and STAFNE: Phys. Rev. **90**, 497 (1953)	O_2-Bi	balanced mirror	-3900 ± 400

FOLDY[1] has investigated the electromagnetic properties of a particle obeying the Dirac equation with the additional term introduced by PAULI in order to take into account the possible existence of an anomalous magnetic moment, and has shown that the neutron-electron interaction can be divided into two parts. The first part is due to the anomalous magnetic moment of the neutron and corresponds to an attraction between neutron and electron which, referred again to the conventional potential well of radius r_e, gives $V_0 = -4080$ ev[2].

This term appears almost sufficient to account for the observed interaction (see Table 4), and therefore leaves little place for the second effect expected according to Eq. (5.4). If the view expressed by this equation is correct, in the immediate vicinity of a neutron one would expect an electric field due to the charge of the bare proton to exist there, for a fraction of the time. This field, of course, would extend only to a very small distance, because it would be screened by the negative charge of the π-meson. Various calculations of this effect indicate a value of the corresponding potential V_0 larger than the difference between the experimental results and the value given above for the contribution arising from the anomalous magnetic moment. According to FRIED[2] the second part of the neutron-electron interaction corresponds to $V_0 = -1300$ ev. However, the uncertainties involved in calculations based on meson theory are so large that the apparent conflict between the experimental results and current ideas about the cloud of virtual mesons surrounding the neutron must be treated with some reservation.

This discussion takes us too far away from the province of the present chapter. Therefore we refer to [*18b*][3] and to Vol. XLIII for more details and we conclude that whatever may be the theoretical interpretation, the above mentioned experiments prove the existence of a short range interaction between electrons and neutrons corresponding to an attractive potential well characterized by the

[1] L. L. FOLDY: Phys. Rev. **87**, 688, 693 (1952).
[2] B. D. FRIED: Phys. Rev. **88**, 1142 (1952).
[3] L.L. FOLDY: Rev. Mod. Phys. **30**, 471 (1958).

following value of the potential well of radius (7.1)

$$V_0 = -(3900 \pm 400)\ \text{ev}. \tag{7.4}$$

In order to appreciate the physical meaning of this figure we can introduce it in (7.3) and compute the electron contribution to scattering in various elements. The results of such a computation are shown in Table 5. Considering the data given in the fourth column of Table 5 and that the form factor $\mathcal{F}(\vartheta)$ is strongly peaked in the forward direction, we feel justified in the following to neglect the neutron-electron interaction and to consider it as a small correction only in experiments in neutron optics.

Table 5. *Relative contribution of electrons to the scattering cross section of various neutral atoms for slow neutrons* ($V_0 = -3900$ ev).

Element	Z	σ_s (barns)	c	cZ
C	6	4 8	$-2\ 3 \times 10^{-3}$	$-1\ 4 \times 10^{-2}$
Xe	54	4 3	-2.4×10^{-3}	-13×10^{-2}
Bi	83	9	-1.67×10^{-3}	-15×10^{-2}

8. The interaction of neutrons with nuclei. The elastic scattering and the (*n*, *α*) processes. In conclusion, in speaking of the interaction of neutrons with matter it is generally the interaction of neutrons with nuclei that is meant. However the field is so wide and complex that only very few historical and general remarks can be given in this section.

Even from the first transmission experiments of CHADWICK it was possible to conclude that the absorption cross section of lead for the neutrons emitted in process (1.1), was of the order of the geometric cross section. This result, although very important, had only a qualitative character because the neutron cross section is a function of the energy, different for every nucleus, and the energy spectrum of the neutrons used in these experiments was complex and extended over a very wide energy interval.

Later it was shown that for neutrons with an energy between 7 and 20 Mev the absorption coefficient varies regularly with the mass number of the nucleus according to the law deduced assuming a nuclear cross section equal to the geometric one (see Sect. 19).

At higher energies "transparency phenomena" begin to reduce the absorption by matter, while at lower energies the individual properties of the various nuclei become evident through a rather complicated and characteristic variation with neutron energy of the corresponding absorption cross sections.

Absorption measurements give only very partial and incomplete information about the nuclear processes produced by neutrons.

The scattering of neutron, the process that had first revealed the existence of this new particle, was further investigated by observing the recoiling atoms[1,2] or the scattered neutrons[3]. But it was also shown, by FEATHER[1] that neutrons can produce nuclear disintegrations, which appear in a cloud chamber as "forks",

[1] N. FEATHER: Proc. Roy. Soc. Lond., Ser. A **136**, 709 (1932).

[2] J. CHADWICK: See footnote 1, p. 9. — M. DE BROGLIE and L. LEPRINCE-RINGUET: C. R. Acad. Sci. Paris **194**, 1616 (1932). — P. AUGER: C. R. Acad. Sci. Paris **194**, 877 (1932); **195**, 234 (1932).

[3] J. R. DUNNING and G. B. PEGRAM: Phys. Rev. **43**, 497 (1933). — J. R. DUNNING: Phys. Rev. **45**, 586 (1934).

i.e. as pairs of tracks starting from the same point in the gas. Of the two tracks, one, usually shorter, is due to the residual nucleus while the other is due to a charged particle emitted in the reaction. Disintegration processes of this type were discovered by FEATHER in O, Ne and F.

From a study of the coplanarity of the forks with the neutron source, a discussion began on what type of process was responsible for the observed forks: namely, whether the incident neutron is captured or not in the reaction. In the first case, the products of the reaction are only two; the two particles giving the fork. In this case the fork must be coplanar with the neutron source. In the second case the reaction products are three and of these the neutron escapes observation; therefore it is to be expected that in general the observed fork will not be coplanar with the neutron source.

The conclusion of this discussion[1], which clearly has a great importance in the study and interpretation of nuclear reactions, was that the overwhelming majority of the observed forks were coplanar with the neutron source: therefore the corresponding reactions were *two-body reactions* in which a residual nucleus and a rather light nuclear particle were emitted.

An examination of the range and ionization of the two tracks allowed FEATHER to conclude that for nitrogen the emitted particle was, at least in a few cases, an α-particle; and that, therefore, the corresponding forks were due to the reaction

$$_7\mathrm{N}^{14}(n, \alpha)\,_5\mathrm{B}^{11} \tag{8.1}$$

which is the reverse of the (α, n) reaction in which neutrons are produced in boron by α-particle bombardment.

From the analysis of the forks FEATHER concluded that, in the majority of cases, the disintegration was of this type, namely a (n, α) reaction, but one case appeared to be in better agreement with a (n, p) process.

These first results extended by FEATHER himself and by other authors[2] to other elements not only led to the discovery of the (n, α) reactions and gave a first indication of the (n, p) processes, but represented the first step in the discovery and investigation of all the reactions produced by neutrons.

IV. The artificial radioactivity. Various reactions produced by neutrons and their general description.

9. The artificial radioactivity produced by α-particles. The discovery of artificial radioactivity was announced by I. CURIE and F. JOLIOT in a note presented to Comptes Rendus in January 1934[3]. They had observed that boron and aluminium when bombarded with Po α-particles, gave a positron emission which did not start immediately when the α-particle source was placed close to them but increased in intensity from that moment onwards, reaching some time later a limiting value. When the Po source was taken away, the positron emission did not cease immediately but started to decrease exponentially with time as does the activity of a radioactive substance. The halftime was 14 min for B, 15 sec for Al and 2.5 min for Mg which showed a similar behaviour with the exception that the emitted electrons were of both signs.

[1] N. FEATHER: Proc. Roy. Soc. Lond., Ser. A **142**, 689 (1933). — W. D. HARKINS, D. M. GANS and H. W. NEWSON: Phys. Rev. **43**, 584 (1933); **44**, 529 (1933). — W. D. HARKINS and D. M. GANS: Phys. Rev. **46**, 397 (1934). — Nature, Lond. **133**, 794 (1934).

[2] L. MEITNER u. K. PHILIPP: Z. Physik **87**, 484 (1934). — R. JAECKEL Z. Physik **96**, 151 (1935).

[3] I. CURIE and F. JOLIOT: C. R. Acad. Sci. Paris **198**, 254, 559 (1934). — J. de Phys. **5**, 153 (1934).

The conclusion was clear. The Po α-particles can produce in these elements nuclear transmutations whose nuclide products are unstable.

The interest of such a discovery lies not only in the fact that it enlarges enormously the number of nuclear reactions that can be produced and investigated, but also in the new possibilities of experiment which it provides. If the nuclide produced has an atomic number Z different from that of the target nucleus, it becomes possible to separate it, using the same chemical procedures already known for the natural radioactive substances. In fact, by detecting the decay electrons, one can easily test the various chemical fractions and recognize in some of them the presence of exceedingly small amounts of the transmutation product.

CURIE and JOLIOT could establish in this way that the active nuclide produced in boron has the same chemical properties as nitrogen and that produced in aluminium has the same properties as phosphorus.

From the fact that aluminium consists of a single isotope ${}_{13}\mathrm{Al}^{27}$, it can be concluded that the observed transmutation is probably due to the (α, n) reaction

$${}_{13}\mathrm{Al}^{27}(\alpha, n)\,{}_{15}\mathrm{P}^{30}.$$

Since the nuclide ${}_{15}\mathrm{P}^{30}$ does not exist in nature, it is very reasonable to consider that it is unstable and that it therefore decays with the emission of a positive electron:

$${}_{15}\mathrm{P}^{30} \rightarrow {}_{14}\mathrm{Si}^{30} + e^{+}.$$

${}_{14}\mathrm{Si}^{30}$ is stable and was already known to be the final product of the reaction

$${}_{13}\mathrm{Al}^{27}(\alpha, p)\,{}_{14}\mathrm{Si}^{30}.$$

10. The artificial radioactivity produced by neutrons. The discovery of artificial radioactivity proved to be of fundamental importance for the investigation of neutrons because less than two months later E. FERMI announced his success in producing artificial radioactivity in aluminium and fluorine by neutron bombardment[1].

During the succeeding months, in collaboration with AMALDI, D'AGOSTINO, RASETTI and SEGRÈ, he published in rapid succession a long series of experimental results on many other elements[2]. In this first series of experiments about sixty elements were irradiated with neutrons and in about forty of them at least one new radioactive product was discovered and often identified[3].

After the discovery of CURIE and JOLIOT it was very natural to think of producing new radioactive nuclides using bombarding particles different from α-particles, as for instance protons or deuterons; experiments of this type were actually made and were successful. The case of neutrons, however, deserves special attention for the wide variety of new phenomena discovered later by a detailed investigation of the artificial radioactivity produced by neutrons. The great efficiency of neutrons in producing nuclear reactions is due to a number of circumstances which more than compensate the disadvantage that (so long as a nuclear reactor is not available) they are only obtainable in rather limited numbers being themselves the products of another nuclear reaction.

---

[1] E. FERMI. Ric. sci. **5** (1), 283 (1934). — Nature, Lond. **133**, 757 (1934).

[2] E. FERMI. Ric. sci. **5** (1), 330 (1934). — E. AMALDI, O. D'AGOSTINO, E. FERMI, F. RASETTI and E. SEGRÈ. Ric. sci. **5** (1), 452, 652 (1934); **5** (2), 21 (1934). — E. FERMI, F RASETTI and O. D'AGOSTINO: Ric. sci. **5** (1), 536 (1934).

[3] E. FERMI, E. AMALDI, O. D'AGOSTINO, F. RASETTI and E. SEGRÈ. Proc. Roy. Soc. Lond., Ser. A **146**, (483 1934).

The factors in favour are all due to the electric neutrality:

(a) They do not lose energy in ionizing matter but only in nuclear reaction. Therefore their efficiency does not change in crossing the target as long as they do not undergo a nuclear collision.

(b) As a consequence of (a) the thickness of the target from which the secondary β-ray can reach the detector is limited by *their* penetration and not by the penetration of the primary particles, as happens for instance in the case of α-particles. This fact alone gives a factor of the order of one hundred in favour of the neutrons.

(c) Finally, their cross section for any reaction is always a rather large fraction of the geometric cross section while in the case of charged particles the cross section is strongly reduced by the so-called Gamow factor representing the penetrability of the electrostatic potential barrier. This last factor has obviously a greater importance if the atomic number Z of the target nucleus is large.

11. (n, p) and (n, γ) processes. The main results of a systematic investigation of the physical and chemical properties of the radioactive nuclides produced by neutron bombardment can be summarized as follows[1]. Almost all elements, whatever their atomic weight, can be activated. The nuclide product is sometimes an isotope of the target nucleus, on other occasions it has an atomic number lower by one or two units. From this point of view there is a marked difference in the behaviour of light and heavy elements: for light elements the active product has in general Z smaller than that of the target nucleus, while for heavy elements (with the exception of uranium and thorium) the active product is always an isotope of the bombarded nucleus.

The results obtained with light elements can, in general, be explained, as due to (n, p) and (n, α) reactions in which the nuclear charge of the nucleus decreases by one and two units respectively.

In these processes it is not the neutron but the outgoing particle which has to cross the electrostatic potential barrier; and the higher this is, the heavier is the residual nucleus. The energy of the neutrons emitted from Be bombarded with α-particles from Po or Rn in equilibrium with their products, (a few Mev, and all neutron sources used at that time were of this type) is such that the penetrability of the electrostatic barrier turns out to be so small for heavy elements that the corresponding cross section is negligible.

The interpretation of the nuclear reactions in which Z does not change met, on the contrary, some difficulty, and the complete solution took some years.

The problem was the following: the fact that in these reactions the atomic number does not change means that passing from the target nucleus to the active product only the number of neutrons present in the nucleus is changed. Therefore, the possible processes are the following: (a) The incident neutron is captured and the corresponding binding energy (7 to 8 Mev) is emitted in the form of one or more photons [(n, γ) process]. (b) The incident neutron re-emerges with a lower energy leaving the target nucleus in an excited state whose mean life for β-decay is shorter than for γ-transition to the ground state [(n, n') process with formation of an isomer of the target nucleus]. (c) The incident neutron re-emerges accompanied by another neutron from the target nucleus [$(n, 2n)$ process].

Process (b) was usually not considered at that time, while against process (c) some obvious objections based on its clearly endoenergetic character were raised.

Process (a) would therefore have remained the more natural interpretation if it had not been for the following fact based on the "one particle model" of the

[1] See footnote 3, p. 14.

reaction universally accepted at that time. The probability of emission of one or more photons of a few Mev energy during the short time that a neutron takes to cross a nucleus, is so small that the calculated cross section for a (n, γ) process turned out to be negligibly small.

This question was soon settled by the direct experimental proof that by far the most frequent process, and the most important one for what follows, was the radiative capture (n, γ), while the processes (n, n') and $(n, 2n)$, discovered only a few years later, are in a certain sense rather exceptional.

The experimental proof mentioned above is based on three main arguments. The first is the possibility of producing the same unstable nuclide of atomic number Z by bombarding elements of the same atomic number Z, as well as elements of atomic number $Z+1$ and $Z+2$. Three classical examples are collected in Table 6. In the first[1,2] example the same vanadium nucleus, identified from its half-life and from its β-ray spectrum, is produced by bombarding vanadium,

Table 6. *Examples of experimental verifications of (n, α) processes.*

$_{25}Mn^{55}(\alpha, n)\,_{23}V^{52}$	$_{24}Cr(n, p)\,_{23}V$	$_{23}V^{51}(n, ?)\,_{23}V^{52}$
$_{13}Al^{27}(n, \alpha)\,_{11}Na^{24}$	$_{12}Mg(n, p)\,_{11}Na$	$_{11}Na^{23}(n, ?)\,_{11}Na^{24}$
$_{12}Mg(\alpha, p)\,_{13}Al$	$_{14}Si^{28}(n, p)\,_{13}Al^{28}$	$_{13}Al^{27}(n, ?)\,_{13}Al^{28}$

chromium or manganese. But manganese consists of a single stable isotope and therefore the product of the (n, α) reaction can be identified. It follows that if this is identical with the unstable nuclide produced by bombarding vanadium, the corresponding process must be of the (n, γ) type.

The second example[3,4] is still more convincing because $_{11}Na^{22}$ was already known to emit positrons instead of negative electrons with a mean life of a few years (2.6y). Similar considerations hold for the third case[3].

The second argument is based on the fact that in all known cases the emitted particles are negative electrons[5], as would be expected for nuclides having an excess of neutrons with respect to the corresponding stable isotope. In the case of the $(n, 2n)$ processes one would expect, at least in general, an emission of positrons.

The third argument tending to exclude the $(n, 2n)$ processes, is based on the conclusion that, as this is an endoenergetic process, one would expect a decrease in the intensity of its product by decreasing the energy of the incident neutrons. After a few attempt in this direction[6] which gave uncertain answers, MEITNER[7] succeeded in proving that the neutrons produced in Be by photoeffect of the γ-rays of Ra ($E \leq 0.670$ Mev, Table 27, p. 160) could produce artificial radioactivity in heavy elements such as I, Ag and Au. This experimental result excludes definitely that in these three cases the involved processes are of the $(n, 2n)$ type. The general validity of such a conclusion was established soon afterwards by the results of experiments with slow neutrons which we consider in the Sects. 13 and 14.

---

[1] See footnote 2, p. 14.
[2] See footnote 3, p. 14.
[3] T. BJERGE and C. H. WESTCOTT Nature, Lond. **134**, 286 (1934).
[4] E. AMALDI, O. D'AGOSTINO and E. SEGRÈ: Ric. sci. **5** (2), 381 (1934).
[5] E. AMALDI and E. SEGRÈ. Nuovo Cim. **11**, 452 (1934), see also footnote 2, p. 14.
[6] T. BJERGE and C. H. WESTCOTT. Nature, Lond. **134**, 177 (1934); see also footnotes 2 and 3, p. 14.
[7] L. MEITNER: Naturwiss. **22**, 759 (1934).

Beside scattering and the (n, α), (n, p), and (n, γ) processes, neutrons can produce a few other types of reactions.

Firstly, there are the two processes mentioned in the previous discussion. The existence of $(n, 2n)$ reactions was proved by HEYN[1] making use of the fact that the reaction product is identical with that of the (γ, n) reaction: the identity of the half-lives and of the β-ray spectra (usually positrons) is in general a convincing argument in favour of this interpretation.

The discovery of nuclear isomerism[2] implies that a special type of radioactive element can be produced by simple excitation of a stable nucleus, provided that this latter has a suitable excited state. This excitation can be produced by exposing the target nucleus to different radiations and in particular to neutrons.

More complicated reactions can be produced by neutrons, as for instance $(n, 3n)$, (n, pn), (n, d), and spallations or stars, two equivalent names used to indicate processes in which many fragments are emitted by the same nucleus. All these processes are endoenergetic and can be produced provided the incident neutron has sufficiently high energy.

Finally an effect of extraordinary importance both for the investigation of the properties of nuclei and for the development of applied nuclear physics, is the fission of heavy nuclei produced by neutrons. To this effect, discovered in 1939 by HAHN and STRASSMANN[3], the article by J. A. WHEELER in Vol. XLI is devoted while D. J. HUGHES' contribution to Vol. XLIV deals with its applications. Therefore we do not treat it here, in spite of its paramount importance.

12. The description of nuclear reactions. It may be useful at this point to recall the terminology usually adopted in the description of nuclear reactions. We shall refer to *two-body reactions*

$$a_1 + \mathrm{A}_1 \rightarrow \mathrm{A}_2 + a_2$$

since in the *low energy region* considered here the great majority of the reactions are of this type. In particular almost all the reactions examined in the previous sections as well as in the following of the present chapter, are of this type.

A few important exceptions do exist of course; for instance in the $(n, 2n)$ processes there are three final products: a residual nucleus and two emitted neutrons. But in most cases, and in particular in the case of $(n, 2n)$ processes, the reaction only apparently is a three-body reaction because, as can be proved, it takes place in two steps. The first step consists of a two-body reaction in which the residual nucleus is left in an exited state. The second step consists of a spontaneous disintegration of the residual nucleus with emission of the second particle.

In general, when the incident particle a_1 gets into contact with a target nucleus A_1, more than one reaction can take place. Therefore we must write, for a complete description of the situation,

$$a_1 + \mathrm{A}_1 \rightarrow \left\{\begin{array}{l} \mathrm{A}_1 + a_1, \\ \mathrm{A}_1^* + a_1', \\ \mathrm{A}_2 + a_2, \\ \mathrm{A}_3 + a_3, \\ \cdots\cdots\cdot \end{array}\right\} \qquad (12.1)$$

[1] F. A. HEYN: Nature, Lond. **138**, 723 (1936); **139**, 842 (1937).

[2] See on this subject the review articles by E. SEGRÈ and A. C. HELMHOLZ [Rev. Mod. Phys. **21**, 271 (1949)] and M. GOLDHABER and R. D. HILL [Rev. Mod. Phys. **24**, 179 (1952)] and Ref. [3], p. 623.

[3] O. HAHN and F. STRASSMANN Naturwiss. **26**, 755 (1938).

The motion of all particles and nuclei appearing in (12.1) may be referred to the *laboratory frame of reference* (l.s. = laboratory system) in which the target nucleus A_1 was at rest before the reaction took place, or to the *center-of-mass frame of reference* (c.m.s. = center-of-mass system) in which the projectile a_1 and the target nucleus A_1 fly one against the other with opposite momenta. In part B (Sect. 33) we shall consider the transformation from one to the other of these two frames of reference. Here let us assume *the motion of all particles to be referred to the c.m.s.*, in which, because of momentum conservation, the two products A_i, a_i of each of the reactions (12.1) fly away one from the other with opposite momenta.

In (12.1) the first reaction represents *elastic scattering* in which the incident particle a_1 reemerges with the same energy it had before the reaction (in the c.m.s.). The second reaction represents *inelastic scattering* in which the incident particle is reemitted with a lower energy and the target nucleus is left in an excited state. In all other reactions the outgoing particle is different from the incident one.

We will include among these also the *radiative capture* of the projectile a_1, in which the outgoing particle is substituted by a photon.

The description of the set of reactions (12.1) may be refined by introducing the concept of reaction channel. In all reactions, except elastic scattering that will be considered later, the two products A_i and a_i can be left in different quantum states due to possible internal excitation and/or to different spin orientations. Any possible pair of products of one of the reactions (12.1) in well defined quantum states is indicated as a *reaction channel*. According to this definition each one of the reactions appearing on the right side of (12.1) is subdivided into a certain number of *outgoing channels*.

Since also the projectile a_1 and the target nucleus A_1 may be in different quantum states, one talks of the *entrance channel* or *initiating channel*, intending with this expression a complete specification of the quantum states of a_1 and A_1.

In principle the initiating channel is completely determined by the chosen experimental conditions: momentum of a_1, spin orientation and internal states of a_1 and A_1. Of all possible outgoing channels only those will be *open*, which respect the conservation laws of energy, angular momentum and parity. In the case of elastic scattering the outgoing channel is identical to the initiating channel and therefore these theorems are always satisfied.

Under actual experimental conditions the reaction is very seldom initiated through a single channel, because, for instance, the spin orientation of the incident particle and/or of the target nucleus are not well defined. The results obtained under these conditions represent averages over all initiating channels present in the actual initial state.

V. The effect of hydrogeneous substances. The mechanism of the slowing down of neutrons.

13. The discovery of the effect of hydrogeneous substances and its qualitative interpretation. In the attempt to establish an empirical scale of the activities produced in various elements when irradiated under standard conditions, the group working at the University of Rome, noticed, in October 1934, some anomalies due to the presence of extraneous bodies in the vicinity of the neutron source and of the bombarded element. It was then decided to study the transmission of lead (due in part, to absorption, in part to scattering) of the neutrons emitted from a Rnα + Be source and detected by means of the activity produced in a small thin-walled cylinder of silver the activity of which was measured by putting the cylinder around a G.M. counter.

Two arrangements shown in Fig. 1 were tried, and it was found that the presence of the layer of lead did not produce any considerable reduction in the activity of the Ag-detector when the irradiation was done with the arrangement (a), while on the contrary a considerable absorption was observed with arrangement (b).

This result was clearly due to the fact that in (b) the layer of Pb absorbs and scatters away from the detector part of the neutrons, while in (a) this effect is almost completely compensated by the scattering of neutrons towards the detector due to the more distant parts of the lead plate.

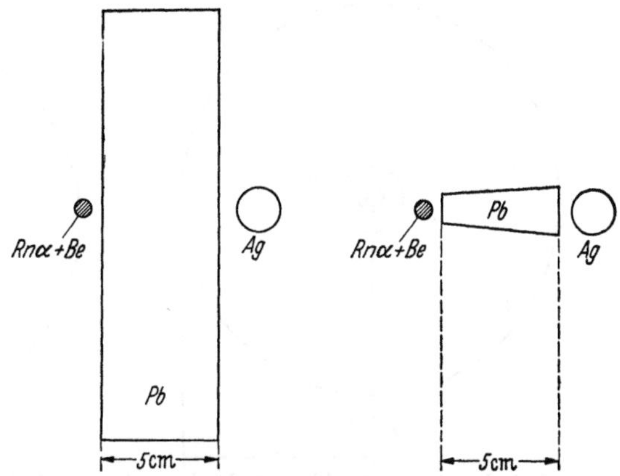

Fig. 1. Arrangement used by FERMI *et al.* for studying neutron transmission through lead and paraffin.

This first result obtained with a heavy element suggested the idea of repeating the experiment, with arrangement (a) using paraffin instead of lead. The new result was indeed surprising, because the presence of the paraffin produced a noticeable increase in the activity of the silver detector.

Blocks of paraffin were next placed all around the neutron source and the detector and it was observed that under these conditions the activity of the silver was enhanced by a large factor which, depending on the geometry, could vary from about ten up to about one hundred.

A preliminary letter of FERMI, AMALDI, PONTECORVO, RASETTI and SEGRÈ[1] sent to the Editor of the Ricerca Scientifica, the 22nd of October 1934, contains the experimental proof of the following points: (a) the effect was due to the neutrons and not to the γ-rays emitted from Rn and its products; (b) it was produced not only by paraffin but also by others hydrogeneous substances, as for instance water; (c) it was observed only if the detector was one of those elements whose reaction product is an isotope of the target nucleus, i.e. according to the arguments given in Sect. 11, only in the case of the (n, γ) processes.

This last rule, later confirmed and extended[2-4], admits a few important exceptions which will be considered below (Sect. 27.).

The existence of this effect was clearly very interesting, not only from the practical point of view of providing a simple means of increasing by a large factor the rate of production of many nuclear reactions, but still more for the

[1] E. FERMI, E. AMALDI, B. PONTECORVO, F. RASETTI and E. SEGRÈ: Ric. sci. **5** (2), 282 (1934).

[2] E. FERMI, B. PONTECORVO and F. RASETTI: Ric. sci. **5** (2), 380 (1934).

[3] E. AMALDI, O. D'AGOSTINO and E. SEGRÈ: Ric. sci. **5** (2), 381 (1934). — E. AMALDI, O. D'AGOSTINO, E. FERMI, B. PONTECORVO, F. RASETTI and E. SEGRÈ: Ric. sci. **5** (2), 467 (1934). — E. AMALDI, O. D'AGOSTINO, E. FERMI, B. PONTECORVO, F. RASETTI and E. SEGRÈ. Ric. sci. **6** (1), 123 (1935). — E. AMALDI, O. D'AGOSTINO, E. FERMI, B. PONTECORVO and E. SEGRÈ: Ric. sci. **6** (1), 435 (1935).

[4] E. AMALDI, O. D'AGOSTINO, E. FERMI, B. PONTECORVO, F. RASETTI and E. SEGRE: Proc. Roy. Soc. Lond., Ser. A **149**, 522 (1935).

various implications involved in its interpretation and the many new effects to which it had opened the door.

When a detector, in the form of a thin-walled cylinder D, is irradiated by a neutron source S inside it and placed at the centre of a large mass of paraffin or water M (Fig. 2), one would expect to observe a certain increase in its activity due to the presence of the hydrogeneous substance. In fact a neutron which has crossed the detector without being captured, can be scattered back by the paraffin and cross the detector a second, a third, and perhaps even more times. But in the first place it is difficult to believe that such an effect can increase the activity by a factor of the order of 30 or 40, and in the second place it can not be invoked in order to interpret the following experiment[1,2].

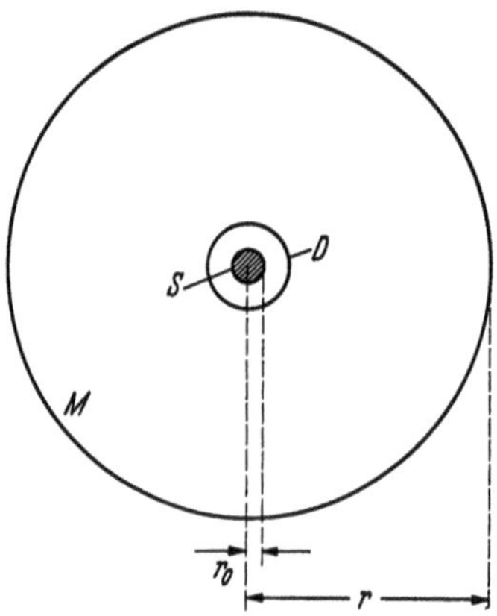

Fig. 2. The hydrogeneous material M when placed around the neutron source S produces a large increase of the activity of the detector D.

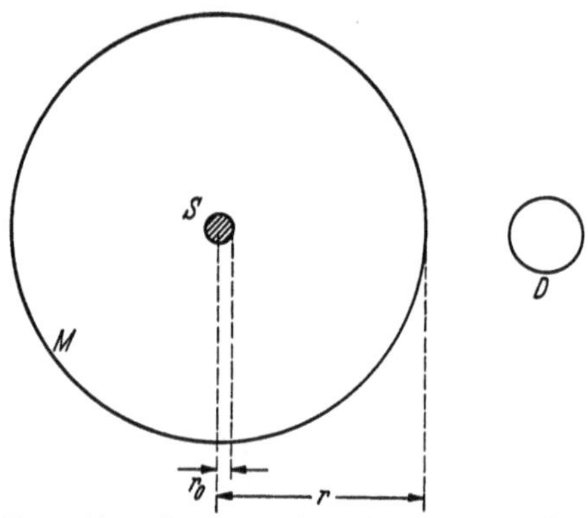

Fig. 3. Also under these conditions (see Fig. 2) the hydrogeneous material M when placed around the neutron source S enhances by a large factor the activity of the detector D.

The silver detector, placed at a distance of 20 cm from the neutron source (Rn α + Be), is irradiated successively in two different ways: firstly in air, secondly surrounding the neutron source with a cylindrical container of $r = 14$ cm diameter, filled with water (Fig. 3). It is clear that, for reasons of symmetry, the number of neutrons crossing the detector is not increased by the presence of the water. But the experimental result is that the presence of water enhances considerably the activity of the detector.

As FERMI realised immediately, the following two assumptions give a plausible explanation of the effect of hydrogeneous substances. It is perhaps the only explanation and certainly the correct one as it became clear from a later and more detailed investigation:

(a) In water or paraffin the velocity of a neutron can be reduced enormously through collisions against protons.

(b) The cross section for certain nuclear reactions, in particular for the (n, γ) processes, increases rapidly with decreasing velocity of the neutron.

Assumption (a) is suggested by the fact that a neutron, in an elastic collision, can confer upon a proton a variable but in general large fraction of its kinetic energy. Therefore, provided the neutron-proton elastic cross section is much larger than the cross section for all processes in which the neutron is absorbed, the neutron can undergo, before being captured in a nucleus, a sufficiently large

[1] See footnote 2, p. 19.
[2] See footnote 4, p. 19.

number of collisions against protons and consequently reduce its velocity to a value enormously smaller than the initial one.

At first sight, assumption (b) appeared rather strange because all excitation curves known at that time, for reactions produced by α-particles, protons and deuterons, had the opposite behaviour. But in all these cases the increase of the cross section with the velocity of the incident particle is mainly due to the increase of the penetrability of the electrostatic potential barrier which does not act in the case of neutrons. This assumption found support in the fact that the increase of activity produced by the presence of hydrogeneous substances, was observed in the case of the (n, γ) processes, but not, at least in general, in the case of (n, p) and (n, α) reactions. In these processes the penetration of the potential barrier by the outgoing particles has to be considered. An increase in the energy of the incident neutron produces a corresponding increase in the energy of the outgoing particle and therefore an increase in the penetrability of the electrostatic barrier.

In the case of (n, γ) processes neither the neutron nor the photons has to cross an electrostatic barrier and therefore no reason exists for expecting an increase of the cross section with increasing velocity of the neutron.

A qualitative analysis of the behaviour of the cross section for these processes was then attempted by FERMI[1,2] as well as by various other authors[3] along the following lines: when the wavelength of the neutron becomes very large with respect to the nuclear dimensions, the probability P that the nucleus absorbs the neutron wave can depend only on the probability density $|\psi|^2$ for the neutron to be found outside the nucleus and no longer on the neutron wavelength, which is practically infinite; in other words it is reasonable to expect that by decreasing the velocity of the neutron the probability of capture P tends to a value independent of the neutron velocity. Expressing P by means of the capture cross section σ_a and the flux-density $n_0 \cdot v$ of the incident neutrons (n_0 = number of neutrons per cm^3) one gets

$$P = n_0 \cdot v \cdot \sigma_a \tag{13.1}$$

i.e.

$$\sigma_a = \frac{P}{n_0} \frac{1}{v} . \tag{13.2}$$

Eq. (13.2) gives, for P = const, the so-called "$1/v$-law", to which we shall return later (Division VI), and more generally reduces the interpretation of the experimental fact that neutrons ot low energy have larger capture cross section than high energy neutrons, to the condition that P must increase, with increasing energy of the incident neutrons, slower than v.

If the preceding interpretation of the effect of hydrogeneous substances is correct, one has to expect a similar effect, although perhaps less pronounced, with other substances.

Experiments with carbon, silicon, iron and lead in place of the paraffin were performed and gave positive results, although smaller than with paraffin, in all cases except that of iron[4,1]. This negative result was interpreted as due to the rather high capture cross section of this element while the positive result

---

[1] See footnote 4, p. 19.

[2] E. FERMI and F. RASETTI· Nuovo Cim. **12**, 201 (1935).

[3] F. PERRIN and W. M. ELSASSER: C. R. Acad. Sci. Paris **200**, 450 (1935). — J. Phys. Radium **6**, 194 (1935). — F PERRIN: C. R. Acad. Sci. Paris **200**, 1749 (1935). — G. BECK and L. H. HORSLEY. Phys. Rev. **47**, 510 (1935). — H. A. BETHE: Phys. Rev. **47**, 747 (1935).

[4] See footnote 3, p. 19.

with lead, which had first revealed the influence of surrounding bodies on the neutron activation, was attributed to inelastic collisions (see Sect. 30).

In conclusion many light elements can be employed to *slow down neutrons* by means of successive elastic collisions, that is to say, using an expression introduced later, they can be employed as *moderators*. The lighter the nucleus, the higher is the "*slowing down power*". From this point of view hydrogen is certainly the best moderator but, as we shall see in the following, once the neutrons have been slowed down it is less efficient than for instance deuterium in maintaining a high density of neutrons, because of the much smaller capture cross section of the latter.

14. Further experimental information on the slowing down of neutrons. In a short time the experiments reported in Sect. 13 were completed by investigating the activation of the detector as a function of the dimensions of the moderator.

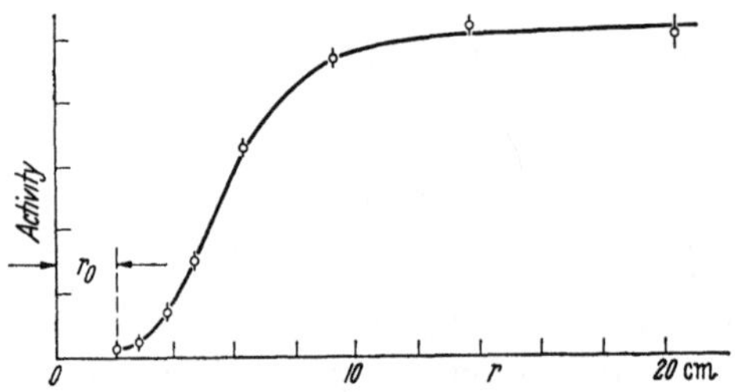

Fig. 4. The activation of a silver detector D as a function of the radius r of the hydrogeneous material (water) in the experimental set-up of Fig. 2 (BJERGE and WESTCOTT).

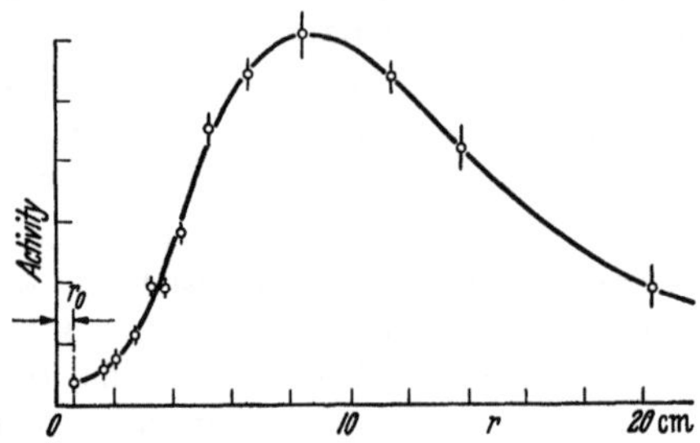

Fig. 5. The counting rate of a small boron ionization chamber D as a function of the radius r of the hydrogeneous material (water) in the experimental set-up of Fig. 3 (BJERGE and WESTCOTT).

Fig. 4 shows the results obtained by BJERGE and WESTCOTT[1] with the experimental set-up of Fig. 2. The detector, in the form of a Ag cylinder, was placed around a Ra α+Be source, and both were contained in a test tube of radius $r_0 = 2$ cm. The activation, measured as a function of the radius r of the surrounding cylinder filled with water, shows a saturation effect when r is larger than 10 cm. This can be interpreted as an indication that fast neutrons, emitted from the source S which undergo their first collisions at distances of more than about 8 cm, can still come back to the detector through the action of the moderator.

The results obtained by the same authors using the experimental set-up of Fig. 3 and by detecting the neutrons by means of a small ionization chamber lined with a thin layer of boron (see Sects. 17 and 26) are given in Fig. 5. The counting rate of the boron chamber as a function of the radius r of the water cylinder surrounding the neutron source S(Raα+Be) shows the general qualitative features of a transition curve; namely it starts from a very low value, when the source S is naked, it increases when the source is surrounded with a water cylinder of radius r reaching a maximum for $r \sim 8$ cm, and then, by a further increase of r, it decreases again.

Two interesting remarks can be made about this curve. The first refers to the part of the curve before its maximum; it shows the gradual generation, from the fast neutrons emitted from the source, of a neutron radiation much more effective in producing the boron reaction (slow neutrons). The increase is very slow in the first centimeters and then gets faster, as one has to expect if the process

[1] T. BJERGE and C. H. WESTCOTT: Proc. Roy. Soc. Lond. **150**, 709 (1935). — Proc. Cambridge Phil. Soc. **31**, 145 (1935).

of slowing down takes place in many collisions. In fact, if the formation of the secondary radiation from the primary one would take place in a single elementary act, the curve had to increase linearly at the origin.

The second remark refers to the second part of the curve; at the maximum the process of degradation in energy of the fast neutrons has reached its full development and a further increase of the thickness of water produces an absorption of the neutrons more effective in producing the boron-reaction (slow neutrons). The question remains open, however, if the decrease observed for $r > 10$ cm represents the absorption law of the fast neutrons emitted from the source or of the slow neutrons produced in the moderator.

The answer depends of course on which of these two radiations is more rapidly absorbed in the moderator. If the slow neutrons are more rapidly absorbed, then for $r \gtrsim 10$ cm, one has a situation that can be described as a kind of equilibrium between the primary radiation (fast neutrons) and the secondary radiation (slow neutrons) and both decrease according to the absorption law of the primary radiation. In the opposite case, the primary radiation would be quickly absorbed so that for large r one would have only slow neutrons. The answer also depends on the energy of fast neutrons emitted from the source, but in general the absorption of the slow neutrons is larger than that of fast neutrons. Therefore the decrease observed in Fig. 5 for large values of r represents the absorption of the fast primary neutrons (Sect. 77 and 80).

These experiments, although very enlightening on the qualitative features of the process of slowing down, were not the most suitable for a quantitative interpretation of these phenomena. The results of measurements of the spatial distribution of slow neutrons in an infinite medium surrounding a fast neutron point source[1], appear to be much more convenient for a comparison with theory, especially when made with neutrons of a well defined energy[2]. This latter point, however, needed the discovery of another important effect, namely the existence in many nuclei of narrow lines for neutron absorption (Sect. 24).

A detailed discussion of the spatial distribution of neutrons will be given in Part C, Divisions III and IV, of this article.

VI. A few important properties of slow neutrons and the quantum mechanical representation of processes produced by neutrons.

15. The anomalously high absorption of slow neutrons in various elements. A simple check of the interpretation given in Sect. 13 of the effect of hydrogeneous substances, and in particular of assumption (b), consists obviously in trying to verify whether the neutrons emerging from the paraffin, which are so efficient in producing artificial radioactivity in certain elements through (n, γ) processes, show, for the same elements, a correspondingly high absorption.

Therefore a systematic investigation of the absorption properties of the neutrons slowed down in paraffin, was started using as detector a small silver or rhodium plate whose activity was measured by means of a G-M counter.

The geometry of these first experiments did not allow the determination of the cross sections, but established qualitatively whether the absorption was very high just in those elements which were strongly activated. It was thus recognized that in general the elements which are strongly activated show an absorption much higher than the other elements and that the same element,

---

[1] See footnote 4, p. 19.
[2] E. Amaldi and E. Fermi: Ric. sci. 7 (1), 310, 454 (1936). — Phys. Rev. **50**, 899 (1936).

strongly activated, say for instance Rh, absorbs "slow neutrons" much more than fast neutrons[1]. In order to give an idea of the orders of magnitude, 0.3 g/cm² of Rh reduced to about one half the intensity of slow neutrons, while the same effect was produced on fast neutrons by a ten times thicker Rh absorber.

Assumption (b) of the preceding section was thus definitely proved.

These first data, although almost of a qualitative nature, were sufficient to show that in many elements the absorption cross sections were enormously larger than the corresponding geometric cross sections.

This result, which would clearly constitute a very serious difficulty from the classical point of view, is perfectly understandable in a quantum mechanical picture of the collision between a low energy neutron and a nucleus. As we shall see in Sect. 18 when the wavelength λ_r of the relative motion is very large with respect to the dimension R of the nucleus, the absorption cross section can be put in the form

$$\sigma_a = \pi \lambdabar_r^2 \, T(E_r) \tag{15.1}$$

where $\lambdabar_r = \lambda_r/2\pi$, E_r the kinetic energy of the relative motion (33.11) and $T(E_r)$ is a number always smaller than 1 but, sometimes, by not too many orders of magnitude. Therefore, if $\lambdabar_r$ is larger than R by a large factor, the cross section (15.1) can be much bigger than πR^2. In particular there are cases in which $T(E_r)$ can be put in the form

$$T(E_r) = \frac{C}{\lambdabar_r} \tag{15.2}$$

where C is a convenient constant. In these cases the combination of Eqs. (15.1) and (15.2) gives again the $1/v$ law already mentioned in Sect. 13.

In order to familiarize ourselves with the orders of magnitude we recall that the wavelength of a neutron of energy E colliding against a nucleus of infinite mass ($M \gg m$) is given by the expression

$$\lambdabar_r \approx \lambdabar = \frac{\hbar}{p} = \frac{\hbar}{\sqrt{2mE}} = \frac{4.55}{E_{\mathrm{ev}}^{\frac{1}{2}}} \cdot 10^{-10}\,\mathrm{cm} \tag{15.3}$$

so that

$$\pi \lambdabar^2 = \frac{65}{E_{\mathrm{ev}}} \cdot 10^{-20}\,\mathrm{cm}^2 = \frac{6.5}{E_{\mathrm{ev}}} \cdot 10^5\,\mathrm{barns}. \tag{15.4}$$

Apart from this general remarks about the absorption cross section, an unxpected result was obtained in the cases of B, Li, and Cd which were found to bsorb very strongly slow neutrons, while not showing any corresponding activity.

16. The emission of γ-rays following the absorption of a neutron. If the strong activity produced by slow neutrons in certain elements, and the corresponding strong absorption of slow neutrons were due to radiative capture processes, the binding energy of the captured neutrons would have to be emitted, in the form of one or more photons, the total energy being about 7 Mev. The detection of these photons was clearly the last step necessary to complete the experimental proof of the picture given in Sect. 13.

Therefore FERMI and coworkers set up an experiment in which the photons emitted by an element under slow neutron irradiation could be converted in a layer of Pb 2 mm thick, and the corresponding secondary electrons detected by means of a G-M counter.

---

[1] See footnotes 2, 3 and 4, p. 19.

They found that in general the elements which show a strong absorption for slow neutrons gave a strong emission of γ-rays. Among these was cadmium, the strong absorption of which did not correspond to any appreciable activation. This behaviour was then attributed to radiative capture of slow neutrons, in one (or more) of the many stable isotopes of this element ($A=106$, 1.22%; $A=108$, 0.92%; $A=110$, 12.8%; $A=111$, 12.80%; $A=112$, 24.0%; $A=113$, 12.30%; $A=114$, 28.80%; $A=116$, 7.66% [*17e*]), since in some of these processes also the reaction products were stable. This interpretation was confirmed a few years later, when using Cd samples enriched in some of the isotopes, it was possible to prove[1] that the strong absorption by this element is due to the process

$$^{113}\mathrm{Cd}\,(n,\gamma)\,^{114}\mathrm{Cd}. \tag{16.1}$$

The strong absorption power for slow neutrons and the good mechanical properties of this element led to its immediate use as a convenient material for screening and canalizing slow neutrons. A further investigation of its properties, that we shall mention later (Sect. 26), showed that it is a tool in neutron experimental physics which is extremely useful and almost without substitute.

These first results on the emission of photons as a consequence of slow neutron capture were soon confirmed and extended by various authors[2,3]. In particular it was shown by FLEISCHMANN[3] that the paraffin in which the neutrons are slowed down gives a weak γ-ray emission: this radiation, as we will see in Sect. 46, is due to the capture of slow neutrons in hydrogen. In fact the production of γ-rays by the passage of fast neutrons through matter and in particular through liquid hydrogen had been observed, before the discovery of slow neutrons, by LEA[4]. The interpretation given by this author in terms of inelastic scattering of fast neutrons with successive deexcitation through γ-ray emission, is correct only in the case of heavy elements. In light elements and in particular in hydrogen the slowing down of neutrons and their successive radiative capture are the prevailing effects.

In more recent times the high intensity beams of slow neutrons provided by neutrons piles, have allowed a careful investigation of the γ-rays emitted by neutron capture[5-7]. We refer to these papers for rather extensive bibliographies on this important subject [*3*].

17. The reactions produced by slow neutrons in boron, lithium and other light elements. No appreciable emission of γ-rays was observed by Fermi and coworkers in the cases of B and Li, the two light elements which absorb slow neutrons very strongly without showing any corresponding activity. The absence of a γ-ray emission clearly excluded a radiative capture process in $_5\mathrm{B}^{10}$ (or $_3\mathrm{Li}^6$) with production of the other stable isotope $_5\mathrm{B}^{11}$ (or $_3\mathrm{Li}^7$).

This gave rise to the suspicion, immediately confirmed by experiment[8], that the strong absorption in these two elements could be due to some instantaneous process with emission of ionizing particles of rather short range.

[1] B. J. MOYER, B. PETERS and E. H. SCHMIDT: Phys. Rev. **69**, 666 (1946).

[2] F. RASETTI: Z. Physik **97**, 64 (1935). — S. KIKUCHI, H. AOKI and K. HUSIMI: Nature, Lond. **137**, 398 (1936). — Proc. Phys.-Math. Soc. Japan **18**, 115 (1936).

[3] R. FLEISCHMANN: Naturwiss. **22**, 839 (1934). — Z. Physik **97**, 242, 265 (1935); see also L. ARZIMOWITSCH, I. KURTSCHATOW, G. LATYSCHEW and W. CHROMOW: Phys. Z. Sowjet. **8**, 472 (1935).

[4] D. E. LEA: Nature, Lond. **133**, 24 (1934). — Proc. Roy. Soc. Lond., Ser. A **150**, 376 (1935).

[5] B. B. KINSEY and G. A. BARTHOLOMEW: Phys. Rev. **93**, 1260 (1954).

[6] L. V. GROSHEV, B. P. ADYASEVICH and A. M. DEMIDOV, p. 39 of Ref. [*15a*].

[7] T. H. BRAID: Phys. Rev. **102**, 1109 (1956).

[8] See footnotes 3 and 4, p. 19.

Similar conclusions were reached by CHADWICK and GOLDHABER[1] who had been guided by the following considerations.

The rule, mentioned in Sect. 13, that slow neutrons can produce only (n, γ) reactions, could fail for some very light element whose electrostatic potential barrier is so small as to make possible processes of the (n, p) or (n, α) type provided they are strongly exoenergetic. These authors in a systematic investigation of light elements found three cases of this type: Li, B and N.

In a first series of experiments, FERMI and coworkers measured the ionization current produced by a neutron source in an ionization chamber, connected to an electrometer, and containing boron, sometimes in the solid state, sometimes as a gas (borontrifluoride). In order to eliminate slow neutrons and observe the background due to the γ-rays emitted by the neutron source (Rn α + Be), the ionization chamber was surrounded with a layer of Cd.

In a second series of experiments a small ionization chamber, connected to a linear amplifier and oscillograph, allowed the measurement of the ionization produced by a single elementary process. From measurements of the range of the emitted particles as well as of the ionization produced by a single disintegration process caused by a slow neutron in a solid layer of boron or in the gas of the ionization chamber (BF_3), these authors concluded[2] that the process responsible for the strong absorption in boron was

$$ {}_5B^{10}(n, \alpha)\, {}_3Li^7; \qquad Q = 2.792 \text{ Mev.} \tag{17.1} $$

The same result was obtained by TAYLOR and GOLDHABER by exposing to slow neutrons nuclear plates loaded with boron[3].

In the case of lithium, CHADWICK and GOLDHABER[1,3] proved that the process responsible for the strong absorption of slow neutrons, was

$$ {}_3Li^6(n, \alpha)\, H^3; \qquad Q = 4.785 \text{ Mev.} \tag{17.2} $$

Some time later BURCHAM and GOLDHABER[4] showed that in nitrogen slow neutrons give rise, with a not too large cross section, to the process

$$ {}_7N^{14}(n, p)\, {}_7C^{14}; \qquad Q = 0.626 \text{ Mev} \tag{17.3} $$

where C^{14} is the well known radioactive isotope of long mean life ($T_{\frac{1}{2}} = 5.568$ y). Two more exoenergetic processes of the (n, p) type, namely

$$ He^3(n, p)\, H^3; \qquad Q = 0.7636 \text{ Mev}, \tag{17.4} $$

$$ Cl^{35}(n, p)\, S^{35}; \qquad Q = 0.62 \text{ Mev} \tag{17.5} $$

were discovered later.

These (n, α) and (n, p) processes produced by slow neutrons are of the highest interest not only for certain special properties that we shall consider in Sect. 27, but also because they offer the possibility of a new technique for detecting slow neutrons (cf. H. H. BARSCHALL's article in Vol. XLV).

18. Outline of the geometrical limitations on scattering and reaction cross sections. In this section we shall summarize a few general theorems due to

[1] J. CHADWICK and M. GOLDHABER: Nature, Lond. **135**, 65 (1935). — Proc. Cambridge Phil. Soc. **31**, 612 (1935).

[2] E. AMALDI: Nuovo Cim. **12**, 223 (1935); see also footnote 4, p. 19.

[3] H. J. TAYLOR and M. GOLDHABER: Nature, Lond. **135**, 341 (1935).

[4] W. E. BURCHAM and M. GOLDHABER: Proc Cambridge Phil. Soc. **32**, 632 (1936).

BOHR, PEIERLS and PLACZEK[1] and BETHE[2] which establish limitations to the scattering and reaction cross sections. An adequate presentation of the considerations sketched here may be found in other articles, especially in Vol. XLI, and in Ref. [3].

These limitations are of geometrical origin and therefore do not depend on the particular model adopted for the description of the nucleus and of its interaction with the incident neutron. They refer to the elastic scattering cross section σ_s and to the reaction cross section σ_r. This latter is defined as the sum of the cross sections for all processes other than elastic scattering, which start with the capture of the incident neutron and end with the emission of other particles, or neutrons of lower energy, or photons. A detailed description of these particular processes requires the adoption of a convenient nuclear model. But this is not necessary for a general discussion of σ_r which is determined by the number of neutrons taken out of the incident beam.

A beam of monoenergetic neutrons moves along the z axis and is partly scattered, partly absorbed by a nucleus placed at the origin. The only assumption that we need is that the radius of the nucleus

$$R = r_0 A^{\frac{1}{3}}; \quad r_0 = 1.4 \times 10^{-13}\,\text{cm} \tag{18.1}$$

is finite and that the nuclear forces fall off so rapidly with increasing distance that for $r \gg R$ the neutrons may be treated as free particles[3]. The two-body problem is transformed into the problem of one particle moving in the field of forces produced by the other, by making use of reduced mass and of relative coordinates (see Sect. 33).

For the sake of simplicity we limit ourselves to considering the case of a target nucleus as well as of an incident particle of *zero spin*. Therefore the considerations given below do not apply to the actual case of neutrons colliding with nuclei. One has to notice, however, that the assumption of zero spin does not affect most of the results, while it simplifies considerably their deduction. The derivation of similar relations made without this assumption may be found in Ref. [3].

The neutral particles of protonic mass and zero spin considered in the following will be indicated as neutrons for the sake of simplicity.

At great distance from the scatterer the wave function has the asymptotic form

$$\psi(r) \rightarrow e^{ikz} + \psi_s(r) \tag{18.2}$$

where $k = 1/\lambda\!\!\!^{-}_r = p_r/\hbar$ refers to the relative motion of the neutron + nucleus system. In (18.2) the first term is a plane wave describing the incident beam of neutrons, the second term is a wave originating at the nucleus which represents the scattered particles. Since in general the latter will depend on the angle of scattering ϑ, we may expand it in Legendre polynomials

$$\psi_s = \sum_{l=0}^{\infty} \frac{c_l}{r}\, e^{ikr}\, P_l(\cos\vartheta). \tag{18.3}$$

From this expression we see that for a complete description of the scattered wave we need the knowledge of the coefficients c_l of the scattered waves of angular momentum l. In principle the complex quantities c_l can be determined by solving the Schrödinger equation of the neutron + nucleus system. This is

[1] N. BOHR, R. PEIERLS and G. PLACZEK: Nature, Lond. **144**, 200 (1939).
[2] H. A. BETHE: Phys. Rev. **57**, 1125 (1939).
[3] Exactly, this is the so-called nuclear *force radius*; cf. e.g. HILL's article in Vol XXXIX.

actually made in simple cases. Here we do not pursue this procedure which would involve the neutron-nucleus interaction; we simply transform (18.2) and (18.3) as follows.

By expanding the plane wave appearing in (18.2) into Legendre polynomials we separate it in pairs of spherical waves: each pair contains an incoming as well as an outgoing wave, both of angular momentum $l\hbar$. For large values of kr this expansion becomes

$$e^{ikz} \to \frac{1}{kr} \sum_{l=0}^{\infty} (2l+1) \frac{i^{l+1}}{2} \left\{ e^{-i\left(kr - l\frac{\pi}{2}\right)} - e^{+i\left(kr - l\frac{\pi}{2}\right)} \right\} P_l(\cos\vartheta). \tag{18.4}$$

Since the nucleus placed at the origin will produce a change only of the outgoing waves, the perturbed wavefunction (18.2) must have the asymptotic general form

$$\psi(r) \to \frac{1}{kr} \sum_{l=0}^{\infty} (2l+1) \frac{i^{l+1}}{2} \left\{ e^{-i\left(kr - l\frac{\pi}{2}\right)} - \eta_l\, e^{+i\left(kr - l\frac{\pi}{2}\right)} \right\} P_l(\cos\vartheta) \tag{18.5}$$

where the η_l are complex numbers. According to (18.2) the scattered wave can be obtained as difference between (18.5) and (18.4)

$$\psi_s \to \psi(r) - e^{ikz} = \frac{1}{kr} \sum_{l=0}^{\infty} (2l+1) \frac{i^{l+1}}{2} (1 - \eta_l)\, e^{i\left(kr - l\frac{\pi}{2}\right)} P_l(\cos\vartheta) \tag{18.6}$$

and therefore the coefficients c_l appearing in (18.3) are given by

$$c_l = \frac{2l+1}{2k}\, i (1 - \eta_l). \tag{18.7}$$

This relation shows how the scattered wave may be described by using the complex numbers η_l instead of the coefficients c_l.

In order to obtain the differential scattering cross section one now has to divide the number of particles scattered per second in the unit solid angle by the number of particles incident per square centimeter per second. The result is

$$\frac{d\sigma_s}{d\omega} = \left| \sum_{l=0}^{\infty} c_l\, P_l(\cos\vartheta) \right|^2. \tag{18.8}$$

The total scattering cross section is obtained by integrating (18.8) over the solid angle and since

$$\int_0^{2\pi} d\varphi \int_{-1}^{+1} P_l(\cos\vartheta)\, P_{l'}(\cos\vartheta)\, d\cos\vartheta = 2\pi \frac{2}{2l+1} \delta_{ll'}$$

one has

$$\sigma_s = \sum_{l=0}^{\infty} \sigma_{s,l} \tag{18.9}$$

with

$$\sigma_{s,l} = \frac{4\pi}{2l+1} |c_l|^2 \tag{18.10}$$

or, making use of (18.7)

$$\sigma_{s,l} = \pi \lambda\!\!\!^{-}{}_r^2 (2l+1)\, |1 - \eta_l|^2. \tag{18.11}$$

The reaction cross section σ_r is determined by computing the number of particles which enters into a sphere of radius $r \gg R$ around the center of the nucleus without leaving it again. Therefore it is equal to the number of particles which enters per second into this sphere as computed from the complete wave function (18.5), divided by the number of particles incident per square centimeter per second. The result of standard computations is

$$\sigma_r = \sum_{l=0}^{\infty} \sigma_{r,l} \tag{18.12}$$

with

$$\sigma_{r,l} = \pi \lambda\!\!\!^{-2}_r (2l+1)(1-|\eta_l|^2). \tag{18.13}$$

A few very important general conclusions may be drawn by inspecting Eqs. (18.9), (18.11), (18.12), (18.13):

(a) The total scattering cross section (18.9) can be subdivided into partial cross sections $\sigma_{s,l}$ corresponding to the scattering of incident particles of angular momentum l. This separation is not possible for the differential scattering cross section (18.8) which includes interference terms between contributions from different values of l. According to (18.12) the reaction cross section σ_r can also be split into partial cross sections each corresponding to a different value of the angular momentum.

(b) Since $\sigma_{r,l}$ can never be negative, it follows from (18.13) that the complex numbers η_l must necessarily satisfy the condition

$$|\eta_l|^2 \leq 1. \tag{18.14}$$

From (18.13) and (18.14) one has that, when $|\eta_l|^2$ varies between 1 and 0, $\sigma_{r,l}$ is bound to vary in the interval

$$0 \leq \sigma_{r,l} \leq \pi \lambda\!\!\!^{-2}_r (2l+1). \tag{18.15}$$

(c) From (6.21) we see that also $\sigma_{s,l}$ must satisfy certain limitations: its minimum value, i.e. zero, is attained for $\eta_l = +1$, its maximum value is attained for $\eta_l = -1$. In general one can write

$$0 \leq \sigma_{s,l} \leq 4\pi \lambda\!\!\!^{-2}_r (2l+1). \tag{18.16}$$

The limitation (18.15) of the l-component of the reaction cross section may be visualized by the following semiclassical consideration [3].

A neutron of momentum p_r and angular momentum $l\hbar$ with respect to the center of the target nucleus, has a classical distance of closest approach given by

$$r_{\min} \approx \frac{l\hbar}{p_r} = l \lambda\!\!\!^{-}_r. \tag{18.17}$$

Since according to quantum mechanics l can take on only integral values, we can divide the incident beam into cylindrical zones around its direction of propagation; the first zone contains the particles of impact parameter between 0 and $\lambda\!\!\!^{-}_r$, the second zone the particles of impact parameter between $\lambda\!\!\!^{-}_r$ and $2\lambda\!\!\!^{-}_r$, the l-zone the particles with impact parameter between $l\lambda\!\!\!^{-}_r$ and $(l+1)\lambda\!\!\!^{-}_r$. Then we have to expect that the reaction cross section $\sigma_{r,l}$ cannot be larger than the cross sectional area of the l-zone, i.e. not larger than $\pi\lambda\!\!\!^{-2}_r(2l+1)$ because no more particles of the considered momentum may be taken out of the incident beam than are in it originally.

Although this estimate agrees with the correct quantum mechanical result (18.15), the argument has to be accepted with some reservations. Since the incident wave corresponds to a beam of particles of well defined velocity, it is impossible according to quantum mechanics, to attribute a well defined value to the impact parameter. It is however approximately correct to assume that the particles of angular momentum $l\hbar$ move in the l-zone.

The fact that according to (18.16) $\sigma_{s,l}$ can attain a value four times larger than that attained by $\sigma_{r,l}$ is due to constructive interference between the incident and the scattered wave.

The semiclassical consideration given above allows the derivation of other important rules when combined with the fact that the nuclear forces fall to zero very rapidly for $r \gtrsim R$.

For *fast neutrons*, defined as neutrons of such an high energy that

$$\lambda\!\!\!^{-}_r < R, \tag{18.18}$$

it follows from (18.17) that they may interact with a target nucleus only for

$$l \lesssim \frac{R}{\lambda\!\!\!^{-}_r} \tag{18.19}$$

and therefore the sums appearing in (18.9) and (18.12) will break, at least in general, at values of l equal to about $R/\lambda\!\!\!^{-}_r$.

On the contrary, *slow neutrons*, defined as neutrons of such a low energy that

$$\lambda\!\!\!^{-}_r \gg R, \tag{18.20}$$

will interact with a nucleus *only* (or almost only) *in S states:* in a P state the classical distance of closest approach is equal to $\lambda\!\!\!^{-}_r$ and therefore neutrons satisfying condition (18.20), remain at such a large distance from the target nucleus that no reaction or scattering process can take place. Recently the Argonne group[1] has found some indication of a weak contribution to the total cross section of U^{238} for slow neutrons due to P states. But in general when discussing the interaction of slow neutrons with nuclei, we shall always refer only to S waves. In particular the asymptotic expression of the wave function (18.2) will reduce to

$$\psi(r) = e^{ikz} + \frac{c_0}{r} e^{ikr} \tag{18.21}$$

with

$$c_0 = \frac{1}{2ik}(\eta_0 - 1). \tag{18.22}$$

Finally we notice that in general it follows from (18.12) and (18.13) that the reaction cross section may be written in the form

$$\sigma_r = \pi \lambda\!\!\!^{-2}_r \sum_{l=0}^{\infty} (2l+1)(1 - |\eta_l|^2) = \pi \lambda\!\!\!^{-2} T(E_r). \tag{18.23}$$

In the case of slow neutrons, in which only S waves are considered, the numerical factor $T(E_r)$ is simply $(1 - |\eta_0|^2)$, and therefore it is always smaller than 1; in certain cases, however, it is smaller than 1 by not too many orders of magnitude (Sect. 15).

19. The case of maximum possible absorption. The example of fast neutrons colliding against large black nuclei. In this section we shall discuss the extreme case of nuclei showing the maximum possible absorption. According to (18.13) this situation is attained, for the l-th partial cross section, when $\eta_l = 0$. Then

[1] L. M. BOLLINGER, R. E. COTÉ, D. A. DAHLBERG and G. E. THOMAS Phys. Rev. **105**, 661 (1957). See the footnote 1 on p. 50 of Sect. 26.

the perturbed wave function (18.5) does not contain the outgoing wave of angular momentum $l\hbar$ and one has

$$\sigma_{s,l} = \sigma_{r,l} = \pi \lambda_r^2 (2l+1). \tag{19.1}$$

We may now combine these expressions with the semiclassical considerations given at the end of the previous section and derive the scattering and absorption cross sections for the ideal extreme case of *fast neutrons* colliding against a *large black* nucleus. The properties of this ideal body are described by the following relations

$$\left.\begin{aligned} &\text{a)} \quad R \gg \lambda_r, \\ &\text{b)} \quad \eta_l = 0 \quad \text{for} \quad l \leq \frac{R}{\lambda_r}, \\ &\text{c)} \quad \eta_l = 1 \quad \text{for} \quad l > \frac{R}{\lambda_r}. \end{aligned}\right\} \tag{19.2}$$

The first refers to the dimensions of the nucleus. Because of condition (a) it is possible to attribute a sharp edge to the nucleus as indicated by conditions (b) and (c); these express that no outgoing wave of angular momentum $l \leq R/\lambda_r$ is present in the wave function (18.5), while the components of higher angular momenta are not affected by the presence of the nucleus. From (19.1) it follows that

$$\sigma_s = \sigma_r = \pi \lambda_r^2 \sum_{l=0}^{R/\lambda_r} (2l+1) = \pi R^2 \tag{19.3}$$

and therefore

$$\sigma_{\text{tot}} = \sigma_s + \sigma_r = 2\pi R^2. \tag{19.4}$$

The fact that the total cross section is twice the geometric cross section is a well known apparent paradox; a black nucleus absorbs all the neutrons falling on its cross sectional area (so that $\sigma_r = \pi R^2$) and therefore it must produce a shadow behind it. This requires the existence of a diffracted wave the intensity of which, integrated over the solid angle, corresponds to a scattering cross section equal to πR^2.

The relations (19.3) and (19.4) have been confirmed experimentally by measuring the absorption and scattering of fast neutrons (10 to 15 Mev) by heavy nuclei (Sect. 30).

20. The case of pure scattering. Phase shift and scattering length. The opposite extreme is that of pure scattering, i.e. $\sigma_{r,l} = 0$. Since this case is attained only when

$$|\eta_l|^2 = 1 \tag{20.1}$$

we can put, without renouncing generality,

$$\eta_l = e^{2i\delta_l} \tag{20.2}$$

with the *phase shift* δ_l real. Introducing (20.2) in (18.7) and (18.11) one has the well known expressions

$$c_l = \frac{2l+1}{k} e^{i\delta_l} \sin \delta_l, \tag{20.3}$$

$$\sigma_{s,l} = 4\pi \lambda_r^2 (2l+1) \sin^2 \delta_l. \tag{20.4}$$

The meaning of the phase shift δ_l may be recognized by introducing (20.2) in (18.5) and transforming the exponentials into trigonometric functions in both the unperturbed wave (18.4) and the perturbed wave (18.5). The result is that the asymptotic behaviours of their l components are respectively

$$\frac{1}{kr}\sin\left(kr - l\frac{\pi}{2}\right) \quad \text{and} \quad \frac{1}{kr}\sin\left(kr - l\frac{\pi}{2} + \delta_l\right). \tag{20.5}$$

If the neutron-nucleus interaction could be switched on and off at will, δ_l would represent the phase difference observed at great distance from the scatterer between the wave with the interaction on and the wave with the interaction off.

At the end of Sect. 18 we have seen that for slow neutrons the asymptotic expression of the wave function is given by (18.21) with the coefficient of the scattered S waves given by (18.22). Introducing (20.2) in (18.22) we have

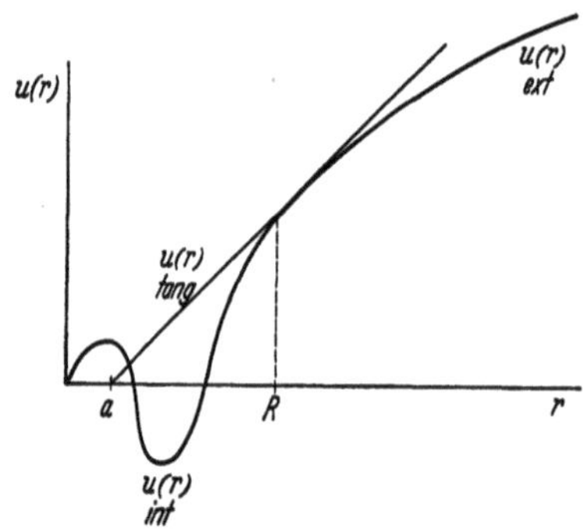

Fig. 6. Geometrical meaning of the scattering length a.

$$c_0 = \frac{1}{k} e^{i\delta_0} \sin\delta_0 \tag{20.6}$$

and from (18.10)

$$\sigma_s = 4\pi \lambda\!\!\!^{-2} \sin^2\delta_0 \tag{20.7}$$

which corresponds to *isotropic distribution* of the scattered neutrons in the c.m.s.

In the limiting case of neutrons of very small energy this expression is very often written in a different form, namely

$$\sigma_s = 4\pi a^2 \tag{20.8}$$

where

$$a = -\lambda\!\!\!^{-}_r \sin\delta_0 \tag{20.9}$$

is called the *scattering length*. Its meaning may be visualized by the following consideration based on the assumption that the neutron-nucleus interaction falls suddenly to zero at $r = R$. Under these conditions the asymptotic expression (20.5) of the perturbed wave function will start to be valid immediately outside of the nucleus so that, for any value of $r \geq R$, we can write

$$u_{\text{ext}}(r) = r\,\psi(r) \propto \sin(kr + \delta_0).$$

Inside the nucleus the behaviour of $u(r)$ is determined by the neutron-nucleus interaction and therefore will be rather complicated (Fig. 6). The equation of the tangent to $u_{\text{ext}}(r)$ for $r = R$ may be written

$$u_{\text{tang}}(r) \propto k\cos(kR + \delta_0)\cdot(r - a')$$

where

$$a' = R - \lambda\!\!\!^{-}_r \tan(kR + \delta_0)$$

represents the intercept on the r axis. According to (18.20) in case of slow neutrons we have

$$kR \ll 1$$

and therefore, if also the phase shift satisfies the condition

$$|\delta_0| \ll 1 \tag{20.10}$$

as it frequently happens, we can write

$$a' \approx R - \lambda\!\!\!^{-}_r (kR + \delta_0) = -\lambda\!\!\!^{-}_r \delta_0 \approx -\lambda\!\!\!^{-}_r \sin\delta_0 = a. \tag{20.11}$$

In conclusion the scattering length a may be interpreted as the intercept on the r axis as shown in Fig. 6, at least as long as

$$\left|\frac{a}{\lambda\!\!\!^{-}_r}\right| = |\sin\delta_0| \ll 1. \tag{20.12}$$

Under the same condition

$$c_0 = \frac{1}{2ik}(e^{2i\delta_0} - 1) \approx \frac{\delta_0}{k} \tag{20.13}$$

and the asymptotic expression of the perturbed wave function (18.21) can be written

$$\psi(r) \to e^{ikz} - \frac{a}{r} e^{ikr}. \tag{20.14}$$

The negative sign has been chosen so that a is positive for a normal scatterer, whose behaviour is in some way similar to that of an impenetrable sphere; one has to keep in mind however that a is in general a function of energy and that in vicinity of a resonance (Sect. 25) it may change sign.

In connection with the remark made above we recall that in the case of an impenetrable sphere of radius R which scatters slow neutrons, one has $-\delta_0 \approx kR \ll 1$ so that (20.7) reduces to

$$\sigma_s = 4\pi \lambda\!\!\!^{-2}_r \sin^2 kR \approx 4\pi R^2. \tag{20.15}$$

The various expressions, from (20.6) on, are sometimes used also when absorption is not negligible, by introducing a complex phase shift

$$\delta_0 = \delta_r + i\,\delta_i. \tag{20.16}$$

Under the assumption that both δ_r and δ_i satisfy conditions similar to (20.10) one has, from (20.13) and (18.10)

$$c_0 = \frac{\delta_r + i\,\delta_i}{k}, \tag{20.17}$$

$$\sigma_s = 4\pi \lambda\!\!\!^{-2}_r (\delta_r^2 + \delta_i^2) = 4\pi (a_r^2 + a_i^2) = 4\pi |a|^2 \tag{20.18}$$

where a_r and a_i are the real and the imaginary parts of the scattering length

$$a = a_r + i\,a_i. \tag{20.19}$$

In this case we have also a reaction cross section which may be computed by means of (18.13). Since

$$|\eta_0|^2 = |e^{2i(\delta_r + i\delta_i)}|^2 = e^{-4\delta_i}$$

one has

$$\sigma_r = \pi \lambda\!\!\!^{-2}_r (1 - e^{-4\delta_i}) = 4\pi \lambda\!\!\!^{-2}_r \delta_i (1 - 2\delta_i) \tag{20.20}$$

which may be written

$$\sigma_r = 4\pi \lambda\!\!\!^{-}_r a_i \left(1 - 2\frac{a_i}{\lambda\!\!\!^{-}_r}\right). \tag{20.21}$$

In agreement with our previous considerations the reaction cross section depends only on the imaginary part of the scattering length.

21. The distinction between potential scattering and resonance scattering. In this section we shall summarize a procedure of transformation of the scattering cross section of neutrons of angular momentum l, Eq. (18.11), which allow a clear distinction between potential scattering and resonance scattering (Sect. 26). The procedure, suggested by WEISSKOPF and coworkers [3] does not involve the adoption of a nuclear model but only the assumption that *the nucleus has a well defined surface* (at $r = R$) which divides space into two regions: the external and the internal region.

In the external region the neutrons move as free particles, and therefore the expansion of their wave function in Legendre polynomials

$$\psi(r) = \sum_{l=0}^{\infty} \frac{u_l(r)}{r} \left(\frac{2l+1}{4\pi}\right)^{\frac{1}{2}} P_l(\cos\vartheta) \tag{21.1}$$

is a known function of the complex numbers η_l (Sect. 18). In fact one can show that

$$u_l(r) = a_l \{u_l^{(-)}(r) - \eta_l u_l^{(+)}(r)\} \tag{21.2}$$

with $u_l^{(-)}(r)$ and $u_l^{(+)}(r)$ two known functions of which the first one corresponds to an incoming, the second one to an outgoing wave. The asymptotic behaviour of $u_l(r)$, as well as the value of the constant a_l, are obtained by comparing (21.2) with (18.5).

In the internal region the wave function has a much more complicated behaviour, the determination of which involves the adoption of a nuclear model for the description of the interaction of the neutrons with nuclear matter.

The condition that the wave function and its derivative are continuous at the surface $r=R$, determines completely the external wave function in terms of the parameters adopted for the description of the structure of the nucleus.

This remark shows that the complex numbers η_l are connected with the condition at the nuclear surface. These conditions may be expressed by imposing that the quantities

$$f_l = R\left(\frac{\frac{d u_l}{d r}}{u_l}\right)_{r=R} \tag{21.3}$$

computed by means of the external wave functions must be equal to the corresponding quantities computed by means of the internal wave functions.

The important remark made by WEISSKOPF and coworkers, in which we are interested here, is that certain very important features of the scattering process may be understood without specifying the internal wave function, but simply by expressing the η_l in terms of the f_l which are then used as the new set of parameters providing a complete description of the behaviour of the nucleus.

If the expression (18.11) is rewritten in the form

$$\sigma_{s,l} = \pi \lambda\!\!\!^{-2}_r (2l+1)\,|A_l|^2, \tag{21.4}$$

the amplitude of the scattered wave

$$A_l = 1 - \eta_l \tag{21.5}$$

is a function of f_l which is easily derived from the explicit forms of $u_l^{(-)}(r)$ and $u_l^{(+)}(r)$.

The general property of A_l in which we are interested here, is that it splits into two parts

$$A_l = A_{\text{pot}}^l + A_{\text{res}}^l \tag{21.6}$$

of which the first *does not depend on* f_l but only on kR, the second one is such a function of f_l that, for $f_l \to \infty$, it goes to zero.

We have noticed before that, because of the conditions regarding continuity of the wave function at the surface of the nucleus, the complex numbers f_l summarize the internal properties of the nucleus. Therefore, it appears justified to indicate the first term, which does not depend on f_l, as *external* (or *potential*) *scattering*, the second term as *internal* (or *resonance*) *scattering*. This distinction is further clarified by noticing that if the nucleus were replaced by a *perfectly impenetrable sphere of radius R*, the condition that the wave function had to fulfill at the nuclear surface would be $u_l(R)=0$, which implies, according to (21.3), $f_l \to \infty$.

In conclusion, combining (21.4) and (21.6) one can write in general

$$\sigma_{s,l} = \pi \lambda\!\!\!^{-2}_r (2l+1)\,|A^l_{\text{pot}} + A^l_{\text{res}}|^2 \tag{21.7}$$

where the potential scattering arises from the small penetrability of the nucleus for the incident neutrons and is usually described in terms of an impenetrable sphere, while the internal or resonance scattering depends on the internal structure of the nucleus. In Sect. 26 we shall come back to this point after having introduced a specific nuclear model.

One has to notice that the separation of the scattering amplitude A_l into a potential and a resonance scattering depends on the assumption of the existence of a well defined nuclear surface and therefore it is only approximate to the extent that R is indetermined.

Furthermore such a separation has only a theoretical meaning since the only quantity that may be measured experimentally is the total scattering amplitude (21.6).

VII. The velocities reached by slow neutrons.

22. Thermal effects. The experiments described up to now give sufficient evidence in favour of the interpretation of the effect of hydrogeneous substances in terms of a considerable reduction of the velocity of the neutrons. The problem of the actual amount of such a reduction was however still completely open. One could think, of course, that the slowing down process goes on until the natural limit imposed by the thermal agitation of the moderator is reached, i.e. until the neutrons have an average kinetic energy equal to $\frac{3}{2}kT$ $(=\frac{3}{2}\times 0.025$ ev $=$ 0.0375 ev for $T=293^\circ$ K). Such an assumption appeared rather disconcerting at first sight, but the quantitative analysis given by FERMI[1], showed that it was a reasonable one. As we will see in Sects. 63 and 64, the energy of the neutron decreases exponentially with the number of collisions and the average number of collisions against protons necessary to reduce it from the initial value E_0 to a value E, is given by the expression

$$\bar{n} = \ln \frac{E_0}{E} + 1. \tag{22.1}$$

From Eq. (22.1) it follows, for instance, that $\bar{n}=19$ for $E_0=2$ Mev and $E=0.04$ ev i.e. a number large but not exceedingly large.

This does not mean of course that the process of slowing down of neutrons in a hydrogeneous moderator goes on until the neutrons have actually reached thermal equilibrium with the medium. As we have already noted in Sect. 13, whether this happens or not will depend on the value of the ratio of the cross section for elastic collision against protons, to the absorption cross section of all nuclei present in the medium, and on its variation with velocity.

In order to check this important point FERMI and coworkers attempted to find a "thermal effect" i.e. they looked for a change in the activation of a slow neutron detector (actually Ag) due to a change in the temperature of the moderator[2]. This experiments, however, gave a negative answer to the question on account of the particular experimental conditions that had been chosen.

[1] See footnote 4, p. 19.

[2] B. PONTECORVO. Nuovo Cim. **12**, 211 (1935); see also footnote 4, p. 19.

The search for the possible existence of a thermal effect was pursued by MOON and TILLMAN[1], who, after a few attempts, succeeded in obtaining a clear result.

The experimental set-up used by these authors is shown in Fig. 7. A Dewar vessel V was placed in a cavity excavated in a large block of paraffin containing a neutron source (Rn α + Be). A cylinder C of paraffin whose external and internal diameters were 3.7 and 1.6 cm respectively was placed inside the Dewar.

A thin cylinder of Ag was placed inside the cylinder C and after irradiation its activity was measured by means of a G-M counter. The experiment consisted in comparing the activity induced by slow neutrons in the silver detector irradiated under two different conditions: firstly when the cylinder of paraffin C was at room temperature ($T = 290°$ K), and secondly, using liquid oxygen, when at a temperature of 90° K. The result obtained by MOON and TILLMAN was that at the lower temperature the activity was higher by a factor

$$J_{90°} = 1.26 \pm 0.04. \qquad (22.2)$$

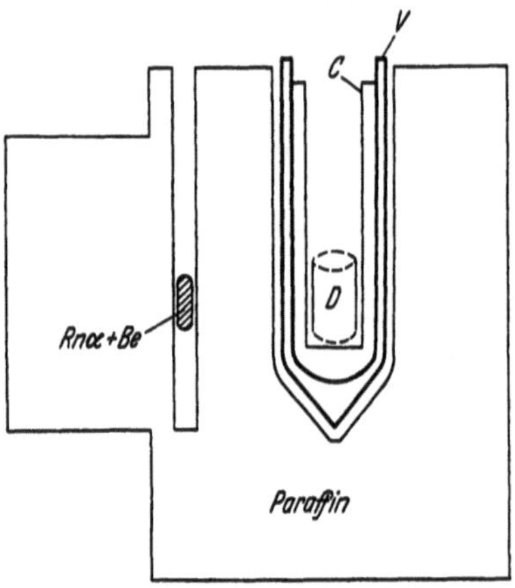

Fig. 7. Experimental set-up used by MOON and TILLMAN to prove the existence of a thermal effect on the activation of various elements by slow neutrons.

An indication of a small thermal effect had also been found independently by other authors[2]; the inconclusive result was due, as in the experiment of FERMI and co-workers, to the ill-chosen experimental conditions.

The first result of MOON and TILLMAN, immediately confirmed by various authors[3] proved that the slowing down process reduces the energy of the neutrons to values close to those corresponding to thermal agitation. Furthermore it opened the way to an understanding of the conditions under which thermal effects can be observed. The activation of a detector, like Ag, can change with a change in the temperature of the paraffin for two different reasons: (a) the capture cross section of Ag depends on the velocity of the neutrons, which depends on the temperature of the moderator; (b) neutron capture by the nuclei of the moderator also depends on the temperature, and therefore a change in temperature affects the intensity of the neutrons reaching the detector.

A detailed consideration of the problem shows that, if the irradiation is done inside a rather extended moderator in which slow neutrons are produced uniformly, the two effects (a) and (b) compensate one another exactly, provided all nuclei present in the moderator, as well as in the detector, have capture cross sections following the $1/v$ law. The experiments mentioned above whose result was negative, or almost negative, had been performed under conditions rather close to this extreme case. In the experiment of MOON and TILLMAN the compensation of the effects (a) and (b) does not occur because the layer of paraffin, whose temper-

[1] P. B. MOON and J. R. TILLMAN: Nature, Lond. **135**, 904 (1935). — Proc. Roy. Soc. Lond., Ser. A **153**, 476 (1936).

[2] J. R. DUNNING, G. B. PEGRAM, D. P. MITCHELL and G. A. FINK: Phys. Rev. **47**, 888 (1935).

[3] E. AMALDI, O. D'AGOSTINO, E. FERMI, B. PONTECORVO and E. SEGRÈ: Ric. sci. **6** (1), 581 (1935).

ature is changed, is rather thin: so it more influences the velocity of the neutrons than their density.

Once the mechanism producing thermal effects is clear, one can choose experimental conditions such that the effect (a) is isolated. This is done by measuring as a function of the temperature of the moderator the absorption coefficients of various substances for slow neutrons.

An effect of this type means that (1) the velocity of the neutrons emerging from the moderator depends on its temperature, and (2) the capture cross section of the substance under investigation has a cross section which depends on the neutron energy at least in the region of thermal velocities.

Measurements of this type were done by various authors[1,2], some of whom thought it possible to use the corresponding results in order to verify the validity in various elements, of the $1/v$ law. According to this law, if the neutrons are in thermal equilibrium with the moderator, the absorption coefficient must vary with the absolute temperature as $T^{-\frac{1}{2}}$. But the interpretation of experiments of this type is rather complicated, because, as we will see in Sect. 107, the problem of whether the neutrons are in a more or less perfect thermal equilibrium with the moderator gives rise to complications, especially at low temperatures.

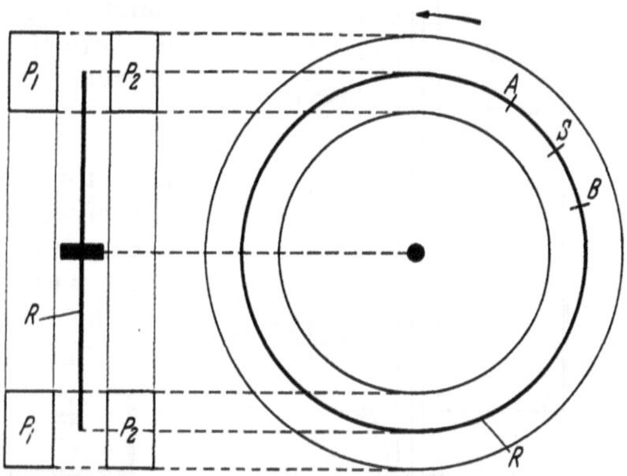

Fig. 8. Scheme of the mechanical experiment used by the Rome group for demonstrating the low velocity reached by slow neutrons.

A very important result of all these early experiments on thermal effects was that they demonstrated the different behaviour of various elements. We shall come back to this point in Sect. 24.

23. Mechanical experiments. As a consequence of the first positive result of Moon and Tillman, the Rome group, was encouraged to attempt to demonstrate the low velocity of slow neutrons by means of a mechanical experiment rather similar to those employed to measure molecular velocities[3].

A source of neutrons (Rnα + Be), S in Fig. 8, was fixed on the edge of a steel wheel R, of 80 cm diameter, together with two slow neutron detectors A and B (Mn or V in different experiments) placed symmetrically with respect to the source at distances of 5 cm from it. The wheel could rotate, at about 1600 turns/min, between two paraffin rings P_1 and P_2, of section 10×20 cm^2, placed 6 cm from the plane of the wheel. Only the relative motion is of importance for the discussion of the experiment. Thus, we can describe the situation as if the source and the detectors were at rest, and the paraffin rings were moving, at a velocity $-\boldsymbol{V}$ of 67 m/sec with respect to them. The neutrons emitted from the source are slowed down in the paraffin and to some extent dragged along by it. The result is an alteration in their spatial distribution with respect to the stationary

[1] See footnote 2, p. 36.
[2] C. H. Westcott and H. Niewodniczanski: Proc. Cambridge Phil. Soc. **31**, 617 (1935). — P. N. Powers, G. A. Fink and G. B. Pegram: Phys. Rev. **49**, 650 (1936). — G. A. Fink. Phys. Rev. **50**, 738 (1936).
[3] See footnote 3, p. 36.

distribution that would be obtained if everything were at rest. One can show[1] that the effect of the movement is the same that one would have by displacing the source a distance $\Delta \boldsymbol{s} = -\boldsymbol{V}\bar{t}$ with respect to its position midway between A and B. In the preceding expression $\bar{t}$ is the average time that neutrons spend in the paraffin before being captured or diffusing out of it.

The result of the experiment was that the activity of the detector following the source in their common motion, was 10% higher than the activity of the detector preceding the source. This fact is a clear proof of the low velocity reached by the neutrons in paraffin and allows the determination of $\bar{t}$. In fact, the same variations of the activities of the two detectors was obtained when everything was at rest, by displacing the source by $\Delta s \approx -0.5$ cm. This value, combined with that of the linear velocity of the source, gives

$$\bar{t} \sim 10^{-4}\,\text{sec}, \tag{23.1}$$

i.e. a time such that a thermal neutron ($v = 2200$ m/sec) would traverse a total path of about 20 cm. But the neutrons are emitted by the source with a much higher velocity and therefore the total path they traverse in 10^{-4} sec is

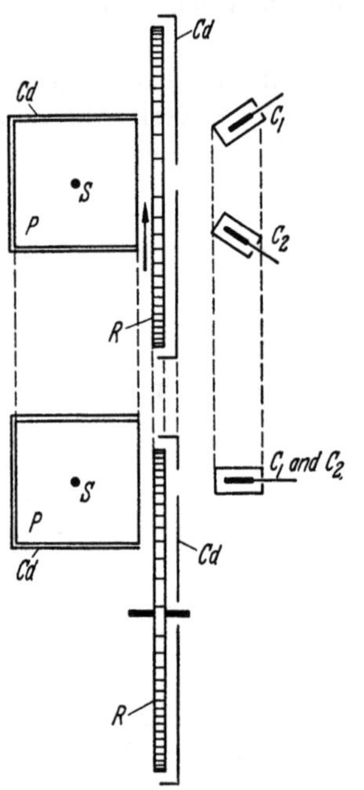

Fig. 9.

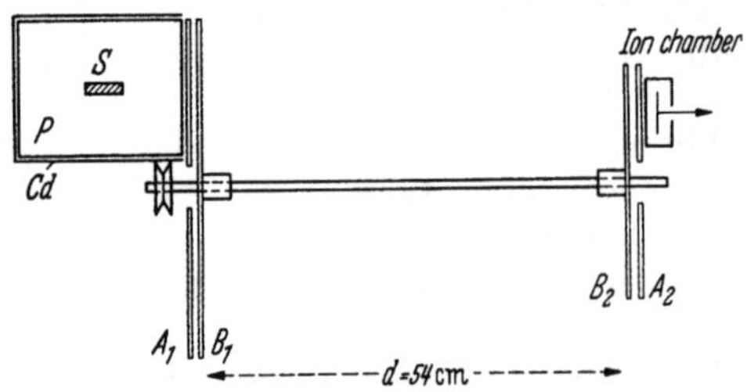

Fig. 10.

Fig. 9. Scheme of the mechanical experiment made by FRISCH and SORENSEN for demonstrating the low velocity reached by slow neutrons.

Fig. 10. The first velocity selector of the Columbia University group.

certainly appreciably longer. If one takes a mean free path for elastic collision of about 1 cm, one can conclude that, according to Eq. (22.1), the average number of collisions is large enough to reduce the velocity of the neutrons to that of thermal agitation[2].

Another very elegant mechanical experiment was done shortly afterwards by FRISCH and SORENSEN[3]. The neutrons emitted by a source S and slowed down by a block of paraffin P, cross the edge of a wooden wheel R, 2 cm thick and 50 cm in diameter, rotating at an angular speed of 4000 revolutions per minute (Fig. 9). A thick layer of Cd with a small hole limits the regions from which slow neutrons emitted from the surface of the wooden wheel can reach two ionization chambers C_1, C_2 filled with BF_3. C_1 and C_2 are in symmetrical positions with respect to the hole in the Cd screen, so that they give exactly the same number

[1] E. AMALDI: Phys. Z. **38**, 692 (1937).
[2] A similar experiment has been done also P. HARTECK and FR. KNAUER: Naturwiss. **24**, 558 (1936).
[3] O. R. FRISCH and E. T. SORENSEN: Nature, Lond. **136**, 258 (1935).

of counts when the wheel is at rest. But when the wheel is rotating, the ionization chamber C_1 displaced in the direction of motion, gives a number of counts 4.2% higher than C_2, proving again that the drift of the neutrons in a moderator moving at a velocity of only 90 m/sec is appreciable.

These two experiments give however only a qualitative indication of the low velocity reached by neutrons.

DUNNING, PEGRAM, FINK, MITCHELL and SEGRÈ[1] built the first "velocity selector" for slow neutrons, based on the same principle used for molecular beams[2] which, in its turn, is an adaptation of the method used by FIZEAU in his classical experiment for the determination of the velocity of light.

Fig. 10 shows schematically the set-up used by the Columbia University group. The neutrons emitted by the source S and slowed down by the paraffin block P are detected by means of an ionization chamber filled with BF_3. The paraffin P, is screened, with a thick layer of Cd, in all directions except towards the velocity selector. This consists of four duraluminium discs of which two (A_1, A_2) are fixed and two (B_1, B_2) rotate, being mounted, at a distance d from each other, on the same shaft. This arrangement rotates at an angular speed ω that can be varied up to 5000 revolutions per minute. Fixed, at regular intervals, on the surface of each of the four discs are n cadmium sectors so that the pair $A_1\, B_1$ and the pair $B_2\, A_2$ behave like Fizeau wheels for slow neutrons, i.e. as shutters which are opened and closed periodically with a frequency $f=\omega n$.

The energy of the neutrons passed or stopped by the velocity selector depends on the separation d of the discs, the frequency f of the bursts of slow neutrons passed by the first shutter $A_1\, B_1$ and the relative phase with which the second shutter $B_2\, A_2$ is opened and closed with respect to $A_1\, B_1$. The velocity v of the neutrons transmitted appears in the theory of the instrument through the corresponding *time of flight*

$$t = \frac{d}{v} \tag{23.2}$$

that they take to cross the distance d from B_1 to B_2.

The details into which a slow neutron spectrum can be analysed by an instrument of this type are determined by its "resolution" in energy $\Delta E/E$ or in velocity $\Delta v/v$, which can be expressed by means of Eq. (23.2) in terms of the relative uncertainty in the determination of the time of flight, $\Delta t/t$. Here Δt is given by the sum of the intervals of time during which the neutrons can pass through the two shutters $A_1 B_1$ and $B_2 A_2$. In spite of the fact that the first velocity selector, built at Columbia University, had a very poor resolution, it gave a few very important results. Furthermore it represented the first step towards the development of other mechanical and electrical neutron spectrometers which can be considered high resolution instruments.

In the following we shall refer very often to experimental results obtained with these neutron spectrometers. A detailed description of them and of their performances is given by J. RAINWATER in Vol. XL; the main points of their behaviour are summarized in Ref. [*12*], [*18b*], [*15b*].

Returning to the problem of the velocity reached by slow neutrons, measurements performed[1] with the instrument of Fig. 10 as well as others made some time later with an improved form of it[3], showed that, at room temperature

---

[1] J. R. DUNNING, G. B. PEGRAM, G. A. FINK, D. P. MITCHELL and E. SEGRÈ: Phys. Rev. **48**, 704 (1935).

[2] I. ESTERMANN, R. FRISCH and O. STERN: Z. Physik **73**, 348 (1931).

[3] G. A. FINK: Phys. Rev. **50**, 738 (1936).

($T = 300°$ K), the slow neutron spectrum emitted from a block of paraffin was represented, at least roughly, by a Maxwellian distribution

$$d\varphi_{\text{th}} = v \cdot d n_{\text{th}} \tag{23.3}$$

where

$$d n_{\text{th}} = \frac{4}{\sqrt{\pi}} N \left(\frac{m}{2kT}\right)^{\frac{3}{2}} e^{-\frac{m v^2}{2kT}} v^2 \, dv = \frac{4}{\sqrt{\pi}} \frac{N}{v_0^3} e^{-\frac{v^2}{v_0^2}} v^2 \, dv, \tag{23.4}$$

and N is the total number of neutrons present in the spectrum. As in the case of molecular beams, the spectrum of the neutrons leaving the surface of the moderator, corresponds to the neutron flux (23.3), not to the neutron density inside the moderator (23.4).

In similar experiments[2] performed with the paraffin cooled to 90° K, the experimental points were found to lie between the Maxwellian distributions calculated for $T = 90°$ K and $T = 300°$ K. This result was interpreted as an indication that slow neutrons reach thermal equilibrium at room temperature but perhaps not at lower temperatures.

How far we can expect that the velocity distribution of the neutrons is well represented by a Maxwellian distribution needs however more detailed considerations which will be given in Sect. 107. For the time being we can assume that at room temperature the velocity distribution of that part of the spectrum of slow neutrons which is absorbed by cadmium, is at least roughy Maxwellian.

A Maxwell spectrum extending over the entire range of velocities, i.e. from $v = 0$ to $v = \infty$, deserves special consideration as a very interesting limiting case. The *most probable velocity* of the neutron density (23.4) is

$$v_0 = \sqrt{\frac{2kT}{m}} = (1.648 \times 10^8 \times T)^{\frac{1}{2}} \text{ m/sec}. \tag{23.5}$$

At room temperature, 20° C,

$$v_0 (293° \text{ K}) = 2198 \text{ m/s} \approx 2200 \text{ m/sec} \tag{23.6}$$

to which there corresponds a neutron energy

$$\tfrac{1}{2} m v_0^2 = kT = 0.0252 \text{ ev} \approx 0.025 \text{ ev}. \tag{23.7}$$

The *average velocity* of the density distribution (23.4) is

$$\bar{v} = \frac{2}{\sqrt{\pi}} v_0 = 1.128 \, v_0 \tag{23.8}$$

and its *root mean square velocity*

$$\sqrt{\overline{v^2}} = \sqrt{\tfrac{3}{2}} \, v_0 = 1.225 \, v_0 \tag{23.9}$$

to which corresponds the well known average kinetic energy

$$\tfrac{1}{2} m \overline{v^2} = \tfrac{3}{2} kT. \tag{23.10}$$

The *most probable energy* of the density distribution is $\frac{1}{2} kT$ and the *most probable velocity* of the neutron flux distribution (23.3) is $(\frac{3}{2})^{\frac{1}{2}} v_0$.

It became costumary to refer to the neutron in a Maxwell distribution to the temperature T, as "kT neutrons" although the average kinetic energy is given, according to the theorem of equipartition of energy, by (23.10).

This convention originates from the following ideal experiment. The average absorption cross section of an extremely thin $1/v$ absorber (Sects. 13 and 15) is measured in a Maxwellian beam of neutrons extending from $v=0$ to $v=\infty$, by means of an extremely thin $1/v$ detector. The condition that the thickness of the absorber and the detector must be both extremely thin is necessary to avoid that the incident spectrum is modified (actually hardened) by the $1/v$ absorption law.

In order to get $\bar{\sigma}_a$ the activation of the detector must be measured without and with the absorber in the beam; here *activation* means a quantity proportional to the number of processes produced in the detector by the incident neutron in a given time. If $\mathfrak{N}_d$ and $\mathfrak{N}_a$ are the numbers of atoms per cm^3 in the detector and in the absorber respectively, and

$$\sigma_d = \frac{k_d}{v}, \qquad \sigma_a = \frac{k_a}{v}$$

the corresponding cross sections, one has the following expressions for the activation without and with absorber

$$a_0 = \mathfrak{N}_d \int \frac{k_d}{v}\, d\varphi_{\text{th}} = \mathfrak{N}_d\, k_d \int d n_{\text{th}}, \tag{23.11}$$

$$a_a = \mathfrak{N}_d \mathfrak{N}_a \int \frac{k_d}{v}\,\frac{k_a}{v}\, d\varphi_{\text{th}} = \mathfrak{N}_d \cdot k_d \mathfrak{N}_a \int \frac{k_a}{v}\, d n_{\text{th}}. \tag{23.12}$$

Introducing (23.3) and (23.4) it follows that

$$\bar{\sigma}_a = \frac{1}{\mathfrak{N}_a}\,\frac{a_a}{a_0} = \frac{2}{\sqrt{\pi}}\,\frac{k_a}{v_0}, \tag{23.13}$$

i.e.

$$\bar{\sigma}_a = \frac{2}{\sqrt{\pi}}\,\sigma_a(v_0) = 1.128\,\sigma_a(v_0). \tag{23.14}$$

The commonly listed values of the *absorption cross section for thermal neutrons* correspond to

$$\sigma_{\text{th}} = \sigma_a(kT) = \sigma_a(v_0) \tag{23.15}$$

with v_0 given by (23.6) [*12*], [*14*]. Eq. (23.14) shows that the average cross section, obtained from transmission experiments of a Maxwellian beam through a thin $1/v$ absorber with a $1/v$ detector, is 1.128 times $\sigma_{a\,\text{th}}$. We will come back to this point in Sect. 135.

Two remarks must be added. The first is that, whenever the experimental conditions correspond to those specified above, from a measurement of $\bar{\sigma}_a$ the value of v_0 can be derived and from this, making use of (23.5), the neutron temperature (Sect. 107).

The second remark is that, according to (23.11), *the activation of a bare $1/v$ detector is proportional to the neutron density integrated over the spectrum.* This remark is of general validity, i.e. independent of the shape of the neutron spectrum, as long as the thickness of the detector is infinitely small; it is due to the fact that the factor v appearing in the expression of the neutron flux (23.3) is compensated by the factor $1/v$ of the detector cross section.

VIII. Selective capture of slow neutrons.

24. The experimental evidence for pronounced selective absorptions. We have already mentioned in Sect. 22, that one of the first indications that different elements have different behaviour with respect to the absorption of slow neutrons was found in the study of thermal effects. This, as well as the results of other experiments that we shall describe below, was the prelude to the discovery of a very important fact: nuclei absorb selectively neutrons of well defined velocities determined by the existence of certain energy levels of the nucleus produced in the capture of the incident neutron. This selective absorption of neutrons has many features in common with the absorption of light of well defined wavelengths by atoms; in particular, the widths of the absorption lines are, for slow neutrons, extremely small indicating an unexpected stability of the corresponding energy levels of the nuclei.

After the discovery that slow neutrons can be absorbed by nuclei with a very high cross section, the considerations indicated in Sect. 22 had been developed and had allowed the prediction of the $1/v$ law. But if this law is valid in general, then all nuclei should show the same behaviour in both thermal experiments and absorption experiments.

In this connection, we must mention a mechanical experiment, different from those described in Sect. 23, done at Columbia University[1], in order to directly verify the $1/v$ law in Cd, Ag and B. According to this law, the absorption probability (13.1) is independent of the relative velocity v_r of the neutron and the nucleus. Therefore one can verify this law by making use of an absorber which moves with a velocity $\boldsymbol{V}$ not negligible with respect to the velocity $\boldsymbol{v}$ of thermal agitation of the neutrons, so that the relative velocity

$$\boldsymbol{v}_r = \boldsymbol{v} + \boldsymbol{V}$$

can be changed at will, by changing $\boldsymbol{V}$. In particular if the absorbing power of the material concerned is measured in two experiments differing only in the sign of the absorber velocity $\boldsymbol{V}$, the same value ought to be obtained if the $1/v$ law is valid, while a variation of absorbing power will directly give the variation of the absorption probability P in the range of velocities $|\boldsymbol{v}-\boldsymbol{V}|$, $|\boldsymbol{v}+\boldsymbol{V}|$.

In the experimental set-up a beam of slow neutrons, emitted from the surface of a paraffin block and detected by a BF_3 ionization chamber, crossed near the edge and at an angle of about 25° with its face, a steel disc carrying on its surface a layer of the absorber. The disc could rotate in both directions at 6000 turns/min corresponding to a linear velocity of 140 m/sec. With Cd (0.02 mm thick) these authors observed a small variation of absorbing power, indicating that the cross section of this element varies less rapidly with velocity than is expected according to the $1/v$ law. On the contrary B and Ag gave results in accordance with the $1/v$ law[2].

But the first to notice an indication of a selective absorption of slow neutrons, were probably BJERGE and WESTCOTT[3] followed immediately by MOON and TILLMAN[4] who, making use of various absorbers (I, Ag, Cu) and detectors (Ag, Cu, I, Rh), gave good evidence for the following rule: the absorbing power of an element depends on the detector used and is a maximum when the same element is used as detector. This is just what is to be expected if the various elements show a preferential absorption for neutrons of certain velocities, different for

[1] F. RASETTI, E. SEGRÈ, G. A. FINK, J. R. DUNNING and G. B. PEGRAM: Phys. Rev. **49**, 104 (1936).

[2] F. RASETTI, D. P. MITCHELL, G. A. FINK and G. B. PEGRAM: Phys. Rev. **49**, 777 (1936).

[3] T. BJERGE and C. H. WESTCOTT: Proc. Roy. Soc. Lond., Ser. A **150**, 709 (1935).

[4] B. P. MOON and J. R. TILLMAN: Nature, Lond. **135**, 904 (1935). — J. R. TILLMAN and P. B. MOON. Nature, Lond. **136**, 106 (1935).

different elements. These results were immediately confirmed by other authors[1], but the observed variation of absorbing power was rather small and did not show any very obvious effect.

The next step was taken by FERMI and AMALDI[2] and independently by SZILARD[3]. The first authors, after verifying the rule of MOON and TILLMAN, started a detailed investigation of the absorption curves of the neutrons emitted from the surface of a block of paraffin.

Their main results were the following:

(a) The absorption curves of Cd with Ag and Rh detectors, although different, show the same general behaviour, namely that the radiation can be analysed in both cases into two components: one strongly absorbed in Cd, designated the "group *C*", the other very weakly absorbed in Cd.

(b) The absorption curve of Ag with a Ag detector shows also a similar behaviour i.e. it shows, as already recognized by MOON and TILLMAN, the existence of a component strongly absorbed by Ag.

(c) In order to verify whether the component strongly absorbed in silver belonged to the spectral region strongly absorbed by Cd, they measured the absorption curve of Ag with a Ag detector, using the neutrons filtered through a layer of Cd thick enough (0.27 g/cm^2) to eliminate almost completely group *C*.

The result of this third experiment, repeated immediately[4] with Rh and In instead of Ag, was that the rule of MOON and TILLMAN was strongly enhanced by the presence of the Cd filter. This obviously means, that the spectral regions selectively absorbed by these elements, do not belong to the spectral region strongly absorbed by Cd (group *C*).

Furthermore, for each element, the thinner the detector the more evident was the selective absorption of the radiation filtered with cadmium, giving rise to unexpectedly large values for the initial absorption coefficients.

SZILARD had the same idea of studying the selective absorption of the neutrons previously filtered through cadmium and showed that In and I had a behaviour similar to that described above for Ag. Some time later a similar behaviour was found also for Au by FRISCH, HEVESY and MCKAY[5].

From these first experimental results it was possible to conclude[4] that while each of the elements mentioned above seems to have a characteristic spectral region of absorption different, at least in general, for each element, all of them have a common spectral region of absorption, namely the component strongly absorbed by cadmium (group *C*). In some cases of course the characteristic spectral regions could in part coincide as, for instance, happens for Rh and In[4].

Once the existence of selective absorptions of slow neutrons by nuclei had been clearly demonstrated, the question of determining the corresponding velocities was clearly the natural and important next step. Today the answer to this question is given by the slow neutron spectrometer mentioned in Sect. 23.

However, perhaps it still is of some interest to show how it was possible to reach a considerable number of conclusions long before such instruments were sufficiently developed to be of practical use in these problems. Let us first

[1] L. ARSIMOVITCH, I. KOURTCHATOV, G. LATYSCHEW and W. CROMOW: Phys. Z. Sowjet. **8**, 472 (1935). — B. PONTECORVO: Ric. sci. **6** (2), 145 (1935). — L. N. RIDENOUR, K. SHINOHARA and D. M. YOST: Phys. Rev. **47**, 318 (1935). — D. P. MITCHELL, J. R. DUNNING, E SEGRÈ and G. B. PEGRAM: Phys. Rev. **48**, 774 (1935).

[2] E. FERMI and E. AMALDI: Ric. sci. **6** (2), 344 (1935).

[3] L. SZILARD: Nature, Lond. **136**, 951 (1935).

[4] L. FERMI and E. AMALDI: Ric. sci. **6** (2) 443 (1935).

[5] O. R. FRISCH, G. HEVESY and H. A. C. MCKAY: Nature, Lond. **137**, 149 (1936).

consider group C. From the simple fact that a velocity distribution roughly Maxwellian was obtained using the velocity selector described in Sect. 23, in which the selection of neutrons is done by means of Cd absorbers, one can conclude[1] that the spectral interval strongly absorbed in Cd covers, at least in part, the thermal energy region. This argument was strengthened and confirmed by the observations of PREISWERK and v. HALBAN[2] in the case of Ag and of RASETTI and FINK[3] in the case of Ag, Rh, In, I that while all these elements show a thermal effect of the type of that investigated by MOON and TILLMAN (see Sect. 22), no effect at all is observed if the same elements are screened with Cd. This means that the characteristic absorption lines of these elements are above the thermal energy region, which, on the contrary, is practically all included in the region absorbed by Cd. This conclusion is correct. We know today that a cadmium layer 0.44 g/cm² (0.02 inch) thick absorbs practically completely the neutrons of a collimated beam whose energy is below 0.4 ev, while the others pass through it almost unaffected (see Sect. 26). The original phenomenological terminology which indicated as group C the spectral region strongly absorbed by Cd, can now be replaced by specifying the thickness [see Eq. (26.20) below] of the Cd absorber used (for a collimated beam): group C covers the thermal region in which the velocity distribution is roughly Maxwellian but it differs from it, because of the capture of neutrons in the moderator, and because of the continuous supply of other neutrons slowed down from higher energies. These effects produce an appreciable change in the energy spectrum, with respect to the Maxwell law, especially in the energy region between 0.1 and 0.4 ev where the energy spectrum tends to decrease as $1/E$ (see Sect. 107).

The problem of determining the energies corresponding to the characteristic absorption lines and the widths of the latter, was also attacked and roughly solved by making use of a technique based on the diffusion properties of neutrons in paraffin[1,4].

From this study it was possible to show that a neutron whose energy is in the characteristic spectral region of Ag, undergoes, in the majority of cases, in a single collision against a proton, a change in energy large enough to remove it from this energy region. Similar results were obtained for other elements. This means that the corresponding absorption lines are very narrow: as we shall see in Sect. 26, their widths in intermediate and heavy nuclei are generally about $^1/_{10}$ ev. On the contrary the neutrons of group C, being mainly thermal neutrons, undergo a number of collisions of the order of 200 without changing their absorption properties: after these collisions they simply disappear, being captured by the nuclei of the moderator. It was also shown by the same authors[1,4], as well as by PREISWERK and v. HALBAN[5], that in some cases it is possible to follow the transformation through collisions in the paraffin of the neutrons lying in the energy range of the characteristic absorption of one element, in neutrons selectively absorbed by another element. Experiments of this type made it possible to order the absorption lines of the various elements in a series of decreasing energy.

---

[1] E. AMALDI and E. FERMI: Ric. sci. **7** (1), 56 (1936).

[2] P. PREISWERK, H. v. HALBAN: Nature, Lond. **136**, 1027 (1935). — H. v. HALBAN and P. PREISWERK: Nature, Lond. **136**, 951 (1935).

[3] F. RASETTI and G. A. FINK: Phys. Rev. **49**, 642 (1936).

[4] E. AMALDI and E. FERMI: Ric. sci. **7** (1), 223, 310, 393, 454 (1936). — Phys. Rev. **50** 899 (1936).

[5] P. PREISWERK and H. v. HALBAN: C. R. Acad. Sci. Paris **202**, 840 (1936). — H. v. HALBAN and P. PREISWERK: Helv. phys. Acta **9**, 318 (1936).

These methods for the determination of the energies and widths of the absorption lines will be briefly mentioned again in Sects. 82α and 90α.

Although very interesting from a methodological point of view, they were soon superseded by others. In particular the determination of the energy of the absorption lines became possible on the basis of an important assumption about the velocity dependence of the absorption in boron which will be mentioned at the end of Sect. 27.

25. The theoretical interpretation of the selective absorption of slow neutrons. In this section we shall very briefly recall the line of ideas about nuclear structure developed in connection with the discovery of selective absorption of slow neutrons in intermediate and heavy nuclei.

These selective absorptions have the character of very narrow absorption lines because their width, of the order of 10^{-1} ev, has not to be compared with the energy of the incident neutron, but with the excitation energy of the nucleus produced in the capture of the neutron. This is of the order of 7 Mev so that the relative width turns out to be about 10^{-8}, i.e. a value at least as small as that found for a narrow line of the optical absorption spectrum.

Another very important experimental result published a few months earlier by DUNNING, PEGRAM, FINK and MITCHELL[1] was that the very large cross section of cadmium for slow neutrons was not due to neutron scattering which was almost undetectable, but practically entirely to radiative capture.

All these experimental results could not be explained in terms of the current ideas, according to which the neutron-nucleus interaction was represented, at least as a first approximation, by means of a potential well. With such a scheme it is possible to find special values of the energy for which the cross section for radiative capture is very large with respect to the normal case. But if the conditions are such that they give a large value of the cross section at very low energy, this would remain very large in an energy interval of at least 10^3 ev, inside which the $1/v$ law was expected to hold.

Furthermore the absorption was expected to be associated with an appreciable scattering.

These difficulties were treated and solved in two independent papers by BREIT and WIGNER[2], and BOHR[3]. The first authors, guided by the analogy with certain molecular resonances, assumed that the neutron, once it had penetrated into the nucleus, could give part of its energy to one of the nuclear particles and thus create an excited metastable state whose mean life was long enough to produce the desired small width of the level. They derived an expression for the cross section for radiative capture which is valid when the neutron is captured in a single resonance level, and which is usually indicated as the one-level formula of BREIT and WIGNER. According to this formula, given by Eq. (26.2) of Sect. 26, the energy dependence of the capture cross section $\sigma_a(E)$ is determined by three parameters[4]. The first is the energy E_R that the incident neutron must have in order to be at resonance, the second is the maximum value σ_{a0} reached by the cross section when the neutron is at resonance, and the third is

[1] J. R. DUNNING, G. B. PEGRAM, G. A. FINK and D. P. MITCHELL: Phys. Rev. **48**, 265 (1935).

[2] G. BREIT and F. WIGNER: Phys. Rev. **49**, 519 (1936).

[3] N. BOHR: Nature, Lond. **137**, 344 (1936). — Naturwiss. **24**, 241 (1936).

[4] We use here the kinetic energy E of the neutron in the l.s. instead of the kinetic energy E_r of the relative motion of the two colliding particles. This approximation is permitted only in the case of neutrons colliding against nuclei of intermediate and heavy mass at rest. see Eq. (25.3) below and Sect. 86.

the halfwidth of the level Γ, which is connected to the corresponding mean life by the usual relation

$$\Gamma = \frac{\hbar}{\tau}. \tag{25.1}$$

The one level formula was soon extended by BETHE and PLACZEK[1] to the case of many levels and since that time many derivations of similar expressions have been given by various authors [5], [3].

The paper of BOHR, presented at the Danish Academy the 27-th of January 1936, is also based on the idea of an intermediate level. But instead of postulating its existence and of deriving the expression for the cross section from it, BOHR develops a new conception of the mechanism of nuclear processes which justifies the existence of many excited levels in nuclei of intermediate and high atomic mass number, whose mean life is sufficiently long to match the experimental results. Introducing numerical values in Eq. (25.1) one gets

$$\tau_{\text{sec}} = \frac{6.6 \times 10^{-16}}{\Gamma_{\text{ev}}} \tag{25.2}$$

which, for $\Gamma \approx 0.1$ ev, gives a mean life of about 10^{-14} sec i.e. a time incomparably longer than that taken by a neutron to cross a nucleus once ($\approx 10^{-21}$ sec).

This means that the intermediate level must have an extraordinary stability, a fact that, according to BOHR'S argument, can be understood on account of the strong nuclear interactions and of the compactness of nuclear matter. As soon as the incident neutron has entered the nucleus it starts to collide with the constituent nucleons. As a consequence its energy is rapidly shared among many particles, no one of which acquires an energy large enough to leave the nucleus. This situation can last for a long time, until one of the particles, through a momentary fluctuation which concentrates in it a sufficient energy, flies out of the nucleus.

According to this point of view any nuclear process A_1 (a_1, a_i) A_i (see Sect. 12) can be described as happening *in two independent steps*: the first is the formation of an *excited compound nucleus* C as a consequence of the capture of particle a_1 by the nucleus A_1; the second is the disintegration of C into a residual nucleus A_i and an emitted particle a_i ($i = 1, 2, \ldots$).

The excited state in which the compound nucleus C is formed, is characterized by certain values of the excitation energy w_c, of the total angular momentum J_c and of parity. The excitation energy is given by the expression[2]

$$w_c = E_r + S_{a_1} = E_1^* + \mathsf{E}_1^* + S_{a_1} = E_1 \frac{M_1}{M_1 + m_1} + S_{a_1} \approx E_1 + S_{a_1} \tag{25.3}$$

where S_{a_1} is the separation energy of the incident particle a_1 from the compound nucleus (for a nucleon $S_{a_i} \approx 7$ Mev), E_1^* and E_1^* the kinetic energies of the incident particle a_1 and of the target nucleus A_1 in the c.m.s. When the target nucleus is at rest, the sum $E_r = E_1^* + \mathsf{E}_1^*$ may be expressed according to Eq. (33.12) in terms of the energy E_1 of the incident particle in the l.s. The last transformation of (25.3) is permitted only in the case of intermediate and heavy nuclei as are those considered here.

The significance of the Bohr conception can be expressed as follows: the mode of disintegration of the compound nucleus does not depend on the specific

[1] H. A. BETHE and G. PLACZEK. Phys. Rev. **51**, 450 (1937).

[2] We assume here that the two colliding particles are both in their ground state. Otherwise their internal excitation energies must be added at the right hand side of Eq. (25.3).

way in which it has been produced, provided it respects the general theorems of conservation of energy, angular momentum and parity.

This assumption is obviously based on the idea that the interval of time during which the compound nucleus exists, i.e. the mean life of the corresponding level, is so long that all the nucleons present have been sufficiently mixed to have completely lost, at the moment in which the decay of C takes place, any memory of the mode of its formation.

The validity of such a picture requires the fulfilment of certain conditions. First the mean free path of the incident particle in nuclear matter must be small with respect to the radius R of the nucleus, and second the excitation energy of the compound nucleus w_c divided by the number of nucleons A present in it, must be small compared to the average separation energy of a nucleon from the compound nucleus. Both these conditions are fulfilled for nuclei and neutron energies satisfying the inequalities

$$A>10 \quad \text{and} \quad E<50 \text{ Mev}. \tag{25.4}$$

These conditions are necessary but not sufficient because we do not know whether some mechanism exists which prevents or reduces the sharing of the energy of the incident neutron among all the nuclear constituents. Deviations from the compound nucleus hypothesis are actually observed in many reactions, even when the inequalities (25.4) are fulfilled (see, for example, Sect. 30). But we will not take further a critical analysis of the Bohr assumption [3][1] nor in general of the various effects which later gave rise to a *shell model*[2] and to a *complex potential model*[3] which combines the main features of the compound nucleus-with those of an optical model.

The Bohr assumption not only represents a fundamental step in the development of our ideas about nuclei, but gives also a satisfactory interpretation of a wide range of effects among which are most of the nuclear reactions produced by slow neutrons.

According to the qualitative description given above, the cross section of a nuclear reaction A_1 (a_1, a_t) A_t can be written as the product of two independent factors

$$\sigma_{(a_1, a_t)} = \sigma_c(a_1) \cdot G_c(a_t). \tag{25.5}$$

The first factor represents the cross section for formation of the compound nucleus C through the considered initiating channel (Sect. 12), the second is the probability that the compound nucleus C decays by emission of the particle a_t in a well defined outgoing channel. $G_c(a_i)$ is referred to as the *branching ratio*, and must satisfy the relation

$$\sum_t G_c(a_t) = 1 \tag{25.6}$$

with the sum extended over all open outgoing channels. According to the Bohr assumption, $G_c(a_t)$ only depends on the energy w_c, the angular momentum J_c and the parity of the compound nucleus, and is independent of the particular initiating channel through which the compound nucleus has been formed.

---

[1] F.L. FRIEDMAN and V.F. WEISSKOPF: p. 134 of Niels Bohr and the development of physics. London: Pergamon Press 1955. — D.C. PEASLEE: Annual Rev. Nucl. Sci. **5**, 99 (1955).

[2] M. G. MAYER and J. H. D. JENSEN: Elementary theory of nuclear shell structure New York: John Wiley & Sons 1955; see also the article by J. P. ELLIOTT and A. M. LANE in Vol. XXXIX of this Encyclopedia.

[3] H. FESHBACH, C.E. PORTER and V.F. WEISSKOPF: Phys. Rev. **96**, 448 (1954). — V.F. WEISSKOPF: Rev. Mod. Phys. **29**, 174 (1957); see also the article by B.B. KINSEY in Vol. XL of this Encyclopedia.

An important exception to this assumption is presented by the decay of the compound nucleus C through the initiating channel, a process that constitutes a part of the elastic scattering. Since the wave describing the elastic scattering is coherent with the incident wave, interference phenomena take place [between A^l_{pot} and A^l_{res} as indicated by Eq. (21.7)]. Therefore the expression (25.5) as well as the following considerations are correct assuming that the particle a_1 can not be reemitted in the entrance channel. In this case the cross section for formation of the compound nucleus $\sigma_c(a_1)$ turns out to be identical with the reaction cross section σ_r considered in Sect. 18.

The treatment of elastic scattering must be made separately and will not be presented here. The result in the particular case of a single isolated level of the compound nucleus, is given by Eq. (26.4) of the next section, for the derivation of which we refer to Ref. [*3*].

The branching ratio $G_c(a_i)$ is usually expressed in terms of the *partial widths* for decay of the compound nucleus into the various open channels,

$$\Gamma_{a_i} = \frac{\hbar}{\tau_{a_i}} \tag{25.7}$$

where τ_{a_i} are the corresponding partial mean lifes. Since $\Gamma_{a_i}/\hbar$ is, by definition, the probability per unit time of the compound nucleus to decay into the considered channel, one can write

$$G_c(a_i) = \frac{\Gamma_{a_i}}{\Gamma} \tag{25.8}$$

where

$$\Gamma = \sum \Gamma_{a_i} \tag{25.9}$$

is the *total width* obtained by extending the sum over all open outgoing channels. From (25.5) and (25.8) it follows that

$$\sigma_{(a_1, a_i)} = \sigma_c(a_1)\,\frac{\Gamma_{a_i}}{\Gamma}\,. \tag{25.10}$$

26. A few important cases of slow neutron resonances (Rh, In, Cd, Au). In this section we shall discuss a few examples of interaction of slow neutrons with nuclei which deserve special consideration not only as illustration of the nuclear resonances mentioned in previous sections, but also as useful tools in experimenting with slow neutrons. As we shall see later they are of current use as detectors or absorbers of neutrons belonging to very narrow energy intervals.

The theoretical treatment of resonance phenomena is considerably simpler in the case of slow neutrons than in the general case, because of the following circumstances. First, as we have seen in Sect. 18, slow neutrons interact with nuclei only in S states. Thus, only the first terms in the expressions (18.9) and (18.12) of the scattering and reaction cross sections have to be considered; and the neutrons are scattered isotropically in the c.m.s. Secondly, the excitation energy w_c of the compound nucleus is almost equal to the neutron separation energy S_n. Therefore, apart from a few exceptional cases observed in light nuclei (Sect. 27), only two decay processes are possible for the compound nucleus; the emission of γ-rays (Sect. 16) resulting in a process of *radiative capture* of the incident neutron, and the reemission of a neutron resulting in what is called *resonance scattering*.

Besides this mechanism of scattering one has to consider the *potential scattering*. As we have shown in Sect. 21 this is the type of scattering that would take place if the nucleus were replaced by a impenetrable sphere. It is responsible for all the observed scattering at large distance from resonances. According to (21.7) the amplitudes, not the intensities, of the wave produced by potential and resonance scattering must be added and therefore an interference term is present in the expression of the scattering cross section.

In conclusion, in the case of slow neutrons colliding against intermediate and heavy nuclei, the total cross section is given by the sum of the radiative capture cross section $\sigma_a(E)$ and the scattering cross section $\sigma_s(E)$

$$\sigma_{\text{tot}} = \sigma_a(E) + \sigma_s(E). \tag{26.1}$$

In order to avoid unnecessary complications we limit ourselves here to considering the case of a *single isolated resonance.* Then the expressions of $\sigma_a(E)$ and $\sigma_s(E)$ are given by the well known *one-level formulas of Breit and Wigner* [5], [3], viz.

$$\sigma_a(E) = 4\pi\,\lambda\!\!\!^-{}^2 g\,\frac{\Gamma_n\Gamma_\gamma}{4(E-E_R)^2+\Gamma^2} \tag{26.2}$$

which reaches its maximum value

$$\sigma_{a0} = 4\pi\,\lambda\!\!\!^-{}_R^2\, g\,\frac{\Gamma_n\Gamma_\gamma}{\Gamma^2} \tag{26.3}$$

for $E=E_R$; and

$$\sigma_s(E) = \pi\,\lambda\!\!\!^-{}^2 g\left|2kR_g + \frac{\Gamma_n}{(E-E_R)^2+i\frac{\Gamma}{2}}\right|^2 + 4\pi(1-g)\,R_{1-g}^2. \tag{26.4}$$

Here Γ_n and Γ_γ are the partial widths for neutron and photon emission by the compound nucleus,

$$\Gamma = \Gamma_n + \Gamma_\gamma \tag{26.5}$$

is the total width, and

$$g = \frac{1}{2}\left[1 \pm \frac{1}{2I+1}\right] \tag{26.6}$$

in the statistical factor due to the two possible orientations of the neutron spin $|\boldsymbol{S}| = \frac{1}{2}$ with respect to the spin $\boldsymbol{I}$ of the target nucleus. The upper (lower) sign must be used when $\boldsymbol{S}$ is parallel (antiparallel) to $\boldsymbol{I}$; the choice between these two possibilities is determined by the value of the total angular momentum $\boldsymbol{J}_c$ of the considered resonant level of the compound nucleus, since, because of total angular momentum conservation, one has $\boldsymbol{J}_c = \boldsymbol{I} + \boldsymbol{S}$.

The potential scattering contributes two terms in (26.4). The first term depends on the effective nuclear radius R_g for neutrons of the same spin orientation as for the resonant level, the second term depends on the effective nuclear radius R_{1-g} for neutrons of opposite spin orientation. Only the first of these two terms corresponds to waves capable of interfering (coherent) with those arising from resonant scattering. Eq. (26.4) may be written

$$\sigma_s(E) = 4\pi\,\lambda\!\!\!^-{}^2 g\,\frac{\Gamma_n^2}{4(E-E_R)^2+\Gamma^2} + 16\pi\,\lambda\!\!\!^-{}^2 g\,\Gamma_n R_g\,\frac{E-E_R}{4(E-E_R)^2+\Gamma^2} + \sigma_{p'} \tag{26.7}$$

which, for $E=E_R$ takes the value

$$\sigma_s(E_R) = \sigma_{s0} + \sigma_{p'}; \qquad \sigma_{s0} = 4\pi\,\lambda\!\!\!^-{}_R^2\, g\,\frac{\Gamma_n^2}{\Gamma^2}. \tag{26.8}$$

In Eq. (26.7) the first term represents the resonance scattering, the second term the interference between resonance and potential scattering. The third term has been indicated by $\sigma_{p'}$ to designate the *constant* (or almost constant) portion of the scattering cross section. This in general may differ from the potential scattering σ_p because of contributions from neighbouring resonances.

It may be interesting to notice here that when two or more resonances of the same class (i.e. with the same l, J_c and parity) are present, the amplitude, not the intensities of the corresponding waves must be added. Therefore they interfere with each other with the result that the analysis of the experimental data is much more involved.

In the present case of a single resonance $\sigma_{p'} = \sigma_p$ and the potential scattering cross section is given by the expression

$$\sigma_p = 4\pi \{g R_g^2 + (1-g) R_{1-g}^2\} = 4\pi \langle R^2 \rangle \tag{26.9}$$

where $\langle\ \rangle$ denotes averaging with respect to spin orientation.

The energy dependence of the neutron width Γ_n and of the radiation width Γ_γ deserves some considerations. From rather general arguments [*5*], [*3*], [*18b*] it follows that the neutron width can be split into two factors. The first factor, indicated as *reduced neutron width*, incorporates all the properties of the interior of the nucleus and depends so little on the neutron energy that it becomes permissible to treat it as constant in a not too wide energy interval. The second factor depends on the conditions outside the nucleus and therefore it is a function of the energy of the incident neutron. In the case of neutrons with $l=0$ one has

$$\Gamma_n = \Gamma_n^0 \cdot E^{\frac{1}{2}}. \tag{26.10}$$

As we shall see later in a few examples, the reduced neutron width Γ_n^0 is usually derived from the experimental data by means of (26.10) and is usually given as a significant feature of the considered resonance.

The particle width Γ_{a_i} for particles a_i with an electric charge and/or an angular momentum $l \neq 0$, is in general smaller than the neutron width Γ_n for neutrons of the same energy and $l=0$, because of the Coulomb potential barrier and/or the centrifugal barrier. The corresponding effects are not incorporated in the reduced width, but in the factor depending on the conditions outside the nucleus. For instance in the case of neutrons of angular momentum $l \neq 0$ and of such a low energy that $\lambda\!\!\!^{-} \gg R$, the effect of the centrifugal barrier consists in introducing an extra factor $\sim (kR)^l$ so that the neutron width for $l \neq 0$ is proportional to $E^{l+\frac{1}{2}}$[1].

The radiation width Γ_γ is not expected to depend on the neutron energy, at least as long as we limit our considerations to slow neutron energies. In fact Γ_γ measures the radiative transition probability from the considered level of the compound nucleus to lower states. Therefore it will depend on the energy, total angular momentum and parity of all these levels. Its dependence on the neutron energy E is determined by its dependence on w_c. But for slow neutrons E is so small with respect to S_n that according to (25.3) an appreciable variation of E gives rise to a negligible variation of w_c and therefore of Γ_γ.

[1] Considerations of this type have been used by the Argonne group [L. M. BOLLINGER, R. E. COTÉ, D. A. DAHLBERG and G. E. THOMAS: Phys. Rev. **105**, 661 (1957)] who has suggested the existence of slow neutron resonances due to P levels. In the curve of the total cross section of U^{238} as a function of energy, a very small resonance has been observed at $E_R = 10.2$ ev the reduced width of which, computed by means of (26.10), is smaller by a factor of 6×10^3 than the average value of the reduced widths found for the many other resonances of the same nucleus in the same energy region. This is what one would espect for a P resonance, because of the influence of the centrifugal barrier factor which, in uranium at 10.2 ev, amounts to 24×10^3.

From these remarks it follows that in the case of nuclei with a single narrow resonance located at an energy E_R large with respect to kT, i.e. such that

$$E_R \gg \Gamma \quad \text{and} \quad E_R \gg kT \tag{26.11}$$

the capture cross section (26.2) will reduce to the following expression for neutron energies of the order of kT

$$\sigma_a(E) \approx \pi g \lambda\!\!\!^{-2} \frac{\Gamma_n \Gamma_\gamma}{E_R^2} = \pi g \lambda\!\!\!^{-2}_R \frac{\Gamma_\gamma}{E_R} \frac{\Gamma_n^0}{E^{\frac{1}{2}}}. \tag{26.12}$$

Since Γ_n^0 and Γ_γ may be treated as constant, at least in a narrow energy interval, this relation expresses the fact that in case of nuclei with a single resonance satisfying conditions (26.11), the radiative capture cross section follows the $1/v$ law in the thermal energy region. This conclusion, valid also in the case of many resonance levels, provided they all satisfy condition (26.11), explains the result of the Columbia University group (Sect. 24) who found that Ag follows approximately the $1/v$ law in the thermal region while other authors showed for this element the existence of at least one resonance at energies larger than 0.4 ev. As we shall see in the following, the same situation is encountered in the cases of Rh, In and Au, not in the case of Cd—the first resonance of which does not satisfy conditions (26.11).

When the first resonant level of a nucleus satisfies conditions (26.11) the resonance scattering is negligible at energies of the order of kT and the scattering cross section (26.7) reduces to

$$\sigma_s \approx 4\pi \langle R^2 \rangle. \tag{26.13}$$

From all these considerations we see that a resonance is entirely described by the parameters E_R, Γ_n, Γ_γ, g, R_g and R_{1-g}. With so many parameters available for adjustment, special methods are needed for correcting the experimental data for Doppler broadening (Sect. 86) and instrument resolution, and for fitting them so as to obtain a unique set of values. These methods may be found in various papers of the Columbia University group[1,2] and of the Brookhaven group[3-5] to which we refer for a wider bibliography on this subject. They are summarized in Refs. [*12*], [*15a*], [*15b*] and in J. RAINWATER'S contribution to Vol. XL.

Here we limit ourselves to noticing that such an analysis is possible because certain regions of the resonance curve primarily depend on certain combinations of the parameters, while other regions are sensitive to different combinations. For instances, the "centre region" of $\sigma_{\text{tot}}(E)$ is dependent primarily on E_R and $\sigma_{\text{tot},0} \cdot \Gamma$ were

$$\sigma_{\text{tot},0} = \sigma_{a0} + \sigma_{s0}, \tag{26.14}$$

the "near wing region" depends on E_R and $\sigma_{\text{tot}\,0} \cdot \Gamma^2$, and the "far wing region" on σ_p. The coefficient of the interference term manifests itself in the asymmetry of the resonance in the near wings. From an experimental determination of σ_{tot} as a function of the energy, all parameters of the one-level formula may be derived with the exception of g and Γ_n which enters into this formula only as

[1] L. J. RAINWATER, W. W. HAVENS, C. S. WU and J. R. DUNNING: Phys. Rev. **71**, 65 (1947).

[2] E. MELKONIAN, W. W. HAVENS and L. J. RAINWATER: Phys. Rev. **92**, 702 (1953).

[3] V. L. SAILOR: Phys. Rev. **91**, 53 (1953).

[4] R. E. WOOD, H H. LANDON and V. L. SAILOR: Phys. Rev. **98**, 639 (1955).

[5] F. G. P. SEIDL, D. J. HUGHES, H. PALEVSKY, J. S. LEVIN, W. Y. KATO and N. G. SJÖSTRAND: Phys. Rev. **95**, 476 (1954).

the product $g \cdot \Gamma_n$. They can, however, be determined separately by measuring the resonance scattering cross section as a function of energy.

The relative importance σ_a and σ_s at resonance is obviously determined by the relative values of Γ_n and Γ_γ and these change with changing atomic weight and neutron energy region [*3*], [*12*], [*18b*]. The various examples considered below all refer to nuclei of atomic weight around or appreciably above one hundred and to neutron energies of a few ev; in all these cases Γ_n is much smaller than Γ_γ and therefore at resonance the radiative capture prevails over scattering. There are cases, however, as for instance the Mn level at 300 ev, in which on the contrary resonance scattering prevails over resonance capture. Such a situation usually occurs for atomic weights around 50 and for neutron energies of the order of several hundred ev.

An extensive list of resonance parameters is given in Ref. [*14*].

α) *Rhodium* is composed of a single isotope: $_{45}Rh^{103}$ which gives rise, by neutron capture, to $_{45}Rh^{104}$ in two isomeric states [*11*]. The dominant scheme of decay may be described as follows[1]: The excited state ($T_{\frac{1}{2}} = 4.3$ min) decays mainly (99.9%) by emission of a γ-ray of 52 kev ($E3$ or $M3$ transition) into the ground state which has a half-life $T_{\frac{1}{2}} = 44$ sec. In 98.5% of the cases the decay of the ground state of Rh^{104} takes place to the ground state of Ru^{104} by emission of a group of electrons of $E_{\max}^{(-)} = 2.5$ Mev. In 1.4 and 0.08% of the cases two other groups are emitted respectively with $E_{\max}^{(-)} \approx 1.9$ and ≈ 0.7 Mev. In 0.1% of the cases the excited state of Rh with $T_{\frac{1}{2}} = 4.3$ min undergoes β-decay. All secondary groups of β-rays are associated with γ-rays of various energies ranging from 0.556 to 1.53 Mev[2]. Both levels of Rh^{104} are produced by slow neutron capture; the activity of short half-life is considerably more intense than the other.

A single resonance at $E_R = 1.26$ ev has been found in the energy interval from 0.01 ev to about 7000 ev. Fig. 11 shows the total cross section as a function of neutron energy obtained by SAILOR[3] using the high resolution Brookhaven crystal spectrometer[4]. Reference to previous work may be found in the paper of this author.

The measurements of σ_{tot} were made for neutron energies ranging from 0.18 to 30 ev with samples of various thicknesses, all very small. The weak resonance at 5.2 ev is due to a small impurity of Ag (0.01%).

The constribution to the total cross section arising from scattering corresponds to $\sigma_{p'} = (5.5 \pm 1.0)$ barns. The observed resonance may be fitted by means of a Breit and Wigner one-level formula [Eqs. (26.2) and (26.4)] as shown by the full line of Fig. 11. The agreement is excellent over the entire range of observations provided small corrections are made near the center of the resonance to account for self-absorption, instrument resolution ($\Delta E/E = 0.026\%$ at resonance) and Doppler effect (Sect. 86). The values of the various parameters appearing in Eqs. (26.2), (26.4) as deduced by the analysis of the data of Fig. 11 are given in

Table 7. *Parameter values of the rhodium resonance* (SAILOR[3]).

$E_R = (1.260 \pm 0.004)$ ev	$J_c = 1$
$\sigma_{tot,0} = (5000 \pm 200)$ barns	$\Gamma_n = (5.1 \pm 0.2) \times 10^{-4}$ ev
$\Gamma = (0.156 \pm 0.005)$ ev	$\mathfrak{S}_{ar} = (980 \pm 70)$ barns
$I = \frac{1}{2}$	$\sigma_p = (5.5 \pm 1.0)$ barns
$g = \frac{3}{4}$	$\sigma_a(2200\ \text{m/s}) = (142 \pm 10)$ barns

[1] M. GOLDHABER and R. D. HILL: Rev. Mod. Phys. **24**, 179 (1952).
[2] M. E. BUNKER, J. P. MIZE and J. W. STARNER: Phys. Rev. **99**, 659 (1955).
[3] See footnote 3, p. 51.
[4] L. B. BORST and V. L. SAILOR: Rev. Sci. Instrum. **24**, 141 (1953).

Table 7, which also contains the absorption resonance integral (Sect. 90)

$$\mathfrak{J}_{ar} = \frac{\pi}{2} \sigma_{a0} \frac{\Gamma}{E_R}. \tag{26.15}$$

Since the spin of $_{45}Rh^{103}$ is $I = \frac{1}{2}$, the possible values of g are $\frac{3}{4}$ and $\frac{1}{2}$; by comparing the values of $g \cdot \Gamma_n$ and $\frac{\Gamma_n}{\Gamma}$ one can conclude that $g = \frac{3}{4}$ is the correct

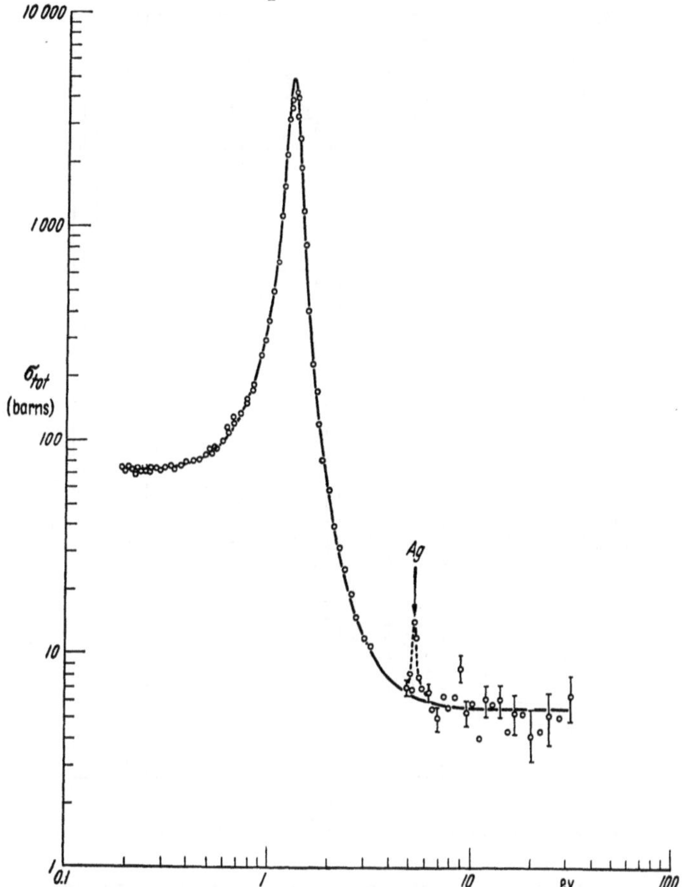

Fig. 11. The total cross section of rhodium as a function of energy (SAILOR).

value. Therefore the resonance corresponds to parallel spins of the incident neutron and of the target nucleus and to a total angular momentum of the compound nucleus $J_c = 1$.

Rhodium has been used, especially at early times of slow neutron physics, as detector of neutrons in the 1 ev energy region. Besides many clear advantages, this element has the disadvantage that the resonance is so low that the effect of chemical binding forces on the slowing down spectrum of neutrons is not negligible (Part D, Divsion IV); furthermore the half-lives of the isomeric states mentioned above are rather short for accurate activation measurements.

β) *Cadmium* ($Z = 48$) is composed of 8 isotopes whose abundances have been given in Sect. 16. No important activity is produced by slow neutrons [*11*] although its absorption cross section is very large.

A very strong resonance is observed at $E_R = 0.178$ ev (Fig. 12) and others much weaker in the energy interval between 10 ev and 10^4 ev (Fig. 13)[1,2] [*14*]. The low energy resonance of Fig. 12 is due to Cd^{113} which, by radiation capture, produces Cd^{114} which also is stable (Sect. 16).

The fit to the one-level formula of BREIT and WIGNER of this low energy resonance is excellent. The values of the resonance parameters are given in Table 8.

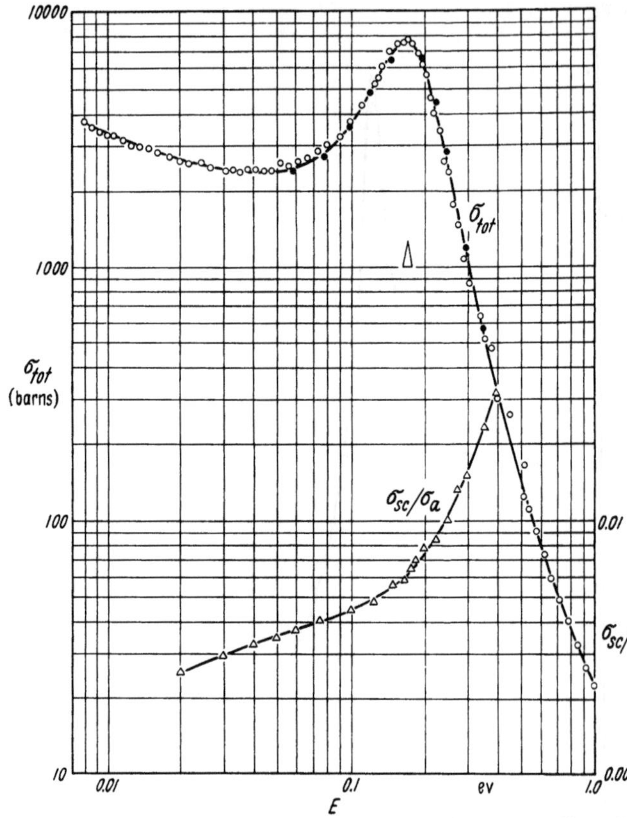

Fig. 12. The energy dependence of the cadmium cross section below 1 ev according to Ref. [*14*].

The cross section at resonance given there refers to the element; when referred to the isotope Cd^{113} it becomes $\sigma_{a0} = 63\,000$ barns.

The ratio of σ_s/σ_a given in Fig. 12 as a function of energy[3] arises from a potential scattering cross section

$$\sigma_{p'} = (4.7 \pm 0.5) \text{ barns}. \tag{26.16}$$

Table 8. *Parameter values of the cadmium resonance* [*14*].

$E_R = 0.178$ ev	$g = \frac{3}{4}$
$\sigma_{\text{tot},0} = (7800 \pm 150)$ barns[4]	$J_c = 1$
$\Gamma = (0.113 \pm 0.005)$ ev	$\Gamma_n = (6.5 \pm 0.2)\ 10^{-4}$ ev
$I = \frac{1}{2}$	$\sigma_{p'} = (4.7 \pm 0.5)$ barns

[1] See footnote 1, p. 51.

[2] See footnote 1, p. 52.

[3] B. N. BROCKHOUSE: Canad. J. Phys. **31**, 432 (1953).

[4] When referred to Cd^{113}: $\sigma_{\text{tot}\,0} = \sigma_{a\,0} = 63\,000$ barns.

The remarkable importance of Cd as a thermal neutron absorber was recognized from early times (Sect. 16); it originates from the existence of the isolated resonance at very low energy. The use of Cd in slow neutron experiments is favoured by the rather good mechanical properties of this metal and by the ease with which it may be alloyed with other metals.

Because of the resonance at $E_R = 0.178$ ev, a Cd layer of convenient thickness may completely absorb thermal neutrons, leaving the neutrons of 1 ev energy or more almost unaffected.

In general, the action of a filter placed in a beam of neutrons depends on the angular distribution of the incident neutrons and may be described as follows.

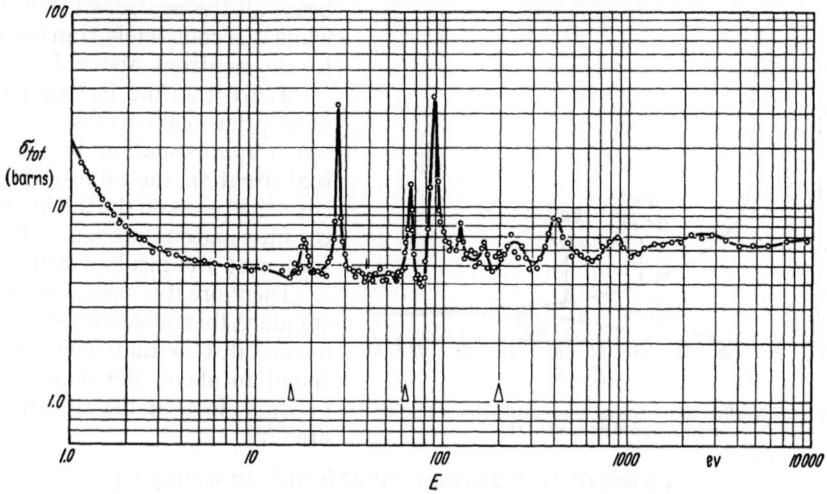

Fig. 13. The total cross section of cadmium between 1 and 10^4 ev according to Ref. [14].

If the incident neutrons constitute a parallel beam perpendicular to the surface of the filter, of thickness x with $\mathfrak{N}$ atoms per cm^3, the incident flux $\varphi_0(E)\,dE$ is transformed into the transmitted flux

$$\varphi_{\text{trans}}(E)\,dE = e^{-\mathfrak{N}\cdot x\cdot\sigma_a(E)}\,\varphi_0(E)\,dE. \tag{26.17}$$

If the incident beam has an angular spread which can be described by a distribution function $f(\vartheta)$ of the angle ϑ measured from the normal to the filter surface, so that

$$\int_{4\pi} f(\vartheta)\,d\omega = 2\pi\int_0^1 f(\vartheta)\,d(\cos\vartheta) = 1,$$

the transmitted flux will be given by the expression

$$\varphi_{\text{trans}}(E)\,dE = \varphi_0(E)\,dE\,2\pi\int_0^1 e^{-\frac{\mathfrak{N}x}{\cos\vartheta}\sigma_a(E)}\,f(\vartheta)\,d(\cos\vartheta) \tag{26.18}$$

which, in the case of isotropic neutron distribution, reduces to

$$\varphi_{\text{trans}}(E)\,dE = \varphi_0(E)\,dE\int_0^1 e^{-\frac{\mathfrak{N}\cdot x}{\cos\vartheta}\sigma_a(E)}\,d(\cos\vartheta). \tag{26.19}$$

The effect of the integration in this case may be roughly described by attributing to the filter an effective thickness almost twice the true thickness x.

In Fig. 14 the results are shown of numerical computations made by HUGHES [12] in the case of Cd-filters placed in an isotropic flux of neutrons under the assumption that the incident neutron flux may be described as composed of a pure Maxwellian distribution and a pure slowing down distribution joined at $E_c = 0.17$ ev (Sect. 107). The full line represents the incident flux, while the broken lines show the neutron fluxes transmitted by $0.01'' = 0.254$ mm $\widehat{=}$ 0.22 g/cm² and $0.03'' = 0.762$ mm $\widehat{=}$ 0.66 g/cm² Cd layers. These two lines rise from zero to a value practically identical to the incident neutron flux in about 0.5 ev. Therefore it is only *approximately correct* to describe the action of a cadmium filter as a *sharp cut-off* E_{Cd} defined as follows: a Cd-filter removes from the incident beam all the neutrons below E_{Cd} while it is completely transparent to the neutrons above E_{Cd}.

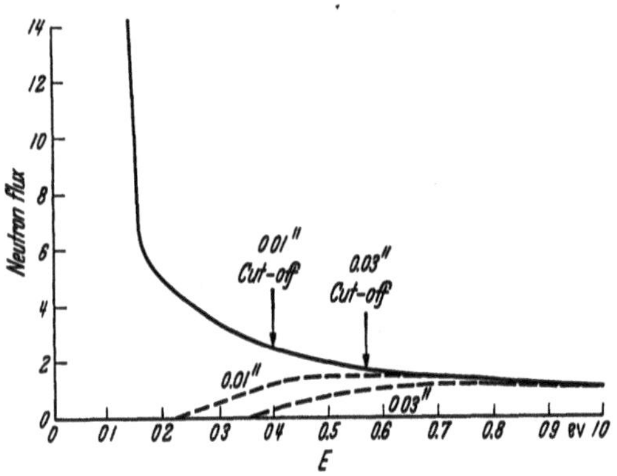

Fig. 14. The effect of a cadmium filter on an isotropic flux of neutrons according to Ref. [12].

Apart from the fact that this description constitutes only a rough approximation to the actual situation, the value of the Cd cut-off depends on the thickness of the filter and on the angular distribution of the incident neutrons.

Therefore it is desirable to use standard thicknesses for different angular distributions which correspond to about the same value of E_{Cd}. Reasonable thicknesses are [12]

$$\text{for } E_{\mathrm{Cd}} = 0.4 \text{ ev} \left\{ \begin{array}{l} x = 0.01'' = 0.254 \text{ mm} \,\widehat{=}\, 0.22 \text{ g/cm}^2 \text{ for isotropic incident neutrons,} \\ x = 0.02'' = 0.51 \text{ mm} \,\widehat{=}\, 0.44 \text{ g/cm}^2 \text{ for parallel incident neutrons.} \end{array} \right\} \quad (26.20)$$

γ) *Indium* is composed of two isotopes [17e]: ${}_{49}\mathrm{In}^{113}$ (4.23%); ${}_{49}\mathrm{In}^{115}$ (95.77%) which both undergo (n, γ) reactions [11].

In^{114} is produced in two isomeric states, the upper state of half-life $T_{\frac{1}{2}} = 50$ d, decays in 96.5% of the cases by emission of a photon of $h\nu = 0.192$ Mev in the lower level which has a half-life $T_{\frac{1}{2}} = 72$ sec, and in 3.5% of the cases by K capture in an excited state of Cd^{114} from which two successive γ-rays are emitted of $h\nu = 0.722$ and $h\nu = 0.556$ Mev[1]. The ground state of In^{114} decays by emitting[2]:

in 98% of the cases, electrons with $E_{\max}^{(-)} = 1.980$ Mev;

in 0.09% of the cases, electrons with $E_{\max}^{(-)} = 0.675$ Mev plus a photon of $h\nu = 1.300$ Mev;

in 0.004% of the cases, positrons of $E_{\max}^{(+)} = 0.400$ Mev;

and

in 1.9% of the cases, by K capture.

Of the two half-lives of this isotope one is too long the other too short to be used for slow neutron detection.

${}_{49}\mathrm{In}^{116}$ also is produced in two isomeric states the upper one of half-live $T_{\frac{1}{2}} =$ 54 min, the lower one of half-life $T_{\frac{1}{2}} = 13$ sec. No transition has been observed

[1] See footnote 1, p. 52.

[2] L. GRODZINS and H. MOTZ. Phys. Rev. **102**, 761 (1956).

between these two states. While the lower level emits electrons of $E_{\max}^{(-)} = 2.95$ Mev, the 54 min level decays by emitting 3 groups of electrons[1]:

in 51% of the cases, electrons of $E_{\max}^{(-)} = 1.00$ Mev plus two photons of $h\nu = 1.085$ Mev and $h\nu = 1.274$ Mev;

in 28% of the cases, electrons of $E_{\max}^{(-)} = 0.87$ Mev plus two photons of $h\nu = 0.137$ Mev and $h\nu = 1.085$ Mev or $h\nu = 0.406$ Mev and $h\nu = 2.09$ Mev;

in 21% of the cases, electrons of $E_{\max}^{(-)} = 0.60$ Mev plus two photons of $h\nu = 1.487$ Mev and $h\nu = 1.274$ Mev

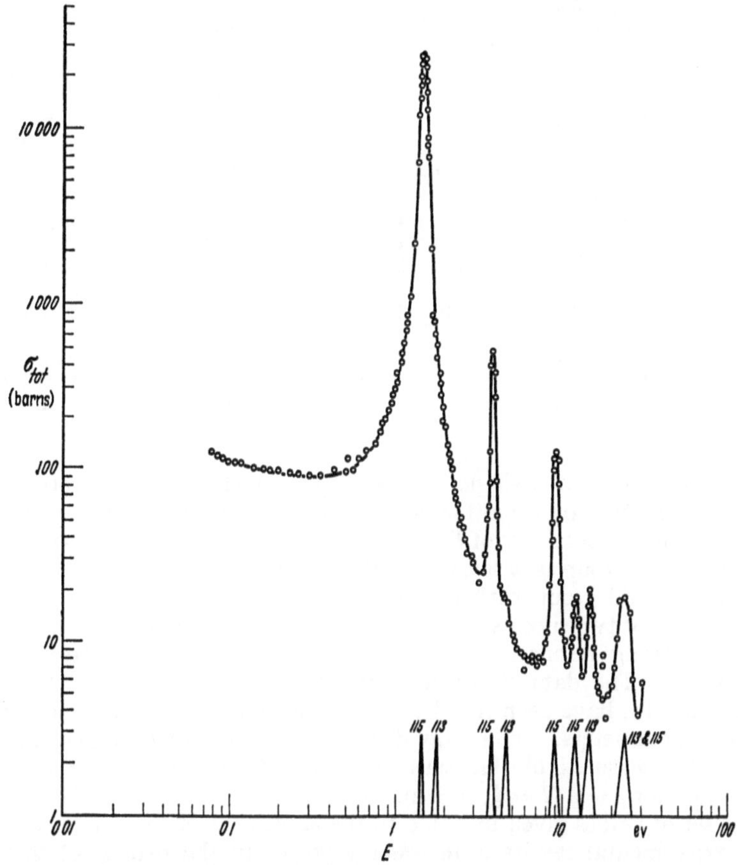

Fig. 15. The total cross section of normal indium as a function of energy (Sailor and Borst).

Fig. 15 shows the total cross section of indium obtained by Sailor and Borst[2] by means of the high resolution crystal spectrometer of the Brookhaven National Laboratory[3]. The assignment of the various resonances to the two isotopes has been made from measurements with enriched isotope samples.

The resonance parameters according to [14] have been collected in Table 9.

The resonance at $E_R = 1.458$ ev is of great practical importance since it gives rise to the 54 min activity providing a very convenient neutron detector sensitive to a narrow energy region around 1.5 ev.

[1] See footnote 1, p. 52.
[2] V.L. Sailor and L.B. Borst: Phys. Rev. 87, 161 (1952).
[3] See footnote 4, p. 52.

Table 9. *Values of resonance parameters of Indium [14].*

Isotope	I	E_R (ev)	Γ_γ (mev)	Γ_n (mev)	Γ_n^0 (mev)$^{\frac{1}{2}}$
$_{49}In^{113}$	$\frac{9}{2}$		80 ± 20[1]		
		1.80 ± 0.03		0.1	
		4.71 ± 0.04		0.104 ± 0.016	0.048 ± 0.007
		14.7 ± 0.1	60 ± 20	7.7 ± 1.0	2.0 ± 0.3
		21.7 ± 0.2		4.4 ± 0.9	0.95 ± 0.20
		25.2 ± 0.2	110 ± 40	9.7 ± 1.6	1.9 ± 0.3
		32.5 ± 0.4		8.5 ± 1.0	1.49 ± 0.18
		45.6 ± 0.6		4.7 ± 1.2	0.69 ± 0.018
		63.6 ± 1.0		1.0 ± 0.4	0.13 ± 0.06
		70.6 ± 1.2		4.7 ± 0.7	0.56 ± 0.08
		82.5 ± 1.5		2.9 ± 1.0	0.32 ± 0.11
		93.2 ± 1.8		40 ± 8	4.1 ± 0.8
		105 ± 2		29 ± 9	2.8 ± 0.9
$_{49}In^{115}$	$\frac{9}{2}$		77 ± 15[1]		
		1.458 ± 0.003	72 ± 2	3.36 ± 0.10	2.78 ± 0.08
		3.86 ± 0.04	81 ± 4	0.318 ± 0.015	0.162 ± 0.008
		9.10 ± 0.09	80 ± 40	1.73 ± 0.17	0.57 ± 0.06
		12.1 ± 0.1	140 ± 60	0.106 ± 0.013	0.031 ± 0.004
		23.0 ± 0.2		1.0 ± 0.2	0.21 ± 0.04
		39.9 ± 0.5	140 ± 50	3.5 ± 0.4	0.55 ± 0.06
		46.3 ± 0.6		0.43 ± 0.09	0.063 ± 0.013
		48.6 ± 0.7		0.65 ± 0.08	0.093 ± 0.0012
		63.2 ± 1.0		1.1 ± 0.3	0.14 ± 0.04
		83.3 ± 1.5		11 ± 2	1.2 ± 0.2
		95 ± 2		2.8 ± 0.6	0.29 ± 0.06

δ) *Gold* is composed of a single isotope $_{79}Au^{197}$ which, by (n, γ) process, gives rise to Au^{198}. This radioisotope has a half-life $T_{\frac{1}{2}} = 2.69$ d [*11*] and emits two groups of electrons respectively of $E_{\max}^{(-)} = 0.96$ Mev (92.4%) and $E_{\max}^{(-)} = 0.295$ Mev (1.76%). The first group is associated with the emission of one photon of $h\nu = 0.41$ Mev, the second group with the emission either (0.33%) of one photon of $h\nu = 1.09$ Mev, or of two successive photons of $h\nu = 0.68$ Mev and $h\nu = 0.41$ Mev[2].

Besides a strong resonance at $E_R = 4.91$ ev (Fig. 16) others are observed above 40 ev [*14*]. The data obtained by WOOD *et al.*[3] with the same technique used for Rh and In, have been fitted to the one-level formula so as to determine the resonance parameters. The energy dependence of the scattering cross section was previously measured by TITTMAN and SHEER[4] and by WOOD[4] who finds clear indication for $g = \frac{5}{8}$ ($I = \frac{3}{2}$; $J_c = 2$).

For the set of values given in Table 10 the agreement of the computed curve with the experimental results is reasonably good. In the region 0.3 to 0.8 ev the observed cross sections all lie one or two barns above the curve. Contribution from resonances at higher energies can be estimated from parameters listed in the paper by SEIDL *et al.*[5] and in Ref. [*14*]. They are negative, because scattering interference terms dominate, and amount to -0.4 barns in the considered range of energy. The fact that the observed points lie above rather than below the curve suggests the existence of levels in Au^{198} lying *below* the binding energy[3].

[1] Average value for the nucleus.
[2] P. E. CAVANAGH: Phys. Rev. **82**, 791 (1951). — D. MAEDER, R. MÜLLER and V. WINTERSTEIGER: Helv. phys. Acta **27**, 3 (1954).
[3] See footnote 4, p. 51.
[4] J. TITTMAN and C. SHEER: Phys. Rev. **83**, 746 (1951). — R. E. WOOD: Phys. Rev. **95**, 644 (1955).
[5] See footnote 5, p. 51.

These would explain also the difference between the value of the absorption cross section at $E=0.025$ ev (2200 m/sec) computed by means of the parameters given in Table 10 [σ_a(2200 m/sec) $= (93.7 \pm 2.3)$ barns] and the results mentioned below [Eqs. (26.21), (26.22)].

The nuclear radii corresponding to each of the two possible spin orientations of the incident neutron, have been derived by combining the values of the coefficient of the interference term and σ_p. The results given in Table 10 are somewhat larger than the radius computed by means of Eq. (18.1), viz. $R=0.82\times10^{-12}$ cm.

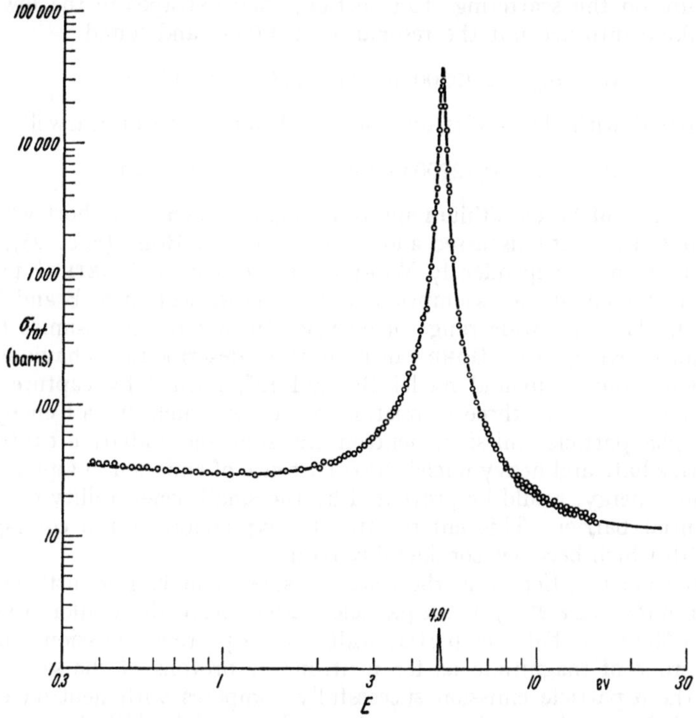

Fig. 16. The total cross section of gold as a function of neutron energy (WOOD *et al.*).

Table 10. *Parameter values of the gold resonance* (WOOD *et al.*)[1].

$E_R = (4.906 \pm 0.010)$ ev	$\Gamma_n = (1.56 \pm 0.04)\times10^{-2}$ ev
$\sigma_{tot,0} = (57000 \pm 2000)$ barns	$\Gamma_n^0 = (7.05 \pm 0.2)\times10^{-3}$ (ev)$^{\frac{1}{2}}$
$\Gamma = (0.140 \pm 0.003)$ ev	$\sigma_{p'} = (11.1 \pm 0.8)$ barns
$I = \frac{3}{2}$	σ_a(2200 m/sec) $= (98.7 \pm 0.6)$ barns
$g = \frac{5}{8}$	$R_g = (0.95 \pm 0.12)\times10^{-12}$ cm
$J_c = 2$	$R_{1-g} = (0.92 \pm 0.15)\times10^{-12}$ cm

In conclusion gold provides a very satisfactory detector of neutrons belonging to a very narrow energy interval around 5 ev with the only disadvantage with respect to other detectors as for instance indium, of a rather long half-life.

The absorption cross section of gold at $E=0.025$ ev has been determined with considerable accuracy by various authors[2,3] since, together with those of

[1] See footnote 4, p. 51.

[2] R. S. CARTER, H. PALEVSKY, V. W. MYERS and D. J. HUGHES: Phys. Rev. **92**, 716 (1953).

[3] P. A. EGELSTAFF: J. Nuclear Energy **1**, 57 (1954).

boron, uranium 235, and uranium 238, it constitutes one of the *standard absorption cross section* which can be used for the determination of other cross sections by relative measurements[1].

CARTER *et al.*[2] at Brookhaven have measured the absorption cross section of gold by means of a "slow chopper" at very low energy (namely below the Au-crystal cut-off: Sect. 102). They found that the cross section follows the $1/v$ law in the 5 to 9 Å region ($E = 0.0033$ to 0.0010 ev). In order to derive the absorption cross section at $v = 2200$ m/s, avoiding errors due to the influence of the crystalline structure on the scattering cross section, they extrapolate the low energy data by taking into account the resonance at 4.9 ev, and found

$$\text{Au:} \quad \sigma_{\text{th}} = \sigma_a(2200\ \text{m/sec}) = (98.7 \pm 0.6)\ \text{barns} \tag{26.21}$$

to be compared with the results obtained by EGELSTAFF[3] at Harwell

$$\text{Au:} \quad \sigma_{\text{th}} = \sigma_a(2200\ \text{m/sec}) = (98.4 \pm 0.9)\ \text{barns}. \tag{26.22}$$

27. The cases of boron, lithium and other light elements. A short while after the preceding ideas were discussed and made known by BOHR (Sect. 25), FRISCH and PLACZEK[4] and independently WEEKS, LIVINGSTON and BETHE[5] presented arguments in favour of the assumption that the cross section of B and Li must follow the $1/v$ law in a wide range of energy. In fact if one assumes that the considerations developed by BOHR can be used to describe the behaviour of very light elements, the compound nuclei B^{11} and Li^7, formed by capture of slow neutrons, may decay by three competing processes which are all energetically possible; alpha particle emission, neutron emission and radiative capture.

In intermediate and heavy nuclei the occurrence of emission of alpha particles of a few Mev energy, would be prevented by the small penetrability of the Coulomb potential barrier. This enters into the expression of the corresponding partial width which becomes considerably reduced.

The situation is different in the cases considered here. For instance, in the case of boron ($Q_0 = 2.8$ Mev), the α-particle passes above the Coulomb potential barrier (2.5 Mev) so that the partial width for α-particle emission will be of the same order of magnitude as for neutrons or protons of the same energy. Therefore the α-particle emission successfully competes with neutron emission and is much larger than radiative capture; the partial width for emission of photons turns out to be 10^{-5} that for emission of α-particles [*3*] which is expected to be of the order of several ten thousands of electron-volts from the experiments on the capture of protons by light nuclei [*5*].

Thus, if one tries to represent the cross section for the (n, α) process in boron by means of the BREIT and WIGNER one-level formula,

$$\sigma_{(n,\alpha)}(E) = \sigma_c(E)\,\frac{\Gamma_\alpha}{\Gamma} = 4\pi\,\lambda\!\!\!^{-2}\,g\,\frac{\Gamma_n\,\Gamma}{4(E-E_R)^2+\Gamma^2}\cdot\frac{\Gamma_\alpha}{\Gamma}, \tag{27.1}$$

$\Gamma \approx \Gamma_\alpha$ is so large that the resonance factor can not affect appreciably the cross section at low energies. Therefore the energy dependence of $\sigma_{(n\,\alpha)}(E)$ will be determined essentially by the two factors $\lambda\!\!\!^{-2}$ and Γ_n with the result that one can write [*5*]

$$\sigma_{(n,\alpha)}(E) \propto E^{-\frac{1}{2}} \propto 1/v \tag{27.2}$$

---

[1] J. A. HARVEY. p. 147 of Ref. [*15b*].
[2] See footnote 2, p. 59.
[3] See footnote 3, p. 59.
[4] O. R. FRISCH and G. PLACZEK: Nature, Lond. **137**, 357 (1936).
[5] D. F. WEEKES, M. S. LIVINGSTON and H. A. BETHE: Phys. Rev. **49**, 471 (1936).

as in the case, considered in Sect. 25, of a narrow resonance located far away from the considered region [see Eqs. (26.11) and (26.12)].

A few years later the remarkable development of the neutron spectrometers allowed the verification of this assumption, whose correctness was established up to about 10^4 ev. The deviation of the total cross section observed in the curves given in Ref. [*14*] around and above 100 ev arises mainly from the constant scattering cross section which is

$$\text{B:}\quad \sigma_s = (4 \pm 1)\ \text{barns} \tag{27.3}$$

and therefore, at these energies, begins to be comparable with the cross section for the (n, α) process. The latter has been measured with considerable care by various authors since, in principle it can be used (together with the capture cross sections of gold, Sect. 26, uranium 235 and uranium 238) as *standard absorption cross section* for the determination of other cross sections by relative measurements[1].

Table 11 contains the results obtained by various authors—by means of transmission experiments—for neutrons of $v = 2200$ m/sec and boron samples containing 19.8% of B^{10} [2].

Table 11. *Boron cross section for (n, α) process at $v = 2200$ m/sec.*

	σ_a (2200 m/sec) (barns)
Argonne group[1]	755 ± 3
Brookhaven group[1]	749 ± 4
Harwell group 1953[3]	772 ± 6
Harwell group 1957[4]	769 ± 4
Columbia group[5]	756 ± 6

Other determinations of σ_a (2200 m/sec) for boron, based on the measurement of the neutron lifetime, will be considered in Sects. 130 and 131.

Some uncertainty in this cross section arises from the variation of the isotopic abundance of B^{10} which is invoked by some authors[4] in order to explain, at least in part, the rather large difference between the results of the Harwell and Argonne-Brookhaven groups. This point will come up again in Sect. 127. The value given in Ref. [*14*] is

$$\text{B:}\quad \sigma_a(2200\ \text{m/sec}) = (755 \pm 4)\ \text{barns}. \tag{27.4}$$

The compounds nucleus of B^{11} emits two groups of α-particles, one corresponding to the ground state, the other to the 0.48 Mev excited state of Li^7 [*1*]. The ratio of the transitions leading to the ground state and to the excited state of Li varies with energy [*2*] reaching the value[6]

$$\frac{\sigma_a(\text{Li}^7)}{\sigma_a(\text{Li}^{7*})} = (6.52 \pm 0.05) \cdot 10^{-2}$$

for thermal neutrons.

Similar considerations hold in the cases of He^3 and Li^6 which also follow the $1/v$ law over rather wide ranges of energy. The corresponding cross sections at $v = 2200$ m/sec are given in Table 12 [*14*].

---

[1] J. A. HARVEY, p. 147 of Ref. [*15b*].

[2] The isotopic composition of boron seems to be different from different sources 18.45 to 18.98% [*17e*]. Previously the B^{10} content of the Argonne-Brookhaven standard boron was quoted as 18.8 atom percent. Recent results have given 19.8% (T.L. COLLINS: Knolls Atomic Power Laboratory, 1958).

[3] A. GREEN, D. J. LITTLER, E. E. LOCKETT, V. G. SMALL, A. H. SPURWAY and E. BOWELL. J. Nuclear Energy **1**, 144 (1955).

[4] P.A. EGELSTAFF: J. Nuclear Energy **5**, 41 (1957).

[5] W.W. HAVENS, E. MELKONIAN and B. RUSTAD: TNCC (US)-25, CU-160, AT 30-1 Gen. 72 (1957).

[6] J. DE JUREN and H. ROSENWASSER: Phys. Rev. **93**, 831 (1954).

The three reactions listed in this table are frequently used for detecting slow neutrons by introducing the corresponding target elements in form of gases (He or BF_3) or of solid layers (Li or B) inside proportional counters or ionization chambers. For details about this techniques we refer to[1] [*17a*] and to the article by BARSCHALL on "Detection of neutrons" in Vol. XLIV.

It may be interesting to recall that the so called "boron assumption" played an important role in the early development of slow neutron physics, because it supplied a very simple method for the determination of the resonance energy E_R

Table 12. *Absorption and scattering cross sections of a few elements of interest for slow neutrons detection. Neutron velocity* $v = 2200$ m/sec

	Isotopic abundance	σ_a (2200 m/sec) (barns)		σ_s (barns)	
		element	isotope	element	isotope
He^3	$1.3\ 10^{-4}$	—	5400 ± 300	0.8 ± 0.2	1.0 ± 0.7
Li^6	$7.52\ 10^{-2}$	71.0 ± 1.0	945	1.4 ± 0.3	
B^{10}	$18.5\ 10^{-2}$	760[2]	4100	4 ± 1	4.0 ± 0.5

of the various elements. Provided each of the elements considered has a single narrow resonance at an energy E_R larger than the cadmium cut-off $\sim$0.4 ev, the mass absorption coefficients in boron of the neutrons filtered by Cd and detected, let us say, by Rh and Au, K_{Rh} and K_{Au}, have a ratio given by the expression

$$\frac{K_{\mathrm{Rh}}}{K_{\mathrm{Au}}} = \sqrt{\frac{E_{R\,\mathrm{Au}}}{E_{R\,\mathrm{Rh}}}}\,. \tag{27.5}$$

As a consequence, the boron assumption gives a method for establishing in a simple way a relative scale of the resonance energies of the various elements, which can even be "calibrated" by measuring the absorption coefficient of boron for the neutrons of group C, K_c, and assuming that it represents, with sufficient accuracy, the absorption coefficient K_{th} for neutrons satisfying the Maxwellian distribution of velocities.

This method, employed by various authors[3] until the use of the neutron spectrometers superseded it, gave fairly good results not only for the resonance energies of many elements, but also for the value of the cadmium cut-off[4].

IX. The main processes involved in the slowing down of neutrons.

28. Qualitative description of the slowing down and diffusion of neutrons and the influence of the molecular structure at very low energies. In order to give a qualitative description of the process of slowing down and diffusion of the neutrons moving through a moderator, we shall refer to the following simple important case.

A source of fast neutrons is embedded in a homogeneous isotropic moderator extending in all directions to such a large distance that one can consider it as infinite. The neutron source is assumed to be isotropic and to emite monoenergetic neutrons of 2 or 3 Mev energy. Furthermore its linear dimensions are supposed

[1] B. ROSSI and H. H. STAUB. Ionization Chambers and Counters. New York. McGraw Hill Book Co. 1949.
[2] See Table 11
[3] H. H. GOLDSMITH and F. RASETTI: Phys. Rev. **49**, 891 (1936); **50**, 328 (1936).
[4] See footnote 4, p. 60

to be very small with respect to the neutron mean free path for the various processes of scattering and absorption (point source).

As we have already noticed in Divisions V and VII, a good moderator mus tbe made up, at least for the most part, of light elements whose scattering cross section is so large as to prevail over the absorption due to all the nuclei present in it.

The neutrons emitted from the source undergo a certain number of elastic collisions against the light nuclei, losing energy and undergoing changes in the direction of their motion.

Through a sufficiently large number of these collisions, which are obviously statistically distributed, the energy is degraded and the spatial distribution extended over a certain region of the moderator.

As long as the energy E of the neutron is large compared to the energies $h\nu_\iota$ of the vibrations of the atoms in the molecules or crystal of the moderator, the target nuclei against which the neutron collides can be treated as free and at rest. This case is considered in detail in part C of the present article.

But when, as a consequence of the slowing down process, the energy of the neutron approaches $h\nu_\iota$, one can no longer neglect the chemical binding forces. The existence of such an effect was recognized by AMALDI and FERMI[1,2] who found that the neutrons of energy corresponding to the resonance of Rh and Ag, i.e. of energies of the order of a few ev, have a mean free path in paraffin about 3 times longer than the neutrons of the group C (thermal neutrons). Such a difference can not be due to a nuclear effect because neutrons of a few electron volts or of $^1/_{40}$ ev both have wavelengths so much larger than nuclear dimensions that their scattering cross sections have already reached the limiting value corresponding to $\lambda\!\!\!^{-} \to \infty$ (see Sect. 20).

But, as was immediately suggested by FERMI [8], the $h\nu_\iota$ due to the chemical forces which bind the protons, or more generally the atoms of the moderator, in molecules or crystals, have values of the order of $^1/_{10}$ ev, and can considerably affect the scattering cross section.

Further complications arise from the thermal agitation and zero point energy of the various degrees of freedom of the moderator, and from the interference between waves scattered by different atoms of the same molecule or micro-crystal. The first of these effects is due to the fact that when the neutron energy starts to approach kT the motion of the moderator atoms can no longer be neglected. On the other hand at these energies the neutron wavelength is comparable with the interatomic distances and therefore interference effects start to be important. These various effects arising from the molecular or crystalline structure of the moderator, will be considered in detail in part D of the present article.

From the point of view of the slowing down process, it is clear that a neutron of energy below about 1 ev colliding against bound atoms can only exchange amounts of energy corresponding to $n=0, \pm 1, \pm 2, \ldots$ times the energies $h\nu_\iota$ of the molecular frequencies of the moderator. In spite of these complications the neutrons reach energies of the order of kT; their spectrum, however does not correspond exactly to the Maxwellian distribution of velocities because, as we have mentioned in Sect. 22, slow neutrons are on the one hand absorbed by the nuclei of the moderator, on the other hand, they are continuously produced by the slowing down of neutrons of higher energy.

---

[1] E. AMALDI and E. FERMI: Ric. sci. **7** (1), 223, 393 (1936).

[2] E. AMALDI and E. FERMI: Ric. sci. **7** (1), 454 (1936). — Phys. Rev. **50**, 899 (1936).

From this qualitative picture of the slowing down process it can be seen that the cross sections of all the nuclei present in the moderator, for all possible neutron-reactions, from the initial energy of a few Mev down to zero energy are involved, and that in the low energy region i.e. around and below 1 ev also the molecular or crystalline structure of the moderator have to be considered in detail.

Different reactions play different roles. The elastic scattering cross section of the light elements present in the moderator determines the slowing down of neutrons; both the differential and the total cross sections as functions of the neutron energy enter the problem.

The absorption of neutrons is determined at low energies by the (n, γ) reactions, at higher energy by the (n, α), (n, p), (n, d) ... reactions which take place with an appreciable cross section only in light and medium elements. The $(n, 2n)$ reactions may have to be considered from two different points of view; they contribute to the absorption of neutrons of energy larger than the corresponding threshold, and they supply a secondary source of neutrons of degraded energy.

The (γ, n) reactions in D and Be (Sects. 46 and 47) produced by γ-rays emitted in the radiative capture of neutrons by other nuclei may also represent appreciable secondary sources of neutrons.

Finally, the inelastic scattering (n, n') may contribute to the slowing down, especially if the moderator contains appreciable amounts of nuclei of intermediate and large mass number.

Accordingly, it appears appropriate to devote the next two sections to a summary of the information available on the more important processes that may influence the slowing down, for neutron energies

$$E \lesssim 15 \text{ Mev}. \tag{28.1}$$

This limitation of the considered energy range is suggested by the fact that the great majority of the phenomena treated in the present article involve neutron energies of the order of a few Mev or less and only rarely energies of the order of 14 or 15 Mev.

In the picture presented here the neutron process of greatest practical importance, the fission of heavy elements, is omitted; because of its paramount importance two other articles in other volumes of this Encyclopedia are completely devoted to it and its applications (see Vols. XLI and XLII).

For a thorough discussion of the theory of the various reactions the reader is referred to Vol. XLI and to Refs. [*9*], [*3*], [*18a*], [*18b*]. A very enlightening comparison between the experimental data and the theoretical predictions, with particular regard to the limits of validity of the Bohr model of the nucleus (Sect. 25) is given by WEISSKOPF[1].

29. Elastic scattering, radiative capture and other neutron induced reactions in a few light elements of interest as moderators (H, D, Be, C, O). In this section we shall review very rapidly the cross sections for elastic scattering and for neutron absorption of those light elements which are generally used as moderators.

The scattering cross sections $\sigma_s(E)$ considered here refer to atoms free and at rest in the l.s. This means that they refer to conditions in which the various effects arising from the molecular structure of the moderator and its thermal agitation (Sect. 28) may be ignored.

This point is of importance only at very low energies, where the equivalent expressions *free atom cross section* and *zero energy cross section* are used to indicate

[1] V. F. WEISSKOPF, p. 23 of Ref. [*15a*]. — Rev. Mod. Phys. **29**, 174 (1957).

the value obtained by extrapolating to zero energy the results of measurements made between a few ev and 100 ev, i.e. in an energy interval where the molecular effects are small and can be easily taken into account (Part D, Sect. 100).

A similar problem does not exist for the absorption cross section $\sigma_a(E)$ as long as it follows the $1/v$ law; and all the nuclei considered in this section are of this type. Under these conditions the value of σ_a does not depend on the molecular structure and on the distribution of the velocities of the target nuclei (Sect. 95). According to Eq. (23.15) the expression "thermal neutron absorption cross section" refers to a neutron velocity equal to 2200 m/sec.

Table 13. *Free atom cross sections* $\sigma_s(0)$ *and thermal neutron absorption cross sections* σ_a(2200 m/sec) *for various light nuclei.*

Nucleus	$\sigma_s(0)$ (barns)	σ_a(2200 m/sec) (barns)
H	20.36 ± 0.10	0.3315 ± 0.0017
D	3.4 ± 0.05	0.00057 ± 0.00001
Be	6.04 ± 0.03	0.010 ± 0.001
C	4.7 ± 0.03	0.0032 ± 0.0002
O	3.75 ± 0.02	<0.002

Table 13 shows the values of $\sigma_s(0)$ and σ_a(2200 m/sec) for the various nuclei considered later in this section in more detail. The value of σ_a for hydrogen is discussed in Sect. 126, that for heavy hydrogen is taken from[1]; all other data given in Table 13 are taken from Ref. [*14*].

α) Hydrogen[2]. The total cross section of neutrons against protons has been measured by many authors for energies varying from 1 ev to a few hundred millions of ev. All these results have been collected critically in Ref. [*14*] and in a review article by FOWLER and BROLLEY[3] to which we refer for extensive literature on this subject. Fig. 13 shows the total cross section as a function of neutron energy in the l.s. between 10^4 ev and 100 Mev [*14*]. Below 10^4 ev, the cross section approaches the value

$$\sigma_{\mathrm{H}}(E) = \sigma_s(0) + \sigma_a(v) \qquad (\text{for } E \to 0) \tag{29.1}$$

where

$$\sigma_s(0) = (20.36 \pm 0.10) \times 10^{-24}\ \mathrm{cm}^2 \tag{29.2}$$

is the zero energy scattering cross section, the determination of which is discussed in Sect. 99, and

$$\sigma_a(v) = (0.3315 \pm 0.0017)\,\frac{2200}{v_{\mathrm{m/sec}}} \times 10^{-24}\ \mathrm{cm}^2 \tag{29.3}$$

is the radiative capture cross section we shall discuss later in this section. At energies of the order of 1000 ev or more the capture cross section is negligibly small with respect to σ_s and therefore the total cross section given in Fig. 17 is practically identical with the scattering cross section $\sigma_s(E)$.

The theoretical curve of Fig. 17 is derived by the procedure described later; namely by introducing into (29.9) and (29.6) the values of the parameters (29.11). These have been obtained from the low energy data and the five high energy points that are plotted in Fig. 17 between 1 and 14 Mev.

[1] L. KAPLAN, G.R. RINGO and K.E. WILZBACH: Phys. Rev. **87**, 785 (1952).

[2] For theoretical detail see also HULTHÉN and SUGAWARA's article in Vol. XXXIX, information on the deuteron photoeffect in BISHOP and WILSON's contribution to Vol. XLII.

[3] J. L. FOWLER and J. E. BROLLEY: Rev. Mod. Phys. **28**, 103 (1956).

Also the angular distribution of the scattered neutrons has been measured by many authors and has been found to be isotropic in the c.m.s. at least for neutron energies around and below 10 Mev. At energies of the order of 30 Mev or more the experimental results on the differential scattering cross section in the c.m.s. can be fitted by an expression of the form

$$\frac{d\sigma(\vartheta^*)}{d\omega^*} = \sigma(90^\circ)\,(1 + A\cos\vartheta^*) \tag{29.4}$$

where the anisotropy parameter A is a monotonically increasing function of the neutron energy. Recent results[1] seem to indicate an appreciable anisotropy

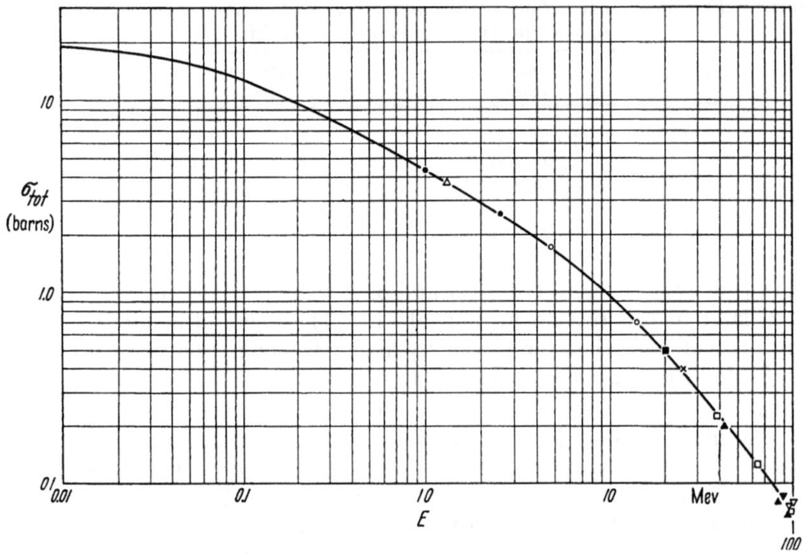

Fig 17. Total cross section of hydrogen atoms for neutron energies between 10^4 ev and 100 Mev [*14*]

($A = 0.05$) already at $E = 14$ Mev. Since most of the considerations given later refer to lower energies, it appears justified to assume in the present article that *the neutron-proton scattering is isotropic in the c.m.s.*

This remark is of considerable importance also for the derivation of the theoretical expression of the total n-p scattering cross section; this derivation can be made by considering *only S waves.*

This was actually done, in early times, by BETHE and PEIERLS [*4*] who used Eq. (20.7) rewritten in the form

$$\sigma_s(E) = \frac{4\pi}{k^2 + k^2\cot g^2\delta_0}; \qquad k = \frac{1}{\lambda\!\!\!^{-}_r} = \frac{p_r}{\hbar} = \left(\frac{mE}{\hbar^2}\right)^{\frac{1}{2}} \tag{29.5}$$

and determined the phase shift δ_0 assuming that the neutron-proton interaction could be represented by a potential well of radius very small compared to $\lambda\!\!\!^{-}_r$. But when, some time later, the n-p cross section was measured for slow neutrons, it was WIGNER [*4*] who realized that the considerable increase of the scattering cross section with decreasing energy of the neutrons could not be represented by such a naive description of the phenomenon. As we shall see later, even taking into account the finite value of the radius of the potential well, the expression (29.5) contains a single adjustable parameter the value of which is fixed by the

[1] See footnote 3, p. 65.

experimental value of the cross section at a conveniently chosen energy. As suggested by WIGNER, the difficulty is solved by taking that the neutron-proton interaction is different for different spin orientations of the incident neutron with respect to the target proton. Then the total scattering cross section of unpolarized neutrons colliding with unpolarized protons is given by the formula

$$\sigma_s = \tfrac{3}{4}\sigma_3 + \tfrac{1}{4}\sigma_1 \tag{29.6}$$

where σ_3 and σ_1 are the scattering cross sections respectively for the triplet ($^3S_1$; parallel spins) and the singlet ($^1S_0$; antiparallel spins) states of the two colliding particles. Each one of these two terms is related to the corresponding phase shift δ_{0i} ($i=3$ or 1) by means of Eq. (29.5) and can be transformed into a very simple explicit function of the neutron energy by means of the *effective range theory*[1] [3], [17d] which neglects the tensor forces and the exchange currents between the two nucleons. If the quantity $k \operatorname{cotg} \delta_{0i}$ is expanded in powers of k^2

$$k \operatorname{cotg} \delta_{0i} = -\frac{1}{a_i} + \frac{1}{2} r_{0i} k^2 - T k^4 \tag{29.7}$$

it is found that, in the energy range considered, the term in k^4 is negligibly small provided that the potential well representing the n-p interaction, does not have a long tail. The approximate relation

$$k \operatorname{cotg} \delta_{0i} = -\frac{1}{a_i} + \frac{1}{2} r_{0i} k^2, \tag{29.8}$$

is known as the *shape independent formula* since any potential well which is specified by two parameters, can be adjusted to give the desired values of a_i and r_{0i}. Introducing (29.8) into (29.5) we obtain

$$\sigma_i(E) = \frac{4\pi}{\left(\frac{1}{a_i} - \frac{1}{2} r_{0i} k^2\right)^2 + k^2}, \qquad i = 3 \text{ or } 1 \tag{29.9}$$

which is a two-parameter formula. For $k=0$, it reduces to the zero energy cross section

$$\sigma_i(0) = 4\pi a_i^2, \qquad i = 3 \text{ or } 1 \tag{29.10}$$

identical with (20.8). Therefore, a_i is the scattering length defined in Sect. 20. The other constant r_{0i}, which also has the dimensions of a length, is called the *effective range*.

Introducing (29.9), with both $i=3$ and $i=1$, into (29.6), one obtains a formula depending on four parameters: a_3, r_{03} and a_1, r_{01}. Of these only three are actually independent since it can be shown that a_3 and r_{03} are bound by the following relation:

$$\gamma = \frac{1}{a_3} + \frac{1}{2} r_{03} \gamma^2, \tag{29.11}$$

where

$$\gamma^{-1} = \left(\frac{\hbar^2}{m\,|E_3|}\right)^{\frac{1}{2}} = \frac{10^{-12}}{(2.317 \pm 0.002)} \approx 4.31 \times 10^{-13}\ \text{cm} \tag{29.12}$$

is the size of the deuteron in its ground state of energy $E_3 = -2.226 \pm 0.003$ Mev [see Sect. 2: Table 1][2].

[1] G. L. SQUIRES: Progress in Nuclear Physics, edited by O. R. FRISCH, Vol. 3, p. 220. London: Pergamon Press 1952.

[2] At distances r larger than the radius b of the potential well, the wave function of the deuteron in its ground state (a pure $^3S_1$ in absence of tensor forces) is proportional to $\exp(-\gamma r)$; and, since $\gamma^{-1} > b$, the linear dimensions of the deuteron are essentially determined by γ^{-1}. We notice that in the approximation $b=0$, one obtains $r_{0i}=0$, and therefore from (29.12), (29.11), (29.9) it follows that σ_3 is completely determined by $|E_3|$. The expression obtained in this approximation corresponds to the original formula of BETHE and PEIERLS in the zero range approximation.

The values

$$\left.\begin{aligned} a_3 &= (5.38 \pm 0.02) \times 10^{-13}\,\text{cm}, \\ a_1 &= (-23.69 \pm 0.06) \times 10^{-13}\,\text{cm}, \\ r_{03} &= (1.704 \pm 0.030) \times 10^{-13}\,\text{cm}, \\ r_{01} &= (2.5 \pm 0.5) \times 10^{-13}\,\text{cm} \end{aligned}\right\} \tag{29.13}$$

have been obtained by means of the following procedure[1]: a_3 and a_1 are determined by combining the zero energy cross section (29.3) with the coherent scattering length as it is deduced from total reflection experiments of slow neutrons at liquid hydrocarbon mirrors[1,2]. The effective range of the triplet state r_{03} is obtained by introducing the values of a_3 and γ into (29.11). Finally r_{01} is adjusted to represent satisfactorily the experimental data on the scattering cross section at neutron energies of a few Mev.

A few remarks may be added to the set of values of the parameters (29.13). First, it has to be noticed that the negative sign of a_1 is necessary in order to match the measured coherent scattering length. Other effects (see later in this section as well as Sect. 46) confirm this conclusion. Secondly, the total scattering cross section at very low energy does not depend on the sign of a_3 and a_1, the largest contribution being provided by the $^1S_0$ state. Finally, the negative sign of a_1, combined with the fact that a_1 is larger than any reasonable value of the range of the nuclear forces ($b \approx 2 \times 10^{-13}$ cm), implies that the $n+p$ system does not have a $^1S_0$ bound state [*5*], [*3*]. In spite of this it is costumary to define the energy $E_1(>0)$ of the $^1S_0$ state responsible for the largest part of the scattering at very low energies, as follows. The relation (29.11), valid for the $^3S_1$ state, is applied to the $^1S_0$ state in order to define the corresponding radius γ_1^{-1} which of course turns out to be negative. The value of γ_1, with opposite sign, is introduced in (29.12) from which a value of E_1 of the order of one hundred kev is deduced. The result of such a computation is however affected by a very large error arising from the very low accuracy of r_{01}. This difficulty does not exist in the zero range approximation which gives

$$E_1 \approx 0.0739\ \text{Mev}. \tag{29.14}$$

A state of this type is called a virtual state [*5*].

The absorption cross section (29.3) is due to radiative capture of neutrons by protons

$${}_0n^1 + {}_1\text{H}^1 \rightarrow {}_1\text{D}^2 + h\nu. \tag{29.15}$$

The γ-rays emitted in this process were observed very early as mentioned in Sect. 16. A few attempts to measure their energy by using absorption methods succeeded in showing that the energy was of the expected order of magnitude[3].

---

[1] M. T. BURGY, G. R. RINGO and D. J. HUGHES: Phys. Rev. **84**, 1160 (1951).

[2] The coherent scattering length (per nucleus)

$$a_{\text{coh}} = \tfrac{3}{4}\,a_3 + \tfrac{1}{4}\,a_1$$

is the average with respect to the spin orientation of the scattering lengths in the triplet and singlet states (see Sect. 98). It can be determined either by measuring the scattering of slow neutrons ($\lambda \to \infty$) by para-hydrogen (i.e. H_2 molecules with antiparallel spins of the two protons) or by measuring the critical glancing angle for reflection on hydrogen mirrors or hydrocarbon mirrors [*12*], [*13*].

[3] R. FLEISCHMANN: Z. Physik **103**, 113 (1936). — S. KIKUCHI, H. AOKI and K. HUSIMI: Nature, Lond. **137**, 186 (1936). — Proc. Phys.-Math. Soc. Japan **17**, 369 (1935).

More recently, BELL and ELLIOT[1] have made a precision measurement of $h\nu$ and have derived from it the binding energy of the deuteron. Their result

$$E_3 = (-2.2230 \pm 0.007)\ \text{Mev} \tag{29.16}$$

agrees with the results of other methods summarized in Table 1 (p. 5). In the experimental set-up of BELL and ELLIOT a slab of paraffin was exposed to the slow neutrons diffusing out of the graphite thermal column of the Chalk River pile. The energy of the γ-radiation produced in paraffin though the process (29.15) was measured in a β-ray spectrometer by comparing the momenta of the photo-electrons ejected from a thin uranium radiator by this radiation, with that of the photo-electrons ejected by the $h\nu = 2.615$ Mev line of ThC″.

The cross section of the process (29.15) has also been measured from the beginning by many authors. Its value (29.3), even for neutrons of thermal energy ($v = 2200$ m/sec), is too small compared to the scattering cross section to be measured by transmission experiments. Therefore various indirect methods—as listed below—have been used.

1. Measurements of the diffusion length of thermal neutrons in water or paraffin[2,3].

2. Measurements of the albedo of thermal neutrons in water or paraffin[2].

3. Comparison of the capture cross sections of H and B by measuring the reflection coefficient for thermal neutrons[4] or the total number of thermal neutrons present in a solution of boric acid in water[5], as a function of the boron concentration.

4. Comparison of the capture cross section of H and other absorbing elements (Ag, Cd) by measuring the activation of solutions of a silver salt as a function of the concentrations of the dissolved elements[6].

5. Direct measurements of the mean life of thermal neutrons in water or paraffin[7].

6. Comparison of the capture cross sections of H and B using the technique of pile oscillation[8].

The first five methods will be discussed in Part E Division II of the present article, while we direct the reader to Refs. [*12*], [*18b*] for the sixth method the detailed discussion of which would take us too far from the scope of this article.

While many measurements have been made of the value of σ_a at thermal energies (see Sect. 23) its velocity dependence has received very little attention In particular no experimental proof has been provided until now of the validity of the $1/v$ law adopted in (29.3). It has only been postulated because of the theoretical considerations given below, which provide a quantitative description of all other effects observed at low energies for the neutron-proton system.

---

[1] R. E. BELL and L. G. ELLIOT: Phys. Rev. **79**, 282 (1950).

[2] E. AMALDI and E. FERMI: Ric. sci. **7** (1), 454 (1956). — Phys. Rev. **50**, 899 (1936).

[3] G. GAMERTSFELDER and M. GOLDHABER: Phys. Rev. **62**, 556 (1946).

[4] E. SEGRÈ: Ric. sci. **7** (1), 389 (1936)

[5] O. FRISCH, H. v. HALBAN and J. KOCH: Danske Vidensk. Selsk. Mat.-Fys. Medd **15**, Nr. 10 (1937). — L. SCHULZ and M. GOLDHABER: Phys. Rev. **67**, 202 (1945). — W. J WHITEHOUSE and G. A. R. GRAHAM: Canad. J. Phys. A **25**, 261 (1947). — B. HAMERMESH, G. R. RINGO and S. WEXLER: Phys. Rev. **90**, 603 (1953).

[6] O. HAXEL and H. VOLZ: Z. Physik **120**, 493 (1942).

[7] J. MANLEY, L. HAWORTH and E. LUEBKE: Phys. Rev. **61**, 752 (1942). — G. v. DARDEL and A. W. WALTNER: Phys. Rev. **91**, 1284 (1953). — F. R. SCOTT D. B. THOMPSON and W. WRIGHT: Phys. Rev. **95**, 582 (1954).

[8] S. P. HARRIS, C. O. MUEHLHAUSE, D. ROSE, H. P. SCHROEDER, G. E. T. THOMAS and S. WEXLER: Phys. Rev. **91**, 125 (1953).

The theory of the process of radiative capture of neutrons by protons (29.14) deserves special consideration because it allows the derivation of direct information on the n-p interaction at very low energies; in particular it allows the determination of the sign of a_1.

The original treatment by FERMI[1], in the zero range approximation, showed that the correct order of magnitude of the cross section σ_a was obtained by attributing the capture to transitions of the $n+p$ system from an initial state $^1S_0$ of the continuum, to the final bound state $^3S_1$ i.e. to the deuteron ground state. The mechanism required for giving rise to such a capture is found in the magnetic dipole radiation in which the spin (and therefore also the magnetic moment) of either the proton or the neutron must change direction. The process is the reverse of the so-called photomagnetic effect discussed in Sect. 46.

The Fermi expression, valid in the approximation of zero range of the nuclear forces, is the following

$$\sigma_{a\,0}(v) = \frac{1}{v}\,\frac{2\pi e^4 \hbar^4}{m^5 c^5}\,(\gamma - a_1^{-1})^2\,\frac{\gamma^2}{a_1^{-2}}\,(\mu_p - \mu_n)^2 \tag{29.17}$$

where μ_n and μ_p are the magnetic moments (5.1) and (5.3) of the neutron and the proton respectively in units of the nuclear magneton (5.2).

The value of the factor $(\gamma - a_1^{-1})^2$ depends on the sign of the scattering length a_1 in the singlet state. Therefore by comparing (29.17) with the measured value of σ_a it is possible to decide if the $^1S_0$ state is real or virtual[1,2].

An approximation better than (29.17) is provided by the effective range theory

$$\sigma_a(v) = \frac{1}{1-\gamma\varrho_3'}\,\sigma_{a0}' \tag{29.18}$$

where σ_{a0}' differs from (29.17) because the factor $(\gamma - a_1^{-1})^2$ is substituted by $(\gamma - a_1^{-1} - \gamma^2 D)^2$ with D denoting a constant which, for several potential shapes, turns out to be

$$D = \tfrac{1}{4}(r_{03} + r_{01})\,. \tag{29.19}$$

This modification, together with the factor

$$\frac{1}{1-\gamma\varrho_3'}, \tag{29.20}$$

represents the correction for the finite range of the forces. In (29.20) ϱ_3' is an abbreviation for

$$\varrho_3' = r_{03} + 4T\gamma^2. \tag{29.21}$$

Thus, ϱ_3' reduces to r_{03} in the shape-independent approximation (29.8) obtained by neglecting the term in k^4 in (29.7). The second term in (29.21) depends on the shape of the potential well: Yukawa, exponential, Gaussian and square wells have been considered with the result that ϱ_3' takes its largest value ($\varrho_3' = 1.768\times10^{-13}$ cm) for the Yukawa potential well, and its smallest value ($\varrho_3' = 1.682\times 10^{-13}$ cm) for the square well, while the exponential and the gaussian wells give intermediate results.

While the effect of tensor forces may be neglected[3], a correction to (29.18) is needed in order to take into account the effect of the currents produced by the exchange of charge between the two nucleons in the deuteron. This cor-

---

[1] E. FERMI· Phys. Rev. **48**, 570 (1935). — See also Ref. [8].
[2] See footnote 2, p. 69.
[3] See footnote 1, p. 67 and footnote 3, p. 65.

rection is the same as that of the photomagnetic effect discussed in Sect. 46; according to (46.11), (46.10) it can be introduced by replacing $(1-\gamma\varrho_3')^{-1}$ in (29.18) with

$$R(0) = \frac{1}{1-\gamma\varrho_3'}\,\frac{1}{1-0\,04}\,. \tag{29.22}$$

In the discussion given in Sect. 46 we shall come back to this point.

β) Heavy hydrogen's abundance in natural hydrogen is 0.0149%. Because of its remarkable importance in applied nuclear physics, this isotope is separated

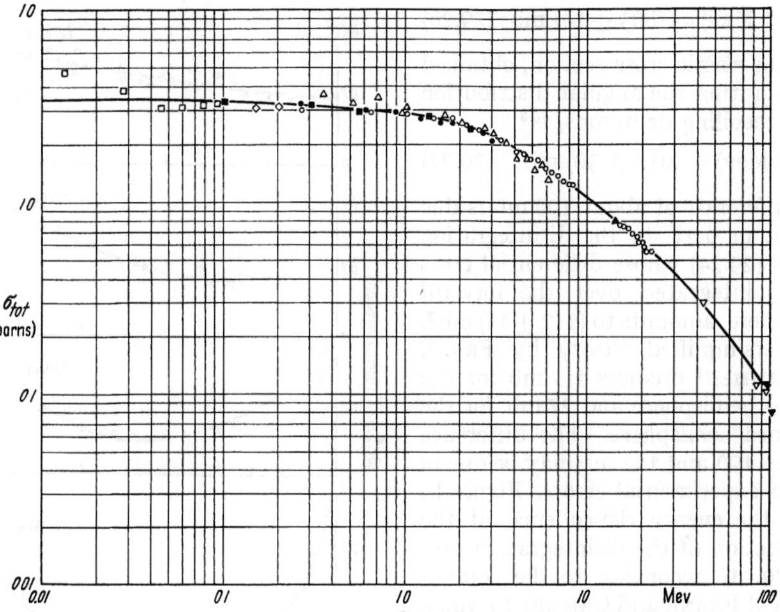

Fig. 18. Total cross section of deuterium for neutron energies between 0.01 and 100 Mev [*14*].

from light hydrogen on a large industrial scale and is available on the market in the form of deuterium gas, heavy water or other compounds at a very high degree of purity.

Three processes take place when a neutron collides against a deuteron: *elastic scattering, radiative capture*

$$_1n^0 + {}_1D^2 \rightarrow {}_1T^3 + h\nu, \qquad Q_0 = 6.251 \text{ Mev} \tag{29.23}$$

and deuteron disintegration

$$_0n^1 + {}_1D^2 \rightarrow {}_1H^1 + {}_1H^1 + {}_0n^1 - 2.226 \text{ Mev} \tag{29.24}$$

for neutron energies above the threshold [see Eq. (34.5)]

$$E_{1\mathrm{T}} = \frac{2+1}{2}\,2.226 = 3.35 \text{ Mev}. \tag{29.25}$$

This latter process is usually described as *inelastic scattering*.

The zero energy scattering cross section and the thermal neutron absorption cross section are given in Table 13. The effect of molecular binding in D_2O at energies of the order of a few ev or less is discussed in Sect. 101.

Figure 18, showing the total cross section between 0.01 and 100 Mev, has been taken from Ref. [*14*] where the reader may find an extensive literature. At energies of the order or below 3 to 4 Mev the total cross section is practically identical to the scattering cross section, since the radiative capture cross section, already negligible at thermal neutron energies, is expected to follow the $1/v$ law. At higher energies the disintegration process (29.24) may give an appreciable contribution[1]; at $E=14.1$ Mev the total cross section is[2]

$$\sigma_{tot}(14.1\text{ Mev}) = (803 \pm 14)\text{ mb} \quad (29.26)$$

while the elastic cross section, obtained by integrating the angular distribution of the recoiling deuterons, is[3]

$$\sigma_s(14.1\text{ Mev}) = (610 \pm 30)\text{ mb}. \quad (29.27)$$

The difference of about 190 mb is due at least in part, to the disintegration process (29.24) whose differential cross section, integrated over the forward hemisphere, amounts to (112 ± 15) mb[3]. An oversimplified theory by FRANK and GAMMEL[4] predicts 117 mb for the forward hemisphere and 27 mb for the backward hemisphere. The difference between 190 and 112 mb may be due in part to experimental errors. Figure 19 shows the energy dependence of the cross section of the disintegration process (29.24) according to the computations of FRANK and GAMMEL to which we shall come back later.

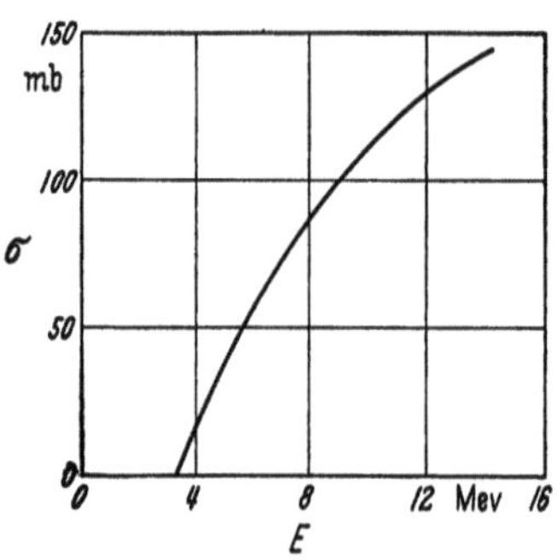

Fig. 19.

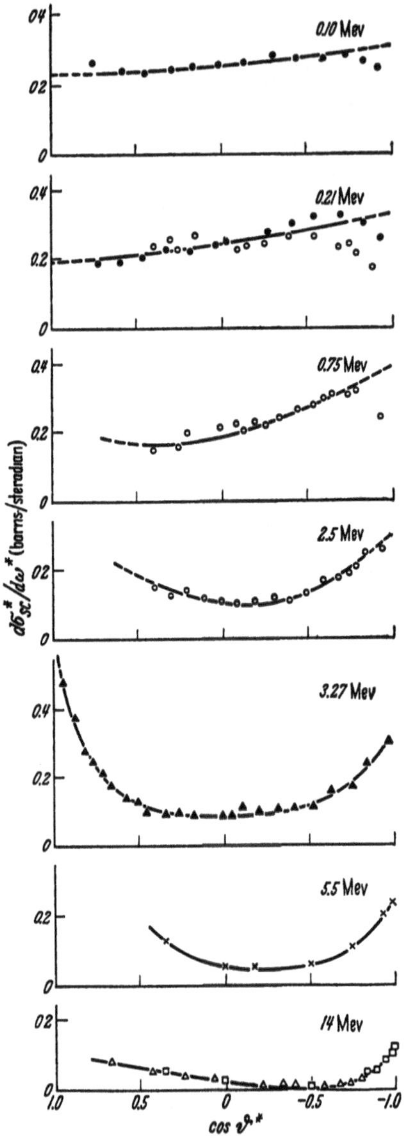

Fig. 20.

Fig. 19. Energy dependence of the cross section of the process of disintegration of the deuteron according to the computations of FRANK and GAMMEL.

Fig. 20. Angular distribution of neutrons scattered elastically by deuterons [14].

[1] M. AGENO, E. AMALDI, D. BOCCIARELLI and G. C. TRABACCHI: Nuovo Cim. 1, 253 (1943). — Phys. Rev. 71, 20 (1947).
[2] H. L. POSS, E. O. SALANT, G. A. SNOW and L. C. L. YUAN: Phys. Rev. 87, 11 (1952).
[3] J. D. SEAGRAVE: Phys. Rev. 97, 757 (1955).
[4] R. M. FRANK and J. L. GAMMEL: Phys. Rev. 93, 463 (1954).

The angular distribution of the scattered neutrons in the c.m.s.[1] is given in Fig. 20 for various neutron energies [*14*]. The anisotropy already evident at $E=0.10$ Mev, increases with increasing energy in a rather complicated way. Figure 21 shows a comparison of the experimental results obtained at $E=14.1$ Mev[2,3] with the results of the computations mentioned below[4].

The radiative capture cross section of the deuteron is so small that it can not be measured by transmission experiments. Its determination is made by various indirect methods similar to those listed before in the case of hydrogen. These methods will be discussed in part E of this article.

The γ-rays emitted in the process (29.23) have been recognized, by means of a pair spectrometer, as a line of $h\nu=(6.244\pm 0.008)$ Mev in the γ-ray spectrum emitted from the reacting vessel of the heavy water Chalk River pile[5,6].

The theoretical investigation of the various $n+\mathrm{D}$ reactions is important as a first step towards the development of a theory of the nuclear reactions in very light elements. It is however rather complicated because it involves the solution of the Schrödinger equation for a three-body problem. And further more the radius of the deuteron (29.12) is so large that *P*- and *D*-waves become important at quite low energies of the relative motion. This point appears evident from the inspection of the angular distributions of the elastically scattered neutrons shown in Fig. 20.

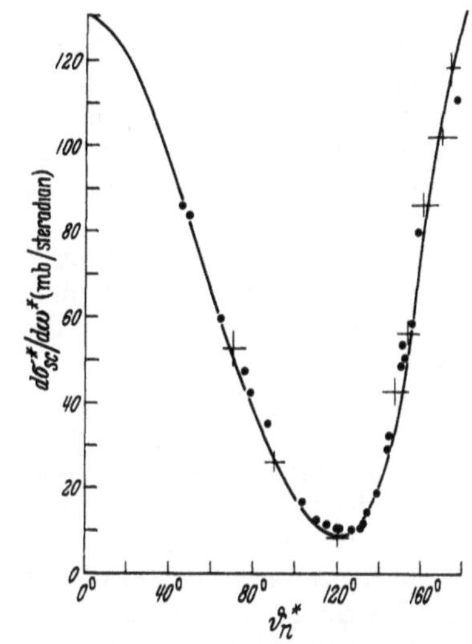

Fig. 21. Elastic scattering of neutrons by deuterons at $E_n=14.1$ Mev. Comparison between the experimental results of ALLRED *et al.* (•) and SEAGRAVE (+) with the computations of CHRISTIAN and GAMMEL.

For a discussion of the theoretical results and their comparison with the experiments made before 1953 we refer the reader to the review article by MASSEY[7].

Since the deuteron in its ground state has spin 1, the total spin of the three-body system $n+\mathrm{D}$ may be $S=\frac{1}{2}$ (doublet state) or $S=\frac{3}{2}$ (quartet state)[8]. Therefore the scattering cross section for unpolarized neutrons colliding against unpolarized deuterons, can be split as follows

$$\sigma_s=\tfrac{1}{3}\sigma_2+\tfrac{2}{3}\sigma_4 \tag{29.28}$$

where σ_2 and σ_4 are respectively the doublet and quartet state cross sections. According to (20.8) at zero energy both σ_2 and σ_4 may be expressed in terms of the corresponding scattering lengths a_2 and a_4.

[1] D. J. HUGHES and R. S. CARTER: Neutron Cross Sections Angular Distributions. BNL 400. 1956.

[2] J. C. ALLRED, A. H. ARMSTRONG and L. ROSEN: Phys. Rev. **91**, 90 (1953).

[3] See footnote 3, p. 72.

[4] R. S. CHRISTIAN and J. L. GAMMEL: Phys. Rev. **91**, 100 (1953).

[5] B. B. KINSEY and G. A. BARTHOLOMEW: Phys. Rev. **80**, 918 (1950).

[6] This and a second line of $h\nu=7.72$ Mev, attributed to neutron capture by aluminium, are the only two strong lines appearing superposed on a continuous unresolved background.

[7] H. S. W. MASSEY: Progress in Nuclear Physics, edited by O. R. FRISCH, Vol. 3, p. 235. 1953.

[8] For theoretical detail see VERDE's article in Vol. XXXIX.

These may be determined, as in the case of n-p collisions, by combining the zero energy cross section, given in Table 17, with the coherent scattering length $(\frac{1}{3}a_2+\frac{2}{3}a_4)$ which is deduced from experiments on the scattering of slow neutrons either by ortho- and para-deuterium[1], or by deuterium-containing crystals such as NaD[2].

Two possible solutions are found by this procedure. From a phase shift analysis of the $p-\mathrm{D}$ (and $n-\mathrm{D}$) scattering data in the energy range 0 to 10 Mev, CHRISTIAN and GAMMEL[3] conclude that the correct set of $n-\mathrm{D}$ scattering lengths is the following

$$\left.\begin{aligned} a_4 &= (6.2 \pm 0.2)\times 10^{-13}\,\mathrm{cm},\\ a_2 &= (0.8 \pm 0.3)\times 10^{-13}\,\mathrm{cm}.\end{aligned}\right\} \tag{29.29}$$

The calculations of these authors are carried out for a general central nuclear potential with either Yukawa or Gauss radial dependence. The potentials are adjusted in range and depth to fit the low energy $n-p$ and $p-p$ data, and, in the case of the Yukawa potential, the high energy $n-p$ data. Tensor forces are not taken into account since they are expected to play a small role in $n-\mathrm{D}$ or $p-\mathrm{D}$ scattering.

As a first step these authors calculate the phase shifts for $l \geqq 1$ using the first Born approximation with symmetrization for equal particles taken into account. This approximation is expected to yield reasonable results for the higher angular momentum states.

The corresponding results are then used as a starting point for the phase shift analysis. The values of the phase shifts obtained by this analysis agree with those calculated in BORN's approximation.

To check these values, the authors felt it necessary to compute the phase shifts in the S states using a variational method with neglect of the polarization of the deuteron. The results of this computation are in close agreement with those of the phase shift analysis, providing a strong *a posteriori* justification for the approximations adopted in these computations.

CHRISTIAN and GAMMEL have also shown that the more important physical process in determining the phase shifts at low energy, especially in the higher momentum states, is the pick-up by one of the particles in the deuteron of the incident particle to form a new deuteron. They are also able to show that only slight differences occur in the phase shifts depending upon the range, shape or tensor nature of the nuclear forces, provided that they are sufficient to bind the deuteron.

In a successive paper FRANK and GAMMEL[4] have treated the inelastic scattering of neutrons (and protons) by deuterons, i.e. the process (29.24) in zero range approximation i.e. assuming that the range of the nuclear forces is small compared to the size of the deuteron and the wavelength of the incident particle. A connection between the elastic and the inelastic scattering cross section is derived, which is independent of the properties of the two-body forces. The result of this simple theory has already been considered before; it agrees much better with the available experimental results than one could expect from the crudeness of the assumptions. We refer to the original paper for more detail.

---

[1] M. HAMERMESH and J. SCHWINGER: Phys. Rev. **69**, 145 (1946). — S. J. NIKITIN, W. T. SMOLYANKIN, W. Z. KOLGANOV, A. W. LEBEDEW and G. S. LOMKAZY, p 81 of Ref. [*15a*]

[2] D G. HURST and N. Z. ALCOCK: Canad. J. Phys. **29**, 36 (1951). — E. O. WOLLAN, C G SHULL and W. C. KEOHLER. Phys. Rev. **83**, 700 (1951).

[3] See footnote 4, p. 73.

[4] See footnote 4, p. 72.

The radiative capture (29.23) may be produced either through magnetic or electric dipole transitions. The contribution from the magnetic moments is the only important one for capture of thermal neutrons, and gives rise to a cross section following the $1/v$ law. VERDE[1] has treated the magnetic dipole contribution from the group theoretical point of view and has shown that it vanishes if the wave function of the triton produced in this process is accurately space symmetric. Since these conditions will be approximately satisfied, VERDE's remark explains the abnormally small capture cross section of the deuteron. A quantitative calculation of the process requires much higher accuracy in the wave functions than that of the wave functions available at present, since a high degree of cancellations is expected to be involved in the computation of the matrix elements.

γ) Beryllium is composed of a single isotope of spin $I = \frac{3}{2}$. Therefore the total spin of the $n + \mathrm{Be}^9$ system can take the two values $S = 2$ (quintet state) and $S = 1$ (triplet state). The following processes take place in beryllium under neutron bombardment [*1*], [*2*]

$$
{}_0n^1 + {}_4\mathrm{Be}^9 \rightarrow \begin{cases} {}_4\mathrm{Be}^9 + {}_0n^1\,, & (29.30)\\ {}_4\mathrm{Be}^{10} + h\nu; \quad Q_0 = 6.811 \text{ Mev}\,, & (29.31)\\ {}_4\mathrm{Be}^{*9} + {}_0n'^1\,, & (29.32)\\ {}_2\mathrm{He}^6 + {}_2\mathrm{He}^4; \quad Q_0 = -0.64 \text{ Mev}\,, & (29.33)\\ {}_4\mathrm{Be}^8 + 2\,{}_0n^1; \quad Q_0 = -1.666 \text{ Mev.} & (29.34)\end{cases}
$$

The total cross section is constant and equal to its zero energy value $\sigma_s(0) = 6.04$ b (Table 17) from 1 ev to about 10 kev and decreases only very little up to about 0.1 Mev [*14*]; it is mainly due to elastic scattering. Deviations from this value due to the crystalline structure of beryllium start to be observed below 0.1 ev (Sect. 101). The spin dependent scattering (Sect. 103) has a very low cross section (< 0.03 mb)[2] and therefore the scattering lengths for the quintet and the triplet states are the same.

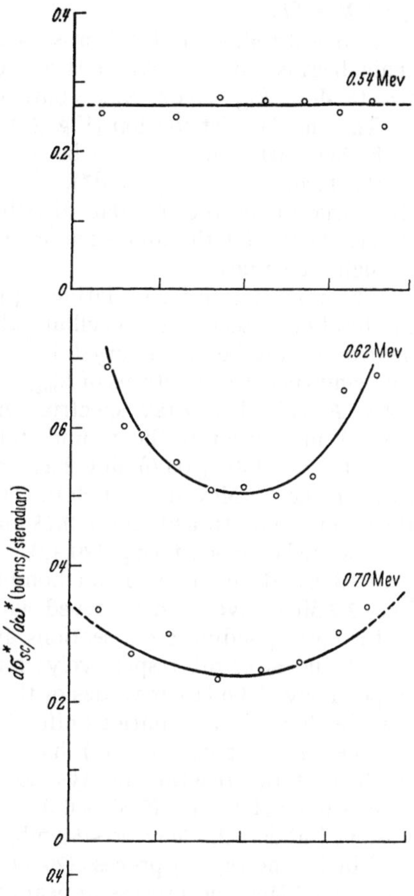

Fig. 22. Angular distribution of neutrons scattered elastically from Be [*14*].

[1] M. VERDE: Helv. phys. Acta **23**, 453 (1950); see also footnote 8, p. 73.
[2] A. PALEVSKI and R. R. SMITH: Phys. Rev. **86**, 604 (1952).

Since they are positive and greater than the nuclear radius, it is concluded that two bound levels with $J_c=2$ and $J_c=1$, odd, may exist with binding energy $\lesssim 1.7$ Mev [1].

Above 0.1 Mev the total cross section begins to decrease and then increases again because of the first of three resonances observed in the region $E=0$ to $E=18$ Mev: $E_{R1}=0.62$, $E_{R2}=0.81$ and $E_{R3}=2.73$ Mev.

The angular distribution (Fig. 22) is isotropic around and below $E=500$ kev; at higher energy it starts to be anisotropic and at the 0.62 Mev resonance, is of the form $1+0.28\,P_2(\cos\vartheta^*)$. We refer to Ref. [2] for the interpretation of the variation of the angular distribution with the energy near the 0.62 Mev resonance and for the literature on the angular distribution measurements made at higher energies.

The cross section for radiative capture (29.31) has been measured by analysing the total cross section of beryllium at various temperatures, as a function of the neutron energy below the crystal cut-off (Sect. 102). The product of the reaction Be^{10} emits negative electrons of $E_{\max}^{(-)}=0.556$ Mev with a half-life $T_{\frac{1}{2}}=(2.7\pm0.4)\times 10^6$ y [1], [2]. The γ-ray spectrum has been investigated with a pair spectrometer[1]; in addition to the ground state line of $h\nu=(6.816\pm0.006)$ Mev, a second line of $h\nu=(3.41\pm0.06)$ Mev has been observed and attributed to a cascade through the 3.4 Mev state of Be^{10}. The intensity of the cascade transition and of the ground state transition are 0.25 and 0.75 photons/capture respectively [1], [2].

The inelastic scattering (29.32) has been investigated by looking for γ-rays emitted by Be under neutron bombardement. Two lines of $h\nu=1.8$ Mev and $h\nu=2.2$ Mev have been observed which might be due to Be. The upper limits of the corresponding cross sections are the following: for $E=2.56$ Mev, $\sigma_{(n,\gamma)}\lesssim 0.3$ mb and $\lesssim 0.2$ mb respectively; for $E=2.74$ Mev $\sigma_{(n,\gamma)}\lesssim 1.8$ mb and $\lesssim 0.3$ mb respectively. The larger values of the cross sections at the higher neutron energy may be due to some spurious effect[2].

The (n,α) process (29.33) has a threshold $E_{1\mathrm{T}}=\frac{10}{9}\,0.64=0.74$ Mev. The product of the reaction decays according to the process $He^6(\beta^-)\,Li^6$, $E_{\max}^{(-)}=(3.50\pm0.05)$ Mev $T_{\frac{1}{2}}=(0.83\pm0.03)$ sec. A resonance for production of He^6 is observed at $E=2.6$ Mev ($\sigma\sim 50$ mb). At 14 Mev $\sigma_{(n,\alpha)}=(10\pm1)$ mb [1], [2].

Finally the $(n,2n)$ process (29.34) has the threshold $E_{1\mathrm{T}}=\frac{10}{9}\,1.666=1.85$ Mev. At $E=3.7$ Mev the process appears to take place in two steps: $Be^9(n,n')\,Be^{9*}$, $Be^{9*}\to Be+n^1$, where Be^{9*} denotes the 2.43 Mev level of beryllium. The cross section for Ra α + Be neutrons is (0.3 ± 0.1) b.

δ) Carbon is composed of two isotopes: C^{12} (98.892%, $I=0$), C^{13} (1.108%, $I=\frac{3}{2}$). The following processes have been observed in the two isotopes:

$$
{}_0n^1+{}_6C^{12}\to\begin{cases}
{}_6C^{12}+{}_0n^1, & & (29.35)\\
C^{13}+h\nu; & Q_0=4.948\ \text{Mev}, & (29.36)\\
{}_6C^{12*}+{}_0n^{1\prime}, & & (29.37)\\
3\alpha+{}_0n^1; & Q_0=-7.277\ \text{Mev}, & (29.38)\\
{}_7C^{11}+2\,{}_0n^1; & Q_0=-18.711\ \text{Mev}, & (29.39)
\end{cases}
$$

$$
{}_0n^1+{}_6C^{13}\to\begin{cases}
{}_6C^{13}+{}_0n^1, & & (29.40)\\
{}_6C^{14}+h\nu; & Q_0=8.169\ \text{Mev}, & (29.41)\\
{}_4Be^{10}+\alpha; & Q_0=-3.845\ \text{Mev}. & (29.42)
\end{cases}
$$

---

[1] B. B. KINSEY, G. A. BARTHOLOMEW and W. H. WALKER: Canad. J. Phys. **29**, 1 (1951). — B. B. KINSEY and G. A. BARTHOLOMEW: Canad. J. Phys. **31**, 537 (1953).

[2] R. B. DAY, p. 9 of Ref. [*15a*]. — Phys. Rev. **102**, 767 (1956).

The total cross section is constant and equal to its zero energy value (Table 17) from about 1 ev to 20 kev and decreases only very little up to 0.15 Mev. Deviations from this value due to the crystalline structure start to be appreciable below 0.1 ev [14]. These deviations are considered in detail in Sects. 101 and 102.

Above 0.15 Mev the cross section decreases monotonically reaching 2.0 b at $E = 1.6$ Mev; various resonances are observed between 2.0 and 7.5 Mev.

Figure 23 shows the angular distribution of the elastically scattered neutrons which is isotropic around and below 1 Mev and starts to deviate appreciably from isotropy around 1.5 Mev. The reader may compare Refs. [1] and [2] for possible interpretations of the resonances observed in the total cross section and of the angular distributions of the scattered neutrons.

The investigation of the γ-rays emitted in process (29.36) has been made by means of a pair spectrometer[1] with the result that in addition to the $h\nu = (4.949 \pm 0.006)$ Mev ground state line, there is a line of $h\nu = (3.68 \pm 0.05)$ Mev with an intensity of 0.3 photons/capture. Other possible lines of 3.1 and 3.9 Mev energy, have intensities lower than 0.10 and 0.06 photons/capture [2].

The γ-ray and α-particle emission as well as the production of inelastically scattered neutrons according to the two processes (29.37) and (29.38) have been investigated mainly at $E = 14$ Mev. The cross section for production of γ-rays is ≈ 0.30 b and 0.19 b according to different authors; that for production of α-particles (0.135 ± 1.04) b and that for inelastically scattered neutrons, in the range $E_{n'} = 0$ to $E_{n'} = 11.5$ Mev, (0.760 ± 0.040) b, in the range $E_{n'} = 0.5$ to $E_{n'} = 12$ Mev, 0.520 ± 0.200 b [2]. At $E = 6.58$ Mev the cross section for production of γ-rays of $h\nu = 4.42$ Mev is 370 mb $(\pm 15\%)$[1].

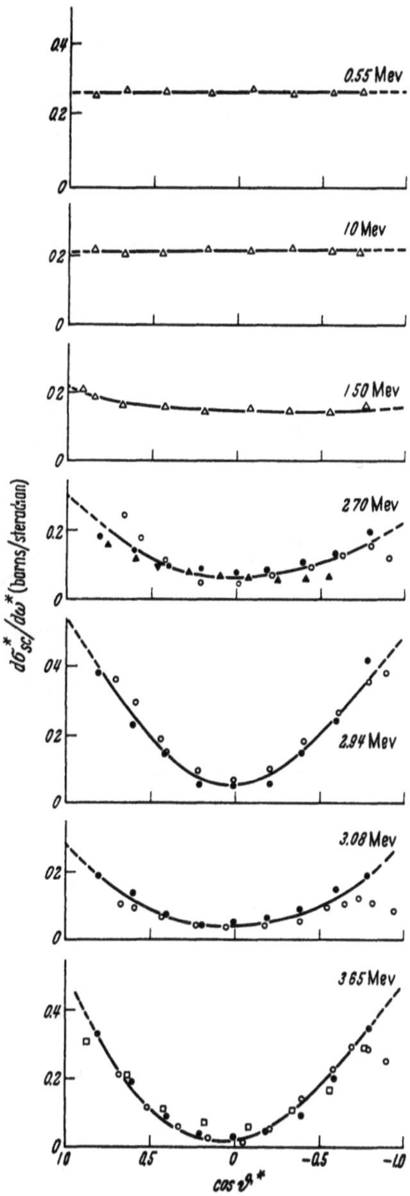

Fig. 23. Angular distribution of neutrons scattered elastically from C [14].

Finally the threshold of the $(n, 2n)$ process (29.39) is $E_{1T} = \frac{13}{12}\, 18.7 = 20.2$ Mev, i.e. above the energy interval considered in this article.

Much less is known about the processes produced by neutrons in C^{13}. The radiative capture of the incident neutron has a thermal energy cross section [14]

$$C^{13}: \sigma_a(2200 \text{ m/sec}) = (0.5 \pm 0.2) \text{ mb}. \qquad (29.43)$$

[1] See footnote 2, p. 76.

The product of the reaction is unstable for emission of electrons: $C^{14}(\beta^-)N^{14}$, $E^{(-)}_{max} = 155$ kev; $T_{\frac{1}{2}} = (5900 \pm 200)$ y. Data and literature about the processes (29.40), (29.42) are found in Ref. [*2*].

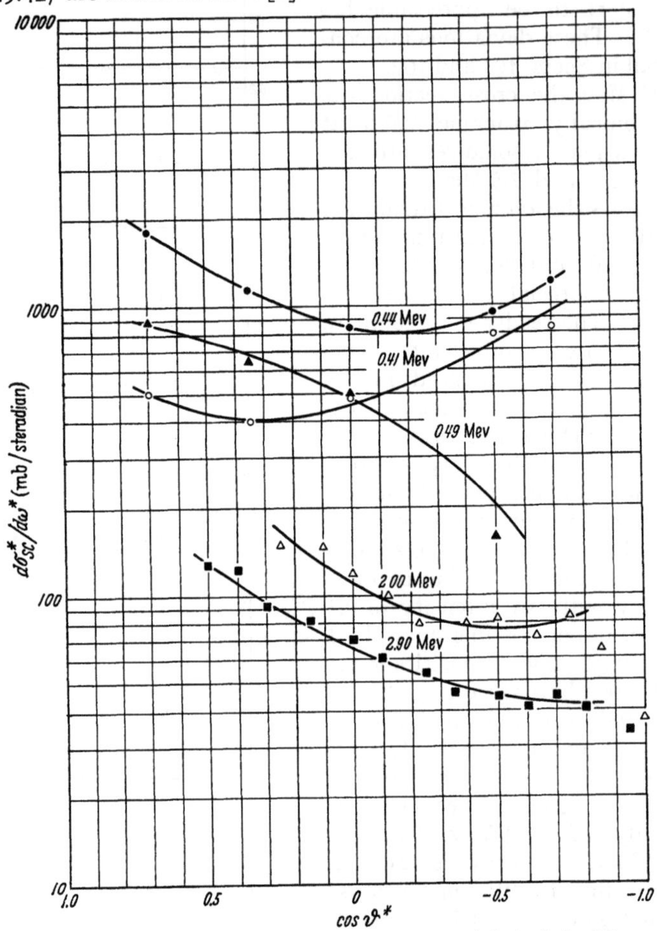

Fig. 24. Angular distribution of neutrons scattered elastically by O^{16}.

ε) *Oxygen* is composed of three isotopes: O^{16} (99.758%, $I=0$), O^{17} (0.0373%, $I=\frac{1}{2}$?), O^{18} (0.2039%, $I=0$). The processes observed in O^{16} are the following

$$
{}_0n^1 + O^{16} \rightarrow \begin{cases}
{}_8O^{16} + {}_0n^1, & & (29.44)\\
{}_8O^{17} + h\nu; & Q_0 = 4.143 \text{ Mev}, & (29.45)\\
{}_8O^{16*} + n', & & (29.46)\\
{}_6C^{13} + \alpha; & Q_0 = -2.201 \text{ Mev}, & (29.47)\\
{}_7N^{16} + p; & Q_0 = -9.62 \text{ Mev}, & (29.48)\\
{}_7N^{15} + d; & Q_0 = -9.885 \text{ Mev}, & (29.49)\\
{}_8O^{15} + 2n; & Q_0 = -15.597 \text{ Mev}. & (29.50)
\end{cases}
$$

The total cross section remains constant and equal to its zero energy value from 1 ev to almost 0.3 Mev where a series of resonances begins, extending to energies of 10 Mev and more [*14*]. The corresponding resonance parameters are given in Ref. [*2*].

The effect of molecular binding begins below about 0.1 ev (see Sect. 100).

From Fig. 24 one sees that the anisotropy, already considerable at $E=0.4$ Mev, changes in a rather complicated way by changing the energy of the neutrons; more data on this point may be found in Ref. [14] and footnote 1 on p. 73.

The cross section for radiative capture (29.45) is given in Table 13. The inelastic scattering (29.46) has a cross section, at $E=14.1$ Mev, of 0.5 b. The neutrons are emitted roughly isotropically. The cross section for production of γ-rays of 6 to 7 Mev is found to be 0.25 b [2].

The following values of the cross sections of the three processes (29.47) to (29.49) have been reported at $E=$ 14.1 Mev [14]:

$$\left.\begin{aligned} \sigma_{(n,\alpha)} &\approx 310 \text{ mb}, \\ \sigma_{(np)} &\approx 35 \text{ mb}, \\ \sigma_{(nd)} &\approx 15 \text{ mb}. \end{aligned}\right\} \quad (29.51)$$

The cross section of the (n, α) process has been measured with considerable accuracy by HUBER and SEITZ[1] for neutrons in the energy range 3.6 to 4.2 Mev (Fig. 25).

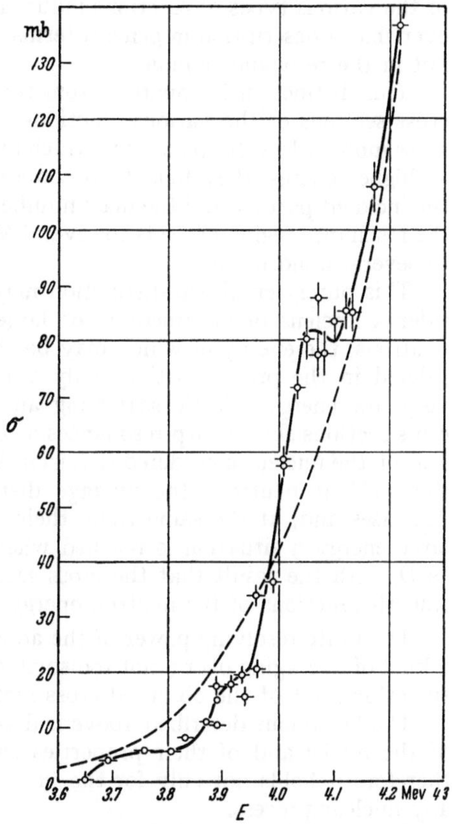

Fig. 25. Cross section for $O^{16}(n, \alpha) C^{13}$; solid curve results of measurements; broken curve penetrability of particles through Coulomb barrier (HUBER and SEITZ).

The threshold of the $(n, 2n)$ process is

$$E_{1T} = \tfrac{17}{16}\, 15.6 = 16.6 \text{ Mev}$$

and therefore it is outside the energy region of interest in the present article.

The following reactions have been observed in O^{17} and O^{18}

$$\left.\begin{aligned} &O^{17}(n,\alpha)\,C^{14}; \quad Q_0 = 1.825 \text{ Mev}, \\ &O^{17}(n,p)\,N^{17}; \quad Q_0 = -8.0 \text{ Mev}, \\ &O^{18}(n,\gamma)\,O^{19}; \quad Q_0 = 3.956 \text{ Mev} \end{aligned}\right\} \quad (29.52)$$

where

$$_7N^{17}(\beta^-)\,_8O^{17**};$$
$$E_{\max}^{(-)} = (3.7 \pm 0.2) \text{ Mev};$$
$$T_{\frac{1}{2}} = (4.14 \pm 0.04) \text{ sec}$$

with $_8O^{17**}$ denoting one (or more) excited states which decay to O^{16} by neutron emission (the neutron spectrum has a maximum at (0.92 ± 0.07) Mev [2]), and

$$_8O^{19}(\beta^-)\,_9F^{19}, \quad E_{\max}^{(-)} = (4.5 \pm 0.3) \text{ Mev} \quad T_{\frac{1}{2}} = 29 \text{ sec}.$$

The cross section of the (n, α) process (29.52) is, at thermal neutron energy

$$O^{17}: \quad \sigma_{(n,\alpha)} = (0.46 \pm 0.11) \text{ b}$$

and that of the (n, γ) process (29.52) [14]

$$O^{18}: \quad \sigma_{(n,\gamma)} = (0.21 \pm 0.04) \text{ mb}.$$

---

[1] P. HUBER and J. SEITZ, p. 121 of Ref. [15a].

30. A few general remarks on the elastic and inelastic scattering in medium and heavy nuclei. The slowing down and diffusion of neutrons is also influenced by the reactions induced by neutrons in medium and heavy elements present in the moderator. Since we can not enter into a complete discussion of the various processes taking place in all the elements of medium and large mass number as a function of energy of the incident neutrons, the reader may compare Refs. [*3*], [*18a*] and [*18b*] [1]. Here, we only can give a summary of the more typical properties of the various types of reactions induced by neutrons. It has to be noticed however that a description in general terms is possible only in the continuum region, not in the resonance region.

This distinction is apparent from the inspection of the empirical data on the cross sections of the various reactions. While at low energies all cross sections show more or less sharp maxima which are characteristic of the particular process, at higher energy they tend to become smooth functions of both the energy of the incident particle and the mass number of the target nucleus. In light elements the resonance region extends to several Mev, while in heavy elements it is limited to several hundred kev.

This purely empirical distinction may be interpreted as follows. Let us consider a nucleus of intermediate or large mass number under bombardment of neutrons, the energy of which may be changed at will. Since fission is not considered in the present article, only two processes are energetically possible at very low energy: elastic scattering and radiative capture. The corresponding cross sections show sharp resonances at certain energies E_{Ri} which are characteristic of the nucleus considered (Division VIII above). By increasing the energy of the incident neutrons the average distance D between successive resonances decreases and, at the same time, their total width Γ increases. At sufficiently large energy a situation is reached where Γ is of the same order of magnitude as D, with the result that the cross sections of these processes tend to become smooth functions of the neutron energy E.

The finite resolving power of the adopted experimental set-up may introduce a kind of averaging over a not too small energy interval which produces a further smoothing out of the observed cross sections.

The behaviour described above is determined by the distribution of the levels of the nuclei and of their properties as functions of the excitation energy w; therefore it holds not only for the (n, n) and (n, γ) reactions, but in general for any nuclear process.

In this section we shall mainly consider the various reactions in the continuum region. In doing this we shall start from the distinction between elastic scattering cross section and reaction cross section as they have been defined in Sect. 19 on very general grounds.

The differential scattering cross section is represented by Eq. (18.8); its actual dependence on the scattering angle ϑ^* (in the c.m.s.) is determined at each energy, by the value of the set of complex numbers η_l. In the case of a large black nucleus with a sharp edge such as that considered in Sect. 19, the values of η_l are given by (19.2) and the differential cross section (18.8) reduces to the well known Fraunhofer diffraction pattern produced by a small absorbing sphere.

This model provides a fairly good qualitative representation of the angular distribution of the neutrons scattered elastically by intermediate and heavy nuclei. At low energy the differential cross section is a rather flat function of the angle; in the limiting case of $\lambda\!\!\!^{-} \to \infty$ the elastic scattering is always isotropic.

[1] B.B. Kinsey: p. 202, Vol. XL of this Encyclopedia.

By increasing the energy, the scattered neutrons start to be peaked in the forward direction, a fact established experimentally for a long time[1]. At still higher energies a very large fraction of the scattered neutrons is concentrated in the forward direction. By increasing the angle of observation the intensity of the scattered neutrons decreases, passes a minimum very close to zero and then increases again showing a second maximum. The angles at which the minimum and the second maximum are observed appear to be related in a simple way to the neutron wavelength and the dimensions of the nucleus[2], as one has to expect in the case of diffraction patterns[3].

Figure 26, taken from the review article by WALT[4], shows the results of observations made at different energies by different authors[5] in Bi, for $E=2.5$ Mev, and Pb, for all other energies. The difference between these two nuclei is expected to be negligible.

While "a black sphere with a sharp edge" is sufficient for a qualitative description of these curves, a quantitative computation imposes a more refined model. In particular the sharp edge at the surface of the nucleus does not provide an accurate picture of the situation. The interaction of the incident neutron with the nucleus must drop smoothly to zero in a surface region of the dimensions of the range of the nuclear forces. Such an effect may be neglected at low energies where $\lambda\!\!\!^{-}$ is very large, but it appears to be important at energies of the order of 10 to 15 Mev or more, if a quantitative interpretation of the curves of Fig. 26 is desired[6].

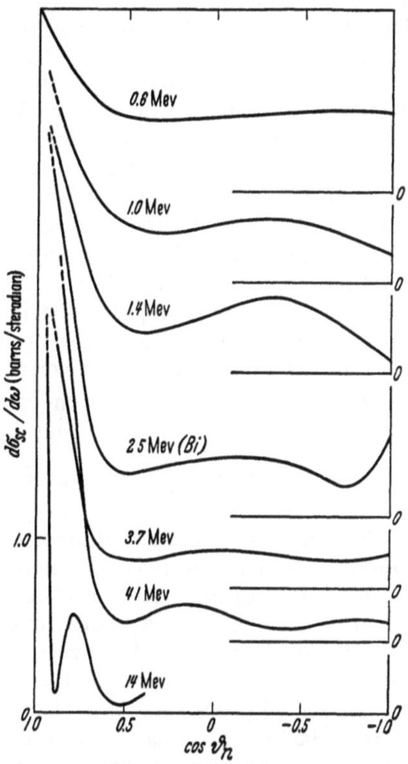

Fig. 26. Differential cross section for elastic scattering of neutrons by heavy elements (Pb or Bi)[4].

At energies larger than those considered here, the interpretation of the differential cross section for elastic scattering, requires the introduction of the assumption that the nucleus has a certain transparency which originates from the increased mean free path of the incident neutron in nuclear matter[7]. We

[1] S. KIKUCHI, H. AOKI and T. WAKATUCHI: Proc. Phys.-Math. Soc. Japan **21**, 410 (1939). Phys. Rev. **55**, 1264 (1939).

[2] E. AMALDI, D. BOCCIARELLI, B. N. CACCIAPUOTI and G. C. TRABACCHI: Nuovo Cim. **3**, 15 (1947). — Physical Society Cambridge Conference, p. 97, 1947.

[3] H. A. BETHE and G. PLACZEK. Phys. Rev. **57**, 1075 (1939).

[4] M. WALT, p. 18 of Ref [*15a*].

[5] M. WALT and H. H. BARSCHALL: Phys. Rev. **93**, 1062 (1954). — W. D. WHITEHEAD and S. C. SNOWDEN. Phys. Rev. **94**, 1267 (1954). — M WALT and J. R. BEYSTER: Phys. Rev. **98**, 677 (1955). — A. LANGSDORF and R. O. LANE: Bull. Amer. Phys. Soc. **30**, No. 3 (1955). — I. L. MORGAN: Bull. Amer. Phys Soc. **30**, No. 3 (1955). — J. H. COON and R. W. DAVIS. Bull. Amer. Phys. Soc. **29**, No. 6 (1954); see also footnote 4.

[6] V. F. WEISSKOPF, p. 23 of Ref. [*15a*].

[7] S. FERNBACH, R. SERBER and B. T. TAYLOR: Phys. Rev. **75**, 1352 (1949).

shall not enter into the details of this interesting problem because it would bring us too far away from the scope of the present article.

The elastic collisions with intermediate and heavy nuclei contribute only very little to the slowing down of neutrons. Under the assumption of isotropic scattering (in the c.m.s.) the average logarithmic energy loss per collision is given by Eq. (62.14); for nuclei of mass number $A \gtrsim 100$, it amounts to 2% or less [see Eq. (62.18)]. When the angular distribution starts to be peaked in the forward direction, as shown by Fig. 26, the average energy loss becomes still smaller because the momentum transferred to the target nucleus is the lower the smaller is the scattering angle of the neutron.

The spatial distribution of the neutrons may be influenced appreciably by the elastic scattering on intermediate and heavy nuclei in the energy region where the differential scattering cross section is a rather flat function of the angle of scattering; but when the energy is so large that the great majority of the collisions deviate the neutron only by a very small angle, the effect of these collisions tends to become negligible. This may be recognized by recalling that the transport mean free path, not the elastic scattering mean free path, enters into the problem. According to Eq. (72.3) l_{tr} contains the factor $(1-\overline{\cos\vartheta})^{-1}$ which, in the case of angular distributions strongly peaked in the forward direction, becomes very large, at the limit, infinite.

The reaction cross section considered in Sect. 18 was measured a long time ago[1-3]. For the purpose of the present discussion we write it as the sum of various terms, some of which may be zero:

$$\sigma_r(E) = \sigma_{(n,\gamma)}(E) + \sigma_{\text{inel}}(E) + \sum_{a_i \neq n} \sigma_{(n,a_i)}(E) + \sum_{a_i, b_i} \sigma_{(n,a_i b_i)} + \cdots. \tag{30.1}$$

The cross section for radiative capture is important only at very low energy where all other reactions are energetically impossible or very improbable. As we have seen in Sect. 26 as a result of the combined effect of resonances and of the $1/v$ factor, $\sigma_{(n,\gamma)}$ may reach extremely large values in the very low energy region. At about 1 Mev, however, it falls to about 0.1 barns and decreases rapidly at higher energies because of the competition of neutron emission.

The (n, a_i) processes, for particles a_i different from neutrons, are much weaker than inelastic scattering, because of the effect of the Coulomb potential barrier. Few exceptional cases, all among very light nuclei, have been considered in Sects. 17 and 27.

The effect of the Coulomb potential barrier is enhanced by the fact that the outgoing particles are emitted preferentially with a low energy so that the residual nucleus is left in a highly excited state [see Eq. (30.7), below].

The $(n, a_i b_i)$ processes like $(n, 2n)$ or (n, pn) are always endoenergetic; for a given target nucleus the Q-values of $(n, 2n)$ and (n, pn) reactions are identical to the Q-values of the (γ, n) (Sect. 48) and (γ, p) reactions respectively. From Table 23 we see that their threshold $E_{1\text{T}} = \frac{1+A}{A}|Q|$ is always larger than ~ 7 Mev with the only exceptions of a few very light elements considered in Sect. 29.

The $(n, a_i b_i)$ processes are described as two-step reactions: the first reaction is a (n, n) or (n, p) reaction which leaves the residual nucleus in a highly excited

[1] D. C. GRAHAM and G. T. SEABORG: Phys. Rev. **53**, 795 (1938). — A. SOLTAN: Nature, Lond. **142**, 252 (1938).

[2] See footnote 2, p. 81.

[3] H. T. GITTINGS, H. H. BARSCHALL and G. G. EVERHART. Phys. Rev. **75**, 610 (1949). — D. D. PHILLIPS, R. W. DAVIS and E. R. GRAVES: Phys. Rev. **88**, 600 (1952).

state from which a second particle may be emitted. Because of the Coulomb barrier the $(n, 2n)$ processes are the strongest among all possible $(n, a_i b_i)$ processes, at least at the energies considered in this article. The corresponding cross section rises rapidly in an interval of a few Mev above the threshold, at the expenses of the (n, n') processes.

In conclusion, provided the energy of the incident neutron is much higher than the first excited state of the residual nucleus and not so high that $(n, 2n)$ processes start to be important, the re-emission of the neutron with a lower energy combined with the residual nucleus left in an excited state, is by far the strongest reaction induced by neutrons.

If the Bohr assumption is adopted and the cross section $\sigma_c(E)$ for the formation of a compound nucleus is defined (Sect. 25), this may be expressed as the sum of the reaction cross section σ_r and of that part of the cross section for elastic scattering which is due to processes going through formation of a compound nucleus. This type of scattering, indicated as internal or resonance scattering, is coherent with the external or potential scattering and therefore it gives rise to the interference effects considered in Sects. 21 and 26.

According to this remark the internal elastic scattering is a process in competition with the inelastic scattering; but in the case of heavy nuclei and neutron energies much larger than the first excited state of the target nucleus, the internal scattering is much smaller than the inelastic scattering. In fact it corresponds to the same physical process with the difference that the target nucleus is left in its ground state and not in one of the many possible excited states.

According to Eq. (25.10) the cross section for that particular type of inelastic scattering in which the product nucleus is left in the excited state w_i, is given by the expression

$$\sigma_{\text{inel}}^{(i)}(E) = \sigma_c(E)\,\frac{\Gamma_n^{(i)}}{\Gamma} \tag{30.2}$$

where $\Gamma_n^{(i)}$ is the partial width for emission of neutrons of energy

$$E_i' = E - w_i \tag{30.3}$$

and Γ the total width for emission either of neutrons of the initial energy (Γ_0, internal elastic scattering), or of neutrons of degraded energy according to (30.3), or of any other particle or photon a_i:

$$\Gamma = \Gamma_0 + \Sigma_i \Gamma_n^{(i)} + \sum_{a_i \neq n} \Gamma_{a_i}. \tag{30.4}$$

If the state w_i considered is the first excited state of the residual nucleus ($w_i = w_1$) and the initial energy (in the c.m.s.) is only slightly higher than w_1, Eq. (30.2) will represent the total inelastic scattering cross section and the spectrum of the inelastically scattered neutrons will consist of a single line the energy of which is given by (30.3) with $i=1$.

In a narrow energy interval immediately above the threshold of the considered inelastic reaction, $\sigma_c(E)$ and Γ may be considered as constants, so that the energy dependence of $\sigma_{\text{inel}}^{(i)}(E)$ will be mainly determined by the energy dependence of $\Gamma_n^{(i)}$. We can conclude, making use of the remark on p. 50 that

$$\sigma_{\text{inel}}^{(i)}(E) \propto E_i'^{\,l+\frac{1}{2}} \tag{30.5}$$

where l is the angular momentum of the emitted neutron. This is determined by the values of the total angular momentum and parity of both the compound nucleus and the residual nucleus in its state w_i. The transition of the residual nucleus from this state to the ground state will take place by emission of one or more photons.

The experimental investigation of the energy spectrum and of the angular distribution of the emitted neutrons and γ-rays as well as of their angular correlations, is a problem of the greatest importance since it allows the derivation of direct information on the quantum numbers characterizing the various levels involved in the process [*18b*].

When the energy of the incident neutrons is sufficiently large to produce the residual nucleus in many different excited states, the inelastic scattering cross section becomes

$$\sigma_{\text{inel}}(E) = \sum_{w_i \leqq w_c} \sigma_{\text{inel}}^{(i)} \tag{30.6}$$

where $\sigma_{\text{inel}}^{(i)}$ is given by (30.1) and the sum has to be extended over all those excited levels w_i of the residual nucleus which lie below the excitation energy w_c of the compound nucleus. To each term in (30.5) there will correspond a different line in the spectrum of the inelastically scattered neutrons.

In the limiting case of a very large number of levels involved in the process, one can make use of a statistical theory developed by WEISSKOPF and coworkers [*3*], in which the sum with respect to the index i characterizing the various excited states of the residual nucleus, is substituted by an integral over the expression (30.2) multiplied by the number of excited states of the residual nucleus which fall between w and $w+dw$. In this schematization, the energy spectrum of the emitted neutrons is a continuum spectrum of the type

$$d\sigma(E, E') = \sigma_c(E)\,\frac{E'}{T^2}\,e^{-\frac{E'}{T}}\,dE' \tag{30.7}$$

where $T = T(E)$ is a function of the energy E of the incident neutron indicated as nuclear temperature; it is determined by the distribution of the levels of the nucleus left after the emission of the neutron.

We shall not enter into more detail on this subject; additional information may be found in Refs. [*3*] and [*18b*]. In the latter book [*18b*] FELD has discussed at length, not only the theoretical aspects but also the experimental data available in 1953.

Here we limit our considerations to recalling the following points. First the qualitative semi-empirical representation of the behaviour of the level density as a function of the excitation energy w and of the mass number A, is still rather uncertain and unexplained[1]. These uncertainties reflect in the nuclear temperature T. Secondly for an excitation energy w_c of the order of 5 Mev T is about 1.5 Mev for $A=55$ and 0.6 Mev for $A=231$; for $w_c \approx 10$ Mev $T=2.2$ Mev for $A=55$ and $T=0.9$ for $A=231$ [*3*]. And finally we notice that according to (30.7) the spectrum of the inelastically scattered neutrons is peaked around $E' \approx T$ and independent of the direction of emission.

The comparison of the predictions of this theory with the experimental data[1] shows that the picture of the phenomenon given by the Bohr assumption and the statistical approximation mentioned above is qualitatively acceptable but *not* quantitatively correct. The more important deviations are the following. The spectrum of the inelastically scattered neutrons observed experimentally, is richer in high energy neutrons than is predicted from (30.7). Furthermore this excess of fast neutrons is mainly concentrated in the forward direction with respect to the direction of the incident neutrons.

---

[1] See footnote 6, p. 81.

These discrepancies between theory and experiments appear to originate from the use of the Bohr assumption which does not give a satisfactory picture of the phenomenon. The observed behaviour of nuclei in these as well as in other reactions [(γ, n), (γ, p), (n, p)] can be explained if one considers a mechanism of direct momentum transfer from the incident neutron to a single nucleon in the nucleus; at rather high energy the mean free path of nucleons in nuclear matter is sufficiently long to allow a non negligible probability for the exit from the nucleus without any further exchange of energy, either of the incident neutron after its first collision, or of the corresponding recoiling nucleon. These arguments are developed in detail in the article by WEISSKOPF mentioned above[1].

Table 14. *Cross sections for inelastic scattering of neutrons* (PASECHNIK *et al.*).

Scatterer	2.5 Mev	2.5 Mev*	3.3 Mev	4.1 Mev	14 Mev
Carbon	—	—	—	—	0.63 ± 0.05
Sodium	0.53 ± 0.26	—	—	0.57	—
Magnesium	0.77 ± 0.25	—	—	0.78 ± 0.20	1.08 ± 0.09
Aluminium	0.96 ± 0.17	—	0.81 ± 0.20	0.78 ± 0.20	1.13 ± 0.07
Phosphorus	0.7 ± 0.2	—	—		—
Sulphur	0.54 ± 0.21	—	—	—	—
Chlorine	0.6 ± 0.3	—	—	0.90 ± 0.3	—
Calcium	0.4 ± 0.2	—	—	—	—
Chromium	1.4 ± 0.3	—	—	—	—
Iron	1.16 ±	—	1.51 ± 0.21	1.58 ± 0.30	1.29 ± 0.07
Cobalt	1.40 ± 0.11	—	—	—	—
Nickel	0.83 ± 0.12	—	1.35 ± 0.25	—	1.38 ± 0.05
Copper	1.58 ± 0.15	—	1.67 ± 0.24	1.68 ± 0.25	1.35 ± 0.05
Zinc	2.18 ± 0.5	—	2.18	1.58 ± 0.3	1.37 ± 0.07
Selenium	1.88 ± 0.17	—	—	—	1.40 ± 0.23
Molybdenum	1.9 ± 0.3	—	—	—	—
Silver	2.13 ± 0.18	—	—	—	—
Cadmium	2.21 ± 0.20	2.1 ± 0.2	2.35 ± 0.35	2.3 ± 0.3	1.83 ± 0.13
Tin	—	1.65 ± 0.3	1.73 ± 0.15	2.11 ± 0.3	1.81 ± 0.13
Antimony	1.88 ± 0.20	2.11 ± 0.20	2.11 ± 0.20	2.13 ± 0.3	1.90 ± 0.16
Tellurium	2.0 ± 0.35	—	—	—	1.97 ± 0.27
Iodine	1.96 ± 0.25	—	—	—	1.88 ± 0.16
Barium	—	0.8 ± 0.5	1.3 ± 0.4	2.2 ± 0.4	—
Tungsten	2.6 ± 0.25	—	—	—	2.35 ± 0.14
Mercury	2.6 ± 0.3	—	2.99 ± 0.3	3.36 ± 0.3	—
Lead	1.69 ± 0.32	—	—	1.72 ± 0.3	2.42 ± 0.09
Bismuth	0.62 ± 0.29	—	0.62 ± 0.5	1.14 ± 0.40	2.44 ± 0.13

* Second run of measurements.

We shall not attempt here to summarize the many experimental results on the (n, n') reactions which have been published since 1953 when this subject was discussed thoroughly by FELD [*18b*]. We limit ourselves to reporting some of the results contained in a few papers which cover wide ranges of neutron energies and of mass numbers.

The report presented by PASECHNIK at the Geneva conference[2] summarizes the results obtained at the Institute of Physics of the Academy of Science of the Ukrainian SSR. The cross section measurements were made by using mono-energetic neutrons of various energies produced by means of the $D(d, n)$ and $D(t, n)$ reactions (Sect. 57). The inelastically scattered neutrons were measured

[1] See footnote 6, p. 81; see also V.F. WEISSKOPF: Rev. Mod. Phys. **29**, 174 (1957) and the article by B.B. KINSEY in Vol. XL of this Encyclopedia.

[2] M. V. PASECHMIK, p. 3 of Ref. [*15a*].

by means of various threshold detectors of increasing reaction threshold[1]: $P^{30}(n,p)$; $Al^{27}(n,p)$; $S^{32}(n,p)$; $Cu^{63}(n,2n)$.

For instance the inelastic scattering cross section at $E=2.5$ Mev was measured by a transmission experiment using as detector the activity induced through the process $P^{30}(n,p)$ the effective threshold of which is 1.7 Mev, and that of neutrons of 14 Mev with the activity induced through the process $Cu^{63}(n,2n)$ the effective threshold of which is at 12.5 Mev.

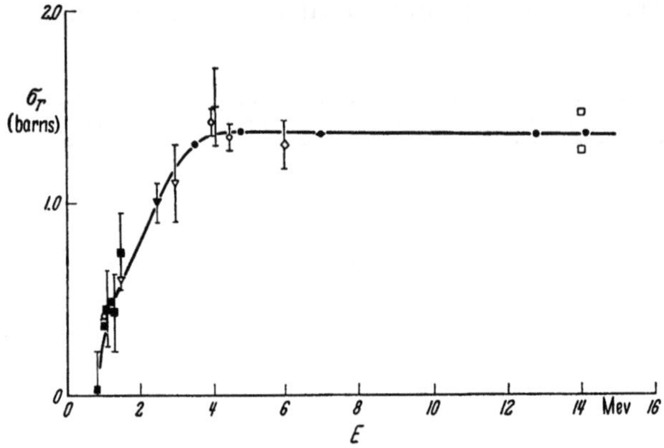

Fig. 27. Reaction cross section of Fe as a function of neutron energy[1,2].

When the effective threshold is only slightly lower than the energy of the incident neutrons, the detector is insensible to the inelastically scattered neutrons. A transmission experiment made under such geometrical conditions that the effect of elastic scattering is exactly compensated, will provide a measurement of σ_r. Since in the range of energies considered here the contributions of all other reactions to (30.1) is negligibly small, σ_r is practically identical to σ_{inel}.

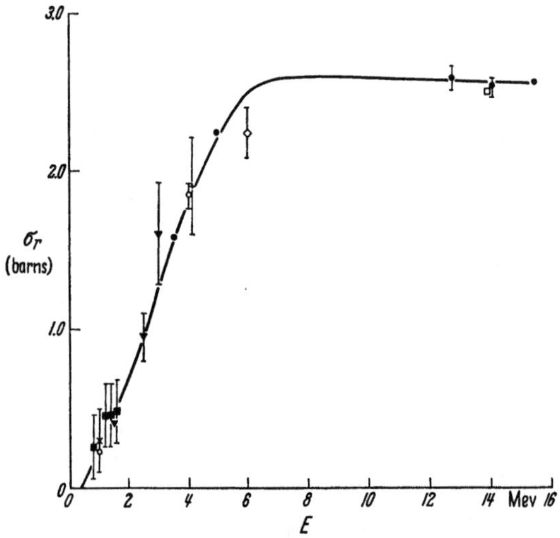

Fig. 28. Reaction cross section of Pb as a function of neutron energy[1,2].

The results of the Russian authors are summarized in Table 14 which shows the dependence of σ_r on E and A.

The energy dependence of the inelastic cross section is illustrated in Fig. 27 and 28 where the results obtained by various American groups have been collected[2]. In the case of Fe the threshold for inelastic scattering may be placed at the first excited state of the principal isotope of iron (850 kev). Above this threshold the cross section rises to a plateau at about 5 Mev where σ_{inel} is about 40% of the total cross section $\sigma_{\text{tot}}=\sigma_s+\sigma_r$.

In order to understand the physical meaning of this result[3,4], we recall that, according to Eq. (19.4), in the case of a black large nucleus with sharp edge σ_r represents 50% of σ_{tot}.

---

[1] H. H. Barschall: Detection of neutrons, in Vol. XLV of this Encyclopedia.

[2] M. Walt and H. H. Barschall: Phys. Rev. **93**, 1062 (1954). — M. Walt and J. R. Beyster: Phys. Rev. **98**, 677 (1955). — E. R. Graves and R. W. Davis: Phys. Rev. **97**, 1205 (1955). — J. R. Beyster, R. L. Henkel, R. A. Nobles and J. M Kister: Phys. Rev. **98**, 1216 (1955).

[3] See footnote 2, p. 81.

[4] See footnote 1, p. 82.

The same behaviour is shown by Pb (Fig. 28). Here the threshold is at about 0.5 Mev and the plateau is reached near 7 Mev where σ_{inel} amounts to about 45% of the total cross section. In both cases the values at 14 Mev are slightly lower than the values at about 5 to 7 Mev.

From Table 14 one recognizes that for neutrons of relatively low energy (2.5 to 4 Mev) σ_{inel} is a smooth function of A following roughly the $A^{\frac{2}{3}}$-law, except for the nuclei containing magic numbers of neutrons (50, 80 and 126: Sn, Ba, Pb and Bi) for which σ_{inel} is considerably lower than for other neighbouring nuclei. Such a behaviour has disappeared at $E=14$ Mev.

The Russian group has also measured in a number of elements the spectrum of the neutrons scattered inelastically at about 90° for $E=2.75$ Mev, by means of a fast neutron spectrometer which consists of an ionization chamber filled with a H_2—A mixture and connected to a pulse height analyser. Various groups of neutrons were observed for which the reader may compare the Pasechnik paper[1] where extensive literature on previous work also may be found.

Many other authors have investigated the inelastic scattering by detecting the γ-rays emitted by the residual nucleus, by means of crystal scintillation spectrometers. Among all these papers we shall limit ourselves to mentioning that by Day[2] which covers, with considerable accuracy, wide ranges of neutron energies and of mass numbers.

B. Reactions with emission of neutrons and neutron sources.

31. Introductory remarks and comparison between neutron sources. This part is devoted to the nuclear processes in which neutrons are emitted. Special consideration is given to those which are of practical interest as sources of neutrons of energy below about 20 Mev. Therefore it overlaps in part with the review article by Fowler and Brolley which deals with monoenergetic techniques in the 10 to 30 Mev range[3]. The production of neutrons of energy in the range of one hundred or more Mev is considered in Vol. XLV by Siegel.

This part will have mainly the character of a collection of data about particular reactions, preceded by a few general considerations about the consequences of the conservation laws. These are presented in a form suitable also for other nuclear reactions for which the energies involved are sufficiently low to neglect relativistic corrections.

A complete investigation of the nuclear reactions involves two types of considerations: the first, mentioned above, consists of the application of the conservation laws while the second deals with the intensity with which the process which is considered takes place under given conditions. This aspect of the various nuclear reactions collected in this part, will be treated usually only from the experimental point of view, with a few theoretical remarks added in some specially interesting cases. For the methods used for calibrating the neutron sources reference is made to Barschall's article[4].

As an introduction to the subject a comparison between neutron sources of different types may be useful.

In general, neutron sources may be divided into *homogeneous sources* from which neutrons of small energy spread are emitted (some times such sources emit more than one group of neutrons) and *heterogeneous sources* which emit

---

[1] See footnote 2, p. 85.
[2] R. D. Day, p. 9 of Ref. [*15a*]. — Phys. Rev. **102**, 767 (1956).
[3] J.L. Fowler and J.E. Brolley: Rev. Mod. Phys. **28**, 103 (1956).
[4] H.H. Barschall: Detection of neutrons, Vol. XLV of this Encyclopedia.

a continuous spectrum of neutrons or, at least, so many groups that they appear to form a continuous spectrum. Homogeneous sources are needed for studies of the interaction of fast neutrons with nuclei, heterogeneous sources are most useful for inducing radioactivity or for experiments with slow neutrons.

There are three principal methods for producing neutrons: neutron producing interactions of radiations from radioactive substances, neutron producing reactions of charged particles accelerated in various machines including secondary processes such as the nuclear photo-effect, and chain reacting devices. There are, in addition, some less important sources of neutrons such as spontaneous fission and cosmic rays. Radioactive sources use either (α, n) or (γ, n) reactions. For reasons of intensity radioactive (α, n) sources have thick targets of light elements and are therefore heterogeneous sources. Even if a thin target were employed, it would be necessary to collimate the α-particles in order to produce a homogeneous source which would result in prohibitively small intensities. The requirement of collimation is not necessary in the case of (γ, n) sources [see Eq. (34.11)] and, therefore, homogeneous (γ, n) sources may be prepared.

All (γ, n) sources and some (α, n) sources have the disadvantage that the intense γ-radiation may interfere with neutron measurements. On the other hand such sources are useful as standards and for some test purposes[1].

For most neutron measurements homogeneous intense sources are needed, and only accelerator sources are satisfactory except at the lowest neutron energies where neutrons of definite velocity may be selected from a continuum. The choice of a reaction for producing homogeneous neutrons is mostly determined by the desired energy. Most experiments in the kev region have been performed with the $\mathrm{Li}(p, n)$ reaction which has a high cross section just above threshold. Because of the small mass of Li, the neutrons emitted at threshold have an energy of 30 kev and the lowest neutron energy at which only one group of neutrons is emitted in the forward direction is 120 kev. It has been suggested that (p, n) reactions of heavier nuclei might be advantageous for obtaining monoenergetic neutrons of energies below 120 kev. For example the $\mathrm{V}^{51}(p, n)$ reaction has been explored for this purpose (Sect. 55), but it produces a very low yield.

Since Be^7 has an excited state at 430 kev, neutrons from the $\mathrm{Li}^7(p, n)\,\mathrm{Be}^7$ reaction are not monoenergetic in the forward direction at energies above 650 kev. Neutrons above this energy are most conveniently obtained from the $\mathrm{T}(p, n)\,\mathrm{He}^3$ reaction. There are no known excited states of He^3 so that this reaction presumably produces monoenergetic neutrons until energies are reached at which He^3 is broken up.

In order to avoid the use of too high bombarding energies exothermic (d, n) reactions instead of endothermic (p, n) reactions frequently serve as sources of neutrons of higher energies. Monoenergetic neutron of energies above 2 Mev may be produced in the $\mathrm{D}(d, n)$ reaction and neutrons above about 12 Mev in $\mathrm{T}(d, n)$ reaction. There is evidence, however, that in the latter reaction break-up of the deuterons above a bombarding energy of 3.7 Mev will produce some low energy neutrons (Sect. 57).

For the spectroscopy of slow neutrons, i.e. neutrons of less than 10 kev energy, one may start with an inhomogeneous fast neutron source and moderate the neutrons in a hydrogenous medium to produce slow neutrons. Neutrons of well-defined velocities are then selected on the basis of their time of flight or their wavelength in a diffraction experiment. The primary inhomogeneous neutron

[1] A.O. HANSON: Chap. IA. — Radioactive neutron sources of fast neutron physics, edit. by J.B. MARION and J.L. FOWLER. New York: Interscience Publisher 1959.

source may be produced either with an accelerator or it may be a reactor. Accelerated particle sources are pulsed for this purpose. If the primary source is a reactor, a rotating aperture ("chopper") interrupts the moderated neutron beam to produce bursts or the neutron beam falls on a crystal which acts as a monochromator.

Most types of accelerators and neutron producing reactions have been used to generate bursts of neutrons: Cockcroft-Walton accelerators with the $D(d, n)$ or $T(d, n)$ reaction, electrostatic accelerators with the $Li(p, n)$ reaction, pulse transformer accelerators with the $D(d, n)$ or $T(d, n)$ reaction, cyclotrons with the $D(d, n)$ or $Be(d, n)$ reaction, a synchrocyclotron with the $W(p, n)$ reactions. Surveys of these pulsed accelerators for slow neutron spectroscopy have been given by HAVENS[1], WIBLIN[2] and CRANBERG[3].

Choppers are more extensively used than crystal monochromators for neutron spectroscopy with reactors although both types of devices are available at most of the larger laboratories at which experiments in slow neutron spectroscopy are carried out with reactor sources. Surveys of neutron velocity selectors used with reactors were given by HUGHES[4], MOSTOVOI *et al.*[5] and WLADIMIRSKI *et al.*[6]. This subject is discussed by HUGHES in Vol. XLIV of the present Encyclopedia.

For experiments in which it is merely desired to induce an activity a wide choice of neutron sources is available. If the desired activity can be induced by thermal neutrons, a moderator may be used to produce the slow neutrons. The highest thermal flux can be generated in reactors. Central thermal neutron fluxes in some typical reactors are listed in Ref. [*12*] and [*18b*].

These introductory remarks may be concluded by recalling that most neutron sources constitute a serious health hazard and that special attention is necessary to keep the average radiation level below a maximum permissible dose[7].

[1] W. W. HAVENS, p. 74 of Ref. [*15b*].
[2] E. R. WIBLIN, p. 35 of Ref. [*15b*].
[3] L. C. CRANBERG, p. 40 of Ref. [*15b*].
[4] D. J. HUGHES, p. 3 of Ref. [*15b*].
[5] V. I. MASTOVOI, M. I. PEVZNER and A. P. TRITOVICH, p. 12 of Ref. [*15b*].
[6] V. V. WLADIMIRSKI, I. A. RADKOWICH and W. W. SOKOLOWSKY, p. 22 of Ref. [*15b*].
[7] The most recent recomandations are those of the NCRP [National Commision on Radiological Protection. Radiology **68**, 260 (1957)]. The maximum permissible accumulated dose (MPD) for occupational radiation exposures at any age $N(\geqq 18$ years) is given by

$$\text{MPD} = 5\,(N - 18)\ \text{rems}$$

provided no annual increment exceeds 15 rems. This relationship corresponds to an average annual increment of 5 rems, i.e. to a weekly MPD of 100 mrems. Such a value of the MPD is attained with an exposure to the following neutron fluxes for 40 hours per week.

Neutron energy	Neutron flux (neutrons/cm² sec)	Neutron energy	Neutron flux (neutrons/cm² sec)
0.025 ev	700	0 5 Mev	30
10 ev	700	1 Mev	20
10 kev	350	2 Mev	12
0.1 Mev	70	3 to 10 Mev	10

These values are obtained by readjusting the data contained in the Report of the International Commission on Radiological Units (ICRU) [Suppl. No. 6 to the Brit. J. Radiol., p. 91, 1956].

It may be recalled that the *absorbed dose* is the amount of energy transferred by the radiation under consideration—in the present case the neutrons—per gram of the medium (tissue). The corresponding unit is the *rad*: 1 rad is the dose corresponding to the absorption of 100 ergs per gram of the medium. The roentgen equivalent physical (i.e. the *rep*, corresponding to

I. The application of the conservation laws to nuclear reactions.

32. General considerations. In all nuclear processes known up to now, the following quantities are conserved: the total electric charge, the total number of nucleons, the classical constants of motion: total energy, linear and angular momentum, and the quantum mechanical constant of motion parity[1].

The conservation of the total electric charge and of the total number of nucleons are thought of as two very similar laws: the conservation of the total electric charge does not forbid the creation of any number of pairs of charged particles of opposite sign. In the same sense the conservation of the total number of nucleons does not forbid the creation of any number of nucleon-antinucleon pairs. In other words what must be conserved in general, is the difference between the numbers of nucleons and of antinucleons. Owing to the very high energy threshold, processes of creation of nucleon-antinucleon pairs do not affect our considerations about low energy nuclear reactions, in which the number of antinucleons is zero and remains zero. Therefore, for the scope of the present chapter, we can express the conservation of the total number of nucleons by the conservation of the total mass number. Similarly, the conservation of the total electric charge can be expressed by the conservation of the total atomic number as long as the creation of electron-positron pairs and beta-decay processes do not intervene.

As a consequence of these two fundamental conservation laws in this simplified form, the sum of the lower as well as of the upper indices of the nuclei and particles involved must remain constant in the equations representing nuclear reactions. For instance, if we write CHADWICK'S reaction in one of the two usual forms

$$_2\mathrm{He}^4 + {}_4\mathrm{Be}^9 \rightarrow {}_6\mathrm{C}^{12} + {}_0n^1 \quad \text{or} \quad {}_4\mathrm{Be}^9\,(\alpha, n)\,{}_6\mathrm{C}^{12}$$

the conservation of electric charge and of the number of nucleons are expressed by the equalities $2+4=6+0$; $4+9=12+1$.

In addition, the constants of motion of classical mechanics are conserved. But the experimental conditions under which a nuclear reaction is investigated are such that in general only the total energy and momentum of the system before the reaction, are well defined, and not its angular momentum.

In fact a well collimated beam of monoenergetic particles impinging on a thin target composed of the nuclei to be bombarded, is represented by a plane wave which contains all possible angular momenta [see Eq. (18.4)]. Therefore, while the application of the conservation of energy and momentum allows the deduction of very useful information, the conservation of angular momentum is in general of no use. As we have seen in Sect. 18 there are cases, however, in which only a few initial states of angular momentum can give rise to the reaction which is considered; then, of course, the corresponding law of conservation will also be of great value.

---

the absorption of 93 ergs per gram of medium) should be abandoned [ICRU Reports: Brit. J. Radiol. **27**, 243 (1954); Nat. Bur. Stand. Handbook, No. 62, 1956].

The *biological effective dose* or *RBE dose* is the dose measured in rems (roentgen-equivalent-man). Its value is obtained by multiplying the *absorbed dose*, expressed in rad, by the "relative biological effectiveness" (RBE) of the radiation. The RBE is taken equal to 1 for X- and γ-rays of any energy; for fast neutrons of up to 10 Mev a RBE of 10 is applicable [G. J. HINE and G.L. BRONWELL: Radiation Dosimetry. New York: Academic Press 1956; B.T. PRICE, C.C. HERTON and K. T. SPINNEY: Radiation Shielding. London: Pergamon Press 1957].

[1] For the non-conservation of parity in β-decay see Vol. XLI, Part 2, and Vol. XLIII of this Encyclopedia.

This is what happens, for instance, in the case of slow neutrons, whose wavelength is much larger than the dimensions of the target nucleus; only the neutrons with zero angular momentum (S waves) can contribute to any observed reaction because the neutrons with angular momentum different from zero ($P, D, \ldots$ waves) have a negligible wave amplitude in the region where the nuclear forces are different from zero.

Going back to the general case, we note that what is conserved is the total angular momentum which includes the intrinsic or spin angular momentum as well as the orbital angular momentum for the relative motion of the reacting particles. What is usually measured experimentally is the statistical angular distribution of the reaction products which depends on the orbital angular momentum of the final states. The analysis of the observed angular distribution in terms of waves of well defined angular momentum is very important for the investigation of nuclear reactions, but it will become practically possible only when we have some reason to limit the number of initial states of given angular momentum that can give rise to the nuclear reaction considered. Only then, as a consequence of the conservation of total angular momentum, is the number of final states limited and the analysis of the experimental results becomes feasible.

Parity is also conserved in nuclear reactions; it expresses the behaviour of the wave function of the system with respect to inversion of all the space coordinates. The conservation of parity means that the wave functions of the initial and final states of a nuclear reaction must have always the same parity. This theorem of conservation can in many cases be very useful in the interpretation of the experimental results [*3*].

33. The transformation from the laboratory to the centre-of-mass frame of reference. In the great majority of cases the experimental investigation of a nuclear process is done by bombarding, with a beam of monoenergetic particles a_1, of mass m_1, velocity $\boldsymbol{v}_1$, momentum $\boldsymbol{p}_1$ and kinetic energy E_1, a thin target containing the bombarded nucleus A_1 (M_1, $\boldsymbol{V}_1$, $\boldsymbol{P}_1$, E_1). The frame of reference fixed to the target, in which the detecting instruments are also at rest, is called the laboratory frame of reference (laboratory system = l.s.). It is, however, well known that this is not the most appropriate frame of reference for the study of a nuclear process. In fact the total energy of the two reacting particles can always be expressed as the sum of two terms: the first represents the translation energy in the l.s. of the centre of mass of the two particles, the second the total energy of the two particles in the frame of reference in which their centre of mass is at rest (centre-of-mass system = c.m.s.).

Obviously only this second frame is of importance for the dynamics of the nuclear process.

Therefore in order to compare the experimental results, obtained in the l.s., with theoretical calculations performed in the c.m.s., one has to follow, at least in principle, a procedure of the following type:

(a) Making use of the energies and momenta of the two reacting particles in the l.s., calculate the velocity of their centre of mass v_m in the l.s., and the total energy available in the c.m.s.

(b) Study the nuclear reaction in the c.m.s., where, by definition, the resultant momentum of all particles is zero, both before and after the reaction.

(c) Making use of v_m, transform the results obtained under (b) in the c.m.s., back to the l.s.

The relativistic treatment of this problem has been given by various authors, in particular by J. BLATON who makes use of an interesting geometrical interpretation[1].

The energies involved in this article are however always so small with respect to the rest energy of the nucleons, that we can make use of the non-relativistic approximation.

Furthermore in almost all cases considered in this article there are only two products of the reaction (see Sect. 12), a residual nucleus A_2 and a particle a_2: therefore we will limit our considerations to the case of reactions of the type

$$a_1 + A_1 \to A_2 + a_2 \tag{33.1}$$

which have also the advantage of being the only ones in which the theorems of conservation of energy and momentum allow the determination of the energy of the emitted particle as a function of the corresponding angle of emission [see Eq. (34.2)]. The theorems of energy and momentum conservation can be written, in the l.s.

$$E_1 + \mathsf{E}_1 = E_2 + \mathsf{E}_2 - Q_S \tag{33.2a}$$

$$\boldsymbol{p}_1 + \boldsymbol{P}_1 = \boldsymbol{p}_2 + \boldsymbol{P}_2 \tag{33.2b}$$

where

$$\left.\begin{aligned} &E_i = \frac{p_i^2}{2m_i}, \qquad \mathsf{E}_i = \frac{P_i^2}{2M_i}, \qquad i = 1, 2 \\ &\boldsymbol{p}_i = m_i \boldsymbol{v}_i, \qquad \boldsymbol{P}_i = M_i \boldsymbol{V}_i, \end{aligned}\right\} \tag{33.3a}$$

and

$$\left.\begin{aligned} &Q_S = Q_0 - w_S \\ &Q_0 = m_1 c^2 + M_1 c^2 - m_2 c^2 - M_2 c^2. \end{aligned}\right\} \tag{33.3b}$$

Q_0 is the "energy release" of the reaction and w_S the sum of the excitation energies of the two products with respect to their ground states.

E_1 represents the kinetic energy of the bombarded nucleus, which is usually only its energy of thermal agitation. One could take into account also the potential energy due to the chemical binding of the bombarded nucleus in the molecules of the target, by including it in Q_S. This term however is always negligible except in the case of elastic scattering ($Q_0 = 0$, $w_S = 0$) of incident particles of extremely low energy (see part D).

In most cases E_1 is so large that one can also put $\mathsf{E}_1 = 0$, $\boldsymbol{P}_1 = 0$ without introducing any appreciable error. We make use, however, of the slightly more general expressions (33.2) as long as they do not give rise to any inconvenience. In some special cases we shall assume, for the sake of simplicity: $\mathsf{E}_1 = 0$, $\boldsymbol{P}_1 = 0$.

Following the procedure sketched above let us start with point (a).

The velocity of the centre of mass of two colliding particles in the l.s., remains constant during the collision. Its classical expression

$$\boldsymbol{v}_m = \frac{m_1 \boldsymbol{v}_1 + M_1 \boldsymbol{V}_1}{m_1 + M_1}, \tag{33.4}$$

remains constant if we substitute the index 2 for the index 1, except for terms of the order of $Q_S/(m_1 + M_1)$, an approximation consistent with the non-relativistic treatment of the problem[2].

---

[1] J. BLATON: Kgl. danske Vidensk. Selsk. **24**, No. 20 (1950).

[2] The relativistic expression

$$\frac{v_m}{c} = \frac{c\,(\boldsymbol{p}_1 + \boldsymbol{P}_1)}{m_1 c^2 + E_1 + M_1 c^2 + \mathsf{E}_1}$$

remains exactly constant during the collision as a consequence of the conservation of momentum and total energy.

If we now introduce, besides v_m, the relative velocity of the two particles

$$\boldsymbol{v}_{ri} = \boldsymbol{v}_i - \boldsymbol{V}_i \qquad i = 1, 2, \tag{33.5}$$

the total kinetic energy can be written

$$E_i + \mathsf{E}_i = \tfrac{1}{2}(m_i + M_i)\, v_m^2 + \tfrac{1}{2}\mu_i v_{ri}^2 \qquad i = 1, 2 \tag{33.6}$$

where

$$\mu_i = \frac{m_i M_i}{m_i + M_i} \qquad i = 1, 2 \tag{33.7}$$

is the reduced mass.

The separation (33.6) is well known, and has already been mentioned at the beginning of this section. It can be completed by calculating the kinetic energies of the two particles in the c.m.s. Their velocities can be obtained by applying the conservation of momentum and Eq. (33.5) to the c.m.s.

$$\left.\begin{aligned} m_i \boldsymbol{v}_i^* &= -M_i \boldsymbol{V}_i^* \\ v_{ri} &= v_i^* + V_i^* \end{aligned}\right\} \quad i = 1, 2 \tag{33.8}$$

from which there follows

$$v_i^* = \frac{M_i}{m_i + M_i} v_{ri}; \qquad V_i^* = \frac{m_i}{m_i + M_i} v_{ri}; \tag{33.9}$$

$$\left.\begin{aligned} E_i^* &= \frac{1}{2} m_i v_i^{*2} = \frac{M_i}{m_i + M_i} \frac{1}{2} \mu_i v_{ri}^2; \\ \mathsf{E}_i^* &= \frac{1}{2} M_i V_i^{*2} = \frac{m_i}{m_i + M_i} \frac{1}{2} \mu_i v_{ri}^2, \end{aligned}\right\} \tag{33.10}$$

$$E_r = \mathsf{E}_i^* + E_i^* = \tfrac{1}{2}\mu_i v_{ri}^2 \qquad i = 1, 2. \tag{33.11}$$

Eqs. (33.10) and (33.11) show the separation of the kinetic energy E_r of the relative motion into the kinetic energies of the two particles in the c.m.s.

If we assume $\boldsymbol{V}_1 = 0$, it follows, from Eqs. (33.4) to (33.6) used with $i = 1$, that

$$E_r = \frac{1}{2}\mu_1 v_{r1}^2 = \frac{M_1}{m_1 + M_1} E_1 \tag{33.12}$$

and from (33.4) and (33.9) that

$$\boldsymbol{V}_1^* = -\boldsymbol{v}_m \qquad v_1^* = v_{r1} - v_m \tag{33.13}$$

two equations of practical importance.

We have thus accomplished part (a) of our procedure. Passing to part (b) we notice that, because of the constancy of the first term on the right of Eq. (33.6), one has

$$\tfrac{1}{2}\mu_2 v_{r2}^2 = \tfrac{1}{2}\mu_1 v_{r1}^2 + Q_S \tag{33.14}$$

which is also of practical importance, especially when combined with (33.12): Eq. (33.14) establishes that for a given reaction and for given initial velocities of the reacting particles, the relative velocity after collision has a well defined value, independent of its direction and therefore of the directions of the velocities of the reaction products.

In order to complete our procedure we have to consider part (c), namely we have to add the velocities $\boldsymbol{v}_2^*$ and $\boldsymbol{V}_2^*$, calculated from v_{r2} by means of Eq. (33.9), to $\boldsymbol{v}_m$ in order to obtain the corresponding velocities in the l.s., $\boldsymbol{v}_2$ and $\boldsymbol{V}_2$.

The whole procedure, and in particular this last step, is shown in Fig. 29, which refers to the reaction

$$ {}_1D^2 + {}_1D^2 \rightarrow {}_2He^3 + {}_0n^1 $$

$$ Q_0 = 3.28 \text{ Mev}, \quad w_S = 0, \quad E_1 = 4 \text{ Mev}, \quad \mathsf{E}_1 = 0. $$

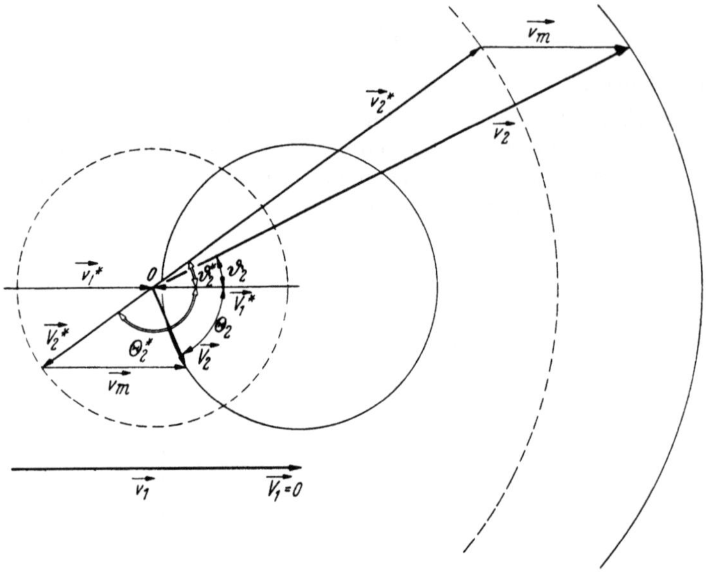

Fig. 29. Velocities of the products of the reaction $D + D \rightarrow He^3 + n$ in the l.s. and in the c.m.s.

By examining this figure one obtains the relations between the angles of emission of the two products ϑ_2 and Θ_2 in the l.s. and the corresponding angles in the c.m.s.

$$ \vartheta_2^* + \Theta_2^* = \pi $$

$$ \tan\vartheta_2 = \frac{\sin\vartheta_2^*}{\frac{v_m}{v_2^*} + \cos\vartheta_2^*}; \qquad \tan\Theta_2 = \frac{\sin\Theta_2^*}{1+\cos\Theta_2^*} = \tan\frac{\Theta_2^*}{2} \tag{33.15} $$

or

$$ \sin(\vartheta_2^* - \vartheta_2) = \frac{v_m}{v_2^*}\sin\vartheta_2; \qquad \sin(\Theta_2^* - \Theta_2) = \sin\Theta_2. \tag{33.16} $$

From Eqs. (33.4), (33.9) one has (in the usual approximation $m_1 + M_1 = m_2 + M_2$)

$$ \left(\frac{v_m}{v_2^*}\right)^2 = \frac{\frac{m_1 m_2}{M_1 M_2}}{1 + \frac{m_1 + M_1}{M_1}\frac{Q_S}{E_1}}. \tag{33.17} $$

Eqs. (33.15) to (33.17) are very useful in the following problems. The intensity $I(\vartheta_2)$ as a function of the angle ϑ_2 in the l.s. has been measured, and we want transform it to the c.m.s.

The intensities and the corresponding solid angles in the two frames of reference must obviously satisfy the relation

$$ I^*(\vartheta_2^*)\, d\omega_2^* = I(\vartheta_2)\, d\omega_2 $$

from which it follows that

$$I^*(\vartheta_2^*) = I(\vartheta_2)\, g(\vartheta_2) \tag{33.18}$$

with

$$g(\vartheta_2) = \frac{d\omega_2}{d\omega_2^*}\,. \tag{33.19}$$

In order to calculate $g(\vartheta_2)$ let us consider the triangle $\boldsymbol{v}_2^*$, $\boldsymbol{v}_m$, $\boldsymbol{v}_2$ of Fig. 30. One has

$$d\omega_2^* = \frac{ds^*}{v_2^{*2}}\,, \qquad d\omega_2 = \frac{\cos(\vartheta_2^* - \vartheta_2)\, ds^*}{v_2^2}$$

where ds^* is an element of the spherical surface of centre O and radius v_2^*. Introducing this expression in (33.19) and making use of (33.16) we obtain

$$g(\vartheta_2) = \frac{\sqrt{1 - \left(\frac{v_m}{v_2^*}\right)^2 \sin^2\vartheta_2}}{\left[\sqrt{1 - \left(\frac{v_m}{v_2^*}\right)^2 \sin^2\vartheta_2} + \frac{v_m}{v_2^*}\cos\vartheta_2\right]^2} \tag{33.20}$$

which completes our rules for transformation from the l.s. to the c.m.s.

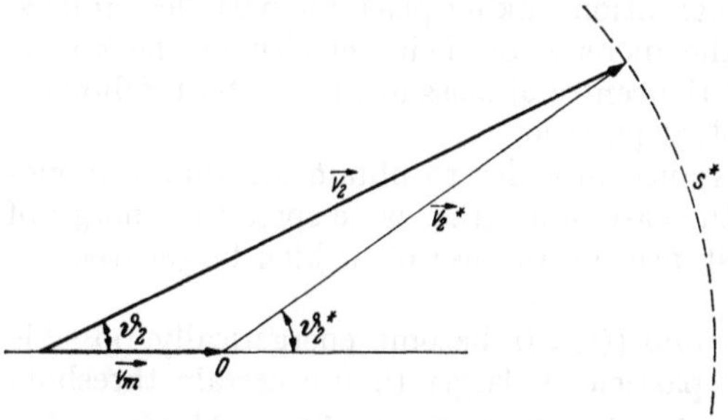

Fig. 30. Transformation of the solid angles from the l s. to the c.m.s.

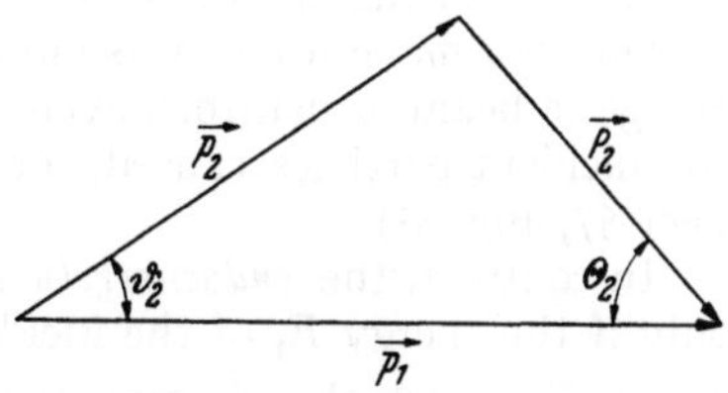

Fig. 31. Momentum conservation.

34. The energy momentum conservation in the laboratory system. It is, however, convenient to have also some relation by means of which one can calculate directly from the initial data the energies of the reaction products in the l.s.

For $P_1 = 0$, Eq. (33.2b) gives (Fig. 31)

$$P_2^2 = p_1^2 + p_2^2 - 2 p_1 p_2 \cos\vartheta_2$$

which, expressing the momenta by means of the corresponding kinetic energies, allows the elimination of E_2 from Eq. (33.2a) ($\mathsf{E}_1 = 0$). One gets

$$Q_S = \frac{m_2 + M_2}{M_2} E_2 - \frac{M_2 - m_1}{M_2} E_1 - 2\frac{\sqrt{m_1 m_2}}{M_2}\sqrt{E_1}\sqrt{E_2}\cos\vartheta_2 \tag{34.1}$$

where $\sqrt{E_1}$, and $\sqrt{E_2}$ have to be taken both with the positive sign because they represent the absolute values of the momenta. This equation can be used to calculate Q_S making use of the kinetic energies of the incident particles E_1 and of the emitted particle E_2 and of the corresponding angle of emission ϑ_2. Solving Eq. (34.1) with respect to $\sqrt{E_2}$ one gets

$$\left.\begin{aligned} \sqrt{E_2} &= \frac{\sqrt{m_1 m_2}}{m_2 + M_2}\sqrt{E_1}\cos\vartheta_2 \pm \\ &\pm \sqrt{\left[\frac{M_2 - m_1}{M_2 + m_2} + \frac{m_1 m_2}{(M_2 + m_2)^2}\cos^2\vartheta_2\right] E_1 + \frac{M_2}{m_2 + M_2} Q_S}\,. \end{aligned}\right\} \tag{34.2}$$

Once we know E_2, we can calculate E_2 directly from (33.2a) and the corresponding angle of emission by means of the following equation:

$$\sin \Theta_2 = \left(\frac{m_2 E_2}{M_2 \mathsf{E}_2}\right)^{\frac{1}{2}} \sin \vartheta_2 \tag{34.3}$$

obtained by considering the momentum triangle of Fig. 31.

In the discussion of Eq. (34.2) one has to distinguish between the case of *exoenergetic reactions*, in which Q_S is positive, and the case of *endoenergetic reactions*, in which Q_S is negative.

In the case of *exoenergetic reactions*, a single real and positive value of $\sqrt{E_2}$ (and therefore a single physically possible value of E_2) exists for any value of $\cos \vartheta_2$ and for any positive value of $\sqrt{E_1}$, from infinity down to zero, provided one considers only the positive sign of the root of Eq. (34.2).

Furthermore an angle of emission $\vartheta_2 > 90°$ in general exists, at which the variation of the first term of Eq. (34.2) compensates approximately the variation of the second term, so that E_2 becomes almost independent of E_1 in a not too narrow interval of this variable.

This fact is clearly due to a compensation, taking place for particles emitted in a backward direction, between the increase of their velocity in the c.m.s., and the increase of the velocity v_m of the centre of mass in the l.s., both following an increase of the energy of the incident particle.

One can make use of this circumstance in order to obtain an almost monoenergetic beam of neutrons even in the case of a rather wide spread in energy of the incident particles caused, for instance, by the use of a thick target (see e.g. Sect. 57, Fig. 85).

In contrast, the *endoenergetic reactions* ($Q_S < 0$) become energetically possible only if the energy E_1 of the incident particles is larger than a certain threshold energy E_{1T} at which $\sqrt{E_2}$ starts to be real and positive. E_{1T} is obtained by imposing the condition that the quantity appearing under the square root of Eq. (34.2) be zero:

$$E_{1T}(\vartheta_2) = -Q_S \frac{M_2}{M_2 - m_1 + \dfrac{m_1 m_2}{m_2 + M_2} \cos^2 \vartheta_2}\,. \tag{34.4}$$

$E_{1T}(\vartheta_2)$ depends on ϑ_2 and is a minimum for $\vartheta_2 = 0$:

$$E_{1T}(0°) = -Q_S \frac{M_2 + m_2}{M_2 + m_2 - m_1} \approx -Q_S \frac{M_1 + m_1}{M_1} \tag{34.5}$$

where the usual approximation $M_2 + m_2 \approx M_1 + m_1$ has been introduced.

When the energy of the incident particles is just above the threshold (34.5), the secondary particles a_2 are emitted only in the forward direction ($\vartheta_2 \approx 0$) with an energy which from Eq. (34.2) turns out to be

$$E_{2T}(0°) = \frac{m_1 m_2}{(m_2 + M_2)^2} E_{1T}(0°) \approx -Q_S \frac{m_1}{M_1} \frac{m_2}{m_2 + M_2}\,. \tag{34.6}$$

Under these conditions the particles a_2 are emitted with a negligible energy in the c.m.s., and the energy (34.6) is that due to the velocity v_m of the centre of mass in the l.s.

As the energy E_1 of the incident particles a_1 is increased above the threshold (34.5), the secondary particles a_2 are emitted in the forward direction in a cone whose angle is obtained from Eq. (34.4) by substituting, in its left-hand side, the value of E_1 concerned. At each angle within this cone, except at the edge, there are two groups of particles whose energies correspond to the $\pm$ signs appearing in Eq. (34.2). They are due to particles emitted in two different directions in the c.m.s., which are observed as moving in the same direction in the l.s., as a consequence of the composition of their velocities with the velocity $\boldsymbol{v}_m$ of the centre of mass.

As the energy of the incident particles a_1 is further increased, the cone widens and the energy of one of the two groups of secondary particles a_2 increases while the energy of the other one decreases and reaches the value zero for

$$E_1 = E_{1m}(90°) = -Q_S \frac{M_2}{M_2 - m_1} \tag{34.7a}$$

at which the cone includes the forward hemisphere. For energies larger than this value only the positive sign in Eq. (34.2) is used and the secondary particles are emitted at any angle in the complete sphere, with a single energy. Therefore Eq. (34.7a) represents, in a certain sense, the threshold for emission of monoenergetic neutrons in the l.s. and as such it is a quantity of practical interest for any neutron source based on an endoenergetic reaction. When the incident particles a_1 have the energy (34.7a) the particles a_2, emitted in the forward direction, have an energy given by

$$E_{2m} = -Q_S \frac{M_2}{M_2 - m_1} \frac{m_1 m_2}{m_2 + M_2} . \tag{34.7b}$$

Finally we want to mention here a point of practical interest in the measurement of the excitation curves of endoenergetic reactions which is connected with the equations given above and with the method of observation of the emitted particles. If the detector of the emitted particles is placed in the forward direction ($\vartheta_2 = 0$) and subtends a very small angle at the target, it can happen that the differential excitation curve shows a maximum immediately above the threshold (34.5) which is not due to the properties of the compound nucleus but to the transformation of the solid angle from the c.m.s. to the l.s. In fact, immediately above threshold the neutrons are emitted within a small forward cone and, therefore, if they move all within the solid angle subtended by the detector the result of the measurement represents the total yield.

By increasing the energy of the incident particles, the cone of emission of particles a_2 widens and only a fraction of the emitted neutrons can be detected; then the result of the measurements is, in a certain sense, closer to a differential yield (differential with respect to the solid angle).

Clear examples of this effect have been observed in various (p, n) reactions (see Division V below).

A special type of endoenergetic reaction, frequently used for the production of neutrons, is the (γ, n) process, the discussion of which differs from the preceding one, only in that in Eq. (33.2) one has obviously to put

$$E_1 = h\nu, \quad p_1 = \frac{h\nu}{c} .$$

Following the same procedure as before one obtains instead of Eqs. (34.1) (34,2), (34,4) and (34.5)

$$Q_S = E_2 \frac{m_2 + M_2}{M_2} - h\nu \left[1 - \frac{h\nu}{2M_2 c^2} + \frac{\sqrt{2m_2 E_2}}{M_2 c} \cos\vartheta_2\right], \tag{34.8}$$

$$\left.\begin{aligned} \sqrt{E_2} &= \frac{1}{2}\frac{h\nu}{c}\frac{\sqrt{2m_2}}{m_2+M_2}\cos\vartheta_2 + \\ &+ \sqrt{(h\nu+Q_S)\frac{M_2}{m_2+M_2} - \frac{1}{2}\left(\frac{h\nu}{c}\right)^2\left[\frac{1}{m_2+M_2} - \frac{m_2\cos^2\vartheta_2}{(m_2+M_2)^2}\right]}, \end{aligned}\right\} \quad (34.9)$$

$$E_{1T}(\vartheta_2) = (h\nu)_{1T} = M_2 c^2 \frac{1-\sqrt{1+\frac{2Q_S}{M_2c^2}\left[1-\frac{m_2}{m_2+M_2}\cos^2\vartheta_2\right]}}{1-\frac{m_2}{m_2+M_2}\cos^2\vartheta_2}. \quad (34.10)$$

For not too large values of $h\nu$ as, for instance, those of the γ-rays emitted by all naturally and artificially radioactive bodies ($h\nu \lesssim 3$ Mev), it is preferable to use approximate expressions instead of Eqs. (34.9) and (24.10). These are obtained by introducing the following simplifications:

$$M_1c^2 \approx (m_2+M_2)\,c^2 \approx A\times 931 \text{ Mev}, \qquad m_2c^2 = 938 \text{ Mev} \approx 931 \text{ Mev},$$
$$M_2c^2 \approx (A-1)\times 931 \text{ Mev},$$

where A is the mass number of the target nucleus.

Taking the square of Eq. (34.9) one has

$$\begin{aligned} E_2 &= \frac{A-1}{A}\left[h\nu + Q_S - \frac{(h\nu)^2}{1862\,(A-1)}\right] + \frac{(h\nu)^2}{931\cdot A}\cos^2\vartheta_2 + \\ &+ h\nu\cos\vartheta_2\left[\frac{2(A-1)\,(h\nu+Q_S)}{A^3\cdot 931} - \frac{(h\nu)^2}{A^3(931)^2}\left(1-\frac{1}{A}\cos^2\vartheta_2\right)\right]^{\frac{1}{2}} \text{ Mev}. \end{aligned}$$

For small values of $h\nu$ one can neglect the second term as well as the second term under the square root so that one obtains

$$E_2 = \frac{A-1}{A}\left[h\nu + Q_S - \frac{(h\nu)^2}{1862\,(A-1)}\right] + \left[\frac{2(A-1)\,(h\nu+Q_S)}{A\cdot 931}\right]^{\frac{1}{2}}\frac{h\nu}{A}\cos\vartheta_2 \quad (34.11)$$

where the term in $\cos\vartheta_2$ is always small.

Various authors have prepared different nomograms[1] or numerical tables[2] in order to represent the more useful among the equations collected in this section and the preceding one, applied to a few reactions of great importance as sources of neutrons.

35. The case of elastic scattering by nuclei at rest. In this case we have

$$m_2 = m_1 = m; \qquad M_2 = M_1 = M; \qquad \mu_2 = \mu_1 = \mu; \qquad w_S = 0, \qquad Q_0 = 0;$$

and therefore, from Eq. (33.14),

$$v_{r2} = v_{r1}, \quad (35.1)$$

i.e., the relative velocity of the two particles does not change in the collision. Therefore, according to Eq. (33.9) which remains unaltered, the absolute values of the velocities of the two particles, in the c.m.s. do not change either; the only changes are in the corresponding directions of motion.

The following equations are valid without any change: (33.9) for the velocities in the c.m.s., (33.10) for the kinetic energies in the c.m.s. and, for $\boldsymbol{V}_1 = 0$, (33.12) and (33.13).

[1] MCKIBLEN: Document MDDC 223 of the Document Division of the AEC, Oak Ridge: these nomograms are reproduced in Ref. [*10*] and [*18b*].

[2] J.L. FOWLER and J.E. BROLLEY: Rev. Mod. Phys. **28**, 103 (1956).

The angular relations (33.15), (33.16), and (33.20) are still valid, and become simplified as a consequence of the fact that (33.17) for $Q_S=0$, is independent of the initial energy E_1:

$$\frac{v_m}{v_2^*} = \frac{m}{M}. \tag{35.2}$$

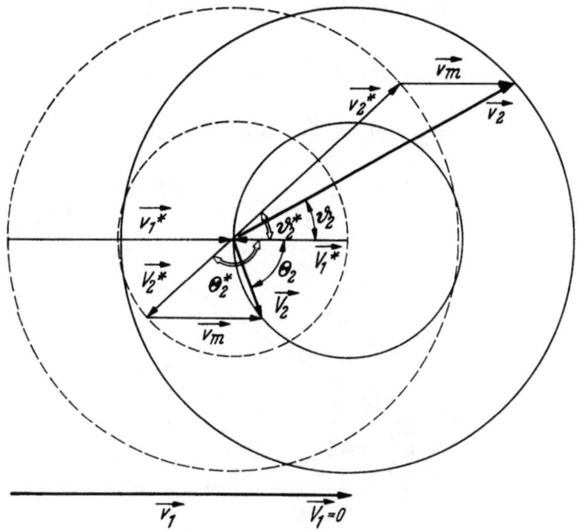

Fig. 32. Example of elastic scattering. $n+\mathrm{D}$.

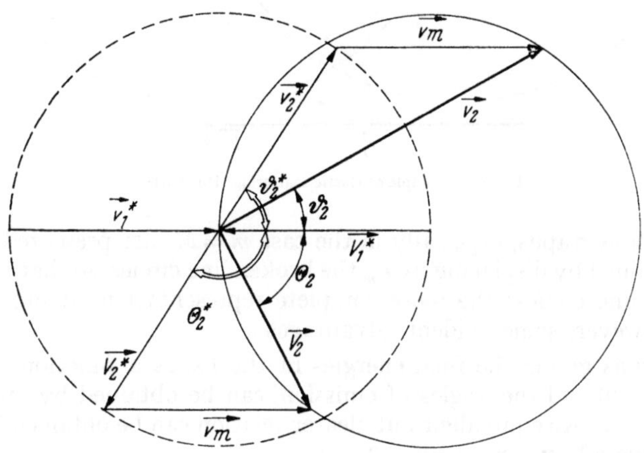

Fig. 33. Example of elastic scattering. $n+\mathrm{H}$.

Figs. 32 to 34, show the velocity diagrams similar to that of Fig. 29 for the following cases:

$$_0n^1 + {}_1\mathrm{D}^2 \rightarrow {}_1\mathrm{D}^2 + {}_0n^1,$$

$$_0n^1 + {}_1p^1 \rightarrow {}_1p^1 + {}_0n^1,$$

$$_2\mathrm{He}^4 + {}_1\mathrm{H}^1 \rightarrow {}_1\mathrm{H}^1 + {}_2\mathrm{He}^4.$$

The graphs have been drawn for given values of the initial energy E_1 which however determines only the scale, and not the angles as a consequence of the fact that (35.2) is independent of E_1.

From these figures one obtains the well known rule

$$\vartheta_2 + \Theta_2 > \frac{\pi}{2} \quad \text{for} \quad m < M,$$

$$\vartheta_2 + \Theta_2 = \frac{\pi}{2} \quad \text{for} \quad m = M,$$

$$\vartheta_2 + \Theta_2 < \frac{\pi}{2} \quad \text{for} \quad m > M.$$

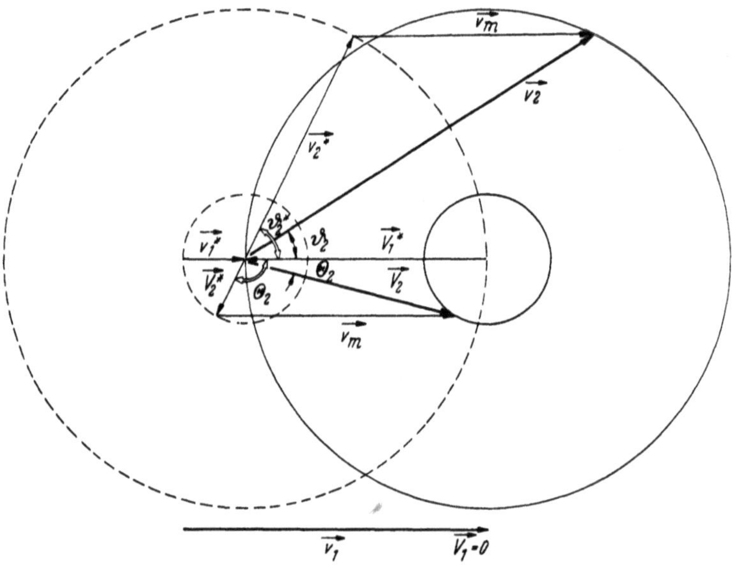

Fig. 34. Example of elastic scattering: $He^4 + {}_1H^1$.

Usually these graphs, especially in the case $m = M$, are presented in a simplified form obtained by displacing by $\boldsymbol{v}_m$ the broken line circles, so that they coincide with the full line circles: the more complete representation adopted in Figs. 29 to 34 has, however, some evident advantages.

The relations giving the final energies in the l.s. as a function of the initial energy E_1 ($V_1 = 0$) and the angles of emission, can be obtained by introducing in Eq. (34.2) $Q_S = 0$. An equivalent but clearer relation can be obtained by considering the triangle $\boldsymbol{v}_2^*$, $\boldsymbol{v}_m$, $\boldsymbol{v}_2$ of Fig. 32:

$$v_2^2 = v_m^2 + v_2^{*2} + 2 v_m v_2^* \cos\vartheta_2^* = v_1 \left[1 - \frac{4mM}{(m+M)^2} \frac{1}{2} (1 - \cos\vartheta_2^*)\right] \quad (35.3)$$

which, by multiplying by $\frac{1}{2}m$ can be written

$$E_2 = E_1 - \mathsf{E}_2, \quad (35.4)$$

$$\mathsf{E}_2 = E_1 \frac{4\,m\,M}{(m+M)^2} \frac{1}{2} (1 - \cos\vartheta_2^*) = E_1 \frac{4\,m\,M}{(m+M)^2} \cos^2\Theta_2 \quad (35.5)$$

where the last angle transformation is based on (33.16).

36. The case of elastic scattering by moving nuclei. Finally the case of elastic scattering of neutrons by nuclei moving with a velocity $\boldsymbol{V}_1 \neq 0$ will be considered. The velocity of the center of mass is deduced from (33.4)

$$v_m^2 = \frac{1}{(m+M)^2}\left[(m v_1)^2 + (M V_1)^2 + 2 m v_1 M V_1 \cos\alpha\right] \tag{36.1}$$

where α is the angle between the initial velocities $\boldsymbol{v}_1$ and $\boldsymbol{V}_1$ (Fig. 35).

The relative velocity v_r, which according to (35.1) remains constant during the collision, is deduced from (33.5)

$$v_r^2 = v_1^2 + V_1^2 - 2 v_1 V_1 \cos\alpha. \tag{36.2}$$

The velocity $\boldsymbol{v}_2$ of the neutron in the l.s. after the collision, is given by a formula similar to (35.3), viz.

$$v_2^2 = v_m^2 + v_2^{*\,2} + 2 v_m v_2^* \cos\beta \tag{36.3}$$

where the angle β between the vectors $\boldsymbol{v}_2^*$ and $\boldsymbol{v}_m$ is different from the angle ϑ_2^* of deflection of the neutron in the c.m.s. It is equal to it only in the case $\boldsymbol{V}_1 = 0$ considered before (Fig. 32).

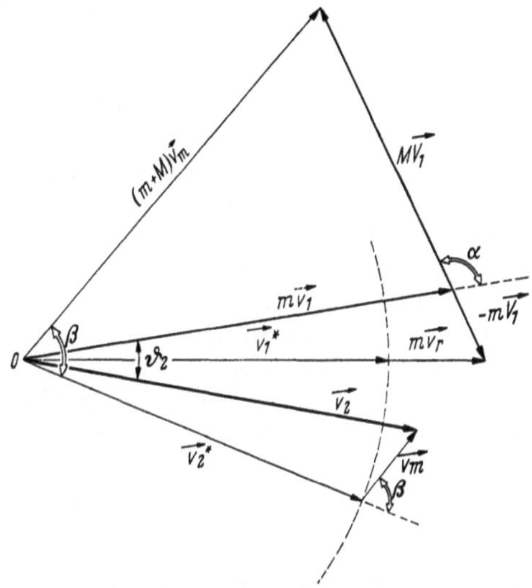

Fig. 35. Example of elastic scattering of a neutron ($m = 1$) of velocity v_1 ($E_1 = 1$ ev) by a nucleus ($M = 4$) moving with velocity V_1 ($E_1 = 1.8$ ev).

In (36.3) $v_2^* = v_1^*$ must be expressed in terms of v_m by means of (33.9) with the subscript i omitted because of the elasticity of the collision.

Therefore, for given values of the initial velocities $\boldsymbol{v}_1$ and $\boldsymbol{V}_1$ (i.e. for given values of v_1, V_1 and $\cos\alpha$) the final velocity v_2 can only vary between the limit speeds

$$\left.\begin{aligned} v_{2\,\max} &= v_m + v_1^* = v_m + \frac{M}{m+M} v_r, \\ v_{2\,\min} &= v_m - v_1^* = v_m - \frac{M}{m+M} v_r. \end{aligned}\right\} \tag{36.4}$$

The ratio of the final to the initial neutron energy in the l.s. is given by

$$\frac{E_2}{E_1} = \frac{v_2^2}{v_1^2} = \frac{v_m^2 + v_2^{*\,2}}{v_1^2} + \frac{2 v_m v_2^*}{v_1^2}\cos\beta, \tag{36.5}$$

an expression linear in $\cos\beta$.

II. The excitation curves of nuclear reactions.

37. Yields and cross sections of nuclear reactions. Under idealized experimental conditions a nuclear reaction would be investigated by allowing a parallel beam of particles a_1 of energy E_1, to be incident on an infinitely thin target containing $\mathcal{N}_1$ nuclei of type A_1 per cm³ in a layer of thickness dx. If a reaction of the type (33.1) takes place, the number of particles a_2 emitted per unit time into the solid angle $d\omega_2$ will be given by

$$dS_2 = S_1 \sigma(E_1, \vartheta_2)\, d\omega_2 \mathcal{N}_1\, dx \tag{37.1}$$

where S_1 is the number of particles a_1 incident per unit time on the target. Eq. (37.1) defines the differential cross section $\sigma(E_1, \vartheta_2)$ for the process (33.1). It depends on ϑ_2, the angle between the incident and emitted particles, but not on the azimuthal angle φ, as long as the polarization of the emitted particles is not taken into account, as we shall do throughout this article.

If E_1 and ϑ_2 are given, the energy E_2 of the emitted particles a_2 can be calculated from Eq. (34.2). Under the idealized conditions assumed so far the emitted particles will be homogeneous in energy provided the reaction products A_2 and a_2 are both left in their ground states. Otherwise there will be several "homogeneous groups" of particles a_2, each corresponding to a different value of the sum w_S of the excitation energies of A_2 and a_2 according to Eq. (33.3b) and (34.2).

In practice, the idealized conditions are never satisfied because (a) the target is not indefinitely thin, (b) the incident beam is never strictly homogeneous, and (c) the detector subtends a finite solid angle. As a consequence, the emitted particles have a finite energy spread which has to be taken into account in the analysis of the experimental data.

The finite target thickness Δx results in an energy loss of the particles a_1 or a_2 or both as they move through the target material. This is important primarily if a_1 or a_2 are charged particles. In the present discussion of neutron sources the case in which only a_1 is charged will be considered. This particle loses an energy

$$dE_1 = \frac{dE_1}{dx}\, dx = k(E_1)\, dx \tag{37.2}$$

in crossing the layer dx, where $k(E_1)$ is the energy loss per unit path length in the target material. $k(E_1)$ is related to the corresponding range $R(E_1)$ by

$$R(E_1) = \int_0^{E_1} \frac{dE_1}{k(E_1)}. \tag{37.3}$$

Values of $R(E_1)$ and $k(E_1)$ may be found for example in Ref. [*17b*] for some standard substances such as aluminium and air under normal conditions. Values for other materials may be obtained from the relation

$$\mathcal{N}_0\, dx_0 = \sum \mathcal{N}_i s_i\, dx \tag{37.4}$$

where $\mathcal{N}_0$ is the number of atoms per cm^3 of the standard substance, dx_0 the corresponding equivalent thickness, s_i the atomic stopping power of atoms of type i relative to the standard substance, $\mathcal{N}_i$ the number of atoms of type i per cm^3. The sum is taken over all chemical elements present in the target.

By integration of Eq. (37.1) the number of particles emitted from a target of finite thickness may be found. To carry out the integration, it is convenient to change variables from x to E_1 according to Eq. (37.2). The integration is to be carried out between the energy E_1 of the incident particles a_1 and the energy E_1' which these particles have after they have passed through the target. E_1' may be found from the range-energy relation (37.3) and the target thickness Δx:

$$R(E_1') = R(E_1) - \Delta x. \tag{37.5}$$

The result of the integration is

$$\left.\begin{aligned} \frac{d S_2}{d \omega_2} &= S_1 \mathcal{N}_1 \int_{E_1'}^{E_1} \frac{\sigma(E_1, \vartheta_2)}{k(E_1)} dE_1 \\ &= S_1 \mathcal{N}_1 \frac{\mathcal{N}_0}{\Sigma \mathcal{N}_i s_i} \int_{E_1'}^{E_1} \frac{\sigma(E_1, \vartheta_2)}{k_0(E_1)} dE_1 \end{aligned}\right\} \tag{37.6}$$

where $k_0(E_1)$ is the energy loss per unit length in the standard substance. The second equality holds only if the relative stopping powers s_i do not vary in the energy interval considered.

A target is called "thin" if $\sigma(E_1, \vartheta_2)$ does not vary appreciably in the energy interval E_1' to E_1. In this case Eq. (37.6) may be written

$$\frac{d S_2}{d \omega_2} = S_1 \mathcal{N}_1 \sigma(E_1, \vartheta_2) \,\Delta x \tag{37.7}$$

which is the same as Eq. (37.1) except that dx is replaced by Δx. The energy interval

$$\Delta_t E_1 = E_1 - E_1' = k(E_1) \,\Delta x \tag{37.8}$$

is usually called the "equivalent thickness" of the target. In practice, for reasons of intensity the target thickness cannot be made arbitrarily small. The finite thickness of the target is one of the factors which produces a spread in energy of the incident particles, as will be discussed in the next section.

A measurement of $d S_2/d \omega_2$ as a function of E_1 is usually referred to as a measurement of the yield of the reaction. Instead of the particles emitted in a fixed direction ϑ_2 sometimes all the particles a_2 emitted are observed. In this case by integration of Eq. (37.7) over $d \omega_2$ one obtains

$$S_2 = S_1 \mathcal{N}_1 \sigma(E_1) \,\Delta x. \tag{37.9}$$

Measurements of reaction yields are normally carried out with a thin target. In the past some measurements have been performed with an "infinitely thick" target, i.e., a target thick enough to stop all the incident particles a_1. For a thick target E_1' defined by Eq. (37.5) is zero $[\Delta x \geqq R(E_1)]$ and, therefore, the lower limit of the integration in Eq. (37.6) is zero. Thick target measurements determine the "integral" yield. Integral yield courves are primarily of interest for some practical applications. In principle, differential yield courves could be obtained by differentiation of integral yield curves, but the results are very inaccurate even if very precise integral curves are available.

If the target is neither very thin nor infinitely thick, the interpretation of the results is complicated. In particular, the differentiation just mentioned cannot be applied.

For reactions which have a threshold E_{1T} the equivalent target thickness defined by Eq. (37.8) will be

$$\Delta_t E_1 = E_1 - E_{1T} \tag{37.10}$$

whenever E_1' is smaller than E_{1T}. In some neutron-producing reactions the cross section rises very rapidly above the threshold so that Eq. (37.7) is not strictly correct unless $\sigma(E_1, \vartheta_2)$ is interpreted as an average value.

Differential yield measurements may give information about energy levels of the compound nucleus and of the product nucleus. Excited states of the product nucleus manifest themselves by the presence of two or more homogeneous groups of particles a_2 for a fixed energy E_1 of the incident particles a_1. As was mentioned at the beginning of this section, the energies E_{2S} of these groups (the index S refers to the energy levels of A_2) vary with E_1 according to Eq. (34.2).

If the ground state of A_2 is stable, it snatural width is zero. The natural width of the excited states of A_2 will, however, contribute to the energy spread of the corresponding groups of a_2. Assuming that the energy spread of the group of a_2 corresponding to the ground state of A_2 is a measure of the resolving power of the apparatus, measurements of the width of the other groups of a_2 enable one to determine the natural widths of excited states of A_2.

Excited states of the compound nucleus are responsible for rapid fluctuations of $d S_2/d \omega_2$ given by Eq. (37.6), when the energy of the incident particle is varied slightly in the vicinity of certain values E_{1R}. These fluctuations appear usually (but not always) as maxima in the intensity of one or more of the groups emitted in the reaction at certain angles with respect to the direction of the incident particles. The observed width of the maximum or, generally speaking, the shape of the fluctuation depends both on the parameters characterizing the level (in particular its total width) and on the resolving power of the experimental arrangement.

Levels in the compound nucleus and the product nucleus may also be observed in measurements of the integral yield curve. In this case the groups of particles a_2 are, however, not homogeneous, but are spread out on the low energy side, because they are produced by incident particles of energies from E_1 to zero. The energy distribution of each group depends for a given E_1 not only on the natural width and the resolving power, but also on the variation of $\sigma(E_1, \vartheta_2)$ with E_1. A well-defined group will be observed only if $\sigma(E_1, \vartheta_2)$ decreases rapidly with decreasing E_1.

While these groups appear at energies E_{2S} which depend on E_1, levels in the compound nucleus give rise to other groups whose energies are independent of E_1. Whenever $E_1 > E_{1R}$, the particles a_1 will at some depth in the target have the energy E_{1R} and there produce these reactions with a higher cross section than at smaller or greater depths. The energy of the particles a_2 corresponding to such a group may be found by inserting E_{1R} into Eq. (34.2). Their energy, therefore, does not depend on E_1. The distribution in energy of the group does not depend on the distribution in energy of the primary particles a_1 provided this primary spectrum does not include particles of energy near E_{1R}.

38. Factors limiting the homogeneity of neutron sources. In Sect. 37 it was mentioned that even if the product of a neutron-producing reaction is left in its ground state, several factors will result in a spread in energy of the neutrons produced in the reaction. The most important of these factors are: (a) the spread in energy of the incident charged particles, (b) the finite thickness of the target, and (c) the finite angle subtended by the detector at the target.

In principle, all three of these sources of neutron-energy spread may be made arbitrarily small, but, in practice, as the energy resolution is improved, the strength of the neutron source decreases rapidly.

For accelerated particle sources a spread in energy of the charged particles may be caused either in the ion source or in the acceleration. The spread introduced in the ion source is of the order of magnitude of the voltage applied to the probe which extracts the ions from the discharge and may be about 1 kv or more. The stability of the accelerating voltage depends on the type of machine. It is best for d.c. accelerators. In Cockcroft-Walton and electrostatic accelerators the accelerating voltage may be stabilized sufficiently well to contribute little to the energy spread.

Whenever a very homogeneous beam of charged particles is needed, it is necessary to select particles of the desired energy by means of magnetic or electrostatic analyzers[1]. The ion beam passes through an appropriately located slit before it enters the magnetic or electric field and through another slit after it has been deflected. Assuming that the field does not vary with time, charged particles of given energy and energy spread will pass through the analyzer. The energy spread is determined by the width of the slits. In this manner beams of charged particles may be selected with energy spread of the order of 100 ev which is less than what is normally used for neutron work.

The effect of the finite target thickness on the energy spread of the neutrons was discussed in Sect. 37. Because of the energy loss of the charged particles passing through the target, neutrons produced at the front and back surfaces of the target will have somewhat different energies. The difference in energy of the neutrons may be greater than the energy loss of the charged particles, particularly at low bombarding energies for exoenergetic reactions, and near threshold for endoenergetic reactions.

Unless the vacuum system in which the target is located is very clean, carbon deposits tend to form on the target. These deposits are caused by decomposition by the beam of organic vapors which may originate in rubber gaskets or pump oil. Since the deposits are never quite uniform they will contribute to the energy spread of the neutrons. The deposits may be greatly reduced by placing a liquid air trap close to the target and by heating the target.

Among the most frequently used target materials for neutron sources are deuterium and tritium. In order to obtain large neutron yields with a given energy spread it is advantageous to use these materials in gaseous form[2]. The gas is usually separated from the vacuum system by a thin foil (see Sect. 53). Straggling in the foil will contribute to the energy spread of the neutrons.

The third important cause of neutron energy spread is the variation of neutron energy with the angle between the incident charged particle and the emitted neutron. Both the size of the target and of the detector determine the range of angles. It is usually possible to reduce the effective target area by focussing the beam of charged particles, but if the focus is too sharp the target may be damaged by the beam. The size of the detector is usually the dominating factor. It is clear that the variation of energy with angle is smallest when the detector is placed at 0°, i.e., in line with the charged particle beam and this direction is normally chosen for experiments.

When experiments are to be carried out with very high energy resolution there are some additional sources of energy spread that have to be taken into account. The thermal motion of the target nuclei produces a Doppler effect. This will amount to a few 100 ev when light nuclei are bombarded with particles of several Mev energy. The Doppler effect could in principle be reduced by cooling the target.

Another possible source of energy spread arises when the bombarding charged particles are molecular ions. Excitation of the molecules results in motion of the nuclei with respect to the center of mass of the molecules and introduces

[1] See Ref. [*17e*]; R.E. Warren, J.L. Powell and R.G. Herb: Rev. Sci. Instrum. **18**, 559 (1947). — W.A. Fowler, C.C. Lauritsen and T. Lauritsen: Rev. Sci. Instrum. **18**, 818 (1947).

[2] E.M. Hafner, W.F. Hornyak, C.E. Falk, G. Snow and T. Coor: Phys. Rev. **89**, 204 (1953). — C.H. Johnson and H.E. Banta: Rev. Sci. Instrum. **27**, 132 (1956).

an additional Doppler broadening. This effect is usually negligible compared to other effects in application to neutron sources.

In Sect. 34 kinematic effects have been discussed which must also be taken into account in evaluating the homogeneity of a neutron source. In particular, the occurrence of two neutron groups just above the threshold of an endoenergetic reaction interferes with neutron measurements at very low neutron energies for accelerated particle sources.

The effects so far mentioned produce a spread in the energy of neutrons produced in the reaction. An additional energy spread may be caused by the interaction of neutrons in surrounding material. Since the neutrons are produced inside the vacuum system of the accelerator they will pass through the walls of the vacuum system and suffer collisions in the material. Neutrons which start out in a direction which is not that of the detector may be scattered into the detector. These neutrons have originally a different energy from those starting towards the detector so that, even if they are elastically scattered by heavy nuclei, they will contribute a component of an undesired energy. Neutrons which start out in the plane of the target have a particularly large probability of being scattered into the detector. Neutrons may also lose energy in elastic or inelastic collisions in the walls of the target chamber. Great care must be taken, therefore, to make the target assembly as light as possible.

Another source of neutrons of energies other than those one wishes to observe, is produced by collisions of neutrons in materials surrounding the source and detector, such as the floor and walls of the laboratory and the air. In order to reduce the effect of such background neutrons the source may be placed at the center of a large laboratory. The effect of background neutrons may be estimated by placing a long shadow bar between the source and detector.

Several methods have been employed to investigate the homogeneity of a neutron beam from an accelerated particle source. The measurement of the apparent width of a sharp resonance in the neutron interaction cross section is frequently used. For example, the total neutron interaction cross section of sulphur has an isolated resonance at 587 kev which is less than 1.5 kev wide.

The contributions to the neutron energy spread made by the effective target thickness and by the energy spread of the incident charged particles may be estimated from measurements on the neutron producing reaction. If this reaction has a narrow resonance, the apparent width of the resonance may serve to measure the effective target thickness and the resolution of the incident beam[1].

For endoenergetic reactions the apparent rise of neutron yield in the forward direction may be used for the same purpose[2]. The yield curve rises at first with increasing slope because of the inhomogeneity of the incident charged particle beam. For sufficiently thick targets the yield curve rises at higher bombarding energies with constant or decreasing slope. The bombarding energy interval over which this rise occurs is a measure of the effective target thickness. This rise is due to the fact that the incident charged particles reach the threshold of the reaction at increasingly greater depth in the target material. Most of the observed yield in the forward direction just above threshold is due to neutrons contained in the forward cone (see Sect. 34). The yield-rise curve may under some conditions lead to an underestimate of the effective target thickness. It is observed occassionally that, after a target has been bombarded for some time, the length of the rise remains essentially unchanged, but the yield curve drops

[1] R.E. PETERSON, H.H. BARSCHALL and C.K. BOCKELMAN: Phys. Rev. **79**, 593 (1950).
[2] R. TASCHEK and A. HEMMENDINGER: Phys. Rev. **74**, 373 (1948).

off more slowly beyond the peak. This may be caused by the build-up of a non-uniform deposit on a part of the target.

Although, from the point of view of yield and variation of energy with angle, observations in the forward direction are preferable, homogeneous neutron beams at very low energies are more readily obtained in the backward hemisphere. The reasons are the following: the second group of neutrons which occurs just above threshold in the forward hemisphere is not present in the backward hemisphere; the neutron energy varies much more slowly with bombarding energy; and neutrons of energy below that given by the motion of the center of mass at threshold are almost impossible to observe in the forward direction. On the other hand, the background caused by scattered neutrons is relatively much larger in the backward hemisphere so that it is advantageous to use, if possible, a collimator which shields the detector from secondary neutron sources.

III. (α, n) reactions.

39. General remarks. To-day more than 50 (α, n) reactions [*11*] are known to occur. Of these only the few which give a relatively high neutron yield are of practical interest as neutron sources. When an accelerator is available, reactions other than (α, n) are usually more convenient for producing intense neutron beams.

(α, n) neutron sources using natural α-particles have, however, several important advantages: (a) in contrast to accelerators and piles they are cheap and easy to prepare and operate; (b) provided they have been carefully enough prepared, they may be used as standards, since their intensity varies with time in a known manner according to the laws of radioactive decay for the parent substance and its daughter products; (c) the region from which neutrons are emitted is small and the overall size of the source is not much larger, a fact of importance in studies of the slowing down and diffusion of neutrons in a moderator; (d) in contrast to accelerator sources, radioactive sources are essentially isotropic, which helps for example in the so-called sphere transmission measurements of absorption and inelastic collision cross sections.

Disadvantages of such sources are: (a) their strength is usually much less than what can be achieved with accelerators or piles; (b) the neutrons are not monoenergetic; (c) depending on the α-emitter used, the neutrons are accompanied by more or less intense γ-radiation which may affect the neutron detector; (d) the sources frequently develop leaks so that hazardous radioactivity may inadvertently be spread.

Sources of this type will be denoted by the symbol Rα+A where R describes the α-emitter (usually in radioactive equilibrium with its daughter products) and A the target element.

The radioactive substances R usually employed are Po^{210}, Ra^{226}, and Pu^{239}. To a lesser extent Rn^{222}, Ac^{227}, Am^{241}, and Cm^{242} have been used. Each of these α-emitters has advantages and disadvantages. The half-life of Rn^{222} is too short for most applications, and Ac^{227}, Am^{241} and Cm^{242} are not availabel in large enough quantities.

Po^{210} has a half-life too short for its use in a standard source, but has the advantage that the intensity of γ-rays is weak[1] so that the γ-rays emitted by a

---

[1] 1.6×10^{-5} photons of 0.8 Mev per α-particle and characteristic X-rays of Pb with comparable intensities: *L*-radiation 2.2×10^{-4}, *K*-radiation 1.6×10^{-6}. R.W. PRINGLE, H.W. TAYLOR and S. STANDIL: Phys. Rev. **87**, 384 (1952). — M. RION: J. Phys. Radium **13**, 244 (1952). — R.J. BREEN and M.R. HERTZ: Phys. Rev. **98**, 599 (1955).

neutron source containing Po are mainly those emitted by the final product of the (α, n) reaction.

The half-life of Ra^{226}, 1620 years, makes this α-emitter a very desirable one for preparing neutron sources of small size and constant strength. 20 days after the source has been prepared the Ra reaches radioactive equilibrium (within 2.7%) with radon and all its α-emitting descendents, except Po^{210} which grows with the half-life of RaD (19.4 years). The slow increase in neutron yield caused by the growth of Po can be corrected by making use of the thick target excitation curve. Since the daughter products of Ra^{226} emit more energetic α-particles than Ra, each of them produces more neutrons than Ra in the normally used neutron sources in which the bombarded light element forms a thick target [Eq. (41.4) and Table 18][1]. The principal disadvantage of a Ra source is that it emits many more γ-rays than neutrons. This γ-radiation offers is hazard in handling the source[2] and strongly affects most neutron detectors.

Table 15. *Neutron yields from* Po α *particles on thick targets.*

Element X	Yield/10^6 Alphas			
	ROBERTS (a)	SEGRÈ and WIEGAND (b)	HALPERN (c)	SZALAY (d)
Li	2.6	4.7		
Be	80	73	50	
B	24	19		
C	0.11			
N	0.01			
O	0.07			
F	12	10		
Na	1.5			
Mg	1.4		0.5	
Al	0.74		0.25	0.22
Si	0.16			
Cl	0.11			
A	0.38			

(a) J. H. ROBERTS: MDDC-731 (1944).
(b) E. SEGRÈ and C. WIEGAND: MDDC-185 (1944).
(c) I. HALPERN: Phys. Rev. **74**, 1234 (A) (1948).
(d) A. SZALAY: Z. Physik **112**, 29 (1939).

As Pu^{239} has become available in recent years, this α-emitter is being used increasingly for the preparation of neutron sources. Its main advantages are the absence of penetrating γ-radiation and the long half-life (24300 years) which results in a constant activity. On the other hand, because of the long half-life and the relatively low α-particle energy, ($E_\alpha = 5.15$, 5.137 and 5.09 Mev) Pu neutron sources have considerably larger sizes for a given source strength than Ra sources. Another difficulty with Pu sources of appreciable size is the neutron multiplication in the Pu caused by the fact that Pu is fissionable. This may result in an increase in source strength when the source is embedded in solid or liquid material, particularly in moderating materials.

Many authors have measured the relative yields of various (α, n) reactions produced in light elements by the α-particles from natural radioactive substances[3]. In Table 15 some results obtained with Poα-particles ($E_\alpha = 5.3$ Mev) incident on

[1] After a source, made with new Ra, has aged a few months, one can use the relation

$$Q_0(t) = Q_0(0)\,[1.12 - 0.12\, e^{-t/28}]$$

where t is the age expressed in years.

[2] 1 mg of Ra in equilibrium with its products and filtered with 0.5 mm of platinum gives a dose (Sievert dose) of 8.25 roentgen per hour at 1 cm distance. It may be recalled that 1 r = 0.93 rem: see footnote 7, p. 89 [Nat. Bur. Stand. Techn. News Bull. **11**, 198 (1957)].

[3] L. MEITNER and K. PHILIPP: Naturwiss. **20**, 929 (1932). — I. CURIE and F. JOLIOT: C. R. Acad. Sci., Paris **196**, 397 (1933); **197**, 237 (1933). — J. Phys. Radium **4**, 21 (1933). — T.W. BONNER: Phys. Rev. **45**, 425 (1934). — R. JACKEL: Z. Physik **91**, 493 (1934). — C.D. ELLIS and W. J. HENDERSON: Proc. Roy. Soc. Lond., Ser. A **146**, 206 (1934). — H. FAHLENBRACH: Z. Physik **94**, 607 (1935). — E. FÜNFER: Phys. Z. **38**, 967 (1937). — A. MEYE: Z. Physik **105**, 232 (1937). — H. BRANDT: Z. Physik **108**, 726 (1938).

a thick target of the indicated element are given[1]. The data of ROBERTS[2] were obtained by slowing down the neutrons in a large block of paraffin in which a number of BF_3 counters were embedded. The distance from the source to the counters was chosen so as to minimize the dependence of the sensitivity of the arrangement to the neutron energy. The experimental set up was calibrated with a Raα + Be source whose intensity was known to about 10%.

The data of SEGRÈ and WIEGAND[3] were obtained from their thick target excitation curves (Fig. 36) for the full energy of the Poα's while those of HALPERN[4] are integrations of thin target excitation curves. All these authors detected the neutrons after they were slowed down, by means of BF_3 counters. SZALAI[5], on the other hand, measured the activity of the product of the $Al^{27}(\alpha, n)\, P^{30}$ reaction.

Table 15 shows that the elements giving a rather high intensity are Li, Be, B and F. Among these Be is the most useful for the construction of an intense neutron source. Fig. 36 shows the integral excitation curves of three of the above mentioned elements.

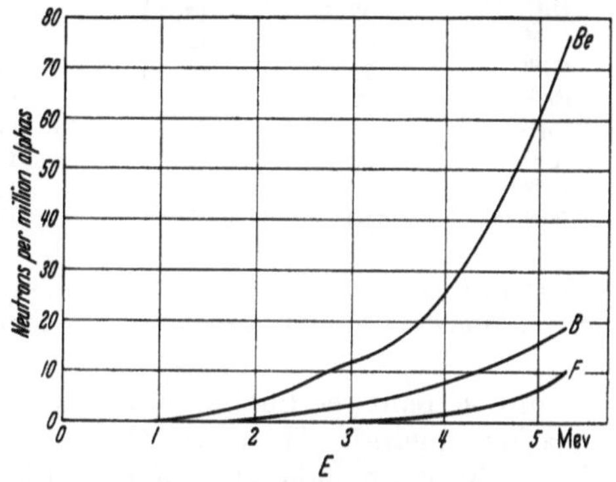

Fig. 36. Thick target excitation functions of Po α + A sources, A = Be, B, F (SEGRÈ and WIEGAND).

An extensive investigation of the γ rays emitted from Po neutron sources has been made by BREEN and HERTZ[6] who quote and discuss the results obtained by other authors.

The thick target yield from 9-Mev α-particles produced in the Princeton cyclotron and incident on a number of elements, was measured by RIDENOUR and HENDERSON[7] by detecting the activity of the reaction products. According to these authors the data given in Table 16 have an uncertainty of about 50%, the principal error arising from the calibration of the Lauritsen electroscope used to measure the β-particles.

More recently ALLEN, NECHAJ, SUN and JENNINGS[8] have measured the angular distribution and the integral yields from 30 Mev α-particles produced in the Pittsburgh cyclotron and incident on a number of elements. The neutrons were detected by means of the threshold detector $S^{32}(n, p)\, P^{32}$ ($T_{\frac{1}{2}} = 14.3$ days). The cross section of this reaction was assumed to be constant over the energy region of the emitted neutrons (probably between 2 and 10 Mev) and to be equal to 0.32 barn[9]. Table 17 contains the neutron yields of a few (d, n) and (α, n) reactions obtained by integrating the corresponding angular distributions. A rather large correction had to be applied to the results of these authors in order to

---

[1] H. L. ANDERSON: Preliminary report No. 3, Nuclear Service Series, National Research Council 1948.
[2] J. H. ROBERTS: MDDC-731 (1944).
[3] E. SEGRÈ and C. WIEGAND: MDDC-185 (1944).
[4] J. HALPERN: Phys. Rev. **74**, 1234 (A) (1948).
[5] A. SZALAI: Z. Physik **112**, 29 (1939).
[6] R. J. BREEN and M. R. HERTZ: Phys. Rev. **98**, 599 (1955).
[7] L. N. RIDENOUR and W. J. HENDERSON: Phys. Rev. **52**, 889 (1937).
[8] A. J. ALLEN, J. F. NECHAJ, K. H. SUN and B. JENNINGS: Phys. Rev. **81**, 537 (1951).
[9] H. H. BARSCHALL: Detection of Neutrons, in Vol. XLV of this Encyclopedia; see also: R. RICAMO: Nuovo Cim. **8**, 383 (1951). — T. HURLIMAN and P. HUBER: Helv. phys. Acta **28**, 33 (1955).

Table 16. *Thick target yields of neutrons for 9 Mev alphas.*

Bombarded isotope	Relative abundance %	Reaction	Radioactive isotope	Half-life	Yield per 10^6 α for pure isotope
B^{10}	20	(α, n)	N^{13}	10.3 min	4.6
N^{14}	99.6	(α, n)	F^{17}	1.07 min	—
Mg^{25}	11.5	(α, p)	Al^{28}	2.56 min	27
Al^{27}	100	(α, n)	P^{30}	2.55 min	16
P^{31}	100	(α, n)	Cl^{34}	32 min	3.4
Cl^{35}*	76	(α, n)	K^{38}	7.65 min	3.0***
Cr^{50}	4.9	(α, n)	Fe^{53} (?)	8.9 min	2.2 (?)
Cr^{53}	10.4	(α, p)	Mn^{56}	160 min	0.30
Ni^{58}	67.3	(α, p)	Cu^{61}	3.4 hrs	0.86
Co^{59}	99.8	(α, n)	Cu^{62}	10.0 min	2.4
Ni^{60}	26.9	(α, n)	Zn^{63}	34 min	0.30
Cu^{63}	70	(α, n)	Ga^{66}	9.4 hrs	0.27
Cu^{65}	30	(α, n)	Ga^{68}	68 min	1.5
As^{75}	100	(α, n)	Br^{78}	6.3 min	0.37
Br^{79}**	50	(α, n)	Rb^{82}	1.5 min	0.20***
Br^{81}**	50		Rb^{84}	9.8 min	0.14

* Rocksalt target.
** NaBr target.
*** Yield multiplied by factor of 2 to take into account Na in target.

take into account the fraction of the neutrons detected by the threshold detector at various angles[1]. Therefore, the uncorrected data given in Table 17 can be taken only as an indication of the order of magnitude. This table is useful, however, for comparing the yields of the (α, n) and (d, n) reactions at high energy.

Table 17. *Fast neutron yield.*

Target nuclei	Fast neutrons/sec/steradian/μamp in forward direction $\times 10^{-8}$		Total fast neutrons/sec/μamp $\times 10^{-8}$		No. bombarding particles required to yield one fast neutron $\times 10^{-3}$	
	15-Mev deuterons	30-Mev alpha	15-Mev deuterons	30-Mev alpha	15-Mev deuterons	30-Mev alpha
$_4Be$	94	22	190	65	0.33	0.48
$_{13}Al$	30	3.1	64	18	0.98	1.6
$_{22}Ti$	12		65		2.4	
$_{24}Cr$	7.2		29		1.8	
$_{25}Mn$	12	1.7	52	14	1.3	
$_{27}Co$	5.5	1.1	26	10	2.0	2.9
$_{29}Cu$	4.8	1.1	29	10	3.0	2.9
$_{41}Cb$	3.0		15		3.3	
$_{42}Mo$	3.1		15		4.0	
$_{47}Ag$	2.2	0.54	14	5.4	4.7	9.6
$_{48}Cd$	2.4	0.56	12	5.2	5.1	5.5
$_{73}Ta$	0.56		3.3		19	
$_{79}Au$	0.37	0.062	2.1	0.65	27	48
$_{82}Pb$	0.36	0.18	2.1	1.6	30	19
$_{83}Bi$	0.23	0.044	1.3	0.47	49	70

ALLEN, NECHAJ, SUN, TENNINGS: footnote 8 on p. 109.

In an (α, n) reaction Z increases by 2 units, while $N = A - Z$ increases by 1 unit. It is to be expected that if no special selection rules intervene, the most exoenergetic (α, n) reactions and therefore the most intense ones, are those in which the bombarded nucleus has even Z and odd N so that the product of the reaction is an even-even nucleus. Among nuclei with $Z \leq 10$ $_4Be^9$, $_6C^{13}$, $_8O^{17}$, and

[1] B. L. COHEN: Phys. Rev. **81**, 632 (1951).

$_{10}Ne^{21}$ satisfy this condition. The relative abundance of the last three nuclei is so small (1.1, 0.04, 0.27%) that in practice only Be has to be considered. The high intensity observed in this element seems to agree with this remark.

For Z larger than 10 the effect of the potential barrier is of importance for not too high energies of the α-particles (potential barrier of Ne ≈ 6 to 7 Mev). At high bombarding energies and for medium and heavy elements other reactions, as, for instance, (d, n) reactions (see Table 17), are, however more useful as neutron sources.

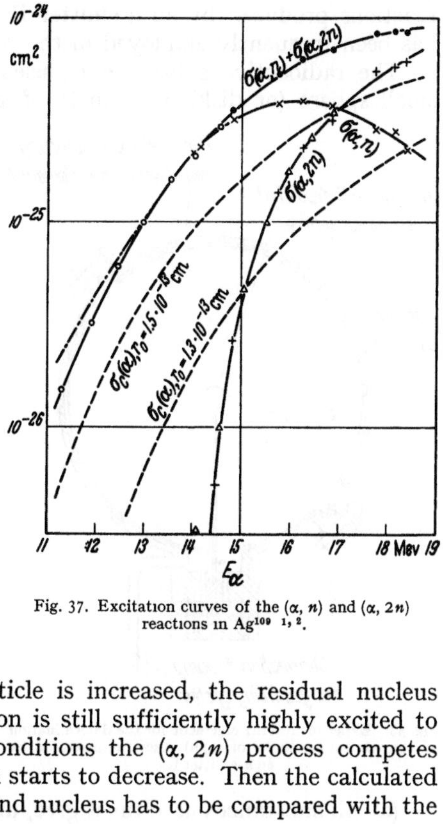

Fig. 37. Excitation curves of the (α, n) and $(\alpha, 2n)$ reactions in Ag^{109} [1, 2].

The experimental results on (α, n) reactions in the high-energy region may be compared with theoretical predictions. The excitation of the compound nucleus produced by the capture of the incident α-particles is so high that the corresponding levels are broadened as a consequence of the high probability of particle emission, in particular of neutron emission. One can then try to apply the continuum theory which gives the cross section for formation of the compound nucleus averaged over many resonances [*3*].

Over a wide energy interval the emission of one neutron is the dominant process, and therefore the cross section for this process is almost equal to the cross section for formation of the compound nucleus

$$\sigma_c(\alpha) \approx \sigma(\alpha, n). \tag{39.1}$$

When the energy of the incident α-particle is increased, the residual nucleus left after the emission of the first neutron is still sufficiently highly excited to emit a second neutron. Under these conditions the $(\alpha, 2n)$ process competes with the (α, n) process whose cross section starts to decrease. Then the calculated cross section for formation of the compound nucleus has to be compared with the sum of the cross sections for both processes

$$\sigma_c(\alpha) \approx \sigma(\alpha, n) + \sigma(\alpha, 2n). \tag{39.2}$$

An example is the (α, n) reaction on Ag^{109} which has been investigated by various authors[1, 2] by measuring the activity of the residual nucleus as a function of the energy of the incident α-particles. Some complication arises from the fact that In^{112} is produced in its ground state ($T_{\frac{1}{2}} = 14.5$ min) as well as in a metastable state ($T_{\frac{1}{2}} = 20.7$ min) which decays in the ground state by emission of a photon ($M3$) of $h\nu = 0.155$ Mev. The ratio of the cross sections for production of In^{112} in these two states has been measured between 14 Mev ($\sigma_m/\sigma_g = 1.3$) and 18 Mev ($\sigma_m/\sigma_g = 2.6$), but not below 14 Mev, where only the 20.7 min period was measured. Therefore there exists some uncertainty in the value of $\sigma(\alpha, n)$ given in Fig. 37 below 14 Mev; for the points indicated by circles a constant ratio has been assumed, while the broken curve indicates an estimate of a reasonable

[1] H. L. Bladt and D. J. Tendam: Phys. Rev. **72**, 1117 (1947).
[2] E. Bleuler, A. K. Stebbins and D. J. Tendam: Phys. Rev. **90**, 460 (1953).

upper limit. Apart from this point the agreement between the continuum theory and the experimental results is very satisfactory, if in the expression for the nuclear radius (18.1) r_0 is taken to be 1.6×10^{-13}. For the determination of the nuclear temperature or level density from the energy dependence of the ratio $\sigma(\alpha, 2n)/\sigma(\alpha, n)$ the reader is referred to the original papers[1,2] and to Ref. [3].

40. A few techniques typical of (α, n) reactions excited by the α-particles of radioactive substances. In the determination of the excitation curves of (α, n) reactions produced by an α-emitter R, the following experimental arrangement has been frequently employed in the past.

The radioactive substance R, used as α-particle source, is deposited on a small sphere (or disk) of diameter d placed at the centre of a hollow sphere (or hemisphere) of radius r (Fig. 38). On the internal surface of the sphere is deposited the target substance A for which the excitation curve has to be measured.

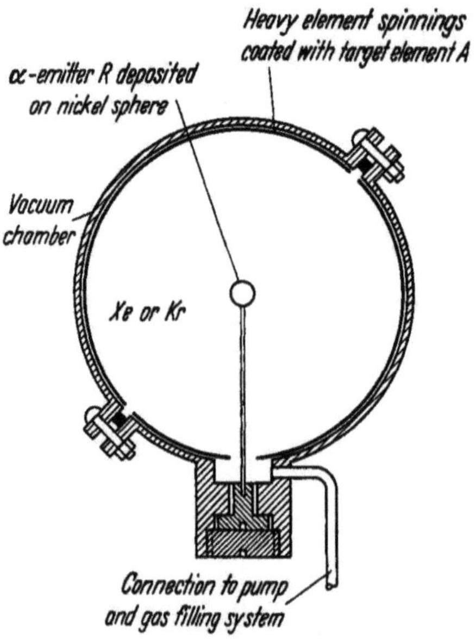

Fig. 38. Experimental arrangement for the determination of the excitation curve of (α, n) reactions produced by an α-emitter R.

By changing the pressure of the gas contained in the sphere, one can change at will the residual range of the α-particles incident on the target. The excitation curve is obtained by measuring with a suitable detector the number of neutrons emitted per second as a function of the pressure of the gas: the maximum energy of the α-particles is obviously obtained when the hollow sphere is evacuated.

The target, as explained in Sect. 37, can be thin or thick and the corresponding excitation curves will be accordingly differential or integral curves.

A few important points have to be considered in the construction of such an experimental set-up:

(a) All substances used must give, under the α-particle bombardment, a much lower (possibly negligible) neutron emission with respect to the target A: as gas filling carbon dioxide or nitrogen may be used or, still better, Kr or Xe; the sphere and the α-particle source support are usually of a medium or heavy element such as Ni ($Z=28$) or Cu ($Z=29$) or better Ag ($Z=47$). In particular the α-particle source must not contain light elements which could undergo (α, n) processes.

(b) If the α-particle source emits γ-rays these can also give a background of neutrons due to (γ, n) processes. Among these, particularly important, is the photodisintegration of the deuteron which is present at a relative abundance of 0.0149%, for instance in the water or paraffin sometimes used for slowing down neutrons.

The background of neutrons can be easily determined by measurements without the target and the contributions due to the various parts of the experimental

[1] See footnote 1, p. 111.
[2] See footnote 2, p. 111.

set-up can be separated by repeating the measurements without and with the gas, at such a pressure that the α-particles do not reach the target.

(c) The neutron detector must be insensitive to the energy of the neutrons. In order to obtain this, the neutrons are usually slowed down and then detected by means of a slow neutron detector, as for instance one or more BF_3 counters embedded in the moderator. If the counters are in fixed positions, their distance from the neutron source must be chosen so as to minimize the sensitivity of the arrangement to the energy of the neutrons.

(d) In the case of differential excitation curves one has to know the overall energy resolving power of the experimental set-up which is determined by: the inhomogeneities of the α-particles emitted from the source, the ratio d/r and the equivalent thickness of the target (Sect. 37).

The same experimental set up of Fig. 38 surrounded by a thin layer of paraffin (1 to 2 cm thick) has been used by various authors[1] in order to determine the existence of levels of the reaction product sufficiently excited to give a negative Q value

$$Q_S = Q_0 - w_S < 0.$$

The method, indicated in the following as "*slow neutron threshold method*", can be applied to any level of the final nucleus if the reaction is endoenergetic, but only to levels with $w_S > |Q_0|$ if the reaction is exoenergetic. The method is based on the fact that when in an endoenergetic reaction the incident particles have an energy immediately above the threshold (34.5), the emitted neutrons have a low energy (34.6) due only to the motion of the centre of gravity in the l.s. Therefore a detector sensitive only to neutrons emitted at low energies, will indicate an intense neutron radiation when the energy of the incident α-particles is equal to or slightly higher than the corresponding threshold.

Below the threshold the process is energetically impossible while for α-particle energies appreciably larger than the threshold, the corresponding neutrons are emitted with rather high energy and are therefore detected with lower efficiency.

41. The Be (α, n) reactions. As was mentioned in Sect. 1, Chadwick discovered the neutron when he bombarded beryllium with α-particles from polonium. Since beryllium is a monoisotopic element, the reaction suggested by Chadwick

$$_2\mathrm{He}^4 + {}_4\mathrm{Be}^9 \rightarrow {}_6\mathrm{C}^{12} + n^1 + 5.708\ \mathrm{Mev}; \tag{41.1}$$

was the simplest assumption and certainly the correct one.

For α-particles of sufficiently high energy there is an indication that the reactions

$$_2\mathrm{He}^4 + {}_4\mathrm{Be}^9 \rightarrow 3\ {}_2\mathrm{He}^4 + n^1 - 1.572\ \mathrm{Mev} \tag{41.2a}$$

$$_2\mathrm{He}^4 + {}_4\mathrm{Be}^9 \rightarrow {}_4\mathrm{Be}^8 + {}_2\mathrm{He}^4 + n^1 - 1.666\ \mathrm{Mev}, \tag{41.2b}$$

also take place with an appreciable intensity. The Be^8 produced in reaction (41.2b) breaks up into two α particles, with the release of 94 kev, in less than 5×10^{-14} sec [2].

It was pointed out in Sect. 39 that the case of Be is the most important one among (α, n) reactions as source of neutrons. Therefore it will be discussed in more detail than other processes of this type.

---

[1] T. Bjerge: Proc. Roy. Soc. Lond., Ser. A **164**, 243 (1938). — O. Haxel and E. Stuhlinger: Z. Physik **114**, 178 (1939).

The excitation curve of the (α, n) process in beryllium has been measured by many authors using the Po α-particles[1]. RASETTI showed that the integral excitation curve for this process was very similar to the excitation curve for emission of γ-rays from Be under α-particle bombardment which had been measured by BECKER and BOTHE[2] by detecting the secondary electrons with a point counter. In particular, both curves showed a hump attributed to a resonance around $E_\alpha = 2$ Mev (see later Figs. 39 and 40). This result shows that the γ-rays observed by BOTHE and BECKER are due to secondary processes in the reaction (41.1).

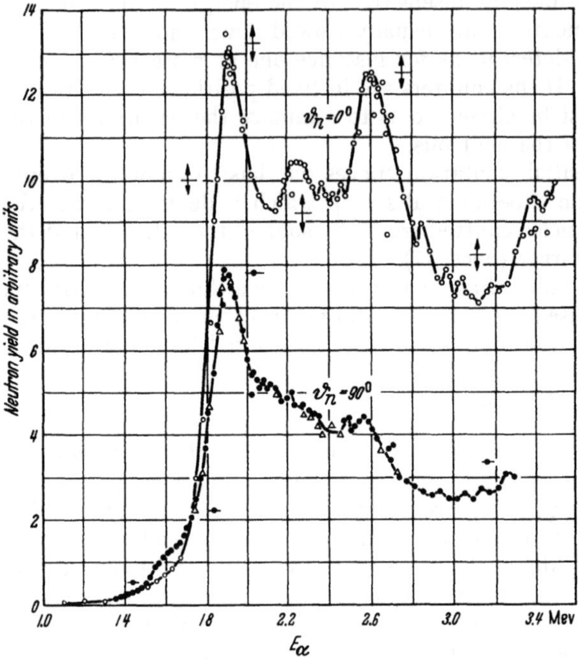

Fig. 39. Differential neutron yield of the $Be^9(\alpha, n) C^{12}$ reaction as a function of the energy of the α-particles (TRUMBLE).

The first differential excitation curve was measured by BERNARDINI, while CRUSSARD and GORODETZKY were the first to attempt to reduce the dependence of the sensitivity of the neutron detector on the energy of the neutrons by slowing them down with paraffin and then using the (n, α) process in boron for detection.

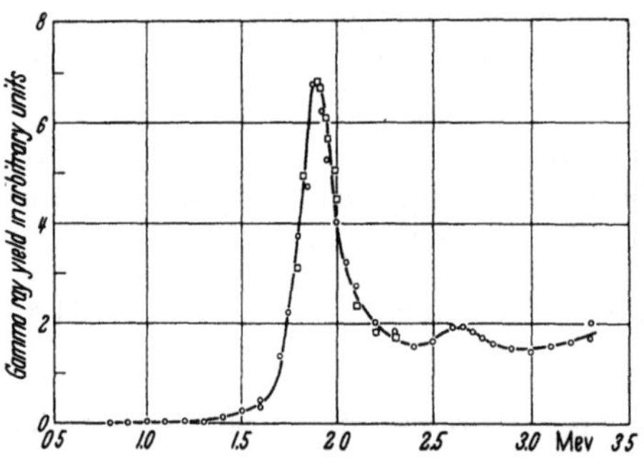

Fig. 40. Gamma ray yield of the $Be(\alpha, n) C^{12}$ reaction (TALBOTT and HEYDENBURG).

The differential yield curve obtained by HALPERN, with an energy resolution slightly better than 20%, represents the best results among those obtained with α-particles from polonium.

More recently various authors[3-5] have measured the excitation curve from a thin target of Be bombarded with α-particles accelerated by means

[1] F. RASETTI: Naturwiss. **20**, 252, 350 (1932). — Z. Physik **78**, 165 (1932). — I. CURIE and F. JOLIOT: C. R. Acad. Sci., Paris **196**, 397 (1933). — J. Phys. Radium **4**, 278 (1933). — J. CHADWICK: Proc. Roy. Soc. Lond., Ser. A **142**, 1 (1933). — G. BERNARDINI: Z. Physik **85**, 555 (1933). — T.W. BONNER: Phys. Rev. **45**, 425 (1934). — J. CRUSSARD and S. GORODETZKY: C. R. Acad. Sci., Paris **205**, 1060 (1937). — T. BJERGE: Proc. Roy. Soc. Lond., Ser. A **164**, 243 (1938). — J. HALPERN: Phys. Rev. **76**, 248 (1949).

[2] H. BECKER and W. BOTHE: Z. Physik **76**, 421 (1932).

[3] R.E. TRUMBLE: Phys. Rev. **94**, 748 (A) (1954). I thank Dr. N.P. HEYDENBURG for making available the given excitation curves.

[4] W.E. BENNETT, P.A. ROYS and B.J. TOPPEL: Phys. Rev. **94**, 924 (A) (1954). I thank Dr. W.E. BENNETT for making available the given graphs.

[5] T.W. BONNER, A.A. KRAUS, J.B. MARION and J.P. SCHIFFER: Phys. Rev. **102**, 1348 (1956).

of an electrostatic accelerator. The high intensity attainable under these conditions allows the use of much thinner targets and of highly monoenergetic incident beams so that the overall energy-resolving power turns out to be 5 to 10 kev.

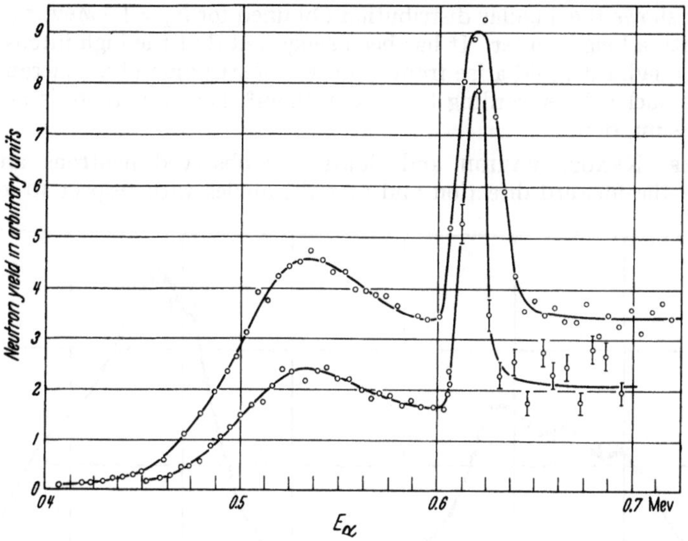

Fig. 41. Differential neutron yield of Be(α, n) C^{12} for two thicknesses of the Be target (BENNETT *et al.*).

Fig. 39 shows the results of TRUMBLE[1]. In this experiment the neutrons were detected by means of a BF_3 "long counter"[2] [*12*], [*18b*] which subtended an angle of 60° at the target. The measurements were done at 0° and 90° with respect

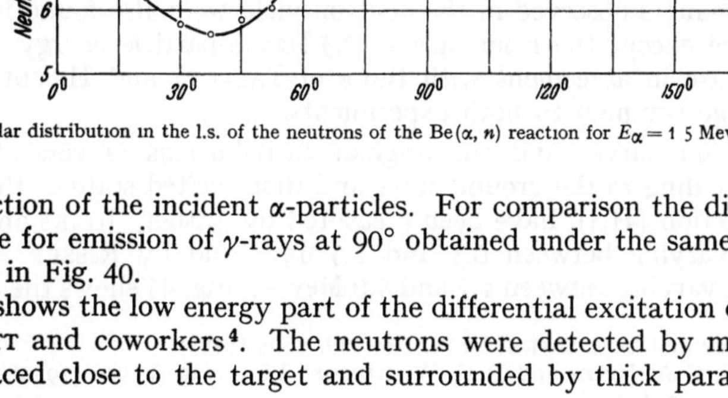

Fig 42. Angular distribution in the l.s. of the neutrons of the Be(α, n) reaction for $E_\alpha = 1.5$ Mev (BENNETT *et al*)

to the direction of the incident α-particles. For comparison the differential excitation curve for emission of γ-rays at 90° obtained under the same conditions[3] is also shown in Fig. 40.

Fig. 41 shows the low energy part of the differential excitation curve obtained by BENNETT and coworkers[4]. The neutrons were detected by means of a BF_3 counter placed close to the target and surrounded by thick paraffin. The two

[1] See footnote 3, p. 114.
[2] H. H. BARSCHALL: Detection of Neutrons, Vol. XLV of this Encyclopedia.
[3] F. L. TALBOTT and N. P. HEYDENBURG: Phys. Rev. **90**, 186 (1953).
[4] See footnote 4, p. 114.

curves are obtained by reducing the thickness of the thin target by a factor of 2. The fact that the height of the resonance at 0.62 Mev remains the same while the width is reduced by a factor two, means that still thinner targets should be used to study it.

Fig. 42 shows the angular distribution obtained for $E_\alpha = 1.5$ Mev, by the same authors with a long counter. It has been suggested that the high intensity in the backward quadrant could arise from a process of stripping of a neutron from the Be^9 target nucleus[1]. According to some authors[2], however, there is no need for such an assumption.

BONNER, KRAUS, MARION and SCHIFFER[3] observed neutrons and γ-rays emitted in the forward direction and at right angles with respect to α-particles

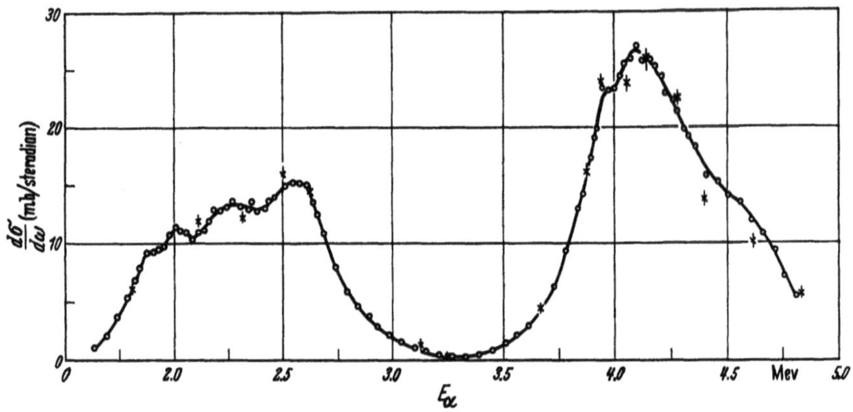

Fig. 43. The differential scattering cross section in the l.s. at 0° for the $Be^9(\alpha, n)\,C^{12}$ reaction leading to C^{12} ground state (RISSER *et al.*).

accelerated in an electrostatic generator and incident on a Be-target, 5 to 10 kev thick. The energy interval investigated extended from 1.8 to 5.3 Mev. Resonances in the Be(α, n) reaction were observed at α-particle energies of 1.9, 2.3, 2.6, 3.98, 4.4 and 5.0 Mev. The first three of these resonances had previously been observed by TRUMBLE[4], the 4.4 Mev resonance had been reported by HALPERN[5]. The differential cross-section measurements are in good agreement with HALPERN's total cross-section determinations.

All the resonances observed in the neutron yield were also found in the γ-ray excitation curve except the resonance at 2.3 Mev α-particle energy. The γ-ray measurements are in agreement with those of TALBOTT and HEYDENBURG[6] in the energy range common to both experiments.

The excitation curves and the angular distributions of resolved neutron groups corresponding to the ground state and first excited state of the final nucleus C^{12} in reaction (41.1), have been measured by JAMES, JONES and WILKINSON[2]—for E_α varying between 0.3 and 1.3 Mev—and by RISSER, PRICE and CLASS[7]—for E_α varying between 1.7 and 5.0 Mev—. Fig. 43 shows the differential

[1] L. MADANSKY and G.E. OWEN: Phys. Rev. **99**, 1608 (1955).
[2] D.B. JAMES, G.A. JONES and D.H. WILKINSON: Phil. Mag. **1**, 949 (1956).
[3] See footnote 5, p. 114.
[4] See footnote 3, p. 114.
[5] See footnote 1, p. 114.
[6] See footnote 3, p. 115.
[7] J.R. RISSER, J.E. PRICE and C.M. CLASS: Phys. Rev. **105**, 1288 (1957).

cross section at 0°, obtained by means of a spectrometer consisting of a polyethylene radiator, a gas proportional counter and a crystal scintillator in coincidence. The corresponding angular distributions (Fig. 44) cannot be fitted—even approximately—by means of a single level of the compound nucleus while a satisfactory agreement can be obtained by the assumption of two overlapping levels. The best fit of the data of Fig. 44 was found from the combination of a $^3/_2{}^+$ level — made by $l=3$ α-particles — and a $^5/_2{}^+$ level — made by $l=1$ α-particles. We recall that the spin of Be^9 ground state is $I={}^3/_2$. We refer to the original paper for the excitation curves and angular distribution of the neutrons corresponding to the first excited state of C^{12}.

Above $E_\alpha=5.3$ Mev the energy of the α-particles of Po, the integral excitation curve has been measured by STUHLINGER[1] using the α-particles emitted from the active deposit of thoron.

The results of this author were obtained by placing the paraffin used to slow down the neutrons, only around the neutron source and the BF_3 counters. Therefore it is to be expected that neutrons emitted from the source with low energy are favoured compared to those emitted with high energy. This fact distorts the excitation curve which becomes steeper for higher values of E_α.

Using a BF_3 counter for detection, BJERGE compared the differential excitation curves obtained for neutrons slowed down in paraffin whose thickness was first large and then only 2 cm, in two successive experiments. He was the first to use the procedure called the slow neutron threshold method (Sect. 40). With the thin paraffin layer the neutrons emitted from the Po α + Be source with a low energy are detected with a higher efficiency than those emitted with higher energy.

He found that, apart from a constant factor, the two curves were pactically identical for $E_\alpha \leqq 4.5$ Mev, while above this energy the "thin paraffin curve" was steeper. BJERGE interpreted this result, confirmed later by STUHLINGER, as owing to the reaction (41.2a) whose threshold is

$$E_{\alpha T}=\frac{4+9}{9}\times 1.57 \approx 2.3 \text{ Mev}.$$

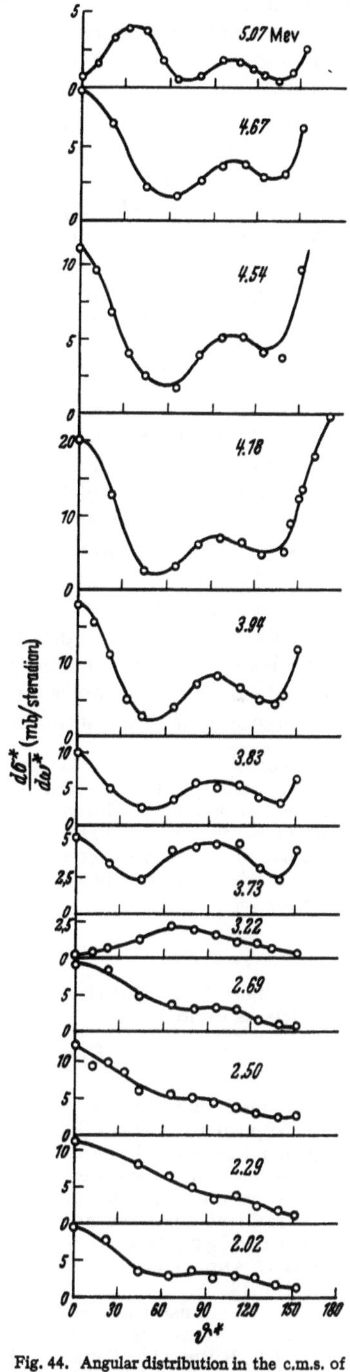

Fig. 44. Angular distribution in the c.m.s. of the neutrons from the $Be^9(\alpha, n)C^{12}$ reaction leading to C^{12} ground state.

[1] E. STUHLINGER: Z. Physik 114, 185 (1939).

As will be discussed later the reactions (41.2) gives very low energy neutrons. This interpretation was further corroborated by BJERGE in a measurement of the excitation curve for γ-rays; apart from a constant factor, the γ-ray excitation curve is almost the same as the neutron excitation curve below about 4.5 Mev, while above this energy it is much less steep than the neutron curve with thick paraffin[1].

The integral yield of sources made of a mixture of Ra, Po or Pu with beryllium is of great practical interest.

In the case of Ra the source is usually prepared of a mixture of $RaBr_2$ and Be powder. The yield depends on the amount of beryllium used in the source because the particles which traverse grains of $RaBr_2$ are absorbed without producing neutrons.

A rough empirical formula which gives the neutron yield as a function of the masses of the constituents of the mixture has been given by ANDERSON[2,3]:

$$Y = 1.7 \times 10^7 \frac{M_{\mathrm{Be}}}{M_{\mathrm{Be}} + M_{\mathrm{RaBr_2}}} \text{ neutrons/sec} \cdot \text{g (Ra)} \tag{41.3}$$

Y refers to young sources, i.e. sources in which all other descendents of Ra are in equilibrium, but no Po in present.

A somewhat higher yield of 1.9×10^7 neutrons/sec · g (Ra) is however regularly attained in small commercial sources prepared using radium containing not more than 50% of the equilibrium amount of Po[4]. From the empirical formula (41.4) given below, one obtains a yield varying from 1.79 to 2.04×10^7 neutrons per second per gram Ra, depending on the Po content of radium[5]. In order to split this yield into the yields produced by each of the α-emitting substances present in the Ra α + Be source, one can make use of the integral yield curve of STUHLINGER[6]. This curve, however, was obtained in an experiment which favoured the detection of low-energy neutrons (produced preferentially by high-energy

[1] According to BJERGE, the processes (41.2) can be described in the following way: the incident α-particle is captured and a highly excited C^{13} (for $E_\alpha = 4.5$ Mev: $w = E_\alpha \frac{9}{13} + S_\alpha = 13.8$ Mev) is produced. This compound state can emit a neutron according to process (41.1) in which some C^{12} nuclei remain excited or it can emit an α-particle leaving a Be^9 nucleus which can also remain either excited or in its ground state. In this last case the α-particle is simply scattered while, when Be^9 is excited, it can disintegrate according to either of the two processes

$$\mathrm{Be}^9 \rightarrow \begin{cases} {}_4\mathrm{Be}^8 + {}_0\mathrm{n}^1 \\ 2\,{}_2\mathrm{He}^4 + {}_0\mathrm{n}^1. \end{cases}$$

For an incident α-particle of energy $E_\alpha = 4.5$ Mev the Be^9 left after the emission of the first α-particle, is excited to

$$E_\alpha \frac{9}{4+9} + Q_0 \approx 3.1 - 1.6 = 1.5 \text{ Mev}.$$

Assuming that the concept of penetration through the electrostatic potential barrier can be applied also to very light elements, at least for obtaining a very rough estimate of the order of magnitude, the probability for an α-particle to pass through the barrier of Be^9 is small for excitation energies smaller than 1.5 Mev. This consideration explains the fact that the reactions (41.2) take place with an appreciable cross section only for energies of the incident α-particle above about $E_\alpha = 4.5$ Mev, while they are already energetically possible at $E_\alpha \gtrsim 2.3$ Mev.

[2] H.L. ANDERSON: Preliminary Report No. 3, Nuclear Science Series, National Research Council 1948.

[3] H.L. ANDERSON and B.T. FELD: Rev. Sci. Instrum. **18**, 186 (1947).

[4] O.J.C. RUNNALLS and R.R. BOUCHER: Canad. J. Phys. **34**, 949 (1956). — G.C. HANNA and O.J.C. RUNNALLS: Canad. J. Phys. **34**, 959 (1956).

[5] See footnote 1, p. 108.

[6] See footnote 1, p. 117.

particles). This fact, according to ANDERSON, explains the discrepancy (about a factor 2) between the results of this author and those of DEMERS and ROBERTS concerning the efficiency ratio of Ra α + Be and Po α + Be. The third column of Table 18 gives the neutron yields deduced by ANDERSON for the various α-emitters of the radium family, by weighting heavily the results of ROBERTS. In the fourth column of the same table are given the results obtained by RUNNALLS and BOUCHER[1] by means of the empirical formula (41.4).

As already mentioned in Sect. 39, a neutron source of the Pu α + Be type possesses some advantages over Po α + Be and Ra α + Be sources and therefore may be expected to replace them where small sizes for a given neutron strength are not the determining factor. The advantages of this type of source are the long half-life of Pu, the absence of penetrating γ-rays and the possibility of preparing intermetallic Pu—Be compounds. Alloy neutron sources rather than pressed mechanical mixtures are expected to have a higher and more reproducible neutron yield (Sect. 44).

The absolute yield of a source composed of 13 g of Pu + 7 g of Be as $PuBe_{13}$ (density 3.7 g/cm³) has been measured by STEWART[2] by observing the proton recoils in nuclear emulsions, and has been found to be 1.2×10^6 neutrons per second for neutrons of energy above 0.5 Mev.

Table 18. *Neutron yields of a* Raα + Be *source.*

α-emitter	α-energy (Mev)	Neutrons/10^6 α	
		(ANDERSON)	(RUNNALLS *et al.*)
Ra	4.777	55	46
Rn	5.486	90	76
RaA	5.998	120	105
RaC′	7.680	200	258
RaF = Po	5.298	80	67

A systematic investigation of the neutron yield from various actinide-beryllium alloys has been made by RUNNALLS and BOUCHER and HANNA and RUNNALLS[1]. In the case of Pu α + Be sources RUNNALLS and BOUCHER have verified that the variation of the neutron yield with the Be—Pu atom ratio ν, may be rather well represented by the formula

$$y = y_{\max} \frac{\nu \cdot s_{\mathrm{Be}}}{\nu \cdot s_{\mathrm{Be}} + s_{\mathrm{Pu}}} \text{ neutrons}/10^6\,\alpha$$

where s_{Be} (= 0.63) and s_{Pu} (= 4.6) are the relative atomic stopping powers of Be and Pu respectively, and the maximum neutron yield $y_{\max}$ has to be adjusted to fit the experimental data. The values of $y_{\max}$ obtained for various actinide-beryllium sources are the following:

$$\begin{aligned} &\mathrm{Pu}^{239} \quad y_{\max} = 60 \pm 2 \quad \text{neutrons}/10^6\,\alpha,\\ &\mathrm{Am}^{241} \quad y_{\max} = 74 \pm 2 \quad \text{neutrons}/10^6\,\alpha,\\ &\mathrm{Cm}^{242} \quad y_{\max} = 112 \pm 3 \quad \text{neutrons}/10^6\,\alpha. \end{aligned}$$

From these data the following empirical expression relating the thick target yield on beryllium, $y_{\max}$, to the particle energy may be derived

$$y_{\max} = 0.152\,(E_{\alpha_{\mathrm{Mev}}})^{3.65} \quad \text{neutrons}/10^6\,\alpha. \tag{41.4}$$

It has to be noticed that the neutron yield for a $PuBe_{13}$ source as that used by STEWART corresponds to $y \approx 38.7$ neutrons per million alphas, or 63% of that obtained for alloys with $\nu \approx 250$.

[1] See footnote 4, p. 118.
[2] L. STEWART: Phys. Rev. **98**, 740 (1955).

The spectrum of the emitted neutrons can be devided in two parts, one of neutrons of a few Mev kinetic energy, from reaction (41.1), the other, first observed by AUGER[1] and attributed to the reactions (41.2), of neutrons of energy of the order of 100 kev.

The high energy part of the spectrum of the Po α + Be source has been investigated by many authors[2,3] who observed the recoil protons by means of different techniques: BONNER and MOTT-SMITH used a cloud chamber filled with H_2 at high pressure, BERNARDINI and BOCCIARELLI and COCHRAN and HENRY[3] a layer of paraffin or polystyrene in front of three proportional counters in coincidence, RICHARDS, DEMERS, and later WHITTMORE and BAKER[3], observed the recoil protons produced in nuclear emulsions, and finally ELLIOT, MCGARRY and FAUST[3] used a coincidence scintillation spectrometer. Fig. 45 shows the results of WHITTMORE and BAKER corrected for the variation of the proton-neutron cross section with energy and for the probability of a track leaving the emulsion.

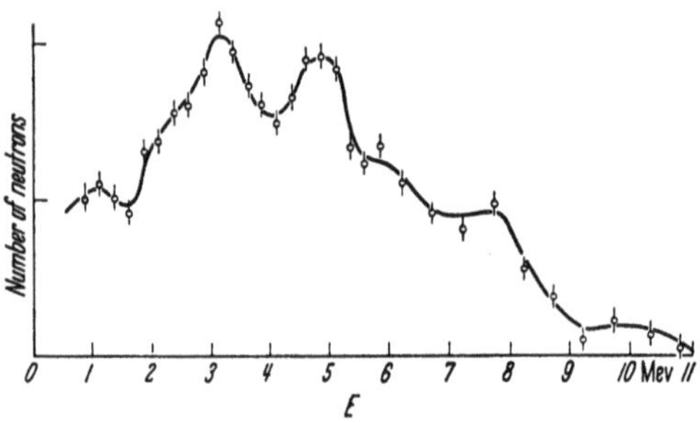

FIG. 45. Energy distribution of neutrons from a Poα + Be source (WHITTMORE and BAKER).

In the interpretation of this curve in terms of levels of C^{12} one has to remember that for a given Q-value the energy of the emitted neutrons varies owing (a) to the variation in the energy of the α-particles caused by the infinite thickness of the target, (b) to the variation in the direction of emission of the neutrons relative to that of the α-partices.

Of the various groups of neutrons appearing in Fig. 45, that at 11 Mev corresponds to the production of C^{12} in its ground state while that at about 8 Mev is probably caused by the resonance in C^{13} at $E_\alpha = 1.9$ Mev (Fig. 39); the other groups are attributed to excited states of the final nucleus C^{12}.

In order to clarify such an interpretation it would be necessary to measure the neutron spectrum under good conditions as it is possible with artificial α-particles. Very few data of this type are available to-day.

BRADFORD and BENNETT[4] have measured with the emulsion technique the neutron spectrum emitted at 90° with respect to a beam of α-particles of 1.4 Mev produced by means of an electrostatic generator and impinging on a thick Be target. Their results (Fig. 46) show two groups, one corresponding to the ground state and the other to the state at 4.4 Mev. The corresponding intensities are in the ratio 1:1.6 while similar measurements at 0° give 1:0.4.

In the interpretation of neutron spectra such as those of Fig. 45 the data on the γ-rays emitted from Be bombarded with α-particles are of great help.

---

[1] P. AUGER: C. R. Acad. Sci., Paris **195**, 234 (1932). — J. Phys. Radium **4**, 719 (1933).

[2] T.W. BONNER and L.M. MOTT-SMITH: Phys. Rev. **45**, 552, 554 (1934); **46**, 258 (1934). — G. BERNARDINI and D. BOCCIARELLI: Ric. Sci. **7** (1), 128 (1936); **7** (2), 89 (1936). — Rend. Acad. Lincei **24**, 59 (1932). — G. BERNARDINI: Ric. Sci. **8** (1), 33 (1937). — H.T. RICHARDS: MDDC, 1504 (1944). — P. DEMERS: Montreal Report MP-74 (1945) and MP-204 (1948).

[3] B.G. WHITTMORE and W.B. BAKER: Phys. Rev. **78**, 799 (1950). — H.W. GUIER, H.W. BERTINI and J.H. ROBERTS: Phys. Rev. **85**, 426 (1952). — H. GURSKY, B. WINNEMORE and D.A. COWAN: Phys. Rev. **91**, 209 (1953). — J.O. ELLIOT, W.I. MCGARRY and W.R. FAUST: Phys. Rev. **93**, 1348 (1954). — R.G. COCHRAN and K.M. HENRY: Rev. Sci. Instrum. **26**, 757 (1955).

[4] C.E. BRADFORD and W.E. BENNETT: Phys. Rev. **78**, 302 (1950).

A detailed discussion of the levels of C^{12} is given in [1] and [2], and only some of the data will be mentioned here. BOTHE[1] showed that the γ-ray spectrum consists of 3 lines of about 3, 4.5 and 7 Mev. The more recent values are (3.25 ± 0.1), (4.425 ± 0.020), and (7.68 ± 0.02) Mev [2].

They can be interpreted with only two excited levels: $w_1 = 4.43$ Mev and $w_2 = 7.65$ Mev, which correspond to the emission of two neutron groups. The corresponding Q-values are $Q_1 = 1.28$ Mev and $Q_2 = -1.97$ Mev respectively.

The direct transition from the excited state w_2 to the ground state has a probability smaller than 1/2500[2]. In the majority of cases the transition goes via the level $w_1 = 4.43$ Mev. This is confirmed by the observation of coincidences

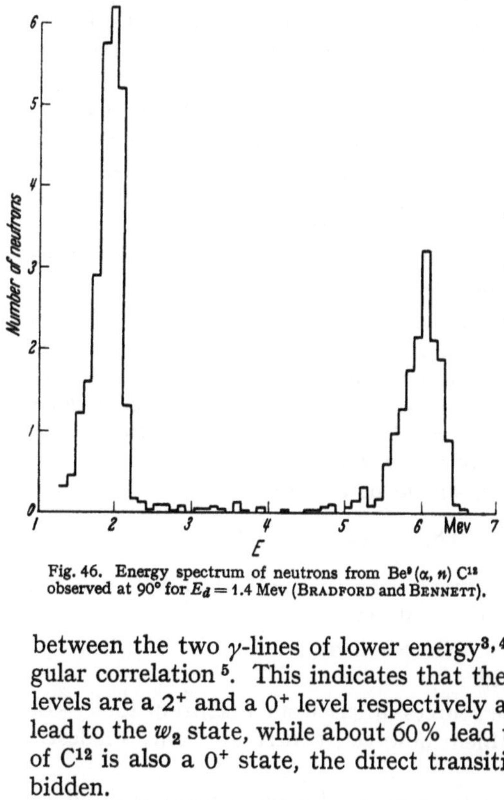

Fig. 46. Energy spectrum of neutrons from $Be^9(\alpha, n)\,C^{12}$ observed at 90° for $E_d = 1.4$ Mev (BRADFORD and BENNETT).

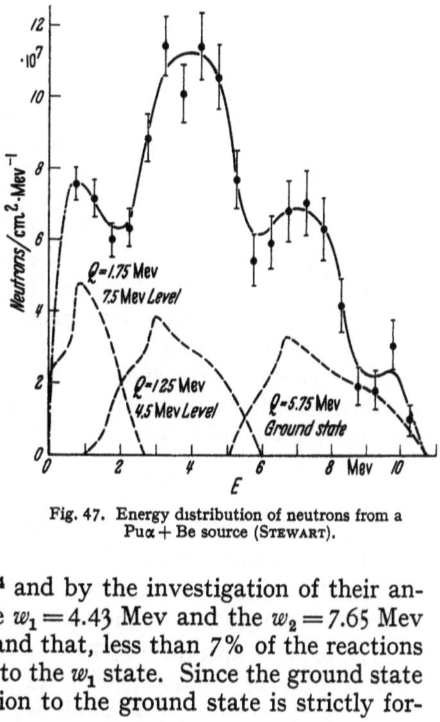

Fig. 47. Energy distribution of neutrons from a Puα + Be source (STEWART).

between the two γ-lines of lower energy[3,4] and by the investigation of their angular correlation[5]. This indicates that the $w_1 = 4.43$ Mev and the $w_2 = 7.65$ Mev levels are a 2^+ and a 0^+ level respectively and that, less than 7% of the reactions lead to the w_2 state, while about 60% lead to the w_1 state. Since the ground state of C^{12} is also a 0^+ state, the direct transition to the ground state is strictly forbidden.

On the other hand the γ-ray line of 3.2 Mev has an intensity relative to the 4.4 Mev line of 3 or 8 to 12% according to various authors[2,4].

One can conclude that the neutron group corresponding to the excited level w_2 must have an intensity of the order of one tenth of the group leading to the level w_1 which is apparently the more intense: this is the group responsible for the maximum observed in Fig. 45 at ≈ 4.8 Mev.

---

[1] W. BOTHE: Z. Physik **100**, 273 (1936).

[2] Y.E. BEGHIAN, H.H. HALBAN, T. HUSAIN and L.G. SANDERS: Phys. Rev. **90**, 1129 (1953).

[3] H. MAIER-LAIBNITZ: Z. Physik **101**, 478 (1936). — K.P. ARTENOV and N.A. VLAZOV: Dokl. Akad. Nauk. SSSR. **77**, 225 (1951).

[4] UEBERGANG: Aust. J. Phys. (unpublished); see [2].

[5] K.G. STEFFEN, O. HEINRICHS and H. NEUERT: Z. Physik **145**, 156 (1956). — K.G. STEFFEN and H. NEUERT: Z. Physik **147**, 125 (1957).

A quantitative analysis of the curve of Fig. 45 has been attempted by HESS[1]. The agreement with the experimental results is rather poor as one may expect because of the uncertainties arising from the fact that it involves the excitation curve and the angular distribution relative to each group of neutrons for energies of the α-particles varying between 0 and 5.3 Mev.

The spectrum of the Pu α+Be source has been measured by STEWART[2]. The result obtained from the observation in nuclear emulsion of 2057 tracks of recoiling protons, is shown in Fig. 47. The dotted curves have been computed, making use of HALPERN'S excitation curve, for the three Q-values corresponding to the ground state and the first two excited states w_1 and w_2 of C^{12}.

The neutron spectrum of sources of the type Rn α+Be or Ra α+Be is obviously much more complicate on account of the complex energy spectrum of the incident α-particles.

The spectrum of the Rn α+Be source has been measured by DUNNING[3] who detected the recoil protons from a layer of paraffin by means of a small ionization chamber connected to a linear amplifier. The spectrum of the Ra α+Be source has been measured by DEMERS[4] and HOUTERMANS and TEUCHER[5] with the emulsion technique, and more recently by SCHMIDT-ROHR[6] who detected the recoil protons produced in high pressure hydrogen contained in three ionization chambers, two in coincidence and one in anticoincidence.

The papers quoted above describe the actual shape of the spectrum which roughly corresponds to what is to be expected from the data deduced from the Po α+Be source, taking into account the energies of the incident α-particles and a few more excited states of C^{12}. Thus, because of the 7.68 Mev α-particles of RaC′ (Table 18), the spectrum of the Ra α+Be sources extends up to 13 Mev. Furthermore, three more excited states of C^{12}, which lie immediately above the two levels previously mentioned [*1*], [*2*], should be considered

$$w_3 = 9.61, \quad w_4 = 10.8, \quad w_5 = 11.1 \text{ Mev}.$$

Their existence was proved by STUHLINGER[7] using the method of slow neutron threshold and was confirmed later by other authors.

Finally it should be noticed that the photoneutron reaction in Be (Sect. 47) contributes about 3% to the total yield.

The neutron spectrum of Rn α+Be sources differs from that from Ra α+Be only because of the absence of Ra which, however, contributes only modestly to the total neutron yield (Table 18).

The neutron spectrum from Ac α+Be sources also has been measured by means of nuclear emulsions[8].

The low energy group of neutrons first observed in a cloud chamber and attributed to the reactions (41.2) by AUGER[9] has an intensity which increases markedly when more energetic α-particles, such as those of RaC′, are used. The results of BJERGE and STUHLINGER quoted above confirm this conclusion and its inter-

---

[1] W.N. HESS: UCRL-3839 (1957).
[2] See footnote 2, p. 119.
[3] J.R. DUNNING: Phys. Rev. **45**, 586 (1934).
[4] See footnote 2, p. 120.
[5] F.G. HOUTERMANS and M. TEUCHER: Z. Physik **129**, 365 (1951).
[6] U. SCHMIDT-ROHR: Z. Naturforsch. **8**, 470 (1953).
[7] See footnote 1, p. 117.
[8] W.R. DIXON, A. BIELESCH and K.W. GEIGER: Canad. J. Phys. **6**, 699 (1957).
[9] See footnote 1, p. 120.

pretation. The low energy group has been investigated by various authors with cloud chambers, proportional counters, and resonance detectors[1].

There is however still some controversy about the relative intensity of this group and that of the high energy spectrum[2]. By comparing the values of $\overline{r^2}$ in water for the neutrons emitted from a Raα + Be source and detected by means of the In resonance, with the $\overline{r^2}$ calculated from the energy spectrum observed in nuclear emulsions by RICHARDS, one can evaluate that the low-energy group contains about 30% of the total number of neutrons.

On the other hand, YALOW, YALOW and GOLDHABER[3] conclude that the intensity of this group is not more than 10% of the total number of neutrons of a Ra α + Be source by comparing the initial slopes of the growth curves of thermal neutrons obtained with a Ra α + Be and with a Ra γ + Be source as functions of the thickness of a paraffin layer.

The Q_0-value of reaction (41.1) is such that the reaction would be a useful one for producing monoenergetic neutrons in the energy range from 6 to 10 Mev. This could be accomplished by accelerating He ions in an electrostatic generator. Neutron energies in the neighborhood of 10 Mev are not readily accessible with other reactions.

The difficulty in using reaction (41.1) is that there are always at least two groups of neutrons produced, the ground state transitions and those to the excited state $w_1 = 4.43$ Mev. Because of the large energy difference between these two groups it might be possible to use detectors which would be sensitive only to the most energetic neutrons. The detection problem is further complicated by the presence of the γ-rays from the deexcitation of the C^{12}-levels. Experiments with monoenergetic neutrons from reaction (41.1) and the difficulties encountered in such experiments have been described[4].

42. The B (α, n) reactions. Normal boron consists of 19.8% B^{10} and 80.2% B^{11}, the isotopic composition depending somewhat on the source of the boron[5]. Under α-particles bombardment the two isotopes undergo the following reactions:

$$\left.\begin{aligned} B^{10} + \alpha \rightarrow N^{13} + n + 1.07\ \text{Mev} \\ N^{13}(\beta^+)\, C^{13} \quad E^{(+)}_{\max} = 1{,}20\ \text{Mev}, \quad T_{\frac{1}{2}} = 10.2\ \text{min}, \end{aligned}\right\} \tag{42.1a}$$

$$B^{11} + \alpha \rightarrow N^{14} + n + 0.5\ \text{Mev}. \tag{42.1b}$$

The product of the first reaction is radioactive, and the reaction may, therefore, be detected by observing the activity of N^{13}.

Reaction (42.1b) is the reverse of

$$N^{14} + n \rightarrow B^{11} + \alpha. \tag{42.2}$$

Reaction (42.2) has been investigated intensively and these results may be used in the analysis of observations on reaction (42.1b) by making use of the reciprocity theorem [*3*].

[1] L. MEITNER and K. PHILIPP: Naturwiss. **20**, 929 (1932). — N. FEATHER: Proc. Roy. Soc. Lond., Ser. A **142**, 689 (1933). — M. BENINI: Rend. Acad. Lincei **29**, 590 (1939). — E. GIEGERL and E. BRODA: Nature, Lond. **167**, 399 (1951).

[2] See footnote 2, p. 118.

[3] R.A. YALOW, R.S. YALOW and M. GOLDHABER: Phys. Rev. **69**, 253 (A) (1946).

[4] A.L. BECKER and H.H. BARSCHALL: Phys. Rev. **102**, 1384 (1956).

[5] See footnote 2, p. 61.

The early work on reactions (42.1a) and (42.1b) was carried out with α-particles from naturally radioactive substances and with normal boron[1]. Because of the limited resolving power available with this technique, the earlier experiments have been superseded by more recent accelerator studies which have shown the presence of many resonances in both reactions. Among the experiments with α-particles from Po, the work of WALKER is of particular interest, because it is an absolute measurement of the total (α, n) cross section for normal boron (Fig. 48), while most of the subsequent accelerator work gives only differential cross sections.

Reactions (42.1a) and (42.1b) have been studied with accelerators by many authors[2]. The measurements cover the energy range from 1 to 5.5 Mev. In their first paper SHIRE *et al.* investigated the $B^{10}(\alpha, n)$ reaction in the energy range from 1 to 2 Mev and found resonances at 1.51, 1.64, and 2.05 Mev. In their second paper they extended the measurements to 2.4 Mev and showed that the resonance at 2.05 Mev is due to the $B^{11}(\alpha, n)$ reaction, but they observed additional resonances in the $B^{10}(\alpha, n)$ reaction at particle energies between 2.1 and 2.3 Mev. In the $B^{11}(\alpha, n)$ reaction resonances were observed at 1,51, 1.58, and 2.05 Mev. In this latter reaction Trumble finds resonances at 1.54, 2.09, 2.63, 3.00 and 3.26 Mev. BONNER *et al.* whose measurements extend from 2 to 5.5 Mev find resonances in the $B^{10}(\alpha, n)$ reaction at 2.16, 2.26, 2.90, 4.53, and 4.85 Mev, and in the $B^{11}(\alpha, n)$ reaction at 2.06, 2.60, 2.93, 2.97, 3.23, 3.54, 3.72, 3.92, 4.25, 4.34, and 5.00 Mev. More detailed information about the energy levels causing these resonances may be found in the review papers by AJZENBERG and LAURITSEN [*1*], [*2*].

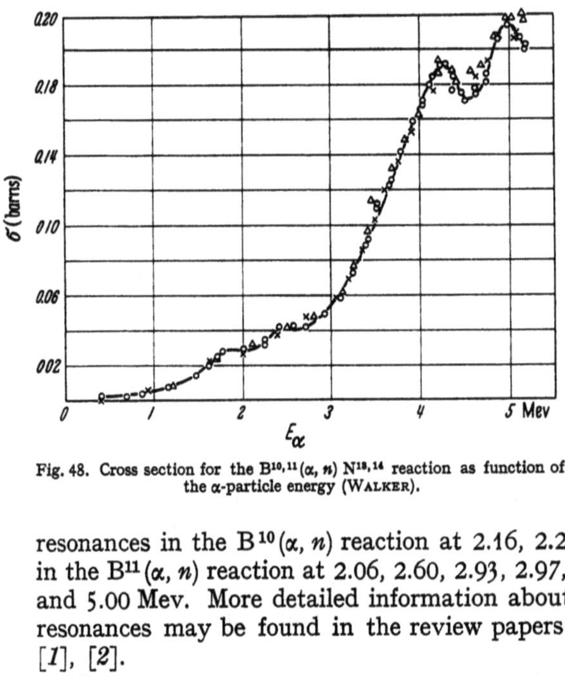

Fig. 48. Cross section for the $B^{10,11}(\alpha, n)\,N^{13,14}$ reaction as function of the α-particle energy (WALKER).

The integral yield of the (α, n) reactions in boron for Po α-particles is given in Table 15 (p. 108). In Fig. 48 the results of WALKER[2] are reproduced which give the cross section with a probable error of about 20% although with rather low resolution.

If one calculates from this curve the integral yield for Po α-particles, one obtaines 19 neutrons per 10^6 α's against 24 given by ROBERTS and 19 given by SEGRÈ (Table 15).

[1] J. CHADWICK: Proc. Roy. Soc. Lond., Ser. A **142**, 1 (1953). — I. CURIE and F. JOLIOT: C. R. Acad. Sci., Paris **196**, 397 (1933). — J. Phys. Radium **4**, 278 (1933). — W. MAURER: Z. Physik **107**, 721 (1937). — F. FÜNFER: Ann. d. Phys. **35**, 147 (1939). — E. STUHLINGER: Z. Physik **114**, 185 (1939). — E. SEGRÈ and C. WIEGAND: MDDC, 185 (1944). — R. L. WALKER. Phys. Rev. **76**, 244 (1949). — A. I. ALICHANOW, A. I. ALICHANIAN and B. S. DZELEPOW: Z. Physik **93**, 350 (1935). — H. FAHLENBRACH: Z. Physik **96**, 503 (1935). — A. SZALAI: Z. Physik **112**, 29 (1939).

[2] R. E. TRUMBLE: Phys. Rev. **94**, 748 (1954). — E. S. SHIRE, J. R. WORMALD, G. LINDSAY-JONES, A. LUNDEN and A. G. STANLEY: Phil. Mag. **44**, 1197 (1953). — E. S. SHIRE and R. D. EDGE: Phil. Mag. **46**, 640 (1955). — T. W. BONNER, A. A. KRAUS, J. B. MARION and J. P. SCHIFFER: Phys. Rev. **102**, 1348 (1956).

B. T. FELD[1] found that a pressed $RaBr_2\, \alpha + B$ source with 2 grams of radium and 5.75 grams of boron has a yield of 4.28×10^6 neutron/second. This result may be expressed by a relation, similar to (41.3),

$$Y = 6.8 \times 10^6 \frac{M_B}{M_B + M_{\mathrm{Ra\,Br_2}}} \quad \text{neutrons/sec} \cdot \mathrm{g\,(Ra)}\,. \tag{42.3}$$

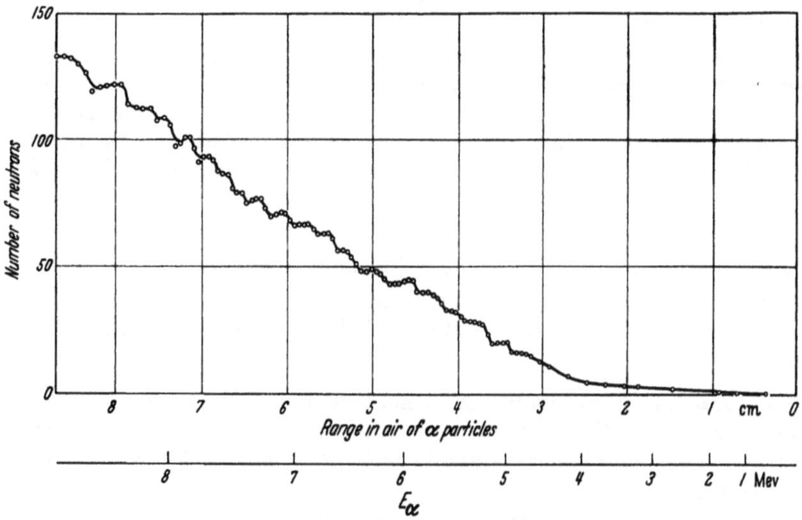

Fig. 49. Integral excitation curve of the $B^{10,11}(\alpha, n)\, N^{13,14}$ reaction as a function of the range and energy of the incident α particles (STUHLINGER).

STUHLINGER's integral excitation curve (Fig. 49) can be used to deduce from this total integral yield the contributions of the various α-emitters in a Ra α + Be source (Table 19). The criticism raised against the measurements of STUHLINGER in the case of Be are not applicable to the case of B, because the neutron spectrum does not contain a low energy group. This is borne out by the good agreement between STUHLINGER's and ROBERTS' results for energies below 5.3 Mev.

Table 19. *Neutron yield of* Raα + B *sources.*

α-emitter	Neutrons/10^6α	α-emitter	Neutrons/10^6 α
Ra	15	RaC	100
Rn	30	RaF	24
RaA	40		

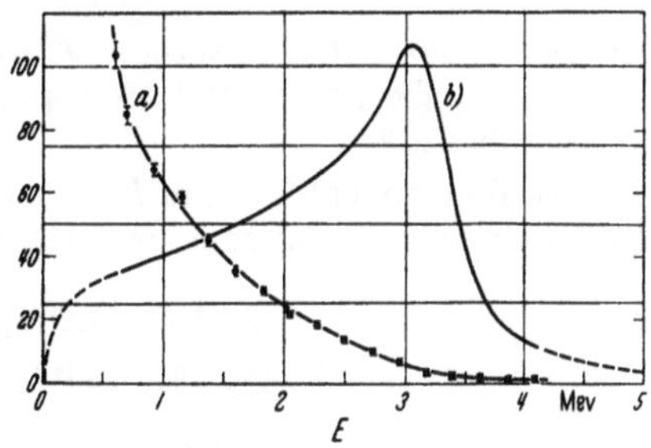

Fig. 50. Experimental proton recoil spectrum (a) and neutron spectrum (b) from a Poα + B source (STAUB).

The neutron spectrum of a Po α + B source appears to be much simpler than that of Be[2]. Fig. 50 shows the results of STAUB[3] who used on ionization chamber

[1] H. L. ANDERSON: Preliminary Report No. 3, Nuclear Service Series. National Research Council, 1948.

[2] J. CHADWICK: Proc. Roy. Soc. Lond., Ser. A **142**, 1 (1933). — I. CURIE and F. JOLIOT: C. R. Acad. Sci., Paris **196**, 397 (1933). — W. MAURER: Z. Physik **107**, 721 (1937). — G. BERNARDINI: Nuovo Cim. **6**, 220 (1938). — I. F. PERLAM, H. T. RICHARDS and L. SPECK: MDDC-39 (1946).

[3] H. STAUB: MDDC-1490 (1947).

similar to the one used by BARSCHALL and KANNER[1]. The chamber was filled with a mixture of 12.0 atm of hydrogen and 3.13 atm of argon so that the correction for recoils leaving the counting volume is large only at energies above 3 Mev. The figure shows the recoil spectrum, and the neutron spectrum corrected for the variation of the neutron-proton cross section with energy and for wall and opacity effects. Similar results have been obtained by other authors with a triple coincidence proton recoil spectrometer[2].

BEGHIAN and coworkers[3] have confirmed STAUB's results using an ionization chamber filled with CH_4 at 20 atm. They observe also a γ-ray line at (2.2 ± 0.2) Mev whose intensity is 9% of the neutron yield and 2% of the intensity of the γ-ray intensity from Po. They attribute this line to a transition in the product of reaction (42.1b) the first two excited states of which are known to be at [*1*], [*2*]

$$w_1 = 2.31, \quad w_2 = 3.95 \text{ Mev}.$$

Of these two levels only the first one can be excited by Po α-particle.

With the method of slow neutron thresholds (Sect. 40) STUHLINGER has concluded that there are 5 excited states of N^{14} for α-particle energies between 5.1 and 8.65 Mev.

43. The (α, n) reactions in a few other light elements. In this section we collect the rather scanty data about the (α, n) reactions in light elements for which neutron emission has been observed directly or at least looked for without success.

The reaction

$$_1D^2(\alpha, n)\,\alpha p - 2.25 \text{ Mev} \tag{43.1}$$

was looked for without success by various authors[4] and observed later by SCHULTZ[5] who estimated the corresponding cross section to be about 1.2×10^{-27} cm² for Ra C′ α-particles, and 3×10^{-27} cm² for those of ThC′.

The reaction

$$_2He^4(\alpha, n)\,Be^7 - 18.98 \text{ Mev} \tag{43.2}$$

was looked for without success at an energy of 39.9 Mev (0.5 Mev above the threshold in the c.m.s.) and an upper limit of the cross section $\sigma < 7 \times 10^{-28}$ cm² was established[6].

According to Table 15, at $E_\alpha = 5.3$ Mev lithium is the fourth element in order of intensity of neutron emission. Its two isotopes undergo the following reactions

$$7.52\% \quad _3Li^6(\alpha, n)\,_5B^9 - 3.983 \text{ Mev}, \tag{43.3}$$

$$92.47\% \quad _3Li^7(\alpha, n)\,_5B^{10} - 2.792 \text{ Mev}. \tag{43.4}$$

$_5B^9$ is unstable against disintegration into a proton and Be^8 or two α-particles, with a mean life of 3×10^{-19} sec (width $\Gamma = 2$ kev) [*1*], [*2*].

Reaction (43.4) is the inverse of the well known reaction produced in boron by slow neutrons (Sect. 17).

---

[1] H.H. BARSCHALL and M.N. KANNER: Phys. Rev. **58**, 590 (1940).

[2] R.G. COCHRAN and K.M. HENRY: Rev. Sci. Instrum. **26**, 757 (1955).

[3] L.E. BEGHIAN, M.A. GRACE and H.H. HALBAN: Proc. Phys. Soc. Lond. **63**, 913 (1950).

[4] Lord RUTHERFORD and A.E. KEMPTON: Proc. Roy. Soc. Lond., Ser. A **143**, 724 (1934). — J.R. DUNNING and G.B. PEGRAM: Phys. Rev. **45**, 295 (1934).

[5] H.L. SCHULTZ· Phys. Rev. **53**, 622 (1938).

[6] D. WALKER, W.T. LINK and W.I.B. SMITH: Proc. Roy. Soc Lond., Ser. A **65**, 861 (1952).

The excitation curve for emission of neutrons by lithium has been measured by various authors[1] with Po α-particles and extended up to 8.78 Mev by HAXEL and STUHLINGER[2] who used the α-particles of the active deposit of thoron. The integral excitation curve obtained by these authors is given in Fig. 51 in arbitrary units, but it can be normalized making use of the data of ROBERTS at 5.3 Mev (Table 15). The arrows in the figure represent the threshold and seven possible resonances suggested by HAXEL and STUHLINGER for the emission of neutrons. BICHSEL and BONNER[3], using monoenergetic α-particles from an electrostatic accelerator, have shown that the reaction (43.4) has its threshold at 4.379 Mev ($Q_0 = -2.788$ Mev) and a resonance at 5.15 Mev. Under α-particle bombardment, lithium gives also rather strong γ-ray emission which is due, at least in great part, to the nuclear excitation produced by the electric field of the incident particle[4].

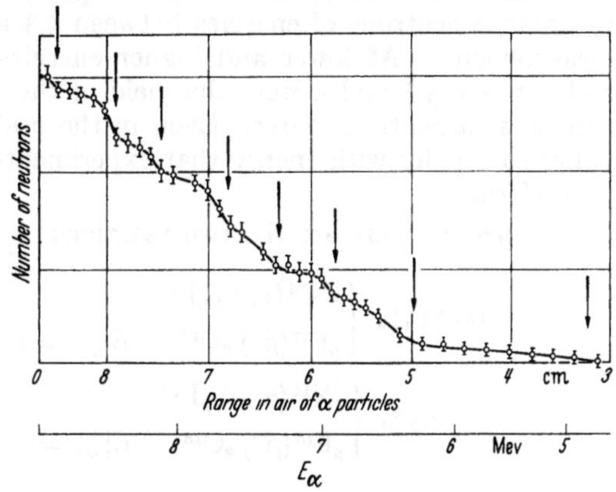

Fig 51. Integral excitation curve of the $Li^{6,7}(\alpha, n)\,B^{9,10}$ reaction as a function of the range and energy of the incident α particles (HAXEL and STUHLINGER).

Very little is known of the neutron spectrum of Poα + Li sources[5]. Some authors[6] find a mean energy of about 250 kev, others[7] of about 500 kev. The low value found by the first group could be explained by assuming that Po remained in sizable granules reducing the effective energy of the α-particles.

We should also recall the results obtained by HAXEL and STUHLINGER[2] with the slow neutron threshold method (Sect. 40). From these the existence of three excited states was deduced which, if attributed to B^{11}, turn out to be at

$$w_1 = (0.8 \pm 0.1); \quad w_2 = (1.4 \pm 0.1); \quad w_3 = (2.2 \pm 0.1)\ \text{Mev}.$$

Under the assumption that the Bohr hypothesis is valid even for very light elements one could try to deduce some information about the spectra of the reactions (43.3) and (43.4) from the experimental results obtained from the reactions ${}_4Be^9(p, n)\,{}_5B^9$ and ${}_4Be^9(d, n)\,{}_5B^{10}$ which have been extensively investigated.

The excitation curve for the reaction[8]

$$C^{13}(\alpha, n)\,O^{16} + 2.20\ \text{Mev} \tag{43.5}$$

has been measured by TRUMBLE and HEYDENBURG, by BONNER *et al.* with the same techniques used by these authors in the case of Be and B, and by BECKER

[1] I. CURIE and F. JOLIOT: J. Phys. Radium **4**, 278 (1953). — P. SAVEL: C. R. Acad. Sci., Paris **198**, 1404 (1934). — K. SCHNETZLER: Z. Physik **95**, 302 (1935).
[2] O. HAXEL and E. STUHLINGER: Z. Physik **114**, 178 (1939).
[3] H. BICHSEL and T.W. BONNER: Phys. Rev. **108**, 1025 (1957).
[4] R. J. BREEN and M.R. HERTZ: Phys. Rev. **98**, 599 (1955).
[5] P. SAVEL: C. R. Acad. Sci., Paris **196**, 1482 (1933).
[6] D.M. BARTON: LA-1609 (1953). — D.S. YOUNG. LA-1938 (1955).
[7] W.N. HESS: UCRL-3839 (1957).
[8] G.A. JONES and D.H. WILKINSON: Proc. Phys. Soc. Lond. **64**, 756 (1951).

and BARSCHALL[1]. The measurements cover the energy range from 1.0 to 5.2 Mev. The yield was measured both in the forward direction and at right angles to the incident α-particle beam. A large number of sharp resonances was observed.

This reaction is of interest as a possible source of monoenergetic neutrons. Since the first excited state of the product of this reaction, O^{16}, lies above 6 Mev, only neutrons leaving O^{16} in its ground state should be produced up to neutron energies above 6 Mev. Although monoenergetic neutrons in this energy range may readily be produced in the (d, d) reaction, the acceleration of deuterons involves background difficulties (see Division VI) which should not be present in the acceleration of α-particles. The $C^{13}(\alpha, n)$ reaction has been used successfully to produce neutrons of energies between 4.3 and 5.8 Mev for total cross section measurements. At lower and higher energies (α-particle energies below 2 Mev and between 3.5 and 4 Mev) the yield of the reaction is too low for most applications as a neutron source. Even in the useful range the yield of the reaction varies so rapidly with energy that experiments with these neutrons are not easy to perform.

In *nitrogen* there are the two reactions

$$99.635\% \left\{ \begin{array}{lll} {}_7N^{14}(\alpha, n)\,{}_9F^{17} & & \\ {}_9F^{17}(\beta^+)\,{}_8O^{17} & E_{\max}^{(+)} = 1.72 \text{ Mev}, & T_{\frac{1}{2}} = 70 \text{ sec} \end{array} \right\} \qquad (43.6)$$

$$0.365\% \left\{ \begin{array}{lll} {}_7N^{15}(\alpha, n)\,{}_9F^{18} & & \\ {}_9F^{18}(\beta^+)\,{}_8O^{18} & E_{\max}^{(+)} = 1.67 \text{ Mev}, & T_{\frac{1}{2}} = 112 \text{ min} \end{array} \right\} \qquad (43.7)$$

of which only the first one has been observed and investigated. The excitation curve of process (43.6) has been measured by various authors[2], in particular by FÜNFER[3] who suggests the existence of 11 resonances of process (43.6) between 6.5 and 8.6 Mev.

BONNER *et al.*[1] have also studied the forward yield of the reaction

$$O^{18} + \alpha \rightarrow Ne^{21} + n - 0.7 \text{ Mev}$$

in the energy range from 2 to 5 Mev and found ten resonances in this interval.

Fluorine is a monoisotopic element which, as appears from Table 15 and Fig. 36, gives the largest neutron emission, after beryllium and boron, under bombardment with Po α-particles. The reaction is

$$\left. \begin{array}{lll} {}_9F^{19}(\alpha, n)\,{}_{11}Na^{22} & Q_0 = -\ 1.928 \text{ Mev} & \\ {}_{11}Na^{22}(\beta^+)\,{}_{10}Ne^{22} & E_{\max}^{(+)} = 0.542 \text{ Mev}, & T_{\frac{1}{2}} = 2.60 \text{ y}. \end{array} \right\} \qquad (43.8)$$

The corresponding excitation curve has been measured by various authors[4]. SAHA[5] finds an indication of 6 resonances, with E_α varying between 3.3 and 5.3 Mev, which do not appear in the curve of SEGRÈ and WIEGAND, perhaps because of the slightly lower resolving power used by the last authors (Fig. 36).

---

[1] R.E. TRUMBLE and N.P. HEYDENBURG: Private communication. — T.W. BONNER, A.A. KRAUS, J.B. MARION and J.P. SCHIFFER: Phys. Rev. **102**, 1348 (1956). — R.L. BECKER and H.H. BARSCHALL: Phys. Rev. **102**, 1384 (1956).

[2] H. FAHLENBRACH: Z. Physik **94**, 607 (1935). — A. MEYE: Z. Physik **105**, 232 (1937).

[3] E. FÜNFER: Phys. Z. **38**, 967 (1937). — Ann. d. Phys. **32**, 313 (1938).

[4] P. SAVEL: C. R. Acad. Sci., Paris **196**, 1482 (1933). — T.W. BONNER: Phys. Rev. **45**, 425 (1934). — E. SEGRÈ and C. WIEGAND: MDDC-185 (1944).

[5] N.K. SAHA: Z. Physik **110**, 473 (1938).

The neutron spectrum of Poα + F sources has not been measured. The only data available are those by RICHARDS[1] for a Poα + BF_3 source, in which the yield from fluorine accounted for about 60% of the neutrons.

Sodium also is a monoisotopic element which undergoes the reaction

$$\left.\begin{array}{l} {}_{11}\mathrm{Na}^{23}(\alpha, n)\,{}_{13}\mathrm{Al}^{26} \\ {}_{13}\mathrm{Al}^{26}(\beta^+)\,{}_{12}\mathrm{Mg}^{26} \qquad E_{\max}^{(+)} = 2.8\ \mathrm{Mev}, \qquad T_{\frac{1}{2}} = 6.5\ \mathrm{sec.} \end{array}\right\} \tag{43.9}$$

The corresponding excitation curve appears to be smooth, probably because of the limited resolving power used in the experiments[2].

In *magnesium* three reactions are possible

$$78.60\% \left\{\begin{array}{l} {}_{12}\mathrm{Mg}^{24}(\alpha, n)\,{}_{14}\mathrm{Si}^{27} - 8.18\ \mathrm{Mev}, \\ {}_{14}\mathrm{Si}^{27}(\beta^+)\,{}_{13}\mathrm{Al}^{27} \qquad E_{\max}^{(+)} = 3.48\ \mathrm{Mev}, \qquad T_{\frac{1}{2}} = 4.9\ \mathrm{sec}; \end{array}\right\} \tag{43.10}$$

$$10.11\% \quad {}_{12}\mathrm{Mg}^{25}(\alpha, n)\,{}_{14}\mathrm{Si}^{28} + (2.14 \pm 0.74)\ \mathrm{Mev}; \tag{43.11}$$

$$11.29\% \quad {}_{12}\mathrm{Mg}^{26}(\alpha, n)\,{}_{14}\mathrm{Si}^{29} - (1.4 \pm 0.84)\ \mathrm{Mev}; \tag{43.12}$$

but only the last two, whose products are stable, can be induced by α-particles of Po. The excitation curve for emission of neutrons has been measured by HALPERN[3] and NAGY[4]; the results of the latter author are given in Fig. 52 in arbitrary units which can be converted to absolute yields making use of the fact that at $E_\alpha = 5.3$ Mev, the absolute integral yield is 3.75×10^{-7} neutrons per α-particle. The reaction prevailing in the observed neutron emission is probably (43.11).

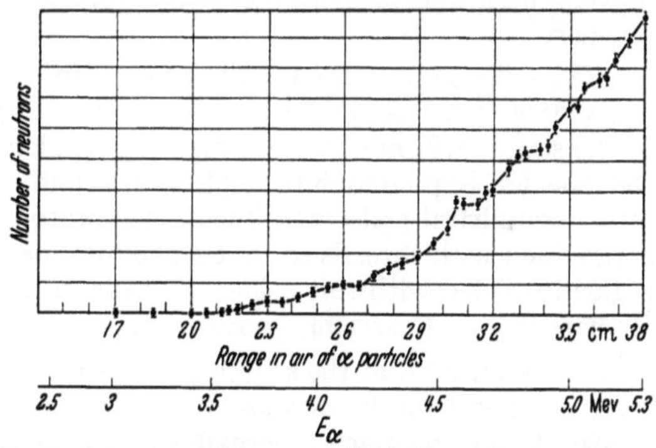

Fig. 52. Integral excitation curve of Mg (α, n) as a function of range and energy of the incident α particles (NAGY).

In *aluminium* we have only the reaction

$$\left.\begin{array}{l} {}_{13}\mathrm{Al}^{27}(\alpha, n)\,{}_{15}\mathrm{P}^{30} \\ {}_{15}\mathrm{P}^{30}(\beta^+)\,{}_{14}\mathrm{Si}^{30} \qquad E_{\max}^{(+)} = 3.5\ \mathrm{Mev}, \qquad T_{\frac{1}{2}} = 2.18\ \mathrm{min} \end{array}\right\} \tag{43.13}$$

which has been investigated for many years.

The excitation curve has been measured by various authors of which some[5] have detected the positrons emitted by P^{30}, some[6] have directly measured the emitted neutrons. SZALAI suggests the existence of 6 resonances between 3.9

[1] H.T. RICHARDS: MDDC-472 (1944).

[2] P. SAVEL: C. R. Acad. Sci., Paris **198**, 1404 (1934). — H. BRANDT: Z. Physik **108**, 726 (1938).

[3] I. HALPERN: Phys. Rev. **76**, 248 (1949).

[4] J. NAGY: Acta phys. Hungar. **3**, 15 (1953).

[5] A.I. ALICHANOW, A.I. ALICHANIAN and B.S. DZELEPOW: Z. Physik **93**, 350 (1935). — H. FAHLENBRACH: Z. Physik **94**, 607 (1935). — R.S. WARING and W.Y. CHANG: Proc. Roy. Soc. Lond., Ser. A **157**, 652 (1936). — A. MEYE: Z. Physik **105**, 232 (1937). — A. SZALAI: Z. Physik **112**, 29 (1939).

[6] E. POLLARD, H.L. SCHULTZ and G. BRUBAKER: Phys. Rev. **53**, 53 (1938). — E. FÜNFER: Ann. d. Phys. **32**, 313 (1939). — I. HALPERN: Phys. Rev. **76**, 248 (1949).

and 5.3 Mev and FÜNFER of 13 resonances between 5.3 and 8.62 Mev. A confirmation of the results of the latter author with modern techniques would be desirable.

A weak emission of neutrons has been observed and the corresponding excitation curves have been measured also in the case of phosphorus[1], chlorine[2] and argon[2,3].

44. Preparation of radioactive (α, n) sources. As was discussed in Sect. 39 Ra is most widely used in the laboratory for preparing neutron sources, and the target element which gives the highest yield is Be. Usually Ra in the compound $RaBr_2$ is mixed with Be metal powder. The yield of the source will depend on the mixing ratio as given in (41.3).

The procedure followed in preparing Ra α + Be sources is discussed by GRAVES *et al.*[1] and by ANDERSON and FELD[2]. Radium bromide is first dissolved in water, and radon gas is permitted to escape. Very fine Be metal powder is added to the solution. A small amount of alcohol decreases the surface tension so that the solution wets the metal. The solution is then evaporated to dryness while it is being stirred to prevent formation of lumps. It is important that the mixture be completely dry because water molecules are dissociated under intense radiation, so that the presence of even small amounts of water may result in the build-up of very high pressure in the source. The dry mixture is pounded to make it powdery.

Ra α + Be sources are usually pressed into a capsule. The purpose of pressing the source is not only to decrease its size, but also to ensure that the strength of the source will not change when it is shaken in use. For pressing the mixture the powder is poured into a steel disc and then pressed in a hydraulic press into a cylindrical pellet which is inserted into the capsule. The top of the capsule is provided with threads so that a cap may be screwed into the capsule. This cap is finally sealed with soft solder. Details of the design of capsules and of devices for handling and carrying such sources are discussed in Refs. [4,5].

It is essential that the source be tightly sealed so that no radon gas will escape from it. Tests for leaks should be performed a few weeks after the source is prepared and regularly thereafter whenever the source is removed from its container. Since the outside of the source is likely to be contaminated during preparation, the freshly prepared source should be thoroughly scrubbed. To test for leaks the source is placed into a closed vessel containing a piece of paper. After a few hours the activity collected on the paper is counted with a Geiger counter. For later routine tests it may be sufficient to wipe the source or the inside of the storage vessel with tissue paper and to check for activity on the paper. Experience has shown that a large fraction of Ra α + Be sources eventually develop at least small leaks.

The danger of unnoticed leakage of a source cannot be overemphasized. In recent years the work of several laboratories has been seriously interrupted for

---

[1] H. BRANDT: Z. Physik **108**, 726 (1938).

[2] E. POLLARD, H.L. SCHULTZ and G. BRUBAKER: Phys. Rev. **51**, 140 (1937); **53**, 53, 351 (1938).

[3] E. FÜNFER: Phys. Z. **38**, 967 (1937). — Ann. d. Phys. **32**, 313 (1938).

[4] A.C. GRAVES and D.K. FROMAN: Miscellaneous Physical and Chemical Techniques of the Los Alamos Project. New York: McGraw-Hill Book Company 1952.

[5] H.L. ANDERSON and B.T. FELD: Rev. Sci. Instrum. **18**, 186 (1947). — H.L. ANDERSON: Neutrons from α-emitters. National Research Council, Committee on Nuclear Science, Preliminary Report No. 3, Washington 1948.

extended periods of time by failure of radioactive neutron sources. Apart from the hazard to personnel, the spreading of radioactivity will make the construction of detectors in the laboratory virtually impossible and will raise the background of all counting devices. For this reason sources should, if possible, be stored outside of the laboratory and they should be checked for leakage each time they are taken out of the storage container.

A definite chemical compound between an α-active element as Ra and a suitable target element such as Be appears more desirable than mechanical mixtures because it will have a more reproducible neutron yield. Radium beryllium fluoride ($RaBeF_4$) has been proposed and details of the technique of preparation have been given[1]. Some experiments by LITTLER[2], however, seems to indicate that such sources have not a better reproducibility than that from mixed sources.

The use of Rn instead of Ra is sometimes convenient since radon is often available from hospitals which do radium therapy. The preparation of a $Rn\alpha + Be$ source is very easy since it amounts to introduction of the radon—after purification from other gases and vapours—in a small glass bulb containing beryllium powder. Being a gas, radon disperses itself uniformly throughout the powder. Because of the short half-life of radon ($T_{\frac{1}{2}} = 3.825$ d) a $Rn\alpha + Be$ source has the disadvantage that its strength varies rapidly with time. It has, however, the advantage that a very small mass of α-emitting material is involved. Thus 100 mg of Be are sufficient to realize the full efficiency of the source, while the same yield from a $RaBr\alpha + Be$ source would require about 13 grams of Be[3].

Occasionally there is some advantage of $Ra\alpha + B$ sources over $Ra\alpha + Be$ sources. Although the latter give a much greater neutron yield, the former have a simpler neutron spectrum which does not extend to as high a neutron energy. $Ra\alpha + B$ sources may be prepared in the manner described for $Ra\alpha + Be$ sources, except that the Be powder is replaced by fine amorphous B powder.

For the preparation of $Po\alpha + Be$ sources, Po is distilled into the inside of the source capsule. The capsule is then filled with Be powder and sealed, usually with nickel. After sealing the capsule is heated to red heat. During this heating the Po vaporizes and diffuses throughout the powdered Be. Again, it is essential to ascertain that the source does not leak. Although Po, unlike Ra, does not have a gaseous radioactive daughter product, its high specific activity results in rapid spreading. Swipe tests with tissue paper reveal even small leaks in the source.

In recent years there has been an increasing interest in the use of the α-emitting elements of the actinide series for the preparation of neutron sources; Ac, Pu, Am, and Cm have been used. The advantage of the actinide elements is that they form alloys with Be. Alloy neutron sources are more desirable than mechanical mixtures because they will have a higher and a reproducible neutron yield. The most widely used sources of this type[4] are $Pu\alpha + Be$. Such sources and their preparation have been described by RUNNALLS and BOUCHER[5]. These authors reduced PuF_3 with powdered Be metal in a BeO crucible in vacuum at 1100 to 1200° C. At this temperature the BeF_2 distilled leaving a fluoride-

[1] E. BRETSCHER, G.B. COOK, G.R. MARTIN and D.H. WILKINSON: Proc. Roy. Soc. Lond., Ser. A **196**, 436 (1949).

[2] D.J. LITTLER: British Report AERE NP/R 1577 (1956).

[3] A.O. HANSON: Chap. IA, Radioactive neutron sources, of: Fast neutron Physics, edit. by J.B. MARION and J.L. FOWLER. New York: Interscience Publisher 1959.

[4] L. STEWART: Phys. Rev. **98**, 740 (1953).

[5] O.J.C. RUNNALLS and R.R. BOUCHER: Canad. J. Phys. **34**, 949 (1956). — A.S. COFFINBERRY and F.H. ELLINGER: The Intermetallic Compounds of Plutonium. International Conference on the peaceful uses of Atomic Energy, Vol. 9, p. 138, Geneva 1955.

free alloy. The product consists of crystals of the intermetallic phase $PuBe_{13}$ in a Be matrix.

The range of Pu α-particles in $PuBe_{13}$ is approximately 20 μ. Sizes of $PuBe_{13}$ crystals large compared to 20 μ would be required in order that the alloy have a neutron yield corresponding to a 13:1 atom mixture of Be and Pu. Most sources prepared so far do not have such large crystals.

IV. The (γ, n) reactions.

45. General remarks. As was mentioned in Sect. 2, the first (γ, n) reaction was discovered by CHADWICK and GOLDHABER[1] in deuterium, using the γ-rays of ThC″. Shortly afterwards a similar process was discovered in beryllium by SZILARD and CHALMERS[2] using the γ-rays of RaC″.

A lengthy series of other elements was examined by the first two authors with the same γ-rays, but without success. Only several years later BOTHE and GENTNER[3] succeeded in observing (γ, n) processes in many elements irradiated with the γ-rays of 17 Mev emitted from lithium under proton bombardment. Since then many more (γ, n) reactions have been established and investigated in detail using high energy γ-rays produced in some nuclear reaction or radiated as bremsstrahlung by high energy electrons accelerated by electronuclear machines.

These results are clearly due to the fact that in most nuclei the binding energy of the neutron is from 6 to 10 Mev, while in deuterium it is only 2.226 Mev and in beryllium 1.666 Mev. The γ-rays of the natural radioactive bodies have all too small an energy to produce (γ, n) processes in nuclei different from the last two exceptional cases.

The same holds for the γ-rays emitted from a few long-lived artificial radioactive nuclei which can be produced today in sufficient amounts to compete in certain cases with the natural radioactive substances.

In the discussion of the (γ, n) reactions as neutron sources one has to treat separately three cases:

(a) The γ-rays of a radioactive body R produce a (γ, n) reaction in D or Be ($R\gamma + D$ or $R\gamma + Be$ sources).

(b) A target element A_2 is irradiated with the γ-rays emitted in some nuclear reaction $a_1 + A_1$ as for instance the $p + Li$ reaction [$(a_1 + A_1)\gamma + A_2$ sources].

(c) A target element A_2 is irradiated with the bremsstrahlung produced by electrons of energy E_0 [$\sigma(E_\gamma, E_0) + A_2$ sources].

The first type of source has some features in common with the $R\alpha + A$ sources, while sources of the second and third types have to be compared with other neutron sources making use of electronuclear accelerator.

A source of the $R\gamma + D$ or $R\gamma + Be$ type has the following advantages: (a) it is cheap and easy to prepare; (b) once prepared it is easy to operate; (c) if prepared with the necessary care its neutron intensity is constant or varies as a well known function of time depending on the half-life of the radioactive substance used; (d) provided the γ-ray emitter is suitably chosen, the neutrons are monoenergetic and located in a low energy region where neutrons can be produced only with some difficulty by means of other processes.

---

[1] J. CHADWICK and M. GOLDHABER: Nature, Lond. **134**, 237 (1934). — Proc. Roy. Soc. Lond., Ser. A **151**, 479 (1935).

[2] L. SZILARD and T. A. CHALMERS: Nature, Lond. **134**, 494 (1934).

[3] W. BOTHE and W. GENTNER: Naturwiss. **25**, 90, 126, 191, 284 (1937). — Z. Physik **106**, 236 (1937).

The disadvantages are: (a) its rather low neutron intensity, at least usually, and (b) the high intensity of γ-radiation superimposed on the neutrons which often imposes awkward limits to the choice of the neutron detector and of the geometry of the experiment.

The more useful sources of this type are the Raγ+Be, MsThγ+Be, MsThγ+D_2O and the $Sb^{124}\gamma$+Be sources. The first and the third of these can also be used as a neutron standard[1]. $_{51}Sb^{124}$ has a half-life of 60 days; it can be prepared in large amounts in a neutron pile, and the neutrons are practically monoenergetic ($E=31$ kev). A Sbγ+Be source with a strength of 1.23×10^3 curies after being irradiated to saturation in a flux of 10^{12} neutrons/cm² sec has been obtained by CROCKER and HENRY[2]; its neutron output amounted to $Q_0=1.3\times10^{10}$ neutrons. Sbγ+Be sources emitting of the order of 10^6-10^7 neutrons per second are commercially available.

The photoneutron sources based on the use of γ-rays produced by bombarding light elements with protons are very weak at least with the proton currents today available in the range of 0.1 to 1.0 Mev (Sect. 50). However, the γ-ray spectra emitted in these reactions are always very simple (one or a few sharp lines of rather high energy), and therefore they are almost irreplaceable for the determination of the cross section for photoneutron production at well-defined energies.

From the point of view of intensity, a photoneutron source based on the use of the bremsstrahlung of high energy electrons produced with a linear accelerator can compete with almost any other source of neutrons especially if one considers the fact that, being a pulsed source of neutrons, it can be used for time-of-flight experiments on neutrons of rather high energy (Sect. 51).

Finally we give[3] a few data which allow an estimate of the health hazard arising from exposures to γ-rays sources. The health hazard arising from exposure to neutrons can be estimated by means of the data given in footnote 7, p. 89.

46. The neutron photoeffect in deuterium. The investigation of the neutron photoeffect in deuterium

$$h\nu + {}_1D^2 \rightarrow {}_1H^1 + {}_0n^1 + E_3 \tag{46.1}$$

provides a number of fundamental results.

First it allows the determination of the binding energy of the deuteron E_3 either by measuring the energy of the recoiling proton produced by γ-rays of well-defined and known $h\nu$, or by determining the photoneutron threshold for X-rays whose energy can be varied at will. Once E_3 is known, one can deduce

[1] A. WATTENBERG: The standardization of neutron measurements, p. 119, Vol. 3 of: Annual Review of Nuclear Science 1953; see also footnote 3, p. 131.

[2] V. S. CROCKER and K. J. HENRY: AERE RP/H 72 (1956).

[3] The maximum permissible weekly dose for occupational exposure (0.1 rem per week: see footnote 7, p. 89) is attained with an exposure to the following photon fluxes for 40 hours per week:

Photon energy (Mev)	Photons/cm² sec	Photon energy (Mev)	Photons/cm² sec
0.01	900	2	750
0.02	3700	3	570
0.05	200000	4	500
0.10	15000	5	400
0.50	2300	10	250
1	1300		

[B. T. PRICE, C. C. HORTON and K. T. SPINNEY: Radiation Shielding. London: Pergamon Press (1957)].

the neutron-proton mass difference, and the mass of the neutron, on the basis of the considerations given in Sect. 2.

Following the original work of CHADWICK and GOLDHABER many measurements of E_3 by these methods have been performed in various laboratories. The review article of STEPHENS[1] contains an extensive bibliography on the subject and a detailed discussion of the various experiments performed up to 1947. More recently E_3 has been measured by MEYER and by HANSON who both used the γ-rays of ThC'' of $h\nu = 2.62$ Mev[2]. The first of these authors determined the energy of the emitted protons [$E_p = (216.8 \pm 0.005)$ kev] in a proportional counter filled with D_2 gas; the second measured the energy of the emitted neutron [$E_n = (197 \pm 10)$ kev] from the pulse height produced by recoil protons in a proportional counter filled with H_2.

A still more precise determination of E_3 has been made by MOBLEY and LAUBENSTEIN[3] who measured the photo-neutron threshold of deuterium (and beryllium: see Sect. 47) by comparison with the $Li^7(p, n)$ threshold (Sect. 54).

This threshold has been determined on the absolute voltage scale ($E_{pT} = 1.882$ Mev $\pm 1\%$). It lies between the ${}_4Be^9(\gamma, n)$ and ${}_1D^2(\gamma, n)$ thresholds so that errors produced by the extrapolation of the energy scale should be small.

The pressure electrostatic generator of the Argonne National Laboratory was operated with the upper tube serving to accelerate positive ions while the lower tube, normally used only for differential pumping, was used, in addition, to accelerate a beam of electrons from ground to an X-ray target placed inside the high voltage electrode. The positive ion beam was separated by a magnetic analyser into an atomic and a molecular beam.

The H_2^+ beam was directed through a cylindrical electrostatic analyser onto a pair of slits where it was used to control the generator voltage. Any inequality in division of the current between these two slits was used to vary the corona current electronically. The heavy water (or beryllium) sample was placed very close to the X-ray target (Fig. 53) and was screened with a layer of lead in order to reduce the intensity of X-rays incident on three proportional counters filled with B^{10} enriched BF_3 at a pressure of 2 atmospheres. A layer of paraffin was placed behind the BF_3 counters on the far side from the sample to give a maximum sensitivity for neutrons in the energy region near threshold.

The photoneutron threshold was obtained by extrapolating the neutron yield versus energy curve to zero yield with the following method. The neutron yield Y (normalized with respect to the intensity of X-rays and after subtraction of the background) is expected to vary, as a function of the energy E_0 of the electron beam immediately above the threshold according to the expression (see Sect. 48)

$$Y = K(E_0 - E_{1T})^m \tag{46.2}$$

where K and m are constants. m depends for various reasons on the experimental set-up (because, for instance, of the decreasing efficiency of the neutron counters with increasing energy), and can be determined by finding by trial, on a logarithmic plot of Y versus $E - E_{1T}$, that value which gives the best straight line. With such a procedure the authors find $m = 2$. Fig. 54 represents the experimental results for deuterium from which one deduces

$$E_3 = (-2.226 \pm 0.003) \text{ Mev} \tag{46.3}$$

[1] W.E. STEPHENS: Rev. Mod. Phys. **19**, 19 (1947).

[2] P. MEYER: Z. Physik **126**, 336 (1949). — A.O. HANSON: Phys. Rev. **75**, 1794 (1949).

[3] R.C. MOBLEY and R.A. LAUBENSTEIN: Phys. Rev. **80**, 309 (1950).

in excellent agreement with results obtained with other methods (Table 1 in Sect. 2). The same procedure was used also by other authors[1] who found

$$E_3 = (-2.227 \pm 0.003)\ \text{Mev}.$$

Apart from the determination of E_3, it is clear that process (46.1) is one of the simplest nuclear reactions for which the comparison between the experimental results and the theoretical predictions provides very direct information about the properties of the neutron-proton system.

Various measurements have been made, from the beginning, of the total cross section[2] and of the angular distribution of particles emitted[3] in process (46.1). More recent results about these quantities are given in Tables 20 and 21 and will be discussed later.

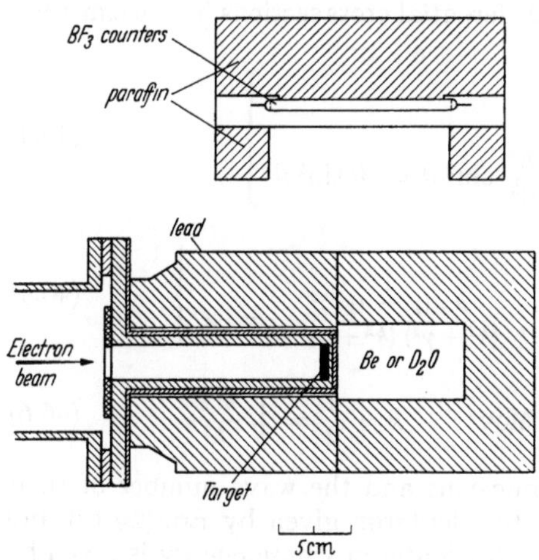

Fig. 53.

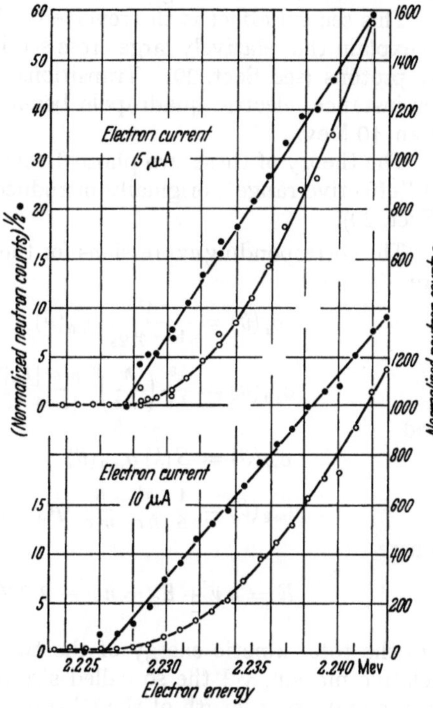

Fig. 54.

Fig. 53. The X-ray target and neutron counters housed in the high voltage electrode for the determination of the threshold of (γ, n) processes in D and Be (Mobley and Laubenstein).

Fig 54. The photoneutron yield of deuterium near threshold (Mobley and Laubenstein).

A first simple theory of process (46.1) can be developed for not too large an energy of the incident photon ($h\nu < 10$ Mev) by neglecting the tensor forces and the exchange currents between the neutron and the proton in the deuteron. The influence of these various simplifying assumptions will be considered later.

[1] J. C. Noyes, J. E. van Hoomissen, W. C. Miller and B. Waldman: Phys. Rev. **95**, 396 (1954).

[2] J. Chadwick and M. Goldhaber: Proc. Roy. Soc. Lond., Ser. A **151**, 479 (1935). — H. Halban: C. R. Acad. Sci., Paris **206**, 1170 (1938). — Nature, Lond. **141**, 644 (1938). — J. A. Van Allen and N. M. Smith: Phys. Rev. **59**, 618 (1941). — H. Wäffler and S. Younis: Helv. phys. Acta **22**, 414 (1949).

[3] J. Chadwick and M. Goldhaber: Proc. Roy. Soc. Lond., Ser. A **151**, 479 (1935). — J. Chadwick, N. Feather and E. Bretscher: Proc. Roy. Soc. Lond., Ser. A **163**, 366 (1937). — J. R. Richardson and L. Emo: Phys. Rev. **51**, 1014 (1937). — H. Halban: Nature, Lond. **141**, 644 (1938). — F. E. Myers and L. C. Van Atta: Phys. Rev. **61**, 19 (1942).

At low energy the transition from the ground state of the deuteron, which in the absence of tensor forces, is a pure $^3S_1$ state, to a state of the continuum in which the two nucleons fly away, can happen through two different mechanisms. The first one is an electric dipole transition from the $^3S_1$ ground state to a $^3P_1$ state and gives the so called *photoelectric effect*; the second is a magnetic dipole transition associated with the flipping of the spins of the particles, from the $^3S_1$ ground state to a $^1S_0$ state and gives the so called *photomagnetic effect.*

This second effect is the reverse of the process introduced by FERMI in order to explain the relatively large cross section for radiative capture of slow neutrons by protons (see Sect. 29). Transitions caused by other multipole radiations as, for instance, electric quadrupole begin to be important only at energies larger than 10 Mev.

The theory of these two photoeffects can be developed[1,2] by using the concept of "effective range" originally introduced in the theory of low-energy scattering (Sect. 29).

The corresponding expressions for the differential cross sections (in the c.m.s.)[3,4] are

$$\left.\begin{aligned}\sigma_e(\vartheta) &= \frac{1}{1-\gamma\varrho_3'}\,\sigma_{e0}(\vartheta)\\ \sigma_{e0}(\vartheta) &= \frac{e^2}{\hbar c}\left(\frac{\hbar}{mc}\right)^2\frac{mc^2|\mathsf{E}_3|^{\frac{1}{2}}E^{\frac{3}{2}}}{[E+|\mathsf{E}_3|]^3}\sin^2\vartheta = \mathcal{B}\sin^2\vartheta\end{aligned}\right\}\tag{46.4}$$

and

$$\left.\begin{aligned}\sigma_m(\vartheta) &= S(k)\,\sigma_{m0}(\vartheta)\\ \sigma_{m0}(\vartheta) &= \frac{1}{6}\frac{e^2}{\hbar c}\frac{\hbar}{mc}(\mu_p-\mu_n)\frac{k\gamma(\gamma-a_1^{-1})^2}{(k^2+\gamma^2)(k^2+a_1^{-1})} = \mathcal{A}\end{aligned}\right\}\tag{46.5}$$

where

$$E = h\nu + \mathsf{E}_3 = h\nu - 2.226\text{ Mev};\qquad k = \frac{1}{\lambda\!\!\!^{-}} = \sqrt{\frac{mE}{\hbar^2}}\tag{46.6}$$

are the total kinetic energy of the two nucleons and the wave number of their relative motion, γ^{-1} the so called size of the deuteron given by Eq. (29.12) and a_1 the scattering length of the ^1S state of the deuteron whose energy is given by Eq. (29.14). σ_{e0} and σ_{m0} are the expressions originally given by BETHE and PEIERLS [4] in the approximation where the range of the nuclear forces is equal to zero.

Although the correction factors for the finite range of nuclear forces are very important in order to try to get an agreement with the experimental results, the general qualitative behaviour of σ_e and σ_m is mainly determined by σ_{e0} and σ_{m0}. By integrating Eqs. (46.4) and (46.5) over the solid angle 4π and introducing the numerical values of the constants one gets

$$\sigma_{e0} = \frac{8\pi}{3}\mathcal{B} = 3.782\times10^{-26}\,\frac{(h\nu-2.226)^{\frac{3}{2}}}{(h\nu)^3}\text{ cm}^2\tag{46.7}$$

$$\sigma_{m0} = 4\pi\mathcal{A} = 6.936\times10^{-28}\,\frac{(h\nu-2.226)^{\frac{1}{2}}}{h\nu(h\nu-2.226+0.0739)}\text{ cm}^2\tag{46.8}$$

where $h\nu$ is expressed in Mev. σ_{e0} has a flat maximum for $h\nu = 2|\mathsf{E}_3| = 4.4452$ Mev which is not due to a resonance but is a result of the peaked momentum

[1] H. A. BETHE and C. LONGMIRE: Phys. Rev. **77**, 647 (1950).
[2] E. E. SALPETER: Phys. Rev. **82**, 60 (1951).
[3] G. L. SQUIRES: Progress in Nuclear Physics, edit. by O. R. FRISCH, Vol. 2, p. 89. 1952.
[4] G. R. BISHOP and R. WILSON: p. 309, Vol. XLII of this Encyclopedia.

distribution in the ground state of the deuteron. σ_{m0} has, however, a resonance peak for $h\nu = |E_3| + E_1 = 2.226 + 0.0739 \approx 2.3$ Mev. Beyond this maximum σ_{m0} drops rapidly so that, for $h\nu$ above 4 Mev it represents less than 5% of σ_{e0}. At high energies, $h\nu \gg 2|E_3|$, both σ_{e0} and σ_{m0} decrease as $(h\nu)^{-\frac{3}{2}}$, while immediately above the threshold $(h\nu - 2.226 \ll E_1)$ σ_{m0} increases as $(h\nu - 2.226)^{\frac{1}{2}}$ and σ_{e0} as $(h\nu - 2.226)^{\frac{3}{2}}$. It is perhaps interesting to note that because of the factor $(\gamma - a_1^{-1})^2$ appearing in Eq. (46.5), σ_{m0} depends on the sign of a_1. Therefore an experimental determination of σ_m, with sufficient accuracy, would allow to recognize whether a_1 is positive or negative, i.e. whether the ^1S state of the deuteron is real or virtual. The experimental data are still too rough to allow such a conclusion to be drawn. We know, however, from other phenomena that the state ^1S is virtual, i.e. that a_1 is negative and E_1 positive (see Sect. 29).

The corrections for the finite range of the nuclear forces are respectively

$$\frac{1}{1 - \gamma \varrho_3'} \tag{46.9}$$

and

$$S(k) = \frac{1}{1 - \gamma \varrho_3'} \frac{\left[1 + \dfrac{k^2 (\frac{1}{2} r_{01} - D) - \gamma^2 D}{\gamma - a_1^{-1}}\right]^2}{1 + \dfrac{\frac{1}{4} r_{01}^2 k^4 - a_1^{-1} r_{01} k^2}{k^2 + a_1^{-2}}} = \frac{1}{1 - \gamma \varrho_3'} s(k) \tag{46.10}$$

where D is given by Eq. (29.19). Both expressions (46.9) and (46.10) reduce to 1 for a range of the nuclear forces equal to zero. For numerical calculation of the expression (46.10) one can use the numerical values of the effective ranges and scattering lengths which give the best fit for the neutron-proton data at low energies [Eq. (29.13)].

The factor (46.9) does not depend on the energy: ϱ_3' is defined by Eq. (29.21), from which we see that it reduces to r_{03} in the shape-independent approximation. If one takes into account the shape of the potential well, ϱ_3' takes its largest value for the Yukawa potential ($\varrho_3' = 1.768 \times 10^{-13}$ cm; see Sect. 29). Introducing this value and (29.12) in Eq. (46.9) one gets a range correction factor for the photoelectric cross section equal to 1.7.

The factor (46.10) can be transformed as follows by neglecting terms in r_{01}^2,

$$S(k) = 1 + \frac{1}{2} \frac{\gamma^2 - 2a_1^{-1}\gamma - k^2}{\gamma - a_1^{-1}} (r_{03} - r_{01}) - \frac{a_1^{-2}(k^2 + a_1^{-1}\gamma)}{(\gamma - a_1)^{-1}(k^2 + a_1^{-2})} r_{01}.$$

From this expression it appears that $S(k)$ does not depend on r_{03} but only on $r_{03} - r_{01}$ and r_{01}, the coefficient of the first term being four times larger than that of the second, for $k = 0$. Therefore one could in principle use the photomagnetic effect in order to obtain an independent experimental determination of $r_{03} - r_{01}$ or, if r_{03} is known, of r_{01} (see Sect. 29).

The effect of the tensor forces on the photodisintegration cross section has been considered in detail and has been shown to be unimportant[1-4].

The currents produced by the exchange of charge between the two nucleons in the deuteron have, however, an influence on the cross section for the photomagnetic effect[5].

[1] W. RARITA and J. SCHWINGER: Phys. Rev. **59**, 436 (1941); **59**, 559 (1941). — H. FESHBACH and J. SCHWINGER: Phys. Rev. **84**, 194 (1951).

[2] See footnote 1, p. 136.

[3] See footnote 2, p. 136.

[4] See footnote 3, p. 136.

[5] N. AUSTERN and R. G. SACHS: Phys. Rev. **81**, 710 (1951).

A rough evaluation has shown that the corresponding correction for a disintegration energy close to zero amounts to an increase of σ_m by about 4%[1]. Assuming that it is energy independent, one can introduce this correction by replacing $S(k)$ in Eq. (46.5), by

$$R(k) = \frac{S(k)}{1 - 0.04}. \tag{46.11}$$

A still better procedure for introducing this correction without making use of the calculated value consists in writing the expressions (46.10) and (46.11) in the form

$$R(k) = R(0)\, s(k) \tag{46.12}$$

where $R(0)$ represents the correction factor for $k=0$. It is the same factor which appears in the expansion for the capture cross section of slow neutrons by protons (Sect. 29). Therefore $R(0)$ is given by the ratio of the experimental value of σ_a to the theoretical value σ_{a0} [Eq. (29.18)]. With such a procedure one obtains[1]

$$R(0) = 1.10. \tag{46.13}$$

The other factor appearing in (46.12) is very close to unity and does not depend very much on the values of r_{01} and D.

From the experimental point of view it is possible to separate the photomagnetic from the photoelectric effect by taking advantage of their different dependence on the angle of emission ϑ with respect to the direction of the incident photon. This different behaviour is due to the fact that the final state is, in the photomagnetic effect, an S state, while in the photoelectric effect it is a P state.

With the notation introduced in Eqs. (46.4) and (46.5), we can write

$$\left.\begin{aligned} \sigma(\vartheta) &= \mathcal{A} + \mathcal{B}\sin^2\vartheta, \\ \sigma &= 4\pi\mathcal{A} + \frac{8\pi}{3}\mathcal{B}, \\ \frac{\sigma_m}{\sigma_e} &= \frac{3}{2}\frac{\mathcal{A}}{\mathcal{B}}. \end{aligned}\right\} \tag{46.14}$$

This last relation shows how one can deduce the ratio σ_m/σ_e from the values of $\mathcal{A}$ and $\mathcal{B}$, which are the quantities directly determined from the measurement of the angular distribution of the emitted protons or neutrons.

Table 20 is a comparison between the experimental values of σ_m/σ_e and the corresponding theoretical results. The data obtained with the procedure described above from the effective range theory refer to ϱ_3' calculated for the Yukawa potential well[1].

In the last column are given also the results of HULTHÉN and NAGEL[2] who have calculated σ_m and σ_e directly, using accurate approximations for the neutron-proton wave functions in the case of a Yukawa potential well.

The calculations were done for energies of the incident photons from threshold to 20 Mev, assuming the same range of the nuclear forces in the ^1S and ^3S states and no interaction in the ^3P state. This last assumption is justified only at low energy[3]. HULTHÉN and NAGEL have also neglected the influence on σ_m of the

---

[1] See footnote 3, p. 136.
[2] L. HULTHÉN and B.C.H. NAGEL: Phys. Rev. 90, 62 (1953).
[3] G.R. BISHOP and R. WILSON: p. 309, Vol. XLII of this Encyclopedia.

Table 20. *Comparison between the experimental and theoretical values of σ_m/σ_e in deuterium.*

Source of photons	$h\nu$ (Mev)	Average of previous experimental results[1]	Results of BISHOP, BEGHIAN and HALBAN[2]	Effective range theory[3]	Yukawa potential well[4]
Na^{24}	2.757	0.30 ± 0.02	0.247 ± 0.007	0.262	0.243
ThC″	2.615	0.39 ± 0.12	0.360 ± 0.008	0.422	0.386
Ga^{72}	2.504	0.61 ± 0.04	0.600 ± 0.02	0.698	0.664

[1] C.A.R. GRAHAM and H. HALBAN: Rev. Mod. Phys. **17**, 297 (1945). — N.O. LASSEN: Phys. Rev. **74**, 1533 (1948); **75**, 1099 (1949). — W.M. WOODWARD and I. HALPERN: Phys. Rev. **76**, 107 (1949). — E.P. MEINERS: Phys. Rev. **76**, 259 (1949). — F. GENEVESE: Phys. Rev. **76**, 1288 (1949). — B. HAMERMESH and H. WATTENBERG: Phys. Rev. **76**, 1408 (1949). — G.R. BISHOP, L.E. BEGHIAN, H. HALBAN, P.F.D. SHAW and R. WILSON: Phys. Rev. **81**, 219 (1951).

[2] G.R. BISHOP, L.E. BEGHIAN and H. HALBAN: Phys. Rev. **83**, 1052 (1951).

[3] See the Yukawa potential case in Table 5 of SQUIRES' review article (cf. footnote 3 on p. 136).

[4] L. HULTHÉN and B.C.H. NAGEL: Phys. Rev. **90**, 62 (1953).

currents due to the charge exchange between the two nucleons. σ_e and σ_m have been calculated for various values of the range of the forces

$$\frac{\hbar}{m_\pi c} = \frac{3.86 \times 10^{-11}}{m_\pi} \text{ cm} \tag{46.15}$$

corresponding to different values of purely phenomenological parameter m_π [in Eq. (46.15) m_π is expressed in electron masses]. The data reported here (in Table 20 and 21 as well as in Fig. 55 and 56) all refer to $m_\pi = 286.4\ m_e$ a value which seems satisfactory from the analysis of the scattering of slow neutrons at zero energy (Sect. 29).

From Table 20 one recognizes that the more accurate experimental values of σ_m/σ_e obtained by BISHOP, BEGHIAN and HALBAN are in better agreement with the theoretical results of HULTHÉN and NAGEL than with those of the effective

Table 21. *Total cross section for neutron production by photoeffect in deuterium.*

$h\nu$ (Mev)	σ_{exp} (10^{-28} cm²)							σ_{theor} (10^{-28} cm²)
	BISHOP and others[1]	PHILLIPS and others[2]	SNELL and others[3]	HOUGH[4]	BARNES and others[5]	WAFFLER and others[6]	MCMURRAY and others[7]	HULTHÉN and NAGEL
2.504	11.9 ± 0.8		10.6 ± 1.1					9.92
2.618	13.9 ± 0.6							12.18
2.757	15.9 ± 0.6		14.3 ± 1.1				15.05 ± 0.5	14.80
4.45 ± 0.04					24.3 ± 1.7			25.36
6.14 ± 0.01		26.9 ± 3.8			21.9 ± 1.0			22.09
7.0		24.2 ± 3.4						19.91
7.39 ± 0.15					18.4 ± 1.5			19.01
8.14 ± 0.08					18.0 ± 1.3			17.33
12.50 ± 0.21					10.4 ± 1.0			10.61
17.6 ± 0.2				7.2 ± 1.5	7.7 ± 0.9	7.1 ± 1.5		6.68

[1] G.R. BISHOP, C.H. COLLIE, H. HALBAN, A. HEDGRAN, K. SIEGBAHN, S. DU TOIT and R. WILSON: Phys. Rev. **80**, 211 (1950).

[2] J.A. PHILLIPS, J.S. LAWSON and P.G. KRUGER: Phys. Rev. **80**, 326 (1950).

[3] A.H. SNELL, E.C. BARKER and L. STERNBERG: Phys. Rev. **80**, 637 (1950).

[4] P.V.C. HOUGH: Phys. Rev. **80**, 1069 (1950).

[5] C.A. BARNES, J.H. CARVER, G.H. STAFFORD and D.H. WILKINSON: Phys. Rev. **86**, 359 (1952).

[6] H. WÄFFLER and S. YOUNIS: Helv. phys. Acta **24**, 483 (1951).

[7] MCMURRAY and C.H. COLLIE: Proc. Phys. Soc. Lond. **68**, 181 (1955).

range theory. This theory gives too small a value of σ_e. In order to bring its results into agreement with the experimental data it would be necessary to increase ϱ_3' from the value used (1.768×10^{-13}) to about 2.1×10^{-13} cm.

This fact supports the results of HULTHÉN and NAGEL which are plotted, for $m_\pi = 286.4\, m_e$, in Fig. 55 in order to show the relative importance and general behaviour of σ_e and σ_m immediately above the threshold, and in Fig. 56 for comparison of $\sigma = \sigma_e + \sigma_m$ with the experimental results.

According to HULTHÉN and NAGEL σ_e may be written

$$\sigma_e = \frac{1}{1-\gamma\varrho_3'}\,\sigma_{e0} F^2 \tag{46.16}$$

similar to (46.4), but where now the factor F^2 appears.

$$F = 1 - 0.640\left(\frac{h\nu}{E_1 + h\nu}\right)^2 \tag{46.17}$$

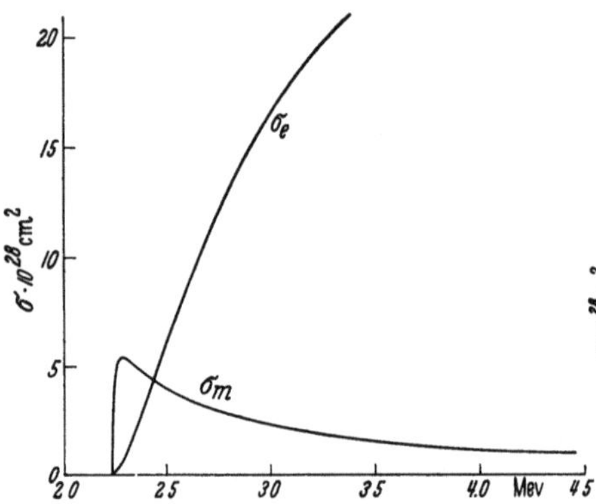

Fig. 55. σ_m and σ_e in D according to calculations of HULTHÉN and NAGEL with a Yukawa potential ($m_\pi = 286.4\, m_e$).

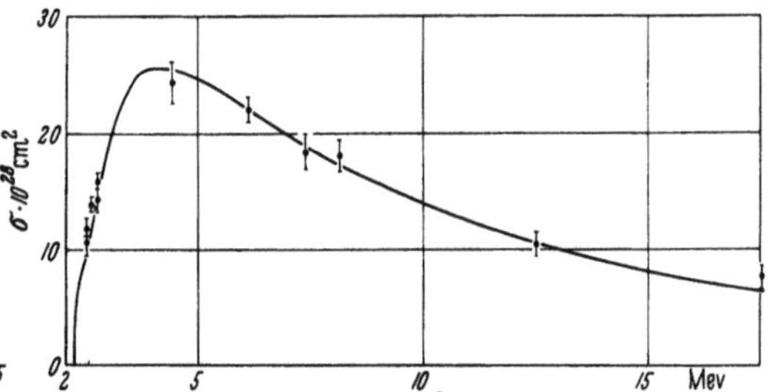

Fig. 56. The total cross section $\sigma_e + \sigma_m$ for neutron photoeffect in deuterium. Experimental points from Table 21. Theoretical curve from HULTÉN and NAGEL ($m_\pi = 286.4\, m_e$).

takes account of the difference between the correct eigenfunction and its asymptotic form used in the effective range theory. E_1 depends on the adopted value for m_π; for $m_\pi = 286.4\, m_e$ one finds $E_1 = 45.67$ Mev.

A few experimental data, appearing in Fig. 56, are collected in Table 21 with the corresponding references. The agreement with the theoretical values of σ is rather satisfactory for $h\nu$ as large as 17 Mev where some of the approximations introduced in the theoretical calculations are certainly questionable[1]. These approximations neglect the phase-shift in the ^3P state, transitions due to absorption of quadrupole radiation, charge exchange currents, relativistic effects, non-central forces and repulsive core. In the effective range theory there is the further approximation of neglecting in Eq. (46.16) the factor F^2. From Eq. (46.17) one sees how the corresponding error increases with the energy $h\nu$.

The relative importance of the various factors appearing in Eq. (46.16) may be seen from WILKINSON'S[2] treatment of the experimental data of BARNES and coworkers[3] for values of $h\nu$ ranging from 4 to 17 Mev. In this energy interval σ_m does not represent more than 5% of the total cross section σ so that the difference $(\sigma)_{\mathrm{exp}} - (\sigma_m)_{\mathrm{theor}}$ can be considered as a good experimental determination of σ_e in spite of some uncertainty in the theoretical value of σ_m. Using Eqs. (46.16) and

[1] See footnote 3, p. 138.
[2] D.H. WILKINSON: Phys. Rev. **86**, 373 (1952).
[3] C.A. BARNES, J.H. CARVER, G.H. STAFFORD and D.H. WILKINSON: Phys. Rev. **86**, 359 (1952).

(46.17) WILKINSON determines that value of ϱ_3' which gives the best fit to the experimental values of σ_e. He finds, for the Yukawa potential, $\varrho_3' = (1.79 \pm 0.09) \times 10^{-13}$ cm to be compared with the value $\varrho_3' = 1.768 \times 10^{-13}$ cm considered previously.

The total cross section and the angular distribution of the emitted protons have been measured with the emulsion technique for $h\nu$ between 4 and 20 Mev by FULLER[1] who used the bremsstrahlung from electrons accelerated to 20.3 Mev in a betatron.

The fact that FULLER finds that σ decreases with increasing $h\nu$ a little more slowly than found by BARNES and others[2] and more slowly than in the theoretical curve of Fig. 56, could be due to his use of a calculated spectrum for the bremsstrahlung which introduces a rather large uncertainty.

The angular distributions are analyzed by comparison with the expression valid in the c.m.s.

$$\sigma(\vartheta^*) \propto A + \sin^2\vartheta^*(1 + 2\beta\cos\vartheta^*); \quad \beta = \frac{v}{c} = \left(\frac{h\nu - |E_3|}{Mc^2}\right)^{\frac{1}{2}} \tag{46.18}$$

deduced by MARSHALL and GUTH[3] under a special assumption about the neutron-proton forces (50% exchange and 50% normal forces). The term $2\beta\cos\vartheta^*$ arises from the interference between the electric dipole and the electric quadrupole term and leads to an asymmetry with respect to the equatorial plane.

The experimental results of FULLER, although in agreement with (46.18), are affected by large errors and uncertainties (mainly because of large background) to allow a definite conclusion to be reached apart from a qualitative indication of the following facts: the isotropic term A seems to go from about 0.05 to zero when $h\nu$ increases from 5 to 7 to about 11 Mev, and then it increases again reaching a rather large value ($A \sim 0.2$) at $h\nu = 20$ Mev. Using the γ-rays emitted from Li under proton bombardment ($h\nu_1 = 14.8$ and $h\nu_2 = 17.6$ Mev; $I_1 : I_2 = 0.45$) WÄFFLER and YOUNIS[4] find $A = 0.12 \pm 0.07$.

The calculations mentioned above[3] are an example of a few similar ones[5,6] carried out for high-energy photons ($h\nu = 10 - 150$ Mev) under various assumptions on the neutron-proton forces. The results are sensitive to the force law and therefore accurate experimental determinations of the cross section and the angular distribution would give important information about the neutron-proton forces. Some interesting conclusions can also be derived by comparing various moments of the experimental cross section for photo-disintegration (integrated up to $h\nu = 155$ Mev, the threshold for meson photo-production) with the corresponding values deduced from the application of general theorems[7,8].

47. The neutron photoeffect in Be. As was mentioned in Sect. 45, emission of neutrons from beryllium under irradiation with the γ-rays of Ra was first observed by SZILARD and CHALMERS[9].

For moderate energies of the incident photons the disintegration process proceeds according to

$$h\nu + {}_4\mathrm{Be}^9 \to {}_4\mathrm{Be}^8 + {}_0n^1 - 1.666\ \mathrm{Mev} \tag{47.1}$$

---

[1] E. G. FULLER: Phys. Rev. 79, 303 (1950).

[2] See footnote 3, p. 140.

[3] J. F. MARSHALL and E. GUTH: Phys. Rev. 76, 1879 (1949).

[4] H. WÄFFLER and S. YOUNIS: Helv. phys. Acta 24, 483 (1951).

[5] See footnote 1, p. 137.

[6] M. E. ROSE and G. GOERTZEL: Phys. Rev. 72, 749 (1947). — J. S. LEVINGER: Phys. Rev. 76, 699 (1949). — L. I. SCHIFF: Phys. Rev. 78, 733 (1950).

[7] J. S. LEVINGER: Phys. Rev. 97, 970 (1955).

[8] See footnote 3, p. 138.

[9] L. SZILARD and T. A. CHALMERS: Nature, Lond. 134, 494 (1934).

where the ${}_4Be^8$ is left in its ground state. The mean life for disintegration in two α-particles

$$ {}_4Be^8 \rightarrow 2\, {}_2He^4 + 0.094 \text{ Mev} \tag{47.2}$$

is shorter than 5×10^{-14} sec [1], [2].

The multibody reaction

$$h\nu + {}_4Be^9 \rightarrow 2\, {}_2He^4 + {}_0n^1 - 1.57 \text{ Mev}$$

has been looked for by ALBURGER with the 1.631 Mev γ-rays from Ne^{20} [1], [2] but no yield was found which could be attributed to this reaction. It was concluded that its cross section can not be larger than 10^{-30} cm² [1].

For sufficiently high γ-ray energies other modes of disintegration become possible. Either the residual nucleus ${}_4Be^8$ may be left in an excited state—the first level of this nucleus is at $w_1 = 2.9$ Mev—or the first step may consist of the emission of a charged particle instead of a neutron such as the following (γ, α) reaction [2]

$$\left.\begin{aligned} h\nu + {}_4Be^9 &\rightarrow {}_2He^5 + {}_2He^4 - 2.57 \text{ Mev}, \\ {}_2He^5 &\rightarrow {}_2He^4 + {}_0n^1 + 1.0 \text{ Mev}. \end{aligned}\right\} \tag{47.3}$$

The given Q values again correspond to the formation of He^5 in its ground state which has a mean life shorter than 2×10^{-21} sec ($\Gamma \geqq 0.32$ Mev), but one could also consider similar processes in which an excited state of He^5 is involved.

The (γ, p) process [1], [2]

$$\left.\begin{aligned} h\nu + {}_4Be^9 \rightarrow {}_3Li^8 + {}_1H^1 - 16.871 \text{ Mev} \\ {}_3Li^8 (\beta^-)\, {}_4Be^{8}*;\ \ {}_4Be^{8}* \rightarrow 2\, {}_2He^4;\ T_{\frac{1}{2}} = 0.88 \text{ sec},\ \ Q_0 = 15.987 \text{ Mev} \end{aligned}\right\} \tag{47.4}$$

becomes possible only at very high energy [3].

Various estimates of the threshold of reaction (47.1) were made by many authors [4] immediately after the discovery of the photoeffect in Be. A good determination of this quantity was made by CHADWICK and GOLDHABER [5] who observed in an ionization chamber the recoiling helium nuclei produced by the photoneutrons of beryllium. MYERS and VAN ATTA [5] measured the number of photoneutrons produced in beryllium by the bremsstrahlung as a function of the energy of the irradiating electrons accelerated with an electrostatic generator.

These values, although correct, have been superseded by more recent results. HANSON [6] measured, in a proportional counter filled with hydrogen, the energy of the recoiling protons from photoneutrons produced in beryllium by the γ-rays of Sb ($h\nu = 1.708$ Mev; see Sect. 49). He found

$$(h\nu)_{1T} = (1.681 \pm 0.013) \text{ Mev} \tag{47.5}$$

---

1 D.E. ALBURGER: Phys. Rev. **88**, 1257 (1952): Ne^{20} is produced in an excited state by β-decay of F^{20}.

2 J.H. CARVER, E. KONDAIAH and B.D. MACDANIEL: Phil. Mag. **45**, 948 (1954).

3 R.N.H. HASLAM, L. KATZ, E.H. CROSBY, R.G. SUMMERS-GILL and A.G.W. CAMERON: Canad. J. Phys. **31**, 210 (1953).

4 A. BRASCH, F. LANGE, A. WALY, T.E. BANKS, T.A. CHALMERS, L. SZILARD and F.L. HOPWOOD: Naturwiss. **22**, 839 (1934). — Nature, Lond. **134**, 880 (1934). — W. GENTNER: C. R. Acad. Sci., Paris **199**, 1211 (1934). — L.N. RIDENAU, K. SHINOHARA and D.N. YOST: Phys. Rev. **47**, 318 (1935). — L. ARSIMONITSCH and P. PALIBIN: Z. Sow. Un. **7**, 245 (1935). — J. ROTBLAT: Nature, Lond. **139**, 963 (1935). — J. Phys. Radium **8**, 477 (1935).

5 J. CHADWICK and M. GOLDHABER: Proc. Roy. Soc. Lond. **151**, 479 (1935). — F.E. MYERS and L.C. VAN ATTA: Phys. Rev. **61**, 19 (1942).

6 A.O. HANSON: Phys. Rev. **75**, 1794 (1949).

which is in good agreement with the more precise value obtained by MOBLEY and LAUBENSTEIN[1] with the method explained in the case of deuterium (Sect. 46)

$$(h\nu)_{1T} = (1.666 \pm 0.002) \text{ Mev}. \tag{47.6}$$

By the same method other authors found (1.662 ± 0.003) Mev[2] and (1.664 ± 0.004) Mev[3].

The total cross section of Be has been measured by many authors[4]. Their results have been discussed and summarized by HOUTERMANS and BARTZ. More recent data are plotted in Fig. 57. The corresponding references are given in the caption.

The curve represents the theoretical results of GUTH and MULLIN[5] who have corrected and improved preceding calculations.

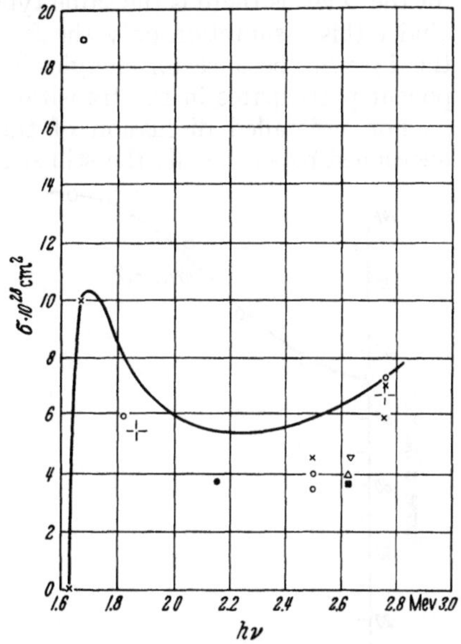

Fig. 57. The photodisintegration cross section of Be^9. ○ B. RUSSEL, D. SACHS, A. WATTENBERG and R. FIDLES: Phys. Rev. **73**, 545 (1948). A. WATTENBERG: Photoneutron sources, Preliminary report No. 6, Nuclear Science Series, Division Mathematics and Physical Science, National Research Council. × A.H. SNELL, E. C. BARKER and L. STERNBERG Phys. Rev. **80**, 637 (1950). △ L.G. ELLIOT: Unpublished. + E. SEGRÈ. Unpublished: for both these last two data see A. WATTENBERG, Photoneutron sources, Preliminary report No. 6, Nuclear Science Series, Division of Mathematics and Physical Science, National Research Council. ● B. HAMERMESH and C KIMBALL· Phys. Rev. **90**, 1063 (1953). ▽ D.L. ALLAN and M. J. POOLE: Nature, Lond. **162**, 373 (1950). ■ R.D. EDGE: Nuclear Phys. **2**, 485 (1956). Theoretical curve: GUTH and MULLIN.

The model adopted by these authors for representing the ground state of Be^9 consists of a Be^8 "core" and an "optical" neutron interacting with a central field (no tensor forces). This, for the sake of simplicity, is represented by a suitable spherical potential well. In order to obtain at least qualitative agreement between the theoretical and the empirical values of the cross section, it seems necessary to assume that the well depends upon the orbital angular momentum (Majorana forces) and upon the orientation of the spin of the neutron.

In such a model the ground state of Be^9 is a $P_{\frac{3}{2}}$ state and the photoeffect, at low energies, is primarily due to electric dipole transitions to the $S_{\frac{1}{2}}$ and $D_{\frac{5}{2}}$, $D_{\frac{3}{2}}$ states of the continuum, while the magnetic dipole transitions do not play an important role. The $P_{\frac{3}{2}} \to S_{\frac{1}{2}}$ transition prevails immediately above the threshold. The $P_{\frac{3}{2}} \to D_{\frac{5}{2}}, D_{\frac{3}{2}}$ transitions give a contribution of the same order of magnitude at 2 Mev and prevail from 3 Mev onward.

According to such a scheme the angular distribution of the emitted neutrons produced by unpolarized γ-rays is expected to be of the type

$$\sigma(\vartheta) = A + B \sin^2 \vartheta$$

[1] R.C. MOBLEY and R.A. LAUBENSTEIN: Phys. Rev. **80**, 309 (1950).

[2] J.C. NOYES, J.E. VAN HOOMISSEN, W.C. MILLER and B. WALDMAN: Phys. Rev. **95**, 396 (1954).

[3] D.R. CONNORS and W.C. MILLER: Bull. Amer. Phys. Soc. **1**, 340 (1956).

[4] L.I. RUSINOW and A.N. SAGAIDAK: Z. Sow. Un. **10**, 203 (1936). — V. DEMENTI, A. LEIPUNSKY and V. MASLOV: Z. Sow. Un. **12**, 764 (1937). — O.R. FRISCH, H. HALBAN and J. KOCH: Kgl. danske Vidensk. Selsk. **15** (1938). — H. HALBAN: C. R. Acad. Sci., Paris **206**, 1170 (1938). — Nature, Lond. **141**, 644 (1938). — F.G. HOUTERMANS and I. BARTZ: Phys. Z. **44**, 167 (1943).

[5] E. GUTH and C. J. MULLIN: Phys. Rev. **76**, 234 (1949).

where, B is due to the $P_{\frac{3}{2}} \to D_{\frac{5}{2}}$ transition, while A is the sum of the contributions of the $P_{\frac{3}{2}} \to S_{\frac{1}{2}}$ and $P_{\frac{3}{2}} \to D_{\frac{3}{2}}$ transitions.

The actual value of the ratio A/B, which is a function of the energy, depends on the detailed specifications about the splitting of the P and D levels[1]. An agreement with the experimental data seems to be obtained by assuming that for the D levels there is the same type of splitting as that adopted for the P level. Under this assumption only the $P_{\frac{3}{2}} \to D_{\frac{5}{2}}$ transitions has to be considered, besides the $P_{\frac{3}{2}} \to S_{\frac{1}{2}}$, because for energies less than 3 Mev only the lower-energy D component participates in the transition.

For a detailed discussion of this problem reference is made to the above mentioned papers[2]. Only the still scanty experimental data will be reported here.

The first measurements of CHADWICK and GOLDHABER[3] made with photons of 2.62 Mev gave an isotropic distribution within the experimental errors of 15%.

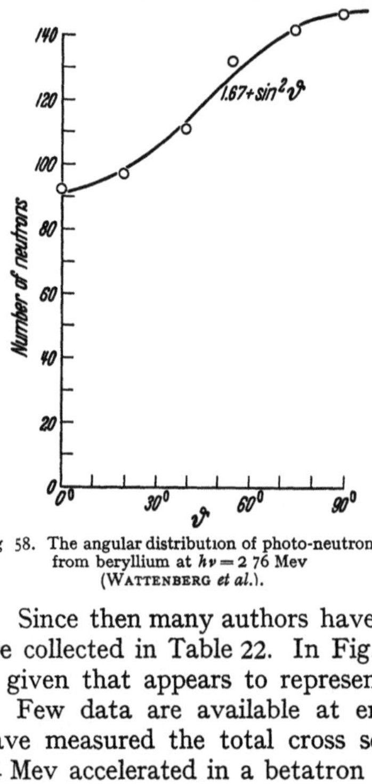

Fig 58. The angular distribution of photo-neutrons from beryllium at $h\nu = 2.76$ Mev (WATTENBERG *et al.*).

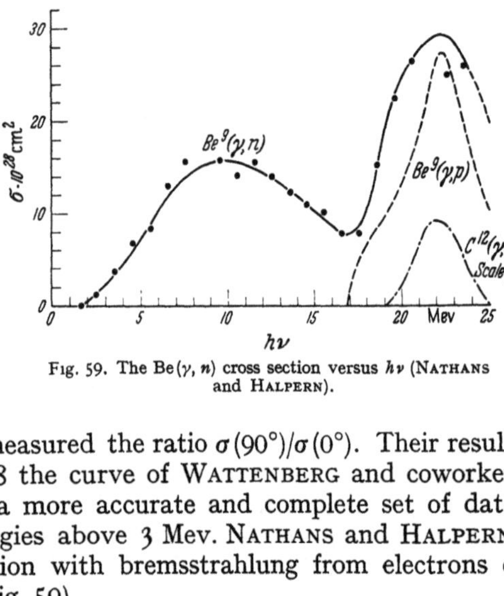

Fig. 59. The Be(γ, n) cross section versus $h\nu$ (NATHANS and HALPERN).

Since then many authors have measured the ratio $\sigma(90°)/\sigma(0°)$. Their results are collected in Table 22. In Fig. 58 the curve of WATTENBERG and coworkers is given that appears to represent a more accurate and complete set of data.

Few data are available at energies above 3 Mev. NATHANS and HALPERN[4] have measured the total cross section with bremsstrahlung from electrons of 24 Mev accelerated in a betatron (Fig. 59).

Their results, although not very accurate immediately above the threshold, give the general trend of the cross section.

Measurements with monoenergetic γ-rays have been made by various authors[5-7]. With the γ-rays of 6.1 Mev produced in the reaction $F^{19}(p, \alpha\gamma)$ CARVER *et al.*[6] find the following value of the total cross section of Be

$$\sigma(6.1 \text{ Mev}) = (12.5 \pm 1.2)\, 10^{-28} \text{ cm}^2 \tag{47.7a}$$

[1] C. J. MULLIN and E. GUTH: Phys. Rev. **76**, 682 (1949). See also A. BORSELLINO: Nuovo Cim. **5**, 263 (1948). — V. DE SABATA, Nuovo Cim. **6**, 368 (1949).

[2] Besides footnotes 5, p. 143 and 1, this page, see also: E. D. COURANT: Phys. Rev. **82** 703 (1951).

[3] See footnote 5, p. 142.

[4] R. NATHANS and J. HALPERN: Phys. Rev. **92**, 940 (1953).

[5] D. ARAKATZU, M. SOMADA, Y. UEMURA, S. YASUMI and Y. SAJI: J. Phys. Soc. Japan **66**, 66 (1951). — V.D. ERIKSEN and C.P. ZALEVSKI: J. Phys. Radium **15**, 492 (1954).

[6] See footnote 2, p. 142.

[7] R.D. EDGE: Nuclear Phys. **2**, 485 (1956).

Table 22. $\sigma(90°)/\sigma(0°)$ *for the photoneutrons from* Be.

Source of γ-rays	$h\nu$ (Mev)	$\frac{\sigma(90°)}{\sigma(0°)} = \frac{A+B}{A}$	Reference	Source of γ-rays	$h\nu$ (Mev)	$\frac{\sigma(90°)}{\sigma(0°)} = \frac{A+B}{A}$	Reference
Ra		1	1	Mn^{56}	1.81	1	4
Ra		1.07 ± 0.09	2	Na^{24}	2.76	1.6	4
ThC″	2.62	1.75 ± 0.04	3	Na^{24}	2.76	1.87 ± 0.05	5
Sb^{124}	1.70	1	4	ThC″	2.62	1.64 ± 0.04	6

[1] L. I. RUSINOW and A. N. SAGAIDAK: Z. Sow. Un. **10**, 203 (1936).
[2] T. A. GOLOBORODKO and L. ROSENKEWITSCH: Z. Sow. Un. **11**, 78 (1937).
[3] T. A. GOLOBORODKO: J. Phys. URSS. **3**, 141 (1940).
[4] B. HAMERMESH, M. HAMERMESH and A. WATTENBERG: Phys. Rev. **76**, 611 (1949).
[5] E. MEINERS: See footnote 5, p. 143.
[6] H. NIEWODNICZANSKI and M. WIELOWIEJSKA: Bull. Acad. Polon. Sci. **2**, 25 (1954).

which fits well on the curve of Fig. 59. Similar measurements were made by EDGE[1] with the γ-rays produced in the reactions $N^{15}(p, \gamma)$, $F^{19}(p, \alpha\gamma)$ and $C^{13}(p, \gamma)$ (Sect. 50). He finds

$$\left.\begin{aligned} \sigma(4.4\text{ Mev}) = (1.8 \pm 3.6), \qquad \sigma(6.1\text{ Mev}) = (11.3 \pm 1.0), \\ \sigma(8.06\text{ Mev}) = (13.8 \pm 1.6) \times 10^{-28}\text{ cm}^2. \end{aligned}\right\} \tag{47.7b}$$

CARVER *et al.*[2] have also investigated the spectrum of the α-particles emitted in coincidence with the neutron, from a thin layer of Be irradiated with photons of 6.1 Mev. Only α-particles of energy larger than 400 kev were recorded.

They found a cross section equal to $(12.8 \pm 2.5) \times 10^{-28}$ cm², i.e. a value equal to (47.7a) representing the total cross section for emission of photo-neutrons. Therefore they conclude that, for $h\nu = 6.1$ Mev, process (47.1) in which the Be^8 is left in its ground state, contributes less than 20% to the value (47.7a). In fact the corresponding α-particles have always energies far below the cut-off of 400 kev of the experimental set-up. From a discussion of the correlated spectra of the two α-particles, they conclude that the cross section (47.7a) is due to process (47.3) and in part to a process proceeding through an excited state of Be^8 (not higher than 2.9 Mev). Both these processes are not considered in the theory of GUTH and MULLIN and therefore any agreement with their data at this energy is fortuitous.

A large polarization of the neutrons emitted in reaction (47.1) has been predicted[3] by using the Guth and Mullin model. No experimental investigation of this effect has yet been made.

48. The neutron photoeffect in medium and heavy elements. Since the pioneering work of BOTHE and GENTNER, mentioned in Sect. 45, the investigation of (γ, n) processes has been extended to elements distributed over the whole periodic system. Some of the experiments were carried through with γ-rays emitted in various reactions (Sect. 50), others with high energy X-rays emitted as bremsstrahlung by electrons accelerated with betatrons, synchrotrons or linear accelerators (Sect. 51).

With this last technique, which will be discussed mainly in this section, the experiments have been extended to energies of a few hundred Mev, where many phenomena take place which, although of great interest, would be outside the scope of the present chapter.

[1] See footnote 7, p. 144.
[2] See footnote 2, p. 142.
[3] W. CZYZ and J. SANRIKI: Nuovo Cim. **3**, 864 (1956).

Therefore in this section and in Sect. 51, only the low energy region will be considered where

$$E_\gamma = h\nu \leqq 30 \text{ Mev} \qquad \lambda\,[\text{cm}] = \frac{123.43}{E_\gamma\,[\text{Mev}]} \cdot 10^{-12} > 4 \times 10^{-12}\,\text{cm}. \qquad (48.1)$$

Data on the photoneutron yields at higher energies are discussed in the review article of STRAUCH[1] and in some original papers[2]. The review article of STRAUCH contains also a detailed discussion of the experimental results in the low energy region as does the review article by TITTERTON[3].

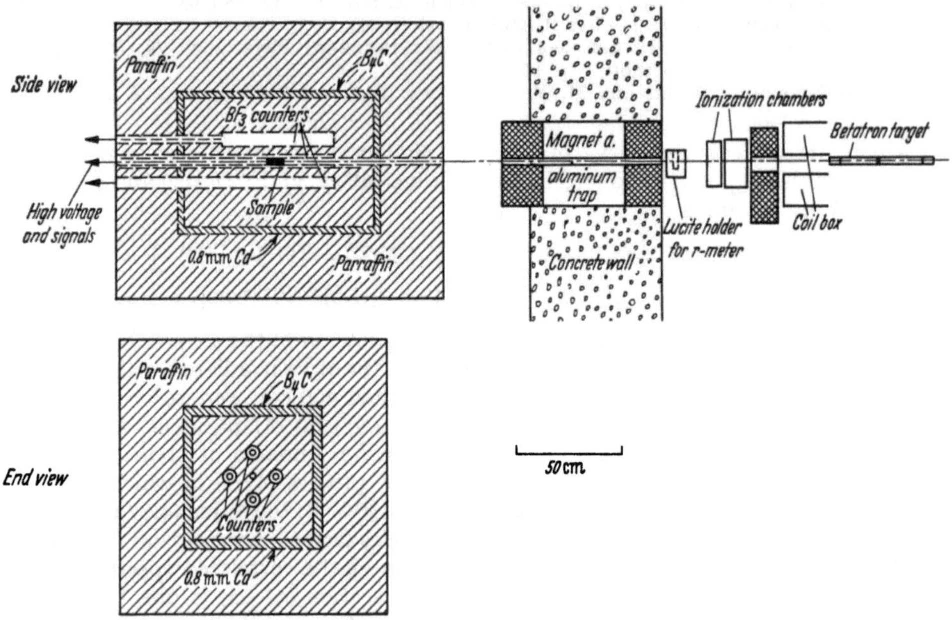

Fig. 60. Experimental set-up used by SHER, HALPERN and MANN for threshold and yield measurements of (γ, n) processes.

The main problem to be discussed in this section is the determination of the cross section $\sigma(E_\gamma)$ of a given (γ, n) process from its threshold E_{1T} to high energies.

When an electron accelerator of sufficiently high energy is available, this is usually done by measuring the yield $Y(E_0)$ of the products of the (γ, n) reaction as a function of the energy E_0 of the electrons. The yield is determined by observation of the emitted neutrons or of the activity of the product, if it is radioactive. A typical experimental set-up for the first type of measurement is shown in Fig. 60. In order to minimize the dependence of the detection efficiency on the energy of the neutrons emitted from the sample, the neutrons are slowed down in paraffin and then detected by BF_3 counters placed at a convenient depth.

The fact that the X-ray beam from electrons produced by a betatron or synchrotron or linear accelerator consists of a series of sharp pulses of a duration of the order of one microsecond repeated at intervals of a few milliseconds, is used to eliminate the piling-up of secondary electrons in the BF_3 counters during the X-ray pulse. By means of an electronic system a gate is usually opened,

[1] K. STRAUCH. Annual Rev. Nucl. Sci. **2**, 105 (1953).

[2] K.M. TERWILLINGER, L.W. JONES and W.N. JARMIE· Phys. Rev. **82**, 820 (1951). — L.W. JONES and K.M. TERWILLINGER· Phys. Rev. **91**, 699 (1953).

[3] E.W. TITTERTON: Progr. Nucl. Phys. **4**, 1 (1955).

say, 20 μsec after the burst and closed some 800 μsec later. The neutrons emitted from the sample during the X-ray pulse need a few microseconds to be slowed down. Then they diffuse through the paraffin for a time of the order of 200 μsec before being captured or before diffusing out of it. A typical curve obtained with the technique described above is shown in Fig. 61 where the yield $Y(E_0)$ is plotted versus the energy E_0 of the irradiating electrons.

Such a curve can be represented by the following equation

$$Y(E_0) = \int_{E_{1T}}^{E_0} \sigma(E_\gamma)\, P(E_\gamma, E_0)\, \eta(E_\gamma)\, dE_\gamma \tag{48.2}$$

where $P(E_\gamma, E_0)\, dE_\gamma$ is the spectrum of the bremsstrahlung emitted by electrons of energy E_0 incident on the target used and $\eta(E_\gamma)$ the efficiency of detection of the photoneutrons whose energy depends on E_γ.

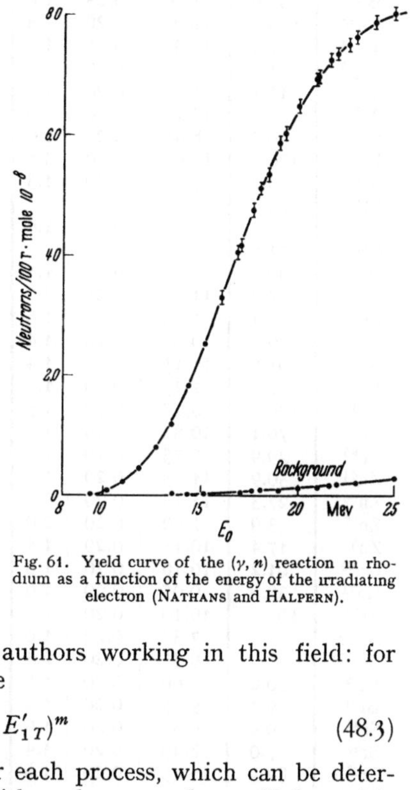

Fig. 61. Yield curve of the (γ, n) reaction in rhodium as a function of the energy of the irradiating electron (Nathans and Halpern).

The derivation of $\sigma(E_\gamma)$ from Eq. (48.2), once $Y(E_0)$ has been accurately measured, gives in general a rather uncertain result, not only because it involves the inversion of an integral equation, but also because the functions $P(E_\gamma, E_0)$ and $\eta(E_\gamma)$ are known only approximately.

In the discussion of the use of Eq. (48.2), one has to distinguish between the case of measurements designed to determine the threshold E_{1T} of the reaction considered and that of mesurements to determine the behaviour of $\sigma(E_\gamma)$ above the threshold.

In the first case one can avoid the use of rather uncertain particular expressions for $P(E_\gamma, E_0)$ and $\eta(E_\gamma)$, and employ the following empirical rule found by various authors working in this field: for values of E_0 very close to E_{1T} one can write

$$Y(E_0) \sim (E_0 - E'_{1T})^m \tag{48.3}$$

where m and E'_{1T} are constants, different for each process, which can be determined by trial. The value of E'_{1T} obtained with such a procedure will depend in general on the adopted value of m. It is, however, usually assumed to represent a good estimate of the threshold E_{1T} of the corresponding process.

It is rather difficult to justify such a procedure in general, and the consideration of some special case will clarify the type of approximations involved. For instance, one may wish to determine the threshold for the (γ, n) processes in D and Be (see Sects. 46 and 47) with a neutron detector for which one can assume $\eta(E_\gamma) = \text{const}$ with good approximation for E_γ close to E_{1T}.

For both these nuclei, immediately above the threshold,

$$\sigma(E_\gamma) \sim (E_\gamma - E_{1T})^{\frac{1}{2}} \tag{48.4}$$

as a consequence of the fact that the photoneutron is emitted in an S state (in the case of the deuteron the photomagnetic effect prevails over the photoelectric effect immediately above threshold).

Tabelle 23. *Thresholds of various nuclei for the (γ, n) reaction* (SHER, HALPERN, MANN[1]).

Isotope	Percent abundance	Present measurements			Class	Isotope	Percent abundance	Present measurements			Class
		Threshold (Mev)	Error (Mev)	m				Threshold (Mev)	Error (Mev)	m	
H^2		2.23	Calib.	1.8	*A*	Cb^{93}	100	8.70	0.20	1.6	*A*
Li^6	7.5	5.35	0.20	1.0	*A*	Mo(?)		6.75	0.25	2.0	*C*
Li^7	92.5	7.15	Calib.	1.0	*A*	Mo(?)		7.95	0.25	2.0	*C*
Be^9	100	1.67	Calib.	0.9	*A*	Ru(?)		7.05	0.20	1.9	*C*
B^{10}	18.4	8.55	0.25	2.1	*A*	Ru(?)		9.50	0.20	1.6	*C*
B^{11}	81.6	11.50	0.25	1.0	*A*	Rh^{103}	100	9.35	0.20	1.6	*A*
Na^{23}	100	12.05	0.20	2.5	*A*	Pd(?)		7.05	0.20	2.1	*C*
Mg^{24}	77.4	16.55	0.25	1.9	*A*	Pd(?)		9.35	0.20	1.6	*C*
Mg^{25}	11.5	7.25	0.20	1.4	*A*	Ag^{109}	48.1	9.05*	0.20	2.5	*B*
Mg^{26}	11.1	11.15	0.20	1.1	*A*	Cd^{113}	12.3	6.55	0.20	2.0	*A*
Al^{27}	100	12.75	0.20	2.0	*A*	In^{115}	95.5	9.05*	0.20	1.9	*B*
Si^{29}	6.2	8.45	0.20	1.0	*A*	Sn^{118}	22.5	9.10	0.20	1.8	*B*
P^{31}	100	12.05	0.20	3.1	*A*	Sn^{119}	9.8	6.60	0.20	1.5	*B*
S^{34}	4.2	10.85	0.20	1.3	*A*	Sb^{121}	56				
Cl^{35}	75.4							8.95	0.25	2.3	*C*
		9.95	0.20	1.6	*C*	Sb^{123}	44				
Cl^{37}	24.6					Te(?)		6.50	0.20	1.9	*C*
V^{51}	100	11.15	0.20	1.4	*A*	Te(?)		8.55	0.20	1.7	*C*
Cr^{52}	83.8	11.80	0.25	2.3	*B*	I^{127}	100	9.10	0.20	1.5	*A*
Cr^{53}	9.4	7.75	0.20	1.6	*B*	Cs^{133}	100	9.05	0.20	1.7	*A*
Mn^{55}	100	10.00	0.20	1.4	*A*	Ba(?)		6.80	0.20	2.0	*C*
Fe^{56}	91.6	11.15	0.25	1.4	*A*	Ba(?)		8.55	0.25	1.7	*C*
Fe^{57}	2.1	7.75	0.20	1.5	*A*	La^{139}	100	8.80	0.20	1.6	*A*
Co^{59}	100	10.25	0.20	1.5	*A*	Ce^{140}	89	9.05	0.20	1.4	*B*
Cu^{63}	70.1	10.85	0.20	1.5	*A*	Ce^{142}	11	7.15	0.20	1.5	*B*
Cu^{65}	29.9	9.75	0.20	1.4	*A*	Ta^{181}	100	7.55	0.20	1.8	*A*
Zn^{64}	50.9	11.65	0.20	1.1	*A*	W(?)		6.25	0.30	1.0	*C*
Zn^{66}	27.3	11.15	0.20	0.7	*B*	W(?)		7.15	0.30	1.6	*C*
Zn^{67}	3.9	7.00	0.20	2.0	*A*	Re^{187}	61.8	7.30*	0.30	1.5	*B*
Zn^{88}	17.4	10.15	0.20	1.4	*B*	Ir^{193}	61.5	7.80*	0.20	1.7	*B*
Ga^{69}	61.2	10.10	0.20	1.8	*B*	Pt^{194}	30.2	9.50	0.20	1.3	*B*
Ga^{71}	38.8	9.05	0.20	1.6	*B*	Pt^{195}	35.3	6.10	0.20	1.6	*A*
As^{75}	100	10.10	0.20	1.6	*A*	Pt^{196}	26.6	8.20	0.20	1.6	*B*
Se(?)		7.30	0.20	1.6	*C*	Au^{197}	100	7.90	0.20	1.7	*A*
Se(?)		9.35	0.20	2.4	*C*	Tl^{203}	21.9	8.80	0.20	1.6	*A*
Br^{79}	50.6	10.60	0.20	1.5	*A*	Tl^{205}	70.9	7.55	0.20	1.1	*A*
Br^{81}	49.4	9.95	0.20	1.4	*A*	Pb^{207}	22.6	6.75	0.20	1.3	*A*
Sr^{86}	9.9	9.50	0.20	1.9	*B*	Pb^{208}	52.3	7.30	0.20	1.0	*A*
Sr^{87}	7.0	8.40	0.20	1.4	*A*	Bi^{209}	100	7.40	0.20	1.7	*A*
Sr^{88}	82.9	11.15	0.20	1.6	*B*						

* These values may be influenced by the presence of a second unresolved isotope.
"Class" refers to isotopic assignment: *A* = certain, *B* = probable, *C* = doubtful or unknown.

On the other hand, the bremsstrahlung spectrum emitted by a thick target, containing $\mathfrak{N}$ atoms /cm³, is given by

$$P(E_\gamma, E_0)\, dE_\gamma = dE_\gamma\, \mathfrak{N} \int\limits_{E_\gamma}^{E_0} \frac{\sigma(E, E_\gamma)}{-\dfrac{dE}{dx}}\, dE \tag{48.5}$$

where $\sigma(E, E_\gamma)$ is the cross section for the production of quanta E_γ by one electron of energy E, and $-dE/dx$ the energy loss that an electron undergoes in the target. The integrand is a slowly varying function of E so that, in a small energy interval,

[1] See footnote 2, p. 149.

it can be assumed to be constant as a first approximation[1]. Then

$$P(E_\gamma, E_0)\, dE_\gamma \sim (E_0 - E_\gamma)\, dE_\gamma \qquad (48.6)$$

which, introduced with (48.4) and $\eta = \text{const}$ in Eq. (48.2), gives

$$Y(E_0) \sim (E_0 - E_{1T})^{\frac{5}{2}}, \qquad (48.7)$$

identical with (48.3) for $m = 2.5$. The theoretical value of m is however of no practical use even in the cases considered here: in the experiments described in Sect. 46 one finds $m = 2$, a fact indicating that some of the approximations involved in the derivation of (48.7) are not correct.

In Table 23 the results of SHER, HALPERN and MANN[2] on the threshold of 77 (γ, n) processes are given. These were obtained with the experimental set up of Fig. 60 and by applying the procedure described above. In the table are also indicated the values of m used in each case.

The more striking phenomenon observed in the low energy region is the existence of the so-called "giant resonance"[3]. A few Mev above the threshold of the (γ, n) process, the cross section increases rapidly with increasing E_γ, reaches a maximum σ_m for a certain value $E_\gamma = E_m$ and then drops to rather low values. As an example Fig. 62 shows the cross section of the process $Cu^{63}(\gamma, n)\, Cu^{62}$ (Cu^{62} is beta-radioactive with $T_{\frac{1}{2}} = 10$ min) as a function of E_γ. From this example one sees that the various results[4,5] agree only qualitatively in the existence and general features of the giant resonance, and not in the details.

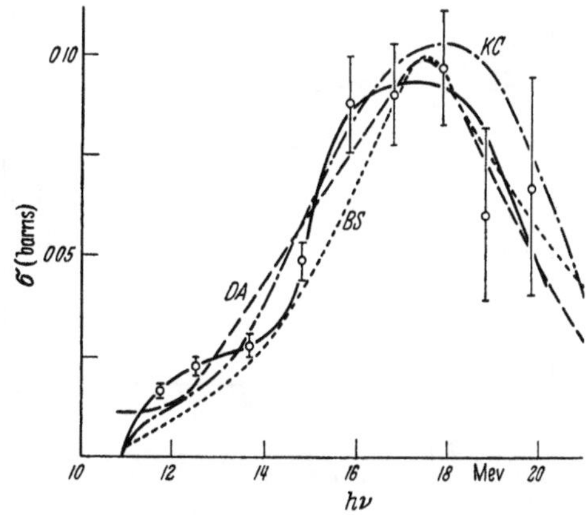

Fig. 62. Cross section of the process $Cu^{63}(\gamma, n)\, Cu^{62}$ as a function of the energy of the photon $h\nu$. Full line = KROHN and SHRADER[4]. DA = DIVEN and ALMY[5]. KC = KATZ and CAMERON[5]. BS = BYERLY and STEPHENS[5].

This is a consequence of the fact, noted above, that, if $Y(E_0)$ has been measured as a function of E_0 with good accuracy (see e.g. Fig. 61), in order to get $\sigma(E_\gamma)$, one has to solve an integral relation and make use of expressions for $P(E_\gamma, E_0)$ and $\eta(E_\gamma)$ which are rather uncertain.

When the measurements of $Y(E_0)$ are performed with an experimental set-up such as that of Fig. 60, so that the integral equation to be solved is given by (48.2), one can apply the so-called photodifference method developed by JOHNS, KATZ, DOUGLAS and HASLAM[6] who substitute for the integral a sum of a few

[1] E. GUTH: Phys. Rev. **59**, 325 (1941). — W.C. MILLER and B. WALDMAN: Phys. Rev. **75**, 425 (1949).
[2] R. SHER, J. HALPERN and A.K. MANN: Phys. Rev. **84**, 387 (1951).
[3] G.C. BALDWIN and G.S. KLAIBER: Phys. Rev. **71**, 3 (1947). — Phys. Rev. **73**, 1156 (1948).
[4] V.E. KROHN and E.F. SHRADER: Phys. Rev. **87**, 685 (1952).
[5] B.C. DIVEN and G.M. ALMY: Phys. Rev. **80**, 407 (1950). — L. KATZ and A.G.W. CAMERON: Canad. J. Res. **29**, 518 (1951). — P.R. BYERLY and W.E. STEPHENS: Phys. Rev. **83**, 54 (1951).
[6] H.E. JOHNS, L. KATZ, R.A. DOUGLAS and R.N.H. HASLAM: Phys. Rev. **80**, 1062 (1950). — J. GOLDEMBERG and L. KATZ: Phys. Rev. **89**, 1300 (1953).

terms calculated for certain values $E_{\gamma i}$ of E_γ conveniently spaced

$$Y(E_{0r}) = \Sigma_i \sigma_i a_{ir} \tag{48.8a}$$

where

$$\left.\begin{aligned} \sigma_i &= \sigma(E_{\gamma i}), & a_{ir} &= P(E_{\gamma i}, E_{0r})\,\eta(E_{\gamma i}), \\ E_{\gamma i} &= E_{0r} - i\,\Delta E & i &= 1, 2, 3, \ldots \end{aligned}\right\} \tag{48.8b}$$

A linear equation of this type can be written for various values of E_{0r} which are also spaced by ΔE (ΔE could e.g. be 1 Mev or 0.5 Mev) starting from a first value E_{01} close to the threshold E_{1T} ($E_{01} - E_{1T} < \Delta E$). One obtains a system of linear equations which allow the determination of the unknowns σ_i in terms of the measured yields and of the coefficients a_{ir}. These are usually calculated

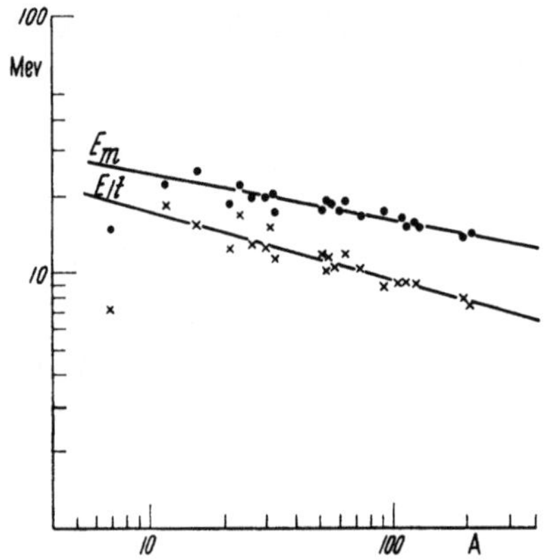

Fig. 63. Plots of E_m and E_{1T} for the (γ, n) reactions as a function of mass number A (MONTALBETTI *et al.*[2]).

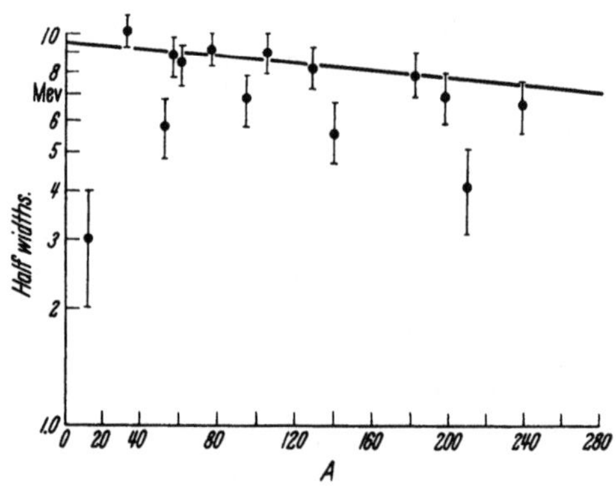

Fig. 64. The width of the giant resonance as function of mass number A (NATHANS and HALPERN).

assuming $\eta(E_{\gamma i}) = \text{const}$ and, using the spectrum given by SHIFF (Sect. 51), corrected for absorption in the various intervening materials (as for instance the walls of the doughnut).

The monitoring of the incident γ-radiation offers some problems because the sensitivity of the monitor (usually a Victoreen r-meter, placed at the centre of a cylinder of lucite) depends considerably on E_0 [1].

Extensive work in this field and by these methods has been done by the Saskatchewan group[2] and the Philadelphia group[3] whose results agree as well as can be expected considering the various uncertainties involved in the procedures described above.

The main features of their common conclusions are the following[4]: (a) all cross section curves that have been investigated to high enough energies show a "resonance behaviour". The shapes of the various curves are not all similar, but it is difficult to estimate how much of this variation is due to experimental uncertainties; (b) the energy E_m of the cross section peak and the threshold E_{1T} are roughly given in Mev as a function of the mass number A by the empirical equations[1]

$$E_m = 37\,A^{-0.186}; \qquad E_{1T} = 32\,A^{-0.270}. \tag{48.9}$$

[1] See footnote 6, p. 149.

[2] R. MONTALBETTI, L. KATZ and J. GOLDEMBERG: Phys. Rev. **91**, 659 (1953).

[3] R. NATHANS and J. HALPERN: Phys. Rev. **92**, 207 (1953); **93**, 437 (1954).

[4] See for a similar discussion the article by G.R. BISHOP and R. WILSON in Vol. XLII of this Encyclopedia.

Therefore the difference $E_m - E_{1T}$ varies only from 7 Mev at $A = 10$, to 6.2 Mev for $A = 200$ (Fig. 63).

(c) The width of the giant resonance, defined as the full width at one half the maximum cross section σ_m, decreases with increasing A as is shown in Fig. 64. Nuclei containing a magic number of neutrons show a narrower resonance than neighbouring nuclei[1].

(d) The peak value of the cross section σ_m increases with increasing A (Fig. 65) as A^α where α has values ranging from $\frac{3}{2}$ [2] to $\frac{5}{3}$ [1] according to different authors.

(e) The cross section integrated between 0 and 25 Mev

$$\sigma_{\text{int}} = \int_0^{25\,\text{Mev}} \sigma(E_\gamma)\, dE_\gamma \qquad (48.10)$$

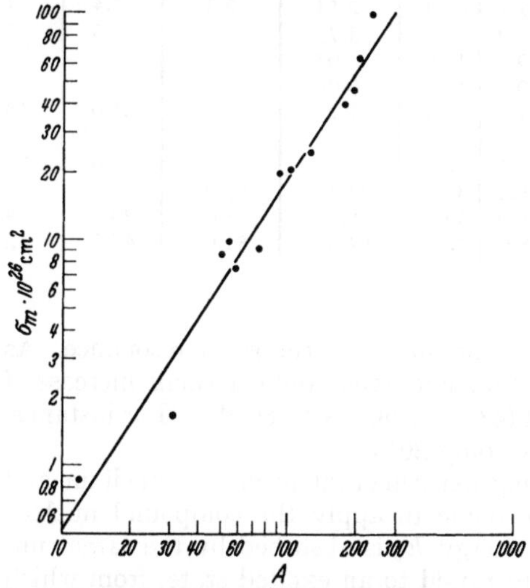

Fig. 65. Double logarithmic plot of the maximum cross section σ_m vs. A (NATHANS and HALPERN).

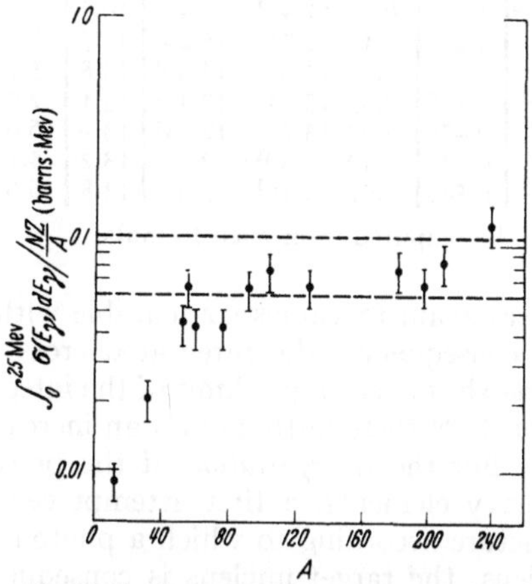

Fig. 66. Total cross section for neutron emission integrated to 25 Mev and divided by NZ/A, vs. mass number A (NATHANS and HALPERN).

is roughly proportional (Fig. 66) for medium and heavy elements to NZ/A ($N = A - Z =$ number of neutrons).

Table 24 gives the results of NATHANS and HALPERN[2] who have investigated 14 monoisotopic elements distributed from carbon to uranium. The yield per mole of the irradiated element and per r of the incident radiation is compared with the results of previous authors[3]. The cross section at 17 Mev, is compared with the results obtained by McDANIEL and coworkers[4] with the γ-rays of the Li(p, γ) reaktion (see Sect. 50). On account of the various uncertainties involved in the derivation of the cross sections from the yield curves, the agreement between the data collected in the two last columns of Table 24 can be considered satisfactory.

A remark which is of interest from the point-of-view of the use of the (γ, n) reactions as neutron sources, as well as for a comparison of the experimental results with the theoretical estimates, is due to JONES and TERWILLINGER[5] who found that, in medium and heavy elements, above the peak of the giant resonance, the cross section falls to a rather low value and then remains constant to energies of about 80 Mev. Above this energy up to 320 Mev the cross section

[1] See footnote 2, p. 150.
[2] See footnote 3, p. 150.
[3] G. A. PRICE and D. W. KERST: Phys. Rev. **77**, 806 (1950).
[4] B. D. McDANIEL, R. L. WALKER and M. B. STEARNS: Phys. Rev. **80**, 807 (1950).
[5] See footnote 2, p. 146.

Table 24. *Summary of data on photoneutron reactions.*

Element	Target thickness g/cm²	Thresholds (γ, n) Mev	Thresholds $(\gamma, 2n)$ Mev	Thresholds (γ, pn) Mev	E_m Mev	Half-width Mev	$\sigma_m \times 10^{26}$ cm²	$\int_0^{25\,\text{Mev}} \sigma\, dE$ Mev-barns	22-Mev yield/mole-r $\times 10^{-6}$ (NATHANS and HALPERN)	22-Mev yield/mole-r $\times 10^{-6}$ (PRICE and KERST)	σ (17 Mev) $\times 10^{26}$ cm² (NATHANS and HALPERN)	σ (17 Mev) $\times 10^{26}$ cm² (Mc DANIEL)
${}_{6}C^{12}$	12.80	18.70	32.60	27.41	22.0	3.0	0.86	0.027	0.028			
${}_{15}P^{31}$	4.41	12.05	24.50	17.90	21.5	10.2	1.67	1.61	0.389	0.33		
${}_{23}V^{51}$	2.68	11.15	19.88	18.80	18.7	5.8	8.69	0.56	2.26	1.6	6.8	3.3
${}_{25}Mn^{55}$	1.57	10.00	19.40	17.72	18.4	8.8	9.69	0.88	3.15	1.9	9.0	4.0
${}_{27}Co^{59}$	2.59	10.25	19.67	17.48	17.3	8.4	7.45	0.63	2.51	2.3	7.4	4.8
${}_{33}As^{75}$	1.34	10.10	17.42[1]	16.82[1]	17.3	9.0	9.03	0.80	3.73	3.9	8.5	7.0
${}_{41}Cb^{93}$	1.36	8.70	17.88[1]	14.08[1]	17.0	6.8	19.5	1.46	6.67	5.8		
${}_{45}Rh^{103}$	1.81	9.35	17.47[1]	15.41[1]	16.5	8.9	20.5	1.94	9.00			
${}_{53}I^{127}$	1.23	9.10	16.72	15.29	15.5	8.3	24.3	2.04	11.2	11.0	21.0	18.0
${}_{57}La^{139}$	...	8.80	14.25[1]	14.90[1]	13.8	5.7	...	...	...	...	...	...
${}_{73}Ta^{181}$	0.468	7.55	13.84[1]	13.47[1]	15.1	7.9	39.7	3.43	21.0	19.0	34.0	25.0
${}_{79}Au^{197}$	0.455	7.90	13.71[1]	12.94[1]	13.9	6.9	45.6	3.19	22.4	19.0		
${}_{83}Bi^{209}$	0.434	7.40	13.30[1]	12.68[1]	13.2	4.1	63.0	4.08	25.0	25.0	20.0	24.0
${}_{92}U^{238}$	0.548	5.97	12.18[1]	13.34[1]	13.8	6.6	98.0	7.15	57.5	51.0	48.0	52.0

[1] Computed from mass formula.

rises again to values comparable with the maximum of the giant resonance. As a consequence, the integrated cross section undergoes only a small increase if one shifts the upper limit of the integral (48.10) from 25 to 80 Mev. For instance, for Ta^{181} these authors find an increase of only 20%.

For the *interpretation* of the preceding experimental results in medium and heavy elements, a first attempt can be made to apply the compound nucleus picture according to which a photon of energy E_γ is absorbed by the target nucleus; the target nucleus is consequently raised to an excited state, from which the emission of one nucleon can occur. According to the Bohr assumption this second step of the reaction depends only on the energy, angular momentum and parity of the compound state, and not on the specific way in which it has been produced (Sect. 25).

As a consequence the cross section for a given process, say for a (γ, n) reaction, can be written as the product of two factors

$$\sigma(\gamma, n) = \sigma_a(E_\gamma)\, G(n) \qquad (48.11)$$

of which the first, $\sigma_a(E_\gamma)$, represents the cross section for absorption of photons of energy E_γ, while the second $G(n)$ is the branching probability for emission of the particle considered, in the present case the neutron.

Below the threshold of the (γ, n) reaction $G(n)$ is equal to zero, while immediately above one has to consider, as competing processes, the elastic and inelastic scattering of photons as well as the (γ, p) process whose threshold is usually not far from that of the (γ, n) reaction. The contribution of the (γ, γ) and (γ, γ') processes is of the order of 10%[1] while the (γ, p) process is hindered by the electrostatic potential barrier. Therefore immediately above the threshold of the (γ, n) reaction, one would expect $G(n)$ to be close to 1 and $\sigma(\gamma, n)$ to be close to $\sigma_a(E_\gamma)$. At higher energies not only the (γ, p) process starts to be of considerable importance but also all other reactions in which the emission of two or more nucleons or complex particles take place: $(\gamma, 2n)$, (γ, np), $(\gamma, 3n)$, (γ, npd), ...

[1] M. B. S. STEARNS: Phys. Rev. **87**, 706 (1952). — J. GOLDEMBERG and L. KATZ: Phys. Rev. **90**, 439 (1953). — C. S. DEL RIO and V. L. TELEGDI: Phys. Rev. **90**, 439 (1953). — E. G. FULLER and R. HAYWARD: Phys. Rev. **94**, 732 (1954).

Therefore, the cross section for photoproduction of neutrons which can be obtained from the neutron yield curve with the procedure described above, has usually a complicated meaning. It represents an average of the corresponding cross sections weighted with their neutron multiplicities. The significance of such an average is to some extent uncertain if the efficiency of the neutron detector depends on the energy of the neutrons. The spectrum of the neutrons emitted from the compound nucleus is different in different processes as one expects also according to the evaporation model of WEISSKOPF and EWING, who express the spectrum in terms of the level density of the residual nucleus (Sect. 30).

In spite of all the above mentioned uncertainties, considerable progress has been made in the theoretical interpretation of these phenomena.

Most theoretical considerations on this subject refer to the absorption cross section $\sigma_a(E_\gamma)$ of the incident photons. Without going into the details, for which reference is made to [3][1], the following represents the main points of the theoretical interpretation of the observations for $E_\gamma \lesssim 30$ Mev where the wavelength (48.1) is large compared to the nuclear size.

One starts with an estimate of a general qualitative nature of the cross section for absorption of electric and magnetic multipole radiations. The results of such an estimate, aiming only to give the general trend, are the following: for a not too large energy of the incident photon, the cross section for absorption of electric multipole radiation is, in general, a rapidly decreasing function of the multipole order l of the radiation. For a given order l, the cross section for absorption of magnetic multipole radiation is smaller than that for the corresponding electric multipole by a factor of the order $\beta^2 = (v/c)^2$ where v is a velocity of the order of those of the nucleons in the nucleus ($\beta^2 \sim 10^{-2}$).

For the interpretation of the absorption of photons in the energy region considered, one needs to take into account, in general, only electric dipole, magnetic dipole and electric quadrupole transitions of which the first one will predominate.

This last remark holds in general, but not if some selection rule intervenes to forbid, or at least to reduce, the probability for electric dipole transitions. This is what one expects to happen at energies below about 15 Mev where the transition probability for electric dipole radiation may be strongly diminished by the correlations between the motion of neutrons and protons in the nucleus. Such a reduction can be understood considering the example of extreme correlation of an even-even nucleus in which neutrons and protons move around bound in subunits, such as deuterons or α-particles, having all the same charge and the same mass. In such a system the electric centre coincides with its centre of mass at any moment independently of the motion of the subunits. Therefore, as long as one neglects the recoil of the nucleus caused by emission or absorption of photons, the corresponding electric dipole moment is always zero. In real nuclei the situation is not as extreme as this; it is believed, however, that there exist strong correlations between the nucleons which considerably diminish the transition probability due to the absorption of electric dipole radiation. Therefore, immediately above the threshold the cross section $\sigma_a(E_\gamma)$ is expected to be rather small and due to the absorption of magnetic dipole and electric quadrupole radiation. At photon energies of about 15 Mev, i.e., when the energy is sufficient to break the correlations of the type mentioned above, an electric dipole absorption sets in and produces a rapid rise of the cross section. From this point on,

[1] G. R. BISHOP and R. WILSON: p. 309, Vol. XLII of this Encyclopedia.

in the low energy region, the absorption of the incident photons is due practically only to electric dipole radiation.

At higher energy the absorption of radiation of poles of higher order as well as mesonic effects cannot be neglected. In particular the rise of the (γ, n) cross section observed between 80 and 320 Mev by JONES and TERWILLINGER is attributed to effects of this type.

According to the qualitative description given above the so-called giant resonance is attributed to the onset of the electric dipole absorption for which a general sum-rule, independent of any particular model of the nucleus, has been derived by LEVINGER and BETHE[1].

This sum rule is similar to the well known Thomas-Reiche-Kuhn sum rule valid for electric dipole absorption by atoms. In the case of a nucleus the situation is different for two reasons: (a) in an atom all the Z electrons, which are the particles contributing to the absorption, have the same charge e. In a nucleus all the A nucleons contribute to the absorption, but the protons have an effective charge $+e\frac{N}{A}$ and the neutrons an effective charge $-e\frac{Z}{A}$; (b) the forces between nucleons are in part exchange forces.

The sum rule derived by LEVINGER and BETHE with some simplifying assumptions in the deduction of the influence of the exchange forces, is

$$\sigma_{\text{int}} = \int_0^\infty \sigma(E_\gamma)\, d E_\gamma = 0.060 \frac{NZ}{A} (1 + 0.8\, x) \text{ Mev} \times 10^{-24} \text{cm}^2 \qquad (48.12)$$

where x is the percentage of exchange forces.

Fig. 66 (p. 151) shows a comparison of Eq. (48.12) with the experimental results obtained from the neutron yield curves. The two horizontal broken lines correspond to $x=0$ and $x=1$ respectively.

The significance of such a comparison for elements of high Z may be understood by considering that the upper limit of the integral of the experimental values of the cross section (48.10) is 25 Mev while in Eq. (48.12) it is infinity, so that the experimental values can be estimated to be about 20% too low. In addition, for high Z, partial reactions involving the emission of charged particles can be neglected; the (γ, γ) and (γ, γ') processes, on the other hand, give a contribution to the integrated absorption cross section of the order of 10 to 20%.

For these two reasons the experimental points have to be raised by 30 to 40%, but the possibility of detection of the two neutrons emitted in the $(\gamma, 2n)$ reaction tends to compensate in part for this error. Therefore, one can conclude that in view of the many uncertainties affecting the experimental results and their comparison with theory, the experimental values of the integrated cross sections are consistent for heavy elements with the predictions of the sum rules.

Uranium is a special case because of the contribution of photofission to the integrated cross section, which gives 2.5 neutrons per process. If one corrects for this effect, the result for uranium is in agreement with those obtained with other heavy elements[2].

For light elements the neutron emission represents only a small contribution to the total photon absorption cross section and therefore one has to expect that the corresponding points fall below the line $x=0$. In fact, for light elements

[1] J. S. LEVINGER and H. A. BETHE: Phys. Rev. **78**, 115 (1950).
[2] See footnote 3, p. 150.

one cannot give simple general rules, as they usually show a characteristic fine structure for the various reactions produced by γ-rays[1,2].

Although the comparison between the sum-rule (48.12) and the experimental results cannot yet be pushed to a quantitative conclusion, one can conclude from this discussion that in medium and heavy elements the largest contribution to the integrated absorption cross section from threshold up to 70 Mev is due to electric dipole radiation. An upper limit of about 70 to 80 Mev constitutes a reasonable choice of energy at which only a small fraction of the electric dipole absorption is lost and no appreciable contribution of mesonic processes takes place.

The advantage of a sum rule such as Eq. (48.12) is that it is rather independent of the adopted nuclear model except for the calculation of the coefficient 0.8 of x, which demands a knowledge of the wave function of the ground state.

In particular, all the preceding discussion is independent of the general behaviour of the cross section as a function of the energy and does not touch the problem of why one observes in all medium and heavy elements the giant resonances nor what is the physical nature of such a peak in the cross section.

Problems of this type can be treated by comparing the experimental results with the predictions of calculations based on some special nuclear model. This has been done by many authors. A first step in this direction is the calculation by sum rules of the various moments of the photoabsorption curve which requires the knowledge of the wave function of the ground state. The more typical examples are the mean energy

$$\bar{E}_\gamma = \int_0^\infty \sigma(E_\gamma)\, E_\gamma \cdot d\,E_\gamma / \sigma_{\text{int}}$$

and the harmonic mean energy

$$E_H = \frac{\sigma_{\text{int}}}{\int\limits_0^\infty \sigma(E_\gamma)\, \frac{d\,E_\gamma}{E_\gamma}}$$

defined by LEVINGER and BETHE[3]. We will not go into the detail of these interesting problems nor will we report the results of calculations based on various models of the nucleus which are summarized in the review article by LEVINGER[4].

Calculations have been done with independent-particle models, subunit and collective models many of which in spite of their apparently large differences give values of E_m in reasonable agreement with each other and not in disagreement with experimental results.

Let us finally consider the data about the energy spectrum and the angular distribution of the neutrons emitted by nuclear photoeffect.

According to the Bohr assumption one expects that the process of decay of a compound nucleus into the various channels is independent of its mode of formation, and that the neutrons are emitted isotropically with a spectrum roughly represented by a Maxwell energy distribution determined by the nuclear temperature of the residual nucleus (Sects. 25 and 30). The protons are expected to have a much lower intensity since their emission is strongly reduced by the potential barrier.

---

[1] See footnote 1, p. 146.
[2] See footnote 3, p. 146.
[3] See footnote 1, p. 154.
[4] J. S. LEVINGER: Annual Rev. Nucl. Sci. 4, 13 (1954); see also footnote 1, p. 153.

HIRZEL and WÄFFLER[1] were the first to point out that in many cases the relative yield of the (γ, p) and (γ, n) reactions did not correspond to this picture: the proton intensity was found to be much stronger than expected.

This first result was confirmed and extended by many authors who have investigated the spectrum and the angular distribution of the particles emitted by various elements exposed to high energy X-rays. It is found that the spectrum contains more high energy particles than from the compound nucleus picture. While the part of the spectrum corresponding to the expected Maxwell distribution is isotropic, the high energy particles are emitted mainly at 90° with respect to the direction of the incident γ-rays.

In order to explain these results there has been suggested[1,2] the existence of an alternative mechanism of absorption of photons besides the formation of the compound nucleus. This mechanism is the absorption of the incident photon by a nucleon at the nuclear surface, which results in its direct ejection before the absorbed energy can be shared with the rest of the nucleus. The existence of such a "direct photoeffect" contradicts the Bohr assumption. One has to notice, however, that it contributes only a minor part to the γ-ray absorption cross section while the major part corresponds to the formation of a compound nucleus.

Because of the potential barrier which reduces the emission of protons at lower energies and higher atomic number Z, one can expect that the direct photoeffect is more important for protons than for neutrons, especially in the case of heavy elements.

COURANT[3] has calculated the direct photoeffect on the basis of an independent-particle model. The cross sections so obtained are an order of magnitude smaller than those measured by HIRZEL and WÄFFLER, but larger than those given by the statistical theory. Such a disagreement may be due to the crudity of the adopted model. Various modifications have been envisaged in order to improve the situation.

The most promising approach to this problem appears to be that made by WILKINSON[4]; he seeks an explanation of the giant resonance in the absorption of γ-rays arising from all the electric dipole transitions of the single nucleons from their initial state (belonging to closed shells as well as to unfilled shells of the ground state of the nucleus) to single-particle excited states. Once the nucleon has been elevated to a sufficiently high single-particle state, it may either interact with the rest of the nucleus and form an ordinary compound nucleus state which subsequently decays by the usual statistical process, or it may be emitted into the continuum without sharing its energy with other nucleons.

Making use of the shell model wave functions, WILKINSON shows that such an absorption is strong enough to account for the experimental observations and that there is such a strong clustering of the considered transitions to give the resonance form to the dipole absorption cross section of γ-rays. This theory explains satisfactorily the width of the giant resonance and its variation with mass number, but meets some difficulty with the absolute value of E_m.

The expected angular distribution resulting from the absorption of electric dipole radiation has the general form

$$A + B \sin^2 \vartheta \tag{48.13}$$

[1] O. HIRZEL and H. WÄFFLER: Helv. phys. Acta **20**, 373 (1947); see also R. K. SHELINE: Phys. Rev. **84**, 1064 (1954).

[2] P. JENSEN: Naturwiss. **35**, 190 (1948). — L. MARQUEZ: Phys. Rev. **81**, 897 (1951).

[3] E. D. COURANT: Phys. Rev. **82**, 703 (1951).

[4] D. H. WILKINSON: Physica, Haag **22**, 1039 (1956); see also footnote 1, p. 153.

which seems to represent satisfactorily the experimental results on the high-energy protons emitted in Rh and Ag[1]. The angular distribution and energy spectrum of photoneutrons has also been investigated by many authors[2].

According to BYERLY and STEPHENS[3] who have investigated the energy distribution of neutrons and the energy and angular distribution of charged particles (p, d, α) emitted by Cu irradiated with 24 Mev X-rays, about 10% of the neutrons and protons have energies greater than expected from the evaporation model.

A similar result has been obtained by PRICE[4] who has investigated the spectrum of the neutrons emitted at 90° by Bi and Ag irradiated with 22 Mev X-rays. Fig. 67 shows the spectrum obtained for Bi with emulsions. Above 4 Mev the observed spectrum is appreciably more intense than that calculated according to the evaporation picture of WEISSKOPF and EWING. It may be noticed here that

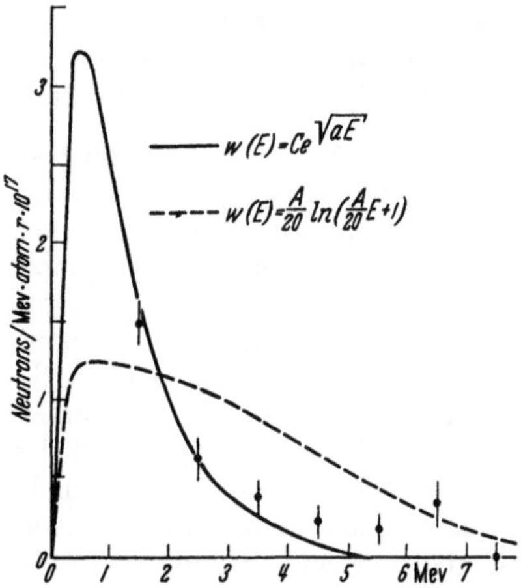

Fig. 67.

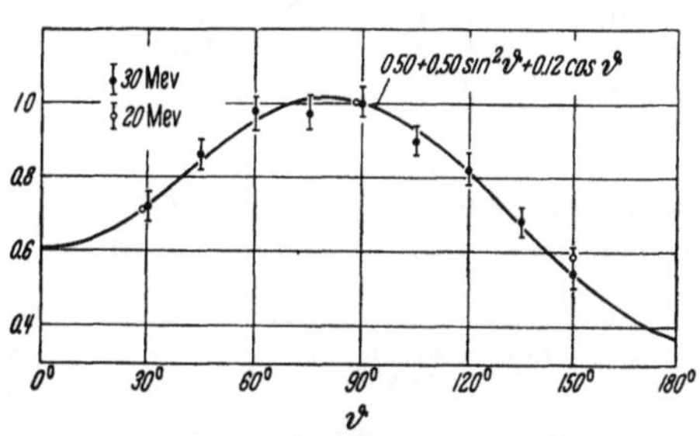

Fig. 68.

Fig. 67. Energy spectrum of photoneutrons emitted from bismuth when irradiated by 22 Mev bremsstrahlung. The solid and dashed curves show the predicted spectra as calculated on the statistical model of the nucleus with the indicated energy level densities (PRICE[4]).

Fig. 68. Angular distributions of photoneutrons from 20 and 30 Mev bremsstrahlung on Bi as observed with a $Si^{28}(n, p)$ threshold detector (Turin group[6]).

according to WILKINSON[5] the spectrum of the directly emitted neutrons decreases by increasing the neutron energy less rapidly than would do a Maxwellian distribution, but it is not concentrated at high energies.

PRICE has also measured the angular distribution of the emitted neutrons using different detectors: a detector of moderated neutrons whose efficiency is energy independent, and two threshold detectors $Al^{27}(n, p)\,Mg^{27}$ and $Si^{28}(n, p)\,Al^{28}$. The first reaction has a threshold at 1.95 Mev but is sensitive primarily to neutrons above 4 Mev; the second reaction has a threshold at 2.69 Mev but is probably most sensitive to neutrons above 5.5 Mev. He has found that the angular distribution can be represented in many cases, as for instance in Ag, Pb, and Bi, by Eq. (48.13), and that the ratio B/A is higher for the heavier elements and he more energetic neutrons. The Turin group[6] finds that the angular distribu-

[1] N.W. CURTIS, J. HORNBOSTEL, D.W. LEE and E.O. SALANT: Phys. Rev. **77**, 290 (1950). — B.C. DIVEN and G.M. ALMY: Phys. Rev. **80**, 407 (1950).

[2] G.A. PRICE and S.W. KERST: Phys. Rev. **77**, 806 (1950). — H.L. POSS: Phys. Rev. **79**, 539 (1950). — P.T. DEMOS, J.D. FOX, I. HALPERN and J. KOCH: Phys. Rev. **86**, 605 (1952).

[3] P.R. BYERLY and W.E. STEPHENS: Phys. Rev. **83**, 54 (1951).

[4] G.A. PRICE: Phys. Rev. **93**, 1279 (1954).

[5] See footnote 4, p. 156.

[6] F. FERRERO, L. GONELLA, A.O. HANSON, R. MALVANO and C. TRIBUNO: Nuovo Cim. **5**, 242 (1957).

tion of fast neutrons emitted from Al, Cu, Ag, Ba, Au and Bi and measured by means of the $Si^{28}(n, p) Al^{28}$ reaction, shows a forward component which needs the addition of a term of the form $C \cdot \cos\vartheta$ to Eq. (48.13) (see Fig. 68).

The existence of such an asymmetry, observed also by PRICE[1] for Au, seems to be well-established, but it is not well-understood in the light of all present theories. On the other hand various experimental results[2] seem to agree with WILKINSON's prediction.

It can be concluded that the spectrum and the angular distribution of photoneutrons in medium and heavy elements need further experimental investigation and theoretical consideration.

Table 25.

Gamma ray spectra ($h\nu > 1.666$ Mev) *of* Ra *and* MsTh *and their decay products.*

Gamma ray spectrum of Ra ($T_{\frac{1}{2}} = 1622$ y) *in equilibrium with its decay products* (RaC).

$h\nu$ (Mev) (a)	Relative intensities in arbitrary units (a)	Photons per disintegration (b)
1.690	0.40	0.0224
1.761	2.42	0.143
1.820	0.41	0.024
2.090	0.37	0.022
2.200	1.00	0.059
2.420	0.50	0.025

Gamma ray spectrum of MsTh ($T_{\frac{1}{2}} = 6.7$ y) *in equilibrium with its decay products* (ThC + C″).

$h\nu$ (Mev) (a)	Relative intensities in arbitrary units (a)	Photons per disintegration (b)
1.800	0.062	
2.200	0.0505	
2.620	1.00	1 × 0.65 (c)
3.20	0.1 (d)	

(a) G. D. LATISHEW: Rev. Mod. Phys. **19**, 132 (1947).
(b) G. D. LATISHEW: J. Exp. Theor. Phys. USSR **10**, 996 (1940).
(c) The factor 0.65 is due to the branching ratio between ThC′ and ThC″.
(d) H. NIEWODNICZANSKI and M. WIELOWIEJSKA: Bull. Acad. Polon. Sci. **1**, 293 (1953).

49. The R γ+D or R γ+Be sources. For a source of this type to be useful, a γ-ray emitter must give photons whose energy exceeds 1.666 Mev, the threshold of Be. In Table 25 are listed the γ-ray lines of radium and mesothorium, in equilibrium with their decay products, satisfying this condition[3]. In the case of Ra the lines reported in Table 25 are all due to RaC; therefore the same data can be used also in the case of Rn in equilibrium with all its products. Similarly in the case of MsTh the reported lines are due to ThC+C″.

As we have already mentioned in Sect. 45 many artificial radioactive nuclei can be used in photoneutron sources.

Table 26, similar to that given by WATTENBERG[4], contains a few radioisotopes chosen for their high intensity γ-radiation and relatively long half-life. The energy and intensity of the emitted photons, the nuclear reactions in which the radioisotopes can be produced, and the cross section for their production through radiative capture of slow neutrons, are also given in the same table. With the exception of a few cases (Zn^{63}, Y^{88}) the radioisotopes listed in Table 26 can be produced by radiative capture of slow neutrons and therefore, can be obtained in large amounts from chain-reacting piles.

Table 27 gives the energies of the emitted neutrons from all the sources obtained by irradiating D or Be with the photons of the various γ-emitters listed in Tables 25 and 26. The theoretical estimate of the energy of the neutrons is obtained from Eq. (34.11) for $\vartheta = 90°$, while in the third column are given the corresponding

[1] See footnote 1, p. 157.
[2] See footnote 6, p 157.
[3] G. D. LATISHEV: Rev. Mod. Phys. **19**, 132 (1947).
[4] A. WATTENBERG: Preliminary Report No. 6, Nuclear Science series. National Research Council, Washington D.C.

Table 26. *Gamma ray emitters useful for photoneutron sources.*

Radio-isotope	Half life (a)	$h\nu > 1.666$ Mev and photons per disintegration (a)	Nuclear reactions for producing isotopes listed in the first column	$\sigma(n,\gamma)\times 10^{24}\,cm^2$
$_{11}Na^{24}$	14.9 h	2.757 1.0	Mg (n, p); Mg (d, α); Na (d, p); Na (n, γ); Al (n, α)	0.41 (c)
$_{13}Al^{28}$	2.27 min	1.782 1.0	Mg (n, d); Mg (α, p); Na (α, n); Al (d, p); Al (n, γ); Si (n, p); daughter Mg^{28}	0.23 (b)
$_{17}Cl^{38}$	37.29 min	2.15 0.47	Cl (d, p); Cl (n, γ); K (n, α)	0.15 (b)
$_{25}Mn^{56}$	2.576 h	1.77; 2.06; 2.88 0.30; 0.20; $\approx$ 0.001	Co (n, d); Cr (α, p); Mn (n, γ); Mn (d, p); Fe (d, α); Fe (n, p)	10.7 (b)
$_{30}Zn^{63}$	38.3 min	1.89; 2.60 0.04; 0.005	Ni (α, n); Cu (p, n); Cu $(d, 2n)$; Zn $(n, 2n)$; Zn (γ, n)	—
$_{31}Ga^{72}$	14.3 h	1.87; 2.21; 2.51 0.08; 0.33; 0.26	Ga (d, p); Ga (n, γ); Ge (n, p); fission Th; fission U; daughter Zn^{72}	1.3 (b)
$_{33}As^{76}$	26.1 h	1.77; 2.06 $\approx$ 0.20; 0.10	Ge (p, n); As (d, p); As (n, γ); Se (n, p); Se (γ, p); Se (d, α); Br (n, α)	4.6 (b)
$_{39}Y^{88}$	104 d	1.853; 2.76 $\approx$ 0.99; $\approx$ 0.01	Sr $(d, 2n)$; Sr (p, n); Y $(n, 2n)$ daughter Zn^{88}	—
$_{49}In^{116}$	53.93 min	2.090 0.25	Cd (p, n); In (n, γ); In (d, p)	150 (b)
$_{51}Sb^{124}$	60 d	1.70 (d) 0.5	Sn (p, n); Sb (d, p); Sb (n, γ); Te (d, α); fission Bi	1.1 (b)
$_{57}La^{140}$	40.0 h	2.51; 3.0 0.04; $\approx$ 0.001	La (d, p); La (n, γ); Ce (n, p); fission Th; fission U; daughter Ba^{140} ($T_{\frac{1}{2}} = 12.8$ d)	7.0 (c)

(a) J. M. HOLLANDER, I. PERLMAN and G. T. SEABORG: Rev. Mod. Phys. **25**, 469 (1953).
(b) L. SEREN, H. N. FRIEDLANDER and S. H. TURKEL: Phys. Rev. **72**, 888 (1947).
(c) A. WATTENBERG: Preliminary Report No. 6, Nuclear Science Series. National Research Council, Washington D.C.
(d) The energy of this line has been reported as (1.708 ± 0.017) Mev by KERN, ZAFFARANO and MITCHELL [Phys. Rev. **73**, 1142 (1948)] and COOK and LANGER [Phys. Rev. **73**, 1149 (1948)], as (1.69 ± 0.02) Mev by FEISTER and CURTISS [J. Res. Nat. Bur. Stand. **40**, 315 (1948)] and as (1.692 ± 0.005) Mev by ALLRIEGER [Phys. Rev. **88**, 1257 (1952)]. Previous authors had found indication of a weak line of 2.04 Mev: see footnote (c).

experimental values obtained by various authors with the methods indicated in the footnotes below the table.

According to Eq. (34.11) the energy of the photoneutrons varies very little, in comparison with other reactions, with the angle of emission. This is due to the fact that the coefficient of $\cos\vartheta$ is small compared with the constant term. From the two following examples

$$\text{Th C}\gamma + \text{D}: \quad h\nu = 2.623\ \text{Mev} \quad Q_0 = -2.226\ \text{Mev}$$
$$E = (196 + 0.27\cos\vartheta)\ \text{kev}$$
$$\text{Na}^{24}\gamma + \text{Be}: \quad h\nu = 2.753\ \text{Mev} \quad Q_0 = -1.666\ \text{Mev}$$
$$E = (966 + 14\cos\vartheta)\ \text{kev}$$

one sees that even if the γ-radiation falls isotropically on the target, the relative spread of the energy of the emitted neutrons corresponds to $\pm 14\%$ in the case of D and $\pm 1.5\%$ in the case of Be. By increasing E the relative spread decreases, so that for almost any monochromatic γ-ray emitter, a Be(γ, n) source is expected to be monoenergetic even if the incident photons are isotropic.

For D the situation is not so good as a consequence of the lighter mass, but the theoretical spectrum of the emitted neutrons is still a rather narrow line.

In principle, this effect can be eliminated, or at least reduced, by designing the experimental arrangement so that the angles subtended by the target at the source and at the neutron detector are both sufficiently small. Such an

Table 27. *Energies and yields of photoneutrons* (a).

(γ,n) sources	Calculated energy (b) (Mev)	Observed energy (Mev)	Neutrons per sec per Curie (1 g of target at 1 cm) $\times 10^{-4}$
Na^{24} + Be	0.966	0.83 (c)	14 (g) 12 (h)
Na^{24} + D	0.261	0.22 (c)	29 (g) 24 (h) 27 (i)
Al^{28} + Be	0.103	—	
Cl^{38} + Be	0.430	—	
Mn^{56} + Be	0.092; 0.350; 1.076	0.15; 0.30 (d)	2.9 (g)
Mn^{56} + D	0.320	0.22 (c)	0.31 (g)
Ga^{72} + Be	0.181; 0.484; 0.750	—	5.9 (g) 3.7 (h)
Ga^{72} + D	0.140	0.13 (c)	6.9 (g) 4.6 (h)
As^{76} + Be	0.089; 0.350	—	
Y^{88} + Be	0.166; 0.972	0.22 (e) 0.16 (f)	10 (i)
Y^{88} + D	0.265	—	0.3
In^{116} + Be	0.377	$>$0.15; 0.30 (d)	0.82 (g)
Sb^{124} + Be	0.031 (m)	0.025 (c), (f) 0.035 (d)	19 (g)
La^{140} + Be	0.747	0.62 (c)	0.23 (g) 0.34 (h)
La + D	0.140	0.13 (c) 0.15 (f)	0.68 (g) 0.97 (h)
MsTh + Be	0.119; 0.475; 0.848 (b)	0.83 (f)	3.5 (j)
MsTh + D	0.196 (b)	0.197 (f)	9.5 (j)
Ra + Be	0.021; 0.084; 0.137 0.377; 0.475; 0.670 (b)	—	1.2 (k)
Ra + D	0.096 (b)	0.12 (e)	0.1 (l)

(a) The intensity of the various groups of neutrons depends on the corresponding number of photons per disintegration (see Table 25 and Table 26).

(b) Differences with similar tables are due to small differences in the adopted values of $h\nu$ and Q_0.

(c) A. WATTENBERG: Phys. Rev. **71**, 497 (1947); from scattering cross section in hydrogen.

(d) D. J. HUGHES and C. EGGLER: Phys. Rev. **72**, 902 (1947): from recoil protons produced in a cloud chamber.

(e) R.D. O'NEAL: Phys. Rev. **70**, 1 (1946). — G. SCHARFF-GOLDHABER: Phys. Rev. **59**, 937 (1941): from spatial distribution of slow neutrons in water.

(f) A.O. HANSON: Phys. Rev. **75**, 1795 (1949): from recoil protons produced in a hydrogen filled proportional counter.

(g) B. RUSSEL, D. SACHS, A. WATTENBERG and R. FIELDS: Phys. Rev. **73**, 545 (1948).

(h) SNELL, BARKER *et al.*: See WATTENBERG, footnote 4 on p. 158.

(i) E. SEGRÈ: See WATTENBERG, footnote 4 on p. 158.

(j) L. G. ELLIOT: See WATTENBERG, footnote 4 on p. 158.

(k) F. G. HOUTERMANS and I. BARTZ: Phys. Z. **44**, 167 (1943). — R.D. O'NEAL and G. SCHARFF-GOLDHABER: Phys. Rev. **69**, 368 (1946). — G.R. GAMERTSFELDER and M. GOLDHABER: Phys. Rev. **69**, 368 (1946).

(l) O.R. FRISCH and R. WALKER: Absolute Calibration of a Ra-Be neutron source. Report of subcommittee on Neutron Measurements and standards.

(m) A second neutron line of 330 kev has been suggested by some authors: see footnote (d) to Table 26.

arrangement, however, greatly reduces the photoneutron emission which, as a consequence of the very low yield of the (γ, n) reactions, is already rather small.

The small yield of these reactions not only requires that in general the source be constructed with a poor geometry as explained above, but also that a rather thick layer of neutron emitter — metallic beryllium or heavy water or heavy paraffin — be used.

It follows that the photoneutrons have an appreciable probability of suffering a collision before escaping from the source. It was pointed out by HUGHES and EGGLER[1] that, because of the low atomic weight of deuterium and beryllium, they are both good moderators and therefore distort the neutron energy spectrum even when present in limited amounts. As a consequence, the resulting energy spectrum will depend on the construction of the source and in general the observed mean energy will be lower and its spread larger than given by Eq. (34.11). With the source described later (Fig. 69), Hughes and Eggler observed a spread in energy of about 25% which they interpreted as due to this effect.

In the last column of Table 27 are given the photoneutron yields of the various sources according to WATTENBERG[2]. Because of the dependence of the neutron emission on the size and shape of the source, the reported data refer to the case of 1 Curie of γ-emitter placed at 1 cm from 1 gram of beryllium metal or 1 gram of heavy water (standard sources). The large variation in the values of the neutron yields is due mainly to the disintegration scheme of the various radioisotopes. Frequently the emission of a useful γ-ray constitutes only a very small percentage of the decay processes (see column 3 of Table 26).

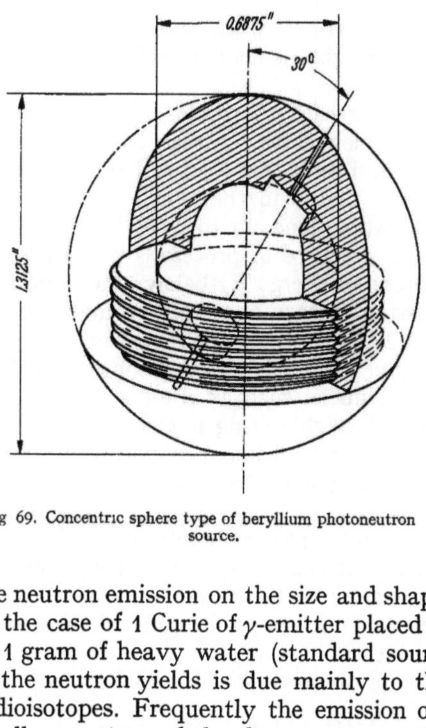

Fig 69. Concentric sphere type of beryllium photoneutron source.

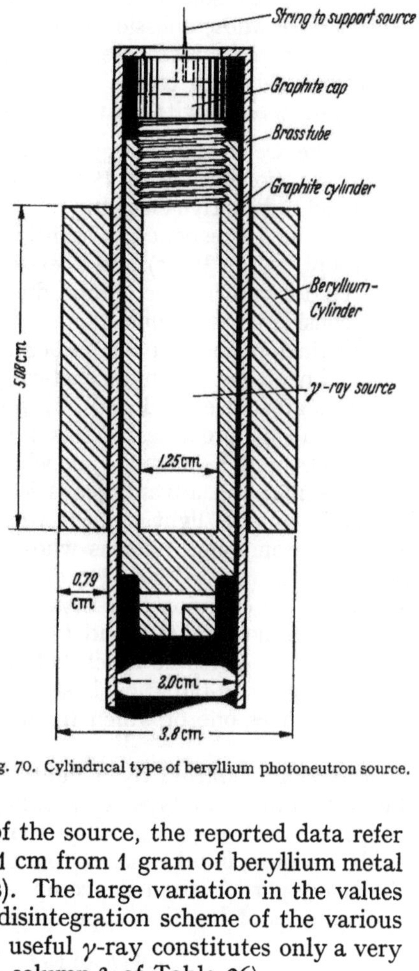

Fig. 70. Cylindrical type of beryllium photoneutron source.

From a practical point of view the source ($_{51}Sb^{124}\,\gamma + Be$) is particularly important because Sb^{124} combines a non negligible (n, γ) cross section with a long halflife ($T_{\frac{1}{2}} = 60$ d) and a good neutron yield. If a pile is available, Sb^{124} can be prepared in amounts of the order of many Curie which is sufficient to give a relatively strong source — of the order of 10^7 neutrons per second over the entire solid angle. As we have seen on p. 133, even strengths of the order of 10^{10} neutrons/sec have been obtained.

[1] D. J. HUGHES and C. EGGLER: Phys. Rev. **72**, 902 (1947).
[2] See footnote 4, p. 158.

Fig. 69 and 70 show the design of two typical (γ, n) sources. The design of such a source is usually a compromise between the necessity of increasing the thickness of the target element to increase the yield, and the desire of keeping it as small as possible not only to reduce the moderation of the photoneutrons in the target itself (see above), but also because the small physical dimensions of the source are important in experiments in which one would like to have available a point-source of neutrons.

The spherical source (Fig. 69) has the advantage of symmetry over the cylindrical source whose emission is anisotropic. This latter has, however, the advantage of easy and quick assembly and disassembly, a feature which reduces the physiological hazard to the operator, especially if one wants to test the effects of the γ-rays alone without the Be target.

50. The use of high-energy γ-rays emitted from excited states of light elements. As was mentioned in Sect. 45 γ-rays with higher energy than those emitted from long-lived radioactive elements can be obtained from a few excited levels of light nuclei. The corresponding half-lives, however, are always very short (since $\Gamma \gtrsim 1$ kev, $\tau \lesssim 10^{-17}$ sec) and therefore the investigation and use of these γ-rays must be done during the bombardment of the light element in which the excited state considered is produced.

In principle one can use as bombarding particles α-particles, deuterons, protons and neutrons. From a practical point of view, however, the more convenient processes are those produced by protons, and only this case will be discussed. More detail on processes of this type can be found in the article by FOWLER, LAURITSEN and LAURITSEN, in which the work done up to 1948 is reviewed[1].

A peculiar and advantageous feature of this type of process is that by bombarding a suitable light element with protons of energy E_p below 1 Mev one can produce a compound nucleus whose excitation energy above its ground state is very high, of the order of 10, or perhaps 20 Mev. Once such a state has been produced, the emission of γ-rays in the same range of energy becomes possible. A situation like this is found for compound nuclei such as ${}_4Be^8$, ${}_6C^{12}$, and ${}_8O^{16}$ whose ground states lie much below those of neighbouring nuclei.

The emitted γ-radiation has also the advantage of being made up of a very few sharp lines one of which frequently has a much greater intensity than the others.

The γ-rays emitted by a light element under proton bombardment can be due to two different types of processes. In some cases they are emitted from the compound nucleus produced by the combination of the target nucleus and the incident proton, in other cases from the residual nucleus produced by the emission of a heavy particle from the compound nucleus.

Examples of the first type of process are observed in C^{12} and Li^7:

$$C^{12}(p,\gamma)\,N^{13} \quad \text{and} \quad Li^7(p,\gamma)\,Be^8,$$

while processes of the second type occur in N^{15} and F^{19}:

$$N^{15}(p,\alpha)\,C^{12*} \quad \text{and} \quad F^{19}(p,\alpha)\,O^{16*}.$$

In these last two examples the γ-rays are emitted from the residual nucleus which, as indicated by the star attached to the corresponding symbols, are frequently left in an excited state.

When the photon is emitted from a compound nucleus its energy depends linearly on the energy E_p of the incident proton. This can be seen by applying

[1] W.A. FOWLER, C.C. LAURITSEN and T. LAURITSEN: Rev. Mod. Phys. **20**, 236 (1948).

the energy and momentum conservation laws to a (p, γ) process. Following the procedure given in Sect. 34 one finds

$$h\nu = Q_S + E_p \frac{M_2 - m_1}{M_2}, \tag{50.1}$$

where m_1, M_2 are the masses of the proton and the residual nucleus and Q_S refers to the state in which the residual nucleus remains after emission of the γ-ray. The term in E_p is small. It has, however, to be considered in connection with the width of the emitted line. With an infinitely thin target and incident protons of infinitely well defined energy E_p, the emitted line would be infinitely sharp, but any spread in energy of the incident protons at the instant of collision will produce through this term a corresponding width of the γ-line. The fact that, as we will see later, the cross section of the processes considered shows sharp resonances, limits the width of the emitted radiation even in the case of thick targets to the width of the corresponding level of the compound nucleus.

When the γ-radiation is emitted from excited states of a residual nucleus, $h\nu$ is not affected by the energy E_p of the incident particle. E_p affects the energy of the primary disintegration (in the examples considered above the emission of the α-particles) but not the subsequent decay (emission of photons) apart from recoil effects which are always negligible. The differential excitation curves for the γ-rays emitted by light elements under proton bombardment show sharp resonances which, after correction for various experimental causes of broadening (mainly spread in energy of the incident protons and finite thickness of the target: Sect. 38) can be represented by the dispersion formula (see Sect. 25 and 26) by adjusting suitably the various parameters appearing in it.

The total width Γ so obtained represents the width of the corresponding state of the compound nucleus which is determined by all its decay processes. If photon emission is among these, $\Gamma \frac{M_2 - m_1}{M_2}$ represents, according to Eq. (50.1), the width of the corresponding γ-line emitted in the case of a thick target.

When the photons are emitted from the residual nucleus, the widths of the corresponding lines are obviously not related to the widths of the resonances of the compound nucleus. They are determined by the mean life of the excited state of the residual nucleus.

Table 28 contains a few data about the more important processes of the type considered here[2] [2].

Table 28. *A few examples of thick target yields for emission of γ-rays by light elements under proton bombardment.*

Reaction	E_p (kev)	$h\nu$ (Mev)	Yield: photon/proton	Target
$N^{15}(p, \alpha)\, C^{12}$	898	4.43	5×10^{-8} [1]	KNO_3 containing 61% N^{15}
$F^{19}(p, \alpha)\, O^{16}$	340	6.14	1.70×10^{-8} [2]	CaF_2
$Be^{9}(p, \gamma)\, B^{10}$	995	7.5	1.9×10^{-8} [3]	Be
$C^{13}(p, \gamma)\, N^{14}$	554	8.1	1.8×10^{-10} [2]	graphite
$B^{11}(p, \gamma)\, C^{12}$	675	16; 12+4	10^{-9} [3]	B_2O_3
$Li^{7}(p, \gamma)\, Be^{8}$	441	17; 14.4; 12	5×10^{-9} [2]	LiOH

[1] C.A. Barnes, J.A. Carver, G.H. Stafford and D.H. Wilkinson: Phys. Rev. 86, 359 (1952).

[2] See footnote 1, p. 162.

[3] A. Schardt, W.A. Fowler and C.C. Lauritsen: Phys. Rev. 86, 527 (1952).

Before making a few comments on the reactions listed in Table 28 it is convenient to compare the intensity of a photoneutron source in which the γ-rays are emitted in one of the reactions listed there with the intensity of other neutron sources, for instance a source of type $R\gamma + A$ (see Sect. 49, Table 27). In order to do this we can use the last column of Table 24 (p. 152) which contains the cross section of various elements for the γ-radiation emitted in the $Li(p, \gamma)$ reaction.

If one irradiates 1 g of uranium metal placed at 1 cm from a thick LiOH target, the total neutron emission corresponds to 3×10^3 neutrons per sec per milliampere ($\triangleq 6.25 \times 10^{15}$ protons/sec) and per gram of irradiated uranium. In spite of the large intensity of the incident protons the neutron intensity is very low as one easily recognizes by comparison of this value with those given in the last column of Table 27 which contains the neutron yields per Curie of activity ($\triangleq 3.7 \times 10^{10}$ disintegrations/sec) of the radioactive substance used and per gram of target element (D and Be).

Table 29. *Resonances of* $N^{15}(p,\alpha)\,C^{12}$ *reaction:* $h\nu = 4.43$ Mev [1].

E_p (kev)	Γ (kev)	$\sigma_{a0} \times 10^{24}$ (cm²)	Thick target yield[1] photon/proton
429	0.9	0.3	4×10^{-9}
898	2.2	0.8	5×10^{-8}
1.210	22.5	0.3	1.8×10^{-7}

[1] Yield for thick KNO_3 target containing 61 atom percent of N^{15}.

Photoneutron sources based on the use of photons produced in some (p, γ) reaction are therefore employed only in special cases; they are more useful for the determination of the cross section for the photoeffect at well defined energies (see Sect. 46 to 48) than for producing an intense flux of neutrons.

The excitation curve of the (p, α) process in N^{15} shows various resonances some of which lead predominantly to C^{12} in its ground state, while others reported in Table 29 leave always or almost always C^{12} in an excited state ($J=2^+$) of 4.43 Mev[1]. No other lower state of C^{12} exists and, therefore, the emitted radiation consists of a single sharp line. The radiative capture of protons takes place in a broad state ($\Gamma \approx 180$ kev) at 1.05 Mev giving γ-rays of 12.1 Mev, but the corresponding intensity ($\sigma \approx 1$ mb) is insignificant relative to that of the 4.4-Mev line at proton energies of about 1 Mev[2] [*1*], [*2*].

The angular distribution of the 4.4 Mev γ-rays is strongly anisotropic [*2*] and roughly represented[2] at 890 kev by $(1 + 0.3 \cos^2 \vartheta)$.

Fluorine bombarded with protons gives 5 groups of α-particles: one, (α_0), leaves O^{16} in its ground state ($J=0^+$), the second (α_π) in a state ($J=0^+$) at 6.06 Mev which decays by emission of electron pairs, while the three remaining groups $(\alpha_1, \alpha_2, \alpha_3)$ lead to γ-ray emitting states at 6.14 ($J=3^-$), 6.91 ($J=2^+$) and 7.12 ($J=1^-$) Mev [*1*], [*2*]. The resonances for $\alpha_1, \alpha_2, \alpha_3$ are different from those for α_0 and α_π. The data relative to resonances of $\alpha_1, \alpha_2, \alpha_3$ located below $E_p = 1$ Mev, are given in Table 30, which contains the resonance energies, the total widths Γ, the cross sections at resonance σ_{a0} and the ratio of the total number (i.e. integrated over the width of the resonance) of α_1 to $\alpha_2 + \alpha_3$[3]. The angular distribution is iotropic only for some resonances, while, for others, it shows a strong anisotropy [*2*]. From these data one sees, for instances, that if one uses the resonance at 340 kev, 97.5% of the particle transitions take place to the 6.14 Mev level and only the remaining 2.5% to states at 6.91 and 7.12 Mev. If under these conditions one observes the α-particles emitted at 90° to the incident protons, the contri-

[1] See footnote 3, p. 163.
[2] See footnote 1, p. 163.
[3] J.M. FREEMAN: Phil. Mag. **41**, 1225 (1950).

bution of $\alpha_0+\alpha_\pi$ is less than 1% of that of $\alpha_1+\alpha_2+\alpha_3$[1]. Furthermore both α-particles and γ-rays are emitted isotropically [2].

These circumstances make possible an accurate measure of the γ-ray flux by counting the α-particles emitted at 90° to the proton beam. This method, first used by VAN ALLEN and SMITH[2], has been further developed by WILKINSON and coworkers[3].

Radiative capture of protons in Be^9 also shows several resonances. The corresponding γ-ray spectra are normally composed of many lines of comparable intensity [2]. But if one uses the broad resonance at $E_p=0.995$ Mev, the compound nucleus B^{10} is formed in a state ($J=2^-$, $T=1$) at 7.48 Mev which decays mainly to the ground state with a thick target yield of $19\times10^{-9}\,\gamma/p$. The angular distribution of the γ-radiation is given by the expression

$$Y(\vartheta) \propto 1 + 0.09 \sin^2\vartheta$$

which suggests a dominant S-wave formation with some D-wave contribution. With protons of $E_p=1.086$ Mev one observes a sharp resonance which decays to the ground state mainly via a state at 0.72 Mev.

With a thick beryllium target the emission of γ-rays is accompanied by a rather intense background of fast neutrons due to second order reactions of the following types: $Be^9(p, d)\,Be^8 - Be^9(d, n)\,B^{10}$ and $Be^9(p, \alpha)\,Li^6 - Be^9(\alpha, n)\,C^{12}$[3].

Table 30. *Resonances below* 1 Mev *of the* $F^{19}(p, \alpha\gamma)\,O^{16}$ *reaction* [2].

E_p (kev)	Γ (kev)	$\sigma_a 0$ (10^{-27} cm²)	$\frac{\alpha_1}{\alpha_2+\alpha_3}$	Angular distribution of γ-rays
224	1.0	>0.2		
340	2.9	160	40	isotropic
483	2.2	>32	3.8	isotropic
598	37	7.1	200	
669	7.5	57	4.6	isotropic
831	8.3	19	—	
873	5.2	540	2.6	
900	4.8	23		
935.3	8.0	180	2.9	isotropic

The radiative capture of protons in C^{13} has many resonances of which only one ($E_p=0.554$ Mev) lies below 1 Mev [2]. Under these conditions the compound nucleus N^{14} is formed in a state ($J=1^-$; $T=1$) which decays mainly into the ground state emitting a single line of 8.1 Mev. Transitions to intermediate states appear to have a much lower intensity[3]. The radiation is isotropic.

The radiative capture of protons by B^{11} comprises two components: one line of 16 Mev corresponding to the ground level transition and a cascade through a state ($J=2^+$) at 4.4 Mev which gives two lines of about 12 and 4 Mev energy. Below 1 Mev there are two resonances whose parameters are collected in Table 31.

The angular distribution at $E_p=0.16$ Mev of the 12 Mev and 16 Mev radiations are given by the expressions

$$h\nu = 12\text{ Mev}, \qquad Y(\vartheta) \propto 1 + 0.25\cos^2\vartheta,$$

$$h\nu = 16\text{ Mev}, \qquad Y(\vartheta) \propto 1 - 0.19\cos\vartheta + 0.21\cos^2\vartheta.$$

By changing the energy of the incident protons, the coefficients of these expressions change in a rather complicated way. In particular a term in $\cos\vartheta$ enters into the expression for $Y(\vartheta)$ at $h\nu=12$ Mev.

[1] J.F. STEIB, W.A. FOWLER and C.C. LAURITSEN: Phys. Rev. 59, 253 (1941).
[2] J. VAN ALLEN and N. SMITH: Phys. Rev. 59, 618 (1941).
[3] See footnote 1, p. 163.

When B^{11} is bombarded with protons the reaction $B^{11}(p, \alpha)\,Be^8$ also takes place. For $E_p \leqq 1$ Mev this process contributes only very little to the γ-radiation, as Be^8 is formed mainly in a state at about 3 Mev which decays by α-particle emission. Only a small number of transitions go to a γ-emitting state of Be^8 at 4.9 Mev. The corresponding line is superimposed on the 4.4 Mev line of C^{12}, and the equality of the intensities of the 16 Mev and 4.4 Mev lines proves that the contribution of the 4.9 Mev line is small. The B^{10} contained in a target of natural boron gives a γ-ray of 0.43 Mev due to the first excited state of Be^7 produced in the reaction $B^{10}(p, \alpha)\,Be^7$.

Furthermore the γ-radiation is accompanied by a rather large background of neutrons due to the second order reactions[1]: $B^{11}(p, \alpha)\,Be^8$—$B^{10,11}(\alpha, n)$.

The radiative capture of protons in Li^7 shows a sharp resonance ($\Gamma = 12$ kev) at $E_p = 0.441$ Mev. The γ-radiation comprises two main components $h\nu = (17.2 \pm 0.2)$ Mev and $h\nu = (14.4 \pm 0.3)$ Mev and probably a third line at $h\nu = 12$ Mev, whose relative intensities vary with the bombarding energy[2,3]. With a thick target and a bombarding energy of 0.510 Mev $I_{17.2}:I_{14.4}:I_{12} = 1:0.55:0.1$[1]. At the 441 kev resonance the 17.2 and 14.4 Mev components are individually isotropic within 6% [2].

The γ-radiation is accompanied by a strong neutron background due to the second order reactions $Li^7(p, \alpha)\,He^4$—$Li(\alpha, n)$.

Table 31. *Resonances in* $B^{11}(p, \gamma)\,C^{12}$ *reaction* [2].

E_p (Mev)	Γ (kev)	σ_{a0} (10^{-30} cm²) $h\nu = 16$ Mev	σ_{a0} (10^{-30} cm²) $h\nu \approx 12$ and 4 Mev
0.163	5	5.5	152
0.675	322	< 2.3	48

51. The use of bremsstrahlung of high energy electrons. The energy and angular distribution of the bremsstrahlung emitted by an electron moving in a pure Coulomb field has been calculated exactly by SOMMERFELD for non-relativistic energies and by BETHE and HEITLER, in Born approximation, for relativistic energies[4]. This last case was reconsidered recently without use of the Born approximation[5].

All these calculations, however, do not take into account the effect of the screening of the electric field of the nucleus by the charge of the outer electrons. This effect was treated for lead by BETHE[6], who represented the atomic field by the Thomas-Fermi model, and gave the results in the form of numerical tables.

A simple formula very convenient for representing the bremsstrahlung spectra produced by high energy electrons incident on a thin target of atomic number Z, was deduced by SCHIFF[7,8].

In the calculation it is assumed that the energies E_0 and E of the incident and scattered electrons, as well as that of the radiated quantum, $E_\gamma = E_0 - E$, are

[1] See footnote 1, p. 163.
[2] R.L. WALKER and B.D. MACDANIEL: Phys. Rev. **74**, 315 (1948).
[3] H. NABHOLZ, P. STOLL and H. WÄFFLER: Helv. phys. Acta **25**, 153 (1952).
[4] W. HEITLER: The quantum theory of radiation, third edit., p. 242. Oxford 1954.
[5] H.A. BETHE and L.C. MAXIMON: Phys. Rev. **93**, 768 (1954).
[6] H.A. BETHE: Proc. Cambridge Phil. Soc. **30**, 524 (1934); and footnote[4].
[7] L.I. SCHIFF: Phys. Rev. **70**, 87 (1946); **83**, 252 (1951).
[8] This author takes into account the effect of the screening by representing the scattering atom by the potential $\frac{Ze}{r} e^{-r/a}$. a is chosen to be inversely proportional to $Z^{\frac{1}{3}}$ according to the Thomas-Fermi model and the constant of proportionality is adjusted to give, in the case of complete screening, the above mentioned results obtained by BETHE with numerical calculations based on the Thomas-Fermi model.

all large in comparison with the rest energy $m_e c^2$ of the electron. This assumption is always reasonably well satisfied under the experimental conditions involved in the photoproduction of neutrons in medium and heavy elements.

The differential cross section as a function of the energy $E_\gamma = h\nu$ and the angle of emission ϑ is given by the expression

$$\left.\begin{aligned}\sigma(E_\gamma, x)\,dx = \frac{4Z^2}{137}\left(\frac{e^2}{m_e c^2}\right)^2 \frac{dE_\gamma}{E_\gamma}\,dx\Big\{\frac{16x^2}{(x^2+1)^4}\,\frac{E}{E_0} - \frac{(E_0+E)^2}{(x^2+1)^2 E_0^2} + \\ + \left[\frac{E_0^2+E^2}{(x^2+1)^2 E_0^2} - \frac{4x^2 E}{(x^2+1)^4 E_0}\right]\ln M(x)\Big\}\end{aligned}\right\} \tag{51.1}$$

where

$$\frac{1}{M(x)} = \left(\frac{m_e c^2 E_\gamma}{2E_0 E}\right)^2 + \left(\frac{Z^{\frac{1}{3}}}{C(x^2+1)}\right)^2; \quad x = \frac{\vartheta}{\vartheta_0}; \quad \vartheta_0 = \frac{m_e c^2}{E_0} \tag{51.2}$$

and C is a constant which in the first paper was taken equal to 191, but was later corrected by Schiff himself to the value $C = 111$[1].

Eq. (51.1) is not correct at angles $\vartheta > \vartheta_0$, an error which is not serious, since most of the radiation is emitted at angles $\vartheta < \vartheta_0 (x < 1)$. Under certain experimental conditions the solid angle subtended by the detector is sufficiently large that it is necessary to consider the spectrum integrated over the angle ϑ. By extending the integration with respect to dx from zero to infinity, which is justified on account of the remark made above, SCHIFF obtains the so-called integrated spectrum

$$\left.\begin{aligned}dE_\gamma \int_0^\infty \sigma(E_\gamma, x)\,dx \\ = \frac{2Z^2}{137}\left(\frac{e^2}{m_e c^2}\right)^2 \frac{dE_\gamma}{E_\gamma}\Big\{\left[\frac{E_0^2+E}{E_0^2} - \frac{2E}{3E_0}\right]\left[\ln M(0) + 1 - \frac{2}{b}\arctan b\right] + \\ + \frac{E}{E_0}\left[\frac{2}{b^2}\ln(1+b^2) + \frac{4(2-b^2)}{3b^3}\arctan b - \frac{8}{3b^2} + \frac{2}{9}\right]\Big\}\end{aligned}\right\} \tag{51.3}$$

where

$$b = \frac{2E_0 E Z^{\frac{1}{3}}}{C m_e c^2 E_\gamma}. \tag{51.4}$$

Eqs. (51.1) and (51.3) are correct for complete screening, i.e. when[2]

$$\gamma = 100\frac{m_e c^2 E_\gamma}{E_0 E}\,\frac{1}{Z^{\frac{1}{3}}} \approx 0$$

and for no screening, i.e., when

$$\gamma = 100\frac{m_e c^2 E_\gamma}{E_0 \cdot E}\,\frac{1}{Z^{\frac{1}{3}}} \gg 1,$$

while they give larger values than they should, though never in error by more than 4%, in the worst case of large Z and energies such that the screening is incomplete (γ of the order of 1—10).

While the significance of Eqs. (51.1) and (51.3) and the importance of the approximations involved in their derivation are very clear, the problem of establishing what formula should be used in order to represent sufficiently well the bremsstrahlung spectrum emitted under certain experimental conditions, is rather involved.

These equations refer to a thin target while, at least in the case of betatrons and synchrotrons, the target is always thick due to the many successive passages of electrons through the target itself.

---

[1] See footnote 7, p. 166.

[2] B. ROSSI: High energy particles, p. 48. New York: Prentice Hall 1952.

Furthermore the multiple scattering of the electrons crossing a thick target gives rise to a spread of their angular distribution which is wider than the angular distribution of the corresponding bremsstrahlung.

If the detector subtends a very small solid angle this last effect tends to make the spectrum observed with a thick target more nearly equal to that with a thin target because the photons emitted in the second and third collisions will be frequently lost because they will be moving at too large angles.

As a conclusion of various qualitative considerations of this type[1], it became customary to represent the spectrum emitted by a betatron or synchrotron by Eq. (51.1) with $x=0$ and $C=191$. In other words, one uses the differential spectrum emitted from a thin target in the forward direction, but with a modified value of the constant C, to obtain in a rough and empirical way

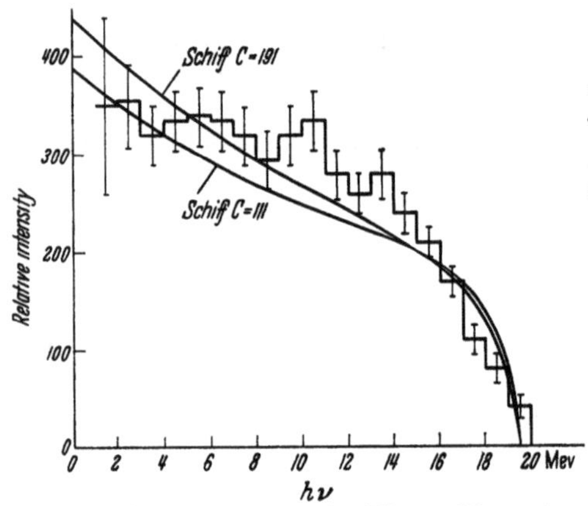

Fig. 71. The experimental results of KOCK and CARTER for the shape of the bremsstrahlung intensity curve. ($E_0 = 19$ Mev).

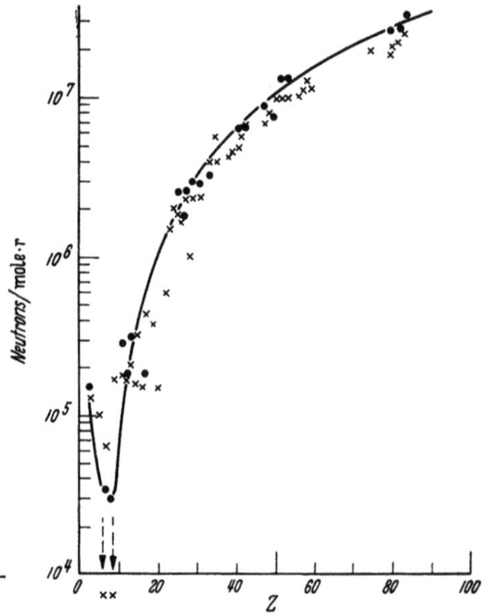

Fig. 72. Neutron yields with betatron operating at 22 Mev: × PRICE and KERST. ● MONTALBETTI *et al.*

an additional contribution to the lower energy portion of the bremsstrahlung curve as one would expect by considerating multiple traversals and scattering of electrons.

A justification of such a procedure can be seen in Fig. 71 taken from the above mentioned paper by KATZ and CAMERON. The experimental data obtained by KOCH and CARTER[2] for electrons of $E_0 = 19.5$ Mev, are compared with Eq. (51.1) for $x=0$ and the two values of the constant C. The two theoretical curves have been fitted to the experimental points so as to have the same number of photons between 13.0 and 19.5 Mev. The curve calculated with $C=191$ seems to represent the experimental points better.

Other experimental work is in agreement with these conclusions[3].

According to JOHNS and others[4] the type of spectrum used in the analysis of the excitation curves (see Sect. 48) does not affect the position of the peak

[1] L. KATZ and A. G. W. CAMERON: Canad. J. Phys. **29**, 518 (1951).

[2] A. W. KOCH and R. E. CARTER: Phys. Rev. **77**, 165 (1950).

[3] C. R. EMIGH: Phys. Rev. **86**, 1028 (1952). — W. M. POWELL, W. HARTSOUGH and M. HILL: Phys. Rev. **81**, 213 (1951). — V. E. KROHN and E. F. SCHODER: Phys. Rev. **87**, 685 (1955). — J. E. LEISS, T. YAMAGATA and A. O. HANSON: Report of the University of Urbana, Illinois.

[4] H. E. JOHNS, L. KATZ and R. A. DOUGLAS: Phys. Rev. **80**, 1062 (1950).

and the integrated cross section of a giant resonance considerably, but only its relative shape.

On the other hand, these same authors take into account the modification of the spectrum (51.1) by absorption in the walls of the doughnut of the electron accelerator, in the monitor, and in other materials placed between the radiating target and the detector. If the detector is not very thin one has to take into account also the self-absorption.

While the above considerations are necessary in order to deduce the photo-effect cross section from the measured neutron yields (see Sect. 48), it is better to make use of the direct empirical data if one is mainly interested in the neutron yield itself.

In Table 24 and in Fig. 72 we give the number of neutrons emitted by various elements per mole and per r as they have been obtained by different groups[1,2].

Similar results have been obtained at 50[3,4], 100, and 320 Mev[5].

It is perhaps noteworthy that for lead the yields at 50 and 22 Mev appear to be about equal[3], and that at 320 Mev the neutron yield depends on Z roughly as at 22 Mev; while a comparison of the absolute scales is of doubtful significance.

The use of bremsstrahlung, produced by a high intensity electron accelerator, as the power source for a pulsed neutron source was suggested by Sir JOHN COCKCROFT[6]. That adequate neutron outputs can be obtained by such means is a direct consequence of the very large currents that can be accelerated by radio-frequency powered, travelling wave focused linear accelerators[7]. A feasibility experiment was carried out by MERRISON[6] in which bremsstrahlung from a 3.2 Mev linear electron accelerator fell onto targets of D_2O or Be metal. The mean neutron outputs were compared with those from a Ra $\alpha+$Be source with the help of an integrating tank of $NaSO_4$ solution (see Sect. 126). This experiment indicates a peak production rate of neutrons, during the 2 μsec electron pulse, of about 2×10^{12} neutrons/sec—roughly in agreement with that calculated from the known photo-neutron cross sections. Owing to the voltage regulation characteristics of the accelerator source, the best compromise was eventually realised with a heavy current (about 120 mA during the pulse) and Be target. It was clear that a very much better output of neutrons would be achieved by increasing the electron energy of the accelerator so as to produce a large proportion of quanta with energies in excess of the photo-emission threshold for heavy elements, and to make full use of the giant resonances of the heavy elements. As an interim measure, dependent on the available radiofrequency power sources, a 15 Mev accelerator[8,9] was installed at Harwell. Tests with Au, W, Cu, Pt, Ag and U targets were made—again using an integrating tank technique. With a pulse electron current of some 35 mA and an electron energy of 13.5 Mev, a uranium target yielded a peak production rate of about 2×10^{14} neutrons/sec in pulses of about 1 μsec; this was some 2 or 3 times better than the best of the other metals. In this case, of course, a separate moderator is necessary for slow neutron work.

---

[1] G.A. PRICE and D.W. KERST: Phys. Rev. **77**, 806 (1950).
[2] R. MONTALBETTI, L. KATZ and J. GOLDEMBERG: Phys. Rev. **91**, 659 (1953).
[3] G.C. BALVIN and F.R. ELDER: Phys. Rev. **78**, 76 (1950).
[4] M.L. PERLMAN and G. FRIEDLANDER: Phys. Rev. **74**, 442 (1948).
[5] D.W. KERST and G.A. PRICE: Phys. Rev. **79**, 725 (1950).
[6] J.D. COCKCROFT, J.C. DUCKWORTH and A.W. MERRISON: Nature, Lond. **163**, 869 (1949).
[7] D.W. FRY *et al.*: Nature, Lond. **162**, 859 (1948).
[8] E.R. WIBLIN, p. 35 of Ref. [*15b*].
[9] C.F. BAREFORD and M.G. KELLIGER: Philips Techn. Rev. **15**, 173 (1954).

The yield is comparable with those calculated by BIRAM[1] — shown in Table 32. From her work (which has been largely verified experimentally by E. J. JONES at Harwell) it is clear that an electron energy in excess of 20 Mev is necessary to achieve still higher yields. After about 30 Mev the increase of yield with electron energy is slower as competing processes set in.

Table 32. *Neutron production by electron bombardment of uranium.*

E_0 (Mev)	40	30	25
Number of neutrons per incident electron	0.020	0.015	0.013

A further accelerator was therefore designed[2,3] for installation at Harwell with a design output of 20 Mev at about 1 ampere peak-amount. Since the maximum pulse repetition rate is 750 per second a fast moving mercury target is being used to cope with the target power of 21 kW. The bremsstrahlung produced in the mercury then interacts with a uranium block for the production of the neutrons. The neutron yield during the pulse is now about 10^{16} neutrons/sec. A suitable amount of moderator surrounds the U-target as is required to produce a neutron spectrum.

Provision has been made for a second similar target which is surrounded by a subcritical mass of pure uranium 235. This assembly will through fission boost the neutron production by a further factor of 10 to 15 and at the same time greatly reduce the γ-ray flash which of course always accompanies the neutron burst. The neutron production rate is therefore now 5×10^{17} neutrons/sec during the pulse which can be as short as $\frac{1}{4}$ μsec. This device has been christened *neutron booster*.

The principal advantage of pulsed neutron sources of this type lies in the possibility to use several flight paths and to carry out many measurements simultaneously. Furthermore, the complete cessation of neutron production during the pulse intervals reduces the background which can be a considerable embarrassment in mechanically interrupted beams from reactors[4].

V. The (p, n) reactions.

52. General remarks. Since the discovery of the first (p, n) reaction in O^{18} by DU BRIDGE, BARNES and BUCK[5], other processes of this type have been investigated over the whole periodic system with protons accelerated by various electronuclear machines.

A few cases among the light and medium elements are of particular interest as neutron sources because they yield monoenergetic neutrons of high intensity and of energy adjustable between wide limits. In the following Sects. 53 to 55 all of the important cases will be discussed in detail, particularly those taking place when targets of T^3, Li^7, and V^{51} are bombarded with protons. The first two reactions are used for producing monoenergetic (or almost monoenergetic) neutrons of energy between a few kev and a few Mev, while the third process may be useful in the region of a few kev.

[1] M.B. BIRAM: A.E.R.E. T/R 1523 (1954).
[2] See footnote 8, p. 169.
[3] C.W MILLER: M.V. Res. Rept. C858.
[4] The data concerning the state of development of linear accelerators for pulsed neutron sources, reported here, are due to the courtesy of Dr. E. BRETSCHER from Harwell.
[5] L.A. DU BRIDGE, S.W. BARNES and J.H. BUCK: Phys. Rev. **51**, 955 (1937).

A large part of the data given in these sections has been taken from the review article of HANSON, TASCHEK, and WILLIAMS [*10*].

A few general remarks may be useful for the understanding of some features common to all the (p, n) processes.

The product of the reaction is the contiguous isobar of the target nucleus shifted from Z to $Z+1$

$$ {}_1p^1 + {}_Z\mathrm{X}^A \rightarrow {}_{Z+1}\mathrm{Y}^A + {}_0n^1 + Q_S . \tag{52.1}$$

Therefore if the target nucleus ${}_Z\mathrm{X}^A$ is stable, the process must be endoenergetic by at least the neutron-hydrogen atom mass difference $\varDelta = m_n - m_H = 0.782$ Mev (see Sect. 2; Table 2, p. 5).

Table 33. *Summary of observed p-n thresholds for $Z \leqq 25$.*

Target and product nucle		Observed E_{pT} (Mev)	$-Q$ (Mev)	$\varDelta E$ (Mev)	$\varDelta E'$ (Mev)	E_{nT} (kev)	E_{nm} at 0° (kev)	Lowest excited level in Mev
H^2	2H	3.339 ± 0.015	2.225	1.442 ± 0.010	1.442 ± 0.00?	371	1979	
H^3	He^3	1.019 ± 0.001	0.764	(used for $n-H$ diff.)	-0.0189 ± 0.0005	63.7	286.5	>2.5
Li^7	Be^7	1.882 ± 0.002	1.646	$+0.864 \pm 0.002$		29.4	120.1	0.435
Be^9	B^9	2.059 ± 0.002	1.852	1.070 ± 0.002		20.6	83.4	>1.5
B^{11}	C^{11}	3.015 ± 0.003	2.762	1.980 ± 0.003	2.003 ± 0.005	20.9	84.5	2.02
C^{12}	N^{12}	20.0 ± 0.1	18.5	17.7 ± 0.1	17.6	118	477	—
C^{13}	N^{13}	3.236 ± 0.003	3.003	2.221 ± 0.003	2.222 ± 0.003	16.5	66.4	2.383
C^{14}	N^{14}	0.664 ± 0.009	0.620	-0.161 ± 0.009	-0.156 ± 0.001	2.9	11.8	2.3
O^{18}	F^{18}	2.590 ± 0.004	2.453	$+1.671 \pm 0.004$	$+1.657 \pm 0.015$	7.2	28.8	
F^{19}	Ne^{19}	4.18 ± 0.25	3.97	3.09 ± 0.25	3.22	10.5	42	
Na^{23}	Mg^{23}	4.78 ± 0.3	4.58	3.80 ± 0.30	3.84	8.3	33	
Cl^{37}	A^{37}	1.640 ± 0.004	1.598	0.816 ± 0.004		1.1	4.6	1.4
A^{40}	K^{40}	$\leqq 2.4$ (?)	2.3 (?)	$\leqq 1.5$ (?)		1.6	5.7	0.81
K^{41}	Ca^{41}	1.25 ± 0.02	1.22	0.44 ± 0.02		0.7	2.8	1.95
Sc^{45}	Ti^{45}	≈ 2.85	2.79	2.01	2.28	1.35	5.4	
V^{51}	Cr^{51}	1.562 ± 0.006	1.532	0.750 ± 0.006		0.58	2.3	
Mn^{55}	Fe^{55}	1.18 ± 0.01	1.16	0.38 ± 0.01		0.38	1.5	

If such a condition were not satisfied the target nucleus would be heavier than the residual nucleus

$$ {}_Z\mathrm{X}^A - {}_{Z+1}\mathrm{Y}^A = \varDelta + Q_S = \varDelta E \tag{52.2}$$

and therefore by emission of a negative electron would transform spontaneously into the residual nucleus of the (p, n) reaction.

Actually the threshold of the (p, n) reactions are almost always higher, namely of the order of 2 to 3 Mev in light nuclei and of the order of 4 to 5 Mev in medium nuclei.

Table 33, given by RICHARDS, SMITH and BROWN[1], summarizes the work—due mainly to the Wisconsin group—on (p, n) threshold of targets up to $Z=25$.

The Q-value of each reaction (column 3) is calculated by means of Eq. (34.5) from the observed thresholds (column 2). The energy difference between the two isobars calculated from Eq. (52.2) is given in column 4, while the same quantity calculated from the beta-ray spectra is given in column 5 for comparison, as explained below. The agreement is in general satisfactory with two exceptions ($B^{11}-C^{11}$, $Sc^{45}-Ti^{45}$) caused by errors in the beta-ray measurements as was shown by later experiments. The last three columns contain information useful

[1] H.T. RICHARDS, R.V. SMITH and C.P. BROWN: Phys. Rev. **80**, 524 (1950).

to those considering these reactions as neutron sources. Column 6 gives the neutron energy at threshold according to Eq. (34.6), column 7 the minimum energy [Eq. (34.7b)] of monochromatic neutrons in the forward direction, which occurs after the cone inside which the neutrons are emitted fills the forward hemisphere, and column 8 the lowest excited state of the residual nucleus which permits a calculation of the maximum neutron energy available from the reaction before the monoenergetic character of the neutrons is lost.

Under the assumption of change independence of nuclear forces, the (negative) Q value of a (p, n) reaction is determined by the following effects (but not by the neutron-hydrogen atom mass difference which appears on both sides of Eq. (52.2) and therefore cancels in the final expression for Q):

(a) The difference between the electrostatic energies of the two isobars of charge Z and $Z+1$ which acts always so as to oppose the introduction of one more charge into a nucleus.

(b) The Pauli principle which frequently imposes the condition that the ground states of the two isobars must be different.

(c) The dependence of the nuclear forces on the spin orientation which gives a difference between the energies of the levels of the two isobars if, as a consequence of the Pauli principle, the same spatial wavefunction combines with different spin orientations.

When the target nucleus is of the type $Z=N-1$, the (p, n) reaction transforms it into its mirror isobar (i.e. a nuclide with all protons replaced by neutrons and viceversa). In this case the threshold is determined only by effect (a), i.e. the difference between the Coulomb energies. This can be calculated by assuming some model for the electric charge distribution in the nucleus.

With the crude assumption that the electric charge is uniformly distributed throughout the nucleus[1], the Coulomb energy difference between the two isobars of atomic number Z and Z' is given by

$$\frac{3}{5}\frac{e^2}{R}\left[Z(Z-1)-Z'(Z'-1)\right]=\frac{Z(Z-1)-Z'(Z'-1)}{A^{\frac{1}{3}}}\,0.596\ \text{Mev}, \qquad (52.3)$$

the nuclear radius R being expressed by the usual formula $R=r_0\,A^{\frac{1}{3}}$ with $r_0=1.45\times10^{-13}$ cm; this term amounts to 1.87 Mev for Be^7—Li^7, 2.77 Mev for C^{10}—B^{10} and 4.39 Mev for Na^{20}—Ne^{20} [2].

When the condition $Z=N-1$ is not satisfied by the target nucleus, one can not invoke symmetry arguments in order to eliminate the effects (b) and (c) whose contribution can be evaluated very roughly by subtracting, in each case, from the observed Q value the Coulomb difference (52.3). The separation between the effects (b) and (c) requires a more detailed knowledge of the isobaric states considered[2] [*2*].

The residual nucleus of the (p, n) reaction is β-unstable and usually transforms back to the target nucleus by emission of a positron or electron-capture:

$$\left.\begin{aligned} {}_{Z+1}Y^A &\to {}_Z X^A + e^+ + e^- + \nu + E^{(+)}_{\max} \\ {}_{Z+1}Y^A &\to {}_Z X^A + \nu + E^{(0)} \end{aligned}\right\} \qquad (52.4)$$

where $E^{(+)}$ and $E^{(0)}$ are the energies released in the two processes[3]. In a few cases the residual nucleus can emit a negative electron and produce by this process

[1] H.A. BETHE: Phys. Rev. **54**, 436 (1938).

[2] More details on this point may be found in Vol. XXXIX of this Encyclopedia: pp. 211 and 417.

[3] The masses are, as usual, those of the neutral atoms: therefore in a β^+ decay one has also to consider in the energy balance, the orbital electron which remains free when Z decreases by 1 unit, while for electron capture the orbital electron does not appear explicitly.

the isobar of charge $Z+2$

$$_{Z+1}Y^A \to {}_{Z+2}Z^A + e^- + \nu + E^{(-)}_{\max}. \tag{52.5}$$

In the case of light elements the behaviour of the cross section above the threshold depends on their individual properties, particularly on the presence of resonances.

For medium and heavy elements and excitation energies of the order of 10 Mev, which are those of the compound nuclei produced by the capture of a proton of a few Mev, the widths of the levels are comparable with their separation.

One can try, therefore, to compare the experimental results with the statistical theory of WEISSKOPF and EWING [3], according to which the cross section of the (p, n) reactions can be split into two factors, the cross section for formation of the compound nucleus [averaged over many levels: $\sigma_c(E_p)$] and the probability of emission of a neutron of any energy. The formulas to be used are similar to those given in Sect. 30 in the discussion of the (n, n') processes, apart from the effect of the Coulomb potential barrier which enters into the expression of $\sigma_c(E_p)$.

The most extensive experimental work with which the theory has been compared, has been performed by the Zürich group[1]. These authors have measured the excitation curves and the cross sections at $E_p = 6.7$ Mev for a number of target nuclei whose Z varies between 28 and 60. As examples in Fig. 73 the excitation curve of $Rb^{87}(p, n)\,Sr^{87}$ and in Fig. 74 the cross section for (p, n) processes as a function of Z are given.

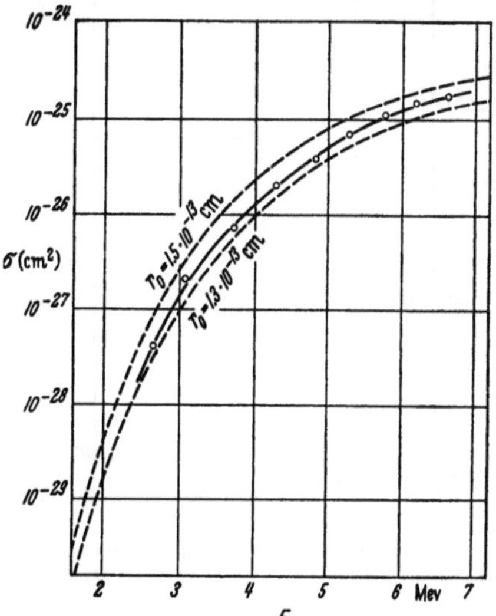

Fig. 73. Excitation curve of the $Rb^{87}(p, n)\,Sb^{87}$ reaction (Zürich group[1]).

The broken curves have been calculated on the basis of the statistical theory for various values of r_0. While individual cases are discussed in the original papers[1], it may be noticed that in general the excitation curves and the absolute values of the cross sections at $E_p = 6.7$ Mev agree with the theory as well as can be expected for values of r_0 between 1.3 and 1.7×10^{-13} cm. The single value of r_0 which on the average gives the best fit for all the data is $r_0 = 1.5 \times 10^{-13}$ cm. Considering the crudity of the theory the agreement can be regarded as generally satisfactory. A few exceptions to the above mentioned rule, i.e. cross sections 10^2 to 10^3 times smaller than those of the neighbouring nuclei, are attributed, in the cases of Ni^{60}, Zr^{90}, and Sn^{116}, to the filling of both proton and neutron shells (or subshells). In other cases the reason for the differences may be that the experimental results obtained by measuring the activity of the reaction product correspond to a partial cross section leading to an isomeric state. For a more detailed investigation of these cases see BOEHM, MARMIER and PREISSWERK[2].

---

[1] J.P. BLASER, F. BOEHM, P. MARMIER and D.C. PEASLEE: Helv. phys. Acta **24**, 3 (1951). — J.P. BLASER, F. BOEHM, P. MARMIER and P. SCHERRER: Helv. phys. Acta **24**, 441 (1951).

[2] F. BOEHM, P. MARMIER and P. PREISSWERK: Helv. phys. Acta **24**, 599 (1951).

Deviations from the statistical theory similar to those observed in (n, p) and (γ, n) reactions (Sects. 30 and 48) are expected also in the case of (p, n) processes. These deviations, however, do not affect appreciably the total cross section, but appear clearly in the angular distribution and energy spectrum of the emitted neutrons[1].

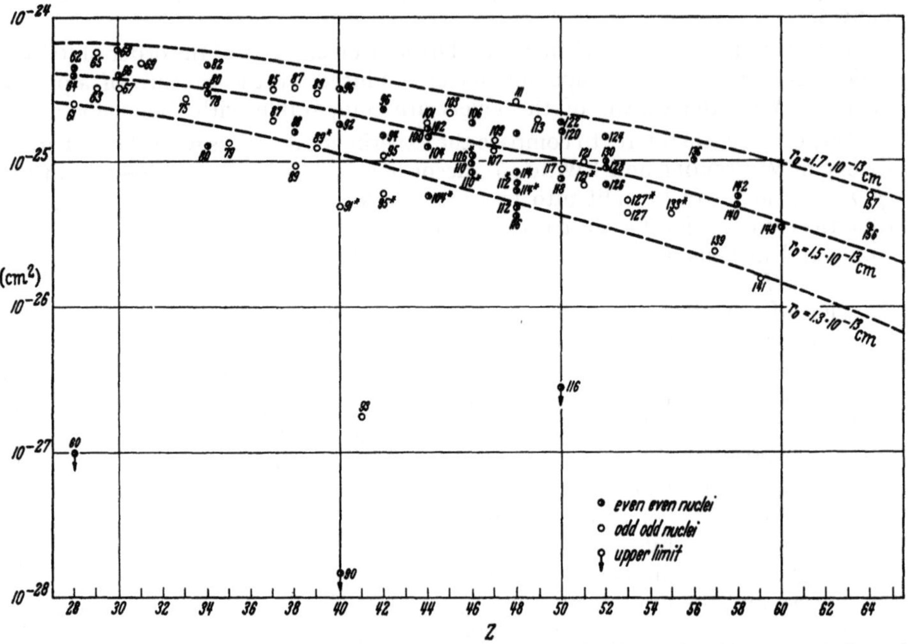

Fig. 74. The cross section of (p, n) processes as a function of Z (Zürich group[2]). Mass number of isotope attached to each point. Star denotes isomeric state.

53. The (p, n) reactions in D and T. The reaction

$$_1H^1 + _1D^2 \rightarrow 2\,_1H^1 + _0n^1 - 2.226 \text{ Mev} \tag{53.1}$$

has been investigated by SMITH and MARTIN[3] using a beam of protons from the Wisconsin electrostatic generator. The energy spread was limited to 0.1% by a cylindrical electrostatic analyser, and the energy was measured against the threshold of $Li^7(p, n)\,Be^7$. By extrapolating the yield curve, they obtained for the threshold

$$E_{pT} = 3.339 \pm 0.015 \text{ Mev}, \tag{53.2}$$

from which

$$Q_0 = -2.225 \pm 0.010 \text{ Mev} \tag{53.3}$$

is deduced. In Sect. 2 this value is compared with the results of other determinations of the binding energy of the deuteron which are used for establishing the neutron mass (Table 1). Other data about reaction (53.1) are given in Table 33 above. It must be noted, however, that in general this reaction cannot compete with others as a source of neutrons, mainly for two reasons: its high threshold (53.2) and the non-homogeneity of its spectrum caused by the presence of three reaction products.

[1] See Sect. 13 of the article by B. B. KINSEY in Vol. XL of this Encyclopedia.
[2] See footnotes 1 and 2, p. 173.
[3] R. V. SMITH and D. H. MARTIN: Phys. Rev. 77, 752 (1950). — R. V. SMITH: Ph. D. Thesis, University of Wisconsin 1950.

One of the more important neutron sources [*10*] is obtained by bombarding tritium with protons

$$_1H^1 + {_1T^3} \rightarrow {_2He^3} + {_0n^1} - 0.764 \text{ Mev}; \tag{53.4}$$

the emitted neutrons are monoenergetic in a wide energy interval, the threshold is low (see Table 33), and the yield is very high immediately above threshold since the electrostatic potential barrier is low.

Neutrons emitted in the c.m.s. in the backward direction are observed in the forward direction in the l.s. for energies of the incident protons between the threshold[1], Eq. (34.5),

$$E_{pT} = 1.019 \text{ Mev} \tag{53.5}$$

and the threshold for emission of monoenergetic neutrons, Eq. (34.7a),

$$E_{pm} = 1.148 \text{ Mev}. \tag{53.6}$$

At an intermediate energy, for example $E_p = 1.1$ Mev, the two groups have the energies

$$E = 23.4 \text{ kev}; \qquad E = 226 \text{ kev}. \tag{53.7}$$

When the energy of the incident protons is decreased to the threshold value (53.5), both values (53.7) tend to $E = 63.7$ kev, while at $E_p = E_{pm}$, the lower energy goes to zero and the higher to

$$E = 286.5 \text{ kev}. \tag{53.8}$$

We notice that the presence of the "low" energy group is not a serious handicap [*10*] because, even for spherical symmetry in the c.m.s., its intensity in the l.s. relative to that of the "fast group" is in the ratio of the corresponding energies (see Sect. 34), i.e. in the preceding example at $E_p = 1.1$ Mev in the ratio $23.4/226 \sim 1/10$.

Furthermore this source should be monoenergetic from the value (53.8) up to neutron energies of about 5 Mev or more because no excited states of He^3 have been observed in the $D(d, n) He^3$ reaction. A numerical table of the values of E as a function of the angle of emission ϑ for E_p ranging between 1.10 and 25 Mev is given in the review article by FOWLER and BROLLEY[2].

Fig. 75 shows the differential cross section at 0° for production of neutrons by the (53.4) reaction[2] obtained by Los Alamos[3,4,5] and Oak Ridge[6] groups. Except for a set of measurements made by PERRY with a stilbene crystal neutron detector, the other data have been obtained with a long counter[7]. The several slight dips in the data of Fig. 75 are attributed to fluctuations of the long counter sensitivity arising from carbon resonances.

Fig. 76 shows again the differential cross section immediately above the thresold, according to the older data [*10*]. The pronounced maximum just above threshold is due to the transformation of the solid angle in the c.m.s. into the l.s. and not to an excited state of the compound nucleus (see Sect. 34).

---

[1] R.F. TASCHEK, G.A. JARVIS, H.V. ARGO and A. HEMMENDINGER: Phys. Rev. **75**, 1268 (1949).

[2] J.L. FOWLER and J.E. BROLLEY: Rev. Mod. Phys. **28**, 103 (1956).

[3] G.A. JARVIS, A. HEMMENDINGER, H.V. ARGO and R.F. TASCHEK: Phys. Rev. **79**, 929 (1950).

[4] G.A. JARVIS: See footnote 2.

[5] J. PERRY: See footnote 2.

[6] H.B. WILLARD, J.K. BAIR and J.D. KINGTON: Phys. Rev. **90**, 865 (1953).

[7] E.M. HAFNER, C.C. LAURITSEN, C.E. FALK, G. SNOW and T. COOR: Phys. Rev. **89**, 204 (1953). — C.H. JOHNSON and H.E. BANTA: Rev. Sci. Instrum. **27**, 132 (1956).

The differential cross section in the l.s. is shown[1,2] in Fig. 77. The total cross section as a function of proton energy given in Fig. 78 is obtained by integrating the data of Fig. 77 over the solid angle.

Only fragmentary information exists with regard to polarization of the T (*p*, *n*) neutrons[1].

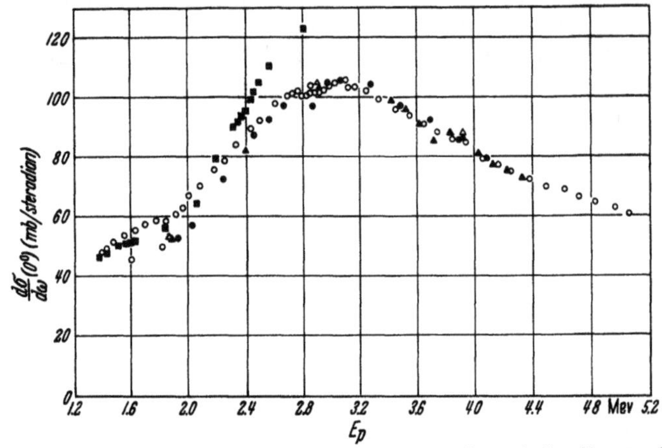

Fig. 75. The differential cross section at 0° of the T(*p*, *n*) He³ reactions according to the Los Alamos and Oak Ridge groups. Solid squares ■ JARVIS *et al.*[3]. Open circles ○ WILLARD *et al.*[4]. Solid circles ● JARVIS *et al.*[1]. Solid triangles ▲ PERRY[1]. Open triangles △ PERRY[1] with stilbene crystal neutron detection.

Targets of tritium have been prepared in gaseous form[5] or the tritium has been absorbed in a layer of a suitable metal[6].

A typical tritium gas target has been described by JOHNSON and BANTA. The gas is separated from the accelerator vacuum system by a thin foil. Nickel foil of about 1 μ thickness covering a circular area of the order of 1 cm in diameter will withstand a pressure difference of an atmosphere. It is necessary to use such thin foils both to reduce the energy loss of the incident protons in the foil and to minimize the straggling in the foil which results in an energy spread of the protons. Such thin foils may be bombarded with several μamp of protons without braking although they frequently leak slowly under bombardment. Care must be taken to reduce tritium contamination of the laboratory from the exhaust of pumps which evacuate the chamber on the vacuum side of the foil. Gaskets exposed to tritium should be made of a non-hydrogeneous material such as teflon to avoid tritium-hydrogen exchange in the gasket material.

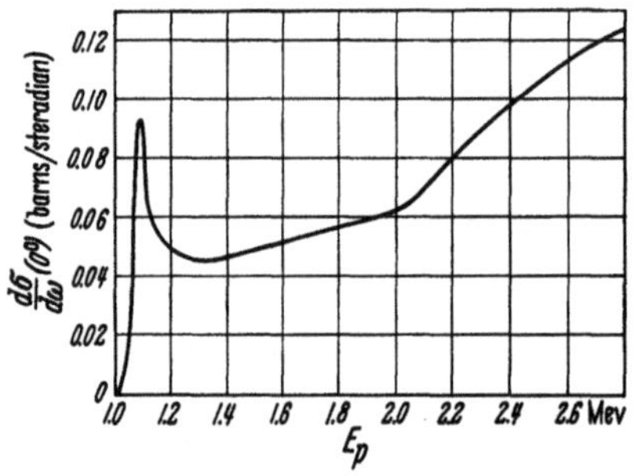

Fig. 76. The differential cross section at 0° of the T(*p*, *n*) He³ reactions shows a maximum immediately above its threshold due to the solid angle transformation from the c.m.s. to the l.s. [*10*].

[1] See footnote 2, p. 175.

[2] In order to avoid the overlapping of the various curves, a shifting ordinate scale has been adopted in Fig. 77. ord-12 means that the reading on the curve must be decreased by 12 mb to get the value of the cross section.

[3] See footnote 3, p. 175.

[4] See footnote 6, p. 175.

[5] See footnote 7, p. 175.

[6] E. R. GRAVES, A. A. RODRIGUES, M. GOLDBLATT and D. I. MEYER: Rev. Sci. Instrum. **20**, 579 (1949). — R. S. RODINI: Rev. Sci. Instrum. **23**, 100 (1952).

The most convenient manner of filling and evacuating the tritium target chamber is with the aid of activated uranium metal. The activated uranium absorbs hydrogen at room temperature to an equilibrium pressure of 25×10^{-3} tor and evolves the hydrogen again around 400° C.

Gas targets cannot be used with low voltage accelerators, because the low energy ions will not penetrate through the foil. In addition, the straggling in the foil limits the attainable energy resolution[1]. Losses of gas because of foil breakage or leaks in the foil are unavoidable. The relative scarcity of tritium and the contamination which it produces have also been deterrents to the use of tritium gas targets.

For these reasons many experiments have been performed with tritium absorbed in thin layers of metal, particularly Zr. The principal disadvantage of such Zr—T targets is that, for a given stopping power, these targets produce a very much smaller neutron yield than gas targets. Zr—T targets were first developed by GRAVES *et al.* for use with a low voltage Cockcroft-Walton accelerator in which deuterons were accelerated[5]. These targets consist of a thin layer of Zr melted onto an 0.8 mm thick W button in vacuum; tritium is then absorbed in the Zr. Zr and W form an alloy at the interface between the two metals so that a strong bond is formed. About one to two atoms of tritium are absorbed per atom of Zr. As long as the temperature of the Zr is below 300° C no thermal outgassing of the target occurs. It is observed, however, that upon prolonged bombardment the yield of neutrons decreases by an amount determined by the charge incident on the target. When a Zr—T target was bombarded with 250 kev deuterons over an area of 0.5 cm², the yield of neutrons produced by the D+T

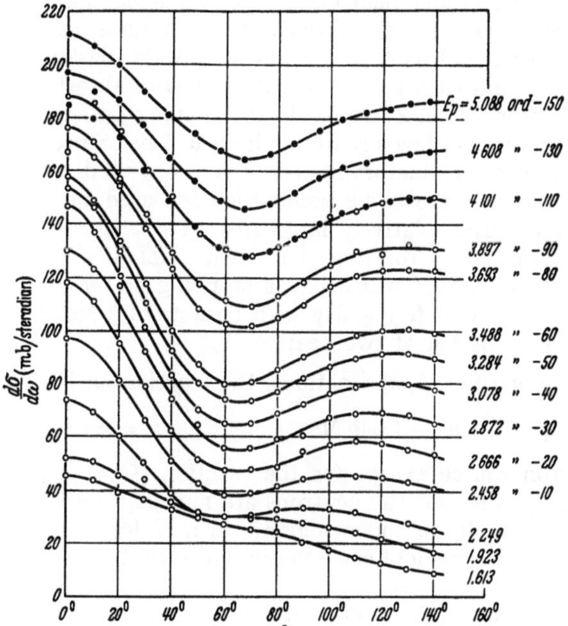

Fig. 77.

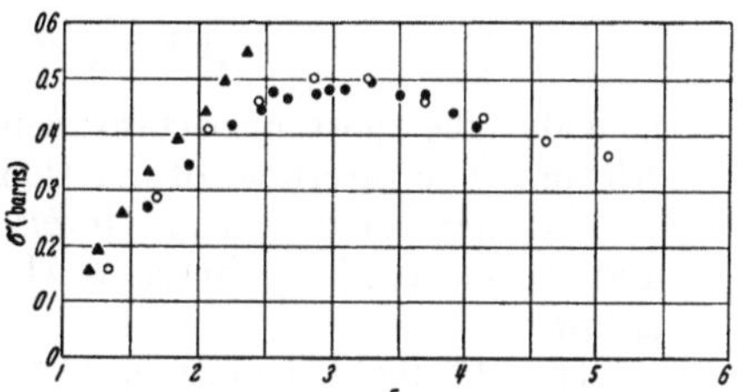

Fig. 78.

Fig. 77. Differential cross section of the $T(p, n) He^3$ reaction as a function of proton energy (l.s.)[1]. Note shifting ordinate scale[2].

Fig. 78. Total cross section of the $T(p, n) He^3$ reaction as a function of proton energy (l.s.). Solid square ■ JARVIS *et al.*[3]. Open circles ○ WILLARD *et al.*[4]. Solid circles ● JARVIS *et al.*[1].

[1] See footnote 2, p. 175.
[2] See footnote 2, p. 176.
[3] See footnote 3, p. 175.
[4] See footnote 6, p. 175.
[5] See footnote 6, p. 176.

reaction decreased to half its initial value after bombardment by 1 coulomb of deuterons. The number of deuterons delivered by this charge to the target is about the same as the number of tritium atoms originally contained in the volume into which the deuterons penetrate so that it is reasonable to assume that deuterium atoms replace tritium atoms in the target. This explanation is corroborated by the observation that neutrons from the D+D reaction appear as the yield of D+T neutrons decreases.

Zr—T targets in which Zr is melted onto W are thick targets for the incident ions and are used primarily for producing 14 Mev neutrons by the D+T reaction. For homogeneous neutron sources employing the p+T reaction thin targets are needed and such targets were developed by LILLIE and CONNOR[1]. A thin layer of Zr is evaporated onto a W backing and then heated in tritium gas to absorb tritium. About one tritium atom is absorbed per Zr atom. In this manner targets may be prepared with stopping powers from a few kev to hundreds of kev. Since cooling of the Zr is important, better heat conductors than W have been used as backings, such as Cu or Ag. If the targets are to be bombarded by large currents, it is desirable that the beam strikes a relatively large area. This may be accomplished by defocussing the beam, or by rotating the target. ALLEN and FERGUSSON[2] report that a rotating Zr—T target which had a stopping power of 50 kev for protons and which was bombarded frequently with 6 μamp of 3 Mev protons showed no deterioration over a period of two years.

54. The (p, n) reaction in Li⁷. Among the (p, n) reactions, the reaction on Li

$$_1H^1 + {}_3Li^7 \rightarrow {}_4Be^7 + {}_0n^1 - 1.646 \text{ Mev} \tag{54.1}$$

provides the most important neutron source except for the reaction (53.4) in tritium. The threshold of reaction (54.1) is low, the neutrons are monoenergetic in a rather narrow energy interval but almost monoenergetic in a wider one, and the yield is large immediately above threshold since the electrostatic barrier is low.

The threshold energy of reaction (54.1) (see Table 34) has been measured with great care by various authors since it is the most convenient and widely

Table 34. *Determination of the threshold of the* $Li^7(p, n)Be^7$ *reaction.*

Reference	Threshold (Mev)	Method
(a)	1.882 ± 0 002	Absolute voltage determination by means of an electrostatic analyser
(b)	1.8819 ± 0.0019	Absolute determination by means of a radio-frequency speed gauge
(c)	1.8813 ± 0.0007	Electrostatic analyser calibrated with RaC′ α-particles
(d)	1.8814 ± 0.0011	Comparison with γ-ray of Au^{198} measured with crystal spectrometer
(d)	1.8797 ± 0.0011	Comparison with γ-ray of Co^{60} measured in terms of proton magnetic moment

(a) R. G. HERB, S. C. SNOWDON and O. SALA: Phys. Rev. **75**, 246 (1949).
(b) W. E. SHOUPP, B. JENNINGS, W. JONES and M. GARBUNY: Phys. Rev. **75**, 336 (1949).
(c) W. J. STURM and V. JOHNSON: Phys. Rev. **83**, 542 (1951).
(d) K. W. JONES, R. A. DOUGLAS, M. T. MCELLISTREM and H. T. RICHARDS: Phys. Rev. **94**, 947 (1954). The gamma ray scales are related to E_{pT} by comparing it, by means of an electrostatic analyser, with the γ-emitting exited state of Mg^{24} produced in the reaction: $Mg^{24}(p, p')Mg^{24}$.

[1] A. B. LILLIE and J. P. CONNOR: Rev. Sci. Instrum. **22**, 210 (1951).
[2] D. V. ALLEN and AT. G. FERGUSSON: AERE NP/R 1720.

used calibration point in the energy scale near 1 Mev. In Table 34 the results of the more important measurements are given.

The threshold for emission of monoenergetic neutrons [Eq. (34.7a)] is

$$E_{pm} = 1.920 \text{ Mev} \tag{54.2}$$

where the neutrons emitted in the forward direction have an energy of 120 kev (Table 33 above). Fig. 79 shows the energy of neutrons emitted at various angles in the l.s. from a thin lithium target, as a function of the difference ΔE_p between the bombarding energy E_p and the threshold E_{pT} in the interval $E_{pm} - E_{pT}$ [10].

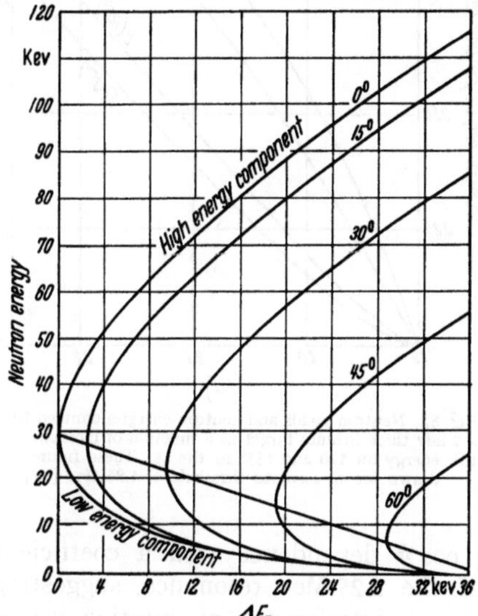

Fig. 79.

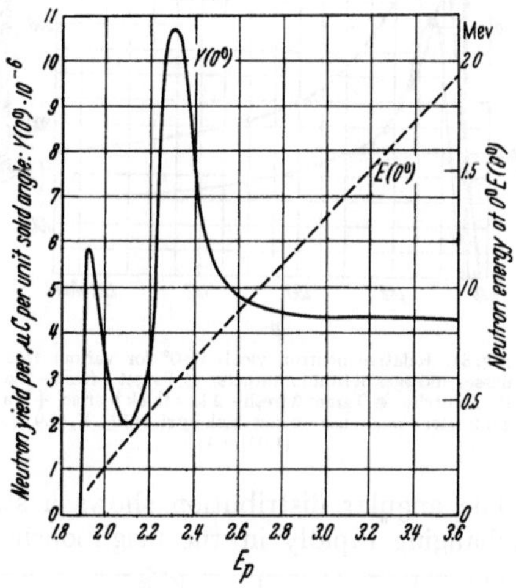

Fig. 80.

Fig. 79. Neutron energies from a thin lithium target as a function of the difference ΔE_p between the bombarding proton energy and the threshold energy, for various angles of emission in the l.s. [10].

Fig. 80. Yield of neutrons at 0° from the Li(p, n) reaction as a function of proton energy. Target thickness 40 kev [10].

For energies of the incident protons larger than $E_p = 2.378$ Mev two groups of neutrons are emitted corresponding to the formation of Be^7 in the ground state and in a (430 ± 5) kev excited state [2]. The intensity of the low energy group with respect to the other is about 3% at $E_p = 2.5$ Mev and increases roughly linearly reaching a value of about 8% at $E_p = 3.0$ Mev[1]; between $E_p = 3.9$ Mev and $E_p = 5.4$ Mev its relative intensity is 10%[2] [1].

Fig. 80 shows the neutron yield of reaction (54.1) in the forward direction obtained with a target 40 kev thick [10]. The pronounced maximum near threshold is due in part to the transformation of the solid angle from the c.m.s. to the l.s. (see Sect. 34 and 53 in the case of tritium) and in part perhaps also to a resonance near threshold[3]. Other resonances are observed at $E_p = 2.25$ ($\Gamma \approx 0.2$ Mev) and $E_p = 4.89$ Mev ($\Gamma \approx 0.4$ Mev)[1].

Fig. 81 shows the influence of the thickness and age of the lithium target on the neutron yield.

[1] R. Batchelor: Proc. Phys. Soc. Lond. **68**, 452 (1955).
[2] L. Cranberg: U.S.A.E. Report L.A. 1654 (1954).
[3] H.W. Newson and J.H. Gibbons: Phys. Rev. **95**, 640 (1954).

Fig. 82 shows the energy of the emitted neutrons and the corresponding yields as a function of E_p at two back angles, 120° and 135°, in the l.s.[1]. These data are useful for planning experiments with neutrons of energy below 80 kev with usable yields.

Fig. 83 gives the total cross section and Fig. 84 the angular distribution in the l.s. for various values of E_p [10][2].

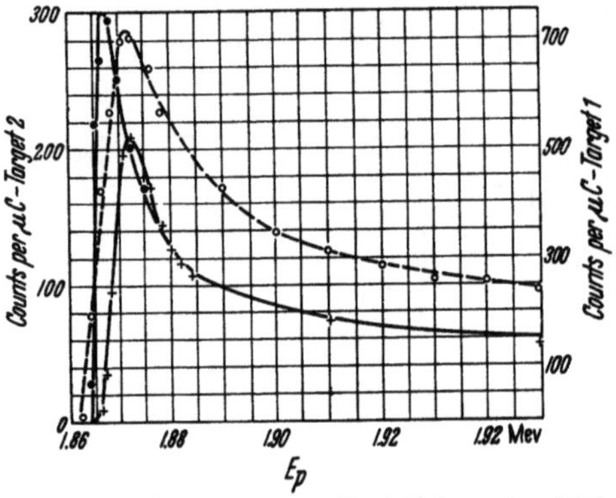

Fig. 81. Relative neutron yields at 0° for various thicknesses and ages of lithium targets. ○ Target 1 fresh—9 kev thick by rise. ● Target 2 fresh—2 kev thick by rise. + Target 2 after 4 weeks use—6 kev thick by rise [10]. $Y_1(1.93)/Y_2(1.93) = 4$.

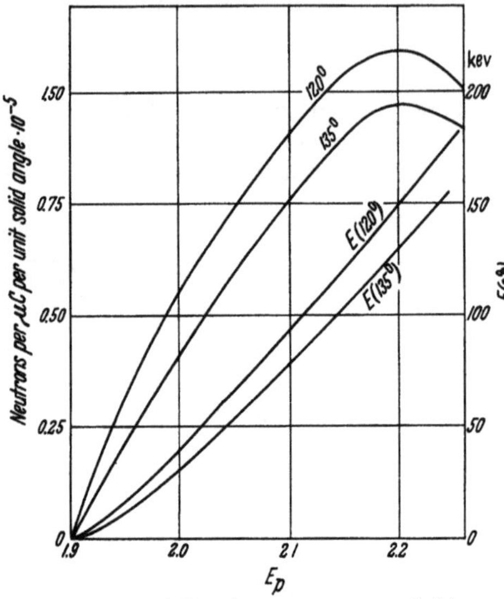

Fig. 82. Neutron yields and neutron energies emitted from a 2 kev thick lithium target as a function of incident proton energy for 120 and 135° in the l.s. These figures are drawn for an assumed threshold of 1.86 Mev [10].

The angular distribution shows a strong cos ϑ dependence with a coefficient changing rapidly in the neighbourhood of the 2.25 Mev resonance, suggesting interference of entering waves of opposite parity[2] [2].

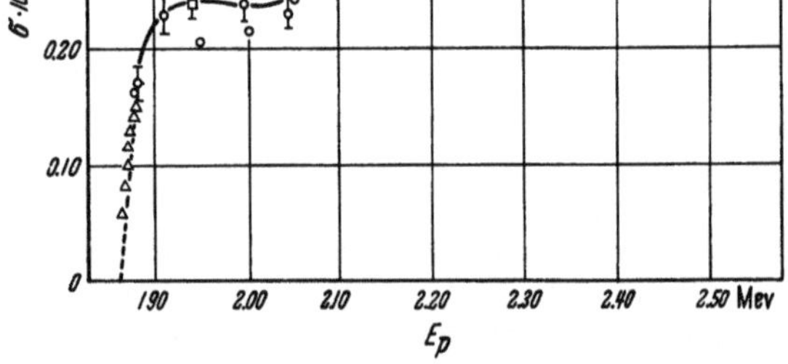

Fig. 83. Total cross section of $Li^7(p,n)$ Be^7 as a function of incident proton energy. The threshold energy is assumed to be 1.86 Mev [10]. --- Extrapolation to threshold assumed linear in E_p; △△△ extrapolation to threshold assumed linear in $\sqrt{E_p}$.

When Li^7 is bombarded with protons, various processes take place, all involving excited states of the compound nucleus Be^8, viz. $Li^7(p,\alpha)\,{}_2He^4$, ${}_3Li^7(p,n)\,{}_4Be^7$, $Li^7(p,p)\,{}_3Li^7$; ${}_3Li^7(p,\gamma)\,{}_4Be^8$.

The first one is strongly exoenergetic ($Q = 17.2$ Mev) so that each one of the two α's gets at least 8.6 Mev. Therefore one would expect that this reaction overcomes the competition of any of the other processes which are however observed with comparable intensity.

Furthermore the excitation curve of the ${}_3Li^7(p,\alpha)\,{}_2He^4$ process shows a continuous rise with increasing energy of the incident protons, while the three other reactions present rather sharp resonances. From a detailed

[1] See footnote 1, p. 179.

[2] R.F. TASCHEK and A. HEMMENDINGER: Phys. Rev. 74, 373 (1948).

study of the situation one can conclude that the compound states of Be^8 which can disintegrate into two α-particles must have even parity and even angular momentum and are different from those which are observed to decay by neutron, proton and γ-ray emission [3].

A qualitative and schematic discussion of the theory of the angular distribution of the neutrons of reaction (54.1) has been given by BREIT and BLOCH[1] who concluded that there is indication of two resonances, one immediately below threshold, the other at ≈ 2.2 Mev.

The problem has been reexamined more recently by ADAIR[2] who finds a good fit to the resonant part of the cross section by attributing the resonance at $E_p = 2.25$ Mev to a $J=3^+$, $T=1$ level of Be^8 at 19.2 Mev above the ground state, while the background intensity is predominantly the result of two states, $J=1^-$, $T=0$ and $J=2^-$, $T=0$, with undetermined location.

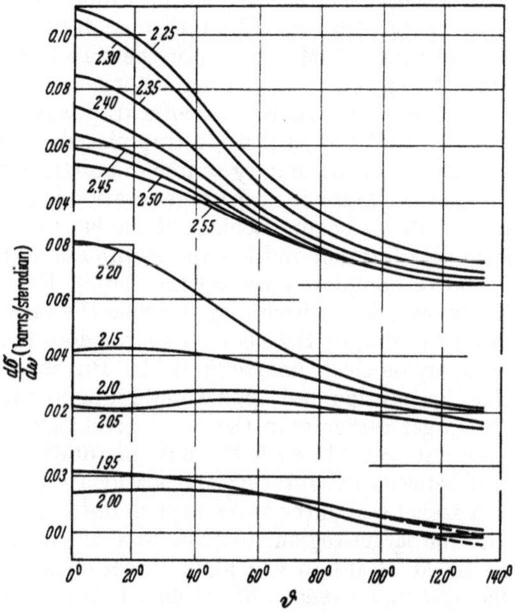

Fig. 84. Differential cross section for the production of neutrons from the $Li^7(p, n)Be^7$ reaction in the l.s. as a function of the angle of the neutron ϑ for various values of E_p (Mev). $E_{p\,T}$ assumed to be 1.86 Mev [10].

This author noticed that the existence of the isolated $J=3^+$ state of Be^8 is an effect of the strong spin-orbit coupling which will result in a polarization of the emitted neutrons. He calculated the polarization of the neutrons emitted at various angles as a function of E_p and found a maximum equal to about 50% near $E_p = 2.25$ Mev and $\vartheta \sim 90°$ in the c.m.s. This conclusion is in satisfactory agreement with the result of measurements done at $E_p = 2.23$ and $\vartheta = 66°$ in the c.m.s. which gave $0.53 \pm 0.06\%$ [3]. Similar results have been obtained by other authors[4].

Difficulties in *maintaining stable targets* under ion bombardment are well known from charged particle scattering and reaction experiments. The problems are enhanced, however, for targets used as neutron sources because of the much larger ion currents normally incident on the target to produce large neutron yields and because it is necessary to keep scattering material away from the neighborhood of the target. The solid metal targets which have been most frequently used for neutron sources are Li targets. Although many authors have studied the behaviour of Li targets very carefully, some of the observations on the behaviour of such targets are not fully understood. Much of this careful work was done by W. D. ALLEN and coworkers[5] at Harwell, and the studies of extremely thin Li targets were performed by H. W. NEWSON and others at Duke University[6].

---

[1] G. BREIT and I. BLOCH: Phys. Rev. **74**, 379 (1954).
[2] R. K. ADAIR: Phys. Rev. **96**, 709 (1954).
[3] R. K. ADAIR, S. E. DARDEN and R. E. FIELDS: Phys. Rev. **96**, 503 (1954).
[4] H. B. WILLARD, J. K. BAIR and J. D. KINGTON: Phys. Rev. **95**, 1359 (1954).
[5] W. D. ALLEN and A. T. G. FERGUSSON: Proc. Phys. Soc. Lond. A **68**, 650 (1955).
[6] See footnote 3, p. 179.

The principal source of deterioration of targets, particularly in charged-particle work, is caused by the build-up of carbon deposits on the targets. Thin layers of organic material which originate in pump oil and gasket materials are reduced to carbon under ion bombardment. This carbon layer is non-uniform in thickness and increases in thickness under continuing bombardment. The build-up of carbon may be prevented by keeping organic vapors from reaching the target. A liquid air trap or a liquid air-charcoal trap placed near the target will accomplish this, provided no rubber gaskets are used in the target chamber between the trap and the target. A trap suitable for this purpose has been described by G. H. MILLER[1]. For the trap to be most effective it has to be placed near the target. It has been reported that an order of magnitude decrease of carbon build-up occurred when a trap was moved from 15 cm to 7 cm from the target. On the other hand, a trap placed so close to the neutron source will form an undesirable secondary source of scattered neutrons.

Another cause of target deterioration is the local heating of the target material by an intense beam. Cooling of the back of the target is not enough to prevent such deterioration and has the additional disadvantage of increasing the condensation of organic vapors on the target. For beam currents above 1 or 2 μamp it has been found necessary to move the target with respect to the beam. If the exact location of the neutron source does not matter, one may move the beam by applying electromagnetic fields. But in most applications the source must be fixed in position and approximate a point source so that it is necessary to move the target rather than the beam. Rotating targets for which the beam position does not coincide with the axis of rotation have been used extensively[2]. The seal between the rotating and stationary parts of the target is either an oil seal or a gasket seal of the Wilson type. Both types of seals introduce organic materials into the target region. In order to avoid this, gyrating targets have been developed by M. WALT in which the target chamber is sealed to the stationary part of the system by means of sylphon bellows. In this arrangement every point on the target moves on a circular path of the same diameter, but the orientation of the target remains fixed, i.e., the target as a whole does not rotate.

Several considerations influence the choice of the target backing. It should scatter as few neutrons as possible, it must be strong enough to withstand a pressure difference of an atmosphere over a large area, it should be a good heat conductor, it should not alloy with the target material although the target metal must adhere firmly to the backing. For Li targets Ta backings, 0.1 to 0.3 mm thick, have been widely used. These backings scatter, however, a too large number of neutrons for some experiments. Less scattering and equal strength may be realized by thin Ag backings. Experience with Ag backings has been unfavourable in that the targets deteriorate for reasons which are not entirely clear. The most satisfactory results have been obtained with Pt backing which may be as thin as 0.05 mm.

Various precautions have been suggested in evaporating Li targets. In order to avoid introducing furnace materials into the vacuum system, heating of the Li by induction has been found convenient. Another method for heating the Li is to have an arrangement by which the crucible containing the Li may be placed into the beam and be heated by the beam. It has been suggested that the Li be redistilled before use, that the target backing be heated in vacuum and be bombarded with a high beam current before evaporation, and that the first material which is given off by the Li should not be used in the final target.

[1] G. H. MILLER: Rev. Sci. Instrum. **24**, 549 (1958).
[2] E. LAMPI: Rev. Sci. Instrum. **22**, 429 (1951).

The target backings are usually cooled on the outside by blowing air against them. This air may be heated above room temperature to reduce deposition of carbon on the target. For high beam current a fine spray of water produces more effective cooling than air.

55. The (*p*, *n*) reactions in other light and medium elements (Be, B, Sc, V). The reaction

$$_1H^1 + Be^9 \rightarrow B^9 + {}_0n^1 - 1.852\ \text{Mev} \tag{55.1}$$

has also a large yield and the properties of the emitted neutrons (see Table 33) are similar to those of reaction (54.1). This latter reaction, however, has the advantage of a lower threshold and ease of target preparation (see Sect. 54) so that it is more frequently used than reaction (55.1).

With a thick Be target and protons of energy below the threshold of reaction (55.1) one observes a neutron emission which is believed to be the result of second order reactions: (*p*, *d*) — (*d*, *n*). The thin target yield below threshold shows no effect above background; therefore it appears that the neutron yield from reaction $Be^9(p, pn)Be^8$ (threshold $E_p = 1.84$ Mev) is negligible at these bombarding energies.

The excitation curve of reaction (55.1) shows a resonance at $E_p = 2.56$ Mev superposed on a general rise to $E_p \sim 4.5$ Mev[1]; a second resonance is observed at $E_p = 4.72$ Mev. The angular distribution is isotropic only very close to threshold, while in the vicinity of the $E_p = 2.56$ Mev resonance it shows maxima at 0^0 and $135°$ in the c.m.s. [*2*]. For the spectrum and angular distribution of the neutrons emitted at $E_p = 6.59$ Mev as well as for their interpretation in terms of levels of B^9 see reference [2].

The reaction

$$_1H^1 + {}_5B^{11} \rightarrow C^{11} + {}_0n^1 - 2.762\ \text{Mev} \tag{55.2}$$

has many properties similar to those of reactions (54.1) and (55.1) with the disadvantage of a still higher threshold (see Table 33). Many resonances have been observed for energies of the bombarding protons between 3.18 and 6.37 Mev [*2*][3]. The (*p*, *n*) reaction in B^{10} has a much higher Q value [$Q_0 = -(4.35 \pm 0.2)$ Mev] which makes it useless as a neutron source; its investigation is however very interesting because it gives information about the level structure of C^{10} [*2*][4].

The excitation curves of a few other light and medium elements can be found in references [1] and [3]. The last authors attribute the rather large cross section variations observed from one element to the other in the low Z region to the competition between the (*p*, *n*) and (*p*, *p*), (*p*, α) processes.

(*p*, *n*) reactions in medium elements such as vanadium and scandium can provide very useful neutron sources [*10*].

The primary interest in such a source is that as a consequence of the rather large mass of the target nucleus the velocity of the centre of mass in the l.s. is very low, at least at bombarding energies near threshold.

It follows that the threshold of these reactions are very close to the corresponding negative Q-values and that the energy of the emitted neutrons near threshold is of the order of a few kev (see Table 33) and approximately independent of the angle of emission.

---

[1] H.T. Richards, R.V. Smith and C.P. Brown: Phys. Rev. **80**, 524 (1950).
[2] F. Ajzenberg and W.W. Buechner: Phys. Rev. **91**, 674 (1953).
[3] J.P. Blaser, F. Boehm, P. Marmier and P. Scherrer: Helv. phys. Acta **24**, 465 (1951).
[4] F Ajzenberg and W. Fronzen: Phys. Rev. **95**, 1531 (1954).

Therefore sources of this type can be used to produce neutrons of 2 to 20 kev.

In order to obtain monoenergetic neutrons one has to use well defined proton energies and very thin targets, conditions which strongly reduce the neutron intensity. But these reactions can still compete with the $Li^7(p, n)Be^7$ reaction used at backward angles in the same energy region.

Their main advantage is due to the more isotropic distribution of the intensity of the emitted neutrons.

A detector placed in the backward hemisphere receives, superposed on the neutrons emitted in that direction, also the neutrons initially emitted in the forward direction and then scattered into the detector; this, of course, makes it

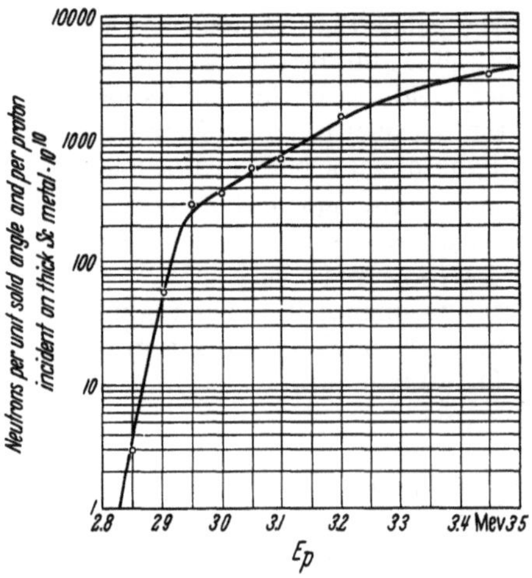

Fig. 85. The $Sc^{45}(p, n)Ti^{45}$ yield curve in the forward direction as a function of the incident proton energy [10].

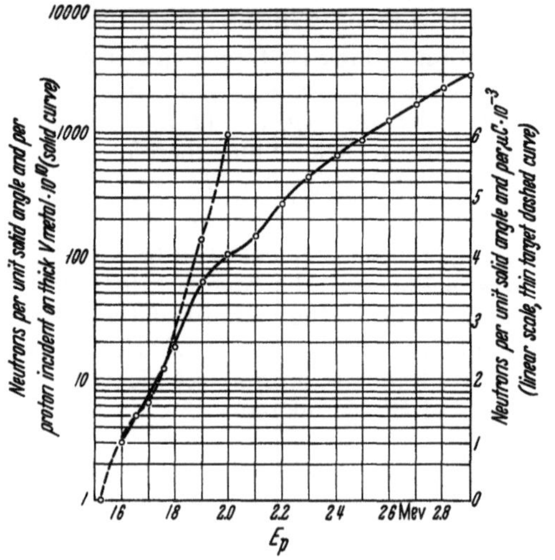

Fig. 86. The $V^{51}(p, n)Cr^{51}$ yield curve in the forward direction as a function of the incident proton energy [10].

difficult to interpret low-energy neutron observations. Such an effect is much more pronounced in the case of lithium than in the case of scandium and vanadium.

The reactions in these two elements are

$$_1H^1 + {}_{21}Sc^{45} \rightarrow {}_{22}Ti^{45} + {}_0n^1 - 2.79\ \text{Mev}, \tag{55.3}$$

$$_1H^1 + {}_{23}V^{51} \rightarrow {}_{24}Cr^{51} + {}_0n^1 - 1.532\ \text{Mev}, \tag{55.4}$$

and the corresponding excitation curves in the forward direction are given in Fig. 85 and 86.

The higher threshold of the scandium reaction makes the neutron yield near threshold much larger than for vanadium. This is due to the smaller influence of the electrostatic potential barrier on the incident proton, which in heavy elements is the most important factor in determining the yield near threshold.

The thick target yield of vanadium although approximately 40 times smaller than that of scandium is still sufficient to render a relatively strong neutron source provided large currents of incident protons are used. The spectrum of the emitted neutrons has been investigated up to $E \approx 400$ kev and no evidence for a slow group of intensity more than 10% of the fast group was observed [10].

VI. The (*d*, *n*) reactions.

56. General remarks. Nuclear reactions of this type were discovered by CRANE, LAURITSEN and SOLTAN[1] who observed an emission of neutrons from Li and Be bombarded with deuterons of energy between 0.4 to 0.9 Mev.

These results were soon confirmed and extended by other authors[2]. To-day processes of this type have been observed in almost all elements throughout the periodic system.

A (d, n) process transforms the target nucleus (Z, N) into the residual nucleus $(Z+1, N)$, i.e., the product of the reaction is the same that one would obtain through the radiative capture of the incident proton by the target nucleus.

The process is usually exoenergetic because the energy released by the binding of the incident proton in a nucleus is almost always much higher than the energy spent to dissociate the incident deuteron. The only exception to this rule among the light elements is observed in C^{12} whose (d, n) process has a small negative Q-value.

The (d, n) processes in light elements are the most convenient neutron sources for accelerators whose maximum energy is around or below 1 Mev. Under these conditions they provide a rather intense neutron flux of energy between 2 and 17 Mev. In the case of deuterium or tritium targets the spectrum of the emitted neutrons is monoenergetic while in other cases (Li, Be, B) the spectrum is complex. The preparation and conservation of metal targets is easier and the intensity is so high that these processes are very useful in many experiments.

By increasing the energy of the incident deuterons one can increase the energy of the emitted neutrons. Table 17 of Sect. 39 allows a comparison of the neutron yield obtained from (α, n) and (d, n) reactions for energies of the incident particles of the order of 20 Mev.

At still higher energy the (d, n) processes are almost the only source of ultra-fast neutrons[3].

The theoretical interpretation of processes produced by deuterons is considerably more complicated than that of other reactions because the deuteron is neither a simple nucleon, nor it is tightly bound, nor symmetric with respect to the electric charge distribution like an α-particle[4].

The very low binding energy of the deuteron $-\mathsf{E}_3 = 2.226$ Mev, its large dimensions with the two constituent nucleons often 3 to 4×10^{-13} cm apart, and the fact that the center of electric charge is displaced with respect to the center of mass by the radius of the deuterons ($\gamma^{-1} = 4.3\times 10^{-13}$ cm), are features which play an important role in determining the course of all processes in which deuterons are involved.

In the interpretation of the reactions induced by deuterons, three main types of mechanism have been suggested which may all occur with appreciable probability. Each one may predominate under particular conditions, i.e., for appropriate values of E_d and Z:

1. Compound nucleus formation. This process can be described as the capture of the incident deuteron as a whole. This produces a compound nucleus which

[1] H. R. CRANE, C. C. LAURITSEN and A. SOLTAN: Phys. Rev. **44**, 629 (1933). — C. R. Acad. Sci., Paris **197**, 913 (1933). — H. R. CRANE and C. C. LAURITSEN: Phys. Rev. **45**, 226 (1934); **45**, 507 (1934).

[2] E. O. LAWRENCE and M. S. LIVINGSTON: Phys. Rev. **45**, 220 (1934). — M. L. E. OLIPHANT, P. HARTECK and Lord RUTHERFORD: Proc. Roy. Soc. Lond. **144**, 692 (1934). — M. C. HENDERSON, M. S. LIVINGSTON and E. O. LAWRENCE: Phys. Rev. **45**, 428 (1934). — H. R. CRANE and C. C. LAURITSEN: Phys. Rev. **45**, 430 (1934).

[3] See the article by R. T. SIEGEL in Vol. XLV of this Encyclopedia.

[4] See the article by B. B. KINSEY in Vol. XL of this Encyclopedia.

then decays in several channels, some of which correspond to the emission of neutrons (Sect. 25).

2. *Electric break-up of the deuteron* in flight by the electric field of the target nucleus. When a deuteron passes by a nucleus without hitting its surface, it is exposed to the Coulomb field of the nucleus. From the point-of-view of the deuteron this field varies with time and can induce a disintegration of the deuteron similar to that produced by a light wave.

3. *Stripping process.* Because of the large dimensions of the deuterons only one of the two nucleons constituting the deuteron may strike the surface of the target nucleus and is stripped off the deuteron, while the other misses the nucleus and continues its flight.

While the first process does not need special considerations for the case of the deuteron, the processes of electric break-up and of stripping are typical of the deuteron as incident particle.

The electric break-up first discussed by OPPENHEIMER[1] has been treated for high energy deuterons by DANCOFF[2] and at low energy by MULLIN and GUTH[3]. The last authors have calculated in the electric dipole approximation the differential cross section for deuteron break-up. In this discussion these authors have included the effect of the Coulomb field on the motion of the center of mass of the deuteron but have neglected the effects of the Coulomb field on the outgoing proton. The procedure is probably adequate for a description of the behaviour of the outgoing neutrons but does not give an adequate description of the outgoing protons, unless the deuteron energy is very high. The angular distribution and total cross section have been calculated numerically for $E_d = 15$ Mev. The angular distribution is peaked in the forward direction with a half-width of $\pm 20°$ roughly independent of the atomic number of the nucleus at least up to $Z = 30$. The procedure of neglecting the finite size of the nucleus yields an upper limit for the total cross section which, for $E_d = 15$ Mev, turns out to be 0.016 barns in Be, 0.22 in Cu and 0.17 in Au. This is of the same order of magnitude or slightly higher than the corresponding cross sections for the stripping process.

In a stripping process neither the incoming deuteron nor the outgoing nucleon need penetrate into the potential barrier and higher angular momenta can participate. For instance in a (d, n) reaction only the proton is captured and therefore has to penetrate the potential barrier while the neutron continues its flight. When protons are captured to form a definite state of the residual nucleus, the corresponding neutrons will constitute a group of definite energy, the angular distribution of which is determined by the conservation of angular momentum[4]

$$\boldsymbol{I}_i + \boldsymbol{l}_d + \boldsymbol{s}_d = \boldsymbol{I}_f + \boldsymbol{l}_n + \boldsymbol{s}_n .$$

Here $\boldsymbol{I}_i$ and $\boldsymbol{I}_f$ are the angular moments of the initial and final nucleus, $\boldsymbol{l}_d$ and $\boldsymbol{l}_n$ the angular momenta of the incident deuteron and outgoing neutron and $\boldsymbol{s}_d$, $\boldsymbol{s}_n$ the corresponding spins.

The difference between $\boldsymbol{I}_f$ and $\boldsymbol{I}_i$ can be expressed in terms of the orbital angular momentum and spin of the captured proton

$$\boldsymbol{I}_f - \boldsymbol{I}_i = \boldsymbol{l}_p + \boldsymbol{s}_p .$$

From these two relations it follows that

$$\boldsymbol{l}_d - \boldsymbol{l}_n = \boldsymbol{l}_p + \boldsymbol{s}_p - (\boldsymbol{s}_d - \boldsymbol{s}_n) .$$

[1] J. R. OPPENHEIMER: Phys. Rev. **47**, 845 (1945).

[2] S.M. DANCOFF: Phys. Rev. **72**, 1017 (1947).

[3] C. J. MULLIN and E. GUTH: Phys. Rev. **82**, 141 (1951).

[4] H. GOEPPERT MAYER and J. H. D. JENSEN: Elementary theory of nuclear shell structure. New York and London: John Wiley & Sons 1955. See also Sect. 42 of the articly by B. B. KINSEY in Vol. XL of this Encyclopedia.

The spin of the neutron will not change because this particle does not interact with the nucleus. Therefore, if $\Delta \boldsymbol{s}_p$ indicates the change of spin of the captured proton from its initial value in the deuteron to its final value in the residual nucleus, one has

$$\boldsymbol{l}_d - \boldsymbol{l}_n = \boldsymbol{l}_p - \Delta \boldsymbol{s}_p .$$

For given $\boldsymbol{l}_d$ and for instance $\Delta \boldsymbol{s}_p = 0$ this equation determines uniquely $\boldsymbol{l}_n$ and therefore the angular distribution of the neutrons from the orbital angular momentum $\boldsymbol{l}_p$ of the captured proton in the residual nucleus.

The first derivation of the angular distribution of the particles emitted in a stripping process of deuterons has been given by Butler[1] who showed that these are unmistakably different for different values of the angular momentum of the captured particle, i.e., in the case of the (*d*, *n*) reactions, the proton.

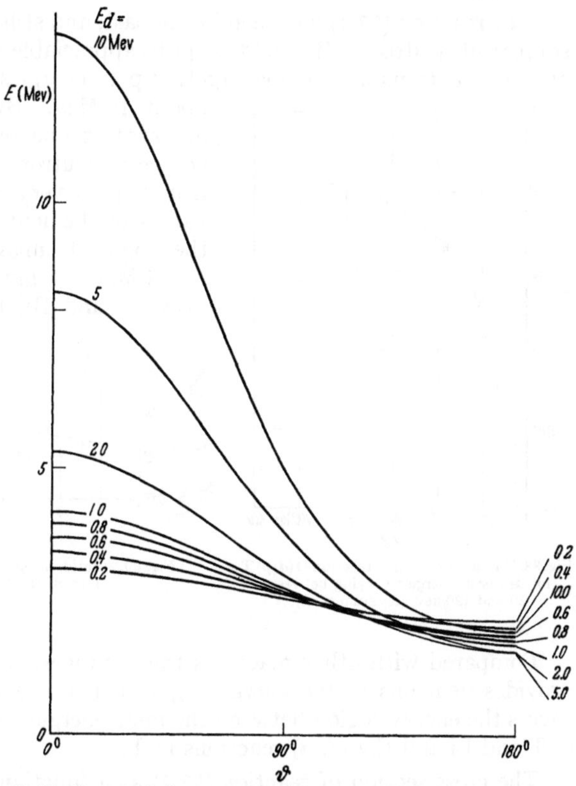

Fig. 87. Energy of the neutrons emitted in the D (*d*, *n*) He³ reaction as a function of the angle of emission in the l.s. for various energies of the incident deuterons.

We refer to Vols. XL to XLII of this Encyclopedia, for details of this theory and for the analytical expression of the angular distributions corresponding to different values of $\boldsymbol{l}_p$. We simply recall that for $\boldsymbol{l}_p = 0$ the angular distribution has its principal maximum at $\vartheta = 0°$ and a secondary one much less pronounced usually at about 47° in the c.m.s. For other values of $\boldsymbol{l}_p$ the angular distribution shows its principal maximum at increasingly larger values of ϑ[2], a feature which can easily be recognized in the experimental results obtained in the case of many groups of neutrons (or protons) emitted in (*d*, *n*) [or (*d*, *p*)] reactions.

57. The (*d*, *n*) reaction in D and T. Oliphant, Harteck and Rutherford[3] discovered that by bombardment of deuterium with deuterons the two following reactions are produced:

$${}_1\mathrm{D}^2 + {}_1\mathrm{D}^2 \rightarrow {}_1\mathrm{He}^3 + {}_0n^1 + 3.27\ \mathrm{Mev}, \tag{57.1}$$

$${}_1\mathrm{D}^2 + {}_1\mathrm{D}^2 \rightarrow {}_1\mathrm{H}^3 + {}_1\mathrm{H}^1 + 4.03\ \mathrm{Mev}. \tag{57.2}$$

---

[1] S. T. Butler: Proc. Roy. Soc. Lond. **208**, 559 (1951).
[2] See footnote 4, p. 186.
[3] M. L. E. Oliphant, P. Harteck and Lord Rutherford: Proc. Roy. Soc. Lond., Ser. A **144**, 692 (1934).

They measured the range of the protons and tritons emitted in reaction (57.2) and the corresponding excitation curve for E_d varying between 20 and 160 kev. They estimated the energy of the neutrons emitted in reaction (57.1) and showed that the yields of the two processes were about the same. The energy of the emitted neutrons was estimated also by other workers[1]. Direct evidence for the two reactions was provided by measuring with a mass spectrograph the H^3 and He^3 percentages present in deuterium exposed to deuteron bombardment[2].

The reaction (57.1) has been in the past, and still is today one of the important sources of neutrons. Its yield is quite appreciable even at very low energies[3,4], and the neutrons are monoenergetic up to energies of the incident deuterons of about 10 Mev. By changing the direction of observation one may obtain a source of monoenergetic neutrons variable between wide limits (Fig. 87). A very useful numerical table of the values of the neutron energy E as a function of the angle of emission, for E_d varying from 0.1 to 25 Mev, is given in the review article by FOWLER and BROLLEY[5].

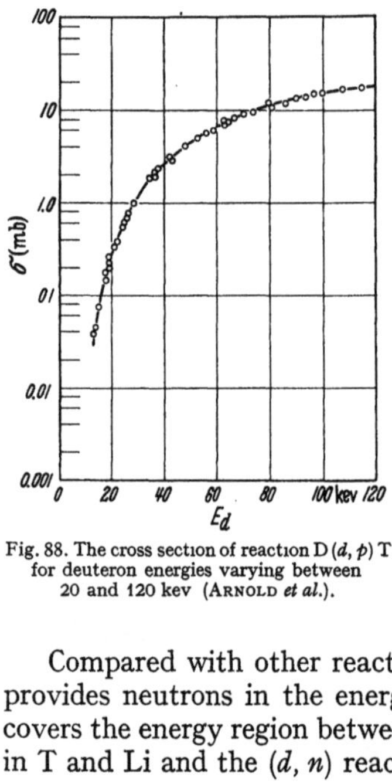

Fig. 88. The cross section of reaction D (d, p) T for deuteron energies varying between 20 and 120 kev (ARNOLD *et al.*).

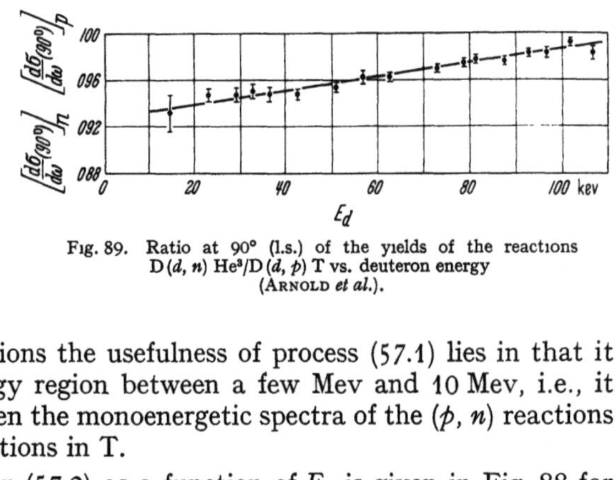

Fig. 89. Ratio at 90° (l.s.) of the yields of the reactions D (d, n) He^3/D (d, p) T vs. deuteron energy (ARNOLD *et al.*).

Compared with other reactions the usefulness of process (57.1) lies in that it provides neutrons in the energy region between a few Mev and 10 Mev, i.e., it covers the energy region between the monoenergetic spectra of the (p, n) reactions in T and Li and the (d, n) reactions in T.

The cross section of reaction (57.2) as a function of E_d is given in Fig. 88 for $E_d \leqq 120$ kev while Fig. 89 gives in the same energy interval the ratio of the cross sections of the processes (57.1) to (57.2) obtained by observing at 90° in the l.s. the He^3 and H^3 nuclei emitted in the two processes[4]. Details of the delicate technique necessary at very low energies are given in the original paper.

The curves of Figs. 88 and 89, combined with the angular distributions of the two reactions allow the determination of the cross section of process (57.1) in the very low energy region ($E_d \leqq 120$ kev).

---

[1] P.I. DEE: Proc. Roy. Soc. Lond., Ser. A **148**, 623 (1935). — P.I. DEE and C.W. GILBERT: Proc. Roy. Soc. Lond., Ser. A **149**, 200 (1935).

[2] W. BLEAKNAY, G.P. HARNWELL, W.W. LOZIER, P.J. SMITH and H.D. SMITH: Phys. Rev. **46**, 81 (1934).

[3] K.D. ALEXOPOULOS: Naturwiss. **23**, 817 (1935). — Helv. phys. Acta **8**, 513 (1935). — R. DÖPEL: Ann. Phys. **28**, 87 (1937). — F.M. PENNING and I.H.A. MOUBIS: Physica, Haag **4**, 1190 (1937). — E. BALDINGER, P. HUBER and H. STAUB: Helv. phys. Acta **11**, 245 (1948).

[4] W.R. ARNOLD, J.A. PHILLIPS, G.A. SAWYER, E.J. STORALL and J.L. TUCK: Phys. Rev. **93**, 483 (1954).

[5] J.L. FOWLER and J.E. BROLLEY: Rev. Mod. Phys. **28**, 103 (1956).

The angular distributions in the c.m.s. can be represented in general by expressions of the type

$$\frac{d\sigma}{d\omega^*} = \left(\frac{d\sigma}{d\omega^*}\right)_{\pi/2} [1 + A(E_d)\cos^2\vartheta^* + B(E_d)\cos^4\vartheta^* + C(E_d)\cos^6\vartheta^*] \quad (57.3)$$

where $A(E_d)$, $B(E_d)$, and $C(E_d)$ are functions of the energy of the incident deuterons. At very low energy, $B(E_d)$ and $C(E_d)$ are negligible and $A(E_d)$ has a different value for the two reactions[1]:

$$\left.\begin{aligned} &\text{for reaction (57.1) and } E_d \leqq 50 \text{ kev:} \quad A(E_d) = (0.31 \pm 0.0058)\cdot E_d, \\ &\text{for reaction (57.2) and } E_d \leqq 50 \text{ kev:} \quad A(E_d) = (0.13 \pm 0.0047)\cdot E_d, \end{aligned}\right\} \quad (57.4)$$

where E_d is expressed in kev.

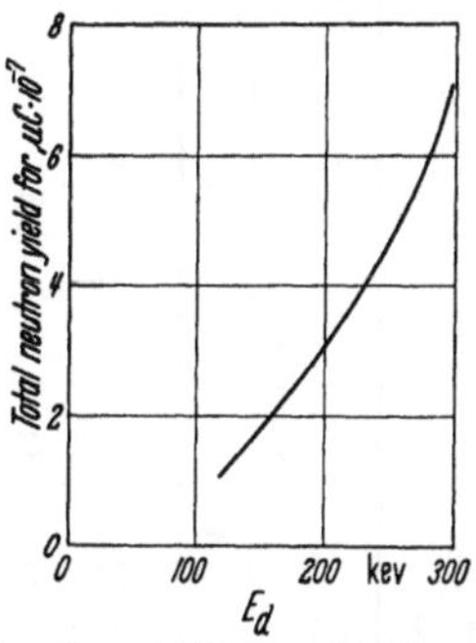

Fig. 90. Total neutron yield from a thick heavy ice target vs. deuteron energy between 100 and 300 kev (COON *et al.* [*10*]).

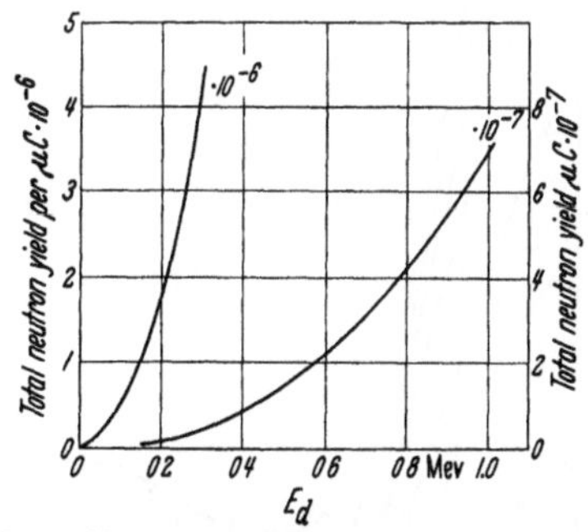

Fig. 91. Total neutron yield from a thick D_2O ice target [*10*].

ARNOLD *et al.*[2] measured the angular distribution by counting the He^3 and H^3 particles emitted in the two reactions at various angles. This technique seems to be more appropriate for a quantitative comparison between the two processes than that of measuring directly the neutrons emitted in process (57.1) and comparing their number with the number of protons emitted in process (57.2). By following this last procedure other authors[3] found that in the same energy region the angular distributions of the two processes do not differ appreciably.

As will be seen later, recent measurements at higher energy appear to confirm the existence of an appreciable difference between the angular distributions of the two reactions.

At these extremely low energies the process (57.1) is a very interesting research subject but not a practically usable neutron source. With high deuteron currents one starts to obtain a usable neutron source for E_d above 100 kev[4]. In Fig. 90 the total yield of neutrons from a thick ice target of D_2O as a function of E_d obtained by COON, DAVIS, GRAVES, MANLEY and NOBLES is given [*10*].

The angular distributions in the c.m.s. as well as the total cross sections of the two processes (57.1) and (57.2) for E_d varying between 156 and 466 kev are given in Table 35[5].

[1] E. A. ELLIOT, D. ROAF and P. F. D. SHAW: Proc. Roy. Soc. Lond., Ser. A **216**, 57 (1953).

[2] See footnote 4, p. 188.

[3] U. TIMM, H. NEUERT and B. ELSNER: Z. Physik **139**, 425 (1954).

[4] R. LADENBURG and R. B. ROBERTS: Phys. Rev. **50**, 1190 (1936). — R. B. ROBERTS: Phys. Rev. **51**, 810 (1937). — R. LADENBURG and M. H. KANNER: Phys. Rev. **52**, 911 (1937). — H. REDDEMANN: Z. Physik **110**, 373 (1938). — J. H. MANLEY, J. H. COON and A. C. GRAVES: Phys. Rev. **70**, 101 (1946). — K. G. MCNEILL and G. M. KEYER: Phys. Rev. **81**, 602 (1951).

[5] G. PRESTON, P. F. D. SHAW and S. A. YOUNG: Proc. Roy. Soc. Lond., Ser. A **226**, 206 (1954). In this paper references to the many previous works on the same subject can be found.

For deuteron energies less than 400 kev the neutrons emitted at 90° in the l.s. are approximately monoenergetic and their energy about 2.5 Mev even for a thick target (Fig. 87) because the decrease in energy of the incident

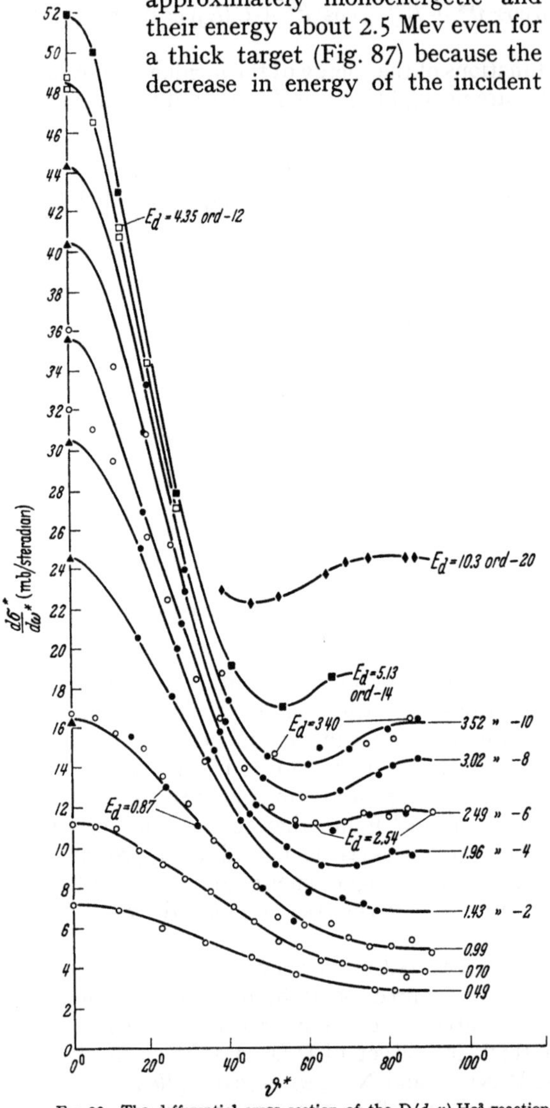

Fig. 92. The differential cross section of the D(d, n)He^3 reaction in the c.m.s. Note shifting ordinate scale. Open circles ○: G. T. HUNTER and H. T. RICHARDS: Phys. Rev. **76**, 1445 (1949). — Solid circles ●: J. H. BLAIR, G. FREIER, E. LAMPI, W. SLEATOR and J. H. WILLIAMS: Phys. Rev. **74**, 1599 (1948). — Open squares □: HENKEL *et al.* — Solid squares ■: SMITH *et al.* — Solid diamonds ◆: K. W. ERICKSON, J. L. FOWLER and E. J. STOVALL: Phys. Rev. **76**, 1141 (1949). — Solid triangles ▲: ROSEN *et al.*, SMITH *et al.* — See also footnote 5, p. 188. — For ordinate scale see footnote 2, p. 176.

deuterons in the target is roughly compensated by the corresponding decrease of the velocity of the center of mass (Sect. 34). Such a compensation does not occur in the forward

Table 35. *Total cross section and angular distribution of the* D$(d, n)^3$ *and* D(d, p)T *reactions for deuteron energies varying between 156 kev and 466 kev* (PRESTON et al. [1]).

Bombarding energy (kev)	Total cross-sections (milli-barns)		Branching ratios σ_n/σ_p		Asymmetry coefficients			
	neutron reaction	proton reaction	for differential cross-sections at $\frac{1}{2}\pi$	for total cross-sections	neutron reaction *A*	neutron reaction *B*	proton reaction *A*	proton reaction *B*
156	29.0 ± 1.8	26.2 ± 2.0	0.99 ± 0.04	1.10 ± 0.04	1.33 ± 0.25	−0.09 ± 0.30	0.77 ± 0.17	0.03 ± 0.18
					(1.25 ± 0.11)	—	(0.80 ± 0.05)	—
206	38.3 ± 2.1	34.7 ± 2.2	0.96 ± 0.03	1.10 ± 0.02	1.27 ± 0.11	0.26 ± 0.20	0.69 ± 0.11	0.30 ± 0.11
					(1.52 ± 0.09)	—	(0.96 ± 0.04)	—
270	51.0 ± 4.6	44.9 ± 3.9	0.96 ± 0.03	1.14 ± 0.07	1.59 ± 0.20	0.31 ± 0.21	0.91 ± 0.14	0.20 ± 0.16
362	60.0 ± 3.7	50.8 ± 4.1	0.98 ± 0.02	1.18 ± 0.05	1.58 ± 0.09	0.56 ± 0.09	0.67 ± 0.14	0.79 ± 0.15
466	73.3 ± 5.4	61.2 ± 4.5	1.00 ± 0.02	1.20 ± 0.05	1.38 ± 0.14	1.23 ± 0.15	0.94 ± 0.09	0.57 ± 0.10

[1] See footnote 5, p. 189.

direction where the neutrons emitted by a thick target show a wide distribution in energy. Their spectrum is, however, strongly peaked toward the highest value corresponding to the initial energy E_d of the deuterons, because the yield increases rapidly with the bombarding energy.

The monitoring of the neutron intensity in this energy region can be done by measuring the protons emitted from the process (57.2) whose energy is sufficiently high (3 Mev) to make their quantitative counting relatively easy. The knowledge of the angular distributions and of the ratio of the cross sections of the two processes (57.1) and (57.2) given in Table 35, can be used to deduce the number of emitted neutrons from the number of counted protons.

For energies of the incident deuterons between 300 and 1000 kev the neutron yield from thick targets has been measured by various authors[1]. In particular AMALDI, HAFSTAD and TUVE compared the neutron yield of reaction (57.1) with the neutron yields of other (d, n) reactions in light elements, by integrating the density of the slowed-down neutrons over a large volume of water surrounding the source.

Their results reduced to a thick heavy ice target agree well with the curve given in Fig. 91. At these bombarding energies, however, the neutrons emitted from a thick target cannot be considered even approximatively monoenergetic.

If monoenergetic neutrons are needed, one must use a thin target. Gas targets have been discussed in Sect. 53.

The preparation of thin D_2O targets may be found in reference[2].

The cross section and angular distribution of reaction (57.1) has been measured around and above 1 Mev by various authors whose results have been reviewed by FOWLER and BROLLEY[3]. The differential cross section at various deuteron bombarding energies is given in Fig. 92 in the c.m.s., and in Fig. 93 in the l.s. Curves of Fig. 93 were integrated over the solid angle to give the total cross section plotted in Fig. 94 versus deuteron energy.

The Los Alamos group[4] has measured by emulsion technique, the angular distribution of reactions (57.1) and (57.2) for deuteron energies varying between 6 and 14 Mev, and has analyzed the corresponding differential cross sections in the c.m.s. in Legendre polynomials. The coefficients obtained by such an analysis are given in Table 36 together with the values of the total cross section obtained by integrating the differential cross section over the solid angle.

It is unfortunate that the monoenergetic character of the neutrons emitted in reaction (57.1) has not been carefully investigated for deuteron energies above 2 Mev and that the background difficulties are so formidable [*10*].

This last difficulty can possibly be overcome[5] by counting the neutrons of process (57.1) emitted in coincidence with a full range He^3: this method would ensure the monoenergetic character of the neutron spectrum by eliminating the effect of background neutrons and furnishing an absolute measure of the neutron flux. The break-up of the incident deuteron in the *d-D* reaction yields a low energy continuous spectrum of neutrons the intensity of which rises with increasing deuteron energy starting from the threshold of the $D(d, np)D$ reaction ($E_{dT} =$ 4.45 Mev)[6].

[1] H.R. CRANE, C.C. LAURITSEN and A. SOLTAN: Phys. Rev. **45**, 507 (1934). — T.W. BONNER and W.M. BRUBAKER: Phys. Rev. **49**, 19 (1936). — E. AMALDI, L.R. HAFSTAD and M.A. TUVE: Phys. Rev. **51**, 896 (1937).

[2] R. RICAMO and W. ZUENTI: Helv. phys. Acta **24**, 419 (1951).

[3] See footnote 5, p. 188.

[4] BROLLEY, PUTNAM and ROSEN: Private communication.

[5] R.R. CURTIS, J.L. FOWLER and L. ROSEN: Rev. Sci. Instrum. **20**, 388 (1949).

[6] L. CRANBERG, A.H. ARMSTRONG and R.L. HENKEL: Phys. Rev. **104**, 1639 (1956).

A theoretical analysis of the processes (57.1) and (57.2) has been attempted by KONOPINSKI and TELLER[1] and extended by BEIDUCK *et al.*[2] who tried to ex-

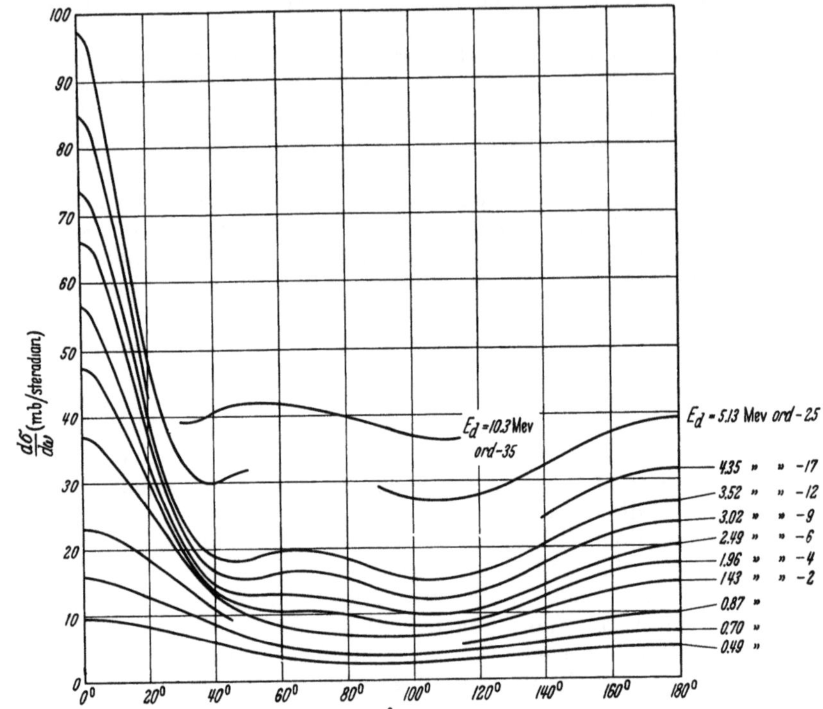

Fig. 93. The differential cross section of the D(d, n) He^3 reaction in the l.s.[3].

plain the behaviour of the coefficients appearing in Eq. (57.3), as they are given by HUNTER and RICHARDS[3], as caused by the different centrifugal barriers met by the various components of the incident deuteron wave. An important result of such an analysis was that, in order to explain the observed angular distribution up to deuteron energies of about 0.5 Mev, it is necessary to introduce strong spin-orbit coupling and P-wave effects. While this theory appears to be inadequate to explain the difference observed between the asymmetry of the two processes (57.1) and (57.2), given in Eq. (57.4) and Table 35, it gave the start to a fruitful line of investigation.

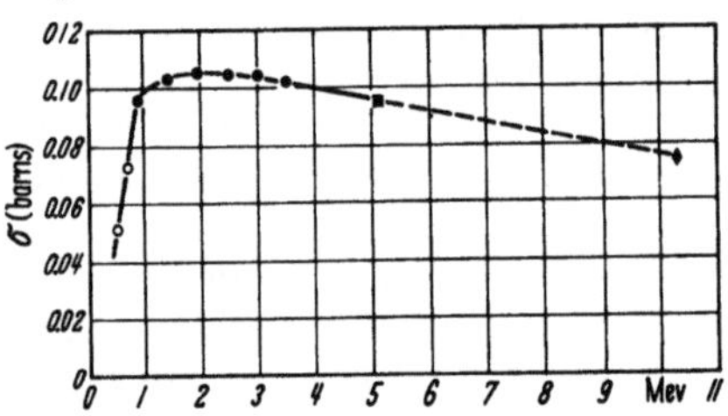

Fig. 94. The total cross section of the D(d, n) He^3 reaction versus deuteron energy (FOWLER and BROLLEY).

It was noticed[4] that strong spin-orbit coupling and P-wave effects are the necessary conditions for the production of polarized particles[5], an effect which

[1] E. J. KONOPINSKI and E. TELLER: Phys. Rev. **73**, 822 (1948).
[2] F. M. BEIDUK, J. R. PRUETT and E. J. KONOPINSKI: Phys. Rev. **77**, 622, 628 (1950).
[3] G. T. HUNTER and H. T. RICHARDS: Phys. Rev. **76**, 1445 (1949); for ordinate scale see footnote 2, p. 176.
[4] R. J. BLIN-STOYLE: Proc. Phys. Soc. Lond. **64**, 700 (1951). — M. CINI: Nuovo Cim. **8**, 1007 (1951). — R. J. BLIN-STOYLE: Proc. Phys. Soc. Lond. **65**, 949 (1952).
[5] L. WOLFENSTEIN: Phys. Rev. **75**, 1664 (1949).

Table 36. *Coefficients of the Legendre polynomials analysis of the differential cross section (mb) of the* D(d, n) *and* D(d, p) *reactions in the* c.m.s. *(Los Alamos group).*

E_d (Mev)	a_0	a_2	a_4	a_6	a_8	a_{10}	a_{12}	$2\int_{0^\circ}^{90^\circ} \sigma(\vartheta)\,d\Omega$ (mb)
				D+D→He³+n				
6	7.40	11.14	15.16	5.28	0.64	0.34		93 ± 5
8	7.03	10.45	16.06	8.02	1.59	0.61		88 ± 4
10	6.43	9.85	14.77	9.22	2.11	0.00		81 ± 4
12	6.04	9.71	14.55	10.94	4.02	1.48		76 ± 4
14	5.93	9.56	13.77	11.26	3.97	1.13	0.26	75 ± 4
				D+D→T+p				
6	6.96	9.45	13.58	5.14	0.85	0.39		87 ± 3
8	6.64	9.36	14.53	7.32	1.73	0.18		84 ± 3
10	6.21	8.42	13.21	8.15	2.15	0.34		78 ± 2
12	5.92	8.77	13.29	9.72	3.06	1.83		74 ± 2
14	5.65	8.42	12.44	10.03	3.58	1.17	0.39	72 ± 2

was successively observed by scattering experiments of the neutrons[1] of the reaction (57.1) and of the protons of reaction (57.2)[2]. According to RICAMO, at an energy of the incident deuterons $E_d = 610$ kev, the neutrons emitted at 45° in the l.s. ($E = 3.32$ Mev) show a polarization of at least 20%. For the theory of the scattering of polarized neutrons see[3].

For deuteron energies of the order of 10 Mev, the stripping process may contribute appreciably to the observed disintegrations. According to FAIRBAIN[4], who has treated the reactions (57.1) and (57.2) ignoring Coulomb effects, under the assumption that they are produced via a stripping process, the stripping becomes the predominant mechanism for bombarding energies greater than 5 Mev in the l.s.

In the collision of deuterons and tritons, monoenergetic neutrons are produced according to the reaction

$$ {}_1\mathrm{T}^3 + {}_1\mathrm{D}^2 \rightarrow {}_2\mathrm{He}^4 + {}_0n^1 + 17.58\ \mathrm{Mev}. \tag{57.5}$$

Details about the construction of gas target are given in Sect. 53; special care has to be used in order to reduce the consumption of tritium which is usually available in limited ammounts.

Fig. 95 shows the energies of the neutrons emitted in process (57.5) for various values of the energy of the bombarding deuterons E_d, as a function of the angle of emission in the l.s. [*10*]. A numerical table of the values of E_n, for ϑ varying between 0 and 180° and E_d from 0.1 to 25 Mev, is given in reference[5].

The cross section of this process is very high at a few hundred kilovolts. Therefore reaction (57.5) is an ideal source for the production of 14 Mev neutrons of large intensity with low voltage accelerating equipment. Furthermore with voltage generators in the range of a few Mev, one can produce neutrons having energies in the range 12 to 30 Mev.

---

[1] H.L. LONGLEY, R.N. LITTLE and J.M. SLYE: Phys. Rev. **86**, 419 (1952). — P. HUBER and E. BAUMGARTNER: Helv. phys. Acta **26**, 420 (1953). — R. RICAMO: Helv. phys. Acta **26**, 423 (1953). — Nuovo Cim. **10**, 1607 (1953). — R.W. MEIER, P. SCHERRER and G. TRUMPY: Helv. phys. Acta **27**, 577 (1954).

[2] G.R. BISHOP, G. PRESTON, J.M. WESTHEAD and H.H. HALBAN: Nature, Lond. **170** 113 (1952).

[3] V. LEPORE: Phys. Rev. **79**, 137 (1950). — M. VERDE: Nuovo Cim. **9**, 376 (1952).

[4] W.M. FAIRBAIN: Proc. Phys. Soc. Lond. A **67**, 990 (1954).

[5] See footnote 5, p. 188.

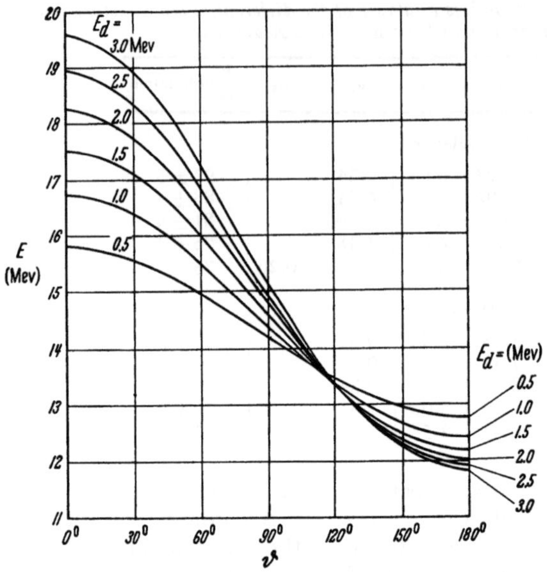

Fig. 95. The energies of the neutrons emitted from the T(*d*, *n*) He^4 reaction for various values of E_d as a function of the angles of emission in the l.s. [*10*].

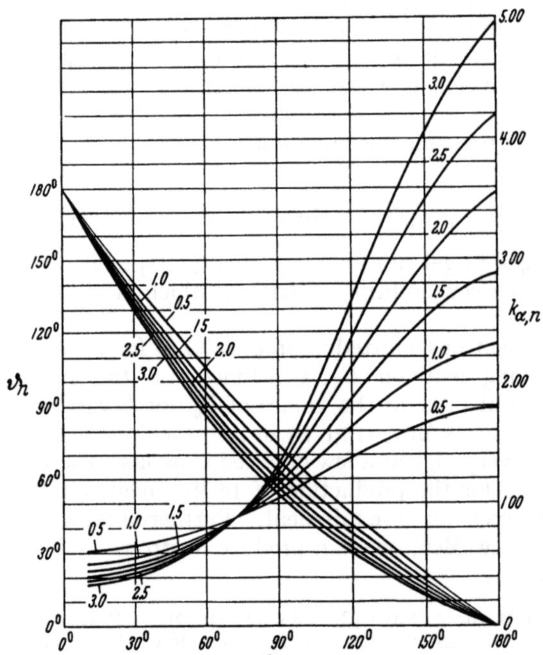

Fig. 96. A solid angle conversion chart for use with the T(*d*, *n*) He^4 reaction. The angle of emission of neutrons in the l.s. ϑ_n, is given as a function of the corresponding angle for α-particles ϑ_α. $k_{\alpha n}$ is the ratio of the cross section for neutron production to the cross section for α-particle production. $k_{\alpha n}$ is shown as a function of ϑ_α. Numbers attached to the various curves indicate deuteron energies in Mev.

The He^4 recoiling nuclei produced in the reaction have a rather high energy so that they can easily be counted, which provides an excellent procedure for absolute calibration of the neutron flux.

Fig. 96 shows the relationship between neutron and α-particle angles and the factor $k_{\alpha n} = \sigma_n/\sigma_\alpha$ needed to multiply the measured α-particle flux to get the corresponding neutron flux for various incident deuteron energies [*10*].

For the results of the first measurements of the excitation curve[1] and of the angular distribution of the particles emitted in the process[2] we refer to the original papers and to the review article [*10*][3].

Fig. 97 shows the cross section of process (57.5) and Fig. 98 the ratio of its cross section to that of the process (57.1), in the very low-energy region ($E_d \leq 120$ Mev)[4]; the latter was obtained by counting simultaneously the α-particles and tritons from the two reactions emitted at 90° by a known mixture of tritium and deuterium as gas target.

Fig. 99 shows the total cross section for bombarding energies of the incident tritons E_t varying between 80 and 1200 kev; these data, having an estimated over-

---

[1] BAKER, HOLLOVAY, SCHREIBER and KING: LAMS **11** (unpublished). See footnote 5, p. 188. — E. BRETSCHER and A. P. FRENCH: Phys. Rev. **75**, 1154 (1949). — D. L. ALLEN and M. J. POOLE: Proc. Roy. Soc. Lond. Ser. A **204**, 488 (1950).

[2] R. F. TASCHEK, A. HEMMENDINGER and G. A. JARVIS: Phys. Rev. **75**, 1464 (1949).

[3] See footnote 5, p. 188.

[4] See footnote 4, p. 188.

all accuracy of $\pm 10\%$, were obtained by counting the He^4 nuclei recoiling at 90° from a gas target[1]. Fig. 100 shows the excitation curve obtained by accelerating the incident deuterons from 10 to 1732 kev and by observing the He^4 recoiling at 90° from thin zirconium-tritium targets[2].

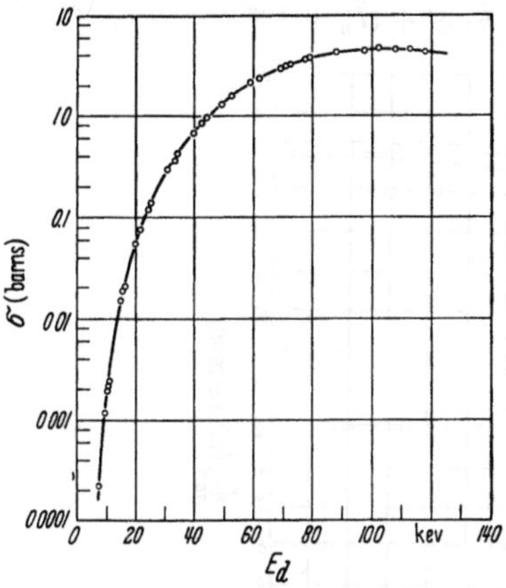

Fig. 97. The $T(d, n) He^3$ cross section vs. deuteron energy (ARNOLD *et al.*).

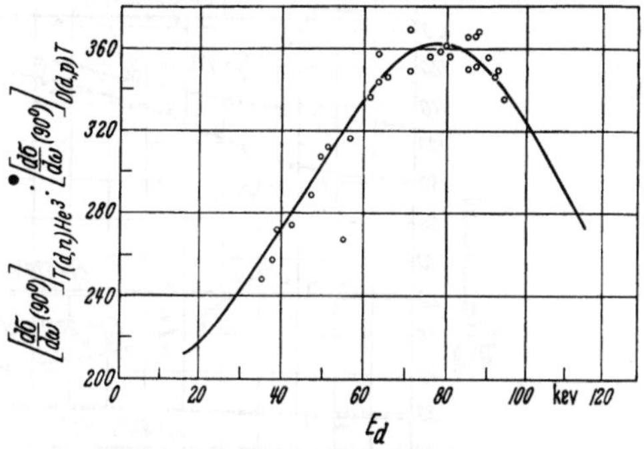

Fig. 98. Ratio at 90° in the l.s. of the yields of the reactions $T(d, n) He^4/D(d, p)T$ vs. deuteron energy (ARNOLD *et al.*).

The angular distribution of the products of reaction (57.5) is isotropic in the c.m.s. up to $E_t = 570$ kev. From Fig. 101 one sees[3] that around $E_t \approx 1$ Mev it starts to be anisotropic, more neutrons being emitted at the backward angles. The same data transformed to the l.s. are given in Fig. 102. The general behaviour of the cross section and the anisotropy of the neutron emission lead one to believe that S-wave interaction takes place between the initial

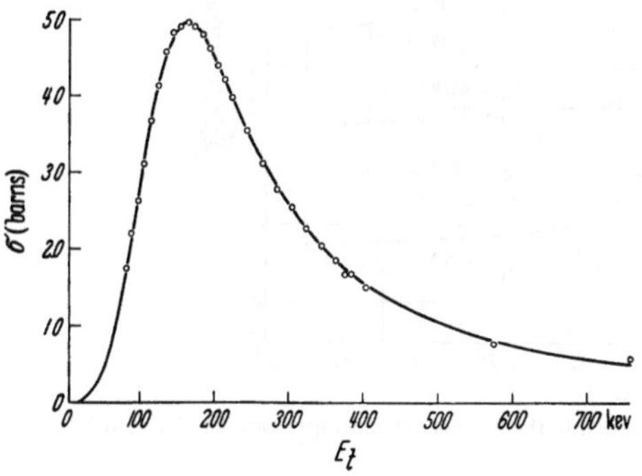

Fig. 99. Cross section of the reaction $D(t, n) He^4$ vs. triton energy (ARGO *et al.*[1]). Curve computed by means of the one level dispersion formula.

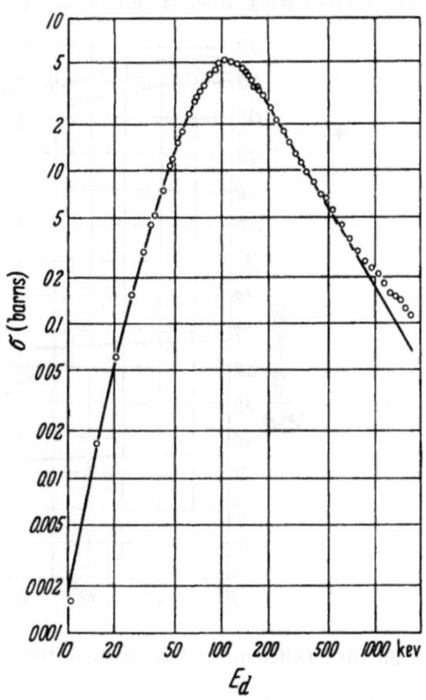

Fig. 100. Cross section of the reaction $T(d, n) He^4$ versus deuteron energy (BONNER and coworkers[2]).

particles forming an intermediate state of the compound nucleus He^5 with $J = \frac{3}{2}^+$ from which the product particles are emitted with relative angular

[1] H. V. ARGO, R. F. TASCHEK, H. M. AGNEW, A. HEMMENDINGER and W. T. LELAND: Phys. Rev. **87**, 612 (1952).

[2] J. P. CONNER, T. W. BONNER and J. R. SMITH: Phys. Rev. **88**, 468 (1952).

[3] See footnote 5, p. 188.

momentum $l=2$. Under this assumption it is possible to fit reasonably well the excitation curves of Fig. 99 (or 100) with the one-level Breit-Wigner dispersion formula[1-3].

$$\sigma = \frac{2J_c+1}{(2S+1)(2I+1)} \frac{4\pi}{k^2} \frac{\Gamma_n \Gamma_t}{4(E_t - E_{Rt})^2 + (\Gamma_n + \Gamma_t)^2} \tag{57.6}$$

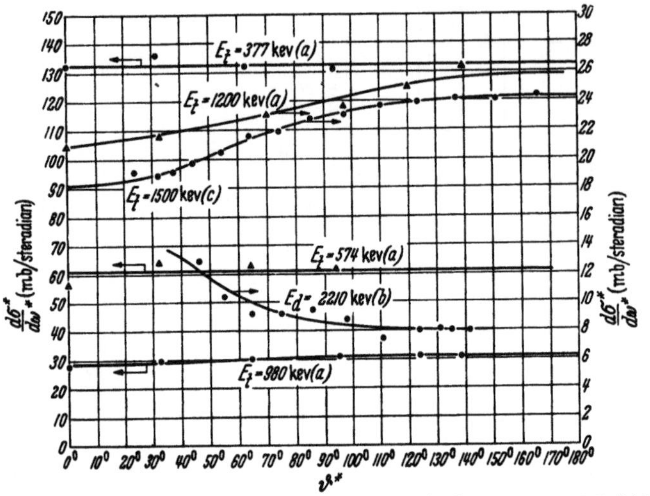

Fig. 101. Differential cross section of the T(d, n) reaction in the c.m.s. (a) footnote 1 on p. 195. (b) T. STRATTON and G.C. FREIER: Phys. Rev. **88**, 261 (1952). (c) A. HEMMENDINGER and H.V. ARGO: Phys. Rev. **98**, 70 (1955).

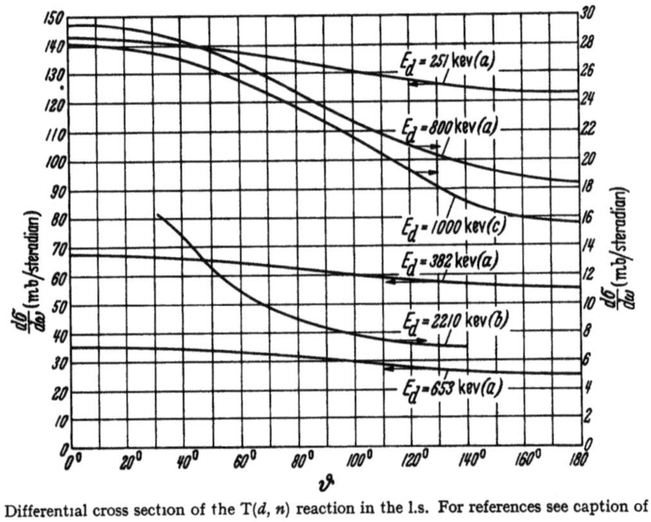

Fig. 102. Differential cross section of the T(d, n) reaction in the l.s. For references see caption of Fig. 101.

where J_c is the total angular momentum of the system, S the spin of the incident particle ($S=1$), I the spin of the target nucleus ($I=\frac{1}{2}$): Γ_n and Γ_t are the widths for emission of the neutron and of the triton, and are energy dependent according to the expression [3]

$$\Gamma = 2k\gamma^2 P$$

[1] See footnote 1, p. 195.

[2] See footnote 2, p. 195.

[3] B.H. FLOWERS: Proc. Roy. Soc. Lond., Ser. A **204**, 503 (1951). — J.L. JOHNSON and H.M. JONES: Phys. Rev. **93**, 1286 (1954).

where k is the wave number of the emitted particle, γ^2 the reduced width assumed to be constant (Sect. 26), and P the penetration factor relative to the Coulomb barrier which depends on the radius a of the interaction. The curve of Fig. 99 is obtained by assigning the following values to the adjustable parameters

$$J_c = \tfrac{3}{2}, \quad E_{Rt} = 194 \text{ kev}, \quad \Gamma_t = 165 \text{ kev}, \quad a = 7 \times 10^{-13} \text{ cm}.$$

E_{Rt} and Γ_t being given in the l.s. when the target deuterons are at rest. The values of the widths at resonance are not uniquely determined because by applying Eq. (57.6) at $E_t = E_{Rt}$ one obtains a second degree equation in the ratio Γ_n/Γ_t. In the present case one notices that at resonance the cross section is nearly equal to the maximum permissible value

$$\frac{2J_c + 1}{(2S+1)(2I+1)} \pi \lambda\!\!\!^{-2}$$

and therefore that both values of Γ_n/Γ_t must be very close to one.

At low energy ($E_d \leqq 120$ kev) the cross section data given in Fig. 97 can be very well fitted by a simple Gamow exponential form[1,2,3]:

$$\sigma = \frac{A}{E_d} e^{-\frac{44.40}{E_d^{\frac{1}{2}}}},$$

E_d being expressed in kev. The logarithmic slope does not agree very well with the value given by others[2].

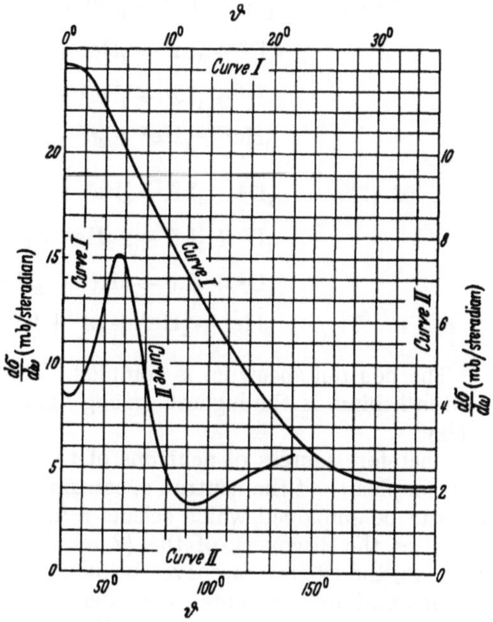

Fig. 103. Differential cross section of the T(d, n) reaction in the l.s. at 10.5 Mev[5].

For energies of the incident deuteron varying between 1.53 and 5.79 Mev, GALONSKY and JOHNSON[4] find that the total cross section deviates from the Breit-Wigner formula (57.6) in the same sense as shown in Fig. 100; it reaches a flat minimum at 3 to 4 Mev above which it increases slightly. This behaviour, combined with measurements at 10.5 Mev[5] may be qualitatively interpreted as evidence of a resonance at about 8 Mev, which would correspond to ≈ 22 Mev excitation in He^5. The angular distribution in this energy interval exhibits forward peaks characteristic of stripping, but there is no detailed agreement between the experimental results and theory[6].

Fig. 103 shows a smoothed curve of the measurements of the differential cross section in the l.s. made at $E_d = 10.5$ Mev by BROLLEY, FOWLER and STOVALL[5] which can be explained by assuming that the predominant process is stripping[6].

[1] See footnote 4, p. 188.
[2] See footnote 3, p. 196.
[3] R. G. JARVIS and D. ROAF: Proc. Phys. Soc. Lond. A **66**, 310 (1953).
[4] H. GALONSKY and C. JOHNSON: Phys. Rev. **104**, 421 (1956).
[5] J. E. BROLLEY, J. L. FOWLER and E. J. STOVALL: Phys. Rev. **82**, 502 (1951).
[6] S. T. BUTLER and J. L. SYMONDS: Phys. Rev. **83**, 858 (1951).

We shall also mention that the break-up of the incident deuteron in the d—T[1,2] reactions yields a low energy continuum spectrum of neutrons the intensity of which rises rapidly with increasing deuteron energy.

Details about deuterium or tritium targets in gaseous form or with these gases absorbed in thin layers of metals, particularly Zr, are given at the end of Sect. 53. In the case of deuterium heavy ice targets are frequently used[3].

58. The (*d, n*) reactions in other light elements (Li, Be, B, C). α) *Lithium.* The two isotopes of lithium when bombarded with deuterons undergo the following processes with emission of neutrons

7.52%:

$$ {}_1D^2 + {}_3Li^6 \rightarrow {}_4Be^7 + {}_0n^1 + 3.374\ \text{Mev}; \tag{58.1}$$

$$ {}_1D^2 + {}_3Li^6 \rightarrow {}_2He^4 + {}_2He^3 + {}_0n^1 + 1.791\ \text{Mev}; \tag{58.2}$$

$$\left.\begin{aligned} &{}_1D^2 + {}_3Li^6 \rightarrow {}_2He^5 + {}_2He^3 + \approx 0.84\ \text{Mev},\\ &{}_2He^5 \rightarrow {}_2He^4 + {}_0n^1 + 0.95\ \text{Mev}. \end{aligned}\right\} \tag{58.3}$$

92.47%:

$$ {}_1D^2 + {}_3Li^7 \rightarrow {}_4Be^8 + {}_0n^1 + 15.017\ \text{Mev}; \tag{58.4}$$

$$ {}_1D^2 + {}_3Li^7 \rightarrow 2\,{}_2He^4 + {}_0n^1 + 15.11\ \text{Mev}; \tag{58.5}$$

$$\left.\begin{aligned} &{}_1D^2 + {}_3Li^7 \rightarrow {}_2He^5 + {}_2He^4 + 14.16\ \text{Mev},\\ &{}_2He^5 \rightarrow {}_2He^4 + {}_0n^1 + 0.95\ \text{Mev}. \end{aligned}\right\} \tag{58.6}$$

The neutron spectrum is complex. The thick target yield of the (*d, n*) reaction in lithium is the highest among light elements for deuteron energies around 1 Mev.

Therefore sources of neutrons of the Li (*d, n*) type are frequently used when the complexity of the neutron spectrum is unimportant or when a neutron-detector can be employed which selects a narrow band of neutron energies.

After the discovery of the (*d, n*) reaction in lithium already mentioned in Sect. 56, the excitation curves below and around 1 Mev have been measured by various authors. In some experiments the absolute yields from a thick lithium target was compared with that of the other light elements[4], in others[5] the excitation curve of reaction (58.5) was investigated[6], in others[7] by means of thin targets of separated isotopes, the differential excitation curve for most of the above mentioned processes in lithium was measured (Fig. 104).

From natural lithium targets one obtains mainly information about the behaviour of Li^7 because of its larger abundance and slightly higher cross section compared to Li^6. For energies of the incident deuteron around 1 Mev one can estimate that Li^6 contributes about 8% of the total number of neutrons[7].

The differential excitation curve for emission of neutrons by Li^6 and Li^7 in the forward direction for $E_d \leq 2.2$ Mev was measured by WHALING *et al.*[8] with a

[1] See footnote 5, p. 188.
[2] R.L. HENKEL, J.E. PERRY and R.M. SMITH: Phys. Rev. **99**, 1050 (1955).
[3] See footnote 2, p. 191.
[4] E. AMALDI, L.R. HAFSTAD and M.A. TUVE: Phys. Rev. **51**, 896 (1937).
[5] J.H.W. WILLIAMS, W.G. SHEPHERD and R.O. HAXBY: Phys. Rev. **51**, 888, 1011 (1937); **52**, 390 (1937).
[6] M.L.E. OLIPHANT, E.S. SHIRE and B.M. CROWTHER: Proc. Roy. Soc. Lond. **146**, 922 (1934).
[7] L.H. RUMBAUGH and L.R. HAFSTAD: Phys. Rev. **50**, 681 (1936). — L.H. RUMBAUGH, R.B. ROBERTS and L.R. HAFSTAD: Phys. Rev. **54**, 657 (1938).
[8] W. WHALING, J.E. EVANS and T.W. BONNER: Phys. Rev. **75**, 688 (1949).

proportional counter filled with argon and biased to count only neutrons of energy larger than 1 Mev.

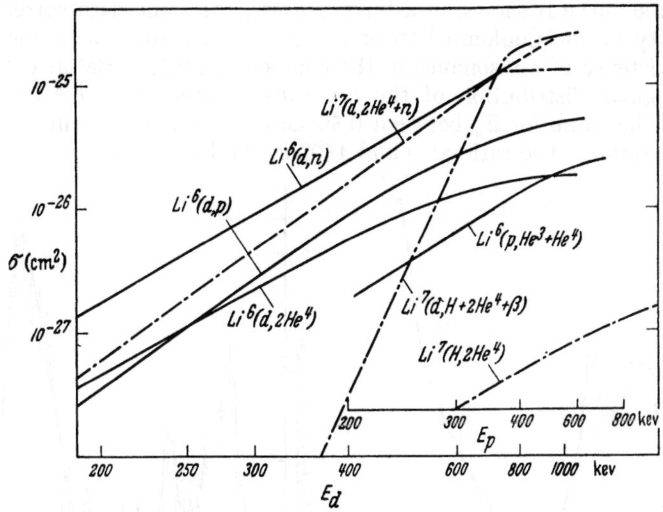

Fig. 104. Cross sections for lithium reactions (RUMBAUGH, ROBERTS and HAFSTAD).

Fig. 105. Cross section for $Li^6(d, n)$ taken at 90° ± 20° to the deuteron beam. Curve A shows the experimental results. Curve B shows the data corrected for Coulomb penetrability (BAGGETT and BAME).

In Figs. 105 and 106 the differential cross section at 90° obtained by BAGGETT and BAME[1] is plotted as a function of E_d in the same energy interval. These

[1] L. M. BAGGETT and S. J. BAME: Phys. Rev. **85**, 434 (1952); see also F. HIRST, I. JOHNSTONE and M. J. POOLE: Phil. Mag. **45**, 762 (1954).

authors detected the neutrons emitted at 90° ± 20° from thin targets of Li^6 and natural lithium, by a long counter. In Fig. 105 curve A represents the cross section while curve B shows, in arbitrary units, the same data corrected for the penetrability of the Coulomb barrier in order to enhance the evidence for the possible existence of a resonance in the compound nucleus Be^8 at 0.4 Mev.

The angular distribution of the neutrons emitted from Li^6 appears to be essentially the same for E_d between 0.59 and 1.7 Mev, with a maximum in the forward direction. The ratio at 0 and 150° in the l.s. is 7:4[1].

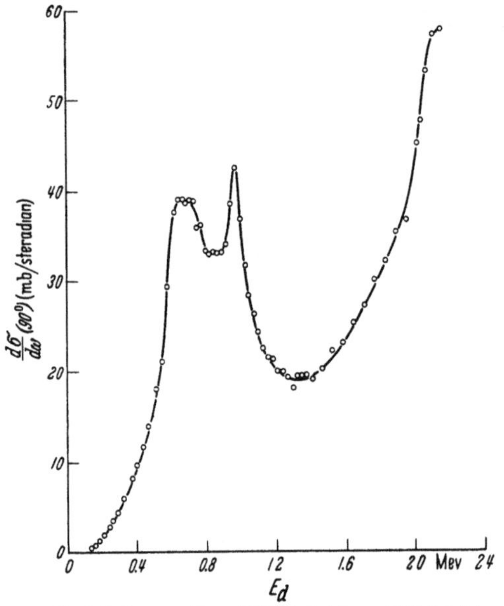

Fig. 106. Cross section of $Li^7(d, n)$ taken at 90° ± 20° to the deuteron beam (Baggett and Bame). The data are uncorrected for Coulomb penetrability.

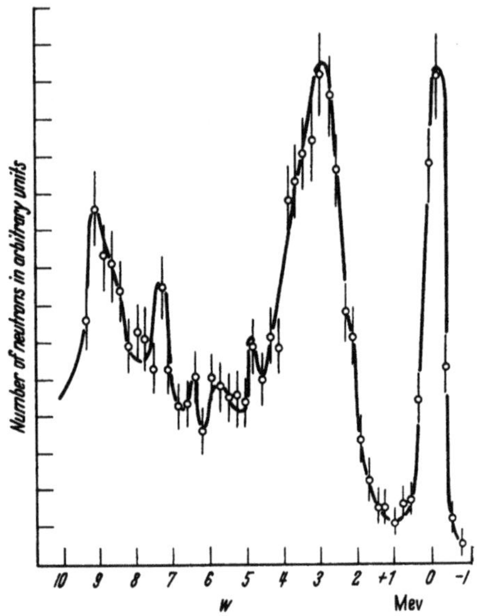

Fig. 107. Neutron spectrum of the $Li^7(d, n)$ reaction. The abscissa indicates the excitation energy of Be^8. The observations have been made for $E_d = 686$ kev at $\vartheta = 90$, 115 and 145° (Ihsan).

The angular distribution of the neutrons from natural lithium was investigated by Trumpy *et al.*[2] who measured the differential excitation curve with a BF_3 counter surrounded by a paraffin cylinder, and the spectrum of the neutrons emitted at different angles in the l.s. (0, 20, 50, 90, 110, 120, 140°) by photographic plates. In the c.m.s. the angular distribution of the neutrons corresponding to the ground state and to the excited states of Be^8 at about 3 Mev appears to show an appreciable increase in intensity going from forward to backward directions with respect to the incident deuterons.

The spectrum of the neutrons was measured by many authors. Some observed recoiling protons or helium nuclei in cloud chambers[3], others recoiling protons in nuclear photographic plates[4,2] or with a spectrometer of counters in coincidence[5].

[1] See footnote 8, p. 198.

[2] B. Trumpy, T. Grotdal and A. Graue: Nature, Lond. **170**, 1118 (1952); Univ. i Bergan Årbok 1953, Naturvit. uk. 8.

[3] T.W. Bonner and W.M. Brubaker: Phys. Rev. **47**, 973 (1935); **48**, 472 (1935). — W.E. Stephens: Phys. Rev. **53**, 212, 223 (1938). — H. Staub and W.E. Stephens: Phys. Rev. **54**, 236, 237 (1938); **55**, 845 (1939).

[4] H.T. Richards: Phys. Rev. **59**, 796 (1941). — L.L. Green and W.M. Gibson: Proc. Phys. Soc. Lond. **62**, 407 (1949). — Cltala, Aquilor and Busquets: An. Real Soc. españ. Fis. y Quim. (Madrid) A **49**, 131 (1953). — G.R. Reid: Proc. Phys. Soc. Lond. **67**, 466 (1954). — M.A. Ihsan: Phys. Rev. **98**, 689 (1955). — Proc. Phys. Soc. Lond. **68**, 393 (1955).

[5] C.C. Trail and C.H. Johnson: Phys. Rev. **95**, 1363 (1954).

In Fig. 107 the results of IHSAN are shown for Li^7. These results were obtained with emulsion technique at $E_d = 686$ kev. The several groups of neutrons correspond, at least in part, to various excited states of Be^8.

Both reactions (58.5) and (58.6) give continuous spectra of neutrons, the first as a consequence of the three-body disintegrations, the latter because the neutrons are emitted by He^5 while still recoiling. The continuous spectrum of neutrons of reaction (58.5) extends to the highest observed energies as a consequence of its high Q-value, while the spectrum of reaction (58.6) contributes to the order of $\frac{3}{4}$ of the neutrons observed below 4 Mev[1-3].

The spectrum of the neutrons observed by TRAIL and JOHNSON[4] to be emitted in the forward direction by Li^7 under bombardment of deuterons of $E_d = 2$ Mev appears to be appreciably simpler than that of Fig. 105 which agrees, however, satisfactorily with all previous observations.

The group of neutrons of higher energy corresponds to the formation of Be^8 from reaction (58.4) in its ground state. But the interpretation of the various other groups in terms of excited levels of Be^8, does not appear completely clear. The paper of TITTERTON[5] and the review articles of AJZENBERG and LAURITSEN [*1*], [*2*] contain a detailed discussion of this point. According to TITTERTON there is evidence for excited states at 4.05, 5.3, 7.5 Mev and possibly at 2.2, 3.4, 4.62 and 4.9 Mev (this last an even state, not a γ-emitter) but not for a 4.9 Mev γ-emitting state (odd state).

Lithium bombarded with deuterons emits γ-rays of 400 kev — arising from the reaction $Li^6(d, p)\,Li^7$ in which two groups of protons are emitted[6] — and of 4.9 Mev whose excitation curve follows very closely that for emission of neutrons[7]. BENNETT *et al.* attributed this radiation to an excited state of Be^8 but according to TITTERTON it seems more reasonable to attribute it to the competing reaction $Li^7(d, p)\,Li^8(\beta^-)$.

The number of photons from this radiation is estimated to be about 3% of the number of neutrons, an intensity which may disturb certain types of experiments.

β) Beryllium. The reaction

$$_4Be^9 + {}_1D^2 \rightarrow {}_5B^{10} + {}_0n^1 + 4.36 \text{ Mev} \tag{58.7}$$

has been used extensively as an intense source of fast neutrons since its discovery and early investigation[8]. The corresponding thick target absolute yield up to $E_d = 1$ Mev has been measured by integrating over a large volume of water surrounding the Be target, the activity induced in a thermal-neutron detector[9]. The neutron intensity from thin targets (10 to 100 kev equivalent thickness) has been measured at 0 and 135° for deuteron energies ranging from $E_d = 0.2$ Mev to $E_d = 1.8$ Mev[10]. The neutrons were detected by means of proportional counters which were filled, in different runs, with argon or with hydrogen.

[1] See footnote 5, p. 198.
[2] See footnote 3, p. 200.
[3] See footnote 4, p. 200.
[4] See footnote 5, p. 200.
[5] E. W. TITTERTON: Phys. Rev. **94**, 206 (1954).
[6] See footnote 7, p. 198.
[7] W. E. BENNET, T. W. BONNER, H. T. RICHARDS and B. E. WATT: Phys. Rev. **71**, 11 (1947).
[8] H. R. CRANE, C. C. LAURITSEN and A. SOLTAN: Phys. Rev. **44**, 692 (1933); **45**, 507 (1934). — M. S. LIVINGSTON, M. C. HENDERSON and E. O. LAWRENCE: Phys. Rev. **44**, 782 (1933). — E. O. LAWRENCE and D. COOKSEY: Phys. Rev. **50**, 1131 (1936).
[9] See footnote 4, p. 198.
[10] J. E. EVANS, C. W. MALICH and J. R. RISSER: Phys. Rev. **75**, 1161 (1949).

Unfortunately the results of these experiments have not been corrected for variations of the cross section and of the efficiency of detection with neutron energy.

The neutron spectrum has been measured by many authors who observed the recoil protons by means of cloud chambers, nuclear emulsions and proportional counter telescopes[1-3]. Fig. 108 shows the recoil proton spectra observed in nuclear plates by PRUITT *et al.*[3] at $\vartheta = 35.8°$ and $\vartheta = 137°1$ with respect to the deuteron beam of $E_d = 0.945$ Mev energy. The group of higher energy corresponds to the ground state of the residual nucleus B^{10} [$Q_0 = (4.43 \pm 0.08)$ Mev] while the four other groups are due to excited levels of B^{10} whose excitation energies are: $w_1 = 0.74 \pm 0.11$, $w_2 = 1.77 \pm 0.10$, $w_3 = 2.20 \pm 0.10$, $w_4 = 3.67 \pm 0.09$ Mev. Spectra similar to those of Fig. 106 have been observed with the same technique by AJZENBERG[2] at $E_d = 3.45$ Mev. From an analysis of these spectra the existence of ten more excited states of B^{10} distributed between 4.78 and 6.77 Mev was proved. For a detailed discussion of all these levels and of the corresponding γ-rays emission[4] we refer to the review articles [*1*], [*2*].

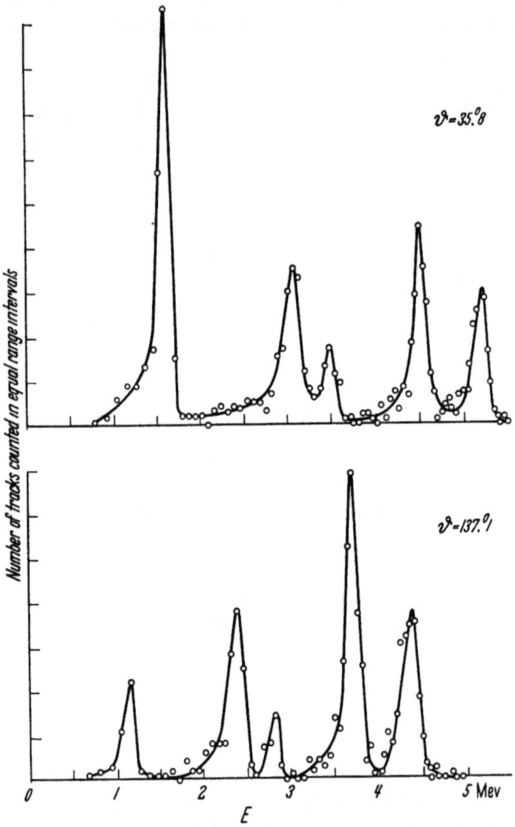

Fig. 108. Recoil proton spectrum of the $Be^9(d, n)$ reaction at $\vartheta = 35.8°$ and $\vartheta = 137.1°$ for $E_d = 0.945$ Mev (PRUITT *et al.*).

An analysis of the angular distribution of the neutrons from the ground state and from the first few excited states at $E_d = 0.86$ Mev[5], $E_d = 0.95$ Mev[3] and $E_d = 3.4$ Mev[2], show evidence for both stripping and compound nucleus formation. At $E_d \approx 0.9$ Mev the neutron group to the excited state $w_4 = 3.67$ Mev shows a strong forward peak with maximum at about 50°, typical of the angular distribution for deuteron stripping[3,6].

The angular distributions of the other groups of neutrons show only some structure in the forward hemisphere suggesting the stripping maximum. One

[1] T.W. BONNER and W.M. BRUBAKER: Phys. Rev. **50**, 308 (1936). — H. STAUB and W.E. STEPHENS: Phys. Rev. **55**, 131 (1939). — C.F. POWELL and G.E.F. FERTEL: Nature, Lond. **144**, 115 (1939). — G.C. REID: Proc. Phys. Soc. Lond. **67**, 466 (1954).

[2] F. AJZENBERG: Phys. Rev. **88**, 298 (1952).

[3] J.S. PRUITT, C.D. SWARTZ and S.S. HANNA: Phys. Rev. **87**, 534 (1952); **92**, 1456 (1953).

[4] V.K. RASMUSSEN, C.C. LAURITSEN and T. LAURITSEN: Phys. Rev. **75**, 199 (1949).

[5] L.L. GREEN, J.P. SCANLON and J.C. WILLMOTT: Proc. Phys. Soc. Lond. **68**, 386 (1955).

[6] See footnote 5, p. 201.

can conclude that at $E_d \approx 0.9$ Mev the $w_4 = 3.67$ Mev level is formed to a large extent by deuteron stripping, while both stripping and compound nucleus formation gives comparable contributions to the production of the other groups.

The contribution of the stripping process disappears at lower deuteron energies while it increases, at least in general, for higher values of E_d[1].

γ) Boron. Normal boron consists of about 19% B^{10} and 81% B^{11}, the isotopic composition depending somewhat on the source of the boron.

Under deuteron bombardment the two isotopes of boron undergo the following reactions[2]:

$$\left.\begin{aligned} {}_5B^{10} + {}_1D^2 &\to {}_6C^{11} + {}_0n^1 + 6.472 \text{ Mev}; \\ {}_6C^{11}(\beta^+)\,{}_5B^{11} &\begin{cases} E^{(+)}_{\max} = (0.985 \pm 0.005) \text{ Mev}, \\ T_{\frac{1}{2}} = (20.39 \pm 0.06) \text{ min}. \end{cases} \end{aligned}\right\} \tag{58.8}$$

$${}_5B^{11} + {}_1D^2 \to {}_6C^{12} + {}_0n^1 + 13.724 \text{ Mev} \tag{58.9}$$

$$\to 3\,{}_2He^4 + {}_0n^1 + 6.5 \text{ Mev}. \tag{58.10}$$

According to the early investigations mentioned above the yields of the two processes (58.9) and (58.10) undergone by B^{11} are of the same order of magnitude (8:11 at $E_d = 375 - 560$ kev).

Figs. 109 and 110 show the excitation curves for emission of neutrons in the forward direction under deuteron bombardment of thin targets of the two separated isotopes[3] (the equivalent thickness of the targets was 30 to 40 kev at $E_d = 1$ Mev).

The curves marked "all neutrons" were obtained with a long-counter subtending a cone of half-angle about 35°. The curves marked "highest Q group" represents the results obtained with a proportional counter filled with helium and biased to respond only to neutrons of the highest Q-values ($Q_0 = 6.5$ for B^{10}, and $Q_0 = 13.7$ Mev and $Q_1 = 9.3$ Mev for B^{11}). The He counter subtended a cone of half-angle about 50°.

The long-counter was calibrated with a Po α + Be neutron source of known flux (±10%). Therefore BURKE *et al.* can estimate that one unit of the ordinate of the "all neutron" curves correspond to

$$\frac{d\sigma}{d\Omega} = 2 \times 10^{-27} \text{ cm}^2/\text{sterad} \quad \text{at} \quad \vartheta = 0^\circ \quad \text{for} \quad B^{10} \quad (\text{Fig. 109}); \tag{58.11}$$

$$\frac{d\sigma}{d\Omega} = 5 \times 10^{-27} \text{ cm}^2/\text{sterad} \quad \text{at} \quad \vartheta = 0^\circ \quad \text{for} \quad B^{11} \quad (\text{Fig. 110}). \tag{58.12}$$

The neutron spectra have been extensively investigated by observing helium and proton recoils with cloud chambers[4] and proton recoils with nuclear emulsions[5-7].

---

[1] See footnote 2, p. 202.

[2] C.C. LAURITSEN and H.R. CRANE: Phys. Rev. **45**, 493 (1934). — J.D. COCKCROFT, C.W. GILBERT and E.T.S. WALTON: Proc. Roy. Soc. Lond. **148**, 225 (1935). — J.D. COCKCROFT and W.B. LEWIS: Proc. Roy. Soc. Lond. **154**, 246 (1936).

[3] W.H. BURKE, J.R. RISSER and G.C. PHILLIPS: Phys. Rev. **93**, 188 (1954).

[4] T.W. BONNER and W.M. BRUBAKER: Phys. Rev. **50**, 308 (1936).

[5] H. STAUB and W.E. STEPHENS: Phys. Rev. **55**, 131 (1936). — C.F. POWELL and G.E.F. FERTEL: Nature, Lond. **144**, 115 (1939). — V.R. JOHNSON: Phys. Rev. **86**, 302 (1952). — M.A. IHSAN: Proc. Phys. Soc. Lond. A **68**, 323 (1955).

[6] G.H. PARIS and P.M. ENDT: Physica, Haag **20**, 585 (1954).

[7] A. GRAUE: Phil. Mag. **1**, 1027 (1956).

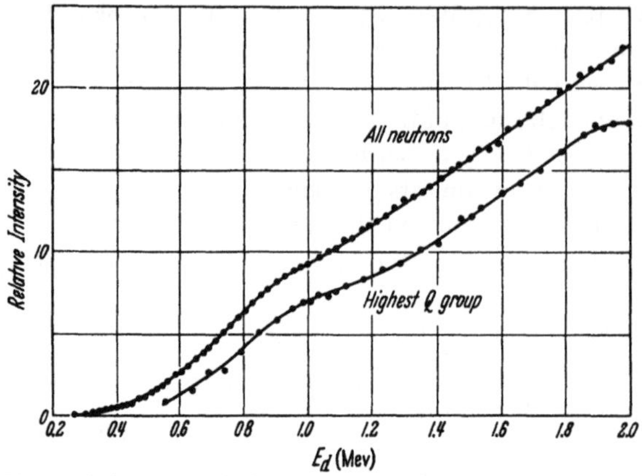

Fig. 109 Thin target excitation curves of the B^{10} (d, n) reaction (BURKE *et al.*).

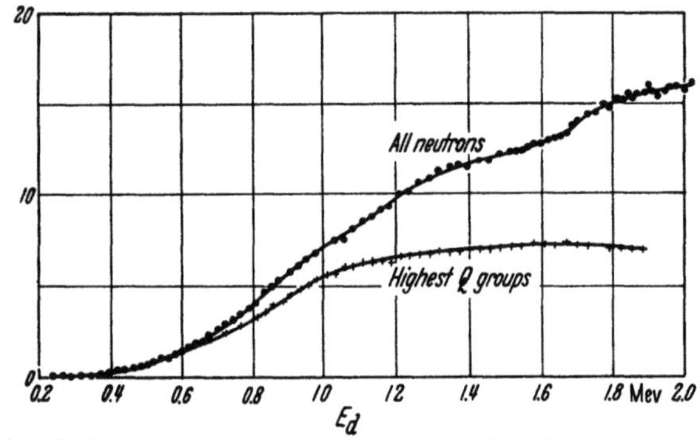

Fig. 110. Thin target excitation curves of the B^{11} (d, n) reaction (BURKE *et al.*).

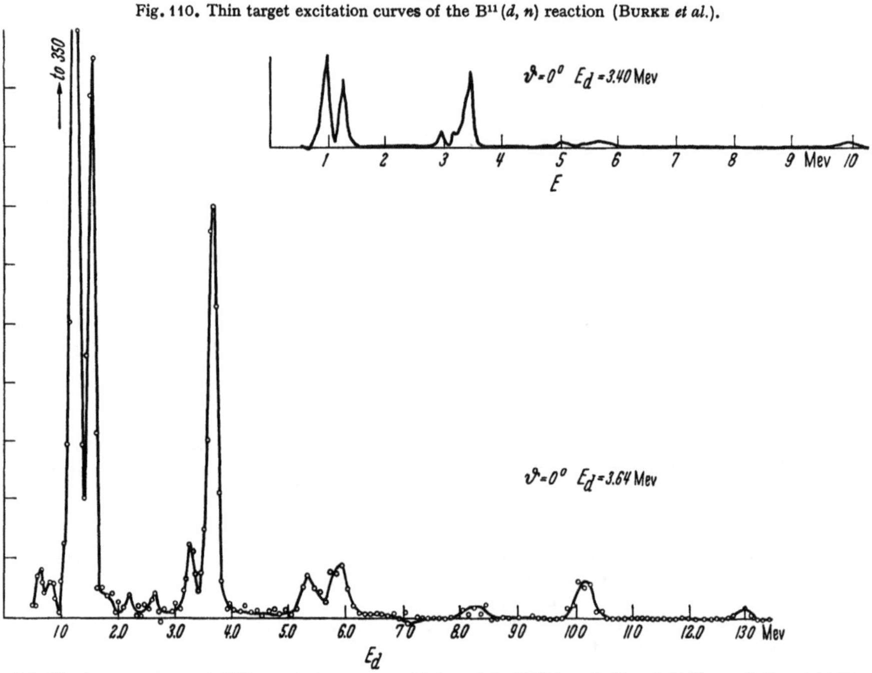

Fig. 111. Neutron spectrum at 0° from deuterons on a 96% enriched B^{10} target, $E_d = 3.40$ Mev and $E_d = 3.64$ Mev (JOHNSON).

Fig. 111 shows the neutron spectrum observed by JOHNSON[1] at 0° from deuterons of $E_d = 3.40$ Mev and $E_d = 3.64$ Mev on 96% enriched B^{10} targets. The data have been corrected for background and for the variations with energy of both the *n*-*p* cross section and the probability of the tracks leaving the emulsion.

The results obtained by PARIS and ENDT[2] at $E_d = 0.58$ Mev and $\vartheta = 0°$ for B^{10} and by IHSAN[2] at $E_d = 0.87$ Mev and $\vartheta = 95°$ for B^{11} are shown in Figs. 112 and 113. The continuous distribution at low energies appearing in this latter figure probably arises from reaction (58.10). The contamination of the target by deuterons producing the D+D neutrons is responsible for the peak marked D in Fig. 112, while in Fig. 113 it is submerged in the continuous distribution mentioned above.

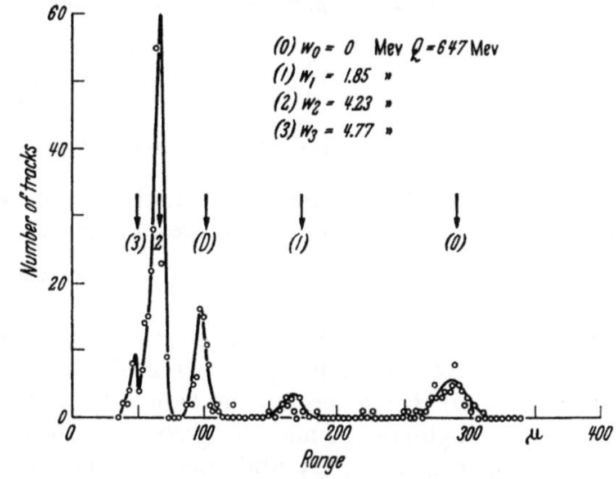

Fig. 112. Neutron spectrum at 0° from a B^{10} target bombarded with deuterons of $E_d = 0.58$ Mev. The neutron group marked D results from the D(*d*, *n*) He^3 reaction (PARIS and ENDT).

For a detailed discussion of the neutron groups appearing in the spectra of Figs. 111 to 113 and the corresponding excited states of C^{11} and C^{12} we refer to the review articles [*1*], [*2*].

The angular distribution of the B^{10} neutrons — studied by PARIS and ENDT[2] at $E_d = 0.58$ Mev, by BURKE *et al.*[3] at $E_d = 0.71$; 1.06 and 1.43 Mev and by GRAUE[4] at 1.08 Mev — shows a contribution of the stripping process even at these low energies. The angular distribution of the B^{11} neutrons has been studied by IHSAN[1] at $E_d = 0.87$ Mev; by BURKE *et al.* at $E_d = 0.71$; 1.0; 1.59 Mev and by GIBSON[5] at $E_d = 8.1$ Mev.

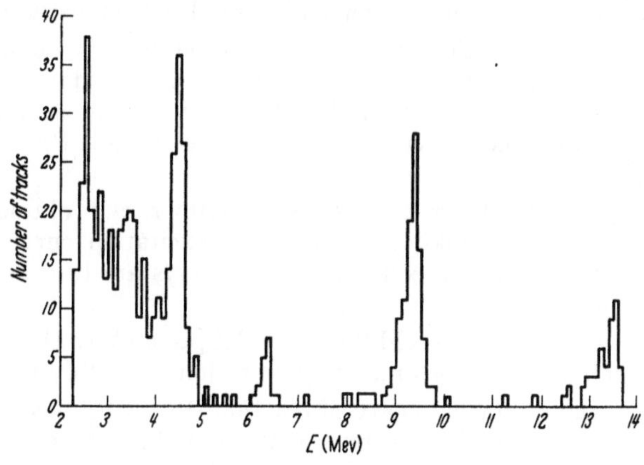

Fig. 113. Neutron spectrum at 95° of the B^{11}(*d*, *n*) reaction for $E_d = 0.87$ Mev (IHSAN).

At the lower energy there is no clear-cut evidence for stripping while this process appears to give the main contribution to the formation of the first excited state (4.47 Mev).

The γ-line corresponding to the transition from this level to the ground state is also observed at low energies of the incident deuteron. The corresponding yield at $E_d = 1.56$ Mev from a thick 100% B^{11} target is 27×10^{-6} γ-rays per deuteron. At higher energies ($E_d > 10.8$ Mev) a 15.2 Mev γ-ray is also observed [*1*], [*2*].

[1] See footnote 5, p. 203.
[2] See footnote 6, p. 203.
[3] See footnote 3, p. 203.
[4] See footnote 7, p. 203.
[5] W.M. GIBSON: Proc. Phys. Soc. Lond. A **62**, 586 (1949). — Phil. Mag. **44**, 297 (1953).

At $E_d = 2$ Mev B^{10} shows two γ-lines with $h\nu = 6.52$ Mev and $h\nu = 7.01$ Mev whose production cross sections are 12 mb and 6 mb respectively [1], [2].

δ) Carbon. Normal carbon consists of two isotopes C^{12} and C^{13} whose abundances are 98.9 and 1.1% respectively. Under deuteron bombardment the following reactions take place

$$\left.\begin{aligned} &{}_6C^{12} + {}_1D^2 \rightarrow {}_7N^{13} + {}_0n^1 - 0.281 \text{ Mev};\\ &\qquad {}_7N^{13}(\beta^+)\,{}_6C^{13} \left\{\begin{aligned} E_{\max}^{(+)} &= (1.202 \pm 0.005 \text{ Mev}),\\ T_{\frac{1}{2}} &= (10.1 \pm 0.1) \text{ min}.\end{aligned}\right.\end{aligned}\right\} \tag{58.13}$$

$${}_6C^{13} + {}_1D^2 \rightarrow {}_7N^{14} + {}_0n^1 + 5.317 \text{ Mev}. \tag{58.14}$$

The first of these processes constitutes the only example of (d, n) reaction with a negative Q-value. The corresponding threshold is

$$E_{1T} = 0.328 \text{ Mev}. \tag{58.15}$$

This reaction is frequently used for producing monoenergetic low-energy neutrons when the maximum energy of the available accelerator is below the threshold of the $T(p, n)He^3$ reaction (see Table 33, p. 171). If one uses targets of normal carbon the high-energy neutron group due to reaction (58.14) represents roughly 1% of the total intensity and therefore one can introduce a corresponding correction of the results. It is better to use a target of the separated isotope C^{12}, but also in this case difficulties arise from the presence of D+D neutrons which may contribute substantially to the emission for very low values of E_d.

Thin carbon targets have been made with thin soot deposit, evaporated paraffin, or India ink on tantalum backing material. Since carbon is always present as a contamination in vacuum systems employing oil diffusion pomps, reaction (58.13) can never be avoided in any deuteron accelerator.

The excitation curve from carbon targets bombarded with deuterons has been measured by many authors: some used a thermal-neutron detector for measuring the neutron distribution in water surrounding the target[1], others[2] measured the activity of N^{13} or made use of a BF_3 counter surrounded with paraffin[3].

Fig. 114 shows the results of BAILY *et al.*[4]. obtained with a long counter at $\vartheta = 0°$ and $\vartheta = 90°$ as well as the excitation curve of γ-rays obtained with a thick-wall Geiger counter. These γ-rays are to a great part due to the process $C^{12}(d, p)C^{13}$.

Between threshold (58.15) and $E_d \approx 0.6$ Mev the neutron cross section is very small and increases regularly so to join smoothly the curves of Fig. 114[1,2].

The excitation curve of reaction (58.14) has been measured by RICHARDSON[5] who detected the neutrons emitted in the forward direction from a C^{13} enriched target (48%) by means of a proportional counter filled with helium at 19 atmospheres for E_d varying between 0.54 and 2.1 Mev, and a long counter for E_d between 0.2 and 0.45 Mev.

---

[1] See footnote 4, p. 198.

[2] H.W. NEWSON: Phys. Rev. **48**, 790 (1935). — R.B. ROBERTS and N.P. HEYDENBURG: Phys. Rev. **53**, 374 (1938). — T.W. BONNER, E. HUDSPETH and W.E. BENNET: Phys. Rev. **57**, 1075 (1940); **58**, 185 (1940).

[3] C.L. BAILEY, M. PHILLIPS and J.H. WILLIAMS: Phys. Rev. **62**, 80 (1942).

[4] C.L. BAILEY, G. FREIER and J.H. WILLIAMS: Phys. Rev. **73**, 274 (1948).

[5] J.E. RICHARDSON: Phys. Rev. **80**, 850 (1950).

The analysis of this excitation curve as well as of that of Fig. 114 shows the existence of many excited levels of the compound nuclei N^{15} and N^{14} for which we refer to the review articles [*1*], [*2*].

For E_d below 2.5 Mev the neutron spectrum of reaction (58.13) consists of a single line corresponding to the indicated Q-value[1], because the first level of the residual nucleus N^{13} lies at $w_1 = (2.29 \pm 0.12)$ Mev. For $E_d = 8$ Mev[1] or

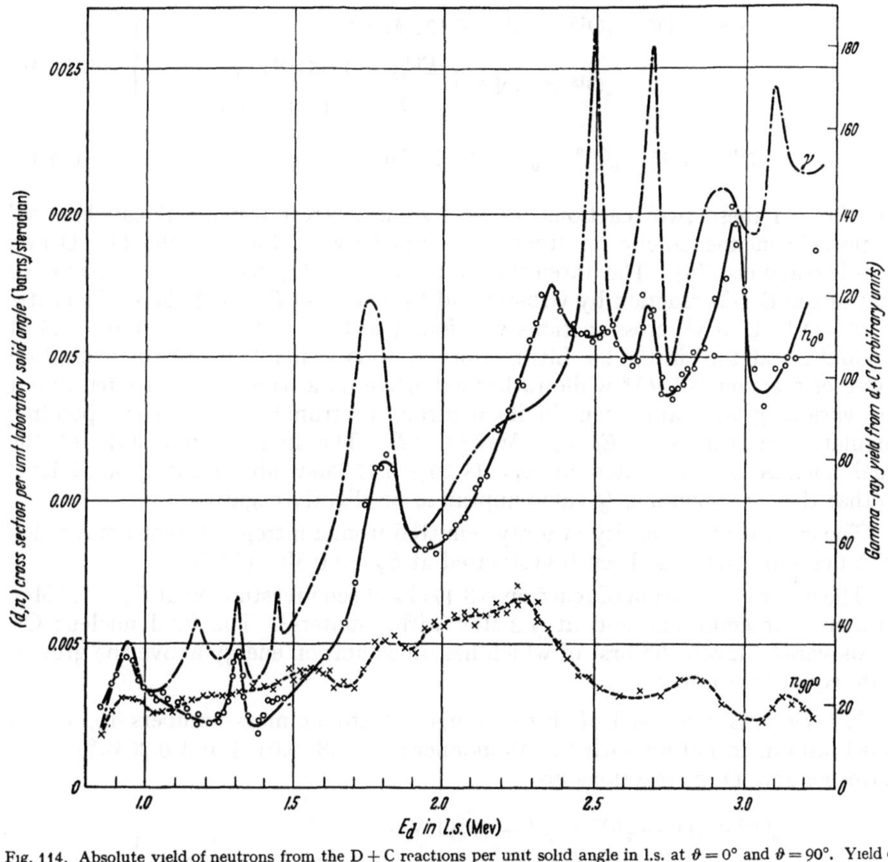

Fig. 114. Absolute yield of neutrons from the D + C reactions per unit solid angle in l.s. at $\vartheta = 0°$ and $\vartheta = 90°$. Yield of gamma-rays to an arbitrary scale (Baily *et al.*).

10 Mev[2] the neutron spectrum shows various groups due to the formation of N^{13} in excited levels. The corresponding angular distributions have been studied and analyzed for stripping process formation[1].

The spectrum of reaction (58.14) has been measured at $E_d = 0.86$ Mev by Green *et al.*[3] and at $E_d = 3.89$ Mev by Benenson[4] for various angles of emission of the neutrons, and the corresponding angular distributions have been analyzed for compound-nucleus formation and stripping process.

For details on the conclusions of this type of analysis as well as for a more extensive bibliography we refer to[3] and to the review articles [*1*], [*2*].

[1] R. Middleton, F.A. El-Bedewi and C.T. Tai: Proc. Phys. Soc. Lond. A **66**, 95 (1953).
[2] J.C. Grosskreutz: Phys. Rev. **76**, 482 (1949).
[3] See footnote 2, p. 202.
[4] R.E. Benenson: Phys. Rev. **90**, 420 (1953).

Evidence has been obtained by HAEBERLI and ROLAND[1] that the neutrons emitted in reaction (58.13) are polarized. The observations were made at 20° for $E_d = 2.95$ Mev.

ε) Nitrogen in nature consists of two isotopes, N^{14} and N^{15} representing respectively 99.635 and 0.365%. The following two reactions take place under deuteron bombardment

$$\left.\begin{array}{l} {}_7N^{14} + {}_1D^2 \rightarrow {}_8O^{15} + {}_0n^1 + 5.122 \text{ Mev}; \\ \qquad {}_8O^{15}(\beta^+)\,{}_7N^{15} \left\{\begin{array}{l} E_{\max}^{(+)} = 1.683 \text{ Mev}, \\ T_{\frac{1}{2}} = (11.80 \pm 0.6) \text{ sec}. \end{array}\right. \end{array}\right\} \quad (58.16)$$

$$ {}_7N^{15} + {}_1D^2 \rightarrow {}_8O^{16} + {}_0n^1 + 9.885 \text{ Mev}. \quad (58.17)$$

The first of these two reactions presents some interest because it can be used to provide monoenergetic neutrons of energy between those of the D+D and D+T reactions [*10*]. The intensity of process (58.16) has been compared to that of the D+D reaction by GIBSON and LIVESEY[2] at $E_d = 0.93$ Mev. The ratio of D+D to D+N^{14} cross sections was found to be 4.4 at $\vartheta = 0$ and 30°, 1.5 at 90° and 2.1 at 150° in the l.s. Little information is available about the excitation curve of reaction (58.16)[3] while rather extensive data have been collected about the various groups appearing in its neutron spectrum[4] and the corresponding angular distributions at $E_d = 7.7$ Mev [*1*], [*2*]. The first excited state of the final nucleus O^{15} is located at $w_1 = (5.29 \pm 0.17)$ Mev above the ground level so that the corresponding Q-value appear to be slightly negative.

The energy and intensity of γ-rays emitted from a nitrogen target bombarded with deuterons has also been investigated at $E_d = 4.0$ Mev [*1*], [*2*].

The neutron spectrum of reaction (58.17) has been investigated at $E_a = 1.2$ Mev with nuclear emulsions and at 3.2 Mev with counters[5]. The final nucleus O^{16} shows various levels the first of which has an excitation energy above the ground state equal to 6.1 Mev.

ζ) Oxygen is composed of three isotopes of atomic mass numbers 16, 17, 18 which appear in nature with the abundancies 99.758, 0.0373 and 0.2039%. The corresponding (d, n) reactions are

$$\left.\begin{array}{l} {}_8O^{16} + {}_1D^2 \rightarrow {}_9F^{17} + {}_0n^1 - 1.631 \text{ Mev}; \\ \qquad {}_9F^{17}(\beta^+)\,{}_8O^{17} \left\{\begin{array}{l} E_{\max}^{(+)} = (1.748 \pm 0.006) \text{ Mev}, \\ T_{\frac{1}{2}} = (66.0 \pm 0.5) \text{ sec}. \end{array}\right. \end{array}\right\} \quad (58.18)$$

$$\left.\begin{array}{l} O^{17} + {}_1D^2 \rightarrow {}_9F^{18} + {}_0n^1 + 3.384 \text{ Mev}; \\ \qquad {}_9F^{18}(\beta^+)\,{}_8O^{18} \left\{\begin{array}{l} E_{\max}^{(+)} = (649 \pm 9) \text{ kev}, \\ T_{\frac{1}{2}} = (112 \pm 1) \text{ min}. \end{array}\right. \end{array}\right\} \quad (58.19)$$

$$ O^{18} + {}_1D^2 \rightarrow {}_9F^{19} + {}_0n^1 + 5.729 \text{ Mev}. \quad (58.20)$$

---

[1] W. HAEBERLI and W.W. ROLAND· Private communication.
[2] W.M. GIBSON and D.L. LIVESEY: Proc. Phys. Soc. Lond. **60**, 523 (1948).
[3] H.W. NEWSON: Phys. Rev. **51**, 620 (1937).
[4] W.E. STEPHENS, K. DJANAB and T.W. BONNER: Phys. Rev. **52**, 1079 (1937). — W.B. ROSE, E.L. HUDSPETH and N.P. HEYDENBURG: Phys. Rev. **87**, 383 (1952).
[5] E.L. HUDSPETH and C.P. SWANN: Phys. Rev. **76**, 464 (1949). — D.C. WORTH. Phys. Rev. **78**, 378 (1950).

Very few data are available about the excitation curve of (58.18), from $E_d=1.8$ to 2.08 Mev[1] and from $E_d=2$ to 5 Mev[2], about the neutron groups[3], their angular distribution and analysis [*1*], [*2*].

Still less is known about process (58.19)[4]. The neutron spectrum of reaction (58.20) at $E_d=2$ Mev has been investigated with nuclear emulsions exposed at 0 and 20°. Four excited levels of F^{18} have been found the first of which lies at 0.9 Mev above ground state[5].

η) Fluorine. The reaction

$$_9F^{19} + {}_1D^2 \rightarrow {}_{10}Ne^{20} + {}_0n^1 + 10.645 \text{ Mev} \tag{58.21}$$

shows various groups[6,4]. The angular distribution of the neutrons corresponding to formation of Ne^{20} in its ground state has been measured and analyzed[7]. For more details about the γ-rays and α-particle emission from excited states of Ne^{20} see Refs.[4], [*1*], [*2*].

VII. Neutrons in nature.

For the sake of completeness brief mention will be made of two natural sources of neutrons, cosmic rays and spontaneous fission.

59. Cosmic rays. The first evidence of the occurrence of neutrons in cosmic radiation was found in 1936 by Rambaugh and Locher[8] from observations of recoil protons in photographic emulsions exposed in a high altitude balloon flight. Observations made independently some time later by Fünfer[9] at sea level and mountain altitudes confirmed this conclusion; the difference between the numbers of counts given by a proportional counter lined with a layer of boron, when recorded without and with a thick Cd absorber, was already appreciable at sea level and increased rapidly with altitude.

In view of the short life-time of neutrons, cosmic-ray neutrons must originate at least within the solar system. Most of them are presumably produced by secondary events in the atmosphere. Primary cosmic-ray particles colliding with nuclei in the air produce, among other products, high-energy nucleons. These high-energy nucleons in turn produce stars which are the source of most of the cosmic-ray neutrons observed[10]. As the neutrons travel through the atmosphere they are moderated by elastic and inelastic collisions. Most of the neutrons are absorbed in nitrogen before they reach thermal energies.

At sea level the intensity of cosmic-ray neutrons is very weak, but it increases markedly with altitude. Only the most sensitive neutron detectors show any appreciable background caused by neutrons at sea level. Usually, the background caused by other radiation, such as radioactive contamination of the counter, is more important than the cosmic-ray neutron background. In addition to its

---

[1] T. W. Bonner and J. W. Butler: Phys. Rev. **83**, 1091 (1951).

[2] See footnote 3, p. 208.

[3] F. Ajzenberg: Phys. Rev. **83**, 693 (1951).

[4] W. F. Hornyak, T. Lauritsen, P. Morrison and W. A. Fowler: Rev. Mod. Phys. **22**, 291 (1950).

[5] R. L. Seal: Phys. Rev. **92**, 389 (1953).

[6] T. W. Bonner: Proc. Roy. Soc. Lond. **174**, 339 (1940). — C. F. Powell: Proc. Roy. Soc. Lond. **181**, 344 (1942).

[7] D. A. Bromley: Phys. Rev. **88**, 566 (1952).

[8] Rambaugh and Locher: Phys. Rev. **44**, 855 (1936).

[9] E. Fünfer: Naturwiss. **25**, 235 (1937).

[10] S. Flügge: Kosmische Strahlung, Editor: W. Heisenberg, p. 132. Berlin: Springer 1943.

dependence on altitude, the flux of cosmic-ray neutrons depends on geomagnetic latitude, on atmospheric pressure, humidity, etc. For practical purposes a more important factor influencing the cosmic-ray neutron background is the material in the neighborhood of the detector, such as the building above it and the ground below. Surrounding materials will not only influence the flux but also the spectrum of neutrons. These changes are caused by scattering and absorption of the incident neutrons and, in addition, by the further production of neutrons in the materials by cosmic rays. For more details on this most interesting subject the reader is sent to Vol. XLVI and to Ref. [1].

60. Spontaneous fission. The process of spontaneous fission of heavy elements was discovered in uranium by PETRZHAK and FLEROV[2]. Although the process was studied extensively in the following years, the first comprehensive report was not published until 1952[3]. As in neutron-induced fission, neutrons are emitted in spontaneous fission, and it is the spontaneous emission of neutrons which is of interest to the present discussion; the general discussion of the process is treated in Vol. XLI.

Among the nuclides of charge $Z \geqq 90$, spontaneous fission is a frequently occurring phenomenon. For the even-even nuclides the logarithm of the half-life for spontaneous fission decreases approximately linearly with Z^2/A, the parameter characteristic of stability against fission, from a half-life of 10^{18} years for $Z^2/A = 35$ (Th^{232}) to 103 years for $Z^2/A = 39$ (Cf^{246})[4,5].

For even-odd nuclides the half-lives for spontaneous fission are of the order of 10^5 times longer than for even-even nuclides.

Nuclides which undergo spontaneous fission are also α-emitters. α-particles interacting with impurities of light elements will produce neutrons. If neutrons are observed to be emitted from a heavy nuclide, it is not always easy to decide whether they are due to (α, n) reactions or to spontaneous fission. Observation of the fission fragments is, therefore, a more reliable method of measuring the spontaneous fission rate than observation of the neutrons, and studies of the neutrons from spontaneous fission are usually carried out by measuring coincidences between neutrons and ionization events produced by fission fragments.

In SEGRÈ's paper[3] the average number of neutrons emitted per spontaneous fission is given as 2.2 ± 0.3 for U^{238}. Subsequently many measurements have been performed both on the average number of neutrons per spontaneous fission, ν, and of the neutron number distributions[6], and a semi-empirical theory has been proposed[7]. For Pu^{240} $\bar{\nu}$ has a value of 2.26 ± 0.05, for Cf^{252} $\bar{\nu} = 3.86 \pm 0.08$. For Pu^{240} the most probable number of neutrons per spontaneous fission is 2, for Cf^{252} it is 4. The distribution in energy of the neutrons is expected to be very similar to that observed for fission induced by thermal neutrons.

The nuclides which occur in appreciable quantities in nature and have measurable spontaneous fission rates are Th^{232}, U^{235} and U^{238} with rates of 4.2×10^{-5}, 3×10^{-4} and 6.9×10^{-3} fissions per gram per second, respectively[3]. These rates are so small that they do not contribute any appreciable neutron background.

[1] W. HEISENBERG (Editor): Kosmische Strahlung. Berlin: Springer 1953.
[2] PETRZHAK and FLEROV: C. R. Acad. Sci. USSR. **28**, 500 (1940).
[3] E. SEGRÈ: Phys. Rev. **86**, 21 (1952).
[4] G.T. SEABORG: Phys. Rev. **85**, 157 (1952).
[5] J.R. HUIZENGA: Phys. Rev. **94**, 158 (1954).
[6] See for example: B.C. DIVEN, H.C. MARTIN, R.F. TASCHEK and J. TERRELL: Phys. Rev. **101**, 1012 (1956). — D.A. HICKS, J. ISE and R.V. PLYLE: Phys. Rev. **101**, 1016 (1956).
[7] R.B. LEACHMAN: Phys. Rev. **101**, 1005 (1956).

The importance of spontaneous fission from the point of view of neutron sources is that some of the heavier nuclides which have high spontaneous fission rates may be used as primary neutron standards. In contrast to other naturally radioactive sources, such sources would have strengths and spectra which would depend only on the amount of material in the source, provided no neutrons are produced by (α, n) reactions in impurities. For certain applications it may be a disadvantage, however, that the neutron emission is not random in time, but that several neutrons are emitted simultaneously.

C. The slowing down of neutrons of energy larger than a few ev.

I. General principles and elementary approach to the problem of slowing down of neutrons.

61. The Boltzmann equation describing the slowing down of neutrons. The experiments described in Sects. 13, 14 and 28 are the simplest examples of experiments on the interaction of neutrons with matter in bulk. As a result of such an interaction the neutrons are slowed down from their initial energy to thermal energy and diffuse through the moderator until they are captured by one of the nuclei present, or escape from the surface of the moderator if this does not extend to infinity in all directions. The simplicity of the ideal experiment described in Sect. 28 lies in the nature and geometry of the moderator and of the fast neutron source.

The theoretical interpretation of experiments of this type, or more generally of experiments on the slowing down of neutrons by a moderator, involves the determination of the neutron distribution, i.e. of a function, say $N(\boldsymbol{r}, \boldsymbol{\Omega}, E, t)$, such that

$$N(\boldsymbol{r}, \boldsymbol{\Omega}, E, t)\, d\boldsymbol{r}\, d\boldsymbol{\Omega}\, dE$$

represents the probable number of neutrons that at a time t are contained in the volume element $d\boldsymbol{r}$ with energy between E and $E+dE$ and with a velocity direction in the solid angle $d\boldsymbol{\Omega}$.

Once the composition of the moderator and the neutron cross sections relative to all processes undergone by the various nuclei when hit by a neutron are known as a function of the energy of the neutrons, one can determine, at least in principle, the function $N(\boldsymbol{r}, \boldsymbol{\Omega}, E, t)$ as a solution of the Boltzmann equation expressing the balance of neutrons; the variation undergone between t and $t+dt$ by the probable number of neutrons belonging to $dE\, d\boldsymbol{\Omega}$ is determined, in each element of volume $d\boldsymbol{r}$, by all processes of absorption and generation in $d\boldsymbol{r}$, of diffusion in and out of $d\boldsymbol{r}$ and $d\boldsymbol{\Omega}$ and of slowing down in and out of dE by elastic as well as by inelastic scattering (and possibly other processes).

Such a general statement of the problem is however of little practical use because of the overwhelming difficulties encountered in the solution of the corresponding integro-differential equation.

As is customary we shall start by writing the Boltzmann equation under certain simplifying assumptions which, although not satisfied in general, represent in many cases a satisfactory description of the actual situation.

The conditions generally adopted in Part C, if not otherwise stated, are the following:

(a) The slowing down of neutrons is due only to elastic scattering.

(b) The energy of the neutrons must be always large compared to the thermal agitation energy ($kT = 0.025$ ev at room temperature) and, if the target atoms are bound in molecules or crystals, large compared to the energies associated with the vibration frequencies due to chemical bonds.

For the sake of simplicity a third condition is sometimes added in the more elementary treatments:

(c) the elastic scattering is isotropic in the center-of-mass frame of reference.

Condition (a) is exactly fulfilled when the energy of the neutron is smaller than the threshold of the (n, n') reactions because all other processes, as (n, np) or $(n, n\alpha)$ reactions which can contribute to the slowing down of neutrons, are always more endoenergetic. The threshold of the (n, n') reaction is related to the energy of the first excited level of the compound nucleus which in light elements is located a few Mev, in heavy elements from 0.1 to 1 Mev, above the ground state (Sect. 30).

Therefore, condition (a) will be fulfilled in the case of neutrons with energy lower than about 6 Mev moving in a moderator composed only of very light elements. As we shall see in Sect. 83, when the energy of the neutrons is of the order of 10 Mev or more, the inelastic scattering contributes in an essential way to the slowing down of neutrons in water (Fig. 147).

In heavy elements, on the contrary, the slowing down by elastic scattering is negligible and the inelastic scattering becomes the main process responsible for the degradation of the neutrons of a few Mev energy; this process will still be effective at energies of the order of a few hundred kev.

Condition (b) imposes a lower limit on the energy of the neutrons, whose value, as we shall see in Part D, where the effect of chemical bond will be considered in detail, is

$$E \gtrsim 1 \text{ ev}. \tag{61.1}$$

Above this value both the chemical bond and the thermal agitation of the nuclei of the moderator can be neglected; the treatment of the collision of a neutron against a nucleus free and at rest becomes elementary (Sect. 35).

As long as condition (61.1) is fulfilled, the de Broglie wavelength (15.3) of the neutron is so short that all interference effects between waves coherently scattered by various nuclei belonging to the same molecule or crystal are negligible; the interaction of the neutron with the medium depends only on the densities and cross sections of the constituent nuclei, and does not depend on its structure.

Condition (c) is only introduced in the more elementary approaches; in more refined treatments of the slowing down process one takes into account the anisotropy of the scattering in the c.m.s. as will be shown in Sect. 79. The condition of isotropy however, is very well satisfied in the case of hydrogen for energies of the incident neutrons lower than about 15 Mev and in the case of carbon for $E < 1.5$ Mev. As we have seen in Sect. 29 appreciable deviations from isotropy are observed in deuterium, beryllium and oxygen already at 0.1, 0.6 and 0.4 Mev respectively.

The Boltzmann equation for the slowing down and diffusion of neutrons, first established by ORNSTEIN and UHLENBECK[1], is the following:

$$\left.\begin{aligned} &\frac{\partial N}{\partial t} + \boldsymbol{v}\cdot\nabla N = -\Sigma_{\text{tot}}(E)\, v\, N + \\ &\quad + \int dE' \int d\boldsymbol{\Omega}'\, \Sigma_s(E'\to E, \boldsymbol{\Omega}'\to\boldsymbol{\Omega})\, v' N(\boldsymbol{r}, \boldsymbol{\Omega}', E', t) + S(\boldsymbol{r}, \boldsymbol{\Omega}, E, t)\,. \end{aligned}\right\} \tag{61.2}$$

[1] L. S. ORNSTEIN and G. E. UHLENBECK: Physica, Haag 4, 478 (1937).

The term $-\boldsymbol{v}\cdot\nabla N$ represents the effect of the diffusion of the neutrons outside the unit element of volume at the considered position $\boldsymbol{r}$; the first term on the right side describes the removal of neutrons due to all interactions, $\Sigma_{\text{tot}}(E)$ being the total macroscopic cross section due to all possible processes produced by neutrons of energy E in the nuclei present in the medium:

$$\Sigma_{\text{tot}}(E) = \sum_i \Sigma_{\text{tot}}^i(E) = \sum_i \mathfrak{N}_i\,\sigma_{\text{tot}}^i(E) = \frac{1}{l_{\text{tot}}(E)}\ \text{cm}^{-1}. \tag{61.3}$$

Here $\mathfrak{N}_i$ is the number of nuclei of type i contained in 1 cm³ of the medium, $\sigma_{\text{tot}}^i(E)$ the total cross section for all possible processes produced by neutrons of energy E in the nuclei of type i, and $l_{\text{tot}}(E)$ the total mean free path of these neutrons in the medium. $\Sigma_{\text{tot}}(E)$ can be split into two parts

$$\Sigma_{\text{tot}}(E) = \Sigma_s(E) + \Sigma_a(E). \tag{61.4}$$

$\Sigma_s(E)$ is the macroscopic scattering cross section, $\Sigma_a(E)$ the macroscopic absorption cross section relative to all processes which result in the absorption of the incident neutron. Therefore it coincides with the macroscopic reaction cross section; the microscopic reaction cross section σ_r has been defined and discussed in Sect. 18. It must be noted that, because of assumption (a), inelastic scattering (n, n') as well as $(n, n\,\alpha)$, $(n, n\,p)$, ... processes, are excluded from our present considerations. If this assumption is dropped, as it becomes necessary at neutron energies around and above 10 Mev, the cross sections of these various processes must be included in $\Sigma_a(E)$, but at the same time the corresponding neutron emission must be introduced in other terms of the Boltzmann equation as we shall explain later.

The second term on the right of Eq. (61.2) represents the effect of elastic scattering in the unit energy interval at E and in the unit solid angle at $\boldsymbol{\Omega}$ from any energy $E'>E$ and from any direction $\boldsymbol{\Omega}'\neq\boldsymbol{\Omega}$. This term contains the macroscopic differential scattering cross section

$$\Sigma_s(E'\to E, \boldsymbol{\Omega}'\to\boldsymbol{\Omega}) = \sum_i \Sigma_s^i(E'\to E, \boldsymbol{\Omega}'\to\boldsymbol{\Omega}) = \sum_i \mathfrak{N}_i\,\sigma_s^i(E'\to E, \boldsymbol{\Omega}'\to\boldsymbol{\Omega}), \tag{61.5a}$$

$$\Sigma_s(E') = \int dE \int_{4\pi} \Sigma_s(E'\to E, \boldsymbol{\Omega}'\to\boldsymbol{\Omega})\,d\boldsymbol{\Omega} \tag{61.5b}$$

where $\sigma_s^i(E'\to E, \boldsymbol{\Omega}'\to\boldsymbol{\Omega})$ represents the differential scattering cross section of the nuclei of type i for the neutrons of energy E. The result of the scattering process is that the energy of the neutron is reduced from E' to E and its direction of motion is changed from $\boldsymbol{\Omega}'$ to $\boldsymbol{\Omega}$; if μ_0 is the cosine of the scattering angle in the l.s., one has

$$\mu_0 = \boldsymbol{\Omega}'\cdot\boldsymbol{\Omega}. \tag{61.6}$$

If one substitutes (61.5a) into the second term on the right-hand side of (61.2), this expression will split into a sum of as many integrals as there are types of nuclei present in the moderator, each of the form

$$J^i(\boldsymbol{r}, \boldsymbol{\Omega}, E, t) = \int dE' \int d\boldsymbol{\Omega}'\, \Sigma_s^i(E'\to E, \boldsymbol{\Omega}'\to\boldsymbol{\Omega})\, v' N(\boldsymbol{r}, \boldsymbol{\Omega}', E', t), \tag{61.7}$$

which are sometimes indicated as *degradation integrals*.

Sometimes one introduces the *collision function* defined as the ratio

$$f(E'\to E, \boldsymbol{\Omega}'\to\boldsymbol{\Omega}) = \frac{\Sigma_s(E'\to E, \boldsymbol{\Omega}'\to\boldsymbol{\Omega})}{\Sigma_s(E')} \tag{61.8a}$$

Its integral, over the whole solid angle and over all possible final energies E, is obviously independent of the initial energy E' and equal to 1.

According to (61.5a) the collision function relative to a moderator containing various types of nuclei will be a weighted average of the collision functions relative to each type of nuclei

$$f(E'\to E, \boldsymbol{\Omega}'\to\boldsymbol{\Omega}) = \sum_i c_i(E')\, f_i(E'\to E, \boldsymbol{\Omega}'\to\boldsymbol{\Omega}), \tag{61.8b}$$

where the weights are given by

$$c_i(E) = \frac{l_s(E)}{l_s^i(E)} = \frac{\Sigma_s^i(E)}{\Sigma_s(E)}. \tag{61.9}$$

By expressing the degradation integrals in terms of the collision function one has

$$\sum_i J^i(\boldsymbol{r},\boldsymbol{\Omega},E,t) = \sum_i \int dE' \int d\boldsymbol{\Omega}'\, \frac{v'\, N(\boldsymbol{r},\boldsymbol{\Omega}',E',t)}{l_s(E')}\, f_i(E'\to E, \boldsymbol{\Omega}'\to\boldsymbol{\Omega}). \tag{61.10}$$

If assumption (a) is dropped, the degradation integrals (61.7) can be modified so as to include the effect of all those processes different from elastic scattering in which the capture of the incident neutron is followed by the (instantaneous) emission of neutrons of degraded energy; examples of processes of this type are the (n, n'), (n, np), ... reactions among which by far the most important one is inelastic scattering. In order to do this, however, one needs the detailed knowledge (usually not available) of the energy spectrum and angular distribution of the emitted neutrons as a function of the energy of the incident neutron, for all nuclei present in the moderator.

In Sect. 74 we shall see, in the case of an approximate treatment frequently used for the slowing down of neutrons (group picture), that sometimes it is convenient to include the degradation integral arising from inelastic scattering in the neutron source term.

The last term on the right-hand side of Eq. (61.2) represents the *neutron source*, i.e. the number of neutrons produced per unit volume and unit time interval in the unit energy interval, per unit solid angle. In this chapter we shall restrict our considerations to *independent and isotropic neutron sources*. By independent source we mean a source the existence and strength of which does not depend on the number of neutrons $N(\boldsymbol{r}, \boldsymbol{\Omega}, E, t)$ present at the point $\boldsymbol{r}$ considered or at other points $\boldsymbol{r}$, in the moderator. This assumption excludes from our considerations: fission sources, when this process is produced by the neutrons belonging to the distribution in which we are interested, (γ, n) sources, when this process is produced by γ-rays emitted in the capture of neutrons by nuclei present in the moderator, $(n, 2n)$ sources distributed in the moderator, etc. It also excludes the treatment of inelastic scattering by the procedure mentioned above and sketched in Sect. 74.

Among all possible isotropic neutron sources a very simple and important type is the isotropic monoenergetic point source located at $\boldsymbol{r}_0$; this is represented by the expression:

$$S(\boldsymbol{r}, \boldsymbol{\Omega}, E, t) = Q_0(t)\, \frac{1}{4\pi}\, \delta(E - E_0)\, \delta(\boldsymbol{r} - \boldsymbol{r}_0). \tag{61.11}$$

where $\delta(x)$ is the Dirac delta function, $\delta(\boldsymbol{r}-\boldsymbol{r}_0) = \delta(x - x_0)\, \delta(y - y_0)\, \delta(z - z_0)$ and $Q_0(t)$ the number of neutrons emitted per second, frequently called the *source strength*. As a consequence of the linearity of the integro-differential equation (61.2), a solution corresponding to an isotropic neutron source distributed in space and in energy, can always be obtained by superimposing solutions of the same equation corresponding to a neutron source of type (61.11) [see Sect. 71, Eq. (71.1)].

Similarly the more general anisotropic and distributed source can be described as the superposition of monoenergetic collimated point sources, each emitting neutrons only in a well defined direction. The analytic expression of an elementary source of this type is obtained by substituting in (61.11) for the factor $(4\pi)^{-1}$ an angular δ-function of $\boldsymbol{\Omega}-\boldsymbol{\Omega}_0$, where $\boldsymbol{\Omega}_0$ represents the direction of emission of neutrons. As already stated above, we shall, however, restrict our considerations to isotropic sources.

In order to make the problem of the solution of the integro-differential equation (61.2) determined, it is necessary to specify boundary conditions; these follow immediately from the physical meaning of $N(\boldsymbol{r}, \boldsymbol{\Omega}, E, t)$ and may be easily formulated in each particular case [7]. For instance, at the interface between two media in direct contact,

$$\left.\begin{array}{c} N\left(\boldsymbol{r}+R\,\boldsymbol{\Omega}, \boldsymbol{\Omega}, E, t+\frac{R}{v}\right) \\ \text{must be a continuous function of } R \text{ for } \boldsymbol{r}+R\,\boldsymbol{\Omega} \text{ at the interface.} \end{array}\right\} \quad (61.12\text{a})$$

On a non-re-entrant[1] free surface separating the moderator from the vacuum

$$\left.\begin{array}{c} N(\boldsymbol{r}, \boldsymbol{\Omega}, E, t)=0 \\ \text{for all } \boldsymbol{\Omega} \text{ entering the moderator and } \boldsymbol{r} \text{ on the surface.} \end{array}\right\} \quad (61.12\text{b})$$

In the case of a moderator extending to infinity with all neutron sources located at finite distances (see Division III)

$$\left.\begin{array}{c} N(\boldsymbol{r}, \boldsymbol{\Omega}, E, t) \\ \text{must tend to zero faster than any power of } r \text{ for } r \text{ going to infinity.} \end{array}\right\} \quad (61.12\text{c})$$

In the case of time dependent problems it is necessary to impose also initial conditions, i.e. to specify

$$N(\boldsymbol{r}, \boldsymbol{\Omega}, E, t) \quad \text{at the initial time } t=t_0. \quad (61.12\text{d})$$

In the discussion of the slowing down of neutrons it is frequently convenient to consider other functions than $N(\boldsymbol{r}, \boldsymbol{\Omega}, E, t)$ for the representation of the neutron distribution or of some of its physical properties. Among these functions the ones more frequently used are the following: The *flux density per unit energy interval and unit solid angle*, sometimes also denoted as *angular number flux*

$$\Phi(\boldsymbol{r}, \boldsymbol{\Omega}, E, t)=v \cdot N(\boldsymbol{r}, \boldsymbol{\Omega}, E, t) \quad (61.13)$$

is defined as the number of neutrons having energies between E and $E+dE$ which per second, at the point $\boldsymbol{r}$ at the time t, cross a unit area normal to $\boldsymbol{\Omega}$. The *total collision density per unit energy interval and unit solid angle*

$$\Psi(\boldsymbol{r}, \boldsymbol{\Omega}, E, t)=\frac{1}{l_{\text{tot}}(E)}\,\Phi(\boldsymbol{r}, \boldsymbol{\Omega}, E, t)=\frac{v}{l_{\text{tot}}(E)}\,N(\boldsymbol{r}, \boldsymbol{\Omega}, E, t) \quad (61.14)$$

represents the number of collisions, per second and cm^3, undergone by neutrons belonging to the unit energy interval at E and to the unit solid angle around $\boldsymbol{\Omega}$, at the point $\boldsymbol{r}$ and time t.

By integration with respect to the direction $\boldsymbol{\Omega}$ one can immediately derive the following integral quantities.

---

[1] A non-re-entrant surface is such a surface that a straight line emerging from it cannot meet it again.

The *neutron density per unit energy interval,*

$$N_0(\boldsymbol{r}, E, t) = \int d\boldsymbol{\Omega}\, N(\boldsymbol{r}, \boldsymbol{\Omega}, E, t), \tag{61.15}$$

representing the number of neutrons per cm³, found in an unit energy interval at E, at the time t at the point $\boldsymbol{r}$.

The *neutron flux per unit energy interval,* sometimes called *number flux,*

$$\Phi_0(\boldsymbol{r}, E, t) = v \cdot N_0(\boldsymbol{r}, E, t), \tag{61.16}$$

representing the number of neutrons per unit energy interval that at the point $\boldsymbol{r}$ and time t cross, in 1 sec, the surface of a sphere of unit cross sectional area.

The *total collision density per unit energy interval,*

$$\Psi_0(\boldsymbol{r}, E, t) = \frac{1}{l_{\text{tot}}(E)}\, \Phi_0(\boldsymbol{r}, E, t) = \frac{v}{l_{\text{tot}}(E)}\, N_0(\boldsymbol{r}, E, t), \tag{61.17}$$

representing the number of collisions per second and per cm³ undergone by the neutrons belonging to the unit energy interval at E, at the point $\boldsymbol{r}$ and time t.

A further fundamental function frequently used is the *slowing down density* $Q(\boldsymbol{r}, E, t)$ defined as the number of neutrons per cm³ and per second that slow down past a given energy E at the point $\boldsymbol{r}$ and time t. Since the number of neutrons in the energy interval from E to $E + dE$ which have collisions near $\boldsymbol{r}$ is $\Psi_0(\boldsymbol{r}, E, t)\, dE$, and the fraction of these that are scattered and not captured is

$$h(E) = \frac{\Sigma_s(E)}{\Sigma_s(E) + \Sigma_a(E)} = \frac{l_{\text{tot}}(E)}{l_s(E)}, \tag{61.18}$$

one has

$$Q(\boldsymbol{r}, E, t) = \int_E^\infty dE'\, h(E')\, \Psi_0(\boldsymbol{r}, E', t) \int_0^E f(E' \to E'')\, dE'' \tag{61.19}$$

where

$$f(E' \to E) = \int d\boldsymbol{\Omega}\, f(E' \to E, \boldsymbol{\Omega}' \to \boldsymbol{\Omega}). \tag{61.20}$$

In certain problems, one considers the quantities (61.15) to (61.17) and (61.19) integrated with respect to the space coordinates $\boldsymbol{r}$ (Division II) or with respect to the energy E (Sect. 81).

The integro-differential equations satisfied by the angular number flux and the total collision density are immediately derived from (61.2) by making use of the corresponding definitions. According to (61.13) one has

$$\left.\begin{aligned} \frac{1}{v}\frac{\partial \Phi}{dt} + \boldsymbol{\Omega} \cdot \nabla \Phi + \frac{1}{l_{\text{tot}}(v)}\, \Phi = \int dE' \int d\boldsymbol{\Omega}'\, \frac{1}{l_s(E')} \times \\ \times \Phi(\boldsymbol{r}, \boldsymbol{\Omega}', E', t)\, f(E' \to E, \boldsymbol{\Omega}' \to \boldsymbol{\Omega}) + S(\boldsymbol{r}, \boldsymbol{\Omega}, E, t) \end{aligned}\right\} \tag{61.21a}$$

and according to (61.14)

$$\left.\begin{aligned} \frac{l_{\text{tot}}(E)}{v}\frac{\partial \Psi}{\partial t} + l_{\text{tot}}(E)\, \boldsymbol{\Omega} \cdot \nabla \Psi + \Psi = \int dE' \int d\boldsymbol{\Omega}' h(E)'\, \Psi(\boldsymbol{r}, \boldsymbol{\Omega}', E', t) \times \\ \times f(E' \to E, \boldsymbol{\Omega}' - \boldsymbol{\Omega}) + S(\boldsymbol{r}, \boldsymbol{\Omega}, E, t). \end{aligned}\right\} \tag{61.21b}$$

Attempts to solve the Boltzmann equation have been made for a long time[1] [*5*], [*16*]; today the methods for the treatment of this equation constitute a wide chapter of mathematical physics [*7*].

In this article we shall not discuss the various cases in which an exact solution of one of the equivalent integro-differential equations (61.2), (61.21a) and

[1] S. FLÜGGE: Z. Physik **111**, 109 (1938).

(61.21 b) has been obtained nor the various approximate methods employed to solve many important special cases. For the treatment of these mathematical problems we refer to the review articles by MARSHAK [16] and by MARSHAK and coworkers[1] and to the books by WEINBERG and NODERER [19] and by DAVISON and SYKES [7].

Since the mathematical problem is essentially the same as that encountered in the study of penetration and diffusion of X rays through matter, we shall frequently refer also to the companion article by FANO, SPENCER and BERGER, in this volume.

In the following, we shall limit ourselves to a discussion of various special cases the solution of which can be derived from elementary procedures. Only occasionally, when necessary in the discussion of experimental data, shall we refer to conclusions derived from the more elaborate theories based on the integro-differential equations given above. A short summary of the most important methods for the approximate treatment of these equations is given in Sects. 79 and 80.

62. The energy loss of neutrons in elastic collisions with nuclei. The lethargy *u* as variable for representing the slowing down of neutrons. The problem of the elastic collision of a neutron against a nucleus of mass $M(\geqq m)$ free and at rest has been discussed in Sect. 35, where two examples are illustrated in Figs. 32 and 33.

The energy E_2 of the neutron after the collision is expressed in terms of its energy E_1, before the collision and of its angle of scattering ϑ_2^* in the c.m.s. by Eqs. (35.4) and (35.5). These, combined together, give

$$E_2 = E_1 \frac{m^2 + M^2 + 2m\,M\cos\vartheta_2^*}{(m+M)^2} \tag{62.1}$$

or

$$\frac{E_2}{E_1} = \frac{1+\alpha}{2} + \frac{1-\alpha}{2}\cos\vartheta_2^* \tag{62.2}$$

having introduced the mass ratio

$$A = \frac{M}{m} \geqq 1 \tag{62.3}$$

and

$$\alpha = \left(\frac{A-1}{A+1}\right)^2. \tag{62.4}$$

Eq. (62.1) or (62.2) shows how the neutron energy after the collision depends on its angle of scattering in the c.m.s. The energy E_2 of the neutron takes its maximum and minimum values respectively for glancing collisions

$$\vartheta_2^* = 0, \qquad \mathsf{E}_2 = 0: \quad E_2 = E_1 \tag{62.5}$$

and head-on collisions

$$\vartheta_2^* = \pi, \qquad \mathsf{E}_2 = \frac{4m\,M}{(m+M)^2} E_1 = (1-\alpha)\,E_1: \quad E_2 = \alpha E_1. \tag{62.6}$$

E_2 is the energy of the recoiling nucleus (35.5). From (62.6) we see that α represents the minimum value of the ratio E_2/E_1.

The probability $f(E_2)\,dE_2$ that E_2 takes a value between E_2 and E_2+dE_2 inside the interval defined by (62.5) and (62.6) can be calculated making use of (62.2) once the corresponding angular distribution in the c.m.s. is known.

[1] R.E. MARSHAK, H. BROOKS and H. HURWITZ: Nucleonics 5 (1), 53 (1949); 5 (2), 59 (1949).

The problem is particularly simple in the case of isotropic distribution in the c.m.s., a circumstance met in a few cases of practical importance; e.g. in the case of collisions of neutrons of energy $E_1 < 15$ Mev against hydrogen (Sect. 29).

In this case the probability that the neutron, after the collision, moves in a direction between ϑ_2^* and $\vartheta_2^* + d\vartheta_2^*$ is proportional to the corresponding solid angle, i.e. it is given by the simple expression

$$\frac{2\pi\, d(-\cos\vartheta_2^*)}{4\pi}.$$

By differentiation of (62.2) one obtains

$$f(E_2)\, dE_2 = \frac{1}{2}\, d(-\cos\vartheta_2^*) = \frac{1}{1-\alpha}\,\frac{dE_2}{E_1}. \tag{62.7}$$

One can conclude that in the case of isotropic elastic scattering in the c.m.s., the distribution of neutron energies after one collision is uniform and is given by

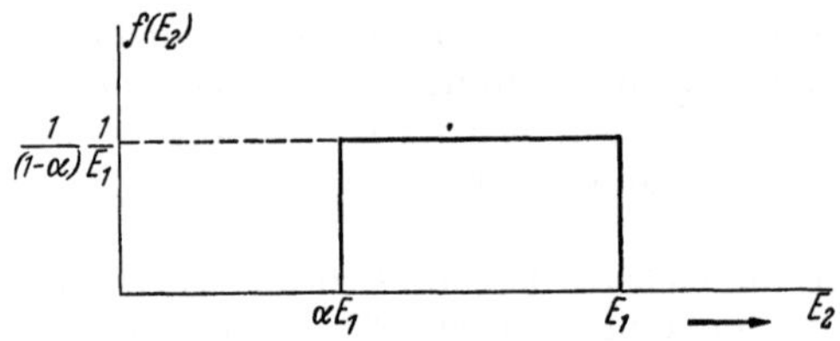

Fig. 115. Distribution of the values of the neutron energy E_2 after a collision in the case of isotropic sattering in the c.m.s.

$$\left.\begin{aligned} f(E_2) &= \frac{1}{(1-\alpha)\,E_1} \\ &= \frac{(A+1)^2}{4A}\,\frac{1}{E_1}, \end{aligned}\right\} \tag{62.8}$$

in the interval

$$\left.\begin{aligned} \alpha E_1 = E_1 \left(\frac{A-1}{A+1}\right)^2 \\ \leqq E_2 \leqq E_1 \end{aligned}\right\} \tag{62.9}$$

as shown in Fig. 115.

The average energy after one collision is given by the formula

$$\overline{E}_2 = E_1 \frac{1+\alpha}{2} = E_1 \frac{A^2+1}{(A+1)^2} \tag{62.10}$$

from which one derives the *average energy decrement per collision*

$$\overline{E_1 - E_2} = E_1 - \overline{E}_2 = \frac{1-\alpha}{2} E_1 = E_1 \frac{2A}{(A+1)^2}. \tag{62.11}$$

In the case of heavy nuclei ($M \gg m$) the following approximate expression may be used

$$\frac{\overline{E_1 - E_2}}{E_1} = \frac{1-\alpha}{2} \approx \frac{1}{\frac{A}{2}+1} \quad \text{for } A \gg 1. \tag{62.12}$$

More useful than the average energy decrement per collision is the *average logarithmic energy decrement per collision* defined as

$$\bar{\xi} = \overline{\ln \frac{E_1}{E_2}} = \int\limits_{\alpha E_1}^{E_1} \ln \frac{E_1}{E_2}\, f(E_2)\, dE_2. \tag{62.13}$$

In the case of isotropic scattering in the c.m.s. one has immediately

$$\bar{\xi} = 1 + \frac{\alpha}{1-\alpha} \ln\alpha = 1 + \frac{(A-1)^2}{2A} \ln \frac{A-1}{A+1}, \tag{62.14}$$

i.e. a value which *does not depend* on the energy of the neutron but only on the mass ratio A.

For $A=1$, i.e. in the case of collisions against protons, Eq. (62.14) gives $\bar{\xi}=1$.

For $A>1$ instead of (62.14) one frequently uses the approximate expressions

$$\bar{\xi}=\frac{1}{2}X-\frac{1}{12}X^2+\cdots \quad \text{for} \quad A>1 \tag{62.15}$$

where

$$X=\ln\frac{1}{\alpha}=2\ln\frac{A+1}{A-1} \quad \text{i.e.} \quad \alpha=e^{-X} \tag{62.16}$$

or better

$$\bar{\xi}\approx\frac{2}{A+\frac{2}{3}}=\frac{1}{\frac{A}{2}+\frac{1}{3}} \quad \text{for} \quad A>1. \tag{62.17}$$

The latter equation is in error by not more than 1% for $A>10$ and by only 3.3% for $A=2$. It differs very little from (62.12); for $A\gg1$ both Eq. (62.12) and Eq. (62.17) may be further simplified to

$$\overline{\frac{E_1-E_2}{E_1}}\approx\bar{\xi}\approx\frac{2}{A} \quad \text{for} \quad A\gg1. \tag{62.18}$$

The use of the average logarithmic energy decrement per collision was originally introduced by FERMI[1] [8] in connection with the following representation of the slowing down process.

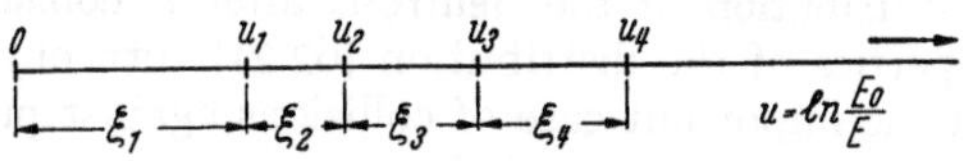

Fig. 116. Representation of the slowing down of a neutron in a logarithmic scale (lethargy).

Let us represent the energy E of the neutron by means of a point on an axis, the abscissa of which — measured from the origin O (Fig. 116) — is given by the quantity

$$u=\ln\frac{E_0}{E} \tag{62.19}$$

frequently called the *lethargy* of the neutron.

If we consider a source of neutrons embedded in an infinite non-absorbing medium, the "history" of each neutron is represented on the axis u by a non interrupted succession of points of abscissa 0, $u_1, u_2, u_3, \ldots$; the point 0 corresponds to the initial energy E_0, with which the neutron is emitted by the source, while the points $u_1, u_2, u_3, \ldots$ represent the energies $E_1, E_2, E_3, \ldots$ taken by the neutron after 1, 2, 3, ... collisions. In this representation the logarithmic decrement of the energy in a collision is given by the distance between two successive points, namely

$$\begin{aligned}\xi_1&=u_1-0,\\ \xi_2&=u_2-u_1,\\ \xi_3&=u_3-u_2,\\ &\cdots\cdots\end{aligned}$$

or in general

$$\xi_r=u_r-u_{r-1}=\ln\frac{E_{r-1}}{E_r}. \tag{62.20}$$

---

[1] E. AMALDI, O. D'AGOSTINO, E. FERMI, B. PONTECORVO, F. RASETTI and E. SEGRÈ: Proc. Roy. Soc. Lond., Ser. A 149, 522 (1935).

The distribution function of the values of ξ_r can be deduced immediately from (62.8), generalized to the case of the r-th collision, and expressed in terms of the lethargy (62.19). Dropping the subscript r, one obtains

$$f(\xi)\,d\xi = \frac{1}{1-\alpha}\,e^{-\xi}\,d\xi\,, \tag{62.21 a}$$

$$0 \leqq \xi \leqq X. \tag{62.21 b}$$

The average logarithmic energy decrement defined by (62.13) and given, in the case of isotropic scattering in the c.m.s., by (62.14), represents the average distance between two successive points of the series $0, u_1, u_2, u_3, \ldots$ of Fig. 116.

The lethargy (62.19) is a very convenient variable for representing the slowing down of neutrons owing to the properties that will be shown in the following sections. These do not depend on the value of the constant E_0 which may be chosen in an arbitrary way. In the case of monoenergetic neutron sources it is very reasonable to chose E_0 equal to the initial energy of the neutrons. But if the source is not monoenergetic, any arbitrarily chosen value of E_0 may be adopted in the definition (62.19). In this case of course, the lethargy u may take positive as well as negative values.

In the following, however, we shall usually refer to monoenergetic neutron sources and therefore E_0 will be usually chosen identical to the initial energy of the neutrons considered.

63. The energy distribution of the neutrons after n collisions in the case of hydrogen. The properties of the distribution (62.21) turn out to be very simple and enlightening in the important case of collisions against protons.

In this case one has

$$A = 1, \quad \bar{\xi} = 1, \quad \alpha = 0, \quad X = \infty$$

and therefore

$$f(\xi)\,d\xi = e^{-\xi}\,d\xi, \quad 0 \leqq \xi = u - u' \leqq \infty, \tag{63.1}$$

similar to the distribution function for the values of the intervals of time t between two successive disintegrations of a radioactive body of mean life τ:

$$e^{-\frac{t}{\tau}}\,\frac{dt}{\tau}, \quad 0 \leqq \frac{t}{\tau} \leqq \infty,$$

whose interpretation in the theory of probability is well known. Eq. (63.1) can also, in a similar way, be interpreted as the product of two factors; the exponential factor represents the probability that no point of our distribution falls in the interval ξ and the factor $d\xi$ represents the probability that the point u falls just in $du = d\xi$. In other words, in the case of (isotropic elastic) collisions against protons, each one of the points u_r is distributed along the axis of Fig. 116 completely independently of the others, with an *average density* $\bar{\xi} = 1$.

From this remark one can deduce a few useful results. For instance, referring to neutrons whose energy has been reduced from the initial value E_0 to a value E, one derives immediately both the average number of collisions $\bar{n}$ as well as the probability $p_n(u)$ that the number of collisions undergone by the neutrons be n.

In order to deduce $\bar{n}$ one notices that in the succession of points $u_1, u_2, u_3, \ldots$ there is one, let us say the last one of subscript n, which falls at the point $u = \ln \frac{E_0}{E}$. Therefore $n-1$ points fall in the interval $(0, u)$ distributed at random with average density $\bar{\xi}=1$. This means[1] that

$$\overline{n-1}=u \tag{63.2}$$

or that

$$\bar{n}=u+1=\ln\frac{E_0}{E}+1. \tag{63.3}$$

From the statistical independence of the successive intervals, it follows that the value of $\bar{n}$ given by (63.3) also represents the average number of collisions necessary to reduce the energy from the initial value E_0 to any value smaller than E: the value of $\overline{n-1}$ does not change if the n-th point instead of being just at u, falls on its right side.

The probability p_n that neutrons, whose energy has been reduced from the initial values E_0 to E, have undergone just n collisions, is simply given by the probability that $n-1$ points fall in the interval $(0, u)$. This is obtained from the Poisson-Bateman formula applied to the case in which the average value is given by (63.2):

$$\left.\begin{aligned} p_n(u) &= \frac{u^{n-1}}{(n-1)!}\,e^{-u} \\ \text{or}\quad p_n(E) &= \frac{\ln\left(\frac{E_0}{E}\right)^{n-1}}{(n-1)!}\,\frac{E}{E_0}. \end{aligned}\right\} \tag{63.4}$$

As an example we give in Fig. 117 the values of $p_n(u)$ as a function of n, for the case $E_0/E=3\times 10^6$. The curve is only slightly asymmetric and very close to a Gaussian distribution. Its maximum lies very close to

$$n=\ln\frac{E}{E_0} \tag{63.5}$$

corresponding to

$$E=E_0 e^{-n} \tag{63.6}$$

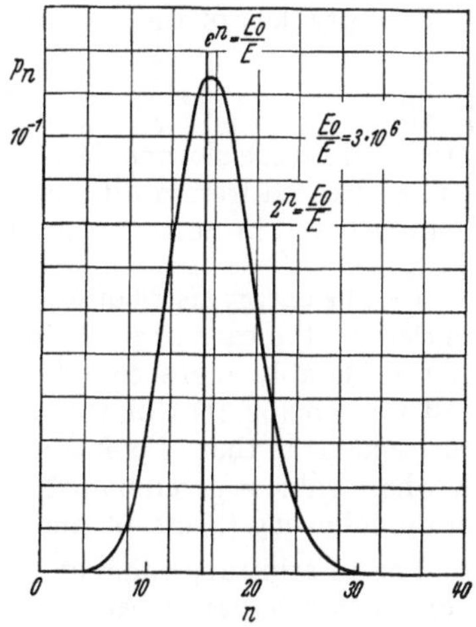

Fig. 117. The probability that a neutron whose energy has been reduced from E_0 to E has undergone n collisions against protons ($E_0/E = 10^6$).

and not to the arithmetic average energy [see (62.10) for $A=1$]

$$E=E_0\cdot 2^{-n}.$$

This remark justifies the practical rule, frequently used, which states that, on the average, the energy of the neutron is reduced in a collision against a proton by a factor $1/e$ (and not $\frac{1}{2}$ as given by the arithmetic average[2] [8]).

Although not very useful in practice, one can also answer immediately questions of the following type: what is the energy distribution of neutrons which have undergone just n collisions[1]? Referring again to the lethargy (62.19), the required probability is simply the probability that u_n falls between u and $u+du$. This is obtained by multiplying the probability (63.4) that just $n-1$ points fall in the interval $(0, u)$, by the probability $du=dE/E$ that the last point falls in du:

$$p_n(u)\,du=\frac{u^{n-1}}{(n-1)!}\,e^{-u}\,du \tag{63.7}$$

[1] E. Lamla: Naturwiss. **24**, 25, 336 (1936).
[2] G.C. Wick: Phys. Rev. **49**, 192 (1936).

or

$$p_n(E)\,dE = \frac{\left(\ln\frac{E_0}{E}\right)^{n-1}}{(n-1)!}\,\frac{dE}{E_0}\,. \tag{63.8}$$

It is clear that once $p_n(u)\,du$ is known, one can immediately derive the total number of neutrons of lethargy between u and $u+du$ for a source of unit strength (i.e. a source emitting one neutron per second) by adding the contributions due to neutrons that have undergone any number of collisions. In the present case one has

$$g(u)\,du = du\sum_{n=1}^{\infty} p_n(u) = \mathrm{e}^{-u}\,du\sum_{n=1}^{\infty}\frac{u^{n-1}}{(n-1)!} = du\,.$$

This result is trivial because it simply states that the probability of finding the neutron emitted by the source in du is just du, as we already knew.

The problem of the derivation of the spectrum

$$n_0(u)\,du = q_0\,g(u)\,du$$

produced by the slowing down of the neutrons emitted by a source of strength q_0, will be treated from a different point of view, in Sects. 67 and 68 for the case of hydrogen and in Sects. 67 and 69 for heavier nuclei.

64. The energy distribution of the neutrons after n collisions against non-hydrogen nuclei. In the case of nuclei of mass $M>m$ the discussion of the slowing down is considerably more complicated[1] than in the case of hydrogen because the distribution (62.21a) of the distances between two successive points no longer corresponds to that of points at random; according to (62.21b) ξ can never exceed X whose value is given by (62.16). This circumstance, as we shall see more clearly in the following (Sects. 67 and 69), gives rise to remarkable complications.

For an arbitrary value of $A(\geqq 1)$ the probability that a neutron, after n collisions, has a lethargy between u and $u+du$, can be deduced from the obvious recurrent relation

$$p_n(u) = \int_{u-X}^{u} p_{n-1}(u')\,f(u-u')\,du' = \frac{1}{1-\alpha}\int_{u-X}^{u} p_{n-1}(u')\,\mathrm{e}^{-(u-u')}\,du' \tag{64.1}$$

obtained by making use of (62.21). The solution of (64.1) is easily found to be [*16*]

$$p_n(u) = \begin{cases} \dfrac{\mathrm{e}^{-u}}{(1-\alpha)^n}\,\dfrac{1}{(n-1)!}\displaystyle\sum_{k=0}^{\bar{k}}(-1)^k\binom{n}{k}(u-kX)^{n-1} & \text{for } 0\leqq u\leqq nX,\\ 0 & \text{for } u>nX \end{cases} \tag{64.2}$$

where $\bar{k}$ is defined as the largest value of k for which the bracket is negative.

Here again as in the case of hydrogen, by summing the $p_n(u)$ with respect to all possible values of n, one can deduce the spectrum of the neutrons, i.e. the total number of neutrons of lethargy between u and $u+du$ present in an infinite non-absorbing medium. This problem, however, will be discussed from a slightly different point of view in Sect. 67, while in Sect. 69 we will take into consideration the effect of capture of neutrons by the nuclei of the medium.

[1] E. U. CONDON and G. BREIT: Phys. Rev. **49**, 229 (1936). — G. C. WICK: Phys. Rev. **49**, 192 (1936). — P. LANGEVIN: Ann. de Phys. **17**, 185 (1942).

Here we are more interested in the derivation of an approximate expression of $p_n(u)$ valid with good accuracy for large values of n.

Let us rewrite (64.2) in the form[1]

$$p_n(u) = \frac{X^{n-1}}{(1-\alpha)^n} e^{-u} f_n(u) \tag{64.3}$$

where

$$f_n(u) = \left\{ \begin{array}{ll} \sum_{k=0}^{\bar{k}} \frac{(-1)^k}{(n-1)!} \binom{n}{k} \frac{(u-kX)^{n-1}}{X^{n-1}} & \text{for } 0 \leqq u \leqq nX, \\ 0 & \text{for } u > nX \end{array} \right\} \tag{64.4}$$

is a function of the ratio u/X symmetric with respect to the value $nX/2$.

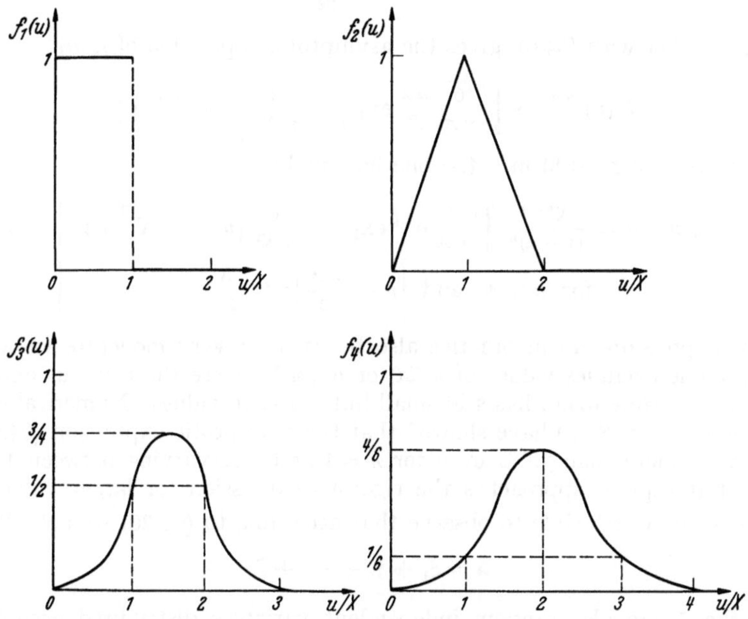

Fig. 118. The distribution of the values of the sum of n independent variables ($x_1, x_2, x_n \leqq X$) for $n = 1, 2, 3, 4$ (GROSJEAN).

From Fig. 118, where $f_n(u)$ is plotted as a function of u/X for the first four values of n, one sees that even for moderate values of n, the function is represented by a bell-shaped curve. The following considerations show that for large values of n it can be approximated by a Gaussian curve; let us consider a variable number x which can take, with equal probability, all values between 0 and X and let us take n of these numbers at random. One can easily recognize that the probability that their sum

$$u = x_1 + x_2 + \cdots + x_n \tag{64.5}$$

belongs to the interval $(u, u+du)$ is given by $f_n(u) \frac{du}{X}$. According to the central limit theorem of the probability calculus, the distribution of the values of the

[1] C.C. GROSJEAN: Medel. Konik. V.L.A. Acad. Wetenscap. Belgie **11**, No. 13 (1949).

sum of n independent random variables tends to a normal distribution as n tends to infinity. Therefore, for large n, we can write

$$f_n(u)\frac{du}{X} \to \frac{du}{U\sqrt{\pi}} \exp\left[-\frac{(u-n\,X/2)^2}{U^2}\right], \quad U^2 = 2\sigma^2, \tag{64.6}$$

where σ is the standard deviation of the distribution of the values (64.5):

$$\sigma^2 = \overline{(u-\bar{u})^2} = \sum_{r=1}^{n} \overline{(x_r - \bar{x}_r)^2} = n\,(\overline{x_r^2} - \bar{x}_r^2),$$

$$\bar{x}_r = \frac{X}{2}, \qquad \overline{x_r^2} = \frac{X^2}{3}.$$

We can conclude

$$\sigma^2 = n\frac{X^2}{12} \tag{64.7}$$

which, together with (64.6), gives the asymptotic expression of $f_n(u)$:

$$f_n(u)\frac{du}{X} \to \sqrt{\frac{6}{n\pi}}\,\frac{du}{X} \exp\left[-\frac{6}{n\,X^2}\left(u-\frac{n\,X}{2}\right)^2\right]. \tag{64.8}$$

By introducing (64.8) in (64.3) one has finally

$$\left.\begin{aligned} p_n(u)\,du &= \frac{X^{n-1}}{(1-\alpha)^n}\sqrt{\frac{6}{n\pi}}\,\mathrm{e}^{-u} \exp\left[-\frac{6}{n\,X^2}\left(u-\frac{n}{2}X\right)^2\right] du \\ &\text{for } n \gg 1 \text{ and } \left|u-\frac{n\,X}{2}\right| \ll \frac{n\,X}{2}. \end{aligned}\right\} \tag{64.9}$$

This expression can in practice also be used for very moderate values of n, provided one excludes values of $u \lesssim 0$ or $u \gtrsim nX$ where the rigorous expression (64.2) is zero while (64.9) has still small but non-zero values. Numerical calculations for $n=4$, 6, 8, 10 have shown[1] that the asymptotic expression (64.9) is in error by no more than 3.5% even for $n=4$ and u/X varying between 1 and 3, and that it rapidly approaches the rigorous expression for larger values of n.

It may be worthwhile to observe that according to (62.20) we actually have

$$u = \xi_1 + \xi_2 + \cdots + \xi_n \tag{64.10}$$

where the ξ_r are also random independent variables distributed according to (62.21). Therefore one can apply the central limit theorem of the probability calculus directly to the $p_n(u)$. From (64.10) one has immediately

$$\bar{u} = \bar{\xi}_1 + \bar{\xi}_2 + \cdots + \bar{\xi}_n = n\,\bar{\xi}, \tag{64.11}$$

at least as long as, according to our assumption of isotropy in the c.m.s., $\bar{\xi}$ is independent of the energy.

The value (64.11) is slightly lower than the value $n\,X/2$ deduced from the consideration of $f_n(u)$, the difference being due mainly to the factor e^{-u}; the average value of u that one deduces from (64.9) is

$$\bar{u} = n\frac{X}{2} - n\frac{X^2}{12}.$$

The remaining difference between this value and the value (64.11) reduces to zero for small values of X [see Eq. (62.15)] as we have to expect because all the

[1] See footnote 1, p. 223.

expressions given here are correct only asymptotically for $n \to \infty$ or, equivalently, for $X \to 0$. A similar remark holds for the expression of the standard deviation relative to the asymptotic expression of $p_n(u)$ which is only asymptotically identical with (64.7). The advantage of the procedure adopted above is that, since the functions $f_n(u)$ are symmetric with respect to their maximum, they can be represented with good accuracy by a Gaussian curve even for small values of n.

The more relevant point for the present discussion is that for $A \gg 1$ the standard deviation (64.7) is rather small as can be seen from the following example.

In the case of carbon, for instance, one has

$$X = 0.3341; \quad \bar{\xi} = 0.158; \quad \frac{\sigma^2}{n} = 0.093$$

and therefore, for a reduction in energy by a factor $E_0/E = 10^6$,

$$n = \frac{\ln 10^6}{\bar{\xi}} = 87; \quad \sigma^2 = 0.80$$

i.e. the fluctuation in energy around the value defined by Eq. (64.11) is of the order of $\delta \ln E \sim \pm\sqrt{0.80} \sim \pm 0.9$ corresponding to a (multiplicating or dividing) factor of about 2.5, which is certainly very small if compared with the reduction in energy: $E_0/E = 10^6$.

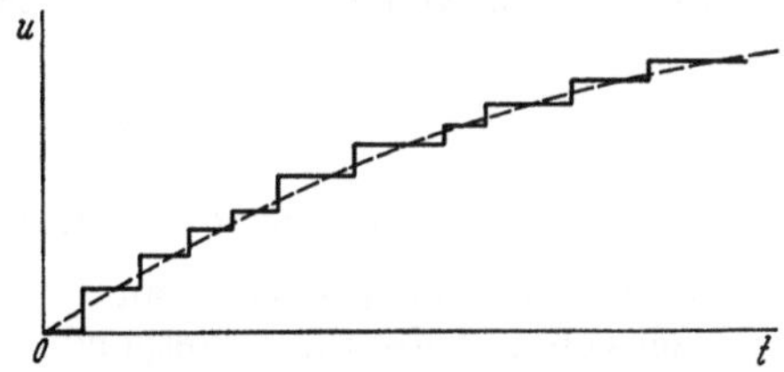

Fig. 119. The lethargy of a neutron as a function of the time elapsed after its emission from the source.

This remark simply shows that for $A \gg 1$ and for a large number of collisions, a given value of n corresponds to a very well defined value of the lethargy, given by (64.11), and, vice-versa, that one can estimate a kind of average number n of collisions necessary to reduce, by elastic isotropic scattering, the energy of the neutrons from the initial energy E_0 to the final value E, by means of Eq. (64.11),

$$n = \frac{u}{\bar{\xi}} = \frac{\ln E_0 - \ln E}{\bar{\xi}}. \tag{64.12}$$

This expression, valid only for $n \gg 1$, can be considered as an extension, to the case of $A > 1$, of (63.3) which is valid for the case of hydrogen.

In column 6 of Table 37 (p. 228), we give the values of n calculated from (64.12) for various materials used as moderators and for an energy reduction by a factor 10^6.

The fact that for $A \gg 1$ and after a large number of collisions, the fluctuations in the lethargy value around its average (64.11) is very small, suggests an approximate description of the slowing down process as a continuous energy loss.

Instead of the discontinuous variable n let us consider, as an independent variable, the time t measured from the instant of emission of the neutron by the source. The corresponding representative point on the axis u of Fig. 116 will remain in the origin O for a certain time which, terminates when the neutron undergoes its first collision. At that moment the representative point jumps to the position u_1, where it remains until the second collision takes place, etc. The lethargy u of the neutron under consideration will be a step function of the time of the type shown in Fig. 119. All discontinuities are less than X and their average value is $\bar{\xi}$. The interval of time that the neutron will spend at a given value of u, will fluctuate around the mean value $l_s(u)/v$, $l_s(u)$ being the scattering mean free path, and

$$v = v_0 \, e^{-u/2} \tag{64.13}$$

the velocity corresponding to u. The behaviour of the step-function of Fig. 119 will depend on the behaviour of $l_s(u)$ as a function of u. But usually l_s does not decrease when the lethargy increases and therefore, considering that with increasing lethargy the neutron moves always slower, we can conclude that the average slope of the step-function $u(t)$ will decrease as t increases, as shown in Fig. 119[1]. The larger the mass ratio A of the moderator the smaller will be X and $\bar{\xi}$, i.e. the discontinuities shown by the curve of Fig. 119.

The plot relative to each individual neutron will of course be different, but for a large A and after a large number of collisions, the deviations from their average behaviour will be small. This remark justifies the substitution, frequently made, of the actual discontinuous relation between lethargy and time with a continuous line, as the broken curve of Fig. 119. In other words, sometimes for $A \gg 1$ and $u \gg X$, the substitution of a continuous energy loss for the actual discontinuous slowing down process, constitutes a satisfactory approximation.

In this approximation u and t are connected by a fixed relation which can be deduced from

$$du = \bar{\xi} \frac{v}{l_s(u)} dt. \tag{64.14}$$

Here $\frac{v}{l_s(u)} dt$ is the number of collisions undergone by the neutron of velocity v in the time interval between t and $t + dt$, $\bar{\xi}$ the average logarithmic energy loss per collision and du the corresponding increase of the lethargy. From (64.14) one derives

$$t = \frac{1}{\bar{\xi}} \int_0^u \frac{l_s(u')}{v(u')} du' = \frac{2}{\bar{\xi}} \int_v^{v_0} \frac{l_s(v')}{v'^2} dv' \tag{64.15}$$

which, for $l_s = \text{const}$, is reduced to

$$t = \frac{2}{\bar{\xi}} l_s \left(\frac{1}{v} - \frac{1}{v_0}\right). \tag{64.16}$$

It has to be noticed that the relations (64.14) and (64.15) hold for any value of the mass number A; but while for very large A, the velocity spread of the neutrons at a certain instant t is so small that their spectrum can be described as an almost monoenergetic line, for light elements the neutron spectrum extends over a very wide energy interval. We shall come back to this point in Sect. 70 where the time dependence of the integral spectrum will be derived. For heavy nuclei ($A \gg 1$) the energy spectrum is with sufficient accuracy given by (70.33) and the average energy is equal to the most probable energy (70.31) which corresponds to the velocity appearing in (64.15). In the case of hydrogen, the velocity spectrum is given by (70.15) according to which the most probable velocity is equal to one half the average velocity, and this (for $v_0 \gg v$) may be obtained by introducing $\bar{\xi} = 1$ in (64.16).

65. The slowing down power and the moderating ratio. According to Eq. (64.12) the greater the value of $\bar{\xi}$ the smaller will be the number, n, of collisions necessary to slow down the neutrons, and therefore the better the moderator. The relative merits of the various materials are, however, determined not only by the corresponding values of $\bar{\xi}$ but also by the corresponding mean free path for scattering and absorption $l_s(E)$, $l_a(E)$ and by their dependence on the energy of the neutrons.

[1] S. GLASSTONE: Principle of nuclear reactor engeneering. New York: Van Nostrand 1955.

Therefore the comparison between the various materials is usually made by considering, besides the values of $\bar{\xi}$, the values of two other quantities called the *slowing down power* and the *moderating ratio*, which give a rough indication of the ability of a material to slow down neutrons and of its effectiveness in producing a large density of slow neutrons.

Before defining these quantities we call attention to the fact that, if the moderator is a compound or a mixture and therefore contains many kinds of nuclei, the corresponding value of the average logarithmic decrement is given by the weighted average (see Sect. 67)

$$\bar{\xi} = \sum_i c_i \bar{\xi}_i = 1 + \sum_i \frac{c_i \alpha_i}{1-\alpha_i} \ln \alpha_i \tag{65.1}$$

where the coefficient c_i represent the relative collision probabilities against a nucleus of kind i defined by Eq. (61.9)

Since, in general, the l_s^i are functions of the energy of the incident neutron, according to (65.1) $\bar{\xi}$ also turns out to be a function of the energy (in spite of the assumption of isotropic scattering in the c.m.s. for all type of nuclei involved) through the energy dependence of the coefficient (61.9).

This complication is sometimes eliminated in the more elementary treatments, by substituting for the coefficients c_i their averages over the lethargy

$$\bar{c}_i = \frac{1}{u_2 - u_1} \int_{u_1}^{u_2} c_i(u)\, du. \tag{65.2}$$

According to (61.9) the coefficients c_i, and therefore $\bar{\xi}$, will be energy independent only if all scattering cross sections σ_s^i of the various nuclei present in the moderator correspond to isotropic scattering in the c.m.s., and furthermore show the same energy dependence.

The *slowing down power* of a material is defined as the ratio of $\bar{\xi}$ to the scattering mean free path for epithermal neutrons,

$$\Sigma_s \cdot \bar{\xi} = \frac{\bar{\xi}}{l_s} \text{ cm}^{-1} \tag{65.3}$$

and represents the average logarithmic decrement of the energy per centimetre of path. Its value would be energy independent if not only $\bar{\xi}$ but also Σ_s were energy independent. Such a condition is certainly not satisfied in the case of hydrogen, while for other nuclei l_s, although not constant, does not vary markedly in a not too wide energy interval (Sect. 29). This inconvenience is overcome by making use in (65.3) of an average value of Σ_s: for instance the values given in column 7 of Table 37 are calculated by averaging Σ_s between 1 ev and 10^5 ev[1].

The slowing down power does not take into account the absorption of the neutrons by the moderator which obviously is an essential aspect of the process of slowing down and diffusion of neutrons.

For instance, materials like boron and lithium can not be used as moderators in spite of their rather large values of $\bar{\xi}$ and $\bar{\xi}/l_s$, because they are too strong absorbers of neutrons.

The *moderating ratio* is defined as the product of the slowing down power (65.3) and the absorption mean free path for thermal neutrons

$$\frac{l_a}{l_s} \bar{\xi} = \frac{\Sigma_s}{\Sigma_a} \bar{\xi}. \tag{65.4}$$

[1] S. GLASSTONE: Principles of nuclear reactor engineering. New York: Van Nostrand 1955.

From the values given in column 8 of Table 37 one sees that heavy water is a far better moderator than ordinary water in spite of the large values of $\bar{\xi}$ and of the slowing down power of this latter material.

Table 37. *Scattering properties of nuclei and slowing down efficiency of moderators (Glasstone).*

Element	A	α	X	$\bar{\xi}$	n $(E_0/E=10^6)$	$\frac{\bar{\xi}}{l_s}$ [1]	$\bar{\xi}\frac{l_a}{l_s}$ [1]
H	1	0	∞	1.000	14.8	1.53 (H_2O)	70
D	2	0.111	2.197	0.725	19.1	0.177 (D_2O)	21000
Be	9	0.640	0.446	0.206	67	0.160	150
C	12	0.716	0.331	0.158	87.5	0.063	170

II. The integral spectrum of the neutrons during the slowing down in an infinite medium.

66. The integral equation of the slowing down spectrum in the stationary case. In this division we shall discuss the following stationary problem: fast monoenergetic neutrons are produced at a constant rate in time and uniformly in space throughout a homogeneous medium of infinite dimensions. We require to find their stationary spectrum which, being space independent, will be called $n_0(E)\,dE$. The data of the problem are the properties of the moderator and the number q_0 of neutrons of energy E_0 emitted by the source per second and per cm³.

At first sight the problem may appear rather fictitious. It has however a considerable heuristic value; furthermore all the equations that we shall derive in the following will also be valid in the case of an arbitrary distribution of mono-energetic neutron sources throughout an infinite medium, provided we limit consideration to the integrals with respect to the space coordinates of the various quantities defined in Sect. 61:

$$n_0(E) = \int d\boldsymbol{r}\, N_0(\boldsymbol{r}, E), \tag{66.1a}$$

$$\varphi_0(E) = \int d\boldsymbol{r}\, \Phi_0(\boldsymbol{r}, E), \tag{66.1b}$$

$$\psi_0(E) = \int d\boldsymbol{r}\, \Psi_0(\boldsymbol{r}, E), \tag{66.1c}$$

$$q(E) = \int d\boldsymbol{r}\, Q(\boldsymbol{r}, E). \tag{66.1d}$$

One has only to pay attention to the fact that, although the equations are the same, the various quantities will have different dimensions in the two cases. So, for instance, the total intensity of the source corresponding to the integrated quantities (66.1) is given by

$$q_0 = \int d\boldsymbol{r}\, Q_0(\boldsymbol{r}) \tag{66.1e}$$

and represents the total number of neutrons of energy E_0 emitted per second throughout the moderator.

The dependence on t of the various quantities has been omitted because of the stationary nature of the problem: the q_0 fast neutrons emitted per second (and per cm³) from the source are continuously slowed down so that a steady-state energy distribution is reached.

In the following, instead of the variable E or v, we shall make frequent use of the lethargy (62.19). The transformation formulas from the variable $x(u)$ ($x=E$ or v in different cases) to the variable u (and viceversa) of any function $F(x)$,

$$F(u) = F(x)\,\frac{dx}{du},$$

[1] Computed with $\Sigma_s = l_s^{-1}$ averaged between 1 ev and 10^5 ev.

are obtained by making use of (62.19) or of its reverse expression

$$E = E_0 \, e^{-u} \tag{66.2}$$

and of (64.13). The results are

$$F(u) = E_0 \, e^{-u} F(E(u)) \tag{66.3 a}$$

$$F(E) = \frac{1}{E} F(u(E)) \tag{66.3 b}$$

$$F(u) = \frac{v_0}{2} e^{-\frac{u}{2}} F(v(u)) \tag{66.3 c}$$

$$F(v) = \frac{2}{v} F(u(v)) \tag{66.3 d}$$

where we have dropped a minus sign originating from the fact that, according to (62.19), an increase du of the lethargy corresponds to a decrease dE (or dv) of the energy (or of the velocity). The change of sign is justified by the fact that the relations (66.3) will be applied to the probable number of neutrons $N(\boldsymbol{r}, \boldsymbol{\Omega}, E, t)$ or to any one of the functions defined by Eq. (61.13) to (61.19), or to their integrals (66.1), which must always be positive because of their physical meaning.

The integral equation for the collision density, relative to the present case, is deduced from Eq. (61.21 b) by the following procedure; first one requires that the distribution be stationary ($\partial\Psi/\partial t = 0$); then one integrates both sides of (61.21b) with respect to $d\Omega$, and finally the dependence of the various quantities on $\boldsymbol{r}$ is dropped (so that $\nabla\psi = 0$). In the case of hydrogen one obtains

$$\psi_{0\delta}(u) = \int_0^u du' \, h(u') \, \psi_{0\delta}(u') \, f(u-u') + q_0 \, \delta(u) \tag{66.4}$$

where, according to (61.20),

$$f(u-u') = f(u' \to u) = \int d\boldsymbol{\Omega} \, f(u' \to u, \boldsymbol{\Omega}' \to \boldsymbol{\Omega}) = \frac{\Sigma_s(u' \to u)}{\Sigma_s(u')}$$

does not depend on $\boldsymbol{\Omega}'$, but only $u - u'$, as it may be recognized from the definition of lethargy (62.19) and from the fact that, according to (62.2), the cosine of the scattering angle in the c.m.s. is always a function of the ratio of the neutron energies after and before the collision. In (66.4) $h(u)$, defined by Eq. (61.18), represents the probability of non-capture in a collision undergone by a neutron of lethargy u; in the case of non-absorbing media one has for any value of u, $h(u) = 1$.

Eq. (66.4) may be transformed by separating out the δ-function part of the solution; if one writes

$$\psi_{0\delta}(u) = q_0 \, \delta(u) + \psi_0(u), \tag{66.5}$$

the first term represents the neutrons experiencing no collision at all, the second term the slowing down spectrum, in which we are mainly interested here. This may be obtained as solution of an integral equation which is derived by introducing (66.5) into (66.4)

$$\psi_0(u) = \int_0^u du' \, h(u') \, \psi_0(u') \, f(u-u') + q_0 \, h(0) \, f(u). \tag{66.6}$$

The only difference with respect to (66.4) is in the source term which in (66.6) may be interpreted as follows. Of the q_0 neutrons emitted per second by the source, $q_0 f(u)$ have their first collision in the unit lethargy interval at u, and of these only the fraction $h(0)$ survives, since the remaining fraction $(1-h(0))$ is captured.

In the case of hydrogen the collision function has non zero values from the initial energy E_0 to E, and therefore the limits of the integral with respect to u' are 0 and u. This fact, as was shown originally by FERMI[1], makes it possible to convert the integral equation (66.6) into a differential equation the solution of which can be derived immediately.

In the case of isotropic scattering in the c.m.s. the collision function is given by Eq. (63.1).

In the case of moderation in media containing a single type of nuclei heavier than hydrogen ($A \geqq 2$) the existence of a maximum value (62.21 b) of the logarithmic energy loss in a single elastic collision, gives rise to complications. The integral equation [similar to (66.6)] valid in the first interval $(0, X)$

$$\psi_0(u) = \int_0^u du'\, h(u')\, \psi_0(u')\, f(u-u') + q_0 h(0) f(u); \quad 0 \leqq u \leqq X \qquad (66.7\text{a})$$

differs from that holding for any value of $u > X$

$$\psi_0(u) = \int_{u-X}^{u} du'\, h(u')\, \psi_0(u')\, f(u-u'); \quad u > X, \qquad (66.7\text{b})$$

because the first collision undergone by the neutrons when emitted from the source, contributes only to the first equation. Furthermore only this can be converted into a differential equation; Eq. (66.7b) cannot be converted because in this case both limits of the integral depend on u.

In the case of isotropic scattering in the c.m.s. the collision function is given by (62.21 a). According to PLACZEK[2] the solution of (66.7a) valid in the first interval $(0, X)$ is immediately derived as in the case of hydrogen. Then the solution in the second interval $(X, 2X)$ is obtained by splitting the integral appearing in (66.7b) into two parts: the first integral is extended from $u' = u - X$ to $u' = X$ and can be computed explicitly since $\psi_0(u)$ is known for $u \leqq X$, the second integral is extended from $u' = X$ to $u' = u$ and is therefore similar to the integral appearing in (66.6) or (66.7a) where u only appears in the upper limit. This new form of the integral equation for the second interval may be transformed into a differential equation which can easily be solved.

By repeating this rather laborious procedure PLACZEK gave the solution in the second, third, fourth, ... interval.

We will not go into the details of the derivation of the solutions of these equations for which we refer to the original papers by various authors[1-4], to various review articles [*16*], lectures [*19*] or books [*7*].

Here we limit ourselves to collecting a few elementary considerations which allow the derivation of the exact solution in the case of hydrogen and of the more useful approximate expressions in the case of heavier nuclei.

The emission by the source of q_0 neutrons per second and per cm^3, all of the same energy E_0, referred to the lethargy (62.19) can be described as the "pouring" on the u-axis of Fig. 116 of q_0 independent series of points $u_1, u_2, u_3, \ldots$ every

[1] E. FERMI: Peter Zeeman Verhandel., Martinus Nijhoff, 's Gravenhage 1935, p. 128.
[2] G. PLACZEK: Phys. Rev. **69**, 423 (1946).
[3] F. T. ADLER: Phys. Rev. **60**, 279 (1941).
[4] I. WALLER: Ark. Mat. Astronom. Fys., Ser. A **34**, 4 (1946).

second. If there is no absorption, each one of these series goes on indefinitely, while in the case of absorption they come to an end at the point where the neutron is captured.

The number of representative points which fall between u and $u+du$, $n_0(u)\,du$, does not change with time (stationary condition) being determined by the statistical equilibrium between the number of points that in each second disappear from du because they are absorbed by the nuclei or because they are slowed down by an elastic collision, and the number of points that in each second jump into du from positions placed on the left of u, having been slowed down by elastic collisions.

67. The slowing down spectrum in non-absorbing media. *α) Hydrogen.* Let us first consider *the case of hydrogen* where we can derive the exact solution by means of the following elementary consideration.

The number $n_0(u)\,du$ of points that at any time lie between u and $u+du$ is given by the product of the number of points each second entering into du, and the average time during which each one of them remains there before jumping out of it as a result of an elastic collision. The first factor is

$$q(u)\,du = q_0\,du \tag{67.1}$$

because: (a) in the case of hydrogen du represents the probability that one point of one of the series $u_1, u_2, u_3, \ldots$ falls in du, and (b) in absence of capture, the slowing down density $q(u)$ defined by (61.19) and (66.1d) is constant and equal to q_0, the number of independent series of points that in each second are poured on the u-axis by the source.

The average time that one of these points spends in du is given by $l_s(u)/v$ and therefore we may write:

$$n_0(u)\,du = q_0\frac{l_s(u)}{v}\,du \tag{67.2}$$

or, taking as independent variable either E or v instead of the lethargy u,

$$n_0(E)\,dE = q_0\frac{l_s(v)}{v}\frac{dE}{E} \tag{67.3}$$

$$n_0(v)\,dv = 2q_0\frac{l_s(v)}{v^2}\,dv \tag{67.4}$$

where we have used the transformation rules (66.3).

From these expressions one also derives immediately the neutron flux (66.1b) and the collision density (66.1d) which, in this case of slowing down without absorption, reduces to

$$\psi_0(u) = \Sigma_s(u)\,\varphi_0(u) = v\,\Sigma_s(u)\,n_0(u). \tag{67.5}$$

It represents the number of elastic collisions undergone by the neutrons per unit lethargy interval at u, per second, per cm^3. From (67.2) and (67.5) one has

$$\psi_0(u)\,du = q_0\,du = q(u)\,du \tag{67.6}$$

or, making use of the transformation rule (66.3b)

$$\psi_0(E)\,dE = q_0\frac{dE}{E}. \tag{67.7}$$

According to (67.1) and (67.6) in absence of absorption, the collision density is constant along the u-axis and, if the moderator is hydrogen, equal to the slowing down density $q(u)=q_0$.

β) In a moderator containing a single type of nuclei different from hydrogen the problem is much more complicated. It can be treated in an elementary way, formally similar to that applied in the case of hydrogen, only if one limits consideration to the asymptotic case in which the neutrons have undergone many collisions

$$n \gg 1 \quad \text{i.e.} \quad u \gg X.$$

In this region, according to the conclusions reached in Sect. 64, the values of u after n collisions, have a normal distribution around the average value $n\bar{\xi}$ with a spread which extends over a few intervals on both sides

$$u = n\bar{\xi} \pm \sqrt{\frac{n}{12}}\,X.$$

Considering that the average distance between two successive points is $\bar{\xi}$, one can describe the distribution of the points of the series $u_n, u_{n+1}, u_{n+2}, \ldots$ for $n \gg 1$, as a uniform distribution of linear density $1/\bar{\xi}$. This is equivalent to saying that the probability that one point of the considered series falls in du is given by

$$\frac{du}{\bar{\xi}}. \tag{67.8}$$

As long as such a description is correct, one can repeat the considerations given previously for hydrogen with the only difference that now one has to use (67.8) instead of du for the probability that one point of the series falls in du. Hence, we may write

$$n_0(u)\,du = q_0 \frac{l_s(v)}{\bar{\xi}} \frac{1}{v}\,du, \tag{67.9}$$

$$n_0(E)\,dE = q_0 \frac{l_s(v)}{\bar{\xi}} \frac{1}{v} \frac{dE}{E}, \tag{67.10}$$

$$n_0(v)\,dv = 2q_0 \frac{l_s(v)}{\bar{\xi}} \frac{dv}{v^2}, \tag{67.11}$$

$$\psi_0(u)\,du = q_0 \frac{du}{\bar{\xi}}; \tag{67.12}$$

these expressions are valid for $u \gg X$ and reduce to (67.2) to (67.4), (67.6) in the case of hydrogen: $\bar{\xi} = 1$.

In these equations the properties of the moderator are summarized by the slowing down power $\bar{\xi}/l_s$ already considered in Sect. 65, Eq. (65.3), the values of which (averaged in the interval 1 to 10^5 ev) have been given for various materials in column 7 of Table 37 (p. 228).

The asymptotic expressions (67.9) to (67.12) are valid however not only for values of $u \gg X$, but also at energies not too small compared with the initial energy E_0. Starting from the corresponding integral equations (66.7a) and (66.7b) applied to non-absorbing media, $h(u) = 1$, PLACZEK[1] has calculated $\psi_0(u)$ as a function of u/X, for $A = 2$, 4, 12, and infinity, by means of the procedure mentioned in Sect. 66.

The results are qualitatively the same in all these cases. In Fig. 120 we give, for $A = 12$ the quantity

$$\frac{\psi_0(u) - \psi_{0\,\text{asy}}(u)}{\psi_{0\,\text{asy}}(u)} = \frac{\psi_0(u)}{\psi_{0\,\text{asy}}(u)} - 1 \tag{67.13}$$

[1] G. PLACZEK: Phys. Rev. **69**, 423 (1946).

as a function of u/X. $\psi_0(u)$ starts from the value

$$\psi_0(0) = \frac{q_0}{1-\alpha}$$

and rises rapidly up to

$$\psi_0(u \to X) = \frac{q_0}{1-\alpha} \, \frac{1}{\alpha^{\frac{\alpha}{1-\alpha}}}$$

for $u = X$, where it drops discontinuously[1,2] by

$$\psi_0(u \to X) - \psi_0(u \leftarrow X) = q_0 \frac{\alpha}{1-\alpha}.$$

This behaviour is due to the fact that in the integral equation the term representing the first collision contribution disappears when u passes from values smaller than X to values larger than X as is shown by (66.7a) and (66.7b). Once past $u = X$, $\psi_0(u)$ goes on rising to a maximum, reaches a minimum with discontinuous tangent at $u = 2X$, and rapidly approaches the asymptotic value (67.12) from which it differs very little even at $u = 3X$.

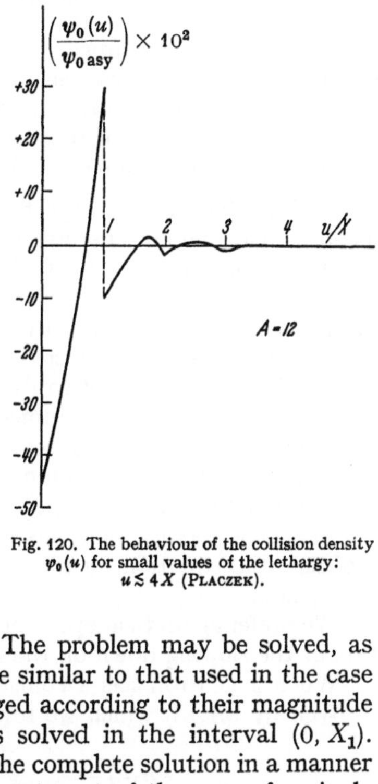

Fig. 120. The behaviour of the collision density $\psi_0(u)$ for small values of the lethargy: $u \lesssim 4X$ (PLACZEK).

In conclusion, we may use the asymptotic expressions (67.9) for any value of the energy satisfying the condition

$$u > 3X \qquad (67.14)$$

or

$$E < \alpha^3 E_0 = \left(\frac{A-1}{A+1}\right)^6 E_0. \qquad (67.15)$$

If the moderator contains nuclei of different atomic weights M_i, the integral equations for the collision density are similar to (66.7a) and (66.7b) with the collision function $f(u-u')$ expressed, according to (61.8b), in terms of the partial collision functions $f_i(u-u')$ relative to the various nuclei present in the moderator, and of the coefficients $c_i(u)$ defined by (61.9). The problem may be solved, as was shown by PLACZEK, by means of a procedure similar to that used in the case of a single nucleus. The quantities X_i are arranged according to their magnitude $X_1 < X_2 < X_3 < \cdots$ and the integral equation is solved in the interval $(0, X_1)$. From this solution one can proceed to construct the complete solution in a manner similar to, though more complicated than the treatment of the case of a single type of nuclei.

If the mean free paths are constant or vary all in the same way with energy so that the coefficients c_i are constant (see Sect. 65), the integral equation (66.7b) is again solved asymptotically by a constant $\psi_{0\,\mathrm{asy}}$ the expression of which is identical to (67.12) with $\bar{\xi}$ given by (65.1).

PLACZEK has derived the complete solution in the case of a mixture of hydrogen with another substance of concentration c defined by (61.9).

---

[1] See footnote 1, p. 232.

[2] S. GLASSTONE and M.C. EDLUND: Nuclear reactor theory, p. 154. New York: Van Nostrand 1956.

In particular, when $1-c \ll 1$ and $1-\alpha \ll 1$, the collision density reaches the asymptotic value that one would have if no hydrogen were present, and then by increasing u, the influence of hydrogen comes into play and $\psi_0(u)$ reaches the constant value (67.12) relative to the mixture considered. The transition takes place, for $1-c \gg \frac{1}{2}(1-\alpha)$, at $u/X \sim \frac{1}{2(1-c)} \ln\left(2\frac{1-c}{1-\alpha}\right)$ which is always a large number when the approximations adopted are valid.

In the more general case when $1-c$ is no longer small compared to 1, the transition will take place earlier. The reader is referred to PLACZEK'S paper for more details.

Before closing this section we notice that by combining (67.1) and (67.12) the following relation is obtained between the asymptotic values of the slowing down density and the collision density

$$\psi_0(u) = \Sigma_s(u)\,\varphi_0(u) = \frac{q(u)}{\xi}. \tag{67.16}$$

In the case of hydrogen this equation is reduced to the equality (67.6) of $\psi_0(u)$ and $q(u)$. One can easily see that in this case the contribution to $q(u)$ due to points located in the interval $(0, u)$ is $q_0(1-e^{-u})$, to which one has to add the contribution $q_0 e^{-u}$ of those series of representative points which do not start on the left of the considered value of u.

68. The slowing down spectrum, in the case of hydrogen, with absorption. Let us start again by considering the case of hydrogen for which one can derive the exact solution by means of elementary considerations. The medium will be composed by hydrogen, whose capture cross section is not very large and follows the $1/v$ law (see Sect. 29), or by a mixture of hydrogen and one or more elements, whose concentrations and cross sections are such as to contribute appreciably to the absorption but not to the scattering. Under these conditions the slowing down of neutrons is determined by hydrogen alone while the other elements only contribute to the absorption. Diluted solutions of strong absorbers of neutrons, as e.g. boron, lithium, silver or uranium, may satisfy these conditions.

In order to determine $n_0(u)\,du$ one could again solve the corresponding integral equation (66.6). This was originally done by FERMI[1] who converted it into a differential equation by the same procedure as that followed in the case of no absorption.

We prefer again to adopt a more elementary approach based on the representation of the slowing down on the u-axis of Fig. 116.

When a neutron has a collision at a point u, the probability of non-capture is given by (61.18) which we rewrite

$$h(u) = \frac{\Sigma_s(u)}{\Sigma_s(u)+\Sigma_a(u)} = \frac{\sigma_s^{\mathrm{H}}}{\sigma_s^{\mathrm{H}}+\bar{\sigma}_a} = \frac{1}{1+\dfrac{l_s}{T v}} \tag{68.1}$$

where σ_s^{H} is the scattering cross section of hydrogen and

$$\bar{\sigma}_a = \frac{1}{\mathfrak{N}_{\mathrm{H}}} \sum \mathfrak{N}_i\, \sigma_a^i = \frac{1}{\mathfrak{N}_{\mathrm{H}}} \Sigma_a \tag{68.2}$$

is the average capture cross section to which possibly not only hydrogen but also other nuclei may contribute, according to our assumptions. The mean life of the neutrons in the medium, is

$$T = \frac{l_a}{v} = \frac{1}{v\,\Sigma_a}. \tag{68.3}$$

[1] See footnote 1, p. 230.

The probability that one particular neutron actually survives to suffer its n-th elastic collision is given by

$$p_n = h(0) \cdot h(u_1) \cdot h(u_2) \cdot \cdots \cdot h(u_{n-1}) \tag{68.4}$$

where $0, u_1, u_2, \ldots, u_{n-1}$ are the values of the lethargy of the neutron under consideration, before the first collision, the second collision, etc.

We are interested in the probability $p(u)$ that the neutron arrives at u without having been captured, independently from the number of collisions n or the particular values $0, u_1, u_2, u_3, \ldots$ at which these collisions take place. Therefore we have to average the expression (68.4) with respect to all possible values of n and of $u_1, u_2, u_3, \ldots, u_{n-1}$ taking into account the corresponding distribution functions. The simplest way of obtaining this is the following. Let us divide the interval $(0, u)$ into many small intervals $\Delta u_1, \Delta u_2, \Delta u_3, \ldots$ and let us write the probability (68.4) in the form

$$p(u) = h(0) \cdot f_1 \cdot f_2 \cdot \cdots \cdot f_i \cdot \cdots \tag{68.5}$$

where each one of the intervals Δu_i contributes a factor f_i; f_i is equal to 1 if the interval does not contain any point of the succession $u_1, u_2, u_3, \ldots$; it is equal to $h(u_i)$ if the interval contains one point. It is now clear that the average value of $p(u)$ can be obtained by averaging each factor f_i independently of all the others because of the independence of the intervals which holds in the case of slowing down by hydrogen (see Sect. 63).

The probability that Δu contains one point of the succession $u_1, u_2, u_3, \ldots$ is equal to Δu, and the probability that it does not contain any point is $1 - \Delta u$ so that

$$\bar{f} = (1 - \Delta u) \cdot 1 + \Delta u\, h(u) = 1 - [1 - h(u)]\, \Delta u.$$

Since Δu is very small, one can write

$$\ln \bar{f} = -g(u)\, \Delta u \tag{68.6}$$

with

$$g(u) = 1 - h(u) = \frac{\Sigma_a}{\Sigma_s + \Sigma_a}\,. \tag{68.7}$$

In the limit, $\sum \ln \bar{f}$ becomes an integral and from (68.6) one gets

$$p(u) = h(0) \exp\left[-\int_0^u g(u)\, du\right] \tag{68.8}$$

or, going back to the variable E and making use of (68.7),

$$p(E) = h(E_0) \exp\left[-\int_E^{E_0} \frac{\Sigma_a(E')}{\Sigma_s(E') + \Sigma_a(E')} \frac{dE'}{E'}\right]. \tag{68.9}$$

Very often the absorption cross section Σ_a is negligible with respect to Σ at the initial energy E_0 so that one can put

$$h(E_0) = 1\,. \tag{68.10}$$

The quantity $p(E)$, representing the probability that a neutron emitted in the medium under consideration with the initial energy E_0, escapes capture while being slowed down to the energy E, is called the *resonance escape probability*. We shall come back later to this point in connection with the same problem in the case of moderators different from hydrogen ($A \geqq 2$) and we then shall see why the term "resonance" is used to describe $p(E)$ (Sect. 69).

In order to determine $n_0(u)\,du$ we can now repeat the argument given in Sect. 67 with the two following modifications.

(a) The slowing down density is no longer independent of u as in the case without absorption [see (67.1)]; as a consequence of capture it is given by the expression

$$q(u) = q_0 \cdot p(u). \tag{68.11}$$

Therefore the number of representative point entering each second the interval du is

$$q(u)\,du = q_0\,p(u)\,du; \tag{68.12}$$

(b) The average time spent by each neutron at u is no more $\frac{l_s(u)}{v}$ but

$$\frac{l_s(u)}{v}\,h(u) = \frac{l_{\text{tot}}(u)}{v}. \tag{68.13}$$

In conclusion we have

$$n_0(u)\,du = q_0\,p(u)\,h(u)\,\frac{l_s}{v}\,du \tag{68.14}$$

or

$$n_0(E)\,dE = q_0\,\frac{\sigma_s(E_0)}{\sigma_s(E_0)+\sigma_a(E_0)}\,\frac{l_{\text{tot}}(E)}{v}\,\exp\left[-\int\limits_E^{E_0}\frac{\sigma_a}{\sigma_s+\sigma_a}\,\frac{dE'}{E'}\right]\frac{dE}{E} \tag{68.15}$$

which may also be written in the form

$$\left.\begin{aligned} n_0(v)\,dv &= C\,\frac{v\,l_s(v)}{v+\dfrac{l_s(v)}{T(v)}}\,\exp\left[-2\int\limits_v^{v_0}\frac{dv'}{v'+\dfrac{l_s(v')}{T(v')}}\right]dv,\\ C &= \frac{2q_0}{v_0\left[v_0+\dfrac{l_s(v_0)}{T(v_0)}\right]};\qquad T=\frac{l_a(v)}{v}\end{aligned}\right\} \tag{68.16}$$

identical with the expression given originally by FERMI[1].

In the important case

$$l_s = \text{const};\qquad \Sigma_a = l_a^{-1} \propto 1/v \tag{68.17}$$

and therefore $T=$const, the spectrum (68.16) is reduced, with respect to the case without absorption (67.4) by the factor

$$p(v)\cdot h(v) \approx \frac{1}{\left[1+\dfrac{l_s}{T v}\right]^3} \tag{68.18}$$

a term $l_s/T v_0$ having been neglected with respect to 1 $(h(v_0)=1)$. Now we have

$$n_0(v)\,dv = 2q_0\,\frac{l_s\cdot v\,dv}{\left[v+\dfrac{l_s}{T}\right]^3} = 2q_0\,l_s\,\frac{v\,dv}{\left[v+\dfrac{\Sigma_a v}{\Sigma_s}\right]^3}. \tag{68.19}$$

In this case the neutron spectrum has a maximum at $v=\frac{l_s}{2T}$ and its intensity is reduced by a factor 2 with respect to the case of no capture, when the velocity reaches the value

$$v_{\frac12} = \frac{1}{2^{\frac13}-1}\,\frac{l_s}{T} \approx 3.86\,\frac{l_s}{T}$$

[1] See footnote 1, p. 230.

while at velocities much larger than this value the effect of absorption can be neglected. For a rough estimate of the situation let us consider the case of pure water; from $l_s = 0.67$ cm, and $T = 203$ μsec it follows $l_s/2T = 1.65 \times 10^3$ cm/sec; $v_{\frac{1}{2}} = 1.27 \times 10^4$ cm/sec i.e. a value much smaller than that ($v = 1.4 \times 10^6$ cm/sec) corresponding to the lower limit (61.1) of applicability of the present considerations. Therefore neutron capture may be neglected in pure water at energies above 1 ev. In a 5% solution of boron ($\mathfrak{N}_B/\mathfrak{N}_H = 5\%$; $\sigma_{th}(B) = 760$ b) $v_{\frac{1}{2}}$ corresponds to an energy of about 1 ev.

The application of (68.19) to a practical case will be correct provided the source emits neutrons of such a low energy that the scattering mean free path in water may be taken as constant: e.g. for a source emitting neutrons of energy E_0 of the order of 10^4 ev (Sect. 29).

Another important case to which one can apply (68.15) is that of a dilute solution containing a nucleus showing resonance absorption. Here again we assume that the slowing down is due to hydrogen while the nucleus under consideration contributes only to the absorption.

For the sake of simplicity let us suppose that the capture cross section may be represented by the simplest formula, namely the one-level formula (26.2) that we write, according to (68.2),

$$\bar{\sigma}_a = \frac{\mathfrak{N}}{\mathfrak{N}_H} \frac{\sigma_{a0}}{1 + \left(\frac{E - E_R}{\Gamma/2}\right)^2} \tag{68.20}$$

where $\mathfrak{N}$ is the concentration of the capturing nucleus. Two approximations are introduced in (68.20): we omit the $1/v$ factor in the resonance formulas (26.2), (26.3); this is legitimate provided

$$\Gamma \ll E_R.$$

Furthermore we neglect the capture by hydrogen which also follows the $1/v$ law, and is always very small in the resonance region.

From (68.20) it follows

$$g(E) = \frac{\bar{\sigma}_a}{\sigma_s + \bar{\sigma}_a} = \frac{\mathfrak{N}\sigma_{a0}}{\mathfrak{N}_H \sigma_s + \mathfrak{N}\sigma_{a0}} \frac{1}{1 + \left(\frac{E - E_R}{\Gamma^*/2}\right)^2} \tag{68.21}$$

where

$$\Gamma^* = \sqrt{1 + \frac{\mathfrak{N}\sigma_{a0}}{\mathfrak{N}_H \sigma_s}}\, \Gamma \tag{68.22}$$

is a kind of effective width of the absorption line when the nuclei are in solution with concentration $\mathfrak{N}$. We will assume that also $\Gamma^* \ll E_R$. Under this condition and assuming $\sigma_s = \text{const}$, we can write

$$\left.\begin{aligned} \ln p(E) &= \int_E^{E_0} g(E') \frac{dE'}{E'} \approx \frac{1}{E_R} \int_E^{E_0} g(E')\, dE' \approx \\ &\approx -\frac{\Gamma^*}{2E_R} \frac{\mathfrak{N}\sigma_{a0}}{\mathfrak{N}_H \sigma_s + \mathfrak{N}\sigma_{a0}} \left\{\frac{\pi}{2} - \arctan\left(\frac{E - E_R}{\Gamma^*/2}\right)\right\} \end{aligned}\right\} \tag{68.23}$$

where we have twice made use of the assumption $E_0 \gg E_R$: once by putting $h(E_0) = 1$, and a second time by extending the integral to infinity instead of to E_0.

The curves given in Fig. 121 have been calculated taking $\mathfrak{N}\sigma_{a0} = \mathfrak{N}_H \sigma_s$, so that $\Gamma^* = \sqrt{2}\,\Gamma$, and $E_R = 5\Gamma^*$. This example, although fictitious because of the oversimplification introduced in the analytical representation of the capture cross section and because of the arbitrariness of the choice of the numerical values of the various constants, clearly illustrates the effect of an absorption line; $h(E)$ has, of course, the shape of an absorption line of half-width Γ^*, while $p(E)$ introduces a permanent decrease of the neutron spectrum $n_0(E)$ passing from energies above to energies below resonance.

From (68.23) one sees that $p(E)$ tends to 1 for $E-E_R \gg \Gamma$ while for $E_R - E \gg \Gamma$ the arctangent tends to $-\frac{\pi}{2}$ and therefore

$$p_0 = \lim_{\frac{E_R-E}{\Gamma}\to\infty} p(E) = \exp\left[-\frac{\pi}{2}\frac{\Gamma^*}{E_R}\frac{\mathfrak{N}\sigma_{a0}}{\mathfrak{N}\sigma_{a0}+\mathfrak{N}_{\mathrm{H}}\sigma_s}\right]. \tag{68.24}$$

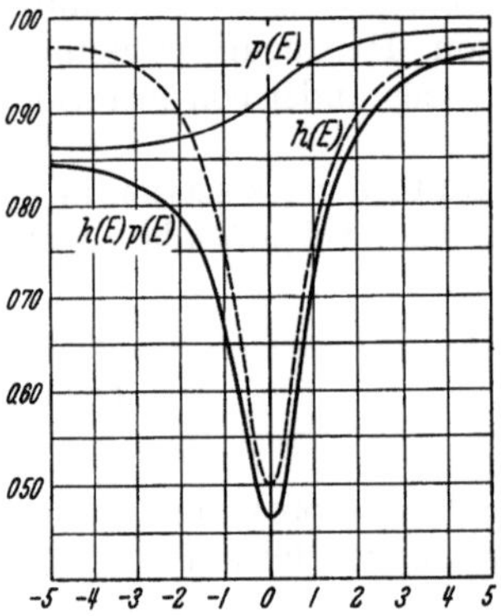

$\dfrac{E-E_R}{\frac{1}{2}\Gamma^*}$

Fig. 121. The resonance escape probability in a solution of a resonant nucleus in water.

This value p_0 is, in a certain sense, the more relevant datum because it represents the fraction of the original number of neutrons which survive far below the resonance. The structure of the exponent in (68.24) is clear: Γ^*/E_R is the width of the absorption line in the logarithmic scale and represents the probability that one of the points of the series $u_1, u_2, \ldots$ falls inside the line itself; the other factor represents the probability that a neutron of the right energy $E_R \pm \frac{\Gamma^*}{2}$ is absorbed before again losing energy through an elastic collision.

By increasing the concentration $\mathfrak{N}$ of the absorbing nuclei, the exponent of (68.24) at first decreases linearly; but later, when $\mathfrak{N}\sigma_{a0} \gg \mathfrak{N}_{\mathrm{H}}\sigma_s$ i.e. when practically all the neutrons which during the slowing down process fall inside the true width of the line Γ are absorbed, the expression (68.24) decreases mainly because Γ^* increases proportionally to the square root of the concentration of the absorbing nuclei.

A similar treatment can also be made in the case of a resonance line whose natural width Γ is much smaller than the Doppler width Δ (see Sect. 87). Then according to (87.6) 87.11 one can write

$$\bar{\sigma}_a = \frac{\mathfrak{N}}{\mathfrak{N}_{\mathrm{H}}}\sigma_{a0}\frac{\sqrt{\pi}}{2}\frac{\Gamma}{\Delta}\exp\left[-\left(\frac{E-E_R}{\Delta}\right)^2\right] \tag{68.25}$$

which is valid only if $\Gamma \ll \Delta$. The case $\Gamma \sim \Delta$ can only be treated numerically.

Then, with the notation

$$\sigma_{ae} = \sqrt{\frac{\pi}{8}}\,\sigma_{a0}\frac{\Gamma}{\Delta},$$

we have

$$\int_E^{E_0}\frac{\bar{\sigma}_a}{\sigma_s+\bar{\sigma}_a}\frac{dE'}{E'} \approx \frac{\Delta}{E_R}\,a\int_{\frac{E-E_R}{\Delta}}^{\infty}\frac{dt}{e^{t^2}+a}; \qquad a = \sqrt{2}\,\frac{\mathfrak{N}\sigma_{ae}}{\mathfrak{N}_{\mathrm{H}}\sigma_s},$$

and we can write

$$p_0 = \lim_{-\frac{E_R-E}{\Delta}\to\infty} p(E) = \exp\left[-\frac{\Delta}{E_R}\varphi(a)\right] \tag{68.26}$$

where

$$\varphi(a) = a \int_{-\infty}^{+\infty} \frac{dt}{e^{t^2} + a}. \tag{68.27}$$

When $a \leqq 1$, one has

$$\varphi(a) = -\sqrt{\pi} \sum_{n=1}^{\infty} \frac{(-a)^n}{\sqrt{n}}.$$

For $a = 1$: $\varphi(1) = 1.073$. For $a \ll 1$, i.e. at low concentration, $\ln p_0$ increases linearly with a, i.e. with $\mathfrak{N}$. Hence, (68.24) and (68.26) yield the same result; this we expect because $g(E) \sim \frac{\sigma_a}{\sigma_s}$ and therefore p_0 only depends on $\int \sigma_a(E)\, dE$ which is independent of the Doppler effect [see Sect. 87].

When $a \gg 1$ one has

$$\varphi(a) = 2\sqrt{\ln a}\left[1 - \frac{\pi^2}{24 (\ln a)^2} - \frac{7\pi^4}{384 (\ln a)^4} + \cdots\right];$$

but for such a high concentration Eq. (68.25) can not be used because it does not correctly represent the line at great distances from its center. In practice this is often the most interesting case because, since $\Delta/E_R \ll 1$, one can not have a large decrease of $p(E)$ without a rather large value of a. Anyhow, in these cases the shape of the line cannot be represented by (68.25). It is more similar to a square well and therefore one can calculate p_0 numerically in each special case without any difficulty.

69. The slowing down spectrum, for $A \geqq 2$, with absorption. The problem is much more complicated than in the case of hydrogen, even if one takes advantage of the fact that in most practical cases the absorption sets in only at energies considerably lower than the initial energy. Under this condition, as was pointed out by PLACZEK[1] who discussed the corresponding integral equations, the collision density has already reached its asymptotic value (67.12) and therefore it is only necessary to investigate how this value is modified by capture.

This is done by considering the integral equation without the direct contribution from the neutron source, Eq. (66.7b), as valid for any value of u. The method of solution depends however on whether the variation of the cross section, and therefore of $h(u)$, in an interval $(u - X, u)$ is slow or fast.

The simplest approach to this problem is by studying the slowing down density $q(u)$ which, for $u > u_0 = 3X$, has the constant value $q(u) = q_0$ as long as the absorption is negligible, but which starts to decrease with increasing u as soon as absorption sets in. The variation of $q(u)$ with the lethargy may also in this case be represented by (68.11), namely we write

$$q(u) = q_0\, p(u) \tag{69.1}$$

where $p(u)$ is the resonance escape probability already considered in the particular case of hydrogen [see Eqs. (68.8) and (68.9), valid in that special case].

Thus the problem is that of determining $p(u)$. Once $p(u)$ is known, the neutron spectrum can be obtained by repeating the argument given in Sect. 68 for hydrogen. In Eq. (67.11) $q(u)$ is substituted for q_0 and $l_{\text{tot}}(u)$ for $l_s(u)$ so that one has

$$n_0(v)\, dv = 2q_0 \frac{l_{\text{tot}}}{\xi} p(v) \frac{dv}{v^2} = \frac{2q_0}{\xi} \frac{l_s}{1 + l_s/v\,T} p(v) \frac{dv}{v^2} = \frac{2q_0}{\xi} l_s\, h(v)\, p(v) \frac{dv}{v^2}. \tag{69.2}$$

[1] See footnote 1, p. 232.

The complications arising in the derivation of $p(u)$ may be understood by the following elementary considerations. Let us suppose that the absorption takes place only in two very narrow regions Δu_1 and Δu_2 placed at u_1 and $u_2 > u_1$. The absorption taking place between u_1 and $u_1 + \Delta u_1$ can be described as a negative neutron source which produces a variation of the slowing down density with respect to its value $q(u \leqq u_1) = q_0$, of an oscillatory type, similar to that shown in Fig. 120 for the case of carbon. According to (67.14) these oscillations will be completely damped at $u > u_1 + 3X$ where $q(u)$ will have reached a new asymptotic value q_1 independent of u.

If the second absorbing region Δu_2 is placed at a point

$$u_2 \geqq u_1 + \Delta u_1 + 4X \approx u_1 + 4X, \tag{69.3}$$

its effect will be that of a second negative neutron source independent of the source placed at u_1; it will produce a new variation of $q(u)$ of an oscillatory type which will be damped again for $u > u_2 + 3X$, where a new constant value q_2 will be reached.

But if the condition (69.3) is not satisfied, the collision density $\psi_0(u)$ has not yet reached a constant value over the lethargy interval X preceding u_2 and therefore the effect of the absorption in the Δu_2 region will depend on its lethargy distance from the first absorption region Δu_1; in this case the value of $q(u)$ for $u > u_2 + 3X$ will also depend on the distance between u_2 and u_1.

α) The case of slowly varying capture is very important because it includes, as a particular example of considerable practical interest, the case of capture cross sections following the $1/v$ law.

Two different procedures have been suggested, the first by PLACZEK[1], the second by GOERTZEL and GREULING[2].

PLACZEK treated this problem starting from the solution of the integral equation (66.7b) valid in the case

$$h(u) = \frac{\Sigma_s(u)}{\Sigma_s(u) + \Sigma_a(u)} = \text{const.} \tag{69.4}$$

It is immediately recognized that in this case $\psi_0(u)$ is represented by an exponential

$$\psi_0(u) = c \exp[-\nu_0 u] \tag{69.5}$$

where c is a constant and ν_0 is the positive root of the transcendental equation

$$1 - \nu = h \frac{1 - \alpha^{1-\nu}}{1 - \alpha} \tag{69.6}$$

obtained by introducing (69.5) into (66.7b). According to the definitions (61.19) (66.1d), the slowing down density $q(u)$ is given by the expression

$$q(u) = \int_{u-X}^{u} h(u')\, \psi_0(u')\, du' \int_{u}^{u'+X} f(u'' - u')\, du'' \tag{69.7}$$

which, using (62.21), (62.16) and the notation

$$\psi_{0s}(u') = h(u')\, \psi_0(u') = v' \Sigma_s(v')\, n_0(u') \tag{69.8}$$

[1] See footnote 1, p. 232.

[2] G. GLASSTONE and M.C. EDLUND: Nuclear Reactor Theory, p. 168. New York: Van Nostrand 1955.

reduces to

$$q(u) = \frac{1}{1-\alpha} \int_{u-X}^{u} \psi_{0s}(u')\,du'\,[e^{u'-u} - \alpha]. \tag{69.9}$$

If (69.4) and (69.5) are introduced in (69.7), (69.8), $q(u)$ also turns out to be exponential so that we may write

$$q = q_0 \exp[-\nu_0(u-u_0)] \quad \text{for } u > u_0 = 3X. \tag{69.10}$$

From Eq. (69.6) we see that the number ν_0 is a function of the cross sections for which an expansion

$$\nu_0 = \frac{w}{\xi}\{1 - c_2 w + c_3 w^2 - \cdots\} \tag{69.11}$$

in powers of

$$w = \frac{g}{h} = \frac{g}{1-g} = \frac{\Sigma_a}{\Sigma_s} \tag{69.12}$$

was given by PLACZEK.

If one now considers the case of a capture cross section varying so slowly that the change of $h(u)$ over one interval X can be neglected, an equation of the type of (69.10) will hold in each interval X and therefore one can write, according to (69.1),

$$p(u) \approx \exp\left[-\int_{u_0}^{u} \nu(u')\,du'\right]. \tag{69.13}$$

If $w(u)$ is small throughout, the expansion (69.11) may be used for $\nu(u)$. Its first term $w(u)/\xi$ will be sufficient if the second one is small; since c_2 is of the order of one, this condition is expressed by the relation

$$\int_{u_0}^{u} w^2(u')\,du' \ll \xi \tag{69.14}$$

which for small ξ, i.e., large masses, is perfectly compatible with the first term being large compared to one:

$$\int_{u_0}^{u} w(u')\,du' \gg \xi.$$

Thus for large masses and small $w(u)$, the expression

$$p(u) = \exp\left[-\frac{1}{\xi}\int_{0}^{u} w(u')\,du'\right] \tag{69.15a}$$

will have a wide range of validity. Going back from the lethargy to the energy as integration variable one has

$$p(E) = \exp\left[-\frac{1}{\xi}\int_{E}^{E_0} \frac{\Sigma_a(E')}{\Sigma_s(E')}\,\frac{dE'}{E'}\right]. \tag{69.15b}$$

In Eq. (69.15a) the lower limit of the integral has been taken 0 instead of u_0; this can be done because, according to our assumptions, capture sets in only for values of $u > u_0$.

For hydrogen, i.e. for $\xi = 1$, Eqs. (69.15a) and (69.15b) reduces to the rigorous expressions (68.8), (68.9) with $h(0) = 1$ and $\Sigma_s \gg \Sigma_a$.

For more accurate expressions of $p(u)$, deduced by PLACZEK by means of a kind of successive approximation procedure, we refer to the original paper[1].

---

[1] See footnote 1, p. 232.

An important example for slowly varying capture is the case of a constant scattering cross section and absorption cross section inversely proportional to the velocity ($1/v$-law). Then, according to (69.12), w is inversely proportional to the velocity

$$w = \frac{v_c}{v}; \quad v_c = \frac{l_s}{T} = \text{const}. \tag{69.16}$$

In this case, however, one can do still better than applying the above procedure, because it is possible to take the variations of the capture cross section $\Sigma_a(v)$ within one interval X not only approximately but exactly into account. For this purpose one uses w as independent variable instead of the lethargy[1] [*16*]

$$u = \ln\left(\frac{w}{w_0}\right)^2. \tag{69.17}$$

Then Eq. (66.7b) is transformed into[1,2]

$$\psi(w) = \frac{2}{1-\alpha}\,\frac{1}{w^2}\int\limits_{rw}^{w} dw'\,\frac{w'}{1+w'}\,\psi(w') \tag{69.18}$$

where

$$r = |\alpha^{\frac{1}{2}}| = \frac{A-1}{A+1} \tag{69.19}$$

and

$$\psi(w)\,dw = \psi_0(u)\,du. \tag{69.20}$$

According to PLACZEK, Eq. (69.18) is satisfied by the power series

$$\psi(w) = \sum_{n=0}^{\infty} \beta_n w^n \tag{69.21a}$$

with

$$\beta_0 = 1, \quad \beta_n = (-1)^n \lambda_n \prod_{\mu=1}^{n} \frac{1}{1-\lambda_\mu} \tag{69.21b}$$

and

$$\lambda_\mu = \frac{2}{\mu+2}\,\frac{1-r^{\mu+2}}{1-r^2}. \tag{69.21c}$$

The expansion (69.21a) converges only for $|w|<1$. The following expansions have been derived by SVARTHOLM[2] by means of Mellin transform methods:

$$\psi(w) = \sum_{n=0}^{\infty} a_n \left(\frac{w}{1+w}\right)^2 \tag{69.22a}$$

with

$$a_n = \sum_{\nu=0}^{\infty} (-1)^\nu \binom{n}{\nu} \Gamma_\nu \tag{69.22b}$$

and

$$\Gamma_0 = \frac{2}{\xi}; \quad \Gamma_\nu = \Gamma_0 \prod_{\mu=1}^{\nu} \frac{1}{1-\lambda_\mu}, \quad \nu \geqq 1, \tag{69.22c}$$

which is valid in the whole range $0 \leqq w \leqq \infty$; and

$$\psi(w) = \frac{1}{w^2}\sum_{n=2}^{\infty} b_n \left(\frac{w}{1+w}\right)^n \tag{69.23a}$$

---

[1] See footnote 1, p. 232.

[2] N. SVARTHOLM: Trans. Chalmers Univ. of Techn., Gothenburg, No. 164, 1955.

with

$$b_n = \sum_{\nu=2}^{n} (-1) \binom{n}{\nu} \Gamma_{\nu-2}, \tag{69.23b}$$

valid for $0 \leqq w < \infty$. Both series (69.22a), (69.23a) represent analytical continuations of a series equivalent to Placzek series (69.21).

The expansions (69.21) to (69.23) are useful for small masses because, for $A \gg 1$, the corresponding coefficients rapidly increase with n[1] [*16*].

For large masses an expansion of $\ln \psi$ rather than of ψ is more convenient. According to PLACZEK the result is

$$\ln p(w) = -\frac{2}{1-r} \left\{ \frac{1+r+r^2}{1+2r} w - \frac{1+3r+2r^3}{2(1+2r)^2} w^2 + \cdots \right\}. \tag{69.24}$$

Turning back to the general case, we recall that (69.13) will be valid only if the change of $h(u)$ in a slowing down interval X is very small. Furthermore expressions such as (69.15) obtained from (69.13) by using the expansion (69.11) will be adequate only if the ratio w of the capture cross section to the scattering cross section is also very small.

If these two conditions are fulfilled, there is another procedure for the derivation of $p(u)$, suggested by GOERTZEL and GREULING[2], which makes use of one of the approximations involved in age theory (Sect. 75). One starts by differentiating Eq. (69.9) with respect to u, so that one obtains

$$\frac{dq}{du} = \psi_{0s}(u) - \frac{1}{1-\alpha} \int_{u-X}^{u} \psi_{0s}(u')\, e^{u'-u}\, du'. \tag{69.25}$$

Then the integrals appearing in (69.9) and (69.25) are computed assuming that $\psi_{0s}(u)$ varies so slowly in the lethargy interval $(u-X, u)$ that $\psi_{0s}(u')$ is adequately given by the first two terms of a Taylor series expansion about u

$$\psi_{0s}(u') \approx \psi_{0s}(u) + (u'-u) \frac{d\psi_{0s}(u)}{du}. \tag{69.26}$$

This approximation, which will be considered in detail in Sect. 79 [see Eq. (79.19) and the discussion on p. 290], is permitted only if in the lethargy interval $(u-X, u)$ $\Sigma_s(u)$ varies very slowly and $\Sigma_a(u)$ is so small that it produces only a small change in the neutron flux $\varphi_0(u)$.

Introducing the expression (69.26) in (69.9) and (69.25) one has

$$q = \bar{\xi} \psi_{0s}(u) + a \frac{d\psi_{0s}}{du} \tag{69.27a}$$

$$\frac{dq}{du} = \bar{\xi} \frac{d\psi_{0s}}{du} \tag{69.27b}$$

where

$$a = \frac{\alpha + \alpha X - \frac{1}{2}\alpha X^2 - 1}{1-\alpha}. \tag{69.28}$$

By setting $d\psi_{0s}/du$ from (69.27b) into (69.27a) we find

$$-a \frac{dq}{du} + \bar{\xi} q = \bar{\xi}^2 \psi_{0s}(u) = \bar{\xi}^2 \Sigma_s \varphi_0(u). \tag{69.29}$$

[1] See footnote 1, p. 232.

[2] G. GLASSTONE and M.C. EDLUND: Nuclear Reactor Theory, p. 168. New York: Van Nostrand 1956.

A second relation between q and the neutron flux $\varphi_0(u)$ is obtained by noticing that when the lethargy increases from u to $u+du$, the slowing down density decreases by dq because of absorption:

$$\frac{dq}{du} = -\Sigma_a(u)\,\varphi_0(u). \tag{69.30}$$

From (69.29) and (69.30) one obtains, by eliminating dq/du,

$$\varphi_0(u) = \frac{q}{\xi\,\Sigma_s(u) + \gamma\,\Sigma_a(u)} \tag{69.31}$$

where

$$\gamma = -\frac{a}{\xi} = \frac{1-\alpha-\alpha X-\frac{1}{2}\alpha X^2}{1-\alpha-\alpha X}. \tag{69.32}$$

Finally introducing (69.31) into (69.29) we obtain

$$\frac{dq}{du} = -\frac{\Sigma_a}{\xi\,\Sigma_s+\gamma\,\Sigma_a}\cdot q \tag{69.33}$$

which, upon integration between the limits $u=0$ and u, gives

$$p(u) = \frac{q(u)}{q_0} = \exp\left[-\int_0^u \frac{\Sigma_a(u')\,du'}{\xi\,\Sigma_s(u')+\gamma\,\Sigma_a(u')}\right]. \tag{69.34}$$

Table 38. *Values of the parameters used in calculation of resonance escape probability*[1].

Element	γ	ξ
Hydrogen . . .	1.000	1.000
Deuterium . .	0.583	0.725
Beryllium . . .	0.149	0.209
Carbon	0.116	0.158

A few values of γ and ξ are collected in Table 38.[1] In the case of hydrogen both ξ and γ are unity and the approximate expression (69.34) of the resonance escape probability $p(u)$ becomes identical with the rigorous expression (68.8)

Although the approximations involved in the derivation of (69.15) and (69.34) are the same, the method of PLACZEK is superior since higher approximations are easily obtained by considering successive terms in the power expansion (69.11). The consideration of successive terms in the series expansion (69.26) would on the contrary change the nature of the differentail equations (69.27) with the result that the problem cannot be solved analytically and has to be integrated numerically [7].

β) The case of rapidly varying capture cross section is also of considerable practical interest because it includes that of resonance absorption which, because of its overwhelming importance, has caused $p(u)$ to be called the resonance escape probability.

If we examine the rigorous procedure followed in the case of hydrogen in order to derive the expression (68.8) of the resonance escape probability, we see that the essential point was the independence of the factors f_i appearing in (68.5), which allowed the averaging of each one of them independently of the others. Such a condition is no longer satisfied in the case of a moderator heavier than hydrogen, and this, as already noticed previously, gives rise to serious complications.

One can, however, investigate what kind of conditions one has to impose in order that the independence of the factors f_i be approximatively satisfied so that one can derive expressions similar to (68.8) for those special cases.

[1] See footnote 2, p. 243.

From the formal point of view the only difference with respect to the case of hydrogen will be in the expression of the probability that the lethargy interval Δu contains one point of the series $u_n, u_{n+1}, u_{n+2}, \ldots$, which now is $\Delta u/\bar{\xi}$.

The individual factors f_i appearing in (68.5) can be taken as approximately independent of the others in two important cases:

(a) *Widely spaced narrow resonances.* The absorber shows one or more resonances separated by intervals where $\sigma_a = 0$. The relative width of each resonance is small compared to the average distance $\bar{\xi}$ between two successive points of the series $u_n, u_{n+1}, \ldots$

$$\frac{\Gamma_s}{E_{Rs}} \ll \bar{\xi} \tag{69.35a}$$

and, in the case of two or more resonances, their spacing is large, i.e. it satisfies condition (69.3), namely

$$u_{s+1} - u_s \geqq 4X, \qquad u_s = \ln \frac{E_0}{E_{Rs}}. \tag{69.35b}$$

If condition (69.35a) were not satisfied because, if e.g. $\Gamma_k/E_{Rk} > X$, for a sufficiently high concentration of the absorber all neutrons would be captured in the k-th resonance and none would be slowed down to lower energies. Condition (69.35b) is a consequence of the considerations given at the beginning of this section which obviously refer to a case of this type.

The result in this case is obviously

$$p(u) = \exp\left[-\frac{1}{\bar{\xi}} \int_0^u \frac{\Sigma_a(u')}{\Sigma_s(u') + \Sigma_a(u')}\, du'\right]. \tag{69.36}$$

(b) *Weak resonance capture.* When the condition

$$\Sigma_a = \sum \mathfrak{N}_i \sigma_a^i \ll \sum \mathfrak{N}_i \sigma_s^i = \Sigma_s \tag{69.37}$$

is satisfied, the fraction of neutrons absorbed in an unit interval at u, will be much smaller than 1:

$$g(u) = \frac{\Sigma_a}{\Sigma_s + \Sigma_a} \ll 1, \tag{69.38}$$

and therefore the influence of one of the factors f_i on the successive ones $f_{i+1}, f_{i+2}, \ldots$ will be negligible without requiring further conditions of the type of those considered in case (a). The argument is equivalent to saying that, if condition (69.37) is satisfied, in order to calculate the slowing down density $q(u)$ one may use (69.30) with $\varphi_0(u)$ expressed in terms of $q(u)$ by means of (67.16) valid for non-absorbing media. In fact for u satisfying condition (67.14) and for such a weak absorption that $\psi_{0s}(u')$ does not change appreciably over the slowing down interval $(u - X, u)$, Eq. (69.9) reduces to

$$q(u) \approx \frac{\varphi_0(u)}{l_s(u)} \bar{\xi} = \frac{\bar{\xi}}{l_s} v n_0(u) \tag{69.39}$$

which is just (67.16)

By integrating the differential equation obtained by combining (69.30) and (69.39), we obtain

$$p(u) = \frac{q(u)}{q_0} = \exp\left[-\frac{1}{\bar{\xi}} \int_0^u \frac{\Sigma_a(u')}{\Sigma_s(u')}\, du'\right] \tag{69.40}$$

which is identical to (69.15a).

In practice one often has to deal with problems which do not correspond to the various cases considered above; for instance one has to derive the slowing down density in a medium showing one or more capture resonances which do not satisfy the conditions (69.35a), (69.35b) and/or (69.37). In such a case the slowing down density may be computed with one of the formulas given above, up to the point where the resonance becomes appreciable. Then the slowing down density $q(u)$ is converted to the slowing down collision $\psi_0(u)$ over one collision interval preceding this point; the values thus obtained are introduced into Eq. (66.7b) in order to solve it numerically from one slowing down interval to the next one as explained in Sect. 66. This process is continued until the resonance has been passed and one of the approximate procedures described above can be used again.

70. The integral spectrum in time dependent problems. In this section we shall discuss the following time dependent problem: a pulsed neutron source emits at the time $t=0$ an infinitely short burst of monoenergetic neutrons

$$q_0(u,t) = q_0 \cdot \delta(u) \cdot \delta(t)$$

which are slowed down in a homogeneous moderator. We may ask how the total number of neutrons of given lethargy

$$n_{0\delta}(u,t) = \int d\boldsymbol{r}\, N_0(\boldsymbol{r},u,t) \tag{70.1}$$

varies as a function of time.

The answer is given by the solution of the following integro-differential equation

$$\frac{\partial n_{0\delta}}{\partial t} + [\Sigma_a(u) + \Sigma_s(u)]\, v\, n_{0\delta} = \int_{u^*}^{u} \frac{v'\, n_{0\delta}(u',t)}{l_s(u')} f(u-u')\, du' + q_0\, \delta(u)\, \delta(t) \tag{70.2}$$

which is obtained from (61.2) by integrating over $d\boldsymbol{\Omega}$ and $d\boldsymbol{r}$ and assuming that the neutron density goes to zero at infinity. The lower limit of the integral u^* is equal to zero for hydrogen as well as for $A \geqq 2$ and $u \leqq X$; it is equal to $u-X$ for $A \geqq 2$ and $u > X$.

As we shall see in Sect. 85, this particular problem is of considerable practical interest. Furthermore, once its solution is known, other problems corresponding to an arbitrary time-dependence of the neutron source strength $q_0(t')$ can be solved by simple integration:

$$n_0(u,t) = \int_0^t dt'\, q_0(t')\, n_{0\delta}(u, t-t'). \tag{70.3}$$

The following remark, of general heuristic value, is sometimes of practical use. If one applies a Laplace transform with respect to time to both sides of Eq. (70.2) one obtains

$$\left[s + (\Sigma_a(u) + \Sigma_s(u))\, v\right] n_{0\delta} = \int_{u^*}^{u} \frac{v'\, n_{0\delta}(u',s)}{l_s(u')} f(u-u')\, du' + q_0\, \delta(u) \tag{70.4}$$

where

$$n_0(u,s) = \int_0^\infty dt\, e^{-st}\, n_{0\delta}(u,t). \tag{70.5}$$

Eq. (70.4) may be regarded as the equation governing the stationary neutron distribution in a fictitious medium in which the scattering cross section is still $\Sigma_s(u) = l_s^{-1}(u)$ and the absorption cross section is given by

$$\Sigma_a^* = \Sigma_a + \frac{s}{v}. \tag{70.6}$$

Therefore we may state that, if the properties of the medium do not change with time, time dependent problems can always be reduced to stationary problems; according to (70.4) to (70.6) the solution of (70.2) is given by the inverse Laplace transform of the solution for the corresponding stationary problem with the absorption cross section Σ_a increased by the amount s/v [*16*] [*7*].

Thus for hydrogen ($A=1$, $u^*=0$) the exact solution for arbitrary variation of the cross sections with lethargy, can be deduced from (68.15); and for heavier nuclei satisfying the two conditions

$$\Sigma_s(u) = \Sigma_s = l_s^{-1} = \text{const} \tag{70.7}$$

and

$$\gamma = v \cdot \Sigma_a(u) = \frac{1}{T} = \text{const} \tag{70.8}$$

the asymptotic expression (i.e. valid for $u > 3X$) of the solution of (70.2) can be deduced, at least in principle, from (69.22). In the latter case the rigorous evaluation of the integral arising from the inversion of the Laplace transform is so difficult, even in the case $\Sigma_a(u)=0$, that other procedures are preferable for practical use.

A second remark refers to the absorption term appearing in (70.2); if the absorption cross section follows the $1/v$ law (70.8), the corresponding term can be eliminated from Eq. (70.2) by replacing $n_{0\delta}(u,t)$ by

$$n_{0\delta}(u,t) \cdot \exp\left[-\frac{t}{T}\right]. \tag{70.9}$$

For the sake of simplicity, we shall in the following only consider Eq. (70.2) without the absorption term $v\,\Sigma_a(u)\, n_{0\delta}(u,t)$ so that the corresponding solution refers either to non-absorbing media, or to media following the $1/v$ law (70.8); in the latter case the final result is obtained by introducing an exponential factor as shown in (70.9). Because of this simplification the fictitius absorption cross section (70.6) will in the following always be

$$\Sigma_a^* = \frac{s}{v}. \tag{70.10}$$

The solution of (70.2) in the *case of hydrogen,* was first given by ORNSTEIN and UHLENBECK[1] who followed a procedure slightly different from that sketched here.

We proceed just as in the discussion of Eq. (66.6); viz. we separate out the δ-function part of the solution of (70.4)

$$n_{0\delta}(u,s) = \frac{q_0\,\delta(u)}{s + \frac{v_0}{l_s(v_0)}} + m_0(u,s). \tag{70.11}$$

By substituting this expression as well as (63.1) into (70.4) one obtains

$$\left[s + \frac{v}{l_s(u)}\right] m_0(u,s) = \int_0^u du'\, \frac{v'\, m_0(u',s)}{l_s(u')}\, e^{-(u-u')} + q_0\, \frac{e^{-u}}{1 + \frac{s\, l_s(0)}{v_0}}$$

the solution of which is essentially (68.15), i.e.

$$m_0(u,s) = C\, \frac{\Sigma_s}{\Sigma_s + \frac{s}{v}}\, \frac{l_s}{v} \exp\left[-s \int_0^u \frac{du'}{s + v'\, l_s(u')}\right]$$

[1] L. S. ORNSTEIN and G. E. UHLENBECK: Physica, Haag **4**, 478 (1937).

with the initial condition

$$m_0(0,s) = q_0 \frac{1}{1+\frac{s\,l_s(0)}{v_0}} \frac{1}{\left(s+\frac{v}{l_s}\right)_{u=0}}.$$

Introducing these expressions in (70.11) and transforming it to the collision density (61.14) one has

$$\psi_0(u,s) = q_0 \left\{ \frac{\delta(u)}{1+s\frac{l_s(0)}{v_0}} + \frac{1}{1+\frac{s\,l_s(0)}{v_0}} \frac{1}{1+\frac{s\,l_s(u)}{v}} \exp\left[-s\int_0^u \frac{du'}{s+v'\,l_s(u')}\right]\right\}. \quad (70.12)$$

The expression of the collision density as a function of lethargy and time, is obtained by taking the Laplace inverse transform of (70.12)

$$\psi_0(u,t) = \frac{1}{2\pi i}\int_{\sigma-i\infty}^{\sigma+i\infty} \psi_0(u,s)\,e^{st}\,ds \quad (70.13)$$

where the contour of integration is chosen to the right of all poles of the integrand [*16*].

The evaluation of the integral (70.13) is in general laborious and can be easily done only when $v/l_s(u)$ can be approximated by certain simple expressions. For instance, the expressions of $\psi_0(u,t)$ for $l_s(v) \propto v$ and $l_s(v) = \text{const}$ are given in MARSHAK's article [*16*]. In the latter case, which is approximately valid for $E \leqq 50$ kev one has

$$\psi_0(u,t) = q_0 \left\{ \frac{v_0}{l_s}\,\delta(u)\,e^{-\frac{v_0 t}{l_s}} + \frac{1}{2}\left(\frac{v}{l_s}\right)^3 t^2 \left[1-\frac{v}{v_0}+\frac{2l_s}{t\,v_0}\right] e^{-\frac{vt}{l_s}} \right\} \quad (70.14)$$

valid for $A=1$, $\Sigma_a=0$, $\Sigma_s = l_s^{-1} = \text{const}$.

The asymptotic expression of the neutron density in hydrogen

$$n_0(u,t)\,du \to \frac{q_0}{2}\left(\frac{vt}{l_s}\right)^2 \exp\left[-\frac{vt}{l_s}\right] du \quad (70.15\text{a})$$

is obtained from (70.14) in the limit

$$v_0 \gg v, \qquad v_0 \gg l_s/t.$$

Transforming (70.15a) from the variable u to the variable v and multiplying and dividing by l_s/t one has

$$n_0(v,t)\,dv \to \frac{q_0}{2}\left(\frac{vt}{l_s}\right)^2 \exp\left[-\frac{vt}{l_s}\right]\frac{2\,dv}{v} = q_0\frac{vt}{l_s}\exp\left[-\frac{vt}{l_s}\right]\cdot d\left(\frac{vt}{l_s}\right). \quad (70.15\text{b})$$

This expression depends on the variables v and t only through the non-dimensional quantity[1]

$$x = \frac{vt}{l_s}. \quad (70.16)$$

We shall come back to this point in the following discussion of non-hydrogeneous moderators.

---

[1] G. PLACZEK: Manhattan Project Report A-25.

For $A \geqq 2$, a rigorous expression of the asymptotic solution of the time dependent problem considered here may be derived in the case specified by the conditions

$$\Sigma_s(u) = \Sigma_s = l_s^{-1} = \text{const}; \quad \Sigma_a = 0. \tag{70.17}$$

One might proceed as in the case of hydrogen, namely derive the solution of the time dependent problem from (69.22); but, as mentioned on p. 247, the evaluation of the integral due to the inverse Laplace transform is too difficult to be of any practical use [*16*].

Therefore other procedures are followed starting from Eq. (70.2) written for $u > 3X$ (so that $u^* = u - X$) without the absorption term and transformed from the variable u to the variable v

$$\frac{\partial n_0}{\partial t} = -\frac{v}{l_s} n_0 + \frac{2}{1-r^2} v \int_v^{v/r} \frac{n_0(v',t)}{l_s} \frac{dv'}{v'}. \tag{70.18}$$

For $l_s = $ const, a further transformation is possible. If one defines[1,2]

$$n(x)\,dx = n_0(v,t)\,dv, \qquad x = \frac{vt}{l_s} \tag{70.19}$$

Eq. (70.18) becomes an integro-differential equation in the single variable x

$$\frac{d}{dx}[x n(x)] = -x n(x) + \frac{2}{1-r^2} x \int_x^{x/r} n(x') \frac{dx'}{x'} \tag{70.20}$$

which has been solved by SVARTHOLM[3] by means of Mellin transform methods. According to this author the two following series may be used

$$n(x) = \frac{1}{x} e^{-x} \sum_{n=0}^{\infty} a_n L_n(x) \tag{70.21}$$

where $L_n(x)$ is an ordinary Laguerre polynomial, and a_n is given by (69.22b); or

$$n(x) = x e^{-x} \sum_{n=2}^{\infty} b_n L_n^{(2)}(x) \tag{70.22}$$

where $L_n^{(2)}(x)$ is an associated Laguerre polynomial of index 2 in the usual normalization, and b_n is given by (69.23b). The series in Eq. (70.21) converges for $0 \leqq x < \infty$ while the series in Eq. (70.22) converges for $0 < x < \infty$ and diverges for $x = 0$.

A finite number of terms of either series represents approximations of correct asymptotic behaviour for large x, with an approximation which may be expressed in terms of the method of moments: an approximation including the first n terms of any of the series (70.21), (70.22) gives exact values for the first n moments.

Another procedure for the derivation of approximate solutions of Eq. (70.20) has been presented by MARSHAK [*16*] and has been modified later by VON DARDEL[2]. It is based on the fact that the exact expression of the moments

$$\overline{x^n} = \int_0^{\infty} x^n\, n(x)\, dx \tag{70.23}$$

---

[1] See footnote 1, p. 248.
[2] G.F. VON DARDEL: Trans. Roy. Inst. Techn. **75** (1954).
[3] N. SVARTHOLM: Trans. Chalmers Univ. Techn., Gothenburg No. **164** (1955).

may be derived directly from Eq. (70.20); for positive and negative values of n, PLACZEK[1] and WALEN[2] find respectively

$$\overline{x^n} = \frac{2}{\xi} \prod_{\nu=1}^{n-1} \frac{\nu}{1-\lambda_\nu}, \tag{70.24a}$$

$$\overline{x^{-n}} = \prod_{\nu=1}^{n} \frac{\lambda_{-\nu}-1}{\nu} \tag{70.24b}$$

where λ_ν is given by (69.21c).

An approximate expression for $n(x)$ is obtained by constructing a trial function $n_{\text{trial}}(x)$ which behaves asymptotically correctly and which yields a reasonable numerical agreement between its moments and those of the exact solution. The following expression has been suggested by MARSHAK

$$n_{\text{trial}}(x) = A\, x^{-1+\frac{2}{1-r^2}} \exp\left[-x - \frac{b}{x}\right] \tag{70.25}$$

where A is a normalization constant and b an adjustable parameter; for $r=0$, i.e. $A=1$, (70.25) reduces to (70.15b).

The procedure followed by MARSHAK may be summarized as follows[3]:

(a) The parameter b is determined by maximizing

$$F_0(x) = x\, n_{\text{trial}}(x)$$

at the point $x = \overline{x^2}/\overline{x}$.

(b) The values of the quantity $2/(1-r^2)$ (column 2 of Table 39) is approximated by the nearest half integer so that the moments of the trial function are given by integrals of the form

$$\int_0^\infty dx \exp\left[-x - \frac{b}{x}\right] \cdot x^{m+\frac{1}{2}} \tag{70.26}$$

(with m integer) which are readily evaluable. The following relation is found to hold between the moments of the exact solution and those of the trial function

$$\overline{x^n_{\text{trial}}} = [1 + a\, n]\, \overline{x^n}$$

where a is a constant to be determined empirically (column 4 of Table 39).

(c) A better approximation $F_1(x)$ is then obtained by writing

$$\int_0^\infty dx\, x^n F_0(x) = [1 + a(n+1)] \int_0^\infty x^n F_1(x)\, dx$$

and solving for $F_1(x)$. This integral equation is equivalent to a differential equation for $F_1(x)$ which can be easily solved

$$F_1(x) = \frac{1}{a}\, x^{\frac{1}{a}} \int_x^\infty \frac{dx'\, F_0(x')}{(x')^{1+\frac{1}{a}}}.$$

[1] See footnote 1, p. 248.

[2] R. WALEN: Rec. trav. Inst. Rech. Struct. Mat. Belgrad 1, 17 (1952).

[3] Attention has to be paid to the fact that the moments (70.23) refer to $n(x)$ while in MARSHAK's article [16] they are defined with respect to $F(x) = x \cdot n(x)$. Therefore our $\overline{x^2}/\overline{x}$ corresponds to MARSHAK's x_{av}.

The results obtained by MARSHAK for $A=2$, 9 and 15, performing a good fit up to the fifteenth moment, are shown as dotted lines in Fig. 122. The values of the adopted parameters are given in Table 39. For $A=9$ and 15 the first thirty moments computed by means of $F_1(x)$ deviate by not more than 5% from the correct values. For $A=2$ the results are less good; the deviation increases from about 5% for $n=16$ to about 26% for $n=30$. This deviation seems to arise mainly from approximation (b); calculations made by VON DARDEL for a hypothetical mass $A=2.618$ for which $2/(1-r^2)$ assumes the value 2.5, showed that a good fit of the moments can be obtained even for low masses, if the exact value of $2/(1-r^2)$ is used in Eq. (70.25).

The full lines given in Fig. 122 represent the results obtained by VON DARDEL with the following procedure. He started again from the trial function (70.25) and approximation (b) for $A=9$ and $A=15$. For $A=2$ the correct value of $2/(1-r^2)$ was used and the corresponding moments were obtained by interpolation on a logarithmic scale between the values computed by means of (70.26) for the five nearest half integers $m+\frac{1}{2}$. The value of the parameter b was finally adjusted so as to give the best fit of the first 20 positive and first 10 negative moments of the exact solution. The curve obtained for $A=2$ is expected to be better than that given by MARSHAK, while the curves corresponding to $A=9$ and 15 are almost as good as those of MARSHAK in spite of the much less laborious derivation. The values of the parameter b determined by VON DARDEL are given in the last column of Table 39.

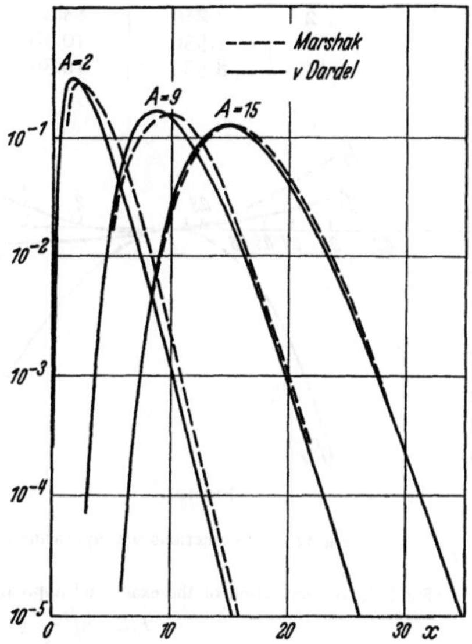

Fig. 122. Approximate velocity distributions as a function of $x=vt/l_s$.

The problem has been further investigated by various Russian authors[1] with special regard for the case $l_s=\text{const}$, $A\gg 1$. KAZARNOVSKY[2,3] starts from Eq. (70.2) without the absorption term and expressed in terms of the collision density, and transforms it into an integral equation in the new variables

$$\psi_0(u,t)=\frac{1}{\eta t}\,\psi(z);\qquad z=\frac{2l_s}{\eta v t} \tag{70.27a}$$

where

$$\eta=1-r=\frac{2}{A+1}. \tag{70.27b}$$

The solution is given in the form

$$\psi(z)\propto\exp\left[\frac{1}{\eta}f_{-1}(z)+f_0(z)+\eta f_1(z)+\cdots\right] \tag{70.28}$$

[1] A.A. BERGMAN, A.I. ISACOFF, I.D. MURIN, F.L. SHAPIRO, I.V. SHTRANIKH, M.V. KAZARNOVSKY, p. 135 of Ref. [*15b*]. In this article a list is given of references to previous Russian papers on this problem.

[2] M.V. KAZARNOVSKY: See the appendix of the preceding reference.

[3] M.V. KAZARNOVSKY: Thesis of the Lebedev Institute of Physics of the Academy of Science, Moscow 1955.

where the functions $f_\iota(z)$ are expressed in parametric form. For instance

$$f_{-1}(z) = -\int_0^y \left(2 - y' \frac{1-e^{-y'}}{e^{-y'}-1+y'}\right) dy'; \qquad z = \frac{2}{y^2}\left(e^{-y} - 1 + y\right). \quad (70.29)$$

Table 39. *Constants for time dependent problems.*

A	$2/(1-r^2)$	$\overline{x^2}/\bar{x}^2$ [1]	a (MARSHAK)	b	b (VON DARDEL)
2	2.250	13.6	0.190	4.86	0.903
9	5.556	10.38	0.0730	50.15	36.0
15	8.533	16.36	0.0526	128.13	105.1

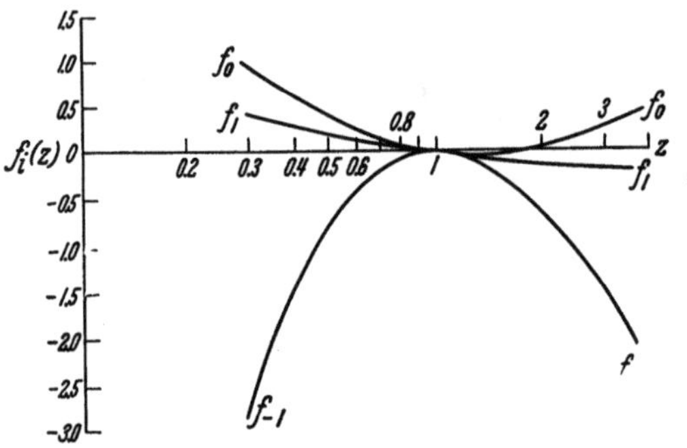

Fig. 123.

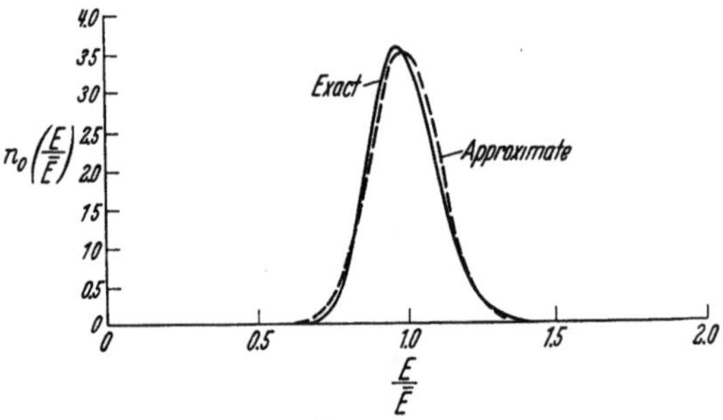

Fig. 124.

Fig. 123. The functions $f_i(z)$ appearing in Eq. (70.28) versus $z = \frac{(A+1)\, l_s}{vt}$ (KAZARNOVSKY).

Fig. 124. A comparison of the exact and approximate expressions of the neutron distribution as a function of $E/\bar{E} = E\Big/\left(\frac{m}{2}(A+1)^2\, l_s^2/t^2\right)$ (KAZARNOVSKY).

Fig. 123 shows graphs of the functions $f_\iota(z)$ the behaviour of which is further illustrated by the following results: for z close to 1

$$f_{-1}(z) = -\tfrac{3}{2}(z-1)^2 + \cdots; \quad f_0(z) = -\tfrac{1}{4}(z-1) + \cdots; \quad f_1(z) = -\tfrac{87}{320}(z-1) + \cdots;$$

for $z \to 0$

$$f_{-1}(z) \sim -\frac{2}{z}; \qquad f_0(z) \sim \frac{3}{2}\ln\frac{2}{z}; \qquad f_1(z) \sim -\frac{1}{4}\ln z;$$

for $z \to \infty$

$$f_{-1}(z) \sim -\frac{y^2}{2}; \qquad f_0(z) \sim \frac{5}{2}|y|; \qquad f_1(z) \sim \frac{y^2}{24}.$$

For $\eta \to 0$, i.e. $A \gg 1$, and z close to one, the following simple approximate expressions may be used for $\psi(z)$:

$$\psi(z) = \sqrt{\frac{3A}{4\pi}}\exp\left[-\frac{3A}{4}(z-1)^2\right]. \qquad (70.30)$$

In Fig. 124 the exact solution (70.28) for $A = 16$ and transformed into $n_0(E/\bar{E})$ is plotted versus $E/\bar{E}$. Here

$$\bar{E} = \tfrac{1}{2} m \bar{v}^2 \qquad (70.31)$$

[1] See footnote 3, p. 250.

is the neutron energy corresponding to $z=1$, i.e. to the velocity

$$\bar{v} = \frac{2 l_s}{\eta t} = (A+1)\frac{l_s}{t} \approx A\frac{l_s}{t}. \tag{70.32}$$

This is identical to (64.16) for $v_0 \gg v$ and $A \gg 1$ so that the approximate expression (62.18) can be used for $\bar{\xi}$.

The broken line given in the same figure represents the Gaussian distribution

$$n_0\left(\frac{E}{\bar{E}}\right) d\left(\frac{E}{\bar{E}}\right) = \sqrt{\frac{3A}{16\pi}} \exp\left[-\frac{3A}{16}\left(\frac{E}{\bar{E}}-1\right)^2\right] d\left(\frac{E}{\bar{E}}\right) \tag{70.33}$$

which has been adjusted to give the correct values of $\bar{E}$ and $\overline{E^2}$. The deviations between the two curves are small enough to justify the use of (70.33) for most calculations of practical interest.

According to (70.33) the neutron energy spread

$$\overline{\left(\frac{\Delta E}{E}\right)^2} = \overline{\left(\frac{E-\bar{E}}{\bar{E}}\right)^2} = \frac{8}{3A} \tag{70.34}$$

arising from the fluctuations of both the neutron path between successive collisions and, to a smaller degree, the energy losses in each collision, decreases by increasing A so that for $A=207$ (Pb) $\sqrt{\overline{(\Delta E/E)^2}}$ reduces to $\sim 11\%$.

For an arbitrary dependence of the mean free path on velocity, the neutron energy distribution can be represented approximately by the expression

$$u_0(v,t) = \frac{C}{t}\exp\left[-\frac{A+1}{2}\left\{\frac{1}{2}k(t)\left[\frac{v}{\bar{v}(t)}-1\right]^2 + \frac{1}{3}L(t)\left[\frac{v}{\bar{v}(t)}-1\right]^3 + \cdots\right\}\right] \tag{70.35}$$

where $k(t)$, $\bar{v}(t)$ and $L(t)$ are functions of time for which analytical expressions have been deduced by KAZARNOVSKY[1].

III. The spatial distribution of neutrons in a medium as a function of the energy.

71. The slowing down kernels and spatial moments. Let us consider a point source which emits isotropically Q_0 neutrons per second all of energy E_0 in an infinite and homogeneous medium. We ask what is the density $N_0(r,u)$ of the neutrons of lethargy u as long as the corresponding energies satisfy condition (61.1). The answer can be obtained in principle by solving the Boltzmann equation (61.2), a problem which in general can be dealt with only under special assumptions about the physical properties of the medium and furthermore by introducing convenient analytical approximations.

The simplest and most frequently used approximate procedure which allows us to solve this problem is the so called age theory, first used by BETHE, KORFF and PLACZEK[2]. It will be considered in Sect. 75; a few elementary improvements to it will be mentioned in Sect. 77. For an extensive treatment and discussion of more refined methods of solution of the Boltzmann equation we refer to various articles [*16*][3,4] and books [*19*], [*7*]. A very short account of these methods will be given in Sects. 79 and 80.

[1] See footnotes 3, p. 251.

[2] H.A. BETHE, S.A. KORFF and G. PLACZEK· Phys. Rev. **57**, 573 (1940).

[3] R.E. MARSHAK, H. BROOKS and H. HURWITZ: Nucleonics **5** (1), 53 and **5** (2), 59 (1949).

[4] V. FANO, L.V. SPENCER and M.J. BERGER: Penetration and Diffusion of X-rays, Article in this volume.

Before turning to the age theory, we will discuss the problem of the second spatial moment of the neutron distribution, the expression for which will be derived in Sects. 73 and 74 by means of kinetic theory methods.

Our considerations will refer to very particular types of neutron sources such as monoenergetic isotropic point sources or monoenergetic isotropic plane δ-sources.

The point source is chosen mainly because it corresponds, with fairly good approximation, to the actual conditions attained in many experiments (see Division IV). Furthermore, as we have noticed in Sect. 61, the flux distribution $\Phi_0(\boldsymbol{r}, u)$ originated by any arbitrary distribution of isotropic sources $Q_0(\boldsymbol{r})$ can always be derived from the one originated by some elementary geometry as, for instance, a point source.

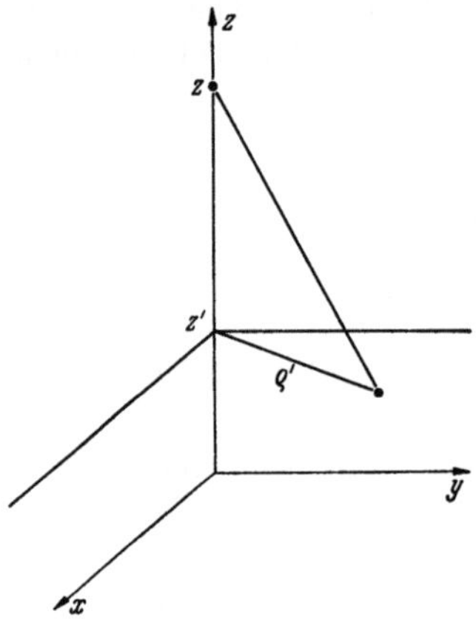

Fig. 125. Point slowing down kernel and plane slowing down kernel are related by Eqs. (71.2a), (71.2b).

The corresponding neutron flux distribution, referred to a source of unit strength, $\Phi_0^{\mathrm{pt}}(\boldsymbol{r}', u)$, is sometimes called the *point slowing down kernel*. One has immediately

$$\Phi_0(\boldsymbol{r}, u) = \int d\boldsymbol{r}'\, Q_0(\boldsymbol{r}')\, \Phi_0^{\mathrm{pt}}(|\boldsymbol{r} - \boldsymbol{r}'|, u). \qquad (71.1)$$

Sometimes, instead of the point slowing down kernel, one uses other kernels as fundamental solution to be introduced in (71.1): for instance the plane slowing down kernel, i.e. the neutron flux distribution $\Phi_0^{\mathrm{pl}}(z, u)$ produced by a monoenergetic isotropic plane δ-source of unit strength.

An isotropic plane δ-source presents some advantages from the theoretical point of view, compared to other elementary geometries, since in the corresponding Boltzmann equation the space coordinates are reduced to the distance z from the plane where the source is located, and the unit vector $\boldsymbol{\Omega}$ defining the direction of motion of the neutrons is reduced to its projection μ on the z axis.

Results obtained with the point slowing down kernel may be transformed into results valid for the plane slowing down kernel and viceversa. This can be seen by making use of (71.1) in order to express the plane slowing down kernel in terms of the point slowing down kernel. From Fig. 125 one has

$$\left.\begin{aligned} \Phi_0^{\mathrm{pl}}(z, u) &= \int_0^{2\pi} d\varphi \int_0^{\infty} \varrho'\, d\varrho' \int_{-\infty}^{+\infty} dz'\, \delta(z')\, \Phi_0^{\mathrm{pt}}\left(\sqrt{(z-z')^2 + \varrho'^2}, u\right) \\ &= 2\pi \int_0^{\infty} \varrho'\, d\varrho'\, \Phi_0^{\mathrm{pt}}\left(\sqrt{z^2 + \varrho'^2}, u\right) = 2\pi \int_{|z|}^{\infty} x\, dx\, \Phi_0^{\mathrm{pt}}(x, u). \end{aligned}\right\} \qquad (71.2\mathrm{a})$$

By differentiation with respect to z one obtains

$$\Phi_0^{\mathrm{pt}}(r, u) = -\frac{1}{2\pi r}\left(\frac{\partial \Phi_0^{\mathrm{pl}}(z, u)}{\partial z}\right)_{z=r}. \qquad (71.2\mathrm{b})$$

These two relationships are general in the sense that they hold whenever Φ_0^{pt} depends only on the distance between $\boldsymbol{r}$ and $\boldsymbol{r}'$.

Other slowing down kernels due to line sources or spherical or cylindrical shell sources are sometimes considered [*19*].

From (71.2) one can derive a relation between the moments of the neutron flux of a point isotropic source defined as

$$\Phi_{0n}^{\mathrm{pt}}(u) = \frac{1}{n!} \int_0^\infty r^n\, \Phi_0^{\mathrm{pt}}(r, u)\, d\boldsymbol{r} \qquad (71.3)$$

and those relative to an isotropic plane δ-source

$$\Phi_{0n}(u) = \Phi_{0n}^{\mathrm{pl}}(n) = \frac{1}{n!} \int_{-\infty}^{+\infty} z^n\, \Phi_0^{\mathrm{pl}}(z, u)\, dz \qquad (71.4)$$

where *only even powers* $n = 2s$ have to be considered because the integral vanishes for symmetry reasons when n is odd. Upon substitution of (71.2b) in (71.3), integrating by parts and using the fact that $z^{2s}\Phi_0^{\mathrm{pl}} \to 0$ as $z \to \infty$, the following relation is derived

$$\Phi_{0,2s}^{\mathrm{pt}}(u) = (2s+1)\, \Phi_{0,2s}(u). \qquad (71.5)$$

A similar relation holds between the averages of the $2s$ powers of the distances traveled by the neutron in the medium considered during its slowing down to the lethargy u; for the point source

$$\overline{r^{2s}} = (2s)!\, \frac{\Phi_{0,2s}^{\mathrm{pt}}(u)}{\Phi_{00}^{\mathrm{pt}}(u)} \qquad (71.6\mathrm{a})$$

and for the plane source

$$\overline{z^{2s}} = (2s)!\, \frac{\Phi_{0,2s}(u)}{\Phi_{00}(u)}\,. \qquad (71.6\mathrm{b})$$

Combining (71.6a) and (71.6b) by means of (71.5), one obtains

$$\overline{r^{2s}} = (2s+1)\, \overline{z^{2s}}. \qquad (71.7)$$

As we shall see later, the quantities $\overline{r^2}$ and $\overline{z^2}$ according to the geometry of the source, deserve particular attention from both the theoretical (Sects. 73, 74, 80) and the experimental point of view (Division IV). Sometimes, instead of $\overline{r^2}$ or $\overline{z^2}$, use is made of the *slowing down length* defined by the relations

$$L_s^2(u) = \frac{\overline{z^2}(u)}{2} = \frac{\overline{r^2}(u)}{6}\,. \qquad (71.8)$$

In Sect. 76 we shall see that in the age theory approximation $L_s^2(u)$ defined by (71.8) is identical to the neutron age τ defined by Eq. (75.10).

When the neutron flux distribution is calculated by other methods than those of age theory, Eq. (71.8) is used as a definition of the neutron age.

The spatial moments defined above refer to the neutron flux distribution $\Phi_0(\boldsymbol{r}, u)$; the same result is obtained if the neutron density $N_0(\boldsymbol{r}, u)$ instead of the neutron flux is used. On the contrary a slightly different result is deduced if the moments are defined using the slowing down density $Q(\boldsymbol{r}, u)$. While the moments $\overline{r_Q^{2s}}$ of $Q(\boldsymbol{r}, u)$ relate to the distances at which the neutrons pass the considered value of the lethargy, the moments $\overline{r^{2s}}$ of $\Phi_0(\boldsymbol{r}, u)$ defined by (71.6a) relate to the distances at which neutrons of the considered lethargy are found. Therefore, for the same lethargy, $\overline{r^{2s}}$ (or $\overline{z^{2s}}$) is larger than $\overline{r_Q^{2s}}$ (or $\overline{z_Q^{2s}}$) by an amount sometimes referred to as the "last flight correction". In Sect. 77 we shall give a rough estimate of such a correction.

Experimental measurements by means of the activation of a thin resonance detector, as those considered in Sects. 82 and 83, give the moments of the neutron density. According to some authors[1] the moments of the slowing down density (in particular the second moment) appear more appropriate in reactor calculations since the last-flight correction is in a sense included in the elementary treatment of the thermal group.

72. The persistence of velocity and the transport mean free path. The collision function. For investigation of the spatial distribution of neutrons in a medium, it is essential to take into account the angular distribution of the neutrons in the l.s. after each collision. This, obviously, is not isotropic even in the case of isotropic scattering in the c.m.s.

The deviation from spherical symmetry is measured by the average value of the cosine of the scattering angle in the l.s.

$$\bar{\mu}_0 = \overline{\cos\vartheta} = \int_{-1}^{+1} \cos\vartheta\, f^*(E, \cos\vartheta^*)\, d(\cos\vartheta^*) \tag{72.1}$$

where, as usual, a star is added to the symbols of all quantities referred to the c.m.s. The quantity $f^*(E, \mu_0^*)\, d\mu_0^*$ $(\mu_0^* = \cos\vartheta^*)$ is the probability that after one collision the direction of motion of the neutrons in the c.m.s. is contained between μ_0^* and $\mu_0^* + d\mu_0^*$. As we shall see later, it is connected with the collision function defined by Eq. (61.8) of Sect. 61. Generally $f^*(E, \mu_0^*)$ will be a function of the energy of the incident neutron, but, even when it does not depend on the energy, $\bar{\mu}_0$ will do. As we shall see later [Eq. (72.10)] $\bar{\mu}_0$ will be energy independent in the case of moderators containing nuclei of a single type which scatter isotropically in the c.m.s.

In the kinetic theory of gases[2], $\bar{\mu}_0$, defined by (72.1), is frequently indicated as the *persistence of velocity* because it measures, in a certain sense, the memory that the neutrons keep, after a collision, of their direction of motion before the collision.

Such a correlation between directions of motions gives rise to the concept of *transport mean free path* l_{tr}, which is defined as the distance which a neutron travels, on the average, in the direction of its initial motion. This distance is equal to the scattering mean free path l_s only if after a collision a neutron has completely forgotten its initial direction; i. e. only if $\bar{\mu}_0 = 0$. But if the neutron has a persistence of velocity $\bar{\mu}_0 \neq 0$, l_{tr} will differ from l_s, being longer, if $\bar{\mu}_0 > 0$ and shorter, if $\bar{\mu}_0 < 0$.

The simplest way of deriving the relation existing between l_{tr} and l_s is that of considering a non-absorbing medium in which both the scattering mean free path l_s and $\bar{\mu}_0$ keep constant values in successive collisions. If we imagine a neutron to be shot into the medium in the z direction, the average distance l_{tr} travelled by the neutron along the z axis will be

$$l_{tr} = l_s + l_s\bar{\mu}_0 + (l_s\bar{\mu}_0)\,\bar{\mu}_0 + (l_s\bar{\mu}_0)\,\bar{\mu}_0\cdot\bar{\mu}_0 + \cdots \tag{72.2}$$

a factor $\bar{\mu}_0$ coming in each time that the path travelled by the neutron between two successive collisions is projected onto the direction of the previous path [see Sect. 73; Eq. (73.4)]. From (72.2) one deduces

$$l_{tr} = l_s \sum_{n=0}^{\infty} \bar{\mu}_0^n = \frac{l_s}{1-\bar{\mu}_0} \tag{72.3}$$

which is the well known formula of the kinetic theory of gases.

[1] H. HURWITZ and P.F. ZWEIFEL: J. Appl. Phys. **26**, 923 (1955).

[2] J.H. JEANS: The dynamical Theory of Gases. Cambridge: Cambridge University Press 1925.

Instead of the transport mean free path l_{tr} one prefers to speak, at least sometimes, of the *transport cross section*, microscopic or macroscopic according to convenience; this is related to l_{tr} by the obvious equations

$$\Sigma_{\text{tr}} = l_{\text{tr}}^{-1} = \mathfrak{N}\,\sigma_{\text{tr}} = \mathfrak{N}\,\sigma_s\,(1-\overline{\mu}_0) = \Sigma_s\,(1-\overline{\mu}_0). \tag{72.4}$$

In the case of an absorbing medium one can define the *diffusion cross section* by adding to the expression (72.4) the absorption cross section

$$\Sigma_d = \Sigma_a + \Sigma_s\,(1-\overline{\mu}_0) = \Sigma_{\text{tot}} - \Sigma_s\,\overline{\mu}_0. \tag{72.5}$$

The reverse of this expression represents the *diffusion mean free path* in an absorbing medium

$$l_d = \frac{1}{\Sigma_a + \Sigma_s(1-\overline{\mu}_0)} \tag{72.6}$$

which is shorter than the transport mean free path (72.3) because in every collision a fraction of the neutrons is captured.

In order to calculate $\overline{\mu}_0$ by means of (72.1) we need expressions for boths $\mu_0(\mu_0^*)$ and $f^*(E, \mu_0^*)$. The first one is given by [see Fig. 32 and Eq. (35.2)]

$$\tan\vartheta_2 = \frac{v_2^*\sin\vartheta_2^*}{v_m + v_2^*\cos\vartheta_2^*} = \frac{\sin\vartheta_2^*}{\cos\vartheta_2^* + \dfrac{1}{A}} \tag{72.7}$$

which, by elementary transformations and dropping the subscript 2, gives

$$\mu_0 = \cos\vartheta = \frac{1 + A\cos\vartheta^*}{\sqrt{1 + A^2 + 2A\cos\vartheta^*}}. \tag{72.8}$$

The quantity $f^*(E, \mu_0^*)$ generally depends on the nature of the forces between the neutron and the target nucleus. Under the assumption of isotropic scattering in the c.m.s. one has

$$f^*(E, \mu_0^*) = \tfrac{1}{2}\,d\mu_0^* \tag{72.9}$$

which, introduced together with (72.8) in (72.1) gives the *energy independent* expression

$$\overline{\mu_0} = \overline{\cos\vartheta} = \tfrac{2}{3}\,A^{-1}. \tag{72.10}$$

For isotropic scattering in the c.m.s. the persistence of velocity is largest in the case of hydrogen, for which $\overline{\mu_0} = \frac{2}{3}$, and decreases with increasing A; for carbon $\overline{\mu_0} = 0.0556$, i.e. the deviation from isotropy is very small.

When the moderator is a compound or a mixture containing a few different types of nuclei, the value of $\overline{\mu_0}$, will be given by a weighted average similar to (61.8b) or (65.1), namely

$$\overline{\mu_0} = \sum_i c_i\,\overline{\mu_{0i}} \tag{72.11}$$

where $\overline{\mu_{0i}}$ is calculated by means of (72.1) for the nuclei of type i, and the coefficient c_i, defined by (61.9), represents the relative probability of collision against a nucleus of type i. The value (72.11) of $\overline{\mu_0}$ will be energy independent only if all the scattering cross sections σ_s^i of the various nuclei present in the moderator correspond to isotropic scattering in the c.m.s., and if furthermore they show the same energy dependence. If this latter condition is not satisfied, the coefficient c_i will be energy dependent and therefore, through them, also $\overline{\mu_0}$.

We shall now derive the expression of the collision function to be introduced in the Boltzmann equations (61.21a) and (61.21b) in terms of the probability

$f^*(E', \mu_0^*)$. This is directly connected with the differential scattering cross section expressed in terms of the initial energy in the l.s. and of the cosine of the scattering angle in the c.m.s.:

$$f^*(u', \mu_0^*) = \frac{2\pi\,\Sigma_s^*(u', \mu_0^*)}{\Sigma_s^*(u')}; \qquad \Sigma_s^*(u') = 2\pi \int_{-1}^{+1} \Sigma_s^*(u', \mu_0^*)\, d\mu_0^* \tag{72.12}$$

where, for the sake of simplicity we assume that a single type of nuclei is present in the moderator, and instead of the energy E' we have used the lethargy u' (62.19).

The probability $f(u', \mu_0)\, d\mu_0$ that after one collision the scattering angle in the l.s. of a neutron of initial lethargy u' is contained between μ_0 and $\mu_0 + d\mu_0$ can be defined by means of an expression similar to (72.12) but with the various quantities referred to the l.s. The following obvious relations equivalent to (33.18) may be useful in the transformation from the c.m.s. to the l.s.

$$\Sigma_s^*(u', \mu_0^*)\, d\mu_0^* = \Sigma_s(u', \mu_0)\, d\mu_0, \tag{72.13}$$

$$\Sigma_s^*(u') = \Sigma_s(u'), \tag{72.14}$$

$$f^*(u', \mu_0^*)\, d\mu_0^* = f(u', \mu_0)\, d\mu_0. \tag{72.15}$$

It will be convenient also to express the cosines of the scattering angles $\mu_0^* = \cos\vartheta^*$ and $\mu_0 = \cos\vartheta$, respectively in the c.m.s. and in the l.s., in terms of u. This is done by introducing in (62.2)

$$\frac{E_2}{E_1} = \left(\frac{v_2}{v_1}\right)^2 = e^{-\xi}; \qquad \xi = u_2 - u_1$$

and solving with respect to μ_0^*

$$\mu_0^*(\xi) = \frac{2}{1-\alpha}\, e^{-\xi} - \frac{1+\alpha}{1-\alpha} = 1 - \frac{2}{1-\alpha}\,(1 - e^{-\xi}). \tag{72.16}$$

In order to derive μ_0 as a function of ξ, we notice that (Fig. 32)

$$v_2 \cos\vartheta_2 = v_2^* \cos\vartheta_2^* + v_m.$$

Expressing v_m by means of (33.4), v_2 by means of (33.9), (33.5) and μ_0^* by means of (72.16), one obtains

$$\mu_0(\xi) = \frac{A+1}{2}\, e^{-\frac{\xi}{2}} - \frac{A-1}{2}\, e^{-\frac{\xi}{2}}. \tag{72.17}$$

Although μ_0 and $\xi = u - u'$ are uniquely related through (72.17) the "macroscopic cross section", appearing in the degradation integral (61.7), is considered to be a function of both μ_0 and ξ as independent variables. Its expression is obtained in terms of $\Sigma_s(u', \mu_0)$ which is directly deduced from measurements of the angular distribution in the l.s. of the scattered neutrons by the nucleus under consideration:

$$\left.\begin{aligned} \Sigma_s(u' \to u, \boldsymbol{\Omega}' \to \boldsymbol{\Omega})\, du\, d\boldsymbol{\Omega} &= \Sigma_s(u', \mu_0)\, \delta(\mu_0(\xi) - \mu_0) \left[-\frac{d\mu_0}{d\xi}\right] du\, d\boldsymbol{\Omega}; \\ \xi = u - u' &\geqq 0 \end{aligned}\right\} \tag{72.18}$$

where $\mu_0(\xi)$ is given by (72.17). One can verify that

$$\int d\boldsymbol{\Omega} \int_{u'}^{u'+X} \Sigma_s(u' \to u, \boldsymbol{\Omega}' \to \boldsymbol{\Omega})\, du = \int \Sigma_s(u', \mu_0)\, d\boldsymbol{\Omega} = \Sigma_s(u'). \tag{72.19}$$

In certain methods for the approximate solution of the Boltzmann equation (61.21a) (which will be mentioned in Sect. 79) it is convenient to expand the "cross section (72.18)" in Legendre polynomials of μ_0

$$\Sigma_s(u' \to u, \boldsymbol{\Omega}' \to \boldsymbol{\Omega}) = \sum_{l=0}^{\infty} \frac{2l+1}{4\pi} B_l(u', \xi)\, P_l(\mu_0); \quad \xi = u - u' \geqq 0. \tag{72.20}$$

The coefficients

$$B_l(u', \xi) = 2\pi \int_{-1}^{+1} \Sigma_s(u' \to u, \boldsymbol{\Omega}' \to \boldsymbol{\Omega})\, P_l(\mu_0)\, d\mu_0 \tag{72.21}$$

can be calculated by introducing (72.18) in (72.21):

$$B_l(u', \xi) = 2\pi\, \Sigma_s(u', \mu_0(\xi))\, P_l(\mu_0(\xi)) \left[-\frac{d\mu_0(\xi)}{d\xi}\right]. \tag{72.22}$$

Sometimes, as a result of analysis of the scattering of neutrons by nuclei, $\Sigma_s^*(u', \mu_0^*)$ is known, rather than $\Sigma_s(u', \mu_0)$, or its expansion in Legendre polynomials of μ_0^*

$$\left.\begin{aligned} \Sigma_s^*(u', \mu_0^*) &= \sum_{l'=0}^{\infty} \frac{2l'+1}{4\pi} B_{l'}^*(u')\, P_l(\mu_0^*); \\ \Sigma_s^*(u') &= 2\pi \int \Sigma_s^*(u', \mu_0^*)\, d\mu_0^* = B_0^*(u'). \end{aligned}\right\} \tag{72.23}$$

The relation existing between $B_l(u', \xi)$ and $B_{l'}^*(u')$ and expressing the transformation rule from the c.m.s. to the l.s., is obtained by introducing (72.23), according to (72.13), into (72.18) and then substituting in (72.21)[1]:

$$B_l(u', \xi) = \sum_{l'=0}^{\infty} T_{ll'}(\xi)\, B_{l'}^*(u') \tag{72.24}$$

with

$$T_{ll'}(\xi) = \frac{2l'+1}{2} P_l(\mu_0(\xi)) P_{l'}(\mu_0^*(\xi)) \left[-\frac{d\mu_0^*(\xi)}{d\xi}\right] \tag{72.25}$$

and $\mu_0^*(\xi)$, $\mu_0(\xi)$ given respectively by (72.16) and (72.17).

If the scattering is isotropic in the c.m.s.

$$B_{l'}^*(u') = \Sigma_s^*(u')\, \delta_{l'0} = \Sigma_s(u')\, \delta_{l'0} \tag{72.26}$$

and therefore, from (72.24) and (72.25),

$$B_l(u', \xi) = T_{l0} \cdot \Sigma_s(u') = \Sigma_s(u') \frac{1}{1-\alpha} e^{-\xi} P_l(\mu_0(\xi)), \quad X \geqq \xi = u - u' \geqq 0. \tag{72.27}$$

In the case of hydrogen ($\alpha = 0$) the expression (72.27) reduces to

$$B_l^{\mathrm{H}}(u', \xi) = \Sigma_s^{\mathrm{H}}(u')\, e^{-\xi} P_l(e^{-\xi/2}); \quad \xi = u - u' \geqq 0. \tag{72.28}$$

73. The second moment of the neutron distribution in the case of hydrogen and hydrogeneous moderators. The spatial distribution of the neutrons in a homogeneous infinite medium may be characterised by the *average square distance* travelled by the neutrons to the lethargy u. This quantity given by (71.6a) with $s = 1$, may be deduced by means of the following kinetic theory methods [*8*][2,3].

[1] P. F. Zweifel and H. Hurwitz jr.: J. Appl. Phys. **25**, 1241 (1954).

[2] E. Fermi: AECD-2664 Oak Ridge 1951.

[3] The same problem was solved by L. S. Ornstein, G. E. Uhlenbeck [Physica, Haag **4**, 478 (1937)] starting from the Boltzmann equation.

If $\boldsymbol{l}_1, \boldsymbol{l}_2, \boldsymbol{l}_3, \ldots, \boldsymbol{l}_n$ are the successive displacements undergone by one of the neutrons emitted from the point source S, its total displacement is (Fig. 126)

$$\boldsymbol{r} = \boldsymbol{l}_1 + \boldsymbol{l}_2 + \boldsymbol{l}_3 + \cdots \boldsymbol{l}_n \tag{73.1}$$

the square of which is

$$r^2 = \boldsymbol{r} \cdot \boldsymbol{r} = l_1^2 + l_2^2 + \cdots l_n^2 + 2(\boldsymbol{l}_1 \cdot \boldsymbol{l}_2 + \boldsymbol{l}_1 \cdot \boldsymbol{l}_3 + \cdots + \boldsymbol{l}_{n-1} \cdot \boldsymbol{l}_n). \tag{73.2}$$

We have now to average this expression, and we shall do this in three steps:

1. keeping constant the length of the displacements l_i and the scattering angles in the l.s. $\vartheta_{i,i+1}$, we average with respect to the azimuthal angles $\varphi_{i,i+1}$;

2. we average over the length of the displacements l_i, and finally;

3. we average over the scattering angles $\vartheta_{i,i+1}$ i.e. over the values of the energies $E_1, E_2, E_3, \ldots E_n$, and over all possible values of n, the number of collisions which reduce the energy to the final value E.

If φ is the angle between the plane $\{\boldsymbol{l}_{n+k-1}, \boldsymbol{l}_n\}$ and the plane $\{\boldsymbol{l}_{n+k-1}, \boldsymbol{l}_{n+k}\}$ we have (Fig. 127)

$$\left.\begin{aligned} \cos\vartheta_{n,n+k} = {} & \cos\vartheta_{n,n+k-1} \times \\ & \times \cos\vartheta_{n+k-1,n+k} + \\ & + \sin\vartheta_{n,n+k-1} \times \\ & \times \sin\vartheta_{n+k-1,n+k} \cos\varphi . \end{aligned}\right\} \tag{73.3}$$

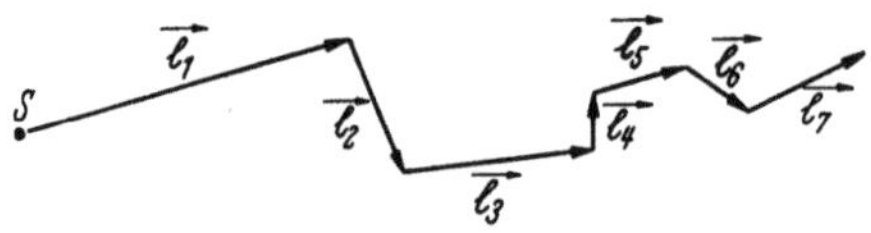

Fig. 126. The path of a neutron in a moderator.

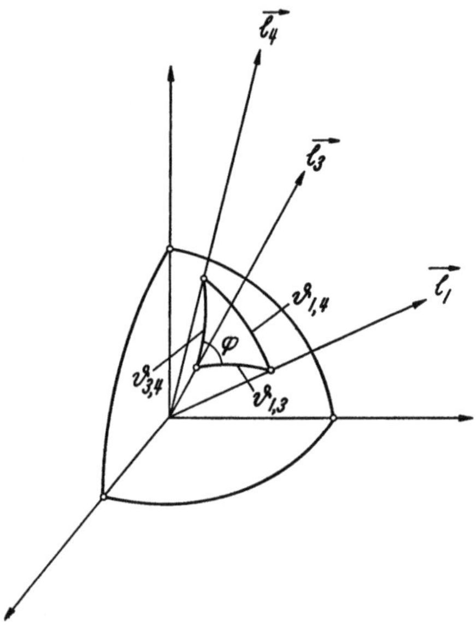

Fig. 127. Illustration of Eq. (73.3) for the particular case $n = 1$, $k = 3$.

The second term vanishes upon averaging over φ and therefore, by successive applications of (73.3) from $n+k-1$ to $n+k-2$, and considering that $\vartheta_{i,i+1}$ does not depend on the azimuthal angle $\varphi_{i,i+1}$, one obtains

$$\langle\cos\vartheta_{n,n+k}\rangle_{\text{av}} = \cos\vartheta_{n,n+1} \cdot \cos\vartheta_{n+1,n+2} \cdot \cos\vartheta_{n+2,n+3} \cdots \cos\vartheta_{n+k-1,n+k} \tag{73.4}$$

where the average refers to all the azimuthal angles.

In the next step, we average over the lengths l_i of the displacements. If $l_s(E_i)$ is the mean free path of a neutron of energy E_i, then the probability that l_i lies between l_i and $l_i + dl_i$ is $\exp(-l_i/l_s(E_i))$ and therefore we have

$$\left.\begin{aligned} \langle l_i\rangle_{\text{av}} &= \frac{\int\limits_0^\infty l \exp[-l/l_s]\, dl}{\int\limits_0^\infty \exp[-l/l_s]\, dl} = l_s(E_i), \\ \langle l_i^2\rangle_{\text{av}} &= \frac{\int\limits_0^\infty l^2 \exp[-l/l_s]\, dl}{\int\limits_0^\infty \exp[-l/l_s]\, dl} = 2 l_s^2(E_i). \end{aligned}\right\} \tag{73.5}$$

By introducing (73.4) and (73.5) in (73.2) we have

$$\left.\begin{aligned}\langle r^2\rangle_{\text{av}} = 2l_s^2(E_0) + 2l_s^2(E_1) + \cdots + 2l_s^2(E_n) + 2[l_s(E_1)\, l_s(E_2)\cos\vartheta_{1,2} + \\ + l_s(E_1)\, l_s(E_3)\cos\vartheta_{1,2}\cos\vartheta_{2,3} + \cdots + l_s(E_2)\, l_3(E_3)\cos\vartheta_{2,3} + \cdots]\end{aligned}\right\} \tag{73.6}$$

where the average refers to the azimuthal angles and to the length of the displacements.

There remains the third and last step to average with respect to the values $E_1, E_2, \ldots, E_n$ and, once the final value E of the energy is fixed, over all possible values of the number n of collisions which can reduce the energy from the initial value E_0 to the final value E. In the following instead of the energy E we will use the lethargy u (62.19).

The case of *pure hydrogen* is relatively simple because according to (72.17) applied to the case $A=1$, one has

$$\cos\vartheta_{r,r-1} = \mathrm{e}^{-\xi_r/2} = \mathrm{e}^{-\frac{1}{2}(u_r - u_{r-1})} \tag{73.7}$$

which, set into (73.4) gives

$$\cos\vartheta_{i,k} = \mathrm{e}^{-\frac{1}{2}(u_k - u_i)}, \qquad k > i. \tag{73.8}$$

Eq. (73.6) can be written

$$\langle r^2\rangle_{\text{av}} = \sum_{i=0}^{n} 2l_s^2(u_i) + 2\sum_{i=0}^{n}\sum_{k=i+1}^{n} l_s(u_i)\, l_s(u_k)\, \mathrm{e}^{-\frac{1}{2}(u_k - u_i)}. \tag{73.9}$$

In averaging this expression one has to treat separately the terms $i=0$ and $i=n$ which both certainly contribute to the value of r^2, while for all other terms one has to take into account that the probability that one point of the series $u_1, u_2, u_3, \ldots u_{n-1}$ falls in du is du. It follows that the average contribution of the interval du to the first sum is $2l_s(u)\,du$ and therefore the average of the first sum turns out to be

$$2l_s^2(0) + 2\int_0^u l_s^2(u)\,du + 2l_s^2(u). \tag{73.10}$$

For the second sum of (73.9) one has, by similar considerations,

$$\left.\begin{aligned}\Big\langle\sum_{k=1}^{n} l_s(0)\, l_s(u_k)\, \mathrm{e}^{-\frac{1}{2}u_k}\Big\rangle_{\text{av}} &= l_s(0)\int_0^u l_s(u)\, \mathrm{e}^{-\frac{u}{2}}\,du + l_s(0)\, l_s(u)\, \mathrm{e}^{-\frac{u}{2}}, \\ \Big\langle\sum_{i=1}^{n-1} l_s(u_i)\, l_s(u)\, \mathrm{e}^{-\frac{1}{2}(u-u_i)}\Big\rangle_{\text{av}} &= l_s(u)\int_0^u l_s(u_i)\, \mathrm{e}^{-\frac{1}{2}(u-u_i)}\,du_i, \\ \Big\langle\sum_{1\le i<k\le n-1} l_s(u_i)\, l_s(u_k)\, \mathrm{e}^{-\frac{1}{2}(u_k-u_i)}\Big\rangle_{\text{av}} &= \int_0^u l_s(u')\,du' \int_0^{u-u'} l_s(u'+u'')\, \mathrm{e}^{-\frac{u''}{2}}\,du'',\end{aligned}\right\} \tag{73.11}$$

and therefore, by combining with (73.10), we finally obtain in the case of slowing down in pure hydrogen

$$\left.\begin{aligned}\overline{r^2}(u) = 2l_s(0) + 2l_s(u) + 2\int_0^u l_s^2(u')\,du' + 2l_s(0)\int_0^u l_s(u')\, \mathrm{e}^{-\frac{u'}{2}}\,du' + \\ + 2l_s(0)\, l_s(u)\, \mathrm{e}^{-\frac{u}{2}} + 2l_s(u)\int_0^u l_s(u')\, \mathrm{e}^{-\frac{1}{2}(u-u')}\,du' + \\ + 2\int_0^u l_s(u')\,du' \int_0^{u-u'} l_s(u'+u'')\, \mathrm{e}^{-\frac{u''}{2}}\,du''\end{aligned}\right\} \tag{73.12}$$

an expression which may be used for numerical computations once $l_s(u)$ is known if the cross section for capture is negligible in the interval of energy considered.

From (73.12) one deduces immediately

$$\left.\begin{aligned}\frac{d\overline{r^2}}{du} &= 4l_s^2(u) + 4l_s(u)\,l_s'(u) + l_s(0)\,[l_s(u) + 2l_s'(u)]\,\mathrm{e}^{-\frac{u}{2}} + \\ &\quad + [l_s(u) + 2l_s'(u)]\int_0^u l_s(u')\,\mathrm{e}^{-\frac{1}{2}(u-u')}\,du'\end{aligned}\right\} \tag{73.13}$$

where

$$l_s'(u) = \frac{d l_s(u)}{du}.$$

We have seen in Sect. 29 that in the energy interval 10^4 to 1 ev, the neutron mean free path in hydrogen is practically constant. Therefore, in this energy interval, which is the most interesting for slowed down neutrons, Eq. (73.13) can be considerably simplified by neglecting all terms containing $l_s'(u)$. Furthermore, if the initial energy E_0 is so high that $\mathrm{e}^{-u/2} = v/v_0$ can be neglected with respect to 1, one can omit the term $2l_s(0)\,l_s(u)\,\mathrm{e}^{-u/2}$ and substitute for the integral, whose value is mainly due to the region close to its upper limit, the simple expression

$$\int_0^u l_s(u')\,\mathrm{e}^{-\frac{1}{2}(u-u')}\,du' \approx 2l_s(u).$$

Under these simplifying assumptions, (73.13) reduces to

$$\frac{d\overline{r^2}}{du} = 6l_s^2(u) \tag{73.14}$$

which, when applied to a finite narrow energy interval $\Delta u = \ln\frac{E_2}{E_1}$ gives

$$\ln\frac{E_2}{E_1} = \frac{\overline{r^2}(E_2) - \overline{r^2}(E_1)}{6l_s^2}. \tag{73.15}$$

The case of pure hydrogen has however a limited practical interest, because all hydrogeneous moderators contain other nuclei which, by absorbing and scattering the neutrons, modify their spatial distribution and therefore the value of $\overline{r^2}$.

The effect of absorption was first considered by ASHKIN [*16*] who took into account the capture by hydrogen itself and/or by other nuclei present in the moderator, provided their contribution to the scattering of neutrons could be completely neglected. The expression derived by this authors is the following

$$\left.\begin{aligned}\overline{r^2}(u) = 2\Bigg\{&l_{\mathrm{tot}}^2(0) + l_{\mathrm{tot}}^2(u) + l_{\mathrm{tot}}(0)\,l_{\mathrm{tot}}(u)\,\mathrm{e}^{-u/2} + l_{\mathrm{tot}}(u)\int_0^u du'\,\frac{l_{\mathrm{tot}}^2(u')}{l_s(u')}\,\mathrm{e}^{-\frac{1}{2}(u-u')} + \\ &+ \int_0^u du'\,\frac{l_{\mathrm{tot}}^3(u')}{l_s(u')} + l_{\mathrm{tot}}(0)\int_0^u du'\,\frac{l_{\mathrm{tot}}^2(u')}{l_s(u')}\,\mathrm{e}^{-u'/2} + \\ &+ \int_0^u du'\,\frac{l_{\mathrm{tot}}^2(u')}{l_s(u')}\int_0^{u'} du''\,\frac{l_{\mathrm{tot}}^2(u'')}{l_s(u'')}\,\mathrm{e}^{-\frac{1}{2}(u'-u'')}\Bigg\}\end{aligned}\right\} \tag{73.16}$$

where

$$\frac{1}{l_{\text{tot}}(u)} = \frac{1}{l_s(u)} + \frac{1}{l_a(u)}$$

is the total mean free path and $l_s(u)$ refers only to hydrogen.

Another generalization of (73.12) was given originally by FERMI [8] who derived a rigorous expression for a non absorbing compound (or mixture) of hydrogen and other nuclei, under the assumption that in the collision against these other nuclei the neutron does not undergo any energy loss and is scattered isotropically in the l.s. This is obviously equivalent to assuming that the masses of the non-hydrogen nuclei are infinite, so that they scatter but do not contribute to the slowing down of the neutrons. The error thus introduced is not too serious as long as the partial scattering mean free path of hydrogen is much shorter than that relative to the heavy nuclei.

The expression derived under these assumptions is[1] [16]

$$\left.\begin{aligned}\overline{r^2}(u) = 2\Bigg\{&\frac{l_s^2(0)}{c(0)} + \frac{l_s^2(u)}{c(u)} + l_s(0)\, l_s(u) \exp\left[-\int_0^u\left[\frac{3}{2} - c(u')\right] du'\right] + \\ &+\int_0^u \frac{l_s^2(u')}{c(u')}\, du' + l_s(u)\int_0^u l_s(u') \exp\left[-\int_{u'}^u\left[\frac{3}{2} - c(u'')\right] du''\right] + \\ &+ l_s(0)\int_0^u l_s(u')\, du' \exp\left[-\int_0^{u'}\left[\frac{3}{2} - c(u'')\right] du''\right] + \\ &+\int_0^u l_s(u')\, du'\int_0^{u'} l_s(u'')\, du'' \exp\left[-\int_{u''}^{u'}\left[\frac{3}{2} - c(u''')\right] du'''\right]\Bigg\}\end{aligned}\right\} \tag{73.17}$$

where, according to (61.9),

$$c(u) = \frac{l_s(u)}{l_s^{\mathrm{H}}(u)},$$

$l_s(u)$ being the total scattering mean free path, and $l_s^{\mathrm{H}}(u)$ the scattering mean free path relative to hydrogen.

This problem has a considerable practical interest because it presents the first attempt to represent the slowing down of neutrons in media such as water. A further improvement with respect to (73.17) has been introduced by VOLKIN[2] who calculated the slowing down length (71.8) taking into account both the capture of neutrons and the anisotropy of scattering in the c.m.s. by the non-hydrogen nuclei.

The simplifying assumptions still remaining in VOLKIN's calculations are: (1) the hydrogen scattering is spherically symmetric in the c.m.s.; (2) the non-hydrogen nuclei do not contribute to the slowing down.

This author makes use of the transport equation (see Sect. 80) with a collision function of the type (61.8b) and represents the scattering cross section by oxygen with an expansion in Legendre functions, in practice reduced to two terms

$$\sigma_s^0(u, \mu_0^*) = \sum_{l=0}^{1} \frac{2l+1}{2}\, \sigma_l(u)\, P_l(\mu_0^*); \qquad \sigma_s^0(u) = \int_{-1}^{+1} \sigma_s^0(u, \mu_0^*)\, d\mu_0^*. \tag{73.18}$$

[1] G. HORWAY: Phys. Rev. 50, 897 (1936).
[2] H.C. VOLKIN: Phys. Rev. 91, 425 (1953). — J. Appl. Phys. 26, 127 (1955).

He obtains

$$\left.\begin{aligned}
L_s^2(u) &= \frac{\overline{r^2}}{6} = \frac{1}{3}\left\{\frac{(l_s^{\mathrm{H}}(0))^2}{[1-b_1(0)]\,[1-b_0(0)]} + H(u) + \int_0^u du'\,\frac{c(u')}{1-b_0(u')}\,H(u')\right\},\\
&\text{where}\\
b_l(u) &= c_0(u)\,\frac{\sigma_l^0(u)}{\sigma_s^0(u)}\,; \qquad c = \frac{l_{\mathrm{tot}}}{l_s^{\mathrm{H}}}\,; \qquad c_0 = \frac{l_{\mathrm{tot}}}{l_s^0}\,,\\
H(u) &= \frac{(l_s^{\mathrm{H}}(u))^2}{[1-b_1(u)]\,[1-b_0(u)]} + \frac{l_s^{\mathrm{H}}(0)\,l_s^{\mathrm{H}}(u)}{[1-b_1(0)]\,[1-b_1(u)]} \times \exp[-G(u)] +\\
&\quad + \frac{l_s^{\mathrm{H}}(u)}{1-b_1(u)}\exp[-G(u)] \times \int_0^u du'\,\frac{l_s^{\mathrm{H}}(u')\,c(u')}{[1-b_1(u')]\,[1-b_0(u')]}\exp[G(u')]\,;\\
G(u) &= \int_0^u du'\left[\frac{1}{2} - \frac{c(u')}{1-b_1(u')} + \frac{c(u')}{1-b_0(u')}\right].
\end{aligned}\right\} \qquad (73.19)$$

Eq. (73.19) has been applied to the calculation of the slowing down length to $E = 1.46$ ev (indium resonance) in water as a function of the initial energy E_0, by using for the cross section of hydrogen the experimental data given in Fig. 17, and for the cross sections and angular distributions of oxygen tables based partly on experimental results, partly on computations[1].

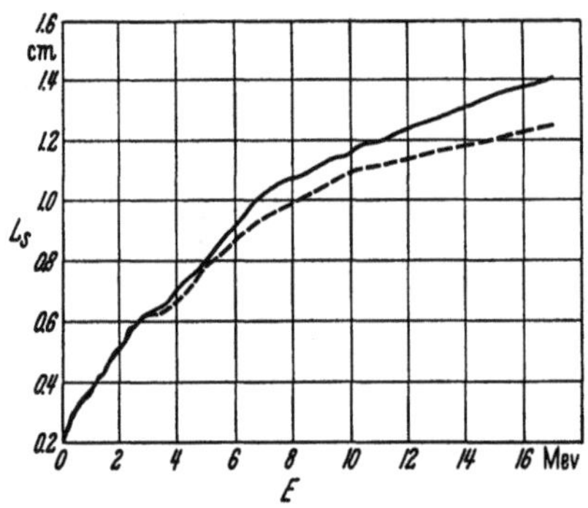

Fig. 128. Slowing down length in water to In-resonance as a function of initial energy when the asymmetry of the oxygen scattering is taken into account (solid curve) and when it is neglected (dashed curve) (Volkin).

Fig. 128 shows $L_s(u)$ as a function of E_0 when the asymmetry of the oxygen scattering is taken into account (solid curve) as well as when the oxygen scattering is assumed to be spherically symmetric ($b_1 = 0$; dashed curve). For source energies E_0 below 3 Mev the difference between the two curves is quite small but increases appreciably at higher energies. The effect of capture, which tends to decrease L_s, is included in both curves so that the curve following from Fermi's Eq. (73.17) should be somewhat above the dashed curve.

The principle of the method used by Volkin for the derivation of (73.19) will be considered in Sect. 80; a comparison with the experimental results will be given in Sect. 85.

74. The second spatial moment of the neutron density in the case of non-hydrogeneous moderators. An elementary derivation of the second spatial moment of the density distribution around a point source embedded in a *medium heavier than hydrogen* can be made by starting from (73.6) and introducing the following approximations[2]:

1. For $A \gg 1$ a large number of collisions is necessary to reduce the energy from the initial value E_0 to the final value E.

2. The scattering mean free path $l_s(E)$ varies very slowly over an energy range of the order of a few intervals X, (62.16).

[1] H. Goldstein: Nuclear Development Associates Inc. Memo 15c-15.
[2] G. Placzek: Phys. Rev. **69**, 423 (1949).

One can reorder the right hand side of (73.6) by writing together first all the terms containing $l_s(u_1)$

$$2l_s(u_1)\,[l_s(u_1) + l_s(u_2)\cos\vartheta_{1,2} + l_s(u_3)\cos\vartheta_{1,2}\cos\vartheta_{2,3} + \cdots], \tag{74.1}$$

then all the remaining terms containing $l_s(u_2)$

$$2l_s(u_2)\,[l_s(u_2) + l_s(u_3)\cos\vartheta_{2,3} + l_s(u_4)\cos\vartheta_{2,3}\cos\vartheta_{3,4} + \cdots],$$

and so on, and average with respect to all possible values of the angles ϑ_{ik}.

The various factors $\cos\vartheta_{ik}$ are independent one from the other, and if we introduce a further assumption, namely that

3. the angular distribution of the scattered neutrons is isotropic in the c.m.s., then the average of each one of them is given by (72.10), a value inversely proportional to the mass ratio A.

Therefore the successive terms in brackets of (74.1) contain increasing powers of $1/A$ and according to assumption (1) will rapidly diminish.

This circumstance, combined with assumption (2), allows us to write

$$l_s(u_1) + l_s(u_2)\,\overline{\cos\vartheta} + l_s(u_3)\,(\overline{\cos\vartheta})^2 + \cdots + l_s(u_2)\,(\overline{\cos\vartheta})^n \approx \frac{l_s(u_1)}{1-\overline{\cos\vartheta}} \tag{74.2}$$

where the various $l_s(u_i)$ occurring in successive terms have all been replaced by $l_s(u_1)$ and the sum, which contains a large number n of terms, has been extended to infinity.

Introducing (74.1) and (74.2) in (73.6) one has

$$\langle r^2(u)\rangle_{\text{av}} \approx 2\left\{\frac{l_s^2(u_1)}{1-\overline{\cos\vartheta}} + \frac{l_s^2(u_2)}{1-\overline{\cos\vartheta}} + \cdots + \frac{l\,(_s^2(u_n)}{1-\overline{\cos\vartheta}}\right\} \tag{74.3}$$

Finally we have to average with respect to u; in Sect. 67 we have seen that if u satisfies condition (67.14) the probability that a point in the series $u_1, u_2, u_3, \ldots$ falls in du is simply du/ξ [see (67.8)]. Therefore, by combining the integration with respect to du, due to the averaging process, with the sum appearing in (74.3), one may conclude

$$\overline{r^2}(u) = \frac{2}{\xi(1-\frac{2}{3}A^{-1})}\int_0^u l_s^2(u')\,du' \tag{74.4}$$

or, going back to the variable E,

$$\overline{r^2}(E) = \frac{2}{\xi(1-\frac{2}{3}A^{-1})}\int_E^{E_0} l_s^2(E')\,\frac{dE'}{E'}\,. \tag{74.5}$$

For $A=1$ the numerical factor reduces to 6 so that, for $l_s=\text{const}$, (74.4) is equivalent to (73.14) deduced for the case of hydrogen. For $A=2$ the numerical coefficient decreases to 4.14 while for $A>2$ it increases again becoming equal to $A+\frac{4}{3}$ for $A\gg 1$.

For a mixture of various nuclei, all satisfying the three assumptions mentioned above, (74.4) must be substituted by

$$\overline{r^2}(u) = 2\int_0^u \frac{l_s^2(u')\,du'}{\xi(u')\left[1-\frac{2}{3}\sum_k c_k(u')\,A_k^{-1}\right]} \tag{74.6}$$

where, according to (65.1) and (72.11) both $\xi/(u)$ and the factor $1-\frac{2}{3}\sum_k c_k(u)\,A_k^{-1}$ are functions of the energy.

As we shall see in Sect. 75 the expressions (74.5) and (74.6) of $\overline{r^2}$ differ by a numerical factor from the neutron age, the derivation of which is based on the same simplifying assumptions.

More elaborate expressions of $\overline{r^2}$ can be found in the review article by MARSHAK [*16*] to which we refer for both the derivation and the analytical expressions of the various results. Here we limit ourself to listing the various cases treated there.

By means of Laplace transform methods MARSHAK derives an equation which is rigorous for any value of A under the assumption of non-capture and isotropic scattering in the c.m.s. In the general case this equation is difficult to apply to nuclei different from hydrogen. It allows however the derivation of useful expressions by introducing further conveniently chosen assumptions about the energy dependence of the scattering mean free path.

The cases considered are the following ones:

1. Exponential theory for a single mass. The moderator is constituted by nuclei all of the same type, and the corresponding scattering mean free path can be represented by means of a sum of exponentials

$$l_s(u) = \sum_{j \geqq 0} A_j\, e^{-a_j u} \tag{74.7}$$

where A_j and a_j are arbitrary constants and $a_0 = 0$. The representation (74.7) is quite general and can be adjusted to the experimental data with a very small number of terms, provided the variation of the mean free path $l_s(u)$ is not too irregular. The result is given by Eq. (84) of MARSHAK's paper.

2. Exponential theory for a mixture. A procedure not very different from the previous one can be applied to a mixture of two nuclei, as for instance hydrogen and deuterium, provided that the scattering mean free paths of the two components vary almost in the same way over the entire energy interval involved in the calculation of $\overline{r^2}$; i.e. one assumes that the total scattering mean free path is represented by (74.7) and that the collision probabilities (61.9) have an energy dependence of the form

$$c_1(u) = \frac{l_s(u)}{l_s^1(u)} = B_0 + \sum_k B_k\, e^{-b_k u},$$

$$c_2(u) = 1 - c_1(u),$$

where B_0 and b_k are arbitrary positive constants and B_k is an arbitrary constant whose absolute value is small compared to B_0

$$|B_k| \ll B_0,$$

i.e. it is required that $c_i(u)$ varies with u more slowly than $l_s(u)$.

This condition is usually fulfilled by the experimental data provided the mean free paths of the component elements do not vary markedly in opposite directions. Eq. (91) of MARSHAK's paper gives the expression of $L_s^2(u)$ obtained for this case by regarding the terms involving B_k as a perturbation with respect to the case $c_1(u) = B_0 =$ const. which is essentially identical to case (1).

3. Interval theory. A mixture of heavy elements of mass ratios

$$2 < A_1 < A_2 < \cdots < A_m$$

and all with scattering mean free path $l_s^1, l_s^2, l_s^3, \ldots l_s^m$ varying slowly with u, is used to slow down neutrons over a lethargy interval $(0, u)$; this interval is very large compared to the maximum logarithmic energy loss X_1 [see (62.16)] relative to the nucleus of lower mass present in the mixture. Under these assumptions one can divide the interval $(0, u)$ into sub-intervals $(0, u_1)(u_1, u_2) \dots (u_{n-1}, u_n)$ such that each of them is large compared to X_1 and $l_s^i(u)$ varies slowly enough inside each sub-interval that it can be substituted with a step-function l_{sk}^i $(k=1, 2, \dots n)$ constant in the k-th interval:

$$l_{sk}^i = \left\{\frac{1}{u_k - u_{k-1}} \int\limits_{u_{k-1}}^{u_k} (l_{sk}^i(u'))^2 \, du'\right\}^{\frac{1}{2}}.$$

The problem is reduced to the treatment of a set of integral equations, one for each sub-interval and therefore each with constant total scattering mean free path. By means of Laplace transform methods, MARSHAK derives the expression of $L_s^2(u)$ given by his Eq. (103). This can be split into a term corresponding to the result of the age theory applied to the case considered, plus correction terms due to the fact that one has made use of the rigorous transport equation.

Tables 40 and 41 contain a few numerical values of L_s obtained by MARSHAK in the case of C and D_2O, as a function of the initial energy and for a final energy $E = 1.46$ ev corresponding to the indium resonance.

Table 40. *Slowing down length L_s of neutrons in carbon. Final energy $E = 1.46$ ev, carbon density 1.6 g/cm³* (MARSHAK 1947).

E_0	u	Slowing down lengths		
		age theory	interval theory	exponential theory
Mev		cm	cm	cm
3.0	14.55	19.2	19.7	19.8
2.0	14.14	17.7	18.2	18.2
1.0	13.45	15.9	16.2	16.2
0.5	12.76	14.7	15.0	15.0
0.25	12.06	13.9	14.1	14.1
0.10	11.15	13.2	13.3	13.3

Table 41. *Slowing down length L_s of neutrons in heavy water. Final energy $E = 1.46$ ev, heavy water density 1.1 g/cm³* (MARSHAK 1947).

E_0	u	Slowing down lengths		
		age theory	interval theory	exponential theory
Mev		cm	cm	cm
3.0	14.55	10.5	11.4	11.9
2.0	14.14	10.1	10.8	10.9
1.0	13.45	9.7	10.1	10.2
0.5	12.76	9.4	9.9	9.8
0.25	12.06	9.1	9.5	9.5
0.10	11.15	8.8	9.2	9.2

In the case of carbon the interval theory was applied so that the variation of $l_s(u)$ in each sub-interval was no more than 10%.

The exponential theory was applied by writing

$$l_s(u) = A_0 + A_1 e^{-a_1 u}$$

and requiring that: (a) the mean free path of neutrons of $E = 1.46$ ev is correctly given, and furthermore that

$$\text{(b)} \qquad \int\limits_0^u (A_0 + A_1 e^{-a_1 u})^2 \, du' = \left\{\int\limits_0^u l_s^2(u) \, du'\right\}_{\text{exper}}$$

and

$$\text{(c)} \qquad \int\limits_0^u (A_0 + A_1 e^{-a_1 u}) \, u' \, du' = \left\{\int\limits_0^u u' \, l_s(u') \, du'\right\}_{\text{exper}}.$$

Condition (b) expresses the fact that the age theory is a good approximation, while condition (c) is chosen to weight the low energy region more than the high energy region.

As one sees from Table 40 in the case of carbon the results obtained with the interval theory and with the exponential theory are in very good agreement and both represent an appreciable improvement with respect to the age theory.

In the case of deuterium the scattering cross section as a function of energy is given in Fig. 18. The results, presented in Table 41 for heavy water, are obtained by applying the interval theory in its "integral form" i.e. in the limiting case when the number of sub-intervals is permitted to become infinite. The exponential theory has been applied by imposing that both the total scattering mean free path and its ratio $c(u)$ to the scattering mean free path in deuterium satisfy conditions of types (a), (b) and (c) already discussed in the case of carbon.

The fact that the experimental angular distribution of the scattered neutrons (Sect. 29) is not taken into account in all these calculations reduces considerably the interest of their comparison with the experimental data.

75. The age theory. As we shall show in Sect. 79, the *age theory*[1] can be derived from the Boltzmann integro-differential equation (61.21a) applied to the stationary case, and represents for $A \gg 1$, a very useful first approximation.

In this section we shall derive it from elementary diffusion theory considerations which refer to the stationary state reached in a homogeneous non-absorbing medium containing only heavy nuclei and one or several neutron sources.

In a region far from the neutron sources and from the boundary of the medium or from absorbing surfaces (as we shall see in Sect. 118 far with respect to l_{tr}), the number of neutron $N_0(\boldsymbol{r}, u)\, d\boldsymbol{r}\, du$ contained in the volume element $d\boldsymbol{r}$ with lethargy between u and $u+du$, does not change with time as a result of a statistical equilibrium between:

1. the neutrons present in $d\boldsymbol{r}$ the energy of which is degraded into or out of the lethargy interval du under consideration, and

2. the neutrons of lethargy u which diffuse into or out of the volume element $d\boldsymbol{r}$.

The first cause of variation of $N_0(\boldsymbol{r}, u)\, d\boldsymbol{r}\, du$ is immediately expressed in terms of the slowing down density $Q(\boldsymbol{r}, u)$, which throughout these considerations is assumed to be a continuous function of the lethargy u. In a unit time interval, the increase of $N_0(\boldsymbol{r}, u)\, d\boldsymbol{r}\, du$ due to neutrons slowed down from higher energies is simply $Q(\boldsymbol{r}, u)\, d\boldsymbol{r}$, and its decrease due to particles which from the lethargy interval du are slowed down to lower energies is $Q(\boldsymbol{r}, u+du)\, d\boldsymbol{r}$. Therefore, the net variation of $N_0(\boldsymbol{r}, u)\, d\boldsymbol{r}\, du$ due to cause 1. is

$$[Q(\boldsymbol{r}, u) - Q(\boldsymbol{r}, u+du)]\, d\boldsymbol{r} = -\frac{\partial Q}{\partial u}\, d\boldsymbol{r}\, du. \tag{75.1}$$

The second cause of variation is derived by recalling that, as in the kinetic theory of gases, the neutron current density $\boldsymbol{J}$ may be related to the neutron density $N_0(\boldsymbol{r}, u)$ by the *Fick's law*

$$\boldsymbol{J} = -D_0 \cdot \nabla N_0(\boldsymbol{r}, u) = -D \cdot \nabla \Phi_0(\boldsymbol{r}, u) \tag{75.2}$$

where $D_0 = D_0(u)$ is the diffusion coefficient (see Sects. 114, 115 and 117)

$$D_0 = D \cdot v = \frac{l_{\mathrm{tr}} \cdot v}{3}. \tag{75.3}$$

The net flux of neutrons of lethargy between u and $u+du$ which diffuse out of the volume $d\boldsymbol{r}$ is given by $d\boldsymbol{r} \cdot du$ times the divergence of $\boldsymbol{J}$, i.e.

$$-\nabla \cdot D_0 \cdot \nabla N_0(\boldsymbol{r}, u)\, d\boldsymbol{r}\, du \tag{75.4}$$

[1] The age theory was for the first time used by H. A. BETHE, S. A. KORFF, G. PLACZEK [Phys. Rev. **57**, 573 (1940)] for treating the slowing down of cosmic ray neutrons through the atmosphere.

which, assuming that the diffusion coefficient D_0 is independent of the spatial coordinates, may be written

$$- D_0 \cdot \nabla^2 N_0(\boldsymbol{r}, u)\, d\boldsymbol{r}\, du. \tag{75.5}$$

The condition of stationarity is obtained by imposing the equality of (75.1) and (75.5), namely

$$D_0 \cdot \nabla^2 N_0(\boldsymbol{r}, u) = \frac{\partial Q}{\partial u}. \tag{75.6}$$

The probable number of neutrons $N_0(\boldsymbol{r}, u)$ is now expressed in terms of the slowing down density $Q(\boldsymbol{r}, u)$ by means of (61.17) and (67.16)

$$N_0(\boldsymbol{r}, u) = \frac{1}{v}\, \Phi_0(\boldsymbol{r}, u) = \frac{l_s}{v}\, \Psi_0(\boldsymbol{r}, u) \approx \frac{l_s}{v}\, \frac{Q(\boldsymbol{r}, u)}{\xi}. \tag{75.7}$$

It has to be noticed, however, that (67.16) holds exactly only asymptotically [i.e. only for values of the lethargy satisfying condition (67.14)] between the space integrals of the collision density and of the slowing down density (66.1 c), (66.1 d) but not between $\Psi_0(\boldsymbol{r}, u)$ and $Q(\boldsymbol{r}, u)$; these two quantities are approximately related by (67.16) only in the limit of extremely slow variation with $\boldsymbol{r}$ (Sect. 66).

Introducing (75.7) into (75.6) we obtain

$$\frac{l_{\text{tr}}\, l_s}{3\xi}\, \nabla^2 Q = \frac{\partial Q}{\partial u} \tag{75.8}$$

which is essentially the differential equation that we were seeking.

Before proceeding, it may be worthwhile noticing, in order to make clearer the significance of Eq. (75.8), that it can also be derived by the following slightly different procedure: one starts from the *continuity equation* of the neutron current

$$\nabla \boldsymbol{J} = - \frac{\partial N_0}{\partial t} \tag{75.9}$$

and one substitutes in it for the current density $\boldsymbol{J}$ its expression (75.2) and for the time differential dt its expression (64.14) in terms of the lethargy differential du.

A useful simplification of Eq. (75.8) can be obtained by substituting for the variable u a new variable called the *age of the neutrons* or simply the *age*, defined as

$$\tau = \int_0^u \frac{l_{\text{tr}}(u')\, l_s(u')}{3\xi}\, du'. \tag{75.10}$$

For media containing nuclei of a single type which scatter the neutrons isotropically in the c.m.s., we can make use of (62.14), (72.3), and (72.10) and write the age in the simplified form

$$\tau = \frac{1}{3\xi\left(1 - \frac{2}{3} A^{-1}\right)} \int_0^u l_s^2(u')\, du'. \tag{75.11}$$

From (75.10) it follows that

$$\frac{\partial Q}{\partial u} = \frac{\partial Q}{\partial \tau}\, \frac{d\tau}{du} = \frac{l_{\text{tr}}\, l_s}{3\xi}\, \frac{\partial Q}{\partial \tau} \tag{75.12}$$

which, combined with (75.8), gives finally[1]

$$\nabla^2 Q = \frac{\partial Q}{\partial \tau}. \tag{75.13}$$

---

[1] E. FERMI: Neutron physics, a revision of I. HALPERN's Notes on E. FERMI's lectures in 1945; AECD-2664 (1951).

This equation, called the *Fermi age equation* or simply the *age equation* is formally identical to the heat conduction equation

$$\nabla^2 T = \frac{c\varrho}{k}\frac{\partial T}{\partial t} \tag{75.14}$$

with the slowing down density $Q(\boldsymbol{r}, \tau)$ corresponding to temperature and the age τ to time (multiplied by the constant factor $\frac{k}{c\varrho}$: k = thermal conductivity, c = specific heat, ϱ = density of the medium). The analogy is however only formal because the age equation refers to the stationary state of the neutron distribution which, at any time is different for different values of the neutron age τ, i.e. of the energy of the neutrons. The heat conduction equation on the other hand, describes how the temperature distribution changes with time, i.e. a transitory phenomenon.

According to the definition (75.10) τ has the dimensions of a square of a length and therefore is sometimes called the *slowing down area* [see Sect. 76, Eq. (76.18)].

One of the two procedures mentioned above for the derivation of the age equation (75.13) makes use of the relation (64.14) between du and dt and therefore, according to (75.10), of a relation between dt and $d\tau$, from which one can derive immediately by integration, the relationship between the time t, defined by (64.15) and the age τ. In the special but interesting case $l_s(u)$ = const over all the energy interval, this relationship may be obtained from (75.11) by expressing the lethargy in terms of the velocity, and this in terms of the slowing down time t by means of (64.16); the result is

$$\tau = \frac{2}{\xi}\,\frac{l_s^2}{3\left(1-\frac{2}{3A}\right)}\ln\left(1+\frac{t}{t_0}\right); \qquad t_0 = \frac{2}{\xi}\,\frac{l_s}{v_0}. \tag{75.15}$$

This procedure makes clear that a description of the slowing down and diffusion processes by means of the age theory corresponds to substituting the discontinuous process of energy degradation undergone by the neutrons in successive elastic collisions, by the continuous energy loss already described qualitatively in Sect. 64 (see e.g. Fig. 119).

The age equation (75.13) has been derived for the case of a non-absorbing medium. We now will show how it can be modified in order to take account of absorption. As we shall see later, in this section as well as in Sect. 79, where the age equation will be derived from the Boltzmann integro-differential equation, the following modification of the age theory is approximately correct only for *very weak capture*.

In the case of an absorbing moderator, the condition of stationarity (75.6) must be substituted by

$$D_0\nabla^2 N_{0a} - \frac{v}{l_a}N_{0a} = \frac{\partial Q_a}{\partial u} \tag{75.16}$$

where the second term on the left hand side obviously corresponds to capture. The subscript a has been attached to N_0 and Q in order to make clear that these quantities now refer to absorbing media.

In order to proceed along the same lines followed for non-absorbing media, we have now to express N_{0a} in terms of Q_a. This can be done by using (75.7) which, because of (69.1), (69.2), will be approximately valid with the same restrictions necessary for non-absorbing media with the weak absorption condition (69.14) added. Introducing (75.7) in (75.16) we obtain

$$\frac{l_{\mathrm{tr}} l_s}{3\xi} \nabla^2 Q_a - \frac{1}{\xi} \frac{l_s}{l_a} Q_a = \frac{\partial Q_a}{\partial u}. \tag{75.17}$$

If now we introduce

$$Q_a(\boldsymbol{r}, u) = Q(\boldsymbol{r}, u)\, p(u) \tag{75.18}$$

where $p(u)$ is the expression (69.15a) of the resonance escape probability valid in the case of weak absorption (69.14), Eq. (75.16) reduces to

$$\frac{l_{\mathrm{tr}} l_s}{3\xi} \nabla^2 Q = \frac{\partial Q}{\partial u} \tag{75.19}$$

formally identical with (75.8).

One concludes that the slowing down density with *weak capture* is approximately equal to the slowing down density without capture (i.e. the solution of the Fermi age equation) for the same age, multiplied by the resonance escape probability (69.15a) at the same energy. In other words the absorption does not affect the space distribution of the neutrons of a given age but merely reduces their density by the space-independent factor (69.15a).

Therefore, in the following, we shall limit ourselves to consider applications of the age theory to non-absorbing moderators.

76. Application of the age theory to a few important cases. We will now apply the age equation (75.13), valid for non-absorbing media, to a few important cases all of which correspond to classical problems of time dependent heat conduction theory[1].

The solution of Eq. (75.13) corresponding to a given distribution of monoenergetic neutron sources $Q_0(\boldsymbol{r})$ in a moderator of given geometry, is uniquely determined by imposing the following conditions:

(a) For $\tau = 0$ the slowing down density $Q(\boldsymbol{r}, \tau)$ must be identical to the assigned neutron source distribution

$$Q(\boldsymbol{r}, \tau = 0) = Q_0(\boldsymbol{r}). \tag{76.1}$$

(b) For any value of $\tau > 0$, $Q(\boldsymbol{r}, \tau)$ must satisfy the boundary conditions relative to the geometry of the moderator. The more naive formulation would be to require the neutron density (and therefore the slowing down density Q) to vanish on all surfaces separating the moderator from vacuum because here the neutrons have an infinite mean free path. In Sect. 117 the solution of the transport equation near a boundary will be discussed in connection with one-velocity group problems and in Sect. 119 it will be shown that, far from boundaries and sources, the solution of the diffusion equation approaches reasonably well the exact solution, provided one stipulates that

$$Q(\boldsymbol{r}, \tau) = 0 \quad \text{on the extrapolated boundary,} \tag{76.2}$$

i.e. at a distance from the actual surface of

$$0.714\, l_{\mathrm{tr}} \tag{76.3}$$

in the direction of the external normal.

The same boundary condition can be applied in the present case of the age equation (75.13); usually one assumes that $l_{\mathrm{tr}}(u)$ is independent of energy, an

[1] H. N. Carslaw: Mathematical theory of the conduction of heat in solids. New York: Dover publications 1945. — E. Persico: Introduzione alle fisica Matematica. Bologna: Zanichelli 1952.

approximation which is unimportant provided the dimensions of the moderator are large compared to the mean free path.

In the case of an infinite non-absorbing medium condition (b) reduces to the condition that Q remains finite over all space for any value of $\tau > 0$.

Before deriving the solutions relative to a few important cases, we note that in an infinite medium the integral over all space of $Q(\boldsymbol{r}, \tau)$ must be equal, for any value of $\tau > 0$, to the rate at which the neutrons are emitted by the source. In fact, since the integral of $\nabla^2 Q$ over all space is zero, by integration of (75.13) one obtains

$$\frac{\partial}{\partial \tau} \int d\boldsymbol{r}\, Q(\boldsymbol{r}, \tau) = 0 \qquad \text{for } \tau > 0, \tag{76.4}$$

i.e.

$$\int Q(\boldsymbol{r}, \tau)\, d\boldsymbol{r} = \text{const} = \int d\boldsymbol{r}\, Q_0(\boldsymbol{r}) = q_0 \qquad \text{for } \tau > 0 \tag{76.5}$$

the value of the constant having been determined by introducing the initial condition (76.1) and the notation (66.1d).

Finally we recall the point mentioned in Sects. 61 and 71: in the case of age theory one also takes advantage of the fact that Eq. (75.13) is linear and homogeneous so that one can limit consideration to the case of such monoenergetic neutron sources the spatial distributions of which are represented by Dirac δ-functions; for the more general case can always be derived from this special one by means of an integration as shown in Eq. (71.1).

α) The case of an isotropic plane δ-source of monoenergetic neutrons in an infinite medium[1]. The slowing down density Q does not depend on x and y but only on the distance z from the plane in which the source is supposed to lie, and, of course, on the age, τ. Therefore, Eq. (75.13) reduces to

$$\frac{\partial^2 Q}{\partial z^2} = \frac{\partial Q}{\partial \tau} \qquad \text{for } z \neq 0 \tag{76.6}$$

whose solution must satisfy condition (a), namely

$$Q(z, 0) = Q_0\, \delta(z) \tag{76.7}$$

where Q_0 is the number of neutrons emitted by the plane source per cm² and per second, and condition (b) which reduces to requiring that $Q(z, \tau)$ remain finite for any value of $\tau > 0$ and for any value of z between $-\infty$ and $+\infty$.

The derivation of particular solutions of (76.6) satisfying conditions (a) and (b) is well known from the corresponding problem of heat conduction: the variables z and τ are separated by letting

$$Q(z, \tau) = Z(z)\, T(\tau) \tag{76.8}$$

which introduced in (76.6) gives rise to the two following differential equations, each in a single variable,

$$\frac{Z''}{Z} = \frac{\dot{T}}{T} = -\omega^2. \tag{76.9}$$

Here ω^2 is a real positive arbitrary constant and the negative sign in front of it is necessary in order to avoid that $T(\tau)$, and therefore $Q(z, \tau)$, increase, for any value of z, with increasing τ. From (76.9) and (76.8) one immediately gets the

[1] S. GLASSTONE and M.C. EDLUND: Nuclear reactor theory. New York: Van Nostrand 1952.

particular solution

$$Q_\omega(z,\tau) = e^{-\omega^2\tau}[A\cos\omega z + B\sin\omega z] \tag{76.10}$$

with ω, A and B as arbitrary constants.

In order to get the required solution satisfying conditions (a) and (b) one takes advantage of the fact that (76.6) is a linear and homogeneous equation so that the expression

$$Q(z,\tau) = \int_0^\infty e^{-\omega^2\tau}[A(\omega)\cos\omega z + B(\omega)\sin\omega z]\,d\omega \tag{76.11}$$

obtained by combining an infinite number of particular solutions of type (76.10), will still be a solution of (76.6); $A(\omega)$ and $B(\omega)$ are two arbitrary functions of ω which have to be determined in such a way that (76.11) satisfies condition (76.7); i.e.

$$Q(z,0) = Q_0\,\delta(z) = Q_0\int_0^\infty [a(\omega)\cos\omega z + b(\omega)\sin\omega z]\,d\omega$$

which, apart from the factor Q_0, is simply the Fourier integral of the δ-function. One obtains

$$a(\omega) = \frac{1}{\pi}\int_{-\infty}^{+\infty}\delta(z)\cos z\omega\,dz = \frac{1}{\pi},$$

$$b(\omega) = \frac{1}{\pi}\int_{-\infty}^{+\infty}\delta(z)\sin z\omega\,dz = 0$$

and therefore

$$Q(z,\tau) = \frac{Q_0}{\pi}\int_0^\infty e^{-\omega^2\tau}\cos\omega z\,d\omega \tag{76.12}$$

from which finally it follows[1]

$$Q(z,\tau) = \frac{Q_0}{(4\pi\tau)^{\frac{1}{2}}}\,e^{-\frac{z^2}{4\tau}}. \tag{76.13}$$

This expression for $\tau=0$ reduces, according to condition (76.7), to the Dirac delta function multiplied by Q_0, while for any value of $\tau>0$, it is a Gaussian becoming wider, and because of (76.5) lower, as the age τ increases (Fig. 129).

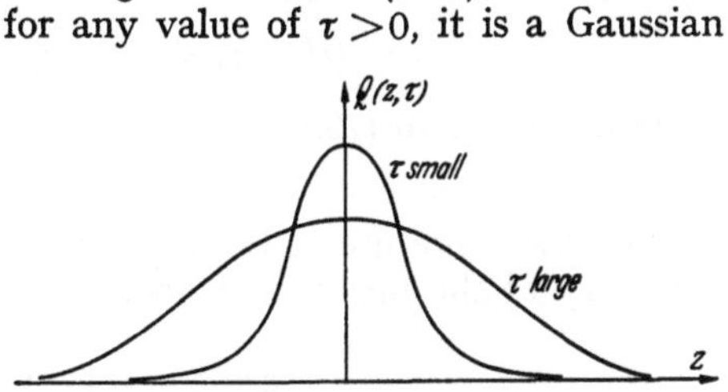

Fig. 129. The slowing down density produced by a plane δ-source in an infinite medium for two values of the age τ.

The corresponding neutron flux is obtained by using the approximate relation (75.7)

$$\left.\begin{aligned}\Phi_0(z,\tau) &\approx \frac{l_s}{\xi}\,Q(z,\tau)\\ &= Q_0\frac{l_s}{\xi}\,\frac{1}{(4\pi\tau)^{\frac{1}{2}}}\,e^{-\frac{z^2}{4\tau}};\end{aligned}\right\} \tag{76.14}$$

according to (71.6b) and (71.4) with $s=1$, the second spatial moment is

$$\overline{z^2} = 2\tau. \tag{76.15}$$

β) The case of an isotropic point source of monoenergetic neutrons in an infinite medium. This case, which is very important from the practical point of view

---

[1] $\int_0^\infty e^{-a^2x^2}\cos(bx)\,dx = \frac{\sqrt{\pi}}{2a}\,e^{-\frac{b^2}{4a^2}}$.

(see Division IV below), can now be treated by applying the relation (71.2b) to (76.13); one gets

$$Q(r,\tau) = \frac{Q_0}{(4\pi\,\tau)^{\frac{3}{2}}}\, e^{-\frac{r^2}{4\tau}} \tag{76.16}$$

with its second spatial moment (71.6a) turning out to be

$$\overline{r^2} = 6\tau. \tag{76.17}$$

By comparing (76.15) and (76.17) with (71.8) one has

$$\tau(u) = L_s^2(u) \tag{76.18}$$

a relationship which justifies the expression "slowing down area" used sometimes to indicate the age τ.

The neutron density $N_0(r, u)$ is obtained by introducing (76.16) in (75.7); its dependence on the lethargy u is best visualized by considering the special case $l_s(u) = \text{const}$ in the energy interval considered. In this case one has from (75.10)

$$\tau = \frac{l_s^2\, u}{3\xi(1-\frac{2}{3}A^{-1})} = \frac{l_{tr}\, l_s}{3\xi}\, u .$$

Introducing this expression in (76.16) one recognizes that, at a given point $\boldsymbol{r}$, the neutron density as a function of u, first increases and then decreases and vanishes; it passes through a maximum at

$$u_{\max} = \frac{\xi\,\overline{r^2}}{2 l_{tr}\, l_s} .$$

γ) The case of an isotropic plane δ-source located in the middle of a moderator of thickness 2c. Let us consider the case of a slab of moderator of infinite dimensions in both x and y directions, but extending from $z=-c$ to $z=+c$ in the z direction. The age equation for this case is again (76.6) which, as we have seen, has particular solutions of type (76.10). The source, located in the middle of the slab, i.e. in the plane $z=0$, is again represented by (76.7).

The difference with respect to the case of an infinite medium occurs when one imposes condition (b) which now is

$$Q(z_0, \tau) = Q(-z_0, \tau) = 0 \quad \text{for} \quad \tau > 0 \tag{76.19}$$

where according to (76.3)

$$z_0 = c + 0.71\, l_{tr} . \tag{76.20}$$

We now make use of (76.10), omitting the sine term in consequence of the symmetry of the present problem. By applying (76.19) to (76.10) one obtains

$$\cos\omega\, z_0 = 0 \quad \text{i.e.} \quad \omega = n\,\frac{\pi}{2 z_0}$$

where n is an odd integer. Therefore the general solution will be

$$Q(z,\tau) = \sum_{n\,\text{odd}} A_n \exp\left[-\frac{\pi^2}{4}\,\frac{n^2}{z_0^2}\,\tau\right] \cos\left[n\,\frac{\pi}{2}\,\frac{z}{z_0}\right] \tag{76.21}$$

where the coefficient A_n is determined by imposing the condition (76.7); this can be done by setting $\tau=0$, multiplying both sides of (76.21) by $\cos\left[n\,\frac{\pi}{2}\,\frac{z}{z_0}\right]$ and integrating with respect to z from $-z_0$ to $+z_0$. The result is

$$A_n = \frac{Q_0}{z_0} .$$

In conclusion one has

$$Q(z,\tau)=\frac{Q_0}{z_0}\sum_{n\,\mathrm{odd}}\exp\left[-n^2\frac{\pi^2}{4}\frac{\tau}{z_0^2}\right]\cos\left[n\frac{\pi}{2}\frac{z}{z_0}\right]. \tag{76.22}$$

For

$$\tau>\frac{4}{\pi^2}z_0^2=\tau_0 \tag{76.23}$$

all terms with $n=3, 5, 7, \ldots$ may be neglected with respect to the first one so that (76.22) can be substituted with sufficient accuracy by

$$Q(z,\tau\gg\tau_0)\approx\frac{Q_0}{z_0}\,\mathrm{e}^{-\frac{\pi^2}{4z_0^2}\tau}\cdot\cos\left[\frac{\pi}{2}\frac{z}{z_0}\right]. \tag{76.24}$$

δ) The case of an isotropic point source placed on the axis of a rectangular prism of sides $2a$, $2b$, $2c$ ($\to\infty$) [19]. This case is important because it corresponds to a geometry frequently used for the determination of the slowing down length of the neutrons in a given material. The point source located at the origin is represented by the expression

$$Q_0(\boldsymbol{r})=Q_0\,\delta(x)\,\delta(y)\,\delta(z). \tag{76.25}$$

Particular solutions of Eq. (75.13) are obtained by separating the four variables x, y, z, and τ in a similar way as shown in (76.8). Then one introduces the boundary conditions (b), i.e. according to (76.3) Q must be zero for any value of $\tau>0$ on the extrapolated boundaries which in this case are the six planes

$$\left.\begin{aligned}x&=\pm x_0=\pm(a+0.71\,l_{\mathrm{tr}}),\\ y&=\pm y_0=\pm(b+0.71\,l_{\mathrm{tr}}),\\ z&=\pm z_0=\pm(c+0.71\,l_{\mathrm{tr}}).\end{aligned}\right\} \tag{76.26}$$

One obtains the particular solution

$$\left.\begin{aligned}Q_{lmn}(x,y,z,\tau)&=\exp\left[-\frac{\pi^2}{4}\left(\frac{l^2}{x_0^2}+\frac{m^2}{y_0^2}+\frac{n^2}{z_0^2}\right)\tau\right]\times\\ &\times\cos\left(l\frac{\pi}{2}\frac{x}{x_0}\right)\cdot\cos\left(m\frac{\pi}{2}\frac{y}{y_0}\right)\cdot\cos\left(n\frac{\pi}{2}\frac{z_0}{z}\right)\end{aligned}\right\} \tag{76.27}$$

(l, m, n, are odd integers), from which one derives the general solution

$$Q(x,y,z,\tau)=\sum_{l,m,n\,\mathrm{odd}}A_{lmn}\,Q_{lmn}(x,y,z,\tau) \tag{76.28}$$

where the coefficients A_{lmn} have to be determined by requiring that for $\tau=0$ (76.28) reduces to (76.25). One obtains

$$A_{lmn}=\frac{Q_0}{x_0\,y_0\,z_0} \tag{76.29}$$

which, combined with (76.27) gives

$$\left.\begin{aligned}Q(x,y,z,\tau)&=\frac{Q_0}{x_0\,y_0\,z_0}\sum_{l,m,n\,\mathrm{odd}}\mathrm{e}^{-\frac{\pi^2}{4}\left(\frac{l^2}{x_0^2}+\frac{m^2}{y_0^2}+\frac{n^2}{z_0^2}\right)\tau}\cos\left(l\frac{\pi}{2}\frac{x_0}{x}\right)\times\\ &\times\cos\left(m\frac{\pi}{2}\frac{y_0}{y}\right)\cdot\cos\left(n\frac{\pi}{2}\frac{z}{z_0}\right).\end{aligned}\right\} \tag{76.30}$$

A very important case is that of a prism of length $2c$ which is very large, let us say

$$2c\to\infty.$$

In this case the z dependent part of (76.30) can be transformed as follows[1]

$$\lim_{z_0\to\infty}\frac{1}{z_0}\sum_{n\,\mathrm{odd}}\exp\left[-\pi^2\frac{n^2}{4z_0^2}\tau\right]\cos\left(n\frac{\pi}{2}\frac{z}{z_0}\right)=\frac{1}{2}\int_0^{\infty}e^{-\frac{\pi^2}{4}\tau k^2}\times$$

$$\times\cos\left(\frac{\pi}{2}zk\right)dk=\frac{e^{-\frac{z^2}{4\tau}}}{(4\pi\tau)^{\frac{1}{2}}}.$$

By introducing this expression in (76.30) one has finally

$$\left.\begin{aligned}Q(x,y,z,\tau)=\frac{Q_0}{x_0y_0}\frac{e^{-\frac{z^2}{4\tau}}}{(4\pi\tau)^{\frac{1}{2}}}\sum_{l,m,\,\mathrm{odd}}\exp\left[-\frac{\pi^2}{4}\left(\frac{l^2}{x_0^2}+\frac{m^2}{y_0^2}\right)\tau\right]\times\\ \times\cos\left(l\frac{\pi}{2}\frac{x}{x_0}\right)\cos\left(m\frac{\pi}{2}\frac{y}{y_0}\right).\end{aligned}\right\}\quad(76.31\,\mathrm{a})$$

The dependence of $Q(x, y, z, \tau)$ along the axis of the prism ($x=0$, $y=0$) on the distance z from the source is identical to (76.13), i.e. it is the same as for an infinite plane source in an infinite medium.

As τ increases the intensity of the distribution falls because of the exponential factor $\exp\left[-\frac{\pi^2}{4}\left(\frac{l^2}{x_0^2}+\frac{m^2}{y_0^2}\right)\tau\right]$ which represents the effect of the leakage of the neutrons out of the lateral surface of the prism.

The series (76.31 a) converges better at large distances from the source than at small distances. It can be transformed into a different form which, as it converges better at short distances, may be useful for numerical computation (at short distances) in special practical cases [*19*].

Passing from these various examples to the *general case*, it is immediately seen that, for a medium of finite dimensions, the solution of the age equation (75.13) can be written in the form

$$Q(\boldsymbol{r},\tau)=\sum_{l,m,n}A_{lmn}R_{lmn}(\boldsymbol{r})\,e^{-\omega_{lmn}^2\tau}\quad(76.31\,\mathrm{b})$$

where A_{lmn} are coefficients to be determined by imposing the initial condition (76.1). $R_{lmn}(\boldsymbol{r})$ and ω_{lmn}^2 are the eigenfunctions and the eigenvalues of the differential equation

$$\nabla^2R(\boldsymbol{r})+\omega^2R(\boldsymbol{r})=0\quad(76.31\,\mathrm{c})$$

which are determined by imposing the boundary conditions (76.2) for the given geometry of the moderator. The eigenvalues ω_{lmn}^2 are sometimes called *geometrical buckling coefficients*.

77. Elementary improvements on the age theory. The intrinsic limitations of the age theory will be discussed in Sect. 79 where we shall indicate its relationship with more refined procedures for the approximate solution of the problem of the slowing down and diffusion of neutrons. The failure of the age theory to describe the neutron distribution at great distances from the source may be recognized, however, from the following qualitative remark.

For the sake of simplicity we assume that, as usually happens, the total mean free path $l_{\mathrm{tot}}(v)$ of the neutrons in the medium has its largest value for $v=v_0$ so that the neutrons which after emission from the source have suffered

[1] See footnote 1, p. 273.

no collision at all, constitute the more penetrating component of the neutronic radiation moving throughout the moderator.

According to the qualitative argument given in Sect. 14, these neutrons may be described as the "primary radiation" and the neutrons of lower energy which are continually generated through collisions against the nuclei of the moderator, as the corresponding "secondary softer components".

The distribution of the "primaries" around a monoenergetic isotropic point source embedded in an infinite medium is proportional to

$$\frac{1}{4\pi r^2}\exp\left[-\frac{r}{l_{\mathrm{tot}}(0)}\right]\frac{1}{l_{\mathrm{tot}}(0)}, \qquad l_{\mathrm{tot}}(0)=l_{\mathrm{tot}}(u=0)=l_{\mathrm{tot}}(E_0), \tag{77.1}$$

an expression which, for r sufficiently large, becomes greater than (76.16) which falls faster than exponentially. As we have noticed in Sect. 14 and as we shall better see in Sect. 80, some kind of *quasi-equilibrium* is established at great distance from the source, between the "primary radiation" and the "secondary radiation" so that the intensity of the latter can not decrease with increasing r, faster than the intensity (77.1) of the primary radiation. This is the problem of "deep penetration" the solution of which needs rather refined methods to be mentioned only very briefly in Sect. 80.

Here we restrict ourselves to noting that an elementary improvement with respect to the age theory may be attained by assuming that the process of slowing down of the neutrons starts only after the first collision[1] [*19*]. In other words, one assumes as source of the slowing down process the spatial distribution of the first collision which is obtained by multiplying (77.1) by $d\boldsymbol{r}$. Recalling (71.1) Flügge obtains—for non-absorbing media—the following expression for the slowing down density "corrected for the first collision"

$$Q(\boldsymbol{r},\tau)=Q_0\int\frac{1}{4\pi r'^2}\,e^{-\frac{r'}{l_s'(0)}}\,\frac{1}{(4\pi\tau_1)^{\frac{3}{2}}}\,e^{-\frac{(\boldsymbol{r}-\boldsymbol{r}')^2}{4\tau_1}}\,\frac{d\boldsymbol{r}'}{l_s(0)} \tag{77.2a}$$

where $l_s'(0)$ is an effective mean free path for the first collision and τ_1 the age computed starting from the average energy of the neutrons after the first collision. If the scattering is isotropic in the c.m.s., this energy is given by

$$E_1=E_0\frac{A^2+1}{(A+1)^2}.$$

By performing the integration one obtains

$$Q(\boldsymbol{r},\tau)=\frac{Q_0}{4\pi l_s'(0)\,r^2}\,x\,e^{-x^2}\{F(a+x)-F(a-x)\} \tag{77.2b}$$

where

$$a=\left(\frac{\tau_1}{l_s'^2(0)}\right)^{\frac{1}{2}},\qquad x=\frac{r}{2\tau_1^{\frac{1}{2}}},\qquad F(z)=\int_0^z t^2\,[1-\mathrm{Erf}(t)]\,dt,\qquad \mathrm{Erf}(t)=\frac{2}{\sqrt{\pi}}\int_0^t e^{-y^2}dy.$$

Flügge gives numerical tables for $F(z)$ which have been extended by Secrest[2]. Eq. (77.2b) has been applied to the analysis of various experimental distributions in hydrogeneous materials[3] [*18b*].

According to (75.7) an approximate expression of the neutron flux distribution $\Phi_0(\boldsymbol{r},u)$ corrected for the first collision is obtained by multiplying (77.2) by $l_s/\bar{\xi}$.

---

[1] S. Flügge: Phys. Z. **44**, 445 (1943).

[2] L. Secrest: Phys. Rev. **77**, 141 (1950).

[3] C.W. Tittle, H. Faul, E.L. Secrest and C. Goodman: Techn. Rep. No. 21 Lab. Nucl. Scien. and Engin. MIT (1949).

In the same spirit and with the same approximations the "last flight correction", mentioned at the end of Sect. 71, may be applied to this expression of $\Phi_0(\boldsymbol{r}, u)$. This correction takes into account the fact that after the neutron has suffered a collision which throws it beyond the final energy $E = E(\tau)$, it experiences a "free ride" without changing energy until the next collision takes place [19]. This correction can be made by introducing another exponential factor containing the mean free path $l_s(u)$ appropriate to the final energy.

In conclusion the neutron flux "corrected" with respect to the age theory for both the first and the last collision is given by

$$\Phi_0(\boldsymbol{r}, \tau) = Q_0 \frac{l_s(u)}{\xi} \int \frac{d\boldsymbol{r}'}{l_s(0)} \int \frac{d\boldsymbol{r}''}{l_s(u)} \frac{e^{-\frac{|\boldsymbol{r}'|}{l_s(0)}}}{4\pi r'^2} \frac{e^{-\frac{(\boldsymbol{r}''-\boldsymbol{r}')^2}{4\tau}}}{(4\pi\tau)^{\frac{3}{2}}} \frac{e^{-\frac{|\boldsymbol{r}-\boldsymbol{r}''|}{l_s(u)}}}{4\pi(\boldsymbol{r}-\boldsymbol{r}'')^2}. \tag{77.3}$$

An expression of this type is called a convolution of three kernels: the point source kernel (76.16) and two transport kernels (77.1).

The corresponding second moment is the sum of the second moments of each kernel:

$$\overline{r^2} = \overline{r_0^2} + \overline{r_{\text{pt}}^2} + \overline{r_u^2}. \tag{77.4}$$

This important theorem can easily be proved by means of Fourier transform methods [19]. It can also be verified by introducing (77.3) into the equation

$$\overline{r^2} = \frac{\int |\boldsymbol{r}|^2 \Phi_0(\boldsymbol{r}, \tau)\, d\boldsymbol{r}}{\int \Phi_0(\boldsymbol{r}, \tau)\, d\boldsymbol{r}} \tag{77.5}$$

and performing the three spatial integrations with

$$\boldsymbol{r}_1 = \boldsymbol{r}, \qquad \boldsymbol{r}_2 = \boldsymbol{r}'' - \boldsymbol{r}', \qquad \boldsymbol{r}_3 = \boldsymbol{r} - \boldsymbol{r}''$$

as independent variables. Of the six terms arising from the expression of

$$|\boldsymbol{r}|^2 = |\boldsymbol{r}_1 + \boldsymbol{r}_2 + \boldsymbol{r}_3|^2,$$

the three containing r_i^2 $(i = 1, 2, 3)$ give rise to the second moments of the corresponding kernels, while the three non-diagonal terms of the form $\boldsymbol{r}_i \cdot \boldsymbol{r}_k$ $(i \neq k)$ vanish identically because of the integration over all possible orientations of the vectors $\boldsymbol{r}_i$.

The second moment of the transport kernel is given by the second equation (73.5) and that of the point source kernel by (76.17), so that we can conclude from (77.4)

$$\overline{r^2} = 2 l_s^2(0) + 6\tau + 2 l_s^2(u). \tag{77.6}$$

Sometimes instead of the expression (77.3), which is rather inconvenient in many practical applications, one prefers to adopt a simple Gaussian distribution

$$\frac{l_s(u)}{\xi} \frac{Q_0}{(4\pi\tau')^{\frac{3}{2}}} \exp\left[-\frac{r^2}{4\tau'}\right] \tag{77.7}$$

where τ' is the "corrected age" adjusted to give the corrected second moment (77.6)

$$\tau' = \frac{1}{6} \overline{r^2}. \tag{77.8}$$

One has to stress here once more that all considerations given in this section provide an approximately correct representation of the slowing down density only in the case of heavy moderators.

78. The group picture. The age theory, corrected for the first and the last collision as described in the previous section, provides a fairly satisfactory description of the neutron distribution in a heavy moderator. But if one tries to use it in problems involving the slowing down of neutrons in an "assembly" composed of two or more large spatial regions (each of them being large compared to the neutron mean free path) filled with different moderators, one faces remarkable difficulties in joining the solutions valid in two neighbouring media on the corresponding interface. This is a consequence of the use of the variable τ which, according to its definition (75.10), does not correspond simply to intrinsic properties of the neutrons, as does the lethargy u, but includes also the various physical features of the moderator.

This type of complication at the boundary is avoided if one makes use, instead of Eq. (75.13), of Eq. (70.8) which is also similar to the heat conduction equation. In this case however the elimination of the difficulty due to the discontinuity of the factor $l_{tr} l_s / 3\bar{\xi}$ on the interfaces, is accompanied by the introduction of complications due to its u-dependence.

These complications may be eliminated by making use of a simplified formulation of the slowing down process, usually called the *multigroup picture* or simply *group picture*, according to which by means of an approximate procedure the age equation is reduced to a finite set of coupled differential equations, each depending only on the spatial variables, not on the lethargy u [*7*], [*19*].

The group picture represents a general class of approximate methods[1-3], differing one from the other in the details of the particular formulation, but all having in common the following general description of the slowing down process.

The lethargy range $(0, U)$ in which we want to know the neutron distribution throughout a composite assembly of moderators, is divided in several discrete sub-intervals called energy groups or simply groups: the first of these, of lethargy range U_1, extends from $u=0$ to $u=u_1$, the second of range U_2, extends from $u=u_1$ to $u=u_2$, ..., and the last, of range U_n, extends from $u=u_{n-1}$ to $u=U_n=U$.

The neutrons in a given energy group i are supposed to diffuse without energy loss, through each spatial homogeneous region $\varrho=1, 2, \ldots$ until they have undergone such a number of collisions that, on the average, the total path length traversed is equal to that which they actually traverse in the true physical case when the lethargy increases from u_{i-1} to u_i. After this they are suddenly removed from the i-th group and pass into the successive group $i+1$, where they diffuse again without energy loss, etc.

The various parameters characterising the media filling the different regions (such as logarithmic energy loss, the cross sections for various processes, etc.) are represented by step functions with respect to u; i.e. they are assumed to be constants and equal to certain appropriate average values over each energy group i, but show a discontinuity at u_i where one passes from the group i to the successive group $i+1$.

We assume that the assembly where the neutrons are produced and slowed down, can be divided into two or more regions, each large enough and sufficiently homogeneous, to justify the use of the age theory for describing the distribution of the neutrons over its volume. We apply to each one of these regions the age equation in the form (75.17), which includes a term representing a weak absorption, and we generalize it: (a) by writing the diffusion term according to (75.4) which is

[1] R. Ehrlich and H. Hurwitz jr.: Nucleonics **12**, No. 2, 23 (1954).
[2] H. Hurwitz jr. and R. Ehrlich: Progress in nuclear energy, Physics and Mathematics, Vol. 1, p. 343. London and New York: Pergamon Press 1956.
[3] M. E. Mandl and J. Howlett: p. 433 of Ref. [*15c*].

valid in the general case of non-homogeneous media, and (b) by adding a source term $S(r, u)$

$$-\nabla \frac{l_{tr} l_s}{3\xi} \nabla Q + \frac{l_s}{\xi l_a} Q = -\frac{\partial Q}{\partial u} + S(\boldsymbol{r}, u). \tag{78.1}$$

This equation, although subject to the limitations discussed in Sect. 79, is of considerable practical interest because, according to the particular problem, the source term may represent not only a true source S_{true}, but also the production of neutrons of lethargy u at the point $\boldsymbol{r}$, due either to inelastic scattering of neutrons of higher energy $u' < u$, or to fission; in general one can write:

$$S(\boldsymbol{r}, u) = S_{\text{true}}(\boldsymbol{r}, u) + S_{\text{inel}}(\boldsymbol{r}, u) + S_f(\boldsymbol{r}, u).$$

In order to take into account the contribution of inelastic scattering to the degradation of the energy of the neutrons, one has to include in the absorption cross section $\Sigma_a = l_a^{-1}$ appearing in the second term on the left-hand side of (78.1), the corresponding cross section $\Sigma_{\text{inel}}(u')$ and one has to write for the source term at a lethargy $u > u'$

$$S_{\text{inel}}(\boldsymbol{r}, u) = \int_0^u \Phi_0(\boldsymbol{r}, u')\, \Sigma_{\text{inel}}(u')\, \chi(u' \to u)\, du' \tag{78.2}$$

where $\chi(u' \to u)$ is the fraction of neutrons of lethargy u' scattered inelastically to lethargy u

$$\int_{u'}^{\infty} \chi(u' \to u)\, du = 1.$$

In the case of fission, the source term may be written

$$S_f(\boldsymbol{r}, u) = \nu \left\{ \int_0^{u_T} \Phi_0(\boldsymbol{r}, u')\, \Sigma_f(u')\, du' + \Phi_{\text{th}}(\boldsymbol{r})\, \Sigma_{f,\text{th}} \right\} S_f(u) \tag{78.3}$$

where ν is the number of neutrons emitted per fission, Σ_f the fission cross section and $S_f(u)$ the normalized fission spectrum (82.6). The thermal neutron flux $\Phi_{\text{th}}(\boldsymbol{r})$ appearing in (78.3) is given by Eq. (109.16); when fission is present, this equation to be considered together with Eq. (78.1).

Consideration of multiplying media is outside the scope of the present article and it has been mentioned here only to show the generality of Eq. (78.1) and the capacity of the set of coupled differential equations that we will now derive from it, to deal with the problem of reactor design calculations[1-3].

By integration of Eq. (78.1) with respect to u over the i-th energy group, one obtains[2]

$$U_i \left(\nabla \frac{l_{tr} l_s}{3\xi} \nabla Q \right)_{i,\text{av}} - U_i \left(\frac{l_s}{\xi l_a} Q \right)_{i,\text{av}} = Q_i - Q_{i-1} - S_i, \tag{78.4}$$

where

$$(A)_{i,\text{av}} = \frac{1}{U_i} \int_{u_{i-1}}^{u_i} A(u)\, du; \qquad S_i = S_i(\boldsymbol{r}) = \int_{u_{i-1}}^{u_i} S(\boldsymbol{r}, u)\, du \tag{78.5}$$

and

$$Q_i = Q_i(\boldsymbol{r}) = Q(\boldsymbol{r}, u_i) \tag{78.6}$$

represents the current density of neutrons, per unit volume at the point $\boldsymbol{r}$, leaving the group i and entering the successive group $i+1$.

---

[1] See footnote 1, p. 279.
[2] See footnote 2, p. 279.
[3] See footnote 3, p. 279.

At this point one makes the approximation of replacing the averages of products appearing in (78.4) by products of averages:

$$\left(\nabla\frac{l_{\mathrm{tr}}\,l_s}{3\xi}\nabla Q\right)_{i,\,\mathrm{av}} \approx \nabla\left(\frac{l_{\mathrm{tr}}\,l_s}{3\xi}\right)_{i,\,\mathrm{av}}\nabla(Q)_{i,\,\mathrm{av}}\,;\;\left(\frac{l_s}{\xi\,l_a}Q\right)_{i,\,\mathrm{av}} \approx \left(\frac{l_s}{\xi\,l_a}\right)_{i,\,\mathrm{av}}\cdot(Q)_{i,\,\mathrm{av}}. \tag{78.7}$$

Such a procedure is justified only if at least one of the two factors does not change appreciably over the i-th group. This point deserves careful consideration while choosing the subdivision in energy groups because these must be adjusted to the actual physical aspects of the particular problem; for instance it will be convenient to choose the group divisions u_i at energy values where some physical threshold or discontinuity takes place.

In this approximation (78.4) can be written

$$\nabla D_i \nabla \Phi_i - \Sigma_{ai}\,\Phi_i = Q_i - Q_{i-1} - S_i \tag{78.8}$$

where

$$\Phi_i = \Phi_i(\boldsymbol{r}) = U_i\left(\frac{l_s}{\xi}\right)_{i,\,\mathrm{av}}(Q)_{i,\,\mathrm{av}} \tag{78.9}$$

is the neutron flux of all the neutrons belonging to the energy group i at the point $\boldsymbol{r}$, and

$$D_i = \frac{(l_{\mathrm{tr}}\,l_s/3\xi)_{i,\,\mathrm{av}}}{(l_s/\xi)_{i,\,\mathrm{av}}}\,;\qquad \Sigma_{ai} = \frac{(l_s/\xi\,l_a)_{i,\,\mathrm{av}}}{(l_s/\xi)_{i,\,\mathrm{av}}} \tag{78.10}$$

are, as well as $S_i(\boldsymbol{r})$, the above mentioned step-functions with respect to the variable u.

In the case of assemblies composed of two or more spatial regions, one gets for each region a set of equations of type (78.8); both the total neutron flux of the i-th group $\Phi_i(\boldsymbol{r})$ and the corresponding "current density" $D_i \nabla \Phi_i$ must be continuous at all interfaces.

In order to solve Eq. (78.8) when Q_{i-1} and S_i are known, one has to postulate an additional relationship, for instance between $(Q)_{i,\,\mathrm{av}}$ and Q_i, Q_{i-1}. This relationship very often takes the form

$$(Q)_{i,\,\mathrm{av}} = \omega_1\,Q_{i-1} + \omega_2\,Q_i \tag{78.11}$$

where ω_1 and ω_2 are appropriately chosen constants. According to the choice of their values one obtains *different schemes*; for instance, the so-called scheme A is obtained assuming $\omega_1 = 0$ and $\omega_2 = f \approx 1$, while scheme B corresponds to $\omega_1 = \omega_2 = \frac{1}{2}$. Other schemes are obtained by postulating a relationship similar to (78.11) but applied to the flux rather than to the slowing down density.

We shall not enter into a detailed discussion aiming at justifying these various procedures and showing their relative merits in the solution of particular problems connected with the calculations of reactor design[1].

Here, from now on, we will limit consideration to the very particular case of a single region filled with a homogeneous non-absorbing non-multiplying medium with negligible inelastic scattering. In this case Eq. (78.8) reduces to

$$D_i \nabla^2 \Phi_i - Q_i + Q_{i-1} + S_i = 0 \tag{78.12}$$

to which we have to add a relationship of the type of (78.11). This may be obtained, for instance, be describing the removal of neutrons from the i-th group as an "absorption process" of mean free path l^*_{ai}. Then, recalling that Q_i is the number of neutrons passing per second and per cm³ from the i-th group to the

[1] See footnotes 1 and 2, p. 279.

successive one, and that Φ_i is the flux of all the neutrons belonging to the i-th group, one has immediately the relation

$$Q_i = \frac{1}{l^*_{a i}} \Phi_i \tag{78.13}$$

which making use of (78.9) reduces to (78.11) in scheme A. The numerical value of the constant factor $f = (Q)_{i,\text{av}}/Q_i$ will depend on the value of $l^*_{a i}$.

If both absorption and generation of neutrons in the i-th group can be neglected, one can define $l^*_{a i}$ as the total path length travelled on the average by a neutron while its energy belongs to the i-th group

$$l^*_{a i} = \int\limits_{u_{i-1}}^{u_i} l_s(u) \frac{du}{\xi} = \left(\frac{l_s}{\xi}\right)_{i,\text{av}} \cdot U_i \tag{78.14}$$

which, combined with (78.9) gives $f = 1$. Ordinarily f will be somewhat larger than unity in a group and region not containing sources but showing a neutron absorption, but may be less than unity in groups and regions in which neutron production prevails.

Introducing (78.13) in (78.12) one obtains

$$D_i \nabla^2 \Phi_i - \frac{1}{l^*_{a i}} \Phi_i + \frac{1}{l^*_{a i-1}} \Phi_{i-1} + S_i = 0. \tag{78.15}$$

Let us now apply the set of Eq. (78.15) to two cases: the first is that of an isotropic point source embedded in an infinite medium treated with a *one-group picture* $n = 1$; the second is that of an isotropic plane δ-source also in an infinite medium which will be treated in general with a *n-group picture.*

In the one-group picture the set of Eq. (78.15) reduces to a single differential equation

$$D_1 \nabla^2 \Phi_1 - \frac{1}{l^*_{a 1}} \Phi_1 + \delta(\boldsymbol{r}) = 0. \tag{78.16}$$

The point source is supposed to be placed at the origin and to have a strength corresponding to the emission of one neutron per second ($\int \delta(\boldsymbol{r})\, d\boldsymbol{r} = 1$). The corresponding solution is immediately obtained

$$Q_1 = \frac{1}{l^*_{a 1}} \Phi_1 = \frac{1}{4\pi r L_1^{*2}} e^{-\frac{r}{L_1^*}} \tag{78.17}$$

where

$$L_1^{*2} = D_1 \cdot l^*_{a 1}. \tag{78.18}$$

The second moment of the distribution (78.17) is

$$\overline{r^2} = 6 L_1^{*2} \tag{78.19}$$

a relation identical to (76.17) valid for the Gaussian distribution (76.16).

We call L_1^*, the slowing down length; according to (78.10) and (78.14) one has

$$L_1^{*2} = \left(\frac{l_{\text{tr}} l_s}{3\xi}\right)_{1,\text{av}} \cdot u_1. \tag{78.20}$$

This expression with the averaging procedure adopted (in a rather arbitrary way) in the present discussion, is identical to (75.10). One can conclude that the one-group picture gives a spatial distribution which is exponential instead of Gaussian but which has the same second moment as in the age theory.

Let us now treat the case of a monoenergetic isotropic plane δ-source of unit strength, embedded in an infinite medium with the *n-group picture*. The set of Eq. (78.15) to be solved involves ordinary instead of partial differential equations:

$$\left.\begin{aligned} & D_1 \frac{d^2 \Phi_1}{dz^2} - \frac{1}{l_{a1}^*} \Phi_1 + \delta(z) = 0, \\ & D_2 \frac{d^2 \Phi_2}{dz^2} - \frac{1}{l_{a2}^*} \Phi_2 + \frac{1}{l_{a1}^*} \Phi_1 = 0, \\ & \dots\dots\dots\dots\dots\dots\dots\dots \\ & D_n \frac{d^2 \Phi_n}{dz^2} - \frac{1}{l_{an}^*} \Phi_n + \frac{1}{l_{an-1}^*} \Phi_{n-1} = 0. \end{aligned}\right\} \tag{78.21}$$

As one sees, the solution of the $(i-1)$-th equation acts as source in the i-th equation. Therefore one can immediately write down the solution of the first equation, representing the one-group picture of our problem

$$Q_1 = \frac{1}{l_{a1}^*} \Phi_1 = \frac{1}{2L_1^*} e^{-\frac{z}{L_1^*}} \tag{78.22}$$

and from this successively deduce the two-group picture, the three-group picture, and so on. The procedure is elementary but tedious if n is large. In this case it is convenient to apply Fourier transform methods by which the following results can easily be deduced [*19*]:

(a) Φ_n is the convolution of all the previous groups and therefore according to the theorem (77.4) of Sect. 77 the square of the corresponding slowing down length is equal to the sum of the squares of the slowing down lengths of each previous group individually

$$\frac{\overline{r^2}}{6} = L_1^{*2} + L_2^{*2} + \cdots + L_n^{*2} \tag{78.23}$$

where

$$L_k^{*2} = D_k l_{ak}^*. \tag{78.24}$$

(b) If the number of groups becomes infinite, but each L_k is reduced so that $\tau = \Sigma L_k^{*2}$ remains finite, the group picture passes over into the continuous age theory.

In conclusion, it is clear that by taking a sufficiently large number of groups it is possible to approximate the age theory to any degree of accuracy and still deal with a much simpler mathematical problem. The approximate slowing down functions which are constructed from group picture exponentials are called *synthetic kernels*.

In the case of assemblies (in particular piles) involving H_2O as a moderator, it is customary to use one or two fast-neutron groups (in addition to the thermal-neutron group). In the case of assemblies moderated with heavy materials as many as six or more groups have been used. One has also to keep in mind that if the neutron source shows an energy spread, as in the case of fission sources, the distribution of the neutrons is not Gaussian even in heavy materials such as graphite. This circumstance favours the group picture with respect to the age theory so that in graphite a three-group picture is only slightly less accurate than a single Gaussian while in H_2O, because of the very long mean free path at high energy, the neutron distribution is better represented by a single group than by a Gaussian.

79. The relation between the age theory and more refined methods for the approximate solution of the Boltzmann equation. Spherical harmonics method:

P_L-approximation and B_L-approximation. In this section we shall very briefly summarize the main features of some of the several methods for the approximate solution of the integro-differential Boltzmann equation in the stationary case, and show under what conditions some of these reduce to the elementary diffusion theory expressed by the age Eq. (75.13). Then the intrinsic limitations of the age theory will be considered in detail and some elementary modification aimed at extending it to the case of media containing hydrogen will be mentioned.

All these procedures deserve a much more complete presentation and thorough discussion. These may be found in various articles[1-5] [*16*] and books [*7*], [*19*] already mentioned in Sect. 61.

Because of the remark made in Sect. 71 we may consider a plane problem without renouncing generality; the angular number flux $\Phi(z, \mu, u)$ will not depend on x and y but only on z, u and μ, the last being the cosine of the angle between the neutron velocity and the z axis.

For the sake of simplicity we shall consider only isotropic sources, a restriction which considerably limits the generality of the treatment sketched here. The general case of anisotropic sources is considered in detail in Sect. 9 of FANO, SPENCER and BERGER's companion article in this volume for the analogous X-ray problem.

Finally in order to avoid complications due to boundary conditions of the type of (61.12a) or (61.12b), we will assume that the moderator is homogeneous and extends to infinity in all directions.

α) The spherical harmonics method. The Boltzmann equation (61.21a) in the one-dimensional stationary case with the various quantities expressed in terms of the lethargy u instead of E and integrated with respect to φ from o to 2π, runs

$$\left.\begin{aligned}&\mu\frac{\partial\Phi(z,\mu,u)}{\partial z}+\Sigma_{\text{tot}}(u)\,\Phi(z,\mu,u)\\&\quad=\int_0^{2\pi}d\varphi\int d\boldsymbol{\Omega}'\int du'\,\Phi(z',\boldsymbol{\Omega}',u')\,\Sigma_s(u'\to u,\boldsymbol{\Omega}'\to\boldsymbol{\Omega})+\tfrac{1}{2}S(z,u)\end{aligned}\right\}\tag{79.1}$$

where

$$\Phi(z,\mu,u)=\int_0^{2\pi}\Phi(z,\boldsymbol{\Omega},u)\,d\varphi=2\pi\,\Phi(z,\boldsymbol{\Omega},u)\tag{79.2}$$

and $S(z, u)$ denotes the strength of the source after integration with respect to μ.

Most methods for solving the integro-differential equation (79.1) involve the expansion in Legendre polynomials of both the scattering cross section and the neutron flux density. This can be done by proceeding as follows. First the expansion (72.20) is introduced into (79.1):

$$\begin{aligned}&\mu\frac{\partial\Phi(z,\mu,u)}{\partial z}+\Sigma_{\text{tot}}(u)\,\Phi(z,\mu,u)\\&\quad=\sum_{l=0}^{\infty}\frac{2l+1}{4\pi}\int_0^{2\pi}d\varphi\int_0^{2\pi}d\varphi'\int d\mu'\int du'\,\Phi(z,\boldsymbol{\Omega}',u')\,B_l(u',u-u')\,P_l(\mu_0)+\frac{1}{2}S(z,u).\end{aligned}$$

---

[1] I. WALLER: Ark. Mat. Astronom. Fys., Ser. A **34**, No. 3 (1946); Ser. A **34**, No. 5 (1946).

[2] M. VERDE and G.C. WICK: Phys. Rev. **71**, 852 (1947).

[3] R.E. MARSHAK, H. BROOKS and H. HURWITZ: Nucleonics **5** (1), 53 (1949); **5** (2), 59 (1949).

[4] H. HURWITZ jr., and P.F. ZWEIFEL: J. Appl. Phys. **26**, 923 (1955).

[5] H. HURWITZ and R.E. EHRLICH: Progress in Nuclear energy. Physics and Mathematics, Vol. 1, p. 343. New York and London: Pergamon Press 1956.

Then the Legendre polynomial $P_l(\mu_0)$ is expressed in terms of μ and μ' by means of the addition theorem for Legendre functions[1] and the integrations with respect to φ and φ' are performed in such a way that the terms in $\cos m(\varphi-\varphi')$ go to zero

$$\left.\begin{aligned}&\mu\frac{\partial\Phi(z,\mu,u)}{\partial z}+\Sigma_{\text{tot}}(u)\,\Phi(z,\mu,u)\\&=\sum_{l=0}^{\infty}\frac{2l+1}{2}\int du'\,B_l(u',u-u')\,P_l(\mu)\int_{-1}^{+1}\Phi(z,\mu',u')\,P_l(\mu')\,d\mu'+\tfrac{1}{2}\,S(z,u).\end{aligned}\right\}\tag{79.3}$$

Finally $\Phi(z,\mu,u)$ is also expanded in Legendre polynomials

$$\Phi(z,\mu,u)=\sum_{l'=0}^{\infty}\frac{2l'+1}{2}\,\Phi_l(z,u)\,P_l(\mu)\tag{79.4}$$

and the integration over μ' is performed. Since

$$\int_{-1}^{+1}P_l(\mu)\,P_{l'}(\mu)\,d\mu=\frac{2}{2l+1}\,\delta_{ll'},$$

the result is

$$\mu\frac{\partial\Phi(z,u,u)}{\partial z}+\Sigma_{\text{tot}}(u)\,\Phi(z,\mu,u)=\sum_{l=0}^{\infty}\frac{2l+1}{2}\,J_l(z,u)\,P_l(\mu)+\frac{1}{2}\,S(z,u)\tag{79.5}$$

where

$$J_l(z,u)=\sum_i J_l^i(z,u)\tag{79.6a}$$

is a sum extending over all types of nuclei present in the moderator, and

$$J_l^i(z,u)=\int_{u_i^*}^{u}\Phi_l(z,u')\,B_l^i(u',u-u')\,du'\tag{79.6b}$$

is the Legendre component of the degradation integral for the nuclei of type i with

$$\left.\begin{aligned}&u_i^*=0 &&\text{for}\quad\begin{cases}A=1\,(\text{hydrogen})\\A>1\ \text{and}\ u\leqq X_i,\end{cases}\\&u_i^*=u-X_i &&\text{for}\quad A>1\ \text{and}\ u>X_i.\end{aligned}\right\}\tag{79.7}$$

Although not strictly necessary in many cases, it is costumary to express the space dependence of the neutron flux density as a Fourier integral

$$\Phi(z,\mu,u)=\frac{1}{2\pi}\int_{-\infty}^{+\infty}\phi(k,\mu,u)\,e^{-ikz}\,dk.\tag{79.8a}$$

The Fourier transform of (79.4)

$$\phi(k,\mu,u)=\int_{-\infty}^{+\infty}\Phi(z,\mu,u)\,e^{ikz}\,dz=\sum_{l=0}^{\infty}\frac{2l+1}{2}\,\phi_l(k,u)\,P_l(\mu)\tag{79.8b}$$

---

[1] See for instance: p. 115 of E. JAHNKE and F. EMDE, Tables of functions. New York: Dover Publications 1945.

$$P_l(\mu_0)=P_l(\mu)\,P_l(\mu')+2\sum_{m=1}^{l}\frac{(l-m)!}{(l+m)!}\,P_l^m(\mu)\,P_l^m(\mu')\cos m(\varphi-\varphi')$$

where

$$\mu_0=\cos\vartheta_0,\quad \mu=\cos\vartheta,\quad \mu'=\cos\vartheta'$$

are such that

$$\cos\vartheta_0=\cos\vartheta\cos\vartheta'+\sin\vartheta\sin\vartheta'\cos(\varphi-\varphi').$$

satisfies the following integral equation obtained by multiplication of (79.5) with $e^{-\imath k z}$ followed by integration over z

$$\left.\begin{aligned}[1-\eta(u)\,\mu]\,\Sigma_{\text{tot}}\sum_{l=0}^{\infty}(2l+1)\,\phi_l(k,u)\,P_l(\mu)\\ =\sum_{l=0}^{\infty}(2l+1)\,J_l(k,u)\,P_l(\mu)+\tfrac{1}{2}\,S(k,u)\end{aligned}\right\}\tag{79.9a}$$

where

$$\eta(u)=\frac{\imath k}{\Sigma_{\text{tot}}(u)}\tag{79.9b}$$

and

$$S(k,u)=\int_{-\infty}^{+\infty}S(z,u)\,e^{\imath k z}\,dz.\tag{79.9c}$$

Eq. (79.9a) is equivalent to an infinite set of coupled Volterra integral equations in the unknown variable $\phi_l(k,u)$. The various methods considered below differ from one another in the procedure adopted to reduce this infinite set to a finite one.

In the majority of the methods, the infinite set of coupled integral equations is obtained by multiplication of Eq. (79.9a) with $P_l(\mu)$ followed by integration over μ

$$\left.\begin{aligned}-\eta(u)\frac{1}{2l+1}\left[(l+1)\,\phi_{l+1}(k,u)+l\,\phi_{l-1}(k,u)\right]+\\ +\phi_l(k,u)=\frac{J_l(k,u)}{\Sigma_{\text{tot}}(u)}+\frac{S(k,u)}{\Sigma_{\text{tot}}(u)}\delta_{l,0}.\end{aligned}\right\}\tag{79.10}$$

The *spherical harmonics P_L-approximation*[1] [*16*] consists in introducing in the equation with $l=L$ of the infinite set (79.10)

$$\phi_{L+1}(k,u)=0\tag{79.11}$$

so that the infinite set is reduced to a set of $L+1$ integral equations which can be used to calculate $\phi_0, \phi_1, \ldots, \phi_L$.

The *spherical harmonics B_L-approximation*[2,3] is obtained if one first divides both sides of (79.9a) by $[1-\eta(u)\,\mu]$ and then multiplies with $P_l(\mu)$ and integrates over the solid angle. The infinite set of coupled equations now becomes

$$\Sigma_{\text{tot}}(u)\,\phi_l(k,u)=\sum_{l'=0}^{\infty}(2l'+1)\,A_{ll'}(\eta)\,[J_{l'}(k,u)+S(k,u)\,\delta_{l'0}]\tag{79.12}$$

where

$$A_{ll'}(n)=\frac{1}{2}\int_{-1}^{+1}\frac{P_l(\mu)\,P_{l'}(\mu)}{1-\eta\mu}\,d\mu.\tag{79.13}$$

The approximation is made that

$$J_l^{\imath}=0\quad\text{for}\quad l>L\tag{79.14}$$

so that one replaces ∞ by L as upper limit in the sum appearing in (79.12).

A discussion of the convergence of these two procedures shows that the B_L-approximation, which neglects J_{L+1}, is superior to the P_L-approximation which neglects ϕ_{L+1}; the P_L procedure may be considered to be an approximation

[1] See footnote 3, p. 284.
[2] See footnote 4, p. 284.
[3] See footnote 5, p. 284.

to the B_L-method because Eq. (79.12) reduces to (79.10) if the integrals (79.13) are evaluated by an $(L+1)$-points Gaussian integration formula[1,2].

The set of $L+1$ coupled integral equations obtained either in the P_L-approximation or in the B_L-approximation may be solved numerically making use of the experimental values of the cross sections as a function of the lethargy u.

In the calculation, the integrals (79.6b) can still be treated differently according to the mass of the corresponding nuclei.

In the case of heavy scatterers the integration interval $(u-X, u)$ is so small that it may become permissible to expand the integrand of (79.6b) in a TAYLOR'S series in $u'-u$ and keep only a few terms

$$J_l^i(k,u) = \sum_{n=0,1,\dots} \int\limits_{u-X_i}^{u} du' \frac{(u'-u)^n}{n!} \frac{\partial^n}{\partial u^n} [\phi_l(k,u)\, B_l^i(u,u'-u)]. \qquad (79.15)$$

When such an approximation is permissible, the degradation integrals are transformed into differential expressions.

As we shall discuss in greater detail later, in the hydrogen case the approximation (79.15) is certainly not allowed. A simplification of the degradation integral is obtained by taking advantage of the isotropy of the scattered neutrons in the c.m.s. at the energies in which we are interested here. Introducing (72.28) in (79.6b) the hydrogen degradation integral turns out to be

$$J_l^{\mathrm{H}}(k,u) = \int\limits_0^u \Sigma_s^{\mathrm{H}}(u')\, \phi_l(k,u')\, P_l\left(\mathrm{e}^{-\frac{\xi}{2}}\right) \mathrm{e}^{-\xi}\, du', \qquad \xi = u-u'. \qquad (79.16)$$

β) The age theory approximation. We can now show under what assumptions the Boltzmann equation (79.1) reduces to the age equation (75.17) in one dimension.

In the following derivation the source term appearing in (79.1) will be omitted because the neutron source has been introduced in the age theory by the initial condition (76.1).

The assumptions are the following:

1. The flux density $\Phi(z,\mu,u)$ is assumed to be almost isotropic so that it is adequately represented by a P_1-approximation. The infinite set of coupled integral equations (79.10) is reduced to two equations

$$i k \phi_1 + \Sigma_{\mathrm{tot}}(u)\, \phi_0 = \sum_i J_0^i = J_0(k,u)$$

$$i\frac{k}{3}\phi_0 + \Sigma_{\mathrm{tot}}(u)\, \phi_1 = \sum_i J_1^i = J_1(k,u)$$

which, by inversion of the Fourier transform, can be written

$$\left.\begin{aligned} \frac{\partial \Phi_1}{\partial z} + \Sigma_a(u)\, \Phi_0 &= -\frac{\partial Q}{\partial u}, \\ \frac{1}{3}\frac{\partial \Phi_0}{\partial z} + [\Sigma_a(u) + \Sigma_s(u)]\, \Phi_1 &= J_1(z,u) \end{aligned}\right\} \qquad (79.17)$$

where $\partial Q/\partial u$ is defined by the relation

$$-\frac{\partial Q}{\partial u} = J_0(z,u) - \Sigma_s(u)\, \Phi_0(z,u) = \int \Sigma_s(u'\to u)\, \Phi_0(z,u')\, du' - \Sigma_s(u)\, \Phi_0(u) \qquad (79.18)$$

---

[1] See footnote 5, p. 284.

[2] G.C. WICK: Z. Physik **121**, 702 (1943). — Phys. Rev. **75**, 738 (1949).

with

$$\Sigma_s(u' \to u) = \Sigma_s(u')\, f(u' \to u) = B_0(u', \xi) = \int \Sigma_s(u' \to u, \boldsymbol{\Omega}' \to \boldsymbol{\Omega})\, d\boldsymbol{\Omega}.$$

2. The integrands of $J_0^i(z, u)$ and $J_1^i(z, u)$ are assumed to vary so slowly over one slowing down interval $(u - X_i, u)$ that in their TAYLOR's series expansions (79.15) it is permissible to keep only two terms for J_0^i and only the first one for J_1^i; thus one obtains

$$\left.\begin{aligned} J_0^i(z, u) &= \Sigma_s^i(u)\, \Phi_0(z, u) - \frac{\partial}{\partial u}\, [\bar{\xi}_i \Sigma_s^i(u)\, \Phi_0(z, u)] + \cdots, \\ J_1^i(z, u) &= \overline{\mu_0^i}\, \Sigma_s^i(u)\, \Phi_1(z, u), \end{aligned}\right\} \tag{79.19}$$

with $\bar{\xi}_i$ and $\overline{\mu_0^i}$ defined by means of (62.15) and (72.10)[1].

By introducing the approximate expression (79.19) in (79.17) and (79.18), these integral equations are transformed into two couples first order differential equations

$$\frac{\partial \Phi_1}{\partial z} + \Sigma_a(u)\, \Phi_0 = -\frac{\partial Q}{\partial u}, \tag{79.20}$$

$$\Phi_1(z, u) = -D \frac{\partial \Phi_0}{\partial z} \tag{79.21}$$

where

$$Q(z, u) = \sum_i \bar{\xi}_i \Sigma_s^i(u)\, \Phi_0(z, u) \tag{79.22}$$

is the usual approximate expression (75.7) for the slowing down density and

$$D = \frac{D_0}{v} = \frac{l_d}{3} = \frac{1}{3}\, \frac{1}{\sum_i [\Sigma_a^i(u) + \Sigma_s^i(u)\, (1 - \overline{\mu_0^i})]} \tag{79.23}$$

---

[1] Introducing (72.24) in (79.15) one obtains (see footnote 4, p. 284.)

$$J_l(k, u) = \sum_{l'=0}^{\infty} \sum_{n=0}^{\infty} \frac{\partial^n}{\partial u^n}\, [\phi_l(k, u)\, B_{l'}^*(u)]\, T_{ll'}^n$$

with $B_l^*(u)$ defined by (72.23) and the constant coefficients $T_{ll'}^n$ defined by

$$T_{ll'}^n = \frac{(-1)^n}{n!} \int_0^X \xi^n\, T_{ll'}(\xi)\, d\xi$$

where $T_{ll'}(\xi)$ is given by (72.25). From (72.23) and (72.14) it follows that

$$\sum_{l'=0}^{\infty} B_{l'}^*(u)\, T_{0l'}^0 = B_0^*(u) = \Sigma_s^*(u) = \Sigma_s(u),$$

$$\sum_{l'=0}^{\infty} B_{l'}^*(u)\, T_{0l'}^1 = -2\pi \int_{-1}^{+1} \xi\, \Sigma_s^*(u, \mu_0^*)\, d\mu_0^* = -\bar{\xi}\, \Sigma_s(u),$$

$$\sum_{l'=0}^{\infty} B_{l'}^*(u)\, T_{0l'}^2 = \frac{1}{2}\, 2\pi \int_{-1}^{+1} \xi^2\, \Sigma_s^*(u, \mu_0^*)\, d\mu_0^* = \frac{\overline{\xi^2}}{2}\, \Sigma_s(u),$$

. . . .

$$\sum_{l'=0}^{\infty} B_{l'}^*(u)\, T_{1l'}^0 = 2\pi \int_{-1}^{+1} \mu_0\, \Sigma_s^*(u\, \mu_0^*)\, d\mu_0^* = \bar{\mu}_0\, \Sigma_s(u),$$

$$\sum_{l'=0}^{\infty} B_{l'}^*(u)\, T_{1l'}^1 = -2\pi \int \xi\, \mu_0\, \Sigma_s^*(u, \mu_0^*)\, d\mu_0^* = -\overline{(\mu_0\, \xi)}\, \Sigma_s(u),$$

. . . .

is the diffusion coefficient appearing in (79.21) which simply expresses FICK's law (75.2) (see also Part E, Division I of this article). By elimination of Φ_1 between (79.20) and (79.21) and making use of (79.23) and (61.16) a second order differential equation is obtained identical to (75.16) in one dimension.

From the short summary given above we see that the derivation of the age equation does not involve the assumption of isotropic scattering in the c.m.s. Therefore the age theory is valid in the general case of any angular distribution (72.23) of the scattered neutrons in the c.m.s.; in this general case both $\bar{\xi}$ and $\bar{\mu}_0 = \overline{\cos\vartheta}$ defined respectively by (62.13) and (72.10) will be energy dependent.

γ) The range of applicability of the age theory. We shall now estimate the range of applicability of the age theory. Assumption (1), i.e. the P_1-approximation, is valid only for almost isotropic neutron distribution, i.e. only if

$$\Phi_1(z, u) \ll \Phi_0(z, u).$$

Introducing the solution (76.14) given by the age theory in (79.21), we find that this condition is satisfied only if

$$z \ll \frac{2\tau}{D} = 6\,\frac{L_s^2(u)}{l_{\mathrm{tr}}(u)}. \tag{79.24a}$$

This inequality refers to the validity of the P_1-approximation for a particular lethargy u. If the age theory is to give a correct answer, the P_1-approximation must have been valid during almost the whole slowing down of the neutrons to the lethargy u. This reduces somewhat the range of applicability of the age theory; it may be shown [7] that, by taking this effect into account, the inequality (79.24a) must be replaced by

$$z \ll 6\,\frac{L_s^2(u)}{(l_{\mathrm{tr}})_{\mathrm{m\,p.}}} \tag{79.24b}$$

where $(l_{\mathrm{tr}})_{\mathrm{m.p.}}$ is the value of l_{tr} for the most penetrating neutrons in the velocity interval (v_0, v). A justification of (79.24b) may be obtained by comparing the distribution of the most penetrating neutrons to the Gaussian distribution given by the age theory. For instance, when l_{tr} takes its largest value for $v = v_0$, the distribution of the most penetrating neutrons is given by (77.1) which becomes of the same order of magnitude as (76.16) for values of r of the order of the right-hand side of (79.24b).

By considering the lethargy dependence of the right-hand side of (79.24b), the failure of the age theory is recognized in the two following extreme cases: (a) for neutron of energy very close to the initial energy E_0 (τ very small) at any distance from the source, and (b) at sufficiently large distance from the source for neutrons of any given energy.

In order to visualize these two extreme cases we may consider an isotropic point source emitting monoenergetic neutrons of energy E_0 in an infinite medium. The neutrons of energy close to E_0 will show an angular distribution strongly peaked in the direction of emission from the source because they have suffered either no collision, or only small angle deflections; therefore their angular distribution will be very poorly represented at any distance from the source, by an interrupted spherical harmonic expansion.

This remark does not affect, however, the distribution of the neutrons of sufficiently large age τ observed at sufficiently small distances from the source, since they originate through a large number of collisions from primary neutrons which have started—immediately after emission—to undergo large angle deflections.

Case (b) is understood by using the qualitative description of neutron deep penetration as sketched in Sects. 14 and 77 and further discussed in Sect. 80. According to this picture at a great distance $\boldsymbol{r}$ from the source, the space distribution of neutrons of any given lower energy is *controlled* by the space distribution of the primary neutrons, i.e. of those among "the most penetrating neutrons", that have travelled from the region of the source to the neighbourhood of $\boldsymbol{r}$, experiencing no collision at all or only small angle deflections. The angular distribution of the primary neutrons is always anisotropic, a feature that is reflected in their space distribution which becomes closer to an exponential than to a Gaussian distribution.

One could object that for any given large value of the distance from the source, there are neutrons of such a low energy (i.e. such a large τ) that condition (79.24b) is fulfilled. But the space distribution even of these neutrons is not given correctly by the age theory for large values of r, because they are in *quasi-equilibrium* with their ancestors the space distribution of which cannot be described in this approximation.

A more detailed description of this picture requires a specification of the energy dependence of the total mean free path of the neutrons in the medium. If, for instance, $l_{\text{tot}}(E)$ has its largest value for $E = E_0$ (Sect. 80) the most penetrating neutrons are those that after emission from the source have suffered either no collision or only small angle deflections. As pointed out above, their angular distribution is strongly anisotropic at any distance from the source.

The predominance of the direction away from the source still exists even when $l_{\text{tot}}(E)$ has its largest value for a neutron energy E_1 such that

$$E_0 > E_1 > E.$$

In this case the most likely fate of the neutrons of energy E observed in the neighbourhood of a point located at a great distance r from the source may be described as follows. The neutron emitted with energy E_0 are slowed down in the vicinity of the source. When they reach values of the energy close to E_1, they have a probability of travelling to great distances r, without undergoing any collision at all or only small angle deflections, larger than the neutrons of higher or lower energies. The slowing down from the energy $\sim E_1$ to the energy E under consideration, takes place in the neighbourhood of r. The angular distribution of the neutrons of energy $\sim E_1$ is strongly anisotropic for values of r much larger than the linear dimensions of the region around the source where the slowing down from E_0 to E_1 takes place.

The conditions of validity of assumption (2) are the following ones: firstly the interval X inside which the broken off TAYLOR's series (79.15) are used must be small, i.e. the atomic weight of all nuclei present in the moderator must be large; as mentioned above, the moderation in hydrogeneous media can for this reason not be handled by means of the age theory. Secondly the collision density must change slowly in an interval $(u - X_\iota, u)$, a condition which can be satisfied only if the scattering mean free path varies slowly in a lethargy range comparable to $\bar{\xi}_\iota$. This condition again excludes application of the age theory to the case of hydrogeneous media.

As was pointed out by WIGNER [*19*] the concomitance of the two conditions —the moderator nuclei must be heavy and the collision density must vary slowly— renders incorrect the extension of the age theory to the case of absorbing media. The heavier is the scattering nucleus, the more collisions a neutron will suffer in any energy interval, and if the moderator has an appreciable capture cross section in the energy interval considered, the greater will be the probability for the neutron to be captured. But an appreciable absorption of neutrons in a slowing down interval X_ι means a rapid variation of the collision density as a function of the lethargy u and therefore the use of the broken off TAYLOR'S expansion (79.15) is not justified. This is the argument mentioned at the end of Sect. 75 where the age theory was extended to the case of weakly absorbing moderators.

δ) The Selengut-Goertzel procedure. A simple improvement which can be made with respect to the age theory when hydrogen is present in the moderator is the Selengut-Goertzel procedure[1-4]. The S.G. method, as it is usually called, consists again in a P_1-approximation with the degradation integrals J_1^ι and J_0^ι evaluated by means of the broken off TAYLOR'S series (79.15), with the exception of $J_0^{\mathrm{H}}(k, u)$ (but not of J_1^{H}) which is left in the integral rigorous form (79.16). By elimination of $\phi_1(k, u)$ between the two integro-differential equations obtained under these assumptions from (79.10), an equation is derived which contains, besides terms in $\phi_0(k, u)$ and its derivatives with respect to u, a term in

$$J_0^{\mathrm{H}}(k, u) = \int_0^u e^{-\xi} \Sigma_s^{\mathrm{H}}(u') \, \phi_0(k, u') \, du' \tag{79.25}$$

which, as one can immediately verify, satisfies the equation

$$J_0^{\mathrm{H}} + \frac{\partial J_0^{\mathrm{H}}}{\partial u} = \Sigma_s^{\mathrm{H}}(u) \, \phi_0(k, u). \tag{79.26}$$

If one assumes that all of the energy degradation takes place in collisions with hydrogen, i.e. that

$$\bar{\xi}_\iota = 0 \quad \text{for} \quad i \neq \mathrm{H}, \tag{79.27}$$

the slowing down density $Q(k, u)$ reduces to the hydrogen contribution, i.e. is given by

$$Q(k, u) = J_0^{\mathrm{H}}(k, u). \tag{79.28}$$

By introducing (79.28) in (79.26) and combining this with the original integro-differential equation, a differential equation in $Q(k, u)$ is obtained; this equation can easily be solved.

This procedure makes use of the correct energy loss suffered by a neutron in a collision with protons (79.25) but neglects the correlation between energy loss and angle of scattering. Because of this approximation the corresponding expression of the neutron age is not correct. This expression is similar to the one given by the age diffusion theory except that it includes a term, representing the first flight correction (Sect. 77), which is however too large[3] compared with the corresponding term appearing in the rigorous expression for the age obtained by the moments method (Sect. 80).

ε) A few numerical results. Table 42 taken from the review article by HURWITZ and EHRLICH[3], shows the values of the age to In-resonance (1.46 ev) of fission neutrons in water computed by various methods. The first flight correction introduced by the S.G. method with respect to the age diffusion theory (applied with $\bar{\xi} = 1$), amounts to 15.3 cm² while the rigorous correction [in the approxima-

[1] G. GOERTZEL: ACE TAB-53 (1950).
[2] See footnote 4, p. 284.
[3] See footnote 5, p. 284.
[4] J.E. WILKINS, R.L. HELLENS and P.F. ZWEIFEL: p. 62 of Ref. [15c].

Table 42. *Computed values of the age τ [cm²] of fission neutrons in water to indium resonance.*

Approximation	Rigorous expression cm²	S.G. method cm²	Age diffusion cm²
No approximation . . . . (a)	25.3	30.9	
Oxygen scattering assumed isotropic (b)	23.6	28.8	
Oxygen assumed not to slow down . (c)	26.6	32.8	17.5
Approximations (b) and (c) both made (d)	25.3		

tion (c)] is 9.1 cm². Therefore the fact that the S.G. value of τ in case (a) turns out to be very close to the experimental results (see Sect. 83; $\tau \approx 30$ cm²) must be regarded as a coincidence.

The so-called modified age diffusion method and modified S.G. method are obtained by introducing the same approximations discussed above on the B_1 rather than on the P_1 approximation[1]. These modified methods may be put in the same form as the corresponding usual methods by conveniently redefining the transport cross section.

A comparison of the numerical results obtained by applying the above mentioned schemes to the slowing down of fission neutrons in water, has been made by WILKINS and others[2]. The slowing down density transform $Q(k, u)$ has been computed numerically for the In-resonance neutrons as a function of k, and for $k=0.1$ cm^{-1} as a function of u, by means of B_1, B_2, B_3, P_1, P_2, P_3, S.G. and modified S.G. methods. The ratio ϱ of the slowing down density computed by means of all other procedures to that computed by means of the B_3-approximation is tabulated and used for comparison. The following conclusions are derived.

Except at extremely large values of k the B_1 approximation is sufficiently accurate for most applications; for $k=0.1$ cm^{-1}, $1.00 \geqq \varrho \geqq 0.99$ for all considered values of u, and for In resonance neutrons ϱ decreases with increasing k, reaching a value $\varrho=0.69$ for $k=1.3$ cm^{-1}.

The P_1 and P_2 calculations are quite poor except for very small k; for In resonance neutrons and $k=0.8$ cm^{-1}, $\varrho=0.048$ for the P_1-approximation and $\varrho=1.56$ for the P_2-approximation.

The P_3 calculation is not much better than the B_1 which involves an amount of labor which may be estimated to be about two orders of magnitude lower, and is significantly poorer than a B_2 calculation.

The S.G. approximation gives rather poor results even for small values of k; for In resonance neutrons $\varrho=0.67$ for $k=0.5$ cm^{-1}, $\varrho=0.94$ for $k=0.1$ cm^{-1}. The modified S.G. method represents an improvement over the ordinary S.G. procedure.

According to these authors the S.G. procedure represents a reasonable method of performing the calculation of slowing down of neutrons in hydrogeneous media in spite of its low accuracy because of the ease of the corresponding computations. The modified S.G. method introduces very little additional complications.

In Fig. 130, there is plotted the age τ ($E=1.46$ ev) of neutrons slowed down to 1.46 ev from an isotropic point source emitting monoenergetic neutrons of energy E_0, as computed by the S.G. procedure. In the same figure there is plotted the product of τ and the fission spectrum (82.6a). From this curve one can recognize the importance of the contribution due to the high energy tail of the spectrum to the value of the age of the In resonance neutrons obtained with a fission source embedded in water.

[1] See footnote 5, p. 284.
[2] See footnote 4, p. 291.

Finally we shall mention the results obtained by HOLTE[1] for the collision density $\Psi_0(\overline{\boldsymbol{r}}, E)$ produced by a monoenergetic point source embedded in a homogeneous non-absorbing medium, in which the neutrons are slowed down only by elastic collisions, assuming isotropy in the c.m.s.

We shall not try to sketch here the analytical procedure followed by this author who starts from the system of difference equations obtained by applying a Laplace transform[2,3] with respect to the lethargy u to Eq. (79.10); the reader is directed for details to the original papers as well as to Chap. 26 of Ref. [*7*]. Here we limit ourselves to recalling that, in his first paper, HOLTE assumes that the scattering mean free path is a constant, but in the second paper he allows l_s to vary as a function of the lethargy according to the convenient and simple expression (74.7) [*16*]. The analytic expressions of $\Psi_0(\boldsymbol{r}, E)$ are valid for all r when E is small; they are rather complicated so that the results must be given by numerical tables for any practical use. The computations have been made for neutrons slowed down to 1 ev. But in order to make the tables useful for other values of E near $E=1$ ev, HOLTE has also computed the first derivative term in the Taylor series for the variable u around $E=1$ ev (i.e. $u=\ln E_0$). Thus, the collision density of neutrons of energy E is written in the form

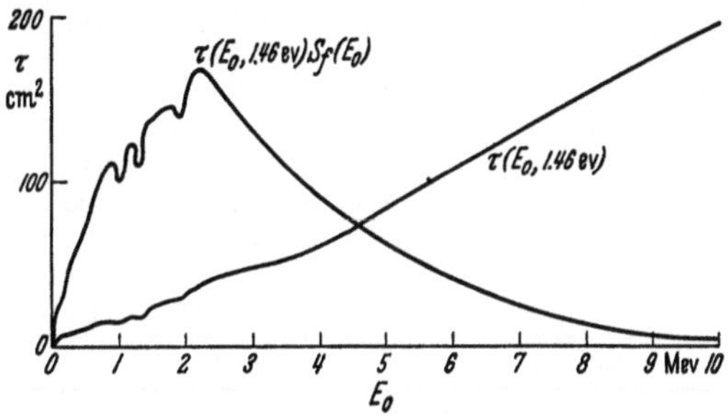

Fig. 130. The age of fission neutrons slowed down in water to the In-resonance computed by the S.G. method (WILKINS *et al.*).

$$\Psi_0(r, E) = \frac{Q_0}{E} 10^{-4} G(r) (1 + \varkappa \Delta u) \tag{79.29}$$

where Q_0 is the number of neutrons emitted by the source per unit of time, $G(r)$ is proportional to the collision density distribution of the neutrons of 1 ev and $\Delta u = -\ln E$ with E in ev. Tables 43 and 44 contain the function $G(r)$ and its logarithmic derivative $\varkappa$ for graphite and light water for a few values of the initial energy E_0 and of the distance r. Also the values of $G_a(r)$, deduced from the age theory, are given for comparison in the same tables.

In the case of carbon (1.60 g/cm³) the scattering mean free path has been taken as constant and equal to $l_s = 2.65$ cm for $E_0 = 0.1$ Mev; for higher values of E_0, l_s is assumed to increase linearly with the neutron energy according to

$$l_s(E)\,[\text{cm}] = 2.65 + 2.35\,E\,[\text{Mev}] \tag{79.30}$$

which constitutes a good approximation up to 2.5 Mev. In Table 43 three figures are always given for $G(r)$, but the last one is uncertain in most cases. Because of the large mass of the carbon nucleus and the moderate variation of l_s with energy, the age theory is a rather good approximation for carbon, except that at very great distances from the source where the neutron distribution given by the age theory decreases too rapidly (Sect. 77).

In the case of water (Table 44) l_s varies very rapidly and was represented by HOLTE by the empirical formula

$$l_s(E)\,[\text{cm}] = 0.65 + 2.09\,E^{\frac{3}{4}}\,[\text{Mev}] \tag{79.31}$$

[1] G. HOLTE: Ark. Fysik **2**, 523 (1951); **3**, 209 (1951); **8**, 165 (1954).

[2] See footnote 1, p. 284.

[3] See footnote 2, p. 284.

Table 43. *Carbon.*

r (cm)	0	10	20	30	50	80	150
$E_0 = 0.1$ Mev:							
$G(r)$	0.623	0.534	0.339	0.163	0.0183	$1.74 \cdot 10^{-4}$	$7.24 \cdot 10^{-11}$
$G_a(r)$	0.587	0.511	0.336	0.168	0.0181	$0.79 \cdot 10^{-4}$	$1.47 \cdot 10^{-14}$
$\varkappa$	−0.13	−0.12	−0.08	−0.01	0.15	0.44	1.05
$E_0 = 0.5$ Mev:							
$G(r)$	0.450	0.398	0.276	0.152	0.0250	$4.85 \cdot 10^{-4}$	$1.32 \cdot 10^{-9}$
$G_a(r)$	0.433	0.386	0.275	0.156	0.0253	$3.01 \cdot 10^{-4}$	$3.44 \cdot 10^{-12}$
$\varkappa$	−0.11	−0.10	−0.07	−0.03	0.09	0.29	0.75
$E_0 = 1.0$ Mev:							
$G(r)$	0.370	0.331	0.241	0.142	0.0288	$8.58 \cdot 10^{-4}$	
$G_a(r)$	0.359	0.324	0.240	0.145	0.0292	$5.86 \cdot 10^{-4}$	
$\varkappa$	−0.10	−0.09	−0.07	−0.03	0.06	0.22	
$E_0 = 2.0$ Mev:							
$G(r)$	0.269	0.246	0.189	0.123	0.0330	$1.88 \cdot 10^{-3}$	$2.33 \cdot 10^{-7}$
$G_a(r)$	0.264	0.243	0.190	0.126	0.0342	$1.42 \cdot 10^{-3}$	$2.77 \cdot 10^{-9}$
$\varkappa$	−0.08	−0.08	−0.06	−0.04	0.03	0.13	0.24

Table 44. *Water.*

r(cm)	0	3	6	12	20	30	50	80
$E_0 = 0.2$ Mev:								
$G(r)$	19.1	11.7	3.40	0.098	$3.2 \cdot 10^{-4}$	$1.7 \cdot 10^{-7}$	$2.5 \cdot 10^{-14}$	
$G_a(r)$	17.7	12.1	3.85	0.040	$7.8 \cdot 10^{-7}$	$4.9 \cdot 10^{-16}$	$1.8 \cdot 10^{-45}$	
$\varkappa$	−0.09	−0.06	0.01	0.13	0.22	0.27	0.29	
$E_0 = 0.5$ Mev:								
$G(r)$	12.3	8.29	3.17	0.198	$2.72 \cdot 10^{-3}$	$1.17 \cdot 10^{-5}$	$2.2 \cdot 10^{-10}$	
$G_a(r)$	11.4	8.56	3.65	0.121	$3.8 \cdot 10^{-5}$	$5.3 \cdot 10^{-12}$	$6.3 \cdot 10^{-34}$	
$\varkappa$	−0.08	−0.05	−0.01	0.07	0.10	0.11	0.12	
$E_0 = 1.0$ Mev:								
$G(r)$	7.59	5.78	2.70	0.312	$1.20 \cdot 10^{-2}$	$2.46 \cdot 10^{-4}$	$1.1 \cdot 10^{-7}$	$1.2 \cdot 10^{-12}$
$G_a(r)$	6.74	5.51	3.02	0.273	$9.2 \cdot 10^{-4}$	$1.3 \cdot 10^{-8}$	$4.6 \cdot 10^{-24}$	$9 \cdot 10^{-62}$
$\varkappa$	−0.06	−0.04	−0.01	0.04	0.05	0.05	0.05	0.05
$E_0 = 2.0$ Mev:								
$G(r)$	4.54	3.54	1.91	0.364	0.0384	$2.48 \cdot 10^{-3}$	$1.3 \cdot 10^{-5}$	$5.9 \cdot 10^{-9}$
$G_a(r)$	3.17	2.81	1.95	0.456	0.0145	$1.7 \cdot 10^{-5}$	$8 \cdot 10^{-15}$	$1.2 \cdot 10^{-37}$
$\varkappa$	−0.04	−0.03	−0.01	0.01	0.02	0.02	0.02	0.02

which takes into account the contribution of oxygen but implies that the scattering cross sections of hydrogen and oxygen vary with energy in the same manner. Except when E is near an oxygen resonance (Sect. 29), this formula is not too bad an approximation in the interval 0 to 2.5 Mev.

HOLTE's results in the case of water are, however, of limited practical use because of the assumption of isotropic scattering in the c.m.s.; Fig. 24 on p. 78 shows that the anisotropy of the neutrons scattered by oxygen is already appreciable at $E = 0.4$ Mev and increases in a rather complicated manner, by increasing the energy of the neutrons. A comparison of the values of $G(r)$ and $G_a(r)$ given in Table 44 shows clearly that the age theory approximation is practically useless in the case of water, as we expected according to our previous considerations.

80. A few remarks about other methods for the approximate solution of the Boltzmann equation: deep penetration problems and method of moments. In the discussion of the limitations inherent in the age theory, as given in the preceding section, we have shown that the assumption of almost isotropy of the angular distribution of the neutrons of given energy, necessary for the validity of the P_1-approximation, is only fulfilled as long as the distance z from the source satisfies condition (79.24b); in fact at distances z of the order of $2\tau/D$ the second term $\Phi_1(z, u)$ in the spherical harmonic expansion of the angular number flux (79.4) becomes of the same order of magnitude as the first term $\Phi_0(z, u)$.

This remark provides a semiquantitative example of the general fact that for sufficiently large values of z, the angular distribution of the neutrons of given energy is so strongly anisotropic, that not only the P_1-approximation, but in general any procedure based on the use of a few terms of the expansion (79.4) becomes inadequate; this may be understood from the qualitative considerations given on p. 289 in the discussion of (79.24).

Therefore the methods for the approximate solution of the Boltzmann equation given in Sect. 79 cannot be used for treating deep penetration problems; these need special procedures which receive a full treatment in Ref. [7]. Monte Carlo methods as well as analytical procedures may be used. The main difficulty involved in the Monte Carlo method is that the number of neutrons found at very large distances in a given energy interval is obviously very small.

The analytical methods however are considerably complicated by the strong anisotropy and its sensitivity to the energy dependence of the cross sections. Therefore only two simple cases have been considered in detail: (a) the total mean free path $l_{\text{tot}}(v)$ of the neutron in the moderator is constant; (b) the total mean free path increases monotonically with increasing neutron energy.

In the following we shall refer to the latter case since it corresponds to the situation actually met in the slowing down of neutrons in the more frequently used moderators (see Sect. 29). In this case the neutrons emitted from the source constitute the most penetrating component among the neutronic radiations moving throughout the moderator.

Therefore one could hope that at great distance from the source a state of radiative equilibrium is reached in which the softer secondary radiations are continually generated from the harder primary radiation (Sects. 29 and 77). According to FANO[1,2] a *radiative equilibrium state* involves the following three features: (1) The intensity ratio of the secondary to the primary radiation approaches asymptotically a maximum value as the distance z from the source increases. (2) The quality of the secondary radiation becomes independent of the depth of penetration because the intensities of its various components of different lethargy bear a constant ratio to the primary intensity. (3) The intensity of the whole radiation is controlled by the attenuation of the primary radiation which, in the plane geometry considered here, follows the exponential law

$$\exp\left[-\frac{z}{l_{\text{tot}}(0)}\right]; \qquad l_{\text{tot}}(0) = l_{\text{tot}}(v_0) = l_{\text{tot}}(E_0). \tag{80.1}$$

A typical radiative equilibrium of this type is reached in the case of a single secondary radiation, the total mean free path of which $l_{\text{tot}}(u)$ differs by a finite amount from the total mean free path of the primaries $l_{\text{tot}}(0)$. The intensity of the secondary radiation, negligibly small at very short distances from the source, increases with increasing z reaching the value

$$\exp\left[-\frac{z}{l_{\text{tot}}(0)}\right]\cdot g(u, \mu) \tag{80.2}$$

[1] U. FANO: J. Res. Nat. Bur. Stand. **51**, 95 (1953).

[2] U. FANO, L. V. SPENCER and M. J. BERGER: Penetration and diffusion of X-rays, in this volume.

for values of z satisfying the condition

$$\left(\frac{1}{l_{\text{tot}}(u)} - \frac{1}{l_{\text{tot}}(0)}\right) \cdot z \gg 1. \tag{80.3}$$

The neutron penetration problem, however, does not lead to a state of equilibrium of this type because neutrons that have suffered collisions only with very small angles of deflection may have an energy (and therefore a total mean free path) which differs by less than any given amount from that of the primary neutrons. This circumstance essentially changes the type of steady state that is actually achieved at great depth; this is sometimes called a *quasi-equilibrium state*. The main difference with respect to an equilibrium state of the type defined above is that the ratio of the intensity of the secondary radiation to that of the primary does not approach a maximum value but grows, with increasing distance, beyond any limit (in contradiction with feature 1); and that at great distances from the source the intensity of the whole radiation is still determined by the attenuation undergone by the primaries but does not follow an exponential law (in contradiction with feature 3). The angular number flux of the neutrons of given lethargy u approaches the asymptotic form

$$\Phi(z, \mu, u) \to f_{\text{pl}}(z) \cdot g(u, \mu) \tag{80.4}$$

where $f_{\text{pl}}(z)$ is different from the expression of the attenuation of the primary radiation (80.1). The validity of the asymptotic expression (80.4) is restricted to those values of z which satisfy condition (80.3) for the value of the lethargy u under consideration.

The actual form of $f_{\text{pl}}(z)$ has been determined by WICK[1] whose work, representing a fundamental step in the understanding of deep penetration problems, is based on the treatment of the order l of the Legendre polynomials as a continuous variable λ. The sum with respect to l appearing in (79.4), is substituted by an integral over λ, and the term in square brackets appearing on the left-hand side of Eq. (79.10) is substituted by a differential operator with respect to λ. Such an approximation gives an adequate description of the actual situation only when the main contribution to $\Phi(z, \mu, u)$ comes from very large values of l.

The derivation of the asymptotic expression of the flux density involves rather complicated analytical developments, for which the reader is directed to the original papers[1] and to Ref. [7]. A detailed discussion of the analogous problem of X-ray deep penetration is given in Sect. 11 of the companion article in this volume.

Assuming that $l_{\text{tot}}(v)$ may be represented by means of a polynomial in the neutron velocity, WICK has shown that

$$f_{\text{pl}}(z) \sim \exp\left[-\frac{z}{l_{\text{tot}}(0)}\right] \cdot \left(\frac{z}{l_{\text{tot}}(0)}\right)^{e} \tag{80.5}$$

where

$$e = -\left(1 + \frac{g_0}{\gamma}\right), \qquad \gamma = \frac{v_0}{l_{\text{tot}}(v_0)} \left(\frac{d l_{\text{tot}}(v)}{d v}\right)_{v=v_0}$$

and g_0 is a number, always negative, obtained as the lowest eigenvalue of a differential equation similar to the Schrödinger equation; the value of g_0 is such that the exponent e usually does not differ much from 1.

[1] G.C. WICK: Phys. Rev. **75**, 738 (1949). — M. VERDE and G.C. WICK: Phys. Rev. **71**, 852 (1947).

The asymptotic behaviour $f_{\text{pt}}(r)$ of the number flux for a point source is obtained by introducing (80.5) in (71.2b)

$$\Phi_0(r, u) \to f_{\text{pt}}(r) \int_{-1}^{+1} g(u, \mu)\, d\mu; \quad f_{\text{pt}}(r) \sim \frac{1}{4\pi r^2}\, e^{-\frac{r}{l_{\text{tot}}(0)}} \cdot \left(\frac{r}{l_{\text{tot}}(0)}\right)^{\varepsilon+1}. \tag{80.6}$$

As we shall see in Division IV, in the interpretation of experimental material most authors do not take account of these theoretical results. They usually omit the factor $(r/l_{\text{tot}})^{\varepsilon+1}$ in (80.6) and use this simplified formula in order to determine that value of the *attenuation length* l'_{tr} which gives the best fit to the empirical data at great distances from the source. Such a procedure does not appear justified in the case of accurate measurements, with monoenergetic neutron sources, extending to values of $r/l_{\text{tot}}(0)$ of the order of ten or more.

Finally the *method of moments* [*16*][1,2] has to be mentioned which allows the calculation of the spatial moments of the neutron distribution to be made directly from the Boltzmann equation.

The spatial moments not only deserve a particular interest because they are frequently used to express the results of experiments on the spatial distribution of neutrons in a moderator. But furthermore, if they are known in sufficient number, the corresponding distribution function can be easily synthetized for instance by means of the polynomial method of SPENCER and FANO[3] (see later).

In order to get the moments of the flux distribution in a medium containing a plane δ-source, one can express the $\phi_l(k, u)$ as power series of k by making use of their definition (79.8):

$$\phi_l(k, u) = \int_{-\infty}^{+\infty} \Phi_l(z, u)\, e^{ikz}\, dz = \int_{-\infty}^{+\infty} \Phi_l(z, u) \sum_{n=0}^{\infty} \frac{(ikz)^n}{n!} = \sum_{n=0}^{\infty} \Phi_{ln}(u)\, i^n k^n \tag{80.7a}$$

where $\Phi_l(z, u)$ is the coefficient of the Legendre polynomial expansion of the angular number flux (79.4) and

$$\Phi_{ln}(u) = \frac{1}{n!} \int_{-\infty}^{+\infty} z^n\, \Phi_l(z, u)\, dz \tag{80.7b}$$

is a generalization of (71.4).

If the expressions (80.7a) are substituted in (79.10) and the coefficients of the various powers of k are equated, a new set of integral equations is obtained in the new unknown variable (80.7b).

The same result however may be obtained without making use of the Fourier transform method. In order to show that, the inverse Fourier transform is applied to (79.10) yielding the following set of coupled integro-differential equations in the Legendre components of the angular number flux $\Phi_l(z, u)$:

$$\left.\begin{aligned} &\frac{1}{2l+1}\left[(l+1)\frac{\partial}{\partial z}\Phi_{l+1}(z, u) + l\frac{\partial}{\partial z}\Phi_{l-1}(z, u)\right] + \Sigma_{\text{tot}}(u)\,\Phi_l(z, u) \\ &\quad = \sum_i \int_{u_i^*}^{u} \Phi_l(z, u')\, B_l^i(u', \xi)\, du' + S(u)\,\delta_{l,0}, \qquad \xi = u - u'. \end{aligned}\right\} \tag{80.8}$$

[1] H. HURWITZ and R. E. EHRLICH: Progress in Nuclear Energy. Physics and Mathematics, Vol. 1, p. 342. London and New York: Pergamon Press 1956.

[2] J.E. WILKINS, R.L. HELLENS and P.E. ZWEIFEL: p. 62 of Ref. [*15c*].

[3] L.V. SPENCER and U. FANO: Phys. Rev. **81**, 464 (1951). — J. Res. Nat. Bur. Stand. **46**, 446 (1951).

The required set of coupled integral equations in the $\Phi_{ln}(u)$ defined by (80.7b) is obtained by multiplying both sides of (80.8) by $z^n/n!$ and integrating with respect to z from $-\infty$ to $+\infty$

$$\left.\begin{aligned}\Sigma_{\text{tot}}(u)\,\Phi_{ln}(u) &= \frac{1}{2l+1}\{(l+1)\,\Phi_{l+1,n-1}(u)+l\,\Phi_{l-1,n-1}(u)\}\\ &= \sum_i \int\limits_{u_i^*}^{u} \Phi_{ln}(u')\,B_l^i(u',\xi)\,du' + S_0\,\delta_{l0}\,\delta_{u0}\end{aligned}\right\} \tag{80.9}$$

where the source has been taken monoenergetic.

The following properties of the moments (80.7b) are of practical importance in looking for solutions of (80.9):

(a) from their definition (80.7b) it follows that

$$\text{for} \quad n<l \quad \Phi_{l,n}(u)=0; \tag{80.10}$$

(b) because of the symmetry of the neutron distribution with respect to the plane $z=0$

$$\Phi(z,\mu,u)=\Phi(-z,-\mu,u),$$

the $\Phi_{ln}(u)$ are identically zero also in the following cases:

$$\left.\begin{aligned}&\text{for } l \text{ even} \quad \Phi_{l,2s+1}(u)=0,\\ &\text{for } l \text{ odd} \quad \Phi_{l,2s}(u)=0,\end{aligned}\right\} \tag{80.11}$$

i.e. the $\Phi_{ln}(u)$ are not identically zero only when n has the same parity character as l.

(c) For $l=0$ the $\Phi_{0,2s}(u)$ represent all the moments of the number flux $\Phi_0(z,u)$ related to $\overline{z^{2s}}$ by Eq. (71.6b).

One of the major advantages of the method of moments lies in the fact that *the moments $\Phi_{0n}(u)$ may be obtained exactly* by *considering only the Eq. (80.9) up to $l=n$.* According to these equations the calculation of $\Phi_{ln}(u)$ requires the knowledge of $\Phi_{l-1,n-1}(u)$ and $\Phi_{l+1,n-1}(u)$ and, because of (80.10), $\Phi_{l+1,n-1}(u)$ is identically zero for $n=l$.

For instance, in order to calculate $\Phi_{02}(u)$ one must calculate $\Phi_{11}(u)$ whose equation involves $\Phi_{00}(u)$ and $\Phi_{20}(u)$ which last vanishes identically because of (80.10). This means that the P_1-approximation of the method of moments gives the correct expression of $\overline{z^2}$ and because of (71.8) of $\tau(u)$.

Similarly the Eqs. (80.9) to determine $\Phi_{04}(u)$ involve $\Phi_{13}(u)$ whose calculation involves in two successive steps $\Phi_{02}(u)$, $\Phi_{22}(u)$, $\Phi_{11}(u)$, $\Phi_{31}(u)=0$, $\Phi_{00}(u)$, $\Phi_{20}(u)=0$.

By means of automatic digital computing machines it has been found possible to calculate moments up to $2s=14$.

Once a number of moments is known, it is possible to calculate the neutron distribution from a plane δ-source (or from a point source) for instance by means of the *polynomial method* developed by SPENCER and FANO[1] for the analogous X-ray problem.

Following these authors, we select a weight function $w(z)$, everywhere positive, such that, as $z\to\pm\infty$, it tends to zero faster than any power of z, and two sets of polynomials

$$p_n(z)=\sum_i c_{ni}\,z^i \tag{80.12a}$$

$$p_n^+(z)=\sum_i c_{ni}^+\,z^i, \tag{80.12b}$$

[1] See footnote 3, p. 297.

such that: (a) the following orthogonality condition is satisfied

$$\int_{-\infty}^{+\infty} w(z)\, p_n(z)\, p_{n'}^{+}(z)\, dz = \delta_{n n'}; \tag{80.13}$$

(b) the products $w(z)\, p_n(z)$ form a complete set of functions which tend to zero as $z \to \pm\infty$; (c) $p_{n+1}^{+}(z)$ always contains one more power (but not necessarily the next higher power) of z than $p_n^{+}(z)$; for instance, instead of all successive powers, $p_n^{+}(z)$ may contain only even (or only odd) powers of z.

Once $w(z)$ and the type of the polynomials $p_n^{+}(z)$ have been chosen according to the criteria mentioned below, the condition (80.13) allows a complete determination of the polynomials in question, provided for given n the same number of coefficients is involved in (80.12a) and (80.12b).

Then the l-component $\Phi_l(z, u)$ of the expansion (79.4) may be written in the form

$$\Phi_l(z, u) = w(z) \sum_n a_{l n}(u)\, p_n(z) \tag{80.14}$$

the coefficients of which are deduced by multiplication by $p_n^{+}(z)$ followed by integration over z:

$$a_{l n}(u) = \int_{-\infty}^{+\infty} p_n^{+}(z)\, \Phi_l(z, u)\, dz = \sum_i c_{n i}^{+} \Phi_{l i} \tag{80.15}$$

expresses $a_{l n}(u)$ as a linear combination of the moments $\Phi_{l i}$ corresponding to the powers of z appearing in (80.12b).

As an example we may consider a case in which only the first s even moments of $\Phi_0(z, u)$ (up to $i = 2s$) are known; $p_n^{+}(z)$ will be chosen to contain only even powers of z so that by applying (80.15) with $l = 0$ successively for increasing powers of z, the coefficients $a_{0n}(u)$ are determined up to $n = s$. For $2n = 0$, Eq. (80.15) provides a relation between $a_{00}(u)$ and $\Phi_{00}(u)$, for $2n = 2$ the same equation provides $a_{01}(u)$ as a linear combination of $\Phi_{00}(u)$ and $\Phi_{02}(u)$, etc. The procedure is applied until there are determined as many coefficients $a_{0n}(u)$ of the expansion (80.14) as there are moments Φ_{0i} available.

The significance of the expansion (80.14) is illustrated by the following remark[1]. The first term is simply a constant $a_{10} p_0$; therefore the weight function $w(z)$ represents "the initial approximation to $\Phi_l(z, u)$" to within the factor $a_{10} p_0$. If $w(z)$ were chosen just equal to $\Phi_l(z, u)$ all other terms of the expansion (80.14) would be zero. Since $\Phi_l(z, u)$ is unknown, $w(z)$ will in general differ from it; but the expansion (80.14) is expected to converge rapidly if the weight function $w(z)$ has been chosen so that it constitutes a sufficiently good approximation to $\Phi_l(z, u)$.

The choice of $w(z)$ must be made in each particular case so that $w(z)$ embodies the available information (different from the moments) on the behaviour of $\Phi_l(z, u)$; as for instance its behaviour for very small and very large values of z. For instance, in the case of an isotropic plane δ-source Fano and Spencer chose: (a) $w(z) = \exp[-z/l_{\text{tot}}(0)]$ and $p_n^{+}(z)$ containing only even powers of z for l even, (b) $w(z) = z/l_{\text{tot}}(0) \cdot \exp[-z/l_{\text{tot}}(0)]$ and $p_n^{+}(z)$ containing only odd powers of z for l odd and (c) $p_n(z)$ a polynomial of order n in $|z|$ for even as well as for odd values of l.

A comparison between the method of moments and the B_3-approximation (Sect. 79) is made by Wilkins and others[2] by considering, as a function of k, the ratio of the number flux transform $\phi_0(k, u)$ calculated by means of these

[1] See footnote 3, p. 297.
[2] See footnote 2, p. 297.

two methods for a neutron energy of 1.46 ev. For k ranging between 0 and 0.1 this ratio remains between 1.000 and 1.007 while for $k>0.1$ it starts to increase appreciably. This behaviour may be understood by considering that the method of moments is expected to be superior to the B_3-approximation for rather large values of r as are encountered in shielding problems, while the B_3-approximation is expected to give better results at very short distances[1].

The rigorous expression of the age $\tau(E, 1.46\text{ ev})$ of a neutron slowed down to 1.46 ev from an isotropic point source emitting monoenergetic neutrons in water (or any other mixture of hydrogen with a non-hydrogenic atom) has been deduced by VOLKIN[2] in the P_1-approximation of the method of moments. In this calculation the capture of neutrons as well as the anisotropy of the scattering in the c.m.s. by oxygen were taken into account. The simplifying assumptions that still remain in the calculation are:

(a) the hydrogen scattering is spherically symmetric in the c.m.s., which is certainly correct in the considered range of energy;

(b) the oxygen nuclei do not contribute to the slowing down;

(c) the inelastic scattering by oxygen is neglected.

In Fig. 128 (p. 264) the solid curve represents the slowing down length $L_s(u)=\sqrt{\tau}$ to $E=1.46$ ev (indium resonance) as a function of the initial energy E_0 computed making use for the cross section of hydrogen of the experimental data given in Fig. 17 and for the cross section and angular distributions of oxygen of the tables of GOLDSTEIN[3] based partly on experimental results, partly on computations. The dashed curve is obtained by assuming the oxygen scattering to be isotropic in the c.m.s. For source energy E_0 below 3 Mev the difference between the two curves is quite small but increases at higher energies. The effect of capture, which tends to decrease $L_s(u)$ is included in both curves so that the curve following from FERMI's Eq. (73.17) would be somewhat above the dashed curve.

Numerical computations by means of fast computers have been made of the flux distribution of 2.03 ev neutrons in water arising from a point isotropic fission source[4]. The following spatial moments have been computed: $\overline{r^2}, \overline{r^4}, \overline{r^6}$ and $\overline{r^8}$ and their values have been used to synthesize the flux distribution of 2.03 ev neutrons up to $r=100$ cm. The comparison of these theoretical results with experiments will be given in Sect. 83 δ (Table 55 and Fig. 145).

The slowing down of neutrons by oxygen scattering was neglected so that the physical assumptions underlying these computations are very similar to those adopted by VOLKIN.

---

[1] The main contribution to the number flux at deep penetration comes from the point (saddle point) of the complex k-plane where

$$(\Sigma_{\text{tot}})_{\text{m.p.}} + i\,k = 0.$$

$(\Sigma_{\text{tot}})_{\text{m.p.}} = (l_{\text{tot}})^{-1}_{\text{m.p.}}$ is the smallest value of $\Sigma_{\text{tot}}(E)$ in the range of energy under consideration and therefore defines the "most penetrating neutrons". A thorough discussion of this point is given for the analogous case of X-rays, in Sects. 9 and 11 of the companion article in this volume. Since the method of moments consists essentially of representing $\phi_0(k, u)$ by means of the broken off power series (80.7a) around the origin, and the origin is the nearest point on the real axis to the saddle point, the method of moments is expected to give better results for large distances than the B_3-approximation for which there exist limitations similar to (79.24).

[2] H.C. VOLKIN: Phys. Rev. **91**, 425 (1953). — J. Appl. Phys. **26**, 127 (1955).

[3] H. GOLDSTEIN: NDA, Memo 15c-15 (1953).

[4] J. CERTAIN and R. ARONSON: NDA 15c-40 (1954).

Other numerical computations have been made[1] with the method of moments for the penetration of fission neutrons in water, whose results are of considerable importance for shielding. Fast neutron doses $D(r)$ and number flux $\Phi_0(r, E)$ in the range 0.33 to 18 Mev have been computed out to $r = 120$ cm for an isotropic point fission source. Three cases have been considered which differ in the method of evaluating the integral term describing the scattering by oxygen in the moment equations (80.9) and which intend to be successively more accurate:

A) The oxygen nucleus is assumed to have infinite mass so that it does not contribute to the energy degradation.

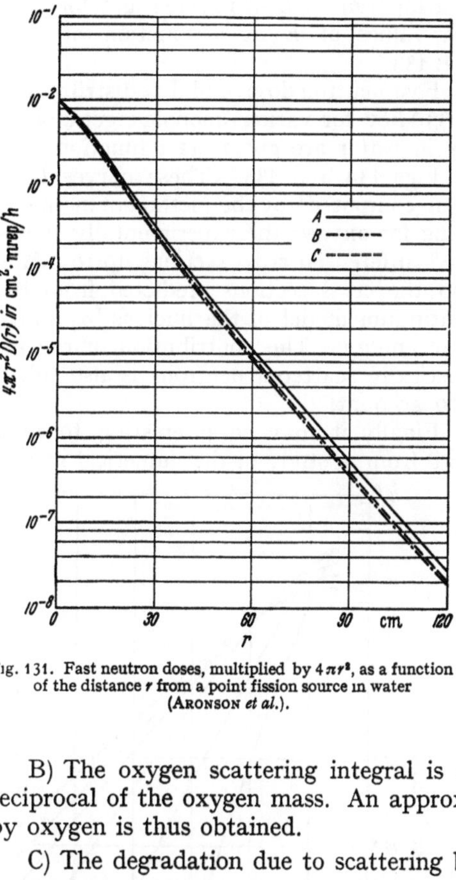

Fig. 131. Fast neutron doses, multiplied by $4\pi r^2$, as a function of the distance r from a point fission source in water (ARONSON *et al.*).

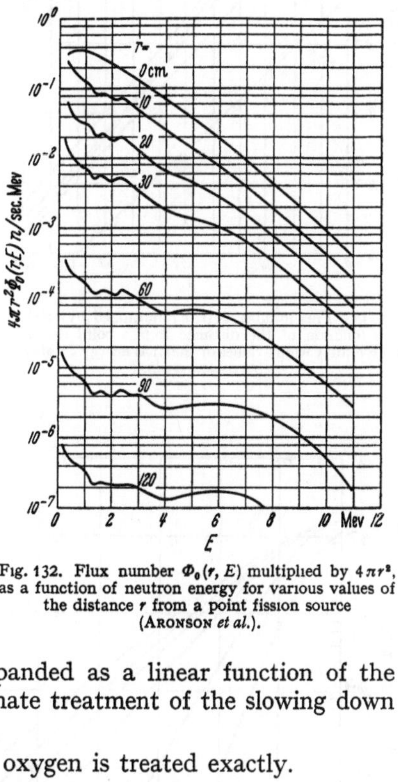

Fig. 132. Flux number $\Phi_0(r, E)$ multiplied by $4\pi r^2$, as a function of neutron energy for various values of the distance r from a point fission source (ARONSON *et al.*).

B) The oxygen scattering integral is expanded as a linear function of the reciprocal of the oxygen mass. An approximate treatment of the slowing down by oxygen is thus obtained.

C) The degradation due to scattering by oxygen is treated exactly.

In all problems the source strength is taken as one neutron per second. In Fig. 131 to 135 the fluxes are expressed in neutrons/cm^2 · sec · Mev and the doses[2] in mrep/h.

Fig. 131 shows $4\pi r^2 D(r)$ as a function of r out to 120 cm in the three cases A, B and C mentioned above. The cases B and C give lower values than case A because of the larger energy degradation due to the contribution of oxygen which is neglected in the latter case. The effect is however small, the dose being reduced by only about 30% at $r = 120$ cm.

The authors conclude that for hydrogeneous media containing in addition to hydrogen only elements heavier than oxygen the procedure used in case A is quite adequate.

[1] R. ARONSON, J. CERTAIN, H. GOLDSTEIN and S. PREISER: NYO-6267 (1954).

[2] See footnote 7, p. 89 and footnote 1, p. 319.

The mean attenuation length defined through the relation — approximately valid for large values of r —

$$4\pi r^2 D(r) \rightarrow e^{-\frac{r}{l'_d}}, \qquad (80.16)$$

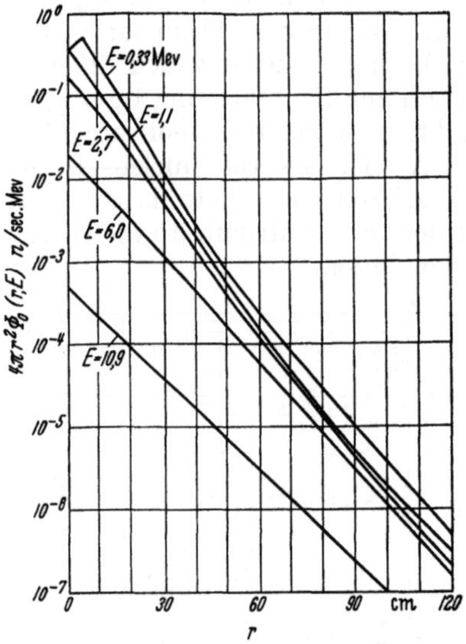

Fig. 133. Flux number $\Phi_0(r, E)$ multiplied by $4\pi r^2$, as a function of the distance from a point fission source for various values of the final energy in Mev (ARONSON *et al.*).

turns out to be 10 cm for r between 60 and 120 cm, either from case A or from case C.

The quantity $4\pi r^2 \Phi_0(r, E)$ calculated in case C is plotted for various values of r as a function of E in Fig. 132 and for various values of E as a function of r in Fig. 133.

Fast neutron doses and flux distributions at 330 kev for various monoenergetic sources in water are given, as a function of r, in Figs. 134 and 135. These curves have been computed[1] by the method of moments using for oxygen the experimentally measured differential cross sections up to 5 Mev and the cross sections predicted from the continuum model of the nucleus [3] above that energy. The contribution of the inelastically scattered neutrons is also taken into account.

Finally it may be interesting to note that from a study of the numerical results

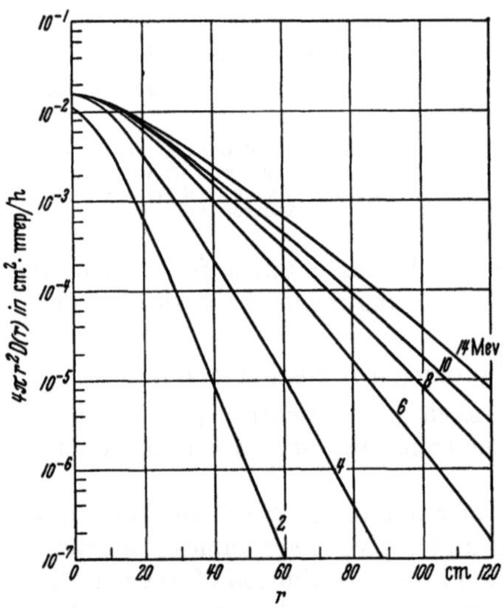

Fig. 134. Fast neutron doses for various monoenergetic sources in water. The parameter is the source energy E_0.

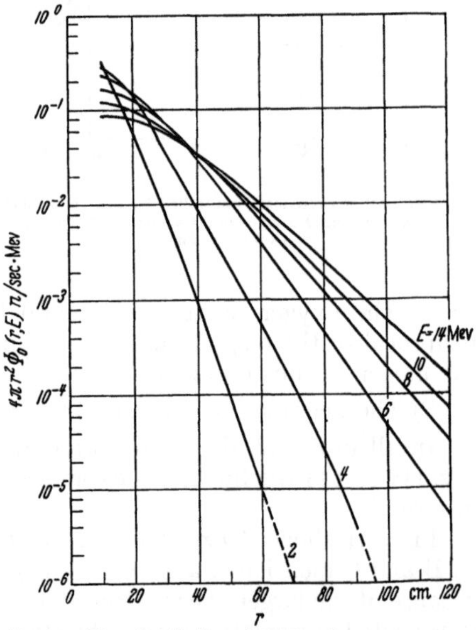

Fig. 135. Flux distribution at 330 kev for various monoenergetic sources in water. The parameter is the source energy E_0.

mentioned above[2] it has been found that the number flux $\Phi_0(r, E)$ in the intermediate energy region (2 ev to 300 kev) for a hydrogeneous moderator

[1] H. GOLDSTEIN, J. CERTAIN, R. ARONSON and J.E. WILKINS: Unpublished; results quoted by WILKINS *et al.*. footnote 2, p. 297.

[2] See footnote 2, p. 297.

may be approximated rather closely in terms of the fast neutron dose $D(r)$ and of the total cross section of hydrogen $\Sigma^H(E)$ by the proportionality relation

$$\Phi_0(r, E) = c \frac{D(r)}{E\,\Sigma^H(E)}. \tag{80.17a}$$

This formula appears to be correct within a few percent in water for distances from the neutron source greater than about 20 cm. For monoenergetic neutron sources of initial energy E_0 embedded in water, the constant of proportionality is roughly given by

$$c = \frac{30}{E_0} \tag{80.17b}$$

where E_0 is expressed in Mev.

81. The slowing down as a function of time in the diffusion theory approximation. The problem of the slowing down and diffusion of a burst of neutrons emitted from an isotropic monoenergetic point source in an homogeneous medium, requires the solution of Eq. (61.2) with

$$S(\mathbf{r}, \boldsymbol{\Omega}, u, t) = \frac{Q_0}{4\pi}\,\delta(\mathbf{r})\,\delta(u)\,\delta(t).$$

We shall point out that the theorem, given in Sect. 70 (p. 246) for the time dependence of the integral spectrum, can be extended to the present case; that is, we can state that in general the solution of the time dependent problem is given by the Laplace inverse transform of the solution of the coresponding stationary problem with the absorption cross section increased by the amount s/v [7]. As in the case of the integral spectrum, this procedure is, however of limited practical importance because of the analytical difficulties usually involved in a rigorous evaluation of the integral arising from the inversion of the Laplace transform.

The problem of the slowing down and diffusion of neutrons as a function of time may be dealt with very easily in the diffusion theory approximation, and this is actually the procedure more frequently used[1,2].

The reader is directed to Sects. 114, 115 and 120 for a detailed discussion of the conditions of validity of this approximate procedure. Here, it may be convenient to recall that the elementary diffusion theory provides a correct picture of the diffusion of neutrons as a function of time in a very weakly absorbing medium (Sects. 114 and 120), only at distances larger than two or three transport mean free paths from any concentrated neutron source, localized absorber or boundary of the moderator (Sect. 113), provided that certain conditions are fulfilled: either the neutron spectrum is independent of position, or the quantities Σ_a, Σ_s and $\bar{\mu}_0$ are independent of energy (Sect. 115).

Above one electron volt, however, Σ_s and $\bar{\mu}_0$ are constant with very good accuracy in a more or less extended energy interval in all the materials which can be used as moderators; for example in water this interval extends from 1 ev up to a few kev, while in graphite, it extends up to 20 kev or more (Sect. 29). On the contrary the capture cross section for all these materials follows the $1/v$ law, but is so small as to be always negligible. Thus we can state that the elementary diffusion theory is expected to provide a fairly good description of the slowing down of neutrons in many cases of practical interest, even without entering the discussion of the constancy of the neutron spectrum with respect to position.

---

[1] G. VON DARDEL: Trans. Roy. Inst. Technol. **75** (1954).

[2] G. VON DARDEL and N. G. SJOSTRAND: Progress in Nuclear Energy. Physics and Mathematics, Vol. 2, p. 183. New York and London: Pergamon Press 1958.

The time dependent diffusion equation (113.46) is obtained in Sect. 114 by imposing the balance of neutrons; the procedure is essentially the one followed in Sect. 75 for deriving the Fermi age equation. In the present case, we write the diffusion equation in the form

$$D_0 \nabla^2 N_{0a}(\boldsymbol{r}, t) - \gamma N_{0a}(\boldsymbol{r}, t) = \frac{\partial N_{0a}(\boldsymbol{r}, t)}{\partial t} \tag{81.1}$$

where

$$N_{0a}(\boldsymbol{r}, t) = \int dv\, N_{0a}(\boldsymbol{r}, v, t) \tag{81.2}$$

and the coefficients

$$\left.\begin{aligned} D_0(t) &= \left(\frac{l_{\mathrm{tr}}\, v}{3}\right)_{\mathrm{av}} = \frac{\int \frac{1}{3} l_{\mathrm{tr}}(v)\, v\, N_{0a}(\boldsymbol{r}, v, t)\, dv}{\int N_{0a}(\boldsymbol{r}, v, t)\, dv} = \frac{\int \frac{1}{3} l_{\mathrm{tr}}(v)\, v\, N_{0a}(\boldsymbol{r}, v, t)\, dv}{N_{0a}(\boldsymbol{r}, t)}; \\ \gamma &= \frac{\int v\, \Sigma_a(v)\, N_{0a}(\boldsymbol{r}, v, t)\, dv}{\int N_{0a}(\boldsymbol{r}, v, t)\, dv} = \frac{\int v\, \Sigma_a(v)\, N_{0a}(\boldsymbol{r}, v, t)\, dv}{N_{0a}(\boldsymbol{r}, t)} \end{aligned}\right\} \tag{81.3}$$

are averages over the neutron spectrum, and have to be considered as functions of the slowing down time t through the relation (64.15) which expresses time in terms of the most probable neutron velocity (Sects. 64 and 70).

Eq. (81.1) turns out to be identical to the age Eq. (75.16) provided that in the latter equation, the derivative of Q with respect to u is transformed, by means of (64.14) and (75.7) into the derivative of N_{0a} with respect to t. Therefore, the final result of the present considerations, i.e. Eq. (81.9) below, could be immediately anticipated. We prefer, however, to derive it again because the present formulation emphasizes those aspects of the problem which are relevant in the discussion of experiments made with pulsed neutron sources (Sect. 84).

If we limit ourselves to the case of the absorption cross section following the $1/v$ law (70.8), $\gamma = 1/T$ is a constant and the corresponding term can be eliminated from (81.1) by means of the substitution (70.9)

$$N_0(\boldsymbol{r}, t) = e^{-\frac{t}{T}} N_{0a}(\boldsymbol{r}, t) \tag{81.4}$$

wich gives

$$D_0(t)\, \nabla^2 N_0(\boldsymbol{r}, t) = \frac{\partial N_0(\boldsymbol{r}, t)}{\partial t}. \tag{81.5}$$

The physical meaning of this equation may be visualized by means of the following examples. Let us suppose that the neutron spectrum is independent of position: then, we can write

$$N_0(\boldsymbol{r}, v, t) = N_0(\boldsymbol{r}, t)\, n_0(v, t). \tag{81.6}$$

Here, $n_0(v, t)$ is the integral spectrum considered in Sect. 70 which, for $l_s = \mathrm{const}$, is represented, with sufficient accuracy, by the Gaussian distribution (70.33). The corresponding half-width (70.34) is inversely proportional to the mass number of the moderator; thus, for very large A, the neutron spectrum can be described as a sharp line of energy (70.31), which, as t increases, shifts towards a lower value which is the same everywhere in the moderator. In this extreme case, the velocity v of the neutron line is related to the slowing down time t by Eq. (64.15) and, thus, the diffusion in space is identical to that of a one velocity model (Sect. 113).

For values of A not very large, the time t appearing as the independent variable in Eq. (81.5) is still related—as long as assumption (81.6) is correct—to the most probable velocity v of the neutron spectrum (70.33) through (64.15). This is true even for hydrogen with the only difference being that in this case the neutron spectrum is given by (70.15).

From the diffusion coefficient (81.3), the corresponding average transport mean free path may be deduced:

$$(l_{\text{tr}})_{\text{av}} = 3\,\frac{D_0(t)}{\bar{v}}\,. \tag{81.7}$$

This expression may be interpreted as the value of $l_{\text{tr}}(v)$ averaged over the neutron flux $\varphi_0(v,t) = v\,n_0(v,t)$.

Eq. (81.5) is considerably simplified if we introduce the neutron age

$$\tau(t) = \int\limits_0^t D_0(t)\,dt \tag{81.8}$$

which is the neutron age expressed as a function of t, instead of u. This expression reduces to (75.10) by expressing t in terms of u by means of (64.15).

If instead of t, the quantity $\tau(t)$ is taken as independent variable, Eq. (81.5) becomes

$$\nabla^2 N_0(\boldsymbol{r},\tau(t)) = \frac{\partial N_0(\boldsymbol{r},\tau(t))}{\partial\tau(t)} \tag{81.9}$$

which is identical with the age Eq. (75.13).

As we shall see in Sect. 111, the same equation may be used for an approximate description of the diffusion of neutrons as a function of time even in both the chemical region and the thermal energy region by a convenient extension of the definitions of $D_0(t)$ and $\tau(t)$.

The solution of Eq. (81.9) in a few important cases is immediately obtained from the results collected in Sect. 76. For example, for the case of an isotropic monoenergetic point source embedded in an infinite medium in which the absorption cross section follows the $1/v$ law, we obtain from (81.9) and (81.4)

$$N_{0a}(\boldsymbol{r},t) = \frac{N_0}{[4\pi\,\tau(t)]^{\frac{3}{2}}}\,e^{-\frac{r^2}{4\tau(t)}}\,e^{-\frac{t}{T}}; \qquad N_0 = \frac{l_s(0)\,Q_0}{\xi\cdot v_0}\,. \tag{81.10}$$

Similarly, for the case of the same source placed on the axis of a rectangular prism with sides $2a$, $2b$, and $2c$, it follows from (76.30), (81.4)

$$\left.\begin{aligned} N_{0a}(\boldsymbol{r},t) &= \frac{N_0}{x_0\,y_0\,z_0}\,e^{-\frac{t}{T}}\sum_{l,m,n\ \text{odd}} e^{-\frac{\pi^2}{4}\left(\frac{l^2}{x_0^2}+\frac{m^2}{y_0^2}+\frac{n^2}{z_0^2}\right)\tau(t)}\times\\ &\quad\times\cos\left(l\,\frac{\pi}{2}\,\frac{x}{x_0}\right)\cos\left(m\,\frac{\pi}{2}\,\frac{y}{y_0}\right)\cos\left(n\,\frac{\pi}{2}\,\frac{z}{z_0}\right)\end{aligned}\right\} \tag{81.11}$$

where x_0, y_0, z_0 are given again by (76.26) with l_{tr} defined by (81.7).

In the general case, the solution of Eq. (81.1) may be written, according to (76.31 b) as

$$N_{0a}(\boldsymbol{r},t) = e^{-\frac{t}{T}}\sum_{l,m,n} A_{lmn}\,e^{-\omega_{lmn}^2\,\tau(t)}\,R_{lmn}(\boldsymbol{r}) \tag{81.12}$$

where $R_{lmn}(\boldsymbol{r})$ and ω_{lmn}^2 are the eigenfunctions and the eigenvalues of Eq. (76.31 c) for the given geometry of the moderator. As can be seen immediately from the particular case (81.11), the eigenvalues ω_{lmn}^2 increase rapidly by increasing the order l, m, n of the successive harmonic modes, so that only the fundamental mode will subsist after a sufficiently high age has been reached. For example in the particular case of Eq. (81.11), the successive terms of the sum become negligibly small with respect to the fundamental mode, after a time such that

$$\tau(t) \gg x_0^2,\ y_0^2,\ z_0^2\,.$$

Thus, in general, after a sufficiently long time has elapsed from the instant of neutron emission from the pulsed source, the neutron distribution is given with good accuracy by the following simple expression:

$$N_{0a}(\boldsymbol{r}, t) \rightarrow e^{-\frac{t}{T}} e^{-\omega^2 \tau(t)} R(\boldsymbol{r}) \tag{81.13}$$

where, for the sake of simplicity, the subscripts of the eigenfunction and eigenvalue of the fundamental mode have been omitted. Under these conditions, the logarithmic derivative of $N_{0a}(\boldsymbol{r}, t)$ with respect to time is given by

$$\frac{1}{N_{0a}(\boldsymbol{r}, t)} \frac{\partial N_{0a}(\boldsymbol{r}, t)}{\partial t} = -\left[\frac{1}{T} + D_0(t)\,\omega^2\right]. \tag{81.14}$$

For $l_s = \text{const}$, $\tau(t)$ is given by (75.15), and the time dependence of (81.13) reduces to

$$N_{0a}(t) \propto \left(\frac{t}{t_0} + 1\right)^{-\alpha} e^{-\frac{t}{T}} \approx \text{const}\, t^{-\alpha} e^{-\frac{t}{T}} \tag{81.15}$$

from which it follows, according to (61.16),

$$\Phi_0(t) \approx \text{const} \cdot t^{-\alpha-1} \cdot e^{-\frac{t}{T}} \tag{81.16}$$

with

$$\alpha = \frac{A\, l_s^2}{3} \omega^2. \tag{81.17}$$

The number of neutrons recorded per unit time by a detector of efficiency proportional to $1/v$, such as a BF_3 counter, is proportional to (81.15) because the factor v, intervening in the definition (61.16) of the flux, is compensated by the factor $1/v$ arising from the detector efficiency (Sects. 23 and 134). We shall come back to this point in Sect. 85.

Finally we should recall the following procedure, suggested by ANTONOV *et al.*[1], for measuring the age $\tau(t)$ for any prefixed value of t. For the sake of definitess, we start by considering a point source embedded in an infinite medium which emits at the time $t = 0$ a burst of fast neutrons. Then, by measuring, at the same time t, the neutron density at two distances r_1 and r_2 from the source, one obtains, from (81.10),

$$\frac{\ln N_0(r_1, t)}{\ln N_0(r_2, t)} = \frac{r_1^2 - r_2^2}{4\tau(t)}. \tag{81.18}$$

Obviously similar relations hold in other geometries. For example, in the case of a prism of finite dimensions in the x and y directions and infinite length, in which the neutron density is measured at two points on the z axis, a relationship is obtained, from (76.31a), which differs from (81.18) only because r_1 and r_2 are replaced by z_1 and z_2.

IV. Experimental data on the space and time distributions of the flux of neutrons of energy ≧1 ev in an infinite medium.

82. General remarks. The experiments on the slowing down of neutrons in a given assembly composed of one or more pieces of moderators and containing neutron sources and possibly neutron absorbers, may be divided into two groups: the first one includes the determinations of the stationary flux distribution as a function of space coordinates inside (and/or outside) the assembly; the second group includes the measurements of the neutron flux at a given position as a function of time. We do not attempt to present here a survey of all experiments

[1] A. V. ANTONOV, A. I. ISAKOFF, I. D. MURIN, B. A. NEUPOCOYEV, I. M. FRANK, E. L SHAPIRO and I. V. SHTRANICH: p. 3 of Ref. [*15c*].

which have been made; we shall limit ourselves to describe only a few typical examples of experiments made inside the moderator under the simplest conditions.

The actual experimental determination of the distribution of neutrons belonging to a narrow energy interval around a chosen value E_R is very simple because it does not involve absolute measurements. As shown in Sect. 89 the activity of a detector of resonance neutrons ($E_R \pm \Gamma_R$) irradiated under a cadmium cover for a given time t at a given position $\boldsymbol{r}$ in a moderator containing a neutron source, is proportional to the density $N_0(\boldsymbol{r}, E_R)$ of the neutrons of energy E_R at the point $\boldsymbol{r}$.

Reliable results from relative measurements of this type can be obtained by careful consideration of a few points, some of which refer to the type and use of the chosen detector and others to the geometry of the experiment.

Because of its finite thickness, the efficiency of the detector is not independent of the direction of the incident neutrons. Therefore, the activity measured on one of its faces depends on the orientation of the foil during irradiation. To a first approximation this inconvenience is eliminated by summing up the activities measured on each face [cf. Eq. (89.14) p. 357].

The activation cross section of the detector generally shows more than one resonance. The effect of all the unwanted resonances (usually those at higher energy) must be eliminated. The correction can be determined experimentally by means of self-absorption and boron absorption. This procedure will be explained later in the discussion of the flux distributions of fission neutrons in water (p. 317) and of photoneutrons from a $Sb^{124}\gamma$+Be source in graphite (p. 337). Eq. (83.5) on p. 322 provides a slightly different form of this correction.

The detector may show other activities different from that produced through the chosen reaction. If these are short-lived one has to start the measurements after they have completely decayed or one has to determine their relative intensitites and correct correspondingly the numbers of counts recorded. If some of these spurious activities have a long life, the "background of the foil" has to be measured after the activity produced through the resonance at E_R has completely decayed and the number of counts recorded has to be correspondingly corrected[1].

Finally it may sometimes be convenient to irradiate the detector inside a tight cover which prevents its contamination. The cover may be made of the same cadmium used to absorb the thermal neutrons or of some inactive material such as aluminium. One has to keep in mind that any type of cover, even if inactive, produces an appreciable change of activation for which a correction has to be made even in the case of relative measurements—if these are not made under exactly the same experimental conditions. The danger of contamination of the foils depends on the type of neutron source and on the moderator, being greater in the case of liquid moderators such as water, heavy water or solutions.

The geometry of the experiments may be different in the different cases. For the determination of the age of the neutrons of a given energy, the ideal geometry is that of an isotropic point source embedded in an infinite medium with a point detector at a distance r. The majority of the experiments reported in Sects. 83 and 84 refer to this case.

[1] For example, neutrons of energy $\gtrsim 0.6$ Mev produce, by inelastic scattering, a metastable state of In^{115} which decays with a half-life $T_{\frac{1}{2}} = 4.5$ h [A. A. EBEL and C. GOODMAN: Phys. Rev. **93**, 197 (1954), H. C. MARTIN, B. C. DIVEN and R. F. TASCHEK: Phys. Rev. **93**, 199 (1954)]. If an In foil is exposed, during an infinite time, to Raα+Be neutrons slowed down in water, the ratio of the activities corresponding to the 4.5 h and 54 m half-lives measured 90 minutes after the end of the irradiation has the following values: at 8 cm from the source ≈ 0.03; at 16 cm from the source ≈ 0.01 [B. C. TITTLE: Nucleonics **8** (6), 5 (1951)].

Deviations from these ideal conditions are due to: (a) the finite dimensions of the source and detector, which in practice can never be neglected with respect to the distance r between their centres; (b) the finite volume of the moderator surrounding both source and detector.

The correction for deviation (a) from the ideal geometry can be introduced as follows. Let

$$\frac{A(r)}{S_1 S_2}$$

be the activity induced per unit area of the detector by the neutrons emitted from unit area of the source, r being the distance between the corresponding surface elements dS_1, dS_2. If the distance r_0 between the centres of the foil and the source is large compared to their linear dimensions, the *activity* $A(r)$ may be expanded in a TAYLOR's series of $|\boldsymbol{r}-\boldsymbol{r}_0|$ and a satisfactory representation of the *measured activity* $A_m(r_0)$ is obtained be keeping only the first two terms:

$$A_m(r_0) = \int dS_1 \int dS_2 \frac{A(r)}{S_1 S_2} \approx A(r_0) + A'(r_0)\,\bar{\varrho} \tag{82.1}$$

with

$$\bar{\varrho} = \int dS_1 \int dS_2 \frac{|\boldsymbol{r}-\boldsymbol{r}_0|}{S_1 S_2}$$

calculated from the shape and dimensions of source and detector. Once $\bar{\varrho}$ is known, the corrected value of the activity can be derived from (82.1) by substituting $A'_m(r_0)$ for $A'(r_0)$:

$$A(r_0) = A_m(r_0) - A'_m(r_0)\cdot\bar{\varrho}. \tag{82.2}$$

In order to determine the effective location of the source centre from which r_0 has to be measured, it is convenient to measure, in a preliminary experiment, the activation of a few identical detectors placed at equal distances from the source on opposite sides. The centre of symmetry of the resulting activity curve can be taken as the effective centre of the neutron source.

Any correction for deviation (b) from the ideal geometry is avoided by choosing the longitudinal and the transverse dimensions of the moderator large enough. The longitudinal dimensions must be so large—even for the greatest values adopted in the actual measurements of the distance r between the source and detector—that both of them are still located at many transport mean free paths of the "most penetrating neutrons" from any boundary of the moderator. The transverse dimensions must be of the same order of magnitude.

Once the activation of the resonance detector has been measured as a function of the distance r from an *isotropic fast neutron source* of small dimensions, and it has been corrected for the various possible deviations from the ideal conditions mentioned above, the spatial moments of the distribution of the neutrons of energy E_R may be computed by means of the formula

$$\overline{r^{2s}} = \frac{4\pi \int_0^\infty r^2 A_{\mathrm{Cd\,o}}(r)\, r^{2s}\, dr}{4\pi \int r^2 A_{\mathrm{Cd\,o}}(r)\, dr}. \tag{82.3}$$

where $A_{\mathrm{Cd\,o}}$ is defined in Sect. 90 (see also Sect. 113).

These experimental values can be compared to the corresponding theoretical predictions deduced by introducing into (71.6a) the quantities $\Phi_{00}^{\mathrm{pt}}(u_R)$ and $\Phi_{02s}^{\mathrm{pt}}(u_R)$ obtained from any one of the methods given in Sects. 76, 79 and 80.

Since the integrals appearing in (82.3) extend to infinity, their computation requires the extrapolation of the experimental values of $A_{\mathrm{Cd\,o}}(\boldsymbol{r})$. A reasonable procedure would be to use WICK's formula (80.6) which contains the factor r^{e+1}. In practice experimentalists make this extrapolation using the expression

$$A_{\mathrm{Cd\,o}}(r) \to \frac{1}{r^2}\, e^{-\frac{r}{l'_{\mathrm{tr}}}} \tag{82.4}$$

which corresponds to the distribution (77.1) of the primary neutrons (Sect. 77). This was found, from early times[1], to allow a satisfactory representation of the empirical data at large distances from the source when the value of the *attenuation length* l'_{tr} was adjusted. Various authors have noticed, however, that the value of l'_{tr} obtained by such a procedure, tends to increase slightly, as r increases; this is what we have to expect according to the correct formula (80.6).

A reliable value of $\overline{r^{2s}}$ is obtained from (82.3) only when the actual measurements of $A_{\mathrm{Cd\,o}}(r)$ extend to such large distances from the source, that the extrapolated part of the integral represents only a small correction with respect to the one directly measured. Such a condition requires the use of strong neutron sources and yields a limit to the maximum order $2s_{\max}$ of the moments that can be computed from a given set of measurements.

For given neutron source and moderator, the value of l'_{tr} appears to be the same for both resonance neutrons of different energies and for thermal neutrons. This is what is expected according to the picture given in Sect. 77, which predicts a quasi-equilibrium state between the most penetrating neutrons and all their descendents of sufficiently low energy. Although two exceptions has been reported in the case of sources emitting neutrons of very low energy (Table 47 on p. 313), the rule that l'_{tr} is independent of the final energy is usually accepted. Sometimes it is even used in the extrapolation of the resonance activation $A_{\mathrm{Cdo}}(r)$ to infinity, which is carried out using the value of l'_{tr} obtained from measurements of the total activity

$$A(r) = A_{\mathrm{Cd\,o}}(r) + A_{\mathrm{th}}(r).$$

Since A is considerably larger than $A_{\mathrm{Cd\,o}}$, it can be measured even at those distances from the source, for which A_{Cdo} is already too small to be accurately determined (Sects. 90 and 113).

When the available amount of moderator is too limited to fulfill the conditions for an "infinite medium", a different geometry may be adopted in order to obtain a good determination of $\overline{r^2} = 6\tau$. For instance, the case of a point source placed on the axis of a rectangular prism has been treated according to age theory in Sect. 76. If the length of the prism is very large, its transverse dimensions are of secondary importance because they only influence the absolute value, not the flux distribution of the neutrons of a given age along the axis of the prism. According to (76.31a) such a distribution is the same as that which we would observe in an infinite medium containing an infinite plane δ-source. From the measured distribution we can derive, making use of (71.2b) and (71.7), the distribution and the spatial moments relative to a point source in an infinite medium.

Other geometries may also be adopted; for example, as we shall see in Sect. 84γ, in certain experiments on the moderation of fission neutrons in graphite, the moderator was in form of a prism and the U^{235} source was a plane isotropic source whose intensity in the transverse (x, y) plane was well represented by the formula

$$Q_0(x, y) = Q_0 \cos\pi\frac{x}{x_0} \cos\pi\frac{y}{y_0}. \tag{82.5}$$

The corresponding solution of the age equation may easily be derived[2].

[1] E. AMALDI and E. FERMI: Ric. Sci. 7 (1), 454 (1936). — Phys. Rev. 50, 899 (1936)

[2] WEINBERG and ROSE: Mon P-240; Mon P-297; see p. 332.

Many of the measurements of the neutron distributions in various moderators reported in the sections to follow have been made with *fission neutron sources*, namely, with the neutrons resulting from thermal neutron fission of U^{235}. The corresponding energy spectrum is well represented by the expression

$$S_f(E) = e^{-\frac{E}{0.965}} \sinh (2.29E)^{\frac{1}{2}} \tag{82.6a}$$

where E is the energy of the emitted neutrons in Mev. This expression, suggested by WATT[1], is derived from simplest considerations of neutron emission and of

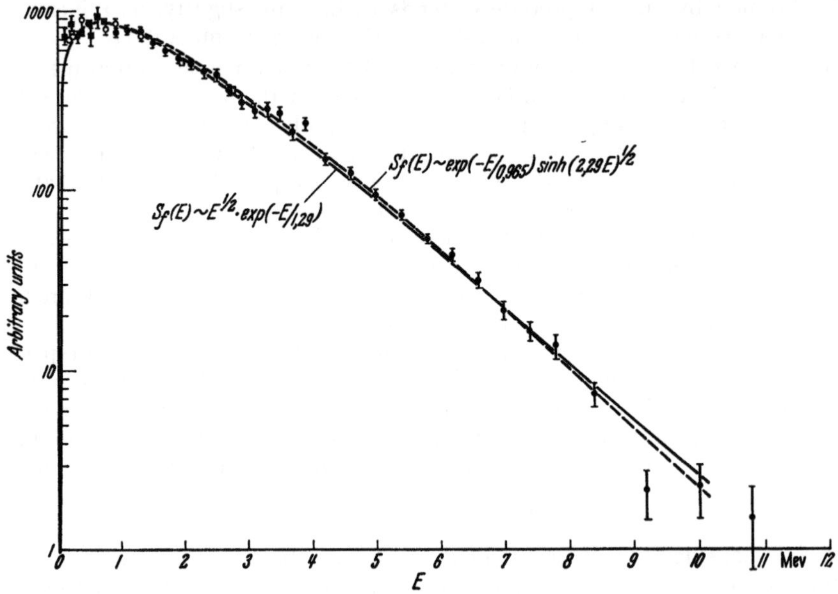

Fig. 136. Comparison of semiempirical expressions of the energy spectrum of fission neutrons of U^{235} with experimental results by various authors (LEACHMAN).

frame of reference transformations, and may be further simplified to the following expression[2]

$$S_f(E) = \sqrt{E}\, e^{-\frac{E}{1.29}}. \tag{82.6b}$$

Fig. 136 shows a comparison of these formulas with the experimental results obtained by various techniques[2].

The neutron spectra emitted by the other sources which will be discussed in the sections to follow are described in Part B of this article.

A comparison of the experimental results of the neutron distribution or of the spatial moments with the corresponding theoretical predictions, can at best be made, in the case of monoenergetic neutrons such as those provided, for example by the Sb γ + Be reaction. It should be noted that the $d+d$ reaction is not an isotropic monoenergetic neutron source because even for rather low incident deuteron energies not only the number but even the emitted neutron energy varies considerably with the angle of emission. In these cases the theoretical derivation of the neutron distribution in a certain direction must be computed by the following procedure; the distribution produced by a collimated neutron source has to be computed first, and then all the distributions corresponding to the emitted

[1] B. E. WATT: Phys. Rev. **87**, 1037 (1952).
[2] R. B. LEACHMANN and others: p. 193 of Ref. [*15a*].

neutrons, with the right energy and intensity in each direction, have to be superimposed. This inconvenience is less pronounced in the case of strongly exoenergetic reactions which give a large intensity already at low incident particle energies; a typical example is provided by the $d+t$ reaction.

Although cadmium-covered iodine, silver, gold and rhodium foils have been used from time to time, the most common resonance detector is provided by cadmium-covered indium foils. The advantages of such a detector are the following:

1. The In (n, γ) In process shows a very sharp resonance at $E_R = 1.46$ ev which definitely prevails over all higher resonances. The corresponding activation integral amounts to 2800 barns (Sect. 91) and the corresponding β-ray spectrum can be easily measured with a thin walled counter or ionization chamber (Sect. 26).

2. Because the energy $E_R = 1.46$ ev is so close to the chemical bond region, the slowing down density derived from the corresponding flux distribution can be introduced—at least in the more elementary treatments—as a source in the diffusion equation of the thermal neutrons.

3. The half-life (54 min) of the interesting process is very convenient for measurements and can easily be separated from those relative to other reactions[1].

4. Indium is a very soft metal and can be easily put in the form of thin foils. During use it needs however some mechanical support, especially when very thin.

Table 45 contains a few typical values of the neutron age

$$\tau_{\text{res}} = \tfrac{1}{6}\,\overline{r_{\text{res}}^2} \qquad (82.7)$$

Table 45. *Age of indium resonance neutrons and migration area of thermal neutrons for various sources and moderators, all in cm².*

	Light water		Heavy water	Beryllium	Graphite
	τ_{In}	M^2	τ_{In}	τ_{In}	τ_{In}
Sb γ+Be					147
Ra γ+Be	~40				
Ra α+Be	45.3	54.7		120 ± 23	
Po α+Be	56.7 ± 1	67.3 ± 1.5		95 ± 11	
Fission source	30.8		109	80.2 ± 2	310
$d+d$ at 0°; $E_n = 3.12$ Mev	43.5 ± 0.8				
$d+t$ at 90°; $E_n = 14.1$ Mev	150 ± 6	156 ± 6			

which will be discussed in detail in Sects. 83 and 84. The same table shows also the values of the migration area of thermal neutrons

$$M^2 = \tfrac{1}{6}\,\overline{r_{\text{th}}^2} \qquad (82.8)$$

which have been obtained by cadmium-difference method (Sect. 133) and which will be considered in Part E.

The experiments on the time dependence of the neutron flux are considered in Sect. 85.

83. The case of water and other hydrogeneous compounds or mixtures. Water or mixtures of water and non-absorbing metals deserve particular consideration because of the effectiveness of hydrogen in slowing down neutrons. Water is moreover a good heat transfer fluid and, therefore, media of this type are of increasing importance in reactor design (Vol. XLIV). On the other hand some

[1] See Sect. 26 and footnote 1, p. 307.

experimental difficulties in measuring neutron distributions become particularly pronounced in hydrogeneous media.

First the large energy loss per n-p collision leads to a short slowing down length L_s. The scale of experiments on slowing down is proportional to L_s and therefore in water the importance of the flux distribution at small distances from the source is accentuated, and the dimensions of source and detector must be smaller than in the case of other moderators.

A second type of difficulty is encountered if the experiment is made with a fission source, as for instance an uranium plate exposed to thermal neutrons. The strong attenuation of thermal neutrons in water requires that the fission plate be placed very close to the source of thermal neutrons; this distorts the slowing down distribution of the neutrons emitted from the plate in the neighborhood of the plate itself.

Finally, because of the rapid variation of the n-p scattering cross section with neutron energy, the slow neutron distribution is rather sensitive to the value of the initial energy. Therefore in hydrogeneous moderators relatively small uncertainties in the energy distribution of the source neutrons lead to rather large variations of the observed slow neutron distribution.

Hydrogeneous media have, however, the advantage with respect to other moderators that, due to the large energy lose per n-p collision, the correction for flux depression can be neglected (Sect. 135).

α) Distribution in water, paraffin and water-metal mixtures of neutrons from $R\alpha + A$ and $R\gamma + A$ sources. The first distributions of resonance neutrons were measured in water with a $Rn\alpha + Be$ source[1]. As neutron detectors I, Ag, and Rh, with and without a 0.5 g/cm² cadmium cover, were used from very short distances up to $r = 20$ cm, where $r^2 A(r)$ already showed the exponential behaviour (82.4).

The fact that the activities $A(r)$ of the various detectors decrease with increasing r according to different laws, was attributed to differences in the energy of the corresponding resonances, a more rapid decrement of $A(r)$ being expected for neutrons of higher energy. These considerations were made quantitative by computing the corresponding $\overline{r^2}$ and trying to estimate, by means of Eq. (73.15), the ratios of the resonance energies E_R of the detectors employed. The values obtained by this method were in reasonable agreement with those derived from boron absorption (see Sect. 27), especially when Eq. (73.15) was corrected to take into account the scattering by oxygen [*5*].

The dependence of the flux distribution on the initial energy of the neutrons was investigated by measuring as a function of r the total activity of a Rh detector for the following (d, n) reactions at $E_d = 740$ kev: $d+d$, $d+$Li, $d+$Be, $d+$C[2].

A few values of $\overline{r^2}$ and l'_{tr} for water obtained by various authors using several $R\alpha + Be$ sources have been collected in Table 46. The values obtained by various authors, for the same neutron source and with the same resonance detector, have a spread of the order of $\pm 5\%$ with respect to their arithmetic average.

Fig. 137 shows a logarithmic plot of $r^2 A(r)$ as obtained by RUSH[3] using a $Ra\alpha + Be$ source and an In detector 1.27×10^{-2} cm thick in a cylindrical galvanized steel tank 120 cm deep and 120 cm in diameter. Curve I represents the total activity of the uncovered detector, whereas curve II represents the resonance activity obtained with the indium foil covered with a cadmium layer 0.101 cm

---

[1] E. AMALDI and E. FERMI: Ric. Sci. **7** (1), 454 (1936). — Phys. Rev. **50**, 899 (1936).
[2] E. AMALDI, L.R. HAFSTAD and M.A. TUVE: Phys. Rev. **51**, 896 (1937).
[3] J.H. RUSH: MDDC-1050. — Phys. Rev. **73**, 157 (1947).

Table 46. *Values of* $\overline{r^2}$ [cm²] *and* l'_{tr} [cm] *for resonance and thermal neutrons in water for various* Rα+Be *sources.*

Source	Rn α+Be		Ra α+Be						Po α+Be				Po α+B
Authors	AMALDI, FERMI[1]		MUNN, PONTECORVO[2]	H. L. ANDERSON[3]	RUSH[3]		FIENO[4]		GRAHAM[5] et al.	BAER[6]		REIER[7] et al.	REIER[8]
	$\overline{r^2}$	l'_{tr}	$\overline{r^2}$	$\overline{r^2}$	$\overline{r^2}$	l'_{tr}	$\overline{r^2}$	l'_{tr}	$\overline{r^2}$	$\overline{r^2}$	l'_{tr}	$\overline{r^2}$	$\overline{r^2}$
Iodine resonance $E_R \sim 20-40$ ev	262	9.45											
Silver resonance $E_R = 5.1$ ev	271	9.45											
Indium resonance $E_R = 1.46$ ev			278	288	272	9.29	291 302	10.0	338			340±7	238±7
Rhodium resonance $E_R = 1.26$ ev	277	9.45											
Thermal neutrons ($T = 290°$ K)	327	9.45	330		328-	9.46			388	418	10.3±0.7	404±9	

thick. Both curves clearly show an exponential behaviour in the range $15 \leq r \leq 40$ cm.

Table 47 shows the values of the relaxation length l'_{tr} for paraffin obtained by BERNSTEIN and FELD [*18b*] using various neutron sources embedded in paraffin. The results appear to be independent of the detector for the two fast neutron sources, but not for the photoneutron sources which emit neutrons of intermediate energy (Sect. 49). A possible explanation of the photoneutron data suggested by FELD is that the exponential character was accidental due to the fact that the measurements were not made over a sufficiently large distance and, therefore, the asymptotic form was not yet reached.

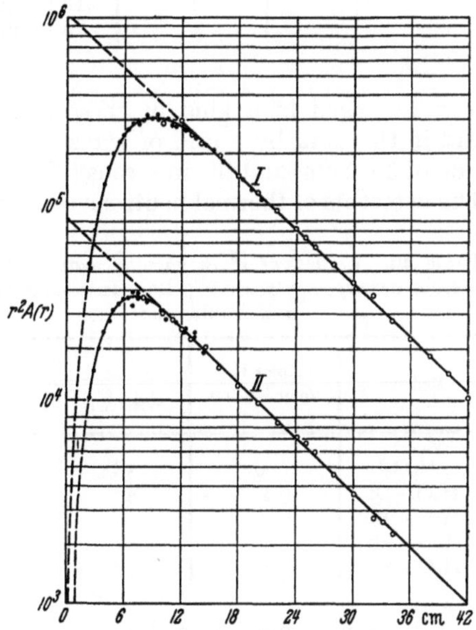

Fig. 137. Logarithmic plot of $r^2 A(r)$ versus r for a Ra α+Be neutron source in water: curve I = uncovered In-detector; curve II = In-detector covered with Cd (RUSH).

Table 47. *Relaxation length* l'_{tr} [cm] *in paraffin for various neutron sources* (*Bernstein and Feld* [*18b*]).

Source	l'_{tr} In-resonance detector	l'_{tr} Thermal neutron detector
Ra α+Be	9.3	9.3
Ra α+B	6.6	6.7
Ra γ+Be	1.8	3.2
Ra γ+D	2.4	3.4

[1] See footnote 1, p. 312.
[2] A. M. MUNN and B. PONTECORVO: Can. J. Res. A **25**, 157 (1947).
[3] See footnote 3, p. 312.
[4] D. FIENO: NACA-RM-E-54HO4 (1954).
[5] R. H. GRAHAM, J. R. DONALDSON, S. H. FITCH, J. W. FLORA and R. E. NATHER: AT(11-1)-74 (MTA-45) (1954).
[6] W. BAER: J. Appl. Phys. **26**, 1235 (1955).
[7] M. REIER, F. OBENSHAIN and R. L. HELLENS: Nucl. Sci. and Engng. **4**, 1 (1958).
[8] M. REIER: AT-11-1-GEN-14 (1957).

The spatial distributions, in water-metal mixtures from Raα + Be and Raγ + Be sources, have been measured by various authors; their results are collected in Tables 48 and 49. In the experiments of MUNN and PONTECORVO[1] the metals

Table 48. *Values of $\overline{r^2}$ and l'_{tr} for various water-metal heterogeneous arrangements in the case of* Ra α + Be *source.*

Medium	Volume H_2O / Volume metal	Detector	MUNN and PONTECORVO[1] $\overline{r^2}$ (cm^2)	FIENO[2] $\overline{r^2}$ (cm^2)	FIENO[2] l'_{tr} (cm)
Water		Dy	330		
		In-resonance	278	291	
				302	~10.0
Water-Bi	1	Dy	562		
	2	In-resonance	398		
	1	In-resonance	477		
	0.5	In-resonance	517		
Water-Pb	1	In-resonance	437		
Water-Fe	1	In-resonance	—	347	~10.0
	1.3	In-resonance	319		
	2	In-resonance		336	~10.0
	3	In-resonance		314	~10.0

were arranged in a kind of chess board pattern of square rods supported on end in the tank by a grid of wires. The lead and bismuth rods had a cross section of 2×2 cm^2 and the iron rods (nickel plated) a cross section of 1.87×1.87 cm^2. The detectors of thermal neutrons consisted of 70 mg each of dysprosium oxide on an aluminium tray 2×2.5 cm. The detectors of resonance neutrons were made of indium, 45 mg each on an aluminium backing 2×2.5 cm. The dysprosium detectors were activated in flat water-tight aluminium boxes, the indium detectors in similar cadmium boxes of wall thickness 0.6 mm. FIENO[2] used In foils 0.17 g/cm^2 thick, covered with 0.88 g/cm^2 cadmium. The two sets of measurements agree as well as can be expected according to the remark made in connection with Table 46.

Table 49. *Values of $\overline{r^2}$ in various water-metal heterogeneous arrangements:* Ra γ + Be — In-*resonance (Munn and Pontecorvo*[1]*).*

Medium	Volume H_2O / Volume metal	$\overline{r^2}$ [cm^2] observed	$\overline{r^2}$ [cm^2] corrected for source dimensions
H_2O	∞	58	~38
H_2O–Bi	2	87	~69
H_2O–Bi	1	95	~77
H_2O–Pb	1	87	~69
H_2O–Fe	1.3	79	~61

According to the Canadian group the relaxation length l'_{tr} for the various media investigated does not change very much; it is smaller, within the experimental error, in mixtures of iron and water than in pure water.

The values of $\overline{r^2}$ for the Raγ + Be source (Table 49) are very uncertain because of the very large correction arising from the considerable dimensions of the source combined with the small value of the age to In-resonance.

[1] See footnote 2, p. 313.
[2] See footnote 4, p. 313.

Table 50 shows the results of similar experiments made with Poα+Be sources. The measurements of BAER[1] were performed in an aluminium tank 50 cm×50 cm×100 cm with the zirconium rods arranged in a kind of lattice. The neutrons were detected with a proportional counter 0.65 cm in diameter and 2.5 cm long filled with BF_3 at a pressure of 1 atmosphere; 96% of the boron was B^{10}. The measurements were made both with and without a 5.1×10^{-2} cm thick layer of Cd surrounding the counter; this layer reduced the counting rate by a factor ~40. No significant differences in the distributions with and without the cadmium cover were observed. The metal used in the experiments was a crystal bar in its rough shape with rod diameters varying from $\frac{1}{2}$ to $\frac{3}{4}$ inch. An

Table 50. *Summary of age measurements in water-metal mixtures from* Po α+Be *sources*[2].

Metal	Volume metal / Volume water	τ_{In} cm²	M^2 cm²	$M^2-\tau_{In}$ cm²	L^2 computed cm²
Fe	2:1	72.1±2.2	75.5±2.3	3.4	2
	1:1	58.8±1.8	66.1±2.0	7.3	2.08
	1:2	56.2±2.2	55.1±2.2	−1.1	2.51
	0	55.4±1.7	68.1±2.0	12.7	
	0	57.9±1.7	66.4±2.0	8.5	
	0[3]	56.3±0.6	64.7±0.6	8.4	
	0[1]		69.6±1.4		
Al	2:1	143.5±4.3	175.7±5.3	32.2	27.8
	1:1	100.2±3	119.6±3.6	19.4	17.7
	1:2	81.2±2.4	94.4±2.8	13.2	12.5
Zr	2:1	98.0±2.9	133.5±4.0	35.5	30.1
	1:1	82.5±2.5	99.2±3.0	16.7	19.2
	1:1[1]		112.8±2.3		
	1:2[1]		89.7±1.8		
	1:4[1]		81.3±1.7		

attempt was made to observe the disorder effect in the array by rearranging the zirconium rods surrounding a fixed position of the counter; no change in counting rate was observed within the statistical probable error of ±2%.

BAER'S results show that the attenuation length has the same value, within the experimental errors, in Zr-water mixtures (for instance $l'_{tr}=11.2\pm0.8$ cm for Zr/water = 1:1) and pure water ($l'_{tr}=10.3\pm0.7$ cm).

The measurements of REIER *et al.*[2] have been made with a foil 0.005 inch (= 1.27×10^{-2} mm) thick of a 50—50 In-Sn alloy with and without a cadmium cover. The tin served only to add rigidity to the foil. A lattice of metal rods of $\frac{5}{8}''$ diameter was placed inside a stainless steel tank of 95 cm×95 cm×158 cm.

A comparison of the results contained in Table 50 with theoretical computations will be given later in this same section.

β) *Distribution in water and metal-water mixtures of In-resonance neutrons from fission sources.* Because of the importance, already mentioned above, of water and hydrogeneous media in reactor design, the flux distribution produced in these moderators by fission neutron sources deserves special consideration.

[1] See footnote 6, p. 313.

[2] The majority of the measurements are due to M. REIER, F. OBENSHAIN and R.L. HELLENS: Nucl. Sci. and Engng. 4, 1 (1958).

[3] R.H. GRAHAM, J.R. DONALDSON, S. H. FITCH, J.W. FLORA and R.E. NATHER: AT (11-1) — 74 (MTA-45) (1954).

In the case of water this problem was investigated earlier by FERMI, ANDERSON and NAGLE[1], and more recently by HILL, ROBERTS and FITCH[2]. These authors placed on the thermal column of the Oak Ridge National Laboratory pile, a tank of aluminium plate 1.25 cm thick filled with water; the horizontal dimensions were $152 \times 152\ \text{cm}^2$ and the height was 183 cm.

Thermal neutrons coming up from the pile entered the tank from the bottom and produced fission in the neutron source. This was a plate 0.2 cm thick of U^{235}—Al alloy (eutectic composition: 18% uranium by weight of which 88% was U^{235}) located at a suitable distance from the bottom.

The fast neutrons emitted in fission and slowed down by water were measured by means of indium foils 0.10 g/cm² thick contained in cadmium boxes of wall thickness 2.74 or 3.07 g/cm² in different sets of measurements.

Source and detectors were held in their position, in planes parallel to the bottom and on the central vertical axis of the tank, by means of aluminium structures so that all relative distances were known to better than 0.03 cm.

By using a rather large number of neutron source and detector sizes, the geometry corrections were kept small for values of r ranging from a few centimeters to 92 cm. They rarely exceeded 10% and were usually much less.

For example, at short distances the neutron source was a disk 1 or 2 cm in diameter, while for intermediate and large distances squares of $10 \times 10\ \text{cm}^2$ or $30.5 \times 30.5\ \text{cm}^2$ area were used; the dimensions of the In foils ranged from $1 \times 1\ \text{cm}^2$ to $4 \times 6.35\ \text{cm}^2$ area.

Four detectors were placed 7.5 cm apart so that they did not perturb one another and were irradiated contemporarely for about an hour. To obtain the distribution between 70 and 92 cm, measurements were made using In foils—without Cd—in aluminium covers which provided mechanical support and prevented contamination.

The best condition for a neutron slowing down experiment would be obtained by placing the neutron source at the centre of a very large volume of the slowing down medium. With a fission source, the thermal neutrons used to produce fission have to diffuse through this slowing down medium. In water the relaxation length of thermal neutrons is about 2.5 cm and therefore a fission source placed 10 cm inward from the water boundary will give less than 2% of the number of fission neutrons which the same source would give at the water surface.

This reduction of intensity would be a serious handicap for measurements at large distances from the source. A scheme to overcome this difficulty, adopted by HILL, ROBERTS and FITCH, consisted in making measurements close to the neutron source with the source 10 cm inward from the water boundary; the measurements at great distances were made with the source placed at the bottom of the tank. This procedure was checked by measuring the neutron distribution for r ranging between 0 and 20 cm with the source located at the two positions explained above. Close to the source the two distributions differed in shape as one would expect, but beyond 10 cm the two curves were proportional. It seems reasonable to conclude that satisfactory results may be obtained by making measurements at distances greater than 10 cm from the source with the source located on the bottom of the tank. And furthermore, since 10 cm corresponds to many transport mean free paths of the indium resonance neutrons, the authors felt that correct results should be obtained for $r < 10$ cm, when the neutron source was located 10 cm inward from the water boundary.

[1] E. FERMI, H.L. ANDERSON and NAGLE: CP-1531 (A 2209).
[2] J.E. HILL, L.D. ROBERTS, T.E. FITCH: ORNL-181. — J. Appl. Phys. **26**, 1013 (1955).

Fig. 138 shows the flux distribution of the In resonance neutrons. The lower curve was obtained with the same In foil covered on both sides with a layer of boron thick enough (0.245 to 0.337 g/cm²) to remove completely the neutrons in the energy range below 10 ev. The results of these measurements are that the boron covered curve is about 3% of the cadmium covered curve, the two curves being almost proportional over a considerable range of distances. One can conclude that in water the distortion of the distribution due to resonances of indium at energies above that at 1.46 ev, is negligible and that the upper curve of Fig. 138 is good enough to be used in calculating the corresponding moments.

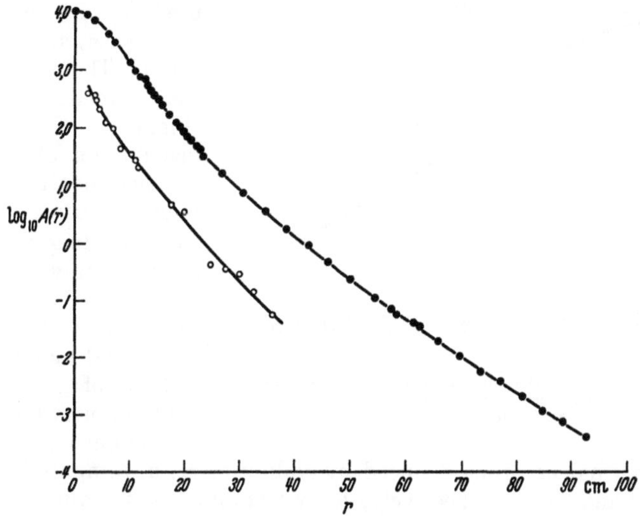

Fig. 138. Logarithmic plot of $A(r)$ versus r for a fission source in water: upper curve Cd covered In-detectors; lower curve B covered In-detectors (Hill, Roberts and Fitch).

Table 51 contains the moments deduced by Hill, Roberts and Fitch from the upper curve of Fig. 138. The fraction of the integrals extrapolated to infinity with

$$l'_{tr} = 8.786 \text{ cm}$$

is negligible in the case of the second, fourth and sixth moments but represents an appreciable part of $\overline{r^6}$ and $\overline{r^8}$. As we have to expect that l'_{tr} is not a constant

Table 51. *The moments of the distribution of ~1.46 ev neutrons from a point fission source in water (Hill, Roberts and Fitch*[1]*).*

$2s$	$\Phi^{pt}_{0,2s} = \int_0^\infty A_{In}(r) r^{2s+2} dr$	Fraction of $\Phi^{pt}_{0,2s}$ extrapolated	$\overline{r^{2s}} = \frac{\Phi^{pt}_{0,2s}}{\Phi^{pt}_{0\,0}}$
0	2.09×10^6	1.40×10^{-5}	
2	3.86×10^8	7.95×10^{-4}	184.7
4	2.56×10^{11}	0.013	1.22×10^5
6	4.37×10^{14}	0.086	2.27×10^8
8	1.36×10^{18}	0.33	8.64×10^{11}

but gradually increases with increasing r due to the hardening of the neutrons, these values are probably underestimated.

[1] See footnote 3, p. 316.

The second moment $\overline{r^2_{\mathrm{In}_2}} = 184.7\ \mathrm{cm}^2$ is appreciably lower than the value previously given ($194\ \mathrm{cm}^2$)[1].

Measurements similar to the previous ones have been made in aluminium-water mixtures[2]. The tank used in the experiment previously described was filled with a stacked grid structure of aluminium plates each 0.63 cm thick separated by spacers of thickness 0.63 cm for the 1:1 mixture and 2×0.63 cm for the 1:2 mixture. The plates lay in horizontal planes parallel to the top of the thermal column of the pile and perpendicular to the axis of the tank along which the measuring foils were located. Fission sources and detectors of various sizes were used as in the case of the measurements in pure water. The measurements were extended up to 68 cm (Fig. 139). From the flux distribution of the In resonance neutrons of this figure one can calculate the corresponding value of $\overline{r^2_{\mathrm{In}}}$ of which only 1% is due to extrapolated area. Table 52 summarizes the values of the second moment, of the age and of l'_{tr} for water and aluminium-water mixtures. In column 5 are given the total age of thermal neutrons deduced by adding to τ_{In} a computed increment $\tau_{\mathrm{th}} - \tau_{\mathrm{In}}$ of 1, 2 and 3 cm² respectively, for water and 0.5, 1.0 aluminium-water mixtures[3].

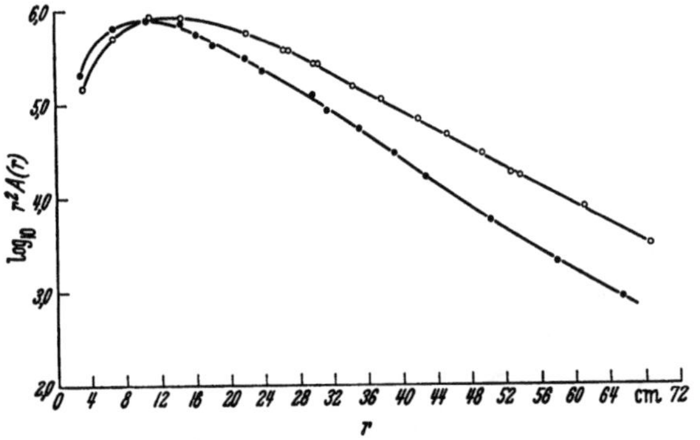

Fig. 139. Logarithmic plot of $r^2A(r)$ versus r for a fission source in water-aluminium heterogeneous arrangements • H_2O: Al = 2:1; ∘ H_2O:Al = 1:1 in volume (ROBERTS *et al.*).

γ) Distribution in water of In-resonance neutrons from monoenergetic sources ($d+t, d+d, N^{17}$). All the experiments described above have the inconvenience that the spectrum of fast neutrons emitted by the source in rather complicated and usually not very well known. Therefore it is difficult to establish how much of the discrepancy between theory and experiments that will be discussed later is due to the energy spectrum adopted and how much to the assumptions or approximations introduced in the calculation of the slowing down process.

Table 52. *Age to indium resonance and to thermal energies from a point fission source in aluminium-water mixtures.*

Volume Al / Volume H_2O	$\overline{r^2_{\mathrm{In}}}$ (cm²)	τ_{In} (cm²)	l'_{tr} (cm)	τ_{th} (cm²)	Reference
0	185	30.8	8.79	31.8	4
0.5	297.4	49.6	8.49	51.6	2
1.0	460.7	76.8	8.98	79.8	2

One of the very few experiments made in good geometrical conditions with a monoenergetic source is due to CASWELL and coworkers[5]. These authors have used neutrons of 14.1 Mev emitted by a tritium + zirconium target, at 90° with respect to the incident beam of deuterons of 150 kev energy. The target, the diameter of which in different runs varied between 0.9 and 1.4 cm was located in a

[1] See footnote 1, p. 316.
[2] L. D. ROBERTS, J. E. HILL and T. E. FITCH: ORNL-294. J. Appl. Phys. **26**, 1018 (1955).
[3] WEINBERG, SOODAK, DISMUKE and ARNETTE: Mon P-219.
[4] See footnote 3, p. 316.
[5] R. S. CASWELL, R. F. GOBHARD, D. W. PADGETT and W. P. DOERING: Nucl. Sci. and Engng. **2**, 143 (1957).

tank of 245 cm diameter and 152 cm deep, at about 76 cm from the bottom and 60 cm from the side wall. The deuteron beam from the accellerator was brought down to the target into the water through a thin-walled aluminium tube 2.5 cm in diameter. The neutrons were detected, for r ranging between 5 and 70 cm from the target, by means of In foils of 2.0×3.4 cm² area and 1.27×10^{-2} cm thick with and without a 5.08×10^{-2} cm thick Cd cover. For r ranging between 45 and 125 cm, a cylindrical B^{10}-lined counter (20 cm long) was used with and without a 5.08×10^{-2} cm thick Cd cover. Foil positions were determined to an accuracy of about 0.1 mm by cathetometer measurements.

The thermal neutron distribution was obtained by "Cd-difference" of the In foil activities for r up to 60 cm and from the B^{10}-counter data for r up to 125 cm; the normalization was obtained from the foil data between 45 and 60 cm. The flux distribution of the resonance neutrons was obtained, as usual, from the activities of the Cd-covered In foils.

Fast neutron dose measurements[1] were also made using a polyethylene-ethylene proportional counter dosimeter[2] for distances from the centre of the target up to 80 cm. The neutron fluxes were calibrated by associate particle counting, namely by counting the He^4 emitted in the $H^3(d, n)$ reaction. Data of this type are of the greatest interest in connection with the problem of neutron shielding from fast neutron sources.

Figs. 140 to 142 show the results obtained by these authors for thermal neutrons, In-resonance neutrons and fast neutron dose.

From the data of Fig. 140 Caswell and others deduce $\overline{r^2_{\mathrm{th}}} = 975$ cm² which, if we correct for the 2.5 cm diameter of the aluminium tube containing the tritium-zirconium target, reduce to $\overline{r^2_{\mathrm{th}}} = 935$ cm². The migration area of thermal neutrons (82.8) is then

$$M^2 = (156 \pm 6)\ \mathrm{cm}^2. \tag{83.1}$$

The chief causes of error are an uncertainty of the order of 0.5 cm in the location of the effective centre of the target source, and the uncertainty in the correction mentioned above for the effect of the evacuated tube containing the target.

From the flux distribution of the In-resonance neutrons of Fig. 141 extrapolated from 70 cm to infinity by means of (82.4) with the same value of l'_{tr} as that deduced for thermal neutrons (Fig. 140) it is found that $\overline{r^2_{\mathrm{In}}} = 900$ cm² and, therefore,

$$\tau_{\mathrm{In}} = (150 \pm 6)\ \mathrm{cm}^2 \tag{83.2}$$

where the largest errors come from the same sources as in the thermal neutron measurements (83.1).

With almost exactly the same technique, a group at the Bureau of Standards[3] has started a systematic investigation of the space distribution of In-resonance —and thermal—neutrons in water produced through the $d+d$ reaction.

In this case, however, some difficulty arises from the contamination of the neutron spectrum by the $d+\mathrm{C}$ reaction whose threshold is at 328 kev (Sect. 58); this inconvenience is avoided in the results reported below since they have been obtained with a deuteron beam of $E_d = 250$ kev impinging upon a deuterated polyethylene foil 1 mg/cm² thick.

As mentioned in Sect. 82 an inconvenient feature of the $d+d$ reaction is the strong variation of the neutron energy with the direction of emission with

---

[1] In Figs. 131 and 134 the absorbed dose is measured in rep, in Fig. 142 it is measured in erg/g. The unit reccomanted by ICRU is the rad: 1 rad $\widehat{=}$ 100 erg/g, 1 rep $\widehat{=}$ 93 erg/g. See footnote 7, p. 89.

[2] G. S. Hurst: Brit. J. Radiol. **27**, 353 (1954).

[3] V. Spiegel, D. W. Oliver and R. S. Caswell: Nucl. Sci. and Engng. **4**, 546 (1958).

respect to the incident beam; in the experimental conditions adopted by these authors the neutron energies are 3.12, 2.52 and 2.00 Mev at 0, 90 and 180°, respectively.

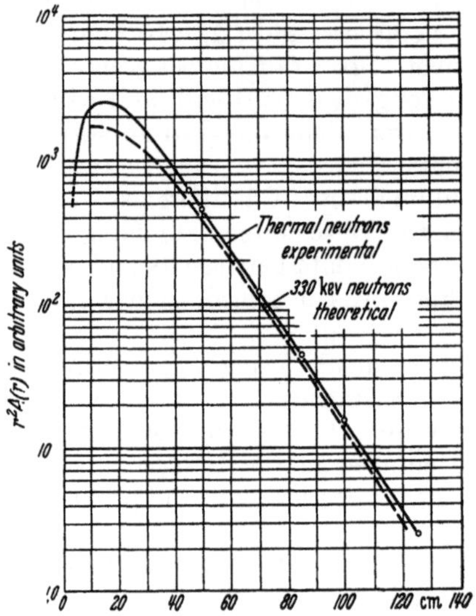

Fig. 140. Thermal neutron space distribution in water due to a point source of $E_0 = 14.1$ Mev: indium foil data 5 to 60 cm, boron lined counter data 45 to 125 cm. (CASWELL *et al.*).

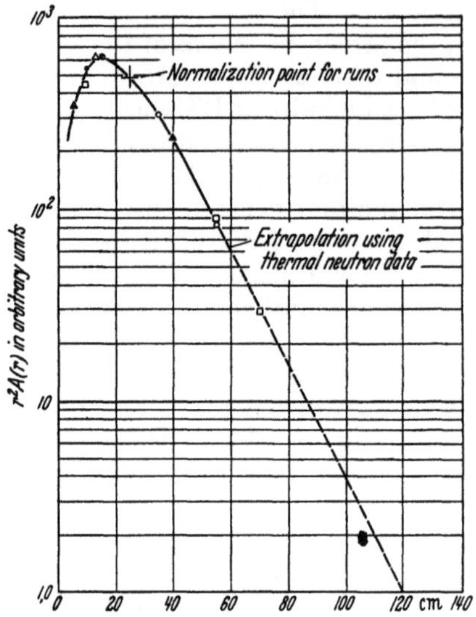

Fig. 141. Indium resonance neutron space distribution in water due to a point source of $E_0 = 14.1$ Mev (CASWELL *et al.*).

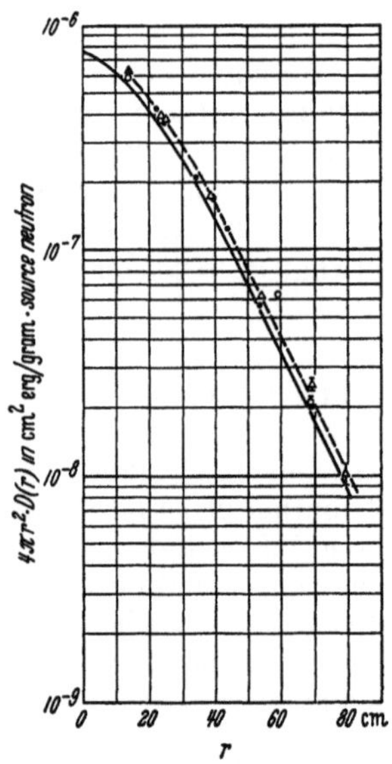

Fig. 142. Space distribution of fast neutron dose in water due to a point source of $E_0 = 14.1$ Mev (Caswell *et al.*).

The measurements were made at various angles: 0, 20.3. 90 and 159.7°. The neutron distribution in the backward direction was considerably distorted by the presence of the duct that brings the deuterons into the water medium. The perturbance of the duct is negligible at 0° and gives rise to a correction of the value of the age of 1.2 cm² at 90°. The results obtained at 0 and 90° are given in column 4 of Table 53. The theoretical values given there will be discussed later (p. 325).

We shall also mention an experiment made by SHURE and ROYS[1] on the spatial distribution in water of thermal neutrons emitted from N^{17} [2]. This radioactive body is produced through the endoenergetic reaction $O^{17}(n, p)N^{17}$ ($Q_0 = -8$ Mev) with a cross section of the order of 10^{-26} cm² [2]. The nuclide N^{17} decays into one or more exited states of O^{17} by beta-ray emission with a half-life of 4.13 sec, and O^{17} promptly emits a neutron in its transition to O^{16}. The spectrum of the delayed neutrons shows a most probable energy of 0.92 Mev, a full width at half-maximum of less than 0.5 Mev and an upper limit of about 2 Mev[3].

---

[1] K. SHURE and P.A. ROYS: Nucl. Sci. and Engng. **2**, 170 (1957).
[2] R.A. CHARPIE, K.H. SUN, B. JENNINGS and J.B. NECHAJ: Phys. Rev. **76**, 1255 (1949).
[3] L.W. ALVAREZ: Phys. Rev. **75**, 1127 (1949).

Table 53. *Age of d+d neutrons to* In-*resonance in water (Spiegel et al.).*

Deuteron energy kev	Angle of neutron emission	Neutron energy Mev	Neutron age [cm²]		
			experimental	theoretical (a)	
				S. G.	rigorous
250	0°	3.12	43.5 ± 1.8	50.4 (b)	39.9 (b)
	90°	2.52	31.9 ± 1.3 (c)	43.9	33.5
150	Averaged			42.1	33.8
100	Averaged			42.1	33.6

(a) P. F. ZWEIFEL: Memo—PFZ—1 (1955).
(b) These two values are determined by a straight line extrapolation of the date given in (a).
(c) A correction for the absence of moderator, displaced by the duct that brings the deuteron beam into the water medium, has been included in this number. The uncorrected value was 33.1 cm².

The N^{17} source is obtained by circulating demineralized water through a thimble located in a high fast flux region of the MTR by means of a constant flow pump. The water is kept about 20 sec in the high flux region and takes about 25 sec to travel from the thimble to the source disk. This consists of a flat piece of lucite, of outer radius $R = 11.5$ cm, in which 14 concentric grooves had been machined. The "active water" enters the disk at the centre, circulates through each groove in turn and leaves at the outer edge. This source may be described as a *flat isotropic source* the strength per unit area of which is

$$S(r) = S(0)\,e^{-a^2\varrho^2} \quad \text{for} \quad 0 \leqq \varrho \leqq R; \quad a^2 = \frac{\pi A}{F b \tau} \tag{83.3}$$

where A is the cross-sectional area of the grooves in the disk, b the distance between the centres of two adjacent grooves, F the flow of water through the disk and τ the mean life of N^{17}. By means of this formula it is possible to convert the actually measured distribution into the distribution that one would observe for a point isotropic source or a plane isotropic δ-source; the procedure is essentially the same followed in Sect. 71 for transforming the neutron flux distribution due to an infinite plane δ-source into that due to a point δ-source.

The thermal neutron distribution was measured along the axis of the flat source, by means of a BF_3 counter $\frac{7}{8}$ inch $= 2.22$ cm in diameter, 9 inch $= 25$ cm long, filled at 55 cm Hg pressure with BF_3 enriched in B^{10} to 96%. The results had to be corrected for the background arising from the $D(\gamma, n)$ reaction produced in water by γ-rays of N^{16} ($T_{\frac{1}{2}} = 7.35$ sec, $h\nu = 6.13$ Mev, $h\nu = 7.10$ Mev). The nuclide N^{16} is produced in water, under neutron bombardment, through the $O^{16}(n, p)\,N^{16}$ reaction.

The thermal neutron distribution, corrected for the background and transformed to an infinite plane δ-source, is shown in Fig. 143. From this the following value of the migration area of thermal neutrons is then deduced:

$$M^2 = 23{,}1\ \text{cm}^2. \tag{83.4}$$

A discussion of this result is given on p. 325.

Finally we should mention the experiments of BARKOV, MAKARIN and MUKHIN[1] on the spatial distribution of the In-resonance neutrons produced bya $Sb^{124}\gamma + Be$ source. According to the data of Table 27 (p. 160), the spectrum of the neutrons emitted in this reaction consists of a line at 31 kev. The measurements were made in a tank 80 cm in diameter and 100 cm high, using In foils 100 mg/cm² thick and of sizes varying from 0.75×1.25 cm² to

[1] L.M. BARKOV, V.K. MAKARIN and K.N. MUKHIN: J. Nucl. Energy 4, 94 (1957).

3.0×2.5 cm² depending upon the distance from the source. During irradiation, the In detectors were contained in boxes of cadmium foil 0.6 cm thick for resonance neutrons and of stainless steal foil 0.2 mm thick for resonance + thermal neutrons (see Sect. 110).

Two neutron sources were employed, both consisting of a beryllium cylinder with an antimony rod placed along its axis; the corresponding geometrical dimensions and neutron strengths are given in Table 54. The small source has obvious geometrical advantages for measurements at short distances, while the large source has a strength large enough to extend the measurements up to $r \approx 35$ cm.

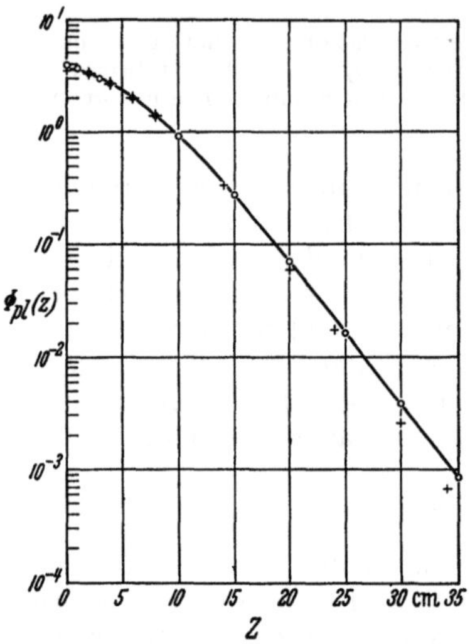

Fig. 143. Thermal neutron flux distribution from a plane source of $N^{1\cdot}$: ○ experiment, + 1 Mev theory (SHURE and ROYS).

The age values derived from the space distribution of the activity of cadmium-covered In detectors are given in column 5 of the same table. Column 6 shows the same quantities after a correction $\delta\tau = (0.2 \pm 0.1)$ cm² has been applied for taking into account the contribution to the In activity arising from neutrons of energy higher than 1.46 ev:

$$\delta\tau = \tau(1.46\text{ ev}) - \tau_{\mathrm{In}} = \frac{a_{3.9}}{a_{\mathrm{Cd}}}\,\tau(3.9\text{ ev} - 1.46\text{ ev}) + \frac{a_{9.1}}{a_{\mathrm{Cd}}}\,\tau(9.1\text{ ev} - 1.46\text{ ev}) + \int_{10\text{ ev}}^{E_0} \tau(E - 1.46\text{ ev})\,\frac{c\,dE}{E\cdot E^{\frac{1}{3}}}. \tag{83.5}$$

Here

$$\tau(E - 1.46\text{ ev}) = \frac{l_{\mathrm{tr}}\,l_s}{3}\ln\frac{E_{\mathrm{ev}}}{1.46},$$

a_{Cd} is the total number of neutrons absorbed in the cadmium covered In foil, and $a_{3.9}$ and $a_{9.1}$ are the numbers of neutrons belonging to the two strong resonances at 3.9 and 9.1 ev (Table 9, Fig. 15) which are absorbed in the In foil. The third term represents the contribution arising from neutrons of energy $E \geqq 1$ ev; c is a constant determined by taking the activation cross section for $E \geqq 10$ ev to be $\sigma = \frac{20}{E^{\frac{1}{3}}} \cdot 10^{-24}$ cm².

A comparison of the results obtained with the two sources shows a difference of 0.4 cm² from which the authors conclude that the age due to a point source should be smaller than 5.48 cm². Since the value computed by means of Eqs. (84.9) and (71.8) for $E_0 = 30$ kev is $\tau(1.46\text{ev}) = 4.9$ cm², the difference between the experimental and the theoretical values of $\tau(1.46$ ev) is in the direction expected; one should keep in mind, however, that it could be due in part to the finite size of the source, in part to the presence of ≈ 330 kev energy neutrons (see footnote (*m*) to Table 27, p. 160).

Table 54. *Age to* 1.46 ev *and migration area to thermal energies of neutrons from* Sb γ + Be *source in water (Barkov et al.).*

Source	Be cylinder Diam. mm	Be cylinder height mm	Sb rod Diam. mm	Sb rod height mm	Source strength sec⁻¹	τ_{In} cm²	τ(1.46 ev) cm²	M^2 cm²	$\Delta\tau$
Large	18	22	8	22	$\approx 10^6$	5.66 ± 0.15	5.86 ± 0.15	14.13 ± 0.25	1.1 ± 0.5
Small	10	20	6	20	$\approx 10^5$	5.28 ± 0.15	5.48 ± 0.15	13.64 ± 0.25	1.0 ± 0.5

The remaining part of Table 54, referring to thermal neutrons, is considered in Sect. 110.

δ) Comparison with theoretical calculations. We shall now try to compare the experimental results reported in this section with the results of calculations made with the methods mentioned in Sect. 79 and 80. As we shall see at the end of this section, the more refined treatments, which take into account the effect of inelastic scattering, give values of the age to the In resonance which agree with the experimental data within the uncertainties of the computation—due to the adopted values of the cross sections—and the estimated standard error of the measurements.

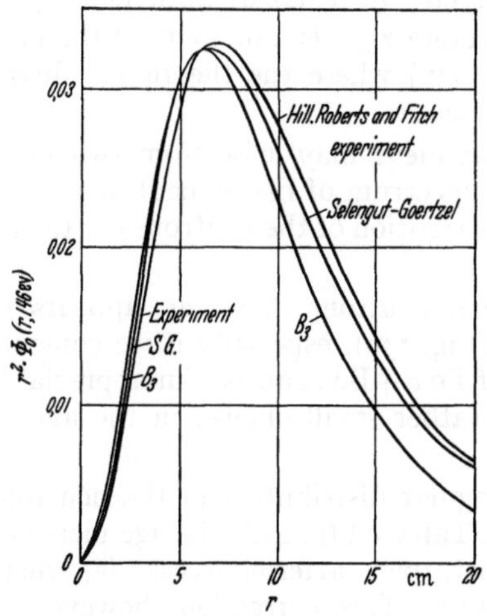

Fig. 144. Comparison of experimental results with theory: space distribution of In-resonance neutrons in water due to a point fission source[1, 2].

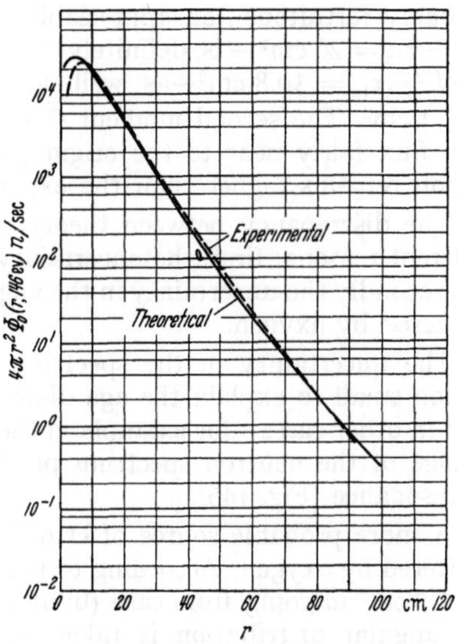

Fig. 145. Comparison of experimental results on flux distribution of In-resonance neutrons with theoretical computations by method of moments[3].

The point fission source flux distribution at $E_R = 1.46$ ev as obtained by the B_3 and S.G. methods[1, 2] are compared with the experimental results of Hill *et al.* in Fig. 144. The B_3 approximation gives a flux distribution which is too large at small distances and too small at large distances. The better agreement of the S.G. procedure is fortuitous and is due to the overestimate of the first flight correction mentioned in Sect. 79.

In Fig. 145 the results of Hill *et al.* (Fig. 138, upper curve) are compared with the flux distribution at 2.03 ev calculated by means of the method of moments for an isotropic fission point source embedded in water[3]. The two curves have been arbitrarily normalized at $r = 20$ cm. The shape of the distribution is expected to be almost the same at 1.46 and 2.03 ev.

The agreement between the two curves is satisfactory except at small distances where there are significant discrepancies. The same situation appears in Table 55 where the values of the corresponding moments are compared. While the agreement is very good for the higher moments (the excellent agreement for $\overline{r^8}$ is

[1] J. E. Wilkins, R. L. Hellens and P. F. Zweifel: p. 62 of Ref. [*15c*].

[2] H. Hurwitz and R. Ehrlich: Progress in nuclear energy. Physics and Mathematics, Vol. 1, p. 343. London: Pergamon Press 1956.

[3] J. Certain and R. Aronson: NDA-15c-40 (1954).

Table 55. *Comparison with experiments of theoretical values of the moments of the flux distribution in water of resonance neutrons produced by a point fission source*[1].

n	$\overline{r^n_{exp}}$	$\overline{r^n_{theor}}$	$(\overline{r^n_{exp}})^{1/n}$ [cm]	$(\overline{r^n_{theor}})^{1/n}$ [cm]
0	1	1	1	1
2	185	154	13.6	12.4
4	1.22×10^5	0.979×10^5	18.7	17.7
6	2.27×10^8	1.934×10^8	24.7	24.1
8	8.64×10^{11}	8.753×10^{11}	31.5	31.4

probably fortuitous), an appreciable discrepancy is found for $\overline{r^2}$. The theoretical age — $\tau_{In}=26$ cm^2 — is definitely lower than the experimental value of HILL *et al.*[2] — $\tau_{In}=30.8$ cm^2 — as well as that given by Russian authors[3] — $\tau_{In}=29\pm1$ cm^2. The second moment $\overline{r^2}$ — and therefore τ_{In} — is a measure of the neutron flux fairly near to the origin ($r\sim5-35$ cm) where the theoretical distribution remains higher than the experimental data.

The discrepancy between theory and experiment may arise from two main different reasons: first, the uncertainty in the spectrum of fission neutrons (82.6) and secondly the uncertainty in the angular distribution of the neutrons elastically scattered by oxygen.

The uncertainty in the spectrum of fission neutrons, however, appears to be too small to explain the age discrepancy (Fig. 136), especially if we consider that in other cases—for example in the case of Po α+Be sources—an appreciable change of the neutron spectrum produces a rather small change in the age to In resonance (Fig. 146).

A more probable source of error is the angular distribution of the neutrons scattered by oxygen. According to the data of Table 42 (p. 292) the age increases by 1.7 cm^2 in going from case (b) of oxygen isotropic scattering to case (a) where the angular distribution is taken into account. This correction, however, is small only because it represents the difference between two large contributions of opposite sign; above 2.5 Mev the oxygen scattering is mainly forward while around 2 Mev it is predominantly backward[4]. Therefore, even a rather small variation of the data relative to the oxygen scattering could produce a considerable change in the corresponding correction of age.

It should be noted that the consideration of inelastic scattering cannot be invoked for explaining this discrepancy since its effect would be in the sense of further reducing the theoretical value.

A discrepancy of this same type is found in the case of a fission source embedded in aluminium-water mixtures; the theoretical values obtained by the method of moments in the P_1-approximation as well as by the S.G. method—under the assumption that both the anisotropy of elastic scattering and the inelastic scattering by the metal can be neglected—turn out to be definitely lower[5] than the experimental values given in Table 52.

The contribution of inelastic scattering to the slowing down is expected to be of limited importance for the case of fission sources because only a rather weak tail of the corresponding spectrum (82.6) extends in the high energy region.

[1] See footnote 3, p. 323.
[2] See footnote 2, p. 316.
[3] S.M. FEINBERG: p. 484 of Ref. [*15c*].
[4] H. GOLDSTEIN: NDA-Memo 15c-15 (1953). — D. J. HUGHES and R. S. CARTER: BNL-400 (1956).
[5] See footnote 1, p. 323.

The situation is quite different for the 14 Mev neutron case; the experimental value (83.2) obtained by CASWELL *et al.*[1], is appreciably lower than the theoretical value given by VOLKIN (Sect. 80): from Fig. 128 we deduce that $\tau_{\text{In}} = 173 \text{cm}^2$. These computations, however, neglect the energy degradation which accompanies oxygen scattering and arises from both elastic as well as inelastic scattering. The experimental point of CASWELL *et al.*, on the contrary, is in reasonable agreement with the theoretical result of REIER *et al.* which takes into account the effect of inelastic scattering as we shall see later in this section (Fig. 147).

The results of fast neutron dose measurements are compared in Fig. 142 with theoretical calculations (Fig. 134) made by GOLDSTEIN *et al.*[2] using the method of moments and taking into account inelastic scattering. Both curves show the same slope as functions of r, but the theoretical one lies about 15% below the experimental data. The agreement is, however, within the estimated standard errors of the experiment (20%) and of the calculation (15%).

Table 53 shows a comparison of the experimental values of τ_{In} for the $d+d$ neutrons with the corresponding theoretical results. These have been obtained by graphical extrapolation of computations made by ZWEIFEL[3], at slightly lower energy (150 and 100 kev, averaged) using the S.G. method and the B_3 approximation. In the second calculation the first four components of the Legendre polynomial expansion of the source distribution were used for deriving the flux as a function of angle and distance, and the ages thus obtained were averaged over the angle. The S.G. method gives age values which are systematically too large for the reasons explained in Sect. 79. In the forward direction the measured age is about ten percent larger than the rigorous theoretical value. The opposite situation is met at 90° where the experimental value is about five percent lower than the theoretical result. A quantitative comparison between theory and experiment is more complicated for the $d+d$ reaction than for all other cases considered in this section because of the dependence of the neutron's energy upon the angle of emission (Sect. 82). Thus, the existence of a disagreement between the experimental result and ZWEIFEL's computation can be stated with certainty only at 0°; its origin is not clear since at neutron energies of the order of 3 to 4 Mev, oxygen inelastic scattering can not be invoked. Probably it arises from errors in the angular distributions of the neutrons elastically scattered by oxygen.

SHURE and ROYS have compared their experimental results obtained with N^{17} neutrons ($E_0 = 920 \pm 200$ kev) with the distribution of thermal neutrons computed by means of the following procedure. They start from the slowing down density of the neutrons of 1 ev obtained by HOLTE[4] for an initial energy of 1 Mev (Table 44) and transform it, by means of (71.2a), into the corresponding distribution in water for a monoenergetic isotropic plane δ-source. This distribution is then assumed to represent a distributed source of thermal neutrons which are allowed to diffuse with $L = 2.73$ cm (Sect. 125, Table 96). The results of such a computation are in good agreement—even on an absolute scale—with the experimental points outwards to 15 cm from the source (Fig. 143). This procedure neglects the slowing down from 1 ev to thermal energies. Therefore, it is interesting to compare the results of the application of HOLTE's theory with these experimental results from a slightly different angle, and compute the difference between the experimental

[1] See footnote 5, p. 318.

[2] H. GOLDSTEIN, J. CERTAINE, R. ARONSON and J.E. WILKINS: Unpublished; results quoted in footnotes 1 and 2, p. 323.

[3] P.F. ZWEIFEL: Memo-PFZ-1 (1955).

[4] G. HOLTE: Ark. Fysik **2**, 523 (1951); **3**, 209 (1951); **8**, 165 (1954).

value (83.4) of the migration area of thermal neutrons and the theoretical value of the age to 1 ev; this gives

$$M^2 - \tau_{1\,\mathrm{ev}} = L^2 + \tau_{1\,\mathrm{ev\text{-}th}} = 8.2\ \mathrm{cm}^2 \tag{83.6}$$

which is compatible with the results reported by other authors.

We shall finally consider the results of computations made by REIER *et al.*[1] in connection with their measurements of the age to In resonance from a Po α+Be neutron source (Table 50, p. 315); these appear to be more realistic than all the computations mentioned above. The Westinghouse group again uses the method of moments in the P_1-approximation, starting from the following data.

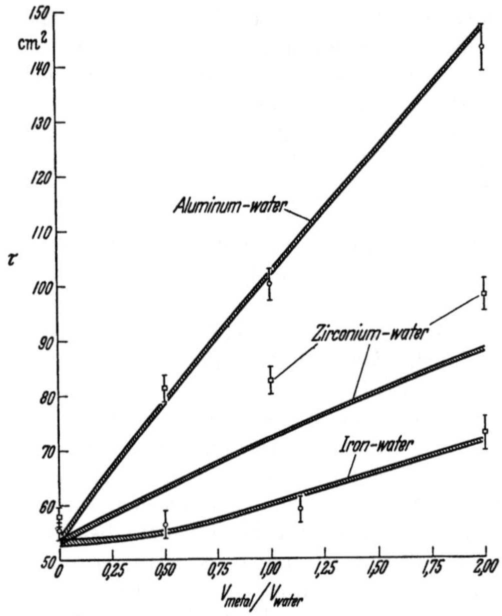

Fig. 146. Neutron age to In-resonance from a Po α + Be source versus metal-water volume ratio (REIER *et al.*).

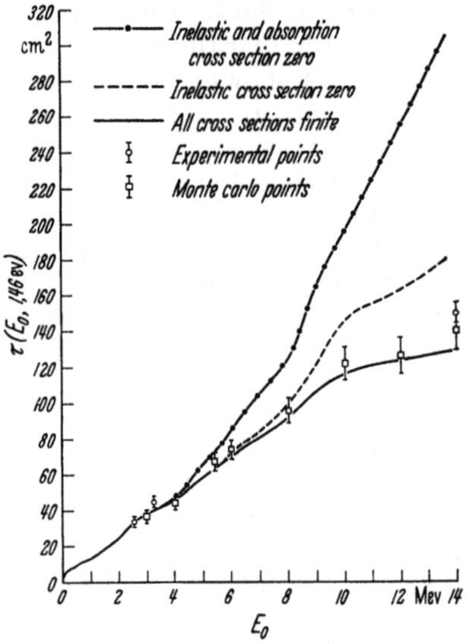

Fig. 147. Neutron age to In-resonance in pure water versus energy of the neutrons emitted from a mono-energetic source (REIER *et al.*).

First, for the neutron spectrum of the source, these authors use the results of WHITMORE and BAKER given in Fig. 45 (p. 104). If the spectra given by other authors[2] are adopted, the final results of the computations change only slightly, as it is indicated by the narrow bands of the age values given in Fig. 146. Secondly, the elastic collision cross sections and angular distributions have been taken from [*14*] and[3]. Whenever the elastic scattering angular distribution data were scanty, an extrapolation was made by fitting the parameters of the complex potential model for neutron scattering at the few energies at which measurements were available. The angular distributions required at the other energies were computed using these parameters. Finally the inelastic scattering cross sections were based on experimental results of the type of those mentioned in Sect. 30. In some cases, notably iron and aluminium, information on the energy spectrum of the inelastically scattered neutrons was available: whenever such information

[1] M. REIER, F. OBENSHAIN and R. L. HELLENS: Nucl. Sci. and Engng. **4**, 1 (1958).

[2] R. G. COCHRAN and K. M. HENRY: ORNL-1479 (1953). — Rev. Sci. Instrum. **26**, 757 (1955). — J. R. BREEN, M. R. HERTZ and D. V. WRIGHT: MLM-1054 (1955). — J. O. ELLIOT, W. I. MCGARRY and W. R. FAUST: Phys. Rev. **93**, 1348 (1954).

[3] D. J. HUGHES and R. S. CARTER. BNL-400 (1956).

was lacking the evaporation model was employed. An attempt has been made to include in the absorption cross section the contributions arising not only from radiative capture, but even from charged particle reactions.

Fig. 146 shows a comparison of the results of measurements of τ_{In} (Table 50) with the computed values. For pure water, iron-water and aluminium-water mixtures the results agree rather well. The discrepancy observed for zirconium-water mixtures are probably due to the considerable uncertainty in the zirconium inelastic cross section.

The sensitivity of the second moment to small variations of the cross sections is illustrated in Fig. 147 where the age to In resonance in pure water is plotted versus the energy E_0 of the neutrons emitted from the source. The effect of suppressing the inelastic scattering cross section or both the inelastic and the capture cross sections of oxygen, increases rapidly with increasing initial energy because of the rapid decrease of the hydrogen cross section with respect to that of oxygen; at 14 Mev the oxygen capture and the inelastic cross sections are 0.4 and 0.5 barns respectively.

A few points obtained by the Monte Carlo method are plotted in the same figure; these are based on the same cross sections, used in the method of moments computation, for all processes except the differential elastic scattering for which the cross sections were taken from the compilation by GOLDSTEIN[1].

The experimental point at 14 Mev is slightly higher than the computed value, but small changes in the oxygen cross sections would bring the two into agreement.

84. The case of moderators with $A > 1$. *α) Heavy water.* The majority of the measurements of flux distribution in heavy water refer to fission source neutrons slowed down to the In resonance. The first determination was made by FRIEDMAN and WATTENBERG[2] and is reported in Vol. II of The Reactor Handbook[3]. The results of measurements of the activity of cadmium covered In detectors as a function of the distance r from a fission point source, were fitted by a function of the form

$$A(r) = \frac{A\,e^{\frac{\tau_1}{\tau_2}}}{8\pi r \tau_2}\left\{\left[1+\operatorname{Erf}\left(\frac{r}{2\sqrt{\tau_1}} - \sqrt{\frac{\tau_1}{\tau_2}}\right)\right] e^{-\frac{r}{\sqrt{\tau_2}}} - \left[1-\operatorname{Erf}\left(\frac{r}{2\sqrt{\tau_1}} + \sqrt{\frac{\tau_1}{\tau_2}}\right)\right] e^{\frac{r}{\sqrt{\tau_2}}}\right\}. \tag{84.1}$$

This expression was obtained by the authors by describing the slowing down process as taking place in two steps. The first is a pure constant velocity diffusion process taking place around the point source such that the diffusion length $L=\sqrt{\tau_2}$; the corresponding flux distribution is given by (78.17). The second step is a slowing down process as described, according to the age theory, by Eq. (76.16) with $\tau=\tau_1$.

This model should not be taken as a correct description of the actual physical situation; it was suggested by the authors because it has the merit of providing the two parameter formula (84.1) which decreases exponentially for large values of r (see Sect. 77).

The best fit of the experimental data is obtained with $\tau_1 = 58\ \text{cm}^2$ and $\tau_2 = 48\ \text{cm}^2$. Since it follows from (84.1) that $\tau = \frac{1}{6}\overline{r^2} = \tau_1 + \tau_2$, the value of the fission

[1] See footnote 4, p. 324.

[2] F.L. FRIEDMAN and A. WATTENBERG: CP 3453 (1946).

[3] The Reactor Handbook, Vol. 2 "Physics", AECD-3645; and McGraw-Hill, New York 1955.

neutron age to In resonance in heavy water, containing 99.8% D_2O, turns out to be

$$\tau_{In} = 106 \text{ cm}^2. \tag{84.2}$$

A slightly higher value has been given by OLCOTT[1] who used, rather arbitrarely, Eq. (76.16) for fitting his experimental points. These were obtained by placing at the centre of a stainless steel spherical shell of 45 cm radius filled with heavy water (99.8% D_2O), a second thin walled spherical shell of 12 cm radius filled with a solution of uranyle fluoride (UO_2F_2)-heavy water. The uranium was 90% U^{235} and the concentration of the solution was such that, when a neutron source was placed at its centre, the assembly had a neutron multiplication of about eight.

By plotting the logarithm of the activity of cadmium covered In-foils versus r^2 (r distance of the detector from the centre of the assembly) the experimental points fall very well into a straight line for r^2 varying between 200 cm² and 1800 cm² ($14 \leqq r \leqq 43$ cm). Assuming that Eq. (76.16) can fit the experimental points obtained under the conditions mentioned above, the slope of this straight line is $(4\tau)^{-1}$, a relation that is used by OLCOTT to deduce $\tau_{In} = (111 \pm 1)$ cm². This procedure would be correct if we could state that (76.16) represents the slowing down in heavy water of the neutrons emitted from a fission point source. In fact, by introducing (76.16) into (71.1) and integrating over the volume of the extended spherical source used in this experiment, it is found that—in a certain interval of distances—the distribution of resonance neutrons still depends upon r according to (76.16). But there is no reason to except that the age theory provides a satisfactory representation of slowing down in heavy water. Furthermore the fission spectrum extends over a wide range of energies (Fig. 136). Therefore—even according to age theory—the actual neutron distribution from a point source should be represented by an integral extending over the fission spectrum of the monoenergetic kernel (76.16). Because of all these reasons the good agreement of this result with those of other authors has to be considered fortuitous.

Measurements of the neutron age in mixtures of light and heavy water ranging from 99.8 to 91.8% D_2O, have been made by WADE[2]. The experiments were performed in an aluminium tank of 76 cm radius and 185 cm height, situated directly over a small pile (Standard Pile: S.P.) of enriched uranium and graphite, which served as a source of neutrons. The source of fission neutrons was a stack of ten foils of highly enriched uranium that was irradiated by thermal neutrons from the S.P.; the foils were 1″ = 2.54 cm in diameter and 0.005″ = 0.127 mm thick. The fission source was placed approximately 35 cm from the bottom of the tank.

The detectors were indium foils 100 mg/cm² thick, $\frac{3}{4}$ inch ≈ 2 cm in diameter with cadmium covers 0.030 inch = 0.76 mm thick.

In order to eliminate the activity produced by the neutrons from the pile, the difference was taken between the activity of the cadmium covered In foils observed with the U^{235} fission source in place and the activity observed with the fission source removed, all other experimental conditions being the same.

Fig. 148 shows the results of the measurements, and Table 56 the values of the age and of the relaxation length as they are obtained from smooth curves drawn through the experimental points, after correction for the finite dimensions

[1] R.N. OLCOTT: Nucl. Sci. and Engng. 1, 327 (1956).
[2] J.W. WADE: DP-163 (1956). — Nucl. Sci. and Engng. 4, 12 (1958).

Table 56. *Age of fission neutron to* In-*resonance in heavy water as a function of the light water contamination (Wade).*

Slowing down media percent D_2O	τ (cm²)	Estimated error (cm²)	l'_{tr} (cm)	Extrapolated portion of area (percent)
99.8	109	±3	8.7	13
99.0	107	±5	8.7	14
98.4	106	±3	9.3	15
95.3	93	±3	9.0	13
94.0	86	±3	7.5	8
91.8	78	±3	6.7	9

of the source and detector [Eq. (82.1)] and for the activity due to high-energy neutrons [see Eq. (83.5)].

Fig. 149 shows τ as a function of the light water contamination. The decrease in τ is very nearly linear in this small range of moderator purity; it amounts to 4 cm² per one percent increase of the light water contamination.

The same author has made a study of the anisotropy of the age in a lattice of aluminium rods in 99.3% D_2O. The aluminium rods,

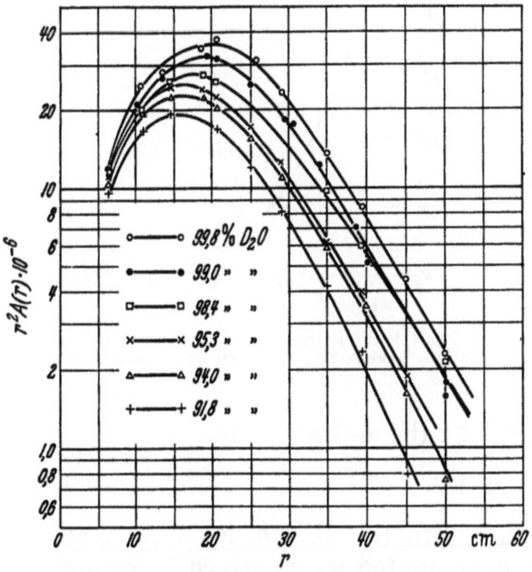

Fig. 148. Logarithmic plot of $r^2A(r)$ in H_2O-D_2O mixtures: point fission source, cadmium covered In-detector (WADE).

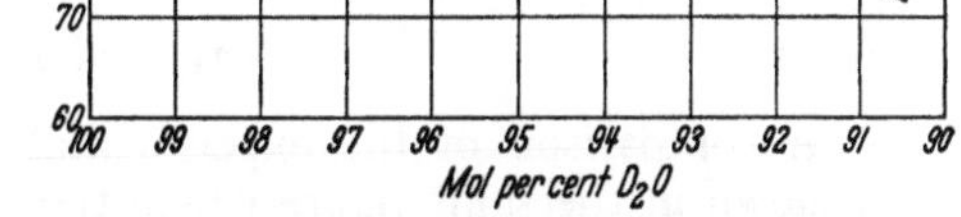

Fig. 149. Age of fission neutrons to In-resonance versus purity of heavy water (WADE).

1 inch in diameter were arranged, in clusters of four, in a kind of chess board pattern so that the volume ratio of aluminium to water was $\frac{1}{10}$. Fig. 150 shows the results of two sets of measurements; the upper curve represents the measurements made in a direction parallel to the axis of the rods, and the lower curve the measurements made in a direction perpendicular to the previous one. The corresponding values of the age and of the relaxation length are given in Table 57.

Table 57. *Age of fission neutrons to* In-*resonance in a lattice of aluminium in heavy water*

$$(99.3\ \%\ D_2O):\frac{\text{Volume Al}}{\text{Volume water}}=\frac{1}{10}\quad(\text{WADE}).$$

	τ (cm²)	Estimated error (cm²)	l'_{tr} (cm)	Extrapolated portion of area (percent)
Parallel to aluminium rods . . .	140	±3	10.2	4
Perpendicular to aluminium rods	108	±5	7.8	20

These results clearly shows the effect of fast neutron streaming along the aluminium rods as predicted by ROSE[1].

The value of 108 cm^2 for the age measured in the direction perpendicular to the aluminium rods is equal to the value for 99.3% D_2O within the precision of the measurements.

β) Beryllium. The flux distribution of In-resonance neutrons for neutrons emitted by a fission point source in Be has been measured by stacking Be-metal blocks in the form of a parallelepiped ($\varrho = 1.85$ g/cm^3) 100 cm × 100 cm × 65 cm high directly on the thermal column located in the top shielding of the Argonne graphite pile[2]. The fission neutron source consisted of a piece of U^{235} metal located on the axis of the parallelepiped. Indium foils wrapped in cadmium, were used as detectors.

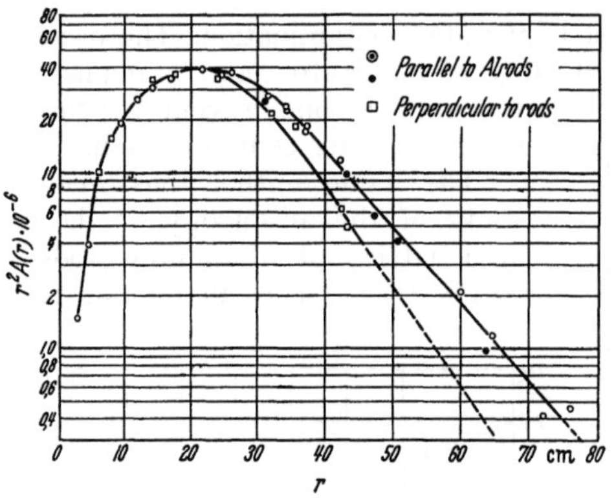

Fig. 150. Logarithmic plot of $r^2 A(r)$ in Al—D_2O lattice (WADE).

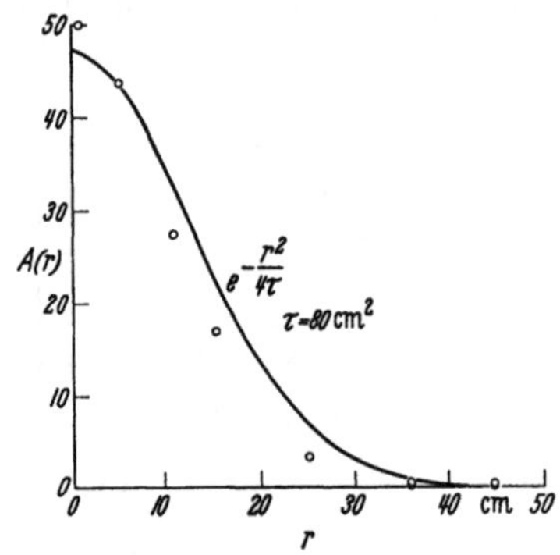

Fig. 151. Flux distribution of In-resonance neutrons produced in beryllium by a fission source (HUGHES *et al.*).

From these measurements, plotted in Fig. 151 the authors derive

$$\tau_{\text{In}} = (80.2 \pm 2)\ \text{cm}^2. \tag{84.3}$$

A comparison of the experimental points with the age theory Eq. (76.16) is shown in the same figure where the curve, computed for $\tau = 80$ cm^2, deviates appreciably from the experimental points, the main reason being probably the continuous spectrum (82.6) of fission neutrons.

According to Eq. (82.4), at large distances from the source (30 cm $\leqq r \leqq$ 50 cm), $r^2 A(r)$ shows an exponential behaviour with $l'_{\text{tr}} = 8.0$ cm. This value corresponds to a scattering cross section for Be of 1×10^{-24} cm^2 and therefore compares well with the scattering cross section of the unmoderated neutrons (1 to 2 barns at several Mev energy).

Other measurements have been made by Russian authors[3] with 1500 kg of Be-metal of the highest grade purity. Beryllium cubes of 4.0 cm edge with a mean density $\varrho = 1.78$ g/cm^2 were stacked in the form of a prism 68 cm × 68 cm × 132 cm. Source and detector were placed in an axial hole 28 mm in diameter. The source was located 50 cm from the bottom.

The detectors were In-foils, 25 mm in diameter, covered with 0.5 mm thick Cd layers. Two series of measurements made with Raα + Be and Poα + Be

[1] M.E. ROSE: Mon P-255.
[2] HUGHES, NOBLES and WALLACE: Unpublished; reported in Ref. [*12*].
[3] L.A. GERASEVA, A.V. KAMAYEV, A.K. KRASIN and I.G. MOROSON: p. 13 of Ref. [*15c*].

sources gave

$$\left.\begin{array}{ll} \text{Ra}\,\alpha + \text{Be}: & \tau = (120 \pm 23)\ \text{cm}^2 \\ \text{Po}\,\alpha + \text{Be}: & \tau = (95 \pm 11)\ \text{cm}^2. \end{array}\right\} \tag{84.4}$$

Figs. 152 and 153 show the experimental points obtained by these authors and a comparison of these with the corresponding curves computed by means of (76.16) using the values (84.4) for τ.

The results obtained with the $\text{Ra}\alpha + \text{Be}$ source have been corrected for the contribution of neutrons due to the $\text{Ra}\gamma + \text{Be}$ process in the moderator by substracting the neutron distribution obtained with a bare Ra source.

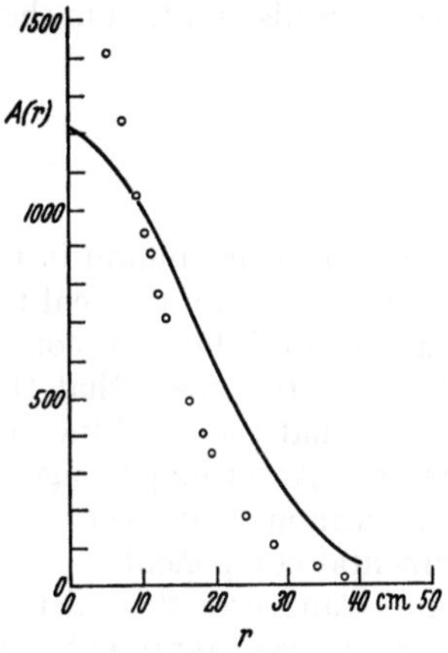

Fig. 152.

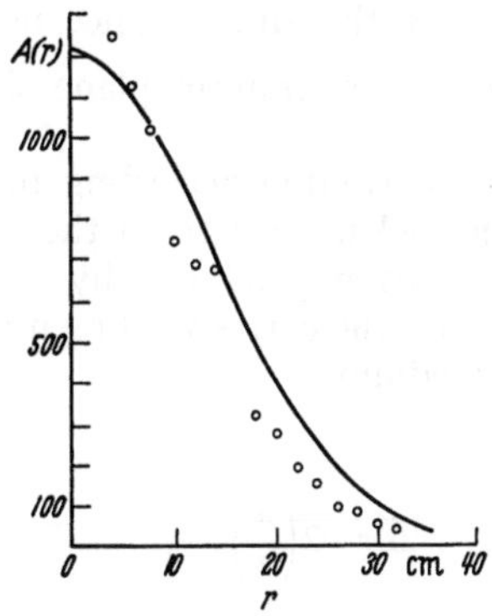

Fig. 153.

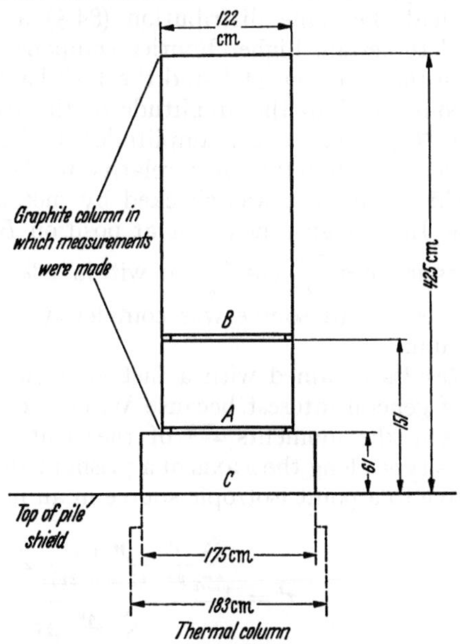

Fig. 154.

Fig. 152. Flux distribution of In-resonance neutrons produced in beryllium by a $\text{Po}\,\alpha + \text{Be}$ source (Geraseva *et al.*).

Fig. 153. Flux distribution of In-resonance neutrons produced in beryllium by a $\text{Ra}\,\alpha + \text{Be}$ source (Geraseva *et al.*).

Fig. 154. Graphite measuring column (Hill *et al.*).

γ) Graphite. The slowing down to In-resonance of fission neutrons in graphite has been investigated by Fermi, Marshall and Marshall[1] and more recently by Hill, Roberts and McCammon[2]. These authors stacked on top of the thermal column (170.5 cm × 170.5 cm) of the O.R.N.L. pile a parallelepiped of graphite henceforward called a graphite measuring column 122 cm × 122 cm in cross section and 366 cm high ($\varrho = 1.652$ g/cm³) as shown in Fig. 154. The thermal flux at the centre of the column in the plane A was 10^6 neutrons/cm² sec. The very large value of the Cd-ratio ($R_{\text{Cd}} = 140000$ at the centre of plane A: see Sect. 90) insured that the contribution of epithermal neutrons diffusing from the thermal column to the graphite measuring column was negligible.

[1] E. Fermi, J. Marshall and L. Marshall: CP-1084.
[2] J.E. Hill, L.D. Roberts and G. McCammon: ORNL-187 (1949).

The fission neutron source consisted of 36 squares of 10 cm side mounted on a graphite sheet in a chess board array the side of which was 122 cm. Each square consisted of a lamination 1 mm thick of an uranium-aluminium alloy (eutectic composition: 18% uranium by weight); 96% of the uranium was U^{235}.

Two experiments were made, the first with the source of fission neutrons in the upper position B, and the second with the source in the lower position A.

With the fission source removed, the thermal neutron distribution in the plane B was represented within 2% by the expression

$$A \cos \pi \frac{x}{a} \cdot \cos \frac{\pi y}{a} \tag{84.5}$$

with $a = 129.3$ cm.

When the source was inserted into the graphite measuring column in the slot at the upper position B, the fast neutron flux generated by the thermal neutron flux had the same distribution (84.5) as its fundamental Fourier component added to certain higher Fourier components. We have to expect that the first two of these are the 11th and the 13th harmonics and that the amplitude of each is about equal to the amplitude of the fundamental. According to age theory (Sect. 76), however, the amplitudes of the higher harmonics are very violently damped, as a function of τ, relative to the fundamental component.

This conclusion was checked by measuring the transverse flux distribution of the In-resonance neutrons at position B; the results were represented by the expression $\cos \frac{\pi x}{a} \cos \frac{\pi y}{a}$ to within 2%. Thus, for the In-resonance neutrons the chess board source was completely equivalent to a uniform plane sheet of uranium.

Results obtained with a fast neutron source distributed according to (84.5) are of general interest because WEINBERG and ROSE[1] have derived the relation between the moments $\overline{z^{2n}}$ of the neutron distribution produced by a source of this type along the z axis of a prism moderator, and the corresponding moments relative to a point isotropic source in an infinite medium:

$$\overline{\tau^n} = \frac{\sum_{\nu=0}^{\infty} \frac{\Delta^\nu}{\nu!} \frac{(n+\nu)!}{(2n+2\nu)!} \overline{z^{(2n+2\nu)}}}{\sum_{\nu=0}^{\infty} \frac{\Delta^\nu}{2\nu!} \overline{z^{2\nu}}}, \qquad \Delta = 2\left(\frac{\pi}{a}\right)^2 \tag{84.6}$$

where

$$\overline{\tau^n} = \int_0^\infty S_f(\tau)\, \tau^n\, d\tau, \qquad \int_0^\infty S_f(\tau)\, d\tau = 1. \tag{84.7}$$

Here $S_f(\tau)\, d\tau$ is the differential fission spectrum (82.6a) with reference to the age τ taken as independent variable.

As we shall see later, the infinite series (84.6) will be approximated by a finite sum up to a small value $\bar{\nu}$.

With the source in position B, measurements were made for z ranging from 5 to 86 cm. At larger distances the activity induced in the foils was too weak to be measured.

Therefore the second experiment was done with the source at the lower position A. Under these conditions, at the expenses of good geometry, a much higher intensity of the source neutrons is obtained which allows the measurements to be made up to $z = 140$ cm.

[1] WEINBERG and ROSE: Mon P-240; Mon P-297.

The results of measurements of the transverse distribution of In-resonance flux at position A could not be represented by a simple cosine of wavelength $2a$ but contained higher harmonics due to the change in the column cross section at A (Fig. 154). Those harmonics, however, were damped out (as a function of τ) so rapidly that there was no difference in shape of the distribution, within experimental error, when the source was at position A or at position B.

As detectors there served In-foils 4.0 cm × 5.35 cm in area and 93.8 mg/cm² thick, irradiated in Cd boxes with a wall thickness 0.34 mm. The irradiation was done by placing these boxes, with their planes parallel to the source plane, inside convenient slots located about 5 cm apart along the vertical axis of the graphite measuring column. During the same irradiation, foils were placed 10 cm apart in order to avoid mutual perturbation.

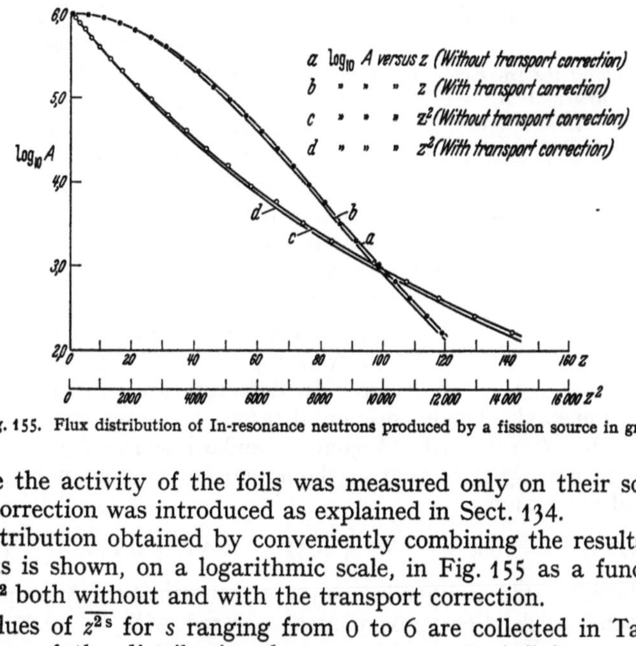

Fig. 155. Flux distribution of In-resonance neutrons produced by a fission source in graphite.

Because the activity of the foils was measured only on their source side, a transport correction was introduced as explained in Sect. 134.

The distribution obtained by conveniently combining the results of the two experiments is shown, on a logarithmic scale, in Fig. 155 as a function of z as well as of z^2 both without and with the transport correction.

The values of $\overline{z^{2s}}$ for s ranging from 0 to 6 are collected in Table 58. The extrapolation of the distribution from $z = 120$ cm to infinity was made with $l'_{tr} = 12.116$ cm. The extrapolation part represents a very small fraction of the total for $s = 2$ but increases in importance with increasing s. As is shown in column 3 of Table 58, for $s = 6$, it represents seventy percent of the total.

Table 58. *The $\overline{z^{2n}}$ moments in a graphite prism* ($\varrho = 1.653$ g/cm³) *(fission neutron source).*

s	$\int_0^{120} A(z)\cdot z^{2s}\,dz$	$\dfrac{\int_{120}^{\infty} -\mathrm{Ei}\left(-\frac{z}{l'_{tr}}\right) z^{2s}\,dz}{\int_0^{120} A(z)\,z^{2s}\,dz + k\int_{120}^{\infty} -\mathrm{Ei}\left(-\frac{z}{l'_{tr}}\right) z^{2s}\,dr}$	$\overline{z^{2n}}$
0	$2.85\cdot 10^{7}$	$5.27\cdot 10^{-5}$	
1	$1.56\cdot 10^{10}$	$1.66\cdot 10^{-3}$	$5.50\cdot 10^{2}$
2	$2.96\cdot 10^{13}$	$1.55\cdot 10^{-2}$	$1.05\cdot 10^{6}$
3	$1.05\cdot 10^{17}$	$7.67\cdot 10^{-2}$	$4.00\cdot 10^{9}$
4	$5.75\cdot 10^{20}$	$2.33\cdot 10^{-1}$	$2.63\cdot 10^{13}$
5	$4.21\cdot 10^{24}$	$4.74\cdot 10^{-1}$	$2.81\cdot 10^{17}$
6	$3.74\cdot 10^{28}$	$7.13\cdot 10^{-1}$	$4.57\cdot 10^{21}$

Table 59 contains the corresponding values of $\overline{\tau^n}$ deduced by applying formula (84.6) for various values of the upper limit $\bar{\nu}$ at which the sum is broken off. These values have been corrected to the graphite standard density $\varrho = 1.600$ g/cm³ by using the well known fact that $\tau \propto \frac{1}{\varrho^2}$. The values of $\overline{\tau^n}$ given on successive lines of Table 59 indicate how formula (84.6) converges for increasing values of $\bar{\nu}$. It will be noted that $\bar{\tau}$ becomes constant well within the experimental accuracy of the measured distribution for values of $\bar{\nu} \geqq 3$ and that $\overline{\tau^2}$ changes negligibly only from $\bar{\nu} = 3$ to $\bar{\nu} = 4$; $\overline{\tau^3}$ on the contrary is still changing, although very slowly, between $\bar{\nu} = 2$, and $\bar{\nu} = 3$. For this reason no $\overline{\tau^n}$ for $n > 3$ were calculated.

The distribution of the In resonance neutrons produced by a point source of Raα+Be type in graphite has been investigated with considerable accuracy in the Argonne standard pile—a measuring graphite parallelepiped (215 cm × 215 cm in section, 305 cm high) used for absolute neutron flux measurements [12]—. It is constructed of 10.2 cm × 10.2 cm graphite bars, several of which are removable and contain small cavities in which foils may be placed for activation; its surface is covered with a layer of cadmium 0.75 mm thick to avoid back-scattering of thermal neutrons into the pile from objects in the room.

Table 59. *Moments of the* In-*resonance neutron distribution in graphite* ($\varrho = 1.600$ g/cm³) *for a fission point source.*

$\bar{\nu}$	$\bar{\tau}$ cm²	$\overline{\tau^2}$ 10⁴ cm⁴	$\overline{\tau^3}$ 10⁴ cm⁷
0	293.0	9.967	4.036
1	304.6	10.90	4.739
2	309.0	11.31	5.096
3	310.3	11.45	5.227
4	310.6	11.48	
5	310.6		

The Raα+Be source (or any other source of small geometrical dimensions) is placed at 92 cm from top of the pile on its z axis. The Cd covered In detectors are irradiated at various positions along the z axis. The experimental results of measurements made with a Raα+Be source (actually the Argonne standard source N# 38) have been fitted, to an accuracy equal to that of the actual foil measurements, by means of the empirical expression

$$q(z) = 0.00178\, A(z) = 16.4\, e^{-\frac{z^2}{(27.12)^2}} + 8.47\, e^{-\frac{z^2}{(41.40)^2}} + 0.13\, e^{-\frac{z^2}{(65.00)^2}}. \quad (84.8)$$

Here $A(z)$ is the saturated activity in counts per minute of the Argonne standard In foils (0.1 g/cm² thick, 3 cm × 10 cm area) covered with Cd (0.02″ = 0.5 mm thick).

Introducing this empirical expression in (71.8), (71.6a), (71.3), the following value of the age of In resonance neutrons is obtained [9]:

$$\text{Ra}\,\alpha + \text{Be} \qquad \tau_{\text{In}} \approx 380 \text{ cm}^2.$$

The slowing down of neutrons from a Raα+Be source in a graphite-iron pile has been investigated by DUGGAL and MARTELLY[1]. The pile, called by the authors a poisoned pile, consisted of graphite plates 1.4 cm thick with a density $\varrho = 1.65$ g/cm³, which alternate with iron sheets 0.1 cm thick of surface density 0.78 g/cm². Under these conditions the iron is responsible for about 10% of the collisions; the medium may be considered as homogeneous because the separation between plates of the same material is substantially smaller than the mean free path. Some of the detectors employed and the results obtained with them are shown in Table 60.

In these experiments the iron is expected to contribute to the slowing down through inelastic collisions as long as the neutron energy is sufficiently high

[1] V.P. DUGGAL and J. MARTELLY: p. 28 of Ref. [15c].

Table 60. *Age of resonance neutrons produced by the slowing down in graphite + iron pile from a Ra α + Be source.*

Detector	τ cm²	l'_{tr} cm
Pd-Cd-covered (24 ev)	307.2 ± 4	—
In-Cd-covered (1.46 ev)	343.6 ± 2.5	14.3 ± 0.3

($\gtrsim$100 kev). At lower energy, when this process becomes energetically impossible and the slowing down takes place only through elastic collisions, the difference between the ages of two groups of neutrons corresponding to two different resonance detectors is expected to differ from the same quantity measured in pure graphite by a factor

$$\eta = \frac{(l_s^m)^2}{\xi_m(1-\overline{\cos\vartheta_m})} : \frac{(l_s^g)^2}{\xi_g(1-\overline{\cos\vartheta_g})}$$

where the indices m and g refer to the graphite-iron mixture and to pure graphite, respectively. DUGGAL and MARTELLY have verified this rule; furthermore, they have found that, by correcting all distances in the ratio $\eta^{\frac{1}{2}}$, the flux distribution of the In-resonance neutrons observed in the poisoned pile overlapped within the experimental errors with the corresponding distribution observed in pure graphite.

The wide energy interval covered by the Raα + Be source neutron spectrum makes a comparison of these experimental results with theory rather uncertain.

Much simpler conditions are attained with a $\mathrm{Sb}^{124}\,\gamma$ + Be source, not only because its neutron spectrum consists of a single line at $E_0 = 31$ kev (Sect. 49), but also because the carbon scattering cross section remains constant ($\sigma_s = 4.59 \times 10^{-24}\,\mathrm{cm}^2$) between 0.1 Mev and 1 ev (see Sect. 29).

Under these conditions one can do better than use the age theory approximation. For example, the computations made by HOLTE and collected in Table 43 of Sect. 79 could be extended to lower values of the energy of the neutrons emitted from the source.

A simpler procedure consists in synthesizing the neutron distribution from the corresponding moments. According to PLACZEK[1], for $u \gg X$ and $l_s(u) = l_s =$ const, the moments of the neutron distribution produced by an isotropic plane δ-source in a medium composed of a single element, are given by the expression [*16*][2]

$$\overline{z^{2s}} = \frac{(2s)!}{s!}\left(\frac{l_s^2\,u}{k}\right)^s\left[1 + \frac{s(s\,B_{11} + B_{10})}{u} + \cdots\right] \tag{84.9}$$

where

$$\left.\begin{aligned} k &= 3\,\xi\,(1-\overline{\cos\vartheta}),\\ B_{11} &= \delta_1 + \delta_2 + \delta_3,\\ B_{10} &= \delta_1 - \delta_3 \end{aligned}\right\} \tag{84.10a}$$

with

$$\left.\begin{aligned} \delta_1 &= \frac{8A-(A-1)^2 X(X+2)}{8A-2(A-1)^2 X},\\ \delta_2 &= \frac{1}{6}\,\frac{12A^2-20-3(A-1)^2(A+2)X}{3A-2},\\ \delta_3 &= \frac{32}{15}\,\frac{4A-(A-1)^2 X}{4A(A^2+1)-(A^2-1)^2 X}. \end{aligned}\right\} \tag{84.10b}$$

[1] G. PLACZEK: Manhattan Project Report A-25.

[2] I. WALLER: Ark. Mat. Astronom. Fys., Ser. A **34**, 5 (1946) has extended these formulas to mixtures.

For $A=1$, Eqs. (84.10) reduce to

$$k=1, \quad B_{11}=\frac{11}{15}, \quad B_{10}=\frac{1}{15}. \tag{84.11a}$$

For $A \gg 1$, Eqs. (84.10) reduce to

$$\begin{aligned} k &= \frac{6}{A}\left(1-\frac{4}{3A}+\cdots\right), \\ B_{11} &= \frac{18}{5A}\left(1-\frac{2}{3A}+\cdots\right), \\ B_{10} &= -\frac{4}{15A}\left(1-\frac{7}{3A}+\cdots\right). \end{aligned} \tag{84.11b}$$

For $s=1$ the terms omitted in (84.9) are of exponentially decreasing order in u/X. For $s>1$ the terms omitted in (84.9) are powers of u; therefore, as s increases, Eq. (84.9) yields increasingly poorer values. It should be noticed that Eq. (84.9) is a considerable improvement over the age theory which under the same assumptions yields

$$\overline{z^{2s}} = \frac{(2s)!}{s!}\left(\frac{l_s^2 u}{k}\right)^s.$$

The distribution function associated with the sequence of moments (84.9) is the following [*16*][1]

$$\Phi_0(z,u) = C\,\mathrm{e}^{-\frac{z^2}{4\tau}}\left\{1+\frac{1}{4u}\left[(B_{11}-2B_{10})-(2B_{11}-B_{10})\frac{z^2}{\tau}+\frac{B_{11}}{4}\left(\frac{z^2}{\tau}\right)^2\right]\right\} \tag{84.12}$$

where C is a normalization constant and, according to our assumptions, $\tau=\frac{l_s^2 u}{k}$.

Eq. (84.12), transformed by means of (71.2b) to an isotropic point source, becomes

$$\Phi_0(r,u) = \frac{C}{(4\pi\tau)^{\frac{3}{2}}}\,\mathrm{e}^{-\frac{r^2}{4\tau}}\left[1+\frac{F(r^2/r)}{4u}\right] \tag{84.13}$$

where

$$F\left(\frac{r^2}{\tau}\right) = 3\,(3B_{11}-2B_{10})-(4B_{11}-B_{10})\frac{r^2}{\tau}+\frac{B_{11}}{\tau}\left(\frac{r^2}{\tau}\right)^2 \tag{84.14}$$

represents the correction term with respect to age theory (76.16).

A source of this type was used by HILL and others[2] in conjunction with a graphite column of the same average density and with the same In detectors as in the fission source experiment described above.

The results of this experiment are plotted on a logarithmic scale versus r^2 in Fig. 156. The solid curve is the best fit of Eq. (84.13) to the experimental points, and it is obtained with

$$\tau_{\mathrm{In}} = 147\ \mathrm{cm}^2. \tag{84.15}$$

The dotted straight line corresponds to the age approximation, i.e. to Eq. (76.16) with the value (84.15) for τ.

From Fig. 156, we see that Eq. (84.13) fits the experimental data very well for r^2 ranging from 400 cm² to 4000 cm², but falls below the experimental points near the origin, just where we would expect the theory to be best. The first point at $r^2=12.6$ cm² is 16% higher than the theoretical value. The reason for this deviation is the activation of In by neutrons of energy higher than the lower resonance as will be shown by the self-absorption experiment described below.

[1] See footnote 1, p. 335.

[2] J.E. HILL, L.D. ROBERTS and G. MCCAMMON: ORNL-201 (1950).

The flux distribution of these "high energy" neutrons is certainly more concentrated towards the source than that of the 1.46 ev neutrons. Therefore it appears reasonable to assume that the contribution to the activation of the In foils by neutrons of energy higher than 1.46 ev becomes negligible for a sufficiently large value of r, for example at $r \approx \sqrt{350}$ cm, where the experimental points fall on the theoretical curve. Thus a self-absorption measurement here would correspond to the 1.46 ev line only. On the contrary, a self-absorption measurement made at very short distances, for example at $r^2 = 12.6$ cm², would consist of two parts, a higher energy part, where the In cross section is so small that the self-absorption is negligible, plus the activity due to the 1.46 ev line which would be self-absorbed in the same way as at $r^2 = 350$ cm².

The saturated activities observed for foils covered with Cd only are

(a) $A_{\text{Cd}}(12.6\ \text{cm}^2)$
$= A_{1.46}(12.6\ \text{cm}^2) +$
$+ A_{\text{high-energy}}(12.6\ \text{cm}^2)$

(b) $A_{\text{Cd}}(350\ \text{cm}^2)$
$= A_{1.46}(350\ \text{cm}^2)$

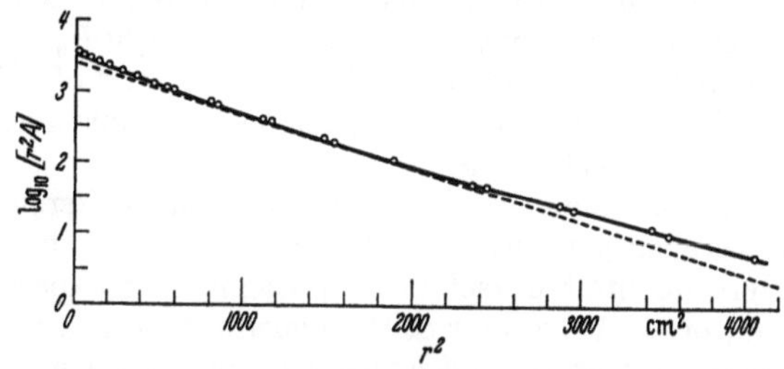

Fig. 156. Flux distribution of In-resonance neutrons produced by a Sbγ + Be source in graphite · theoretical curve → solid; plane gaussian curve $\tau = 147$ cm² → dotted (Hill *et al.*).

and the saturated activities for foils covered with Cd and In (Sect. 88) are

(c) $A_{\text{Cd+In}}(12.6\ \text{cm}^2) = A_{1.46}(12.6\ \text{cm}^2) \cdot F + A_{\text{high-energy}}(12.6\ \text{cm}^2)$

(d) $A_{\text{Cd+In}}(350\ \text{cm}^2) = A_{1.46}(350\ \text{cm}^2) \cdot F$

where F is the self-absorption factor of the 1.46 ev line, which is independent of r. By introducing the measured values of $A_{\text{Cd}}(350\ \text{cm}^2)$ and $A_{\text{Cd+In}}(350\ \text{cm}^2)$ into Eqs. (b) and (d) the value of F is derived. This, together with the measured values of $A_{\text{Cd}}(12.6\ \text{cm}^2)$ and $A_{\text{Cd+In}}(12.6\ \text{cm}^2)$ allow the determination of $A_{1.46}(12.6\ \text{cm}^2)$ and $A_{\text{high-energy}}(12.6\ \text{cm}^2)$ and, therefore, of the correction of the activation of In foils due to neutrons of energy higher than 1.46 ev. Measurements of this type were made by Hill and others at 10 foil positions and for three thicknesses of the In absorber. The good agreement for the three thicknesses confirms the correctness of the procedure. After correction for this effect, the experimental points of Fig. 156 turn out to be in very good agreement with the theoretical curve even for small values of the distance r from the source.

Since in the present case one has $l_s = \text{const}$, Eq. (74.4) reduces to

$$\overline{r^2} = 6\tau = \frac{l_s^2}{\xi\left(1 - \frac{2}{3A}\right)} \ln \frac{E_0}{E} \tag{84.16}$$

which allows a determination of E_0; from the value (84.15) of the age together with $E = 1.46$ ev and $l_s = 2.63$ cm (corresponding to $\varrho = 1.653$ g/cm² and $\sigma_s = 4.59$ barns) Hill *et al.* derive

$$E_0 = (20 \pm 4)\ \text{kev}$$

in not to bad agreement with the results of other measurements of the energy of the neutrons emitted from the $Sb^{124}\gamma + Be$ source (Sect. 49, Table 27).

The same graphite measuring column with the same $Sb^{124}\gamma + Be$ neutron source was used for measuring the age of Au-resonance neutrons ($E_R = 4.8$ ev)[1].

[1] W. L. Carlson: ORNL-784.

The detectors used in this experiment were gold foils $4\ \text{cm}\times 6.35$ cm in area and 103.1 mg/cm² thick covered with Cd. The result,

$$\tau_{\text{tot}} = (128 \pm 2)\ \text{cm}^2$$

may be combined with (84.15) in order to check Eq. (84.16). The difference $\tau_{\text{In}} - \tau_{\text{An}} = (19 \pm 2)\ \text{cm}^2$ compares well with the theoretical value $(18.6 \pm 0.3)\ \text{cm}^2$ deduced by applying (84.16) to the age difference between 4.8 and 1.46 ev.

85. A few experiments on the slowing down of neutrons as a function of time. The slowing down process may be investigated by measuring, as a function of time, the flux of neutrons of given energy E, which are produced in a moderator by the slowing down of neutrons of energy $E_0 > E$ emitted from a pulsed source. The techniques encountered in experiments of this type are very similar to those involved in time-of-flight spectrometers considered by J. RAINWATER in Vol. XL of this Encyclopedia.

A discussion of the similarities and differences between these two types of experiments is given in the review article by VON DARDEL and SJÖSTRAND[1]. Here we limit ourselves to noticing that experiments of this type may provide the correct behaviour of the neutron flux as a function of time only if the experimental set-up has been designed in such a way that the slowing down time is long compared to: (a) the time length of the burst of neutrons emitted from the source, (b) the fluctuations of the time response of the detector, and (c) the channel width of the time analyser connected to the detector. The following types of detectors have been used: boron trifluoride counters[2], scintillation counters (ZnS + boric compound[3] or Eu-activated LiI crystals[4]) and proportional counters arranged to detect the gamma rays produced in the capture of resonance neutrons by various elements[5] as we shall explain later in this section.

The slowing down time may be computed by means of (64.15), or, for $l_s =$ const, by means of (64.16) which, for $v_0 \gg v$, reduces to

$$t = \frac{2}{\xi}\frac{l_s}{v}. \tag{85.1}$$

By combining this expression with

$$v\ [\text{cm/sec}] = 1.38\times 10^6\, E^{\frac{1}{2}}\ [\text{ev}] \tag{85.2}$$

the values given in column 3 of Table 61 are obtained. From these data we recognize that an adequate investigation of the slowing down process in water requires very short time intervals for both the neutron burst emitted from the source and the channels of the time analyser. This means, not only a rather daring technique, but also a low neutron intensity. These difficulties decrease when the mass number of the moderator is increased so that they are strongly reduced in the case of elements as heavy as lead.

The first experiment of this type has been made by VON DARDEL[2] who studied the time dependence of the transmission through boron and cadmium absorbers of different thickness, for the neutrons emitted from a d+d source ($E_d = 150$ kev) and slowed down in light and heavy water.

---

[1] G. VON DARDEL and N. G. SJÖSTRAND: Progress in nuclear energy. Physics and Mathematics, Vol. 2, p. 183. New York and London: Pergamon Press 1958.

[2] G. VON DARDEL: Trans. Roy. Inst. Technol. **75** (1954).

[3] A. BRACCI and C. COCEVA: Nuovo Cim. **4**, 59 (1956).

[4] E.C. CAMPBELL and P.H. STELSON: Nucl. Sci. Engng. **1957**.

[5] A.A. BERGMAN, A.I. ISAKOFF, I.D. MURIN, F.L. SHAPIRO, I.V. SHTRANIKH and M.V. KAZARNOVSKY: p. 135 of Ref. [*15b*].

The moderator was a cube, of approximately 19 cm side, with the target of the accelerating tube located near its center. The detector was a BF_3 counter placed at a distance of about 1 cm from the surface of the moderator which could be covered with absorbing layers.

The experimental results were consistent with theoretical predictions based on the theory of slowing down against free protons and deuterons at rest (see Sects. 70 and 81) corrected for: (a) the distortion due to the time-of-flight delay from moderator surface to detector, and (b) the angular distribution of the neutrons emerging from the moderator surface [see Eq. (122.20)]. A quantitative comparison with theory was not possible because of the limited resolving power of the experimental set-up; both the duration of the neutron pulse and the channel width of the time analyzer were not less than 2 μsec to be compared to a slowing-down time to 1 ev of 1.3 μsec in light water and of 10 μsec in heavy water (column 3 of Table 61).

Table 61. *Slowing down time and thermalization time constant (von Dardel and Sjöstrand).*

Moderator	Density g/cm³	Slowing-down time to 1 ev μsec	Thermalization time constant: t_{th}^0 (d) μsec
Water	1.0	1.3 (a)	6 (b)
Heavy water . . .	1.1	10 (a)	58 (b)
Beryllium	1.8	10	22
Beryllium oxide .	2.95	—	29 (c)
Graphite	1.6	24	35
Lead	11.3	430	910

(a) Presence of oxygen neglected.
(b) Molecule assumed to be rigid.
(c) Be and O assumed to be free.
(d) For a discussion of these data see Sects. 111, 129 and 131.

A more detailed investigation of the slowing down process was made by Russian authors[1], who have developed a neutron spectrometer based on the measurement of the slowing-down time in a moderator composed of heavy elements.

The principle of the method is the following: a burst of neutrons of energy E_0 is emitted from a source at the center of a prismatic moderator in which it is slowed down. According to Eq. (81.6) its velocity distribution is given by (70.31) and its space distribution by (81.11).

A sample containing a particular type of nuclei with a resonance for neutron capture at $E = E_R$ is placed inside the moderator close to the neutron source. When the neutrons, during the slowing-down process, pass the energy $E \approx E_R$ their intensive capture by the sample appears as a strong emission of gamma rays (Sect. 16).

If the intensity of the gamma rays emitted by the sample is recorded as a function of time, the resonances in the neutron capture cross section appear as intensity maxima.

We shall not enter into a detailed comparison of this type of spectrometer with those based on time-of-flight measurements. Here we only notice that the slowing-down time (85.1) corresponds to a time-of-flight over a distance

$$L = \frac{2}{\xi} l_s$$

which for lead ($A = 207$, $l_s = 2.9$ cm) turns out to be 6 meters.

Furthermore the resolving power of a slowing-down time spectrometer is determined by the spread in velocity which arises from three different causes:

[1] See footnote 5, p. 338.

(a) The velocity spread of the neutrons emitted from the source. This contribution decreases during the slowing down process because of a peculiar type of focussing arising for $l_s = \text{const}$ from the fact that the probability per unit time interval, for a neutron of experiencing a collision v/l_s, is proportional to the velocity. Therefore, in a given time interval, neutrons with a larger velocity undergo more collisions and thus lose more energy than neutrons of lower velocity.

(b) The velocity spread arising from the fluctuations of both the neutron path between successive collisions, and the energy losses in each collision. This contribution is zero for $E = E_0$ and increases as E decreases reaching, for $E \ll E_0$ the asymptotic value: $\frac{\overline{\Delta v^2}}{v^2} = \frac{2}{3A}$ [see Eq. (70.30)].

(c) For values of the energy E of the order of a few ev or less, the thermal motion of the atoms can not be neglected (see Part D of this article). Assuming that the temperature T_0 of the moderator is much larger than the corresponding Debye temperature Θ_D, the contribution of this effect turns out to be $\frac{kT_0}{4E}$.

By combining these three effects one has

$$\frac{\overline{\Delta v^2}}{v^2} = \frac{\overline{\Delta v_0^2}}{v_0^2}\frac{E}{E_0} + \left(1 - \frac{E}{E_0}\right)\frac{2}{3A} + \frac{kT_0}{4E}. \tag{85.3}$$

In the case of lead the second and third terms combined together give at room temperature ($kT_0 \sim \frac{1}{40}$ ev)

$$\sqrt{\frac{\overline{\Delta v^2}}{v^2}} = 5.7 \times 10^{-2}\left(1 + \frac{1}{E\,[\text{ev}]}\right)$$

while the first term depends on the nature of the neutron source.

The experiment of the Russian authors was made with neutrons (of about 14 Mev) emitted from a Zr+T target bombarded with deuterons of 300 kev.

The target was placed at the center of a lead prism of 1.3 m × 1.3 m × 2 m or 2.0 m × 2.0 m × 2.3 m sides in different runs.

The neutrons emitted from the source were slowed down mainly by inelastic collisions, in a negligibly short time, until their energy fell below the first excited state of lead: $w_1 = 0.57$ Mev. From this time on, the slowing down took place through elastic collisions and, therefore, can be described by the theoretical considerations given in Sects. 70 and 81. Thus, for the initial neutron spectrum, we have to take that of the neutrons when their energy passes below w_1; the Russian authors estimated $E_0 \approx 100$ kev and $\frac{\overline{\Delta v_0^2}}{v_0^2} \sim 0.3$.

The duration of the neutron burst was varied between 0.5 and 2.0 μsec. The counters used as detectors were placed close to the target and were connected to a 50-channel time analyzer with channel widths of 2 μsec. Two types of counter were used; the first was a BF_3 proportional counter used for measuring the time dependence of the neutron density (81.15); the second a glass walled proportional counter filled with a mixture of argon and isopentane, which was used for measuring the intensity of the gamma rays due to the capture of neutrons by the sample. The latter consisted of metal foils of convenient thickness wrapped around the proportional counter.

As we have shown in Sect. 81, the neutron density at a given position inside the moderator is given, as a function of time, by Eq. (81.15); from thermal neutron cross section measurements, the neutron mean life in lead is computed to be $T = 900$ μsec and the parameter α computed by means of (81.17) is expected to be equal to 1 for a lead cube of 1.2 m sides, and to 0.35 for a lead block of 2.0 m × 2.0 m × 2.3 m sides. Therefore the time dependence of the neutron density at the position of the detector is expected to be determined mainly by the factor t^α for $t \ll 900$ µsec, and mainly by the exponential factor, for $t \gtrsim 900$ µsec. We shall come back to this point at the end of this section.

The counting rates of the boron counter and of the γ-rays counter may be written as follows:

$$\left.\begin{aligned} I_B(t) &= C_1 \Phi_0(t) \cdot t, \\ I_\gamma(t) &= C_2 \Phi_0(t)\, \sigma_\gamma(t) \end{aligned}\right\} \tag{85.4}$$

where C_1 and C_2 are two constants, and $\sigma_\gamma(t)$ is the capture cross section of the sample, averaged over the neutron spectrum (70.33) and expressed in terms of t by means of (70.31), (70.32). By combining with Eq. (85.4) one obtains

$$\sigma_\gamma(t) = \frac{C_2}{C_1} \frac{I_\gamma(t)}{I_B(t)}\, t \tag{85.5}$$

which allows a determination of the relative variation of the cross section as a function of the neutron energy (70.31).

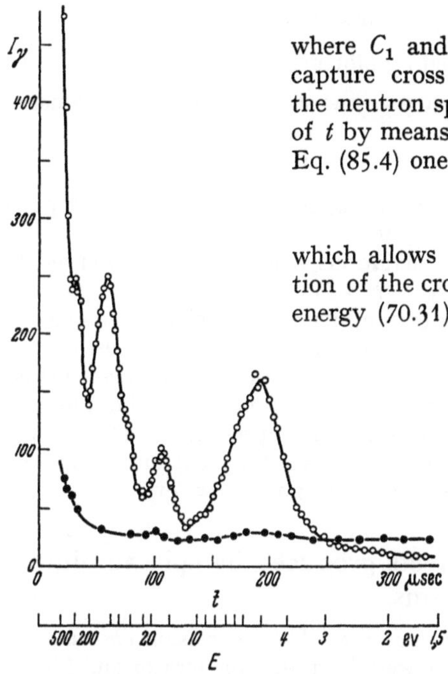

Fig. 157. Capture gamma-ray intensity curve obtained from Ag sample 0.1 mm thick (BERGMAN *et al.*).

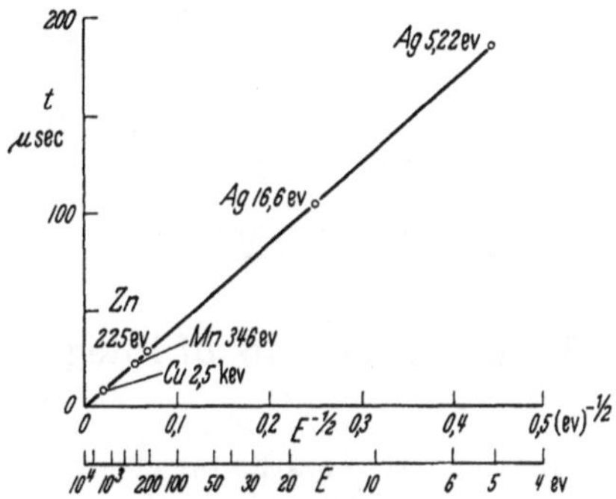

Fig. 158. Plot of moderation time t versus mean energy of neutrons at time t (BERGMAN *et al.*).

The absolute value of $\sigma_\gamma(t)$ could also be found by determining the coefficient C_2/C_1; this can be done by measuring the value of I_γ/I_B in the case of a capture cross section known with high accuracy.

By means of the experimental set-up described above, BERGMAN *et al.*[1] have measured $I_\gamma(t)$ for a number of elements (Ag, Zn, Mn, Cd, Cu). The upper curve of Fig. 157 shows the results obtained with a sample of Ag 0.1 min thick placed in a lead block 1.3 m × 1.3 m × 2.0 m sides. The lower curve represents the gamma ray background measured without sample.

The levels of silver at 5.22, 16.6 and 31.1 ev are clearly shown. Similar results were obtained with the other elements.

In Fig. 158 a plot is given of the slowing-down time obtained by the Russian authors for various well-separated resonances versus the corresponding values of $E_R^{-\frac{1}{2}}$ as given by direct spectrometer measurements. The linear behaviour of the curve agrees with (70.31) and (70.32).

According to (85.4) the counting rate of the boron trifluoride counter is expected to be proportional to the neutron density (81.15). Therefore,

$$\ln I_B(t) = \text{const} - \alpha \ln t - \frac{t}{T}. \tag{85.6}$$

[1] See footnote 5, p. 338.

Measurements of $I_B(t)$ in a lead block of $2\,\mathrm{m}\times 2\,\mathrm{m}\times 2.3\,\mathrm{m}$ sides gave the following results. From 1000 to 7000 μsec where the neutrons have already reached thermal energies, $I_B(t)$ falls exponentially with a mean life that, after correction for leakage of the neutrons from the block was found to be

$$T = (890 \pm 40)\ \mu\mathrm{sec}$$

in very good agreement with the computed value given above.

Once T is known, Eq. (85.6) is verified by plotting $\mathrm{Log}\, I_B(t) + 0.43\,\frac{t}{T}$ versus $\mathrm{Log}\, t$. From Fig. 159 we see that, for $t > 10$ μsec the slope of the curve is constant and equal to

$$\alpha = 0.35$$

in very good agreement with the computed value $\alpha = 0.36$.

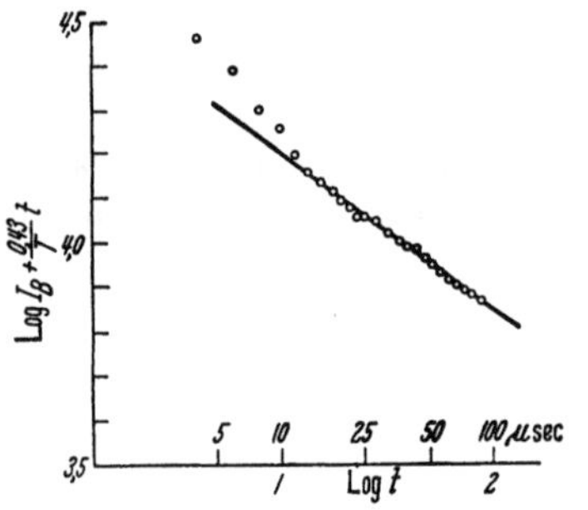

Fig. 159. Dependence of neutron intensity on moderation time (BERGMAN *et al.*).

The deviations from linearity observed for $t < 10$ μsec ($E > 2$ kev) may be due to various causes: influence of higher harmonics in Eq. (81.11), variations of l_s in lead, etc.

Thus we can conclude that these experiments provide a direct check of the adequacy of Eq. (81.15) for representing the slowing down process in lead for neutrons with energies between 2 kev and a few ev.

V. Measurements of integral effects produced by epi-cadmium neutrons.

86. General remarks on epi-cadmium neutrons. The expression *epi-cadmium neutrons* is used to indicate the neutrons slowed down in a moderator and filtered by a cadmium layer 0.2 to 0.6 mm thick. Their spectrum is shown in Fig. 14 (p. 56): the cadmium filter completely eliminates the neutrons of energy below 0.2 to 0.3 ev, and leaves practically unaffected those of energy higher than 0.7 to 0.8 ev. A more detailed description of the transmitted flux requires a specification of the thickness of the cadmium filter and of the angular distribution of the incident neutrons; it may be noticed, however, that in general the *transparency* of the cadmium filter increases from zero to one within about 0.5 ev. This energy interval is sufficiently narrow to justify the description—very rough indeed—of the effect of the cadmium filter in terms of a *cut-off energy* E_{Cd} such that all the neutrons of energy below E_{Cd} are completely absorbed, while those of energy larger than E_{Cd} pass through it unaffected.

The value of E_{Cd} varies, according to the experimental conditions, between 0.3 to 0.5 ev; Eq. (26.20) indicates the thicknesses suggested by HUGHES [*12*] in order to have

$$E_{\mathrm{Cd}} \approx 0.4\ \mathrm{ev} \tag{86.1}$$

for two extreme angular distributions of the neutrons incident on a flat filter: for an isotropic angular distribution 0.25 mm = 0.22 g/cm² Cd, and for a parallel beam moving perpendicularly to the filter's surface 0.50 mm = 0.44 g/cm² Cd.

When the description of the absorption of a cadmium filter in terms of a cut-off energy E_{Cd} is inadequate, the spectrum of epi-cadmium neutrons must be measured by means of a spectrometer, or it must be computed by introducing into (26.18) the unfiltered flux and the capture cross section (26.2) with the values of the resonance parameters of cadmium given in Table 8 (p. 54).

In the considerations given in the following, the oversimplified description of the absorption by cadmium filters in terms of the cut-off energy (86.1) has been generally adopted, with a few exceptions in which the more refined procedure sketched above has been followed; for example, Figs. 161 and 162 of Sect. 88.

It may be noticed that, instead of cadmium, any other material with a strong isolated capture resonance located below 1 ev could be used as a filter in order to eliminate the thermal neutron component. The only difference would be a change in the value of the cut-off energy. A material sometimes suggested as a possible substitute for cadmium is gadolinium—$E_R = 0.030$ ev, $\Gamma_\gamma = (100 \pm 30)$ mev, $\Gamma_n = (0.65 \pm 0.10)$ mev; $\sigma_{a0} = 44000$ barns [*14*]—which gives $E_{\text{Gd}} \approx 0.2$ ev for an 0.061 g/cm² Gd-filter in a collimated beam.

In spite of the fact that neutron resonances are observed below the cadmium cut-off as well as above it, the epi-cadmium neutrons are very often improperly indicated also as *resonance neutrons*; this expression is used even when the experimental set-up does not select, out of the continuous spectrum of cadmium filtered neutrons, those belonging to more or less narrow bands.

The techniques used for the experimental investigation of epi-cadmium neutrons—or more generally of slow neutrons—can be divided into two different categories according to the type of information—differential or integral information—which they may provide for the investigated quantities.

To the first group, there belong: time of flight spectrometers and crystal spectrometers of all types, i.e. every experimental set-up which, by means of a convenient analysis of the directly measured quantities, allows the derivation of the energy dependence of the incident neutron flux. These techniques can not be substituted for a detailed investigation of the energy dependence of the various cross sections of the nuclei. The description of these instruments and of the corresponding methods of analysis of the experimental data is outside the scope of the present article; therefore, the reader interested in such details is directed to the article by Rainwater in Vol. XL of this Encyclopedia and to Ref. [*12*], [*15b*], [*18b*].

Here we limit ourselves to a discussion of the second group of techniques, namely, those techniques which provide information only on *integral effects*, i.e. on effects which involve neutrons extending over large energy regions—usually from the cadmium cut-off up to energies of the order of a few Mev.

As we shall see below, these procedures are characterised by a remarkable instrumental simplicity, being essentially based only on the use of convenient combinations of detectors and absorbers. These were the only methods at the disposal of experimentalists in the early days of neutron physics for investigating the interaction of slow neutrons with matter; they are, still to-day, of considerable practical interest—in spite of the great development reached by the neutron spectrometers—as can be seen from the following remarks.

First of all, the investigation of integral effects—when made under conveniently chosen conditions—provides subsidiary information on the absorption and scattering cross sections which are sometimes very useful even in the analysis of spectrometer data. As we shall see in Sects. 91 and 94, in the case of nuclei with a single isolated resonance, the results of integral experiments provide the values of certain simple combinations of the resonance parameters.

Secondly, in many cases the integral effect in itself is of considerable practical interest and a directly measured value is much more reliable than the value obtained by computing the integral of expressions containing the cross sections measured by means of neutron spectrometers. Examples of situations of this type are met in the case of the resonance escape probability (Sects. 92 and 93) and of the effective resonance integral. As we shall see in Sect. 93 in the case of heterogeneous systems the experimental approach is the only method which can provide a reliable value of the effective resonance integral.

Finally the techniques based on the simple use of detectors and filters are practically the only methods which can be employed in experiments made inside an assembly composed of one or more media without perturbing appreciably the pre-existing neutron distribution. A few examples of experiments of this type have already been reviewed in Division IV: for instance, the flux distribution is measured as a function of the space coordinates inside an infinite medium containing a fast neutron source by means of a resonance detector consisting of a cadmium covered In foil (Sects. 83 and 84).

The latter remark permits to classify the slow neutron experiments into two categories: *external experiments*, i.e. experiments made with beams of neutrons coming out of an assembly—such as a moderator containing a fast neutron source or a pile—, and *internal experiments*, i.e. experiments made inside the assembly. The spectrometers clearly are the only instruments which can be used in external experiments for the determination of the energy dependence of neutron fluxes, while the detector and filter techniques discussed in the following sections provide the only simple approach to internal experiments. These techniques are frequently used also for the determination of integral effects in external experiments; in this case the same result might be obtained—at least in principle—by integrating the data provided by a spectrometer.

Thus, in the following sections, we will discuss in detail both external and internal experiments made with resonance detectors and/or filters. The main difference between the two cases arises from the fact that a beam of neutrons may have such a small angular spread that it can be described as a parallel beam for most practical purposes, while in the internal experiments the neutron flux is spread over the total solid angle according to a law—usually unknown—which is determined by the composition and geometry of the medium and which turns out to be different at different points.

Neutron detectors used inside a moderator are called *neutron probes*. Their properties will be discussed from a general point of view in Sect. 134 where most of the considerations refer to the case of thermal neutrons. A few remarks given there are of importance also for the case of resonance neutrons.

As we shall see in the discussion to follow the interpretation of the results of the integral effects requires, at least sometimes, a knowledge of the energy dependence of the neutron flux. The neutron density in a weakly absorbing medium is given by Eq. (67.10) down to about 1 ev. This, converted into the neutron flux,

$$\varphi_0(E)\,dE = \frac{q_0\,l_s(E)}{\xi}\,\frac{dE}{E} \tag{86.2a}$$

and with

$$l_s(v) = l_s = \text{const}, \tag{86.2b}$$

leads to the well-known $1/E$ energy dependence. As it has been stressed in Sect. 66, Eq. (67.10) either refers to the density integrated over the entire volume of an infinite homogeneous moderator containing monoenergetic sources emitting, all together, q_0 neutrons per second, or to the neutron density found at each point of an infinite moderator in which the neutrons are produced by a distributed source emitting q_0 neutrons per second and per cm^3.

A situation corresponding to the first case may be obtained in an internal experiment by repeating the measurements of the quantity under investigation at many different positions conveniently distributed throughout the moderator and then computing its integral with respect to space coordinates. A few procedures for directly obtaining the value of such a volume integral are mentioned in Sect. 134.

A situation corresponding to the second case is expected to be found in a region of constant, or almost constant neutron density. Various experiments have confirmed this expectation and actually it has been found that Eq. (86.2) fits rather well the measured flux at a given position inside a moderator, even if the neutron density varies appreciably from point to point. For example, Fig. 212 (p. 470) shows a plot of the neutron flux measured at four different distances from the source in water at 291° K. From these curves we see that the dE/E formula not only represents fairly well the experimental data at all the considered distances, but that it extends even below 1 ev, i.e. into the energy region where the influence of the chemical bond of the atoms of the moderator is not negligible (Part D of this article).

Fig. 223 (p. 484) shows the results of similar measurements made at a given position in the case of a graphite pile. The experimental points are very well represented by the empirical expression (108.14) from which it follows that, slightly above

$$E_C = 0.12\ \text{ev},$$

the neutron flux reduces to (86.2).

Thus, we may conclude that in many cases of practical interest, the neutron flux at a given position inside a moderator, like for example water or graphite, is given with sufficient accuracy by (86.2) down to the cadmium cut-off (86.1). This assumption is adopted throughout the following Sects. 87 to 94; its correctness should always be checked before using the various relations given in the following section for the interpretation of results obtained under new experimental conditions.

The presence of an absorbing material will produce a change of the neutron spectrum (86.2). An important distinction must be made between homogeneous and heterogeneous arrangements. The expression *homogeneous system* refers to assemblies containing an absorbing material uniformly distributed throughout the moderator; for the discussion of problems of this type we shall make use of the considerations given in Sects. 68 and 69. By *heterogeneous system* we mean a non absorbing—or weakly absorbing—medium containing one or more lumps of absorbing materials. This case—of considerable importance in connection with reactor design—gives rise to rather complicated analytical problems so that an empirical approach appears to be much simpler and more reliable than all present theoretical treatments (Sect. 93).

The detectors and the absorbers of epi-cadmium neutrons show a different behaviour according to the energy dependence of their cross section for neutron capture: the main distinction can be established between materials following the $1/v$-law and materials with one or more resonances located above the cadmium cut-off

$$E_R > E_{\text{Cd}}. \tag{86.3}$$

The ideal case is that of a material with a single sharp resonance satisfying the inequality (86.3). In Sect. 94 we shall discuss the general behaviour of the various absorption curves obtained by combining in various ways $1/v$ or resonance absorbers with $1/v$ or resonance detectors.

Finally it may be noticed that there are two different methods for selecting out of the continuous spectrum (86.2) the neutrons belonging to a narrow energy band. The first procedure consists of using a resonance detector with a single isolated resonance; the second procedure is based on the use of a $1/v$ detector and a filter with one single isolated resonance. The difference between the responses of the $1/v$ detector when exposed to epi-cadmium neutrons without and with the resonance filter in front of it, corresponds to the neutrons absorbed in the filter. These two procedures will be discussed in detail for the case of In in Sects. 88 and 89.

87. Natural width and Doppler broadening. The derivation of the theoretical expressions for the various integral effects discussed in the following sections involves the energy dependence of the absorption and scattering cross sections, which, in the simple case of a single isolated resonance, are given by the one-level formula of BREIT and WIGNER (Sect. 26).

These formulas provide a correct description of the effective cross sections only when the target nuclei are at rest. In many cases, however, the velocities of the target nuclei due to thermal agitation are not negligible compared to neutron velocity, with the result that the energy dependence of the various cross sections is distorted by Doppler effect.

For the computation of such a distorsion it is convenient to consider separately a few ideal examples, a convenient combination of which allows the understanding of much more general cases. The four simplest cases are: (a) The capture cross section follows the $1/v$ law. (b) The capture cross section shows a single sharp resonance. (c) The scattering cross section is constant. (d) The scattering cross section shows a single resonance line.

Case (a) is discussed in Sect. 96 with the result that when $\sigma_a(v) \propto 1/v$ its behaviour is not affected at all by the thermal agitation of the target nuclei. Case (c) is treated in Sect. 97 for a monoatomic gas; in that section the constant scattering cross section is obtained as a particular case of a slowly varying cross section (97.2). Case (c) for atoms bound in molecules or crystals is discussed in Sects. 100 to 103.

In this section we shall limit our consideration to case (b) i.e. to the absorption, by a nucleus with a single sharp resonance line, of neutrons of energy very close to resonance. The same procedure can be applied to computing the Doppler distorsion of the resonance term (the first term) appearing in the scattering cross section (26.7) (Sect. 92).

For the computation of the Doppler effect we shall use for the capture cross section at resonance the expression

$$\sigma_a(E) = \frac{\sigma_{a0}}{1 + \left(\frac{E - E_R}{\Gamma/2}\right)^2} \tag{87.1}$$

which differs from (26.2), (26.3) because the variable factor $\lambda\!\!\!^{-2}\,\Gamma_n$—which is proportional to $1/v$—has been replaced by its constant value at $E = E_R$; thus Eq. (87.1) gives the correct shape of σ_a in the vicinity of E_R and decreases monotonically tending to zero for both $E \gg E_R$ and $E \ll E_R$.

An estimate of the distorsion of (87.1) due to the Doppler effect requires the velocity distribution of the capturing atoms and this depends on the molecular structure of the corresponding material (see Part D of this article).

BETHE and PLACZEK[1] were the first to treat this problem in the case of a gaseous sample. The results obtained for such a case agree with an accuracy sufficient

[1] H. A. BETHE and G. PLACZEK: Phys. Rev. 51, 450 (1937).

for the present purpose with those obtained in the case of a Debye crystal provided that the sample temperature T is much larger than the Debye temperature Θ_D. This point, already recognized by BETHE and PLACZEK, can be justified by means of considerations similar to those given in Sect. 104 in connection with the problem of the scattering of neutrons by nuclei bound in crystals, and was clearly proved by LAMB as we shall see below.

Following BETHE and PLACZEK, we adopt the Maxwellian distribution of velocities

$$M_{MT}(V_x)\,dV_x = \left(\frac{M}{2kT}\right)^{\frac{1}{2}} e^{-\frac{MV_x^2}{2kT}}\,dV_x \tag{87.2}$$

for the component V_x of the velocity of the capturing nuclei in the direction of motion of the neutron. The energy appearing in (87.1) is the kinetic energy E_r of the relative motion of the neutron moving against the nucleus, not the kinetic energy E of the neutron in the l.s. According to (33.11) one has

$$E_r = \tfrac{1}{2}\mu_1 v_{r1}^2 = \tfrac{1}{2}\mu_1 (v - V_x)^2.$$

In the case considered here, of heavy nuclei ($M \gg m$) with a narrow resonance ($\Gamma \ll E_R$), belonging to a sample at a temperature T such that $kT \ll E_R$ one can write

$$E_r \approx \frac{1}{2} m (v - V_x)^2 = E\left[1 + \left(\frac{V_x}{v}\right)^2\right] - (2mE)^{\frac{1}{2}} V_x \approx E - (2mE)^{\frac{1}{2}} V_x. \tag{87.3}$$

Inserting the value of V_x derived from (87.3) into (87.2), we obtain for the probability of a given relative energy E_r:

$$w(E_r)\,dE_r = \frac{1}{\pi^{\frac{1}{2}}}\, e^{-\frac{(E_r - E)^2}{\Delta^2}}\,\frac{dE_r}{\Delta} \tag{87.4}$$

where

$$\Delta = 2\left(\frac{m}{M} E_R kT\right)^{\frac{1}{2}} \tag{87.5}$$

is the *Doppler width*.

The effective capture cross section, i.e. the effective shape of the resonance line, for neutrons of energy E, is obtained by combining (87.5) and (87.1) and integrating over E_r

$$\sigma_a(\eta, E) = \int \sigma_a(E_r)\, w(E_r)\, dE_r = \sigma_{a0}\, \psi(\eta, z) \tag{87.6}$$

where

$$z = \frac{E - E_R}{\Gamma/2}, \tag{87.7}$$

$$\eta = \frac{\Gamma}{\Delta}, \tag{87.8}$$

$$\psi(\eta, z) = \frac{\eta}{2\pi^{\frac{1}{2}}} \int_{-\infty}^{+\infty} \frac{e^{-\frac{1}{4}\eta^2 (z-y)^2}}{1 + y^2}\, dy, \qquad y = \frac{E - E_R}{\Gamma/2}. \tag{87.9}$$

In the general case $\psi(\eta, z)$ is a complicated function of z; its behaviour is determined by the value of the parameter η which, according to (87.8) provides a measure of the Doppler effect: when $\eta \gg 1$ the Doppler width is negligible with respect to the natural width, while the opposite is true when $\eta \ll 1$. In heavy elements the natural width of low lying levels is usually larger than the Doppler width at room temperature by a factor varying between 2 and 5 (Table 62).

Table 62. *A few examples of Doppler broadening at room temperature* ($T = 293°$ K).

Capturing nucleus	E_R ev	Γ ev	Δ ev	η	Θ_D °K	$\frac{\Gamma+\Delta}{2k\Theta_D}$	$\sqrt{\frac{E}{kT}}$
Rh^{103}	1.26	0.156	0.035	4.5	370	3	1.11
Ag^{109}	5.1	0.149	0.068	2.2	215	6	1.03
Cd^{113}	0.178	0.113	0.040	2.8	160	5.5	1
In^{115}	1.46	0.075	0.036	2.1	100	6.5	1
Ir^{193}	1.3	0.087	0.026	3.3	283	2.35	1 07
Au^{197}	4.9	0.140	0.050	2.8	180	6.8	1

Numerical tables of $\psi(\eta, z)$ have been published by various authors[1,2]; the following properties may easily be shown[1,3].

(a) For $\eta \gg 1$ (pure natural width)

$$\psi(\eta, z) = \frac{1}{1+z^2}. \tag{87.10}$$

(b) For $\eta \ll 1$ (pure Doppler width) and $z \ll \eta^{-2}$

$$\psi(\eta, z) = \tfrac{1}{2}\pi^{\frac{1}{2}}\eta\, e^{-\frac{1}{4}\eta^2 z^2}. \tag{87.11}$$

(c) For $z \gg \eta^{-2}$ and any η

$$\psi(\eta, z) = \frac{1}{1+z^2}\left(1 - \frac{2}{\eta^2}\,\frac{1-3z^2}{(1+z)^2} + \frac{12}{\eta^4}\,\frac{1-10z^2+5z^4}{(1+z^2)^4} + \cdots\right) \approx \frac{1}{1+z^2}. \tag{87.12}$$

(d) For $z = 0$ and any η

$$\psi(\eta, 0) = \tfrac{1}{2}\pi^{\frac{1}{2}}\eta\, e^{-\frac{1}{4}\eta^2}\left[1 - \mathrm{Erf}\left(\tfrac{1}{2}\eta\right)\right] \tag{87.13}$$

from which it follows that

$$\text{for}\quad \eta \gg 1 \quad \psi(\eta, 0) = 1, \qquad \sigma_a(\eta, 0) = \sigma_{a0} \tag{87.14a}$$

$$\text{for}\quad \eta \ll 1 \quad \psi(\eta, 0) = \frac{\pi^{\frac{1}{2}}}{2}\eta, \qquad \sigma_a(\eta, 0) = \frac{\pi^{\frac{1}{2}}}{2}\sigma_{a0}\frac{\Gamma}{\Delta}. \tag{87.14b}$$

(e) From the definition (87.9) it follows that, for any value of η,

$$\int_{-\infty}^{+\infty} \psi(\eta, z)\, dz = \pi. \tag{87.15}$$

This relation is well known from the corresponding classical case of optics[1]: *the total absorption of a spectral line* of a very thin sample—in our case the integral of the cross section (87.6) with respect to E from $-\infty$ to $+\infty$—*does not depend on the Doppler effect, and is in any case proportional to the product* $\Gamma \cdot \sigma_{a0}$. We shall come back to this point in Sect. 91, Eq. (91.5).

The problem has been re-examined by LAMB[4] under the assumption that the capturing nuclei are bound in a Debye crystal (Sect. 99). The procedure, in some way similar to that sketched in Sect. 103 for the computation of the inelastic scattering of a neutron by a crystal, involves the derivation of the probability for the capture of the neutron of momentum p, by a definite lattice atom A to form a new atom B with emission of a γ-ray and the simultaneous transition

[1] M. BORN: Optik. Berlin: Springer 1933.

[2] V.L. SAILOR: BNL-257 (T-40) (September 1953). — M.E. ROSE, W. MIRONKER, P. LEAK, L. ROSENTHAL and J.K. HENDRICKSON: WAPD-SR-506 (October 1954).

[3] See footnote 1, p. 346.

[4] W.E. LAMB: Phys. Rev. 55, 190 (1937).

of the crystal from the state $\{n_s\}$ to the state $\{m_s\}$. Here the set of numbers denoted by $\{n_s\}$ (and $\{m_s\}$) gives the numbers $n_1, n_2, n_3, \ldots, n_s$, of phonons in the various modes of oscillation of the lattice. The final state is reached through an intermediate state in which there exists neither a neutron nor a γ-ray but a compound nucleus C with the lattice in a state $\{m'_s\}$. The shape of the line is then obtained by summing such a probability over all possible final states satisfying the conservation theorems and averaging it with respect to the occupation distribution of the initial states of the crystal at the temperature T.

The final result, in the form of an integral in the complex plane, is rather difficult to evaluate exactly in the general case. But, if the condition of *weak binding*

$$\Gamma + \Delta \gg 2k\Theta_D \tag{87.16}$$

is met, the effective shape of the line is given by the same expressions (87.6) to (87.9) valid for a gas, with the only difference with respect to (87.5) being that in the expression for the Doppler width, the factor kT relative to the temperature T of the sample is replaced by the average energy per vibrational degree of freedom of the crystal including the zero point energy:

$$\Delta_c = \Delta \sqrt{\frac{\bar{E}}{kT}} \tag{87.17}$$

where

$$\bar{E} = \tfrac{1}{3}(\bar{E}_l + 2\bar{E}_t) \tag{87.18a}$$

$$\left.\begin{aligned} \bar{E}_j(kT) &= 3\left(\frac{T}{\Theta_{Dj}}\right)^3 \times \\ &\times kT \int_0^{\Theta_{Dj}/T} \left\{\frac{1}{e^x - 1} + \frac{1}{2}\right\} x^2\, dx \end{aligned}\right\} \tag{87.18b}$$

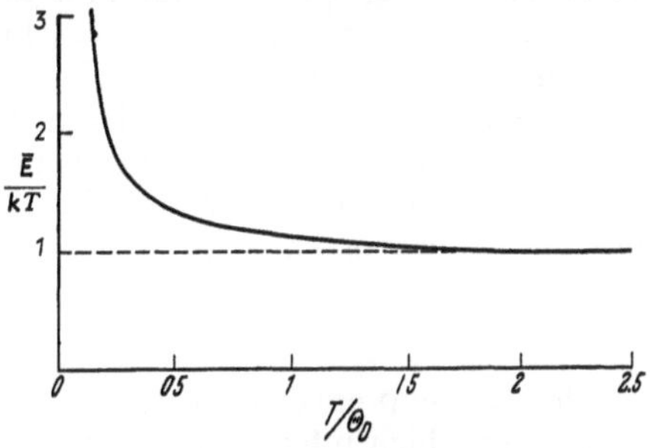

Fig. 160. Plot of $\bar{E}/kT$ for a Debye crystal versus T/Θ_D (LAMB).

and the subscripts $j=l$ or $j=t$ refer to the longitudinal and the transverse waves. In Fig. 160 the values of $\bar{E}/kT$ computed by means of (87.18), are plotted versus T/Θ_D under the usual simplifying assumption that $\Theta_{Dl}=\Theta_{Dt}=\Theta_D$; the figure shows that even for so low a temperature as for $T=1.5\,\Theta_D$ the Doppler width of the absorption line is almost the same in a crystal as in a gas.

In the opposite extreme case of very *strong binding* of the atoms in the crystal ($\Theta_D \to \infty$), the shape of the line simply reduces to (87.1) as we have to expect.

Numerical computations made by the same author for a few cases of intermediate binding, show that in a crystal there is, in general, not only a broadening of the line with respect to the gas, but even a shift and a rudimentary fine structure.

Table 62 shows a few examples of interest for the detection and absorption of resonance neutrons. In the case of the first resonance of Ag, Cd, In and Au the condition of weak binding (87.16) is rather well satisfied and the correction factor $\sqrt{\bar{E}/kT}$ appearing in (87.17) is practically equal to 1. In the case of the first resonance of Rh and Ir the condition of weak binding is only rather poorly satisfied so that the correction factor read from the graph of Fig. 160 must be taken only as an indication of the order of magnitude.

For E_R of the order of 200 ev or more, the Doppler width becomes comparable with or even larger than Γ in nuclei of mass number $A>100$. At still higher energies—for heavy nuclei above several thousands of ev—the width of the absorption lines becomes comparable with their average distance because of neutron width broadening [Eq. (26.10)] and Doppler effect, so that the various resonances merge into each other giving a continuous absorption cross section.

88. The use of filters. In order to illustrate the use of filters in experiments on epi-cadmium neutrons, we start by considering a parallel beam of slow neutrons, such as may be obtained from a hole opened in a pile according to the criteria discussed in Sect. 90. A cadmium layer, for example, 0.03 inch (=0.76 mm = 0.66 g/cm²) thick eliminates the thermal neutron component so that, the flux $\varphi_{\mathrm{Cd}}(E)\,dE$ of the cadmium filtered neutrons will be similar to that of the lower of the two broken curves in Fig. 14 (p. 56).

The counting rate due to neutrons of energy between E and $E+dE$ of a thin $1/v$ detector—for example a BF_3 counter—placed in the beam, will be proportional to

$$\frac{1}{E^{\frac{1}{2}}}\,\varphi_{\mathrm{Cd}}(E)\,dE. \tag{88.1}$$

According to Fig. 14 the cadmium absorption becomes negligible for neutron energies larger than 0.8 ev; above this energy we can make use of Eq. (86.2) because the energy dependence of the neutron flux in the beam is the same as that inside the pile in the vicinity of the bottom of the hole (Sect. 90).

The insertion into the beam of a filter of macroscopic absorption cross section $\Sigma_a(E)$ and thickness x modifies the counting rate of the $1/v$ detector which will now be proportional to

$$\mathrm{e}^{-\Sigma_a(E)\cdot x}\,\frac{1}{E^{\frac{1}{2}}}\,\varphi_{\mathrm{Cd}}(E)\,dE. \tag{88.2}$$

The two expressions (88.1) and (88.2) are illustrated in Figs. 161 and 162, the first of which refers to a resonance filter, the second to a $1/v$ filter. In these figures, taken from HUGHES' book [*12*], the counting rate *per unit logarithmic energy interval*

$$\Delta\,\mathrm{Log}\,E \propto \frac{\Delta E}{E}$$

is plotted—in arbitrary units—versus Log$E$. The advantage of a graph of this type is that the area between two arbitrarily chosen values of the abscissa is proportional to the counting rate arising from neutrons belonging to the corresponding energy interval.

Of the three curves given in Fig. 161 that marked 0.76 g/cm² Cd represents the epi-cadmium counting rate (88.1), and therefore corresponds to the 0.03″-curve of Fig. 14. It may be noticed that in the plot adopted here, for neutron energies larger than about 0.8 ev, this curve is a simple semilogarithmic plot of $E^{-\frac{1}{2}}$.

The other two curves represent the counting rates measured behind In-filters of two different thicknesses. The smaller thickness has been chosen so that practically only the neutrons corresponding to the 1.46 ev In-resonance are absorbed in the filter. This condition is easily attained because the first In-resonance is much larger than the resonances at higher energies. The curve marked 0.1 g/cm² In shows that for thicknesses of this order of magnitude, the absorption of neutrons corresponding to the 3.9 and 9.1 ev resonances (Table 9) are appreciable.

We may conclude that an In-filter placed in a parallel epi-cadmium beam behaves like a filter with a single resonance at 1.46 ev only if its thickness does not exceed about

$$20\ \mathrm{mg/cm^2} \triangleq \frac{20\times10^{-3}\ \mathrm{g}}{7.3\ \mathrm{g/cm^3}} = 2.74\times10^{-3}\ \mathrm{cm}. \tag{88.3}$$

Fig. 162, similar to Fig. 161, refers to the case of boron-filters and serves to illustrate the effect of a $1/v$ absorber on the counting rate of epi-cadmium neutrons. The curve marked 0.76 g/cm² Cd is identical to that given in Fig. 161 and represents the epi-cadmium counting rate without boron filters (88.1). The other three curves show the effect of boron filters of increasing thicknesses; a layer of natural boron 0.2 g/cm² thick reduces to about one half the counting rate of epi-cadmium neutrons of energy between the cadmium cut-off and about 10 ev.

We have already mentioned at the end of Sect. 86 that filters combined with, for example, a $1/v$ detector, may be used for selecting from the continuous spectrum of epi-cadmium neutrons those belonging to a more or less narrow energy band. The procedure is similar to that used for selecting from the continuous

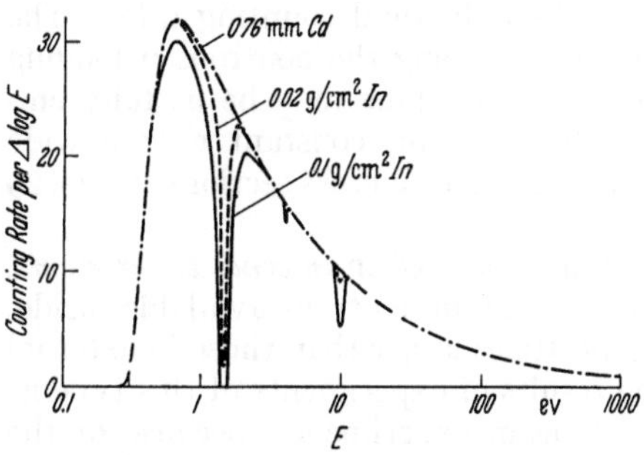

Fig. 161. Effect of In-filters on the counting rate—per unit logarithmic energy interval—of a thin $1/v$ detector exposed to a parallel beam of epi-cadmium neutrons.

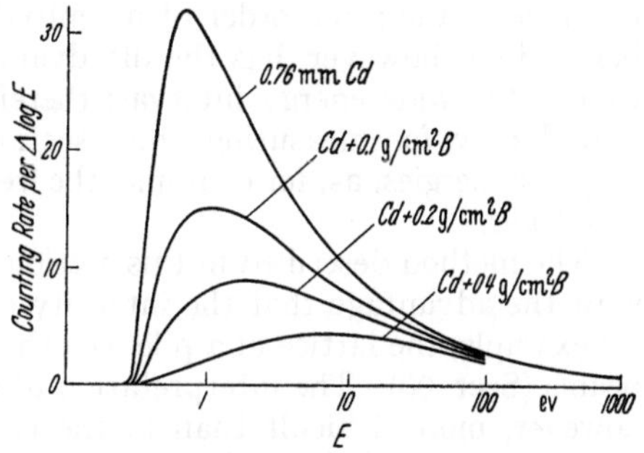

Fig. 162. Effect of B-filters on the counting rate—per unit logarithmic energy interval—of a thin $1/v$ detector exposed to a parallel beam of epi-cadmium neutrons.

spectrum of slow neutrons the thermal neutron component by taking the difference of the counting rates given, for example, by a BF_3 counter, without and with a cadmium filter in front of it. This procedure, called the *cadmium-difference* method, has already been mentioned in Sect. 23 and will be discussed in detail in Sect. 133.

Similarly, in order to select from the continuous spectrum of epi-cadmium neutrons those belonging to the In resonance at 1.46 ev, one can take the difference of the counting rates given by a BF_3 counter placed in a Cd-filtered beam without and with an In-filter 0.02 g/cm² thick.

The main disadvantage of the *difference method* when applied to resonance filters such as indium, arises from the fact that the result is usually given by a small difference between large counting rates. For example the indium-difference for a filter 0.02 g/cm² thick can be estimated from the areas of the curves in Fig. 161; since it amounts to about 4% of the total counting rate N, its relative statistical error is $\sqrt{2}/0.04 \sim 35$ times larger than $1/\sqrt{N}$. Therefore accurate results of measurements made by means of the indium difference method with thicknesses of the filter of the order of (88.3), can be obtained only with high intensity epi-cadmium beams.

An increase in In-foil thickness would of course produce an increase in In-difference, but this would arise only in part from a broadening of the 1.46 ev line while the rest would originate from the resonances at 3.9 and 9.1 ev.

This inconvenience is much less pronounced for other resonance filters; for example rhodium has up to about 1000 ev no other resonance of intensity comparable to that at 1.28 ev (Fig. 11) and gold shows only six very weak resonances, between 40 and 300 ev [*14*] besides that at 4.9 ev (Fig. 16). Thus, in these cases, a rather large filter thickness can be employed since it simply produces a broadening of the dip in the counting rate curve arising from the absorption of neutrons in the resonance wings.

For resonances of energy larger than a few ev, the fractional part of the neutron flux removed by the filter is much smaller than that for resonances at a few ev, and therefore the inaccuracy of the results is correspondingly larger. This is the reason why the use of the difference procedure with resonance filters can be conveniently employed, when a very intense epi-cadmium beam is available only for resonances of a few ev.

The difference procedure may also be used with a *boron-filter*. Since in this case there is no reason for keeping the thickness of the filter very small, the difference between the counting rates without and with the boron absorber can be made of the same order of magnitude as the individual counting rates. The boron filter, however, has the disadvantage of absorbing the neutrons belonging to a rather wide energy interval; therefore, this technique may be usefully employed only for measuring cross sections which remain constant over a wide range of energies, as, for example, the free atom scattering cross sections (Sects. 29 and 100).

The method described in this section may also be used in *internal experiments*, with the advantage that the intensity of epi-cadmium neutrons available inside, for example, the lattice of a pile, is a thousand times larger than those in external beams (Sect. 90). The interpretation of the results of experiments of this type is, however, more difficult than in the case of beam experiments, because of the angular spread of the incident neutrons. Even assuming that the sensitivity of the $1/v$ detector is rigorously isotropic, Eq. (88.2) must be modified replacing by $x/\cos\vartheta$ the thickness of the filter, and integrating the expression thus obtained over the solid angle. We shall not enter here into a discussion of this point since it will come up again at the end of Sect. 94 in connection with the problem of the absorption curves obtained under different experimental conditions. Here, the obvious remark may suffice that in internal experiments a filter behaves as if its effective thickness were larger than its true thickness x by a factor of the order of 2.

89. A few remarks about the use of detectors. We shall not enter into a discussion of the methods employed for detecting neutrons, since these may be found in BARSCHALL's article in Vol. XLIV of this Encyclopedia. We assume that the reader is familiar with all the procedures described there—in particular with BF_3 counters and foil techniques—and proceed to consider those features of these instruments which are relevant to experiments of the type described in this article. For details on the construction and the various practical aspects of the use of these detectors, the reader is directed to various books[1,2] [*12*] and articles [*17a*], [*18b*].

All slow neutron detectors are based on the use of materials whose nuclei absorb the incident neutrons giving rise—either directly or through the radioactive product of a nuclear reaction—to the emission of radiations which can be easily measured. In the first case, i.e. in direct processes, the delay of emission of the radiation with respect to the neutron capture, is so short ($\sim 10^{-14}$ sec) that the process can be considered instantaneous. In the second case the half-

[1] B. B. ROSSI and H. H. STAUB: National Nuclear Science Series: Experimental techniques. New York: McGraw-Hill 1949.

[2] A. C. GRAVES and D. K. FROMAN: National Nuclear Science Series: Experimental techniques. New York: McGraw-Hill 1952.

life of the radioactive product is much longer; usually, for convenience of measurement, only those processes are used which give products of half-life not shorter than about half a minute and not longer than a few days.

The instantaneous radiation may consist of heavily ionizing particles as in the case of boron and a few other light elements (Sects. 27) or in the case of U^{235} fission. It may consist of γ-rays emitted in neutron capture (Sect. 16); an example of the latter technique—first used by FLEISCHMANN[1] for detecting thermal neutrons captured by cadmium—has been mentioned in Sect. 85 in connection with experiments on the slowing down of neutrons as a function of time[2].

The delayed radiations emitted by the radioactive product of the reaction produced by slow neutrons may be β- and γ-rays or even X-rays arising from electron-capture processes.

In the simplest cases, the essential part of the detector can be described as a flat layer of absorbing material of thickness x and area S. The total number of neutrons absorbed by the detector is given by [Eq. (134.3)]

$$a = S \int |\cos\Theta|\, d\Omega \int_0 dE\, \Phi(E, \Theta, \varphi) \left(1 - e^{-\bar{\Sigma}_a(E) \frac{x}{\cos\Theta}}\right). \tag{89.1 a}$$

If we assume, for the sake of simplicity, that the angular number flux Φ can be written as the product of a function of the energy $\varphi_0(E)$ and a (normalized) function of the angles $f(\Theta, \varphi)$, Eq. (89.1 a) becomes

$$a = S \int |\cos\Theta| f(\Theta,\varphi)\, d\Omega \int_0^{E_0} \left(1 - e^{-\bar{\Sigma}_a(E) \frac{x}{\cos\Theta}}\right) \Theta_0(E)\, dE \tag{89.1 b}$$

where $\bar{\Sigma}_a(E)$ is the effective macroscopic absorption cross section—i.e. the macroscopic absorption cross section corrected for thermal agitation of the capturing nuclei (Sect. 87)—and Θ the angle of incidence of the neutrons measured from the normal to the detector surface.

If $\bar{\Sigma}_a(E)$ arises only from the process whose radiation is used for detecting the captured neutrons, the quantity (89.1) may be called the *total activation* of the detector.

The expression (89.1) is correct only if the scattering cross section is so small that

$$\bar{\Sigma}_s(E) \frac{x}{\cos\vartheta} \ll 1 \tag{89.2}$$

for any value of E and Θ: otherwise the scattering would produce an increase with respect to $x/\cos\Theta$ of the path length crossed by the neutrons inside the detector.

The absorbing layer is called *thin* if for any value of E and Θ one has

$$\bar{\Sigma}_a(E) \frac{x}{\cos\Theta} \ll 1 \tag{89.3}$$

so that Eq. (89.1) reduces to

$$a = x\, S \int_0^{E_0} \bar{\Sigma}_a(E)\, \varphi_0(E)\, dE = x\, S\, \mathfrak{N} \int_0^{E_0} \bar{\sigma}_a(E)\, \varphi_0(E)\, dE \tag{89.4}$$

---

[1] R. FLEISCHMANN: Z. Physik **97**, 242, 265 (1935); **111**, 281 (1938). — Phys. Z. **36**, 806 (1935); **39**, 879 (1938).

[2] A.A. BERGMAN, A.I. ISAKOFF, I.D. MURIN, F.L. SHAPIRO, I.V. SHTRANIKH and M.V. KAZARNOVSKY: p. 135 of Ref. [*15b*].

where, according to (61.3), (61.4), $\mathfrak{N}$ is the number of capturing atoms/cm³ present in the detector.

The inequality (89.3) is indicated sometimes as the condition of *no self-protection* since it imposes that all the atoms present in the sample are exposed to the same neutron flux. This condition may be very severe in the vicinity of a resonance E_R and for values of Θ not far from 90°. This happens in internal experiments with resonance detectors, the thickness of which must be less than a few mg/cm² (Sect. 90); as we shall see there, in certain cases, Au and U^{238} detectors have been used with thicknesses X of the order of 1 or 2×10^{-5} g/cm², i.e.

$$x = \frac{X}{\varrho} = \frac{2\times 10^{-5}\,\text{g/cm}^2}{\sim 20\,\text{g/cm}^3} \cong 10^{-6}\,\text{cm} = 100\,\text{Å}. \tag{89.5}$$

The activation (89.1) of a detector does not represent the quantity actually measured. The *counting rate* of an instantaneous detector, such as a counter or an ionization chamber, differs from the activation because of wall effects, if it contains the neutron-absorbing material in gaseous form; it differs from (89.1) because of geometrical effects and particle self-absorption if it contains the absorbing material in form of solid layers.

If the detector is a foil in which a radioactive nucleus is produced through a (n, γ) reaction, the activation (89.1) represents the initial value of the number of atoms disintegrating per second after an exposure t to neutrons much longer than the half-life $T_{\frac{1}{2}}$ of the reaction product $(t/T_{\frac{1}{2}}\to\infty)$. The *measured activity* A, i.e., the initial number of counts per second recorded by the adopted counting system, differs from a because of various effects, the more important of which are the following: (a) the solid angle subtended by the counter of β (or γ) rays with respect to the foil introduces a geometrical factor ω; (b) the absorption of the radiation in the walls of the counter introduces another factor which we call w, and (c) the self-absorption of the radiation in the foil itself gives rise to a correction which will be discussed in detail below.

The corrections for these various effects must be made, in each case by appropriate procedures, so that it does not make much sense to discuss them in general. The only point that we shall stress here is that the quantity in which one is usually interested is the activation of the detector, not its measured activity (or its counting rate). Therefore the behaviour of a detector, is satisfactory only if its measured activity A is proportional to the activation a, i.e. only if one can write

$$A = \varepsilon\, a \tag{89.6}$$

where the efficiency of detection ε is a constant independent of the conditions of irradiation, i.e. independent of the position of the detector or of the presence of filters in front of it. While for absolute measurements the value of the constant ε must be known, it is not necessary in making relative measurements; these only require that ε is a constant.

The importance of this point may be illustrated by considering the case of a foil activated uniformly with respect to its surface through a (n, γ) process and whose β-rays are measured, for example, by means of an end-window Geiger-Müller counter; while the effects (a) and (b) do not depend on the conditions of irradiation of the foil, the self-absorption does depend upon these conditions. Let us call $\alpha(x')\,dx'$ the activation of the layer between x' and $x'+dx'$ inside the foil so that the total activation of the foil is

$$a = \int_0^x \alpha(x')\,dx'. \tag{89.7}$$

The self-absorption undergone by the β-rays may be easily taken into account by making use of the empirical exponential law

$$e^{-\mu x} \tag{89.8}$$

which is known to be approximately correct for moderate thickness x, provided that the mass absorption coefficient μ/ϱ is measured in the material of the foil with the same counter and under the same geometry adopted in the actual measurements of the foil activity. Then the activity of the foil, measured on its face $x=0$ (Sect. 134) can be written

$$A=\omega w \int_0^x \alpha(x')\, e^{-\mu x'} d x'. \tag{89.9}$$

The value of this integral depends on the distribution of the activation $\alpha(x)$ inside the detector. This is usually different under different conditions of irradiation and may be considered known whenever the corresponding self-absorption curve is known (Sect. 94 and 136). It can always be represented with sufficient accuracy by an expression of the type

$$\alpha(x')=\sum_i \alpha_i\, e^{\bar{k}_i x'} \tag{89.10}$$

where the sum includes a few terms, each with two adjustable constants α_i and $\bar{k}_i$.

When the detector is thin according to the definition (89.3), $\alpha(x')$ is constant $\alpha(x')=\alpha$ so that (89.9) reduces to (89.6) with

$$\varepsilon=\omega w \frac{1-e^{-\mu x}}{\mu x} \tag{89.11}$$

which is independent of the conditions of irradiation. Thus we can conclude that in the case of a thin detector, i.e. a detector satisfying condition (89.3), the measured activity is proportional to the activation, because the absorption of the recorded β-rays inside the detector gives rise simply to a constant factor. The computation of this factor is made in different ways according to the penetrations of the recorded radiation; for μx so small with respect to 1 that terms of the order of $(\mu x)^2$ are negligible, the correction factor is practically equal to one. When only terms of order $(\mu x)^3$ or higher can be neglected, one can write

$$\frac{1-e^{-\mu x}}{\mu x} \approx 1-\frac{\mu x}{2} \approx e^{-\frac{\mu x}{2}}$$

while in the opposite case of $\mu x \gg 1$ the correction factor turns out to be $\frac{1}{\mu x}$ i.e. the detector behaves as a thin detector of thickness $1/\mu$ instead of x. In the latter case the detector can be considered as thin even if condition (89.3) is not fulfilled provided

$$\Sigma_a(E) \frac{\mu^{-1}}{\cos\Theta} \ll 1 \tag{89.12}$$

for any value of E and Θ. In this case the measured activity is due only to the superficial layers of effective thickness $1/\mu$ and the use of a detector thickness larger than 2 or 3 times μ^{-1} is useless since the activation of the deeper layers cannot be observed.

We shall not give here further details on the determination of the various factors entering into the detection efficiency ε which may be found elsewhere[1,2,3] [*12*], [*17a*].

For intensity reasons, one is obliged very often, to use thick detectors, i.e. detectors which do not fulfill the condition of no self-protection (89.3).

This circumstance must be considered with special attention in the case of *internal experiments* since then the two following effects may play an important role: (a) the activation of a non-spherical detector—usually a flat foil or a cylindrical counter—placed at a given position *depends on the detector orientation*, and (b) the detector produces a *depression* of the neutron flux in its vicinity.

These effects are considered in detail in Sect. 134 and 135 respectively. Here we limit ourselves to recalling that the first one can be eliminated by averaging the response of the detector with respect to its orientation; for a flat foil this can be made with sufficient accuracy by averaging the activities measured over both detector sides after exposure with a fixed orientation [see Eq. (134.10) and the corresponding discussion].

The depression of the neutron flux is of considerable importance in the case of thermal neutrons (Sect. 135) while it is negligible for detectors of epi-cadmium neutrons with a single sharp resonance embedded in a medium containing only very light elements. Under these conditions a resonance neutron which has crossed the detector without being absorbed and which is scattered back to cross it again, has always undergone an energy loss sufficiently large to be no more at resonance[4]. The flux depression may on the contrary become appreciable even in the case of epi-cadmium neutrons if the detector absorption extends over a large energy interval and the surrounding medium has a small slowing-down power, since then many collisions are required to appreciably reduce the energy of the neutrons.

The consideration given above can now be applied to *detectors of epi-cadmium neutrons*; these are obtained by exposing to slow neutrons a thin detector—as defined above by Eq. (89.3)—covered with a cadmium filter.

In the approximation in which the effect of the cadmium filter can be described in terms of the cadmium cut-off (86.1) one has

$$A_{\mathrm{Cd}} = F_{\mathrm{Cd}}^{-1}\,\varepsilon\,a_{\mathrm{Cd}} \tag{89.13a}$$

$$a_{\mathrm{Cd}} = \varkappa\, S\,\mathfrak{N} \int_{E_{\mathrm{Cd}}}^{E_0} \bar{\sigma}_a(E)\,\varphi_0(E)\,dE \tag{89.13b}$$

where A_{Cd} is the activity measured behind the cadmium filter and F_{Cd} the correction factor arising from the absorption of epi-cadmium neutrons in the Cd filter. As we shall see below, it depends on the thicknesses of both detector and filter.

---

[1] See footnote 2, p. 352.

[2] H. H. BARSCHALL: Detection of neutrons, Vol. XLV of this Encyclopedia.

[3] Conference on absolute β-counting. (National Research Council, U.S.A., Preliminary Report, No. 8, October 1950.) — Meeting of the Subcommittee on Beta and Gamma Measurements and Standards. (National Research Council U.S.A.); (a) February 1950; (b) October 1952; (c) January 1954. — Standardization of Radioactive Sources. G. G. MANOV: Annual Rev. Nucl. Sci. **4**, 51—68 (1954). — Measurements of Disintegration Rate. J. L. PUTNAM: Beta and Gamma-Ray Spectroscopy (Editor K. SIEGBAHN). Amsterdam: North Holland Publishing Co. 1955. — J. STEYN: Proc. Phys. Soc. Lond. A **69**, 865 (1956). — B. GRIMBERG et Y. LE GALLIC: J. Phys. Radium **17**, 35A (1956).

[4] E. AMALDI and E. FERMI: Ric. Sci. **7** (1), 454 (1936). — Phys. Rev. **50**, 899 (1936).

We have mentioned in Sect. 86 that epi-cadmium neutron detectors may be distinguished in $1/v$ detectors and resonance detectors. Examples of detectors of the first type are BF_3 counters or foils of materials such as Na, and Al, which are activated through (n, γ) processes which do not show resonances of appreciable intensity (see Table 65); examples of resonance detectors are provided by foils of Rh, In, Ag, I, Au etc., a few of which have been discussed in detail in Sect. 26.

The same remark holds here as in the case of filters: in order to be an *ideal resonance detector* of epi-cadmium neutrons, a material should have a single sharp resonance satisfying the inequality (86.3). In this case $\bar{\sigma}_a(E)$ is given by Eq. (87.6) which shows a sharp maximum at $E = E_R$, while $\varphi_0(E)$ is a slowly varying function of the neutron energy. Therefore we may replace $\varphi_0(E)$ by $\varphi_0(E_R)$ in (89.13) and state that the activity of an infinitely thin resonance detector with a single sharp resonance is proportional to the flux of the neutrons of energy E_R at the position of irradiation: $A_{\text{Cd}} \propto \varphi_0(E_R)$ [see Eq. (134.16)].

A relation of this type holds even when the ideal resonance detector is thick; then, according to (134.10), the proportionality exists between the neutron flux and the *arithmetic average* $\bar{A}_{\text{Cd}}$ of the activities measured on the two sides of the foil. Therefore, making use of (75.7) we can state that *in the case of an epi-cadmium detector with a single sharp resonance satisfying condition (86.3), even if its thickness is finite,* one has,

$$\bar{A}_{\text{Cd}}(\boldsymbol{r}) = K_{\text{res}}\, Q(E_R, \boldsymbol{r}) \tag{89.14}$$

where $\bar{A}_{\text{Cd}}(\boldsymbol{r})$ is defined by Eq. (134.10), $Q(E_R, \boldsymbol{r})$ is the slowing down density at energy E_R and space coordinates $\boldsymbol{r}$, and K_{res} is a constant which depends on the absorption coefficients for neutrons and for electrons (or γ-rays) of the detector, its thickness and area, the properties of the moderator and the type of counter used for measuring the activities.

This ideal behaviour—i.e. an activation cross section with a single sharp resonance—is shown, however, only by very few resonance detectors, perhaps even by none. As we have pointed out in Sect. 88, rhodium appears to approach rather well these specifications since no other resonances of intensity comparable to that of Fig. 11 has been observed up to about 1000 ev. Gold has only six weak resonances, besides that at 4.9 ev (Fig. 16), between 40 and 300 ev, while the capture cross section of each one of the other elements mentioned above shows a number of appreciably strong maxima [*14*].

This inconvenience, however, can be avoided by choosing the thickness x of a resonance detector so small that only the resonance of largest intensity—usually that at lower energy—is responsible for almost all the activation. In the case of In foils, for example, the activation produced by a parallel beam of neutrons is practically due only to the 1.46 ev resonance, if the foil thickness does not exceed about 20 mg/cm². This may be recognized from Fig. 161 which refers to the case of an In filter placed in front of a $1/v$ detector, but which may serve also as a qualitative illustration of the behaviour of an In detector.

When the thickness of the In foil is larger than the value (88.3) and/or the incident neutron beam has an appreciable angular spread—as it happens, for example, in the case of internal experiments—, the contribution to the activation of the foil arising from the resonances at higher energy may become appreciable. An estimate of this effect and a computation of the corresponding correction can be made by repeating the activity measurement of the In detector with a boron or In cover of convenient thickness added to the cadmium filter. The boron filter eliminates mainly the neutrons of energy below 10 ev (Fig. 162, p. 351) and therefore allows for an estimate of the activity due to higher energy neutrons. The In filter, when sufficiently thin, eliminates only the neutrons corresponding to the 1.46 ev resonance so that the activity of a rather thin In detector exposed behind a (Cd+In) filter is mainly due to capture of neutrons of energy larger than 1.46 ev.

In internal experiments, as those reported in Division IV, one can take advantage of the fact that the activation due to the "higher energy neutrons" decreases by increasing the distance of detector from source faster than the activation due to lower energy neutrons. A few examples of this type of correction have been given in Sects. 83 and 84. The lower curve of Fig. 138 shows the effect of a boron filter thick enough (0.245 to 0.337 g/cm²) to eliminate a very large fraction of the neutrons of energy smaller than 10 ev, while in Sect. 84 the effect of the addition of an In filter has been discussed in detail in connection with measurements of the distribution in graphite of In-resonance neutrons produced by a $Sb^{124}\gamma+Be$ source (Fig. 156). A description of the same procedure applied to the case of fission neutrons slowed down in heavy water can be found in WADE's paper[1].

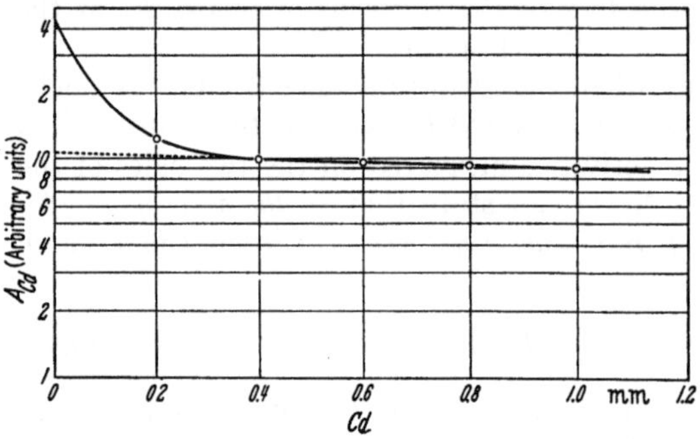

FIG. 163. Absorption curve in cadmium for neutrons slowed down in water as used by BRACCI *et al.* for computing the correction factor F_{Cd} for an In detector 7.5 mg/cm² thick.

The correction factor F_{Cd} appearing in (89.13a) is introduced in order to take into account the absorption of epi-cadmium neutrons by the Cd filter. It can be determined by measuring the activity of the resonance detector—for example an In foil—with various thicknesses x_{Cd} of Cd around it, plotting the results as a function of x_{Cd} and extrapolating to $x_{Cd}=0$ the part of the curve obtained with cadmium layers thicker than about 0.5 mm. (Fig. 163). It may be recognized that the value of F_{Cd} obtained by such a procedure depends also on the detector thickness especially if the detector is sensitive to neutrons of energy close to the cadmium cut-off. For example, in the case of In, the 1.46 ev resonance is so close to E_{Cd} that thick foils pick up an appreciable amount of activation on the resonance wing extending into the half-ev energy region, thus giving a large value of F_{Cd}[2].

Table 63. *Correction factor* F_{Cd} *for absorption of* In-*resonance neutrons by* Cd *filters of thickness* x_{Cd}.

Authors[3]	x_{Cd} mm	In foil thickness mg/cm²	F_{Cd}
WALKER	1.04	3.59	1.077
BRACCI, FACCHINI, GERMAGNOLI	0.6	7.5	1.108±0.020
DACAY, PAINE and GOODMAN	1.04	58.8	1.20
RUSH	1.04	93.5	1.23
KUNSTADTER	1.04	94.5	1.50

[1] J.W. WADE: DP-163 (1956). — Nucl. Sci. and Engng. **4**, 12 (1958).
[2] C.W. TITTLE: Nucleonics **9**, No. 1, 60 (1951).
[3] R.L. WALKER: MDDC-414 (1940). — J. DACEY, H.W. PAINE and C. GOODMAN: Tech. Rep. No. 23 M.I.T. (October 1949). — J.H. RUSH. Phys. Rev. **73**, 271 (1948). — J. KUNSTADTER: Phys. Rev. **78**, 484 (1950). — A. BRACCI, U. FACCHINI and E. GERMAGNOLI: Nuovo Cim. **7**, 881 (1950).

Table 63 shows the results obtained by various authors[1] using cadmium filters 0.6 and 1.04 mm thick and various thicknesses of the In detector. The measurements of KUNDSTADTER were made in graphite, while all other authors used water. A very rough estimate of F_{Cd} for other values of x_{Cd} is shown in Fig. 164 which was deduced by TITTLE[2] by means of a graphical interpolation procedure of a semilogarithmic plot of the data of Table 63 and the point at $x_{\mathrm{Cd}}=0$.

Fig. 165 shows the results obtained by MEISTER[3] in a cavity opened inside a graphite block arranged to as to give an isotropic distribution of thermal neutrons (Sect. 135). The values of F_{Cd} found by the same author for paraffin are slightly lower than those in graphite.

The absorption in cadmium is not expected to depend on the medium except in so far as the medium might affect the isotropy of the neutrons. The effective absorption however, may depend on whether the Cd displaces a medium, like for example water, or whether it takes up more or less space in a cavity.

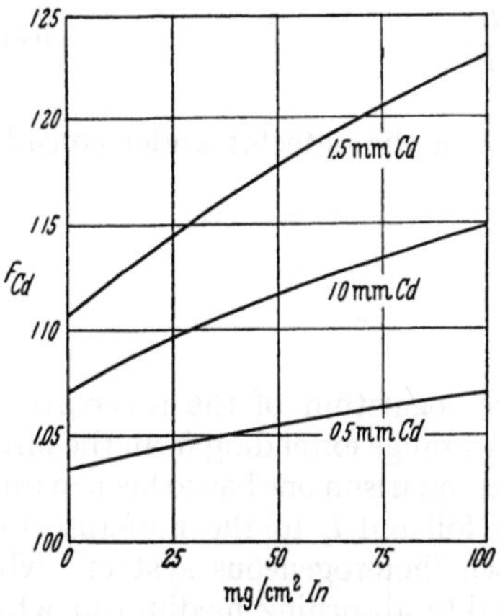

Fig. 164. An estimate of F_{Cd} as a function of In-foil thicknesses, for various values of x_{Cd} deduced by TITTLE from the data obtained by various authors (Table 63).

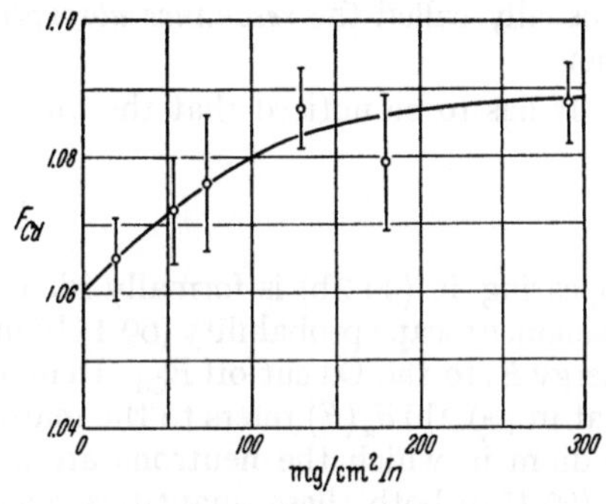

Fig. 165. F_{Cd} as a function of In-foil thicknesses in graphite (MEISTER).

Some authors[3,4,2] have attempted to compute F_{Cd} assuming that the absorption of epi-cadmium neutrons in cadmium can be described as the absorption of an one velocity group (Sect. 136α) for which the absorption coefficient is given by the absorption coefficient of cadmium averaged over the spectrum of epi-cadmium neutrons which activate the detector. The usefulness of theoretical expressions of this type is very doubtful, even for the extrapolation to $x_{\mathrm{Cd}}=0$, because $\overline{K}_{\mathrm{Cd}}=\frac{\Sigma_{\mathrm{Cd}}}{\varrho}$ depends on x_{Cd} as well as on the detector thickness, and furthermore the scattering undergone by epi-cadmium neutrons in the cadmium filter is ignored. Therefore, the results of direct measurements extrapolated to $x_{\mathrm{Cd}}=0$ by means of a logarithmic plot, appear to be simpler and more reliable.

Finally we shall mention that if the foil is exposed inside a holder of some very weakly absorbing materials, such as lucite or aluminium, another correction factor has to be introduced which also has to be determined experimentally by varying the holder thickness[1,2].

---

[1] See footnote 3, p. 358.
[2] See footnote 2, p. 358.
[3] H. MEISTER: Z. Naturfrosch. **10**a, 669 (1955).
[4] G.M. BRANCH: AEC MDDC 747 (1946).

90. The activation of a thin detector and its cadmium ratio. The activity of a thin detector exposed to epi-cadmium neutrons, after correction for the absorption of epi-cadmium neutrons by the Cd-filter,

$$A_{\mathrm{Cd}\,0} = F_{\mathrm{Cd}}\, A_{\mathrm{Cd}} \tag{90.1}$$

can be expressed in a very convenient form whenever we know that the incident neutron flux is given, with sufficient accuracy, by Eqs. (86.2a), (86.2b). By introducing these expressions into (89.13) we obtain

$$A_{\mathrm{Cd}\,0} = \varepsilon\, a_{\mathrm{Cd}} \tag{90.2a}$$

$$a_{\mathrm{Cd}} = q_0 \frac{l_s}{\xi}\, x\, S\, \mathfrak{N}\, \mathfrak{J}_{a\,E_{\mathrm{Cd}}} \tag{90.2b}$$

where

$$\mathfrak{J}_{a\,E} = \int_E^{E_0} \bar{\sigma}_a(E)\, \frac{dE}{E} \tag{90.3}$$

is usually called the *resonance absorption integral* of the detector under consideration.

It has to be noticed that the quantity

$$\frac{l_s}{\xi}\, \mathfrak{J}_{a\,E_{\mathrm{Cd}}}$$

appearing in (90.2b) is formally identical to the logarithm of the reverse of the resonance escape probability (69.15b) in the energy range extending from the initial energy E_0 to the Cd cut-off E_{Cd}. In making this comparison one has to keep in mind that in (90.2b) $\bar{\sigma}_a(E)$ refers to the material of the foil and l_s to the non-absorbing medium in which the neutrons are slowed down (heterogeneous system), while in (69.15b) both these quantities refer to a weakly absorbing medium in which the neutrons are moderated (homogeneous system).

Since the main contribution to the resonance absorption integral usually comes from energy regions far below about 1 Mev, its value (90.3) is very insensitive to the value adopted for its upper limit, as long as it is of the order of 1 Mev or more. Thus, many authors take $E_0 = 1$ Mev and write

$$\mathfrak{J}_{a\,E_{\mathrm{Cd}}} = \int_{0.4\,\mathrm{ev}}^{1\,\mathrm{Mev}} \bar{\sigma}_a(E)\, \frac{dE}{E}\,. \tag{90.4}$$

The *cadmium ratio* $R_{\mathrm{Cd}}(X)$ of a detector X placed at a given position, is defined as the ratio of the measured activity $A(X)$ of the uncovered detector, to its measured activity when it is covered with a cadmium filter 0.5 to 0.8 mm (i.e. 0.4 to 0.7 g/cm²) thick

$$R_{\mathrm{Cd}}(X) = \frac{A(X)}{A_{\mathrm{Cd}\,0}(X)} = \frac{1}{F_{\mathrm{Cd}}}\,\frac{A}{A_{\mathrm{Cd}}} \approx \frac{A}{A_{\mathrm{Cd}}}\,. \tag{90.5}$$

If the detector is so thin that self-protection with respect to the neutrons can be neglected, A and A_{Cd} can be expressed by means of (89.6), (89.4) and (89.13), respectively. The result

$$R_{\mathrm{Cd}}(X) = 1 + \frac{\int_0^{E_{\mathrm{Cd}}} \bar{\sigma}_a(E)\, \varphi_0(E)\, dE}{\int_{E_{\mathrm{Cd}}}^{E_0} \bar{\sigma}_a(E)\, \varphi_0(E)\, dE} \tag{90.6}$$

is independent of both the detection efficiency of the counting system and the total number $x\,S\,\mathfrak{N}$ of atoms constituting the detector. Making use of (90.2b), (133.6) and of the remark made at the beginning of Sect. 87 about the independence of $\bar{\sigma}_a$ of thermal agitation in the case of $1/v$ absorbers[1], we can write

$$R_{\mathrm{Cd}}(X) - 1 = \frac{\sigma_{a\,\mathrm{th}}\,\varphi_{\mathrm{th}}}{\frac{q_0\,l_s}{\xi}\,\mathfrak{J}_{a\,E_{\mathrm{Cd}}}} \tag{90.7}$$

which expresses the cadmium ratio in terms of the absorption resonance integral $\mathfrak{J}_{a\,E_{\mathrm{Cd}}}$, the thermal capture cross section $\sigma_{a\,\mathrm{th}}$ defined by (23.15), the thermal neutron flux φ_{th} and the factor $q_0 l_s/\xi$.

For the experimental determination of $R_{\mathrm{Cd}}(X)$ two procedures may be adopted: (a) The same sample—or two identical samples—is activated successively at the same position, once uncovered, the other time covered by cadmium. For accurate results great care must be devoted to the monitoring system whose readings enter as normalisation factors. (b) Two identical samples—one uncovered, the other covered by cadmium—are activated simultaneously at two different positions. Then the ratio of the activities obtained at the two positions for uncovered as well for cadmium covered foils, must be measured with great accuracy. Furthermore, if the irradiation is made inside the moderator, attention must be paid in order to avoid that the flux depression of thermal neutrons (Sect. 135) produced by the Cd-filter, reduces the activity of the uncovered foil. This inconvenience can be avoided by separating the foils 10 cm or more.

For a $1/v$ detector, $\sigma_{a\,\mathrm{th}}$ as well as the resonance absorption integral can easily be computed:

$$\frac{\sigma_{a\,\mathrm{th}}}{\mathfrak{J}_{a\,E_{\mathrm{Cd}}}} = \frac{(0.025)^{\frac{1}{2}}}{\int\limits_{E_{\mathrm{Cd}}}^{\infty} E^{-\frac{3}{2}}\,dE} = \frac{(0.025)^{\frac{1}{2}}}{2\,(E_{\mathrm{Cd}})^{\frac{1}{2}}} = \frac{1}{2}\left(\frac{0.025}{0.40}\right)^{\frac{1}{2}} = 2\,. \tag{90.8}$$

In the following we shall use this numerical value, which, however, is expected to be correct only within about 10% since it derives from the description of the absorption by cadmium in terms of the Cd cut-off and on the adoption of the numerical value (86.1).

Introducing (90.8) into (90.7) we have

$$\frac{q_0\,l_s}{\xi} : \varphi_{\mathrm{th}} = \frac{R_{\mathrm{Cd}}(1/v) - 1}{2}\,. \tag{90.9}$$

This equation is of practical interest since it allows for the computation of the ratio of the epi-cadmium flux to the thermal flux from the experimental value of R_{Cd} for a $1/v$ detector. It may be noticed that, according to (86.2), the quantity $q_0 l_s/\xi$ is the neutron flux in any $\ln E_1/E_2 = \ln e = 1$ energy interval such as 1 to 2.72 ev, 2.72 to 7.4 ev, ... or 100 to 272 ev. Thus, for example, from the value

$$R_{\mathrm{Cd}}\left(\frac{1}{v}\right) \approx 33 \tag{90.10}$$

found within the lattice of both a graphite and a heavy water pile, one deduces [*12*]

$$\frac{q_0\,l_s}{\xi} : \varphi_{\mathrm{th}} = 1:16\,. \tag{90.11}$$

[1] See also Eq. (96.5) and the corresponding footnote on p. 396.

This means that at the considered locations the flux of epi-cadmium neutrons in any energy decade as 1 to 10 ev, 10 to 100 ev etc., is $\frac{\ln 10}{16} = \frac{2.30}{16} = \frac{1}{7}$ of the thermal flux.

The value (90.10) represents a very small value for the cadmium ratio of a $1/v$ detector; much larger values are found, for example, inside the reflector or inside the thermal column of reactors.

Fig. 166, taken from the book by HUGHES [12], shows the variation with distance from the core, of the thermal flux and the resonance flux observed with a $1/v$ detector inside the reflector of the Brookhaven pile. The epi-cadmium flux

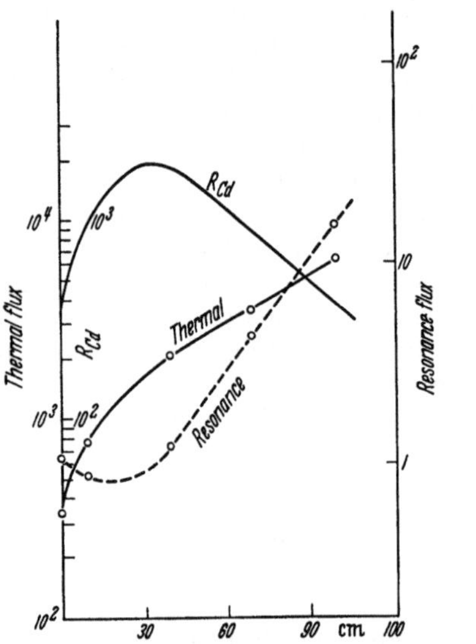

Fig. 166. The experimental thermal and resonance fluxes observed with a $1/v$ detector in the reflector of the Brookhaven graphite pile. Distance measured from outer face of reflector (HUGHES *et al.*).

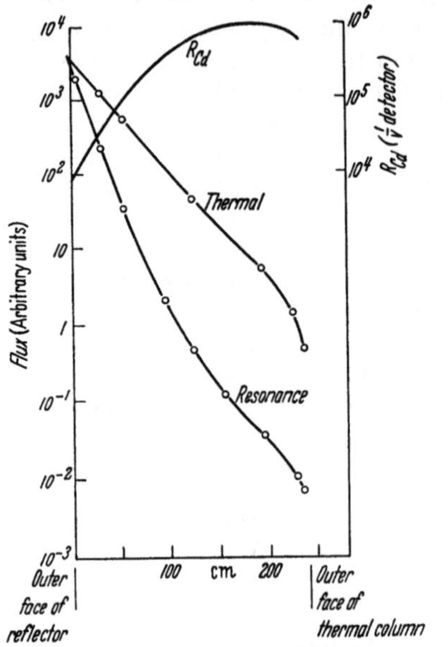

Fig. 167. The experimental thermal and resonance fluxes observed with a $1/v$ detector in the thermal column of the Argonne heavy water pile (SEREN [12]).

decreases faster than the thermal flux with distance from the core so that the cadmium ratio increases moving outward through the reflector. The maximum observed in the $R_{\text{Cd}}(1/v)$ curve, which occurs at about 30 cm from the external surface of the reflector, is due to the increase of the resonance flux observed in this region, caused by fast neutrons which leak out of the center of the pile through air-cooling channels, stream along the cooling gap and finally migrate into the outer region of the reflector. The large value of the cadmium ratio

$$R_{\text{Cd}}\left(\frac{1}{v}\right) \approx 3000 \tag{90.12}$$

reached at about 30 cm from the external surface of the reflector combined with a still considerable value of the neutron flux (Table 64) makes this region of the reflector very appropriate for thermal neutron experiments (see later in this section).

Fig. 167, also taken from HUGHES's book, shows in arbitrary units the results of similar measurements made with In-foils, inside the thermal column of the Argonne heavy water pile. The resonance flux decreases approximately exponentially with a relaxation length of 13 cm, a value which corresponds reasonably well to what one has to expect from the penetration of the fast primaries (Sect. 80) in graphite (for $E \sim 2$ Mev, $l_s \sim 8$ cm). The thermal flux decreases exponentially and the observed relaxation length of 30 cm is shorter than the diffusion length (50 cm; Sect. 128) because of the neutrons leaking out of the sides of the finite thermal column. The cadmium ratio for a $1/v$ detector, reaches values of the order of 10^6 towards the outer face of the thermal column, but in that region the thermal flux has rather low values.

It is clear that for experiments with resonance neutrons it is convenient to choose locations where the value of $R_{\text{Cd}}(1/v)$ is low, while for experiments with thermal neutrons in which the contamination of epi-cadmium neutrons must be kept as low as possible, locations with a large value of $R_{\text{Cd}}(1/v)$ are preferable; a limitation in the choice of such a location is imposed by the fact that large values of $R_{\text{Cd}}(1/v)$ are found in the region of low thermal neutron flux; this circumstance is clearly shown by the few examples collected in Figs. 166 and 167 and in Table 64.

Table 64. *Examples of thermal neutron flux φ_{th} and cadmium ratio $R_{\text{Cd}}(1/v)$ values in Brookhaven and Argonne piles* [12].

	φ_{th}	$R_{\text{Cd}}(1/v)$
Brookhaven graphite pile		
At the center of the lattice	5×10^{12}	~ 33
At ~ 30 cm inside the reflector	4×10^{11}	~ 3000
Argonne heavy water pile		
At the center of the lattice	10^{12}	~ 33
At the inner face of the thermal column	10^{11}	~ 10000

For experiments made outside the pile, holes are opened into the pile to the region where the desired values of φ_{th} and $R_{\text{Cd}}(1/v)$ are found. Thus in the Brookhaven pile the epi-cadmium neutrons are investigated at best through an experimental hole opened into the pile lattice, while thermal neutrons are better investigated through a hole that opens about 30 cm into the reflector. If the hole opened into the pile lattice has 10 cm$\times$10 cm sides and the neutrons are observed at a 6 meter distance from its inner end, the fluxes available are, according to Table 64 and Eq. (90.9)

$$\varphi_{\text{th}} = 5\times10^{12}\frac{10^2}{4\pi\,(600)^2} = 1.1\times10^8 \text{ neutrons/cm}^2\text{ sec};$$

$$\varphi_{\text{epi-Cd}}(E)\,dE = \frac{1.1\times10^8}{16}\,\frac{dE}{E} = 6.9\times10^6\,\frac{dE}{E} \text{ neutrons/cm}^2\text{ sec}.$$

At the bottom of the hole in the reflector, the thermal flux and the epi-cadmium flux are respectively 12.5 times and and $12.5\times1500/16 \sim 1200$ times smaller than the corresponding values at the center of the pile.

In the case of resonance detectors with strong resonances, the cadmium ratio can reach very low values. In this case the condition of thin detector is very stringent since the inequality (89.3) must be fulfilled even at the center of the resonance line where the cross section may be exceedingly large. The samples used by Harris, Muehlhause and Thomas[1] were usually less than 5.0 mg/cm² thick and, when R_{Cd} was smaller than 5, they were 1 mg/cm² or even 0.2 mg/cm². These very thin detectors were obtained either by evaporating

[1] S. P. Harris, C. O. Muehlhause and G. E. Thomas: Phys. Rev. **79**, 11 (1950).

or by fixing metals or oxide powders with organic adhesive (thinned zapon) on a 1-mil thick puron (electrolytic iron) foil. MACKLIN and POMERANCE[1] used 0.015 mg/cm² of Au, 0.07 mg/cm² of U^{238} and 0.3 mg/cm² of Th.

Table 65. *Cadmium ratio values according to Argonne group*[2].

Target isotope (X)	10^{24} nuclei per cm² (isotopic)	Chemical form	$R_{Cd}(X)$	Target isotope (X)	10^{24} nuclei per cm² (isotopic)	Chemical form	$R_{Cd}(X)$
$_{11}Na^{23}$	7.4×10^{-5}	NaF	70.3	$_{49}In^{115}$	1.5×10^{-6}	In	3.26
$_{13}Al^{27}$	2.9×10^{-4}	Al	43.1	$_{51}Sb^{121}$	1.7×10^{-5}	Sb	2.27
$_{15}P^{31}$	3.1×10^{-4}	P	63.8	$_{51}Sb^{123}$	1.3×10^{-5}	Sb	1.55
$_{21}Sc^{45}$	2.6×10^{-5}	Sc_2O_3	69.0	$_{53}I^{127}$	1.3×10^{-5}	PbI_2	2.45
$_{23}V^{51}$	2.1×10^{-5}	V_2O_5	70.6	$_{59}Pr^{141}$	1.8×10^{-5}	PrO_2	30.9
$_{25}Mn^{55}$	6.5×10^{-5}	Mn	35	$_{62}Sm^{152}$	4.1×10^{-6}	Sm_2O_3	3.43
$_{27}Co^{59}$	6.1×10^{-5}	Co	23.8	$_{63}Eu^{151}$	8.3×10^{-6}	Eu_2O_3	45.7
$_{29}Cu^{63}$	1.4×10^{-4}	Cu	30.8	$_{72}Hg^{180}$	8.0×10^{-6}	HgO_2	14.9
$_{29}Cu^{65}$	6.0×10^{-5}	Cu	30.5	$_{72}Hg^{all}$	2.3×10^{-5}	HgO_2	3.0
$_{31}Ga^{69}$	3.1×10^{-5}	Ga_2O_3	5.79	$_{73}Ta^{181}$	2.7×10^{-5}	Ta	25.1
$_{31}Ga^{71}$	22.0×10^{-5}	Ga_2O_3	7.96	$_{74}W^{186}$	3.1×10^{-7}	W	3.93
$_{33}As^{75}$	6.0×10^{-5}	As	4.36	$_{75}Re^{185}$	6.6×10^{-6}	Re	3.61
$_{35}Br^{79}$	1.0×10^{-5}	$PbBr_2$	3.25	$_{75}Re^{187}$	1.1×10^{-5}	Re	8.50
$_{39}Y^{89}$	4.2×10^{-5}	Y_2O_3	47.9	$_{77}Ir^{191}$	6.9×10^{-6}	Ir	9.37
$_{41}Nb^{93}$	2.5×10^{-5}	Nb_2O_5	9.59	$_{77}Ir^{193}$	1.1×10^{-5}	Ir	3.89
$_{45}Rh^{103}$	2.6×10^{-5}	Rh	7.94	$_{79}Au^{197}$	5.0×10^{-7}	Au	2.91
$_{47}Ag^{107}$	1.5×10^{-5}	Ag	13.34	$_{81}Tl^{203}$	3.0×10^{-5}	Tl	3.59
$_{47}Ag^{109}$	zero extrapolation	Ag	3.20	$_{81}Tl^{205}$	7.3×10^{-5}	Tl	7.12
$_{49}In^{113}$	6.8×10^{-7}	In	2.68				

Table 65 shows the values of R_{Cd} obtained by the first group of authors[2] in a "rabbit tube" consisting of a pneumatic tube which contained a pair of samples of the same material, one uncovered, the other covered with cadmium, and bombarded in the pile at a position next to the heavy water tank. The cadmium ratio for the $1/v$ detector

$$R_{Cd}\left(\frac{1}{v}\right) \approx 70$$

was obtained from the data of $_{11}Na^{22}$, $_{21}Sc^{45}$ and $_{22}V^{51}$ which do not have any important neutron capture levels in the low energy region. The smallest value was obtained with $_{51}Sb^{123}$ ($R_{Cd} \sim 1.5$) followed by $_{51}Sb^{121}$ ($R_{Cd} \sim 2.27$), $_{53}I^{127}$ ($R_{Cd} \sim 2.45$), $_{49}In^{113}$ ($R_{Cd} \sim 2.68$) and $_{79}Au^{197}$ ($R_{Cd} \sim 2.91$). The largest value was obtained with $_{66}Dy^{164}$ ($R_{Cd} \simeq 215$) because of the existence of a resonance with $E_R < E_{Cd}$ (actually E_R negative).

The influence of the thickness of the sample on its cadmium ratio may be illustrated by the following example [*12*]: an indium foil 0.1 g/cm² thick has a cadmium ratio about five times that obtained with a very thin indium sample (0.6 mg/cm²). While the latter absorbs less than 1% of the neutrons of exact resonance energy $E_R = 1.46$ ev ($\sigma_{a0} \approx 30000$ barns) and therefore does not distort the incident spectrum, the self-protection of resonance neutrons in the 0.1 g/cm² foil enhances that part of the activation which originates from thermal neutrons.

91. The resonance integrals. α) *Definitions and theoretical considerations.* The *resonance absorption integral*, defined by Eq. (90.3), can be used for deriving

[1] R.L. MACKLIN and H.S. POMERANCE: Progress in nuclear energy. Physics and Mathematics, Vol. 1, p. 179. London and New York: Pergamon Press 1956, and p. 96 of Ref. [*15c*].
[2] See footnote 1, p. 363.

useful information on the behaviour of $\sigma_a(E)$ above the cadmium cut-off. To such an end it has been found convenient to split $\mathfrak{I}_{aE_{\text{Cd}}}(X)$ into the sum of two terms, the first of which $\mathfrak{I}_{ar}(X)$ arises from the capture resonances of the material X, while the second $\mathfrak{I}_{1/v\,\text{Cd}}(X)$ represents that part of the $1/v$ absorption which extends beyond the cadmium cut-off

$$\mathfrak{I}_{aE_{\text{Cd}}}(X) = \mathfrak{I}_{ar}(X) + \mathfrak{I}_{1/v\,\text{Cd}}(X). \tag{91.1}$$

Here

$$\mathfrak{I}_{1/v\,\text{Cd}}(x) = \int\limits_{E_{\text{Cd}}}^{E_0} \sigma_a(0.025\text{ ev}) \left(\frac{0.025\text{ ev}}{E}\right)^{\frac{1}{2}} \frac{dE}{E} \tag{91.2a}$$

represents the value which one would obtain for $\mathfrak{I}_{aE_{\text{Cd}}}(X)$ if the sample X would absorb according to the $1/v$ law starting from the value of $\sigma_a(E)$ at $E = 0.025$ ev (Sects. 90, Eq. (90.3)). Thus, according to the definition (91.1), the *pure resonance absorption integral* $\mathfrak{I}_{ar}(X)$ is zero whenever the sample follows the $1/v$ law. Eq. (91.2a) has been written by taking into account that the behaviour of a $1/v$ absorber is not affected by the Doppler effect (Sects. 87 and 96).

If the value (86.1) of the cadmium cut-off is introduced in (91.2a) as we have done in (90.8), we obtain

$$\mathfrak{I}_{1/v\,\text{Cd}}(X) \approx 0.50\,\sigma_{a\,\text{th}}. \tag{91.2b}$$

According to MACKLIN and POMERANCE[1] a better approximation is given by

$$\mathfrak{I}_{1/v\,\text{Cd}}(X) \approx 0.44\,\sigma_{a\,\text{th}}. \tag{91.2c}$$

This value holds strictly for gold and has been obtained by correcting the experimental results of previous authors[2] according to the more recent parameters of the gold resonance (Table 10, Sect. 26); it is usually also adopted for other elements.

Before describing the methods employed for measuring $\mathfrak{I}_{ar}(X)$, we shall derive its theoretical expression in a few important cases. First we shall consider a sample with a single isolated resonance satisfying condition (86.3). We then have to compute

$$\mathfrak{I}_{ra}(X) = \int\limits_{E_{\text{Cd}}}^{E_0} \left\{\bar{\sigma}_a(E) - \sigma_{a\,\text{th}} \left(\frac{0.025\text{ ev}}{E}\right)^{\frac{1}{2}}\right\} \frac{dE}{E} \tag{91.3}$$

where $\bar{\sigma}_a(E)$ is obtained by correcting the one-level formula, Eq. (26.2), for the Doppler effect. The same result may be obtained with sufficient accuracy when the resonance is narrow and well above the cadmium cut-off—i.e. when the two inequalities

$$E_R \gg \Gamma, \quad E_R \gg E_{\text{Cd}} \tag{91.4}$$

are both fulfilled—by substituting for the quantity appearing in the curled brackets in (91.3), the approximate expression (87.6); this was derived from (87.1) by taking into account the Doppler effect. According to (87.6), (87.15) the result is *independent of the Doppler broadening*, so that we can write in general

$$\mathfrak{I}_{ar}(X) = \frac{\pi}{2}\sigma_{a0}\frac{\Gamma}{E_R} \tag{91.5}$$

[1] R.L. MACKLIN and H.S. POMERANCE: p. 96 of Ref. [*15c*].
[2] S.P. HARRIS, C.O. MUEHLHAUSE and G.E. THOMAS: Phys. Rev. **79**, 11 (1950).

or, making use of (26.12),

$$\mathfrak{J}_{ar}(X) = \frac{\pi}{2}\left(4\pi\lambda_R^2 g \frac{\Gamma}{E_R}\right)\frac{\Gamma_n \Gamma_\gamma}{\Gamma^2}. \tag{91.6}$$

A third form, equivalent to (91.5), is obtained by expressing σ_{a0} in terms of $\sigma_{a\,\mathrm{th}}$; introducing $E = kT$ into (26.12), we have

$$\mathfrak{J}_{ar}(X) = 2\pi\sigma_{a\,\mathrm{th}}\frac{(E_R kT)^{\frac{1}{2}}}{\Gamma}. \tag{91.7}$$

Finally, combining (91.5) with (90.2), (90.3), (91.1), (91.2c) we obtain

$$A_{\mathrm{Cd}0} = \varepsilon\, x\, S\, \mathfrak{N}\, q_0 \frac{l_s}{\xi}\left[\frac{\pi}{2}\sigma_{a0}\frac{\Gamma}{E_R} + 0.44\sigma_{a\,\mathrm{th}}\right]. \tag{91.8}$$

Apart from the correction term $0.44\sigma_{a\,\mathrm{th}}$, this relation was derived by BETHE and PLACZEK[1] [5] who improved a similar expression used by AMALDI and FERMI[2] in the first estimate of the relative width Γ/E_R of the resonances observed in various (n, γ) processes.

Some authors[3] have defined—and measured—the *resonance scattering integral*

$$\mathfrak{J}_{sr}(X) = \int_{E_{\mathrm{Cd}}}^{E_0} [\sigma_s(E) - \sigma_p]\frac{dE}{E} \tag{91.9}$$

where $\sigma_s(E)$ and σ_p are the scattering cross section and the cross section for potential scattering (Sects. 21 and 26). If the sample has a single scattering resonance, the integral can be computed by introducing into (91.9) the one-level formula (26.7); it may be useful to recall that, according to (26.10), the factor $\lambda^2\Gamma_n^2$ is constant.

If the resonance satisfies the two conditions (91.4), a constant factor $1/E_R$ may be substituted for $1/E$ in (91.9). Then the contribution of the interference term appearing in (26.7) goes to zero and (91.9) reduces to

$$\mathfrak{J}_{sr}(X) = \frac{\pi}{2}\sigma_{s0}\frac{\Gamma}{E_R} \tag{91.10}$$

which, making use of (26.8), may be written

$$\mathfrak{J}_{sr}(X) = \frac{\pi}{2}\left(4\pi\lambda_R^2 g \frac{\Gamma}{E_R}\right)\frac{\Gamma_n^2}{\Gamma^2}. \tag{91.11}$$

Combining (91.6) with (91.11) we obtain

$$\frac{\mathfrak{J}_{sr}}{\mathfrak{J}_{sr} + \mathfrak{J}_{ar}} = \frac{\Gamma_n}{\Gamma}, \qquad \frac{\mathfrak{J}_{ar}}{\mathfrak{J}_{sr} + \mathfrak{J}_{ar}} = \frac{\Gamma_\gamma}{\Gamma}. \tag{91.12}$$

Returning to the resonance absorption integral, we notice that Eq. (91.5) may easily be extended to the case of two or more well-separated resonances, all satisfying the two conditions (91.4),

$$\mathfrak{J}_{ar}(X) = \frac{\pi}{2}\sum_i \sigma_{a0i}\frac{\Gamma_i}{E_{Ri}}.$$

A rough estimate of the resonance absorption integral in the case of many separated resonances can be made by means of the following expression derived

---

[1] H. A. BETHE and G. PLACZEK: Phys. Rev. **51**, 450 (1937).
[2] E. AMALDI and E. FERMI: Ric. Sci. **7** (1), 454 (1936). — Phys. Rev. **50**, 899 (1936).
[3] See footnote 2, p. 365.

by DRESNER[1];

$$\mathfrak{J}_{ar}(X)=\int_{E_{R1}}^{\infty}\sigma_a(E)\frac{dE}{E}=\frac{1.3\times 10^6}{E_{R1}}\,\frac{2\pi\bar{\Gamma}_\gamma}{D^*}\,\frac{\beta-\ln(1+\beta)}{\beta^2}\text{ barns},\quad \beta=\frac{\bar{\Gamma}_\gamma}{\bar{\Gamma}_n^0 E_{R1}^{\frac{1}{2}}} \qquad (91.13)$$

where E_{R1} is the energy of the first resonance in ev, $\bar{\Gamma}\gamma$ is the average of the radiative width of resonances for neutrons of zero orbital momentum, $\bar{\Gamma}_n^0$ is the average of the reduced width (26.10) and D^* is an energy of the order of the average level distance D [2].

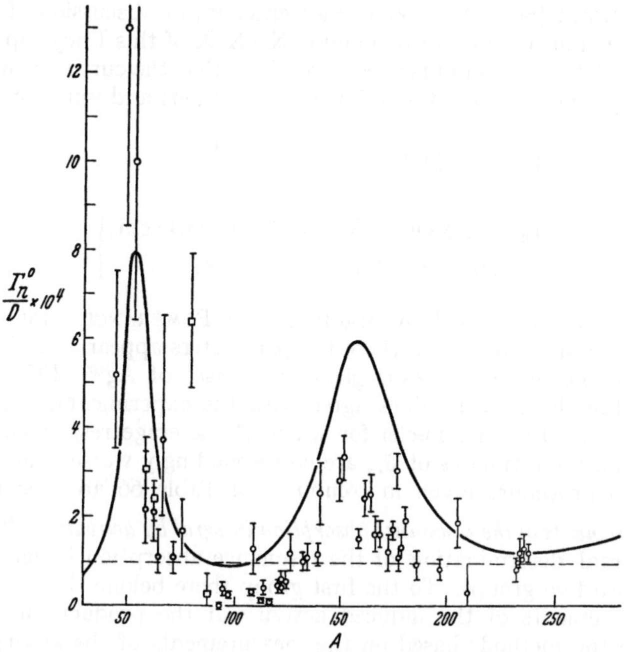

Fig. 168. The average value of Γ_n^0/D as a function of A. Continuum theory – – –; cloudy crystal-ball model —— (WEISSKOPF).

The computation of $\mathfrak{J}_{ar}(X)$ by means of (91.13) involves the values of $\bar{\Gamma}_\gamma/D^*$, $\bar{\Gamma}_n^0/D^*$, and E_{R1}. The two parameters $\bar{\Gamma}_\gamma/D^*$ and $\bar{\Gamma}_n^0/D^*$ are not expected to depend on the neutron energy as long as E varies between 0 and 1 Mev [3]. Their dependence on the mass number A may be derived from the analysis of the resonance parameters obtained by means of slow neutron spectrometers.

The parameter $2\pi\bar{\Gamma}_\gamma/D^*$ is small from small values of A to $A=100$, and is approximately 0.02 for A varying between 100 and 200, except for sharp dips in the vicinity of magic number nuclei. The corresponding variations of $\mathfrak{J}_{ar}(X)$ are unimportant for $A>100$[1].

Fig. 168 shows a plot of the experimental values of $\bar{\Gamma}_n^0/D$ versus A and a comparison with the corresponding theoretical predictions[3]. The continuum theory of nuclear reactions (broken line) [3] gives a constant value for $\bar{\Gamma}_n^0/D$ in disagreement with the experimental data. These are rather well represented by the curve

[1] See footnote 1, p. 364.

[2] The definition of D^* is given on p. 402 of Ref. [3]; a table of experimental values of Γ_γ, Γ_n^0, D^* and D is given on p. 474 of Ref. [3].

[3] R. S. CARTER, J. A. HARVEY, D. J. HUGHES and V. E. PILCHER: Phys. Rev. 96, 113 1954).

computed from the cloudy crystal-ball model (full line) which predicts the existence of maxima for $A = 11$, 55 and 155.

This theory[1], based on the use of a complex potential[2]

$$V = V_1 + i V_2, \quad V_2 = \zeta V_1 \tag{91.14a}$$

for representing the interaction between the incident neutron and the nucleus, constitutes an attempt—very successful in many respects—of describing those aspects of nuclear reactions which do not fall in the two-stage scheme originally suggested by BOHR (Sect. 25). We cannot enter into a discussion of this theory for which the reader is directed to Volume XXXIX of this Encyclopedia and to the references[3,1,4]. We limit ourselves to recalling that the curve given in Fig. 168 has been computed assuming the following expressions and numerical values for the potential[3];

$$V_1 = -V_0\left((1 + \exp\left[\frac{r-R}{d}\right]\right)^{-1} \tag{91.14b}$$

with

$$\left.\begin{aligned} V_0 &= 42\,\text{Mev}, \quad R = 1.35\times 10^{-13} A^{\frac{1}{3}}\,\text{cm},\\ d &= 0.5\times 10^{-13}\,\text{cm}, \quad \zeta = 0.08.\end{aligned}\right\} \tag{91.14c}$$

Eq. (91.13) has been used by MACKLIN and POMERANCE[5], for an estimate of the resonance integrals; even where the parameters appearing in it are experimentally well known—as for example in the cases of Ag^{107}, I^{127}, Ta^{181}, Re^{187}, Th^{232}, U^{238}—the theoretical values agree with the experimental determinations within 50% only. In other nuclei for which the average resonance parameters are less certain, the estimates of $\mathfrak{J}_{ra}$ are correspondingly worse. The values computed by this procedure given in column 2 of Table 66 are designated by *s*.

β) Measurements of the resonance absorption integral by activation. The methods for experimental determination of the resonance absorption integral (91.1) can by divided into two groups. To the first group there belong the methods based, on the measurements of the induced activity of the product nucleus; to the second group the methods based on the measurements of the absorption of the incident neutrons.

Procedures of the first type were used from early times in the study of slow neutron physics[6] and were based on the use of (90.2) which expresses the measured activity A_{Cd} of a cadmium covered foil in terms of $\mathfrak{J}_{a\,Cd}$. These procedures—aiming to determine Γ/E_R by means of (91.8)—have the inconvenience, even when considerably improved[7], of requiring a knowledge of the efficiency of detection ε of the counting system.

The need of such an absolute calibration is avoided in the procedure based on the determination of the cadmium ratio[8,9,5]. If $\mathfrak{J}_{a\,Cd}(X)$ and $\mathfrak{J}_{a\,Cd}(S)$ denote the resonance absorption integrals of a sample X and a standard substance S, and

---

[1] H. FESHBACH, C.E. PORTER and V.F. WEISSKOPF: Phys. Rev. **96**, 448 (1956).
[2] H.A. BETHE: Phys. Rev. **57**, 1125 (1940).
[3] V.F. WEISSKOPF: Rev. Mod. Phys. **29**, 174 (1957).
[4] F.L. FRIEDMAN and V.F. WEISSKOPF, in: NIELS BOHR and the Development of Physics, edit. by PAULI, ROSENFELD, WEISSKOPF. London: Pergamon Press 1955.
[5] See footnote 1, p. 364.
[6] See footnote 2, p. 366.
[7] H.L. ANDERSON: Phys. Rev. **80**, 499 (1950). The work contained in this article was made in 1940.
[8] L. SEREN, M. FRIEDLANDER and S. TURKEL: Phys. Rev. **72**, 888 (1947).
[9] See footnote 2, p. 365.

$R_{\mathrm{Cd}}(X)$ and $R_{\mathrm{Cd}}(S)$ denote the corresponding cadmium ratios, it follows from (90.7) that

$$\mathfrak{J}_{a\,\mathrm{Cd}}(X) = k\,\frac{\sigma_{a\,\mathrm{th}}(X)}{R_{\mathrm{Cd}}(X)-1} \tag{91.15}$$

where

$$k = \frac{R_{\mathrm{Cd}}(S)-1}{\sigma_{a\,\mathrm{th}}(S)}\,\mathfrak{J}_{a\,\mathrm{Cd}}(S) \tag{91.16}$$

is a constant which can be computed from the measured values of $R_{\mathrm{Cd}}(S)$ and $\sigma_{a\,\mathrm{th}}(S)$ and the values of $\mathfrak{J}_{a\,\mathrm{Cd}}(S)$ obtained by performing the integral (90.3) over the absorption cross section deduced from direct measurements made by means of slow neutron spectrometers.

While in much of the early activation work In was used as a standard [*12*], more recently Au was adopted[1,2]. The first resonance of this element (Fig. 16) gives the main contribution to the resonance integral; the next six weak resonances known from time-of-flight studies (Sect. 86) give only a very small contribution. From the parameters given by SEIDL *et al.*[3], one obtains[2]

$$\mathfrak{J}_{a\,r}(\mathrm{Au}) = 1513 \text{ barns} \pm 4\% \tag{91.17a}$$

to which one has to add

$$\mathfrak{J}_{1/v\,\mathrm{Cd}}(\mathrm{Au}) = 45 \text{ barns} \tag{91.17b}$$

for a 0.44 g/cm² Cd filter in a collimated beam [see Eq. (26.20)].

Table 66, taken from the review article of MACKLIN and POMERANCE contains a summary of measurements made by the ARGONNE[1] and the Oak Ridge groups[2]. The results of activation measurements given in column 4 all refer to gold as a standard with

$$\mathfrak{J}_{a\,\mathrm{Cd}}(\mathrm{Au}) = 1558 \text{ b}; \qquad \sigma_{a\,\mathrm{th}}(\mathrm{Au}) = 98 \text{ b}. \tag{91.18}$$

Finally we shall mention the work of MYASISHCHEVA *et al.*[4] who measured by means of an activation method in a heavy water pile, the resonance absorption integral of Th by comparison with In, Au and U standards. The more characteristic feature of this experiment is that the samples were irradiated in the form of highly diluted solutions of the nitrates of the materials. The results differed according to the conditions of irradiation which in different runs was made inside the lattice or inside the reflector. These differences are probably due to deviations of the neutron spectrum from the simple law (86.2). We refer to the original paper for more detail as well as for a comparison with the data collected in Table 66.

γ) Measurements of the resonance integral by absorption. The methods for the experimental determination of $\mathfrak{J}_{a\,\mathrm{Cd}}$ based on the measurement of the absorption of the incident neutrons give, at least in principle, the same results as the methods based on the activation, only when the sample is composed of a single isotope which, by neutron capture, gives rise to a single radioactive product. If the sample is composed of two or more isotopes, the resonance integral obtained by absorption measurements will represent the average of the resonance integrals of each isotope weighted according to its relative abundances;

[1] See footnote 2, p. 365.
[2] See footnote 1, p. 364.
[3] F. G. P. SEIDL, D. J. HUGHES, H. PALEVSKY, J. S. LEVIN, W. Y. KATO and N. G. SJÖSTRAND: Phys. Rev. **95**, 476 (1954).
[4] G. G. MYASISHCHEVA, M. P. ANIKINA, L. L. GOL'DIN and B. V. ERSHLER: J. Nucl. Energy **5**, 230 (1957).

Table 66. *Resonance capture integrals* (10^{-24} cm²)[1].

Nuclide or natural element	Calculated		Measured	
	Resonance	1/v	Activation	Absorption[2]
$_3$Li		31		28
$_{11}$Na^{23}	0.12 (th)	0.25	~0.24	0.27
$_{12}$Mg		0.027		0.9
$_{13}$Al^{27}		0.095	~0.16	0.18
$_{14}$Si		0.06		0.5
$_{15}$P^{31}		0.084	~0.092	<2
$_{16}$S		0.22		0.6
$_{17}$Cl		14.3		12
$_{19}$K		0.89		1.1
$_{20}$Ca		2.0		2
$_{21}$Sc^{45}		10.6	~10.7	
$_{22}$Ti		2.5		3.0
$_{23}$V^{51}		2.2	~2.2	
$_{24}$Cr		1.3		1.9
$_{25}$Mn^{55}	12 (th)	5.9	~11.8	10.8
$_{26}$Fe		1.1		2.1; 2.3 (0.5 ev)
$_{27}$Co^{59}	90 (th)	16.3	49.3	48
$_{28}$Ni		2.0		4
$_{29}$Cu		1.6	3.7	4; 3.3 (0.5 ev)
$_{29}$Cu^{63}	0.8 (s)	1.9	4.4	
$_{29}$Cu^{65}	0.8 (s)	0.92	2.2	
$_{30}$Zn		0.48		2
$_{31}$Ga^{69}	20 (s)	0.62	9.2	
$_{31}$Ga^{71}		1.58	15	
$_{33}$As^{75}	170 (th)	1.8	36.8	33
$_{35}$Br^{79}	62 (s)	4.8	147	
$_{38}$Sr		0.53		16
$_{39}$Y^{89}		0.62	0.91	
$_{40}$Zr				3
$_{41}$Nb^{93}		0.47	3.87	8.3
$_{42}$Mo		1.1		13
$_{45}$Rh^{103}	1080 (th)	66	656	575
$_{46}$Pd		3.6		23
$_{47}$Ag		27		>650
$_{47}$Ag^{107}	82 (s)	13	74	
$_{47}$Ag^{109}	1206 (th)	37	1160	
$_{49}$In^{113}		26	1050	
$_{49}$In^{115} (a)	1752 (th)	87	2640	
$_{50}$Sn		0.26		4.3
$_{51}$Sb				115 (0.5 ev)
$_{51}$Sb^{121}		3.0	162	
$_{51}$Sb^{123}		1.1	~138	
$_{52}$Te		2.0		36
$_{53}$I^{127}	150 (s)	2.9	140	130 (0.5 ev)
$_{56}$Ba		0.53		7.5
$_{57}$La		4.0		11; 5.5 (0.5 ev)
$_{59}$Pr^{141}		4.9	11.3	7 (0.5 ev)
$_{60}$Nd				40 (0.5 ev)
$_{62}$Sm^{152} (a)		61	>1750	
$_{63}$Eu^{153} (a)		1400 (b)	950	
$_{71}$Lu				720 (0.5 ev)
$_{72}$Hf	2778 (th)	51	~1750 (c)	2800 (0.5 ev)
$_{72}$Hf^{180}		4.4	21.8	
$_{73}$Ta^{181}	555 (s)	9.4	590	
$_{74}$W^{186}	180 (t)	15	355	170 (0.5 ev)
$_{75}$Re^{185}	1630 (th)	44	1160	

[1] R. L. MACKLIN and H. S. POMERANCE: Footnote 1, p. 364.

[2] Most of the measurements correspond to $E_{Cd} = 0.4$ ev; a few cases, corresponding to a higher value of E_{Cd}, are indicated as (0.5 ev).

Table 66. (Continued.)

Nuclide or natural element	Calculated		Measured	
	Resonance	$1/v$	Activation	Absorption
$_{75}Re^{187}$	310 (s)	33	305	
$_{77}Ir^{191}$ (a)	7680 (th)	420	3500	
$_{77}Ir^{193}$ (a)	976 (th)	57	1370	
$_{78}Pt$		3.7		69
$_{79}Au^{197}$	1390 (th)	45	1558 (d)	
$_{80}Hg$		(b)		31; 73 (0.5 ev)
$_{81}Tl^{203}$		4.8	129	
$_{81}Tl^{205}$		0.04	0.5	
$_{82}Pb$		0.08		0.1
$_{83}Bi$		0.014		0.5
$_{90}Th^{232}$	83 (s)	3.2	69.8	
$_{92}U^{233}$ (f)	650 (g)	234		
$_{92}U^{235}$ (f)	>260 (s)	255 (b)	271	
$_{92}U^{238}$	290 (s)	1.2	282, 297	
$_{94}Pu^{239}$ (f)	250 (g)	(b)		

(a) A near thermal resonance leads to considerable dependence on the details of the cadmium absorber that was used.
(b) The capture cross section is not $1/v$ near thermal energies.
(c) Capture gamma radiation used to determine the cross section; some self-protection in the sample.
(d) Adopted as the standard, using $\sigma_{\text{th}} = 98 \times 10^{-24}$ cm².
(f) Fission integral.
(g) Graphical estimate. For Pu^{239}, a cutoff at 2.0 ev was used and no separate $1/v$ contribution.
(s) Estimated from average level parameters: Eq. (91.13).
(th) Estimated from parameters of the first large resonance and the thermal cross section.

and even in the case of a single isotope it may happen, as for example in the case of rhodium, that the resonance integral obtained by measuring the activity of a certain period (in the case of Rh the 4.3 min period) differs appreciably from the resonance integral obtained from absorption measurements, since the latter includes the contribution of neutron capture leading directly to other isomeric states decaying with different periods (in the case of Rh the state of Rh^{104} with $T_{\frac{1}{2}} = 44$ sec).

A first procedure for the experimental determination of $\mathfrak{J}_{ar}(X)$ by absorption measurements, essentially due to VON HALBAN and PREISSWERK[1], consists of a direct application of the considerations on the resonance escape probability given in Sect. 68 for hydrogeneous moderators, and in Sect. 69 for moderators with $A \geqq 2^2$ [12].

We recall that in an infinite medium containing neutron sources emitting q_0 fast neutrons per second, the slowing down density integrated over the whole volume $q(E)$ depends on the absorption properties of the medium as indicated by (68.12) or (69.1); these equations reduce to (67.12) in the case of negligible absorption. Therefore, the resonance escape probability of an absorbing material X can be obtained from the ratio

$$p(E_R) = \frac{q_{\text{mod}+X}(E_R)}{q_{\text{mod}}(E_R)} \tag{91.19}$$

where $q_{\text{mod}}(E)$ and $q_{\text{mod}+X}(E)$ are the slowing down densities integrated over the whole volume of two infinite media containing, in two successive experiments,

[1] H.V. HALBAN and P. PREISSWERK: Nature, Lond. 137, 905 (1936). — J. Phys. Radium 8, 29 (1937).
[2] F.L. FRIEDMAN: p. 135 of the Science and Engineering of Nuclear Power. Cambridge: Addison Wesley 1952.

the same fast neutron source; $q_{\rm mod}(E)$ refers to the pure moderator, i.e. to a medium with negligible absorption; $q_{{\rm mod}+X}(E)$ refers to the moderator poisoned with the absorbing material X, i.e. to a medium obtained by dispersing the material X, uniformly over the entire moderator volume.

Since according to (89.14) the integrated slowing down density q at an energy E_R is proportional to the volume integral of the activity of a resonance detector $\bar{A}_{\rm Cd}(\boldsymbol{r})$ with a single sharp resonance at $E = E_R$, we deduce from (91.19) that

$$p(E_r) = \frac{\int\limits_{{\rm mod}+X} \bar{A}_{\rm Cd}(\boldsymbol{r})\, d\boldsymbol{r}}{\int\limits_{\rm mod} \bar{A}_{\rm Cd}(\boldsymbol{r})\, d\boldsymbol{r}} . \tag{91.20}$$

We must now recall the various expressions derived in Sects. 68 and 69 for $p(E)$. In the case of hydrogeneous media—such as solutions of an absorbing material in water—the resonance escape probability is given exactly by (68.9) which reduces to (68.24) or (68.26) when the material X has a single isolated resonance. For media with $A \geqq 2$, as for example graphite, various approximate expressions—such as (69.15), (69.24), (69.34), (69.36) and (69.40)—have been derived for different conditions specified in Sect. 69. By inspecting all these expressions we see that in order to derive from $p(E_R)$ the resonance integral defined by Eq. (90.3) we should use expression (69.40) which holds for very weak absorbing media. Thus in the poisoned medium the material X must be distributed at such a small concentration that

$$\Sigma_a(E) \ll \Sigma_s(E) \tag{91.21}$$

for any value of E. If $\Sigma_s(E) = \text{const}$, as it happens in graphite over a wide energy interval (Sect. 29), we can write Eq. (69.40) in the form

$$p(E_R) = e^{-\frac{l_s}{\bar{\xi}} \mathfrak{N}_X \int\limits_{E_R}^{E_0} \sigma_a(E) \frac{dE}{E}} \tag{91.22}$$

where $\mathfrak{N}_X$ is the number of atoms of the material X per cm^3 of the poisoned medium; it should be noted that the values of l_s and $\bar{\xi}$ appearing in (91.22) correspond to the poisoned medium and therefore the contribution to the scattering of the neutrons by the material X must be taken into account.

The resonance integral (90.3) may be also deduced from measurements in water solutions of the material X provided the macroscopic absorption cross section satisfies the inequality (91.21) for any value of E, and for E of the order of 20 kev or more it becomes negligibly small; under these conditions $\Sigma_s^{\rm H}$ can be considered as approximately constant (Fig. 17), and Eq. (68.9) turns out to be identical with (91.22) with $\bar{\xi} = 1$.

Two final remarks must be added; first the absorption integral obtained from (91.22) differs from (90.3) because the lower limit is E_R—for example, the In resonance is at $E_R = 1.46$ ev—and not the cadmium cut-off $E_{\rm Cd} = 0.4$ ev. Secondly, the thickness of the detector to be used for measuring $A_{\rm Cd}(\boldsymbol{r})$ is not limited in this method by the condition of no self-protection (89.3), but by the less stringent condition that the contribution to the measured activity arising from higher resonances is negligible. As we have seen in Sects. 88 and 89, this condition is fulfilled rather satisfactorily in the case of an indium foil placed in a parallel neutron beam for thicknesses smaller than about 20 mg/cm^2. Since in an internal experiment, as that considered here, the neutrons have all possible directions, the thickness of the In detector should not exceed about 10 mg/cm^2.

A different method has been used by SPIVAK *et al.*[1] whose experimental arrangement is shown schematically in Fig. 169. A beam of neutrons comes out of a pile through a hole opened in its shielding, crosses a monitor (*7*) and a cadmium filter (*6*), is further collimated by a cylindrical collimator of boron carbide 60 cm long and 8 mm in diameter and finally impinges on the sample X (*4*). This is placed at the centre of a cubic cavity of 50 cm side opened inside a graphite cube of 190 cm external side. Twelve BF_3 proportional counters, connected in parallel, are located inside the graphite and constitute the detecting system. The intensity of the fast neutron component ($E>100$ kev) was reduced by placing an aluminium filter 40 cm thick in the pile channel; such a filter causes a 5 to 10 fold relative attenuation of the high energy portion of the spectrum because

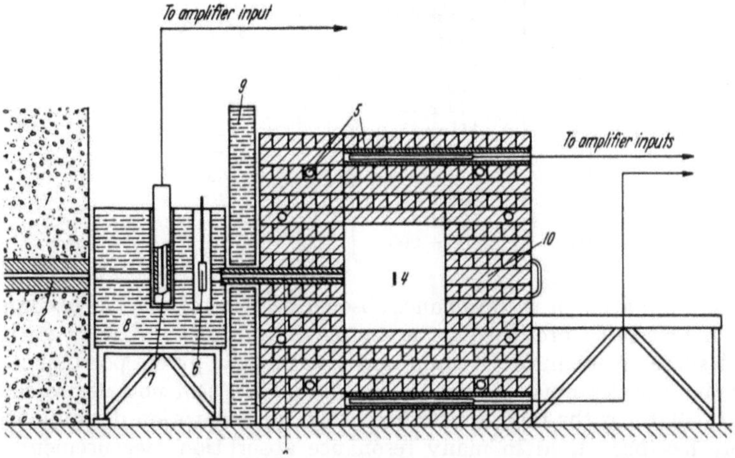

Fig. 169. Schematic section of the arrangement used by SPIVAK *et al.* for measuring the absorption integrals: *1* Pile shielding; *2* first collimator (steel); *3* second inlet collimator (boron carbide); *4* investigated sample at the cavity centre; *5* proportional boron counters (12 pieces); *6* cassette with filter; *7* fission chamber with U^{235} layer; *8* shield (paraffin-boron) *9* front wall shielding of the prism (paraffin-boron), *10* plug pulled out when replacing the sample.

the total cross section of Al for neutron energies between 100 kev and 2 Mev ($\sim$4.5 b) is three times larger than for energies under 10 kev.

The principle of the method consists in measuring the counting rates N_0 and N_X recorded by the detecting system, without and with the sample X inside the cavity. The difference $N_0 - N_X$ is directly related to the change in the number of neutrons in the cavity due to the neutron absorption in the sample[2]. This is correct only if the detector efficiency is insensitive to neutron scattering in the sample and is independent of the neutron energy. The first point was checked with the help of thick scatterers of graphite and bismuth, the second point was verified by means of photoneutron sources of various energies. The apparatus was calibrated using boron as a standard.

The results obtained by the Russian authors are in good agreement with those obtained by activation by the Argonne group, as one can see from Table 67. Some of their results have also been included in Table 66.

When a pile is at one's disposal, the absorption resonance integral can be determinede by procedures which consist of minor modifications of the classical methods used for absorption measurements in piles: the methods of the *danger*

[1] P.E. SPIVAK, B.G. EROZOLIMSKY, V.I. LAVRENCHIK and G.E. DOROFEYER: p. 91 of Ref. [*15c*].

[2] The same method applied to fissionable materials has allowed a determination of the number of neutrons emitted per fission.

coefficient and of the *pile oscillator*. The reader is directed to the book by HUGHES [*12*] for details of these procedures; the principle of the danger coefficient method is summarized in Sect. 108 in connection with the problem of measuring the effective neutron temperature from thermal neutron absorption. The use of the pile oscillator for measuring the resonance absorption integrals is briefly explained below.

Table 67. *Resonance absorption integrals for some elements.*

No.	Element	Results of SPIVAK et al.	Values of the resonance activation integrals computed from the data of Refs. [*1*] and [*2*]
1	Fe	2.3 ± 0.25	—
2	Cu	3.3 ± 0.3	3.35[1,2]
3	W	170 ± 20	—
4	Hg	73 ± 5	—
5	I	130 ± 18	119[1]
6	Sb	115 ± 12	138[1]
7	Nd	40 ± 6	—
8	Pr	7 ± 3	9.3[1]
9	La	5.5 ± 2	
10	Lu	720 ± 70	
11	Hf	2800 ± 600	

For the determination of the resonance absorption integral, one has to compare the change in reactivity produced by a sample X and a standard material S when placed successively inside a cadmium sheath at a given position in the pile lattice; the samples must satisfy the condition (89.3) of no self-protection.

The pile oscillator is the most sensitive method for absorption determinations and therefore has been used in many resonance absorption measurements. It consists of a mechanical apparatus that moves a sample in and out of the nuclear reactor by means of a tube extending through the center of the core. In the Argonne heavy water pile, the period of the oscillator is 20 sec, i.e., the sample is kept 10 sec in the pile and 10 sec outside the pile alternatively.

The amplitude of the oscillating component of the neutron flux induced by the movement of the absorbing sample, is proportional to the neutron absorption in the sample itself. The resonance absorption is measured by placing a cadmium tube into the pile which screens the sample during its motion from the thermal flux.

The calibration of the apparatus is made by means of a standard material, usually In or Au. The samples must be very thin so that the dE/E flux is not distorted by resonance absorption; this, however is not a serious disadvantage because of the very high sensitivity of the method which allows for the measurement of samples of total cross sections ($= x S \mathfrak{N} \sigma_a$) as low as 0.01 cm².

The method presents however a few disadvantages; first of all the high neutron absorption of the cadmium tube protecting the sample from thermal neutrons, restricts its use to piles with a large amount of excess reactivity. Thus, for example, when natural uranium is used as fuel, only heavy water piles can be employed. And even in this case some difficulty may be met with samples of very high resonance absorption cross section.

Furthermore the scattering by the sample may contribute to the moderation of the neutrons with an increase of the reactivity, for which a correction has to

[1] See footnote 2, p. 365.

[2] H.V. LICHTENBERGER, R. NOBLES, J.D. MONK, K. KUBITSCHEK and S.M. DANCOFF: Phys. Rev. **72**, 164 (1947).

be made. We will not go into further details of the actual procedure, since these may be found in [12].

Most of the work on the resonance absorption integral has been made with the Argonne heavy water pile[1]. The results obtained at this laboratory, revised for the gold standard (91.18) and corrected for the $1/v$ component[2,3], are given in column 5 of Table 66. The accuracy can best be judged by comparison with those cases for which the activation value is also known. The elements Na, Al, Cl, K, Sc and V are expected to be $1/v$ absorbers and their resonance absorption integrals are about $\frac{1}{2}$ their thermal absorption which is in agreement with (91.2b) or (91.2c).

δ) Measurements of the resonance scattering integral. This quantity, defined by (91.9) has been measured by the Argonne group[4] by means of the experimental set-up schematically shown in Fig. 170. The neutrons coming out through a

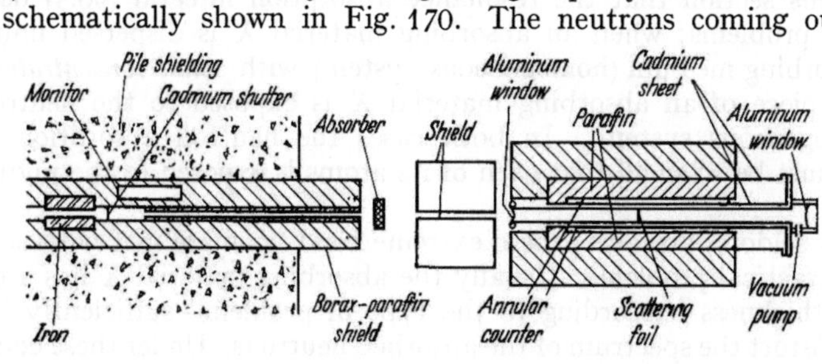

Fig. 170. Schematic diagram of the apparatus used by the Argonne group for measuring resonance scattering integrals.

hole opened into the shield of the Argonne heavy water pile, cross a cadmium absorber and, after further collimation, impinge on the scattering foil which is placed at the center of an evacuated cylindrical tube surrounded by a 4π-annular proportional neutron counter filled with enriched BF_3. The counter is surrounded by a paraffin reflector and borax-paraffin shield.

In spite of the paraffin reflector which markedly increases the sensitivity of the BF_3 counter at high energies, the efficiency of the detector was a slowly decreasing function $S(E)$ of the neutron energy, the expression for which was determined empirically.

The measurement consists in comparing the counting rate C_{C} of a flat scatterer (carbon) of scattering cross section $\sigma_s(\mathrm{C})$ containing $\mathfrak{N}_{\mathrm{C}}$ atoms/cm², with the counting rate C_X of a sample X containing $\mathfrak{N}_X$ atoms/cm²:

$$\left.\begin{aligned} C_{\mathrm{C}} &= \mathfrak{N}_{\mathrm{C}}\,\sigma_s(\mathrm{C}) \int_{E_{\mathrm{Cd}}}^{E_0} S(E)\,dE = \mathfrak{N}_{\mathrm{C}}\, s_0\, \sigma_s(\mathrm{C}), \\ C_X &= \mathfrak{N}_X \int_{E_{\mathrm{Cd}}}^{E_0} \sigma_s'(E)\, S(E)\, dE + \mathfrak{N}_X\, s_0\, \sigma_p(X) \end{aligned}\right\} \tag{91.23}$$

where $\sigma_s'(E)$ represents the first term of Eq. (26.7) and $\sigma_p(X)$ the corresponding potential scattering cross section. For samples with a single isolated narrow resonance, there holds the approximate relation

$$\int_{E_{\mathrm{Cd}}}^{E_0} \sigma_s'(E)\, S(E)\, dE \approx \mathfrak{J}_{sr}(X)\, S(E_R). \tag{91.24}$$

[1] Langsdorf, Harris and Seild: Quoted in Ref. [12].
[2] See footnote 1, p. 365.
[3] See footnote 1, p. 364.
[4] See footnote 2, p. 365.

Combining the latter with (91.23) one finds

$$\mathfrak{J}_{s\,r}(X) = \frac{s_0}{S(E_R)}\left[\frac{C_X}{C_C}\,\frac{\mathfrak{N}_C}{\mathfrak{N}_X}\,\sigma_s(C) - \sigma_p(X)\right]. \tag{91.25}$$

A rather wide series of measurements has been made by the Argonne group. Unfortunately because of the use of Eq. (91.24), the results obtained by means of (91.25) are correct only for samples with a single isolated resonance. As examples we recall the following results: for Rh, $E_R = 1.26$ ev, $\frac{\Gamma_n}{\Gamma} = 0.043$; for Au, $E_R = 4.8$ ev, $\frac{\Gamma_n}{\Gamma} = 0.14$, $E_R > 345$ ev, $\frac{\Gamma_n}{\Gamma} \sim 0.9$.

92. The effective resonance integral in homogeneous systems. We have seen in the previous section that the resonance absorption integral (90.3) occurs in two types of problems: when an absorbing material X is dispersed uniformily in a non-absorbing medium (homogeneous system) with *small concentration*, and when a *thin* piece of an absorbing material X is exposed to the neutron flux (86.2) (heterogeneous system). In both cases the neutron absorption by the material X must be so small that each of its atoms is exposed to the undistorted flux (86.2).

Only very seldom, however, these extreme conditions are met with sufficient accuracy in practical problems. Usually the absorbing material X has a concentration or a thickness—according to the type of problem—sufficiently large to appreciably distort the spectrum of the absorbed neutrons. Under these conditions the resonance absorption integral is replaced by the *effective resonance integral*

$$[\mathfrak{J}_{a\,E_{Cd}}(X)]_{\text{eff}} = \int\limits_{E_{Cd}}^{E_0} [\sigma_a(E)]_{\text{eff}} \frac{dE}{E} \tag{92.1}$$

where $[\sigma_a(E)]_{\text{eff}}$ is called the *effective capture cross section* per atom of type X in the system under consideration.

Because of the finite absorption, the neutron spectrum is depleted in the vicinity of the resonances of the material X, with the result that $[\sigma_a(E)]_{\text{eff}}$ is always smaller than $\sigma_a(E)$ and tends to the latter only in the extreme case of infinite dilution or infinitely small thickness. Therefore the ratio

$$\mathfrak{A} = \frac{\sigma_a(E)}{[\sigma_a(E)]_{\text{eff}}} \tag{92.2}$$

usually called the *volume advantage factor*[1,2], will never be smaller than unity.

In this section we shall discuss the resonance escape probability in a homogeneous system containing an absorbing material X of finite concentration. The experimental determination of this quantity can be made using the first of the three methods presented in Sect. 91γ which is based on Eqs. (91.19), (91.20). As we have pointed out in Sect. 69, if the moderator is non-hydrogeneous the resonance escape probability cannot be calculated rigorously except in the trivial case of absorption and scattering cross sections independent of energy. Various approximate expressions have been derived in that section, the validity of which is limited by certain restrictive conditions. Here we shall start from (69.36) which holds for widely spaced narrow resonances and which reduces,

---

[1] S. GLASSTONE and M.C. EDLUND: The elements of nuclear reactor theory. New York: Van Nostrand 1952.

[2] S. GLASSTONE: Principles of nuclear reactor engineering. New York: Van Nostrand 1955.

for $\bar{\xi}=1$, to (68.9). This holds exactly for the case of an hydrogeneous medium containing an absorbing material X, which, because of its large mass, does not contribute appreciably to the slowing down of neutrons. When $\Sigma_s(E)=l_s^{-1}$ is constant over the energy interval considered, Eq. (69.36) can be written in the form

$$p(E)=\exp\left\{-\frac{l_s}{\bar{\xi}}\,\mathfrak{N}_X\,[\mathfrak{J}_{aE}(X)]_{\text{eff}}\right\} \tag{92.3}$$

formally identical to (91.22). In this expression $\mathfrak{N}_X$ is the number of atoms of type X per cm^3 present in the poisoned medium, and $[\mathfrak{J}_{aE}(X)]_{\text{eff}}$ the effective resonance integral (92.1) with

$$[\sigma_a(E)]_{\text{eff}}=\frac{\bar{\sigma}_a(E)}{1+\mathfrak{N}_X\dfrac{\bar{\sigma}_a(E)}{\Sigma_s(E)}} \tag{92.4}$$

and

$$\bar{\sigma}_a(E)=\frac{\Sigma_a(E)}{\mathfrak{N}_X} \tag{92.5}$$

representing the capture cross section per atom of type X. The volume advantage factor (92.2) turns out to be

$$\mathfrak{A}=\frac{1}{h(E)}=\frac{\Sigma_s+\Sigma_a}{\Sigma_s}=1+\mathfrak{N}_X\,\frac{\bar{\sigma}_a(E)}{\Sigma_s(E)}\,. \tag{92.6}$$

The conditions of validity of the expression obtained by introducing (92.4) in (92.1) are the same as those which have been specified above for the validity of (68.9) and (69.36). Thus, in the case of a non-hydrogeneous poisoned moderator, $\bar{\sigma}_a$ should originate only from capture by the material X whose resonances are narrow and widely spaced; in the case of hydrogeneous moderators, $\bar{\sigma}_a$ will include the contribution to capture by atoms of any type present in the medium.

In Sect. 68 we have discussed in detail the dependence of the resonance escape probability on the concentration of an absorbing material with a single isolated resonance. The following remarks can be added here.

(a) The effective cross section (92.4) is always smaller than $\bar{\sigma}_a(E)$ and takes the value $\sigma_a^X(E)$ only for $\mathfrak{N}_X=0$ ($\mathfrak{A}=1$). Therefore the effective resonance integral is always smaller than the resonance absorption integral and tends to the latter for infinite dilution of the absorbing material

$$\lim_{\frac{\Sigma_s}{\mathfrak{N}_X}\to\infty}[\mathfrak{J}_{aE}(X)]_{\text{eff}}=\mathfrak{J}_{aE}(X)\,. \tag{92.7}$$

Because of the factor $\mathfrak{N}_X/\Sigma_s$ in front of $[\mathfrak{J}_{aE}]_{\text{eff}}$ in the exponent of (92.3), the actual effect of a change of $\mathfrak{N}_X$ on $p(E)$ will depend on which of the two opposing factors predominates. For $\mathfrak{N}_X$ equal to zero the effective resonance integral reaches its maximum value, but then $p(E)$ is always unity.

(b) For increasing values of $\mathfrak{N}_X$, the scattering cross section per atom of type X, $\Sigma_s/\mathfrak{N}_X$, tends to the value σ_s^X of the pure material X without moderator and $[\sigma_a(E)]_{\text{eff}}$ reaches the value

$$\sigma_s^X=\frac{\sigma_s^X}{\sigma_s^X+\sigma_a^X}\,.$$

The volume advantage factor then attains its maximum value

$$\mathfrak{A}=1+\frac{\sigma_a^X(E)}{\sigma_s^X(E)}\,.$$

(c) The physical reason for the decrease of the effective resonance integral in a concentrated mixture of an absorbing material X with a non-absorbing moderator, originates from the decrease of the neutron spectrum in the vicinity of a capture cross section resonance. This is clearly shown by Eq. (68.14) for hydrogeneous media and by Eq. (69.2) for non-hydrogeneous media. Both these expressions contain the factor

$$h(E) = \frac{\Sigma_s(E)}{\Sigma_s(E) + \Sigma_a(E)}$$

which has a minimum roughly where $\Sigma_a(E)$ has a maximum; its reverse represents the volume advantage factor (92.2). Fig. 121 (p. 238) illustrates the behaviour of $h(E)$, $p(E)$ and their product in the case of a water solution of a material X with a single isolated resonance.

(d) While $\mathfrak{J}_a(X)$ is not affected by the Doppler effect, so that it is independent of the medium temperature, the effective resonance integral (92.1) is *temperature dependent*. This fact emerges, for example, from the computation done in Sect. 68 of the resonance escape probability of a water solution of an absorbing material with a single resonance; for a finite concentration of the absorbing material, the result was different according to the value of the ratio of the natural width to the Doppler width of the resonance. We should recall here that the temperature dependence of the effective resonance integral is an effect of considerable importance in connection with the transient behaviour of reactors[1].

The problem of the temperature dependence of the effective resonance integral in a homogeneous system has been treated by DRESNER[2] who rewrites Eq. (92.4) in the more general form

$$[\sigma_a(E)]_{\text{eff}} = \frac{\sigma_a(\eta, E)}{1 + \dfrac{\sigma_a(\eta, E) + \sigma_s'(\eta, E)}{\sigma_{sn}}} \tag{92.8}$$

where $\sigma_a(E)$ and $\sigma_s'(E) + \sigma_p$ are the absorption and scattering cross section of the atoms of the material X, and

$$\sigma_{sn} = \frac{\mathfrak{N}_m}{\mathfrak{N}_X}\,\sigma_s^m + \sigma_p \tag{92.9}$$

is the sum of the scattering cross section of the nuclei of the moderator σ_s^m —computed per atom of type X—and the potential scattering cross section σ_p of the atoms of type X. It may be useful to notice that, according to the definition (92.8), the scattering mean free path to be used in the exponent of (92.3) is

$$l_s^{-1} = \mathfrak{N}_X\,\sigma_{sn} = \mathfrak{N}_m\,\sigma_s^m + \mathfrak{N}_X\,\sigma_p. \tag{92.10}$$

The effect of thermal agitation is taken into account in (92.8) by correcting for the Doppler effect both the absorption cross section $\sigma_a(E)$ and the resonant part of the scattering cross section $\sigma_s'(E)$.

In the simple case of a single narrow resonance, $\sigma_a(E)$ is given by (26.2), (26.3) and therefore $\sigma_a(\eta, E)$ by (87.6). Similarly, by neglecting the interference term in (26.7), we obtain

$$\sigma_s'(\eta, E) = \sigma_{s0}\,\psi(\eta, z). \tag{92.11}$$

[1] The effective multiplication factor of a reactor depends, among other quantities, on the resonance escape probability. Therefore a change of the power level, and therefore of the reactor temperature affects the effective multiplication factor.

[2] L. DRESNER: ORNL-CF-55-9-74 (1955).

The effective resonance integral is now obtained by introducing (87.6) and (92.11) in (92.8) and this in (92.1). Assuming that the resonance at E_R is very narrow, the factor $1/E$ can be replaced by $1/E_R$ with the result

$$[\mathfrak{J}_{aE}(X)]_{\text{eff}} = \frac{\pi}{2}\,\frac{\Gamma}{E_R}\,\sigma_{a0}\,\frac{2\beta}{\pi}\,J(\eta,\beta) \tag{92.12}$$

where

$$\beta = \frac{\sigma_{sn}}{\sigma_{a0}}\,\frac{\Gamma_\gamma}{\Gamma} \tag{92.13}$$

and

$$J(\eta,\beta) = \int_0^\infty \frac{\psi(\eta,z)}{\beta+\psi(\eta,z)}\,dz. \tag{92.14}$$

By comparing (92.12) with (91.5) we see that the effects of finite dilution and Doppler broadening are taken into account by the factor

$$\frac{2\beta}{\pi}\,J(\eta,\beta).$$

The following properties of $J(\eta,\beta)$ have been pointed out by Dresner.

(a) The asymptotic behaviour of $J(\eta,\beta)$ for large β always satisfies the inequalities

$$\frac{\pi}{2}\,\frac{1}{\beta} > J(\eta,\beta) > \frac{\pi}{2}\,\frac{1}{\beta}\left(1-\frac{1}{\beta}\right). \tag{92.15}$$

(b) The asymptotic behaviour of $J(\eta,\beta)$ for small β is obtained by noting that, as long as $\psi(\eta,z)\gg\beta$, one has $\frac{\psi}{\beta+\psi}\sim 1$. Thus, the integrand of (92.14) can be replaced by 1 until $\psi(\eta,\beta)\sim\beta$. If the value of z at which the approximate equality is fulfilled is sufficiently large so that, according to (87.12) $\psi(\eta,z)$ is essentially given by $(1+z^2)^{-1}$, then we may replace $\psi(\eta,z)$ by $(1+z^2)^{-1}$ not only for large z but even for small z, since the integrand is practically unity for small z whether we put $\psi(\eta,z)$ or $(1+z^2)^{-1}$ into it. In this case we have

$$J(\eta,\beta) = \frac{\pi}{2}\,\frac{1}{[\beta(1+\beta)]^{\frac{1}{2}}}. \tag{92.16}$$

If, when $\psi(\eta,z)\approx\beta$, $\psi(\eta,z)\approx\frac{1}{1+z^2}$, then the value of z determined by the first equality is $z\approx\beta^{-\frac{1}{2}}$; the approximation of replacing $\psi(\eta,z)$ by $(1+z^2)^{-1}$ will be a good one only if $\frac{6\beta}{\eta^2}\ll 1$ as we can easily see from (87.12).

(c) The asymptotic behaviour of $J(\eta,\beta)$ for large values of η is obtained by making use of (87.12) with the result

$$J(\eta,\beta) = \frac{\pi}{2}\,\frac{1}{[\beta(1+\beta)]^{\frac{1}{2}}}. \tag{92.17}$$

(d) For the computation of $J(\eta,\beta)$ the following relation—which holds for fairly large c—may be useful:

$$\int_c^\infty \frac{\psi}{\beta+\psi}\,dz = \frac{\frac{\pi}{2}-\arctan\left[c\sqrt{\frac{\beta}{1+\beta}}\right]}{[\beta(1+\beta)]^{\frac{1}{2}}}. \tag{92.18}$$

The results of the numerical computations made by DRESNER are given in Figs. 171 and 172 where the abscissa k is defined by the relation

$$\beta = 10^{-5} \cdot 2^k. \qquad (92.19)$$

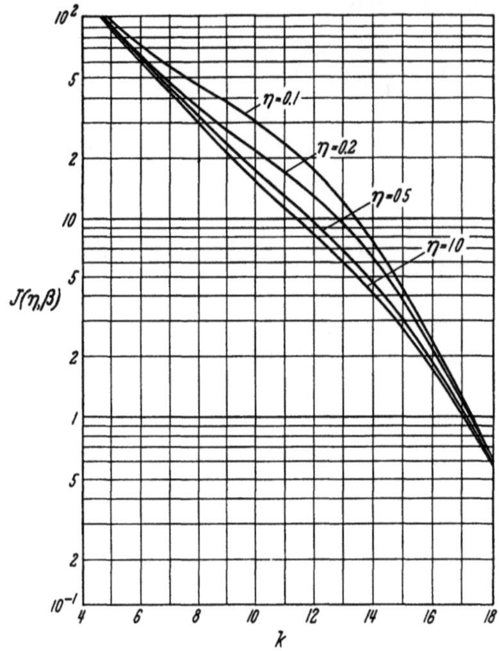

Fig. 171. $J(\eta, \beta)$ versus k. — k defined by Eq. (92.19) (DRESNER).

Fig. 171 shows $J(\eta, \beta)$ for $\eta = 0.1$; 0.2; 0.5; 1.0 as a function of k when this quantity varies between 4 and 18. According to one of its general properties, $J(\eta, \beta)$ decreases as η increases. Fig. 172 shows $J(\eta, \beta)$ for $\eta = 0.1$ and 1 for a large range of values of k; for

$$\beta \leqq 1.6 \times 10^{-4} \quad \text{and} \quad \beta \geqq 2.6$$

all curves have essentially the same value. For $\eta = 1.0$ the curve is practically indistinguishable from (92.17), which is correct for $\eta \to \infty$.

Thus the Doppler effect is only important in computing effective resonance integrals if $2.6 \geqq \beta \geqq 1.6 \times 10^{-4}$. It is also clear from this result that $J(\eta, \beta)$ has the predicted asymptotic form in the limit of both large and small β.

These results, extended to the case of many narrow resonances separated by regions without absorption, were used by the same author for the computation of the effective resonance integrals of U^{238} and Th^{232} as functions of σ_{sn} defined

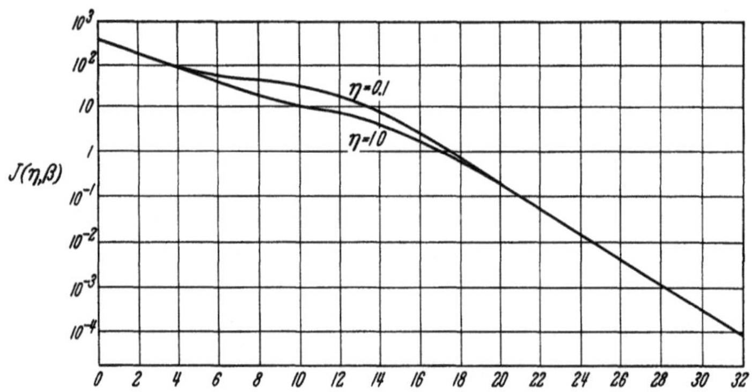

Fig. 172. $J(\eta, \beta)$ versus k. — k defined by Eq. (92.19).

by (92.9). The curves obtained by such a procedure appear to be in satisfactory agreement with the experimental results given by many authors[1] (Fig. 173). More details on this problem can be found in SAMPSON and CHERNICK's review article[2].

[1] R. L. MACKLIN and H. S. POMERANCE: Progress in nuclear energy. Physics and Mathematics, Vol. 1, p. 179. New York and London: Pergamon Press 1956.

[2] J. B. SAMPSON and J. CHERNICK: Progress in nuclear energy. Physics and Mathematics, Vol. 2, p. 223. New York and London: Pergamon Press 1958.

93. The effective resonance integral in heterogeneous systems. The investigation of the effective resonance integral in heterogeneous systems was started in connection with the problem of reactor design, since it was realized from early times[1,2] [*18b*] that a considerable increase of the resonance escape probability with respect to capture by U^{238}, can be obtained by disposing the absorbing material—in this case natural uranium—in lumps embedded in the moderator. The neutrons which are slowed down in the moderator show very little depletion at energies close to the U^{238} resonances. But when they reach the surface of a lump, those with energy close to the center of the most intensive adsorption lines, are quickly absorbed so that at a certain depth inside the lump only the

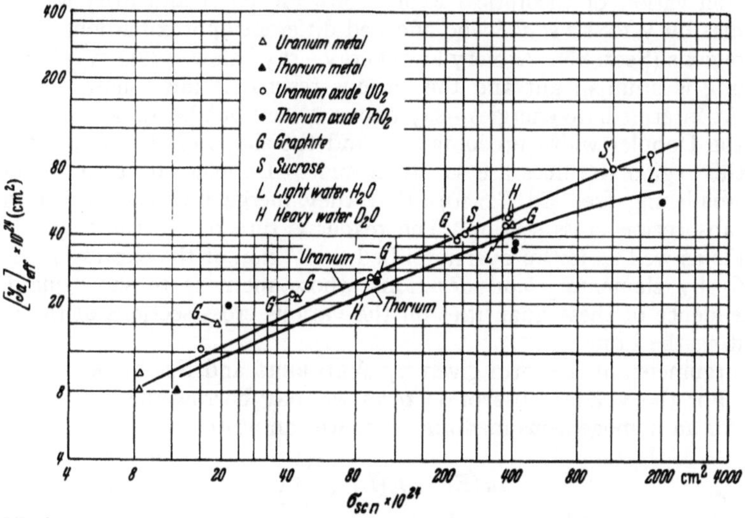

FIG. 173. Effective resonance integral of U^{238} and T^{232} diluted with various materials versus scattering cross section per absorbing atom. Theoretical curves computed by DRESNER.

neutrons belonging to the wings of strong resonances or to weak resonances may be found. These neutrons are absorbed with a much smaller absorption coefficient, and therefore the total neutron absorption taking place in the lump can be roughly divided into a surface and a volume effect with much higher specific absorption taking place at the surface.

The pioneering experimental work on spheres of uranium oxide made by FERMI and ANDERSON together with R. R. WILSON and others at Princeton University, verified this prediction. Therefore, the effective cross section $[\sigma_a(E)]_{\text{eff}}$ used to describe the absorption of the neutrons per absorbing atom of type X, averaged over the whole volume of the lump, can be written in the form

$$[\sigma_a(E)]_{\text{eff}} = a(E) + b(E)\,\frac{1}{\gamma} \tag{93.1}$$

with

$$\gamma = \frac{M_l}{S_l}\ \text{g/cm}^2. \tag{93.2}$$

The first term $a(E)$ corresponds to the volume absorption, the second term $b(E)\,\frac{1}{\gamma}$ to the surface absorption and M_l and S_l represent the mass and the surface of the lump which contains $\mathfrak{N}$ absorbing atoms/cm³.

[1] E. P. WIGNER: J. Appl. Phys. **17**, 857 (1946). — E. FERMI: Science **105**, 27 (1947).
[2] H. HALBAN, F. JOLIOT and L. KOWARSKI: C. R. Acad. Sci., Paris **229**, 909 (1949).

According to (92.1) the effective resonance integral is obtained by multiplying (93.1) by dE/E, followed by integration over E. The result can be written

$$[\mathfrak{J}_{aE}(X)]_{\text{eff}} = A\left(1+\frac{\mu}{r}\right) \tag{93.3}$$

where

$$A = \int_E^{E_0} a(E)\,\frac{dE}{E}; \qquad \mu = \frac{1}{A}\int_E^{E_0} b(E)\,\frac{dE}{E}$$

are two constants typical of the material constituting the lump.

The derivation of theoretical expressions for $a(E)$ and $b(E)$ was made in early times by WIGNER *et al.*[1] who adopted different approximations in different energy regions (Sect. 87). At very low energy—say below ~200 ev—the Doppler broadening is unimportant and the various resonance lines show their natural shape. At intermediate energies—say above 200 ev but below several thousands of ev—the Doppler width predominates and the shape of the lines is Gaussian. Above this region the lines are so broad due to Doppler effect and to neutron-width broadening [see Eq. (26.10)] that they merge into each other giving a smooth absorption cross section. The reader is directed to the original papers for details as well as for an oversimplified derivation of the expressions for $a(E)$ and $b(E)$ by elementary considerations whose merit is to show qualitatively the dependence of these quantities on the various cross sections of the material constituting the lump.

The argument, in the form given by WEINBERG and NODERER [*19*][2], runs as follows. For the volume absorption term we may borrow the expression (92.4) derived for an homogeneous medium; thus we can write

$$a(E) = \sigma_a(E)\,\frac{\Sigma_s}{\Sigma_s+\Sigma_a} \tag{93.4}$$

where, now, the various cross sections refer to the material composing the lump; the neutron flux by which one has to multiply (93.4) in order to obtain the average number of captures per absorbing atom, is the average of $\varphi_0(\boldsymbol{r}, E)$ over the volume V_l of the lump:

$$\overline{\varphi_l(E)} = \frac{1}{V_l}\int_{V_l} \varphi_0(E,\boldsymbol{r})\,d\boldsymbol{r}. \tag{93.5}$$

In fact the rate of volume absorption is given by

$$\mathfrak{N}\, V_l\,\overline{\varphi_l(E)}\,a(E). \tag{93.6}$$

For the surface absorption term the expression

$$b(E) = \frac{1}{4}\left(\frac{\Sigma_a}{\Sigma_s+\Sigma_a}\right)^2 \frac{\varrho}{\mathfrak{N}}\,\frac{\overline{\varphi}_s}{\overline{\varphi}_l} \tag{93.7}$$

is obtained by multiplying the following factors: (a) the resonance flux

$$\varphi_{\text{res}} = \varphi_0(E) - \varphi_0(E)\,h(E) = \varphi_0(E)\,\frac{\Sigma_a}{\Sigma_s+\Sigma_a}$$

i.e. the difference between the flux $\varphi_0(E)$ in the moderator very close to the surface of the lump and the depleted flux $\varphi_0(E)\,h(E)$ present in the interior of the lump,

[1] E.P. WIGNER, E. CREUTZ, H. JUPNIK and T. SNYDER: J. Appl. Phys. **26**, 260 (1955).

[2] S. GLASSTONE and EDLUND: The elements of nuclear reactor theory. New York: Van Nostrand 1952.

integrated over the surface S_l of the lump; (b) a geometrical factor — taken equal to $\frac{1}{4}$ — representing the fraction of the flux $\varphi_{\text{res}}(E)$ that actually impinges on the lump surface; (c) the probability

$$\frac{\Sigma_a}{\Sigma_s+\Sigma_a}$$

that a neutron of energy E is absorbed during its first collision in the lump; and finally (d) the ratio of the density ϱ of the material constituting the lump to the factor $\mathfrak{N}\, V_l\,\overline{\varphi}_l$ appearing in (93.6)

$$\frac{S_l}{\mathfrak{N}\, V_l\,\overline{\varphi}_l}=\frac{\varrho}{\mathfrak{N}}\,\frac{S_l}{M_l}\,\frac{1}{\overline{\varphi}_l}.$$

Among other rough approximations, the derivation of (93.4) and (93.7) contains the tacit assumption that a resonance neutron suffering an elastic collision in the outer layer of the lump will be thrown out of the resonance region; this is not true in the case of heavy elements such as uranium since the average energy loss per collision $\xi\sim\frac{2}{A}\sim 10^{-2}$ is of the order of the natural width of the resonances at about 10 ev.

In spite of these remarks, the expressions (93.4) and (93.7) are useful for a quantitative discussion of the dependence of $[\sigma_a(E)]_{\text{eff}}$ on the absorption and scattering cross sections as well as the number of absorbing atoms present in the lump.

The effective resonance integral of uranium has been computed also by VAN DER HELD[1] who starts from the theoretical expression for the absorption curve (Sect. 94) of a thick layer of material with resonances represented by Eq. (87.6).

In practice it is more convenient to follow a more empirical approach. The effective resonance integral $[\mathfrak{J}_a]_{\text{eff}}$ is measured for a given material and for a given geometrical shape of the lump as a function of the corresponding dimensions, and the two constants A and μ appearing in (93.3) are adjusted to fit the experimental results. This was done in early times and the corresponding results are usually quoted[2-4] as follows:

for uranium metal: $$[\mathfrak{J}_a]_{\text{eff}}=9.25\left[1+2.67\frac{S_l}{M_l}\right]\text{ barns},$$

for uranium oxide: $$[\mathfrak{J}_a]_{\text{eff}}=9.25\left[1+1.67\frac{S_l}{M_l}\right]\text{ barns}.$$

More recently UNTERMEYER of the Argonne National Laboratory[5], found that a better representation of the empirical data was obtained by using the formula

$$[\mathfrak{J}_{aE}(X)]_{\text{eff}}=A\left(1+\frac{\mu}{r+r_0}\right)\tag{93.8}$$

which is obtained by introducing into (93.3) a third parameter to be adjusted.

The values of the constants which give a satisfactory fit of the experimental results of many authors in Th^{232} and U^{238} metals are reproduced in Table 68.

---

[1] E.F.M. VAN DER HELD: p. 107 of Ref. [*15c*].

[2] See footnote 2, p. 382.

[3] S. GLASSTONE: Principles of Nuclear reactors engineering. New York: Van Nostrand 1955.

[4] See, for example, Phys. Today 5, 32 (May 1952).

[5] See footnote 1, p. 380.

It can be verified that with these values, the effective resonance integral reduces for $r=0$ — i.e. for infinitely diluted absorber — to the resonance absorption integrals given in column 4 of Table 66 (p. 370).

A thorough discussion of the role played by the effective resonance integral (93.8) in reactor design may be found in various books on applied nuclear physics[1,2] as well as in the article by HUGHES in Vol. XLIV of this Encyclopedia. We shall limit ourselves to recalling that a regular arrangement of fuel lumps in a moderator may be described as a lattice of equivalent cells. The advantage of this picture is that each cell may be treated as a unit whose behaviour is described by means of the various quantities averaged over the cell volume. In particular, as we shall see below, the resonance escape probability may be defined for the cell as a whole, so that the complications arising from its dependence on the spatial distribution of resonance neutrons is avoided.

Table 68. *Parameters for the effective resonance integral*[3].

	A (10^{-24} cm^2)	μ g/cm^2	r_0 g/cm^2
Th^{232} metal	8.37	0.55	0.08
U^{238} metal	10.2	2.95	0.11

According to the definition of the effective cross section, the rate of absorption of neutrons in the fuel lump per second per unit energy interval at E is given by

$$\bar{\varphi}_l(E)\, V_l\, \mathfrak{N}\, [\sigma_a(E)]_{\text{eff}}. \tag{93.9}$$

Since slowing down takes place mainly in the moderator, the following expression of the slowing down density integrated over the volume V_m of the moderator will constitute—according to (66.3b), (69.39)—a sufficiently good approximation:

$$q(E) = \bar{\varphi}_m(E)\, V_m\, \xi^m\, \Sigma_s^m \cdot E \tag{93.10}$$

where

$$\bar{\varphi}_m(E) = \frac{1}{V_m} \int\limits_{V_m} \varphi_0(\boldsymbol{r}, E)\, d\boldsymbol{r}. \tag{93.11}$$

Since $q(E)$ decreases as E decreases because of the absorption in the lump, we can write[2]

$$\frac{dq(E)}{dE} = -\,\bar{\varphi}_l(E)\, V_l\, \mathfrak{N}\, [\sigma_a(E)]_{\text{eff}} \tag{93.12}$$

which, divided by (93.10) and integrated with respect to E, yield

$$q = q_0\, p(E) \tag{93.13}$$

where $p(E)$ is the resonance escape probability of the cell under consideration:

$$p(E) = \exp\left[-\frac{\mathfrak{N}}{\xi^m \Sigma_s^m} \frac{V_l}{V_m} \int [\sigma_a(E)]_{\text{eff}} \frac{\bar{\varphi}_l(E)}{\bar{\varphi}_m(E)} \frac{dE}{E}\right]. \tag{93.14}$$

This expression reduces to (92.3) in the case of a homogeneous system, since in that case both V_l/V_m and $\bar{\varphi}_l(E)/\bar{\varphi}_m(E)$ turn out to be unity. The factor $\bar{\varphi}_m(E)/\bar{\varphi}_l(E)$ is called the *disadvantage factor* for resonance neutrons. If in the resonance region $\bar{\varphi}_l/\bar{\varphi}_m$ is assumed to be independent of energy, the expression (93.14) reduces to

$$p(E) = \exp\left[-\frac{\mathfrak{N}}{\xi^m \Sigma_s^m} \frac{V_l \bar{\varphi}_l}{V_m \bar{\varphi}_m} [\mathfrak{J}_a(X)]_{\text{eff}}\right] \tag{93.15}$$

[1] See footnote 2, p. 382.
[2] See footnote 3, p. 383.
[3] R. L. MACKLIN and H. S. POMERANCE: Progress in nuclear energy. Physics and Mathematics, Vol. 1, p. 179. New York and London: Pergamon Press 1956.

where for $[\mathfrak{J}_a(X)]_{\text{eff}}$ use can be made of the effective resonance integral (93.8). We shall not go into further detail of the limits of validity of such an approximation which may be found in Refs. [1, 2].

Experimentally, the effective resonance integral can be determined essentially by the same methods as used for the determination of $\mathfrak{J}_{a\,\text{Cd}}$ which have been described in Sect. 91.

For example the methods used by CREUTZ *et al.*[3] for determining the effective resonance integral of spheres of uranium and uranium oxide embedded in graphite, were based on the measurement of the activation of a lump after irradiation behind a cadmium filter. A sphere of the chosen radius, filled with U_3O_8 powder and protected by a cadmium layer, was irradiated inside a graphite block with the neutrons produced by a cyclotron through the Li(p, n) reaction. After irradiation the powder was removed, placed in a large bottle and thoroughly shaken as to be mixed. Then a portion of the U_3O_8 so shaken was dissolved in HNO_3, and by means of classical chemical procedures the uranium fraction was separated from fission products and UX. The solvant (ether) containing the uranium fraction was then blown off and the residue dryed and ground. Finally the activity of the $T_{\frac{1}{2}} = 23.5$ min period—due to U^{239} produced by neutron capture of U^{238}—was measured.

More recently the effective resonance integral of uranium metal has been measured by ERICSON *et al.*[4] and by MUEHLHAUSE *et al.*[5] (for $r = 0.07$ to $8.9\ \text{g/cm}^2$) using the pile oscillator method and by RISSER *et al.*[6] by means of the activation induced in U^{238} metal foils mounted as a part of a cylinder of natural uranium.

CROCKER[7] at Harwell measured the transmission curve for thick uranium aborbers in a well collimated beam of neutrons filtered with cadmium. The detectors were foils of U^{238} (containing 1 atom of U^{235}/6000 atoms of U^{238}) in the form of U_3O_8—3 mg/cm² thick—fixed by means of zapon lacquer on thin inactive disks. The detector foils were backed by 0.5 cm of uranium so that they were essentially measuring the activation at a given position in a mass of uranium.

The transmission curves, obtained by measuring the activity of the 23.5 min period, can be fitted to a three exponential formula like (89.10) which then is used for computing the absorption in lumps of uranium of different shapes —slabs, spheres, cylinders—as a function of S_l/M_l. The results do not depend very much on the adopted shape and appear to be in agreement with the values computed by means of (93.8) with the parameters reproduced in Table 68.

The absorption of Th rods with r ranging between 2.1 and 9.8 g/cm² was measured by the danger coefficient method by UNTERMEYER and EGGER[8] who estimated the accuracy of the results to be 20%.

The temperature dependence of the effective resonance integral was measured by CREUTZ *et al.*[9] for uranium and uranium oxide and was found to be of the order of 10^{-2}% per °C. MACKLIN and POMERANCE[8] report some result of RODEBACK for Th^{232} [(0.020 ± 0.003)% per °C] and U^{238} [(0.008 ± 0.002)% per °C] and of DAVIS per U^{238} [(0.0174 ± 0.002)% per °C].

[1] See footnote 2, p. 382 and footnote 3, p. 383.

[2] J.B. SAMPSON and J. CHERNICK: Progress in Nuclear Energy. Physics and Mathematics, Vol. 2, p. 233. New York and London: Pergamon Press 1958.

[3] E. CREUTZ, H. JUPNIK, T. SNYDER and E.P. WIGNER: J. Appl. Phys. **26**, 271 (1955).

[4] V.O. ERIKSON, R.W. MEIER and P. SCHMID: p. 105 of Ref. [*15c*].

[5] MUEHLHAUSE, RADKOWSKY and UNTERMEYER: Quoted in footnote 3, p. 384.

[6] RISSER, ARFKEN, CUYKENDALL, STEPHENSON and CALDWELL: Quoted in footnote 3, p. 384.

[7] V.S. CROCKER: p. 102 of Ref. [*15c*].

[8] See footnote 3, p. 384.

[9] E. CREUTZ, H. JUPNIK and E.P. WIGNER: J. Appl. Phys. **26**, 276 (1955).

94. The absorption curves of epi-cadmium neutrons. A great variety of absorption curves for slow neutrons can be obtained from transmission experiments, since their behaviour depends on many factors: the spectrum and the angular distribution of the incident neutrons, the energy dependence of the cross sections of both absorber and detector, the thickness and the absorption coefficient of the detector for the β- or γ-radiation used for detecting the neutrons, etc.

The case of epi-cadmium neutrons is considered in this section, while the case of thermal neutrons is considered in Sect. 136.

In order to simplify the discussion, we shall first consider the case of a parallel beam impinging perpendicularly on the surface of a flat absorber of thickness x behind which is placed a detector very thin with respect to neutron absorption. We rewrite the condition of no self-protection (89.3) in the form

$$X_d K_d(E) \ll 1 \tag{94.1}$$

where

$$K_d(E) = \frac{\Sigma_a^d(E)}{\varrho_d} = \frac{\sigma_a^d(E)}{A_d} \mathfrak{N}_{\text{av}} \tag{94.2}$$

$$X_d = x_d \varrho_d \tag{94.3}$$

are the neutron mass absorption coefficient in cm²/g and the thickness of the detector in g/cm².

We do not need impose the condition that also the β- or γ-ray absorption in the detector itself $X_d \mu/\varrho_d$ is small, since the determination of an absorption curve does not involve absolute measurements (Sect. 89), but only the variation of the activity of the detector as a function of the thickness X of the absorber.

Under these assumptions the absorption curve is, in general, determined by

$$T(X) = C \int_{E_{\text{Cd}}}^{E_0} e^{-X K_a(E)} K_d(E)\, \varphi_0(E)\, dE \tag{94.4}$$

where

$$C^{-1} = \int_{E_{\text{Cd}}}^{E_0} K_d(E)\, \varphi_0(E)\, dE$$

is a normalization factor such that $T(0) = 1$. Typical cases are obtained by combining in various ways $1/v$ and resonance absorbers with $1/v$ and resonance detectors.

The case of a $1/v$ absorber

$$K_a(E) = \frac{k_a}{E^{\frac{1}{2}}} \tag{94.5}$$

combined with a resonance detector with a single narrow resonance [see Eq. (87.1)]

$$K_d(E) = \frac{K_0}{1 + \left(\frac{E - E_R}{\Gamma/2}\right)^2}, \qquad K_0 = \frac{\sigma_{a0}}{A_d} \mathfrak{N}_{\text{av}} \tag{94.6}$$

well above the cadmium cut-off, has already been considered in Sect. 27. The detector is activated only by the neutrons belonging to a very narrow energy interval—from about $E_R - \frac{1}{2}\Gamma$ to about $E_R + \frac{1}{2}\Gamma$—in which the incident neutron flux $\varphi_0(E)$ and the absorption coefficient of the absorber (94.5) can be considered as constant and equal to their values at $E = E_R$. For a thin In-detector, for example, one has $\Gamma/E_R = 0.06$ (Table 62) so that $K_a(E)$ of a $1/v$ absorber such as boron varies within the first In resonance between the limits: $K_a(E_R)\left[1 \mp \frac{0.03}{2}\right]$.

Therefore the absorption curve is given by the exponential

$$T(X) = \exp\left[-X K_a(E_R)\right]$$

and the ratio of the absorption coefficients of the same $1/v$ absorber obtained from measurements made with two thin resonance detectors $K_a(E_{R1})$ and $K_a(E_{R2})$ is given by (27.5). As we have mentioned in Sect. 27 this relation has been extensively used for measuring the resonance energies E_R in many elements[1].

The case of a $1/v$ detector combined either with a resonance absorber or with a $1/v$ absorber has already been considered in Sect. 88. The integrand of (94.4) is shown in Fig. 161 for the case of an In-absorber. For very small thicknesses ($X_{\mathrm{In}} \leqq 20$ mg/cm²) only the first resonance contributes to the absorption, and the corresponding dip in the neutron spectrum is so narrow that within its width the factor $\varphi_0(E)\, K_d(E)$ remains practically constant. When the absorber thickness increases, the contribution to absorption due to the resonance wings increases faster than the contribution arising from the center of the line, so that the dip becomes wider and wider and the variation of the factor $\varphi_0(E)\, K_d(E)$ has to be taken into account.

As a result of this behaviour the absorption curve is not exponential. It is further complicated when, as in the case of In, higher resonances start to contribute to absorption as the thickness X of the absorber increases.

The integrand of Eq. (94.4) for the case of $1/v$ absorber and $1/v$ detector is illustrated in Fig. 162. The corresponding absorption curve appreciably deviates from an exponential curve because the absorbed radiation is strongly inhomogeneous; this is obviously due to the fact that by increasing the absorber thickness the average energy of the transmitted neutrons increases and therefore the corresponding average absorption decreases.

By far the most important case is that of *self-absorption* or *self-indication* curves for resonance materials, i.e. the case of a resonance absorber combined with a resonance detector of the same material at the same temperature.

For the sake of simplicity we shall refer to a material with a single resonance so narrow that the factor $\varphi_0(E)$ does not vary appreciably within the line width. If the Doppler width is negligible with respect to the natural width ($\eta \gg 1$) the absorption coefficients of both absorber and detector are given by (94.6), so that we can write, for an infinitely thin detector,

$$R(y) = \frac{1}{\pi} \int_{-\infty}^{+\infty} e^{-\frac{2y}{1+z^2}} \frac{dz}{1+z^2} \tag{94.7a}$$

where

$$2y = K_0 X \tag{94.7b}$$

and the limits of the integral E_{Cd} and E_0 are replaced by $\mp\infty$, since the integrand becomes negligibly small as soon as z appreciably differs from one. With $z = \tan\frac{\vartheta}{2}$, we obtain[2]

$$R(y) = \frac{e^{-y}}{2\pi} \int_{-\pi}^{+\pi} e^{-y\cos\vartheta}\, d\vartheta = e^{-y} I_0(y) \tag{94.8}$$

[1] H.H. Goldsmith and F. Rasetti: Phys. Rev. **49**, 891 (1936); **50**, 328 (1936). — H.V. Halban and P. Preiswerk: Nature, Lond. **137**, 905 (1936). — J. Phys. Radium **8**, 29 (1937). E. Amaldi and E. Fermi: Ric. Sci. **7** (1), 454 (1936). — Phys. Rev. **50**, 899 (1936). — O.R. Frisch: Kgl. danske Vidensk. Selsk. **14**, 12 (1937). — H.V. Lichtenberger, R.G. Nobles, G.D. Monk, H. Kubitschek and S.M. Dancoff: Phys. Rev. **72**, 164 (1947).

[2] O.R. Frisch: See footnote 1.

where $I_0(y) = J_0(i\,y)$ is the Bessel function with imaginary argument of order zero. For small values of y one has

$$R(y) = 1 - y + \cdots = 1 - \tfrac{1}{2} K_0 X \qquad (\text{for } y \ll 1) \tag{94.9}$$

as one can see directly from (94.8). Eq. (94.9) expresses the following theorem established by BETHE and PLACZEK[1] [5]; the initial self-absorption coefficient K_s is equal to $\frac{1}{2}$ the absorption coefficient K_0 corresponding to the center of the resonance line.

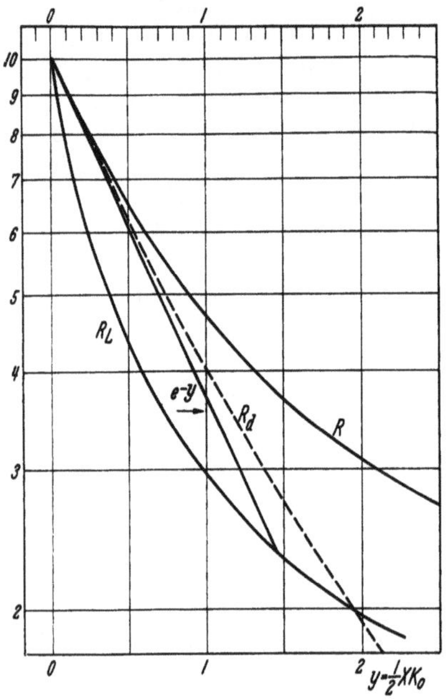

Fig. 174. Self-absorption curves of a parallel beam of resonance neutrons.

A semilogarithmic plot of $R(y)$ is shown in Fig. 174 from which we see how the slope of the absorption curve decreases as the thickness of the absorber increases. A few numerical values are given in Table 69. For large values of y, the following asymptotic expansion can be deduced from those valid for the Bessel functions[2]

$$R(y) = \frac{1}{(2\pi y)^{\frac{1}{2}}} \left\{1 + \sum_{n=1}^{\infty} \frac{1^2 \cdot 3^2 \cdot \cdots \cdot (2n-1)^2}{n!\,(8y)^n}\right\} \qquad (\text{for } y \gg 1). \tag{94.10}$$

When the Doppler width is much larger than the natural width—i.e. when $\eta \ll 1$—, the absorption coefficient is given, according to (87.11), by

$$K_a(E) = K_0 \frac{\pi^{\frac{1}{2}}}{2} \eta \exp\left[-\frac{1}{4}\eta^2 z^2\right], \qquad K_0 = \frac{\sigma_{a0}}{A} \mathfrak{N}_{\text{av}}. \tag{94.11}$$

The corresponding absorption curve turns out to be

$$\begin{aligned} R_d(y) &= \frac{1}{\sqrt{\pi}} \int_{-\infty}^{+\infty} e^{-t^2 + \sqrt{2}\, y e^{-t^2}} dt = \frac{1}{\sqrt{\pi}} \sum_{n=0}^{\infty} \frac{(-\sqrt{2}\,y)^n}{n!} \int_{-\infty}^{+\infty} e^{-(n+1)t^2} dt \\ &= 1 - y + \frac{y^2}{\sqrt{3}} - \frac{y^3}{3\sqrt{2}} + \cdots + \frac{(-\sqrt{2}\,y)^n}{n!\,\sqrt{n+1}} + \cdots \end{aligned} \tag{94.12a}$$

where

$$y = \frac{1}{2} K_0 \frac{\pi^{\frac{1}{2}}}{2} X. \tag{94.12b}$$

For $y \gg 1$ one finds

$$R_d(y) \longrightarrow \frac{1}{y} [2\pi \ln (y\sqrt{2})]^{-\frac{1}{2}},$$

but for large values of the absorber thickness the assumption that the line has the Gaussian shape (94.11) is not correct, since, according to (87.12), at large distances from the resonance, the cross section behaviour is always determined

[1] H. A. BETHE and G. PLACZEK: Phys. Rev. 51, 450 (1937).

[2] This, as well as other asymptotic expansions given in this section, has been derived by G. C. WICK, who has also computed Tables 69 and 70.

Table 69. *Self-absorption curves of a parallel beam of resonance neutrons.*

y	$R(y)$	$\mathfrak{R}(y)$	y	$R(y)$	$\mathfrak{R}(y)$	y	$R(y)$	$\mathfrak{R}(y)$
0.00	1.0000	0.0000	0.48	0.6549	0.3877	0.96	0.4763	0.6548
0.02	0.9803	0.019802	0.50	0.6450	0.4007	0.98	0.4710	0.6643
0.04	0.9612	0.039226	0.52	0.6354	0.4135	1.00	0.4658	0.6737
0.06	0.9426	0.058253	0.54	0.6260	0.4261			
0.08	0.9246	0.076923	0.56	0.6169	0.4386	1.1	0.4414	0.7190
0.10	0.9071	0.095240	0.58	0.6080	0.4508	1.2	0.4198	0.7620
0.12	0.8901	0.11324	0.60	0.5993	0.4629	1.3	0.4004	0.8030
0.14	0.8736	0.13085	0.62	0.5909	0.4748	1.4	0.3831	0.8422
0.16	0.8576	0.14826	0.64	0.5827	0.4865	1.5	0.3674	0.8797
0.18	0.8420	0.16526	0.66	0.5747	0.4981	1.6	0.3533	0.9157
0.20	0.8269	0.18184	0.68	0.5669	0.5095	1.7	0.3405	0.9504
0.22	0.8123	0.1982	0.70	0.5593	0.5208	1.8	0.3289	0.9839
0.24	0.7980	0.2143	0.72	0.5519	0.5319	1.9	0.3182	1 0162
0.26	0.7841	0.2302	0.74	0.5447	0.5429	2.0	0.3085	1.0476
0.28	0.7707	0.2457	0 76	0.5377	0.5537			
0.30	0.7576	0.2610	0.78	0.5308	0.5644	2.5	0.2700	1.1916
0.32	0.7449	0.2760	0.80	0.5241	0.5749	3.0	0.2430	1.3195
0.34	0.7325	0.2908	0.82	0.5176	0.5853	3.5	0 2228	1.4357
0.36	0.7205	0.3053	0 84	0.5113	0.5956	4.0	0.2070	1.5430
0.38	0.7088	0.3196	0.86	0.5051	0.6058	4.5	0.1942	1.6432
0.40	0.6974	0.3337	0.88	0.4991	0.6158	5.0	0.1835	1.7375
0.42	0 6863	0.3475	0.90	0.4932	0.6258			
0.44	0.6756	0.3611	0.92	0.4874	0.6356	5.5	0.1745	1.8270
0.46	0.6651	0.3745	0.94	0.4818	0.6452	6.0	0.1667	1.9123

by the natural width. A semilogarithmic plot of (94.12a) is given in Fig. 174. Attention should be paid to the fact that, according to (94.7b) and (94.12b), the same value of the abscissa does not correspond to the same value of the absorber thickness for the two curves $R(y)$ and $R_d(y)$. It may be noticed that the curve $R_d(y)$ is closer to an exponential than $R(y)$.

The derivation of similar expressions for intermediate cases, i.e. for $\eta \sim 1$ becomes too laborious and will not be made here. We may simply compute the initial absorption coefficient for the general case

$$K_s = \frac{\int\limits_{-\infty}^{+\infty} K_a(\eta_a, E)\, K_d(\eta_d, E)\, \varphi_0(E)\, dE}{\int\limits_{-\infty}^{+\infty} K_d(\eta_d, E)\, \varphi_0(E)\, dE} \tag{94.13}$$

where $K_a(\eta_a, E)$ and $K_d(\eta_d, E)$ are obtained by introducing (87.6) in (94.2); η_d and η_a indicate the values of the parameter (87.8) for the detector and the absorber respectively, since they may be at different temperatures. Assuming as before, that the resonance is sufficiently narrow to take $\varphi_0(E) \approx \varphi_0(E_R)$, one finds[1] [5]

$$K_s = \tfrac{1}{2} K_0 \psi(\bar{\eta}, 0) \tag{94.14}$$

where

$$\bar{\eta} = \frac{2\,\eta_a \eta_d}{[\eta_a^2 + \eta_d^2]^{\frac{1}{2}}}\,. \tag{94.15}$$

When both absorber and detector are at the same temperature, so that $\eta_a = \eta_d = \eta$ one has

$$\bar{\eta} = \sqrt{2}\,\eta$$

[1] See footnote 1, p. 388.

and Eq. (94.14) states that the cross section for self-indication is $\frac{1}{2}$ the resonance cross section at one half the temperature of absorber and detector.

It may be interesting to recall at this point that by measuring the activity of a thin resonance detector (91.8), its initial coefficient for self-indication (94.14), its resonance energy E_R, for example, by means of its absorption in boron, and its capture cross section for thermal neutrons, one obtains a set of relationships between the parameters of the resonance which allow for the experimental determination of Γ[1,2] [*5*].

Other types of absorption curves may be considered; for example the absorption curve for an absorber in which the resonance level *overlaps in part* the resonance level of the detector. The mutual absorption coefficient found under these conditions depends on the difference between the energies and the widths of the two resonances. We shall not discuss these cases here; they may be found in BETHE's review article [*5*].

Here we limit ourselves to notice that when the temperature of the absorber is different from that of the detector, the self-absorption curve mentioned above constitutes an interesting particular example of overlapping levels. If the absorber temperature is higher than that of the detector, the initial absorption coefficient (94.14) will be smaller than that obtained when they have both the same temperature, since, according to (87.13), an increase in Doppler broadening corresponds to a decrease in the absorption coefficient at the center of the line. This effect was clearly shown by VON HALBAN and PAXTON[3] for the case of Ag[4].

We can now extend these considerations to the case of a detector of finite thickness. The procedure given below is essentially the same as described in Sect. 136β for the case of monoenergetic neutrons. Referring, as an example, to the case of self-absorption of a resonance with negligible Doppler width, we have the following expression for the activation of the detector when irradiated behind an absorber of thickness $X=2y/K_0$:

$$\int_y^{y+y_d} R(y)\,dy = \mathfrak{R}(y+y_d) - \mathfrak{R}(y) \tag{94.16}$$

where

$$y_d = \tfrac{1}{2} K_0 X_d$$

and

$$\mathfrak{R}(y) = \int_0^y R(y)\,dy = \int_0^y \mathrm{e}^{-y} I_0(y)\,dy = y\,\mathrm{e}^{-y}\{I_0(y) + I_1(y)\} \tag{94.17}$$

is the total activation—i.e. the total number of neutrons absorbed in a layer of thickness y (Table 69). In writing (94.17), the well-known relations $I_0'=I_1$, $I_1'=-I_1+I_0$ have been used.

For $y\gg 1$ the following asymptotic expression may be used

$$\mathfrak{R}(y) = \left(\frac{2y}{\pi}\right)^{\frac{1}{2}}\left[1 - \frac{1}{8y} - \frac{3}{128y^2} - \frac{15}{1024y^3} - \cdots\right]. \tag{94.18}$$

We shall now discuss the effect of an angular spread of the incident neutrons, referring again to the case of self-absorption with a resonance detector. If the angular distribution follows the LAMBERT's law (136.2), one has, for an infinitely

---

[1] E. AMALDI and E. FERMI: Ric. Sci. **7** (1), 454 (1936). — Phys. Rev. **50**, 899 (1936)

[2] See footnote 1, p. 388.

[3] H. VON HALBAN and H. PAXTON: Nature, Lond. **141**, 116 (1938).

[4] The same effect was previously observed by ARSENJEWA-HEIL, HEIL and WESTCOTT: Nature, Lond. **138**, 462 (1936), but the interpretation was not correct.

thin detector,

$$R_L(y) = \int_0^{\pi/2} R\left(\frac{y}{\cos\Theta}\right) \sin\Theta \, d\Theta \tag{94.19}$$

where the factor $\cos\Theta$ due to the LAMBERT's law disappears because of the factor $1/\cos\Theta$ arising from the probability of detection $\frac{K_d\, dX_d}{\cos\Theta}$. With $t = y/\cos\Theta$, one has

$$\left.\begin{aligned} R_L(y) = y \int_y^{\infty} \frac{e^{-t}}{t^2} I_0(t)\, dt &= 1 - y \ln \frac{\gamma y}{2} - 3\left(\frac{1}{2} y\right)^2 + \\ &+ \frac{5}{3}\left(\frac{1}{2} y\right)^3 - \frac{35}{36}\left(\frac{1}{2} y\right)^4 + \frac{21}{40}\left(\frac{1}{2} y\right)^5 - \frac{77}{500}\left(\frac{1}{2} y\right)^6 + \\ &+ \frac{143}{1260}\left(\frac{1}{2} y\right)^7 - \frac{143}{3136}\left(\frac{1}{2} y\right)^8 + \cdots \end{aligned}\right\} \tag{94.20}$$

where $\ln\gamma = 0.5772\ldots$ is the Euler-Mascheroni constant which enters when $I_0(y)\cdot e^{-y}$ is expanded into a series of powers of $y/2$ and the additive constant of the integral is determined. For $y \gg 1$, one has

$$R_L(y) \to \frac{1}{(2\pi\, y)^{\frac{1}{2}}} \left\{\frac{2}{3} + \frac{1}{20y} + \frac{9}{448\cdot y^2} + \cdots\right\} \tag{94.21}$$

while for intermediate values of y the behaviour of $R_L(y)$ is shown in Fig. 174.

Table 70. *Self-absorption curves of resonance neutrons when their angular distribution is given by the Lambert law.*

y	$R_L(y)$	$\mathfrak{R}_L(y)$	y	$R_L(y)$	$\mathfrak{R}_L(y)$
0.00	1.0000	0.0000	0.30	0.5417	0.4235
0.01	0.9527	0.01948	0.40	0.4790	0.5253
0.02	0.9191	0.03818	0.50	0.4307	0.6160
0.03	0.8907	0.0563	0.60	0.3922	0.6982
0.04	0.8654	0.0738	0.70	0.3609	0.7734
			0.80	0.3350	0.8429
0.06	0.8216	0.1076	1.00	0.2948	0.9685
0.08	0.7840	0.1396			
0.10	0.7508	0.1703	2.00	0.1975	1.443
			4.00	0.1358	2.086
0.15	0.6818	0.2418	8.00	0.0950	2.980
0.20	0.6265	0.3071	16.00	0.0668	4.235

In order to obtain the activation of a detector of finite thickness, we need know the total number of neutrons absorbed in a layer of thickness $X = 2y/K_0$. This is given by

$$\mathfrak{R}_L(y) = \int_0^{\pi/2} \mathfrak{R}\left(\frac{y}{\cos\Theta}\right) 2\cos\Theta \sin\Theta \, d\Theta. \tag{94.22}$$

By changing the variable $t = y/\cos\Theta$, and integrating by parts, one obtains

$$\mathfrak{R}_L(y) = \mathfrak{R}(y) + y\, R_L(y) \tag{94.23}$$

from which it follows that

$$\mathfrak{R}'_L(y) = 2 R_L(y). \tag{94.24}$$

Similarly, if the angular spread of the incident neutrons is given by the generalized Fermi law (136.3), one obtains, for the activation of a thin detector,

$$R_G(y) = \frac{1}{1+\frac{1}{2}a}\left\{\left(1-\frac{3}{4}a\,y\right)R_L(y) + \frac{1}{2}a\left[R(y) + \frac{1}{2}\,\mathfrak{R}(y)\right]\right\} \qquad (94.25)$$

and for the total number of neutrons absorbed in the layer y

$$\mathfrak{R}_G(y) = \mathfrak{R}(y) + \frac{2+a}{3+2a}\,y\,R_G(y) + \frac{y}{3+2a}\,R_L(y). \qquad (94.26)$$

A diagram of $R_G(y)$ for $a=\sqrt{3}$ [FERMI's angular distribution (122.20)] is given in Ref. [5].

When the detector does not fulfill the condition of no self-protection (89.3), the absorption of the β- (or γ-) rays in the detector itself should be calculated. This can be done by means of the procedure described in Sect. 136β.

D. Slowing down and diffusion of neutrons in the chemical region, i. e. E less than or of the order of 1 ev.

I. Survey of the main features of the problem.

95. Various effects influencing the scattering of neutrons. The slowing down of neutrons below a few ev needs special consideration because in this energy region the following phenomena will play a role: (a) the thermal agitation of the atoms of the moderator; (b) the chemical forces binding these atoms in crystals or molecules and the quantization of the corresponding vibrational and rotational levels; (c) the interference between waves scattered from various atoms of the same crystal or molecule.

In practice all, or almost all of these effects are present together so that it is usually rather difficult and even sometimes impossible to disentangle the role played by each one of them from that of the others. The following general remarks may be useful in order to clarify their relative importance under different circumstances.

(a) Various experiments (Sects. 22 and 23) prove that the neutron can be slowed down to energies comparable or even much smaller than those of the thermal agitation of the atoms of the moderator. The finite velocity of the target nuclei produces a change in the scattering cross section and furthermore it makes possible collisions in which the kinetic energy of the neutron increases at the expense of the energy of the atoms of the moderator.

If the velocity of the neutron is large compared to that of the atoms, the change in cross section may be described as a Doppler effect. It is indeed customary to refer to a Doppler effect even if the velocity of the neutron is lower than that of the atoms; but in this case the phenomenon is better described by saying that the atoms hit the neutron rather than that the neutron hits the atoms.

The velocity of thermal agitation of atoms depends not only on the temperature of the moderator but also on its structure: translational, rotational and vibrational degrees of freedom may play a more or less important role in different cases.

A "pure Doppler effect" due to thermal agitation could be observed in the ideal case of a monatomic gas moderator. As we shall see below, it gives rise to a change in the scattering cross section which obviously depends strongly on the temperature of the moderator. There are however cases in which a Doppler

MIX
Papier aus verantwortungsvollen Quellen
Paper from responsible sources
FSC® C105338

If you have any concerns about our products,
you can contact us on
ProductSafety@springernature.com

In case Publisher is established outside the EU,
the EU authorized representative is:
Springer Nature Customer Service Center GmbH
Europaplatz 3, 69115 Heidelberg, Germany

Printed by Libri Plureos GmbH
in Hamburg, Germany

ENCYCLOPEDIA OF PHYSICS

EDITED BY

S. FLÜGGE

VOLUME XXXVIII/2

NEUTRONS AND RELATED GAMMA RAY PROBLEMS

WITH 338 FIGURES

SPRINGER-VERLAG BERLIN HEIDELBERG GMBH

1959

HANDBUCH DER PHYSIK

HERAUSGEGEBEN VON

S. FLÜGGE

BAND XXXVIII/2

NEUTRONEN UND VERWANDTE GAMMASTRAHLPROBLEME

MIT 338 FIGUREN

SPRINGER-VERLAG BERLIN HEIDELBERG GMBH

1959

ISBN 978-3-642-45922-1 ISBN 978-3-642-45920-7 (eBook)
DOI 10.1007/978-3-642-45920-7

Ursprünglich erschienen bei Springer-Verlag OHG.Berlin
Göttingen . Heidelberg 1959

Contents.

Similarly, if the angular spread of the incident neutrons is given by the generalized Fermi law (136.3), one obtains, for the activation of a thin detector,

$$R_G(y) = \frac{1}{1+\frac{1}{2}a}\left\{\left(1-\frac{3}{4}a\,y\right)R_L(y) + \frac{1}{2}a\left[R(y) + \frac{1}{2}\,\mathfrak{R}(y)\right]\right\} \tag{94.25}$$

and for the total number of neutrons absorbed in the layer y

$$\mathfrak{R}_G(y) = \mathfrak{R}(y) + \frac{2+a}{3+2a}\,y\,R_G(y) + \frac{y}{3+2a}\,R_L(y)\,. \tag{94.26}$$

A diagram of $R_G(y)$ for $a=\sqrt{3}$ [FERMI's angular distribution (122.20)] is given in Ref. [5].

When the detector does not fulfill the condition of no self-protection (89.3), the absorption of the β- (or γ-) rays in the detector itself should be calculated. This can be done by means of the procedure described in Sect. 136β.

D. Slowing down and diffusion of neutrons in the chemical region, i. e. E less than or of the order of 1 ev.

I. Survey of the main features of the problem.

95. Various effects influencing the scattering of neutrons. The slowing down of neutrons below a few ev needs special consideration because in this energy region the following phenomena will play a role: (a) the thermal agitation of the atoms of the moderator; (b) the chemical forces binding these atoms in crystals or molecules and the quantization of the corresponding vibrational and rotational levels; (c) the interference between waves scattered from various atoms of the same crystal or molecule.

In practice all, or almost all of these effects are present together so that it is usually rather difficult and even sometimes impossible to disentangle the role played by each one of them from that of the others. The following general remarks may be useful in order to clarify their relative importance under different circumstances.

(a) Various experiments (Sects. 22 and 23) prove that the neutron can be slowed down to energies comparable or even much smaller than those of the thermal agitation of the atoms of the moderator. The finite velocity of the target nuclei produces a change in the scattering cross section and furthermore it makes possible collisions in which the kinetic energy of the neutron increases at the expense of the energy of the atoms of the moderator.

If the velocity of the neutron is large compared to that of the atoms, the change in cross section may be described as a Doppler effect. It is indeed customary to refer to a Doppler effect even if the velocity of the neutron is lower than that of the atoms; but in this case the phenomenon is better described by saying that the atoms hit the neutron rather than that the neutron hits the atoms.

The velocity of thermal agitation of atoms depends not only on the temperature of the moderator but also on its structure: translational, rotational and vibrational degrees of freedom may play a more or less important role in different cases.

A "pure Doppler effect" due to thermal agitation could be observed in the ideal case of a monatomic gas moderator. As we shall see below, it gives rise to a change in the scattering cross section which obviously depends strongly on the temperature of the moderator. There are however cases in which a Doppler effect arises not from thermal agitation but from the chemical bond and its quantization. As we shall see later (Sect. 100) such an effect does not depend on the temperature of the moderator.

(b) When the neutron reaches energies comparable to the chemical binding energy of the atoms in molecules or crystals of the moderator (a few tenths of an ev), the target nuclei no longer can be considered free, and both the scattering cross section and the energy loss per collision show a dependence on neutron energy which is more or less complicated according to the structure and temperature of the moderator.

Because of the quantization of the vibrational and rotational degrees of freedom a sharp distinction can be made between elastic and inelastic collisions; elastic collisions are those which leave the target molecule or crystal in its initial state i and the neutron with energy unchanged. Inelastic collisions, on the contrary, take place with an energy exchange between neutron and molecule or crystal; if the molecule is excited to a higher state the neutron loses energy; if the molecule is deexcited i.e., if it undergoes a transition from a higher to a lower level, the neutron energy is correspondingly increased. A cross section $\sigma_{ik}(E)$ can be defined corresponding to the scattering of the incident neutron of energy E accompanied by the transition of the target molecule or crystal from the initial state i to the final state k.

If all the scattering cross sections, $\sigma_{ik}(E)$, elastic ($k=i$) and inelastic ($k\neq i$), were known, the total scattering cross section $\sigma_{s\,\mathrm{tot}}$ could be obtained by the following procedure. First, for a given initial state i the summation of the $\sigma_{ik}(E)$ is done with respect to all possible final states k (including $k=i$), and afterwards the expression so obtained is averaged with respect to the occupation distribution of the initial molecular states i at the temperature T_0 of the moderator.

The total scattering cross section $\sigma_{s\,\mathrm{tot}}$ obtained by such a procedure depends both on the structure and on the temperature of the moderator.

For values of E much larger than the spacing Δ of the molecular levels, $\sigma_{s\,\mathrm{tot}}(E)$ approaches a limit independent of the structure and of the temperature of the moderator; this value is usually called σ_{free} (abbreviated with σ_f) because it represents the elastic cross section of slow neutrons against free atoms. Some details on the law followed by $\sigma_{s\,\mathrm{tot}}(E)$ in approaching σ_f will be given in Sect. 100.

If, on the contrary, the moderator is at such a low temperature that all its molecules are in their ground state and the energy of the neutron is so small as to be insufficient to excite the lowest molecular level, only elastic collisions will be possible. Under these conditions for $E\to 0$ the limit reached by $\sigma_{s\,\mathrm{tot}}(E)$ is equal to the limit of $\sigma_{00}(E)$. This limit will be called in the following σ_{bound} abbreviated with σ_b.

A very simple relation was established by FERMI [8] between σ_f and σ_b. As will be shown in Sect. 98 the ratio σ_b/σ_f is equal to the square of the ratio of the corresponding reduced masses. If one considers for instance the collision of a neutron against a single nucleus of mass M bound to a fixed centre, one has

$$\sigma_b = \sigma_{\mathrm{bound}} = \left(\frac{m}{\mu}\right)^2 \sigma_{\mathrm{free}} = \left(\frac{m}{\mu}\right)^2 \sigma_f; \qquad \mu = \frac{m M}{m+M}. \tag{95.1}$$

Eq. (98.10) is a generalization of (95.1) to the case of collisions of a neutron against a nucleus of mass M bound to a molecule of finite mass M_{mol}.

The relation (95.1), or more generally (98.10) is usually called the *reduced mass effect*. The lighter the target nucleus the larger is this effect. The maximum value, 4, is reached by σ_b/σ_f in the case of hydrogen bound to very heavy molecules ($M_{\mathrm{mol}}/m\to\infty$).

In the case of slow neutrons with wavelengths much larger than the dimensions of the scatterer, only S waves need to be considered. The corresponding amplitude a has the dimension of a length and is usually called the *scattering length* (Sect. 19). In the absence of absorption, a is real; even in the case of relatively strong absorption its imaginary part is always very small. The scattering cross section and scattering length are related by Eq. (20.18) which we rewrite here:

$$\sigma_s = 4\pi \, |a|^2. \tag{95.2}$$

This is usually applied to the case of free atoms. If by chemical forces the atoms are bound to a crystal or to a very heavy molecule, and the incident neutrons have energy $E \to 0$, one can define a scattering length for bound atoms a_{bound} (abbreviated a_b) which is again related to σ_{bound} by (95.2). This relation, combined with (95.1), gives

$$a_b = \frac{A+1}{A} a_f, \qquad A = \frac{M}{m}. \tag{95.3}$$

(c) Interference effects between the neutron waves scattered from the various nuclei constituting a molecule or a crystal start to be important when the neutron wavelength

$$\lambda = \frac{h}{m\,v} = \frac{h}{\sqrt{2mE}} = \frac{2.87 \times 10^{-9}}{E_{\text{ev}}^{\frac{1}{2}}} \text{ cm} \tag{95.4}$$

becomes comparable with the interatomic distances.

Under these conditions the passage of neutrons through a moderator is complicated by diffraction phenomena which are similar, in many respects, to those observed in the case of X-rays.

Preisswerk and von Halban[1] were the first to prove experimentally that slow neutrons can be diffracted by crystals. Their conclusion was immediately confirmed by Mitchell and Powers[2], while the basic principles of the theory were established independently at the same time by Elsasser[3] and Wick[4]. Since then the investigation of the scattering of slow neutrons by molecules and crystals has become a problem in itself of paramount importance; for it provides new information about the interaction of neutrons with nuclei and a new tool for the study of the structure of solid, liquid and gaseous matter.

These very important problems, however, are outside the scope of the present chapter. Therefore we make reference to G. R. Ringo's article in Vol. XXXII of this Encyclopedia and to various articles and books[5] [*13*], [*18b*] for their detailed treatment while we restrict ourselves to recalling in Sect. 99 and the following chapters those results which appear to be directly connected with the problem of thermalization and diffusion of neutrons through a moderator.

96. The integral equation of the slowing down spectrum in the stationary case. In this section we shall discuss, for neutron energies below 1 ev, the same problem as was already considered in Sect. 66 for higher energies. Neutrons produced at a constant rate in time and uniformly in space throughout a homogeneous

---

[1] P. Preisswerk and H. von Halban: C. R. Acad. Sci., Paris **203**, 73 (1936). — J. Phys. Radium **8**, 29 (1937). — P. Preisswerk: Helv. phys. Acta **10**, 400 (1937).

[2] D. P. Mitchell and P. N. Powers: Phys. Rev. **50**, 486 (1936).

[3] W. M. Elsasser: C. R. Acad. Sci., Paris **202**, 1029 (1936).

[4] G. C. Wick: Ric. Sci. **8** (1), 400 (1937). — Phys. Z. **38**, 403, 689 (1937).

[5] E. O. Wollan and C. G. Shull: Nucleonics **3**, 8 (July 1948); **3**, 17 (August 1948). — G. E. Bacon: Neutron diffraction. Oxford: Clarendon Press 1955. — C. G. Shull and E. O. Wollan: Applications of neutron diffraction to solid state problems, Vol. 2, p. 138. New York: Academic Press 1956.

medium of infinite dimensions, are slowed down and enter the chemical bond energy region; we have to find their stationary spectrum $n_0(v)\,dv$.

This problem, in the energy region considered here, is considerably more involved than the case treated in Sect. 66; because of the effects mentioned in Sect. 95, the scattering cross section and the rate of energy exchange between the neutrons and the medium depend on the temperature and molecular structure of the moderator.

The spectrum $n_0(v)$ can be obtained as a solution of the following integral equation

$$[\gamma(v) + V(v)]\, n_0(v) = \int_0^\infty G_s(v' \to v)\, n_0(v')\, dv' \tag{96.1}$$

which was first introduced and discussed by WIGNER and WILKINS[1]. The significance of its various terms may be recognized by comparing it to Eq. (61.2). The two terms on the left-hand side of (61.2) do not appear in (96.1) because of the condition of stationarity ($\partial/\partial t = 0$) and of uniform distribution in space ($\nabla = 0$); furthermore the variable $\boldsymbol{\Omega}$ has disappeared because of the integration over the solid angle which leads to the definition (61.15), (66.1a) of $n_0(v)$. These three operations are just those used in Sect. 66 to derive Eq. (66.4) from Eq. (61.21b).

In the present case the source term appearing in (61.2) has been omitted because sources emitting neutrons in the energy interval considered here do not exist. Thus, only three terms of Eq. (61.2) survive corresponding to the three terms of Eq. (96.1). The difference between the two cases is embodied in the form of the coefficients of the two equations. The coefficients $\gamma(v)$ and $V(v)$ represent the absorption and scattering probability per unit time, and the kernel $G_S(v' \to v)\,dv$ represents the probability per unit time for a neutron of velocity v' to be scattered into the velocity interval between v and $v + dv$; one has

$$\int G_S(v \to v')\, dv' = V(v)\,. \tag{96.2}$$

These various quantities are referred to the l.s. where the moderator is at rest but its molecules or atoms are in motion because of thermal agitation and chemical binding (zero-point energy). They may be expressed in terms of the corresponding cross sections in the frame of reference in which the target nucleus is at rest and the neutron moves with relative velocity v_r, which is given by Eq. (36.2) as a function of the velocities in the l.s. of the neutron and the target nucleus.

For instance, the coefficient of the first term on the left-hand side of Eq. (96.1), representing the absorption probability per unit time, is given by the formula

$$\gamma(v) = \int v_r\, \Sigma_a(v_r)\, \mathfrak{N}(\boldsymbol{V}_1)\, d\boldsymbol{V}_1 \tag{96.3}$$

where $\mathfrak{N}(\boldsymbol{V}_1)\, d\boldsymbol{V}_1$ is the normalized velocity distribution of the atoms of the moderator and $\Sigma_a(v_r)$ is the macroscopic absorption cross section in the frame of reference of the target nucleus. The result of the integration appearing in (96.3) depends in general on the molecular structure and thermal agitation of the moderator because of the factor $\mathfrak{N}(\boldsymbol{V}_1)\, d\boldsymbol{V}_1$. But if all the nuclei present in the moderator follow the $1/v$ law, i.e. if the condition

$$v_r\, \Sigma_a(v_r) = \frac{1}{T} = \text{const} \tag{96.4}$$

[1] E.P. WIGNER and J.E. WILKINS: AECD 2275 (1948).

is satisfied over the energy interval considered here, the expression (96.3) will reduce to[1]

$$\gamma(v) = \gamma = \frac{1}{T} = \text{const}. \tag{96.5}$$

In the following we will restrict consideration to this simple case because it corresponds very satisfactorily to the actual behaviour of the more frequently used moderators in the low energy region ($E \lesssim 1$ ev).

For the other two quantities appearing in (96.1), namely $V(v)$ and $G_S(v' \to v)$, there do not exist simple expressions of wide validity even in the case of a constant scattering cross section in the frame of reference where the target nucleus is at rest $[\Sigma_s(v_r) = \text{const}]$. They always need the specification of the structure of the moderator and therefore we must refer to the following sections where simple idealized cases are treated in detail.

An important property of Eq. (96.1) valid if the moderator is in thermal equilibrium at the temperature T_0, is the following[2]: if there is no absorption, thermal equilibrium will be established between the neutron and the moderator, and the neutron will have the Maxwell distribution of velocities corresponding to the temperature T_0[3]. This means that Eq. (96.1) is satisfied, for $\gamma = 0$, by

$$M_{T_0 m}(v)\, dv \frac{1}{2} d(\cos\alpha) = \frac{4}{\sqrt{\pi}} \left(\frac{m}{k T_0}\right)^{\frac{3}{2}} e^{-\frac{m v^2}{2kT_0}} v^2\, dv \frac{1}{2} d(\cos\alpha). \tag{96.6}$$

The principle of detailed balance states that this equilibrium will be established between any two velocities, i.e.

$$G_S(v' \to v)\, M_{T_0 m}(v') = G_S(v \to v')\, M_{T_0 m}(v). \tag{96.7}$$

This equality expresses a general property of the kernel $G_S(v' \to v)$ because, by definition, it does not depend on the absorption properties of the moderator.

WIGNER and WILKINS have used the equality (96.7) for transforming Eq. (96.1) into an integral equation with the symmetric kernel

$$g_S(v' \to v) = G_S(v' \to v) \left[\frac{M_{T_0 m}(v')}{M_{T_0 m}(v)}\right]^{\frac{1}{2}}. \tag{96.8}$$

The equation obtained by these authors is

$$\nu(v)\,[\gamma + V(v)] = \int_0^\infty g_S(v' \to v)\, \nu(v')\, dv' \tag{96.9}$$

where

$$\nu(v) = \frac{n_0(v)}{[M_{T_0 m}(v)]^{\frac{1}{2}}} \tag{96.10}$$

is the new unknown function.

---

[1] W. E. LAMB [Phys. Rev. **51**, 187 (1937)] has given the quantum mechanical derivation of this same equation, taking into account the chemical potential binding the capturing atom in a crystal or molecule, according to the procedure sketched in Sect. 98.

[2] See footnote 1, p. 395.

[3] This statement is obvious in the ideal case of a monatomic gaseous moderator with $\Sigma_a(v_r) = 0$; the thermodynamic system of neutrons + moderator may be described as a mixture of two gases in thermal equilibrium at the temperature T_0. It can be extended to the case of a real moderator with a velocity distribution which may differ from the Maxwellian one, due to quantization of energy levels, and for which inelastic collisions are possible (provided the absorption of neutrons is negligibly small). To do this one considers a volume, half of which is filled with the real moderator, and the other half with the ideal moderator in equilibrium at the same temperature T_0. From the principle of detailed balance it follows that the velocity distribution of the neutrons is the same and that it is Maxwellian in both halves of the volume [H. D. BROWN and D. S. ST. JOHN: DP 33 (1954)].

Some authors[1] make use, instead of Eq. (96.1), of the integral equation for the number flux (61.16), (66.1b) expressed in terms of the neutron energy

$$[\bar{\Sigma}_a(E) + \bar{\Sigma}_s(E)]\,\varphi_0(E) = \int_0^\infty dE'\,\bar{\Sigma}_s(E' \to E)\,\varphi_0(E'). \tag{96.11}$$

The coefficients appearing in this equation represent the actually observed macroscopic cross sections in the l.s. and will be called in the following *effective cross sections*. They are connected to the coefficients of Eq. (96.1) by the relations

$$\left.\begin{aligned} \bar{\Sigma}_a(v) &= \frac{1}{v}\gamma(v), \\ \bar{\Sigma}_s(v) &= \frac{1}{v}V(v), \\ \bar{\Sigma}_s(v' \to v) &= \frac{1}{v'}G_S(v' \to v). \end{aligned}\right\} \tag{96.12}$$

Making use of the equality (96.7), Eq. (96.11) may be transformed into the convenient form

$$\bar{\Sigma}_a(E)\,\bar{\varphi}(E) = \int_0^\infty dE'\,\bar{\Sigma}_s(E \to E')\,[\bar{\varphi}(E') - \bar{\varphi}(E)] \tag{96.13}$$

where the unknown function

$$\bar{\varphi}(E) = \frac{\varphi_0(E)}{v\,M_{T_0 m}} = \frac{\varphi_0(E)}{\varphi_M(E)} \tag{96.14}$$

represents the deviation of the actual number flux from the Maxwell distribution at the temperature T_0. For $\Sigma_a = 0$ the solution of Eq. (96.12) is $\bar{\varphi} = \text{const}$.

97. The Doppler effect for a monatomic gaseous moderator. For monatomic gaseous moderators the effects (b) and (c) mentioned in Sect. 95 — namely the effects due to chemical bond and wave interference — do not exist. Only the velocity distribution $\mathfrak{N}(\boldsymbol{V}_1)\,d\boldsymbol{V}_1$ of the atoms has to be taken into account. Assuming that the gaseous moderator is in thermal equilibrium, the velocity distribution reduces to the Maxwell distribution

$$M_{T_0 M}(V)\,dV\,d\left(\frac{\cos\alpha}{2}\right) = \frac{4}{\sqrt{\pi}}\left(\frac{M}{kT_0}\right)^{\frac{3}{2}} e^{-\frac{MV^2}{2kT_0}}\,V^2\,dV\,\frac{1}{2}\,d(\cos d) \tag{97.1}$$

which allows the derivation of analytical expressions for the various terms appearing in (96.1).

The coefficient of the first term on the left-hand side of (96.1) is given by (96.5), provided that the absorption cross section of the nuclei of the moderator follows the $1/v$ law (96.4). In order to derive $V(v)$ and $G_S(v' \to v)$ the dependence of the scattering cross section on the relative velocity v_r must be known. The formula

$$\Sigma_s(v_r) = \sum_i \Sigma_i\, e^{-\alpha_i v_r^2} \tag{97.2}$$

is convenient because it gives analytical expressions for $V(v)$ and $G_S(v' \to v)$ and has sufficient generality to fit the actual behaviour of many moderators by an appropriate choice of the constants Σ_i, α_i[2].

The slowing-down kernel could be calculated first and then $V(v)$ could be deduced by integration according to (96.2). The coefficient $V(v)$ may also be calculated directly from

$$V(v) = \int v_r\,\Sigma_s(v_r)\,M_{T_0 M}(V)\,dV\,d\left(\frac{\cos\alpha}{2}\right). \tag{97.3}$$

[1] H. HURWITZ, M. S. NELKIN and G. T. HABETLER: Nucl. Sci. and Engng. 1, 280 (1956).
[2] H. D. BROWN and D. S. ST. JOHN: DP-33 (1954).

Introducing (97.1) and (97.2) into (97.3) BROWN and ST. JOHN[1,2] found

$$\left.\begin{aligned} V(v) &= \sum_i \Sigma_i W_i(v) = v\,\overline{\Sigma}_s(v)\,. \\ W_i(v) &= \tau_i^3\left[\left(\tau_i^2 v + \frac{1}{2A\beta^2 v}\right)\operatorname{Erf}\left(\sqrt{A}\,\beta\,\tau_i v\right)\exp\left[-\alpha_i\tau_i^2 v^2\right] + \right. \\ &\quad \left. + \frac{\tau_i}{\beta\sqrt{\pi A}}\exp\left[-A\beta^2 v^2\right]\right] \end{aligned}\right\} \tag{97.4}$$

where

$$\tau_i^2 = \frac{A\beta^2}{A\beta^2+\alpha_i}, \qquad \beta^2 = \frac{m}{2kT_0} = \frac{1}{v_0^2}; \qquad \operatorname{Erf}(x) = \frac{2}{\sqrt{\pi}}\int_0^x e^{-y^2}\,dy. \tag{97.5}$$

The case

$$\Sigma_s(v_r) = \Sigma_{s0} = \text{const} \tag{97.6}$$

had been previously considered by WIGNER and WILKINS[3] and by VON DARDEL[4] and can be obtained by reducing (97.2) and (97.4) to a single term with $\alpha_0 = 0$

$$\left.\begin{aligned} \overline{\Sigma}_s(v) &= \frac{V(v)}{v} = \Sigma_{s0}\left\{\left(1+\frac{1}{2\delta^2}\right)\operatorname{Erf}(\delta) + \frac{1}{\sqrt{\pi}\,\delta}\exp\left[-\delta^2\right]\right\} \\ \text{with}\qquad \delta &= \sqrt{A}\,\beta\,v = \sqrt{A}\,\frac{v}{v_0}\,. \end{aligned}\right\} \tag{97.7}$$

The kernel $G_S(v'\to v)$ appearing in (96.1) represents the probability per unit time that a neutron of velocity v' will be scattered into the range between v and $v+dv$ and is related to $V(v)$ by (96.2). It is given by

$$G_S(v'\to v) = \int_{-1}^{+1} d\left(\frac{\cos\alpha}{2}\right)\int_0^\infty v_r\,\Sigma_s(v_r)\,M_{T_0 M}(V)\,P(v'\to v;\,V,\cos\alpha)\,dV. \tag{97.8}$$

Here $P(v_1\to v_2;\,V_1,\cos\alpha)$ is the probability that in a collision of a neutron and an atom of velocities v_1 and V_1 forming an angle α (see Fig. 35 of Sect. 36) the neutron is left with velocity v_2. In order to derive $P(v_1\to v_2;\,V_1,\cos\alpha)$ use can be made of the following remark. According to (36.5) the final energy of the neutron varies linearly with $\cos\beta$ between a minimum and maximum value, β being the angle between the final velocity $\boldsymbol{v}_2^*$ of the neutron in the c.m.s. and the velocity $\boldsymbol{v}_m$ of the centre of mass in the l.s. (Fig. 36). On the other hand in the case of slow neutrons the only waves which are scattered appreciably by nuclei are the S waves which correspond to isotropic scattering in the c.m.s., i.e. to a probability of scattering into the solid angle $2\pi d(\cos\beta)$ given by $d\left(\frac{\cos\beta}{2}\right)$.

This means that the probability that the final energy of the neutron lies between E and $E+dE$ is constant and different from zero only for $E_{\min}\leqq E\leqq E_{\max}$; i.e.

$$P(v'\to v;\,V,\cos\alpha) = \begin{cases} \dfrac{2v\,dv}{v_{\max}^2 - v_{\min}^2} & \text{for}\quad v_{\min}\leqq v\leqq v_{\max} \\ 0 & \text{for}\quad \begin{cases} v < v_{\min} \\ v > v_{\max} \end{cases} \end{cases} \tag{97.9}$$

---

[1] H.D. BROWN and D.S. ST. JOHN: DP 33 (1954).
[2] E.R. COHEN: p. 405 of Ref. [*15c*].
[3] E.P. WIGNER and J.E. WILKINS: AECD 2275 (1948).
[4] G.F. VON DARDEL: Phys. Rev. **94**, 1272 (1954).

where $v_{\min}$ and $v_{\max}$ are given by (36.4). Introducing (97.1), (97.2) and (97.9) in (97.8) the following expression is obtained[1,2]

$$\left.\begin{aligned} G_S(v'\to v) &= \sum_i \Sigma_i\, G(v', v; \alpha_i),\\ G(v', v, \alpha_i) &= \frac{(A+1)^2}{4A}\,\frac{\tau_i^3}{\lambda_i}\,\frac{v}{v'}\times\\ &\times\left\{\exp\left[-\alpha_i\tau_i^2 v'^2\right]\left[\operatorname{Erf}(\beta\vartheta_i v-\beta\zeta_i v')\pm\operatorname{Erf}(\beta\vartheta_i v+\beta\zeta_i v')\right]+\right.\\ &\left.+\exp\left[\frac{\beta^2}{A}(v'^2-\lambda_i v^2)\right]\left[\operatorname{Erf}(\beta\vartheta_i v'-\beta\zeta_i v)\mp\operatorname{Erf}(\beta\vartheta_i v'+\beta\zeta_i v)\right]\right\} \end{aligned}\right\} \tag{97.10}$$

where the upper signs refer to $v<v'$ and the lower signs to $v>v'$, and

$$\vartheta_i=\frac{A+1}{2\tau_i\sqrt{A}};\qquad \zeta_i=\tau_i\sqrt{A}-\vartheta_i;\qquad \lambda_i=1+A(1-\tau_i^2). \tag{97.11}$$

If the scattering cross section (97.2) reduces to (97.6), Eq. (97.10) becomes[3]

$$\left.\begin{aligned} G_S(v'\to v) &= \Sigma_{s0}\frac{A+1}{4A}\,\frac{v}{v'}\left\{\operatorname{Erf}(\beta\vartheta v-\beta\zeta v')\pm\operatorname{Erf}(\beta\vartheta v+\beta\zeta v')+\right.\\ &\left.+\exp\left[\frac{\beta^2}{A}(v'^2-v^2)\right]\left[\operatorname{Erf}(\beta\vartheta v'-\beta\zeta v)\mp\operatorname{Erf}(\beta\vartheta v'+\beta\zeta v)\right]\right\}. \end{aligned}\right\} \tag{97.12}$$

In order to give an approximate description of the slowing-down process when the neutron energy starts to approach the energy of thermal agitation of the moderator, it is convenient to derive the expression for the mean energy loss per unit time[4]

$$\overline{(E-E')}=\int \Sigma_s v_r\left[\frac{1}{2}mv^2-\frac{1}{2}mv'^2\right]M_{T_0M}(V_1)\,dV_1\,d\left(\frac{\cos\alpha}{2}\right) \tag{97.13}$$

where

$$E=\tfrac{1}{2}mv^2 \quad\text{and}\quad E'=\tfrac{1}{2}mv'^2$$

are the initial and final energy of the neutron under consideration. Under the assumption (97.6) the integral appearing in (97.13) can easily be evaluated

$$\left.\begin{aligned} \overline{(E-E')} &= \frac{A v^3}{(A+1)^2}\Sigma_{s0}\left\{\operatorname{Erf}(\sqrt{A}\beta v)\left[1-\frac{2A-1}{A\beta^2v^2}-\frac{4A+1}{4A^2+\beta^2v^4}\right]+\right.\\ &\left.+\frac{\exp[-\sqrt{A}\beta v]}{\sqrt{\pi}A\beta^2v^2}\left[1-\frac{4A-1}{2A\beta^2v^2}\right]\right\}. \end{aligned}\right\} \tag{97.14}$$

If the velocity of the neutron is large compared to the velocity of the atoms of the moderator, i.e. if

$$E\gg\frac{m}{M}kT_0 \quad\text{or}\quad A\beta^2v^2=A\frac{mv^2}{2kT_0}\gg 1,$$

one can substitute for (97.14) the approximate expression

$$\overline{(E-E')}=\frac{2A}{(A+1)^2}v\Sigma_{s0}\left\{E-\frac{2A-1}{A}kT_0-\frac{4A+1}{4A^2}\frac{(kT_0)^2}{E}\right\}. \tag{97.15}$$

---

[1] See footnote 2, p. 397.
[2] See footnote 2, p. 398.
[3] See footnote 3, p. 398.
[4] See footnote 4, p. 398.

In the case of heavy moderators ($A \gg 1$) a further approximation can be made to calculate the relative mean energy loss rate

$$\frac{\overline{(E-E')}}{E} \approx v\,\Sigma_{s0}\,\frac{2}{A}\left(1-\frac{2kT_0}{E}\right). \tag{97.16}$$

Similar considerations, neglecting the thermal motion of the moderator, would give, according to (62.12),

$$\frac{\overline{(E-E')}}{E} = v\,\Sigma_{s0}\,\frac{1-\alpha}{2} \approx v\,\Sigma_{s0}\,\frac{2}{A}. \tag{97.17}$$

By comparing this with (97.16) one can conclude that in the case of a gaseous moderator the consideration of thermal agitation gives rise to the correction factor $(1-2kT_0/E)$, which takes into account the fact that, if E approaches thermal energy, the slowing-down power of the moderator decreases.

In the case of heavy moderators ($A \gg 1$) the expressions for the relative energy loss (62.12) and for the average logarithmic energy loss (62.17) valid for neutron energies above 1 ev, are approximately equal. Therefore, in the same approximation, one can apply the correction factor derived for the mean energy loss rate to the average logarithmic energy loss[1]

$$\bar{\xi}_{\text{cor}} \approx \bar{\xi}\left(1-\frac{2kT_0}{E}\right). \tag{97.18}$$

The corrected average logarithmic energy loss (97.18) will be substituted for $\bar{\xi}$ in the following discussion in order to extend the validity of a few relations holding for $E > 1$ ev down to lower values of the neutron energy.

By combining (67.16) with (66.3d) the following relation between slowing down density $q(v)$ and neutron density $n_0(v)$ is obtained

$$q(v) = \bar{\xi}\,\Sigma_{s0}\,\tfrac{1}{2}\,v^2\,n_0(v).$$

Substituting for $\bar{\xi}$ the expression (97.18), one has

$$q(v) = \Sigma_{s0}\bar{\xi}\left(1-\frac{2kT_0}{E}\right)\frac{1}{2}\,v^2\,n_0(v) = \frac{1}{2}\,\Sigma_{s0}\bar{\xi}\,(v^2-2v_0^2)\,n_0(v) \tag{97.19}$$

which is expected to hold approximately for

$$A \gg 1 \quad\text{and}\quad v^2 \gg v_0^2. \tag{97.20}$$

A second relation between $q(v)$ and $n_0(v)$ is given by (69.30) which states that the slowing-down density is reduced as the neutron energy decreases, by the absorption of neutrons:

$$dq = \overline{\Sigma}_a \cdot v \cdot n_0(v)\,dv = \gamma\,n_0(v)\,dv. \tag{97.21}$$

Combining this with (97.19) one obtains

$$\frac{1}{q}\,\frac{dq}{dv} = \frac{2\gamma}{\bar{\xi}\,\Sigma_{s0}}\,\frac{1}{v^2-2v_0^2} \tag{97.22}$$

from which it follows that

$$q(v) = q_0\left(\frac{v-\sqrt{2}\,v_0}{v+\sqrt{2}\,v_0}\right)^{\omega}, \qquad \omega = \frac{A\,\gamma}{2\sqrt{2}\,v_0\,\Sigma_{s0}} \tag{97.23}$$

and

$$n_0(v) = q_0\,\frac{A}{\Sigma_{s0}}\,\frac{(v-\sqrt{2}\,v_0)^{\omega-1}}{(v+\sqrt{2}\,v_0)^{\omega+1}}. \tag{97.24}$$

[1] E.R. COHEN: NAA, SR 1127 (1955).

This expression, as will be shown in Sect. 107, can be used to extend the neutron spectrum valid above 1 ev, to velocities v which are much larger than v_0, but at the same time sufficiently small to join with continuity to a Maxwellian distribution of velocities, which is expected to hold approximately in the case of weak absorption in the energy region of thermal agitation.

98. The effect of chemical binding forces. The theory of the scattering of neutrons by nuclei bound in molecules cannot be established by the usual methods because the motion of the centre of mass of the two colliding particles can not be eliminated; for the problem is complicated by the intervention of a third body, the molecule, which can take up momentum.

A special approximate method was developed by FERMI [*8*]; later the problem was stated by BREIT[1] in terms of an exact integral equation by means of which it is possible to calculate corrections to FERMI's result. This has to be considered as the first approximation to the more accurate equations [*3*]; as will be shown later the higher approximations give corrections sufficiently small to be neglected.

Two types of forces intervene in the problem: the nuclear forces between the neutron and the nucleus, and the chemical binding forces between the corresponding atom and its molecule. The strength of the nuclear forces is millions of times greater than that of the chemical binding forces. Therefore it appears unjustified, at least at first sight, to introduce the neutron-nucleus interaction as a perturbation of the system consisting of a free neutron and a bound atom.

However, the following remark can be made. First the range b of the nuclear forces is extremely small with respect to the dimensions $\mathfrak{A}$ of the region where the chemical binding forces have to be considered

$$b \ll \mathfrak{A}. \tag{98.1}$$

As molecular dimensions $\mathfrak{A}$ one can use the amplitude of the vibrations of the atoms in the molecule which are of the order of 10^{-9} cm. The range b of the neutron-nucleus interaction is not larger than 10^{-12} cm and therefore condition (98.1) is amply satisfied.

Because of condition (98.1) the neutron-nucleus interaction can be described to a very good approximation as a *contact interaction* which is different from zero only when the neutron is at the same position as the nucleus [see Eq. (98.4) below].

In the case of slow neutrons the range correction is negligible. Therefore the only parameter which enters into the description of the neutron-nucleus interaction is the scattering length, a_f, which is connected to the cross section for scattering by free nuclei by the expression

$$\sigma_f = \int a_f^2\, d\omega^* = a_f^2 \int d\omega^* = 4\pi\, a_f^2 \tag{98.2}$$

where $d\omega^*$ is the solid angle in the c.m.s.

With this in mind a second remark about the scattering of slow neutrons by nuclei bound in molecules may be added. Provided the two following inequalities are fulfilled,

$$a_f \ll \lambda\!\!\!^{-}, \qquad a_f \ll \mathfrak{A}, \tag{98.3}$$

the effect of the neutron-nucleus interaction on both the plane wave describing the incident neutron (or better the relative motion of the neutron and the molecule) and the wave function describing the motion of the atom in the molecule, is sufficiently small to justify the application of perturbation theory.

[1] G. BREIT: Phys. Rev. **71**, 215 (1947).

In conclusion, provided conditions (98.1) and (98.3) are fulfilled, the cross section for elastic as well as inelastic collisions of neutrons with bound nuclei are given, with good accuracy, by the first Born approximation with the actual neutron-nucleus interaction replaced by the Fermi pseudopotential

$$V = 4\pi a_f \frac{\hbar^2}{2\mu} \delta(\boldsymbol{r}_n - \boldsymbol{r}_M) \tag{98.4}$$

where μ is the reduced mass of the two colliding particles the positions of which are defined respectively by the vectors $\boldsymbol{r}_n$ and $\boldsymbol{r}_M$, and $\delta(\varrho)$ is the three-dimensional Dirac δ-function.

The higher approximations calculated by BREIT give a contribution of only 0.3% [3]. If one considers that the chemical binding forces are always introduced by adopting some simplified model of the molecular structure of the moderator, one can conclude that FERMI's approximate procedure is sufficiently accurate for all practical purposes.

The scattering length appearing in (98.4) is a_f. In many cases, as for instance that of the scattering of neutrons by crystals (Sects. 99, 102, 103), it is more convenient to express the pseudopotential (98.4) in terms of a_b. According to (95.3) one has

$$V = 4\pi a_b \frac{\hbar^2}{2m} \delta(\boldsymbol{r}_n - \boldsymbol{r}_M). \tag{98.5}$$

In order to illustrate the use of FERMI's method and the origin of the *reduced mass effect* mentioned in Sect. 95 and used in the transformation of (98.4) into (98.5), let us consider the collision of a slow neutron with a nucleus of mass M belonging to a molecule of mass M_{mol}. The problem is that of deriving the differential scattering cross section for a process in which, because of the neutron-nucleus interaction, the internal motion of the molecule is excited from the initial state of energy w_1 to a final state of energy w_2 while the neutron is scattered from its initial momentum $\boldsymbol{p}_1$ to a final momentum $\boldsymbol{p}_2$ belonging to the solid angle $d\omega^*$. The momenta $\boldsymbol{p}_1$ and $\boldsymbol{p}_2$ refer to the relative motion of the neutron with respect to the centre of mass of the molecule, and $d\omega^*$ is the solid angle in the c.m.s. of neutron and molecule. The unperturbed wave function of the system in its initial state is given by the product of two factors

$$\exp[i\boldsymbol{K}_1 \cdot (\boldsymbol{r}_n - \varrho)] \cdot u_1(\boldsymbol{r}_M, \ldots). \tag{98.6}$$

The first factor is a plane wave representing the relative motion of the neutron and the molecule. The position of the neutron is $\boldsymbol{r}_n$, that of the centre of mass of the molecule is ϱ and the momentum of their relative motion is

$$\hbar|\boldsymbol{K}_1| = |\boldsymbol{p}_1| = \mu' v_1, \quad \mu' = \frac{m + M_{\text{mol}}}{m \cdot M_{\text{mol}}}.$$

The second factor appearing in (98.6) is the internal wave function of the molecule, which will depend on all the coordinates necessary to describe its internal motions. The position $\boldsymbol{r}_M$ of the atom of mass M with respect to the centre of mass of the molecule, is the only coordinate which has been indicated explicitly.

The final state of the system will be described by an expression similar to (98.6) with the subscript 2 instead of 1.

The standard Born approximation applied to the pseudo-potential (98.4) leads, after integration with respect to the neutron position, to the following expression for the differential scattering cross section

$$d\sigma_{12} = \left(\frac{\mu'}{\mu}\right)^2 a_f^2 \frac{v_2}{v_1} \left| \int e^{i\varkappa\cdot\boldsymbol{r}_M} u_2^*(\boldsymbol{r}_M, \ldots)\, u_1(\boldsymbol{r}_M, \ldots)\, d\boldsymbol{r}_M\, d\tau \right|^2 d\omega^*. \tag{98.7}$$

Here the integration runs over all the coordinates of the molecular wave function and

$$\boldsymbol{\varkappa} = \boldsymbol{K}_2 - \boldsymbol{K}_1 \tag{98.8}$$

is the momentum (in units of $\hbar$) transferred by the neutron to the molecule which must satisfy the conservation of energy law:

$$\left.\begin{aligned} \frac{p_1^2}{2\mu'} + w_1 - \left(\frac{p_2^2}{2\mu'} + w_2\right) &= \frac{\hbar^2}{2\mu'}\left(K_1^2 - K_2^2\right) + w_1 - w_2 \\ &= \frac{\hbar^2}{2\mu'}\left(\varkappa^2 + 2\boldsymbol{K}_1 \cdot \boldsymbol{\varkappa}\right) + w_1 - w_2 = 0. \end{aligned}\right\} \tag{98.9}$$

In order to derive an explicit expression for the cross section by means of (98.7) the molecular wave functions need to be known. A few models of the molecular structure, which in spite of their simplicity still reflect the more relevant features of the actual moderators, will be considered later.

A few general conclusions, however, can be drawn without specifying the form of the molecular wave functions in the two limiting cases of "very high" and "very low" energy of the incident neutron.

"Very high" energy means here that the energy of the incident neutron is so large (of the order of a few ev) that the binding energy of the nucleus in the molecule can be neglected. We then return to the case of elastic collisions against free nuclei; therefore $w_1 = w_2 = 0$, $\mu' = \mu$, and according to (98.9), $v_2 = v_1$.

Furthermore the internal molecular wave functions both reduce to the corresponding normalization factors so that the integral in (98.7) is different from zero and equal to 1 only when the collision takes place with conservation of momentum $\boldsymbol{K}_2 = \boldsymbol{K}_1$. When this condition is satisfied the cross section (98.7) turns out to be identical to (98.2).

The other limiting case of "very low" energy is that of neutrons whose energy is not sufficient to excite any molecular state. If the assumption is added that the molecules are all in their ground states, as would happen at very low temperature, energy exchanges will be impossible. Therefore the neutron will only undergo elastic collisions for which one has $v_2 = v_1$ and $u_2(\boldsymbol{r}_M, \ldots) = u_1(\boldsymbol{r}_M, \ldots)$. Furthermore, at such a low energy the wave-length is long compared to the molecular dimensions ($\boldsymbol{\varkappa} \cdot \boldsymbol{r}_M \sim \varkappa \mathfrak{A} \approx 0$) so that the exponential factor appearing in (98.7) can be replaced by unity. In conclusion the integral appearing in this expression is unity and the cross section is given by

$$d\sigma_b = \left(\frac{\mu'}{\mu}\right)^2 a_f^2\, d\omega^*, \qquad \sigma_b = 4\pi\, a_f^2 \left(\frac{\mu'}{\mu}\right)^2 = \sigma_f \left(\frac{\mu'}{\mu}\right)^2 \tag{98.10}$$

which exceeds the cross section with free nuclei by the factor $(\mu'/\mu)^2$.

For instance, in the case of collisions with protons ($M = m$) belonging to a molecule of mass M_{mol}, one has

$$\left(\frac{\mu'}{\mu}\right)^2 = \frac{4}{\left[1 + \frac{m}{M_{\mathrm{mol}}}\right]^2}. \tag{98.11}$$

For $M_{\mathrm{mol}}/m \gg 1$ the limit for $E \to 0$ of the neutron cross section is 4 times larger than that for free protons.

The factor $(\mu'/\mu)^2$ represents the limiting value for $E \to 0$ of the *chemical factor* which is defined, in a rather empirical way, as the ratio of the cross section of "thermal neutrons" colliding with bound atoms of a moderator to the cross section of slow neutrons colliding with free atoms of the same type. It can be calculated from (98.2) and (98.7) by averaging the latter expression over all possible types of collisions and over the actual spectrum of the incident neutrons. Such a calculation requires the adoption of a model of the chemical bond of the nucleus in the corresponding molecule.

The simplest model is the one originally used by FERMI [8] who represented the hydrogen atoms of an hydrogenous moderator as harmonic isotropic oscillators vibrating about fixed equilibrium positions. Under this assumption the wave function of the internal molecular motion is given by the product of the wave functions of three linear harmonic oscillators of the same frequency. The integration appearing in (98.7) can easily be done and yields for the elastic cross section

$$\left.\begin{aligned} d\sigma_{00} &= 4a_f^2\, e^{-y(1-\cos\vartheta)}\, d\omega \\ \sigma_{00} &= 4\pi a_f^2 \frac{1-e^{-4y}}{y} \end{aligned}\right\} \qquad y = \frac{E}{h\nu} \tag{98.12}$$

where ϑ is the angle of scattering in the l.s. of the incident neutron of energy E; and for the inelastic cross section

$$d\sigma_{0n} = 4a_f^2 \left(\frac{y-n}{y}\right)^{\frac{1}{2}} \frac{[2y-n-2y^{\frac{1}{2}}(y-n)^{\frac{1}{2}}\cos\vartheta]^n}{n!}\, e^{-(2y-n-2y^{\frac{1}{2}})(y-n)^{\frac{1}{2}}\cos\vartheta}\, d\omega. \tag{98.13}$$

This expression refer to neutrons which undergo an energy loss

$$w_n - w_0 = n\,h\,\nu$$

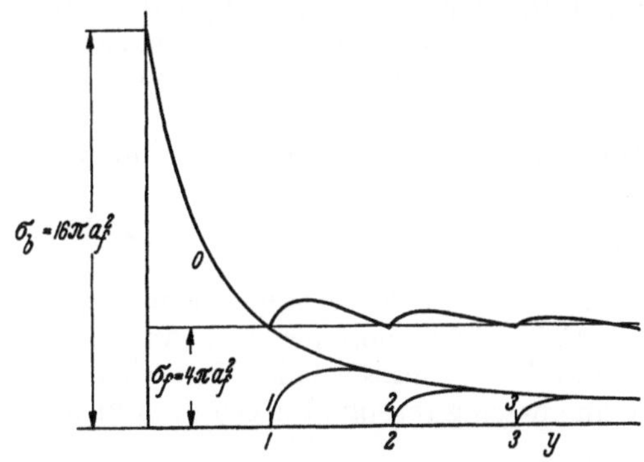

Fig. 175. Elastic and inelastic scattering cross sections for a neutron interacting with a harmonic isotropic oscillator of protonic mass (FERMI).

while the oscillator is excited from the ground state of energy $w_0 = \frac{3}{2}h\nu$ to the state of energy

$$w_n = n\,h\,\nu + w_0,$$
$$n = n_1 + n_2 + n_3 = 1, 2, \ldots.$$

In Fig. 175, there are plotted the elastic (marked 0) as well as the inelastic cross sections for $n = 1, 2, 3, \ldots$ as a function of $y = \frac{E}{h\nu}$. The total cross section

$$\sigma_{s\,\mathrm{tot}}(E) = \sum_{n=0}^{\infty} \sigma_{0n}(E),$$

given in the same figure, shows clearly the typical behaviour previously discussed in the general case; i.e., for $E=0$, the cross section $\sigma_{s\,\mathrm{tot}}(E)$ reaches the value (98.10) while for large E ($\gg h\nu$) it approaches the value (98.2) corresponding to the scattering by free nuclei (in the present case protons).

Attempts to obtain a better representation of the actual behaviour of a hydrogeneous moderator like paraffin have been made by BETHE [5] and ARLEY[1]. The model adopted by ARLEY is based on the following two assumptions only the first of which was used by BETHE:

(a) Each proton oscillates in a harmonic potential which is assumed to be anisotropic, since it can be derived from molecular spectra that the protons oscillate at a higher frequency in the direction of the valency-bond than in the

[1] N. ARLEY: Kgl. danske Vidensk. Selsk., math.-fys. 16, 1 (1938).

perpendicular direction. For paraffin ARLEY used $\nu_z = 3000\ \text{cm}^{-1} \rightarrow h\nu_z = 0.37$ ev; $\nu_x = \nu_y = \eta\, \nu_z$ with $\eta = 0.4$ i.e. $h\nu_x = h\nu_y = 0.148$ ev.

(b) In order to take into account the fact that the spectrum of the frequencies of the molecular motions in solid matter extends to quite small values, ARLEY assumes that the protons and their potentials can more freely like gas molecules with a Maxwell velocity distribution, so that the kinetic energy exchanges with these "molecules" are substituted for the energy exchanges between the neutrons and the small frequencies. The effective mass ascribed to these "molecules" is that of the group CH_2.

This latter assumption gives an oversimplified picture of the actual situation; it may, however, be adopted as long as the neutron energy is large compared to the energies corresponding to the low frequencies.

The theoretical results obtained from the models of BETHE and ARLEY are compared with the experimental data in the case of cetane in Sect. 101 (see Fig. 188).

SACHS and TELLER[1] have considered the scattering of neutrons by polyatomic gases for the special case where the energy of the neutron is at the same time: (a) so low that the bonds within the molecule may be considered rigid, and (b) so high that the neutron can excite high rotational levels[2].

The results of the collision of the incident neutron with a nucleus of the molecule, e.g. a proton, will generally consist of a change of the rotational and translational motion of the rigid molecule. Within the restricted energy range defined by the conditions mentioned above, the collision may be treated according to classical mechanics by associating with the struck nucleus a "hypothetical mass point" which moves at the same velocity $\boldsymbol{V}$ as the nucleus, has the same momentum $\boldsymbol{P}$ as the molecule, but whose mass is a tensor which depends on the mass and structure of the molecule.

This tensor expresses the linear relationship between the velocity $\boldsymbol{V}$ and the momentum $\boldsymbol{P}$

$$\boldsymbol{V} = \frac{1}{\mathsf{M}} \boldsymbol{P}. \tag{98.14}$$

Such a description is physically plausible because, for example, a neutron in a head-on collision with a proton of a CH_4 (rigid) molecule, in a direction perpendicular to the C—H bond, will feel an effective mass smaller than when it strikes the same proton moving along the line passing through the centre of mass of the molecule.

In order to derive the tensor $1/\mathsf{M}$, SACHS and TELLER note that the more general infinitesimal displacement of a rigid molecule (translation + rotation) may be separated into the sum of two orthogonal displacements of different kinds. The first is such that the struck nucleus remains fixed in space while in the displacements of the second type the angular momentum about the struck nucleus is zero. Because the incident slow neutron is always represented by an S wave, only displacements (translation + rotation) of the second type have to be considered.

The energy transferred to the translational + rotational motion of the molecule becomes

$$E_{TR} = \boldsymbol{V}_1 \cdot \hbar \boldsymbol{\varkappa} + \frac{1}{2}\, \hbar \boldsymbol{\varkappa}\, \frac{1}{\mathsf{M}}\, \hbar \boldsymbol{\varkappa} \tag{98.15}$$

[1] R. G. SACHS and E. TELLER: Phys. Rev. 60, 18 (1941).

[2] The spacing of the rotational levels of a molecule is determined by the constants $\hbar^2/2I_k$ where I_k are the principle moments of inertia. Therefore condition (b) is equivalent to requiring that the wavelength of the neutron is small compared to the linear dimensions of the molecule. It follows that under condition (b) interference between the waves scattered from different nuclei in the same molecule can be neglected.

where V_1 is the initial velocity of the scattering nucleus, $\hbar\varkappa$ the momentum transferred to the molecule, and $1/M$ is the *inverse mass tensor*

$$\frac{1}{M_{ii}} = \left(\frac{r_j^2}{I_k} + \frac{r_k^2}{I_j} + \frac{1}{M_{\text{mol}}}\right), \qquad \frac{1}{M_{ij}} = -\frac{r_i r_j}{I_k} \tag{98.16}$$

where I_i, I_j, I_k are the principle moments of inertia and r_i, r_j, r_k are the coordinates of the struck nucleus with respect to the principle axes of inertia of the molecule.

As will be shown in Sect. 106 the slowing down of neutrons in light and heavy water has been calculated following a simplified procedure related to the inverse mass tensor (98.16).

Because of the two conditions (a) and (b) the region of applicability of the procedure of SACHS and TELLER is rather limited.

MESSIAH[1] has extended the mass tensor treatment to higher energies by taking into account the vibrations of the molecule under the assumption that: (a) both kT_0 and the initial neutron energy E are large with respect to the separation of the rotational levels, and (b) E is below the threshold for excitation of higher vibrational levels. The general procedure for deriving the scattering cross sections consists in applying the first Born approximation with the pseudopotential (98.4). Under the assumptions mentioned above it is possible to separate from the matrix element, corresponding to the transition from the initial to the final state, the part involving the translational and rotational states of the molecule. The only role of this part is to guarantee the conservation of linear and angular momentum, i.e. to require that the energy transferred to the translational + rotational motion is given by the classical expression (98.15). Thus, the transition probability for the neutron to be scattered from the initial momentum $\boldsymbol{p}_1$ to the final momentum $\boldsymbol{p}_2$ within a given solid angle $d\omega$, with the molecule excited from the vibrational state $w_1 = w_0$ to the vibrational state w_2, is equal to the product of the square of the matrix element relative to the vibrational transition and to the density of final states satisfying the conservation of energy

$$\frac{p_1^2}{2m} + w_1 = \frac{p_2^2}{2m} + w_2 + E_{TR}.$$

The expression obtained by such a procedure for a given orientation of the molecule and a given value of the initial velocity V_1 of the target nucleus has to be averaged with respect to all possible orientations and over all the initial values of V_1 the distribution of which depends on the temperature T_0 of the scattering medium. Finally it has to be summed with respect to all possible vibrational transitions and elastic scattering.

The general expressions obtained by means of this procedure have been applied by MESSIAH to the H_2 and CH_4 molecules at room temperature ($T_0 = 300°$ K) for neutron energies below the threshold for inelastic collisions, where only elastic scattering has to be considered. A comparison with the experimental results of the Columbia group (see Sect. 101) in the case of CH_4 is given in Fig. 176.

The agreement is very satisfactory for energies between the threshold for inelastic collisions (first point on the left) and an energy of about 0.01 ev where the theoretical curve starts to deviate from the experimental points. Such a behaviour is expected because a neutron of 0.01 ev energy is able to transfer to only half of the molecules (at room temperature) the energy necessary for a transition to the next higher rotational level, and therefore the mass tensor approximation becomes questionable.

[1] A.M.L. MESSIAH: Phys. Rev. **84**, 204 (1951).

The disagreement is increased by decreasing the neutron energy and becomes considerable at about 0.004 ev where other approximations introduced in the actual derivation of the elastic cross section are no longer justified.

The scattering of neutrons by free molecules has been reexamined in great generality by ZEMACH and GLAUBER[1] who applied FERMI's pseudopotential (98.5) to molecules composed of N different atomes

$$V = 4\pi \frac{\hbar^2}{2m} \sum_{\nu=1}^{N} a_{b\nu}\, \delta(\boldsymbol{r}_n - \boldsymbol{r}_\nu).$$

Then, the differential cross section (98.7) generalized to the case of the transition from the initial state i to the final state f with the energy exchange

$$\varepsilon = w_i - w_f, \tag{98.17}$$

becomes

$$\frac{d\sigma_{if}(\vartheta)}{d\omega} = \sum_{\nu,\nu'=1}^{N} a_{b\nu} a_{b\nu'} \frac{K_f}{K_i} \langle \psi_i | e^{i\boldsymbol{\varkappa}\cdot\boldsymbol{r}_\nu} | \psi_f \rangle \langle \psi_f | e^{-i\boldsymbol{\varkappa}\cdot\boldsymbol{r}_{\nu'}} | \psi_i \rangle. \tag{98.18}$$

This expression should be summed with respect to all possible final states f, and averaged with respect to the occupation distribution of the initial molecular states i at the temperature T_0 of the medium (Sect. 95, p. 393). The sum over the final states should be restricted to transitions which fulfil the conservation of energy. This condition can be introduced by making use of the δ-function

$$\frac{d\sigma_i(\vartheta)}{d\omega} = \sum_f \int_{-\varepsilon_0}^{\infty} d\varepsilon\, \delta(w_i - w_f - \varepsilon) \frac{d\sigma_{if}}{d\omega}.$$

The lower limit of the integral is such that the final neutron energy $\hbar^2 K_f^2/2m$ is always non-negative.

ZEMACH and GLAUBER obtain the following expression

$$\left.\begin{aligned} d\sigma_i(\vartheta) &= \int_{-\varepsilon_0}^{\infty} d\varepsilon \frac{d\sigma_i(\varepsilon,\vartheta)}{d\omega} = \int_{-\varepsilon_0}^{\infty} d\varepsilon \cdot \sum_{\nu,\nu'=1}^{N} a_{b\nu} a_{b\nu'} \int_{-\infty}^{+\infty} e^{-i\frac{\varepsilon}{\hbar}t} \times \\ &\times \langle \psi_i | e^{i\frac{H}{\hbar}t} e^{i\boldsymbol{\varkappa}\cdot\boldsymbol{r}_\nu} e^{-i\frac{H}{\hbar}t} e^{-i\boldsymbol{\varkappa}\cdot\boldsymbol{r}_{\nu'}} | \psi_i \rangle dt \end{aligned}\right\} \tag{98.19}$$

by making use of the Fourier representation of the δ-function, introducing the molecular Hamiltonian H, and assuming that the final states ψ_f form a complete set. The eigenvalues are eliminated by using the relationships

$$\exp\left(-i\frac{w_f}{\hbar}t\right)\psi_f = \exp\left(-i\frac{H}{\hbar}t\right)\psi_f, \quad \text{etc.}$$

The expression $d\sigma_i(\varepsilon,\vartheta)/d\omega$ appearing under the integral with respect to ε, represents the differential cross section of the molecule in its initial state i for scattering of neutrons at the angle ϑ with an energy transfer ε.

Eq. (98.19) expresses the cross section for any energy transfer, in terms of the expectation values in the initial state of the molecule, and constitutes a convenient starting point for the development of various techniques to be used in treating the various molecular degrees of freedom. The corresponding Hamiltonians are sufficiently simple to permit exact or approximate computation of the operator products by elementary methods. Thus the explicit use of the molecular wavefunctions is avoided.

[1] A. C. ZEMACH and R. J. GLAUBER: Phys. Rev. **101**, 118 (1956).

We do not discuss here the devices used by the authors for the computation of the operator products appearing in (98.19) for which the reader is referred to the original papers[1,2].

The general procedure sketched above was illustrated by the same authors by treating a few simplified molecular systems; vibrating molecules (i.e. coupled harmonic oscillators), rotators (i.e. single point of mass M constrained to move on the surface of a sphere)and monoatomic gases.

As an example we recall that in the case of vibrating molecules, by performing the thermal averaging before computing the individual transition probabilities, ZEMACH and GLAUBER obtain[3]

$$\left\langle\frac{d\sigma(\varepsilon,\vartheta)}{d\omega}\right\rangle_{T_0}=\frac{K_f}{K_i}\sum_{\nu,\nu'=1}^{N}a_{b_\nu}a_{b_{\nu'}}e^{i\boldsymbol{\varkappa}\cdot(\boldsymbol{d}_\nu-\boldsymbol{d}_{\nu'})}W_{\nu\nu'},\qquad \varepsilon=\Sigma n_\lambda\hbar\omega_\lambda$$

where the symbol $\langle\ \rangle_{T_0}$ denotes averaging with respect to the occupation distribution of the initial states at the temperature T_0, $\boldsymbol{d}_\nu$ $(\nu=1,2,\ldots,N)$ represents the equilibrium position of the ν-th nucleus with respect to the centre of mass of the molecule and

$$W_{\nu\nu'}=\prod_\lambda\exp\left\{-\hbar\,\frac{2(\boldsymbol{\varkappa}\cdot\boldsymbol{c}_\nu^\lambda)^2+(\boldsymbol{\varkappa}\cdot\boldsymbol{c}_\nu^\lambda)^2}{4\hbar\omega_\lambda}\coth\left(\frac{\hbar\omega_\lambda}{2kT_0}\right)\right\}\times$$
$$\times\exp\left(-\frac{n_\lambda\hbar\omega_\lambda}{2kT_0}\right)I_{n_\lambda}\left(\frac{\hbar^2(\boldsymbol{\varkappa}\cdot\boldsymbol{c}_\nu^\lambda)\,(\boldsymbol{\varkappa}\cdot\boldsymbol{c}_{\nu'}^\lambda)}{2\hbar\omega_\lambda\sinh\left(\frac{\hbar\omega_\lambda}{2kT_0}\right)}\right).$$

Here $I_n(x)$ is the modified Bessel function and $\boldsymbol{c}_\nu^\lambda$ the amplitude vector of the displacement of the ν-th nucleus due to the λ-th vibrational mode of the molecule; it is defined by the relationship

$$\boldsymbol{u}_\nu(t)=\sum_\lambda\boldsymbol{c}_\nu^\lambda q_\nu(t)$$

where $\boldsymbol{u}_\nu(t)$ is the more general displacement of the ν-th nucleus from its equilibrium position, and $q_\lambda(t)$ is a set of normal coordinates ($\lambda=1,2,\ldots,2N-6$; or $\ldots,2N-5$ for linear molecules).

These procedures have been extended by the same authors to include the spin dependence of neutron scattering and the spin correlations within the same molecule; they have been applied to the scattering of neutrons by CH_4 molecules[4]. In this case the total cross section was found to be in very good agreement with both the experimental results and the theoretical computations by MESSIAH (Fig. 176). The corresponding differential scattering cross section represents very well the experimental results of ALCOCK and HURST[5] and therefore constitute an appreciable improvement with respect to the theoretical computations made by other authors[5,3].

We shall come back to this point at the end of Sect. 101 where the *rigid molecule* and the *vibrating molecule* models will be considered for discussing the case of molecules with large momenta of inertia, like CF_4 (Fig. 190).

The application of the procedure of ZEMACH and GLAUBER to more complicated molecules, like for example H_2O, is however considerably involved. There-

[1] G. C. WICK: Phys. Rev. **94**, 1228 (1954).
[2] See footnote 1, p. 407.
[3] Equivalent results had been obtained by a more involved procedure by N. K. POPE, Canad. J. Phys. **30**, 597 (1952).
[4] A. C. ZEMACH and R. J. GLAUBER: Phys. Rev. **101**, 129 (1956).
[5] N. Z. ALCOCK and D. G. HURST: Phys. Rev. **75**, 1609 (1949); **83**, 1100 (1951).

fore KRIEGER and NELKIN[1] have tried to use the formalism of the previous authors as starting point for approximate calculations of the scattering by a gas of molecules, for neutron energies below the vibrational threshold, but large compared with the rotational level separation; these are just the assumptions adopted by MESSIAH in above mentioned works. The method of KRIEGER and NELKIN gives results which are even in better agreement with experiments than

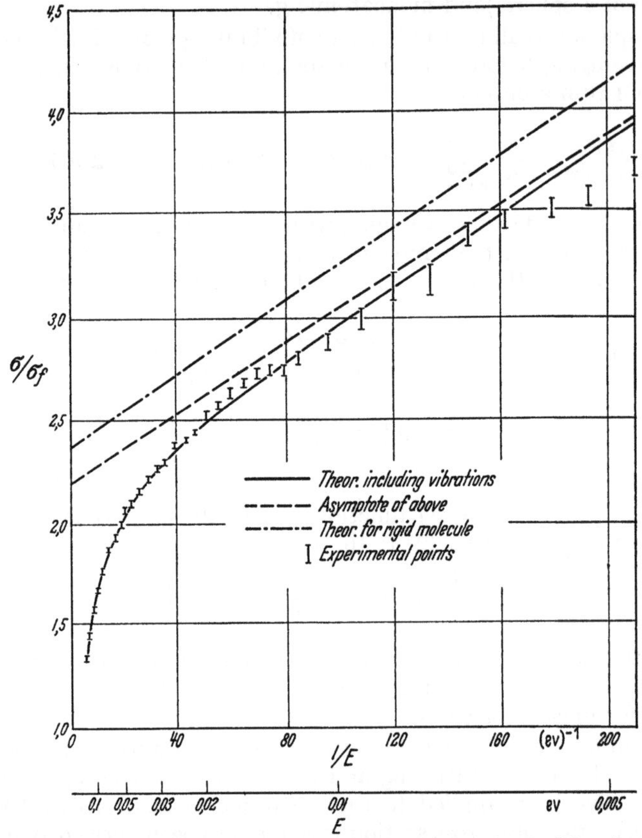

Fig. 176. Energy dependence of the slow neutron scattering cross section per proton by CH_4 molecules at room temperature ($T = 300°$K). First point on the left indicates the energy threshold for inelastic collision. Experimental points by Columbia University group; theoretical curves by MESSIAH.

those of MESSIAH when they are applied to the computation of the total cross section of the CF_4 molecule (Fig. 176); but they deviate violently from experiments in the case of H_2O molecule (Fig. 189). In this case the models considered in Sect. 101α appear to allow a better representation of the experimental data on the total cross section, especially if a judicious choice of the parameters is made (Fig. 188, p. 427). The paper of KRIEGER and NELKIN contains a derivation, from the expression of ZEMACH and GLAUBER, of the above mentioned approximate classical procedure of SACHS and TELLER and of the oversimplified model of BROWN and ST. JOHN (Sect. 106).

99. The separation of the scattering cross section of an individual nucleus into a coherent and an incoherent part. The temperature diffuse scattering. As

[1] T. J. KRIEGER and M. S. NELKIN: KAPL-1597 (1956); KAPL-1597 Addendum No. 1 (1957).

we have already seen in Sect. 95, when a neutron of wavelength (95.4) comparable to interatomic distances, moves through a medium composed of arrays of atoms such as molecules or crystals, interference effects start to become important. Since these effects are particularly pronounced for crystals, we shall refer in the following to the collision of a slow neutron with a non-absorbing micro-cystal which later (Sect. 102) will be considered as part of a polycrystalline substance. If not all, at least the majority of the considerations given below also apply with only minor changes, to the case of molecules.

Since slow neutrons have wavelengths comparable to those of X-rays, it is to be expected that the diffraction effects observed when neutrons fall on a crystal will be similar to those observed with X-rays[1]. Important differences do exist however between these two cases; these arise mainly from the following two facts. While X-rays are scattered by atomic electrons, neutrons are scattered by nuclei. Furthermore while the mass of the scattered particle is zero in the case of X-rays, it has a non-negligible value in the case of neutrons. Some consequences of the first remark will appear in the course of the present discussion, while the second point will be illustrated in Sect. 103.

In this section we shall discuss the distinction between coherent and incoherent scattering which is well known from the classical cases of light, X-rays and electrons.

Such a distinction may be clarified by considering a system of scatterers, possibly with a complex internal structure, whose centres are located at certain positions $\boldsymbol{r}_1, \boldsymbol{r}_2, \ldots \boldsymbol{r}_N$[2]. When a monocromatic incident wave hits the system, the intensity of the scattered wave is obtained by squaring the total amplitude. Since this may be expressed as the sum of the amplitudes of the partial waves originating from the single scatterers, the total intensity will generally differ from the sum of the intensities of the partial waves because of *cross terms* which give rise to interference effects. The importance of these effects depends on the structure of the system as it may be seen from the following examples.

If the scattered waves bear all a definite phase relation to the incident wave, so that they can interfere with it, the scattering is described as *absolutely coherent*. A situation of this type is met, for example, in the case of a system satisfying the following conditions: the scatterers of the system are all identical, they are all located at strictly fixed positions and no change of the internal state of the system (or of part of it) takes place during the scattering process. The scattered waves are absolutely coherent even if the scatterers are not all identical, provided that the locations of the different types of scatterers are known in detail.

In the case of actual crystals, however, one has always to deal with statistical distributions of properties as well as of locations of the scatterers, so that certain averages have to be taken. Incoherent scattered waves arise from any random feature of locations or of properties of the scatterers. As an example of the first type of incoherence we recall that of slow neutron scattering by a monatomic gas; in this case the coherent scattering does not exist at all because of the completely random nature of the distribution of the locations of the scattering centres. An example of incoherent scattered waves arising from a random distribution of a property of the scattering centres will be shown below in the case of crystals containing an element composed of different isotopes.

---

[1] Cf. the treatment of neutron diffraction and X-ray diffraction in Vol. XXXII of this Encyclopedia.

[2] M. LAX: Rev. Mod. Phys. 23, 287 (1951).

An essentially different type of incoherent radiation, called *strictly incoherent*[1,2], is observed when a change in the *internal state of a single scatterer* is involved in the collision process. The most obvious example is that of waves scattered inelastically by an individual scatterer, because of the different frequency, which makes any interference impossible not only with the incident wave but even with the waves scattered (elastically or inelastically) by other scatterers.

Another example of strictly incoherent radiation will be shown below in the case of a crystal composed of a monoisotopic element with nuclear spin $\boldsymbol{I} \neq 0$, if the spin of the neutron flips in the collision and the spin $\boldsymbol{I}$ of a *single nucleus* undergoes a change of one unit due to conservation of total angular momentum.

The example mentioned above of waves scattered inelastically by a single scatterer must not lead us to believe that all waves scattered inelastically by a crystal are incoherent. As we shall see in Sect. 103 there are processes in which the incident neutron exchanges energy not with a single nucleus, but with one of the elastic waves of the lattice; in such a case the inelastically scattered wave is described as *relatively coherent* because it can not interfere with the incident radiation in consequence of its different frequency, but still gives rise to interference effects.

Thus we see that the coherent or incoherent character of the scattered waves arises from the more or less regular structure of the crystal as a whole, and not from the properties of the individual nuclei.

In dealing with phenomena of this type it became customary, however, to split the scattering cross section of a neutron with an individual nucleus[3] belonging to a crystal, into two parts

$$\sigma_s = \sigma_{\text{coh}} + \sigma_{\text{incoh}}. \tag{99.1}$$

The *coherent scattering cross section* σ_{coh} is defined as the cross section for that type of scattering which produces waves capable of interfering; the *incoherent scattering cross section* σ_{incoh} corresponds on the contrary to waves which cannot give rise to interference phenomena. Such a distinction is unnecessary and in some way arbitrary but it has some merits from the practical point of view and, therefore, it has been adopted in the interpretation and discussion of almost all the experimental results on this subject. The arbitrariness of the separation (99.1) lies mainly in the fact that from the elastic scattering cross section σ_s which has a well defined physical meaning, one separates the coherent cross section σ_{coh}. The latter represents the cross section which one has to attribute to the nucleus, if bound in an ideal crystal, in order that it should give the same amount of interference which one observes with the real crystal. The incoherent cross section σ_{incoh}, defined as the difference between σ_s and σ_{coh}, is responsible for the rest of the scattered radiation which constitutes a kind of diffuse background.

The ideal crystal is defined by the three following conditions:

(a) it contains only monoisotopic elements;

(b) all the nuclei of these elements have spin zero;

(c) the crystal is rigid so that the displacements of the nuclei from their equilibrium positions are zero both before as well as after the collision.

---

[1] G. Breit: Rev. Mod. Phys. **5**, 91 (1933).

[2] See footnote 2, p. 410.

[3] Because of the magnetic interaction of the neutron with the atomic magnetic moment, it would be more general to refer the definitions of the various scattering cross sections to an atom rather than to a nucleus. The definitions are here referred to a nucleus because in the following discussion only nuclear interactions are considered.

With this picture in mind one usually says that three different sources of incoherent scattering arise from deviations from the three conditions (a) to (c) listed above[1]. Deviations from conditions (a) and (b) are called respectively isotopic incoherence and spin incoherence. These effects are typical of the case of neutrons since they originate from neutron-nucleus interactions[2]. If condition (c) is dropped, inelastic collisions between the neutron and the crystal become possible, and give rise to a radiation, which is well known from the classical case of X-rays as the *temperature diffuse scattering*. As we shall see later, the corresponding waves may be incoherent or relatively coherent according to the structure of the system of scatters.

Isotopic incoherence can be illustrated by considering the ideal case of a crystal satisfying conditions (b) and (c), but deviating from condition (a). The simplest case is that of a crystal containing various elements, all monoisotopic with the exception of one, which is composed of two isotopes of relative abundances c_1 and $c_2 (c_1 + c_2 = 1)$. In these circumstances the wave scattered coherently by an individual nucleus of this element has an amplitude equal to the isotopic average of the amplitudes a_{b1} and a_{b2} of the two isotopes, and therefore the corresponding cross section is

$$\sigma_{\text{coh}} = 4\pi \, |c_1 a_{b1} + c_2 a_{b2}|^2. \tag{99.2}$$

Since the two isotopes are not distributed regularly, the crystal cannot be considered as perfect for neutron scattering. The disorder introduced by the random distribution of the isotopes gives rise to an additional scattering of incoherent type whose cross section in the present ideal case is obtained, according to (99.1), as the difference between the elastic scattering cross section σ_s and the coherent elastic cross section (99.2).

Independently of the structure of the sample (i.e. if it is a gas, an amorphous solid or a crystal) the total elastic scattering cross section σ_s per nucleus of the element under consideration is given by the isotopic average of the scattering cross sections of the individual isotopes

$$\sigma_s = 4\pi \left\{ c_1 |a_{b1}|^2 + c_2 |a_{b2}|^2 \right\}. \tag{99.3}$$

Introducing (99.2) and (99.3) in (99.1) we obtain

$$\sigma_{\text{incoh}} = 4\pi \, c_1 c_2 \, |a_{b1} - a_{b2}|^2. \tag{99.4}$$

From this expression we see that the scattering cross section due to pure isotopic incoherence is proportional to the square of the modulus of $a_{b1} - a_{b2}$ and to the product of the abundances of the two isotopes. If both of these have the same scattering length $a_{b1} = a_{b2}$, σ_{incoh} is equal to zero. In order to have an appreciable scattering due to this cause, not only must the difference $a_{b1} - a_{b2}$ be large, but also the relative abundances c_1 and c_2 of the two isotopes must be of the same order of magnitude.

Spin incoherence corresponds to deviations from condition (b). It arises from the fact that the interaction of a slow neutron with a nucleus of spin $\boldsymbol{I}$ in general depends on the orientation of the neutron spin $|\boldsymbol{S}| = \frac{1}{2}$ with respect to $\boldsymbol{I}$. This phenomenon may be illustrated by considering the ideal case of a crystal satisfying

[1] We do not consider other types of incoherence as for example that arising from random orientation of the atomic moments in a paramagnetic material. These phenomena are outside the scope of the present chapter. Cf. the article by R. J. BLIN-STOYLE and M. A. GRACE in Vol. XLII of this Encyclopedia. See also footnote 3 on p. 412.

[2] G. C. WICK: Phys. Z. 38, 403 (1937).

condition (c) and composed of various monoisotopic elements all with zero nuclear spin, with the exception of one which has spin $\boldsymbol{I}$. Indicating with a_{b+} and a_{b-} the scattering lengths corresponding to the scattering of neutrons having the spin $\boldsymbol{S}$ parallel or antiparallel with respect to $\boldsymbol{I}$, and assuming a random distribution of the orientations of both $\boldsymbol{S}$ and $\boldsymbol{I}$, the probabilities of the two possible states are respectively

$$f_+ = \frac{I+1}{2I+1}, \qquad f_- = \frac{I}{2I+1}. \tag{99.5}$$

By substituting f_+ and f_- for c_1 and c_2 and at the same time a_{b+} and a_{b-} for a_{b1} and a_{b2} in (99.2), (99.4), and (99.3), we obtain the various cross sections per nucleus of the element under consideration

$$\left.\begin{aligned} \sigma_{\text{coh}} &= 4\pi\,|f_+ a_{b+} + f_- a_{b-}|^2, \\ \sigma_{\text{incoh}} &= 4\pi f_+ f_- \,|a_{b+} - a_{b-}|^2, \\ \sigma_s &= 4\pi\,\{f_+ |a_{b+}|^2 + f_- |a_{b-}|^2\}. \end{aligned}\right\} \tag{99.6}$$

Although the expressions obtained by this procedure are correct, attention must be paid to the fact that spin incoherence is essentially different from isotopic incoherence because a certain fraction of the spin dependent scattering takes place with a reversal of the neutron spin $\boldsymbol{S}$ and a simultaneous change of one unit in the spin $\boldsymbol{I}$ of the struck nucleus. According to the definition given above, the waves scattered in this process are strictly incoherent. The spin flip probability P is given by [*9*], [*18b*]

$$P = \frac{2}{3}\,\frac{\sigma_{\text{incoh}}}{\sigma_s}$$

and can reach its maximum value of $\frac{2}{3}$ for completely incoherent scattering. This process has been investigated experimentally in various cases and, in particular, in the case of hydrogen[1] which shows a strong spin-dependent interaction [Sect. 29, Eq. (29.13)]:

$$a_3 = a_{f+} = 0.54 \times 10^{-12}\,\text{cm}, \qquad a_1 = a_{f-} = -\,2.40 \times 10^{-12}\,\text{cm}.$$

In the general case when both assumptions (a) and (b) are not fulfilled and therefore isotopic and spin incoherence are both present, σ_s, σ_{coh}, and σ_{incoh} are given by the expressions

$$\left.\begin{aligned} \sigma_s &= 4\pi\,\langle a_b^2\rangle, \\ \sigma_{\text{coh}} &= 4\pi\,\langle a_b\rangle^2, \\ \sigma_{\text{incoh}} &= 4\pi\,\{\langle a_b^2\rangle - \langle a_b\rangle^2\}, \end{aligned}\right\} \tag{99.7}$$

with the symbol $\langle\cdots\rangle$ denoting averages over spin orientations and isotopic composition.

Finally we shall consider the case of a crystal satisfying condition (a) and (b) but not (c). In other words, we allow the atoms of the ideal crystal (composed of monoisotopic elements with zero nuclear spin) to move around their equilibrium positions in consequence of thermal agitation and zero-point energy. Under these conditions two new effects take place: The first one may be described as a reduction of the intensity of the Laue-Bragg coherent scattering. The second effect is the appearance of inelastic scattering in which the neutron exchanges energy with a more or less extended part of the crystal. This radiation constitutes the so-called *temperature diffuse scattering*.

1 W. E. MEYERHOF and D. B. NICODEMUS: Phys. Rev. **82**, 5 (1951).

A quantitative discussion of the problem involves the frequency spectrum of the lattice vibrations

$$2\pi\nu_j = \omega_j = \omega_j(\boldsymbol{q}) \tag{99.8}$$

as a function of the corresponding propagation vector $\boldsymbol{q}$ and polarization j ($j = 1, 2, 3$ corresponds to two transversal and one longitudinal vibration).

The various approximate procedures usually adopted for the description of crystal properties, differ one from the other mainly in the assumptions made about the function (99.8).

In the Einstein model of a crystal the frequency ω_j is assumed to be a constant independent of $\boldsymbol{q}$ and j. In this very rough model a crystal composed of N atoms is described as $(3N-6) \approx 3N$ quantized harmonic oscillators all of the same frequency ω_0; the characteristic Einstein temperature is given by $\Theta_E = \hbar\omega_0/k$.

A better approximation is provided by the Debye model which replaces: (a) the dispersion relation (99.8) by a continuous spectrum of frequencies whose velocity of propagation is a constant independent of $\boldsymbol{q}$ and j, equal to its limiting value for $\lambda \to \infty$ (i.e. equal to the velocity of sound c)

$$\omega_j = c\,|\boldsymbol{q}| \tag{99.9}$$

and (b) the cell of the recoprocal lattice (see Sect. 102) by a sphere of radius

$$\frac{q_0}{2\pi} = \left(\frac{3}{4\pi v_0}\right)^{\frac{1}{3}} \tag{99.10}$$

where v_0 is the volume of the unit cell of the crystal.

The continuous spectrum extends from $\omega = 0$ to a maximum value $\omega_{\max}$ determined by imposing the condition which gives the correct number of degrees of freedom of the crystal. This is characterized by the Debye temperature $\Theta_D = \hbar\omega_{\max}/k$.

Finally, more realistic models can be obtained by intoducing simple expressions for the interatomic forces in the crystal depending on a few constants which have to be determined. These are adjusted to give the best fit to the experimentally determined values of the elastic constants of the crystal. This procedure has been envisaged by PLACZEK and VAN HOVE[1] and adopted by SQUIRES[2]. For instance in the case of scattering of slow neutrons by aluminium crystals (face-centered cubic lattice) Squires assumes that the interatomic forces are harmonic and different from zero only between first and second nearest neighbours. On these assumptions the most general set of forces is specified by five force constants which are determined, as explained above, from the experimental values of elastic constants relative to a single crystal of aluminium. We shall come back to this rather refined procedure in connection with the discussion of inelastic scattering by crystals (Sect. 103, p. 456).

The problem of the coherent scattering by a crystal satisfying conditions (a) and (b) has been treated by HALPERN, HAMERMESH and JOHNSON[3] and by WEINSTOCK[4] in the Debye approximation, which, in this case, is expected to give reliable results[1,2] (Sect. 103). The result may be described as a reduction of the amplitude a_b of the wave scattered coherently by an individual nucleus of the crystal

$$a_b \exp\left[-W(T_0)\,\frac{\sin^2\vartheta}{\lambda^2}\right]. \tag{99.11}$$

[1] G. PLACZEK and L. VAN HOVE: Nuovo Cim. **1**, 233 (1955).
[2] G.L. SQUIRES: Phys. Rev. **103**, 304 (1956).
[3] O. HALPERN, M. HAMERMESH and M.H. JOHNSON: Phys. Rev. **59**, 981 (1941).
[4] R. WEINSTOCK: Phys. Rev. **65**, 1 (1944).

Here λ is the wavelength of the neutron, 2ϑ the angle of scattering and the exponential factor is the *Debye-Waller factor*[1,2] known from the case of X-rays. It depends on the monotonically increasing function of T_0/Θ_D

$$W(T) = \frac{6h^2}{M k \Theta_D} \Phi\left(\frac{T_0}{\Theta_D}\right) \tag{99.12}$$

where

$$\Phi\left(\frac{T_0}{\Theta_D}\right) = \frac{1}{4} + \left(\frac{T_0}{\Theta_D}\right)^2 \int\limits_0^{\Theta_D/T_0} \frac{x\,dx}{e^x - 1}. \tag{99.13}$$

Tables and graphs of the function (99.13) may be found in various X-ray textbooks and articles[1,2]. It varies from 0.25 to 0.6 for T_0/Θ_D increasing from 0 to 0.55, and then goes on tending asymptotically to T_0/Θ_D.

When the unit cell of the crystal contains two or more atoms, one has to introduce the *structure amplitude* F defined by Eq. (102.12) as the amplitude of the wave scattered coherently by the unit cell of the crystal. In this case many authors attach the Debye-Waller factor to F, rather than to the scattering lengths of the individual nuclei appearing in (102.12), and write the coherent cross section per unit cell in the form

$$(d\sigma_{\text{u.c.}})_{\text{coh}} = |F|^2 \exp\left[-2W(T_0)\,\frac{\sin^2\vartheta}{\lambda^2}\right] d\omega. \tag{99.14}$$

Since (99.12) contains the mass M of the nucleus, such a procedure is permissible only if all the nuclei contained in the unit cell have at least approximately the same mass.

Besides the reduction of intensity of the Laue-Bragg coherent elastic scattering, the thermal agitation of the crystal gives rise to inelastic scattering, the angular dependence of which is determined by the existence of more or less pronounced correlations between the displacements of the various nuclei present in the crystal[3,4]. If one assumes that the thermal vibrations are distributed at random so that there is no correlation at all between the displacements of different nuclei, even when neighbours, the temperature diffuse scattering per unit cell is given by the expression

$$(d\sigma_{\text{u.c.}})_{\text{temp. diff.}} = |F|^2 \left\{1 - e^{-2W(T_0)\frac{\sin^2\vartheta}{\lambda^2}}\right\} d\omega. \tag{99.15}$$

The intensity of this radiation is not isotropic as were the spin and isotropic incoherent radiations considered above; it increases with increasing 2ϑ, the angle between the direction of observation and that of the incident neutrons.

The extreme assumption of displacements distributed at random is certainly not correct; it is more reasonable to assume the existence of a correlation between the displacements of the nuclei located at positions $\boldsymbol{r}_i$ and $\boldsymbol{r}_j$, which is a function smoothly decreasing as the distance $(\boldsymbol{r}_i - \boldsymbol{r}_j)$ increases. Under these conditions the temperature diffuse scattering is expected to exhibit broad interference maxima in the vicinity of the Laue-Bragg directions. These maxima are broad because of the small "effective crystal size" due to the decrease of the correlation with distance; they are expected to be in the vicinity of the Bragg directions

---

[1] Cf. M. Blackman: The specific heat of solids. This Encyclopedia, Vol. VII, part 1, especially Sect. 27, p. 377.

[2] S.K. Allison and A.H. Compton: X-rays in theory and experiment, p. 437. London: McMillan 1935. — F.C. Blake: Rev. Mod. Phys. **5**, 169 (1933).

[3] M. Born and V. Kármán: Phys. Z. **14**, 15, 65 (1913).

[4] See footnote 2, p. 410.

because the propagation vector $\boldsymbol{q}$ of the elastic waves, which contribute more intensely to inelastic scattering, is very small with respect to the propagation vector of the neutron [see Eq. (103.20)].

The behaviour described by (99.14) and (99.15) is shown in Fig. 177 where a plott is given of the results obtained by SHULL and WOLLAN[1] in the case of lead (face-centered cubic lattice) using the powder method. The neutron diffraction pattern can be divided into a series of Bragg peaks due to the interference of the coherent scattered waves (Sect. 102), superimposed on a diffuse background increasing with the angle ϑ. Since in lead the coherent scattering cross section per nucleus has been found to differ negligibly from the total cross section per nucleus, both isotopic and spin incoherence can be ignored. Therefore the background shown in Fig. 177 may be attributed to temperature diffuse scattering superimposed on multiple scattering effects. By determining empirically these effects, SHULL and WOLLAN succeeded in comparing, on an absolute

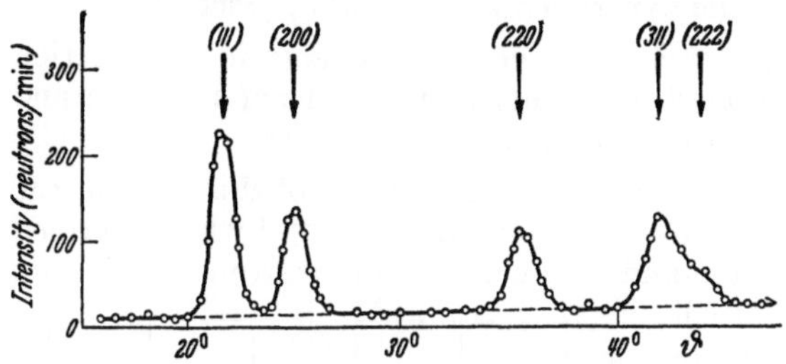

Fig. 177. Neutron diffraction pattern for powdered Pb (SHULL and WOLLAN).

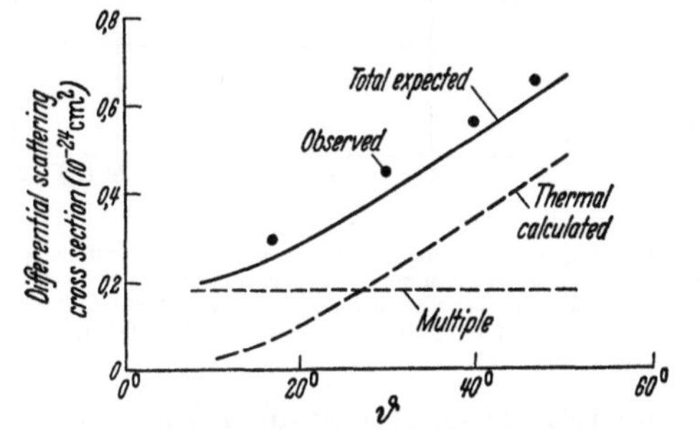

Fig. 178. Comparison between calculated and observed thermal diffuse scattering for powdered Pb (SHULL and WOLLAN).

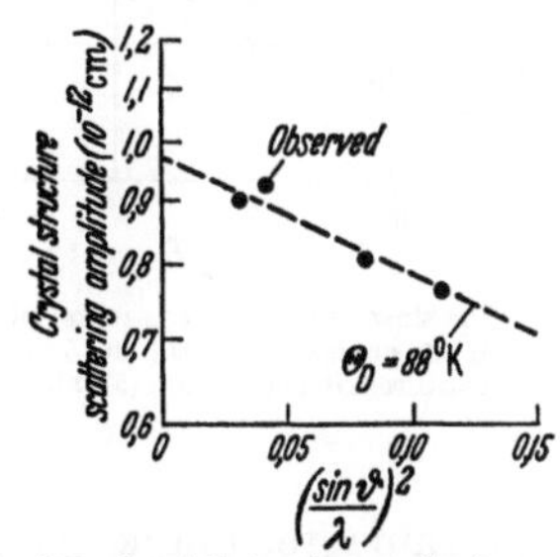

Fig. 179. Crystal structure scattering amplitude for the Pb diffraction peaks of Fig. 177 (SHULL and WOLLAN).

scale, the temperature diffuse scattering computed by means of (99.15) with the corresponding experimental results (Fig. 178). Furthermore, by comparing the measured intensities of the successive Bragg peaks of Fig. 177 with the corresponding values computed for an ideal rigid crystal [i.e. a perfect crystal satisfying conditions (a), (b) and (c)] these same authors succeeded in measuring the Debye-Waller factor as a function of $\left(\frac{\sin\vartheta}{\lambda}\right)^2$ and compare it with the corresponding theoretical value deduced from (99.14) and (99.12) (Fig. 179). The very satisfactory agreement between the theory and the experimental results shown in both Fig. 178 and 179 was obtained by making use, for the Debye temperature of lead, of the value $\Theta_D = 88°$ K as derived from specific heat measurements.

II. The scattering of slow neutrons by molecules.

100. The influence of molecular structure on the scattering cross section for neutron energies above 1 ev. PLACZEK[2] was the first to point out that the scatter-

[1] C. G. SHULL and E. O. WOLLAN: Phys. Rev. **81**, 527 (1951).
[2] G. PLACZEK: Phys. Rev. **86**, 377 (1952).

ing cross section of neutrons of energy larger than the spacing of molecular levels still deviates appreciably from the cross section for free nuclei σ_f (98.2). As will be shown later [see Eq. (100.1)] this deviation is mainly due to a Doppler effect whose importance is expected to depend on the mass of the scatterer: the lighter the scatterer, the more important is the effect. This is the reason why this effect is appreciable in collisions against bound protons, a case which deserves special consideration because of the direct information on the $^1S_0$ state of the n-p system which is derived from the experimental value of σ_f (Sect. 29).

In the case of heavier scatterers, such as O and N, the change in cross section due to this molecular effect is expected to be only a few tenths of a percent and is therefore not detectable with existing experimental techniques.

The first experimental evidence of an energy dependence of this type for the neutron-proton scattering cross section was presented by JONES[1] who performed transmission measurements in water and cyclohexane for neutron energies up to 100 ev by means of the Cornell University time-of-flight velocity spectrometer.

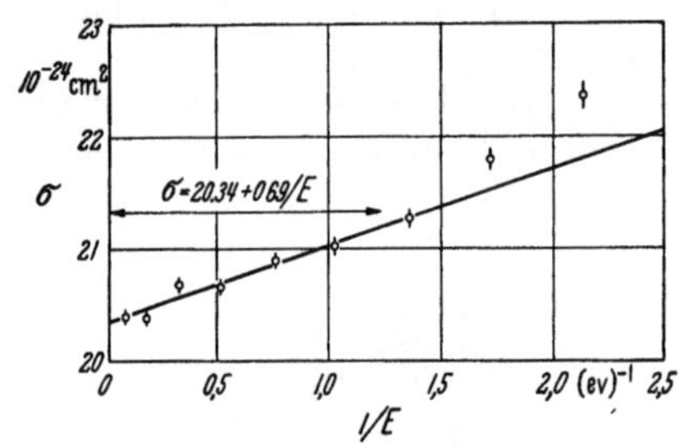

Fig. 180. The slow neutron cross section of protons bound in H_2 gas in the energy region 0.5 to 15 ev. Correction for capture has been made (MELKONIAN).

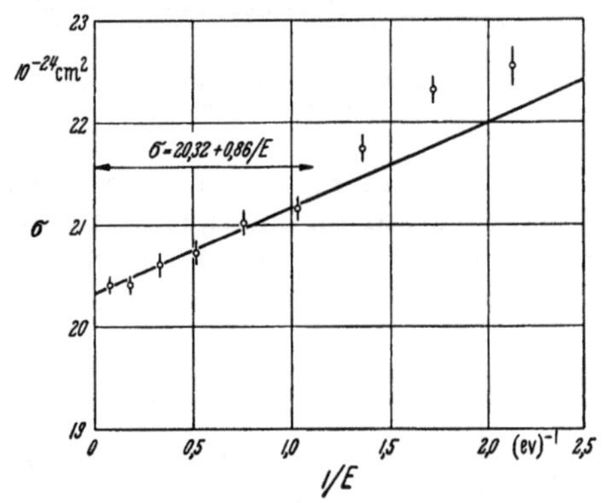

Fig. 181. The slow neutron cross section of protons bound in gaseous n-butane in the energy region 0.5 to 15 ev. The carbon cross section was taken as 4.70 barns. Correction for capture has been made (MELKONIAN).

The same technique was used some time later at Columbia University by MELKONIAN[2], who measured the transmission of gas samples of hydrogen and n-butane and of liquid samples of water. In Figs. 180 to 182 the hydrogen cross section deduced in these three cases has been plotted against $1/E$ for E varying between about 0.5 and 15 ev.

In the case of water and n-butane, the cross sections of O and C were taken as constants and equal to $\sigma_0 = 3.73$ barns, $\sigma_c = 4.70$ barns. Correction has been made for capture of neutrons by protons assuming a $1/v$-law with the capture cross section $\sigma_a = 0.3$ barns for $E = 0.025$ ev. The straight lines shown in these figures have been obtained by means of the method of least squares, using the data above 0.8 ev. The good fit of the experimental points obtained in this energy region is in agreement with PLACZEK's theoretical prediction and allow a determination of σ_f as the extrapolated value corresponding to $1/E = 0$. In Figs. 180 and 181 a few points at energies below 0.8 ev have been included in order to show the beginning of the deviation from the $1/E$ behaviour (see Sect. 101).

A similar plot of the cross section in the case of H_2 is given in Fig. 183 for neutron energies greater than 15 ev in order to show that the proton cross section remains practically constant between 15 and a few thousand electronvolts.

[1] W.B. JONES: Phys. Rev. **74**, 364 (1948).
[2] E. MELKONIAN: Phys. Rev. **76**, 1744 (1949).

As we have mentioned before, the linear behaviour of the scattering cross section shown in Figs. 180 to 182 had been predicted by PLACZEK[1,2] who treated the scattering of neutrons by systems of N nuclei like molecules or crystals; he made use of the first Born approximation with the Fermi pseudopotential (98.5) describing the interaction between the neutron and the nuclei. As has been shown in the special case of a proton bound by a harmonic isotropic potential (Sect. 98) the cross section is in general a complicated function of the neutron energy (as an example, see Fig. 175). In particular the slope of this function changes abruptly whenever E attains the threshold for the excitation of a new level of the molecule. PLACZEK has shown however that with the exception of the case of a harmonically and isotropically bound proton, which will be considered later, for E large compared to the level spacing Δ of the scattering system, these

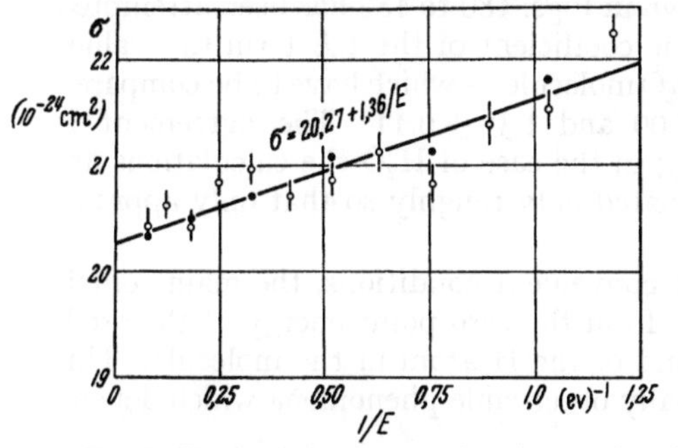

Fig. 182. The slow neutron cross section of protons bound in H_2O in the energy region 1 to 15 ev. The oxygen cross section was taken as 3.73 barns. Correction for capture has been made (MELKONIAN).

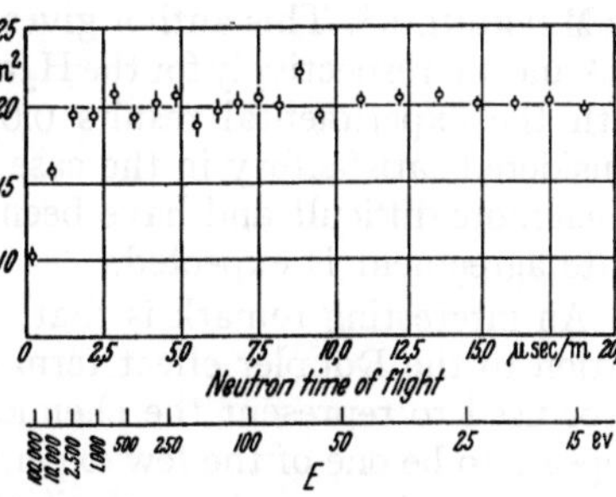

Fig. 183. The slow neutron cross section of protons bound in H_2 for neutron energy greater than 15 ev (MELKONIAN).

changes of slope become negligible. Thus σ_s behaves as a smooth function of E which can be represented by means of simple asymptotic formulas describing the approach to the limiting value (98.2). These formulas derived by PLACZEK for an array of molecules (or crystal) of random orientation, under the assumption that the nuclei involved are heavy compared to the neutron, give the asymptotic behaviour of both the total and differential cross sections as a series in powers of $1/E$. The actual calculation of the coefficients needs the specification of a model for the chemical binding forces of the atoms in the molecule.

The results of PLACZEK have been extended by WICK[3] with only minor modifications, to molecules containing light nuclei, by making use of what is called the "short collision time approximation", i.e. under the assumption that the duration of the collision is short compared to the natural periods of the molecule.

In the case of molecules composed of a single nucleus bound by a fixed potential $U(r)$, the Placzek-Wick asymptotic expression of the total scattering cross section is

$$\sigma_s(E) = \sigma_f \left\{1 + \frac{1}{3}\,\frac{m}{M}\,\frac{K_{\mathrm{av}}}{E} - \frac{1}{32}\left(\frac{M+m}{M}\right)^2 \frac{m}{M}\,\frac{C_{\mathrm{av}}}{E^3} + \cdots\right\} \tag{100.1}$$

where K is the kinetic energy of the nucleus,

$$C = \frac{1}{3}\,\frac{\hbar^2}{M}\,(\mathrm{grad}\,U)^2,$$

[1] See footnote 2, p. 416.
[2] See also MESSIAH: J. Phys. Radium 12, 670 (1951). — Phys. Rev. 84, 204 (1951)
[3] G.C. WICK: Phys. Rev. 94, 1228 (1954).

and the averages have to be taken (1) over the configuration of the initial state of the molecule, (2) over the distribution of the initial states present in the array of molecules with which the neutron collides.

The second term in (100.1) is of the order of the squared ratio of the velocities and represents the Doppler effect; the third term, usually negligible with respect to the other two for E between 1 and 10 ev, represents the direct effect of chemical binding.

The authors quoted have given special attention to the three-dimensional isotropic and non-isotropic harmonic oscillators at a temperature T as well as to systems composed of several nuclei, for which the interference of waves scattered by different centres is taken into account[1,2] (see Sect. 103).

A quantitative comparison between the theoretical expression (100.1) and the linear behaviour of the cross section shown in Figs. 180 to 182 has been attempted by MELKONIAN[3]. This author gives for the coefficient of the $1/E$ term the values 0.65 and 1.0 respectively for the H_2 and H_2O molecules—which have to be compared with the experimental results 0.69 ± 0.09 and 1.36 ± 0.11. The agreement is considered satisfactory in the case of H_2; in the case of H_2O the calculations are much more difficult and have been evaluated only roughly so that only approximate agreement is expected.

An interesting remark is that, under convenient conditions, the main contribution to the Doppler effect term arises from the zero point energy of the oscillator used to represent the chemical bond of the H atom in the molecule. This appears to be one of the few experimentally observable phenomena which depend on this quantum mechanical effect.

As mentioned above, PLACZEK discovered that in the case of a proton bound harmonically and isotropically to a fixed centre, the asymptotic behaviour of the cross section is not normal. Writing the cross section in the form

$$\frac{\sigma}{\sigma_f} = 1 + \frac{c(y)}{y}, \qquad y = \frac{E}{\hbar\omega}, \tag{100.2}$$

one would expect, according to (100.1), that for $y \to \infty$, $c(y)$ should approach the value

$$c(y) = \frac{1}{3}\,\frac{K_{\text{av}}}{\hbar\omega} = \frac{1}{3\hbar\omega}\cdot\frac{1}{2}\cdot\frac{3}{2}\hbar\omega = \frac{1}{4}.$$

PLACZEK found instead that $c(y)$ approaches asymptotically a periodic function having cusps at integral values of y. Thus the cusps which exist in the low energy region of the cross section *vs.* energy curve are preserved in the asymptotic representation of the cross section. The average of the function $c(y)$ over one period is however equal to $\frac{1}{4}$, the value expected from the general formula.

In principle the oscillations of the $1/E$ term in (100.2) should be observable, provided a sufficiently high energy resolution were available. A thorough discussion of the transition from the "normal" behaviour of the asymptotic cross section to the behaviour peculiar to a harmonic and isotropic oscillator of protonic mass, has been given by WICK, who presents the following picture of the phenomenon.

For y close to an integer, a head-on collision may leave the neutron almost at rest and the proton, originally in its ground state, swinging with a large amplitude. After a certain number of periods the proton will again pass near the

[1] See footnote 2, p. 416.
[2] See footnote 3, p. 418.
[3] See footnote 3, p. 417.

initial position where the neutron was left at rest and will collide with it a second time. Such behaviour is however peculiar to the harmonic and isotropic oscillator of protonic mass and even a small anharmonicity or anisotropy is sufficient to change $c(y)$ to its normal behaviour.

101. The influence of the molecular structure on the cross section for neutron energies below 1 ev. An influence of the molecular structure on the scattering cross section of neutrons in this energy region had been recognized since long (Sect. 28). A thorough investigation, however, became possible only when high intensity well collimated beams of monoenergetic slow neutrons became available.

In this section, we first discuss the measurements of the total scattering cross section by transmission experiments in gases and liquids (subsection α), and later a few measurements of the angular distribution of the scattered neutrons and their interpretation in terms of oversimplified models: the rigid molecule and the vibrating molecule models (subsection β, p. 427).

α) Total scattering cross section from transmission experiments. By means of transmission experiments FERMI and MARSHALL[1] measured the total cross section of gas molecules for neutrons of wavelength $\lambda = 5.51$ Å ($E = 0.003$ ev). These neutrons were obtained by filtering, through a prism of BeO microcrystals 40 cm long (100 g/cm²) and 10 cm × 10 cm side, the neutrons emitted from the graphite thermal column of the Argonne pile. The filtering action of microcrystalline substances[2] is discussed in Sect. 103 to which we refer for details.

In the case of BeO the limiting wavelength is 4.4. Å. An analysis of the beam under these conditions[3] showed (Fig. 192) that its intensity is negligible for short wavelengths, rises abruptly to a maximum value at wavelengths of approximately 4.5 Å and then decreases gradually, reaching about $\frac{1}{6}$ of the maximum value at about 6 Å. In conclusion the neutrons belong to a fairly narrow band of average wavelength $\bar{\lambda} \approx 5.5$ Å.

The results obtained by these authors for various molecules, in the form of gases at room temperature, are collected in Table 71 from which one sees a marked

Table 71. *Cross sections per molecule of various gases at room temperature for $\lambda = 5.1$ Å* (FERMI and MARSHALL).

Molecule	σ barns (observed)	Sum of the total cross sections of constituent atoms	σ barns calculated from classical interference theory
CO_2	24.5	13.0	24.8 or 4.1 for opposite phase
N_2O	57.8	34	55 or 41 for opposite phase
O_2	16.2	8.2	13.2
N_2	47.4	30	44.4
CF_4	41.5	21	38 or 7.5 for opposite phase
H_2	170	42	

difference between the experimental data (column 2) and the sum of the total cross sections of the constituent atoms (column 3). The energy of the neutrons and the temperature of the gas are sufficiently low to allow us to disregard the complications due to excitation or deexcitation of vibrational levels of the molecules. This circumstance considerably simplifies the theoretical calculation

[1] E. FERMI and L. MARSHALL: Phys. Rev. **71**, 666 (1947).
[2] G.C. WICK: Ric. Sci. **8** (1), 400 (1937). — Phys. Z. **38**, 403, 689 (1937).
[3] E. FERMI and L. MARSHALL: Phys. Rev. **70**, 815 (1946).

which must take into account the processes of excitation and deexcitation of rotational levels, the Doppler effect due to translational thermal motion of the molecule and the interference of the waves scattered by the constituent nuclei.

Except the case of H_2, which will be discussed later, the differences can only to a small extent be explained as Doppler effect and are mostly due to interference. An estimate of this effect can be made by neglecting the neutron mass in comparison with the atomic masses in the molecule. In this approximation one obtains results identical with those of the classical theory of interference, which is quite good in the case of X rays where the scattered particle is a photon. In the case of neutrons scattered by molecules such as N_2, O_2, etc., the approximation is not so good [see below, in this section, Eqs. (101.2) and (101.3)].

The results of such a calculation are given in column 4 of Table 71. The agreement between theory and observation may be considered satisfactory provided that, in the case of CO_2, CF_4 and N_2O, the scattering lengths a_{bi} of the component nuclei are assumed to have the same sign (Sect. 20).

Table 72. *Cross sections of* CF_4 *at room temperature* (FERMI and MARSHALL).

λ Å	Velocity of neutrons (m/sec)	σ_s barns
0.881	4590	19.5
1.099	3600	19.0
1.316	3000	21.5
1.492	2650	20.8
5.1	775	41.5

The effect of interference is clearly shown by the results obtained by the same authors in the case of CF_4 at room temperature. The cross section per molecule was measured as a function of the velocity of the neutrons by means of a flourite crystal monochromator. Table 72 shows that the cross section of the CF_4 molecule remains almost constant and very close to the sum of the cross sections of the constituent nuclei (21 barns) up to $\lambda = 1.5$ Å and then increases by about a factor 2 between 1.5 and 5.1 Å.

Table 73. *Cross section of* H_2 *at room temperature* (FERMI and MARSHALL).

λ Å	Velocity (m/sec)	σ_s barns	Factor for Maxwell distribution	σ_s/factor barns
0.76	5212	50.9	1.048	48.6
0.93	4256	53.2	1.071	49.7
1.11	3546	56.2	1.099	51.1
1.46	2714	61.2	1.172	52.2
1.68	2347	64.4	1.231	52.3
2.24	1765	68.0	1.407	48.3
5.1	775	169.7	2.514	67.5

In the case of H_2 molecules the Doppler effect is important because the thermal agitation velocity of the molecules is larger than the velocity of the neutrons ($E = 0.003$ ev corresponds to $v = 775$ m/sec). Under these conditions a neutron-H_2-molecule collision may be better described by saying that the molecules hit the neutron rather than that the neutron hits the molecules.

A correction factor for the Doppler effect has been calculated by FERMI and MARSHALL by means of a relation similar to (97.7), assuming that condition (97.6) is fulfilled, and is given in column 4 of Table 73. Column 3 shows the cross section of H_2 molecules at room temperature, as a function of the neutron velocity, which was obtained from transmission experiments with monoenergetic neutrons produced by BRAGG reflection on a fluorite crystal. The last column shows the H_2-molecule cross section after correction for the Doppler effect.

The results of a similar experiment made at liquid air temperature (83° K) are shown in Table 74. The Doppler correction factor is closer to 1 than in the previous case even at $\lambda = 5.1$ Å. Comparison of the observed and corrected values of σ_s at $\lambda = 5.1$ Å, at liquid air temperature, with the corresponding values at room temperature, shows clearly the increase of the cross section due to the increase of thermal agitation of the molecules.

At liquid air temperature, practically only the rotational levels $J=0$ and $J=1$ of the H_2 molecule are involved. Therefore, the experimental results may be compared with the theoretical calculation of the cross section carried out by TELLER and SCHWINGER[1] in connection with the problem of neutron scattering by ortho- and para-hydrogen.

Neutrons with energy less than 0.068 ev colliding with para-hydrogen (anti-parallel proton spins: $S=0$) in its ground state ($J=0$; vibrational quantum number $v=0$) can be scattered either elastically or inelastically with excitation of the molecule to its first rotational level, i.e. to the ground state of ortho-hydrogen ($S=1$, $J=1$, $v=0$).

This latter process, requiring 0.023 ev, will take place only exceptionally under the experimental conditions of Table 74 and therefore it may be disregarded in the present case.

When the neutron energy is less than 0.045 ev, the scattering of neutrons by ortho-hydrogen (parallel proton spins: $S=1$) can be either an elastic or an inelastic process in which the molecule is converted to the ground state of parahydrogen while the neutron takes up the excess energy.

Table 74. *Cross section of* H_2 *at 83° K; ortho-para ratio unchanged from room temperature* (FERMI and MARSHALL).

λ Å	Velocity (m/sec)	σ_s barns observed	σ_s (barns) corrected for absorption and Doppler effect	σ_s (barns) Teller-Schwinger theory
0.836	4730	44	43	—
1.202	3290	51	49	—
1.638	2420	55	52	56
1.884	2100	56	51	56
5.1	775	81	52	80

The cross section corresponding to these various processes have been calculated by SCHWINGER and TELLER by means of (98.7) making use of the rotational wave functions of the H_2 molecule. For a more detailed discussion of this problem we refer to L. HULTHÉN and M. SUGAWARA's article in Vol. XXXIX of this Encyclopedia, Sect. 18.

The theoretical data given by FERMI and MARSHALL in the last column of Table 74 have been obtained by applying the Schwinger-Teller theory to an ortho-parahydrogen mixture corresponding to room temperature equilibrium ($\frac{3}{4}:\frac{1}{4}$) assuming

$$a_1 = -2.40\times 10^{-12}\text{ cm} \quad\text{and}\quad a_3 = 0.54\times 10^{-12}\text{ cm}$$

for the singlet and triplet state of the neutron-proton system (Sect. 29).

The agreement is not too good especially for the lowest velocity, where the comparison may be vitiated by the large Doppler correction.

Similar experiments with deuterium gas at liquid air temperature were performed by the same authors[2]. The cross section per molecule is found to be 21.3 barns which has to be compared with a cross section per free deuterium atom of 3.44 barns. The large value of the molecular cross section again gives clear experimental evidence of positive interference between waves scattered by the two constituent deuterons. The reader is referred to the original paper and to M. VERDE's article in Vol. XXXIX for a detailed discussion of the spin dependence of the scattering length in the neutron-deuteron collision.

Extensive experimental work has been done on the problem of slow neutron scattering by molecules by means of time-of-flight velocity neutron spectro-

[1] J. SCHWINGER and E. TELLER: Phys. Rev. **52**, 286 (1937).
[2] E. FERMI and L. MARSHALL: Phys. Rev. **75**, 578 (1949).

meters[1-4]. A list of the more important materials investigated is given in Table 75. Some of MELKONIAN's results[3] are plotted in Figs. 184 to 189.

The agreement with the results of FERMI and MARSHALL discussed above for N_2 and O_2 (Table 71) and H_2 (Tables 73 and 74) is very satisfactory provided that, as is suggested in a later paper by the same authors[5], the corresponding neutron wavelength is taken equal to 5.43 Å instead of 5.1 Å.

Table 75. *List of materials investigated by means of time-of-flight neutron spectrometers.*

		Reference
Paraffin ($C_{22}H_{46}$)	solid	Columbia Univ. group[1]
D_2O	liquid	Columbia Univ. group[1]
H_2O	liquid	JONES[2]
H_2O	liquid	MELKONIAN[3]
H_2O	liquid	KELLER[4]
Cyclohexane (C_6H_{12})	liquid	JONES[2]
O_2	gas	MELKONIAN[3]
N_2	gas	MELKONIAN[3]
Methane (CH_4)	gas	MELKONIAN[3]
Ethane (C_2H_6)	gas	MELKONIAN[3]
Propane (C_3H_8)	gas	MELKONIAN[3]
n-butane (C_4H_{10})	gas	MELKONIAN[3]
Ethyline (C_2H_4)	gas	MELKONIAN[3]
1,3-butadiene (C_4H_6)	gas	MELKONIAN[3]
Cetane ($C_{16}H_{34}$)	liquid	MELKONIAN[3]
Butane	liquid	MELKONIAN[3]
Oxalique acide (HCOOH)	microcrystal powder	KELLER[4]
Sodium formiate (HCOONa)	microcrystal powder	KELLER[4]
Copper sulfate $SO_4Cu + 5H_2O$	microcrystal powder	KELLER[4]

The points due to CARROLL[6] refer to values measured by means of the Cd-difference method (Sect. 133) applied to the neutrons which were emitted by a Rn α + Be source embedded in a paraffine howitzer, and detected with a BF_3 ionization chamber.

The agreement with the results of MELKONIAN is very satisfactory in all cases provided CARROLL's values are plotted at $E = 0.04$ ev instead of at $E = 0.025$ ev. This shows clearly that the effective energy of the C-neutrons does not correspond to that calculated by assuming that their velocity distribution is within good approximation a Maxwellian distribution at room temperature (see Sect. 108).

The following remarks may be added to the results obtained in a few typical cases.

O_2 molecule (Fig. 184). Oxygen capture is negligible in this energy region and the Doppler effect due to thermal agitation of the molecule is not important.

[1] L. J. RAINWATER, W. W. HAVENS, J. R. DUNNING and C. S. WU: Phys. Rev. **73**, 733 (1947).
[2] W.B. JONES: Phys. Rev. **74**, 364 (1948).
[3] E. MELKONIAN: Phys. Rev. **76**, 1750 (1950).
[4] A. GILBERT, R. KELLER and J. ROSSEL: Helv. phys. Acta **19**, 493 (1946). — R. KELLER: Helv. phys. Acta **23**, 627 (1950). — W. GUBER: Z. Naturforsch. **7**, 725 (1952). — J. A. JANIK: p. 54 of Ref. [*15a*].
[5] See footnote 2, p. 422.
[6] H. CARROLL: Phys. Rev. **60**, 702 (1941).

Therefore almost the entire increase in cross section shown in Fig. 184 is due to molecular effects. According to (98.10), the reduced mass effect gives an increase of 6%, while the interference effect tends to double the cross section at very low energies. This is due to the fact that oxygen is predominantly composed of a single isotope of spin zero, so that the main contributions to incoherent scattering, namely those due to isotopic composition and nuclear spin orientation, are not present in appreciable amount. The paramagnetism of O_2 is expected to give an appreciable contribution to the cross section arising from the interaction between the magnetic moment of the molecule and that of the neutron[1].

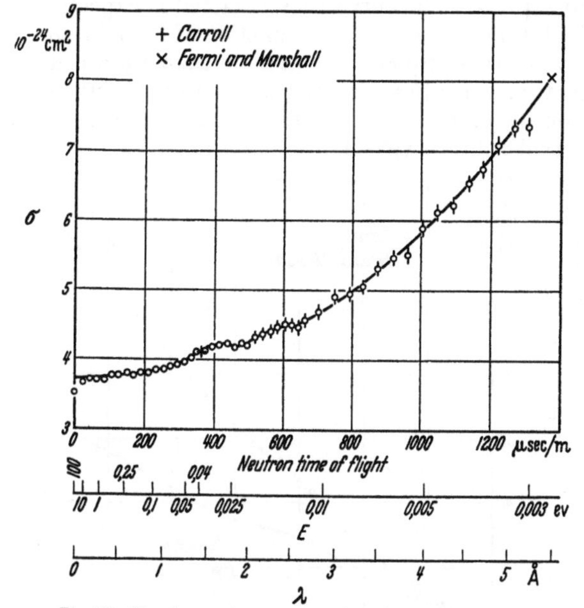

Fig. 184. The slow neutron cross section of oxygen bound in O_2 gas down to 0.003 ev (MELKONIAN).

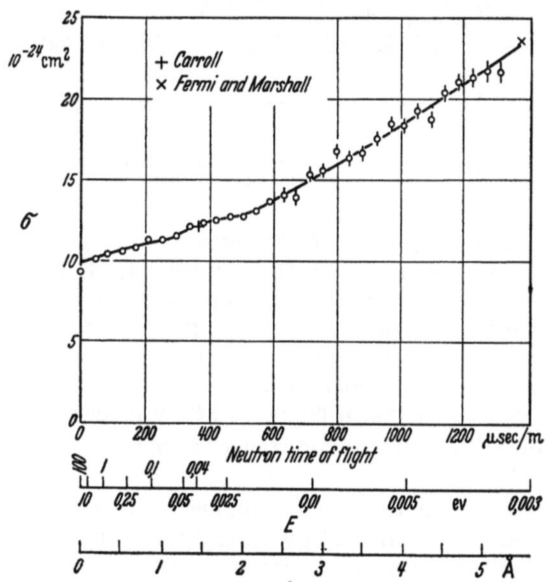

Fig. 185. The slow neutron cross section of nitrogen bound in N_2 down to 0.003 ev (MELKONIAN).

N_2 molecule (Fig. 185). Neutron capture contributes appreciably to the total cross section. Using the $1/E^{\frac{1}{2}}$ slope based on measurements above 1 ev, MELKONIAN deduces the following values of the capture cross section: $\sigma_a = 2.15$ barns and $\sigma_a = 7.2$ barns respectively at $E = 0.025$ ev and $E = 0.003$ ev. The interference effects are complicated by the spin 1 of the predominant isotope N^{14}, which gives rise to an appreciable amount of incoherent scattering. The reduced mass effect amounts to 7%.

H_2 molecule (Fig. 186). The main contribution to the rapid increase of the cross section is due to the Doppler effect from the thermal motion of the molecules. If the H_2 cross section were independent of energy, the Doppler effect would produce an increase by a factor 2.4 at $E = 0.003$ ev. Actually neutron capture has also to be considered. This follows the $1/v$ law (96.4). The "corrected" curve shown in Fig. 186 is obtained by MELKONIAN from the experimental curve, by ap-

[1] O. HALPERN and M.H. JOHNSON: Phys. Rev. 55, 898 (1939). — O. HALPERN: Phys. Rev. 72, 746 (1947).

proximate elimination of the effect of thermal translational motion, on the assumption that the resulting velocity dependence should be of the form of a constant plus a $1/v$ term due to capture. The curve thus obtained is expected to be a good approximation to the total cross section as a function of the relative velocity as long as the neutron velocity is equal to or greater than the mean molecular velocity. For lower neutron velocities it should become progressively poorer because in this energy region the cross section rather depends on the mean molecular velocity than on the actual neutron velocity.

The reduced mass effect increases the scattering cross section by a factor $(\frac{3}{2}:\frac{1}{2})^2 = 1.78$.

The influence of vibrational and rotational levels and the interference between the waves scattered by the two protons of a molecule are rather complicated at room temperature.

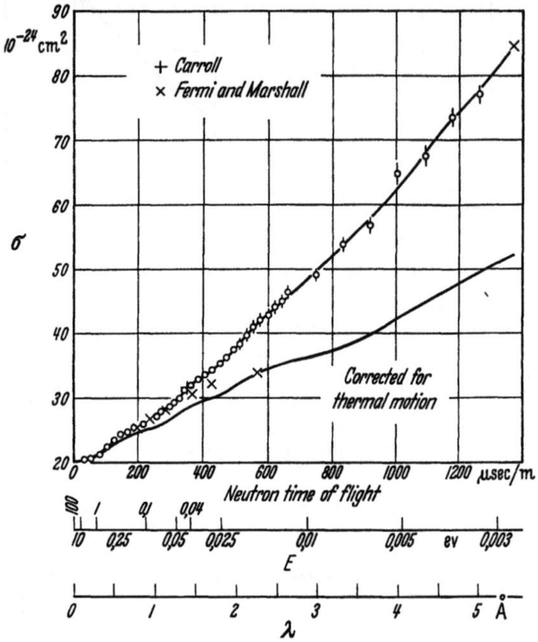

Fig. 186. The slow neutron cross section of protons bound in H_2 gas down to 0.003 ev (Melkonian).

Hydrocarbons. The results obtained by Melkonian for various hydrocarbons show some common features which may be described as follows (Fig. 187). The cross section per hydrogen atom starts from the common value σ_f at about 1 ev, and remains almost the same for all the hydrocarbons down to energies of about 0.1 ev. At about 0.5 ev the cross section undergoes a rapid increase with decreasing neutron energy which is probably to be interpreted as the setting-in of the effect of the C-H binding. Between 0.1 and 0.01 ev the cross section for a given neutron energy increases in the order of increasing molecular weight. Below 0.01 ev the hydrocarbon curves continue to increase and appear somewhat to converge. Some of the irregularities shown in this energy region are probably due to inaccuracy of the experimental results. At very low energies the value $4\sigma_f$ is readily reached according to the prediction of the reduced mass effect (95.1), (98.10) applied to the case of hydrogen bound in heavy molecules.

The same curve was obtained over the entire energy range from 1 ev down to 0.003 ev for liquid and gaseous n-butane.

In Fig. 188 a comparison is presented of the results obtained with cetane with the theoretical curves calculated by Bethe and by Arley (Sect. 98). In order to describe the effect of the C-H chemical bond, Bethe adopts an anisotropic potential well with vibrational numbers $\nu_z = 3200$ cm^{-1} and $\nu_x = \nu_y = 800$ cm^{-1}. The agreement, which is rather poor below 0.01 ev, is slightly improved by changing the lower frequency from 800 to 1200 cm^{-1}. The general trend of the experimental curve seems to be better described by Arley's results who takes into account, although with an oversimplified model, the very low frequencies of the molecular structure (see Sect. 98).

H_2O molecule. The curve of Fig. 189 is obtained by adopting $\sigma_0 = 3.73$ barns. The normal vibrations[1,2] of the non-linear molecule H_2O (106°) are $\hbar\omega_1 = 0.45$, $\hbar\omega_2 = 0.20$, $\hbar\omega_3 = 0.47$ ev and the rotational levels have a separation of 0.006 ev. In the liquid and solid state[3] the vibrational degrees of freedom are left relatively unchanged while the rotational degrees of freedom degenerate to an optical band of hindered rotation at about 0.1 ev and the translational degrees of freedom

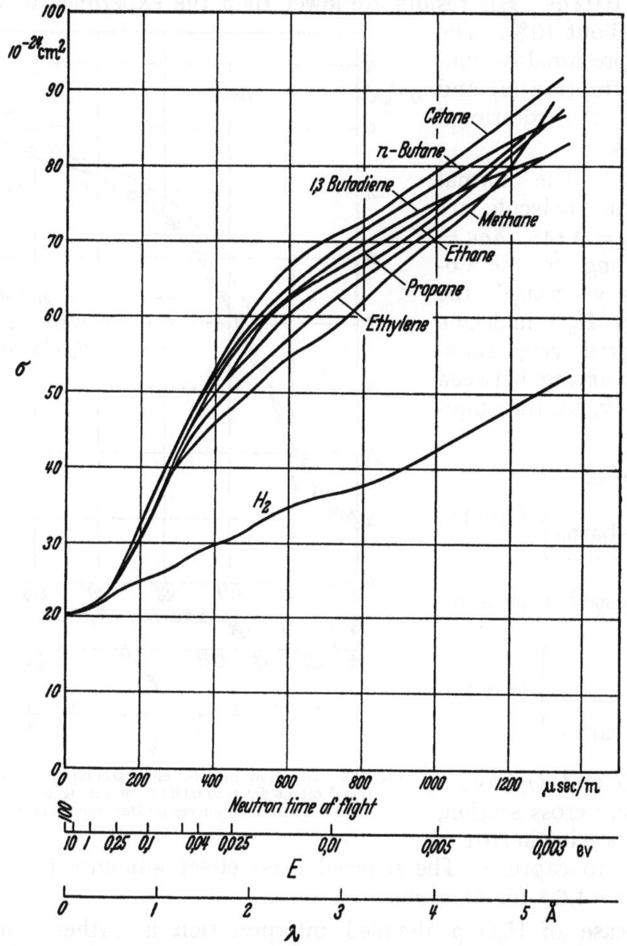

Fig. 187. The slow neutron cross section of protons bound in various hydrocarbons and in H_2 gas. Approximate corrections for the effect of thermal translational motion have been made.

give rise to a band of acoustical frequencies extending from zero to a Debye frequency corresponding to about $k\Theta_D = 0.026$ ev. The experimental curve of Fig. 189 looks similar to those obtained for the heavier hydrocarbons.

The total scattering cross section (elastic + inelastic) of the H_2O molecule at room temperature has been computed by GORYUNOV[4] under the following

---

[1] For the molecular data on vibrational levels cf. J. LECOMTE's paper in Vol. XXVI of this Encyclopedia.

[2] G. HERZBERG: Infrared and Raman spectra of polyatomic molecules. New York: Van Nostrand 1947.

[3] G.F. v. DARDEL: Trans. Roy. Inst. Technical 75 (1954).

[4] A.F. GORYUNOV: J. Nucl. Energy 4, 109 (1957).

simplifying assumptions: (a) the water molecules do not interact with one another; (b) the molecules are in their vibrational ground states and excitation to higher vibrational levels does not occur. Since $kT_0 \approx 0.025\,\text{ev} \ll \hbar\omega_2 = 0.2\,\text{ev}$, this assumption is certainly correct; thus the molecules can be treated as rigid rotators. Using the Fermi pseudo-potential and taking into account the rotational states with $J \leqq 3$, GORYUNOV computes the total cross section for E varying between 0.05 and 0.02 ev. His results are lower than the experimental values of MELKONIAN by about 10%. The discrepancy is presumably connected with assumption (a) and the omission of the rotational levels with $J \geqq 4$.

D_2O molecule. The normal vibrations of this molecule are $\hbar\omega_1 = 0.33$, $\hbar\omega_2 = 0.146$, $\hbar\omega_3 = 0.34$ ev. According to the Columbia University group[1] the cross section per D_2O molecule can be represented very satisfactorily, for E varying between 10 ev and 0.005 ev, by the empirical formula

$$\left.\begin{aligned}\sigma_{D_2O} &= \sigma_f(D_2O) + \\ &+ \frac{0.72}{E^{\frac{1}{2}}}\ \text{barns}\end{aligned}\right\} \qquad (101.1\,\text{a})$$

where E is expressed in ev and

$$\left.\begin{aligned}&\sigma_f(D_2O) \\ &= 2\sigma_f(D) + \sigma(O) \\ &= (10.5 \pm 0.03)\ \text{barns}.\end{aligned}\right\} \qquad (101.1\,\text{b})$$

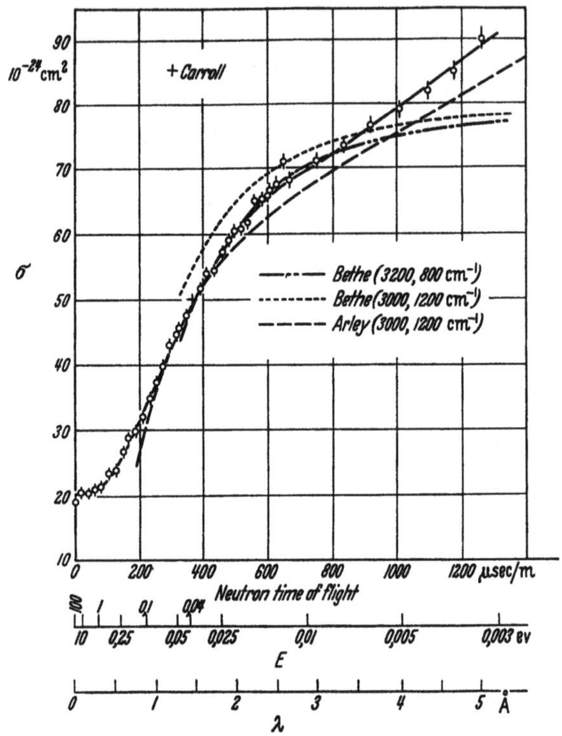

Fig. 188. The slow neutron cross section of protons bound in liquid cetane down to 0.003 ev according to MELKONIAN. Theoretical curves by BETHE and ARLEY.

Because both D and O have a very small capture cross section, the $1/E^{\frac{1}{2}}$ term appearing in (101.1) can not be due to capture. The reduced mass effect amounts to $(\frac{3}{2})^2 = 2.25$ for D and $(\frac{17}{16})^2 = 1.06$ for O atoms.

As in the case of H_2O a detailed interpretation is rather complicated vibrational and rotational states have to be taken into account as well as the Doppler effect due to thermal translational motion. The interference of waves scattered by the component nuclei is complicated by various possible orientations of protons or deuterons and neutron spins.

β) Differential scattering cross section from angular distribution measurements with spectrometers. The diffraction pattern obtained from molecular gases has been investigated by ALCOCK and HURST[2] at Chalk River, who used a NaCl crystal spectrometer which automatically recorded the scattered neutrons every 8° from ~8 to 90°. The gases studied (D_2, O_2, CO_2, N_2, CF_4, CH_4) were contained in

[1] See footnote 1, p. 423.

[2] G.D. HURST and N.Z. ALCOCK: Canad. J. Phys. **89**, 36 (1950). — N.Z. ALCOCK and D.G. HURST: Phys. Rev. **75**, 1609 (1949); **83**, 1100 (1951).

a steel vessel at room temperature and at about 60 atmospheres pressure. Since the scattering from the gas was not very intense, precautions were necessary to reduce the background.

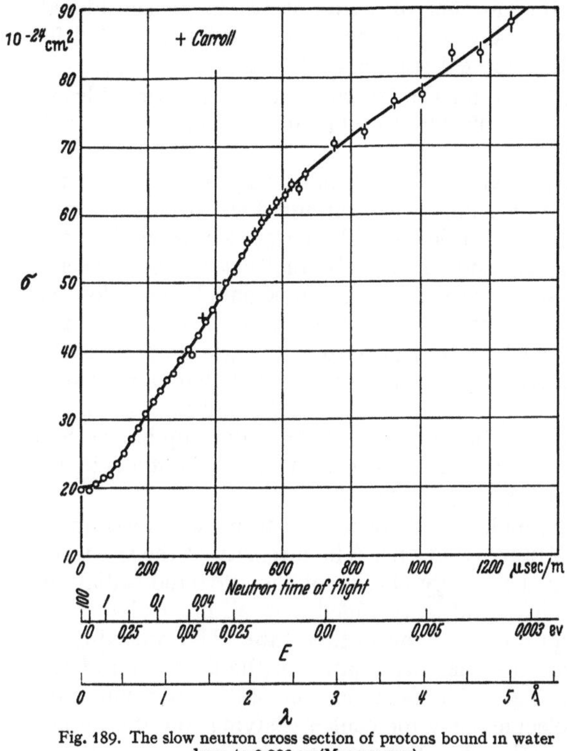

Fig. 189. The slow neutron cross section of protons bound in water down to 0.003 ev (MELKONIAN).

Fig. 190 shows the results obtained for CF_4 with neutrons of $\lambda = 1.063$ Å ($E = 0.072$ ev). The first point was corrected for intermolecular interference following the procedure developed by DEBYE[1] in case of X-ray diffraction (the so-called "outer effect" correction). The two curves correspond to models discussed below.

The main features of the quantum mechanical treatment of the problem of scattering of slow neutrons by molecules have been presented in Sect. 95 and 98. Eq. (98.7) gives the differential cross section $d\sigma_{ik}$ in the c.m.s., which has to be summed over all final states k and averaged with respect to the occupation distribution of the initial molecular states i at the temperature T_0 of the scatterer (Sect. 95). The procedure is rather complicated and in practice it has been applied only to very simple cases such as those of hydrogen[2] and deuterium[3] molecules. The results of these computations refer to the c.m.s. and for comparison with experiments must be transformed to the l.s. for a Maxwellian distribution of molecular translational velocities[4].

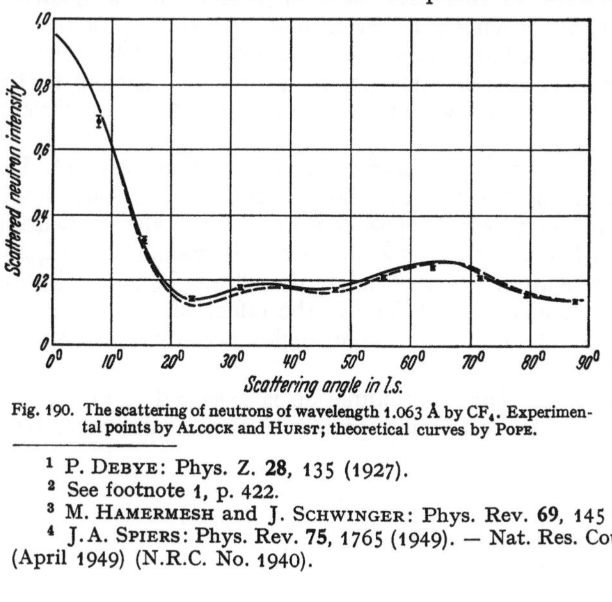

Fig. 190. The scattering of neutrons of wavelength 1.063 Å by CF_4. Experimental points by ALCOCK and HURST; theoretical curves by POPE.

The application of such a procedure, even to other simple cases, becomes, however, very cumbersome. For instance the formulae derived for D_2 can in principle be used also for N_2, since 99.2% of these molecules are N^{14}—N^{14}, i.e. of the same

[1] P. DEBYE: Phys. Z. **28**, 135 (1927).
[2] See footnote 1, p. 422.
[3] M. HAMERMESH and J. SCHWINGER: Phys. Rev. **69**, 145 (1946).
[4] J.A. SPIERS: Phys. Rev. **75**, 1765 (1949). — Nat. Res. Counc. (Canada) Report CRT-417 (April 1949) (N.R.C. No. 1940).

type as D_2. But the greater mass and size of the nitrogen molecule produce a decrease in the spacing of rotational levels with a consequent increase of both the number of transitions and the number of initial states to be considered. For neutrons of $\lambda = 1.063$ Å ten times as many transitions must be computed for nitrogen at room temperature than for deuterium at 77° K[1].

The procedure of SACHS and TELLER for the treatment of the rotational levels of rigid moleules and its extension by MESSIAH to vibrating molecules (Sect. 98), constitute steps towards simplification.

A schematization of the problem which at first sight may even appear too drastic, was made by FERMI and MARSHALL[2], who computed the total scattering cross section of the molecules listed in Table 71 with the procedure used in the theory of coherent scattering of X-rays and electrons by molecules (rigid molecule approximation). The same procedure was used by ALCOCK and HURST[5] for the computation fo the differential cross section of the molecules listed above, and was discussed by SPIERS[3] and POPE[4] with regard to its relationship with the general quantum mechanical treatment. It has to be noticed that it represent a further simplification of the "static approximation" which constitutes the point of departure for the considerations that brought PLACZEK[5] to develop his method of computation of the (incoherent) scattering cross section by means of an expansion in inverse powers of $A = M/m$ (Sects. 100 and 103).

In the static approximation the molecule is considered to have a fixed orientation with all its atoms fixed in definite positions. The intensity of the scattered wave is then computed, neglecting the energy change of the scattered radiation, by means of the HUYGHENS' principle, and is averaged over all molecular orientations and over the actual configuration of the initial state. The procedure is usually applied in a simpified form in some way related to the discussion of the ideal scatterer (crystal or molecule) given in Sect. 99: the static approximation is first applied to an ideal rigid molecule, i.e. a molecule satisfying condition (c) of Sect. 99, and then the effect of molecular vibrations is introduced in a second step by means of a convenient factor similar to the Debye-Waller factor (99.11).

When applied to X-ray scattering by molecules the *rigid molecule model* gives very satisfactory results; in this case the energy of an X-ray quantum is always large compared to the spacing of the electronic levels so that the relative change of momentum of the scattered radiation is exceedingly small. Furthermore the atomic form factor decreases so steeply with increasing angle ϑ that it makes the reduction due to molecular vibrations [see Eq. (101.2 below] difficult to observe.

In the case of neutrons these approximations are certainly not so good. The changes in neutron momentum due to transitions between vibrational levels cannot be neglected in the energy region considered here. Even the momentum changes due to transitions between rotational levels are appreciable in the case of light molecules and for neutron energies below the threshold for excitation of a vibrational degree of freedom. And finally, the effect of the vibrational degree of freedom must be observable at large angles because the scattered neutron wave is isotropic (S wave).

It is therefore concluded that this approximation is expected to give a sufficiently good description of the diffraction pattern for molecules with large

[1] See footnote 2, p. 427.
[2] See footnote 1, p. 420.
[3] See footnote 4, p. 428.
[4] N. K. POPE: Canad. J. Phys. **30**, 587 (1952).
[5] G. PLACZEK: Phys. Rev. **86**, 377 (1952).

moments of inertia so that the spacing of the rotational levels is very small, and for neutron energies sufficiently small to make impossible or at least very improbable the excitation of vibrational levels.

The scattering cross section of a *rigid molecule* in the c.m.s. is given by the expression[1,2]

$$\left.\begin{aligned}\frac{d\sigma(\vartheta^*)}{d\omega^*} &= \left(\frac{A_{\text{mol}}}{A_{\text{mol}}+1}\right)^2 \sum_{ij} \frac{[\sigma_{bi}\sigma_{bj}]^{\frac{1}{2}}}{4\pi}\,\frac{\sin X_{ij}}{X_{ij}},\\ \text{with}\qquad A_{\text{mol}} &= \frac{M_{\text{mol}}}{m}, \quad X_{ij} = \frac{2r_{ij}}{\lambda}\sin\frac{\vartheta^*}{2}\end{aligned}\right\} \tag{101.2}$$

which is very similar to the classical expression valid for X-rays[3]. The sum is extended over all the atoms present in the molecule whose relative distances are r_{ij}. If one denotes with σ_{bi} the total bound scattering cross section of the i-th nucleus when $i=j$, and its coherent part only, when $i\neq j$, then Eq. (101.2) represents the total scattering cross section: coherent + incoherent.

If the rigidity of the molecule is relaxed, each term in the sum appearing in (101.2), is multiplied by the factor

$$\mathrm{e}^{-2\overline{\delta r_{ji}^2}\left(\frac{\sin\frac{\vartheta^*}{2}}{\lambda}\right)^2}, \tag{101.3}$$

where $\overline{\delta r_{ij}^2}$ is the mean square variation of the distance r_{ij} between atoms i and j. The model obtained by such a procedure is called the *vibrating molecule* model. The quantities $\overline{\delta r_{ij}^2}$ may be computed by representing the vibrational levels of the molecule by means of a set of uncoupled anisotropic oscillators placed at the mean nuclear sites.

These theoretical considerations have been developed in detail for spherical top molecules of the type YX_4 and have been compared[2] with the experimental results of ALCOCK and HURST[1] for CF_4 and CH_4.

In the case of CF_4 molecules one of the four fundamental vibrational frequencies (doubly degenerated) corresponds to $\hbar\omega = 0.054$ ev which is comparable with kT. Therefore there is an appreciable population in the second vibrational level at room temperature and furthermore neutrons of 0.072 ev can induce transitions in this mode of vibration. These effects have been neglected in the computation of the two curves of Fig. 190. The dotted curve corresponds to a rigid molecule with fixed centre of mass. It has been computed by means of (101.2) neglecting the transformation from the c.m.s. to the l.s., which represents a small correction in case of heavy molecules like CF_4. The curve was computed taking $r_{\text{C-F}} = 1.33$ Å, $\sigma_{\text{C}} = 5.52$ b, $\sigma_{\text{F}} = 3.99$ b, $\sigma_{\text{F, coh}} = 3.46$ b.

The continuous curve of Fig. 190 has been fitted to the experimental curve by taking into account, by means of (101.3), the effect of zero point motion of the vibrational modes of the molecule. Making use of the detailed description of the interatomic forces known from spectroscopy, POPE was able to establish that the C-atom is equivalent to an isotropic oscillator with a mean amplitude of 0.028 Å perpendicular to the bond direction and 0.030 Å along the bond direction.

---

[1] See footnote 2, p. 427.

[2] See footnote 4, p. 429.

[3] M. H. PIRENNE: The diffraction of x-rays and electrons by free molecules. Cambridge: Cambridge University Press 1946. — R. W. JAMES: The optical principle of the diffractions of x-rays, Chap. IX. London: G. Bell & Sons Ltd. 1948.

An attempt to apply the same procedure to the case of methane was unsuccessful; this is not surprising since the various approximations involved by the models considered here are certainly not valid.

An excellent agreement with the experimental data is obtained, on the contrary, by ZEMACH and GLAUBER by applying to the CH_4 molecule the procedure described at the end of Sect. 98.

The scattering of neutrons by liquids and its interpretation in terms of correlation functions would take us too far from the province of the present article and therefore we refer the reader to [*13*] and to Vol. XXXII.

III. The scattering of slow neutrons by polycrystalline materials.

102. Elastic scattering. In this section we shall discuss the elastic scattering of slow neutrons by a crystal as a whole under the assumptions that all its atoms have a negligible cross section for neutron capture and that the dimensions of the crystal are sufficiently small so that we can neglect the attenuation undergone by the incident wave when crossing the crystal whatever its orientation may be [see Eq. (102.19) below]. Afterwards, we shall consider the transmission of slow neutrons through a polycrystalline moderator composed of microcrystals whose orientations are distributed at random.

The expressions for the elastic (and inelastic) cross sections have been derived in the Debye approximation by WICK[1], HALPERN, HAMERMESH and JOHNSON[2] and by WEINSTOCK[3]. Their results refer to the case of a crystal composed of a single element, satisfying conditions (a) and (b) of Sect. 99, whose atoms are arranged in a simple Bravais lattice[4] the unit cell of which contains a single atom. The work of these authors has been extended by CASSELS[5] to crystals showing both isotopic and spin incoherence.

The general procedure followed in the derivation of the elastic scattering cross section is similar to that found in classical text books on X-ray diffraction[6] and need only be outlined here[7, 5, 8].

At large distance from a microcrystal of given orientation the wave function of the neutron has the form

$$\psi = e^{iK_1 z} - \frac{e^{iK_2 r}}{r} \sum_i a_{b\,i}\, e^{i\boldsymbol{r}_i\cdot(\boldsymbol{K}_1-\boldsymbol{K}_2)} \tag{102.1}$$

similar to (20.14). The first term represents the incident wave of propagation vector $\boldsymbol{K}_1$ in the z direction; the second term represents the scattered wave of propagation vector $\boldsymbol{K}_2$.

The sum is to be extended over all the nuclei contained in the crystal, the scattering length and position of which are denoted respectively by $a_{b\,i}$ and

$$\boldsymbol{r}_i = \boldsymbol{a}\, n_1 + \boldsymbol{b}\, n_2 + \boldsymbol{c}\, n_3 \tag{102.2}$$

where $\boldsymbol{a}, \boldsymbol{b}, \boldsymbol{c}$ are the fundamental translation vectors of the lattice under consideration and n_1, n_2, n_3 three integers running from zero to N_1, N_2, N_3, respectiv-

[1] G.C. WICK: Ric. Sci. **8** (1), 400 (1937). — Phys. Z. **38**, 403, 689 (1937).

[2] O. HALPERN, M. HAMERMESH and M.H. JOHNSON: Phys. Rev. **59**, 981 (1941).

[3] R. WEINSTOCK: Phys. Rev. **65**, 1 (1944).

[4] P.P. EWALD: Handbuch der Physik, Vol. 24, p. 191. 1927.

[5] J.M. CASSELS: Progress Nuclear Physic, Vol. 1, p. 185. London: Butterworth-Springer 1950.

[6] A.H. COMPTON and S.K. ALLISON: X-rays in theory and experiment. London: McMillan & Co. Ltd. 1935.

[7] E. FERMI and L. MARSHALL: Phys. Rev. **71**, 589 (1947).

[8] E.O. WOLLAN and G.G. SHULL: Nucleonics **3**, 8 (July 1948); **3**, 17 (August 1948).

ely. It is well known that (102.1) is valid only at a distance from the origin, r, so large compared to the crystal dimensions, that the approximation

$$\frac{1}{|\boldsymbol{r}-\boldsymbol{r}_i|} \approx \frac{1}{r}$$

can be made for any value of i. In the case of elastic scattering considered here, no energy exchange takes place between the neutron and the crystal and therefore

$$|\boldsymbol{K}_2| = |\boldsymbol{K}_1|. \tag{102.3}$$

In order to calculate cross sections, let us first confine consideration to the case of an ideal rigid crystal satisfying the three conditions (a) to (c) of Sect. 99. The waves scattered by each of the atoms in the microcrystal are coherent and therefore the intensity of the scattered neutrons is obtained by taking the square of the modulus of the sum of the amplitudes of the waves scattered by the single nuclei. Hence one derives the differential cross section per nucleus due to coherent elastic scattering in the $\boldsymbol{K}_2$ direction

$$\frac{d\sigma_{\text{el, coh}}(\boldsymbol{K}_2)}{d\omega} = \frac{1}{N}\left|\sum_i a_{b\,i}\, e^{i\,\boldsymbol{r}_i\cdot(\boldsymbol{K}_1-\boldsymbol{K}_2)}\right|^2 \tag{102.4}$$

where $N = N_1 \cdot N_2 \cdot N_3$ is the total number of atoms present in the crystal.

For crystals containing a single element whose atoms are arranged in a Bravais lattice with one atom per unit cell, the scattering length $a_{b\,i}$ may be factorized and the square of the sum can be transformed by taking advantage of the fact that N_1, N_2, N_3 are very large integers. The result of such a transformation[1] is

$$\frac{d\sigma^{\tau}_{\text{el, coh}}(\boldsymbol{K}_2)}{d\omega} = \frac{(2\pi)^3}{v_0}\, a_b^2\, \delta(\boldsymbol{K}_1-\boldsymbol{K}_2+2\pi\boldsymbol{\tau}) \tag{102.5}$$

where v_0 is the volume of the unit cell (in the present case this is equal to the volume per atom) and $\boldsymbol{\tau}$ a vector of the reciprocal lattice[2].

---

[1] See footnote 5, p. 431.

[2] The treatment of the phase difference $\boldsymbol{r}_i \cdot (\boldsymbol{K}_1 - \boldsymbol{K}_2)$ is most conveniently carried out by use of the *reciprocal lattice* (see footnote 4, p. 431). This is defined as the ensemble of points the positions of which are given by

$$\boldsymbol{\tau}(h', k', l') = h'\boldsymbol{a}^* + k'\boldsymbol{b}^* + l'\boldsymbol{c}^* \tag{a}$$

where h', k', l' are integers and $\boldsymbol{a}^*, \boldsymbol{b}^*, \boldsymbol{c}^*$ are the primitive translations of the reciprocal lattice defined by the relations

$$\boldsymbol{a}^*\cdot\boldsymbol{a} = \boldsymbol{b}^*\cdot\boldsymbol{b} = \boldsymbol{c}^*\cdot\boldsymbol{c} = 1; \quad \boldsymbol{a}^*\cdot\boldsymbol{b} = \boldsymbol{a}^*\cdot\boldsymbol{c} = \boldsymbol{b}^*\cdot\boldsymbol{a} = \boldsymbol{b}^*\cdot\boldsymbol{c} = \boldsymbol{c}^*\cdot\boldsymbol{a} = \boldsymbol{c}^*\cdot\boldsymbol{b} = 0. \tag{b}$$

The following properties of $\boldsymbol{\tau}(h', k', l')$ may be verified: (1) $\boldsymbol{\tau}(h', k', l')$ is normal to the plane of ordinary space which intersects the axis of the crystal at intervals a/h', b/k', c/l'; (2) the length of $\boldsymbol{\tau}$ is equal to the reciprocal of the spacing of these planes

$$|\boldsymbol{\tau}(h', k', l')| = \frac{1}{d(h', k', l')}. \tag{c}$$

If one divides h', k', l' by their largest common integral factor m the three integers

$$h = \frac{h'}{m}, \quad k = \frac{k'}{m}, \quad l = \frac{l'}{m} \tag{d}$$

are the Miller indices of the set of lattice planes whose normal is parallel to $\boldsymbol{\tau}(h', k', l')$ and whose spacing is

$$d = \frac{d(h', k', l')}{m}. \tag{e}$$

The Bragg conditions have the following simple significance in $\boldsymbol{K}$ space, obtained by multiplying all distances of reciprocal space by 2π. The propagation vector $\boldsymbol{K}_1$ of the incident

Because of the three-dimensional Dirac δ-function appearing in (102.5) the differential elastic coherent cross section will be negligibly small unless the condition

$$\boldsymbol{K}_1 - \boldsymbol{K}_2 + 2\pi\boldsymbol{\tau} = 0 \tag{102.6}$$

is fulfilled. Eq. (102.6) is the Bragg condition: the vector $\boldsymbol{K}_1 - \boldsymbol{K}_2$ must be parallel to $\boldsymbol{\tau}$, i.e. perpendicular to the set of lattice planes (h, k, l) and the angle of scattering 2ϑ between $\boldsymbol{K}_2$ and $\boldsymbol{K}_1$ must satisfy the relation

$$|\boldsymbol{K}_1 - \boldsymbol{K}_2| = \frac{2\pi}{\lambda} 2\sin\vartheta = 2\pi\tau = 2\pi\frac{m}{d} \tag{102.7}$$

or

$$2d\sin\vartheta = m\lambda. \tag{102.8}$$

An upper index τ has been attached to the symbol of the differential cross section (102.5) in order to recall that it represents the contribution arising from $\boldsymbol{\tau}$-reflection, i.e. from the reflection of order m at the set of lattice planes of Miller indices (h, k, l).

In order to pass to the case of polycrystalline materials we must average (102.5) over all crystal orientations, an operation that in the following will be denoted by the symbol $(\ldots)_{\text{av}}$.

This is done by averaging over all directions of $\boldsymbol{\tau}$ which, for any given value, has a fixed orientation relative to the crystal axes. We choose a polar axis in K-space along the direction of $\boldsymbol{K}_1 - \boldsymbol{K}_2$ and denote the polar coordinates of $\boldsymbol{\tau}$ by τ, ϑ_τ and φ_τ, so that

$$\left(\frac{d\sigma^{\tau}_{\text{el. coh}}}{d\omega}\right)_{\text{av}} = \frac{1}{4\pi}\int \frac{d\sigma^{\tau}_{\text{el. coh}}}{d\omega} \times \\ \times d\varphi_\tau \sin\vartheta_\tau\, d\vartheta_\tau .$$

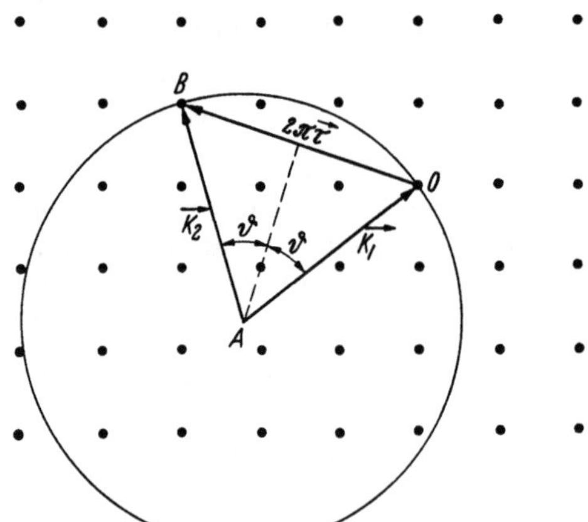

Fig. 191. Ewald's construction in K-space.

If τ_x, τ_y, τ_z are the corresponding Cartesian coordinates taken with the z direction coincident with the polar axis, the Jacobian[1] becomes

$$D(\tau_x, \tau_y | \vartheta_\tau, \varphi_\tau) = \tau^2 \sin\vartheta_\tau \cos\vartheta_\tau$$

and therefore

$$\int_0^{2\pi} d\varphi_\tau \int_0^{\pi} \delta(\boldsymbol{K}_1 - \boldsymbol{K}_2 + 2\pi\boldsymbol{\tau}) \sin\vartheta_\tau\, d\vartheta_\tau$$
$$= \frac{1}{(2\pi)^2\tau^2}\int \delta(\tau_x)\,\delta(\tau_y)\,\delta(|\boldsymbol{K}_1 - \boldsymbol{K}_2| - 2\pi\tau_z)\,\frac{1}{\cos\vartheta_\tau} = \frac{1}{(2\pi)^2\tau^2}\,\delta(2K_1\sin\vartheta - 2\pi\tau).$$

radiation is drawn in Fig. 191 so that it terminates at the origin O of reciprocal space. If the sphere of radius $|\boldsymbol{K}_1|$ is drawn about A, the possible directions of diffracted rays are those determined by points like B which lie on the sphere. Condition (102.6) states that the vector $\boldsymbol{K}_2$ must close the triangle AOB. When two or more atoms are present in the unit cell of the crystall, the positions of which are given in terms of $\boldsymbol{a}$, $\boldsymbol{b}$ and $\boldsymbol{c}$ by the expression

$$\boldsymbol{r}_j = \boldsymbol{a}u_j + \boldsymbol{b}v_j + \boldsymbol{c}w_j, \tag{f}$$

the corresponding phase difference is obtained from (102.6), (a), (f) and (d)

$$|\boldsymbol{r}_j\cdot(\boldsymbol{K}_1 - \boldsymbol{K}_2)| = 2\pi(h'u_j + k'v_j + l'w_j) = 2\pi m(hu_j + kv_j + lw_j). \tag{g}$$

[1] See footnote 3, p. 431.

Introducing this expression in (102.5) one obtains

$$\left(\frac{d\sigma^{\tau}_{\mathrm{el,coh}}}{d\omega}\right)_{\mathrm{av}} = \frac{1}{2}\,\frac{a_b^2}{v_0}\,\frac{1}{\tau^2}\,\delta(2K_1\sin\vartheta - 2\pi\tau)\,. \tag{102.9}$$

Thus, for the neutrons scattered elastically we find the familiar powder scattering of X-rays along certain circular cones about the line of incidence, the so-called *Debye-Scherrer rings*. Each ring only depends on the magnitude of $\boldsymbol{\tau}$, but in general there are several $\boldsymbol{\tau}$ in the reciprocal lattice having the same absolute value τ. Therefore the correct intensity of a particular ring, is obtained by multiplying Eq. (102.9) by the multiplycity M of the reflection considered. We shall come back to this point in the discussion of Eq. (102.17).

In order to obtain the total elastic coherent scattering cross section per nucleus originating from a particular τ reflection, we must now integrate (102.9) over all directions of the scattered neutrons. Recalling that $d\omega = 2\pi\sin(2\vartheta)\,d(2\vartheta)$ with ϑ varying between 0 and $\pi/2$, one obtains

$$\left(\sigma^{\tau}_{\mathrm{el,coh}}\right)_{\mathrm{av}} = \frac{\lambda^2}{2}\,\frac{a_b^2}{v_0}\,\frac{1}{\tau(h',k',l')} \tag{102.10}$$

which represents the total elastic coherent scattering cross section per nucleus of a polycrystalline material arising from m-th order reflection on the set of lattice planes (h, k, l) of spacing d.

The *total elastic coherent scattering cross section* for any type of reflection is finally obtained by summing over all sets of crystal planes and orders of reflection which are consistent with Bragg condition (102.7)

$$\left(\sigma_{\mathrm{el,coh}}\right)_{\mathrm{av}} = \frac{\lambda^2}{2}\,\frac{a_b^2}{v_0}\sum_{|\tau|\le\frac{2}{\lambda}}\frac{1}{\tau} \tag{102.11}$$

where the sum has to be extended over all points of the reciprocal lattice contained in the sphere of radius $2/\lambda$.

Eq. (102.11) holds for a very special ideal case but can be generalized as follows.

First it can be extended to crystals with two or more atoms per unit cell. The procedure to be followed for the derivation of the cross section is almost identical to that given above with the difference that passing from (102.4) to (102.5) instead of the scattering length a_b, the *structure amplitude*

$$F = \sum_{\substack{j\\ \text{atoms of}\\ \text{unit cell}}} a_{b_j}\,e^{i\boldsymbol{r}_j\cdot(\boldsymbol{K}_1-\boldsymbol{K}_2)} \tag{102.12}$$

has to be factorized. In this expression the summation is over the atoms of the unit cell which have scattering amplitude a_{bj} and coordinates u_j, v_j, w_j in terms of the fundamental translation vectors of the lattice [see Eq. (f) of footnote 2 on p. 432]. Because of Eq. (g) of that note we may write

$$\left.\begin{aligned} F(h',k',l') &= \sum_{\substack{j\\ \text{atoms of}\\ \text{unit cell}}} a_{bj}\exp\left[2\pi i\,(h'u_j + k'v_j + l'w_j)\right],\\ &\qquad h' = m\,h,\; k' = m\,k,\; l' = m\,l.\end{aligned}\right\} \tag{102.13}$$

In the case of crystals containing nuclei of a single type the scattering amplitude (102.13) may be written

$$F(h',k',l') = a_b\,\mathrm{S}(h',k',l') \tag{102.14}$$

where

$$\mathsf{S}(h', k', l') = \sum_{\substack{\text{atoms of} \\ \text{unit cell}}} e^{2\pi i (h' n_1 + k' n_2 + l' n_3)} \tag{102.15}$$

is called the *geometrical amplitude factor*. In the expression of the cross section [see Eq. (102.17) below] there enters the square of the modulus of (102.13) which is called the *structure factor*. As an example we recall that for hexagonal close-packed lattices of identical atoms (as for instance Be) one has

$$|\mathsf{S}|^2 = 1 + \cos \pi \frac{2h' + 4k' + 3l'}{3}. \tag{102.16}$$

The second step in the generalization of Eq. (102.11) consists of allowing the atoms of the crystal to oscillate around their equilibrium positions because of thermal agitation and zero point energy. This introduces the Debye-Waller factor as shown in Eqs. (99.11).

In conclusion, we can write the cross section per unit cell for transmission of neutrons through polycrystalline materials, due to coherent elastic scattering of any order on all lattice planes, in the form

$$(\sigma_{\text{el, coh}})_{\substack{\text{av} \\ \text{unit cell}}} = \frac{\lambda^2}{2} \frac{1}{v_0} \sum_{\tau \leqq \frac{2}{\lambda}} \left(|F|^2 M \frac{1}{\tau}\right)_{h'k'l'} e^{-2W(T_0)\left(\frac{\tau}{2}\right)^2} \tag{102.17}$$

where the Debye-Waller factor has been inserted taking into account the fact that according to (102.7) one has

$$\frac{\sin\vartheta}{\lambda} = \frac{\tau}{2},$$

and where M denotes the *multiplicity*, i.e. the number of possible orientations of the unit cell for the particular reflection. It arises from the fact that the sum appearing in (102.17) extends over the values of $\tau = \frac{m}{d} \leqq \frac{2}{\lambda}$. It did not appear in (102.11) because there the sum was extended over all points of the reciprocal lattice which fall inside the sphere of radius $2/\lambda$.

In order to illustrate the dependence of the coherent elastic scattering cross section (102.17) on the neutron wavelength λ, we may start by considering the transmission of a beam of neutrons through a polycrystalline material whose wavelength is longer than the largest spacing $d_{\max}$ of the crystal lattice

$$\lambda \geqq \lambda_c = 2\, d_{\max}. \tag{102.18}$$

Under these conditions no real value of ϑ exists for which the Bragg condition (102.8) is fulfilled. Therefore, as pointed out by Wick[1], the elastic transmission cross section is negligible for all possible orientations of the microcrystals and the beam of neutrons is transmitted through finite thicknesses of polycrystalline materials with almost unreduced intensity. As the wavelength is decreased the transmission cross section (102.17) of the polycrystalline material suddenly increases for $\lambda = \lambda_c$ because the lattice planes which are spaced $2d_{\max}$ start to reflect the incident neutrons. By further decreasing the wavelength below the *cut-off value* λ_c, the cross section decreases as λ^2 until a second sudden increase takes place when λ reaches the next largest value $2d(h', k', l')$. Other discontinuities of the same type will occur each time λ reaches a value $2d(h', k', l')$

[1] See footnote 1, p. 431.

relative to a new set of lattice planes in a suitable order of reflection. For neutrons of sufficiently large energy the individual discontinuities become less and less pronounced so that the cross section (102.17) tends to a smooth function of λ.

Table 76 contains the cut-off wavelength λ_c and the corresponding values of the neutron energy E_c for various light polycrystalline materials of interest as moderators. A more complete collection of data of this type may be found in Ref. [*18b*].

The existence of a cut-off wavelength beyond which a non-absorbing polycrystalline material is almost transparent allows the production of beams of *cold neutrons*[1]. If an intense beam of thermal neutrons from a pile impinges on the base of a long thin column of a polycrystalline non-absorbing material the neutrons with $\lambda < \lambda_c$ will suffer Bragg-reflections, change direction of motion and eventually leave the column, while the neutrons with $\lambda > \lambda_c$ will suffer only a very weak attenuation due to incoherent scattering. Fig. 192 shows the spectrum obtained by FERMI and MARSHALL by filtering the neutrons from the Argonne pile through a beryllium oxide filter of this type (see Sect. 101).

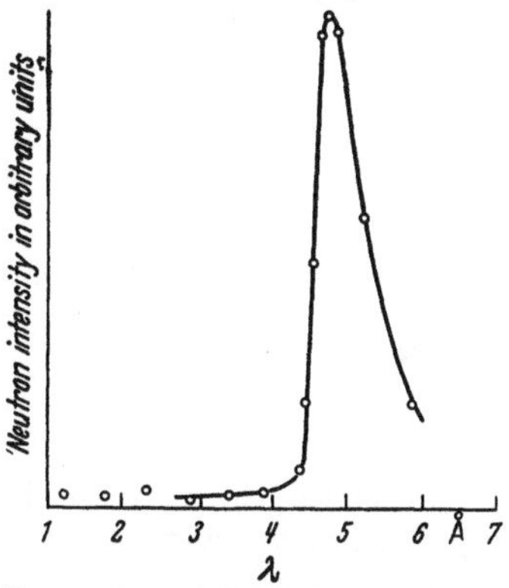

Fig. 192. Spectral distribution of cold neutrons from a beryllium oxide filter (FERMI and MARSHALL).

Table 76. *Cut-off wavelength λ_c and corresponding energy E_c for various polycrystalline materials.*

	λ_c (Å)	E_c (ev)
Be	3.95	0.052
BeO	4.4	0.0042
C	6.69	0.00183
Al	4.67	0.00375

The typical behaviour described by (102.17) is shown by the experimental results of FERMI, STURM and SACHS[2] who were the first to investigate this phenomenon. Figs. 193 and 194 show the transmission cross section obtained by these authors in the case of microcrystalline Be metal reduced to chips of linear dimensions smaller than $5 \cdot 10^{-5}$ cm [see Eq. (102.19)], and in the case of a sintered block of BeO prepared from very fine powder crystals. The measurements were made with a mechanical velocity selector between 0.004 and 0.020 ev and with a crystal spectrometer between 0.04 and 1 ev.

The curves given in Figs. 193 and 194, calculated by means of (102.17), agree very well with the experimental results. The agreement however, is determined by a number of conditions which are not satisfied in the general case of polycrystalline moderators. Among various reasons for deviation from the behaviour discussed above, the most important one appears to arise from the occurrence in the sample of crystals so large that the scattering peaks are wiped out by "extinction". This expression refers to the fact that the incident beam may be depleted of neutrons of a given energy because of scattering by an individual crystal for which the Bragg condition (102.8) is fulfilled.

Such an effect will be avoided if the linear dimensions of the microcrystals are so small that the intensity of the neutrons of a given energy, scattered out of the beam by a single crystal at the Bragg angle, represents a small fraction

[1] See footnote 7, p. 431.

[2] E. FERMI, N. J. STURM and R. G. SACHS: Phys. Rev. **71**, 589 (1957); see also L. WINSBERG, D. MENEGHETTI and S. S. SIDHU: Phys. Rev. **75**, 975 (1949).

of the intensity of the incident neutrons: for instance not more than 1%. This fraction can be obtained by means of the following semiqualitative consideration. If l defines the linear dimensions of a microcrystal, at the Bragg angle, the intensity of the scattered beam is given, apart from a numerical factor, by the expression

$$I_{\text{scatt}} \sim \left(\frac{F\,l^3}{v_0}\right)^2 \cdot \left(\frac{\lambda^2}{l^2 \sin\vartheta}\right)$$

where the first factor represents the intensity of the scattered beam per unit solid angle, and the second factor the solid angle in which the scattered beam is concentrated; its value is determined by the finite dimensions of the crystal. In the same units, the intensity of the incident beam is given by the projected cross sectional area of the crystal $l^2 \sin\vartheta$. By introducing the correct numerical factor[1,2] and making use of (102.7) one obtains

$$\frac{I_{\text{scatt}}}{I_{\text{incid}}} = \frac{1}{\pi}\,\frac{l^2}{v_0^2}\,\frac{|F|^2}{\tau^2}.$$

If the condition

$$\frac{I_{\text{scatt}}}{I_{\text{incid}}} \lesssim 10^{-2}$$

is imposed, one obtains

$$l \leqq \frac{\pi^{\frac{1}{2}}}{10}\,\frac{\tau\,v_0}{|F|}. \tag{102.19}$$

With $\tau = 3\times 10^7\ \text{cm}^{-1}$, $|F|^2 = 10^{-23}\ \text{cm}^2$ and $1/v_0 = 4\times 10^{22}$ one has

$$l \leqq 5\times 10^{-5}\ \text{cm}.$$

This condition refers to the "perfect microcrystals" of which the material is composed, not to the crystallites consisting of many perfect microcrystals orientated at small angles with respect to each other.

Other sources of deviation from (102.17) are: some regularity in the orientations of the microcrystals, and the presence of films of impurities, as for instance water, absorbed on the surface of the grains constituting the powder.

Finally the various causes of incoherence discussed in Sect. 99 produce a reduction of the elastic coherent scattering (102.17) and introduce a corresponding background of diffuse scattering.

In fact, if isotopic and spin incoherence are present, the derivation of the elastic scattering cross section (coherent + incoherent) must be modified as follows. For the sake of simplicity we confine ourselves to the case of crystals composed of a single element and containing only one single atom per unit cell. We again start from Eq. (102.4) which in this case must be averaged with respect to all possible distributions of the isotopes and spin orientations among the lattice points. The result of this operation is that the cross section can be split into two parts, proportional respectively to σ_{coh} and σ_{incoh} as defined by Eq. (99.7). From this point on, one proceeds exactly as in the derivation of (102.11) from (102.5) given above. The result is[3]

$$(\sigma_{\text{el, coh}})_{\text{av}} = \frac{\lambda^2}{2v_0}\,\frac{\sigma_{\text{coh}}}{4\pi} \sum_{\tau \leqq \frac{2}{\lambda}} \frac{e^{-W(T_0)\,(\tau/2)^2}}{\tau}, \tag{102.20}$$

$$(\sigma_{\text{el, incoh}})_{\text{av}} = \sigma_{\text{el, incoh}} = \frac{\sigma_{\text{incoh}}}{2W(T_0)}\,\lambda^2 \left\{1 - e^{-2W(T_0)\frac{1}{\lambda^2}}\right\}, \tag{102.21}$$

with σ_{coh} and σ_{incoh} referring to bound nuclei.

[1] See footnote 7, p. 431.
[2] See footnote 2, p. 436.
[3] See footnote 5, p. 431.

We notice that (102.17) reduces to (102.20) if applied to the case of polycrystalline materials composed of a single element ($F = a_b\,\mathsf{S}$) with a single atom

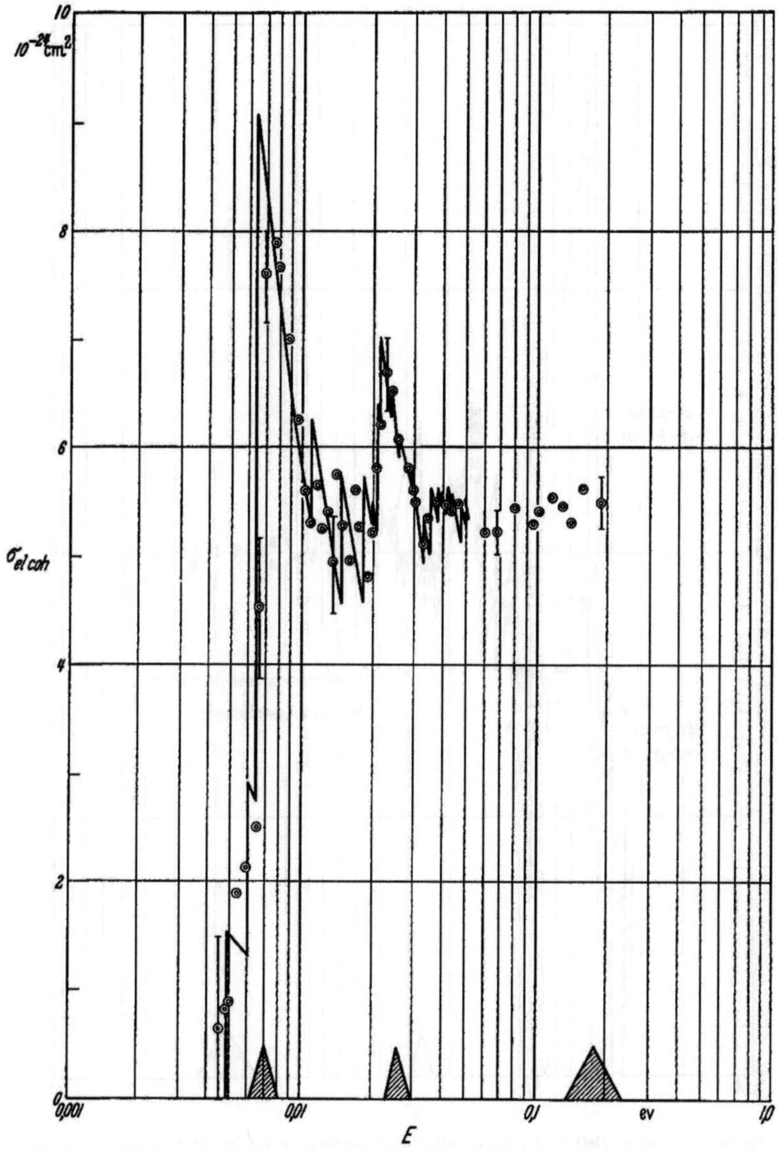

Fig. 193. The microcrystalline beryllium cross section. The solid line is calculated by means of (102.17). Triangles indicate the resolution of the instrument in the various ranges of measurements (FERMI, STURM and SACHS).

per unit cell ($\mathsf{S} = 1$), provided that the amplitude a_b is replaced by its coherent part

$$a_{\text{coh}} = \left(\frac{\sigma_{\text{coh}}}{4\pi}\right)^{\frac{1}{2}}$$

where σ_{coh} is defined by (99.7).

Concerning Eq. (102.21) we have to notice that, whenever

$$\frac{2W(T_0)}{\lambda^2} \ll 1, \tag{102.22}$$

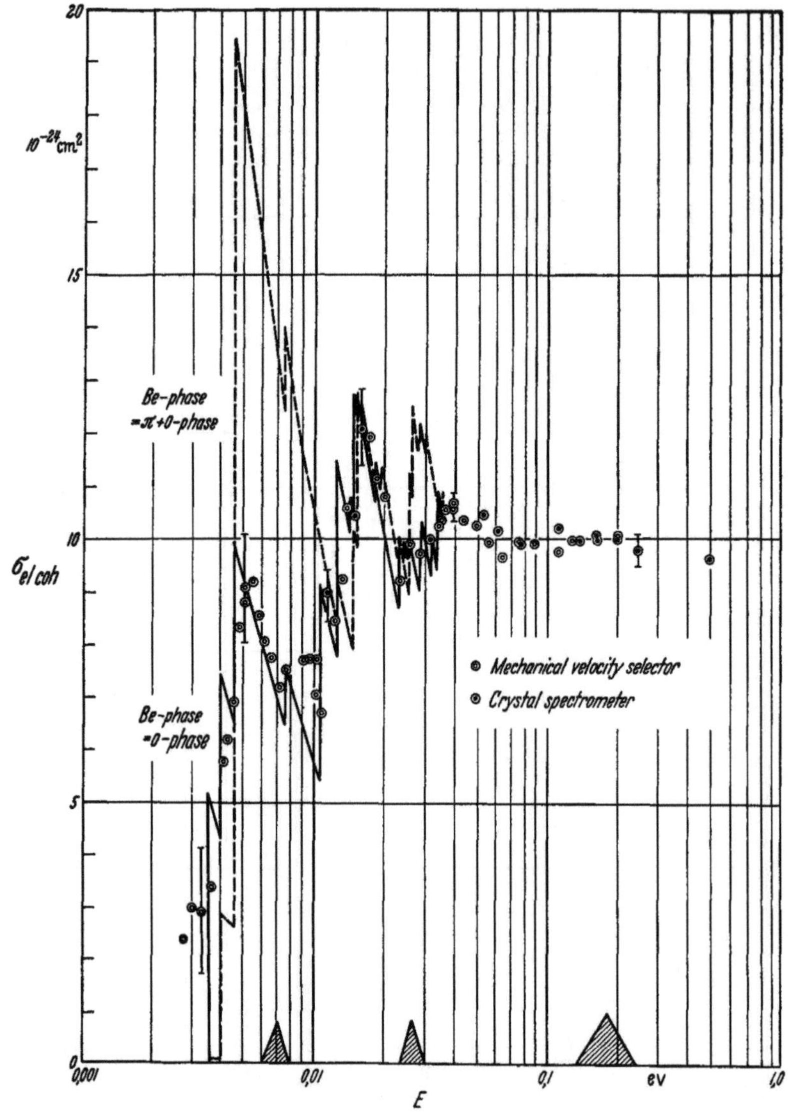

Fig. 194. The microcrystalline BeO cross section. Alternate theoretical curves for the same and the opposite phase of neutron scattering are shown as solid and dotted lines respectively. Results indicate that the Be phase is the same as the oxygen phase (Fermi, Sturm and Sachs).

it is permitted to expand the exponential in powers of $2W(T)/\lambda^2$ and to break the series off after the linear term with the result

$$\sigma_{\text{el, incoh}} = \sigma_{\text{incoh}}. \tag{102.23}$$

Therefore, if condition (102.22) is fulfilled, as frequently happens, the elastic incoherent cross section will neither depend on the crystal temperature nor on the neutron wavelength. In Sect. 103 we shall refer to this result in the discussion of Eq. (103.15).

At the end of Sect. 103, after a discussion of inelastic scattering, we shall reexamine the elastic scattering from a more general point of view [see Eq. (103.31) and the discussion which follows it].

103. Inelastic scattering. The inelastic scattering cross section σ_{inel} arises from scattering processes in which energy is exchanged between the neutron and the crystal. If the neutron transfers energy to the lattice vibrations the process is described as phonon emission by the neutron; alternatively if the neutron absorbs energy from the lattice vibrations the process is described as phonon absorption.

This remark suggests the decomposition of σ_{inel} into partial cross sections, σ_l, for the emission or absorption of l phonons

$$\sigma_{\text{inel}} = \sum_{l \geqq 1} \sigma_l . \tag{103.1}$$

Sometimes each term of the sum is split into two parts

$$\sigma_l = \sigma_{+l} + \sigma_{-l} , \tag{103.2}$$

where the first represents the cross section for emission of l phonons, the second the cross section for absorption of l phonons. The term with $l=0$ has been omitted in (103.1) since it corresponds to elastic scattering which has been treated in Sect. 102.

As we shall see below, the convergence of (103.1) depends on the energy of the neutron and the temperature and structure of the crystal. Before discussing this point it is convenient to express the total cross section for inelastic scattering as a sum of two terms, corresponding to the separation (99.1) of the scattering cross section per nucleus into a coherent and an incoherent part

$$\sigma_{\text{inel}} = \sigma_{\text{inel, coh}} + \sigma_{\text{inel, incoh}} \tag{103.3}$$

with the first term proportional to σ_{coh}, the second term to σ_{incoh} as defined for bound nuclei by (99.7). Combining (103.1) with (103.3) we have

$$\sigma_{\text{inel, coh}} = \sum_{l \geqq 1} \sigma_{l\,\text{coh}} , \tag{103.4}$$

$$\sigma_{\text{inel, incoh}} = \sum_{l \geqq 1} \sigma_{l\,\text{incoh}} . \tag{103.5}$$

Theoretical expressions for the various cross sections defined above have been derived by many authors following the procedure mentioned in Sect. 98, namely in the first Born approximation with the neutron-nucleus interaction described by the Fermi pseudopotential (98.5).

The theory of processes in which a single phonon is exchanged has been worked out by many authors in the Debye approximation for crystals containing a single element arranged in a Bravais lattice with a single nucleus per unit cell[1-4]. The expressions obtained for this case can be generalized to any Bravais lattice containing a single element, by multiplication by the square of the geometrical structure factor as defined by (102.15).

---

[1] R. WEINSTOCK: Phys. Rev. **65**, 1 (1944).
[2] A. AKHEISER and J. POMERANCHUK: J. Exp. Theor. Phys. USSR. **17**, 770 (1947).
[3] J.M. CASSELS: Progress in nuclear physics, Vol. 1, p. 185. 1950.
[4] D. KLEINMAN: Phys. Rev. **81**, 326 (1951).

All papers mentioned above may be considered extensions of that by WEINSTOCK[1] who derived the expression for

$$\sigma_{1\,\mathrm{coh}} = \sigma_{+1\,\mathrm{coh}} + \sigma_{-1\,\mathrm{coh}}$$

for crystals satisfying conditions (a) and (b) of Sect. 99 (p. 411) by making use of harmonic oscillator wave functions to represent the lattice vibrations. CASSELS[2] established the expressions for $\sigma_{+1\,\mathrm{incoh}}$ and $\sigma_{-1\,\mathrm{incoh}}$ as they arise from isotopic and/or spin incoherence.

As long as neutron energy and crystal temperature are low, the one-phonon processes constitute the main contribution to σ_{inel}. The multiphonon processes become appreciable at high neutron energy and/or for high temperature of the crystal, since in the former case the neutron can emit and absorb, while in the latter case it can absorb, many phonons.

α) Total inelastic cross section for Debye crystals. The incoherent approximation. Multiphonon processes have been considered by FILKENSTEIN[3] for an Einstein crystal, and by SQUIRES[4], PLACZEK[5,6], SINGWI and KOTHARI[7] and HELPIN and JACROT[8] in Debye approximation.

SQUIRES makes use of the decomposition of $\sigma_{\mathrm{inel,\,incoh}}$ into partial cross sections and computes the term corresponding to the emission of m phonons and the simultaneous absorption of n phonons by starting from the expressions for the probabilities P^f_{+1}, P^f_{-1} that the neutron gives or receives from the f-th oscillator one single phonon[9]. The probability that the neutron gives one phonon to each of the $f_1, f_2, \ldots, f_m$-th oscillators and receives one from each of the $g_1, g_2, \ldots, g_n$-th oscillators is immediately obtained in terms of P^f_{+1} and P^f_{-1} since the $m+n$ processes considered are independent one from the other. It may be mentioned here that the probability for the absorption of two phonons of the same type, is N^{-1} times the probability for the absorption of two different types of phonons, N being the total number of nuclei in the crystal.

The differential cross section for the particular set of excitation and deexcitation processes is obtained by multiplying the corresponding probability by a convenient factor. By averaging over all possible distributions of the various isotopes and spin orientations among the lattice points, the expression for the cross section may be divided into two parts, proportional respectively to σ_{coh} and σ_{incoh}. The total inelastic incoherent cross section $\sigma_{\mathrm{inel,\,incoh}}$ is finally obtained by performing the following operations: (a) summation over the three directions of polarization for each phonon, f and g, (b) integration over all possible directions of the scattered neutrons, and (c) summation over all combinations of the oscillators.

This procedure has the inconvenience that the convergence of (103.5) becomes progressively poorer with increasing neutron energy and/or temperature of the crystal, and that the structure of the higher terms is exceedingly complicated.

A new method of calculation of the total cross section for incoherent scattering has been introduced by PLACZEK[5] who used essentially the same procedure that he had previously developed[10] for the derivation of Eq. (100.1). The neutron

[1] See footnote 1, p. 440.
[2] See footnote 3, p. 440.
[3] R. J. FILKENSTEIN: Phys. Rev. **72**, 907 (1947).
[4] G.L. SQUIRES: Proc. Roy. Soc. Lond., Ser. A **212**, 192 (1952).
[5] G. PLACZEK: Phys. Rev. **93**, 895 (1954).
[6] G. PLACZEK: Phys. Rev. **105**, 1240 (1957).
[7] K.S. SINGWI and L.S. KOTHARI: p. 56 of Ref. [*15a*].
[8] A. HELPIN and B. JACROT: J. Phys. Radium **16**, 35 (1955).
[9] See footnote 1, p. 440.
[10] G. PLACZEK: Phys. Rev. **86**, 377 (1952).

variables, such as wavelength and momentum, are expressed in terms of the neutron energy E and the cross section is expanded in powers of the inverse of the mass ratio A

$$\sigma_{\text{el+inel, incoh}} = \sigma_{\text{incoh}} \left\{ 1 + \sum_{n \geq 1} \frac{A^{(n)}}{A^n} \right\}. \tag{103.6}$$

This expansion converges rapidly for all energies of the neutron provided A satisfies the two conditions

$$A = \frac{M}{m} \gg 1, \quad A > \frac{T_0}{\Theta_D}. \tag{103.7}$$

Under these conditions, which are well satisfied in most practical cases for temperatures right up to the melting point of crystals, the term in the third power of A^{-1} may be neglected so that (103.6) reduces to

$$\sigma_{\text{el+inel, incoh}} = \sigma_{\text{incoh}} \left\{ 1 + \frac{A^{(1)}}{A} + \frac{A^{(2)}}{A^2} \right\}. \tag{103.8}$$

Analytic expressions of the coefficients $A^{(1)}$ and $A^{(2)}$ in terms of $E/k\Theta_D$ and T_0/Θ_D suited for numerical computation are given in the second paper by PLACZEK[1] while the first paper contains a table of numerical values of the coefficients of expansions of $A^{(1)}$ and $A^{(2)}$ in powers of T_0 and λ, which are valid in the case $E/k\Theta_D \ll 1$, $T_0/\Theta_D \gg 1$.

The same expressions of the coefficients $A^{(1)}$ and $A^{(2)}$ have been derived by SINGWI and KOTHARI[2] who have extended the method of PLACZEK to the calculation of $\sigma_{\text{inel, coh}}$.

These authors also derive the cross section in the first Born approximation by making use of the matrix element of the Fermi pseudo-potential (98.5) between the initial state, with the neutron having wave vector $\boldsymbol{K}_1$ and the lattice vibrations in the states $n_1, n_2, \ldots$, and the final state characterized by $\boldsymbol{K}_2$ and $n'_1, n'_2, \ldots$ A considerable gain in simplicity in the calculation of this matrix element is obtained by SINGWI and KOTHARI by adopting the field theory method based on the use of creation and destruction operators for the description of processes of excitation and deexcitation of lattice vibrations. By such a procedure they obtain the following expressions for the terms appearing in the expansion (103.5)

$$\sigma_{l\,\text{incoh}} = \frac{\sigma_{\text{incoh}}}{2K_1^2\, l!} \left(\frac{3}{2A}\right)^l \prod_{j=1}^{l} \int_{-1}^{+1} X_j \left(\coth \frac{X_j}{2T_0} - 1\right) dX_j \int_{K_2-K_1}^{K_2+K_1} \mathrm{e}^{-\frac{6\Phi}{A} t^2} \cdot t^{2l+1}\, dt, \tag{103.9}$$

where

$$X_j = \frac{\hbar\,\omega_j}{k\,\Theta_D},$$

$\Phi(T_0)$ is given by (99.13), T_0 is expressed in terms of the Debye temperature and the unit of length (in which $1/K_1$ and $1/K_2$ are measured) is $\left[\frac{\hbar^2}{2m\,k\,\Theta_D}\right]^{\frac{1}{2}}$.

The l values of X_j appearing in (103.9) must be chosen in such a way as to satisfy conservation of energy, which in the units used by the authors, can be written

$$K_2^2 = K_1^2 + \sum_{j=1}^{l} X_j. \tag{103.10}$$

---

[1] See footnote 5, p. 442.
[2] See footnote 7, p. 442.

It must be noted that in order to take into account both emission as well as absorption processes, the integral with respect to X_j had to be extended from -1 to $+1$ so that X_j takes on both positive and negative values.

Eq. (103.9) had also been derived by PLACZEK[1] and was the starting point of his expansion in negative powers of A (103.6).

The coherent part of the inelastic cross section is split by SINGWI and KOTHARI into two terms

$$\sigma_{\text{inel, coh}} = \sum_{l=1}^{\infty} \sigma_{l\,\text{coh}} = \sum_{l=1}^{\infty} [\sigma_{l\,\text{coh}}^{(0)} + \sigma_{l\,\text{coh}}^{(1)}] = \sigma_{\text{inel, coh}}^{(0)} + \sigma_{\text{inel, coh}}^{(1)} \tag{103.11}$$

where the second term may be considered as a correction with respect to the first one. The first term is given by the exact relation

$$\frac{\sigma_{l\,\text{coh}}^{(0)}}{\sigma_{\text{coh}}} = \frac{\sigma_{l\,\text{incoh}}}{\sigma_{\text{incoh}}}, \qquad l \geqq 1. \tag{103.12}$$

The so-called *incoherent approximation* suggested by PLACZEK, in which it is assumed that

$$\frac{\sigma_{\text{inel, coh}}}{\sigma_{\text{coh}}} = \frac{\sigma_{\text{inel, incoh}}}{\sigma_{\text{incoh}}}$$

is not always justified, since in many cases $\sigma_{\text{inel, coh}}^{(1)}$ represents a non-negligible fraction of $\sigma_{\text{inel, coh}}^{(0)}$.

By expanding the correction term $\sigma_{\text{inel, coh}}^{(1)}$ according to (103.4), SINGWI and KOTHARI obtain

$$\sigma_{l\,\text{coh}}^{(1)} = \frac{\sigma_{\text{coh}}}{2K_1^2\, l!} \left(\frac{3}{2A}\right)^l \sum_i \frac{Z_{r_i}}{r_i} \prod_{j=1}^{l} \int_{-1}^{+1} X_j\, dX_j \left(\coth \frac{X_j}{2T} - 1\right) \int_{K_2-K_1}^{K_2+K_1} e^{-\frac{6\Phi}{A} t^2}\, t^{2l} \sin r_i t \cdot dt \tag{103.13}$$

where r_i is the modulus of the lattice vector (102.2) defining the equilibrium position of the i-th nucleus in the crystal, and Z_{r_i} the number of nuclei at the lattice distance $|\boldsymbol{r}_i|$. After summation over l, an expansion into inverse powers of A can also be obtained for $\sigma_{\text{coh}}^{(1)}$,

$$\sigma_{\text{coh}}^{(1)} = \sum_{n \geqq 1} \frac{B^{(n)}}{A^n}, \tag{103.14}$$

and expressions suitable for numerical computation in terms of K_1 and T_0 can be derived for the coefficients $B^{(n)}$.

These formulas have been applied by the same authors to calculating the inelastic cross section as a function of both neutron wavelength λ_1 and sample temperature T_0 for the following polycrystalline materials: Be, Mg, Al, Fe and Pb. In the numerical computation of $\sigma_{\text{inel, incoh}}$ for all cases except that of Be, the term $A^{(2)}/A^2$ in (103.8) was neglected since it is about 0.5% of $A^{(1)}/A$. In the computation of the correction term $\sigma_{\text{inel, coh}}^{(1)}$ appearing in (103.11), all terms beyond $B^{(1)}/A$ in the expansion (103.14) have been neglected for all polycrystalline materials considered. In the case of Be, on account of its small mass, the term $A^{(2)}/A^2$ was taken into account; it amounted to about 10% of the main term but with negative sign. Higher terms have not been computed since they are expected to contribute only a few percent. To these results for Be we shall again refer in Sect. 107.

The best conditions for a comparison with the experimental results are obtained at neutron wavelengths greater than the cut-off (see Table 76) since for $\lambda \leqq \lambda_c$ the main contribution to the cross section arises from the elastic coherent scattering discussed in Sect. 102.

[1] See footnote 6, p. 441.

For $\lambda > \lambda_c$ the remaining cross section may generally be divided into three parts

$$\sigma_{\text{tot}} = \sigma_a + \sigma_{\text{el, incoh}} + \sigma_{\text{inel}} \tag{103.15}$$

The first term represents the absorption cross section which usually follows the $1/v$ law (i.e. $\sigma_a \propto \lambda$) and is independent of the sample temperature. The elastic incoherent cross section is given by (102.21) which reduces to (102.23) when (102.22) is satisfied, as occurs in most practical cases. It is therefore independent of both sample temperature and neutron wavelength. The last term may be calculated by means of (103.3), (103.5), (103.9), (103.11), (103.12), (103.13). Its dependence on λ is in general rather complicated. However, if we only consider neutrons of very low energy (such as those satisfying the condition $\lambda \gg \lambda_c$) and to polycrystalline samples at very low temperature, the main contribution to σ_{inel} will be given by σ_{-1} i.e. by processes in which the neutron absorbs a single phonon from the lattice vibration.

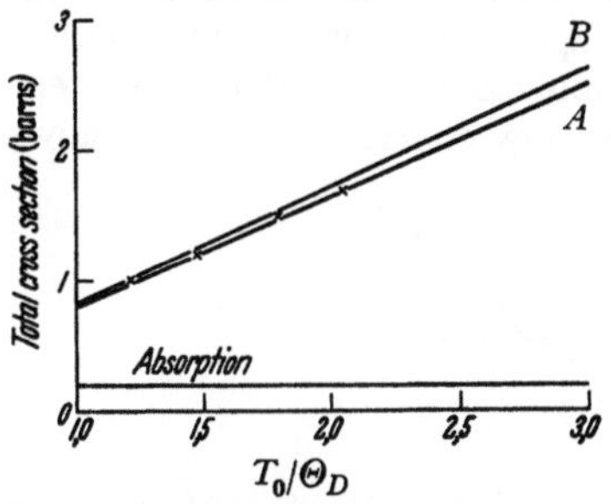

Fig. 195. The total cross section of Mg for λ = 6.25 Å, versus temperature. Experimental points are taken from SQUIRES; curve B is PLACZEK's incoherent approximation; curve A is the same as B, with SINGWI and KOTHARI's correction included.

For K_1 very small this term follows the $1/v$ law ($\sigma_{-1} \propto \lambda$) as well as σ_a, but can be distinguished from the latter because it increases rapidly with sample temperature[1].

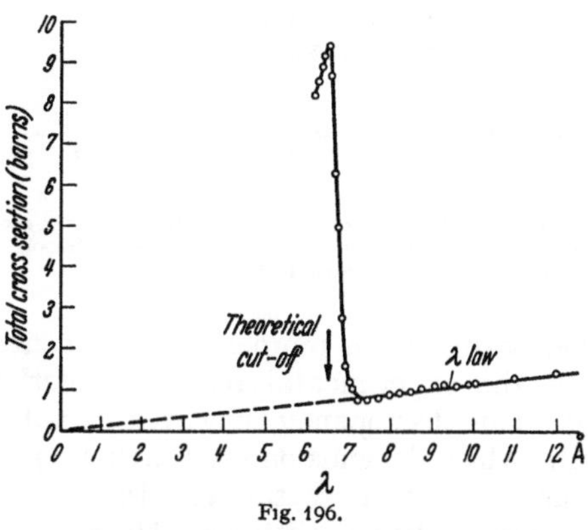

Fig. 196.

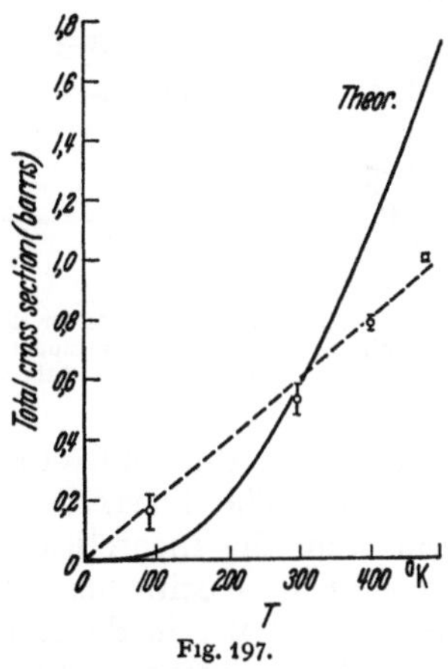

Fig. 197.

Fig. 196. The total cross section of graphite versus λ at room temperature (HUGHES *et al.*).

Fig. 197. The total cross section of graphite at $\lambda = 7$ Å as a function of temperature. Experimental points were obtained by the Brookhaven group, the theoretical curve by KLEINMAN.

Fig. 195 taken from SINGWI and KOTHARI's paper, shows the variation with temperature of the total cross section (103.15) in the case of Mg (hexagonal close-packed, $\lambda_c = 5.5$ Å, $\Theta_D = 305°$ K) for neutrons of $\lambda = 6.25$ Å. Curve B represents

[1] This behaviour of σ_{-1} can be understood by means of arguments similar to those given in Sects. 13, 25 and 27, where the conditions for having $\sigma_a \propto \lambda$ have been discussed. In the limit of very long wavelength of the incident neutron, the probability P of absorption of a phonon does not depend on the velocity of the incident neutron and therefore, according to (13.2), $\sigma_{-1} \propto \lambda$. Also in the present case the constancy of P is due to the fact that the outgoing particle is fast so that the corresponding width Γ_n does not depend on the velocity of the incident neutron (Sect. 27) [*12*].

the incoherent approximation in which $\sigma^{(1)}_{\text{inel, coh}}$ is neglected, while curve A includes this term which gives a negative contribution of the order of 4% of the total scattering cross section. The agreement of curve A with the experimental results by SQUIRES[1] is fairly satisfactory.

The total cross section for graphite obtained by HUGHES and coworkers [*12*] at room temperature is shown in Fig. 196. The linear behaviour of the total cross section with λ, observed beyond the cut-off $\lambda_c = 6.69$ Å, is due to coherent inelastic scattering since graphite is essentially monoisotopic and of spin zero (so that $\sigma_{\text{incoh}} = 0$) and σ_a is negligible even at very low energies. Fig. 197 shows

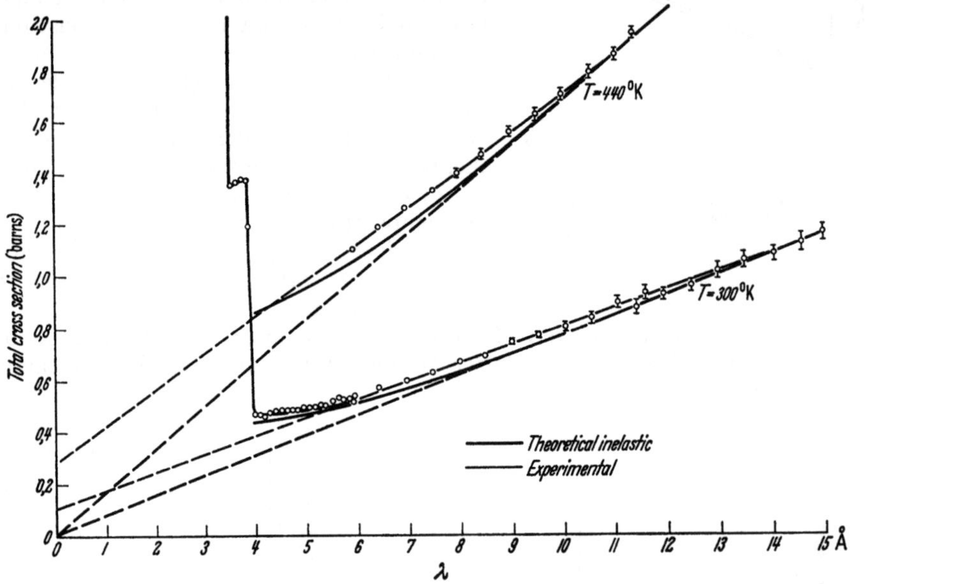

Fig. 198. The total cross section of Be for slow neutrons at 300 and 440° K, measured with the Brookhaven chopper, compared with the theoretical curves computed by KLEINMAN. The broken lines represent the linear extrapolation of the experimental and theoretical cross sections to zero wavelength (HUGHES).

the variation of the total cross section with sample temperature for neutrons of $\lambda = 7.0$ Å. The linear behaviour of the experimental results is markedly different from the theoretical curve[2,3]. This discrepancy, however, is not surprising since the specific heat of graphite, which depends on the lattice frequency spectrum essentially in the same way as the inelastic scattering, also varies less rapidly with temperature than that of a typical Debye substance. The anomalous behaviour of both the inelastic neutron scattering and the specific heat is attributed by HUGHES [*12*] to the layer structure of graphite which suppresses some of the lattice vibrations.

The case of Be (hexagonal close-packed: $\lambda_c = 3.95$ Å, $\Theta_D = 1000°$ K) is very interesting because this monoisotopic element has spin $\frac{3}{2}$ and small capture cross section ($\sigma_a = 0.056$ barns at $\lambda = 10$ Å).

The results obtained by the Brookhaven group[4] [*12*] with a neutron chopper, are compared with the theoretical curve computed by KLEINMAN[3] in Figs. 198

[1] See footnote 4, p. 441.

[2] See footnote 2, p. 440.

[3] See footnote 4, p. 440.

[4] JOHNSON, H. PALEVSKY and D. J. HUGHES: Phys. Rev. **82**, 345 (1951). — H. PALEVSKY and R. R. SMITH: Phys. Rev. **86**, 604 (1952).

and 199. The total cross section varies linearly with λ only beyond 10 Å, where neutron capture by Be and absorption of a single phonon by the neutron are the only processes to be considered, apart from a possible contribution, independent of λ, arising from spin-dependent incoherent scattering. For shorter wavelengths possible energy transfer from the neutron to the lattice is suggested by HUGHES [12] as a cause of departure from the "λ law". Since beryllium has more than one atom per unit cell, its lattice vibrations contain acoustical as well as optical branches whose behaviour is difficult to predict.

An estimate of the spin-dependent incoherent cross section can be made by extrapolating to $\lambda=0$ the linear part observed for $\lambda > 10$ Å [12]. The fact that at all three investigated temperatures ($T_0=300°$ K, $T_0=440°$ K in Fig. 198 and

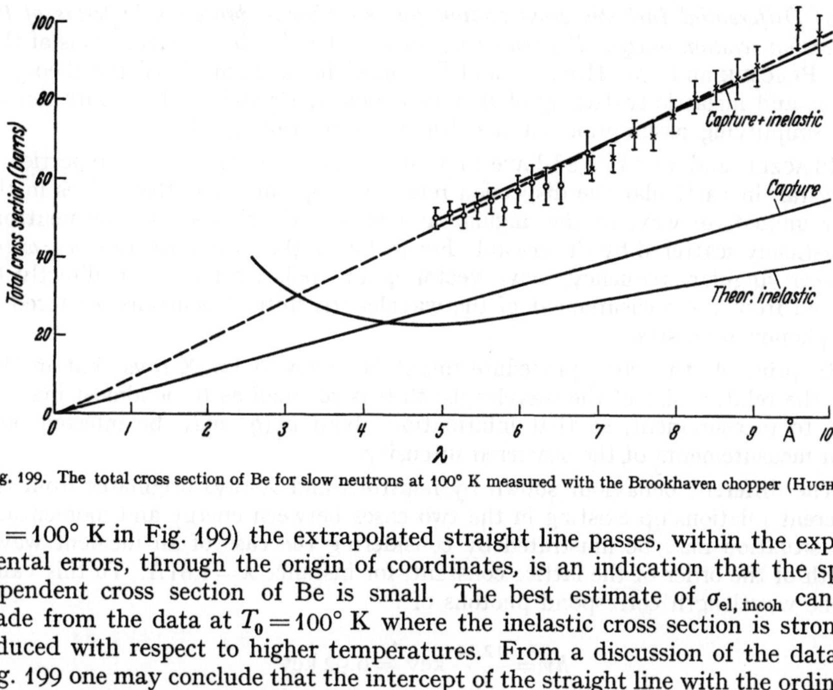

Fig. 199. The total cross section of Be for slow neutrons at 100° K measured with the Brookhaven chopper (HUGHES).

$T_0=100°$ K in Fig. 199) the extrapolated straight line passes, within the experimental errors, through the origin of coordinates, is an indication that the spin-dependent cross section of Be is small. The best estimate of $\sigma_{\text{el, incoh}}$ can be made from the data at $T_0=100°$ K where the inelastic cross section is strongly reduced with respect to higher temperatures. From a discussion of the data of Fig. 199 one may conclude that the intercept of the straight line with the ordinate axis can be placed at less than 5 mb.

From the same data the Brookhaven group succeeded to give an estimate of the absorption cross section of Be by substracting from the experimental curve the calculated cross section for one-phonon absorption. From this procedure they find $\sigma_a=(10\pm 3)$ mb at $v=2200$ m/sec (assuming that the $1/v$ law extends to this velocity) in good agreement with the value obtained from diffusion of slow neutrons (Sect. 128γ).

The variation of the elastic cross section with sample temperature at $\lambda=10.0$ Å calculated by SINGWI and KOTHARI agrees only moderately well with the experimental results of the Brookhaven group.

Similar measurements[1] made with aluminium at temperatures varying between 100° and 900° K show deviations of about 10% from the prediction of the incoherent approximation[2] and of about 20% from the theory of SINGWI and KOTHARI. This discrepancy may be an indication of the failure of the Debye

[1] R.L. ZIMMERMANN and H. PALEVSKY: Phys. Rev. 98, 1162 (1955).
[2] See footnote 5, p. 441.

approximation. Because of its success in describing specific heats of solids, this model is expected, however, to be a fair approximation for effects involving the average of certain functions over the frequency spectrum $\omega_j(\boldsymbol{q})$ of the crystal. For instance, one would expect that it should be good enough to calculate the total incoherent cross section, and its energy and temperature dependence, and the reduction in intensity of the Bragg reflection of neutrons (Debye-Waller factor).

For the other effects, however, in particular for the angular and energy distribution of inelastically scattered neutrons, the details of the vibration spectrum play a much more important role so that the Debye approximation is no longer expected to give reliable results.

β) Differential inelastic cross section for one-phonon processes in terms of the dispersion relation $\omega_j(\boldsymbol{q})$. The scattering surface. Guided by considerations of this type PLACZEK and VAN HOVE[1-3] and SQUIRES[4] have reconsidered the theory of elastic and inelastic scattering of slow neutrons by crystals without introducing any simplifying assumption on the dispersion relation (99.8).

PLACZEK and VAN HOVE[1] have first shown that the dynamical properties of a crystal, in particular the dispersion relation $\omega_j(\boldsymbol{q})$, manifest themselves in the most immediate way, in the angular and energy distribution of the neutrons inelastically scattered by the crystal. For instance, the correct relation $\omega = \omega_j(\boldsymbol{q})$ between angular frequency, wave vector $\boldsymbol{q}$ and polarization j, can directly be derived from the measurement of the wavelength shift of neutrons scattered in one-phonon processes.

In principle the same procedure might be followed for X-rays; but in this case the relative size of the wavelength shift is so small as to be almost inaccessible to measurement, so that information about $\omega_j(\boldsymbol{q})$ may be inferred only from measurements of the scattered intensity.

The different behaviour shown by neutrons and X-rays originates from the different relationship existing in the two cases between energy and momentum. The situation may be illustrated by considering the case of an incident wavelength of the order of the lattice constant, for instance $\lambda = 1.81$ Å. To this value of the wavelength correspond photons of

$$h\nu = \frac{12.34}{\lambda_{\text{Å}}}\,\text{kev} = 6.92\,\text{kev}$$

and neutrons of

$$E = 0.025\,\text{ev}.$$

Therefore the relative energy change undergone by the incidental radiation in a process of absorption or emission of one phonon (the energy of which is of the order of kT) will be exceedingly small for X-rays and of the order 1 for neutrons.

In the case of one phonon processes energy conservation is expressed by

$$K_2^2 - K_1^2 = \pm \frac{2m}{\hbar}\,\omega_j(\boldsymbol{q}), \tag{103.16}$$

the upper sign corresponding to the absorption and the lower sign to the emission of one phonon by the neutron.

---

[1] G. PLACZEK and L. VAN HOVE: Phys. Rev. **93**, 1207 (1954).

[2] G. PLACZEK and L. VAN HOVE: Nuovo Cim. **1**, 233 (1955).

[3] G. PLACZEK, B. R. A. NIJBOER and L. VAN HOVE: Phys. Rev. **82**, 392 (1951)

[4] G. L. SQUIRES: Phys. Rev. **103**, 304 (1956).

Eq. (103.16) holds for both coherent and incoherent scattering. In the latter case, no further condition relates the phonon variables to the momentum transferred by the neutron $-\hbar\boldsymbol{\varkappa}$, where

$$\boldsymbol{\varkappa} = \boldsymbol{K}_2 - \boldsymbol{K}_1. \tag{103.17}$$

Therefore neutrons scattered incoherently in one-phonon processes may have any value of the final energy between the following limits: for one-phonon absorption

$$E_1 = \frac{\hbar^2}{2m} K_1^2 \leqq E_2 \leqq \frac{\hbar^2}{2m} K_1^2 + \hbar\omega_{\max} = E_1 + \hbar\omega_{\max} \tag{103.18}$$

where $\omega_{\max}$ is the maximum of $\omega_j(\boldsymbol{q})$ for all $\boldsymbol{q}$ and j; for one-phonon emission

$$E_1 = \frac{\hbar^2}{2m} K_1^2 \geqq E_2 \geqq \begin{cases} 0 & \text{if } \hbar\omega_{\max} > E_1, \\ E_1 - \hbar\omega_{\max} & \text{if } h\omega_{\max} < E_1. \end{cases} \tag{103.19}$$

For coherent scattering the interference between waves scattered by the various nuclei imposes a further restriction

$$\boldsymbol{K}_1 - \boldsymbol{K}_2 + \boldsymbol{q} + 2\pi\boldsymbol{\tau} = 0 \tag{103.20}$$

where $\boldsymbol{\tau}$ can be any vector of the reciprocal lattice (see footnote 2 on p. 432). If $\boldsymbol{q} = 0$ and $\omega_j(\boldsymbol{q}) = 0$ are inserted into (103.20) and (103.16) these two relations respectively reduce to the Bragg condition (102.6) and to condition (102.3) that the scattering is elastic.

If $\boldsymbol{K}_1$ is fixed relative to the crystal, only those values of $\boldsymbol{K}_2$ are allowed in one-phonon coherent scattering, which satisfy Eqs. (103.16) and (103.20) for the same value of $\boldsymbol{q}$. These values may be found for a given crystal by means of the following geometrical construction in $\boldsymbol{K}$ space, similar to that used in Sect. 102 (Fig. 191) for elastic reflection.

Fig. 200 shows the wave vector $\boldsymbol{K}_1$ of the incident radiation terminating at the origin, O, of $\boldsymbol{K}$-space (i.e. the space of the reciprocal lattice with all distances multiplied by 2π). If the set of all admissible propagation vectors $\boldsymbol{q}$ is erected at any lattice point $2\pi\boldsymbol{\tau}$, the wave vector $\boldsymbol{K}_2$ satisfying condition (103.20) is obtained by completing the quadrangle $AOBC$. The locus of end-points of vectors $\boldsymbol{K}_2$ obtained by such a construction and satisfying condition (103.16) is termed a *scattering surface*. Since $\omega_j(\boldsymbol{q})$ is a periodic function of $\boldsymbol{q}$,

$$\omega_j(\boldsymbol{q} + 2\pi\boldsymbol{\tau}) = \omega_j(\boldsymbol{q}), \tag{103.21}$$

the equation of the scattering surface, obtained by combining (103.16), (103.17) and (103.20) is

$$K_2^2 - K_1^2 = \pm \frac{2m}{\hbar} \omega_j(\boldsymbol{\varkappa}) \tag{103.22}$$

with the upper (lower) sign corresponding to scattering with energy gain (loss) by the neutron.

A discussion of the properties of the scattering surface may be found in the references[1-4]. Here let us only note that:

(a) When taken with the positive sign (phonon absorption), Eq. (103.22) has solutions for any value of $|\boldsymbol{K}_1|$ varying between zero and infinity, while with the

[1] R.D. LOWDE: Proc. Roy. Soc. Lond., Ser. A **221**, 206 (1954).
[2] See footnote 1, p. 447.
[3] See footnote 2, p. 447.
[4] See footnote 3, p. 447.

negative sign, $|\boldsymbol{K}_1|$ must in general satisfy certain conditions discussed below if solution are to exist [see Eqs. (103.23) to (103.25)].

(b) For a given $\boldsymbol{K}_1$ the surface S depends only on the dispersion relation (99.9). And finally:

(c) Since there are three such relations, one corresponding to each value of the polarization j, for each of the two signs appearing in (103.22) S will consist of three surfaces S_1, S_2 and S_3 which cross each other.

As an example we give in Fig. 201 the section through the scattering surface at the plane $q_3=0$ calculated by SQUIRES[1] (see below in this section) for one-phonon absorption processes in aluminium with $\lambda=6.74$ Å and $\boldsymbol{K}_1$ parallel to the q_1 axis. We recall that aluminium has a face-centered cubic lattice (with the side of the cube: $a=4.04$ Å) and therefore its reciprocal lattice is body-centered cubic (with the side of the cube: $2b=2/a$).

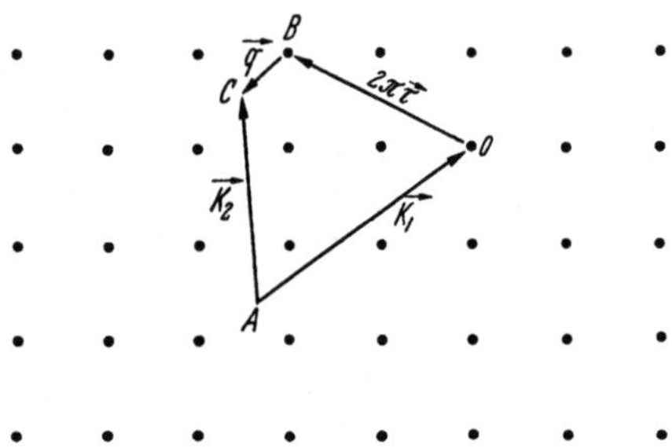

Fig. 200.

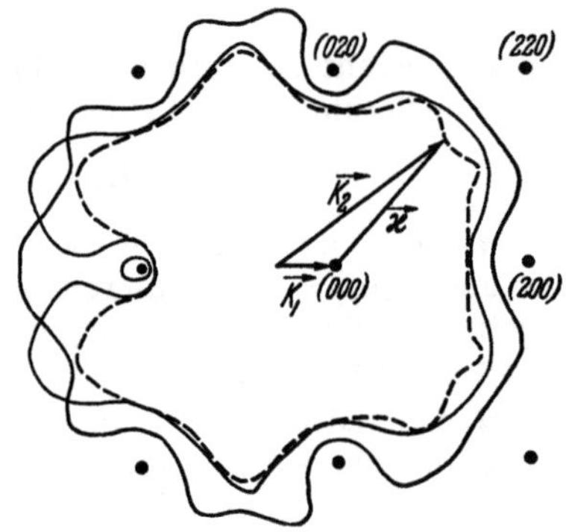

Fig. 201.

Fig. 200. K-space construction in one-phonon processes.

Fig. 201. Section through the scattering surface at the plane $q_3=0$ for neutrons with $\lambda=6.74$ Å (K_1 along q_1 axis) (SQUIRES).

As mentioned above, a solution of Eq. (103.22) in case of the negative sign (one-phonon emission processes) can only exist if $|\boldsymbol{K}_1|$ is larger than a value $K_1^{(1)}$

$$|\boldsymbol{K}_1| > K_1^{(1)}. \tag{103.23}$$

It is only for

$$|\boldsymbol{K}_1| > K_1^{(2)}, \qquad K_1^{(2)} > K_1^{(1)} \tag{103.24}$$

that the neutrons are scattered in every direction through this type of process. The actual values of $K_1^{(1)}$ and $K_1^{(2)}$ depend on the details of $\omega_j(\boldsymbol{q})$. It is, however, generally true that for a simple Bravais lattice (i.e. a lattice containing one atom per unit cell)

$$K_1^{(1)} \leqq \pi\,\tau_0 = \frac{2\pi}{\lambda_c}. \tag{103.25}$$

Combining (103.23) with (103.25), we see that one-phonon emission processes will in general only be possible if $\lambda_1 \leqq \lambda_c$ [2].

When the energy of the incident neutron is large enough to allow phonon emission processes, there are six or more scattering surfaces corresponding to a fixed value of $\boldsymbol{K}_1$: three of them corresponding to phonon absorption ($K_2^2 > K_1^2$) and three to phonon emission ($K_2^2 < K_1^2$). This situation is in general met with λ being of the order of 1 Å.

[1] See footnote 4, p. 447.
[2] G. C. WICK: Physik Z. 38, 403 (1937).

A detailed discussion of the scattering surface in the limiting case of very long wavelength ($K_1=0$) is particularly instructive[1,2]. An important role in determining its general shape is played by the parameter

$$F = \frac{\frac{4\pi}{3} K_{2\,\max}^3}{(2\pi)^3 v_0^{-1}} = \frac{\frac{4\pi}{3}\left(\frac{2m}{\hbar}\omega_{\max}\right)^{\frac{3}{2}}}{(2\pi)^3 v_0^{-1}}$$

defined as the ratio of the volume of the sphere attainable through one phonon absorption process and the volume of the lattice cell in $\boldsymbol{K}$-space.

The physical meaning of the scattering surface and its practical uses are obvious. For given $\boldsymbol{K}_1$ neutrons scattered coherently in one-phonon processes may be emitted in every direction and for each direction the outgoing energy has a finite number of discrete values. Therefore, measurements of the outgoing energy as a function of direction determine the surface S and thus yield $\omega_j(\boldsymbol{q})$.

It may be instructive to compare the shape of the scattering surface deduced from (103.22) with that derived for a Debye crystal. Since in this model all possible values of $\boldsymbol{q}$ fill a sphere of radius $q_{\max}$ around each lattice point of $\boldsymbol{K}$ space, one recognizes that the corresponding scattering surface is in general markedly different from the correct one. This remark has been stressed in different ways by PLACZEK and VAN HOVE[2] and by SQUIRES[3] with the common conclusion that the Debye approximation is inadequate to predict the detail of the angular and energy distributions of inelastically scattered neutrons.

Turning to the cross sections, one has for incoherent one-phonon processes[4,1]

$$\left.\begin{aligned} \frac{d\sigma_{1\,\mathrm{incoh}}}{d\boldsymbol{K}_2} &= \frac{2}{A K_1} \mathrm{e}^{-\langle(\boldsymbol{\varkappa}\cdot\boldsymbol{u})^2\rangle_{\mathrm{av}}} \frac{v_0}{(2\pi)^3} \sum_j \int d\boldsymbol{q} \frac{[\boldsymbol{\varkappa}\cdot\boldsymbol{e}_j(\boldsymbol{q})]^2}{|K_2^2-K_1^2|} \times \\ &\times\left[\frac{1}{\exp[\hbar\omega_j(\boldsymbol{q})/kT_0]-1}+\frac{1}{2}(1\mp 1)\right]\cdot\delta\left(K_2^2-K_1^2\mp\frac{2m}{\hbar}\omega_j(\boldsymbol{q})\right) \end{aligned}\right\} \qquad (103.26)$$

where $\exp[-\langle\boldsymbol{\varkappa}\cdot\boldsymbol{u}\rangle_{\mathrm{av}}]$ is the Debye-Waller factor, $\boldsymbol{u}$ the displacement vector of any nucleus in the crystal from its equilibrium position and the average $\langle\ldots\rangle_{\mathrm{av}}$ has to be taken over the thermal equilibrium distribution. The vector $\boldsymbol{e}_j(\boldsymbol{q})$ is the polarization vector of the phonon defined by j and $\boldsymbol{q}$. The upper (lower) sign corresponds to scattering with energy gain (loss) by the neutron. The integration is extended over one cell of $\boldsymbol{K}$ space. The argument of the δ-function directly corresponds to the energy conservation condition (103.16).

Since

$$d\boldsymbol{K}_2 = d\Omega_2 K_2^2 dK_2 \quad \text{and} \quad E_2 = \frac{\hbar^2 K_2^2}{2m},$$

Eq. (103.26) multiplied by $\frac{m K_2}{\hbar^2}$ gives directly the energy distribution of the scattered neutrons in a given direction (that of $\boldsymbol{K}_2$) per unit solid angle.

Through a detailed discussion of (103.26) PLACZEK and VAN HOVE[1] have shown that the energy distribution of scattered neutrons varies in general with direction but in each direction it has singularities occuring at energies independent of the direction:

$$\frac{\hbar^2 K_2^2}{2m} = \frac{\hbar^2 K_1^2}{2m} \pm \hbar\omega_c \qquad (103.27)$$

[1] See footnote 1, p. 447.
[2] See footnote 2, p. 447.
[3] See footnote 4, p. 447.
[4] See footnote 3, p. 447.

where ω_c are the singular frequencies of the crystal[1]. Apart from exceptional cases they are the values of $\omega_j(\boldsymbol{q})$ at the point where $\operatorname{grad}\omega_j(\boldsymbol{q})=0$.

Multiphonon incoherent scattering does not produce such singularities.

The foregoing discussion was concerned with single crystals. For powders the cross section (103.26) must be averaged over all possible orientations. Since the singularities in the energy distribution occur at energies independent of direction, they are retained in the averaging and the qualitative behaviour of the energy distribution of incoherent inelastic scattered neutrons remains unchanged.

The coherent scattering cross section is given by the expression[2]

$$\left.\begin{aligned}\frac{d\sigma_{1\,\mathrm{coh}}}{d\boldsymbol{K}_2}=\frac{2}{A\,K_1}\,\mathrm{e}^{-\langle(\boldsymbol{\varkappa}\cdot\boldsymbol{u})^2\rangle_{\mathrm{av}}}\sum_j\frac{[\boldsymbol{\varkappa}\cdot\boldsymbol{e}_j(\boldsymbol{\varkappa})]^2}{|K_2^2-K_1^2|}\left[\frac{1}{\exp\left[\frac{\hbar\,\omega_j(\boldsymbol{\varkappa})}{k\,T_0}\right]-1}+\frac{1}{2}\,(1\mp1)\right]\times\\ \times\,\delta\Big(K_2^2-K_1^2\mp\frac{2m}{\hbar}\,\omega_j(\boldsymbol{\varkappa})\Big).\end{aligned}\right\}\qquad(103.28)$$

The argument of the δ-function corresponds directly to Eq. (103.22) of the scattering surface and both $\boldsymbol{e}_j(\boldsymbol{\varkappa})$ and $\omega_j(\boldsymbol{\varkappa})$ are defined for all values of $\boldsymbol{\varkappa}$ through periodicity conditions as (103.21).

According to (103.22) the neutrons scattered in a given direction have a discrete spectrum corresponding to outgoing vectors $\boldsymbol{K}_{21}, \boldsymbol{K}_{22}, \boldsymbol{K}_{23}, \ldots$. The cross section per unit solid angle relative to the neutron group of wave vector $\boldsymbol{K}_{2i}$ is obtained by multiplication of (103.28) with $K_2^2\,dK_2$ and integration over dK_2 in the region around $\boldsymbol{K}_{2i}$. The result is particularly simple in the limit of long incident wavelengths so that

$$K_2^2-K_1^2\approx K_2^2.$$

In this case one obtains

$$\left(\frac{d\sigma_{1\,\mathrm{coh}}}{d\Omega_2}\right)_{K_{2i}}=\frac{2}{A\,K_1}\,\mathrm{e}^{-\langle(\boldsymbol{\varkappa}\cdot\boldsymbol{u})^2\rangle_{\mathrm{av}}}\,\frac{[\boldsymbol{K}_{2i}\cdot\boldsymbol{e}_j(\boldsymbol{K}_{2i})]^2}{\left\{\exp\left[\frac{\hbar\,\omega_j(\boldsymbol{K}_{2i})}{kT_0}\right]-1\right\}\cdot\left|2K_{2i}-\frac{2m}{\hbar}\,\frac{d\omega_j}{d\varkappa}\right|}\,.\qquad(103.29)$$

For powders this expression must be averaged over all crystal orientations producing in each direction an outgoing energy distribution continuous over finite intervals.

Multiphonon coherent scattering by single crystals gives a continuous energy distribution in each direction. Only two-phonon processes produce singularities of the same type as those observed in one-phonon incoherent scattering; these occur however at energies unrelated to the singular frequencies of the crystal and vary with the outgoing direction. They can, therefore, be distinguished from the singularities resulting from one-phonon incoherent scattering, and for powders they disappear by averaging over orientations.

Finally one has to notice that the behaviour of the elastic coherent scattering cross section when the Bragg condition is approached i.e. when

$$|\boldsymbol{K}_1+2\pi\boldsymbol{\tau}|^2\approx K_1^2$$

may be derived from Eq. (103.29) by applying it to scattering with $\boldsymbol{\varkappa}$ nearly equal to $2\pi\boldsymbol{\tau}$ and with a very small energy transfer ($K_2^2\approx K_1^2$). Since

$$\exp\left[-\frac{\hbar\,\omega_j(\boldsymbol{\varkappa})}{kT_0}\right]-1\approx\frac{\hbar\,\omega_j(\boldsymbol{\varkappa})}{kT_0}=\frac{\hbar^2}{2m\,kT_0}\,|K_2^2-K_1^2|,$$

[1] L. van Hove: Phys. Rev. **89**, 1189 (1953).

[2] See footnote 1, p. 447.

the cross section will vary proportionally to $(K_2^2 - K_1^2)^{-2}$. This special case was considered in detail by various authors[1,2].

γ) Further discussion of the total elastic + inelastic cross section. The static approximation. The approximation adopted by various authors in calculating the interference effects in the scattering of slow neutrons by crystals have been reexamined and compared in a succeeding paper by PLACZEK and VAN HOVE[3].

In the previous more conventional approach to this problem, the total cross section for scattering (elastic + inelastic) per nucleus of a crystal containing atoms of a single element, is written as

$$\sigma_s = \sigma_{\text{coh}} S_c + \sigma_{\text{incoh}} S_i \tag{103.30}$$

where σ_{coh} and σ_{incoh} are defined by (99.7) and the numerical factors S_c and S_i depend on the wavelength of the incident neutrons, the crystal temperature and, in general, on the orientation of the crystal. PLACZEK and VAN HOVE point out that Eq. (103.30) can more conveniently be rewritten in the form

$$\sigma_s = 4\pi \langle a_b^2 \rangle S_i + 4\pi \langle a_b \rangle^2 \delta S \tag{103.31}$$

with

$$\delta S = S_c - S_i. \tag{103.32}$$

Here the distinction between coherent and incoherent scattering disappears since the first term represents the scattering from a single nucleus in the crystal and the second the interference effect.

The factor S_i may be expressed as a sum of terms S_{il} for the emission or absorption of l-phonons. This expansion however converges very poorly, particularly at high temperatures and/or short wavelengths, as has already been stated while discussing Eq. (103.5) at the beginning of this section. By extending the considerations given there one finds that it is a more convenient procedure for the evaluation of S_i to expand it in negative powers of the mass ratio $A = M/m$. This type of expansion does not apply in general to the interference term δS which is a highly singular function of the wave vector $\boldsymbol{K}_1$ of the incident neutrons. The main singularities occur in both the elastic and the inelastic part, at the values of K_1 satisfying the Bragg condition (102.6).

The expansion of δS in terms representing the emission or absorption of l phonons

$$\delta S = \sum_{l \geq 0} \delta S_l \tag{103.33}$$

converges far more rapidly than the corresponding expansion of S_i because of the condition of interference of the waves scattered by the various nuclei

$$\boldsymbol{K}_1 - \boldsymbol{K}_2 + \sum_{i=1}^{l} \mathsf{E}_i \boldsymbol{q}_i + 2\pi \boldsymbol{\tau} = 0 \tag{103.34}$$

with $\mathsf{E}_i = \pm 1$ according to whether the i-th phonon has been absorbed or emitted by the neutron. Condition (103.34) is an extension of (103.20) which was valid for one-phonon processes; it introduces in the expression of δS_l a δ-function of its left-hand side which, combined with the integration with respect to $dq_1\, dq_2 \ldots dq_l$ produces a rapid decrease with increasing l of the contribution to the interference term of l-phonon processes.

[1] R. SEEGER and E. TELLER: Phys. Rev. 62, 37 (1942). — I. WALLER and P.O. FRÖMAN: Ark. Fysik 4, 183 (1951).
[2] See footnote 1, p. 448.
[3] See footnote 2, p. 447.

In the limiting case of very short wavelengths ($\lambda \to 0$) and polycrystalline materials, $(S_i)_{av}$ is obtained by expressing σ_f in terms of σ_b according to (95.1) in the asymptotic Placzek-Wick formula (100.1) so that

$$(S_i)_{av} = \left(\frac{M}{m+M}\right)^2 \left(1 + \frac{1}{3}\frac{m}{M}\frac{K_{av}}{E}\right) - \frac{1}{32}\frac{m}{M}\frac{C_{av}}{E^3}. \tag{103.35}$$

For Debye crystals containing a single element, the coefficients appearing in (103.35) are given by the following expressions

$$K_{av} = \frac{36 k T_0^4}{\Theta_D^3} \int_0^{\frac{\Theta_D}{2T_0}} x^3 \cdot \coth x \cdot dx, \quad C_{av} = \frac{96 k^3 T_0^6}{\Theta_D^3} \int_0^{\frac{\Theta_D}{2T_0}} x^5 \cdot \coth x \, dx$$

which may be computed by means of the following expansions

for $T_0 \ll \Theta_D$:

$$K_{av} = \frac{9}{16} k\Theta_D \left\{1 + \frac{8}{15}\left(\frac{\pi T_0}{\Theta_D}\right)^4\right\}, \quad C_{av} = \frac{[k\Theta_D]^3}{4}\left\{1 + \frac{32}{21}\left(\frac{\pi T_0}{\Theta_D}\right)^6\right\};$$

for $T_0 \gg \Theta_D$:

$$K_{av} = \frac{3}{2} kT_0 \left\{1 + \frac{1}{20}\left(\frac{\Theta_D}{T_0}\right)^2\right\}, \qquad C_{av} = \frac{3}{5} kT_0 k^2 \Theta_D^2 \left\{1 + \frac{5}{84}\left(\frac{\Theta}{T_0}\right)^2\right\}.$$

For heavy nuclei Eq. (103.35) begins to hold at a neutron energy slightly larger than $k\Theta_D$ and for lighter nuclei at somewhat higher neutron energies.

The asymptotic expression for short wavelengths of the interference term for a polycrystalline material is given by

$$(\delta S)_{av} = \frac{\lambda^2}{8\pi v_0^{\frac{2}{3}}} I \tag{103.36}$$

where v_0 is the volume of the unit cell of the crystal and I a numerical coefficient. This was first calculated in the "*static approximation*"[1] well known as the basis of the theory of X-ray scattering by molecules, crystals and liquids. The calculation essentially consists in neglecting the neutron energy changes associated with the transitions undergone by the crystal, and averaging the cross section computed for nuclei fixed in definite positions over the actual configurations of the initial state. In this approximation one obtains

$$I = I^{(0)} - I^{(1)} \frac{u_0^2}{v_0^{\frac{2}{3}}} - I^{(2)} \frac{u_0^4}{v_0^{\frac{4}{3}}} \tag{103.37}$$

where u_0 represents the root-mean-square displacement of a nucleus from its equilibrium position. The first term is given by the expression[2,1]

$$I^{(0)} = 3 - v_0^{\frac{2}{3}} \sum_{r_i}{}' \frac{\exp[-\pi v_0^{-\frac{2}{3}} r_i^2]}{\pi r_i^2} - \frac{2\pi}{v_0^{\frac{1}{3}}} \sum_{\tau}{}' \frac{\operatorname{Erf}(\tau v_0^{\frac{1}{3}}/2\pi)}{\tau} \tag{103.38}$$

where the sum extends over the lattice vector $\boldsymbol{r}_i$ with omission of $\boldsymbol{r}_i = 0$. The remaining terms in (103.37) represent the effect of thermal and zero point motion and amount in general to a correction too small to be of practical importance.

---

[1] See footnote 3, p. 447.
[2] See footnote 2, p. 447.

The phonon expansion of the numerical factor I

$$I = \sum_{l \geqq 0} I_l$$

has been investigated by the same authors[1] with the result that I_1 is found to be of the order $(u_0/v_0^{\frac{1}{3}})^2$, I_2 of the order $(u_0/v_0^{\frac{1}{3}})^4$ and in general

$$\sum_{l_0}^{\infty} I_l = O\left(\frac{u_0}{v_0^{\frac{1}{3}}}\right)^{2l_0}$$

[but not $I_l \sim (u_0/v_0^{\frac{1}{3}})^{2l}$]. Therefore I_0 differs from both I and $I^{(0)}$ only by terms of order $(u_0/v_0^{\frac{1}{3}})^2$ so that $(\delta S)_{\text{av}}$ is given almost entirely by its elastic part $(\delta S_0)_{\text{av}}$ and in the same approximation it may be computed by means of (103.36) to (103.38) neglecting terms in $u_0^2/v_0^{\frac{2}{3}}$ and $u_0^4/v_0^{\frac{4}{3}}$.

The size of $(\delta S)_{\text{av}}$, but not its finer structure, is well represented by $(\delta S_0)_{\text{av}}$ even at somewhat longer wavelengths provided that $K_1 v_0^{\frac{1}{3}} \gg 1$. In this wavelength region however, the asymptotic expression (103.36) for the computation of $(\delta S_0)_{\text{av}}$ must be replaced by the expression

$$(\delta S_0)_{\text{av}} = \frac{1}{K_1^2}\left\{\frac{\pi}{2v_0}\sum_{\tau < 2/\lambda_c}\frac{1}{\tau}\,\mathrm{e}^{-2W(T_0)\left(\frac{\tau}{2}\right)^2} - \frac{1}{2W(T_0)}\left(1 - \mathrm{e}^{-2W(T_0)\frac{1}{\lambda^2}}\right)\right\} \quad (103.39)$$

obtained by combining (102.20), (102.21) with (103.31) and (103.32).

At still longer wavelengths this approximation breaks down; this can be seen from the fact that, for $\lambda < \lambda_c$, $(\delta S_0)_{\text{av}}$ tends to -1 while the inelastic part of $(\delta S)_{\text{av}}$ increases, for all finite temperatures, as $K_1^{-1} = \lambda\!\!\!^{-}_1$ [see Eqs. (103.26), (103.28)].

Finally PLACZEK and VAN HOVE discuss in detail the Debye approximation, concluding that, when quantitative calculations are concerned, this model is much less adequate to calculate the interference effect than the incoherent cross section.

δ) Some experiments on the differential cross section for inelastic scattering. A first attempt at measuring the angular and energy distribution of neutrons scattered inelastically by crystals has been made by EGELSTAFF[2] who used the neutrons of the Harwell pile filtered through 18 inches of Pb (see Sect. 102). The maximum energy of the neutrons in the beam was 0.026 ev and their mean energy was 0.017 ev. A sample of polycrystalline material was placed in the path of the beam and the neutrons scattered at 30° ± 5° were counted with a BF_3 counter. A rough estimate of the energy change undergone by the neutrons in the scattering process was obtained by comparing the transmission of a lithium fluoride plate (0.14 g/cm² thick) for the scattered beam with that obtained for the incident neutrons.

The experiments by BROCKHOUSE and HURST[3] are essentially based on the same principle; these authors selected neutrons of 0.35 ev emitted from the Chalk River pile by means of a NaCl crystal spectrometer. The neutrons scattered at 90° ± 30° by polycrystalline samples of diamond, graphite, aluminium and lead were detected by a ring of BF_3 proportional counters. By means of cylindrical absorbers of Cd, of thickness ranging up to 0.7 g/cm², these authors measured the cadmium transmission of the scattered neutrons. Between neutron energies of 0.2 and 0.5 ev the Cd cross section falls rapidly from over 6000 barns

[1] See footnote 2, p. 447.
[2] P. A. EGELSTAFF: Nature, Lond. **168**, 290 (1951).
[3] B. N. BROCKHOUSE and D. G. HURST: Phys. Rev. **88**, 542 (1952).

to less than 150 barns. Therefore the Cd cross section is a sensitive indicator of the energy near 0.35 ev even when averaged over a rather wide spectrum. The results of these experiments show that the change in energy undergone by the neutrons is about that expected for an Einstein crystal[1].

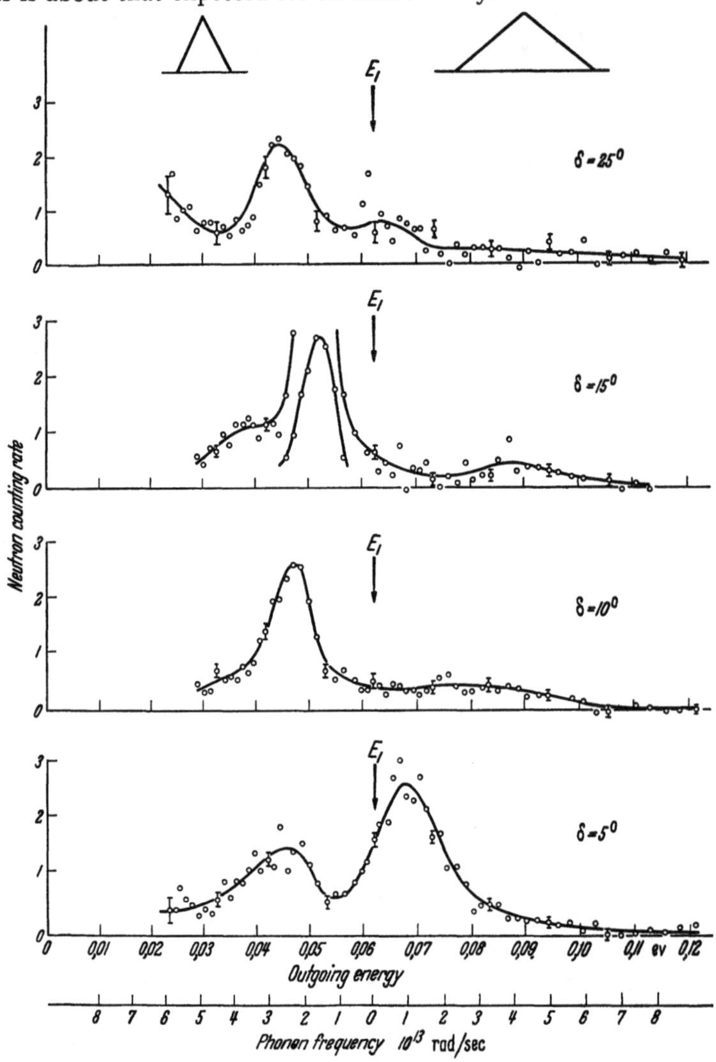

Fig. 202. Typical energy distribution of neutrons inelastically scattered by an aluminium single crystal; E_l = energy of incident neutrons (BROCKHOUSE and STEWART).

Much more detailed information has been obtained by BROCKHOUSE and STEWART[2] who investigated the scattering by an aluminium monocrystal making use of two crystal spectrometers: the first one to select a monoenergetic beam of neutrons of $\lambda = 1.15$ Å ($E = 0.062$ ev), the second one to analyse the spectrum of the neutrons scattered at an angle of 95.1° by a single aluminium crystal. This was oriented so that its (1, 1, 1) axis was in the plane of the spectrometer

[1] See footnote 3, p. 441.

[2] B.N. BROCKHOUSE and A.T. STEWART: Phys. Rev. **100**, 756 (1955).

and its (0, 1, 1) axis vertical. The angle and wavelength used satisfy the Bragg condition for the (3, 3, 3) reflection when the (1, 1, 1) direction bisects the scattering angle. Measurements were made with the crystal rotated about the vertical direction by an angle δ from the (3, 3, 3) Bragg position, keeping the angle of scattering constant. Energy distributions were obtained for eleven angular settings of the crystal at room temperature of which a few are shown in Fig. 202. Groups of neutrons are seen which represent both energy losses and gains corresponding to phonon emission and absorption.

Similar experiments have also been made by other authors[1,2].

The energy resolution in these experiments has been limited mainly by the intensity available and it is not sufficient to show the details anticipated by theory. In spite of this, SQUIRES[3] has attempted a comparison with the results of detailed calculations based on the use of the dispersion relation (and frequency distribution function) deduced from the elastic constants directly measured on single aluminium crystals, under the assumption that the crystal forces are harmonic and that they exist only between first and second nearest neighbours (see Sect. 99, p. 414). We do not however enter into a discussion of such a comparison and of the possible origin of the rather poor agreement between theory and experiments because it would take us too far away from the scope of the present article.

IV. The slowing down and diffusion of neutrons.

104. Qualitative remarks on the slowing-down problem. Below a few electron-volts the problem of the slowing down of neutrons becomes much more involved than at higher energy because the various cross sections entering the problem show a considerably more complicated energy and angle dependence. These complications arise from the effects mentioned in Division I and considered in some detail in Divisions II and III of this part.

Even in the simplest stationary case such as that of neutrons slowed down uniformly throughout a homogeneous moderator filling all the space, one faces the problem of solving the integral equation (96.1), or one of the equivalent equations (96.9), (96.11), (96.13), the coefficients of which show an energy dependence which reflects the details of the molecular structure of the moderator. More general problems, whose solution involves the neutron density as a function of space variables, are still more complicated because the dependence of the cross section on the angle of scattering enters into the problem through the kernel of the corresponding integral equation (Sect. 108).

As we shall see later in Sects. 105, 107, 108, if the neutron absorption is weak and the gradient of the neutron density small, the low-energy part of the neutron spectrum is roughly represented by a Maxwellian distribution relating to a temperature not lower than that of the moderator. When the absorption is strong the low-energy tail of the neutron spectrum goes to zero even faster. Therefore the slowing down of neutrons by a medium at room temperature will be affected by the behaviour of the cross section above, around and somewhat below $kT_0 \sim$ 0.025 ev; at energies much lower than this value the behaviour of the cross section has much less influence. If, e.g., neutrons moving through a polycrystalline moderator at room temperature are elastically scattered, the behaviour of the

[1] B. JACROT: C. R. Acad. Sci., Paris **240**, 745 (1955).

[2] R. S. CARTER, H. R. MUETHER, D. J. HUGHES and H. PALEVSKI: Bull. Amer. Phys. Soc. II, No. **1**, 55 (1956).

[3] G. L. SQUIRES: Phys. Rev. **103**, 304 (1956).

cross section around and below the cut-off (Table 76, p. 436) is expected to have a considerable influence on the very-low-energy tail and only a minor influence on the bulk of the spectral distribution. From Figs. 193 and 194 (Sect. 102) one sees, however, that the discontinuities of the elastic cross section are still present in the region around kT_0 and disappear only at higher energies.

As mentioned in Sect. 102 the sharpness of these discontinuities depends on various conditions which in many practical cases are not all under control, such as for instance the size of the microcrystals, the presence of impurity films adsorbed on their surfaces etc.

Similar remarks can be made about the inelastic scattering cross section which determines the slowing down of neutrons from higher energies down to energies of the order kT_0, and the establishment of conditions of almost thermal equilibrium between the moderator and the neutron gas. Its coherent part shows maxima in some way similar to those shown by the coherent elastic cross section, its incoherent part shows discontinuities which depend on the detail of the dispersion relation of the crystal (Sect. 103).

All these difficulties and uncertainties justify the fact that the ideal case of monatomic gaseous moderators is the only one investigated thoroughly until now and that very little has been done on more realistic media such as liquids and crystals (Sects. 106 and 107).

It has to be noticed, however, that besides its heuristic value, a monatomic gaseous moderator, when conveniently adjusted, may constitute a very rough first approximation to both molecular moderators and heavy polycrystalline moderators.

For instance, as we shall see in Sect. 106, the behaviour of light and heavy water, has been roughly described by means of a model based on a monatomic gaseous moderator of convenient mass.

The following simple argument by BROCKHOUSE and HURST[1,2] shows how the behaviour of polycrystalline moderators compares with that of monatomic gases. For the sake of simplicity the crystal is described by means of the Einstein model so that the kinetic energy of an oscillator at the temperature T_0 will be $\frac{1}{2}kT_0$ whenever kT_0 is large compared to the level spacing $\hbar\omega_0 = k\Theta_E$ of the crystal. Therefore, the energy transferred in a collision between a neutron of energy E and such an oscillator will have a distribution with a spread in energy of the order of

$$\left(\sqrt{E} + \sqrt{\frac{kT_0}{2A}}\right)^2 - \left(\sqrt{E} - \sqrt{\frac{kT_0}{2A}}\right)^2 = \sqrt{\frac{8kT_0 E}{A}}.$$

If this energy spread is large compared to the spacing $\hbar\omega_0$ of the energy levels of the crystal, the probability of collisions with an exchange of n phonons will be appreciable even for values of $n > 1$. One can conclude that, if the inequality

$$\sqrt{\frac{8kT_0 E}{A}} \gg k\Theta_E$$

or

$$\frac{E}{kT_0} \gg \frac{A}{8}\left(\frac{\Theta_E}{T_0}\right)^2 \tag{104.1}$$

is satisfied, the relative probabilities for scattering with exchange of 0, 1, 2, 3, ... phonons of energy $\hbar\omega_0$ calculated for an Einstein crystal, will form a histogram

[1] B.N. BROCKHOUSE and D.G. HURST: Phys. Rev. **88**, 542 (1952).
[2] E.R. COHEN: p. 405 of Ref. [*15c*].

which approximates a continuous curve representing the energy exchange distribution calculated for a Maxwellian monatomic gas of the same mass A. This remark is illustrated by Fig. 203 on which there are plotted the results of calculations made for neutrons of $E = 0.35$ ev scattered by C, Al and Pb samples at $T_0 = 293°$ K[1]. The histograms have been calculated by means of FINKELSTEIN's[2] formulae taking $k\Theta_E = 0.150$ ev, $k\Theta_E = 0.0259$ ev, $k\Theta_E = 0.00565$ ev in the three cases as suggested from specific heat measurements[3]. While for C no resemblance exists between the behaviour of the Einstein crystal and that of the Maxwell gas, already for Al a monatomic gas appears to be not too bad an approximation.

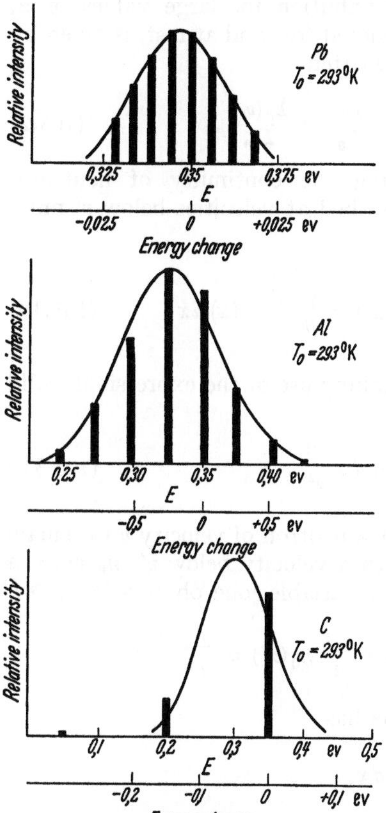

Fig. 203. The theoretical energy distributions of initially monoenergetic neutrons (0.35 ev) after scattering from a gas (solid line) and an Einstein crystal (BROCKHOUSE and HURST).

Similar considerations may be applied to the Debye model (Sect. 99) with the difference that the free gas approximation is better since in this case there is a continuum of oscillators[4].

In general the approximation of a Maxwell gas tends to be better for heavier elements where the Einstein and Debye temperature are low and breaks down for crystals in which the Einstein temperature is high and hence the level spacing large. The important cases of graphite and beryllium belong to this latter class and, therefore, for these the gas approximation is expected to be rather poor. In Sect. 107 we shall see, however, that the deviations between the results of the heavy gas approximation and the heavy Debye crystal approximation are very small when applied to the case of Be.

Many details on the neutron spectra can be found in the review article by POOLE *et al.*[5]

105. The integral spectrum in the case of a monatomic gaseous moderator. The integral equation obtained by introducing the expressions (96.5), (97.4), (97.10) into (96.1) cannot be solved analytically in the general case. Numerical methods of the solution have been used in a few interesting cases; the results will be reported later in this and the following section.

In the case of a hypothetical *monatomic hydrogen moderator* with absorption cross section following the $1/v$ law (96.4) and with constant scattering cross section (97.6), WIGNER and WILKINS[6] have reduced the integral equation with

[1] See footnote 1, p. 457.
[2] R. J. FINKELSTEIN: Phys. Rev. **72**, 907 (1947).
[3] N. F. MOTT and H. JONES: The theory of the properties of metals and alloys. London: Oxford University Press 1936.
[4] See footnote 2, p. 457.
[5] M. J. POOLE, M. S. NELKIN and R. S. STONE: Progress in Nuclear Energy, Physics and Mathematics, Vol. 2, p. 91. New York and London: Pergamon Press 1958.
[6] E. P. WIGNER and J. E. WILKINS: AECD-2275 (1948).

symmetric kernel (96.9) to a differential equation and have solved it by numerical methods. These authors use the independent variable

$$x = \beta v = \sqrt{\frac{m v^2}{2 k T_0}} = \frac{v}{v_0} \tag{105.1}$$

where v_0 is the most probable neutron velocity of the Maxwell distribution at temperature T_0. The asymptotic neutron distribution for large values of x, where the atoms of the moderator can be considered free and at rest, is given by (68.19) which in the present notation, may be written

$$n_0(x)\,dx = \frac{C\,x\,dx}{(x+\Gamma)^3}; \qquad \Gamma = \frac{\beta\gamma}{\Sigma_{s0}} = \frac{\Sigma_a(v_0)}{\Sigma_{s0}}. \tag{105.2}$$

The constant C was determined by requiring the continuity of neutrons, namely that the total number of neutrons absorbed at velocities below v' must be equal to the slowing down density at v':

$$q(v') = \int_0^{v'} v\,\Sigma_a(v)\,n_0(v)\,dv = \gamma \int_0^{v'} n_0(v)\,dv = \frac{\gamma}{\beta}\int_0^{x'} n_0(x)\,dx. \tag{105.3a}$$

When v' is large, $q(v')$ may be calculated by making use of the expressions valid in the case of free nuclei at rest:

$$q(v') = \int_{v'}^{\infty} v\,\Sigma_s(v)\,n_0(v)\,dv\,\frac{v'^2}{v^2}. \tag{105.3b}$$

According to (62.7) v'^2/v^2 is the probability for a neutron of velocity v to remain after a collision with a free proton at rest, with a velocity below v'; $n_0(v)\,dv$ is given by (68.19). By changing the integration variable, one obtains for $x' \gg \Gamma$

$$q(v') = \Sigma_{s0}\beta^{-2} x'^2 \int_{x'}^{\infty} n_0(x)\,\frac{dx}{x} \approx \Sigma_{s0}\beta^{-2}\frac{x'^2}{2}\,n_0(x') \approx \frac{\Sigma_{s0}}{\beta^2}\,\frac{C}{2}.$$

By substituting this expression into (105.3) one has

$$C \cong 2\Gamma \int_0^{\infty} n_0(x)\,dx.$$

Numerical calculations were made for the following three values of the constant

$$\Gamma = 0.226,\ 0.451,\ 2.257.$$

The results of WIGNER and WILKINS are plotted in Fig. 204 together with the Maxwell distribution, marked $\Gamma = 0$, and the asymptotic distributions in absence of absorption

$$n_0(x) = \frac{C}{x^2} \tag{105.4}$$

which are represented by the straight lines on the right. The latter are useful in order to visualize the deviation from the dv/v^2 law, which, for x larger than 1 but still comparable to Γ, do not arise from the thermal motion of the atoms of the moderator, but rather from neglect of the absorption term in (105.2). For x sufficiently large, however, Eq. (105.2) always reduces to (105.4) and therefore the neutron spectra always approach the straight lines on the right of Fig. 204.

As we shall see later, the problem of the slowing down of neutrons by a gaseous moderator has been treated by means of the Monte Carlo method by BATE,

COVEYOU and OSBORN[1] for various values of the mass ratio $A = M/m$ and under the assumption of constant scattering cross section (97.6) and $1/v$ absorption cross section (96.4). The results obtained in the case of monatomic hydrogen gas for $\Gamma=0.222$ and 0.451 are in good agreement with those of WIGNER and WILKINS. The results obtained in the case of heavier moderators will be considered later in this section and in Sect. 107.

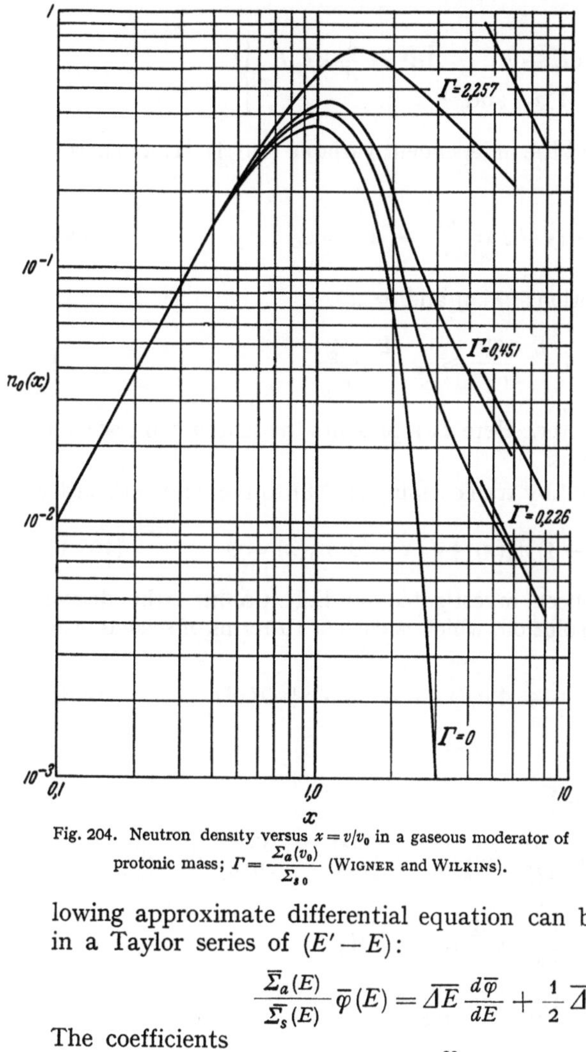

Fig. 204. Neutron density versus $x = v/v_0$ in a gaseous moderator of protonic mass; $\Gamma = \frac{\Sigma_a(v_0)}{\Sigma_{s0}}$ (WIGNER and WILKINS).

When the atomic mass of the monatomic gaseous moderator is different from 1, the integral equation (96.1) is much more difficult to handle even under the assumptions (96.4) and (97.6) on the absorption and scattering cross sections. In the limiting case of $A \gg 1$—i.e. in the case of very heavy moderators—WILKINS[2] succeeded in reducing the integral equation (96.1) to a second order differential equation which is accurate to order $1/A$. The same equation was later derived by HURWITZ, NELKIN and HABETLER[3] by means of a different procedure which is summarized below.

These authors start from the integral equation (96.13) and notice that if the kernel $\overline{\Sigma}_s(v' \to v)$ is narrow compared to the range in which the function $\overline{\varphi}(E')$ defined by (96.14) has large variations, the following approximate differential equation can be obtained by expanding $\overline{\varphi}(E')$ in a Taylor series of $(E'-E)$:

$$\frac{\overline{\Sigma}_a(E)}{\overline{\Sigma}_s(E)}\,\overline{\varphi}(E) = \overline{\Delta E}\,\frac{d\overline{\varphi}}{dE} + \frac{1}{2}\,\overline{\Delta E^2}\,\frac{d^2\overline{\varphi}}{dE^2} + \cdots. \tag{105.5}$$

The coefficients

$$\overline{\Delta E^n} = \frac{1}{\overline{\Sigma}_s(E)} \int\limits_0^\infty dE' (E'-E)^n\, \overline{\Sigma}_s(E \to E') \tag{105.6}$$

are the moments of the kernel which can be evaluated for any moderator if the differential energy transfer cross section $\overline{\Sigma}_s(E' \to E)$ is known. A consistent approximation is obtained in the case of a heavy gaseous moderator by considering only first order terms in the inverse mass ratio $1/A$. The expansion of the

[1] R.F. COVEYOU, R.R. BATE and R.K. OSBORN: ORNL-1958 (1955).
[2] E.R. COHEN: p. 405 of Ref. [15c].
[3] H. HURWITZ, M.S. NELKIN and G.J. HABETLER: Nucl. Sci. and Engng. 1, 280 (1956).

right side of (105.5) needs to be taken only unto the second derivative since all higher terms will have coefficients of order $1/A^2$ or less. In conclusion, Eq. (105.5) reduces to a second order differential equation whose coefficients (105.6) are obtained by following the same procedure, with the same approximations as were used to derive (97.16):

$$\left.\begin{aligned}\overline{\Delta E} = \frac{2}{A}(2kT_0 - E), \quad \overline{\Delta E^2} = \frac{4}{A} kT_0 E, \\ \text{for } A \gg 1 \text{ and } v^2 \gg v_0^2.\end{aligned}\right\} \tag{105.7}$$

Introducing these expressions and the independent variable (105.1), HURWITZ *et al.* find the differential equation

$$x\frac{d^2\overline{\varphi}}{dx^2} + (3 - 2x^2)\frac{d\overline{\varphi}}{dx} - \Delta\overline{\varphi} = 0 \tag{105.8}$$

where, according to the assumptions (96.4) and (97.6),

$$\Delta = 2A\frac{\bar{\Sigma}_a(kT_0)}{\Sigma_b} \approx \frac{4}{\xi}\frac{\bar{\Sigma}_a(kT_0)}{\Sigma_{s0}} \tag{105.9}$$

is an energy independent parameter. Here Σ_b represents the macroscopic scattering cross section for bound atoms.

WILKINS' equation is obtained by transforming (105.8) to the neutron density $n_0(x)$

$$x\,n_0''(x) + (2x^2 - 1)\,n_0'(x) + (4x - \Delta)\,n_0(x) = 0. \tag{105.10}$$

This differential equation has been investigated by E. R. COHEN[1] who derives the following expression for the solution, which is correct to terms of second order in Δ:

$$n_0(x) = x^2 e^{-x^2}\left\{\frac{4}{\sqrt{\pi}} + \Delta\,\mu_1(x) + \Delta^2\mu_2(x) + O(\Delta^3)\right\} \tag{105.11}$$

with

$$\begin{aligned}\mu_1(x) &= \sum_{n=0}^{\infty}\frac{x^{2n+1}}{(2n+1)\,\Gamma(n+\frac{5}{2})} \\ &= \frac{4x}{3\sqrt{\pi}}\left[1 + \frac{2x^2}{15} + \frac{4x^4}{175} + \frac{8x^6}{2205} + \frac{16x^8}{31185} + \frac{32x^{12}}{49495} + \cdots\right], \\ \mu_2(x) &= \frac{1}{4}\sum_{n=1}^{\infty}\frac{x^{2n}}{n(n+1)!}\left[\sum_{K=0}^{n-1}\frac{(K+1)!}{(2K+1)\,\Gamma(K+\frac{5}{2})}\right] \\ &= \frac{x^2}{6\sqrt{\pi}}\left[1 + \frac{19x^2}{90} + \frac{157x^4}{3780} + \frac{1263x^6}{176400} + \cdots\right].\end{aligned}$$

Fig. 205 shows the results of numerical computations for various values of the parameter Δ. For values of x large enough, the neutron density $n_0(x)$ approaches the expression (105.4) of the elementary solution valid, in absence of absorption, above 1 ev. The normalization of the curves of Fig. 205 is such that the constant C appearing in (105.4) is given by

$$C = \frac{\Delta}{2}\exp\left[\Delta \cdot a(\Delta)\right] \tag{105.12}$$

and may be obtained from the semilogarithmic plot of $a(\Delta)$ given in Fig. 206.

[1] E. R. COHEN: Nucl. Sci. and Engng. 2, 227 (1957).

A comparison of the flux distribution $\varphi_0(E)$ for a heavy gaseous moderator and for a monatomic hydrogen gas[1] has been made by Hurwitz *et al.*[2]. Such a comparison is possible because in both cases the absorption and scattering cross sections are assumed to satisfy the same conditions (96.4) and (97.6) so that both Γ and Δ turn out to be energy independent parameters. According to the definitions (105.2) of Γ and (105.9) of Δ, the comparison has to be made by chosing the cross sections so that

$$\Delta = 4\Gamma$$

and by matching the asymptotic behaviours of $\varphi_0(x)$. Hurwitz *et al.* find that, for moderate absorption, the two distributions are very similar. This is mainly due to the fact that both distributions show the same asymptotic behaviour at high energy, both satisfy the condition of conservation of neutrons and in the low energy region must both approach a Maxwell spectrum. Similar remarks will be made at the end of the Sects. 106 and 107.

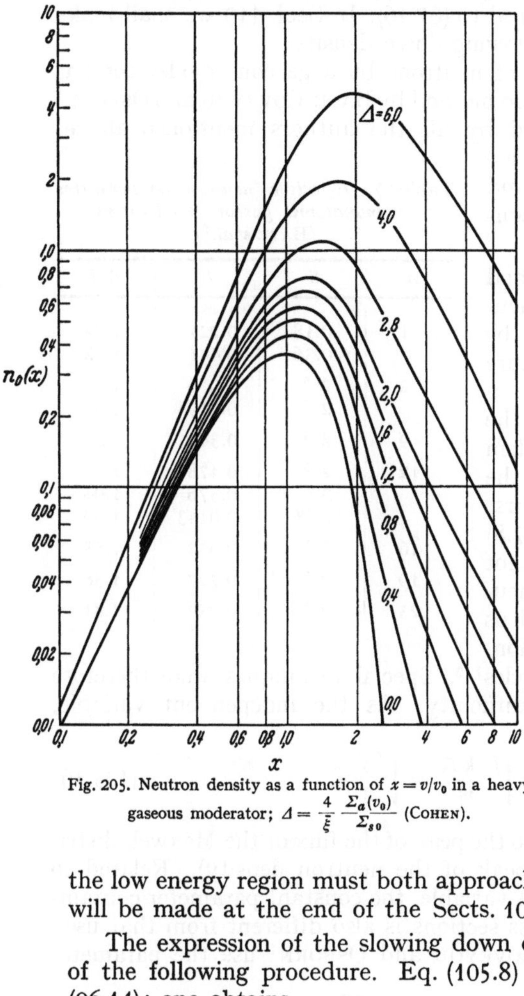

Fig. 205. Neutron density as a function of $x = v/v_0$ in a heavy gaseous moderator; $\Delta = \frac{4}{\xi}\,\frac{\Sigma_a(v_0)}{\Sigma_{s0}}$ (Cohen).

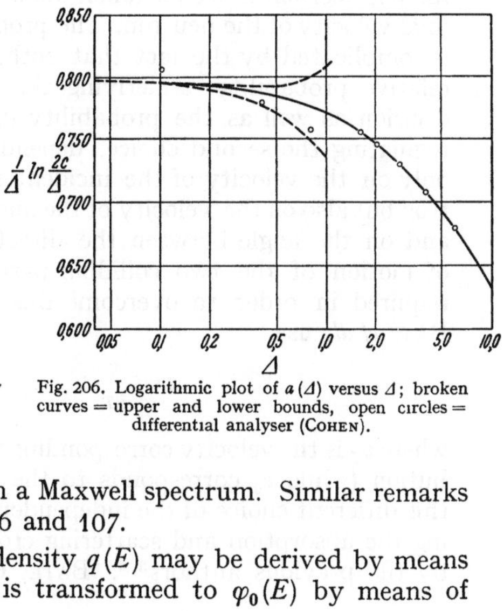

Fig. 206. Logarithmic plot of $a(\Delta)$ versus Δ; broken curves = upper and lower bounds, open circles = differential analyser (Cohen).

The expression of the slowing down density $q(E)$ may be derived by means of the following procedure. Eq. (105.8) is transformed to $\varphi_0(E)$ by means of (96.14); one obtains

$$\overline{\Sigma}_a(E)\,\varphi_0(E) = \frac{2\Sigma_b}{A}\left[\varphi_0(E) + E\,\frac{d\varphi_0(E)}{dE} + E\,kT_0\,\frac{d^2\varphi_0(E)}{dE^2}\right]. \qquad (105.13)$$

The right-hand side of this equation is then substituted for $\overline{\Sigma}_a(E)\,\varphi_0(E)$ in (69.30) and the differential equation so obtained is integrated from $E=0$ to E with the condition $q(0)=0$. The result is

$$q(E) = \frac{2\Sigma_b}{A}\left[(E - kT_0)\,\varphi_0(E) + E\,kT_0\,\frac{d\varphi_0}{dE}\right]. \qquad (105.14)$$

[1] See footnote 6, p. 458.
[2] See footnote 3, p. 460.

For $E \gg kT_0$ this expression for $q(E)$ approaches the asymptotic value

$$q(E) \to \frac{2\Sigma_b}{A} E\,\varphi_0(E) \tag{105.15}$$

as one can recognize, for example by computing $d\varphi_0/dE$ by means of (86.2). The expression (105.15), for $A \gg 1$, is identical to (67.16). In Sect. 110 we shall make use of the expression (105.14) of the slowing down density.

The problem of the slowing down of neutrons by a gaseous moderator has been treated by means of the Monte Carlo method by BATE, COVEYOU and OSBORN[1] under the same assumptions adopted by all the authors mentioned above: namely the absorption cross section is assumed to follow the $1/v$ law (96.4) and the scattering cross section to be constant (97.6).

Table 77. *Effective neutron temperatures in monatomic gaseous moderators* (BATE *et al.*[2]).

A	K	Γ	T_n/T_0
1	0.18	0.221	1.24
	0.36	0.441	1.42
2	2^{-5}	0.076	1.09
	2^{-7}	0.0194	1.02
9	2^{-5}	0.344	1.28
12	2^{-5}	0.478	1.35
	2^{-8}	0.576	1.04
	2^{-10}	0.0143	1.00
16	2^{-5}	0.612	1.55
19	2^{-5}	0.727	1.58
25	2^{-5}	0.957	1.81

The use of the Monte Carlo method implies that in each collision of a neutron with a nucleus a decision must be made as to whether the event is a capture or a scattering. In this second case a further decision must be taken about the final velocity of the neutron. The problem is complicated by the fact that both, the relative probability underlying the first decision as well as the probability (97.8) regulating the second choice, depend not only on the velocity of the incident neutron but also on the velocity of the nucleus and on the angle between the directions of motion of the two colliding particles[1,2]. Special techniques were therefore required in order to overcome this difficulty. As the independent variable, BATE *et al.* use

$$z = \frac{v}{v_T}\,; \qquad v_T = \sqrt{\frac{3kT_0}{m}} = \sqrt{\frac{3}{2}}\,v_0\,, \tag{105.16}$$

where v_T is the velocity corresponding to the peak of the flux of the Maxwell distribution (while v_0 corresponds to the peak of the neutron density). Related to this different choice of the independent variable, the constant parameter containing the absorption and scattering cross sections is also different from that used by the previous authors[3,4]. BATE, COVEYOU and OSBORN[1] use the parameter

$$K = \frac{\Sigma_a(v_T)}{\Sigma_{s0}} = \sqrt{\frac{2}{3}}\,\frac{\Sigma_a(v_0)}{\Sigma_{s0}} = \sqrt{\frac{2}{3}}\,\Gamma. \tag{105.17}$$

The only other parameter which needs be specified in order to characterize a particular moderator is the mass ratio A. Table 77 contains the values of K and A; the flux distribution for these parameter values was obtained by following as many neutron histories as were necessary to provide a total of at least 65000 scattering events. The velocity interval corresponding to the energy range from

[1] See footnote 1, p. 460.
[2] See footnote 2, p. 460.
[3] See footnote 6, p. 458.
[4] See footnote 3, p. 460.

0 to $1024 \cdot \frac{3}{2} kT_0$ was divided into 512 equal parts each of which corresponds at room temperature to $\Delta v = 169$ m/sec.

As already mentioned above, the results obtained in the case of hydrogen are in good agreement with the results of WIGNER and WILKINS (Fig. 204).

As examples of the results obtained for larger values of A, Figs. 207 to 209 show the flux distribution obtained for $A = 9$, $K = 2^{-5}$; $A = 9$, $K = 2^{-3}$; and $A = 12$, $K = 2^{-5}$. In each figure the Maxwell flux distribution and the $1/v \times$ flux distribution are also given for comparison. For strong absorption (K large) the flux distribution definitely deviates from the Maxwell flux distribution even

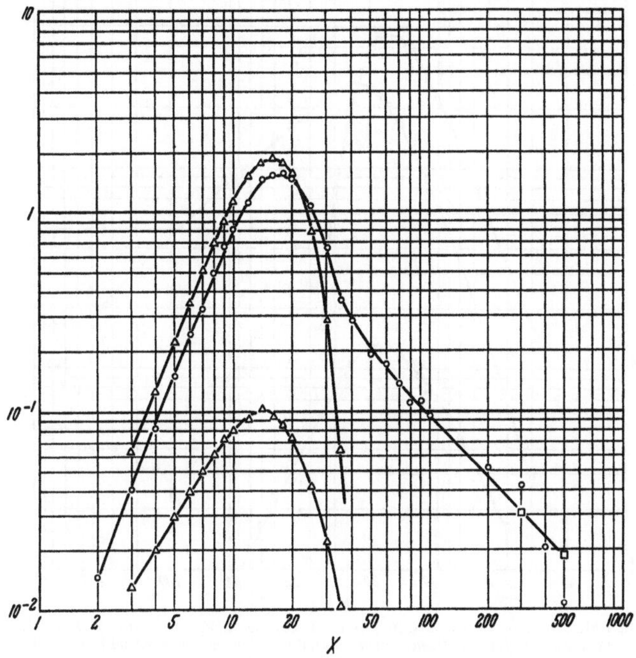

Fig. 207. Neutron flux distribution versus $X = 16v/v_T$ in a gaseous moderator: $A = 9$, $K = 2^{-5}$. ○ Actual flux; △ (upper curve) Maxwell flux; △ (lower curve) actual number density = actual flux/X; □ $1/v$ flux with same normalization (BATE *et al.*).

for neutron energies around kT_0, while for weak absorption the deviations are moderate. In the latter case a measure of these deviations is given by the neutron temperature as defined in Sect. 108.

The examples of Figs. 207 to 209 do not correspond to the actual cases of beryllium and graphite mixed with convenient amounts of heavy $1/v$ absorbers, because no effect of the crystalline structure of these materials is contained in the treatment of the slowing down presented in this section. It has to be noticed however that, whatever may be the molecular structure of the moderator, in the case of weak absorption, the neutron spectrum must approach a Maxwell distribution in the region below and around kT_0, and a slowing down spectrum (68.19) valid for chemically unbound nuclei in the high energy region. At intermediate energies on the contrary the behaviour of the spectrum will depend on the molecular structure, being different in the case of crystalline moderators from the case of gaseous moderators. A certain limitation to the difference between the spectra valid in these two cases is however imposed even in this intermediate energy region by the conservation of neutrons (105.3).

In Sect. 108 we shall see that in the case of Be the heavy Debye crystal approximation deviates very little from the heavy gas approximation (Fig. 218). There are, however, no good reasons to anticipate a similar result for other polycrystalline moderators.

106. The slowing down integral spectrum in light and heavy water. The neutron spectrum in light and heavy water has been numerically calculated by BROWN and ST. JOHN[1] who have adopted the following simplified model:

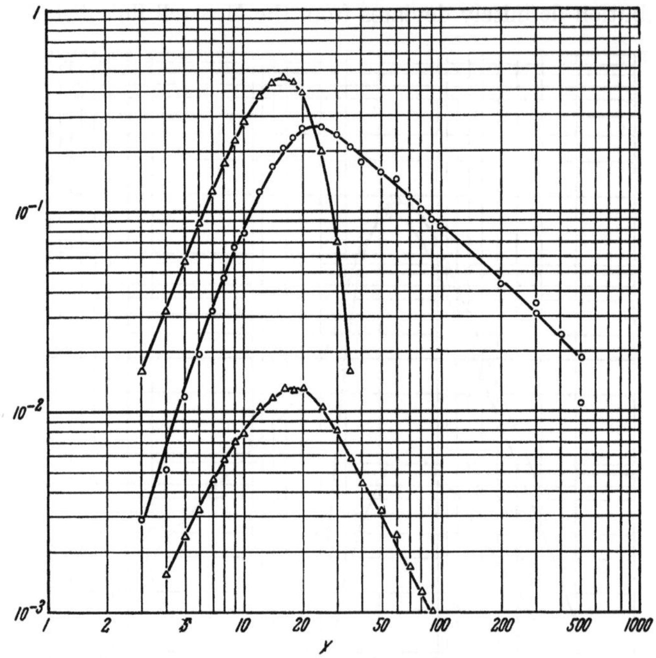

Fig. 208. Neutron flux distribution versus $X = 16 v/v_T$ in a gaseous moderator: $A = 9$, $K = 2^{-3}$. ○ Actual flux; △ (upper curve) Maxwell flux; △ (lower curve) actual number density = actual flux/X (BATE *et al.*).

(a) The slowing down is due only to hydrogen and not to oxygen or other heavy nuclei present in the moderator. The proton (or deuteron) scattering cross section is represented by (97.2) with only two terms, namely

$$\Sigma_s(v_r) = \Sigma_{s0} + \Sigma_{s1} \exp[-\alpha_1 v_r^2]. \tag{106.1}$$

The values of the three constants Σ_{s0}, Σ_{s1} and α_1 (Table 78) are adjusted so that (97.4) fits well the experimental data obtained by the Columbia University group (Sect. 101).

(b) The absorption cross section, including the contribution of heavy nuclei which may be dissolved in the water, follows the $1/v$ law (96.4), (96.5).

(c) The calculation begins at an energy E_2 large compared to the energy required to excite the vibrational modes of the molecules. For energies above E_2 the absorption can be neglected and the scattering cross section (106.1) reduces to its constant term Σ_{s0}, so that the neutron spectrum is given by (67.4), i.e. by

$$n_0(v') = \frac{C}{v'^2}, \qquad C = \text{const for } v' > v_2. \tag{106.2}$$

[1] H.D. BROWN and D.S. ST. JOHN: DP-33 (1954). — H.D. BROWN: DP-64 (1956).

This high energy region provides, through a further scattering process, the source of neutrons $S(v)$ at velocities $v \leqq v_2$. According to this picture the integral equation (96.1) is replaced by the following one:

$$[\gamma + V(v)]\, n_0(v) = \int_0^{v_2} G_s(v' \to v)\, n_0(v')\, dv' + S(v). \qquad (106.3)$$

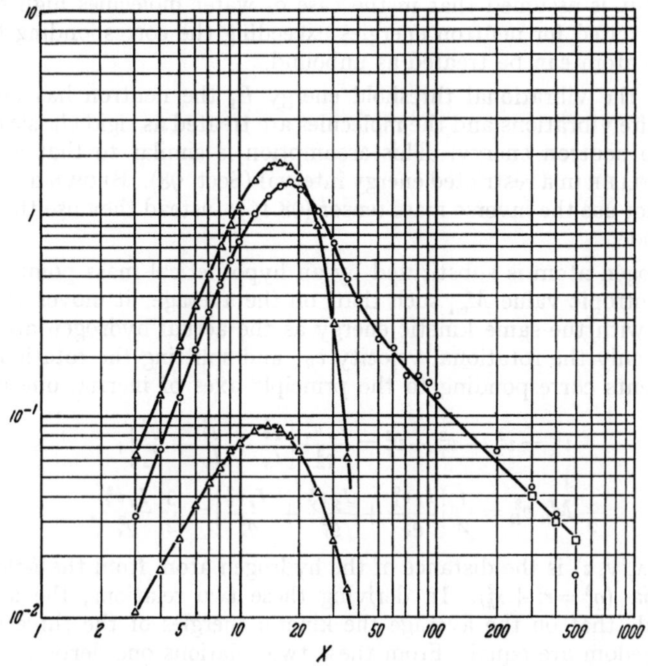

Fig. 209. Neutron flux distribution versus $X = 16v/v_T$ in a gaseous moderator: $A = 12$, $K = 2^{-5}$. ○ Actual flux; △ (upper curve) Maxwell flux; △ (lower curve) actual number density = actual flux/X, □ $1/v$ flux with same normalization (BATE *et al.*).

The source term is given, according to (62.7), (106.2), by

$$\left.\begin{aligned} S(v) &= 0 && \text{for } v < \alpha^{\frac{1}{2}} v_2, \\ S(v) &= \int_{v_2}^{v/\alpha^{\frac{1}{2}}} v'\, \Sigma_{s0}\, n_0(v')\, \frac{2v\, dv'}{v'^2(1-\alpha)} = \frac{C\, \Sigma_{s0}\, v}{1-\alpha} \left[\frac{1}{v_2^2} - \frac{\alpha}{v^2}\right] && \text{for } \alpha^{\frac{1}{2}} v_2 \leqq v \leqq v_2. \end{aligned}\right\} \qquad (106.4)$$

The constant C appearing in (106.4) is a kind of normalization constant. Because of the conservation of neutrons

$$\int_0^{v_2} S(v)\, dv = \int_0^{v_2} \gamma\, n_0(v)\, dv,$$

C can be determined by requiring that $n_0(v)$ takes an arbitrary value for a conveniently chosen velocity.

(d) Between the energy E_2 and an energy E_1 called the rotational threshold energy, the hydrogen (or deuterium) atoms are considered free and at rest. In this energy region the vibrational modes of the molecule play an important role. The justification given by the authors for this oversimplifying assumption is that according to (98.12), (98.13), valid for an isotropic harmonic oscillator, the ratio

$$f(y) = \frac{1}{4\pi} \int \frac{\sigma_1(\cos\vartheta)}{\sigma_0(\cos\vartheta)}\, d\omega, \qquad y = \frac{E}{\hbar\omega}$$

increases rapidly immediately above the threshold for excitation of the first vibrational level, $y=1$. For $y=1.5$, $f(y)$ is already about 3, showing that inelastic scattering starts to prevail over elastic collisions immediately above threshold. If it is assumed that in the case of water molecules high modes are also rapidly excited for neutron energies exceeding the corresponding threshold, the hydrogen atom can be treated as unbound.

(e) Below the vibrational threshold energy E_1 the neutron has not enough energy to excite vibrations and the molecules are treated as rigid classical rotators down to zero neutron energy. This assumption is similar to that adopted by SACHS and TELLER in a restricted energy interval (Sect. 98). BROWN and ST. JOHN however do not use the inverse mass tensor (98.16); instead they use the following simplified model.

The hydrogen atom is substituted by an hypothetical mass point which has a constant isotropic value M_{eff} such that, on the average, it moves at the same velocity and with the same kinetic energy as the actual hydrogen atoms. Considering first only the rotational velocity v_R, and splitting the rotational energy into components corresponding to the principle axes of inertia, one has

$$v_R^2 = v_x^2 + v_y^2 + v_z^2 = \frac{I_x v_x^2}{\varrho_x^2}\left[\frac{\varrho_x^2}{I_x} + \frac{\varrho_y^2}{I_y} + \frac{\varrho_z^2}{I_z}\right],$$

$$\frac{1}{2} M_R v_R^2 = \frac{1}{2}\left[\frac{I_x v_x^2}{\varrho_x^2} + \frac{I_y v_y^2}{\varrho_y^2} + \frac{I_z v_z^2}{\varrho_z^2}\right] = \frac{3}{2}\,\frac{I_x v_x^2}{\varrho_x^2},$$

where $\varrho_i\,(i=x, y, z)$ is the distance of the hydrogen atom from the i-th principle axis of inertia $(\varrho_i^2 = r_j^2 + r_k^2)$. In deriving these two relations, the assumption has been made that on the average the kinetic energies of the three rotational degrees of freedom are equal. From these two relations one derives

$$\frac{1}{M_R} = \frac{1}{3}\left[\frac{\varrho_x^2}{I_x} + \frac{\varrho_y^2}{I_y} + \frac{\varrho_z^2}{I_z}\right]. \tag{106.5}$$

The effective mass M_{eff} is obtained by considering also the translational motion. If M_{mol} is the mass of the molecule one has

$$\frac{1}{M_{\text{eff}}} = \frac{1}{M_R} + \frac{1}{M_{\text{mol}}}. \tag{106.6}$$

This expression, which is one third of the trace of the inverse mass tensor (98.16), can be obtained by requiring that: (1) the velocity of the hypothetical mass point is equal to the sum of the rotational and translational velocities averaged over all directions of v_{TR} with respect to v_R.

The result is:

$$v^2 = v_R^2 + v_{TR}^2.$$

(2) The kinetic energy of the rotational motion is equal, on the average, to the kinetic energy of the translational motion

$$\tfrac{1}{2} M_R v_R^2 = \tfrac{1}{2} M_{\text{mol}} v_{TR}^2;$$

and finally (3) the kinetic energy of the hypothetical particle of mass M_{eff} is equal to one half of the sum of the translational and rotational kinetic energies

$$\tfrac{1}{2} M_{\text{eff}} v^2 = \tfrac{1}{2}\left[\tfrac{1}{2} M_R v_R^2 + \tfrac{1}{2} M_{\text{mol}} v_{TR}^2\right].$$

The factor $\frac{1}{2}$ appearing in the latter equality is introduced by the authors because they use the Maxwell distribution (97.1) for free particles having three degrees of freedom, while the description of the water molecules as classical rotators and translators corresponds to six degrees of freedom.

A critical discussion of this oversimplified procedure and of its relationship to more refined methods such as the mass tensor method, has been made by KRIEGER and NELKIN[1] (see end of Sect. 98).

Table 78 contains the values of the constants adopted by BROWN and ST. JOHN in order to specify their model.

Table 78[2]. *Parameter values used in the molecular model of light and heavy water by* BROWN *and* ST. JOHN.

	Σ_{s0} (barns)	Σ_{s1} (barns)	α_1 (a.m.u./ev)$^{\frac{1}{2}}$	M_{eff}/m	E_2 (ev)	E_1 (ev)
H_2O	20.7	13.1	4.5	1.9	2.87	0.226
D_2O	3.3	1.6	2.88	3.6	2.87	0.143

The integral equation (106.3) was converted into a form suitable for numerical computation by replacing the integral by a sum

$$n_i = \frac{S_i + \sum_{j=1}^{n} G_{ij} \Delta v_j n_j}{V_i + \gamma}. \qquad i = 1, 2, \ldots n;$$

and the system of linear equations thus obtained was solved by standard iteration procedure with an IBM machine.

The neutron flux distribution obtained by such a procedure for light and heavy water at 80° C ($kT_0 = 0.0304$ ev) and for different values of the absorption probability γ (Table 79) are shown in Figs. 210 and 211.

Table 79. *Values of absorption cross sections of D_2O used by Brown and St. John.*

γ	0.0426	0.064	0.1	0.2 sec^{-1}
Σ_a (2200 m/sec)	0.0111	0.0169	0.0264	0.0528 cm^{-1}

All curves have been arbitrarily normalized to the same value for $E = 0.005$ ev. The curve marked $\gamma = 0$ corresponds to the Maxwell flux distribution.

In order to investigate the influence of the chosen model on the results and to obtain an indication of the magnitude of possible errors, the neutron flux distribution was computed again for heavy water at 80° C, by changing one or the other of the parameters characterizing the model:

1. For $\Sigma_a(2200) = 0.0111$ cm^{-1} the computation was repeated with effective mass ratios $A_{eff} = 2$ and 5. The deviations of these spectra from the spectrum obtained with the effective mass $A_{eff} = 3.6$ are less than 5% for all neutron energies between 0.005 and 1 ev.

2. The flux distribution was computed for various values of γ assuming that the scattering cross section was constant and equal to Σ_{s0} from above 1 ev down to low energy. Under these conditions the effect of the chemical bond, introduced empirically by means of the velocity dependence of the scattering cross section

[1] T. J. KRIEGER and M. S. NELKIN: KAPL-1597 (1956); KAPL-1597 Addendum No. 1 (1957).

[2] For the data on vibrational and rotational levels of the H_2O and D_2O molecules see Sect. 101, p. 426.

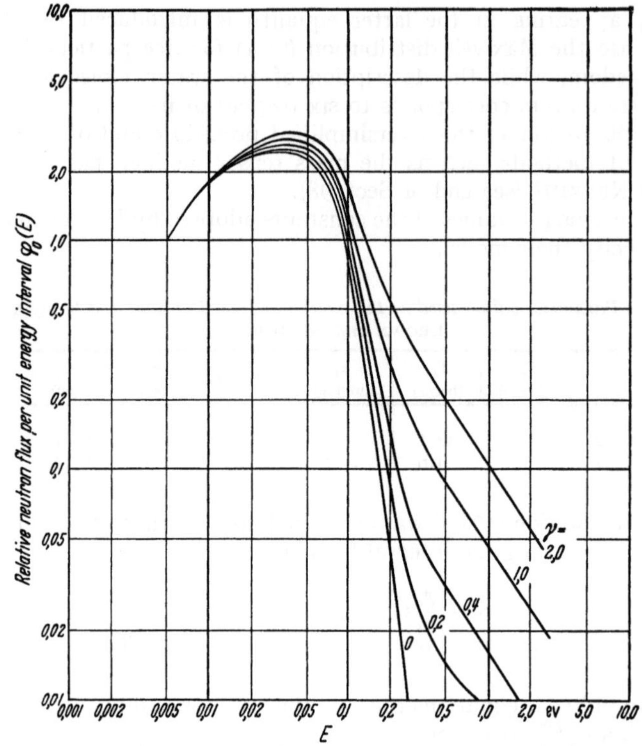

Fig. 210. Neutron flux distributions in H_2O at 80° C for different amounts of absorption (BROWN and ST. JOHN).

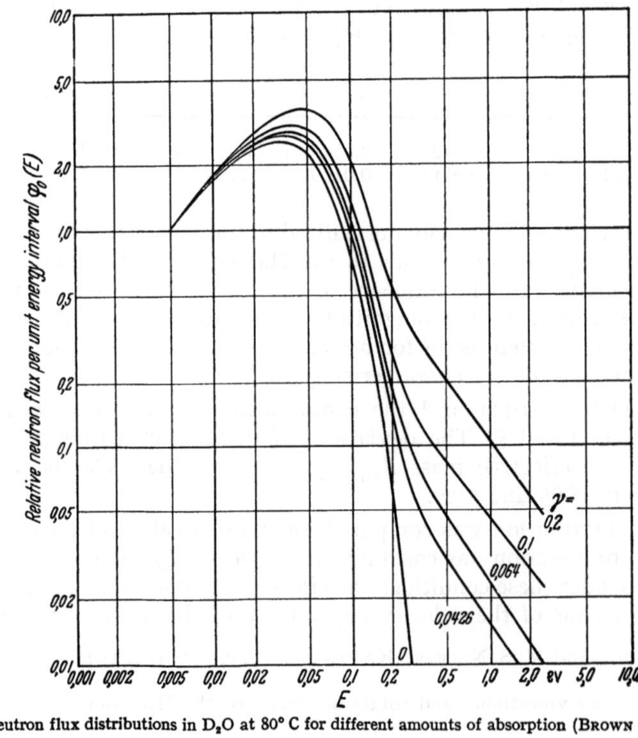

Fig. 211. Neutron flux distributions in D_2O at 80° C for different amounts of absorption (BROWN and ST. JOHN).

(106.1), is disregarded. The distortion of the spectrum is small for low absorption but increases with increasing γ and is about 10% for $\gamma = 0.2$; i.e. the correction is about 10% and negative at 0.005 ev, decreases to zero at 0.05 ev, reaches a

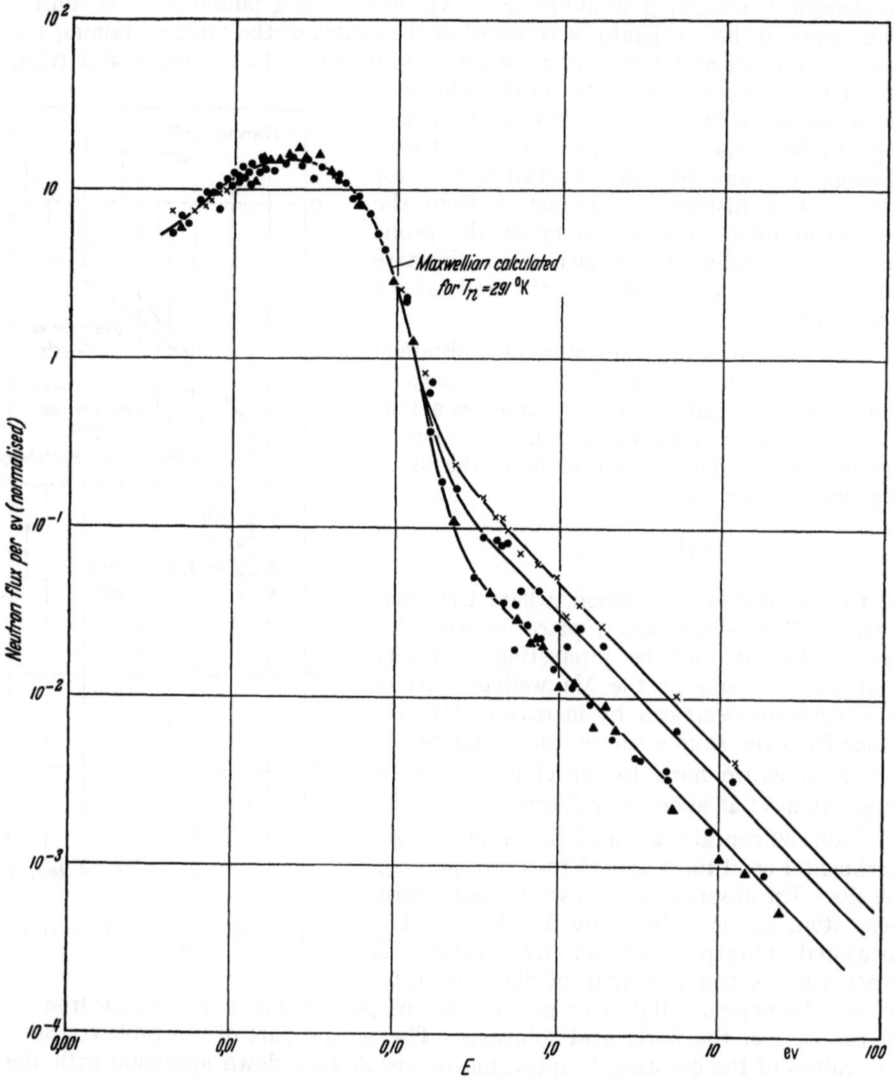

Fig. 212. Neutron spectra in water at 291° K. × Spectrum at 2 cm from source; ● (upper curve) spectrum at 5 cm from source; ▲ spectrum at 10 cm from source; ● (lower curve) spectrum at 12.5 cm from source (Poole).

positive broad maximum around 0.2 ev, and then decreases again becoming zero at 1 ev.

3. The flux distribution was computed for different values of the vibrational threshold energy: $E_1 = 0.090$ ev and $E_1 = 0.226$ ev. The deviations of these spectra with respect to that computed with $E_1 = 0.143$ ev and given in Fig. 211 is, below 0.05 ev, not more than 2%. In a narrow region between 0.1 and 0.2 ev the deviation reaches +8% at $E_1 = 0.090$ ev and −20% at $E_1 = 0.226$ ev. It drops to negligible values at neutron energies larger than 0.5 ev.

A systematic experimental investigation of the neutron spectrum in water and in solutions of boric acid in water has been made by POOLE[1] using the method of pulsed neutron sources (Sects. 85, 111 and 131). An electron linear accelerator bombarding uranium (Sect. 51) is used as a pulsed neutron source. The target of the accelerator was placed at the centre of the tank containing the moderator and a probe tube, at right angle with respect to the accelerator tube, ended at a distance x from the target. The neutrons, slowed down in the moderator and entering into the probe tube from the bottom, passed through a collimator and reached a detector placed at a distance of 12 meters from the moderator tank. The geometry of the probe tube and the collimator is such as to eliminate all the neutrons emerging from the walls of the probe tube.

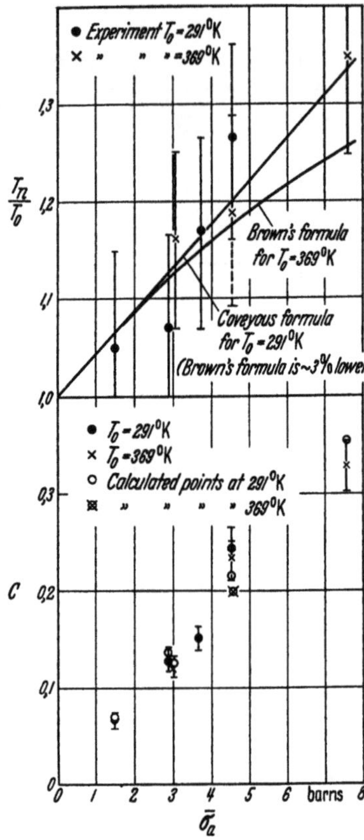

Fig. 213. Plot of T_n/T_0 and of C versus $\bar{\sigma}_a$ (POOLE).

Fig. 212 shows the neutron spectra obtained with pure water at 291° K for four values of the distance x. All the spectra have been normalized so that they overlap in the thermal energy region. The transition from the slowing down region, where

$$n_0(E) = \frac{C}{E},$$

to the thermal energy region, where the spectrum is Maxwellian, takes place between 0.5 and 0.1 ev. It may be interesting to notice that the intensity of the Maxwellian part of the spectrum decreases by increasing the distance from the source slower than that of the $1/E$ part, as we have to expect because $\overline{r_{th}^2}$ is larger than $\overline{r^2}$ at about 1 ev (Sects. 83 and 84).

Table 80 contains a list of the employed experimental conditions and of the corresponding results. The absorption cross section per hydrogen atom $\bar{\sigma}_a$ is defined by Eq. (68.2). The measured effective neutron temperature T_n (Sect. 108) is equal to that of the moderator within the experimental error in the case of pure water and deviates from it in the case of the boric acid solutions. The second part of the table contains the values of the constant C appearing in the slowing down spectrum with the normalization explained above. Fig. 213 shows, in graphical form, the same comparison of the experimental values with the corresponding theoretical predictions for T_n/T_0 and C as a function of $\bar{\sigma}_a$.

Figs. 214 and 215 show a comparison of some of the results of POOLE with the curves, computed by AMSTER[2] by means of the Wigner-Wilkins formula (Sect. 105). The agreement is much better than one would have expected; some deviations, however, do appear. In Fig. 215 between 0.1 and 0.3 ev, i.e. in the transition region between the Maxwellian and the $1/E$ flux distributions, the

[1] M. J. POOLE: J. Nucl. Energy 5, 325 (1956).
[2] H. J. AMSTER: Nucl. Sci. and Engng. 2, 394 (1957).

Wigner-Wilkins spectra remain below the experimental results. In the $1/E$ region, on the contrary, the theoretical curves tend to remain slightly above the experimental data.

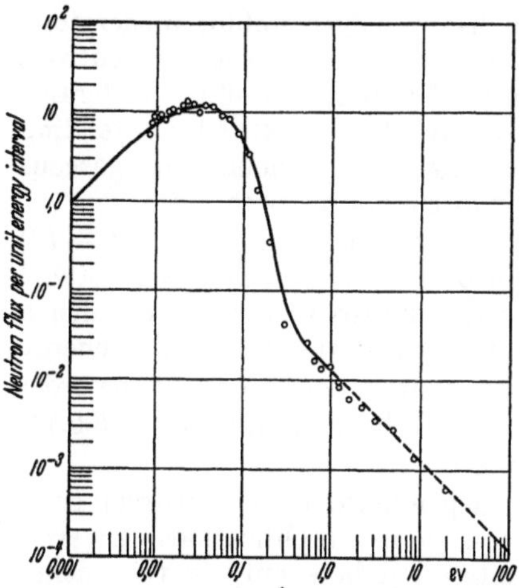

Fig. 214. Comparison of neutron spectrum in water at 98° C obtained by POOLE with curves computed by means of Wigner and Wilkins formula (AMSTER).

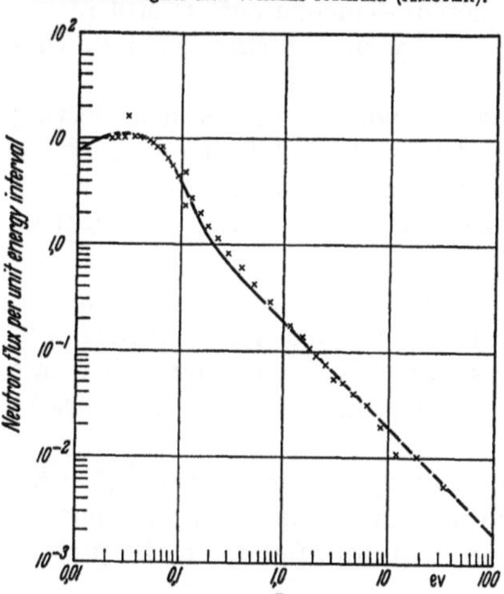

Fig. 215. Comparison of neutron spectrum in boric acid in water at 97° C ($\bar{\sigma}_a = 4.55$ barns per H atom) obtained by POOLE, with curves computed by means of Wigner and Wilkins formula (AMSTER).

These deviations may be qualitatively explained in terms of the remarkable difference existing between the scattering cross section of the hydrogeneous gas moderator used by WIGNER and WILKINS and that of the water molecules.

Table 80. *Summarized results on T_n and C in water and solutions of boric acid in water* (POOLE).

Moderator	$\bar{\sigma}_a$ (barns)	T_0 (°K)	Position of probe (cm from source)	Measured T_n (°K)	Averages	Calculated T_n COVEYOU	Calculated T_n BROWN	Calculated T_n AMSTER	Measured C (All ± 10%)	Calculated C POOLE	Calculated C AMSTER	Calculated C BROWN
Tap water	0.3	291	2	308 ± 30					0.048			
Tap water	0.3	291	5	314 ± 30					0.033			
Tap water	0.3	291	10	283 ± 30	299 ± 15				0.016			
Tap water	0.3	291	12.5	297 ± 30					0.015 (0.016 ± 0.001)			
Demineralized water	0.3	291	10	304 ± 30					0.018			
Demineralized water	0.3	291	12.5	314 ± 30					0.016			
Boric-acid solution	1.5	291	10	306 ± 30		313	313	303	0.067	0.069	0.072	
Boric-acid solution	2.9	291	10	310 ± 30		332	330	320	0.130	0.137	0.139	
Boric-acid solution	3.73	291	10	340 ± 30		345	342		0.13	0.19	0.180	
Boric-acid solution	4.55	291	10	368 ± 35		353	348	341	0.214	0.214	0.215	0 240
Boric-acid solution	3.03	369	10	430 ± 40					0.121	0.125		
Boric-acid solution	4.54	370	10	432 ± 40				417	0.236	0.19		
Boric-acid solution	7.6	371	10	501 ± 50			466	452	0.328	0.356	0.34	0.35
Boric-acid solution	9.9	371	10	515 ± 50					0.44			

The same reasons as those mentioned at the end of Sect. 105 in connection with the discussion of the curves plotted in Fig. 207, may be invoked here in order to explain the insensitivity of the flux distribution to the type of model adopted for the computation of the coefficients appearing in the integral equation of the neutron spectrum[1].

107. The integral spectrum in the case of heavy polycrystalline moderators. Comparison with experiments in beryllium. The energy distribution of neutrons in an infinite homogeneous medium composed of a polycrystalline moderator and a $1/v$ absorbing material has been investigated by NELKIN[2] by extending the procedure developed by HURWITZ *et al.*[3] for heavy monatomic gaseous moderators [see Eqs. (105.5), (105.8) and their discussion in Sect. 105].

We have seen that for heavy gases the energy change kernel $\bar{\Sigma}_s(E\to E')$ appearing in the integral equation (96.13) can be expanded into powers of A^{-1} keeping only the first order terms, and that in this approximation the expansion (105.5) can be broken off after the second term, because the moments of energy transfer $\overline{\Delta E^n}$ of order n higher than the second, contain coefficients of the order of $1/A^2$ or less. By such a procedure we have reduced the integral equation (96.13) to the second order differential equation (105.8).

For polycrystalline moderators, the energy dependence of the scattering cross section is much more complicated than for gases. Some simplifications, however, may be obtained by noticing that we are interested here only in the energy distribution, not in the space distribution of the neutrons, so that the elastic part (coherent + incoherent) of the cross section does not enter the problem.

The inelastic part of the cross section (coherent + incoherent) is still rather complicated, but can be simplified considerably by introducing the following approximations.

First one adopts the incoherent approximation, which consists of neglecting the term $\sigma^{(1)}_{\text{inel, coh}}$ in Eq. (103.11). According to SINGWI and KOTHARI such an approximation is not always justified (Sect. 103); for instance for Be, to which the numerical computations given below in this section refer, $\sigma^{(1)}_{\text{inel, coh}}$ is negative and represents about 10% of the total cross section.

Secondly the approximation is made of describing the crystal in terms of the Debye model (Sect. 99) neglecting the polarization of phonons. As we have seen in Sect. 103, PLACZEK and VAN HOVE, as well as SQUIRES, have shown that this approximation is questionable if the energy distribution of the inelastically scattered neutrons is wanted in detail.

NELKIN starts from a convenient form[4, 5] of the energy change kernel $\bar{\Sigma}_s(E\to E')$ valid under the above mentioned approximations and expands it into powers of A^{-1}. In two respects, however, the situation is quite different in the present case from that of gases.

While the expressions (105.7) of the moments of energy transfer $\Delta E = E' - E$ are correct only when the velocity v of the neutron is larger than that of the atoms of the moderator, a limitation of this type does not exist for crystals. Therefore the expression of the energy change kernel derived by such a procedure can be used in the low energy region, in particular below the crystal cut-off

---

[1] M. J. POOLE, M. S. NELKIN and R. S. STONE: Progress in Nuclear Energy. Physics and Mathematics, Vol. 2, p. 91. New York and London: Pergamon Press 1958.

[2] M. NELKIN: Nucl. Sci. and Engng. **2**, 199 (1957).

[3] H. HURWITZ, M. S. NELKIN and G. J. HABETLER: Nucl. Sci. and Engng. **1**, 280 (1956).

[4] O. KLEINMANN: Brown University thesis 1951; see footnotes 2 and 3.

[5] R. J. GLAUBER: Phys. Rev. **98**, 1692 (1955).

($E<E_c$, Table 76, p. 436) where, according to the discussion of Eq. (103.15) the comparison of theoretical expressions for σ_{inel} with experiments can very easily be made.

Furthermore in the case of crystals all moments of the energy change, $\overline{\Delta E^n}$, defined by (105.6), give contributions of order A^{-1} with the result that the integral equation describing the slowing down cannot be transformed into a differential equation by neglecting terms in some small power of A^{-1}.

In conclusion, the energy change kernel introduced by NELKIN in Eq. (96.13) is given by the following expression valid to the first order in $1/A$:

$$\left.\begin{aligned}&\overline{\Sigma}_s(E\to E')\,dE'\\&\quad=\begin{cases}\Sigma_b\dfrac{3}{A}\dfrac{kT_0}{k\Theta_D}\left(\dfrac{E'}{E}\right)^{\frac{1}{2}}\dfrac{E'+E}{k\Theta_D}\dfrac{\Delta E/kT_0}{e^{\Delta E/kT_0}-1}\dfrac{dE'}{k\Theta_D} & \text{for } |\Delta E|\leqq k\Theta_D,\\ 0 & \text{for } |\Delta E|>k\Theta_D,\end{cases}\end{aligned}\right\}\tag{107.1}$$

where Σ_b is the macroscopic scattering cross section for bound atoms and $\Delta E = E-E'$.

It may be checked that this expression satisfies the detailed balance condition (96.7). A comparison of (107.1) with experimental results in the case of Be will be given below (Figs. 216 and 217).

Once a convenient approximate expression of the energy-change kernel has been derived, we can turn to the integral equation (96.13) describing the slowing down of neutrons in an infinite homogeneous medium if no actual source is present in the energy interval under consideration.

The assumption is made that for $E\geqq E_0$ the flux $\varphi(E)$ is given by some known asymptotic expression $\varphi_0(E)$ while in the region, $E<E_0$, $\varphi(E)$ satisfies the integral equation

$$[\overline{\Sigma}_a(E)+\overline{\Sigma}'_s(E)]\,\varphi(E)=\int_0^{E_0}\overline{\Sigma}_s(E'\to E)\,\varphi(E')\,dE'+S(E)\tag{107.2}$$

where

$$S(E)=\int_{E_0}^{\infty}\overline{\Sigma}_s(E'\to E)\,\varphi_0(E')\,dE'-\varphi_0(E)\int_{E_0}^{\infty}\overline{\Sigma}_s(E\to E')\,dE'\tag{107.3}$$

and

$$\Sigma'_s(E)=\int_0^{E_0}\overline{\Sigma}_s(E\to E')\,dE'.\tag{107.4}$$

It is rather arbitrary to treat the term

$$\varphi_0(E)\int_{E_0}^{\infty}\overline{\Sigma}_s(E\to E')\,dE'$$

as part of the source but, together with the definition (107.4), it yields the neutron conservation condition in the very simple form

$$\int_0^{E_0}S(E)\,dE=\int_0^{E_0}\overline{\Sigma}_a(E)\,\varphi(E)\,dE.$$

A check of the choice of $\varphi_0(E)$ is obtained by verifying that it agrees with $\varphi(E)$ in the region below E_0 where the source term does not vanish.

In order to obtain a convenient asymptotic method for determining $\varphi_0(E)$, NELKIN rewrites the integral equation for $E>E_0$ using the lethargy (Sect. 62)

$$u=\ln\frac{k\Theta_D}{E}$$

as the new variable. He obtains

$$\bar{\Sigma}_a(u)\,\varphi_0(u) = \int_{-\infty}^{+\infty} [\bar{\Sigma}_s(u\to u')\,\varphi_0(u') - \bar{\Sigma}_s(u\to u')\,\varphi_0(u)]\,du'$$

with

$$\bar{\Sigma}_s(u\to u') = E'\,\bar{\Sigma}_s(E\to E'); \qquad \varphi_0(u) = E\,\varphi_0(E).$$

By expanding $\varphi_0(u')$ in a Taylor series about $u'=u$ and neglecting all terms beyond the first derivative, he obtains the following approximate expression:

$$\varphi_0(u) = \varphi_0(u_0)\exp\left\{\int_{u_0}^{u} [\bar{\Sigma}_a(u') - \delta\bar{\Sigma}_s(u')]\,\frac{du'}{\bar{\xi}'\,\bar{\Sigma}_s(u')}\right\}$$

where

$$\delta\bar{\Sigma}_s(u) = \int_{-\infty}^{+\infty} [\bar{\Sigma}_s(u'\to u) - \bar{\Sigma}_s(u-u')]\,du'; \quad \bar{\xi}'\,\bar{\Sigma}_s(u) = \int_{-\infty}^{+\infty} (u'-u)\,\bar{\Sigma}_s(u'\to u)\,du'$$

may easily be calculated by making use of the mass expansion kernel (107.1). One can verify that, in the high energy limit $\frac{k\Theta_D}{E} \ll 1$,

$$\delta\bar{\Sigma}_s \to 0, \qquad \bar{\xi}'\,\bar{\Sigma}_s \to \frac{2m}{M}\,\Sigma_{s0} = \bar{\xi}\,\Sigma_{s0},$$

i.e. one returns to the result of the Fermi age theory.

This procedure has been applied to the computation of the slowing down of neutrons in Be. Before doing this, a check has been made of the mass expansion kernel (107.1) by comparing it with the experimental results of the Brookhaven group. The total inelastic cross section of Be[1] is compared with (107.1) after integration with respect to E'. For the Debye temperature the value $\Theta_D = 1000°$ K is used as obtained from specific heat measurements. Fig. 216 shows that the agreement is as good as one would expect (about 10%) because the error due to the use of the incoherent approximation is negative and about 10%, and the error due to neglecting terms of order $1/A^2$ is also roughly 10% and positive.

Fig. 217 shows a comparison between the energy distribution of cold incident neutrons scattered at 90° by beryllium at room temperature[2] with the corresponding spectrum computed by means of (107.1); the agreement is even better than could have been expected. One can, therefore, conclude that even if the agreement between Eq. (107.1) and experiments does not rest on a very sound physical basis and may be in part fortuitous, the mass expansion kernel (107.1) can be used with some confidence for the computation of slowing down in Be.

The actual computation has been made by replacing the integral equation (107.2) by an approximate set of linear algebraic equations which are solved by an iterative procedure with an IBM machine.

The results are plotted as continuous curves in Fig. 218 for three values of the parameter (105.9), i.e.

$$\Delta = 2A\,\frac{\bar{\Sigma}_a(kT_0)}{\Sigma_{s0}} \approx \frac{4}{\xi}\,\frac{\bar{\Sigma}_a(kT_0)}{\Sigma_{sb}}.$$

The dotted curves represent the results obtained for a heavy gaseous moderator (Sect. 105). The deviations between the crystalline and the gaseous moderator are very small even for $\Delta = 5$ where the Maxwellian peak has completely

[1] H. Palevsky, D. J. Hughes and R. L. Zimmermann: Phys. Rev. **99**, 611 A (1955). — BNL-325 and p. 68 of Ref. [*14*].

[2] R. S. Carter, D. J. Hughes and H. Palevsky: Phys. Rev. **99**, 611 (1955).

disappeared. NELKIN notices that in considering the computation procedures presented in this section and in Sect. 105 one does not recognize any good a priori reason for expecting the influence of crystalline structure to be so small.

As mentioned in the discussion of Figs. 208 to 209 of Sect. 105, detailed balance, neutron conservation and the use of the same asymptotic expression in the high energy region, are all conditions which constrain the slowing down

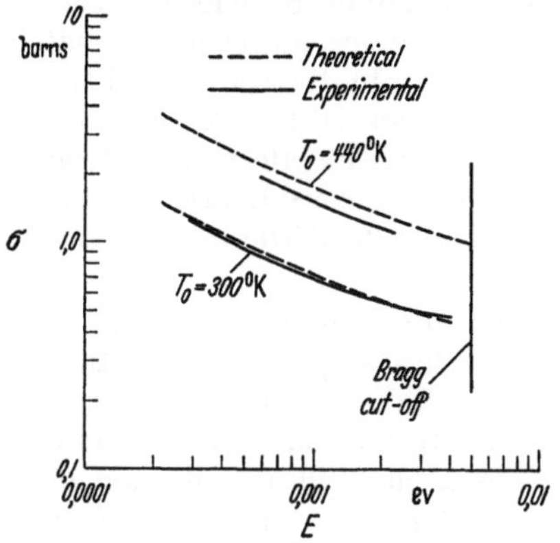

Fig. 216.

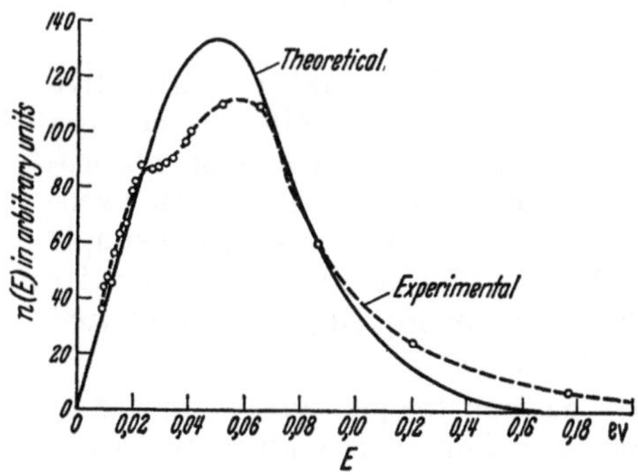

Fig. 217.

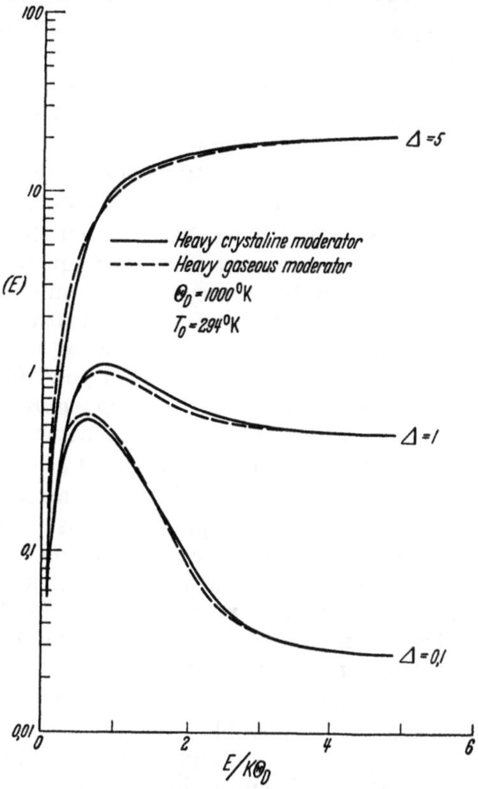

Fig. 218.

Fig. 216. The total inelastic cross section of polycrystalline beryllium as a function of neutron energy (NELKIN).

Fig. 217. The energy distribution of initially cold neutrons scattered through 90° by polycrystalline beryllium (NELKIN).

Fig. 218. The thermal spectra for $1/v$ absorption in a heavy crystalline moderator compared to those for a heavy gaseous moderator (NELKIN).

spectra computed for different moderators to remain rather close to each other. According to NELKIN, in the comparison between a Debye crystal and a gas two further similarities may play an appreciable role. The first is the continuous spectrum of vibrational frequencies of the Debye crystal which permits a continuous energy transfer from the neutron to the moderator and viceversa (see Fig. 203 and its discussion at the end of Sect. 104). The second similarity is the heavy mass limit ($A \gg 1$) which emphasizes the approximate constancy of the total scattering cross section.

One has, however, to keep in mind that there are no reasons for expecting that the great similarity between the gas and the crystal model found for Be must hold a priori for other polycrystalline moderators.

108. The effective neutron temperature. In Sect. 96 it has been pointed out that in absence of both neutron source and absorption, the Maxwell distribution of velocities (96.6) corresponding to the temperature T_0 of the moderator, must be a solution of the integral equation (96.1). In other words, if there is neither source nor absorption, nor leakage of neutrons at the surface, thermal equilibrium is established between the neutrons and the moderator; the neutrons thus have, for purely thermodynamical reasons, the Maxwell distribution of velocities corresponding to the temperature T_0 of the moderator. In practice, however, thermal equilibrium is not established because of neutron emission by the source and the constant drain of neutrons due to absorption and leakage. In this section only the effect of absorption will be taken into account, since the effect of neutron leakage at the moderator surface is eliminated by referring consideration to the stationary case of neutrons slowed down uniformly throughout the entire space filled with a homogeneous moderator.

In Sect. 105 it has been shown that for strong absorption the distortion of the neutron spectrum in the energy region around a few $k T_0$ is so severe that the spectrum bears very little resemblance to a Maxwell distribution. For weak absorption, on the contrary, the distorsion is moderate and one can therefore ask what temperature T_n must be introduced into a Maxwell distribution of velocities, in order to obtain a good fit to the actual neutron spectrum; T_n is called the *effective neutron temperature.* Its deviation from the moderator temperature T_0 is a measure of the deviation of the actual spectrum from that corresponding to thermal equilibrium.

Such a description of the neutron spectrum, although qualitatively attractive does not have a sound physical basis. In fact the definition of an effective neutron temperature implies that the neutron spectrum for a certain value of the absorption parameter Δ defined by (105.9), may be brought to coincide with the Maxwellian distribution — rigorously valid for the case of no absorption — by a simple change of velocity scale. Now, it can easily be verified that this is not true; for instance the spectra computed by COHEN by means of Eq. (105.11) and plotted in Fig. 205, deviate considerably from such a behaviour.

Thus, it is possible to define an effective neutron temperature by means of some appropriate analytical recipe, but such a definition is ambiguous in the sense that different recipes will give different effective temperatures. In spite of this remark, the use of the effective neutron temperature as a convenient parameter for the description of the neutron spectrum in a moderator or in a pile is so widely spread, that we fell obliged to report in detail the various theoretical and experimental approaches to this problem.

In order to determine T_n, COHEN[1] makes the assumption that the neutron spectrum may be represented, with adequate accuracy, by a Maxwell distribution with neutron temperature T_n

$$M_{T_n m}(v) = \frac{4}{\sqrt{\pi}} \frac{1}{v_n^3} e^{-\frac{v^2}{v_n^2}} v^2, \qquad v_n = \sqrt{\frac{2kT_n}{m}} \tag{108.1}$$

in the region where v is smaller than a certain velocity $y v_n$ and by the approximate expression (97.24) for $v > y v_n$. Considering that this approximate expression is valid only in the limit $v^2 \gg v_0^2$, one can conclude that a representation of this type will be adequate only if the condition $(y v_n)^2 \gg v_0^2$ is fulfilled. As will be shown later this condition is satisfied (Fig. 219).

[1] E. R. COHEN: NAA-SR-1127 (1955).

The condition of continuity of the two expressions (108.1) and (97.24) at $v = y v_n$ may be imposed by introducing (108.1) in (97.19) with $v = y v_n$:

$$q(y v_n) = \frac{4}{\sqrt{\pi}} \frac{\Sigma_{s0} v_n}{A} y^2 (y^2 - 2t) e^{-y^2} \tag{108.2}$$

where

$$t = \left(\frac{v_0}{v_n}\right)^2 = \frac{T_0}{T_n}. \tag{108.3}$$

On the other hand the conservation of neutrons (105.3a) demands that $q(y v_n)$ is equal to the total number of neutrons absorbed at velocities below $y v_n$,

$$q(y v_n) = \gamma \int_0^{y v_n} n_0(v)\, dv = \gamma \frac{4}{\sqrt{\pi}} \int_0^{y} e^{-u^2} u^2\, du = \gamma \left[\operatorname{Erf}(y) - \frac{2y}{\sqrt{\pi}} e^{-y^2}\right]. \tag{108.4}$$

Equating (108.2) to (108.4), the following relationship between the two adjustable parameters t and z is obtained

$$\frac{\Sigma_a(kT_0)}{\xi \Sigma_{s0}} = \frac{A}{2} \frac{\Sigma_a(kT_0)}{\Sigma_{s0}} = \frac{2y^2(y^2 - 2t) e^{-y^2}}{\sqrt{\pi t}\left[\operatorname{Erf}(y) - \frac{2y}{\sqrt{\pi}} e^{-y^2}\right]}. \tag{108.5}$$

In order to completely determine the neutron spectrum a further condition must be imposed, e.g., that the neutron distribution calculated by the present procedure should fit the neutron spectrum derived by solving the differential equation (105.8) which is valid in the case of heavy moderators.

A convenient procedure for achieving this fit is to equate the moments

$$M_k = \int_0^\infty v^k n_0(v)\, dv$$

relative to the two spectra. One has to notice however that according to (108.1) and (97.24) the neutron spectrum given by the present procedure varies as v^2 for $v \ll v_n$ and as v^{-2} for $v \gg v_n$. It is therefore impossible to define a mean velocity or a mean energy for this distribution. One can, however, compute a characteristic neutron energy for the distribution from the ratio of the moments M_0 and M_{-2}:

$$\langle E \rangle = \frac{2}{m} \frac{M_0}{M_{-2}} = \frac{2}{m} \frac{\int_0^\infty n_0(v)\, dv}{\int_0^\infty v^{-2} u_0(v)\, dv}.$$

By requiring that under the same conditions [i.e. the same values of A, Σ_{s0}, $\Sigma_a(kT_0)$] this quantity should have the same value when computed from the present distribution and from Wilkin's spectrum, Cohen derives[1,2] an implicit relationship between z and t which, when combined with (108.5), allows the determination of both them.

In Fig. 219, y^2 is plotted as a function of $\frac{\Sigma_a(kT_n)}{\xi \Sigma_{s0}} = \frac{A}{2} \frac{\Sigma_a(kT_n)}{\Sigma_{s0}}$. The curves marked A, B, C have been computed by means of (108.5). Curve A is calculated including the correction for neutron temperature, i.e. with T_n given by (108.7). Curve B is calculated for a moderator at $T_0 = 0$ (constant slowing down power) and curve C by assuming that the temperature T_n of the neutrons

[1] See footnote 1, p. 477.

[2] R.E. Cohen: p. 405 of Ref. [*15c*].

is identical to that of the moderator ($t=1$). From these curves one recognizes that y is always sufficiently large to justify the use of (97.24) for $v \geqq y v_n$. The broken curves are computed by means of (110.17) with $t=1$. This expression takes into account the effect of diffusion of neutrons in space-dependent problems (Sects. 109 and 110).

In Fig. 220, $t^{-1} = T_n/T_0$ is plotted as a function of $\Sigma_a(kT_0)/\xi\,\Sigma_{s0}$. The curves marked 1 and 2 correspond to two different approximations adopted in the actual computation and represent, respectively, upper and lower limits. The dotted line represents the tangent at the origin whose equation is

$$T_n = T_0\left[1 + \frac{2}{\sqrt{\pi}} A \frac{\Sigma_a(kT_0)}{\Sigma_{s0}}\right] \tag{108.6}$$

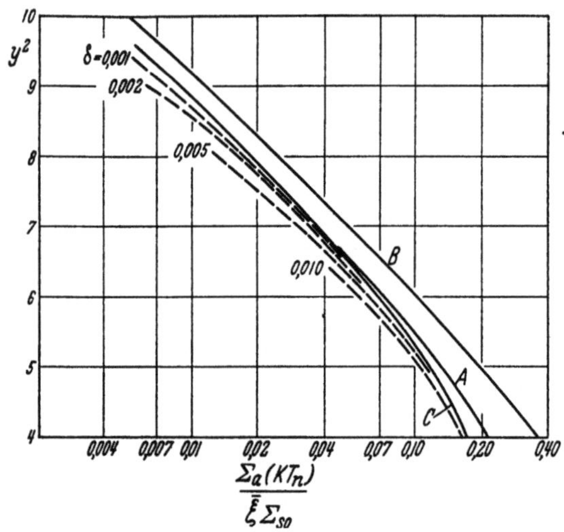

Fig. 219.

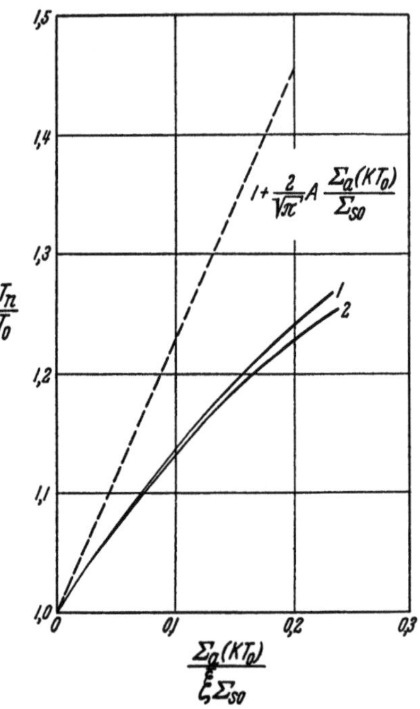

Fig. 220.

Fig. 219. Transition energy as a function of $\frac{\Sigma_a(kT_n)}{\xi\,\Sigma_{s0}}$. Curve A: calculated including correction for neutron temperature, curve B: constant slowing down power: moderator at $T_0=0$ ($t=0$). Curve C: Neutron temperature equal to moderator temperature ($t=1$). The dashed curves represent the effect of finite leakage in the case $t=1$ (Cohen).

Fig. 220. Neutron temperature as a function of $\Sigma_a(kT_0)/\xi\,\Sigma_{s0}$ (Cohen).

which can be used for

$$A \gg 1 \quad \text{and} \quad A\,\frac{\Sigma_a(kT_0)}{\Sigma_{s0}} \ll 1\,.$$

The curves of Fig. 220 can be represented by the approximate linear relationship

$$T_n = T_0\left[1 + 0.60\,A\,\frac{\Sigma_a(kT_0)}{\Sigma_{s0}}\right] \tag{108.7}$$

valid for

$$A \gg 1 \quad \text{and} \quad 0 < A\,\frac{\Sigma_a(kT_0)}{\Sigma_{s0}} < 0.5\,.$$

These results may be compared with those obtained by Bate, Coveyou and Osborn[1] using the Monte Carlo method for various values of the mass ratio A as indicated in Table 77 (p. 463). These authors determined the effective neutron

[1] R.R. Bate, R.R. Coveyou and R.K. Osborn: ORNL 1958 (1955).

temperature T_n for not too strong absorption by a least-square fitting of a Maxwell neutron distribution to the computed flux in the region

$$0 \leqq X = 16z \leqq 35$$

i.e.

$$0 \leqq E \leqq 7.2\,kT_0. \tag{108.8}$$

Their results, given in Table 77, may be represented, within 5% in all cases, by the linear expression

$$T_n = T_0[1 + 1.11 A \cdot K] = T_0\left[1 + 0.91 \frac{\Sigma_a(kT_0)}{\Sigma_{s0}}\right]. \tag{108.9}$$

The value of the constant is intermediate between those appearing in (108.6) and (108.7) which refer to heavy moderators.

The case of water has been investigated by BROWN[1] who determined T_n by a least-square fitting of a Maxwell flux distribution to the flux distributions of Figs. 210 and 211. In the case of heavy water the effective neutron temperature satisfies within 1% the relationship

$$T_n = T_0\left[1 + 1.92 \frac{\Sigma_a(kT_0)}{\Sigma_{sf}}\right], \tag{108.10}$$

where Σ_{sf} is the macroscopic scattering cross section of unbound atoms $[\sigma_{sf}(D) = 3.5$ barns$]$. In the case of light water it is not possible to fit the values of T_n by a linear formula because the coefficient of $\Sigma_a(kT_0)/\Sigma_{sf}$ decreases with increasing absorption and increasing moderator temperature as shown in Table 81.

α) *Various procedures for the experimental determination of T_n.* There are two experimental procedures which are frequently used for the determination of

Table 81. *Coefficients of* $\frac{\Sigma_a(kT_0)}{\Sigma_{sf}}$ *as a function of* $\Sigma_a(kT_0)$ *and* kT_0 *for the case of* H_2O (BROWN).

kT_0	$\Sigma_a(0.0257)=$				
	0.026	0.052	0.104	0.260	0.520
0.0257	0.876	0.985	0.876	0.854	0.810
0.0304	0.805	0.805	0.805	0.745	0.685
0.035	—	0.657	0.704	0.657	0.591
0.040	—	0.614	0.614	0.562	0.509

the effective neutron temperature reached by the neutrons in a medium. Both of them make use of very thin $1/v$ neutron detectors such as BF_3 counters of very small absorption or thin $1/v$ foils (Sect. 133). We recall that the number of counts given by such instruments is proportional to the neutron density because the factor v appearing in the flux distribution is cancelled by the factor $1/v$ due to the efficiency of the detector.

In the case of reactors a third method can be applied based on the comparison of the neutron absorption by a $1/v$ and a non $1/v$ absorber.

The first method consists in measuring the velocity distribution $n_0(v)$ of the neutrons by means of a convenient spectrometer. The most probable velocity v_n and the effective temperature T_n are derived by fitting a Maxwell distribution (108.1) to the experimental results. Before, doing this, however, the actual measurements need to be corrected for the efficiency and resolution of the spectrometer which both vary with neutron velocity, and for deviations from the $1/v$

[1] H.D. BROWN: DP-64 (1956).

behaviour of the detector due to its finite thickness. The instrumental corrections make it difficult to obtain velocity distributions sufficiently accurate to allow the derivation of a reliable value of T_n.

The principle of the second procedure consists in a transmission measurement through a very thin layer of a $1/v$ absorber (for instance boron) of the neutrons, the temperature of which has to be measured (Sect. 23). Under ideal conditions, i.e. infinitely thin $1/v$ absorber and detector and parallel beam of neutrons, the average cross section of the absorber is given by

$$\bar{\sigma}_a = 1 - \frac{C_a}{C_0} \tag{108.11}$$

where C_a and C_0 are the Cd-differences of the numbers of counts given by the detector with and without the absorber. Introducing the cross sections $\sigma_a(v)$ and $\sigma_d(v)$ of respectively absorber and detector one has

$$\bar{\sigma}_a = \frac{\int\limits_0^{v_{\text{Cd}}} n_0(v) \cdot v\,\sigma_a(v) \cdot \sigma_d(v)\,dv}{\int\limits_0^{v_{\text{Cd}}} n_0(v)\,v\,\sigma_d(v)\,dv}$$

with the integrals extended from zero to the Cd cut-off v_{Cd}. If both $\sigma_a(v)$ and $\sigma_d(v)$ follow the $1/v$ law,

$$\bar{\sigma}_a = \frac{\int\limits_0^{v_{\text{Cd}}} M_{T_n m}(v)\,\frac{k_a}{v}\,dv + \int\limits_0^{v_{\text{Cd}}} [n_0(v) - M_{T_n m}(v)]\,\frac{k_a}{v}\,dv}{\int\limits_0^{v_{\text{Cd}}} M_{T_n m}(v)\,dv + \int\limits_0^{v_{\text{Cd}}} [n_0(v) - M_{T_n m}(v)]\,dv} = \frac{\frac{2}{\sqrt{\pi}}\,\frac{k_a}{v_n} + A}{1+B} \tag{108.12}$$

where $M_{T_n m}(v)$ is the Maxwell distribution of velocities, which is assumed to be practically zero for $v \geqq v_{\text{Cd}}$, and A and B are correction terms which may be taken into account by means of a successive approximation procedure. In the first approximation one can write

$$\bar{\sigma}_a = \frac{2}{\sqrt{\pi}}\,\frac{k_a}{v_n} = 1.128\,\sigma_a(v_n). \tag{108.13}$$

This relationship allows the determination of v_n by introducing on the left-hand side the experimental value (108.11) for $\bar{\sigma}_a$. From v_n one can calculate T_n by means of (108.1).

Once T_n has been determined in the first approximation, one can calculate A and B and derive second approximation values of v_n and T_n. One has to notice that A and B are small because $n_0(v) - M_{T_n m}(v)$ is different from zero only in a narrow velocity interval immediately below v_{Cd}. Their calculation is, however, to a certain extent arbitrary because a definite assumption must be made about the actual behaviour of the neutron density $n_0(v)$. This may be suggested by the results of theoretical calculations. Unfortunately, the velocity interval of interest in calculating A and B is that in which the approximate representation of the molecular structure of the moderator has most influence on the behaviour of $n_0(v)$.

Since in practice both an absorber and a detector of finite thickness must be used in order to have statistically significant results, a further correction must be introduced for the hardening of the beam (Sect. 136, Fig. 280).

Both the spectrometer technique and the boron absorption technique as described above, make use of a parallel beam of neutrons. This can be obtained by performing the experiment on a beam of neutrons coming out through a hole penetrating inside the medium to the point where the effective neutron temperature is to be measured. The geometry must be arranged so that only the neutrons coming directly from the bottom of the hole, not those emitted by its walls, can reach the detector.

The boron absorption technique can also be used by placing inside the moderator, where the value of T_n has to be measured, a thin $1/v$ foil between two thin $1/v$ absorbers, provided adequate corrections are introduced for obliquity, flux depression and scattering (Sects. 135 and 136).

Finally, in the case of reactors, one can determine the effective neutron temperature by comparing the reductions of reactivity produced by a $1/v$ absorber and a non-$1/v$ absorber when placed successively in the pile.

In fact the change of the pile reactivity ΔK caused by a particular sample is a direct measure of its neutron absorption; it can be determined by finding what displacement of a control rod is necessary in order to maintain a constant power level with and without the sample inserted. Details of this procedure, indicated in the literature as the *danger coefficient method*, may be found in reference [*12*].

In the case of a $1/v$ absorber the number of neutrons absorbed by a thin layer is proportional to the neutron density and is independent of the neutron velocity distribution, while for a non-$1/v$ absorber the absorption depends also on the velocity distribution and therefore, if the latter is Maxwellian, on the effective temperature T_n of the neutrons.

In the following, various measurements are described in some detail. As we shall see different methods, and even the same method, often give different results, even when applied to similar assemblies. The poor agreement may be understood by considering the many corrections involved in the derivation of the effective neutron temperature from the actual measured quantities; there are in addition a few secondary effects discussed below in this section which depend on the detail of the geometry and of the polycrystalline structure of the moderator.

β) Effective temperature measurements by means of spectrometers. Early experiments with mechanical velocity selectors indicated that the velocity distribution of the *C*-group (i.e. the slow neutrons obtained by the Cd difference procedure: Sect. 133) from a paraffin block or howitzer containing a Rn α+Be source, was roughly in agreement with a Maxwellian distribution at room temperature[1]. Later, more accurate experiments with time-of-flight velocity spectrometers[2,3] seemed to prove that the neutron spectrum of the *C*-group does not correspond to thermal equilibrium with the moderator. For instante MANLEY *et al.*[2] found that the neutron spectrum from a 14 cm side cube of paraffin with a source of 2.5 Mev neutrons at the centre, is approximately Maxwellian with $T_n \sim 400°$ K, a value about 100° C higher than that of the moderator. This result, as well as some others mentioned below, cannot be compared with the theoretical considerations presented above which neglect the leakage of neutrons at the surface of the moderator. The importance of this effect is shown by the fact that with a larger amount of paraffin Manley found $T_n \sim 340°$ K.

The same technique was used by RAINWATER and HAVENS[4] who investigated the neutron spectrum emitted from a paraffin slab 2.6 cm thick contained in a

[1] J.R. DUNNING, G.B. PEGRAM, G.A. FINK, D.P. MITCHELL and E. SEGRÈ: Phys. Rev. **48**, 704 (1935). — G.A. FINK: Phys. Rev. **50**, 738 (1936).

[2] J.H. MANLEY, L.J. HAWORTH and E.A. LUEBKE: Phys. Rev. **69**, 405 (1946).

[3] R.F. BACHER, C.P. BAKER and B.D. MCDANIEL: Phys. Rev. **69**, 443 (1946).

[4] J. RAINWATER and W.W. HAVENS: Phys. Rev. **70**, 146 (1946).

box of plywood 8 mm thick, in which the neutrons from a Be target bombarded with 8 Mev deuterons were slowed down. The experimental results obtained under these conditions are well represented by a Maxwell distribution with $T_n \approx 430°$ K for neutron energies between 0.005 and 0.1 ev $\approx 4\,kT_0$. By increasing the thickness of the paraffin slab to 6.2 cm, satisfactory agreement was obtained with $T_n = 390°$ K up to about 0.2 ev $\approx 8\,kT_0$.

The velocity distribution of the neutrons from the Argonne heavy water pile was investigated by ZINN[1] with a calcite crystal spectrometer. For neutron energies between 0.004 and 0.30 ev this author found $T_n - T_0 \approx 90°$C from a point inside the reactor as well as from the graphite thermal column. From measurements with a LiF crystal spectrometer made on the neutron beam coming out from a side hole of the same pile, STURM[2] found $T_n = 400°$K. The data of this author are plotted in Fig. 221 where the curve has been calculated by means of the expression of

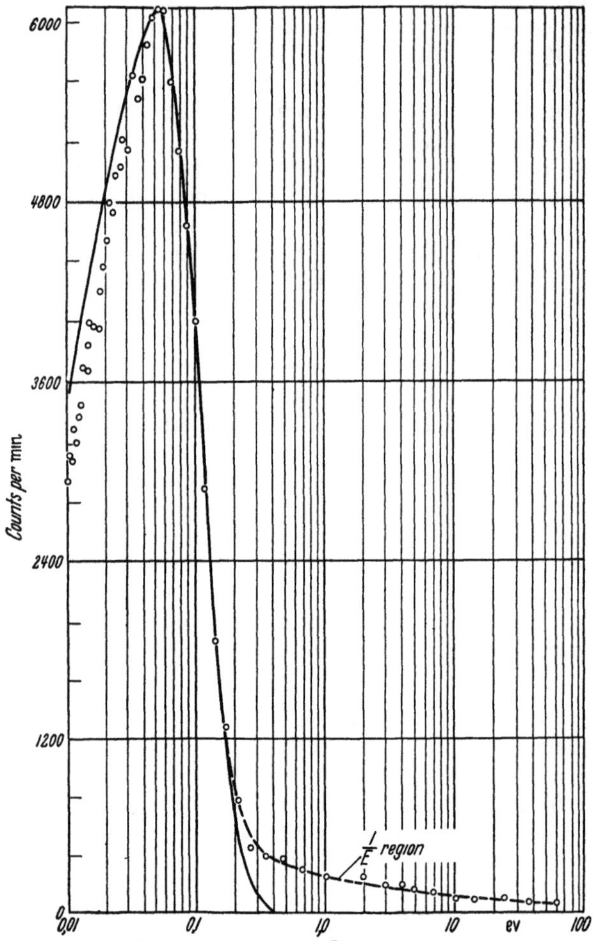

Fig. 221. Neutron spectrum in the Argonne heavy water pile measured with a crystal spectrometer. The solid line represents the theoretical Maxwellian distribution at $T = 400°$ K (STURM).

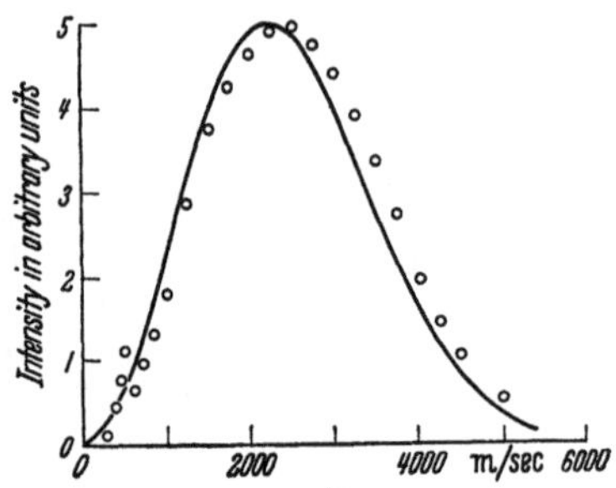

Fig. 222. The neutron spectrum of the Argonne heavy water pile measured with a chopper (RINGO).

GOLDBERGER and SEITZ[3] relating the crystal reflectivity to the glancing angle for a $1/v$ detector. BERNSTEIN *et al.* [*12*], also using a crystal, found a temperature 155° C above the moderator temperature of the Oak Ridge graphite pile.

With a mechanical velocity selector (chopper) RINGO [*12*] obtained the results given in Fig. 222 for the thermal column of the Argonne heavy water pile. The experimental points, which have been corrected for the instrumental effects

[1] W.H. ZINN: Phys. Rev. **71**, 752 (1947).
[2] W.J. STURM: Phys. Rev. **71**, 757 (1947).
[3] M.L. GOLDBERGER and F. SEITZ: Phys. Rev. **71**, 294 (1947).

mentioned above, indicate a temperature somewhat higher than $T_0 = 293°$ K for which the theoretical curve is calculated. The intensity peak shown at 500 m/sec is due to the transparency of graphite for neutrons of wavelength larger than the cut-off λ_c (Sect. 102, Table 76). This effect is discussed at the end of Sect. 108γ.

A very small difference between the neutrons and moderator temperature was found by TAYLOR[1] who, by means of a crystal spectrometer, studied the velocity distribution of the neutrons emitted from one of the thermal column holes in the Harwell graphite pile. The results obtained by this author, after correction for: (a) higher order reflections, (b) detection efficiency (BF_3 counter of finite thickness), and (c) reflectivity of the crystal[2], are plotted versus λ in Fig. 223. The best fit of the low energy part of the spectrum is obtained with $kT_n = 0.028$ ev corresponding to $T_n = 326°$ K. The average temperature of the graphite during the period of the experiment at a point near the centre of the pile was $T_0 = 325°$ K. The part of the spectrum between 0.1 and 0.5 ev has been analyzed by fitting the difference between the observed spectrum and the Maxwellian distribution with $kT_n = 0.028$ ev. One finds that the spectrum changes from a pure Maxwellian shape below 0.1 ev to a Maxwellian shape plus an $E^{-\frac{3}{2}}$ term above 0.15 ev. The behaviour may be most simply expressed by the following semi-empirical formula

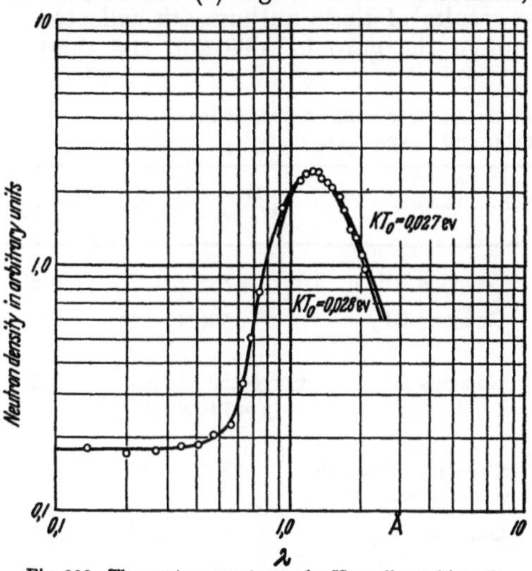

Fig. 223. The neutron spectrum of a Harwell graphite pile measured with a crystal spectrometer (TAYLOR).

$$\left.\begin{aligned} n_0(E)\,dE &= \left\{E^{\frac{1}{2}}\, e^{-\frac{E}{E_0}} + \frac{C}{E^{\frac{3}{2}}}\left[1 - \frac{1}{1+(E/E_c)^n}\right]\right\} dE\,, \\ E_0 = 0.028\text{ ev},\quad & E_c = 0.12\text{ ev},\quad C = 3.2\times 10^{-5},\quad n = 16. \end{aligned}\right\} \qquad (108.14)$$

The experimental result $T_n - T_0 = (1 \pm 12)°$K must be compared with the theoretical value derived from Fig. 220 according to which for $\Sigma_a(kT_0)/\xi\, \Sigma_{s0} = 0.52$, is $T_n/T_0 = 1.075$ or, with $T_0 = 325°$K, $T_n - T_0 = 24°$K[3,4].

The results of a systematic investigation of the effective neutron temperature in light water and in solutions of boric acid in water, made by POOLE, have been reported in Sect. 106, to which we refer for details. A comparison of the experimental values of T_n with the corresponding theoretical predictions, is given both in Table 80 (p. 472) as well as in Fig. 213.

γ) *Effective neutron temperature measurements from absorption.* The results obtained with the boron absorption method have also varied over a range of about 100° C in spite of the simplicity of the method.

[1] B. T. TAYLOR: AERE-NR 1005 (1952).
[2] G. E. BACON and R. D. LOWDE: Acta crystallogr. 1, 303 (1948).
[3] See footnote 1, p. 477.
[4] See footnote 2, p. 478.

WATTENBERG and JANKOWSKI [*12*] obtained $T_n = 330°$ K for a beam from the thermal column of the Argonne heavy water pile. HUGHES *et al.*[1], at the same place, found practically room temperature by measuring the transmission of the beam through a piece of calibrated Pyrex glass using a $1/v$ detector. After correction for the hardening of the beam they found $v_n = 2180$ m/sec, in excellent agreement with $v_0 = 2200$ m/sec.

A systematic investigation with the boron absorption method of the effective temperature of the neutrons diffusing through the thermal column of the Argonne graphite pile has been made by FERMI, MARSHALL and MARSHALL[2]. The temperature of the thermal column during the experiments was in the neighbourhood of 30° C (i.e. $T_0 = 303°$ K). The results of these authors are collected in Table 82. In case 1 the filtering action of the graphite pile and the thermal

Table 82. *Summary of results on effective neutron temperature* (FERMI, MARSHALL and MARSHALL)*.

Source of neutrons	Absorber	Cross section for kT_n neutrons (cm²)	Effective temperature (°K)
1. Beam from surface of thermal column	Gaseous BF_3	$\sigma_B = 855 \times 10^{-24}$	198
2. Beam passed through a 3.7-cm slab of paraffin	Gaseous BF_3	598×10^{-24}	408
3. Beam passed through 7.6 cm of heavy water at 33.7° C in a container 18-in. diameter	Gaseous BF_3	Corrected to 20.4° C $\sigma_B = 710 \times 10^{-24}$	288
4. Beam passed through a 22 cm column of graphite 10 cm square	Pyrex plate calibrated in velocity selector	2800×10^{-24}	18.4
5. Beam from hole in thermal column 125 cm deep, 10 cm square	Gaseous BF_3	701×10^{-24}	293
6. Beam from a "black hole" in thermal column, a hole 10 cm × 10 cm × 22 cm high connected to surface of thermal column by a 42-cm tube of cadmium of internal diameter 2.5 cm	Gaseous BF_3	755×10^{-24}	255

* The fact that these authors used σ_B(2200 m/sec) = 703 barns instead of the more recent value (27.4), does not affect their conclusions.

column produces "cold neutrons" (Sect. 102). In case 2 the slower neutrons are removed preferentially, because both absorption and scattering cross sections of hydrogen are large for lower velocities and also because at higher energy the scattering is preferentially in the forward direction. In case 3 heavy water acts somewhat in the same way as paraffin and therefore the effective temperature of the neutrons is raised from the initial value $T_n = 198$ to 288° K. The fact that the latter value is close to the actual temperature of the heavy water is probably accidental. In case 4 the filtering effect of graphite is clearly shown. Most scattered neutrons are removed from the beam and the graphite column is so long that hardly any of the warm neutrons can travel the whole distance without being scattered. In case 5 the neutrons in the beam should be a fair sample of the neutrons present at the bottom of the hole. Their temperature is very close to the temperature of the medium as one has to expect because the experimental conditions are similar to those of black body radiation from a hole in the

[1] D. G. HUGHES, J. R. WALLACE and R. H. HOLTZMAN: Phys. Rev. **73**, 1277 (1948).
[2] E. FERMI, J. MARSHALL and L. MARSHALL: Phys. Rev. **72**, 193 (1947).

wall of a furnace. Finally case 6 was expected to give a good temperature value but failed to do so probably because the hole was not deep enough.

Measurements of the effective neutron temperature inside a moderator have been made by BRANCH[1] and KÜCHLE[2]. The first of these authors used as $1/v$ detectors In-foils 0.1 mm thick and as $1/v$ absorbers calibrated Pyrex plates; he found $T_n - T_0 = 140°$ K inside the Oak Ridge graphite pile.

Table 83. *Summary of results on effective neutron temperature* (KÜCHLE).

Arrangement	δ' [mg/cm²]	s [cm]	$\frac{\Phi_{th}}{\Phi_{epi}}$	$\frac{A_a}{A_0}$	T_n [°K]
1. Graphite pile 189×189×197 cm, two 500 mC Ra-Be sources in the vertical axis of the pile, at about 20 cm distance from top and base respectively					
(a) Detector in the middle of the pile in the center of a hollow sphere of radius 10 cm	359.5	79	137	0.7529 ± 0.0026	311 ± 10
	535.5	79	137	0.6759 ± 0.0035	304 ± 10
(b) Detector in the middle of the pile without cavity	359.5	79	164	0.7441 ± 0.0020	
(c) Detector in a distance of 54 cm from the vertical axis of the pile	359.5	96	328	0.7452 ± 0.0037	285 ± 20
2. Graphite pile 99×108×134 cm, two 500 mC Ra-Be sources in the vertical axis of the pile, at about 16 cm from top and base, respectively. Detector in the center of a sphere of radius 10 cm in the middle of the pile	359.5	51	28	0.7591 ± 0.0011	336 ± 5
3. Graphite cube with 91 cm length of side. A 500 mC Ra-Be source in the center of the cube, detector 25 cm above, no cavity					
(a) Graphite cube surrounded by 30 cm paraffin	359.5	25	18	0.7585 ± 0.0015	334 ± 16
(b) Between graphite and paraffin a cadmium foil of 1 mm thickness	359.5	25	13	0.7640 ± 0.0028	360 ± 19
(c) Cubical, thermal plane source obtained from the difference of (3a) and (3b)	359.5	—	∞	0.7443 ± 0.0040	282 ± 22

KÜCHLE, using very thin In foils (65.6 mg/cm²; 9 mm diameter) as $1/v$ detectors, and Au foils (12 mm diameter) as $1/v$ absorbers, measured the effective neutron temperature inside various graphite columns containing one or two Raα+Be sources (Table 83). The calculation of the activation of the foil, with and without the absorber, was made for a neutron field of rotational symmetry as given by elementary diffusion theory (Sect. 134). The neutron current term was eliminated by taking the mean value of the activation measurements made on both sides of the irradiated foils (Sect. 134). Therefore only the density term $\frac{1}{4\pi} n_0(v)\, dv$ had to be considered in calculating the activation of the In detectors.

The ratio A_a/A_0 of the activations with and without absorber was computed by means of the procedure described in Sect. 136 assuming for $n_0(v)\, dv$ an isotropic Maxwellian distribution. Two corrections were applied to A_a/A_0: the first, of 0.2%, was due to β-ray self-absorption (Sect. 136β); the second, due

[1] G.M. BRANCH: MDDC-747 (1946).
[2] M. KÜCHLE: Nucl. Sci. and Engng. 2, 87 (1957).

to scattering of neutrons inside the absorbers, produced a change of 10% in the absorption cross section of gold. Epithermal activation was eliminated by means of Eq. (108.12). Finally the correction for neutron flux depression was reduced to less than 0.1% by placing the detector at the centre of a hollow sphere of about 10 cm radius. Under these conditions the probability that a neutron passing through the foil surface should be scattered back, is so small that the diminution of the backscattering due to foil absorption is negligible.

Table 83 shows the results obtained by KÜCHLE under various conditions with $T_0 = 290°$ K; here δ' is the thickness of the Au absorber and s the average source distance. The indicated errors represent standard deviations from the actual measurements, to which must be added $\pm 20°$ C resulting mainly from uncertainties in the cross sections employed. The two measurements (1a) differ in the thickness of the gold absorber. The measurement (1b) was made in order to test the accuracy of the correction for neutron flux depression. The difference of 1% with respect to (1a) in A_a/A_0 (which leads to a change of the neutron temperature of 20° C), indicates that the correction is not sufficiently accurate. This result was taken into account in measurements (1c) and (3). No dependence of the activity on the orientation of the foil was observed (Sect. 134).

KÜCHLE tries to generalize his experimental results by the following semi-qualitative argument. Since the epithermal flux Φ_{epi} can be considered as the source of thermal neutrons, the ratio $\Phi_{\text{th}}/\Phi_{\text{epi}}$ is proportional to the effective lifetime of thermal neutrons due to both leakage and absorption. On the other hand, the difference $T_n - T_0$ caused by incomplete thermal equilibrium is expected to be inversely proportional to the effective lifetime. Therefore the relation

$$\frac{\Phi_{\text{th}}}{\Phi_{\text{epi}}}(T_n - T_0) \approx \text{const}, \qquad (108.15)$$

is expected to be approximately valid. Before estimating the constant appearing in (108.15), KÜCHLE points out that there are two effects to be considered which may both produce an appreciable decrease of T_n—whenever the measurements are made in an assembly of limited dimensions or close to a surface.

The first effect originates from accumulation of cold neutrons ($\lambda < \lambda_c$, Sect. 102) while the second effect arises from the depression of fast thermal neutrons (Sect. 129).

In graphite the incoherent scattering mean free path $(\Sigma_{\text{incoh}}^{-1})_{\text{cold}}$ of cold neutrons ($E \leqq 0.00183$ ev) is about four times the incoherent mean free path of other thermal neutrons $(\Sigma_{\text{incoh}}^{-1})_{\text{th}}$. With Φ_{cold} denoting the flux of cold neutrons, one has (Sect. 133)

$$\Phi_{\text{th}}(\boldsymbol{r}) = \int \frac{e^{-\Sigma_s |\boldsymbol{r}-\boldsymbol{r}'|}}{4\pi(\boldsymbol{r}-\boldsymbol{r}')^2} \Sigma_s \Phi_{\text{th}}(\boldsymbol{r})\, d\boldsymbol{r}' \quad \text{for } \Sigma_a \ll \Sigma_s \qquad (108.16)$$

and

$$\Phi_{\text{cold}} = p \frac{(\Sigma_{\text{incoh}})_{\text{th}}}{(\Sigma_{\text{incoh}})_{\text{cold}}} \int \frac{e^{-(\Sigma_{\text{incoh}})_{\text{cold}} |\boldsymbol{r}-\boldsymbol{r}'|}}{4\pi(\boldsymbol{r}-\boldsymbol{r}')^2} (\Sigma_{\text{incoh}})_{\text{cold}} \Phi_{\text{th}}(\boldsymbol{r})\, d\boldsymbol{r}' \qquad (108.17)$$

with

$$p = \frac{\int_0^{v_c} v\, n_0(v)\, dv}{\int_0^{\infty} v\, n_0(v)\, dv},$$

where v_c indicates the neutron velocity corresponding to λ_c. Since $\Phi_{\text{th}}(\boldsymbol{r})$ is determined by the geometry of the assembly and by the diffusion length of thermal neutrons, comparison of (108.16) with (108.17) yields that there is no appreciable accumulation of cold neutrons in the interior of assemblies of dimensions large compared to $(\Sigma_{\text{incoh}}^{-1})_{\text{cold}}$. In the immediate region of the boundaries, however, the effect becomes noticeable because of the different coefficient appearing in the exponential of Eqs. (108.16) and (108.17). On the surface of an assembly the cold neutrons may be accumulated so that their intensity is increased by a factor two or three and therefore a determination of the effective neutron temperature by transmission measurements may lead to temperature values which are too low by about 30° C.

The depression of fast thermal neutrons is due to the fact that the leakage of neutrons is proportional to their velocity (Sect. 129). According to KÜCHLE, this effect produces a depression of respectively 10, 16 and 22° for the measurements (2), (3a) and (3b) of Table 83. After the introduction of these corrections the constant appearing on the right-hand side of (108.15) can be estimated as $(1400 \pm 500)°$ K. Taking $\Phi_{\text{th}}/\Phi_{\text{epi}} \sim 16$ for the core of a graphite reactor [*12*] one obtains $T_n - T_0 = (87 \pm 31)°$ K which is in agreement with the result of ANDERSON *et al.*[1] but in disagreement with that of BRANCH[2]; this disagreement could be explained, at least in part, as due to the finite thickness of the detector employed by BRANCH.

δ) Effective temperature measurements inside reactors by the danger coefficient method. ANDERSON *et al.*[3] measured the effective neutron temperature inside the Argonne graphite pile by comparing the reduction of the reactivity produced by a thin layer of boron and a thick cadmium absorber when placed at the centre of the reactor. The boron absorbers were prepared by depositing known amounts of a solution of borax in many aluminium dishes each containing a disk of filter paper. By evaporation the boron was deposited uniformly on the filter-paper disks in such a thin layer that in spite of obliquity the effect of self-absorption was not more than 2%.

The cadmium absorber, in form of wires of $D = 1.006$ mm diameter and $l = 128.25$ cm length, are thick enough to absorb all the neutrons with energy less than the Cd cut-off which strike their surface (black cadmium absorber). Hence their absorption depends on the average neutron velocity.

For isotropically distributed neutrons the cross section of such a wire is

$$a_{\text{Cd}} = \tfrac{1}{4} \pi D l \text{ cm}^2.$$

A comparison of the effect of the absorption of boron and cadmium when successively introduced into the pile gives information about the average velocity of the neutrons. In making this comparison it is necessary to subtract that part of the effect of boron which is due to the absorption of neutrons above the Cd cut-off. This correction was found to be 2.08% by measuring, with a BF_3 counter, the transmission through a thick Cd layer of a beam of neutrons coming out from a hole extending into the centre of the pile. With this correction the authors found that an absorber, following the $1/v$ law below the Cd cut-off and not absorbing at all above it, must have a total cross section $(a_{\text{B}})_{\text{corr}} = N_{\text{B}} \sigma_{\text{B}} = 2.095$ cm² in order to produce the same change of reactivity as a thick Cd absorber of total cross section $a_{\text{Cd}} = 1.625$ cm². From these data they conclude that in the centre of the pile the average velocity of the neutrons of energy below the Cd cut-off is 2836 m/sec which corresponds to an effective neutron temperature $T_n = 382°$ K, i.e. $T_n - T_0 = 90°$ C.

[1] H.L. ANDERSON, E. FERMI, A. WATTEMBERG, G.L. WEIL and W.H. ZINN: Phys. Rev. **72**, 16 (1947).
[2] See footnote 1, p. 486.

GAWIN[1], using a similar procedure did not observe any temperature difference ($T_n = T_0 = 300°$ K; Table 84) inside the Thermal Test Reactor (TTR). This experiment differs from that of ANDERSON *et al.* mainly because, instead of a black cadmium absorber, a very thin Cd layer (~ 30 mg/20 cm²) was used by GAWIN in order to reduce the distortion of the neutron flux. He measured by the danger coefficient method the ratio $(\Delta K)_{\text{Cd}}/(\Delta K)_{1/v}$ of the reduction in reactivity produced by placing successively a thin Cd-absorber and a $1/v$-absorber in a slot at the centre of the internal graphite thermal column of the TTR. In case of weak absorption ΔK is proportional to the neutron absorption of the sample. Therefore $(\Delta K)_{\text{Cd}}/(\Delta K)_{1/v}$ is a measure of the ratio of the corresponding neutron absorption rates; this was computed for several values of kT_n under the assumption that the neutron distribution is isotropic and Maxwellian. As $1/v$ absorbers GAVIN used, in successive sets of measurements, thin foils of Au (1.64 g/19 cm²), Mn (1.49 g per 19 cm²), Cu (15.5 g/19 cm²), Fe (13.9 g/19 cm²) and Co (2.27 g/19 cm²). The thickness of each sample was such as to absorb about 3% of the incident neutrons so that the relative change of reactivity $\Delta K/K$ was of the order of 10^{-4}.

Table 84. *Summary of results on effective neutron temperature* (GAVIN).

Element	$(\Delta K)_{\text{Cd}}/(\Delta K)_{1/v}$	E_0 (ev)
Au	1.11	0.0262
Mn	2.62	0.0276
Cu	1.01	0.0256
Fe	1.50	0.0276
Co	0.675	0.0256

Since all the $1/v$ detectors listed above, as well as Cd, show an appreciable epithermal absorption, the experimental values of $\Delta K/K$ have to be corrected before they are used for the determination of kT_n from the theoretical value of the ratio $(\Delta K)_{\text{Cd}}/(\Delta K)_{1/v}$. The epicadmium component of $\Delta K/K$ for each of the $1/v$ absorbers was measured by inserting the samples into a small cadmium sheath located at the normal position in the pile. Attention was paid to the fact that, because of the large absorption of the cadmium shield, the constant of proportionality between $(\Delta K/K)_{\text{epi-Cd}}$ and the neutron absorption was changed by about 10%.

The epithermal-epicadmium contribution [corresponding to A in Eq. (108.12)] is computed by taking $kT_n = 0.026$ ev, the cadmium cut-off $E_{\text{Cd}} = 0.5$ ev, and the transition energy between a pure Maxwellian distribution and a slowing down spectrum $E_c = 0.12$ to 0.13 ev as suggested by TAYLOR's result (108.14). This correction amounted to about 2% and was not too sensitive to the values of E_{Cd} and E_c selected. The correction of $(\Delta K/K)_{\text{Cd}}$ [corresponding to constant B in Eq. (108.12)] was calculated using $E_{\text{Cd}} = 0.5$ ev and amounted to about 7%.

The results of 5 independent measurements are shown in Table 84 and give $T_n = 300°$ K with a standard deviation of 2% to be compared with a reactor temperature measured with a thermometer $T_0 = 300°$ K.

109. Space dependent problems in the stationary case. The Boltzmann equation describing the slowing down of neutrons in the low energy region is formally similar to Eq. (61.2) which is valid at neutron energies larger than 1 ev. The only difference is that the absorption and scattering cross sections $\Sigma_a(u)$ and $\Sigma_s(u)$ and the kernel $\Sigma_s(u' \to u, \boldsymbol{\Omega}' \to \boldsymbol{\Omega})$ must now be substituted with the corresponding effective values $\overline{\Sigma}_a(u)$, $\overline{\Sigma}_s(u)$ and $\overline{\Sigma}_s(u' \to u, \boldsymbol{\Omega}' \to \boldsymbol{\Omega})$. According to (96.12) these quantities are kinds of averages of the cross sections with respect to the thermal agitation of the target nuclei. Their explicit expressions can be derived from those given in Sect. 97 for the simple case of a gaseous moderator. In this section we shall again restrict consideration to this simple case.

[1] G.B. GAVIN: KAPL-1142 (1954). — Nucl. Sci. and Engng. 2, 1 (1957).

In fact the case of polycrystalline moderators is very complicated; the coefficients derived in Sect. 107 for a heavy Debye crystal, cannot be transformed into the form necessary for space dependent problems since they have been derived by ignoring the elastic scattering cross section. And this depends in such a complicated way on neutron energy, angle of scattering and on the temperature and structure of the crystal (Sect. 102) that the treatment of space dependent problems becomes exceedingly involved.

One has to notice that while $\Sigma_s(u'\to u, \boldsymbol{\Omega}'\to\boldsymbol{\Omega})$ is different from zero only for values of u' belonging to the interval $(u-X, u)$, it is not generally possible to assign the limits of u' between which $\overline{\Sigma}_s(u'\to u, \boldsymbol{\Omega}'\to\boldsymbol{\Omega})$ is different from zero. For instance for sufficiently low energy (sufficiently large u), $\overline{\Sigma}_s(u'\to u, \boldsymbol{\Omega}'\to\boldsymbol{\Omega})$ is different from zero even for $u'>u$ because of collisions in which the neutron gains energy at the expenses of the thermal agitation of the moderator.

The similarity of the mathematical problems allows the extension to the present case of some of the methods for the approximate solution of the Boltzmann equation summarized in Sect. 79.

Since in the following sections we need approximate expressions of the neutron age which may be used in the low energy region, we can start by treating the one-dimensional transport equation in the P_1-approximation. As has been shown in Sect. 79 this procedure, which gives the correct value of τ, substitutes for (79.1) the Eqs. (79.17). These, rewritten with the notations appropriate in the present case and for a moderator containing nuclei of a single type, are

$$\left.\begin{aligned} &\frac{\partial \Phi_1}{\partial z} + \overline{\Sigma}_a(u)\,\Phi_0(z,u) = -\frac{\partial Q}{\partial u}, \\ &\frac{1}{3}\frac{\partial \Phi_0}{\partial z} + [\overline{\Sigma}_a(u) + \overline{\Sigma}_s(u)]\,\Phi_1(z,u) = \int \overline{B}_1(u',\xi)\,\Phi_1(z,u')\,du' \end{aligned}\right\} \tag{109.1}$$

with

$$-\frac{\partial Q}{\partial u} = \int \overline{B}_0(u'\to u)\,\Phi_0(z,u')\,du' - \overline{\Sigma}_s(u)\,\Phi_0(z,u). \tag{109.2}$$

The second Eq. (109.1) reduces to FICK's law (79.21) i.e. to

$$\Phi_1 = -D\frac{\partial \Phi_0}{\partial z} \tag{109.3}$$

only if the diffusion coefficient defined by the relation

$$\frac{1}{3D} = \frac{1}{l_d} = \overline{\Sigma}_a + \overline{\Sigma}_s - \frac{1}{\Phi_1(z,u)}\int du'\,\overline{B}_1(u'\to u)\,\Phi_1(z,u') = \overline{\Sigma}_a + \overline{\Sigma}_s(1-\bar{\mu}_0) \tag{109.4}$$

does not depend on z. This condition is fulfilled in the high-energy region if $\Phi_1(z,u')$ varies so slowly with u' in the integration interval $(u-X, u)$ that it becomes permitted to expand it in a Taylor's series in $(u'-u)$ and retain only the first term. By applying the same approximate procedure to the integral appearing in (109.2) the neutron age equation is derived, the validity limits of which have been discussed in Sect. 79.

In order to follow as much as possible the general scheme of the age theory also in the low energy region, HURWITZ *et al.*[1] proceed as follows. They expand the kernel $\overline{\Sigma}_s(u'\to u)$ in Legendre polynomials and obtain from Eq. (96.7)

$$\overline{B}_1(u'\to u) = \overline{B}_1(u\to u')\,\frac{\Phi_M(u)}{\Phi_M(u')}$$

[1] H. HURWITZ, M. S. NELKIN and G. J. HABETLER: Nucl. Sci. and Engng. 1, 280 (1956).

which may now be introduced into the integral of Eq. (109.4) yielding

$$\bar{\mu}_0 = \overline{\cos\vartheta} = \frac{1}{\bar{\Sigma}_s} \int du'\, \bar{B}_1(u \to u')\, \frac{\bar{\varphi}_1(z, u')}{\bar{\varphi}_1(z, u)} \tag{109.5}$$

where $\bar{\varphi}_1(z, u) = \frac{\Phi_1(z, u)}{\Phi_M(u)}$. If $\Phi(z, u)$ is Maxwellian for the P_1-component, the last factor becomes unity and the integral can be computed once the kernel $\bar{\Sigma}_s(u \to u')$ is known. In the case of a heavy gaseous moderator with constant scattering cross section (97.6), $\bar{\Sigma}_s(u \to u')$ is given by (96.12), (97.7). From this, neglecting terms in $1/A^2$, one obtains

$$\bar{\mu}_0 = \overline{\cos\vartheta} \sim \frac{2}{3A}. \tag{109.6}$$

The validity conditions for this expression are: (a) the neutron absorption is so small, and the flux distribution varies so slowly with z throughout an infinitely extended gaseous moderator, that the neutron spectrum at each point very closely approaches the Maxwell distribution of velocities; (b) the mass of the nuclei of the moderator is so large that terms in $1/A^2$ can be neglected.

In conclusion, provided the absorption is so small and A so large that

$$\bar{\Sigma}_a \ll \Sigma_{s0}, \qquad \Sigma_{s0}\bar{\mu}_0 \ll \Sigma_{s0} \tag{109.7}$$

one can assume that Fick's law (109.3) will hold with a diffusion coefficient given by

$$D = \frac{1}{3[\bar{\Sigma}_a + \bar{\Sigma}_{s0}(1 - \bar{\mu}_0)]} \approx \frac{1}{3\bar{\Sigma}_{s0}}. \tag{109.8}$$

Then Eq. (109.3) and the first Eq. (109.1) may be combined to give

$$-D\nabla^2 \Phi_0(\boldsymbol{r}, u) + \bar{\Sigma}_a \Phi_0(\boldsymbol{r}, u) = -\frac{\partial Q}{\partial u}, \tag{109.9}$$

where the two terms on the left hand side are of the same order of magnitude.

This equation, though it looks very much like an age-diffusion equation of the usual type, yet is quite another type of equation. This may be seen in the following way: In order to obtain a consistent first-order approximation in $1/A$, we had to introduce on its right-hand side the expression for dQ/du derived from (105.11). With the variables

$$u = \ln\frac{kT_0}{E} = \ln\left(\frac{v_0}{v}\right)^2, \qquad \Phi_0(\boldsymbol{r}, u) = E\,\Phi_0(r, E), \tag{109.10}$$

Eq. (105.14) becomes

$$Q(\boldsymbol{r}, u) = \frac{2\Sigma_{s0}}{A}\left[(1 - 2\,\mathrm{e}^{-u})\,\Phi_0(\boldsymbol{r}, u) - \mathrm{e}^{u}\,\Phi'(\boldsymbol{r}, u)\right] \tag{109.11}$$

from which it follows that

$$\frac{dQ}{du} = \frac{2\Sigma_{s0}}{A}\left[-2\,\mathrm{e}^{u}\Phi_0 + (1 - 3\,\mathrm{e}^{u})\,\Phi_0' - \mathrm{e}^{u}\Phi_0''\right]. \tag{109.12}$$

Introducing this expression on the right-hand side of (109.9) a differential equation is obtained which has a structure essentially different from the age equation because of the presence of a term containing the second derivative of Φ_0 with respect to the lethargy u.

Various approximate procedures have been suggested in order to maintain the essential form of the age-diffusion equation; these procedures are all equivalent to assuming that $\Phi_0(\boldsymbol{r}, u)$ is a slowly varying function of u. E. R. Cohen[1,2]

[1] E. R. Cohen: NAA-SR-1127 (1955) and p. 405 of Ref. [*15c*].

[2] See footnote 1, p. 490.

suggests that the term in Φ_0'' in (109.11), should be neglected while HURWITZ *et al.*[1] substitute $\frac{1}{2}\Phi_0'$ for Φ_0'' in (109.12). In both cases the result is that (109.12) becomes

$$\frac{dQ}{du} = \frac{2\Sigma_{s\,0}}{A}\left[2\frac{kT_0}{E}\Phi_0(\boldsymbol{r},u) - \left(1 - a\frac{kT_0}{E}\right)\frac{\partial\Phi_0}{\partial u}\right] \tag{109.13}$$

with $a=2$ in the case of COHEN, and $a=3.5$ in the case of HURWITZ *et al.*

Although the second method appears to give slightly more accurate results, we shall here follow COHEN's procedure which has the advantage of a simpler physical interpretation. This can be recognized transforming Eq. (109.11) into the variable v by means of (109.10) and

$$\Phi_0(E) = v\,n_0(v)\frac{dv}{dE} = \frac{1}{m}n_0(v).$$

One obtains for $q(v)$ the expression (97.19), which was derived in Sect. 97 by introducing into the expression for $q(v)$, valid for neutron energies larger than 1 ev, the average logarithmic energy loss corrected for the thermal agitation of the nuclei of a gaseous heavy moderator, Eq. (97.18).

Since (97.19) is approximately valid only when the conditions (97.20) are satisfied, the differential equation obtained from (109.9) with COHEN's procedure will provide an approximate description of the slowing down of neutrons only by a weakly absorbing heavy moderator in the epithermal region. As the lower limit of the velocity interval in which this procedure can be used, one may take $v = y v_n$, where v_n and y have been defined in Sect. 108 [see Eq. (108.2) and Fig. 219].

At lower energies the solution of (109.9) is usually obtained by a different procedure which consists of regarding all the neutrons of velocity $v \leqq y v_n$ as belonging to a single group called the "thermal group". This is possible if the assumption is made that throughout the moderator the flux distribution can be taken as the product of a function of $\boldsymbol{r}$ and a function of u

$$\Phi_0(\boldsymbol{r},u) = \Phi_{\text{th}}(\boldsymbol{r})\cdot\varphi_{\text{th}}(u) \quad \text{for} \quad u \geqq u_{\text{th}} \tag{109.14}$$

where

$$u_{\text{th}} = \ln\frac{kT_0}{\frac{1}{2}m\,y^2 v_n^2} \quad \text{and} \quad \int_{u_{\text{th}}}^{\infty}\varphi_{\text{th}}(u)\,du = 1. \tag{109.15}$$

The lower the neutron absorption and the smaller the variation of $\Phi_0(\boldsymbol{r},u)$ with $\boldsymbol{r}$ are, the more accurate will be the description of the actual situation given by the separation of variables (109.14). In the limiting case of uniform slowing down of neutrons throughout a non-absorbing medium filling all the space, $\varphi_{\text{th}}(u)$ is the same everywhere and corresponds to the Maxwell distribution of velocities at the temperature of the moderator.

Introducing (109.14) into (109.9) and integrating with respect to u from u_{th} to infinity, one obtains

$$-\overline{D}\,\nabla^2\Phi_{\text{th}}(\boldsymbol{r}) + \overline{\overline{\Sigma}}_a\,\Phi_{\text{th}}(\boldsymbol{r}) = Q(\boldsymbol{r},u_{\text{th}}) \tag{109.16}$$

with

$$\left.\begin{aligned} \overline{D} &= \int_{u_{\text{th}}}^{\infty} du\,D(u)\,\varphi_{\text{th}}(u), \\ \overline{\overline{\Sigma}}_a &= \int_{u_{\text{th}}}^{\infty} du\,\overline{\Sigma}_a(u)\,\varphi_{\text{th}}(u). \end{aligned}\right\} \tag{109.17}$$

---

[1] See footnote 1, p. 490.

We refer to Sects. 115 to 117 for a detailed discussion of the conditions under which the behaviour of thermal neutrons is correctly described by means of the diffusion equation (109.16).

110. The neutron age correction. The approximate formulation of the space dependent problems given in the preceding section for the stationary case, has been applied in computing the migration area of thermal neutrons

$$M^2 = \tfrac{1}{6}\,\overline{r_{\rm th}^2}\,, \tag{110.1}$$

where $\overline{r_{\rm th}^2}$ is the average of the square of the distance travelled by thermal neutrons from a monoenergetic point source embedded in an infinite homogeneous moderator and the point at which they are observed.

It is customary and physically plausible to split the migration area (110.1) into two terms

$$M^2 = \tau(E_m) + L^2. \tag{110.2}$$

Here L is the *diffusion length* of thermal neutrons i.e. the average distance travelled by the neutrons once they have entered the thermal group, and $\tau(E_m)$ is the neutron age from the initial energy E_1, to a certain final energy E_m called *migration energy*, at which the slowing down process can be considered to be finished and the neutrons become thermalized. In most elementary calculations E_m is taken arbitrarily equal to kT_0. As we shall see in the following it appears more reasonable to take for E_m a value of about $4kT_0$.

One has to notice, however, that the only quantity which has a well-defined physical meaning is M^2, while the separation (110.2) of the migration area into a thermal diffusion area plus a slowing down age, is an arbitrary procedure; it is possible to give some appropriate analytical recipe for making such a separation, but the result will be different with different recipes.

Instead of $\overline{r_{\rm th}^2}$ it is more convenient to compute $\overline{z_{\rm th}^2}$ i.e. the average of the square of the distance at which thermal neutrons are observed from a monoenergetic plane δ-source. According to (71.8) and (110.1) one has

$$M^2 = \tfrac{1}{2}\,\overline{z_{\rm th}^2}\,. \tag{110.3}$$

α) Theoretical considerations. The equations describing the slowing down and diffusion of neutrons from the initial velocity v_1 to thermal velocity can be written as follows[1]

$$v \geqq y v_n \begin{cases} D v \dfrac{\partial^2 N_0(z,v)}{\partial z^2} - \gamma N_0(z,v) = \dfrac{\partial Q(z,v)}{\partial v} + \delta(z)\,\delta(v - v_1)\,, & (110.4)\\[2ex] Q(z,v) = \tfrac{1}{2}\xi\,\Sigma_{s0}(v^2 - 2v_0^2)\,N_0(z,v)\,, & (110.5)\end{cases}$$

$$v \leqq y v_n \begin{cases} \varPhi_1 D \dfrac{\partial^2 \varPhi_{\rm th}(z)}{\partial z^2} - \gamma\,\varPhi_0\,\varPhi_{\rm th}(z) + Q(y v_n, z) = 0\,, & (110.6)\\[2ex] \left.\begin{aligned} \varPhi_1 &= \frac{4}{\sqrt{\pi}}\int\limits_0^{y v_n} \frac{v^3}{v_n^3}\,e^{-\frac{v^2}{v_n^2}}\,dv = \frac{2 v_n}{\pi^{\frac{1}{2}}}\,e^{-y^2}\left[e^{y^2} - (1 + y^2)\right];\\ \varPhi_0 &= \frac{4}{\sqrt{\pi}}\int\limits_0^{y v_n} \frac{v^2}{v_n^3}\,e^{-\frac{v^2}{v_n^2}}\,dv = \operatorname{Erf}(y) - \frac{2y}{\pi^{\frac{1}{2}}}\,e^{-y^2}\,,\end{aligned}\right\} & (110.7)\\[2ex] N_0(z,v) = \dfrac{4}{\sqrt{\pi}}\,\dfrac{v^2}{v_n^3}\,e^{-\frac{v^2}{v_n^2}}\,\varPhi_{\rm th}(z)\,. & (110.8)\end{cases}$$

[1] E. R. Cohen: NAA-SR-1127 (1955).

Eq. (110.4) is simply Eq. (109.9) expressed in terms of the velocity v and the neutron density $N_0(z, v) = \frac{1}{v}\,\Phi_0(z, v)$ and with the source term added. The second equation is (97.19); the third is (109.9) expressed in terms of the neutron density; finally (110.8) is obtained by introducing (108.1) into (109.14).

Taking the Fourier transform of (110.4) and (110.6) we obtain

$$-\left[D k^2 + \gamma\right] Q(k, v) + \tfrac{1}{2}\,\xi\,\Sigma_{s0}(v^2 - 2v_0^2)\left\{\frac{\partial Q}{\partial v} + \frac{1}{\sqrt{2\pi}}\,\delta(v - v_1)\right\} = 0, \tag{110.9}$$

$$-\left[\Phi_1 D k^2 + \Phi_0 \gamma\right] \phi_{\text{th}}(k) + Q(k, y v_n) = 0 \tag{110.10}$$

with

$$\left.\begin{aligned} Q(k, v) &= \frac{1}{\sqrt{2\pi}} \int_{-\infty}^{+\infty} e^{ikz}\, Q(z, v)\, dz, \\ \phi_{\text{th}}(k) &= \frac{1}{\sqrt{2\pi}} \int_{-\infty}^{+\infty} e^{ikz}\, \Phi_{\text{th}}(z)\, dz. \end{aligned}\right\} \tag{110.11}$$

Eq. (110.9) is similar to (97.22) which is valid in the corresponding space independent problem (Sect. 97). Its solution can be written

$$Q(k, v) = Q(k, y v_n)\left(\frac{v - \sqrt{2}v_0}{y v_n - \sqrt{2}v_0}\right)^{\delta+\omega}\left(\frac{v + \sqrt{2}v_0}{y v_n + \sqrt{2}v_0}\right)^{\delta-\omega} \tag{110.12}$$

with

$$\delta = \frac{D k^2}{\xi\,\Sigma_{s0}}, \qquad \omega = \frac{\Sigma_a(kT_0)}{\xi\,\sqrt{2}\,\Sigma_{s0}}. \tag{110.13}$$

The asymptotic behaviour of $Q(k, v)$ for $v \gg \sqrt{2}\,v_0$ is

$$Q(k, v) \to Q(k, y v_n)\,(y^2 - 2t)^{-\delta}\left(\frac{y + \sqrt{2t}}{y - \sqrt{2t}}\right)^{\omega}\left(\frac{v}{v_n}\right)^{2\delta}$$

which may be used to express $Q(k, y v_n)$ in terms of the value of $Q(k, v)$ for $v = v_1$ provided $v_1 \gg \sqrt{2}\,v_0$

$$Q(k, y v_n) = Q(k, v_1)\,(y^2 - 2t)^{\delta}\left(\frac{y - \sqrt{2t}}{y + \sqrt{2t}}\right)^{\omega}\left(\frac{v_n}{v_1}\right)^{2\delta}. \tag{110.14}$$

We now impose the condition of continuity at $v = y v_n$ on the expressions of $Q(k, v)$ derived from (110.5) and (110.8) and obtain the following relation similar to (108.2)

$$Q(k, y v_n) = \frac{2}{\sqrt{\pi}}\,\xi\,\Sigma_{s0}\,v_n\,y^2(y^2 - 2t)\,e^{-y^2}\,\phi_{\text{th}}(k). \tag{110.15}$$

From this relation and Eq. (110.14) it follows that

$$\phi_{\text{th}}(k) = \frac{\sqrt{\pi}}{2}\,\frac{Q(k, v_1)}{\xi\,\Sigma_{s0}\,v_n}\,e^{y^2} y^{-2}\,(y^2 - 2t)^{\delta-1}\left(\frac{y - \sqrt{2t}}{y + \sqrt{2t}}\right)^{\omega}\left(\frac{v_n}{v_1}\right)^{2\delta} \tag{110.16}$$

which expresses the Fourier transform of the spatial distribution of thermal neutrons as a function of k, y and t. One has to notice however that y is not an independent variable and can be expressed as a function of k and t. This can be done by using (110.6) or its Fourier transform (110.10) which expresses the conservation of neutrons

$$Q(k, y v_n) = \left[\Phi_1 D k^2 + \Phi_0 \gamma\right] \phi(k).$$

Equating the right-hand side of this equation to the right-hand side of (110.15) we obtain

$$\frac{\Sigma_a(kT_0)}{\xi\,\Sigma_{s0}} = \frac{2y^2(y-2t)\,e^{-y^2}}{\sqrt{\pi t}\left[\mathrm{Erf}\,(y) - \frac{2y}{\sqrt{\pi}}\,e^{-y^2}\right]}\left[1-\delta\,\frac{e^{y^2}-1-y^2}{y^2(y^2-2t)}\right] \tag{110.17}$$

which reduces to (108.5) for $\delta=0$, i.e. in the case of space independent problems. The broken curves of Fig. 219, giving y^2 as a function of $\frac{\Sigma_a(kT_0)}{\xi\,\Sigma_{s0}}$ have been computed by means of this expression for various values of δ and for $t=1$ ($T_n=T_0$). As long as δ is small with respect to ω, the difference between these dotted curves and the heavy line C valid for $t=1$ in the case of space-independent problems, is small enough to justify the use of curve A even in the case of space-dependent problems. In this approximation the relation between T_n and T_0, which is given in Fig. 220 and by Eqs. (108.6), (108.7) obtained for $\delta=0$, may also be used in the present case of space-dependent problems.

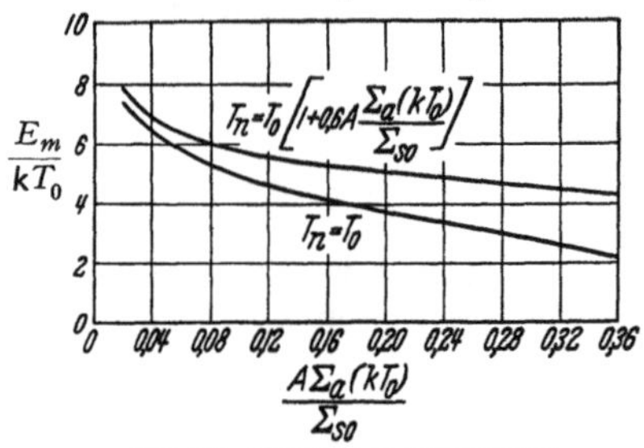

Fig. 224. Migration energy (Cohen).

In conclusion the Fourier transform of the spatial distribution of thermal neutrons is given by (110.16) as a function of k with y expressed in terms of k and t by means of (110.17) and t computed by means of (108.7).

We can now compute the migration area M^2 without inverting the Fourier transform since, according to (80.7)

$$\overline{z^2} = \left[\frac{1}{\Phi_{\mathrm{th}}(k)}\,\frac{d^2\Phi_{\mathrm{th}}(k)}{dk^2}\right]_{k=0}.$$

From this relationship Cohen obtains

$$M^2 = \frac{D}{\xi\,\Sigma_{s0}}\ln\frac{v_1^2}{y^2v_n^2-2v_0^2} + L^2\,\frac{1-(1+y^2)\,e^{-y^2}}{\mathrm{Erf}\,(y)-\frac{2y}{\sqrt{\pi}}\,e^{-y^2}} \tag{110.18}$$

with

$$L^2 = \frac{2D}{\sqrt{\pi t}\,\Sigma_a(kT_0)}.$$

If we define the *migration energy* E_m as the lower limit of an equivalent age integral for migration, viz.

$$M^2 = \frac{D}{\xi\,\Sigma_{s0}}\ln\frac{E_1}{E_m} + L^2 \approx \frac{l_{\mathrm{tr}}\,l_s}{3\xi}\ln\frac{E_1}{E_m} + L^2 \tag{110.19}$$

it follows[1] that

$$E_m = kT_n(y^2-2t)\exp\left[\frac{1-\frac{2}{\sqrt{\pi}}\,y^2+y^2-e^{-y^2}[1-\mathrm{Erf}\,(y)]}{y^2(y^2-2t)}\right]. \tag{110.20}$$

The ratio E_m/kT is plotted in Fig. 224 as a function of $\frac{A\,\Sigma_a(kT_0)}{\Sigma_{s0}}$ for both $T_n=T_0$ and T_n given by (108.7). From these curves we see that in order to compute correctly the migration area of thermal neutrons for $0.01<\frac{A\,\Sigma_a(kT_0)}{\Sigma_{s0}}<0.5$, one has to assume that they enter the thermal group at about 4 or 5 times kT_0.

[1] E. R. Cohen: p. 405 of Ref. [15c].

Eq. (110.19) can be used to calculate the increase in age when the neutrons are slowed down, in a heavy gaseous moderator, from In resonance (1.46 ev) to thermal energy (for $T_0 = 273° + 20°$ K, $E_m \approx 0.1$ ev)

$$\Delta\tau = \tau_{\text{th}} - \tau_{\text{In}} = \frac{D}{\xi \Sigma_{s0}} \ln \frac{1.46}{0.1} = \frac{l_s^2}{\xi} \frac{2.68}{3 \cdot 0.984} \sim \frac{l_s^2}{\xi} \tag{110.21}$$

which applied, for example, to the case of $A = 12$, gives $\Delta\tau = \frac{(2.65)^2}{0.158} = 45$ cm², which can be compared with the experimental result (110.35).

In Sect. 111 we shall discuss a different procedure to separate the migration area (110.1) into a slowing down age and a diffusion area.

Another quantity which is of considerable practical interest—especially in connection with thermal reactor design—is the *age to absorption* defined as $\frac{1}{6}$ of the average of the square of the distance travelled by a neutron between its emission from a monoenergetic point source embedded in an infinite homogeneous medium and its absorption.

Assuming that the absorption cross section $\Sigma_a(v_r)$ follows the $1/v$ law (96.4) in the whole velocity interval from $v = v_1$ down to $v = 0$, the absorption probability per unit time, $\gamma = v\Sigma_a(v)$, is independent of the neutron velocity, and therefore the *absorption density* $A(z)$ is proportional to the total neutron density rather than to the density of the thermal group only:

$$A(z) = \int_0^{v_1} \gamma N_0(z, v)\, dv = \gamma \int_0^{v_1} N_0(z, v)\, dv. \tag{110.22}$$

According to (80.7) the age to absorption is given by

$$\tau_{\text{abs}} = \frac{1}{2}\,\overline{z^2_{\text{abs}}} = \frac{\frac{1}{2}\int_0^\infty z^2 A(z)\, dz}{\int_0^\infty A(z)\, dz} = \frac{1}{2}\left[\frac{1}{A(k)} \frac{d^2 A(k)}{dk^2}\right]_{k=0} \tag{110.23}$$

where $A(k)$ is the Fourier transform of (110.22). Once the expression for τ_{abs} has been obtained, it is possible to define an *effective energy* E_{eff} as the lower limit of the age to absorption, i.e. such that

$$\tau_{\text{abs}} = \frac{l_a l_s}{3\xi} \ln \frac{E_1}{E_{\text{eff}}}. \tag{110.24}$$

A first calculation of E_{eff} was made by introducing into (110.22) the approximate expressions of the neutron density to be derived from (110.8), (110.11) for $0 \leqq v \leqq y v_n$ and from (110.5), (110.12) for $v > y v_n$ [1,2]. The result is a slowly varying function of $\frac{A\,\Sigma_a(kT_0)}{\Sigma_{s0}}$ which remains very close to $15\,kT_0$ for $\frac{A\,\Sigma_a(kT_0)}{\Sigma_{s0}}$ varying between 0.01 and 0.5. This value, however, is certainly much too high because of the fact that in the transition energy region the approximate spectrum adopted in this calculation remains considerably below the spectrum plotted in Fig. 205.

A better value of E_{eff} was later derived by COHEN with the following procedure[3].

[1] E.R. COHEN: NAA-SR-1127 (1955).
[2] See footnote 1, p. 495.
[3] E. R. COHEN: Nucl. Sci. and Engng. 2, 227 (1957).

He starts from the remark that, for constant scattering cross section, the total number of scattering collisions which occur in the medium below the velocity x_0, is given by

$$R_s = \int_0^{x_0} \Sigma_{s0}\, x\, n_0(x)\, dx = \Sigma_{s0} \int_0^{x_0} x\, n_0(x)\, dx, \tag{110.25}$$

where x is defined by (105.1) and $n_0(x)$ is the solution of WILKINS's equation (105.10). If the absorption cross section follows the $1/v$ law the number of neutrons absorbed in the moderator at a velocity less than x_0 is

$$R_a = \int_0^{x_0} \Sigma_a(x)\, x\, n_0(x)\, dx = \Sigma_a(kT_0) \int_0^{x_0} n_0(x)\, dx. \tag{110.26}$$

The ratio of these two rates is the average number of scattering collisions which the neutron makes before it is absorbed. Therefore the mean-square distance which the neutron travels from the velocity x_0 to absorption, is given by

$$\overline{r^2_{\text{abs}}} = 2 l_s^2 \frac{R_s}{R_a} = \frac{2}{\Sigma_{s0}\, \Sigma_a(kT_0)} \frac{\int_0^{x_0} x\, n_0(x)\, dx}{\int_0^{x_0} n_0(x)\, dx}, \tag{110.27}$$

from which according to (71.8) τ_{abs} is derived by dividing by 6.

The integrations are made by COHEN using the expression (105.11) for $n_0(x)$. The result is given again by (110.24) with

$$E_{\text{eff}}(\text{ev}) = 5.167\, \mathrm{k}\, T_0\, \mathrm{e}^{-1{,}127 \cdot \Delta}. \tag{110.28}$$

This expression gives, for example for $\Delta = 0.4$, $E_{\text{eff}} \approx 0.08$ ev, i.e. a value much smaller than the value obtained by means of the approximate procedure mentioned above.

β) Experimental results. We now turn to the experimental determination of the age increase $\Delta\tau$ undergone by the neutrons when they are slowed down from the In-resonance at 1.46 ev to thermal energies. According to the definition of migration energy given above, one can define

$$\Delta\tau = \Delta\tau\,(1.46\ \text{ev} - E_m) = \tau(1.46\ \text{ev}) - \tau(E_m). \tag{110.29}$$

Combining this relation with (110.2) one finds

$$\Delta\tau = M^2 - \tau(1.46\ \text{ev}) - L^2. \tag{110.30}$$

Thus, $\Delta\tau$ can be derived from the measured values of M^2, $\tau(1.46\ \text{ev})$ and L^2; it should be noticed, however, that since $\Delta\tau$ is small while M^2, $\tau(1.46\ \text{ev})$ and L^2 are large, the value of $\Delta\tau$ derived from (110.30) is usually affected by large errors.

This remark is true especially in the case of *light water*. The value of $M^2 - \tau_{\text{In}}$ may be readily deduced from the data collected in Tables 46, 48 and 50. The more accurate results are those contained in the latter table which gives, for pure water,

$$M^2 - \tau_{\text{In}} \approx 8.4\ \text{cm}^2.$$

Taking $L^2 = 7.2\ \text{cm}^2$ (Sects. 128 and 131) one obtains

$$\Delta\tau \approx 1.2\ \text{cm}^2 \quad (\text{for } H_2O \text{ at } 22^\circ\,\text{C}).$$

The problem of determining $\Delta\tau$ in water has been treated with considerable care by BARKOV *et al.*[1] who used neutrons of the $Sb^{124}\gamma + Be$ source in order to deal with relatively small values of M^2 and $\tau(1.46 \text{ ev})$. The experimental arrangement used by these authors has been described in Sect. 83; the corresponding results are summarized in Table 54 (p. 322). Here we should add a few details about the determination of M^2 made by means of the Cd-difference procedure applied to In detectors. The values of M^2 given in column 7 of Table 54 have been obtained by correcting the experimental values for the contribution to the activation of the In foils arising from neutrons of energy between the cadmium cut-off $E_{Cd} = 0.5$ ev and thermal energies ($E = 0.025$ ev). The calculation—very rough indeed—is based on the assumption that the expression for the neutron flux

$$d\varphi_0(E) = \frac{l_s}{\xi}\,\frac{dE}{E}$$

valid above 1 ev, can be used through the chemical bond region down to kT_0. A lower limit of the correction is obtained by assuming that the slowing down from 0.5 ev to thermal energies is due to free protons ($\xi = 1$, $\sigma_s = 20$ barns), an upper limit is obtained by assuming that only free protons are responsible for the slowing down from 0.5 to 0.1 ev and only free monatomic "water molecules" ($A = 18$, $\sigma_s = 80$ barns) from 0.1 to 0.025 ev. Thus the following two expressions for the activation of the In foils due to intermediate neutrons are obtained by BARKOV *et al.*

$$\text{lower limit:}\quad A_{\text{int}} = 0.77\,A_{\text{Cd}},$$
$$\text{upper limit:}\quad A_{\text{int}} = 1.43\,A_{\text{Cd}}.$$

The "pure thermal neutrons activation" is then given by

$$A_{\text{th}}(\boldsymbol{r}) = A(\boldsymbol{r}) - A_{\text{Cd}}(\boldsymbol{r}) - A_{\text{int}}(\boldsymbol{r}). \tag{110.31}$$

The migration area M^2 must now be computed from the two distributions

$$\text{lower limit:}\quad A_{\text{th}}(\boldsymbol{r}) = A(\boldsymbol{r}) - 1.77\,A_{\text{Cd}}(\boldsymbol{r}),$$
$$\text{upper limit:}\quad A_{\text{th}}(\boldsymbol{r}) = A(\boldsymbol{r}) - 2.43\,A_{\text{Cd}}(\boldsymbol{r}).$$

These equations, applied to the results obtained with the large source, (Sect. 83, p. 322) give

$$\text{lower limit:}\quad M^2_{(l)} = (13.92 \pm 0.05)\ \text{cm}^2,$$
$$\text{upper limit:}\quad M^2_{(u)} = (14.35 \pm 0.05)\ \text{cm}^2.$$

Since the true value of M^2 must lie between these two, BARKOV *et al.* take

$$M^2 = \frac{M^2_{(l)} + M^2_{(u)}}{2} = 14.13\ \text{cm}^2$$

with an uncertainty of

$$\frac{M^2_{(u)} - M^2_{(l)}}{2} = 0.22\ \text{cm}^2.$$

The same procedure was applied for deriving the value of M^2 corresponding to the measurements made with the small source.

Combining the values of $\tau(1.46 \text{ ev})$ and M^2 given in Table 54 with the value (128.10) of the diffusion length obtained by the same authors [Sect. 128;

[1] L.M. BARKOV, V.K. MAKARIN and K.N. MUKLIN: J. Nucl. Energy 4, 94 (1957).

$L^2 = (7.18 \pm 0.11)$ cm²] one obtains

$$\left.\begin{array}{ll}\text{with the large source:} & \Delta\tau\,(1.46\text{ ev} - 0.025\text{ ev}) = (1.1 \pm 0.5)\text{ cm}^2 \\ \text{with the small source:} & \Delta\tau\,(1.46\text{ ev} - 0.025\text{ ev}) = (1.0 \pm 0.5)\text{ cm}^2 \\ & \qquad\qquad (\text{in } H_2O \text{ at } 22^\circ\text{ C}).\end{array}\right\} \quad (110.32)$$

If the correction due to intermediate neutrons is computed by considering only the energy interval $E_{\text{Cd}} = 0.5$ ev, $E_m = 0.1$ ev, the result is

$$\Delta\tau = \Delta\tau\,(1.46\text{ ev} - 0.1\text{ ev}) = (0.6 \pm 0.3)\text{ cm}^2 \quad (\text{in } H_2O \text{ at } 22^\circ\text{ C}), \quad (110.33)$$

which probably provides the best estimate of this quantity available today. This value is lower than that computed by age theory based upon the assumption that the slowing down is produced by free hydrogen and oxygen nuclei

$$\Delta\tau = \Delta\tau\,(1.46\text{ ev} - 0.1\text{ ev}) = 1.1\text{ cm}^2. \quad (110.34)$$

By similar considerations DUGGAL and MARTELLY[1] obtain

$$\Delta\tau\,(1.46\text{ ev} - 0.025\text{ ev}) = (56 \pm 6.5)\text{ cm}^2 \quad \text{in graphite } (\varrho = 1.60\text{ g/cm}^3). \quad (110.35)$$

When the slowing down from 1.46 ev to thermal energies is described by means of the procedure illustrated in Fig. 225, the corresponding *age correction* should be defined as follows

$$\Delta\tau_m = \tau_m - \tau\,(1.46\text{ ev}) \quad (110.36)$$

where τ_m is given by (111.2). According to DUGGAL and MARTELLY one has

$$\Delta\tau_m = (38 \pm 9)\text{ cm}^2 \quad \text{in graphite } (\varrho = 1.60\text{ g/cm}^3). \quad (110.37)$$

111. The slowing down from 1.46 ev to thermal energies through the chemical bond region, as a function of time. We shall close this part with a discussion of the time dependence of the slowing down through the chemical bond region, paying special attention to the approach to thermal equilibrium between the neutron gas and the moderator.

We start from the considerations given in Sect. 81, which refer to the case of a burst of fast monoenergetic neutrons emitted at the time $t = 0$, by a pulsed source in a medium in which the absorption cross section follows the $1/v$ law. As we have shown there, the diffusion of neutrons above 1 ev is fairly well described in many cases by the elementary time-dependent diffusion equation (81.1). In Sect. 85 (Figs. 158 and 159), it has been shown that the solution of this equation fits the experimental data very well down to energies of the order of 1 ev.

On the other hand, we shall see in Sect. 115 that the diffusion of neutrons in thermal equilibrium with a weakly absorbing medium can also be described, in good approximation, by a diffusion equation of the same type, the only difference between the two cases being that for neutron energies above 1 ev, the diffusion coefficient $D_0(t)$ is a function of time, while, in the thermal energy region, it has a constant value; this is simply a consequence of the fact that in the thermal energy region the spectrum is Maxwellian—or almost Maxwellian—and, therefore, does not change with time.

Thus, the question arises as to whether the same diffusion equation can, or cannot, be used to describe the slowing down even in the intermediate energy region of the chemical bond. It is rather difficult to give an answer to this question since in this energy interval, the scattering cross section is definitely energy

[1] V.P. DUGGAL and J. MARTELLY: p. 28 of Ref. [*15c*].

dependent because of the various molecular effects which have been discussed in Part D of the present article. Therefore, the only possibility for a correct use of the diffusion theory approximation, rests upon the independence of the neutron spectrum from the space coordinates (cf. Sect. 115 below). But this could be established only by a thorough investigation of the behaviour of the solution of the Boltzmann equation in the chemical bond region; and unfortunately, such an investigation has not yet been done.

In spite of these difficulties, it has become customary to assume that, the *diffusion equation (81.1) with a diffusion coefficient $D_0(t)$ independent of space coordinates is valid even in the chemical bond region.*

If the absorption cross section follows the $1/v$ law, as usually occurs, Eq. (81.1) can be transformed into (81.5) and this into Eq. (81.9) by defining the *generalized neutron age*[1]. This is related to $D_0(t)$ by (81.8) and differs from it, only because now it is defined for any neutron energy, — starting from the initial energy down to thermal energies — as

$$\tau(t) = \int\limits_0^t D_0(t)\,dt = \int\limits_0^t \frac{l_{tr}(t)}{3} v(t)\,dt. \tag{111.1}$$

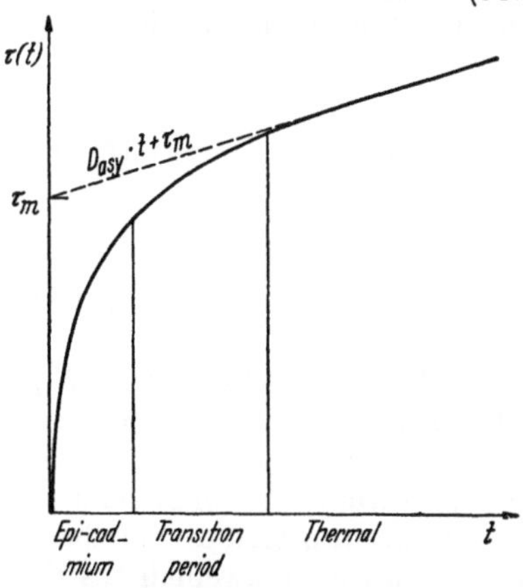

Fig. 225. The generalized neutron age as a function of time (V. DARDEL).

Fig. 225 shows the qualitative behaviour of the generalized age as a function of time as given by VON DARDEL[2]. At high energy (i.e. t very small) the diffusion coefficient $D_0(t)$, defined by Eq. (81.3), is large so that $\tau(t)$ increases rapidly. As the slowing down proceeds, the slope of the curve $\tau(t)$ decreases and during the *transition period* corresponding to the chemical bond energy region, $\tau(t)$ approaches a straight line the angular coefficient of which is the asymptotic value $D_{asy}(t)$. This represents the constant diffusion coefficient for thermal neutrons.

The zero time intercept of the asymptote in Fig. 225 may be defined as the age for slowing down to thermal energies:

$$\tau_m = \lim_{t\to\infty} \int\limits_0^t \{D(t) - D_{asy}\}\,dt. \tag{111.2}$$

The diffusion length for thermal neutron diffusion is then given by the expression

$$L = \sqrt{T D_{asy}}\,. \tag{111.3}$$

The definition (111.2) for the slowing down age to thermal energies, has the advantage over other definitions [Sect. 110; Eq. (110.19), Fig. 224] of being independent of the absorption properties of the medium.

Once the age has been generalized to any energy value, the solution of the age equation (81.9) can also be applied to any energy. Therefore, either Eqs. (81.10), (81.11) according to the geometry of the moderator or, in general, Eq. (81.12), can be used during the transition period and in the thermal energy region.

This approach to the problem is of considerable practical importance in the thermal energy region where the generalized age $\tau(t)$ increases linearly with time.

[1] G. VON DARDEL: Trans. Roy. Inst. Technol. **75** (1954).

[2] G. VON DARDEL and N. G. SJÖSTRAND: Progress in Nuclear Energy, Physics and Mathematics, Vol. 2, p. 183. New York and London: Pergamon Press 1958.

Such a behaviour can easily be verified experimentally, and, as we shall see in Sects. 130 and 131, leads to the experimental determination of the various parameters involved in the description of the diffusion of thermal neutrons.

In the transition period, the measured time dependence of the neutron density could usefully be compared with the theoretical predictions. So far, such a procedure has found very little application so that in this energy region, not only the theoretical predictions, but even the experimental information about the behaviour of $D_0(t)$ are very scanty.

As we shall see in Sect. 131, even the value of τ_m has been determined only for the case of neutrons from the D + Be reaction ($E_d = 300$ kev) slowed down in beryllium oxide. The result of this experiment is given by Eq. (131.7).

A different approach to the problem has deen suggested by VON DARDEL, who expresses the rate of energy loss of the neutrons at a particular time t by means of the following equation:

$$\frac{d\bar{E}(t)}{dt} = \int \Sigma_s(E)\, \overline{(E-E')} \cdot v \cdot n_0(v, t)\, dE \tag{111.4}$$

with

$$\bar{E}(t) = \frac{\int\limits_0^\infty E\, n_0(v, t)\, dv}{\int n_0(v, t)\, dv}. \tag{111.5}$$

In Eq. (111.4) $\Sigma_s(E)$ as well as the average energy loss per collision $\overline{(E-E')}$ depends upon the structure of the moderator and can be calculated in terms of some conveniently chosen model. The calculation of the integral appearing in (111.4) requires, however, a knowledge of the neutron spectrum; and, unfortunately, we do not know much about $n_0(v, t)$ during the transition period except that, for sufficiently small absorption, it gradually approaches a Maxwellian distribution in thermal equilibrium with the moderator.

This difficulty was overcome by VON DARDEL who introduced the assumption that when the neutron gas is sufficiently close to thermal equilibrium, its velocity spectrum is approximately represented by a Maxwellian distribution with a temperature T different from the temperature T_0 of the moderator. This instantaneous temperature T of the neutron gas must not be confused with the effective neutron temperature T_n defined in Sect. 108. While T_n provides a measure of the deviations due to absorption of the stationary neutron spectrum from thermal equilibrium, $T(t)$ is here thought of as a function of time which, for t going to infinity, tends to the temperature T_0 of the moderator.

Under this assumption, the integral appearing in (111.4) is easily computed for any conveniently chosen model of the moderator by substituting for $n_0(v, t)$ the Maxwell distribution (96.6) with T_0 replaced by $T(t)$.

The actual calculations have been carried out for a montaomic gaseous moderator[1] as well as for a crystal moderator[2,3,4].

For a monatomic gaseous moderator of constant scattering cross section, the quantities $\Sigma_s(E)$ and $\overline{(E-E')}$ are given by (97.7) and (97.14) respectively; by making use of these expressions, VON DARDEL[1] obtains

$$\frac{d\bar{E}}{dt} = -\frac{8}{\sqrt{\pi}} \left(\frac{2kT_0}{m}\right)^{\frac{1}{2}} \Sigma_{s0} \frac{A^{\frac{1}{2}}}{(A+1)^{\frac{3}{2}}} \left\{1 + \frac{A}{A+1}\, \frac{T-T_0}{T_0}\right\}^{\frac{1}{2}} k(T-T_0) \tag{111.6a}$$

[1] See footnote 1, p. 500.

[2] R. GURNEY: AECD, 2449 (1949).

[3] L. S. KOTHARI and K. S. SINGWI: J. Nucl. Energy 5, 342 (1957); 6, 59 (1958).

[4] R.C. BHANDARI: J. Nucl. Energy 6, 104 (1958).

which, for $T(t) - T_0 \ll T_0$, reduces to[1]

$$\frac{d\bar{E}}{dt} = -\frac{1}{t^0_{\rm th}} \frac{3}{2} k\left(T(t) - T_0\right). \tag{111.6b}$$

Here

$$\bar{E}(t) = \tfrac{3}{2} kT(t), \tag{111.7}$$

and

$$t^0_{\rm th} = \frac{3\sqrt{\pi}}{16} \frac{1}{v_0 \Sigma_{s0}} \frac{(A+1)^{\frac{3}{2}}}{A^{\frac{1}{2}}} \tag{111.8}$$

is called the *thermalization time*. The data given in Column 4 of Table 61 (p. 339) have been computed by means of this equation. From (111.6b) it follows that

$$T(t) - T_0 = \text{const}\, e^{-t/t^0_{\rm th}} \tag{111.9}$$

a law which can be easily verified experimentally, as we shall see below.

Two attempts have been made to treat the same problem in the case of polycrystalline moderators[2,3]; the numerical results refer to beryllium in both cases. The first attempt was made by GURNEY[2] who calculated the slowing down from 1.46 ev to thermal energies taking into account the lattice binding by means of the Einstein model of the crystal (Sect. 99). This model, however, breaks down for neutron energies less than a few $k\Theta_E$ (for beryllium $\Theta_E = 740°$ K, $k\Theta_E \approx 0.064$ ev) because, as soon as the neutron has reached an energy even slightly lower than $k\Theta_E$, it can no longer lose any more energy to the lattice and, therefore, the thermalization of the neutron becomes impossible.

This inconvenience was avoided by KOTHARI and SINGWI[3] who used the following slightly improved model of a beryllium crystal. (a) For neutron energies

$$E < k\Theta_D$$

(for beryllium $\Theta_D = 1000°$ K, $k\Theta_D = 0.085$ ev) the crystal is described by means of the Debye model, and the cross sections for the elastic collisions as well as for the one- and two-phonon exchanges are computed in the incoherent approximation (Sect. 103) by introducing into Eq. (103.9) successively $l = 0$, 1 and 2. (b) For neutron energies

$$E > 2\, k\Theta_D$$

where multiphonon processes are important, the crystal is described by means of the Einstein model. The corresponding cross section for exchange of l phonons is readily derived from the expression valid for a Debye crystal by introducing the condition that all phonons have the same energy. Because of this condition the calculations are much simpler. (c) In the intermediate energy region, the various quantities are graphically interpolated.

Once the cross sections for elastic as well as for inelastic scattering have been obtained, the average energy loss $\overline{(E - E')}$ is computed. Finally, this quantity as well as $\Sigma_s(E)$ are introduced into (111.4) and following VON DARDEL, a Maxwellian distribution at the instantaneous temperature $T(t)$ is substituted for the neutron spectrum $n_0(v, t)$.

---

[1] See footnote 2, p. 500.
[2] See footnote 2, p. 501.
[3] See footnote 3, p. 501.

The results of KOTHARI and SINGWI agree, in their main qualitative lines, with those obtained by VON DARDEL for a monatomic gaseous moderator, and can be summarized as follows. For a moderator temperature $T_0 = 300°$ K, the equilibrium temperature T_{eq} defined by Eq. (129.19), turns out to be 304° K and the total slowing-down time for neutrons from a fission source to $T = 310°$ K amounts to 143 μsec; of this, only 13 μsec are spent by the neutron in reaching the In-resonance at $E = 1.46$ ev.

The fact that the slope of a plot of $\log[T(t) - T_0]$ versus time decreases with increasing time, can be interpreted as an increase of the thermalization time constant t_{th}^0 which, starting from a rather small value, increases with decreasing temperature and, for neutron energies below $0.5\, k\, \Theta_D$, attains the value[1]

$$t_{th}^0 = 29 \ \mu\text{sec} \tag{111.10}$$

which should be compared with the value obtained by VON DARDEL for a gaseous medium (Table 61, p. 339).

Finally, SINGWI and KOTHARI make use of the definition of the generalized age (111.1) in order to numerically compute the increase in age from the In-resonance to the thermal energy; the results show that

$$\tau_{th} - \tau_{In} = 25 \ \text{cm}^2. \tag{111.11}$$

The experimental check of Eq. (111.9) and the determination of the thermalization time as well as its possible time dependence is made by means of the same technique employed in the investigation of the time variation of the neutron density produced in a moderator by a pulsed neutron source. This problem has been discussed in Sect. 85 for energies higher than 1 ev, and it will be treated in Sects. 130 and 131 for thermal energies.

In the present case, the quantity measured as a function of time by means of a $1/v$ detector, is the neutron transmission through a thin $1/v$ absorber; from this transmission the instantaneous temperature $T(t)$ can readily be computed by means of Eqs. (108.11), (108.12), (108.13) and (108.1).

The first experiments of this type have been carried out by VON DARDEL[2,3] who studied, in "bad geometry", the boron transmission of the neutrons emitted from a cube of water containing a pulsed source (Sect. 85). The main results of this author can be summarized as follows. For $T_0 = 273°$ K the neutron temperature $T(t)$ approaches equilibrium with the moderator in a thermalization time of 7 μsec, a value which is in very good agreement with the prediction of Eq. (111.8) which gives 6.3 μsec for a mass $A = 18$ of the scattering centres. Similar measurements made at different water temperature ($T_0 = 98°$ K and $T_0 = 354°$ K) show that the neutron energy falls with time along the same curve, independent of T_0, during the slowing down period corresponding to energies above 1 ev. But when the neutron temperature becomes of the order of that of the moderator, the various curves deviate one from the other and approach (roughly exponentially) the corresponding T_0 value. At liquid air temperature ($T_0 = 98°$ K), the value of the thermalization time is found to be 20 μsec, i.e., twice the value (10 μsec) predicted by Eq. (111.8) at this temperature. The disagreement is probably due to the fact that we are far below the Debye temperature of ice so that the monatomic gas picture becomes inadequate[2,3].

[1] See footnote 4, p. 501.
[2] See footnote 1, p. 500.
[3] See footnote 2, p. 500.

Similar measurements have been made in graphite by Antonov *et al.*[1] and by Beckurts[2], and in beryllium oxide by Iyengar *et al.*[3]. Since all these authors used "good geometry", corrections arising from the angular distribution of the neutrons emerging from the medium were avoided.

The agreement between the results of Antonov *et al.* and Beckurts is very satisfactory. Fig. 226 shows the results obtained by Beckurts who used a silver foil as a $1/v$ absorber in the energy region below 1 ev. The experimental points fit Eq. (111.9) with

$$t_{\mathrm{th}}^{0} = (185 \pm 45)\ \mu\mathrm{sec} \tag{111.12}$$

i.e., with a value by a factor five larger than the prediction of Eq. (111.8) for a monatomic gaseous medium (Table 61, $t_{\mathrm{th}}^{0} \approx 35\ \mu\mathrm{sec}$).

The results of Mani *et al.* on beryllium oxide and of Antonov *et al.* on graphite are discussed in Sect. 131, where they are compared with the results of an indirect determination of t_{th}^{0} based on the measurement of the diffusion cooling coefficient C (Sect. 129).

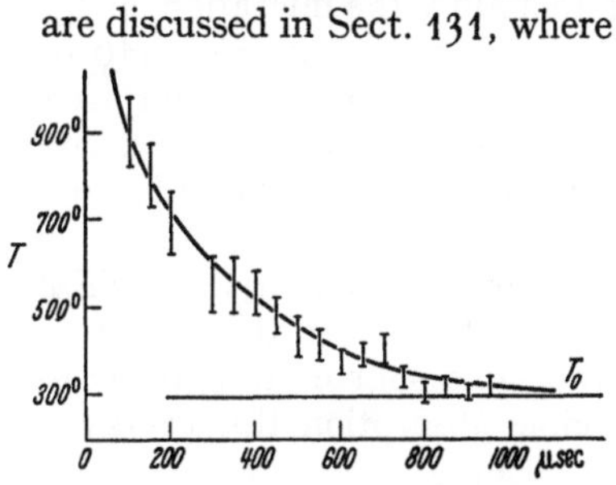

Fig. 226. Neutron temperature as a function of time in graphite (Beckurts).

We may leave the subject with the final statement that the thermalization time constant t_{th}^{0} turns out to be of the order of 150 to 200 μsec in all polycrystalline materials so far investigated (Be, BeO and C) and that it therefore appears to be larger by a factor between five and seven with respect to the theoretical values obtained with Debye crystal models.

E. The diffusion of thermal neutrons.

I. General features and mathematical aspects of the diffusion of thermal neutrons.

112. A few remarks on thermal neutrons and their treatment as an one-velocity group. The expression "thermal neutrons" is used to indicate the low energy tail of the neutron spectrum present in a moderator, which under certain conditions, deviates moderately from a Maxwellian distribution. As we have seen in Sect. 96, the stationary neutron spectrum is exactly Maxwellian at the temperature T_0 of the moderator only in the ideal limiting case of uniform slowing down in a moderator which satisfies the following conditions: it fills all the space, it does not contain neutron sources and does not absorb neutrons. In the slightly more realistic case in which both neutron sources and weak absorption are present and distributed uniformly throughout the moderator, the neutron spectrum can be described as a Maxwellian distribution joined smoothly into a slowing-down spectrum. This is clearly shown by the curves of Figs. 205, 207—211 and 218 where the results of various theoretical calculations have been plotted. Experiments made in water (Fig. 212), and graphite (Fig. 223) very satisfactorily confirm this conclusion.

If the absorption is increased, the deviations of the low-energy tail of the spectrum from a Maxwellian distribution at the temperature T_0 become appre-

[1] A.V. Antonov, A.I. Isakoff, I.D. Murin, B.A. Neupocoyev, I.M. Frank, F.L. Shapiro and I.V. Shtranich: p. 83 of Ref. [*15c*].

[2] K. Beckurts: Nucl. Sci. and Engng. **2**, 516 (1957).

[3] S. B. D. Iyengar, G. S. Mani, R. Ramanna and N. Umakanth: Proc. Ind. Acad. Sci. **45**, 215, 224 (1957).

ciable; for moderate absorption these deviations can be represented by introducing an effective neutron temperature T_n larger than T_0; but for strong absorption the neutron spectrum is so distorted that even this rather arbitrary procedure turns out to be inadequate.

Further distorsions of the low-energy tail of the neutron spectrum will occur in the vicinity of a moderator boundary or of a strong localized absorber.

Thus, we see that the expression "thermal neutrons" refers to neutrons of energy of the order of kT_0 and that it may indicate, in different cases, quite different shapes for the low-energy tail of the neutron spectrum.

But even if we limit our consideration to an infinite moderator with a very small capture cross section, it is rather arbitrary to fix an energy which separates the thermal neutrons from those which are still undergoing the slowing down process. A reasonable value of this energy is represented by the migration energy E_m, various definitions of which are given in Part D: namely, in Sect. 110 by means of Eqs. (110.19), (110.20) (Fig. 224), and in Sect. 111 by means of the graphical procedure indicated in Fig. 225. Its value depends somewhat on the recipe adopted for its definition, but, in all cases, it turns out to be of the order of a few kT_0, i.e. about 0.1 ev at room temperature.

In some elementary treatments the chemical region is ignored and the dE/E spectrum is directly joined onto the Maxwellian distribution at the energy at which the corresponding curves, normalized to the same source strength, cross one another; at room temperature and for negligible absorption this energy is about 0.17 ev [*12*].

From the experimental point of view, the separation between the low energy tail of the spectrum and the neutrons of higher energy is usually made at the Cd-cut-off E_{Cd} which may be placed at about 0.4 ev if the recipe (26.20) is followed.

From the inspection of Figs. 212 and 223, both of which refer to media with very small capture, we see that the neutron spectrum with energy below E_{Cd} includes a considerable excess of neutrons of energy between $\sim$0.1 and $\sim$0.4 ev with respect to the Maxwellian distribution. Thus, the comparison of experimental results obtained by means of the Cd-difference method (Sect. 133) with theoretical results obtained for a Maxwellian distribution, requires always the computation of a convenient correction: see, for example, the procedures described in Sect. 108 [Eq. (108.12)] and Sect. 110 [Eq. (110.31)].

A correct treatment of the diffusion of thermal neutrons through a moderator requires the solution of the Boltzmann integro-differential equation which is obtained by substituting in Eq. (61.2a) for $\Sigma_a(v)$, $\Sigma_s(v)$ and $\Sigma_s(v'\to v, \boldsymbol{\Omega}'\to\boldsymbol{\Omega})$ the corresponding effective cross sections $\overline{\Sigma}_a(v)$ $\overline{\Sigma}_s(v)$ and $\overline{\Sigma}_s(v'\to v, \boldsymbol{\Omega}'\to\boldsymbol{\Omega})$. These quantities, according to their definition (96.12), should embody the molecular structure of the moderator; in practice they are computed by substituting for the actual moderator some simplified model, a few examples of which have been given in Part D of this article.

But even in this case, the solution of the transport equation involves considerable analytical difficulties and, therefore, one substitutes the actual problem with an oversimplified case which nevertheless approximates, at least in certain cases, the main features of thermal neutron diffusion.

If the conditions are such that thermal equilibrium is reached—at least approximately—with the atoms of the moderator, the neutrons diffuse with an energy which, on the average, remains constant since energy gains are as probable as energy losses. The process goes on for a very large number of collisions before capture brings it to an end.

Such a behaviour is expected to be similar to that of a group of neutrons, all having the same velocity, which diffuse without exchanging energy with the moderator, a case which has the advantage that the corresponding transport equation is much easier to solve than the Boltzmann equation relative to any one of the models considered in Part D (Sects. 105 to 107).

The transport equation without energy loss can be derived from Eq. (61.21 a) by imposing the condition that $\Sigma_s(E' \to E, \boldsymbol{\Omega}' \to \boldsymbol{\Omega})$ be zero unless $E = E'$; this means that $\Sigma_s(E' \to E, \boldsymbol{\Omega}' \to \boldsymbol{\Omega})$ behaves as a δ-function with respect to E so that

$$\int dE'\, \Phi(\boldsymbol{r}, \boldsymbol{\Omega}', E')\, \Sigma_s(E' \to E, \boldsymbol{\Omega}' \to \boldsymbol{\Omega}) = \Phi(\boldsymbol{r}, \boldsymbol{\Omega}', E)\, \Sigma_s(E, \boldsymbol{\Omega}' \to \boldsymbol{\Omega}). \quad (112.1)$$

Introducing (112.1) and $\partial\Phi/\partial t = 0$ into (61.21 a), we obtain the stationary *transport equation for an one-velocity group:*

$$\boldsymbol{\Omega} \cdot \nabla \Phi(\boldsymbol{r}, \boldsymbol{\Omega}) + \Sigma_{\text{tot}} \cdot \Phi(\boldsymbol{r}, \boldsymbol{\Omega}) = \int d\boldsymbol{\Omega}'\, \Phi(\boldsymbol{r}, \boldsymbol{\Omega}')\, \Sigma_s(\boldsymbol{\Omega}' \to \boldsymbol{\Omega}) + S(\boldsymbol{r}, \boldsymbol{\Omega}) \quad (112.2)$$

where the variable E has been dropped since, in the present case, it is a constant.

The boundary conditions for the solution of (112.2) are derived by applying (61.12) to the stationary case. At the interface between two media in direct contact

$$\left.\begin{array}{l}\Phi(\boldsymbol{r} + R\boldsymbol{\Omega}, \boldsymbol{\Omega}) \text{ must be a continuous function} \\ \qquad \text{of } R \text{ for } \boldsymbol{r} + R\boldsymbol{\Omega} \text{ at the interface.}\end{array}\right\} \quad (112.3\text{a})$$

On a non re-entrant free surface separating the medium from a vacuum

$$\Phi(\boldsymbol{r}, \boldsymbol{\Omega}) = 0 \quad \text{for all } \boldsymbol{\Omega} \text{ entering the medium and for } \boldsymbol{r} \text{ on the surface.} \quad (112.3\text{b})$$

When the medium extends to infinity and all the neutron sources are located at a finite distance,

$$\left.\begin{array}{l}\Phi(\boldsymbol{r}, \boldsymbol{\Omega}) \text{ must tend to zero faster than any} \\ \qquad \text{power of } r \text{ for } \boldsymbol{r} \text{ going to infinity.}\end{array}\right\} \quad (112.3\text{c})$$

At first sight the application of Eq. (112.2) to the diffusion of thermal neutrons implies a neglect of their velocity spread. It may be noticed, however, that when physical conditions are met such that one can state "a priori" that the energy spectrum is independent of the position and direction of motion of the neutrons, one can write

$$\Phi(\boldsymbol{r}, \boldsymbol{\Omega}, E) = \Phi(\boldsymbol{r}, \boldsymbol{\Omega}) \cdot \varphi(E). \quad (112.4)$$

Then by introducing (112.4) into (61.21 a) and integrating with respect to dE, one obtains an equation of the same form as (112.2) [7] with

$$\left.\begin{aligned}\Sigma_{\text{tot}} &= \frac{\int \Sigma_{\text{tot}}(E')\, \varphi(E')\, dE'}{\int \varphi(E')\, dE'}, \\ \Sigma_s(\boldsymbol{\Omega}' \to \boldsymbol{\Omega}) &= \frac{\int \varphi(E)\, dE \int \Sigma_s(E' \to E, \boldsymbol{\Omega}' \to \boldsymbol{\Omega})\, \varphi(E')\, dE'}{\int \varphi(E')\, dE'}, \\ S(\boldsymbol{r}, \boldsymbol{\Omega}) &= \frac{\int S(\boldsymbol{r}, \boldsymbol{\Omega}, E')\, \varphi(E')\, dE'}{\int \varphi(E')\, dE'}.\end{aligned}\right\} \quad (112.5)$$

According to the discussion given above, there are cases in which thermal neutrons fulfill—at least approximately—condition (112.4) with $\varphi(E)$ given by a Maxwellian flux distribution.

This or equivalent arguments [19] are brought forward by various authors to justify the use of the one-velocity group equation (112.2) in describing the behaviour of thermal neutrons. Here we shall not push further this discussion which will be taken up again, from different points of view, at the end of Sect. 115 and in Sect. 117. We shall only recall that, whenever the absorption of thermal neutrons and/or their escape from the moderator is appreciable, there is no longer thermal equilibrium with the medium so that the neutron spectrum is no longer Maxwellian and its shape depends on the position and direction of the motion. Thus we can conclude that the one-velocity group transport equation will provide an adequate description of thermal neutron diffusion only in those regions of weakly absorbing media where their flux varies very slowly with position. In particular we should expect that Eq. (112.2) will not describe correctly the behaviour of thermal neutrons near a strong absorber or a boundary.

In spite of these remarks, the one-velocity group approximation is usually adopted for the treatment of thermal neutron diffusion; in particular it will be used in this part of the present article.

The one-velocity group theory has been widely investigated by many authors, in particular by PLACZEK and his coworkers. The most systematic and general presentations of this theory and its applications are given in Refs. [*6*] and [*7*].

A detailed development of the one-velocity theory is also justified by other applications different from that of thermal neutron diffusion. Firstly, it gives a complete description of the diffusion of neutrons if both the total cross section and the differential scattering cross section are constant over the considered energy interval. Secondly, the one-velocity problems must be solved before one can hope to include the effect of velocity changes. Moreover, the solution of one-velocity problems can be used for solving more general problems; for example the problem of the slowing down in a non-absorbing medium can be reduced to a one-velocity problem by taking the Laplace transform of the corresponding equation with respect to energy [*16*].

Finally, the multigroup picture of the slowing-down process mentioned in Sect. 78 is based on dividing the neutrons into velocity groups. The neutrons of each group can be described by the one-velocity equation (112.2) with sources depending on the neutrons of the other groups.

As we shall see in the following, the transport equation of an one-velocity group (112.2)—or an equivalent integral equation which is derived in Sect. 113—can be solved rigorously in a few simple cases (Sects. 116 to 118). The problem of neutron diffusion at a constant velocity can, however, be tackled from a more elementary point of view based on the analogy with other diffusion phenomena (Sect. 114). Such an elementary approach constitutes the first step (Sect. 115) of a successive approximation method (Sect. 123) and may provide, by a convenient choice of boundary conditions (Sect. 119), a satisfactory approximate description of the asymptotic part of the rigorous solution. Sects. 120 and 121 contain a few applications of elementary diffusion theory to a few interesting cases, and Sect. 122 illustrates an over-simplified one-dimensional model. Finally, time dependent problems are very briefly considered in Sect. 124.

113. The integral equation for one-velocity group diffusion in a medium with isotropic scattering in the laboratory system. In this section we shall show that the integro-differential equation (112.2) with the boundary conditions (112.3) can be transformed into an integral equation which has a very simple physical meaning [*6*]. Such a transformation could have been done in the most general case, i.e. starting, for example, from Eq. (61.2a) with the boundary conditions (61.12) [*7*]. We give the derivation for the very special case of an one-velocity group of neutrons scattered isotropically in the l.s., because in this case the form of the integral equation is so simple as to allow the derivation of a few important theorems some of which are mentioned in Sect. 116 and Sect. 118.

The condition of isotropic scattering in the l.s. is introduced into Eq. (112.2) by writing

$$\Sigma_s(\boldsymbol{\Omega}' \to \boldsymbol{\Omega}) = \frac{\Sigma_s}{4\pi} = \frac{1}{4\pi\, l_s} = \frac{h}{4\pi\, l_{\text{tot}}} \tag{113.1a}$$

where h is defined by (61.18):

$$h = \frac{l_{\text{tot}}}{l_s} = \frac{\Sigma_s}{\Sigma_s + \Sigma_a}.$$

In the one-velocity theory h is a constant, the physical meaning of which can be extended as will be shown below.

If the medium consists of a single non re-entrant homogeneous body and the neutron source is isotropic,

$$S(\boldsymbol{r}, \boldsymbol{\Omega}) = \frac{s(\boldsymbol{r})}{4\pi}, \tag{113.1b}$$

then as a consequence of (112.2), (112.3) the neutron flux

$$\Phi_0(\boldsymbol{r}) = \int \Phi(\boldsymbol{r}, \boldsymbol{\Omega})\, d\boldsymbol{\Omega} \tag{113.2}$$

must satisfy the following integral equation:

$$\Phi_0(\boldsymbol{r}) = \frac{1}{4\pi} \int \left[\frac{h}{l_{\text{tot}}} \Phi_0(\boldsymbol{r}') + s(\boldsymbol{r})\right] \frac{e^{-\frac{|\boldsymbol{r}-\boldsymbol{r}'|}{l_{\text{tot}}}}}{|\boldsymbol{r}-\boldsymbol{r}'|^2}\, d\boldsymbol{r}' \tag{113.3}$$

with the integral extended over all the space.

Eq. (113.3) can be written directly as will be shown later for the special case of isotropic sources with plane symmetry [see Eq. (113.11)]; the first term on the right-hand side represents the flux contribution at $\boldsymbol{r}$ which arises from the neutrons which have undergone their last collision in the volume element $d\boldsymbol{r}'$ at $\boldsymbol{r}'$, and the second term represents the contribution of neutrons emitted by the source in the volume element $d\boldsymbol{r}'$ at $\boldsymbol{r}'$ which reach the point $\boldsymbol{r}$ without undergoing any collision.

It may be interesting to notice that Eq. (113.3) holds even in the case of multiplying media, provided that the numerical coefficient h is interpreted as *the average number of secondary neutrons emitted per collision.* Then, an absorbing medium will correspond, according to the original definition (61.18), to $h<1$ and a medium with pure scattering—i.e. without capture or production of secondary neutrons—to $h=1$; a multiplying medium will give $h>1$ whenever the emission of secondary neutrons prevails over absorption.

The derivation of (113.3) from (112.2) can be made as follows [*6*], [*7*]. Because of the simplifying assumptions of isotropy for both scattering and neutron source (113.1), the transport equation (112.2) can be written in the form

$$\boldsymbol{\Omega} \cdot \nabla \Phi(\boldsymbol{r}, \boldsymbol{\Omega}) + \frac{1}{l_{\text{tot}}} \Phi(\boldsymbol{r}, \boldsymbol{\Omega}) = \frac{h}{4\pi\, l_{\text{tot}}} \int \Phi(\boldsymbol{r}, \boldsymbol{\Omega}')\, d\boldsymbol{\Omega}' + \frac{s(\boldsymbol{r})}{4\pi}. \tag{113.4}$$

Since $\boldsymbol{\Omega} \cdot \nabla$ is simply the derivative along the direction of $\boldsymbol{\Omega}$, by replacing $\boldsymbol{r}$ by $\boldsymbol{r} - R\boldsymbol{\Omega}$, we can rewrite (113.4) in the form

$$-\frac{d}{dR} \Phi(\boldsymbol{r} - R\boldsymbol{\Omega}, \boldsymbol{\Omega}) + \frac{1}{l_{\text{tot}}} \Phi(\boldsymbol{r} - R\boldsymbol{\Omega}, \boldsymbol{\Omega}) = Q(\boldsymbol{r} - R\boldsymbol{\Omega}) \tag{113.5}$$

where

$$Q(\boldsymbol{r}) = \frac{1}{4\pi}\left[\frac{h}{l_{\text{tot}}} \int \Phi(\boldsymbol{r}, \boldsymbol{\Omega}')\, d\boldsymbol{\Omega}' + s(\boldsymbol{r})\right] = \frac{1}{4\pi}\left[\frac{h}{l_{\text{tot}}} \Phi_0(\boldsymbol{r}) + s(\boldsymbol{r})\right] \tag{113.6}$$

may be considered as an *effective neutron source.*

If $Q(r)$ is supposed to be known, Eq. (113.5) can be solved for $\boldsymbol{\Phi}(\boldsymbol{r}-R\boldsymbol{\Omega},\boldsymbol{\Omega})$ with the result

$$\Phi(\boldsymbol{r}-R\boldsymbol{\Omega},\boldsymbol{\Omega})=\Phi(\boldsymbol{r}-R_0\boldsymbol{\Omega},\boldsymbol{\Omega})\,e^{\frac{R-R_0}{l_{tot}}}+\int\limits_R^{R_0} Q(\boldsymbol{r}-R'\boldsymbol{\Omega})\,e^{\frac{R-R'}{l_{tot}}}\,dR' \tag{113.7}$$

which, for $R=0$ and $R_0\to\infty$ reduces to

$$\Phi(r,\boldsymbol{\Omega})=\int\limits_0^\infty Q(\boldsymbol{r}-R'\boldsymbol{\Omega})\,e^{-\frac{R'}{l_{tot}}}\,dR'. \tag{113.8}$$

In order to prove the last transformation, two cases have to be distinguished according to the extension of the medium. The first case is met if the line $\boldsymbol{r}-R_0\boldsymbol{\Omega}$, defined by certain values of $\boldsymbol{r}$ and $\boldsymbol{\Omega}$, cuts the surface of the medium at some value of $R_0=R_0'>0$; then $\Phi(\boldsymbol{r}-R_0'\boldsymbol{\Omega},\boldsymbol{r})$ vanishes in consequence of the boundary condition (112.3b). Thus, on the right-hand side of (113.7), the first term is zero and the integral can be extended to infinity since $Q(r)$ is zero in the vacuum surrounding the medium.

The second case to be considered is that of a medium extending to infinity; then for $R_0\to\infty$, the first term on the right-hand side of (113.7) goes to zero in consequence of the boundary condition (112.3c), and (113.7) again reduces to (113.8). This is identical to (113.3), as one can see by substituting (113.6) into (113.8), putting $\boldsymbol{r}'=\boldsymbol{r}-R'\boldsymbol{\Omega}$, noticing that $dR'\,d\boldsymbol{\Omega}'=\frac{d\boldsymbol{r}'}{R'^2}=\frac{d\boldsymbol{r}'}{|\boldsymbol{r}-\boldsymbol{r}'|^2}$ and then integrating over $d\boldsymbol{\Omega}$ [7].

The integral equation (113.3) is considerably simpler than the transport equation (113.4) since it only involves a function of $\boldsymbol{r}$. Once the solution of (113.3) is known, the effective source is also known—because of (113.6)—and $\Phi(\boldsymbol{r},\boldsymbol{\Omega})$ may be calculated by inserting it into (113.8) [6]. An example of the application of this procedure is given in Sect. 116 for the case of a source-free homogeneous infinite medium.

Before closing the discussion of the integral equation valid in the case of isotropic scattering in the l.s., we may derive it from first principles for the case of an arbitrary distribution of isotropic sources having plane symmetry.

The neutrons that cross the plane at z in the direction defined by a certain value of $\mu=\cos\vartheta$ are those that started to move in that direction either after their last collision or at the moment in which they have been emitted by the source provided that they have suffered no collisions at all. Therefore, we can write [19]

$$\left.\begin{aligned}\Phi(z,\mu)&=\frac{h}{2l_{tot}}\int\limits_{-\infty}^{z}\Phi_0(z')\,e^{-\frac{z-z'}{l_{tot}\mu}}\frac{dz'}{\mu}+\int\limits_{-\infty}^{z}s(z')\,e^{-\frac{z-z'}{l_{tot}\mu}}\frac{dz'}{\mu}\quad\text{for }\mu>0,\\ \Phi(z,\mu)&=\frac{h}{2l_{tot}}\int\limits_{\infty}^{z}\Phi_0(z')\,e^{-\frac{z'-z}{l_{tot}\mu}}\frac{dz'}{\mu}+\int\limits_{\infty}^{z}s(z')\,e^{-\frac{z'-z}{l_{tot}\mu}}\frac{dz'}{\mu}\quad\text{for }\mu<0\end{aligned}\right\} \tag{113.9}$$

where

$$\Phi_0(z)=\int\limits_{-1}^{+1}\Phi(z,\mu)\,d\mu. \tag{113.10}$$

By integration over μ between -1 and $+1$, we obtain

$$\Phi_0(z)=\int\limits_{-\infty}^{+\infty}\left[\frac{h}{2l_{tot}}\Phi_0(z')+s(z')\right]E_1\left(\frac{|z-z'|}{l_{tot}}\right)dz' \tag{113.11}$$

where

$$E_n(z) = \int_1^\infty e^{-uz} u^{-n} du \tag{113.12}$$

are functions which frequently appear in problems of this type. They are tabuated for n up to 4 in Ref. [6].

The integral equation (113.11) is simply Eq. (113.3) applied to the case of (isotropic) sources with plane symmetry. The kernels appearing in these two equations are called the *plane transport kernel* and *the point transport kernel*, respectively. Table 85 contains the transport kernels for a few other simple source geometries; they can be derived by applying the relations (71.2) to the point transport kernel (Sect. 71).

So far, in this section, we have considered only the case of isotropic scattering in the l.s.

It should be noticed however that the procedure sketched above for deriving a rigorous solution of the transport equation can be generalized to the case of anisotropic scattering whenever $\Sigma_s(\boldsymbol{\Omega}' \to \boldsymbol{\Omega})$ can be represented by a polynomial of degree n in $\mu_0 = \boldsymbol{\Omega}' \cdot \boldsymbol{\Omega}$. In this case, the expression (113.6) of the effective source involves angular moments of $\Phi(\boldsymbol{r}, \boldsymbol{\Omega})$ up to the n-th order. This is easily seen, for example, in the case of almost isotropic scattering in the l.s., i.e. if the expansion (115.2a) of $\Sigma_s(\boldsymbol{\Omega}' \to \boldsymbol{\Omega})$ in Legendre polynomials of μ_0 breaks off at the second term (Sect. 117 and 120). Then, by inserting Eq. (120.5) into (113.6), an expression for $Q(r)$ is obtained which depends upon both $\Phi_0(\boldsymbol{r})$ and $\boldsymbol{J}(\boldsymbol{r})$.

Table 85. *The transport kernels*[1].

Geometry	Notation	Source normalization	Kernel
Point	$K_p(\boldsymbol{r}, \boldsymbol{r}')$	$\frac{1\text{ neutron}}{1\text{ sec}}$	$\dfrac{e^{-\frac{\lvert \boldsymbol{r}-\boldsymbol{r}'\rvert}{l_{\text{tot}}}}}{4\pi \lvert \boldsymbol{r}-\boldsymbol{r}'\rvert^2}$
Plane	$K_{pl}(z, z')$	$\frac{1\text{ neutron}}{1\text{ sec} \times 1\text{ cm}^2}$	$\frac{1}{2} E_1\left(\frac{\lvert z-z'\rvert}{l_{\text{tot}}}\right)$
Spherical shell	$K_s(r, r')$	$\frac{1\text{ neutron}}{1\text{ sec} \times \text{shell}}$	$\frac{1}{8\pi r r'}\left[E_1\left(\frac{\lvert r-r'\rvert}{l_{\text{tot}}}\right) - E_1\left(\frac{\lvert r+r'\rvert}{l_{\text{tot}}}\right)\right]$
Line	$K_l(r, \varphi, r', \varphi')$	$\frac{1\text{ neutron}}{1\text{ sec} \times 1\text{ cm}}$	$\frac{1}{2\pi l_{\text{tot}}} \int_1^\infty K_0\left(\frac{\varrho}{l_{\text{tot}}} y\right) dy$; $\varrho = \sqrt{r^2 + r'^2 - 2 r r' \cos(\varphi' - \varphi)}$
Cylindrical shell	$K_c(r, r')$	$\frac{1\text{ neutron}}{1\text{ sec} \times 1\text{ cm}}$	$\frac{1}{2\pi l_{\text{tot}}} \int_1^\infty K_0\left(\frac{r}{l_{\text{tot}}} y\right) I_0\left(\frac{r'}{l_{\text{tot}}} y\right) dy \quad r > r'$; $\frac{1}{2\pi l_{\text{tot}}} \int_1^\infty K_0\left(\frac{r'}{l_{\text{tot}}} y\right) I_0\left(\frac{r}{l_{\text{tot}}} y\right) dy \quad r < r'$

[1] I_0 and K_0 are modified Bessel functions of the first and second kind. — Cf. e g. N.W. MCLACHLAN: Bessel functions for Engineers. Oxford: Clarendon Press 1934.

Returning to the general case, we see that the angular moments of $\Phi(\boldsymbol{r}, \boldsymbol{\Omega})$ will also appear—through $Q(r)$—in the integral equation (113.8), with the result that its treatment becomes a problem considerably more involved than in the case of isotropic scattering in the l.s.

In order to solve this equation, one can multiply it by various powers of the components of $\boldsymbol{\Omega}$ and integrate over $d\boldsymbol{\Omega}$. Thus, a set of integral equations is obtained for the moments involved; after solving these equations, the moments are introduced in the expression for $Q(r)$ and finally $\Phi(\boldsymbol{r}, \boldsymbol{\Omega})$ is computed by means of (113.8) [6].

Instead of this procedure—which replaces the original equation by a set of integral equations in the angular moments of $\Phi(\boldsymbol{r}, \boldsymbol{\Omega})$—it is also possible to reshuffle (113.8) so that the part of its right-hand side containing $\Phi(\boldsymbol{r}, \boldsymbol{\Omega}')\,\Sigma_s(\boldsymbol{\Omega}' \to \boldsymbol{\Omega})$ splits into a linear combination of integrals, the coefficients of which depend on the B_l of the expansion (115.2a) broken off at the term with $l = n$. The integrals, in the plane symmetry case, have the form

$$\int_{-\infty}^{+\infty} \Phi_0(z')\, E_{2l+1}\left(\frac{|z-z'|}{l_{\text{tot}}}\right) dz' \qquad (l \leq n).$$

It is easily seen that for isotropic scattering in the l.s. all the coefficients mentioned above—except the one with $l=0$—are zero, so that the equation for anisotropic scattering reduces to (113.11).

Thus, even for anisotropic scattering in the l.s. the transport equation can be replaced by an integral equation as long as the mean free path is the same everywhere and $\Sigma_s(\boldsymbol{\Omega}' \to \boldsymbol{\Omega})$ is represented by a polynomial of degree n in μ_0. The solution of such an equation, however, is sometimes a very difficult problem and is always at least rather cumbersome. Therefore, very often it becomes more convenient to deal with the original transport equation; this will be shown in Sect. 117 where the problem of a source-free homogeneous medium is treated by imposing upon the scattering in the l.s. less restrictive conditions than those adopted in Sect. 116.

Finally we want to mention that the procedures described in this section can be extended also to the case of anisotropic neutron sources [7][1].

114. The elementary diffusion theory. The picture presented in Sect. 112 for the diffusion of thermal neutrons in a large portion of a weakly absorbing moderator very closely resembles other diffusion phenomena well known from the investigation of the properties of gases or liquids. Neutron diffusion, however, is a simpler process than gas diffusion because the neutron density is so small in all physical systems considered in this article, that neutron-neutron collisions can always be neglected. Even at the centre of a high flux reactor the neutron flux density does not exceed 10^{17} thermal neutrons per cm^2 per sec; since the n-n collision cross section is expected to be smaller than the n-p cross section (29.2), the time that will elapse on the average, between two successive n-n collisions undergone by the same neutron will be longer than $5 \cdot 10^5$ sec. This has to be compared with a neutron half-life for decay (Table 3, p. 8) of 730 sec, and a neutron half-life for nuclear capture of 203 μsec in water (Table 96, p. 583), and 12.2 msec in graphite (Table 100, p. 585).

In this respect the diffusion of neutrons is more like the diffusion of electrons in a metal, of molecules in a foreign gas, or of ions in a diluted solution, rather than like gaseous diffusion. The fact that the neutron-neutron collisions can be

[1] C.C. Grosjean: Nuovo Cim. **11**, 11 (1954).

neglected implies that the fundamental equations describing the diffusion of neutrons are linear; and, therefore, these equations can be integrated in many more cases than can the non-linear equations encountered in the dynamical theory of gases [*19*].

In classical diffusion phenomena, such as the diffusion of molecules or of ions in a foreign gas or liquid, it is found that the diffusing "substance" tends to move from regions of high density to those of low density. The same is expected to happen in the case of neutrons since more collisions per unit volume take place in a region of high density, and after colliding the neutrons move away from the collision centres.

This remark has suggested, since early times[1], the attempt to describing the diffusion of thermal neutrons by means of the so-called *elementary diffusion theory*. This is based on the *Fick's law of diffusion* which may be stated as follows: let us consider the neutrons diffusing, in a stationary state and having a constant velocity, inside a large portion of a non-absorbing moderator. At a point $\boldsymbol{r}$ far from any neutron source, boundary, or localized absorber, *the net current* $\boldsymbol{J}$ —defined as the net number of neutrons flowing per unit time through a unit area normal to the direction of flow—*is proportional to the gradient of the neutron density:*

$$\boldsymbol{J}(\boldsymbol{r}) = -D_0 \nabla N(\boldsymbol{r}). \tag{114.1 a}$$

The constant D_0 is called the *diffusion coefficient*. Very often in this article we make use of the FICK's law in the form

$$\boldsymbol{J}(\boldsymbol{r}) = -D \cdot \nabla \Phi_0(\boldsymbol{r}) \tag{114.1 b}$$

where D is the diffusion coefficient for the flux:

$$D = \frac{D_0}{v}. \tag{114.1 c}$$

The relationship (114.1 b) can be derived by means of the following elementary consideration based on the assumption that in each collision the neutrons are scattered isotropically in the l.s.[2]. We have also to keep in mind that, since the neutrons are monoenergetic, the scattering cross section is a constant: $\Sigma_s = \text{const}$.

The number of neutrons flowing in the negative z-direction per unit time through an area dS placed at the origin in the xy plane can be computed by integrating over the upper half space the contribution arising from each element of volume dV', i.e.

$$J_{-z}\, dS = \frac{dS}{4\pi} \Sigma_s \int_0^\infty dr' \int_0^{2\pi} d\varphi' \int_0^{\pi/2} \Phi_0(r')\, e^{-\frac{r'}{l_s}} \cos\vartheta' \sin\vartheta'\, d\vartheta'. \tag{114.2}$$

It may be noticed that, apart from the fact that this equation refers to a system without sources and without capture, its right-hand side differs from that of Eq. (113.3) because the integral is extended only over one half space and contains a factor $\cos\vartheta'$ as is necessary for obtaining, on the left-hand side, the neutron current flowing in the negative z-direction, instead of the neutron flux.

Thus, under the assumptions specified above, Eq. (114.2) is rigorous; the presence of the exponential factor, which reduces heavily the contributions arising from points placed far from the origin, justifies the adoption of the follow-

[1] E. AMALDI and E. FERMI: Ric. Sci. 7 (1), 56, 454 (1936). — Phys. Rev. 50, 899 (1936).
[2] S. GLASSTONE and M.C. EDLUND: Nuclear reactor theory. New York: Van Nostrand 1952.

ing approximate procedure: the flux density $\Phi_0(\boldsymbol{r})$ is expanded in a Taylor series:

$$\Phi_0(\boldsymbol{r}) = \Phi_0(0) + x\left(\frac{\partial\Phi_0}{\partial x}\right)_{r=0} + y\left(\frac{\partial\Phi_0}{\partial y}\right)_{r=0} + z\left(\frac{\partial\Phi_0}{\partial z}\right)_{r=0} + \cdots \qquad (114.3)$$

and only the first-order terms are kept. Since the integration over φ is between the limits of zero and 2π, the terms containing x and y do not contribute. The integration with respect to ϑ' and r' of the other two terms is immediately carried out with the result

$$J_{-z} = \frac{1}{4}\,\Phi_0(0) + \frac{l_s}{6}\left(\frac{\partial\Phi_0}{\partial z}\right)_{r=0} + \cdots. \qquad (114.4\text{a})$$

Following the same procedure, the neutron density current in the positive z-direction is found to be

$$J_{+z} = \frac{1}{4}\,\Phi_0(0) - \frac{l_s}{6}\left(\frac{\partial\Phi_0}{\partial z}\right)_{r=0} + \cdots. \qquad (114.4\text{b})$$

Since the point chosen as the origin of the coordinates is arbitrary we can conclude that the net current density along the z-direction at the point $\boldsymbol{r}$ is given by

$$J_z(\boldsymbol{r}) = J_{+z}(\boldsymbol{r}) - J_{-z}(\boldsymbol{r}) = -\frac{l_s}{3}\,\frac{\partial\Phi_0}{\partial z} \qquad (114.4\text{c})$$

where the derivative with respect to z must be taken at the same point $\boldsymbol{r}$. Two other relations similar to (114.4c) may be found for the x- and y-components of $\boldsymbol{J}$, and, therefore, FICK's law in the form (114.1b) is proved and the diffusion coefficient D is given by the expression

$$D = \frac{l_s}{3} \qquad (114.5)$$

which refers to the case of a non-absorbing medium whose nuclei scatter the neutrons isotropically in the l.s.

In Sect. 117β we shall derive an expression for the diffusion coefficient in the more realistic case of media showing capture and anisotropic scattering in the l.s.

The following remarks may be added to the above derivation of the elementary FICK's law (114.1b).

First, one recognizes immediately that the net current

$$\boldsymbol{J} = \boldsymbol{J}_+ - \boldsymbol{J}_-$$

is correct up to second-order terms since these are identical in both $\boldsymbol{J}_+$ and $\boldsymbol{J}_-$ and therefore, cancel one another.

Secondly, the condition of the validity of (114.1) is that the flux density can be expressed with sufficient accuracy by three terms of the expansion (113.11) over a distance of the order of two or three mean free paths. This condition excludes the use of FICK's law in the vicinity of concentrated neutron sources, boundaries or strong absorbers which all produce fast variations of $\Phi_0(\boldsymbol{r})$ with $\boldsymbol{r}$. This limitation, however, concerns regions of linear dimensions of the order of two or three mean free paths because of the exponential factor appearing under the integral of Eq. (114.2).

The *elementary diffusion equation* can now be written down by expressing the continuity of neutrons: the time derivative of the probable number of neutrons contained in the volume element dV

$$\frac{N_0\,\partial(\boldsymbol{r})}{\partial t}\,dV$$

is equal to the sum of the following contributions: (a) the number of neutrons diffusing into dV:

$$-\nabla \boldsymbol{J} \cdot dV = \nabla (D \nabla \Phi_0)\, dV = D \nabla^2 \Phi_0\, dV;$$

(b) the number of neutrons captured per second in dV:

$$-\Sigma_a \Phi_0\, dV$$

and (c) the number

$$S(\boldsymbol{r})\, dV$$

of neutrons produced per second, through some convenient process, inside the volume element dV. In conclusion, we can write

$$D \nabla^2 \Phi_0 - \Sigma_a \Phi_0 + S = \frac{\partial N_0}{\partial t}. \tag{114.6}$$

This equation, which is similar to the Fourier heat equation, is widely used in problems of neutron diffusion and in particular in reactor theory[1,2]. One has to keep in mind that its use is limited by the conditions of validity of the elementary FICK's law; therefore it refers to monoenergetic neutrons, diffusing with constant scattering cross section in a weakly absorbing moderator, at distances greater than two or three mean free paths from strong localized sources and absorbers, or boundaries of the moderator.

In Sects. 115 and 124, we shall show under which conditions the diffusion equation (114.6) can be deduced from the transport equation, while in Sects. 120 and 121 we shall discuss its solution in a few classical cases.

In the stationary case, Eq. (114.6) is usually written in the slightly different form

$$\nabla^2 \Phi_0 - \frac{1}{L^2} \Phi_0 + \frac{S}{D} = 0 \tag{114.7}$$

where

$$L = \sqrt{\frac{D}{\Sigma_a}} \tag{114.8}$$

is a length called the *diffusion length*. According to (114.5), in a medium which scatters the neutrons isotropically, one has

$$L = \left(\frac{l_s l_a}{3}\right)^{\frac{1}{2}}. \tag{114.9}$$

More approximate expressions for the diffusion length will be derived later [see Eqs. (116.12) and (117.21)].

115. The elementary diffusion equation as the P_1-approximation to transport theory. In this section we shall show how the stationary diffusion equation may be derived from the transport equation of a one-velocity group (112.2) by applying to it the spherical harmonics method. For the sake of simplicity we shall refer to the one-dimensional case. Then, following the same procedure as in Sect. 79, Eq. (112.2) reduces to

$$\left.\begin{aligned} &\mu \frac{\partial \Phi(z,\mu)}{\partial z} + \Sigma_{\text{tot}} \Phi(z,\mu) \\ &\quad = \sum_{l=0}^{\infty} \frac{2l+1}{2} B_l P_l(\mu) \int \Phi(z,\mu') P_l(\mu')\, d\mu' + \frac{1}{2} S(z,\mu) \end{aligned}\right\} \tag{115.1}$$

[1] See footnote 2, p. 512.

[2] S. GLASSTONE: Principles of Nuclear reactor engineering. New York: Van Nostrand 1955.

where the B_l represent the coefficients of the expansion of $\Sigma_s(\boldsymbol{\Omega}'\to\boldsymbol{\Omega})$ in Legendre polynomials in terms of the cosine of the angle of scattering $\mu_0=\boldsymbol{\Omega}'\cdot\boldsymbol{\Omega}$

$$\Sigma_s(\boldsymbol{\Omega}'\to\boldsymbol{\Omega})=\sum_{l=0}^{\infty}\frac{2l+1}{4\pi}B_l\,P_l(\mu_0). \tag{115.2a}$$

It may be noticed that Eqs. (115.1) and (115.2a) can be derived from (79.3) and (72.20), respectively, by imposing that the coefficients $B_l(u', u-u')$ behave as δ-functions with respect to $u-u'$:

$$\int B_l(u', u-u')\,\Phi(z,\mu',u')\,du' = B_l(u,0)\,\Phi(z,\mu',u) = B_l\cdot\Phi(z,\mu')$$

where the lethargy u has been dropped since, in the present case, it is a constant. The latter set of relations—equivalent to (112.1)—shows that the neutrons diffuse without exchanging energy with the moderator.

In order to solve Eq. (115.1), we follow the same procedure as that of Sect. 79; we substitute for $\Phi(z,\mu,u)$ the expansion in Legendre polynomials (79.4), and for the source term the similar expansion

$$S(z,\mu,u)=\sum_{l=0}^{\infty}\frac{2l+1}{2}S_l(z,u)\,P_l(\mu). \tag{115.2b}$$

The equation

$$\left.\begin{aligned}&\sum_{l=0}^{\infty}(2l+1)\,P_l(\mu)\left[\mu\frac{\partial\Phi_l(z,u)}{\partial z}+\Sigma_{\text{tot}}(u)\,\Phi_l(z,u)\right]\\&\quad=\sum_{l=0}^{\infty}(2l+1)\,P_l(\mu)\left[B_l(u)\,\Phi_l(z,u)+S_l(z,u)\right]\end{aligned}\right\} \tag{115.3}$$

is then multiplied by $P_l(\mu)$ and integrated over μ. One then obtains the following infinite set of linear differential equations

$$\frac{l+1}{2l+1}\frac{d\Phi_{l+1}(z)}{dz}+\frac{l}{2l+1}\frac{d\Phi_{l-1}(z)}{dz}+\Sigma_{\text{tot}}\,\Phi_l(z)=B_l\,\Phi_l(z)+S_l(z). \tag{115.4}$$

The spherical harmonics P_L-approximation (Sect. 79) consists of introducing the "characteristic equation"

$$\Phi_{L+1}(z)=0 \tag{115.5}$$

into the equation with $l=L$.

A few general remarks about this procedure are given in Sect. 123. Here we limit ourselves to the simplest possible case: under the assumption that the source is isotropic,

$$S_l(z)=\delta_{l0}\,S_0(z), \tag{115.6}$$

the set of infinite differential equations (115.4) in the P_1-approximation reduces to the following two equations:

$$\frac{d\Phi_1(z)}{dz}+\Sigma_{\text{tot}}\,\Phi_0(z)=\Sigma_s\,\Phi_0(z)+S_0(z), \tag{115.7a}$$

$$\frac{1}{3}\frac{d\Phi_0(z)}{dz}+\Sigma_{\text{tot}}\,\Phi_1(z)=\Sigma_s\bar{\mu}_0\,\Phi_1(z), \tag{115.7b}$$

where use has been made of the obvious identities

$$B_0=\int_{-1}^{+1}\Sigma_s(\mu_0)\,d\mu_0=\Sigma_s;\qquad B_1=\int_{-1}^{+1}\Sigma_s(\mu_0)\,P_1(\mu_0)\,d\mu_0=\Sigma_s\bar{\mu}_0 \tag{115.8}$$

with $\bar{\mu}_0$ defined by (72.1).

Eq. (115.7b) may be written

$$\Phi_1(z) = -D\frac{d\Phi_0(z)}{dz} \tag{115.9a}$$

where

$$D = \frac{l_d}{3} = \frac{1}{3}\,\frac{1}{\Sigma_a + \Sigma_s(1-\bar{\mu}_0)}, \tag{115.9b}$$

$$\Phi_1(z) = \int_{-1}^{+1} \mu\,\Phi(z,\mu)\,d\mu = J(z) = J_+(z) - J_-(z) \tag{115.9c}$$

and

$$J_+(z) = \int_0^1 \mu\,\Phi(z,\mu)\,d\mu, \quad J_-(z) = \int_0^{-1} \mu\,\Phi(z,\mu)\,d\mu. \tag{115.10}$$

Thus, Eq. (115.9a) differs from the elementary FICK's law (114.1b) only in the diffusion coefficient which is now expressed in terms of the diffusion mean free path (72.6). It has to be noticed that Eq. (115.9b) reduces to (114.5) in the case of a non-absorbing medium ($\Sigma_a = 0$) whose nuclei scatter isotropically in the l.s. ($\bar{\mu}_0 = 0$).

By elimination of Φ_1 between (115.9a) and (115.7a), one obtains

$$D\frac{d^2\Phi_0(z)}{dz^2} - \Sigma_a\,\Phi_0(z) = S_0(z) \tag{115.11}$$

which is identical with the elementary diffusion equation (114.6) for the one-dimensional stationary case.

We may conclude that, whenever in a stationary problem the flux density $\Phi(z,\mu)$ is correctly represented by the first two terms of the expansion in Legendre polynomials and the neutron source is isotropic, the transport equation for one velocity group is equivalent to the diffusion equation (114.7) with the diffusion coefficient given by (115.9b). As we shall see in Sect. 124 the validity of the time dependent diffusion equation (114.6) requires the fulfilment of further conditions: its form is correct only in the limit of zero mean free path and zero absorption cross section, but infinite velocity, so that both the diffusion coefficient $D_0 = \frac{1}{3} l_a v$ and the neutron lifetime $T = (\Sigma_a v)^{-1}$ remain finite.

Before closing this section, we should recall the considerations given in Sect. 109 which lead to Eq. (109.16) for describing the diffusion of thermal neutrons in a heavy monatomic gaseous moderator. This equation differs from the one-velocity group diffusion equation (114.7) only because: (a) the diffusion coefficient D and the absorption cross section Σ_a are replaced by the corresponding averages over the neutron flux distribution (109.17). (b) The source term S_0 is given by the slowing down density at a certain energy E_m—of the order of a few kT—at which the neutrons are considered to enter into the thermal group.

Both these remarks are of considerable practical importance since they define which numerical values of the coefficients appearing in (114.7) should be used for comparing the results deduced from the elementary diffusion theory with experiments. According to (109.17), when transforming from the variable u to the variable v, one obtains

$$\bar{D} = \frac{\frac{1}{3}\int l_d(v)\,\varphi_{\text{th}}(v)\,dv}{\int \varphi_{\text{th}}(v)\,dv}, \tag{115.12a}$$

$$\bar{\bar{\Sigma}}_a = \frac{\int \bar{\Sigma}_a(v)\,\varphi_{\text{th}}(v)\,dv}{\int \varphi_{\text{th}}(v)\,dv}. \tag{115.12b}$$

In the more elementary treatments one takes for $\varphi_{th}(v)$ the Maxwellian flux distribution (101.1), (101.2)[1]. In more realistic treatments, the Wigner and Wilkins spectrum (Sect. 105)[2], or semiempirical expressions of the type of (108.14)[3], are used.

The source term in (109.16) is given by the slowing-down density at a certain energy E_m simply because that equation describes the diffusion of thermal neutrons in the same region where they have been slowed down. There are cases, however, in which the slowing down takes place in a region different from that where the diffusion of thermal neutrons is studied; a typical example is provided by a medium placed in contact with the external surface of the thermal column of a reactor.

For this reason, throughout Part E of this article a generic notation S is used for denoting the sources, in spite of the fact that sources of thermal neutrons do not exist in nature; they are always produced by the slowing down of neutrons of higher energy.

Finally, it should be noticed once more that the validity of Eq. (109.16) requires that condition (109.14) is fulfilled; this can be deduced from (112.4) by integration over the full solid angle. In Sect. 117 we shall come back to this point from a different point of view: thermal neutrons may be treated as a one-velocity group whenever their spectrum is independent of the space coordinates and of the direction of motion.

116. The rigorous solution of the integral equation in a homogeneous source-free medium with isotropic scattering in the laboratory system. In this section we are going to show that, in the case of a homogeneous source-free medium, the rigorous solution of the integral equation (113.3) is identical with the solution of the elementary diffusion equation [6]. We start from (113.3), without the source term, so that

$$\Phi_0(\boldsymbol{r}) = \frac{h}{4\pi l_{tot}} \int \Phi_0(\boldsymbol{r}')\, e^{-\frac{|\boldsymbol{r}-\boldsymbol{r}'|}{l_{tot}}} \frac{d\boldsymbol{r}'}{|\boldsymbol{r}-\boldsymbol{r}'|^2}, \tag{116.1}$$

and we ask if there exist particular solutions of the form

$$\Phi_0(\boldsymbol{r}) = e^{i\boldsymbol{K}\cdot\boldsymbol{r}}, \qquad \boldsymbol{K} = K\boldsymbol{w} \tag{116.2}$$

where K is a complex quantity of physical dimensions l^{-1}, and $\boldsymbol{w}$ is a unit vector: $\boldsymbol{w}\cdot\boldsymbol{w} = w^2 = 1$. By inserting (116.2) into (116.1), we obtain

$$e^{i\boldsymbol{K}\cdot\boldsymbol{r}} = h\,\frac{\arctan(K l_{tot})}{K l_{tot}}\, e^{i\boldsymbol{K}\cdot\boldsymbol{r}}. \tag{116.3}$$

Thus, the expression (116.2) is a solution of (116.1) for arbitrary $\boldsymbol{w}$ if K is a solution K_0 of the transcendental equation

$$1 = h\,\frac{\arctan(K_0 l_{tot})}{K_0 l_{tot}}. \tag{116.4}$$

A more general solution of (116.1) is given by

$$\Phi_0(\boldsymbol{r}) = \int f(\boldsymbol{w})\, e^{iK_0\boldsymbol{w}\cdot\boldsymbol{r}}\, d\boldsymbol{w} \tag{116.5}$$

where $f(\boldsymbol{w})$ is an arbitrary function of $\boldsymbol{w}$.

[1] S. Glasstone and M. C. Edlund: Nuclear reactor Theory. New York: Van Nostrand 1952.

[2] See, for example, the tables of thermal neutron cross sections computed by H. Amster and R. Suarez [WAPD + TM + 39 (1957)—WAPD — 185 (1958)] for various mixtures of H, U^{235}, Pu^{234} and $1/v$ absorbers.

[3] C. H. Westcott, W. H. Walker and T. K. Alexander: Internat. Conference on peaceful uses of Atomic Energy, Geneva 1958: A Conf. 15 P 202. — C. H. Westcott: A. E. C. L.-670 (1958).

It can be proved that (116.5) constitutes the more general solution of (116.1) if the restriction is added that $\Phi_0(\boldsymbol{r})\, e^{-\frac{r}{l_{\text{tot}}}}$ vanishes as r goes to infinity[1]. This condition corresponds to the boundary condition (112.3c) for an infinite medium.

By differentiating (116.5), we obtain

$$\nabla^2 \Phi_0(\boldsymbol{r}) = \int f(\boldsymbol{w})\, \nabla^2 e^{iK_0 \boldsymbol{w}\cdot\boldsymbol{r}} d\boldsymbol{w} = -K_0^2 \int f(\boldsymbol{w})\, e^{iK_0 \boldsymbol{w}\cdot\boldsymbol{r}} d\boldsymbol{w}$$

or

$$\nabla^2 \Phi_0(\boldsymbol{r}) + K_0^2 \Phi_0(\boldsymbol{r}) = 0. \tag{116.6}$$

Thus, if $\Phi_0(\boldsymbol{r})$ is a solution of the integral equation (116.1), it satisfies the *wave equation* (116.4), where K_0 is a solution of (116.4).

From a discussion of Eq. (116.3), one finds that

$$\left.\begin{array}{lll} \text{for } h > 1 & K_0 \text{ is real,} \\ \text{for } h = 1 & K_0 = 0, \\ \text{for } h < 1 & K_0 \text{ is imaginary.} \end{array}\right\} \tag{116.7}$$

In the present article we are mainly interested in the latter case. Thus we put

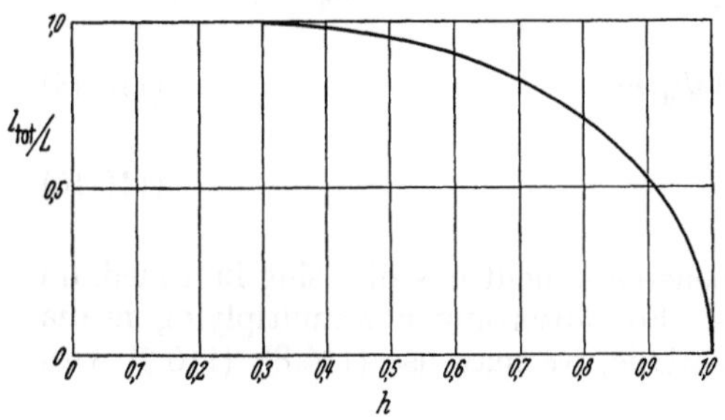

Fig. 227. l_{tot}/L versus h [6].

$$K_0 = i\varkappa_0. \tag{116.8}$$

If we call *diffusion length* the inverse of $\varkappa_0$:

$$L = \frac{1}{\varkappa_0}, \tag{116.9}$$

Eq. (116.6) is transformed into

$$\nabla^2 \Phi_0(\boldsymbol{r}) - \frac{1}{L^2} \Phi_0(\boldsymbol{r}) = 0 \tag{116.10}$$

which is identical with (114.7) for $S(\boldsymbol{r}) = 0$, apart from a difference in the expression for the diffusion length L. Fig. 227—taken from the book by CASE, DE HOFFMANN and PLACZEK [6],—shows $l_{\text{tot}}\,\varkappa_0 = \frac{l_{\text{tot}}}{L}$ versus h for $0 \leq h \leq 1$, in accordance with (116.4). For

$$1 - h = \frac{\Sigma_a}{\Sigma_s + \Sigma_a} \ll 1, \tag{116.11}$$

i.e. in weakly absorbing media, one obtains, from (116.4),

$$\left.\begin{aligned} L &= \frac{l_{\text{tot}}}{\sqrt{3(1-h)}} \left\{1 - \frac{2}{5}(1-h) + \frac{12}{175}(1-h)^2 - \cdots\right\} \\ &= \left(\frac{l_{\text{tot}}\, l_a}{3}\right)^{\frac{1}{2}} \left\{1 - \frac{2}{5}\frac{\Sigma_a}{\Sigma_s + \Sigma_a} + \frac{12}{175}\left(\frac{\Sigma_a}{\Sigma_s + \Sigma_a}\right)^2 - \cdots\right\} \end{aligned}\right\} \tag{116.12}$$

which should be compared to (114.9).

The procedure may be completed by computing $\Phi(\boldsymbol{r}, \boldsymbol{\Omega})$; this is done—according to the remark given on p. 509 of Sect. 113—by introducing the general solution (116.5) of the integral equation (116.1) into (113.6) with $s(\boldsymbol{r}) = 0$, and then by inserting the expression so obtained into (113.8):

$$\Phi(\boldsymbol{r}, \boldsymbol{\Omega}) = \frac{h}{4\pi l_{\text{tot}}} \int \frac{f(\boldsymbol{w})\, e^{iK_0 \boldsymbol{w}\cdot\boldsymbol{r}} d\boldsymbol{w}}{1 + iK_0\, l_{\text{tot}}\, \boldsymbol{w}\cdot\boldsymbol{\Omega}}. \tag{116.13}$$

[1] The reader may find a proof of this theorem for the one-dimensional case in Ref. [6].

The angular moments of $\Phi(\boldsymbol{r},\boldsymbol{\Omega})$ can be computed from this equation without specifying the arbitrary function $f(\boldsymbol{w})$. We limit ourselves to calculate the first angular moment which represents the net current density vector:

$$J_i = \int \Omega_i\, \Phi(\boldsymbol{r},\boldsymbol{\Omega})\, d\boldsymbol{\Omega} \qquad (i = x, y, z). \tag{116.14}$$

From (116.13) it follows that

$$J_i = \frac{h}{4\pi l_{\text{tot}}} \int f(\boldsymbol{w})\, e^{i K_0 \boldsymbol{w}\cdot\boldsymbol{r}}\, v_i\, d\boldsymbol{w}$$

where

$$v_i = \int \frac{\Omega_i\, d\boldsymbol{\Omega}}{1 + i K_0 l_{\text{tot}}\, \boldsymbol{w}\cdot\boldsymbol{\Omega}} \qquad (i = x, y, z)$$

are the components of a vector $\boldsymbol{v}$, the values of which are derived in the footnote[1]

$$v_i = w_i \frac{4\pi}{i K_0 l_{\text{tot}}} \frac{h-1}{h}.$$

Since

$$\int f(\boldsymbol{w})\, e^{i K_0 \boldsymbol{w}\cdot\boldsymbol{r}}\, w_i\, d\boldsymbol{w} = \frac{1}{i K_0} \frac{\partial}{\partial x_i} \int f(\boldsymbol{w})\, e^{i K_0 \boldsymbol{w}\cdot\boldsymbol{r}}\, d\boldsymbol{w} = \frac{1}{i K_0} \frac{\partial}{\partial x_i} \Phi_0(\boldsymbol{r})$$

we obtain

$$J(\boldsymbol{r}) = -D \operatorname{grad} \Phi_0(\boldsymbol{r}) \tag{116.15}$$

with

$$D = \frac{h-1}{l_{\text{tot}} K_0^2}. \tag{116.16}$$

Eq. (116.15) is FICK's law for monoenergetic neutrons diffusing in a medium which scatters isotropically in the l.s. For absorbing non-multiplying media ($h < 1$), as they are considered in this article, we may use (116.8), (116.9) with the result:

$$D = \frac{1-h}{l_{\text{tot}}} L^2 = L^2 \Sigma_a \tag{116.17}$$

which is identical with (114.8) apart from the value of L which, now, is given by the solution (116.8), (116.9) of (116.4).

Similar results also might be derived for the case of anisotropic scattering in the l.s. by applying the procedure mentioned on p. 510 of Sect. 113 to the corresponding integral equation. The simplicity of the derivation is, however, lost even in cases slightly more complicated than those considered here. Therefore, in the next section, we shall treat essentially the same problem—without imposing the condition of isotropic scattering in the l.s.—by means of a different procedure which starts directly from the transport equation.

---

[1] We start from the quantity

$$\boldsymbol{w}\cdot\boldsymbol{v} = w_x v_x + w_y v_y + w_z v_z = \int \frac{\boldsymbol{w}\cdot\boldsymbol{\Omega}\, d\boldsymbol{\Omega}}{1 + i K_0 l_{\text{tot}}\, \boldsymbol{w}\cdot\boldsymbol{\Omega}}. \tag{a}$$

Making use of (116.4), one finds

$$\boldsymbol{w}\cdot\boldsymbol{v} = \frac{4\pi}{i K_0 l_{\text{tot}}} \frac{h-1}{h}, \tag{b}$$

i.e. a value independent of the orientation of $\boldsymbol{w}$ as we should expect from symmetry reasons. This means that the vectors $\boldsymbol{w}$ and $\boldsymbol{v}$ should be proportional to one another. From this remark, combined with $w^2 = 1$, it follows that

$$v_i = w_i \frac{4\pi}{i K_0 l_{\text{tot}}} \frac{h-1}{h}. \tag{c}$$

Eq. (116.6) permits the derivation of the infinite medium solutions corresponding to particular symmetries. For spherical symmetry, (116.6) becomes:

$$\frac{1}{r}\frac{d^2}{dr^2}\left(r\,\Phi_0(r)\right)+K_0^2\,\Phi_0(r)=0 \tag{116.18}$$

with the general solution

$$\Phi_0(r)=A\frac{\sin K_0 r}{r}+B\frac{\cos K_0 r}{r}, \tag{116.19a}$$

where A and B are two arbitrary constants. By adding the condition that the flux must remain finite for any value of r, one obtains:

$$\Phi_0(r)=A\frac{\sin K_0 r}{r}. \tag{116.19b}$$

For $h<1$ we may use (116.8), (116.9) which give the result

$$\Phi_0(r)=A\frac{e^{-r/L}}{r}. \tag{116.19c}$$

In the plane symmetry case, Eq. (116.6) becomes

$$\frac{d^2\Phi_0(z)}{dz^2}+K_0^2\,\Phi(z)=0 \tag{116.20}$$

with the general solution

$$\Phi_0(z)=A\sin K_0 z+B\cos K_0 z \tag{116.21a}$$

where A and B are two arbitrary constants.

For $h<1$, Eq. (116.21a) becomes

$$\Phi_0(z)=C_1\,e^{-\varkappa_0 z}+C_2\,e^{+\varkappa_0 z} \tag{116.21b}$$

with C_1 and C_2 two new arbitrary constants.

117. The derivation of the FICK's law from the transport equation. In this section we shall try to determine under what conditions FICK's law is a rigorous consequence of the transport equation for a source-free medium. We do this without introducing the restriction that the scattering in the l.s. should be isotropic.

Let us first consider the case of monoenergetic neutrons diffusing without energy loss (one-velocity theory). Then let us examine the case of polyenergetic neutrons without energy loss, and, finally, the more general case of neutrons which diffuse by exchanging energy with the moderator.

In view of these extensions, even in the discussion of the one-velocity case, we shall explicitely indicate the dependence of the various quantities upon the neutron energy.

In the case of *monoenergetic neutrons*, we start from the one-dimensional transport equation (115.1) without the source term; we multiply it by μ and integrate with respect to μ from -1 to $+1$. Recalling (115.8), (115.9c) and (115.9b), the result can be written as

$$J(z,u)=-\,l_d\int_{-1}^{+1}\mu^2\frac{\partial\Phi(z,\mu,u)}{\partial z}\,d\mu. \tag{117.1}$$

Since μ and z are independent variables, it is permitted to interchange the order of differentiation and integration so that one can write

$$J(z, u) = -l_d \frac{\partial}{\partial z} [\overline{\mu^2}\, \Phi_0(z, u)] \tag{117.2}$$

where

$$\overline{\mu^2} = \frac{\int_{-1}^{+1} \mu^2 \Phi(z, \mu, u)\, d\mu}{\int_{-1}^{+1} \Phi(z, \mu, u)\, d\mu} = \frac{\int_{-1}^{+1} \mu^2 \Phi(z, \mu, u)\, d\mu}{\Phi_0(z, u)} \tag{117.3}$$

denotes the average of μ^2 with respect to the angle. Eq. (117.2) is called the *generalized Fick's law for monoenergetic neutrons*. By comparing it with (114.1b), we can conclude that the elementary FICK's law holds exactly for monoenergetic neutrons diffusing without energy loss, if $\overline{\mu^2}$ is independent of position; for, in that case, Eq. (117.2) can as well be written [*19*]

$$J(z, u) = -D \frac{\partial \Phi_0(z, u)}{\partial z} \tag{117.4}$$

with

$$D = \overline{\mu^2}\, l_d. \tag{117.5}$$

Thus, we are led to discussing in which cases $\overline{\mu^2}$ is independent of z. Such cases may be found by introducing the expansion (79.4) into (117.3); since

$$\mu^2 = \tfrac{1}{3} P_0(\mu) + \tfrac{2}{3} P_1(\mu)$$

one obtains

$$\overline{\mu^2} = \frac{1}{3}\left[1 + 2\frac{\Phi_2(z, u)}{\Phi_0(z, u)}\right] \tag{117.6}$$

from which it follows that $\overline{\mu^2}$ will be a constant whenever

$$\Phi_2(z, u) = 0 \tag{117.7a}$$

or

$$\frac{\Phi_2(z, u)}{\Phi_0(z, u)} = \text{const.} \tag{117.7b}$$

These conditions are *exactly* fulfilled in two important cases: condition (117.7a) is satisfied by the asymptotic distribution in a non-capturing medium, and condition (117.7b) is satisfied by the asymptotic distribution in a capturing medium [*19*]. Here by asymptotic distribution we mean the distribution found at a large distance with respect to l_{tr} from any source (of lethargy u), boundary or localized absorber (Sect. 118).

α) Asymptotic distribution of monoenergetic neutrons in non-capturing media. Condition (117.7a) is certainly fulfilled when the expansion (79.4) breaks off with the second term; i.e. when

$$\Phi(z, \mu, u) = \tfrac{1}{2} \Phi_0(z, u) + \tfrac{3}{2} \Phi_1(z, u)\, P_1(\mu). \tag{117.8}$$

This expression satisfies the transport equation (115.1) when: (a) the source term is dropped from the equation; (b) the angular number flux $\Phi(z, \mu, u)$, is linear in z and the number current $\Phi_1(z, u)$ is constant with respect to z; (c) the absorption cross section is zero so that $\Sigma_{tot}(u) = \Sigma_s(u)$. These statements may be verified by expanding the coefficient of (117.8) in powers of z and μ:

$$\Phi(z, \mu, u) = A + Bz + C\mu + Dz\mu + Ez^2 + \cdots$$

($A, B, C, D, E, \ldots$ are constant with respect to both z and μ) and then substituting this expression into (115.1) without the source term. By equating the coefficients of equal powers of z and μ on both sides, one finds

$$\cdots = E = D = 0; \quad C = -l_{tr} B \quad \text{and} \quad \Sigma_{tot}(u) = \Sigma_s(u).$$

Thus, the flux distribution in a non-capturing homogeneous medium, at great distances from any source or boundary, is given by the expression

$$\Phi(z, \mu, u) = \tfrac{1}{2} \Phi_0(z, u) - \tfrac{1}{2} l_{tr} \Phi_0'(z, u) P_1(\mu) \tag{117.9a}$$

where

$$\Phi_0(z, u) = 2A + 2Bz, \quad \Phi_0'(z, u) = \frac{\partial \Phi_0(z, u)}{\partial z} = \int_{-1}^{+1} \frac{\partial}{\partial z} \Phi(z, \mu, u)\, d\mu = 2B \tag{117.9b}$$

with A and B being arbitrary constants.

It should be noticed that the angular distribution (117.8) or (117.9a) corresponds to the P_1-approximation considered in Sect. 114; it is usually indicated as the angular distribution of elementary diffusion theory. As we shall see below, there is another angular distribution which holds rigorously for a capturing medium and which also leads to the elementary diffusion theory. This angular distribution takes on the same form as (117.8) in the limit of very weak capture.

According to (117.6), (117.7a), in this case the diffusion coefficient (117.5) is given by the expression

$$D = \tfrac{1}{3} l_{tr} \tag{117.10}$$

which is identical with (115.9b) if $\Sigma_a = 0$.

β) Asymptotic distribution of monoenergetic neutrons in capturing media. Condition (117.7b) is fulfilled when $\Phi(z, \mu, u)$ can be expressed as the product of a function of z and a function of μ, since then all harmonics $\Phi_l(z, u)$ with $l \geqq 1$ are in a constant ratio to $\Phi_0(z, u)$ for all z.

In order to compute $\overline{\mu^2}$, we start again from the Boltzmann equation (115.1), without the source term, written for the case of *almost isotropic scattering*, i.e. under the assumption that the expansion (115.2a) of the scattering cross section in the l.s. breaks off with the second term ($l = 1$); thus we have

$$\left.\begin{aligned} &\mu \frac{\partial \Phi(z, \mu, u)}{\partial z} + \Sigma_{tot}(u)\, \Phi(z, \mu, u) \\ &\quad = \frac{\Sigma_s(u)}{2} \int_{-1}^{+1} \Phi(z, \mu', u)\, d\mu' + \frac{3}{2} \Sigma_s \bar{\mu}_0 \mu \int_{-1}^{+1} \mu' \Phi(z \mu', u)\, d\mu'. \end{aligned}\right\} \tag{117.11}$$

We try to determine a constant $\varkappa$ and a function $g(\mu)$ such that the expression

$$\Phi(z, \mu, u) = e^{\pm \varkappa z} g(\mu) \tag{117.12}$$

is a solution of (117.11). By inserting (117.12) into (117.11), we obtain

$$[\Sigma_{tot}(u) \pm \varkappa \mu]\, g(\mu) = \tfrac{1}{2} \Sigma_s(u) [g_0 + 3 g_1 \bar{\mu}_0 \mu] \tag{117.13}$$

where

$$g_l = \int_{-1}^{+1} g(\mu') P_l(\mu')\, d\mu'. \tag{117.14}$$

Eq. (117.13) permits the determination of $g(\mu)$; if one takes

$$g(\mu) = \text{const} \frac{A + \bar{\mu}_0 \mu}{\Sigma_{tot}(\mu) \pm \varkappa \mu}, \tag{117.15}$$

the integrals (117.14) with $l=0$ and $l=1$ may readily be calculated and give the result

$$A+\bar{\mu}_0\mu=\frac{\Sigma_s}{2}\frac{A}{\varkappa}\ln\frac{\Sigma_{\rm tot}+\varkappa}{\Sigma_{\rm tot}-\varkappa}\pm\frac{\Sigma_s\bar{\mu}_0}{2}\left(\frac{2}{\varkappa}-\frac{\Sigma_{\rm tot}}{\varkappa^2}\ln\frac{\Sigma_{\rm tot}+\varkappa}{\Sigma_{\rm tot}-\varkappa}\right)\pm$$
$$\pm\frac{3}{2}\Sigma_s\bar{\mu}_0\mu A\left(\frac{2A}{\varkappa}-\frac{\Sigma_{\rm tot}}{\varkappa^2}\ln\frac{\Sigma_{\rm tot}+\varkappa}{\Sigma_{\rm tot}-\varkappa}\right)+\frac{3}{2}\Sigma_s\bar{\mu}_0^2\mu\left(-\frac{2\Sigma_{\rm tot}}{\varkappa^2}+\frac{\Sigma_{\rm tot}^2}{\varkappa^3}\ln\frac{\Sigma_{\rm tot}+\varkappa}{\Sigma_{\rm tot}-\varkappa}\right).$$

By equating the coefficients of equal powers of μ, one obtains the two following transcendental equations which allow the determination of A and $\varkappa$:

$$A=\pm\frac{\frac{\Sigma_s\bar{\mu}_0}{2}\left(\frac{2}{\varkappa}-\frac{\Sigma_{\rm tot}}{\varkappa^2}\ln\frac{\Sigma_{\rm tot}+\varkappa}{\Sigma_{\rm tot}-\varkappa}\right)}{1-\frac{\Sigma_s}{2\varkappa}\ln\frac{\Sigma_{\rm tot}+\varkappa}{\Sigma_{\rm tot}-\varkappa}}, \tag{117.16a}$$

$$\frac{\Sigma_s}{2\varkappa}\ln\frac{\Sigma_{\rm tot}+\varkappa}{\Sigma_{\rm tot}-\varkappa}=\frac{1+\frac{2\Sigma_s\Sigma_a\bar{\mu}_0}{\varkappa^2}}{1+3\Sigma_{\rm tot}\Sigma_a\bar{\mu}_0/\varkappa^2}. \tag{117.16b}$$

Eq. (117.16b) does not contain A. Therefore, one can first determine those among its roots which do not contradict physics, and then compute—by means of (117.16a)—the corresponding values of A.

Thus, we are led to discussing the roots of (117.16b). First we notice that —in agreement with (117.12)—we are always concerned with pairs of roots $\pm\varkappa_0$. Secondly, only those roots are permitted from the physical viewpoint which fulfill the inequalities

$$-\frac{1}{l_{\rm tot}}=-\Sigma_{\rm tot}<\mathrm{Re}\,(\varkappa_0)<\Sigma_{\rm tot}=\frac{1}{l_{\rm tot}}. \tag{117.17}$$

In fact, if roots of (117.16b) which do not belong to the interval defined by (117.17) would be used, solutions of the form (117.12) which vary *faster* than $e^{\pm z/l_{\rm tot}}$ as z goes to infinity would be obtained. As we have pointed out on various occasions (for example in Sects. 77 and 80), this is physically unacceptable.

The discussion becomes very simple for the case of *isotropic scattering in the l.s.*—already treated by a different procedure in Sect. 116. Then $\bar{\mu}_0=0$ and Eq. (117.16b) reduces to:

$$\frac{\Sigma_s}{2\varkappa_0}\ln\frac{\Sigma_{\rm tot}+\varkappa_0}{\Sigma_{\rm tot}-\varkappa_0}=1 \tag{117.18}$$

which is equivalent to (116.4)[1].

Eq. (117.18) has a single pair of real roots[2] plus an infinite set of imaginary ones. We are only interested in the real roots; these satisfy the condition (117.17) which ensures that no singularity of $g(\mu)$ exists for μ varying between -1 and $+1$.

The constant A in (117.15) need not be computed in the present case, since it can be absorbed in the normalization constant.

Thus we have

$$\Phi(z,\mu,u)=\text{const}\cdot\frac{e^{\pm\varkappa_0 z}}{\Sigma_{\rm tot}\pm\varkappa_0 l_{\rm tot}\mu} \tag{117.19}$$

---

[1] $\arctan x=-\frac{i}{2}\ln\frac{1+ix}{1-ix}$.

[2] It may be noticed that the double sign of $\varkappa_0$ is irrelevant in the considerations of Sect. 116, since there $\varkappa_0$ always appears multiplied by the unit vector $\boldsymbol{w}$ of arbitrary direction.

where $\varkappa_0$ are the real roots of (117.18). The two solutions (117.19) of the one-dimensional source-free transport equation (115.1) for isotropic scattering in the l.s. correspond to the solution

$$\frac{e^{i\varkappa_0 \boldsymbol{w}\cdot\boldsymbol{r}}}{1+i\varkappa_0 l_{\text{tot}}\,\boldsymbol{u}\cdot\boldsymbol{\Omega}}$$

which appears under the integral sign of (116.13). This is immediately seen by making use of (116.8) and noticing that—in the plane geometry considered here—only two directions of propagation need be considered: $\boldsymbol{w}\cdot\boldsymbol{r}=\pm z$.

The value of $\overline{\mu^2}$ corresponding to (117.19) is obtained by introducing this equation into (117.6); the result is

$$\overline{\mu^2}=\frac{\Sigma_{\text{tot}}\,\Sigma_a}{\varkappa_0^2}=\Sigma_{\text{tot}}\,\Sigma_a\,L^2 \qquad (\text{for } \bar{\mu}_0=0) \tag{117.20a}$$

where use has been made of the definition (116.9) of the diffusion length. According to (117.5)

$$D=\frac{\overline{\mu^2}}{\Sigma_{\text{tot}}}=\frac{\Sigma_a}{\varkappa_0^2}=\Sigma_a\cdot L^2 \tag{117.20b}$$

which is identical with (114.8), apart from the different definition of L.

We now return to the case of *almost isotropic scattering in the l.s.*, i.e. to the case in which the expansion of the scattering cross section (115.2a) reduces to a linear expression in $\cos\vartheta$. Since in the present article only non-multiplying media are considered, we restrict our considerations to the case $h\leqq 1$. Then, it can be shown that Eq. (117.16b) can have only one pair of roots and this is true even if $\Sigma_s(\boldsymbol{\Omega}'\to\boldsymbol{\Omega})$ is a quadratic form in $\cos\vartheta$. This, however, is not true in general. Examples, with $h\gg 1$, have been constructed in which two pairs of such roots exist [7]. In the present article only non-multiplying media are considered. We may, therefore, restrict our discussion to the case of a single pair of roots of Eq. (117.16b).

If the capture cross section is small compared to the scattering cross section, the transcendental equation (117.16b) can be solved by a series expansion; the result, correct to terms of first degree in $\frac{\Sigma_a(u)}{\Sigma_s(u)}$, is

$$\frac{1}{L^2}=\varkappa_0^2=3\,\Sigma_{\text{tot}}\,(1-\bar{\mu}_0)\,\Sigma_a\left[1-\frac{4}{5}\,\frac{\Sigma_a}{\Sigma_{\text{tot}}}+\frac{\Sigma_a\bar{\mu}_0}{\Sigma_{\text{tot}}(1-\bar{\mu}_0)}-\cdots\right]. \tag{117.21a}$$

For isotropic scattering Eq. (117.21a) becomes

$$\frac{1}{L^2}=\varkappa_0^2=3\,\Sigma_{\text{tot}}\,\Sigma_a\left[1-\frac{4}{5}\,\frac{\Sigma_a}{\Sigma_{\text{tot}}}\right] \qquad (\text{for } \bar{\mu}_0=0) \tag{117.21b}$$

which is identical to (116.12).

From this expression, we see that for $\Sigma_a\ll\Sigma_s$ one has $\varkappa_0\ll\Sigma_{\text{tot}}$ so that it becomes possible to expand the denominator of (117.15) in powers of $\frac{\varkappa_0\mu}{\Sigma_{\text{tot}}}$ and keep only the first two terms. Thus we have

$$\Phi(z,\mu,u)=\text{const}\;e^{\pm\varkappa_0 z}\left[1\pm\frac{\varkappa_0}{\Sigma_{\text{tot}}}\,\mu\right] \tag{117.22a}$$

which shows the same angular distribution as the one given in absence of absorption by (117.9a) the only difference being that now

$$\Phi_0(z, u) = \text{const}\, e^{\pm \varkappa_0 z}, \quad l_{tr}\, \Phi(z, u) = \pm\, \text{const}\, e^{\pm \varkappa_0 z} \cdot \frac{\varkappa_0}{\Sigma_{tot}}. \tag{117.22b}$$

The procedure given above for almost isotropic scattering can be extended to the case in which the expansion (115.2a) breaks off with a term of comparatively low order $l = n > 1$. In this case the transport equation is similar to (117.11) with the only difference being that now $n+1$ terms appear on its right-hand side. Again, we look for a solution of the form (117.12); if this expression is inserted into the transport equation, one obtains

$$[\Sigma_{tot}(u) \pm \varkappa\mu]\, g(\mu) = \frac{h}{2} \Sigma_{tot} \sum_{l=0}^{n} (2l+1)\, B_l\, g_l\, P_l(\mu) \tag{117.23}$$

which is an obvious generalization of Eq. (117.13). By multiplying both sides with $P_r(\mu)$ $(r \leqq n)$ and integrating over μ from -1 to $+1$, a set of $n+1$ homogeneous linear equations in the constants g_l, defined by (117.14), is obtained. The determinant of coefficients of these equations must vanish; its roots provide all possible values of $\varkappa$. For each root $\varkappa_\iota$, the corresponding set of values of the constants g_l can be computed and from these the function $g(\mu)$ can be deduced by means of (117.23) with $\varkappa = \varkappa_\iota$.

Thus one is led to consider the roots of the determinantal equation. A detailed discussion may be found in Ref. [7]; here we restrict our consideration to the following few remarks.

Firstly—in agreement with (117.12)—the roots always occur in pairs $\pm\varkappa_\iota$. Secondly, the roots permitted by physics are only those satisfying the inequalities (117.17).

Thirdly, for very weak absorbers, i.e. $1-h \ll 1$, a pair of real roots $\pm\varkappa_0 = \pm\frac{1}{L}$ always exist which tend to zero $(L \to \infty)$ as $h \to 1$; irrespective of the values of the B_l, one always has [7]

$$L = \left(\frac{l_{tr}\, l_a}{3}\right)^{\frac{1}{2}} [1 + O(1-h)] \tag{117.24}$$

where the first term depends only on B_0 and B_1—i.e., according to (115.8), only on Σ_s and $\bar{\mu}_0$—, while $B_2, B_3, \ldots, B_n$ enter only the correction term. For strongly absorbing media, no simple general rule can be given for computing the roots $\pm\frac{1}{L}$.

Finally, the determinental equation may have more than one single pair of real roots satisfying the inequalities (117.17). It has been shown that, for any h, if n is sufficiently large, the constants $B_1, B_2, \ldots, B_n$ can be chosen so that the determinantal equation has any prescribed number of pairs of roots permitted by physics[1] [7]. These roots, however, are always larger in absolute value than $|\varkappa_0|$. This statement may be understood by means of the following argument. Let us start from a situation in which $B_1, B_2, \ldots, B_n$ are so small that a single pair of roots $\pm\varkappa_0 = \pm\frac{1}{L}$ exist, and let them gradually increase towards the set of values for which there is a second pair of roots $\pm\varkappa_1 = \pm\frac{1}{L_1}$. As the B_l increase, a point is reached where the new pair of roots enter the interval (117.17) at its ends—i.e., at $\pm\frac{1}{L_1} = \pm\frac{1}{l_{tot}}$—and begin to move, along the real axis, towards the origin remaining always outside the pair of roots $\pm\frac{1}{L}$.

[1] B. DAVISON: Nat. Res. Coun. Canada LT 18 (1947).

Thus, if other pairs of physically permitted roots exist, they always fulfill the inequalities

$$\frac{1}{L^2} < \frac{1}{L_1^2} < \frac{1}{L_2^2}. \tag{117.25}$$

However, these situations in which two or more pairs of roots of the determinantal equation exist have not yet been met with in actual applications, so that the discussion given above has more an heuristic value rather than a practical one.

If the determinantal equation has only one pair of roots, the more general solution of a source-free homogeneous infinite medium problem is given by a superposition of waves of the form (117.12) all corresponding to the same value of the diffusion length L. The neutron flux is obtained, according to (113.10), by integrating $\Phi(z,\mu)$ over μ; therefore, it will show the same dependence on space coordinates, with the result that it will be a solution of the wave equation (116.6). This is no longer true if two or more pairs of roots are physically admissible.

γ) Diffusion of polyenergetic neutrons without energy loss. In the preceding sub-section we have shown under what conditions FICK's law may be derived from the transport equation in the case of monoenergetic neutrons diffusing without energy loss. We shall now drop the assumption that the neutrons are monoenergetic and proceed to consider the case in which the neutron spectrum extends over a finite energy interval. If the assumption is kept that scattering takes place without energy loss, the diffusion of the neutrons is just a superposition of the diffusion of many groups of neutrons, each with a fixed energy. Then the expression for the total net current is obtained by integrating Eq. (117.2) with respect to u. The result, called the *generalized Fick's law for polyenergetic neutrons without energy loss,*

$$J(z) = \int J(z,u)\,du = -\int du\, l_d(u)\frac{\partial}{\partial z}\left[\overline{\mu^2}\,\Phi_0(z,u)\right] \tag{117.26}$$

can be written as follows

$$J(z) = -\frac{\partial}{\partial z}\left[\left(l_d(u)\,\overline{\mu^2}(u)\right)_{\text{av}}\,\Phi_0(z)\right] \tag{117.27a}$$

where

$$\left(l_d(u)\,\overline{\mu^2}(u)\right)_{\text{av}} = \frac{\int du\, l_d(u)\,\overline{\mu^2}(u)\,\Phi_0(z,u)}{\int du\,\Phi_0(z,u)} = \frac{\int du\, l_d(u)\,\overline{\mu^2}(u)\,\Phi_0(z,u)}{\Phi_0(z)} \tag{117.27b}$$

denotes the average with respect to the neutron spectrum [actually with respect to the number flux $\Phi_0(z,u)$]. In deriving (117.27a) the following two remarks have been used: firstly $l_a(u)$ does not depend on z, and secondly, since z and u are independent variables, it is permitted to exchange the order of the differentiation with respect to z and of the integration with respect to u.

The relationship (117.27a) reduces to the FICK's law with

$$D = \left(l_d(u)\,\overline{\mu^2}\right)_{\text{av}} \tag{117.28}$$

whenever (117.27b) is independent of z.

We may conclude that, even if one of the two conditions (117.7a, b) is fulfilled, so that $\overline{\mu^2}$ does not depend on z, the expression (117.27b) may depend on z. This will not be the case—and the FICK's law will be rigorously valid—*only if either the neutron spectrum does not depend on z, or $\overline{\mu^2}\, l_d(u)$ does not depend on u.*

δ) Diffusion of polyenergetic neutrons with energy loss. This is the more general case; therefore we start from Eq. (79.3) without the source term, we multiply it by μ and integrate it with respect to both μ and $\boldsymbol{u}$. According to (115.8) the result is

$$\frac{d}{dz}\left[(\overline{\mu^2})_{\text{av}}\,\Phi_0(z)\right] = -\int\left[\Sigma_{\text{tot}}(u) - \Sigma_s(u)\,\bar{\mu}_0(u)\right] J(z,u)\,du \tag{117.29a}$$

where

$$(\overline{\mu^2})_{\text{av}} = \frac{\int du \int \mu^2\,\Phi(z,\mu,u)\,d\mu}{\int du \int d\mu\,\Phi(z,\mu,u)} = \frac{\int du \int \mu^2\,\Phi(z,\mu,u)\,d\mu}{\Phi_0(z)} \tag{117.29b}$$

will in general be a function of z. One has to notice, however, that, *if one of the conditions (117.7a, b) is fulfilled, as it occurs in the two cases considered above, $\overline{\mu^2}$ is constant even with respect to u.* Then $(\overline{\mu^2})_{\text{av}}$ turns out to be independent of z and Eq. (117.29a) may be written

$$(\overline{\mu^2})_{\text{av}}\frac{d\Phi_0(z)}{dz} = -\int\left[\Sigma_{\text{tot}}(u) - \Sigma_s(u)\,\bar{\mu}_0(u)\right] J(z,u)\,du\,. \tag{117.30}$$

The right-hand side may be expressed as the product of

$$J(z) = \int J(z,u)\,du\,,$$

times a factor independent of z in the two following cases: (a) $J(z,u)$ can be separated into two factors, one depending only on z, the other depending only on u; (b) the three quantities $\Sigma_{\text{tot}}(u)$, $\Sigma_s(u)$ and $\bar{\mu}_0(u)$ are constants with respect to u.

In these two cases, Eq. (117.30) reduces to the elementary FICK's law (114.1b). Therefore, we can state that the elementary FICK's law holds exactly, even in the general case, provided the value of $\overline{\mu^2}$ is constant with respect to $\boldsymbol{r}$, for neutrons of each energy, and, furthermore, either condition (a) or condition (b) is fulfilled.

The actual example which better approaches case (a) is that of thermal neutrons diffusing in a very weakly absorbing moderator so that they are in thermal equilibrium—or almost thermal equilibrium—with the moderator at each point $\boldsymbol{r}$; then their spectrum is with rather good accuracy Maxwellian at each point and, therefore, it is independent of $\boldsymbol{r}$ (Sect. 112).

An example of case (b) is provided by the diffusion of neutrons in a non-capturing, non-crystalline, non-hydrogeneous medium, as for instance graphite, for neutron energies larger than 0.1 ev and smaller than 20 kev. In this energy interval, the scattering cross section as well as $\bar{\mu}_0$ are practically constant (Sect. 29) and the capture cross section, which follows the $1/v$ law, is sufficiently small to be neglected.

118. A few results of the one-velocity transport theory. In this section we shall summarize a few results of the one-velocity transport theory without giving the corresponding proofs, which would bring us too far away from the general scheme of the present article. The great majority of the corresponding analytical derivations may be found in Refs. [6] and [7] which this section is mainly based upon. The results reported here will be illustrated as much as possible by means of graphs taken from Ref. [6] where the reader may find extensive numerical tables.

Since the formulas can be simplified by measuring all lengths in total mean free paths l_{tot}, we shall frequently use, instead of r or z, the dimensionless variables

$$\varrho = \frac{r}{l_{\text{tot}}} \tag{118.1a}$$

and

$$\zeta = \frac{z}{l_{\text{tot}}} \tag{118.1b}$$

which measure the so-called *optical distance*.

α) *Isotropic sources.* We may start from the solution $\Phi_0^{\mathrm{pt}}(r)$ of the integral equatio

$$\Phi_0^{\mathrm{pt}}(\boldsymbol{r}) = \int \left[\frac{h}{l_{\mathrm{tot}}}\,\Phi_0^{\mathrm{pt}}(\boldsymbol{r}') + \delta(\boldsymbol{r}')\right] \frac{e^{-\frac{|\boldsymbol{r}-\boldsymbol{r}'|}{l_{\mathrm{tot}}}}}{4\pi|\boldsymbol{r}-\boldsymbol{r}'|^2}\,d\boldsymbol{r}' \tag{118.2}$$

corresponding to an isotropic point source of unit strength

$$s(\boldsymbol{r}) = \delta(\boldsymbol{r}).$$

This particular choice finds its justification in Eq. (71.1) which states that the neutron flux $\Phi_0(\boldsymbol{r})$ due to an arbitrary distribution of isotropic sources $s(\boldsymbol{r})$ is given by

$$\Phi_0(\boldsymbol{r}) = \int \Phi_0^{\mathrm{pt}}(\boldsymbol{r}-\boldsymbol{r}')\,s(\boldsymbol{r}')\,d\boldsymbol{r}'. \tag{118.3}$$

Thus we are led to look for the general solution of (118.2) in an infinite medium, with the boundary conditions (112.3c), for the case $h \leqq 1$.

The case of multiplying media ($h > 1$) is outside the scope of the present article. It should be noticed, however, that in an infinite medium with $h > 1$, there is no steady-state solution when sources are present, because the neutron flux is everywhere continuously increasing. This case, however, can be treated by procedures similar to those followed for $h < 1$ and its solution is of considerable interest in connection with both time dependent problems and problems involving finite media.

The general solution of (118.2) is found by adding to the general solution (116.5) of the homogeneous equation (116.1) a particular solution of (118.2). This was first derived by BOTHE[1] by means of Fourier transform methods and was further investigated by PLACZEK, WIGNER, DAVISON, PEIERLS and VOLKOFF[2].

The result is that $\Phi_0^{\mathrm{pt}}(\boldsymbol{r})$ can be decomposed into two parts

$$\Phi_0^{\mathrm{pt}}(\boldsymbol{r}) = p^{\mathrm{pt}}(\boldsymbol{r}) + \Phi_{0\,\mathrm{asy}}(\boldsymbol{r}) \tag{118.4a}$$

where

$$\Phi_{0\,\mathrm{asy}}(\boldsymbol{r}) = C\,l_{\mathrm{tot}}^{-1}\,\frac{e^{-\varkappa_0 r}}{4\pi r} \tag{118.4b}$$

$$C = \frac{2\,l_{\mathrm{tot}}^2\,(L^2 - l_{\mathrm{tot}}^2)}{h\,L^2\,(h\,L^2 + l_{\mathrm{tot}}^2 - L^2)} \tag{118.4c}$$

represents the so-called *asymptotic solution*, and

$$p^{\mathrm{pt}}(\boldsymbol{r}) = \frac{e^{-r/l_{\mathrm{tot}}}}{4\pi r^2}\,\varepsilon\!\left(\frac{r}{l_{\mathrm{tot}}}\right) \tag{118.5}$$

is a function which predominates over $\Phi_{0\,\mathrm{asy}}(\boldsymbol{r})$ at short distances from the source but becomes much smaller than $\Phi_{0\,\mathrm{asy}}(\boldsymbol{r})$ for r sufficiently large. The behaviour of the function $\varepsilon(\varrho)$ depends on the value of h as shown in Fig. 228 which also includes a few curves corresponding to $h > 1$. Tables of numerical values of $\varepsilon(\varrho)$ are given in Ref. [*6*].

The asymptotic solution (118.4b) satisfies the wave equation (116.18). Therefore, we can state that the solution of the elementary diffusion theory in an infinite medium containing an isotropic point source, will provide a correct description of the actual flux distribution only at so large a distance from the source that $p^{\mathrm{pt}}(\boldsymbol{r})$ is negligibly small.

---

[1] W. BOTHE: Z. Physik **118**, 401 (1941); **119**, 493 (1942).

[2] G. PLACZEK: Phys. Rev. **60**, 166A (1941). — E.P. WIGNER: U.S.A. Rep. A-20 (1941). — B. DAVISON and R. PEIERLS: British Rep. MS-76 (1943). — G. PLACZEK and G.M. VOLKOFF: Can. rep. MT-4-NRC-1548 (1943). — E.P. WIGNER: U.S.A. Rep. CP-1120 (1943).

This remark brings us to a detailed comparison of the two terms appearing in (118.4a). The ratio

$$\frac{\int \Phi_{0\,\mathrm{asy}}(\boldsymbol{r})\,d\boldsymbol{r}}{\int \Phi_0^{\mathrm{pt}}(\boldsymbol{r})\,d\boldsymbol{r}} = K \qquad (118.6\mathrm{a})$$

provides a measure of the fraction of the total number of neutrons present in the asymptotic distribution. It may be shown that

$$\left.\begin{aligned} K &= D\,C\,l_{\mathrm{tot}}^{-1} \\ &= C(1-h)\frac{L^2}{l_{\mathrm{tot}}^2} \end{aligned}\right\} \quad (118.6\mathrm{b})$$

where D is the diffusion coefficient (116.17) and C is the constant (118.4c). Fig. 229 shows a plot of K versus h; for

$$1-h = \frac{\Sigma_a}{\Sigma_s+\Sigma_a} \ll 1$$

one has

$$\left.\begin{aligned} K &= 1 - \tfrac{4}{5}(1-h) - \\ &- \tfrac{104}{175}(-h)^2 - 1 \\ &- \tfrac{68}{175}(1-h)^3 - \cdots . \end{aligned}\right\} \quad (118.7)$$

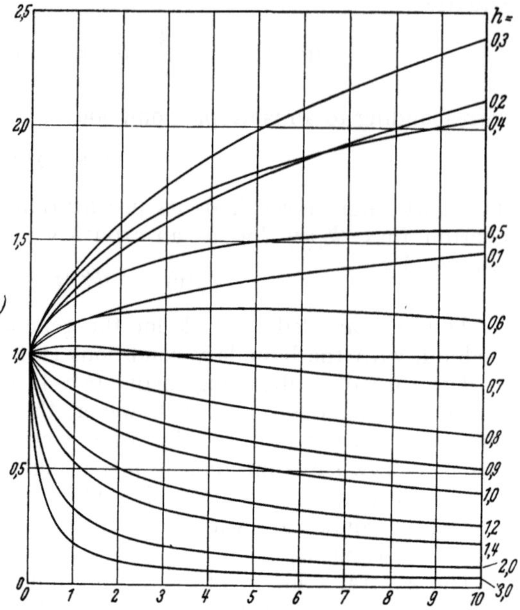

Fig. 228. $\varepsilon(\varrho)$ versus $\varrho = r/l_{\mathrm{tot}}$: see Eqs. (118.4), (118.5) [6].

In the case of pure scattering ($h=1$) one has $K=1$ so that, at first sight, one would conclude that all neutrons are in the asymptotic distribution. In this case, however, one has

$$\Phi_{0\,\mathrm{asy}}(r) = \frac{1}{2\pi\, l_{\mathrm{tot}}\, r}$$

which corresponds to an infinite number of neutrons present in the medium; the result $K=1$ is due to the fact that the number of neutrons present in the non-asymptotic distribution is finite, while the number of neutrons present in the asymptotic distribution is infinite. This is what we should expect, since for $h=1$, the continuous emission of neutrons by the source is not compensated by any absorption.

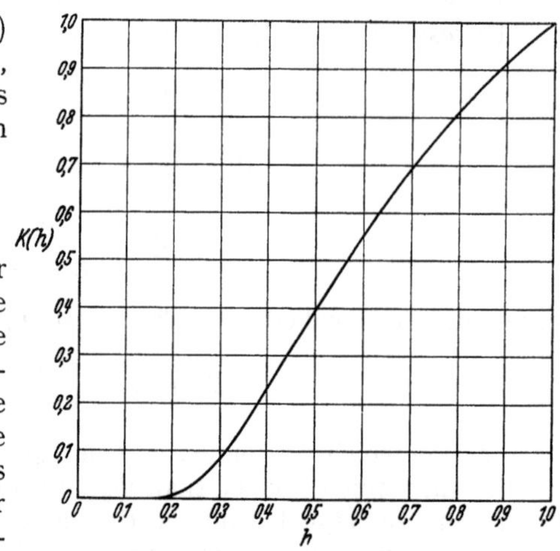

Fig. 229. $K(h)$ versus h see Eq. (118.6) [6].

The mean square distances from the source corresponding to $\Phi_0^{\mathrm{pt}}(\varrho)$, $\Phi_{0\,\mathrm{asy}}(\varrho)$ and $p^{\mathrm{pt}}(\varrho)$ also may serve for an estimate of the relative importance of the various terms in (118.4):

$$\left.\begin{aligned} \overline{\varrho^2} &= \frac{2}{1-h}, \\ \overline{\varrho^2_{\mathrm{asy}}} &= \frac{6}{\varkappa_0^2\, l_{\mathrm{tot}}^2} = 6\frac{L^2}{l_{\mathrm{tot}}^2} \\ \overline{\varrho^2_{\text{non-asy}}} &= \frac{2}{1-h}\;\frac{1-3DK\,l_{\mathrm{tot}}^{-1}}{1-K}. \end{aligned}\right\} \qquad (118.8)$$

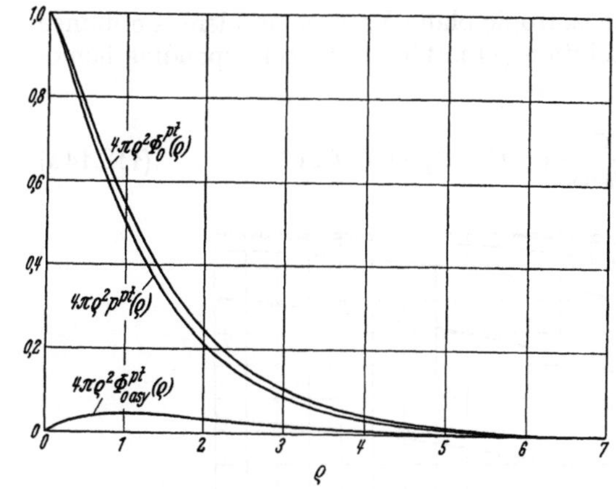

Fig. 230. The flux distribution around a point source embedded in an infinite medium for $h = 0.3$; $\varrho = r/l_{\text{tot}}$ [6].

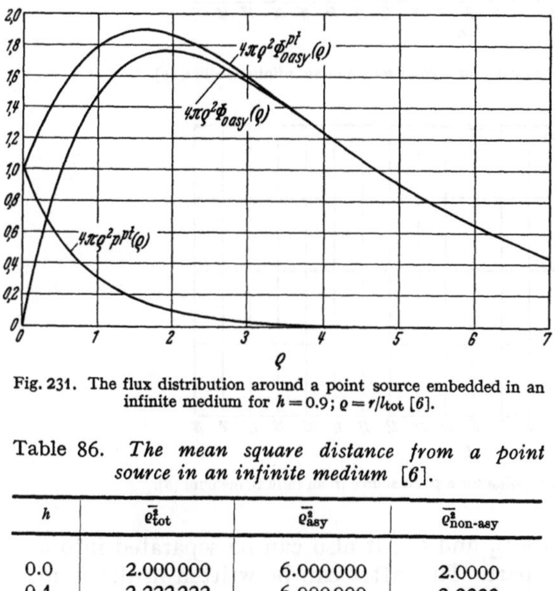

Fig. 231. The flux distribution around a point source embedded in an infinite medium for $h = 0.9$; $\varrho = r/l_{\text{tot}}$ [6].

Table 86. *The mean square distance from a point source in an infinite medium* [6].

h	$\overline{\varrho^2_{\text{tot}}}$	$\overline{\varrho^2_{\text{asy}}}$	$\overline{\varrho^2_{\text{non-asy}}}$
0.0	2.000000	6.000000	2.0000
0.1	2.222222	6.000000	2.2222
0.2	2.500000	6.001091	2.4743
0.3	2.857143	6.031155	2.5745
0.4	3.333333	6.176306	2.4815
0.5	4.000000	6.544403	2.3097
0.6	5.000000	7.288170	2.1254
0.7	6.666667	8.738259	1.9527
0.8	10.000000	11.888606	1.7982
0.9	20.000000	21.733132	1.6622
0.92	25.000000	26.704866	1.6370
0.94	33.333333	35.010787	1.6126
0.96	50.000000	51.650855	1.5891
0.98	100.000000	101.625052	1.5658
1.00	∞	∞	1.5429

Table 86 (Sect. 113) shows the numerical values of these quantities as a function of h.

The total flux as well as the corresponding asymptotic and non-asymptotic parts, multiplied by $4\pi\varrho^2$ are plotted versus ϱ in Figs. 230 and 231 for two values of h. The neutron flux has a very different behaviour in the vicinity of the origin for the two cases; for $h = 0.3$, $4\pi\varrho^2\,\Phi_0^{\text{pt}}(\varrho)$ decreases monotonously, while for $h = 0.9$, it first increases to a maximum and then decreases.

Figs. 232 and 233 show the ratio $\Phi_{0\,\text{asy}}(\varrho)/\Phi_0^{\text{pt}}(\varrho)$ versus ϱ for a few values of h. The value ϱ^* of ϱ for which

$$\Phi_{0\,\text{asy}}(\varrho^*) = p^{\text{pt}}(\varrho^*) \quad (118.9)$$

is given by

$$\varrho^* = \frac{r^*}{l_{\text{tot}}} = 0.031\, e^{2/h}. \quad (118.10)$$

Thus $\varrho^* = 0.031 \cdot e^2 = 0.223$ for $h = 1$, and $\varrho^* \to \infty$ for $h \to 0$, and is already $\varrho^* = 1.5 \cdot 10^7$ for $h = 0.1$.

Problems with isotropic plane sources can be treated essentially by the same procedure. The solution $\Phi_0^{\text{pl}}(z)$ of Eq. (113.11), with

$$s(z) = \delta(z) \quad (118.11)$$

can be divided into two parts:

$$\Phi_0^{\text{pl}}(z) = p^{\text{pl}}(z) + \Phi_{0\,\text{asy}}(z) \quad (118.12)$$

where

$$\Phi_{0\,\text{asy}}(z) = \frac{C}{2}\,\frac{L}{l_{\text{tot}}}\,e^{-\frac{z}{L}} \quad (118.13\text{a})$$

and

$$p^{\text{pl}}(z) = \frac{1}{2}\,E_1\left(\frac{|z|}{l_{\text{tot}}}\right)\eta\left(\frac{z}{l_{\text{tot}}}\right). \quad (118.13$$

The constant C is given by (118.4b), $E_1(\zeta)$ is given by (113.12) with $n = 1$, and $\eta(\zeta)$ is a function, in some way similar to $\varepsilon(\varrho)$, given for a few values of h in Fig. 234. Figs. 235 and 236 are similar to Figs. 232 and 233 apart from the different geometry of the two cases.

The more general solution of the isotropic plane δ-source problem is obtained by adding (118.12) to the general solution (116.21 b) of the corresponding homogeneous equation. The result

$$\Phi_0^{\text{pl}}(z) = p^{\text{pl}}(z) + \frac{C}{2\varkappa_0 l_{\text{tot}}} e^{-\varkappa_0 z} + C_1 e^{\varkappa_0 z} + C_2 e^{-\varkappa_0 z} \tag{118.14a}$$

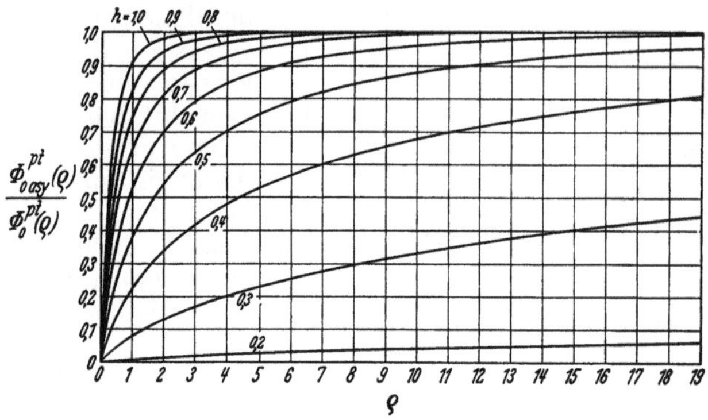

Fig. 232. The ratio $\Phi_{0\,\text{asy}}/\Phi_0^{\text{pt}}$ versus $\varrho = r/l_{\text{tot}}$ for a point source in an infinite medium [6].

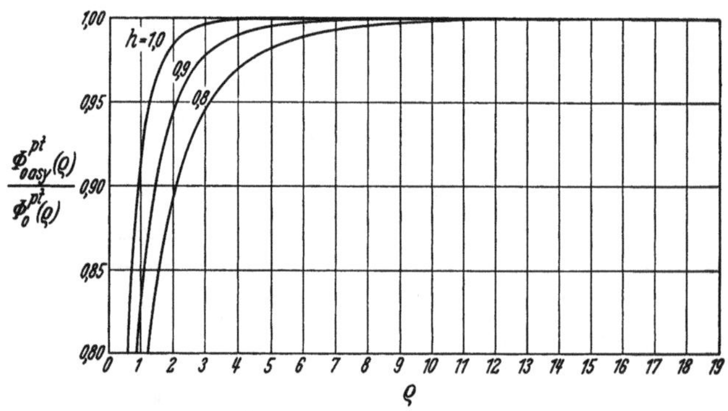

Fig. 233. The ratio $\Phi_{0\,\text{asy}}/\Phi_0^{\text{pt}}$ versus $\varrho = r/l_{\text{tot}}$ for a point source in an infinite medium [6].

contains the two arbitrary constants C_1 and C_2; it also can be separated into a non-asymptotic and an asymptotic part. The latter can be written in the form

$$\Phi_{0\,\text{asy}}^{\text{pl}}(z) = \left\{ \begin{array}{ll} \left(C_2 + \frac{C}{2\varkappa_0 l_{\text{tot}}}\right) e^{-\varkappa_0 z} + C_1 e^{\varkappa_0 z} & \text{for} \quad z > 0 \\ C_2 e^{-\varkappa_0 z} + \left(C_1 + \frac{C}{2\varkappa_0 l_{\text{tot}}}\right) e^{\varkappa_0 z} & \text{for} \quad z < 0. \end{array} \right\} \tag{118.14b}$$

The asymptotic solution for negative z is obtained from that for positive z by adding

$$\frac{C}{2\varkappa_0 l_{\text{tot}}} \sinh \varkappa_0 z. \tag{118.15}$$

β) Anisotropic sources. The results given above may be extended to the case of anisotropic sources. For the sake of simplicity, we shall refer to the case of plane anisotropic sources

$$S(\boldsymbol{r}, \boldsymbol{\Omega}) = \delta(z) f(\mu). \tag{118.16}$$

The neutron flux $\Phi_0^{\text{pl, anis}}(z)$ can be divided into two parts:

$$\Phi_0^{\text{pl, anis}}(z) = \Phi_0^S(z) + \Phi_0^m(z). \tag{118.17}$$

The first term $\Phi_0^S(z)$ represents the flux of the neutrons which have come from the source without having suffered a collision, the second term $\Phi_0^m(z)$ represents the flux due to a distributed isotropic plane source of strength $h\,\Phi_0^S(z')$. Thus we have

$$\Phi_0^{\text{pl, anis}}(z) = \Phi_0^S(z) + h\int_{-\infty}^{+\infty} \Phi_0^S(z')\,\Phi_0^{\text{pl}}\left(\frac{|z-z'|}{l_{\text{tot}}}\right) dz' \tag{118.18}$$

where $\Phi_0^{\text{pl}}(z)$ is the isotropic plane source solution determined in subsection 118α. For $\Phi_0^S(z)$, we have [Eq. (113.9)]

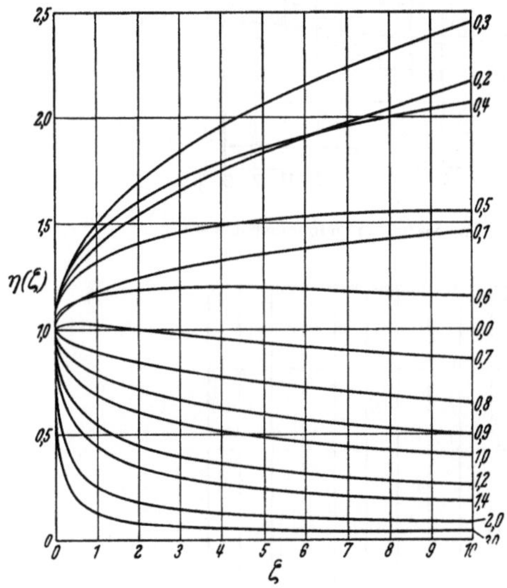

Fig. 234. $\eta(\zeta)$ versus ζ: see Eqs. (118.12), (118.13b).

$$\Phi_0^S(z) = \begin{Bmatrix} \displaystyle\int_0^1 f(\mu')\exp\left[-\frac{z}{\mu' l_{\text{tot}}}\right]\frac{d\mu'}{\mu'} \\ \text{for } z>0 \\ \displaystyle\int_0^{-1} f(\mu')\exp\left[-\frac{z}{\mu' l_{\text{tot}}}\right]\frac{d\mu'}{\mu'} \\ \text{for } z<0 \end{Bmatrix} \tag{118.19}$$

and, therefore, $\Phi_0^{\text{pl, anis}}(z)$ can be evaluated by elementary integration. The calculation of the integral carried out by introducing (118.12), (118.13) into (118.18) leads to

$$\Phi_0^{\text{pl, anis}}(z) = p^{\text{pl, anis}}(z) + \Phi_{0\,\text{asy}}^{\text{anis}}(z) \tag{118.20}$$

where

$$\Phi_{0\,\text{asy}}^{\text{anis}}(z) = h\,C_{\mp}\,e^{\mp\varkappa_0 z} \tag{118.21a}$$

with

$$C_{\pm} = -h\,l_{\text{tot}}\frac{\partial\varkappa_0}{\partial h}\int_{-1}^{+1}\frac{f(\mu)\,d\mu}{1\pm\varkappa_0 l_{\text{tot}}\mu} \tag{118.21b}$$

and $p^{\text{pl, anis}}(z)$ is the non-asymptotic solution whose form depends on $f(\mu)$ but whose general properties are, in some way similar to those of $p^{\text{pl}}(z)$: in particular it is always of the order of $e^{-z/l_{\text{tot}}}$.

The general anisotropic plane source solution of plane symmetry is obtained by adding (116.21b) to (118.20). Thus we have

$$\Phi_{0\,\text{asy}}^{\text{anis}}(z) = \begin{Bmatrix} (C_1 + C_-)\cosh\varkappa_0 z + (C_2 - C_-)\sinh\varkappa_0 z & \text{for} \quad z>0 \\ (C_1 + C_+)\cosh\varkappa_0 z + (C_2 + C_+)\sinh\varkappa_0 z & \text{for} \quad z>0 \end{Bmatrix} \tag{118.22}$$

where C_1 and C_2 are arbitrary constants.

Similar results may be obtained with point anisotropic sources.

It should be noted that the considerations given above hold regardless of the sign of $f(\mu)$, i.e., of the *sign of the source.* Thus, they may be used for dealing with *localized absorbers which may be treated as neutron sinks or negative sources.* An example of this type of problem is given in Sect. 135.

γ) *Reduction of finite-medium problems to infinite-medium problems.* The extension of (118.12) to the case of anisotropic sources (118.20) is of considerable importance in treating the flux distribution in finite media, since it opens the way to the following theorem [6], [7].

A homogeneous medium extending over a volume V is bounded by a surface S; outside S the medium is absent or has different properties. We indicate with $\boldsymbol{\Phi}(\boldsymbol{r}, \boldsymbol{\Omega})$ the angular number flux actually present in the medium and with $\boldsymbol{\Phi}^{\infty}(\boldsymbol{r}, \boldsymbol{\Omega})$ a possible angular number flux in an infinite medium having the same values of l_{tot} and h found in V. *The angular number flux $\boldsymbol{\Phi}^{\text{ex}}(\boldsymbol{r}, \boldsymbol{\Omega})$ in the infinite*

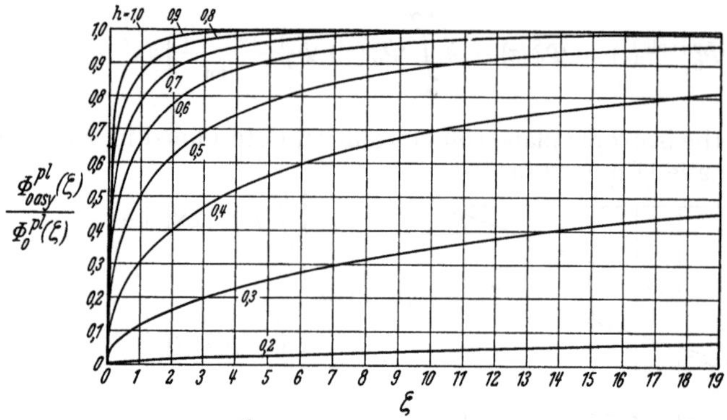

Fig. 235. The ratio $\Phi_{0\,\text{asy}}/\Phi_0^{\text{pl}}$ versus $\zeta = z/l_{\text{tot}}$ for a plane δ-source in an infinite medium.

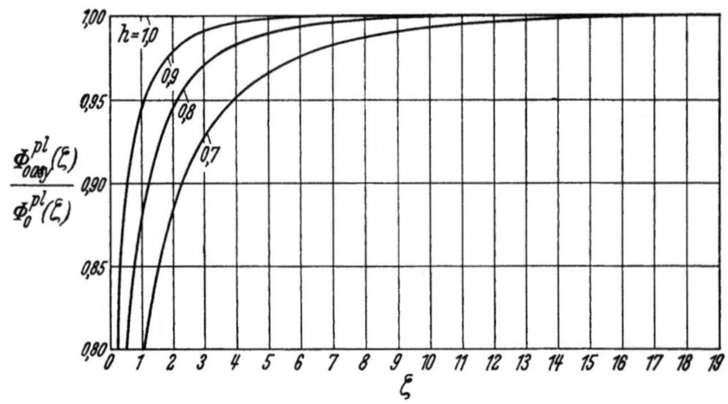

Fig. 236. The ratio $\Phi_{0\,\text{asy}}/\Phi_0^{\text{pl}}$ versus $\zeta = z/l_{\text{tot}}$ for a plane δ-source in an infinite medium.

medium which satisfies the same conditions at infinity as $\boldsymbol{\Phi}^{\infty}(\boldsymbol{r}, \boldsymbol{\Omega})$, but in presence of the independent neutron source

$$S(\boldsymbol{r}, \boldsymbol{\Omega}) = [\boldsymbol{\Phi}(\boldsymbol{r}, \boldsymbol{\Omega}) - \boldsymbol{\Phi}^{\infty}(\boldsymbol{r}, \boldsymbol{\Omega})]\, n(\boldsymbol{r}, \boldsymbol{\Omega}), \tag{118.23a}$$

will coincide with $\boldsymbol{\Phi}(\boldsymbol{r}, \boldsymbol{\Omega})$ over V and with $\boldsymbol{\Phi}^{\infty}(\boldsymbol{r}, \boldsymbol{\Omega})$ outside V. In Eq. (118.23a), $n(\boldsymbol{r}, \boldsymbol{\Omega})$ is defined by the following relations

$$n(\boldsymbol{r}, \boldsymbol{\Omega}) = \left\{\begin{array}{l} 1 \text{ for } \boldsymbol{r} \text{ on } S \text{ and } \boldsymbol{\Omega} \text{ pointing into } V, \\ -1 \text{ for } \boldsymbol{r} \text{ on } S \text{ and } \boldsymbol{\Omega} \text{ pointing outside } V, \\ 0 \text{ otherwise.} \end{array}\right\} \tag{118.23b}$$

The theorem may be understood by noticing that, with $\boldsymbol{\Phi}^{\text{ex}} = \boldsymbol{\Phi}^{\infty}$ outside V, the number of neutrons entering V from outside, added to the contribution of the source (118.23), is just equal to $\boldsymbol{\Phi}^{\text{ex}}(\boldsymbol{r}, \boldsymbol{\Omega})$ on the surface; i.e., it is equal to the number of neutrons entering V in the actual case of a finite medium. Similarly, with $\Phi^{\text{ex}} = \Phi$ inside V, the number of neutrons leaving V added to the contribution of the source (118.23) is equal to that leaving V if, in absence of sources, the distribution inside V is Φ^{∞} [7].

Thus *the angular number flux in any finite region bounded by a surface S is equal to that in an infinite region when suitable sources are added on the surface S; these sources are given by* (118.23).

The formal reduction of a finite medium problem to an infinite medium problem permits the derivation of important conclusions even without explicit knowledge of $\Phi(\boldsymbol{r}, \boldsymbol{\Omega})$ for $\boldsymbol{r}$ on the medium surface.

Thus, for example, one can state [7], that since the source (118.23) for fixed $\boldsymbol{r}$ is not a δ-function of $\boldsymbol{\Omega}$, its expansion in spherical harmonics will converge. Therefore, due to the theorem stated above, the spherical harmonics procedure (Sect. 123) can be applied to finite medium problems.

Of still wider use—because it provides the basis for a correct application of elementary diffusion theory—is the following conclusion [7]. *In any region, finite or not, in which l_{tot} and h are constant, the neutron flux in the one-velocity theory can be represented in the form*

$$\Phi_0(\boldsymbol{r}) = p(\boldsymbol{r}) + \Phi_{0\,\text{asy}}(\boldsymbol{r}) \tag{118.24a}$$

where $\Phi_{\text{asy}}(\boldsymbol{r})$ satisfies the wave equation (116.6) *and $p(\boldsymbol{r})$ is of the order of* $e^{-\frac{R}{l_{\text{tot}}}}$:

$$p(\boldsymbol{r}) = O\left(e^{-\frac{R}{l_{\text{tot}}}}\right) \tag{118.24b}$$

where R is the distance from $\boldsymbol{r}$ to the nearest source or boundary.

Thus, if there is a region far removed from boundaries and from all sources, so that $p(\boldsymbol{r})$ is negligible, the neutron flux $\Phi_0(\boldsymbol{r})$ satisfies the homogeneous equation (116.6) of elementary diffusion theory.

It may be useful to add two remarks [6]. The first is that, in general, one has to distinguish between the asymptotic solution and the infinite-medium solution of the wave equation (116.6). The latter is given by (116.19b), since it should be finite for any value of $\boldsymbol{r}$; the asymptotic solution is, on the contrary, given by (116.19a), and reduces to (116.19b) only if the considered region includes the origin.

The second remark refers to a very simple extension of the previous conclusions. So far the region, where the exact solution reduces to the asymptotic solution, was required to be source-free. The introduction of a constant isotropic distribution of sources

$$S_0 = \frac{s_0}{4\pi} = \text{const} \tag{118.25}$$

does not modify the conclusion. If $\Phi(\boldsymbol{r}, \boldsymbol{\Omega})$ is the solution of the transport equation containing the source (118.25), then

$$\chi(\boldsymbol{r}, \boldsymbol{\Omega}) = \Phi(\boldsymbol{r}, \boldsymbol{\Omega}) - \frac{1}{4\pi}\,\frac{s_0}{1-h} \tag{118.26}$$

satisfies the homogeneous transport equation. Thus, in the region where

$$\chi(\boldsymbol{r}) = \int \chi(\boldsymbol{r}, \boldsymbol{\Omega})\, d\boldsymbol{\Omega}$$

satisfies (116.6), the neutron flux

$$\Phi_0(\boldsymbol{r}) = \int \Phi(\boldsymbol{r}, \boldsymbol{\Omega})\, d\boldsymbol{\Omega}$$

is a solution of the equation

$$(\nabla^2 - \varkappa_0^2)\, \Phi_0(\boldsymbol{r}) = -\frac{\varkappa_0^2\, s_0}{1-h}. \tag{118.27}$$

δ) Standard problems. The exact solution of the integral equation (113.3)—valid for isotropic sources and isotropic scattering in the l.s.—has been derived in a few simple cases of bounded media with such a high symmetry as to reduce the space dependence to a single—or at most—to two independent scalar coordinates. These problems are sometimes indicated as *standard problems*. A detailed discussion of the more interesting cases may be found in Chap. VI and VII of DAVISON's book [7]. The most important standard problem is MILNE's *problem*, i.e., *the determination of the neutron flux distribution in a source-free homogeneous medium occupying a halfspace*[1]. From its solution, the solution of other standard problems may be constructed. Thus, DAVISON and PLACZEK[2] have obtained the solution of the half space problem with arbitrary source distribution.

Detailed treatment of the following simple cases may be found in [7][3,4]: (a) An isotropic neutron source distributed uniformly throughout the medium.

(b) A plane isotropic source parallel to the medium surface placed at a certain depth inside the medium.

(c) An isotropic point source placed at a certain depth inside the medium[5].

(d) A plane isotropic source placed at a certain distance outside the medium, parallel to its surface. In this case, the neutrons emitted by the source enter the medium and give rise to a continuous distribution of isotropic sources corresponding to the neutron distribution after the first collision multiplied by h. From the solution of this problem one can compute the *albedo* of the medium. This is defined (Sect. 120) as the ratio of the current emerging from the medium surface to the incident current.

Finally, the half-space solution has been generalized[2] to the case of two adjacent half-spaces with different l_{tot} and h and without and with neutron sources of the types (a) and (b) mentioned above.

The importance of the solutions of these standard problems lies in the fact that they may be used for constructing the solution of much more involved problems. Let us consider a system consisting of several homogeneous media of such a symmetry—plane, spherical or cylindrical—that the neutron flux depends only on one scalar coordinate, and let us decompose it into a number of "asymptotic regions"—each filled with a source-free homogeneous medium—separated, and possibly bounded, by regions containing sources and boundaries. The solution of this problem may be constructed from the solutions of independent standard problems defined as follows [6].

One considers a single separating or bounding region and extends the asymptotic regions adjacent to it over all space. Thus one obtains a standard problem, the solution of which is assumed to be known. The complete solution for a bounding region contains one arbitrary constant and for a separating region, two arbitrary constants.

---

[1] G. PLACZEK and W. SEIDEL: Phys. Rev. **72**, 550 (1947). — G. PLACZEK: Phys. Rev. **72**, 556 (1947). — C. MARK: Phys. Rev. **72**, 558 (1947). — J. LE CAINE: Phys. Rev. **72**, 564 (1947). — S. FRANKEL and S. GOLDBERG: U.S.A. Rep. AECD-2056. This problem is identical with the problem of the diffusion of radiation through the sun mass towards its surface (E. MILNE: Handbuch der Astrophysik, Vol. 3, p. 1) which had been solved—explicitly—as far as the angular distribution of the emerging radiation is concerned—by N. WIENER and E. HOPF: Berl. Berichte, Math.-phys. Kl. p. 696 (1931).

[2] B. DAVISON and G. PLACZEK: Res. Coun. Canada. Rep. MT-118.

[3] B. DAVISON: Canad. J. Res. A **28**, 303 (1950). — S. CHANDRASEKAR: Canad. J. Res. A **29**, 14 (1951).

[4] B. DAVISON: Report CTR-358 (1946). — R.E. MARSHAK: Phys. Rev. **71**, 443 (1947).

[5] J.P. ELLIOT: AERE T/R 972 (1952).

The solution of the original problem is then obtained by joining in the asymptotic regions the solutions of the various standard problems and, thereby, determining the arbitrary constants.

A further simplification occurs if one is only interested in the behaviour of the neutron flux $\Phi_0(\boldsymbol{r})$ in the asymptotic regions. In such cases, all we need is to join the asymptotic solutions relative to each region with the asymptotic solutions relative to adjacent regions and thus determine the arbitrary constants. These conditions may be written in the form of boundary conditions imposed upon (116.6). An estimate of the distance at which the non-asymptotic part of the solution becomes negligible should be added.

ε) The solution of Milne's problem. As an example let us review here the results of the investigation of MILNE's problem. The derivation of the expressions given below can be found in [6] and [7]. The problem consists of the determination of the neutron distribution in an infinite half-space ($z > 0$) filled with a source-free homogeneous medium which scatters isotropically in the l.s. It represents an idealization of the actual case of neutron sources with plane symmetry placed at very large distance from the plane $z=0$; then the neutrons diffuse through the medium until they reach its surface and flow out of it.

According to the general theorem (118.24), the neutron flux splits into two terms:

$$\Phi_0(z) = \Phi_0(0) \left[\frac{2(L^2 - l_{\text{tot}}^2)}{h L^2 - L^2 + l_{\text{tot}}^2}\right]^{\frac{1}{2}} \sinh\frac{z+d}{L} + p^M(z) \tag{118.28a}$$

which, in the case of non-absorbing media, reduces to

$$\Phi_0(z) = \sqrt{3}\, \Phi_0(0)\, \frac{z+d}{l_s} + p^M_{h=1}(z). \tag{118.28b}$$

The first term in (118.28) is the asymptotic solution which vanishes at a distance d beyond the medium surface:

$$\Phi_{\text{asy}}(-d) = 0. \tag{118.29}$$

The length d is usually called the *extrapolated end point*. Eq. (118.29a) can be regarded as the boundary condition satisfied by $\Phi_{0\,\text{asy}}(z)$ in MILNE's problem. Fig. 237 shows a graph of hd/l_{tot} versus h for h varying between zero and infinity. It starts from 1 at $h=0$, decreases monotonically to a minimum at $h=1$:

$$\frac{d(1)}{l_s} = 0.710446 \tag{118.30a}$$

and then rises slowly as h tends to infinity:

$$\lim_{h\to\infty} \frac{h\,d(h)}{l_{\text{tot}}} = \frac{3}{4}. \tag{118.30b}$$

A table of numerical values of d/l_{tot} is given in [6]. For $0.6 \leqq h \leqq 2$, the value of the extrapolated-end point may be approximated within 0.7% by

$$\frac{d}{l_{\text{tot}}} = \frac{0.7104}{h}. \tag{118.30c}$$

The behaviour of the asymptotic part of the solution of MILNE's problem at the medium-surface is sometimes alternatively described by means of the *extrapolation length*

$$\lambda = \frac{\Phi_{0\,\text{asy}}(0)}{\Phi'_{0\,\text{asy}}(0)} \tag{118.31}$$

where the dash denotes differentiation with respect to z. From (118.28) one finds

$$\lambda = L \tanh \frac{d}{L} \tag{118.32a}$$

which is plotted as a function of h in Fig. 238; from (118.32a) (118.30c), if $h \approx 1$, there follows

$$\lambda \cong \frac{0.7104}{h^{\frac{1}{2}}} l_{\text{tot}} . \tag{118.32b}$$

Starting from the flux distribution (118.28a), the neutron current may be readily computed:

$$J(z) = \frac{L^2(h-1)}{l_{\text{tot}}} \Phi_0'(z) \left[1 + O\left(e^{-\frac{z}{l_{\text{tot}}}}\right)\right] \tag{118.33a}$$

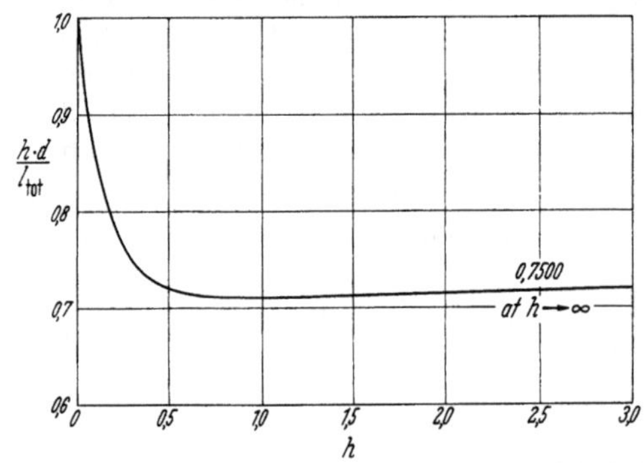

Fig. 237. The extrapolated end-point multiplied by h in units of l_{tot} versus the number of secondary neutrons per collision h [6].

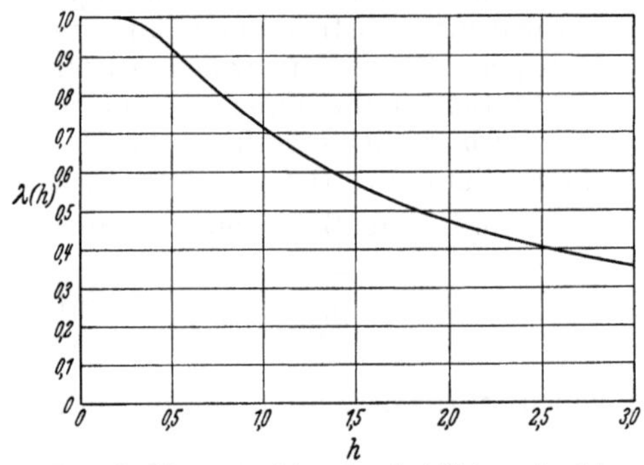

Fig. 238. The extrapolation length $\lambda(h)$ in units of l_{tot} versus the number of secondary neutrons per collision h [6].

which, in the limit $h \to 1$, becomes

$$[J(z)]_{h=1} = -\frac{1}{3} l_s [\Phi_0'(z)]_{h=1} \left[1 + O\left(e^{-\frac{z}{l_{\text{tot}}}}\right)\right]. \tag{118.33b}$$

At the medium surface ($z = 0$) the outgoing current is

$$J(0) = -\frac{L\sqrt{1-h}}{l_{\text{tot}}} \Phi_0(0) \tag{118.34a}$$

and, for $h = 1$,

$$[J(0)]_{h=1} = -\frac{1}{\sqrt{3}} [\Phi_0(0)]_{h=1} . \tag{118.34b}$$

The non-asymptotic term appearing in (118.28a):

$$p^M(z) = O\left(e^{-\frac{z}{l_{\text{tot}}}}\right)$$

is always negative so that near the surface

$$\Phi_0(z) < \Phi_{0\,\text{asy}}(z) .$$

Fig. 239 shows a plot of

$$\frac{\Phi_{0\,\text{asy}}(0)}{\Phi_0(0)} - 1$$

versus h, and Fig. 240 shows a plot of

$$\frac{J_{\text{asy}}(0)}{J(0)} .$$

The ratio

$$-\frac{p^M(z)}{\Phi_0(z)}$$

is plotted in Fig. 241 for a few values of h. Finally Fig. 242 shows the angular number flux $\Phi(0,\mu)$ for z on the medium surface, normalized to 1:

$$\int_{-1}^{0} \Phi(0,\mu)\, d\mu = 1 .$$

According to Eq. (123.34) the angular distribution of the neutrons emerging from the medium surface is given by

$$\Phi(0,\mu)\,\mu\, d\mu .$$

Since the medium is assumed to fill all half-space $z > 0$, the emerging neutrons correspond to μ varying between -1 and 0.

ζ) The extension of the solutions of the isotropic point source problem and of Milne's problem to the case of anisotropic scattering in the l.s. So far in this section we have restricted our considerations to the solution of the integral equation for the case of isotropic scattering in the l.s. We will now outline very briefly the extension of some of the previous results to the case of moderate anisotropic scattering. The latter expression is used for indicating that the expansion (115.2a) of the differential scattering cross section breaks off with a term of comparatively low order $l = n$. This is the assumption also adopted in Sect. 117β for the discussion of the problem of a homogeneous source-free infinite medium.

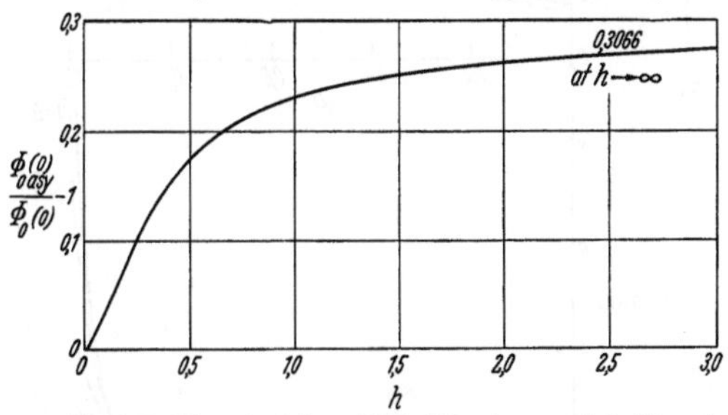

Fig. 239. The ratio $[\Phi_{0\,\text{asy}}(0)/\Phi_0(0)] - 1$ versus h in MILNE's problem [6].

We first consider the problem of a point isotropic source embedded in an infinite medium which scatters anisotropically. If the transcendental equation considered in Sect. 117β has *only one pair of roots*—as is normally the case—then the expression for the neutron flux has the same structure as in the case of isotropic scattering. The result can be written in the form—similar to (118.4)

$$\Phi_0(r) = A\,\frac{e^{-r/L}}{r} + O\left(e^{-r/l_{\text{tot}}}\right). \tag{118.35a}$$

The first term represents the asymptotic part $\Phi_{0\,\text{asy}}(r)$ which satisfies the wave equation (116.6), the second term represents the non-asymptotic part which becomes negligibly small as soon as r is larger than a few l_{tot}.

If two pairs of roots exist—as may occur, at least in principle, in some special case—then the rigorous solution of the problem takes the form

$$\Phi_0(r) = A\,\frac{e^{-r/L}}{r} + A_1\,\frac{e^{-r/L_1}}{r} + O\left(e^{-r/l_{\text{tot}}}\right) \tag{118.35b}$$

where A and A_1 are two constants. The first two terms in (118.35b) should be regarded as $\Phi_{0\,\text{asy}}$, which now is no longer a solution of the wave equation (116.6). However because of the inequality (117.25), the second term decreases with increasing r faster than the first one. Therefore, one can regard it as part of the non-asymptotic term and consider only the first term in (118.35b) as $\Phi_{0\,\text{asy}}(r)$.

This satisfies the wave equation (116.6) which thus can be used even in this case, provided one keeps in mind that now the non-asymptotic term is of the order of e^{-r/L_1} rather than of the order $e^{-r/l_{tot}}$ [7].

The MILNE's problem has been treated for almost isotropic scattering ($n=1$) with the result that in a weakly capturing medium one has [7]

$$\frac{d}{l_{tr}} = \frac{0.7104}{h} - \frac{1-h}{h}\left[0.508\,\bar{\mu}_0 - 0.156\,\bar{\mu}_0^2\right] + O\left(\frac{(1-h)^2}{1-\bar{\mu}_0}\right). \qquad (118.36)$$

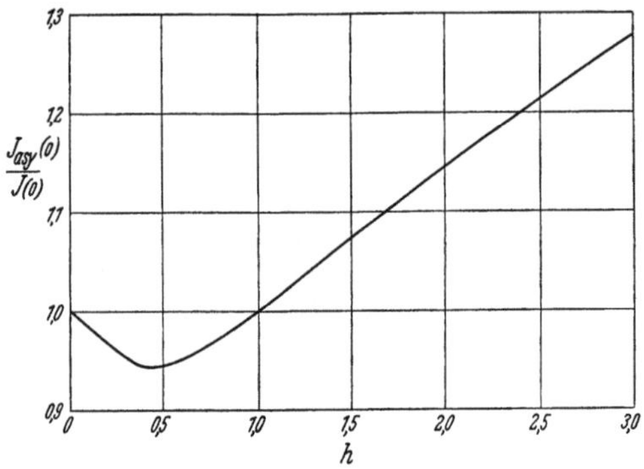

Fig. 240.

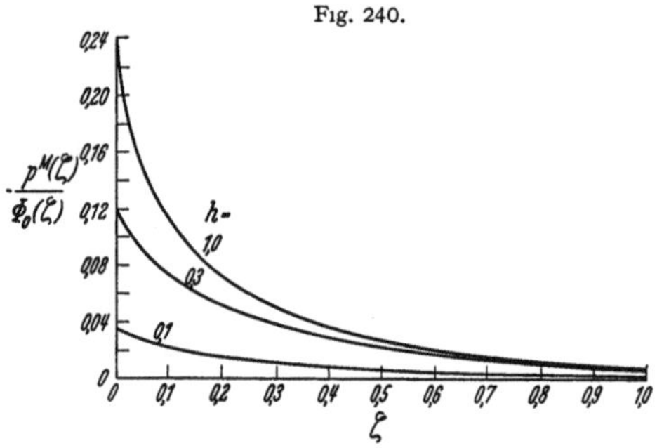

Fig. 241.

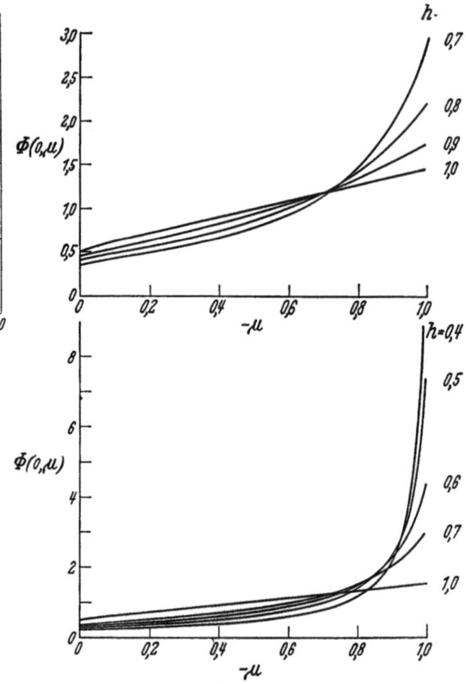

Fig. 242.

Fig. 240. The ratio $J_{asy}(0)/J(0)$ versus h in MILNE's problem [6].

Fig. 241. The ratio $p^M(\zeta)/\Phi_0(\zeta)$ versus ζ in MILNE's problem.

Fig. 242. The angular number flux $\Phi(0,\mu)$ at the free surface in MILNE's problem [6]. The angular distribution of the emerging neutrons is given by $\Phi(0,\mu)\,\mu\,d\mu$.

Thus, to a first approximation, the effect of a slight anisotropy of the differential scattering cross section in the l.s., consists of replacing l_s by l_{tr} in (118.30a).

119. The diffusion approximation: boundary conditions and effect of sources. As we have seen in Sects. 116 and 118, the exact solution of the integral equation describing the diffusion of neutrons in the one-velocity approximation can be derived in a few very simple cases under the assumption that the medium scatters isotropically in the l.s. In more complex cases the exact solution, even if it could be obtained at all, would be too complicated for any practical use.

Therefore, recourse should be made to approximate procedures. Among these, the simplest approach is the *diffusion approximation*; it shows considerable similarities to elementary diffusion theory but differs from it because boundary conditions and neutron sources are introduced according to rules obtained by generalizing the results of the exact treatment of a few standard problems (Sect. 118). These rules will be given below in this section.

Higher approximations—the results of which approach more closely the exact neutron distribution—will be considered very briefly in Sect. 123.

The diffusion approximation is based on the theorem (118.24) which states that: in a homogeneous medium far from any source or boundary, the exact solution reduces to its asymptotic part and this satisfies the wave equation (116.6). Thus, if the dimensions of the medium are much larger than the total mean free path l_{tot}, broad regions will, in general, exist in which the exact solution can be replaced by its asymptotic part. Therefore, we may content ourselves with the determination of $\Phi_{0\,asy}(\boldsymbol{r})$ or even of a solution $\Phi_0(\boldsymbol{r})$ of the equation

$$\nabla^2 \Phi_0(\boldsymbol{r}) - \frac{1}{L^2}\Phi_0(\boldsymbol{r}) = 0 \tag{119.1}$$

which approaches as nearly as possible $\Phi_{0\,asy}(\boldsymbol{r})$.

The solution $\Phi_0(\boldsymbol{r})$ would be identical to $\Phi_{0\,asy}$ if we could impose to it the same boundary conditions. This, however, can only be done if $\Phi_{0\,asy}$ is already known, i.e., just if the problem of determining Φ_0 by means of Eq. (119.1) has no practical interest.

When $\Phi_{0\,asy}(\boldsymbol{r})$ is unknown—so that its exact behaviour at the boundaries is not at one's disposal—one can hope to obtain a solution $\Phi_0(\boldsymbol{r})$ of (119.1) which approaches reasonably well $\Phi_{0\,asy}(\boldsymbol{r})$ by postulating it to satisfy the boundary conditions which hold exactly for $\Phi_{0\,asy}(\boldsymbol{r})$ in some simpler problem already solved. As we shall see below in a few cases, the choice of these boundary conditions is not free from any ambiguity.

Before entering this discussion we recall that in Eq. (119.1), the properties of the medium are synthesized by the corresponding diffusion length L. If $h \leqq 1$—as always occurs in the media considered in this article—L is expressed in terms of the scattering and absorption cross sections of the medium by means of (116.12) whenever the scattering is isotropic in the l.s. In Sect. 117, it has been shown that essentially the same procedure may be used even if the scattering is slightly anisotropic in the l.s.; then the diffusion length should be calculated by means of (117.21a).

The following property of a spherical system will turn out to be very useful: whenever the scattering is isotropic in the l.s. and the mean free path l_{tot} is independent of the position, the determination of the neutron distribution in a spherically symmetrical system can be reduced to the determination of the neutron distribution in a certain system with plane symmetry [7].

One starts from Eq. (116.1) with the integral extending only to $r' \leqq a$; by a we indicate the outer radius of the spherical system which may be finite or infinite. Since

$$d\boldsymbol{r}' = r'^2\,dr'\,d\mu\,d\varphi, \qquad \mu = \frac{\boldsymbol{r}\cdot\boldsymbol{r}}{r r'}, \qquad r r'\,d\mu = 2\,|\boldsymbol{r}-\boldsymbol{r}'|\,d\,|\boldsymbol{r}-\boldsymbol{r}'|,$$

on integrating over angles, one finds

$$\Phi_0(\boldsymbol{r}) = \frac{1}{2l_{tot}}\,\frac{1}{r}\int_0^a h(r')\,\Phi_0(r')\left[E_1\left(\frac{|\boldsymbol{r}-\boldsymbol{r}'|}{l_{tot}}\right) - E_1\left(\frac{r+r'}{l_{tot}}\right)\right]dr'$$

where the slight generalization has been made of assuming the mean number of secondaries h to be a function of r. This allows the treatment of media with certain types of inhomogeneities (see the example at the end of Subsection β). If we extend $\Phi_0(r)$ and $h(r)$ to $r<0$ by putting

$$\Phi_0(-r) = \Phi_0(r), \qquad h(-r) = h(r), \tag{119.2a}$$

we obtain

$$r\,\Phi_0(r) = \frac{1}{2l_{\text{tot}}}\int_{-a}^{+a} h(r')\cdot r'\,\Phi_0(r')\,E_1\!\left(\frac{|\boldsymbol{r}-\boldsymbol{r}'|}{l_{\text{tot}}}\right)d\boldsymbol{r}' \tag{119.2b}$$

which, by comparison with (113.11), can be identified with the equation for a source-free slab of thickness $2a$ having infinite transversal dimensions. In the equivalent plane system, the neutron flux is given by $r\,\Phi_0(r)$ which—because of (119.2a)—must be an odd function of r [7].

α) Boundary conditions at a free surface. The following rule may be adopted for any non-re-entrant surface. The boundary condition is given with sufficient accuracy by Eq. (118.29); i.e.

$$\left.\begin{array}{r}\Phi_0(\boldsymbol{r}+\boldsymbol{n}_s\,d)=0 \text{ for } \boldsymbol{r} \text{ on any surface } S \text{ separating}\\ \text{the medium from the vacuum.}\end{array}\right\} \tag{119.3}$$

Here $\boldsymbol{n}_s$ is the external normal to S and d is the extrapolation end-point whose value is given—for a plane infinite surface—by Eq. (118.30) and Fig. 237 for the case of isotropic scattering in the l.s. and by Eq. (118.36) for the case of a weakly absorbing medium which scatters the neutrons slightly anisotropically.

A few remarks should be added. Eq. (118.29) is rigorously satisfied by the asymptotic part of the exact solution of MILNE's problem: an infinite plane surface separates a half-space filled with a homogeneous source-free medium from the vacuum. Therefore, the application of Eq. (119.3), for example, to the two opposite faces of a slab of infinite transversal dimensions and finite thickness b, involves the tacit assumption that the mutual interference of the two boundaries can be neglected. This will be correct in the limit $b/l_{\text{tot}}\to\infty$.

Furthermore, if Eq. (119.3) is applied to a curved (non-re-entrant) surface, the value of d is expected to depend upon the curvature of the surface. In the limit of a radius of curvature much larger than l_{tot} the error introduced by using the extrapolated end-point valid for a plane surface is, however, negligibly small.

Thus, for example, in the case of a parallelepiped of linear dimensions much larger than l_{tot}, we should expect that Eq. (119.3) provides a rule for the boundary conditions which is correct in regions near the central part of each face, but fails near the edges and corners. In spite of this, the solution obtained by solving Eq. (119.1) with the boundary conditions (119.3) is expected to be sufficiently correct, for most purposes, everywhere except in a superficial layer of thickness of the order of 3 or 4 times l_{tot}.

The ambiguity mentioned above in the choice of the approximate boundary conditions (119.3) can be shown by noticing that the boundary conditions rigorously valid in MILNE's problem can be put into two equivalent alternative forms given by (118.29) and (118.31) respectively.

A different result is obtained according to the choice made between these two equations for the generalization to curved surfaces. For example, the boundary condition on a spherical surface of radius a, obtained by combining (118.31) with (119.2), is the following:

$$a\,\Phi_0(a) = -\lambda\left[\frac{d\,r\,\Phi_0(r)}{dr}\right]_{r=a}. \tag{119.4}$$

Thus, while this equation involves the radius of curvature a, Eq. (119.3) is independent of it.

β) Boundary conditions at an interface. These are given in general by Eq. (112.3a), which, for the sake of simplicity, may be re-stated in the following way for the one-dimensional case:

$$\Phi^{(-)}(z_s, \mu) = \Phi^{(+)}(z_s, \mu). \tag{119.5}$$

Here the superscripts $^{(-)}$ and $^{(+)}$ denote quantities referring to the two media separated by the plane located at $z = z_s$. The physical meaning of Eq. (119.5) is that the number of neutrons reaching the interface from the left in the direction μ should be the same as the number of neutrons leaving the interface towards the right in the same direction.

If both $\Phi^{(-)}(z,\mu)$ and $\Phi^{(+)}(z,\mu)$ are expanded in spherical harmonics according to (79.4), Eq. (119.5) is equivalent to stating that at the interface

$$\Phi_l^{(-)}(z_s) = \Phi_l^{(+)}(z_s). \tag{119.6}$$

This is the form of the boundary conditions appropriate for the infinite set of linear equations (115.4).

We have seen in Sect. 115 that the use of the diffusion equation (119.1) corresponds to solving the infinite set of linear equations (115.4) in the P_1-approximation in which the $\Phi_l(z)$ are assumed to be zero for $l \geqq 2$.

In this approximation, the conditions (119.6) reduce to

$$\left.\begin{aligned} \Phi_0^{(-)}(z_s) &= \Phi_0^{(+)}(z_s), \\ \Phi_1^{(-)}(z_s) &= \Phi_1^{(+)}(z_s). \end{aligned}\right\} \tag{119.7}$$

The first of these equations states the continuity of the neutron flux and the second the continuity of the net neutron current (115.9c). According to (115.9a), Φ_1 can be expressed in terms of $d\Phi_0/dz$ so that we may rewrite Eq. (119.7) in the form

$$\Phi_0^{(-)}(z_s) = \Phi_0^{(+)}(z_s), \tag{119.8a}$$

$$D^{(-)}\left(\frac{d\Phi_0^{(-)}}{dz}\right)_{z_s} = D^{(+)}\left(\frac{d\Phi_0^{(+)}}{dz}\right)_{z_s}, \tag{119.8b}$$

thus specifying the values of Φ_0 and its first derivative at the interface, as it is possible for the solution of a second order differential equation.

The boundary conditions (119.8) lead or do not lead to a rigorous asymptotic solution, depending upon the importance of neutron capture.

When both media do not capture, the solution of the transport equation —far from any neutron source—is rigorously given in each medium by an expression of the type (117.9); i.e. by expressions which contain only two spherical harmonics (P_0 and P_1) and which are linear in z. This is true even at the interface, where the asymptotic flux and the net current are given by equations similar to (118.28b), (118.33b). In this case, the diffusion coefficient is given by (117.10) so that we may write

$$\Phi_0^{(-)}(z_s) = \Phi_0^{(+)}(z_s), \tag{119.9a}$$

$$l_{\text{tr}}^{(-)}\left(\frac{d\Phi_0^{(-)}}{dz}\right)_{z_s} = l_{\text{tr}}^{(+)}\left(\frac{d\Phi_0^{(+)}}{dz}\right)_{z_s}, \tag{119.9b}$$

equations which are *rigorously correct*.

When one or both of the media capture neutrons, the asymptotic solution of the transport equation is given by (117.12), (117.15) which contain all the spherical harmonics. Then the boundary conditions (119.7), requiring only continuity of the neutron flux and of the neutron net current, are sufficient to determine the solution in the diffusion approximation, but not that of the transport equation.

Thus we may conclude that the boundary conditions (119.7) or (119.8) are only approximate, and that they tend to become the boundary conditions rigorously satisfied by $\Phi_{0\,\mathrm{asy}}(z)$ only in the limit of negligible capture; then they reduce to (119.9).

For more detailed discussions, the reader is directed to Refs. [*19*] and [*7*].

In the case of thermal neutrons, the values of D (or l_{tr}) to be inserted into (119.8) or (119.9) are averages over the neutron spectrum as shown by (115.12a).

The considerations given above actually refer to an infinite plane interface. The y may be extended to spherical interfaces by means of the theorem (119.2), the application of which, however, requires isotropic scattering in the l.s. and that l_{tot} be the same in both media. The same boundary conditions are usually assumed to be approximately valid even for other shapes of interfaces for which there exist equal mean free paths and nearly equal mean numbers h of secondaries in both media [*7*].

γ) Black bodies. This expression is used (Sect. 108) for indicating localized strong absorbers such as, for example, lumps or thick sheets of cadmium embedded in a medium. Since the scattering cross section of cadmium is very small (Table 8), practically all thermal neutrons incident on its surface are absorbed.

The boundary conditions to be applied to the surface of a black body depend upon its dimensions. If the black body has linear dimensions which are much larger than l_{tot} in the surrounding medium, then the situation is similar to that encountered in the case of a free surface and the boundary conditions discussed in subsection α may be applied.

The extreme opposite case of a black body, of linear dimensions a which are much smaller than l_{tot}, embedded in a non-capturing medium with a constant flux distribution at large distances $\Phi_0(\infty) = \mathrm{const}$, can be treated in the diffusion approximation and yields the result [*7*]

$$\left.\begin{aligned} \tfrac{1}{3} l_s \operatorname{grad} \Phi_0(\boldsymbol{r}) &= \tfrac{1}{4} \Phi_0(\infty) \\ \Phi_0(\boldsymbol{r}) &= \Phi_0(\infty)\left[1 + O\left(\frac{a^2}{r\, l_{\mathrm{tot}}}\right)\right]. \end{aligned}\right\} \tag{119.10}$$

Thus, according to the definition (118.31) of the extrapolation length, we can write

$$\lambda = \tfrac{4}{3} l_s \qquad \text{for } h = 1 \text{ and } a \ll l_s. \tag{119.11a}$$

This relation may even be used if one of the three dimensions of the black body is infinite as it occurs in the case of an infinite cylinder of small radius a.

We may also expect that, replacing l_s by l_{tot}, Eq. (119.11a) will be approximately correct even in the case of weakly absorbing media:

$$\lambda \approx \tfrac{4}{3} l_{\mathrm{tot}} \qquad \text{for } 1 - h \ll 1 \text{ and } a \ll l_{\mathrm{tot}}. \tag{119.11b}$$

A comparison of (119.11) and (118.32b) shows that for black bodies, of any linear dimensions, embedded in a non-capturing medium, the extrapolated length should remain between the limits

$$0.71\, l_s \leqq \lambda \leqq \tfrac{4}{3} l_s. \tag{119.12}$$

Fig. 243 shows the actual dependence of λ upon a for a black sphere and a black cylinder of radius a embedded in a non-capturing medium [*7*][1].

[1] B. Davison: Proc. Phys. Soc. Lond. A **64**, 881 (1951).

δ) Cavities and gaps [7]. The problem of establishing the boundary conditions at the surface of a gap or of a cavity opened in a homogeneous medium is much more involved. We shall limit ourselves to considering *two cases in which the presence of the cavity does not affect the neutron distribution.*

The first case is met if a cavity of any shape is opened in a non-capturing medium and the source distribution and boundary conditions at infinity—or at the outer surface—are such that

$$\Phi_0(\boldsymbol{r}) = \text{const}. \tag{119.13}$$

This theorem can be proved by rewriting Eq. (113.8) generalized to the case of non-homogeneous media:

$$\Phi(\boldsymbol{r}, \boldsymbol{\Omega}) = \int_0^\infty Q(\boldsymbol{r} - R'\,\boldsymbol{\Omega})\, e^{-\varrho'}\, dR'. \tag{119.14a}$$

Here

$$\varrho' = \int_0^{R'} \frac{dR''}{l_{\text{tot}}(\boldsymbol{r} - R''\,\boldsymbol{\Omega})} \tag{119.14b}$$

represents the generalized optical distance (118.1a) and $Q(r)$ is given by (113.6). Because of the assumptions specified above,

$$Q(r) = Q = \text{const}$$

and, therefore,

$$\Phi(r, \boldsymbol{\Omega}) = Q \int_0^\infty e^{-\varrho'}\, dR' \tag{119.15}$$

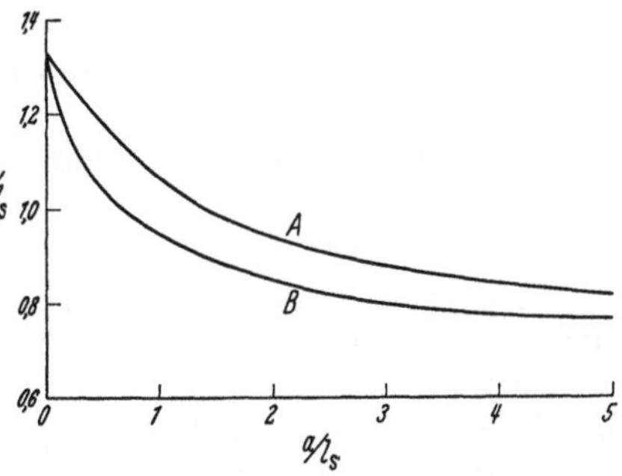

Fig. 243. The extrapolation length for a black sphere (A) and a black cylinder (B) as a function of the radius a [7].

where the value of the integral is not affected by the presence of cavities. In Sect. 135, we shall make use of this theorem in the discussion of experiments on the perturbation of the neutron distribution produced by an absorbing foil.

The second case in which the presence of a gap does not modify the conclusions reached in the previous sections is that of a plane symmetry system with a plane slit perpendicular to the symmetry axis z. Since the solution depends only upon the optical distance ζ (118.1b) and the mean free path in the gap is infinity—so that its optical thickness is zero—the formulas derived for a homogeneous medium will still hold, provided that the geometrical thickness of the slit is ignored.

The discussion of more involved problems—such as plane slits with a lateral flux variation or cavities of other shapes—may be found in Ref. [7].

ε) The effect of sources. In order to take into account the effect of sources in the diffusion approximation, the following two cases should be discussed separately.

The first case is that of sources situated at isolated points. Then the rigorous flux distribution is obtained by adding a certain number of solutions of the type (118.4) each multiplied by the corresponding source strength

$$s_0 = 4\pi\, S_0. \tag{119.16}$$

The diffusion approximation gives a correct description of the actual situation only at distances from all sources larger than a few mean free paths where all the non-asymptotic terms start to be negligibly small. The asymptotic distribution, however, contains only the fraction (118.6) of the total number of neutrons present in the medium. Therefore, the solution of the diffusion equation

$$\nabla^2 \Phi_0 - \frac{1}{L^2}\,\Phi_0 = -\,\overline{S}_0\,\delta(r) \tag{119.17a}$$

will be identical with $\Phi_{0\,\mathrm{asy}}(\boldsymbol{r})$—for $r \gg l_{\mathrm{tot}}$—only if we take

$$\overline{S}_0 = K\, S_0 \tag{119.17b}$$

where K is given by (118.6b) and S_0 is the actual source strength.

In the case of weakly absorbing media, K is given by (118.7) which corresponds to a correction of 0.8% in the case of water and almost ten times less in the case of graphite.

The second case to be discussed is that of the neutron distribution inside a medium of volume V in which the sources are distributed uniformly. The answer to this question has already been given at the end of Sect. 118γ; according to (118.27), one may use the diffusion approximation provided that the source strength per unit volume is replaced by

$$\overline{s}_0 = \frac{1}{L^2(1-h)}\, s_0 . \tag{119.18}$$

The same conclusion holds when the neutron strength per unit volume s_0 is not a constant but varies linearly with the Cartesian coordinates.

120. The albedo. The expression albedo was introduced[1] [8] for indicating the reflection power of a medium. Although it may be defined for neutrons of any energy, in particular for epi-cadmium neutrons, the usefulness of this quantity is clearly apparent only in the case of thermal neutrons.

Two kinds of albedo should be distinguished [8][2]: the *current albedo*, or simply albedo β[2-4] and the *density albedo*, or experimental albedo β_e [8][1,2,5]. The first definition is of considerable interest in connection with boundary conditions between two media, the second one occurs in a relationship between the average number of collisions undergone by a thermal neutron before being captured and two measurements of the neutron density under convenient experimental conditions.

α) *The albedo* is defined as the ratio of the number of neutrons reflected from a medium surface to the number of neutrons incident upon it. Its value depends, of course, upon the angular distribution of the incident neutrons as well as on the size and geometry of the reflector; a thin sheet of scatterer will reflect less neutrons than a thick one.

The usefulness of the albedo may be understood by considering the following situation: a medium A, in which the neutrons are produced, is bounded by a source-free medium B having different properties. The question arises whether it is possible to describe the medium B simply by a reflection coefficient β and to determine the neutron distribution in medium A from the albedo of medium B without paying any further attention to this latter medium. Such a description will be useful only if the albedo of medium B is *independent* of the properties and of the source distribution in the medium A[6].

This, in general, is not the case. In the one-velocity transport theory, the albedo depends upon the properties of the media in contact and, furthermore,

[1] E. AMALDI and E. FERMI: Ric. Sci. **7** (1), 454 (1936). — Phys. Rev. **50**, 899 (1936). — E. FERMI· AECD 2664 (1951). It should be kept in mind that the word albedo is used for the coefficient of diffused reflection of white light by various surfaces: in astronomy the albedo of the moon, of the earth, etc., in technology the albedo of paper, cardboard, etc.

[2] G.C. WICK: Rend. Linc. **23**, 774 (1936).

[3] O. HALPERN, R. LUENEBURG and O. CLARK: Phys. Rev. **51**, 1020 (1937).

[4] W. DE GROOT and K.F. NIESSEN: Physica, Haag **7**, 199 (1940).

[5] G.C. WICK: Z. Physik **121**, 702 (1943).

[6] G. PLACZEK: The concept of albedo in elementary diffusion theory, p. 77 in: The Science and Engineering of Nuclear Power. Edit. by C. GOODMAN, Vol. II, p. 77. Cambridge, Mass.: Addison-Wesley Press Inc. 1949.

its knowledge is not sufficient to determine the behaviour of the angular number flux in the medium A near its surface [see Eq. (119.6)].

This situation, however, is different whenever the diffusion theory is valid because, as we shall see below, in this approximation and for problems of high symmetry, the albedo of a source-free medium is an intrinsic property, —i.e., independent of the adjacent medium—and its knowledge is sufficient for the determination of the neutron distribution in the adjacent medium[1].

The simplest case, which will be considered first, is that of two media filling the two half-spaces on the left and on the right of the plane $z=0$ respectively.

The albedo of the right-hand medium B is then, by definition, given by

$$\beta = \frac{J_{\text{out}}}{J_{\text{inc}}} = \frac{J_-}{J_+} = \frac{\int_0^{-1} \Phi(0,\mu)\,\mu\,d\mu}{\int_0^{1} \Phi(0,\mu)\,\mu\,d\mu}. \tag{120.1}$$

Since we assume that the diffusion approximation is valid, we may use the results of Sect. 117β; for isotropic scattering in the l.s. we have, according to (117.19),

$$\Phi(0,\mu) = \frac{1}{\Sigma_{\text{tot}} - \varkappa_0\,\mu}$$

and, therefore,

$$\beta = \frac{1 - \frac{L}{l_{\text{tot}}} \ln\left(1 - \frac{l_{\text{tot}}}{L}\right)}{-1 - \frac{L}{l_{\text{tot}}} \ln\left(1 - \frac{l_{\text{tot}}}{L}\right)}. \tag{120.2}$$

If the absorption is weak ($\varkappa_0 = 1/L \ll \Sigma_{\text{tot}}$), we may expand the logarithm and obtain the result[1-3]

$$\beta = 1 - \frac{4}{\sqrt{3}} \sqrt{\frac{\Sigma_a}{\Sigma_{\text{tot}}}} = 1 - \frac{2.31}{\sqrt{N}} \tag{120.3}$$

where

$$N = \frac{\Sigma_{\text{tot}}}{\Sigma_a} = \frac{l_a}{l_{\text{tot}}} \tag{120.4}$$

is the average number of collisions undergone by a neutron before capture.

From Eq. (120.3), we see that for $N \to \infty$ (i.e. $\Sigma_a \to 0$), the albedo tends to unity, as we should have expected.

The same result (120.3) may be obtained by means of a slightly different procedure which allows the derivation of an expression for the albedo without specifying the geometry of the reflector; this is embodied by the corresponding extrapolation length which can easily be calculated afterwards for any chosen shape of the reflector[4].

If the absorption is so small that the two-harmonics angular distribution can be used, we may write

$$\Phi(\boldsymbol{r}, \boldsymbol{\Omega}) = \frac{1}{2\pi}\left\{\frac{1}{2}\,\Phi_0(\boldsymbol{r}) + \frac{3}{2}\,\boldsymbol{J}(\boldsymbol{r}) \cdot \boldsymbol{\Omega}\right\} \tag{120.5}$$

which is obtained by generalizing (117.8), (115.9c) to the case in which grad $\Phi_0(\boldsymbol{r})$ has an arbitrary direction instead of that of the z axis. We now may consider a

[1] See footnote 2, p. 545.
[2] See footnote 3, p. 545.
[3] See footnote 4, p. 545.
[4] See footnote 6, p. 545.

surface element dS at the interface between the two media A and B. If $\boldsymbol{n}_s$ is the inward normal to dS in medium B, we can decompose the vector $\boldsymbol{J}$ into a tangential and a normal component. Thus, we may write:

$$\boldsymbol{J}\cdot\boldsymbol{\Omega} = J_n \cos\vartheta + J_t \sin\vartheta\cos\varphi \tag{120.6}$$

where ϑ is the angle between $\boldsymbol{\Omega}$ and $\boldsymbol{n}_s$ and φ is the angle between $\boldsymbol{J}_t$ and the projection of $\boldsymbol{\Omega}$ upon the surface element.

The incident and reflected currents are given by

$$\left.\begin{aligned} J_{\text{out}} &= \int_0^{2\pi} d\varphi \int_0^{-1} \Phi(\boldsymbol{r},\boldsymbol{\Omega})\,\mu\,d\mu = \tfrac{1}{4}\Phi_0 - \tfrac{1}{2}J_n,\\ J_{\text{inc}} &= \int_0^{2\pi} d\varphi \int_0^{1} \Phi(\boldsymbol{r},\boldsymbol{\Omega})\,\mu\,d\mu = \tfrac{1}{4}\Phi_0 + \tfrac{1}{2}J_n,\end{aligned}\right\} \tag{120.7}$$

where $\mu = \cos\vartheta$, and J_t has disappeared because of the integration with respect to φ. From (120.1) and (120.7), we obtain the following expression for the albedo

$$\beta = \frac{\frac{1}{4}\Phi_0 - \frac{1}{2}J_n}{\frac{1}{4}\Phi_0 + \frac{1}{2}J_n}. \tag{120.8}$$

Since, according to (117.4), (117.5), (117.6), (117.7a)

$$J_n = -\frac{1}{3}\,l_d\left(\frac{\partial\Phi_0}{\partial n_s}\right)_{r_s} \tag{120.9}$$

we obtain

$$\beta = \frac{1-\dfrac{2}{3}\dfrac{l_d}{\lambda}}{1+\dfrac{2}{3}\dfrac{l_d}{\lambda}} = \frac{\lambda-\dfrac{2}{3}l_d}{\lambda+\dfrac{2}{3}l_d} = 1-\frac{4}{3}\,\frac{l_d}{\lambda+\dfrac{2}{3}l_d} \tag{120.10}$$

where

$$\lambda = -\frac{\Phi_0(r_s)}{\left(\dfrac{\partial\Phi_0}{\partial n_s}\right)_{r_s}} \tag{120.11}$$

is the *extrapolation length* already defined in Sect. 118 [Eq. (118.31)] for the special case of an infinite plane free-surface with grad Φ_0 perpendicular to it. The difference in sign appearing in Eqs. (120.11) and (118.31) originates from the opposite directions of flow of the neutrons: in MILNE's problem the neutrons diffuse from right (medium) to left (vacuum) while here they diffuse from left (medium A) to right (medium B).

Since we started from Eq. (120.5)—which is correct only for very weak absorption (Sect. 117β)—, the validity of (120.10) is subject to the following conditions

$$\Sigma_a \ll \Sigma_s \quad \text{i.e.} \quad l_d \approx l_{\text{tr}}. \tag{120.12}$$

Thus, the albedo is determined by the ratio l_d/λ. As we shall see from subsequent examples λ, is independent of the properties and the source distribution of medium A if the neutron flux $\Phi_0(\boldsymbol{r})$ depends only upon one coordinate. This is the case with plane, spherical or cylindrical symmetry[1].

Before entering the discussion of a few special cases, it should be noticed that the solution of the problem mentioned at the beginning of this section can be carried out in two steps: (1) The calculation of the albedo β_B for the source-free medium B. (2) The determination of the neutron flux distribution in medium A for a given source distribution, geometry and albedo at the interface.

[1] See footnote 6, p. 545.

In order to see how the second step is carried out, we solve Eq. (120.10) with respect to l_d/λ:

$$\left(\frac{l_d}{\lambda}\right)_B = \frac{3}{2}\,\frac{\beta_B - 1}{\beta_B + 1}. \tag{120.13}$$

A subscript B is added to the various quantities in order to stress that they all refer to medium B. If now we recall the definition of λ (120.11), we see that —apart from an unimportant constant factor—the boundary conditions (113.7), generalized to the case of an interface of normal $\boldsymbol{n}_s$, can be written in the new form:

$$\left[\frac{l_d\left(\dfrac{\partial \Phi_0}{\partial n_s}\right)_{r_s}}{\Phi_0(r_s)}\right]_A = \frac{3}{2}\,\frac{\beta_B - 1}{\beta_B + 1}. \tag{120.14}$$

From Eq. (120.13), it follows that for $\beta = 1$, $\lambda = \infty$; and for $\beta_B = 0$

$$\lambda = \tfrac{2}{3}\, l_d. \tag{120.15}$$

The latter situation corresponds to the case of an infinite plane surface separating medium A from a half-space evacuated or filled with a very strong absorber (black body). Therefore, it should be compared to (118.32b) with $h \to 1$, from which it differs in the numerical factor.

The boundary condition (120.15) could have been obtained directly by assuming that the flux distribution in medium A is given by (117.9); it constitutes the boundary condition at the free surface of a non-capturing medium in the diffusion approximation. In this approximation the extrapolated end point is

$$d = \tfrac{2}{3}\, l_{\mathrm{tr}} \tag{120.16}$$

which should be compared to the exact value (118.30a).

We may turn to computing the albedo in a few simple cases. Since all the considerations given here are subject to condition (120.12), the albedo will always be very close to unity, and, therefore, it is more practical to consider the *albedo defect*

$$1 - \beta = \frac{4}{3}\,\frac{l_{\mathrm{tr}}}{\lambda + \frac{2}{3} l_{\mathrm{tr}}} \tag{120.17}$$

which, for $\lambda \gg l_{\mathrm{tr}}$, as it occurs frequently, reduces to

$$1 - \beta = \frac{4}{3}\,\frac{l_{\mathrm{tr}}}{\lambda} \qquad \text{for } l_{\mathrm{tr}} \ll \lambda. \tag{120.18}$$

Since the term $\frac{2}{3} l_{\mathrm{tr}}$ appearing in the denominator of (120.17) originates from the current term in (120.5), the expression (120.18) represents the albedo defect in the case of isotropic angular distribution of the incident neutrons.

Table 87 contains the values of the albedo defect for a few simple geometries of the reflector.

In the case of an infinite half-space, the albedo defect, calculated from (120.17), is

$$1 - \beta = \frac{4}{\sqrt{3}}\,\frac{1}{\sqrt{N}}\left(1 - \frac{2}{\sqrt{3}}\,\frac{1}{\sqrt{N}} + \cdots\right). \tag{120.19}$$

Therefore, the relative error derived from the use of the approximate expression (120.18) is $\frac{2}{\sqrt{3N}}$; it amounts to 8% in water and paraffin ($N \sim 180$) and to 3% in graphite ($N \sim 1000$).

Table 87. *Flux distribution and albedo defect for a few characteristic cases in diffusion approximation.*

Geometry of the reflecting medium	Flux distribution Φ_0	Extrapolation length λ	Albedo defect $1-\beta$ for $\lambda \gg l_{tr}$
Infinite half space	$e^{-z/L}$	L	$\frac{4}{\sqrt{3}} \frac{1}{\sqrt{N}}$
Slab of thickness a ($\bar{a} = a + d$)	$\sinh \frac{\bar{a} - z}{L}$	$L \cdot \tanh \frac{\bar{a}}{L}$	$\frac{4}{3} \frac{l_{tr}}{L} \coth \frac{\bar{a}}{L}$ $\frac{4}{3} \frac{l_{tr}}{a + \frac{4}{3} l_{tr}}$ for $a \ll L$ $\frac{4}{3} \frac{l_{tr}}{L}$ for $a \gg L$
Sphere of radius a	$\frac{\sin r/L}{r}$	$\frac{L}{\left(\cot \frac{a}{L} - \frac{L}{a}\right)}$	$\frac{4}{3} \frac{l_{tr}}{L} \left(\coth \frac{a}{L} - \frac{L}{a}\right)$ $\frac{4}{9} \frac{a l_{tr}}{L^2}$ for $a \ll L$ $\frac{4}{3} l_{tr} \left(\frac{1}{L} - \frac{1}{a}\right)$ for $a \gg L$
Infinite medium surrounding sphere of radius a	$\frac{e^{-r/L}}{r}$	$\frac{L}{1 + \frac{L}{a}}$	$\frac{4}{3} l_{tr} \left(\frac{1}{L} + \frac{1}{a}\right)$
Spherical shell inner radius a, outer radius b, neutrons incident from inside ($\bar{b} = b + d$; $d = \bar{b} - a = b - a + d$)	$\frac{\sinh \frac{\bar{b} - r}{L}}{r}$	$\frac{L}{\coth \frac{\bar{b} - a}{L} + \frac{L}{a}}$	$\frac{4}{3} \frac{l_{tr}}{L} \left(\coth \frac{\bar{b} - a}{L} + \frac{L}{a}\right)$ $\frac{4}{3} l_{tr} \left(\frac{1}{d} + \frac{1}{a}\right)$ for $d \ll L$ $\frac{4}{3} l_{tr} \left(\frac{1}{L} + \frac{1}{a}\right)$ for $d \gg L$

β) The experimental albedo. The determination of the albedo (120.1) requires measuring of the currents J_- and J_+ which in principle, can be done by means of current probes (Sect. 134β). These have the disadvantage, however, of producing an appreciable perturbation of the neutron distribution so that the result of such measurements would not correspond correctly to the definition (120.1). An experimental determination of the reflecting power of a medium for thermal neutrons can be done very conveniently by the following procedure which is based upon the use of density probes (Sect. 134α) which produce perturbations of the neutron distribution which can be made—at least in principle—as small as desired.

As we shall see below, this procedure leads to a different definition of the albedo: the *experimental albedo*[1] [8][2] representing the ratio

$$\beta_e = \frac{\int_0^{-1} \Phi(0, \mu)\, d\mu}{\int_0^{1} \Phi(0, \mu)\, d\mu}. \tag{120.20}$$

[1] See footnote 1, p. 545.
[2] See footnote 2, p. 545

The experimental set-up used for this type of measurements is schematically shown in Fig. 244. The medium under investigation in the shape of a block of very large dimensions with respect to the diffusion length, is cut into two parts A and B along the x, y-plane. The source or the sources, are placed symmetrically with respect to this plane so that the distribution of neutrons of about 1 ev energy is also symmetric and almost independent of z. Under these conditions, the distribution of epi-cadmium neutrons—i.e. the distribution of the source of thermal neutrons—may be considered as uniform and equal to s_0 neutrons/cm³ sec in a not too narrow region surrounding the origin O. For the following considerations, this condition should be fulfilled over a region of the order of three or four diffusion lengths.

An infinitely thin $1/v$ detector (i.e. an ideal density probe: Sect. 134α) is inserted between the two blocks and its activity due to thermal neutrons (Sect. 133) is determined, once with both sides uncovered

$$A_{\text{th}}\,(\text{detector uncovered}) = A_{\text{th}}(\infty) \tag{120.21a}$$

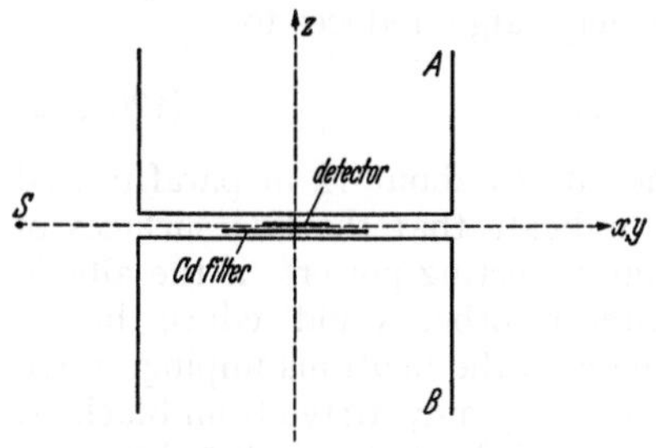

Fig. 244. Experimental set-up for albedo measurements.

and again with one of its faces, let us say the lower one, covered by a thick cadmium absorber (black surface for thermal neutrons):

$$\left.\begin{array}{r} A_{\text{th}}\,(\text{detector covered on} \\ \text{one side}) = A_{\text{th}}(0)\,. \end{array}\right\} \tag{120.21b}$$

The adopted notations $A_{\text{th}}(\infty)$ and $A_{\text{th}}(0)$ will be explained below.

According to Eqs. (133.1), (133.6) and (133.8), the measured activities are proportional to the values of the neutron fluxes or neutron densities present, under the considered experimental conditions, at the location of the detector. Thus the ratio

$$\eta = \frac{A_{\text{th}}(\infty)}{A_{\text{th}}(0)} = \frac{\Phi_0(\infty)}{\Phi_0(0)} = \frac{N_0(\infty)}{N_0(0)} \tag{120.22}$$

is independent of the detector properties.

Two expressions of η may now be derived: the first one gives η in terms of parameters—such as N and $\bar{\mu}_0$—describing the properties of the medium, and the other is a relationship between η and the experimental albedo (120.20).

In order to derive the theoretical expression for η in terms of the medium parameters, we notice that on the right-hand side of (120.22) we may use the densities calculated for a medium filling one half-space in which the neutrons are uniformly produced at a source strength density of s_0 thermal neutrons per cm³ sec. In fact, in absence of the Cd absorber, the neutron density will be identical to that observed at very large distance ($z\to\infty$) from the free surface of the medium, i.e.,

$$A_{\text{th}}(\infty) \sim N_0(\infty) = s_0\, T \quad \text{neutrons/cm}^3 \tag{120.23a}$$

where T is the mean life of thermal neutrons for capture in the medium.

The presence of the cadmium absorber transforms the x, y-plane into a free surface for thermal neutrons without disturbing their source distribution at all.

Thus we are led to calculating the neutron density distribution in a homogeneous medium filling one-half space. The result is immediately obtained by noticing that the neutron density at a depth z is given by the difference between its value (120.23a) and the number of neutrons which, starting from the considered position, leak out of the medium through its free surface. Thus, we may write

$$N_0(z) = s_0 T - s_0 T P(z) \tag{120.23b}$$

where $P(z)$ is the probability of a neutron at depth z to escape from the medium through its surface. Making use for $P(z)$ of the expression (122.7) derived in the one-dimensional approximation with isotropic scattering in the l.s. (Sect. 122), we obtain

$$N_0(z) = s_0 T \left(1 - \frac{\sqrt{N}}{\sqrt{N}+1} e^{-\frac{z}{L}}\right) \tag{120.24}$$

from which it follows that

$$\eta = \sqrt{N} + 1. \tag{120.25a}$$

This differs very little—for $N \gg 1$—from the rigorous expression valid for isotropic scattering in the l.s., (123.38a). If the scattering is not isotropic, the ratio η is given by Eq. (123.38b) which, for N very large, reduces to

$$\eta = \sqrt{N(1-\bar{\mu}_0)} + 1. \tag{120.25b}$$

The experimental value of the ratio η amounts to about 12 in paraffin and to about 50 in graphite. These large values indicate that the two surfaces of the blocks A and B of Fig. 244 have very large reflecting powers; if the albedo were close to zero, the insertion of the cadmium absorber would reduce the activity of the detector by a factor 2 since in one case the neutrons impinge upon the detector on both sides, while in the other case they only arrive from block A.

Let us now derive the relationship between η and β_e. Let us call ζ the probability of a thermal neutron, leaving A (or B) and crossing the detector, to be captured in the detector; its value has to be understood as an average over the angular distribution of the neutrons[1]. The activity $A_{\text{th}}(0)$ is due to neutrons coming out only from block A and crossing the detector just once; it is obviously proportional to ζ. The activity $A_{\text{th}}(\infty)$, on the contrary, arises from neutrons coming out of both blocks and crossing the detector a great number of times. It can be calculated by means of the following reasoning: a neutron coming out of one of the two blocks, let us say block B, has a probability ζ of being captured and a probability $1-\zeta$ of not being captured. In the second case it has a probability β_e of returning to the detector and cross it a second time.

At its second crossing of the detector, the neutron has again the probability ζ of being captured and the probability $1-\zeta$ of passing through it. If it has not been captured, there is once more a probability β_e that it returns to the detector a second time, etc. Thus the activity $A_{\text{th}}(\infty)$ turns out to be given, apart from unessential factors, by the geometrical series

$$A_{\text{th}}(\infty) = 2\zeta\left[1 + (1-\zeta)\beta_e \cdot (1-\zeta)^2\beta_e^2 + \cdots\right] = \frac{2\zeta}{1-(1-\zeta)\beta_e}. \tag{120.26}$$

The factor 2 arises from the fact that neutrons coming out of both blocks A and B contribute equally to $A_{\text{th}}(\infty)$.

Thus we may write

$$\eta_\zeta = \frac{2}{1-\beta_e+\zeta\beta_e} \tag{120.27a}$$

---

[1] If the angular distribution of the incident neutrons is taken to be (122.20), one finds, for $X\Sigma_a \ll 1$,

$$\xi = \sqrt{3}\, X \Sigma_a$$

where X is the thickness of the detector and Σ_a is its macroscopic activation cross section.

which, for $\zeta \to 0$—i.e., in the case of an infinitely thin detector—reduces to

$$\eta = \frac{2}{1-\beta_e}. \tag{120.27b}$$

Solving this equation with respect to β_e we obtain

$$\beta_e = 1 - \frac{2}{\eta} = 1 - 2\,\frac{A_{\text{th}}(\infty)}{A_{\text{th}}(0)} \tag{120.28}$$

which actually may be used as a definition of the experimental albedo. Media such as water or graphite have very large experimental albedos, as we recognize by introducing the above mentioned values of η into (120.28). The result ist $\beta_e \approx 0.82$ for water, and $\beta_e \approx 0.93$ for graphite (see Table 104, p. 596).

121. The solution of the stationary diffusion equation in various geometries. In Sect. 119 we have discussed the boundary conditions and the source strength reduction that should be adopted in order that the solution of the stationary diffusion equation (114.7) approximates as nearly as possible the rigorous solution of the transport theory (Sects. 116 to 118).

In this section, we shall derive the solution of the diffusion equation in various geometries of interest in many practical problems; a wider collection on solutions may be found in various articles[1] and books[2].

We rewrite, for convenience, the diffusion equation

$$\nabla^2 \Phi_0 - \frac{1}{L^2}\,\Phi_0 + \frac{\bar{S}}{D} = 0 \tag{121.1}$$

and recall that $\bar{S}$ is the source strength corrected according to the criteria discussed in Sect. 119ε, D is the diffusion coefficient and L is the diffusion length. Various expressions for these two quantities have been derived in the previous sections. The diffusion coefficient is given by Eq. (115.9b) in the P_1-approximation and by Eq. (117.10) in the rigorous treatment of the source-free non-capturing medium with anisotropic scattering, and is in general expressed in terms of L and Σ_a by Eq. (117.20b). The diffusion length L is given: in elementary diffusion theory with isotropic scattering by (114.9), and in transport theory for a weakly absorbing medium with almost isotropic scattering by Eq. (117.21a) which reduces to (117.21b) in the case of isotropic scattering. For isotropic scattering and any value of the capture cross section, L is given in graphical form in Fig. 227. In all other cases the diffusion length may be calculated as the inverse of the absolute value of the first—and usually the only—real root, physically permitted according to (117.17), of a determinant equation considered in Sect. 117.

In order to find the solution of Eq. (121.1) for various geometries of the source and the surrounding medium, we may take advantage of the fact that we already know its solution for an isotropic point source embedded in an infinite homogeneous medium. This solution, normalized to unit source strength, can be used as the GREEN's function $G_p(\boldsymbol{r}, \boldsymbol{r}')$ which allows the construction of the solution of Eq. (121.1) corresponding to any arbitrary distribution $s(\boldsymbol{r})$ of isotropic sources in an infinite medium; according to (71.1) or (118.3), we have

$$\Phi_0(\boldsymbol{r}) = \int G_p(\boldsymbol{r}' - \boldsymbol{r})\,\bar{s}(\boldsymbol{r}')\,d\boldsymbol{r}'. \tag{121.2}$$

Thus, instead of looking directly for the solutions of Eq. (121.1) corresponding to various geometries of the neutron source, we can construct them by means of

[1] P. R. WALLACE: Nucleonics **4** (2) 30 (1949); **4** (3) 48 (1949).

[2] S. GLASSTONE and M. C. EDLUND: Nuclear reactor theory. New York: Van Nostrand 1952.

(121.2). The same procedure may usually be applied even to the case of finite media by a convenient superposition of images.

The expression for $G_p(\boldsymbol{r}, \boldsymbol{r}')$ is readily obtained from (116.19c) by adapting it to just 1 neutron leaving the source per second. Thus we stipulate that

$$\lim_{R\to 0} 4\pi R^2 D \frac{dG_p}{dR} = -1 \qquad (R = |\boldsymbol{r} - \boldsymbol{r}'|), \tag{121.3}$$

from which it follows

$$G_p(R) = G_p(\boldsymbol{r}, \boldsymbol{r}') = \frac{e^{-\frac{|\boldsymbol{r}-\boldsymbol{r}'|}{L}}}{4\pi D|\boldsymbol{r}-\boldsymbol{r}'|}. \tag{121.4}$$

This expression is the GREEN'S function of the elementary diffusion theory. It is sometimes called the *point diffusion kernel* because of the following remark.

Eq. (121.2) can be interpreted in two different ways; if the source distribution $\bar{s}(\boldsymbol{r})$ is known, it simply allows the calculation of the flux distribution by superposition of solutions corresponding to a point source. It can, however, also be viewed as an integral equation for the source density $\bar{s}(\boldsymbol{r})$ [*19*]. Such a point of view can be adopted for determining the neutron distribution in a medium with regions where the neutron source density $\bar{s}(\boldsymbol{r})$ is proportional to the neutron flux.

An example of this kind of problems is met in calculating the flux depression produced by an absorbing foil placed in an infinite medium where neutrons are produced at a constant and uniform rate per sec and per cm³ (Sect. 135). If the foil is very thin, the effect of its absorption may be described as a negative source extending over the foil volume of strength

$$s = -\Sigma_a \Phi_0 \text{ neutrons/sec cm}^3.$$

Since s is proportional to Φ_0, Eq. (121.2) becomes an integral equation in the unknown flux distribution. Another example of problems of the same kind is the determination of the flux distribution in a multiplying medium—actually rather unrealistic—in which k secondary thermal neutrons are produced directly per each captured thermal neutron [*19*]. Also in this case the source strength at a given point is proportional to the local neutron flux, so that Eq. (121.2) becomes an integral equation in the unknown function $\Phi_0(\boldsymbol{r})$.

In this section, we use Eq. (121.2) only to calculate the flux distribution produced in an infinite homogeneous medium by a few very simple sources which may be used in turn as *diffusion kernels:* the plane source, the spherical shell source, the linear source and the cylindrical shell source.

The details of the integrals calculation may be found in Ref. [*19*] and are not reported here. The results are summarized in Table 88 (p. 554) to which we add the following remarks.

The spherical shell diffusion kernel of radius r' can be written

$$\left.\begin{aligned} G_S(r, r') &= \frac{1}{8\pi\varkappa_0 D r r'}\left[e^{-\varkappa_0|r-r'|} - e^{-\varkappa_0|r+r'|}\right] \\ &= \frac{\sinh \varkappa_0 r'}{\varkappa_0 r'}\,\frac{e^{-\varkappa_0 r}}{4\pi D r} \qquad r > r' \\ &= \frac{e^{-\varkappa_0 r'}}{4\pi D r'}\,\frac{\sinh \varkappa_0 r}{\varkappa_0 r} \qquad r < r'. \end{aligned}\right\} \tag{121.5}$$

Thus, at points outside the shell ($r > r'$), the neutron flux is the same as if the shell were concentrated at its center with a total strength $\frac{\sinh \varkappa_0 r'}{\varkappa_0 r'}$. Inside ($r < r'$), the shell cannot be replaced by a single point [*19*].

The same remark holds in the case of the cylindrical diffusion kernel. This result is a generalization of well-known theorems of electrostatic potential theory, as we should expect by noticing that Eq. (121.1) reduces to the potential equation as $\varkappa_0 = \frac{1}{L} \to 0$.

Therefore this property holds only in the diffusion theory approximation and not in the rigorous transport theory.

Finally, it can readily be seen that the plane diffusion kernel G_{pl} and the plane transport kernel K_{pl} considered in Sect. 113 (Table 85, p. 510) are related by the formula [19]:

$$K_{\text{pl}}(z, z') = D \int_{\Sigma_{\text{tot}}}^{\infty} G_{\text{pl}}(\varkappa_0, z, z')\, d\varkappa_0. \tag{121.6}$$

Relationships of the same type also hold in other geometries.

Table 88. *The diffusion kernels* [19] $\varkappa_0 = \frac{1}{L}$.

Geometry	Notation	Source normalization per sec	Flux
Point	$G_p(\boldsymbol{r} - \boldsymbol{r}')$	1 neutron	$\frac{e^{-\varkappa_0 \lvert \boldsymbol{r} - \boldsymbol{r}' \rvert}}{4\pi D \lvert \boldsymbol{r} - \boldsymbol{r}' \rvert}$
Plane	$G_{\text{pl}}(z, z')$	1 neutron/cm²	$\frac{e^{-\varkappa_0 \lvert z - z' \rvert}}{2\varkappa_0 D}$
Spherical shell	$G_s(r, r')$	1 neutron per shell of radius r'	$\frac{1}{8\pi r r' \varkappa_0 D} \left[e^{-\varkappa_0 \lvert r - r' \rvert} - e^{-\varkappa_0 \lvert r + r' \rvert}\right]$
Line	$G_l(r, \varphi, r', \varphi')$	1 neutron per unit length	$\frac{1}{2\pi D} K_0(\varkappa_0 \varrho)$ $\varrho^2 = r^2 + r'^2 - 2 r r' \cos(\varphi - \varphi')$
Cylindrical shell	$G_c(r, r')$	1 neutron per shell of radius z' and unit length	$\frac{1}{2\pi D} \begin{cases} K_0(\varkappa_0 r)\, I_0(\varkappa_0 r') & r > r' \\ K_0(\varkappa_0 r')\, I_0(\varkappa_0 r) & r < r' \end{cases}$

I_0 and K_0 are modified Bessel functions of the first and second kind of zero order (N. W. McLACHLAN: Bessel functions for engineers. Oxford. Clarendon Press 1954; A. ERDÉLYI, W. MAGNUS, F. OBERHETTINGER and F. G. TRICOMI. Higher transcendental functions II. New York: McGraw Hill 1953).

We turn now to the case of finite media of particular shapes, such as a rectangular parallelepiped or a cylinder with a plane neutron source distributed over one of its bases. The results of the present discussion will be used in Sect. 128γ for the interpretation of a few experiments designed for the determination of the diffusion length.

The solutions of these problems can readily be found by writing Eq. (121.2) in a coordinate system appropriated to the geometry of the medium and then separating the variables. As coordinates one uses rectangular Cartesian coordinates in the case of the parallelepiped, and cylindrical coordinates in the case of the cylinder. The procedure is very similar to the one followed in Sect. 76 and, therefore, it is not repeated here in all details.

In the case of a rectangular parallelepiped of sides $2a$, $2b$, and $2c$ with a source $S(x, y)$ distributed over its base $z=0$, we should find the solution $\Phi_0(x, y, z)$ of the equation

$$\frac{\partial^2 \Phi_0}{\partial x^2} + \frac{\partial^2 \Phi_0}{\partial y^2} + \frac{\partial^2 \Phi_0}{\partial z^2} - \varkappa^2 \Phi_0 = 0, \tag{121.7}$$

satisfying (a) the initial condition

$$D\left(\frac{\partial \Phi_0}{\partial z}\right)_{z=0} = -\frac{1}{2} S(x, y), \tag{121.8a}$$

and (b) the boundary conditions, similar to (76.26), that the flux should be zero on the extrapolated boundary defined by the planes

$$\left.\begin{aligned} x &= \pm x_0 = \pm (a + 0.71\, l_{\text{tr}}), \\ y &= \pm y_0 = \pm (b + 0.71\, l_{\text{tr}}), \\ z &= \pm z_0 = \pm (c + 0.71\, l_{\text{tr}}). \end{aligned}\right\} \tag{121.8b}$$

We assume that the variables are separable

$$\Phi_0(x, y, z) = X(x)\, Y(y)\, Z(z)$$

so that Eq. (121.7) is transformed into

$$\frac{1}{X}\frac{d^2 X}{d x^2} + \frac{1}{Y}\frac{d^2 Y}{d y^2} + \frac{1}{Z}\frac{d^2 Z}{d z^2} - \varkappa_0^2 = 0. \tag{121.9}$$

Since each of the first three terms depends upon a different variable, this equation can be satisfied only if each one of them is a constant

$$\frac{1}{X}\frac{d^2 X}{d x^2} = -\alpha^2; \quad \frac{1}{Y}\frac{d^2 Y}{d y^2} = -\beta^2; \quad \frac{1}{Z}\frac{d^2 Z}{d z^2} = \omega^2, \tag{121.10}$$

where

$$-\alpha^2 - \beta^2 + \omega^2 - \varkappa^2 = 0. \tag{121.11}$$

The choice of the signs for the constants α^2, β^2 and ω^2 anticipates, in part, the consequences of stipulating the boundary conditions (121.8b) which can be fulfilled only by solutions of the Eqs. (121.10) of the form

$$\left.\begin{aligned} X_l &= A_l \cos l \frac{\pi}{2}\frac{x}{x_0}, \\ Y_m &= B_m \cos m \frac{\pi}{2}\frac{y}{y_0}, \\ Z_\Omega &= C_\omega \sinh \omega (z_0 - z) = C'_\omega\, \mathrm{e}^{-\omega z}(1 - \mathrm{e}^{-2\omega(z_0 - z)}) \end{aligned}\right\} (l, m = \text{odd integers}) \tag{121.12}$$

with A_l, B_m and C_ω arbitrary constants and l and m odd positive integers. Because of (121.11), the constant ω^2 can only take on the discrete set of values

$$\omega_{lm}^2 = \varkappa^2 + \frac{l^2}{4}\frac{\pi^2}{x_0^2} + \frac{m^2}{4}\frac{\pi^2}{y_0^2}. \tag{121.13}$$

Thus, any expression of the form

$$S_{lm} \cos\left(l \frac{\pi}{2}\frac{x}{x_0}\right) \cdot \cos\left(m \frac{\pi}{2}\frac{y}{y_0}\right) \mathrm{e}^{-\omega_{lm} z}\left(1 - \mathrm{e}^{-2\omega_{lm}(z_0 - z)}\right) \tag{121.14}$$

provides a particular solution of Eq. (121.7). The solution which satisfies the initial condition (121.8a) is obtained by a superposition of an infinite number of particular solutions of this type. Thus we have

$$\left.\begin{aligned}\Phi_0(x,y,z) = \frac{2}{x_0\, y_0 \cdot D} \sum_{l,\, m\,\mathrm{odd}} S_{lm} L_{lm} \cos\left(l\,\frac{\pi}{2}\,\frac{x}{x_0}\right) \times \\ \times \cos\left(m\,\frac{\pi}{2}\,\frac{y}{y_0}\right) \mathrm{e}^{-\frac{z}{L_{lm}}}\left(1-\mathrm{e}^{-\frac{2}{L_{lm}}(z_0-z)}\right)\end{aligned}\right\} \tag{121.15}$$

where

$$\frac{1}{L_{lm}^2} = \omega_{lm}^2 = \varkappa_0^2 + \frac{\pi^2}{4}\left(\frac{l^2}{x_0^2} + \frac{m^2}{y_0^2}\right) \tag{121.16}$$

and

$$S_{lm} = \int_{-x_0}^{+x_0} dx \int_{-y_0}^{+y_0} dy\, S(x,y) \cos\left(l\,\frac{\pi}{2}\,\frac{x}{x_0}\right) \cos\left(m\,\frac{\pi}{2}\,\frac{y}{y_0}\right). \tag{121.17}$$

If the parallelepiped is very long ($z_0 \to \infty$), the exponential $\exp\left[-\frac{2}{L_{lm}}(z_0 - z)\right]$ tends to zero, and the various terms in (121.15) decrease exponentially along the z axis with a relaxation length L_{lm}:

$$\left.\begin{aligned}\Phi_0(x,y,z) = \frac{2}{x_0\, y_0\, D} \sum_{l,\, m\,\mathrm{odd}} S_{lm} L_{lm} \cos\left(l\,\frac{\pi}{2}\,\frac{x}{x_0}\right) \cos\left(m\,\frac{\pi}{2}\,\frac{y}{y_0}\right) \mathrm{e}^{-\frac{z}{L_{lm}}} \\ (\text{for } z_0 \to \infty).\end{aligned}\right\} \tag{121.18}$$

From Eq. (121.18), we see that the relaxation length of the l, m-harmonic decreases as the corresponding orders l and m increase; thus, far from the source, only the (0, 0) harmonic remains, i.e., one has

$$\Phi_0(x,y,z) \to S_{00} L_{00} \cos\left(\frac{\pi}{2}\,\frac{x}{x_0}\right) \cos\left(\frac{\pi}{2}\,\frac{y}{y_0}\right) \mathrm{e}^{-\frac{z}{L_{00}}} \quad (\text{for } c \to \infty,\ z \gg L_{00}) \tag{121.19a}$$

where

$$\frac{1}{L_{00}^2} = \frac{1}{L^2} + \left(\frac{\pi}{2}\right)^2 \left[\frac{1}{x_0^2} + \frac{1}{y_0^2}\right]. \tag{121.19b}$$

The problem of a cylinder of radius R and height h with a neutron source distributed over its base $z=0$ may be treated by means of essentially the same procedure with the only difference being that now one starts from Eq. (121.1) written in cylindrical coordinates. The result, similar to (121.15), is

$$\Phi_0(r,z,\varphi) = \sum_{l=0,1,2\ldots} \sum_{m=1,2,3} (A_{lm} \cos l\varphi + B_{lm} \cos m\varphi)\, J_l\left(\lambda_{lm} \frac{r}{\overline{R}}\right) \frac{\sinh \frac{\overline{h}-z}{b_{lm}}}{\sinh \frac{\overline{h}}{b_{lm}}} \tag{121.20}$$

where $J_l(x)$ are the Bessel functions, λ_{lm} are the roots of J_l ($m=1, 2, \ldots$; $\lambda_{01} = 2.4048$; $\lambda_{02} = 5.5201$; ...) and b_{lm} is the relaxation length of the l, m-harmonic

$$\frac{1}{b_{lm}^2} = \frac{1}{L^2} + \frac{\lambda_{lm}^2}{\overline{R}^2} \tag{121.21}$$

where

$$\overline{R} = R + 0.71\, l_{\mathrm{tr}}, \qquad \overline{h} = h + 0.71\, l_{\mathrm{tr}}$$

and, finally, A_{lm}, B_{lm} are constants to be determined by imposing upon $\Phi_0(r,z,\varphi)$ an initial condition similar to (121.8a). The same remarks hold here as in the case of (121.15); for $h \to \infty$ the z dependence along the cylinder axis becomes purely exponential: $\exp[-z/b_{lm}]$, and since b_{lm} decreases as l and m increase, at very large distance from the source, the flux distribution (121.20) reduces to its fundamental mode. Thus we have

$$\Phi_0(r,z,\varphi) \to A_{01} J_0\left(\lambda_{01}\frac{r}{\bar{R}}\right)\frac{1}{\sinh\frac{\bar{h}}{b_{01}}} e^{-\frac{z}{b_{01}}} \quad (\text{for } h \to \infty,\ z \gg b_{01}) \quad (121.22\text{a})$$

with

$$\frac{1}{b_{01}^2} = \frac{1}{L^2} + \left(\frac{2.4048}{\bar{R}}\right)^2. \quad (121.22\text{b})$$

122. The one-dimensional medium model. In this section we are going to mention an oversimplified model, introduced by FERMI [8][1] at a very early stage of neutron physics, which allows a rather quick estimate of the various features of thermal neutron diffusion. As we shall see below, in spite of its extreme crudeness, this model provides predictions which are in many cases in reasonable agreement with those derived from more elaborate or even rigorous treatments. It actually provides the first step of a successive approximation procedure which will be considered in Sect. 123 ε.

The neutron is described as a point which moves in a fictitious one-dimensional medium of scattering mean free path l_f and a probability $1/N$ of being captured at each collision. In other words, the actual "random walk" of the neutron in space is replaced by the "zitter" movement of its projection upon a straight line z.

The fictitious point has a probability $(N-1)/N$ of not being captured at each collision and—if the actual scattering of neutrons in space is assumed to be isotropic in the l.s.—it has, a probability $\frac{1}{2}$ of moving after the collision, either in the same direction or in the opposite direction it had immediately before. The same type of reasoning may be followed if the scattering is anisotropic in the l.s. $(\bar{\mu}_0 \neq 0)$, the only difference being that the neutron in this case has the probabilities $\frac{1+\bar{\mu}_0}{2}$ and $\frac{1-\bar{\mu}_0}{2}$ respectively of keeping and reversing the direction of its motion at each collision.

In order to transform the formulas obtained from such a model to the corresponding actual three-dimensional case, one can take for N the ratio (120.4) of the total cross section to the absorption cross section, and for l_f the actual scattering mean free path l_s divided by $\sqrt{3}$:

$$l_f = \frac{l_s}{\sqrt{3}}. \quad (122.1\text{a})$$

Since the time interval between two successive collisions should obviously be the same for the neutron and for its projection on the z axis, the velocity v_f of the fictitious point is given by

$$v_f = v\frac{l_f}{l_s} = \frac{v}{\sqrt{3}}. \quad (122.1\text{b})$$

The relation (122.1 a) may be justified by various arguments. For example, one can notice that in the three-dimensional case, the average of the square of the displacement undergone by a neutron in m successive steps of its random walk is given by

$$\overline{r^2} = 2 l_s^2 m.$$

[1] E. FERMI: AECD-2664 (1951).

The factor 2 appearing in this equation arises, according to (73.5), from the fact that the average of the square of many successive free paths crossed by a neutron, is twice the square of its mean free path. If we denote by $\overline{z^2}$ the projection of $\overline{r^2}$ upon the z direction, one has

$$\overline{z^2} = \frac{1}{3}\,\overline{r^2} = \frac{2}{3}\,l_s^2\, m. \tag{122.2}$$

On the other hand, the value of $\overline{z_f^2}$ calculated for the point moving in the fictitious one-dimensional medium is given by

$$\overline{z_f^2} = 2\,l_f^2\, m \tag{122.3}$$

which provides the correct value (122.2), if l_f is assumed to be related to l_s by (122.1a).

As a typical example how to use this model, we may consider the following case: a neutron is placed at a depth z inside a slab of infinite transverse dimensions and finite thickness a. We want to know the probability $P(z)$ for the neutron to get out of the slab through its surface at $z=0$. The problem is obviously equivalent to asking for the fraction of neutrons escaping from the slab through its surface at $z=0$ if an isotropic point source, emitting Q_0 thermal neutrons per second, is placed at a depth z inside the medium.

The probability $P_f(z)$ relative to the corresponding one-dimensional model can be separated into the sum of two terms. The first one is the probability that the neutron starts to move from its initial position z towards the surface $z=0$ and gets out of it without having undergone any collision: $\frac{1}{2}\,\mathrm{e}^{-z/l_f}$. The second term is the product of three probabilities: (a) the probability that the neutron undergoes its first collision between z' and $z'+dz'$: $\frac{1}{2}\,\mathrm{e}^{-\frac{|z-z'|}{l_f}}\,dz'/l_f$; (b) the probability that this collision is elastic: $(N-1)/N$; and (c) the probability that the neutron gets out of the slab through its surface at $z=0$ from the new position z': $P_f(z')$. By integration over z' between $z=0$ and $z=a$, we obtain

$$P_f(z) = \frac{1}{2}\,\mathrm{e}^{-\frac{z}{l_f}} + \frac{N-1}{2N\,l_f}\int\limits_0^a P_f(z')\,\mathrm{e}^{-\frac{|z-z'|}{l_f}}\,dz'. \tag{122.4}$$

This is a very simple integral equation, the solution of which can be put into the form

$$P_f(z) = A\,\mathrm{e}^{\varkappa z} + B\,\mathrm{e}^{-\varkappa z}.$$

The constants A, B and $\varkappa$ may be found by inserting the latter expression into (122.4). The result is

$$\left.\begin{aligned} A &= \frac{(N-\sqrt{N})\,\mathrm{e}^{-\varkappa a}}{(\sqrt{N}+1)^2\,\mathrm{e}^{\varkappa a} - (\sqrt{N}-1)^2\,\mathrm{e}^{-\varkappa a}}; \qquad B = \frac{(N+\sqrt{N})\,\mathrm{e}^{\varkappa a}}{(\sqrt{N}+1)^2\,\mathrm{e}^{\varkappa a} - (\sqrt{N}-1)^2\,\mathrm{e}^{-\varkappa a}};\\ \varkappa &= \frac{1}{l_f\sqrt{N}}\,. \end{aligned}\right\} \tag{122.5}$$

The expressions (122.5) become much simpler if the slab is assumed to extend to infinity on its right-hand side ($a\to\infty$). Then one has

$$A=0;\qquad B = \frac{\sqrt{N}}{\sqrt{N}+1},\qquad \varkappa = \frac{1}{l_f\sqrt{N}} \tag{122.6a}$$

from which it follows that

$$P_f(z) = \frac{\sqrt{N}}{\sqrt{N}+1}\,\mathrm{e}^{-\frac{z}{l_f\sqrt{N}}}. \tag{122.6b}$$

Making use of (122.1a), we obtain the corresponding probability of the actual three-dimensional problem:

$$P(z) = \frac{\sqrt{N}}{\sqrt{N}+1}\,e^{-\frac{z}{L}} \tag{122.7}$$

where

$$L = \left(\frac{N l_s^2}{3}\right)^{\frac{1}{2}}$$

is the diffusion length. In the approximation

$$N = \frac{\Sigma_{\text{tot}}}{\Sigma_a} = \frac{l_a}{l_{\text{tot}}} \approx \frac{l_a}{l_s}$$

—which is correct in weakly absorbing media—we obtain

$$L = \left(\frac{l_s l_a}{3}\right)^{\frac{1}{2}}, \tag{122.8}$$

a result identical with (114.9).

We can now use the various expressions derived above in order to compute the *albedo* (Sect. 120).

In the case of the fictitious one-dimensional medium, we may proceed as follows: a neutron moving from the left to the right enters the medium through its boundary at $z=0$. The probability that it suffers its first collision between z and $z+dz$, without being captured, is given by

$$\frac{N-1}{N}\,e^{-\frac{z}{l_f}}\,\frac{dz}{l_f}. \tag{122.9a}$$

The one-dimensional albedo is obtained by the multiplication of (122.9a) by $P_f(z)$ followed by the integration with respect to z over the slab thickness:

$$\beta_f(a) = \frac{N-1}{l_f N}\int_0^a P_f(z)\,e^{-\frac{z}{l_f}}\,dz. \tag{122.9b}$$

Making use of (122.6b), one obtains

$$\beta_f(a) = \frac{(N-1)\,[e^{\varkappa a} - e^{-\varkappa a}]}{(\sqrt{N}+1)^2\,e^{\varkappa a} - (\sqrt{N}-1)^2\,e^{-\varkappa a}}. \tag{122.10a}$$

For a tending to infinity, the previous equation reduces to

$$\beta_f(\infty) = \frac{\sqrt{N}-1}{\sqrt{N}+1} = 1 - \frac{2}{\sqrt{N}} + \cdots. \tag{122.10b}$$

The same type of reasoning leads to the *transmission coefficient* of a one-dimensional slab of thickness a:

$$T(a) = \frac{4\sqrt{N}}{(\sqrt{N}+1)^2\,e^{\varkappa a} - (\sqrt{N}-1)^2\,e^{-\varkappa a}}. \tag{122.11}$$

In the case of a non-capturing medium, i.e., in the limit $N\to\infty$, this expression becomes

$$T(a) = \frac{2l_f}{a+2l_f} \qquad (N\to\infty). \tag{122.12}$$

To calculate the albedo in the three-dimensional case we should keep in mind that its value clearly depends upon the angular distribution of the incident neutrons on the medium surface. Therefore, we should first compute the albedo for neutrons incident upon the medium surface in a direction forming an angle ϑ with the normal to the surface. This is obtained by repeating the reasoning followed in the one-dimensional case with the probability (122.9a) replaced by

$$\frac{N-1}{N}\,\mathrm{e}^{-\frac{z}{l_s\cos\vartheta}}\,\frac{dz}{l_s\cos\vartheta}$$

and $P_f(z)$ replaced by $P(z)$. Thus, now we have, instead of (122.9b),

$$\beta(a,\vartheta)=\frac{N-1}{l_s N}\int\limits_0^a P(z)\,\mathrm{e}^{-\frac{z}{l_s\cos\vartheta}}\,\frac{dz}{\cos\vartheta}\,. \tag{122.13a}$$

The integration is readily done; for the interesting case $a\to\infty$, one obtains

$$\beta(\vartheta)=\frac{\sqrt{N}-1}{\sqrt{N}+\sqrt{3}\cos\vartheta}=1-\frac{1+\sqrt{3}\cos\vartheta}{\sqrt{N}}+\cdots. \tag{122.13b}$$

We may now use this expression for computing the albedo corresponding to any assigned angular distribution of the incident neutrons. Thus, for example, if we assume the Lambert law, i.e., that the neutrons have the probability

$$\cos\vartheta\,\frac{d\omega}{\pi}=2\cos\vartheta\sin\vartheta\,d\vartheta$$

to move within the solid angle between ϑ and $\vartheta+d\vartheta$, we obtain

$$\beta=\int\limits_0^{\pi/2}\beta(\vartheta)\,2\cos\vartheta\sin\vartheta\,d\vartheta=\frac{2}{\sqrt{3}}(N-1)\left[1-\frac{\sqrt{N}}{3}\ln\left(1+\sqrt{\frac{3}{N}}\right)\right] \tag{122.14a}$$

which, if N is very large, reduces to

$$\beta=1-\frac{1+\frac{2}{\sqrt{3}}}{\sqrt{N}}+\cdots. \tag{122.14b}$$

The numerical coefficient $1+2/\sqrt{3}=2.15$ has to be compared with the correct value given by Wick[1] (120.19), from which it differs by 7%.

The same procedure can be used to calculate the transmission coefficient for neutrons incident in directions between ϑ and $\vartheta+d\vartheta$ upon a slab of thickness a. The result is, for $N\to\infty$,

$$T(\cos\vartheta)=\frac{l_s}{a+\frac{2l_s}{\sqrt{3}}}\left[\cos\vartheta+\frac{1}{\sqrt{3}}-\left(\cos\vartheta-\frac{1}{\sqrt{3}}\right)\mathrm{e}^{-\frac{a}{l_s\cos\vartheta}}\right]. \tag{122.15}$$

We may now derive the solution of Milne's problem (Sect. 118ε) by means of the one-dimensional model. A non-capturing medium fills the half-space at the right side of the plane $z=0$ bounding it towards vacuum. Neutron sources emitting Q_0 thermal neutrons per second are located inside the medium at a very large distance (∞) from its surface. These neutrons diffuse from the right to the left towards the free surface at $z=0$. In the absence of capture, the diffusion length is infinite so that the one-dimensional diffusion equation reduces to $d^2\Phi_0/dz^2=0$. This means that we should have (Sect. 117α)

$$\Phi_0(z)=A+Bz \tag{122.16}$$

[1] G.C. Wick: Rend. Linc. **23**, 774 (1936).

where A and B are two constants. These can be determined by stipulating that the number of neutrons crossing each plane z per second should be equal to Q_0; i.e., we postulate that

$$Q_0 = \frac{1}{2}\frac{1}{l_f}\int_0^\infty \Phi_0(z+z')\,\mathrm{e}^{-\frac{z'}{l_f}}\,dz' - \frac{1}{2l_f}\int_0^z \Phi_0(z-z')\,\mathrm{e}^{-\frac{z'}{l_f}}\,dz'. \tag{122.17}$$

By inserting (122.16) into (122.17) and computing the integrals, one obtains

$$Q_0 = l_f B; \qquad A = l_f B$$

from which it follows that

$$\Phi_0(z) = B(z + l_f). \tag{122.18a}$$

Making use of (122.1), we finally obtain

$$\Phi_0(z) = B\left(z + \frac{l_s}{\sqrt{3}}\right). \tag{122.18b}$$

Comparing this expression with (118.28b), we see that the one-dimensional model gives the correct behaviour of the asymptotic solution in a non-capturing medium, with the only difference that the extrapolated end point in the present case has the value $d/l_s = \frac{1}{\sqrt{3}} = 0.58$, while elementary diffusion theory gives $d/l_s = \frac{2}{3} = 0.75$, and the rigorous transport theory $d/l_s = 0.71$.

The neutron distribution (122.18b) has been used by Fermi to calculate the angular distribution of thermal neutrons diffusing out of a plane infinite free surface bounding a medium containing sources, which are all located at very great distances.

The number of collisions taking place per second in an element of volume dV located at depth z inside the medium is given by

$$\Phi_0(z)\,\frac{dV}{l_s}.$$

If the scattering is isotropic in the l.s., the probability that one of the colliding neutrons gets out of the medium surface $z=0$ with a direction of motion between ϑ and $\vartheta + d\vartheta$ without suffering any further collisions is

$$\mathrm{e}^{-\frac{z}{l_s\cos\vartheta}}\,\frac{\sin\vartheta\,d\vartheta}{2}.$$

Therefore, the total number of neutrons diffusing out of the medium surface in the direction between ϑ and $\vartheta + d\vartheta$ is given by

$$\Phi(0,\cos\vartheta)\cos\vartheta\,d(\cos\vartheta) = \text{const}\times\sin\vartheta\,d\vartheta\int_0^\infty \Phi_0(z)\,\mathrm{e}^{-\frac{z}{l_s\cos\vartheta}}\,dz \tag{122.19}$$

where dV has been taken equal to $a\,b\,dz$, with a and b being two arbitrary constant lengths in the two directions perpendicular to the z direction.

Introducing (122.18b) into (122.19) and normalizing to 1 neutron leaving per second in the solid angle 2π, one obtains the *Fermi's angular distribution*

$$\frac{2\sqrt{3}}{2+\sqrt{3}}\,(1+\sqrt{3}\cos\vartheta)\cos\vartheta\,d(\cos\vartheta)\,\frac{d\varphi}{2\pi}. \tag{122.20}$$

This expression has been frequently used in the past and is still used sometimes today. According to rigorous computations made by Placzek[1] already mentioned

[1] G. Placzek: Phys. Rev. **72**, 556 (1956).

in Sect. 118ε (Fig. 242), FERMI's law (122.20) is in error by 7.2% at $\cos\vartheta=0$ ($\vartheta=90°$), but for $\cos\vartheta>0.1$ ($\vartheta<84°$) its error is below 1% throughout. We shall come back to this point at the end of Sect. 123ε [cf. Eq. (123.36)].

123. A few remarks on second-order diffusion theory. As we have seen in Sect. 118, the exact solution of the transport equation can be derived only in a few simple cases. More complex problems are usually treated by means of the elementary diffusion theory (Sects. 114, 120 and 121) which—when supplemented by convenient boundary conditions and source corrections (Sect. 119)—can provide a not too bad description of the asymptotic part of the rigorous solution (Sects. 116 to 118). When a better approximation is required, recourse should be made to successive approximation procedures which are able to give results of arbitrarily high accuracy provided that a sufficient amount of labour is expended in their calculation. These methods, sometimes indicated as *second order diffusion theories*, are described in detail in various articles[1] and books [*19*], [*7*]; here we limit ourselves to outlining, very briefly, their main features and to report a few of the corresponding results.

Two main procedures may be distinguished: *Wick's method*[2] and *spherical harmonics method*. The latter was first suggested by JEANS, in connection with certain problems of transfer of radiation in stars, and has been developed later for neutron transport problems by the Chicago group [*19*] and the Montreal group[3,4] [*19*] of the Plutonium Project. The most systematic investigation has been made by PLACZEK and co-workers at Montreal.

For the sake of simplicity, we shall refer to the one-dimensional transport equation written in the form (115.1).

α) The spherical harmonics method. As we have already shown in Sect. 115, the spherical harmonics method proceeds as follows: the Boltzmann transport equation (115.1) is reduced to the infinite set of linear differential equations (115.4) in which the unknowns are the harmonic components $\Phi_l(z)$ of the angular number flux (79.4). This set of equation is then cut off at a finite number $L+1$ by means of the approximation (115.5).

The accuracy of the result as well as the amount of labour required to solve this finite set of equations, rapidly increase as the order of the approximation L increases.

The method may be applied even if the scattering is anisotropic and the cross sections are functions of position, so that any constant velocity problem can be solved—in principle—to any required degree of accuracy.

In the presence of neutron sources, the linear equations are non-homogeneous; the corresponding general solution is obtained by adding to any particular solution of the inhomogeneous equations, the general solution of the homogeneous set. E.g. in the case of isotropic scattering in the l.s., this set turns out to be

$$\left.\begin{aligned} \Phi_1' + \Sigma_a \Phi_0 &= 0,\\ \tfrac{2}{3}\Phi_2' + \Sigma_{\text{tot}}\Phi_1 + \tfrac{1}{3}\Phi_0' &= 0,\\ \tfrac{3}{5}\Phi_3' + \Sigma_{\text{tot}}\Phi_2 + \tfrac{2}{5}\Phi_1 &= 0,\\ \tfrac{4}{7}\Phi_4' + \Sigma_{\text{tot}}\Phi_3 + \tfrac{3}{7}\Phi_2' &= 0 \text{ etc.}, \end{aligned}\right\} \tag{123.1}$$

[1] R.E. MARSHAK, H. BROOKS and H. HURWITZ: Nucleonics **4** (6), 43 (1949).
[2] G.C. WICK: Z. Physik **121**, 702 (1943).
[3] J.C. MARK: Nat. Res. Coun. Can. Report MT 92 (1944); Report MT 97 (1945).
[4] R.E. MARSHAK: Phys. Rev. **71**, 443 (1947).

where

$$\Phi_l' = \frac{d\Phi_l}{dz}.$$

The following discussion refers to this set of equations. The P_1-approximation is obtained by considering only the first two equations with $\Phi_2(z)=0$; as we have shown in Sect. 115, by elimination of Φ_1 the elementary diffusion equation is obtained (first order diffusion theory).

Similarly, the P_2-approximation consists of considering the first three equations (123.1) with $\Phi_3=0$. These involve the three unknown functions Φ_0, Φ_1, Φ_2; it is readily seen that, by elimination of Φ_2 and Φ_1, a second order differential equation is obtained which differs from that obtained in the P_1-approximation only in the expression for the coefficients. Thus, passing from the P_1 to the P_2 approximation, the structure of the solution does not change.

The P_3-approximation involves the first four equations of the set (123.1) with $\Phi_4=0$. The elimination of Φ_3, Φ_2 and Φ_1 leads to a fourth order differential equation in Φ_0. This means that new terms with respect to the P_1- and P_2-approximations are included in the P_3-solution; as we shall see below, they correspond to the non-asymptotic part of the rigorous solution considered in Sect. 118.

The P_4-approximation does not change the structure of the analytical problem with respect to the P_3-approximation, while an essential change again takes place in passing from the P_4- to the P_5-approximation.

This is a general property of the set (123.1): *the successive approximations proceed in steps of two spherical harmonics*; each one of the odd approximations P_1, P_3, P_5, ... introduces new terms in the expression for the neutron flux with respect to the preceding even approximation. Each one of the even approximations P_2, P_4, P_6, does not produce any change in the structure of the solution with respect to the preceding odd approximation, but simply changes the expressions for the various parameters appearing in it [*19*], [*7*].

We shall come back later, from a different point of view, to this same remark. It is clear now, however, that a substantial increase of accuracy is always obtained in passing from any even approximation to the successive odd approximation, while only a minor advantage is obtained in passing from odd approximations to the successive even ones.

This is the reason why usually only odd-order approximations are used.

We may now proceed to find the actual structure of these solutions. Instead of reducing the set of $L+1$ homogeneous equations to a differential equation in $\Phi_0(z)$, it is more convenient to look for a solution of the original set of linear equations (123.1) of the form

$$\Phi_l(z) = \sum_j A_{lj}\, e^{\varkappa_j z}, \qquad l=0, 1, 2, \ldots L. \tag{123.2}$$

Inserting (123.2) into (123.1) and equating the coefficients of each $e^{\varkappa_j z}$ to zero, one obtains a set of linear homogeneous algebraic equations in the A_{lj} $(l=0, 1, 2, \ldots L)$. For example, in the P_3-approximation, this set is

$$\left.\begin{aligned}
\Sigma_a A_{0j} + \varkappa_j A_{1j} \qquad\qquad\qquad\qquad &= 0,\\
\varkappa_j A_{0j} + \Sigma_{\text{tot}} A_{1j} + \tfrac{2}{3}\varkappa_j A_{2j} \qquad\quad &= 0,\\
\tfrac{2}{5}\varkappa_j A_{1j} + \Sigma_{\text{tot}} A_{2j} + \tfrac{3}{5}\varkappa_j A_{3j} &= 0,\\
\tfrac{3}{7}\varkappa_j A_{2j} + \Sigma_{\text{tot}} A_{3j} &= 0.
\end{aligned}\right\} \tag{123.3}$$

If these equations have to be consistent, the determinant of the coefficients should be zero; an equation of degree $L+1$ in the constant $\varkappa_j$ is thus obtained. For example, in the P_3-approximation one has

$$\frac{9}{35}\frac{1}{\Sigma_{\mathrm{tot}}^2}\varkappa_j^4 - \left(1+\frac{11}{7}\frac{\Sigma_a}{\Sigma_{\mathrm{tot}}}\right)\varkappa_j^2 + 3\Sigma_a\Sigma_{\mathrm{tot}} = 0 \tag{123.4}$$

with the roots

$$\left.\begin{aligned}\varkappa_1 &= \pm(3\Sigma_{\mathrm{tot}}\Sigma_a)^{\frac{1}{2}}\left(1-\frac{2}{5}\frac{\Sigma_a}{\Sigma_{\mathrm{tot}}}\right),\\ \varkappa_2 &= \pm\frac{\sqrt{35}}{3}\Sigma_{\mathrm{tot}}\left(1+\frac{2}{5}\frac{\Sigma_a}{\Sigma_{\mathrm{tot}}}\right).\end{aligned}\right\} \tag{123.5}$$

A detailed discussion [7] shows that, in general, the determinantal equation of the odd approximations (L odd: P_1, P_3, P_5 ...) has always $L+1$ roots, i.e., the same number of roots as the number of spherical harmonics moments retained. For $h<1$ they have $\frac{L+1}{2}$ different pairs of real non-zero values of opposite sign. The pairs of smaller absolute values go to zero as h tends to unity. A similar behaviour is shown by the roots of the determinantal equation for even approximations (L even: P_2, P_4, P_6, ...) with the difference that now the number of roots is L.

Table 89 shows the expressions for these roots for both odd and even approximations up to $L=7$ [19].

Table 89. *The roots of the determinantal equation of the spherical harmonics method for isotropic scattering in the l. s.* [19].

Odd approximation	Even approximations
P_1: $\varkappa_1 = \pm\sqrt{3\Sigma_{\mathrm{tot}}\Sigma_a}$	P_2: $\varkappa_1 = \pm\dfrac{\sqrt{3\Sigma_{\mathrm{tot}}\Sigma_a}}{\sqrt{1+\frac{4}{5}\frac{\Sigma_a}{\Sigma_{\mathrm{tot}}}}}$
P_3: $\varkappa_1 = \pm\sqrt{3\Sigma_{\mathrm{tot}}\Sigma_a}\left(1-\frac{2}{5}\frac{\Sigma_a}{\Sigma_{\mathrm{tot}}}\right)$ $\varkappa_2 = \pm\frac{\sqrt{35}}{3}\Sigma_{\mathrm{tot}}\left(1+\frac{2}{5}\frac{\Sigma_a}{\Sigma_{\mathrm{tot}}}\right)$	P_4: $\varkappa_1 = \pm\sqrt{3\Sigma_{\mathrm{tot}}\Sigma_a}\left(1-\frac{2}{5}\frac{\Sigma_a}{\Sigma_{\mathrm{tot}}}\right)$ $\varkappa_2 = \pm\sqrt{\frac{45}{23}}\Sigma_{\mathrm{tot}}\left(1+0.2012\frac{\Sigma_a}{\Sigma_{\mathrm{tot}}}\right)$
P_5: $\varkappa_1 = \pm\sqrt{3\Sigma_{\mathrm{tot}}\Sigma_a}\left(1-\frac{2}{5}\frac{\Sigma_a}{\Sigma_{\mathrm{tot}}}\right)$ $\varkappa_2 = \pm 1.225\,\Sigma_{\mathrm{tot}}\left(1+0.1123\frac{\Sigma_a}{\Sigma_{\mathrm{tot}}}\right)$ $\varkappa_3 = \pm 3.203\,\Sigma_{\mathrm{tot}}\left(1+0.2877\frac{\Sigma_a}{\Sigma_{\mathrm{tot}}}\right)$	P_6: $\varkappa_1 = \pm\sqrt{3\Sigma_{\mathrm{tot}}\Sigma_a}\left(1-\frac{2}{5}\frac{\Sigma_a}{\Sigma_{\mathrm{tot}}}\right)$ $\varkappa_2 = \pm 1.146\,\Sigma_{\mathrm{tot}}\left(1+0.0685\frac{\Sigma_a}{\Sigma_{\mathrm{tot}}}\right)$ $\varkappa_3 = \pm 1.9825\,\Sigma_{\mathrm{tot}}\left(1+0.1994\frac{\Sigma_a}{\Sigma_{\mathrm{tot}}}\right)$
P_7: $\varkappa_1 = \pm\sqrt{3\Sigma_{\mathrm{tot}}\Sigma_a}\left(1-\frac{2}{5}\frac{\Sigma_a}{\Sigma_{\mathrm{tot}}}\right)$ $\varkappa_2 = \pm 1.1032\,\Sigma_{\mathrm{tot}}\left(1+0.0448\frac{\Sigma_a}{\Sigma_{\mathrm{tot}}}\right)$ $\varkappa_3 = \pm 1.592\,\Sigma_{\mathrm{tot}}\left(1+0.1400\frac{\Sigma_a}{\Sigma_{\mathrm{tot}}}\right)$ $\varkappa_4 = \pm 4.458\,\Sigma_{\mathrm{tot}}\left(1+0.2153\frac{\Sigma_a}{\Sigma_{\mathrm{tot}}}\right)$	

It should be noticed that $\varkappa_1$ is essentially the reciprocal of the diffusion length[1]. It starts, in the P_1-approximation, with the value corresponding to elementary diffusion theory and changes, by increasing the order of approximation, tending to approach the solution of the transcendental equation (117.18). The other roots $\pm\varkappa_2$, $\pm\varkappa_3$ etc., are always of the order of l_{tot}^{-1} so that the corresponding terms in the general solution will correspond to the non-asymptotic part of the rigorous solution (Sect. 118).

The fact that the even P_L-approximations involve the same number of roots as the odd P_{L-1}-approximations corresponds to the remark made above that the structure of the solution does not change in passing from an odd approximation to the successive one, while it changes essentially in passing from an even approximation to the successive odd approximation. Therefore in the following we shall only refer to odd approximations.

Table 90 contains the numberical values of the roots of the determinantal equation [7] in units of Σ_{tot}

$$\nu_j = \frac{|\varkappa_j|}{\Sigma_{\text{tot}}} \tag{123.6}$$

as computed by MARK[2] as a function of h for $L = 1, 3, 5, 7$.

Table 90. *Values of ν_j for $L = 1, 3, 5, 7$ [7].*

h	L = 1	3		5			7			
	j = 1	1	2	1	2	3	1	2	3	4
1.0	0	0	1.972027	0	1.225211	3.202945	0	1.103185	1.591779	4.458085
0.9	0.547723	0.525560	2.055185	0.525430	1.240875	3.296688	0.525430	1.108937	1.615640	4.554851
0.8	0.774597	0.711876	2.145773	0.710456	1.260956	3.393117	0.710413	1.116799	1.642629	4.652966
0.7	0.948683	0.834508	2.241834	0.829085	1.286086	3.491504	0.828671	1.127655	1.672473	4.752078
0.6	1.095445	0.922637	2 341383	0.909562	1.316073	3.591190	0.907696	1.142345	1.704602	4.851871
0.5	1.224745	0.988751	2.442706	0.964672	1.349619	3.691603	0.959481	1.160900	1.738235	4.952060
0.4	1.341641	1.039801	2.544478	1.002603	1.384742	3.792265	0.992327	1.181880	1.772515	5.052401
0.3	1.449138	1.080124	2.645751	1.029086	1.419574	3.892788	1.012964	1.203058	1.806656	5.152683
0.2	1.549193	1.112590	2.745892	1.048012	1.452833	3 992862	1.026230	1.222733	1 840027	5.252727
0.1	1.643168	1.139170	2.844500	1.061917	1.483848	4.092252	1.035120	1.240155	1.872179	5.352384
0	1.732051	1.161256	2.941340	1.072421	1.512380	4.190778	1.041352	1.255230	1.902832	5.451533

The solution of our problem in a given P_L-approximation (L odd) can now be written as follows

$$\Phi_0(z) = A_{01}^{+}\,e^{\varkappa_1 z} + A_{01}^{-}\,e^{-\varkappa_1 z} + A_{02}^{+}\,e^{\varkappa_2 z} + A_{02}^{-}\,e^{-\varkappa_2 z} + \cdots + A_{0\frac{L+1}{2}}^{-}\,e^{-\varkappa_{\frac{L+1}{2}} z} \tag{123.7a}$$

with A_{0j}^{+} and A_{0j}^{-} being arbitrary constants. The coefficients A_{lj}^{+}, A_{lj}^{-} $\left(0 < l \leqq L;\ 1 \leqq j \leqq \frac{L+1}{2}\right)$ appearing in the expressions for the other harmonic components $\Phi_l(z)$ are readily obtained: by insertion of (123.7a) into the first one of Eqs. (123.3),

[1] While in other sections this quantity is indicated by $\varkappa_0$, here it appears more convenient to use the notation $\pm\varkappa_1$ for the first pair of roots of the determinantal equation of the set of linear homogeneous algebraic equations (123.3).

[2] See footnote 3, p. 562.

one obtains A_{1j}^+, A_{1j}^-. Once these are known, the second one of these equations is used for expressing A_{2j}^+, A_{2j}^- in terms of A_{0j}^+, A_{0j}^-, A_{1j}^+, A_{1j}^-; etc.

Thus, for example, in the P_3-approximation, one has

$$\left.\begin{aligned}
\Phi_1(z) &= -\frac{\Sigma_a}{|\varkappa_1|}\left[A_1^+ e^{\varkappa_1 z} - A_1^- e^{-\varkappa_1 z}\right] - \frac{\Sigma_a}{|\varkappa_2|}\left[A_2^+ e^{\varkappa_2 z} - A_2^- e^{-\varkappa_2 z}\right],\\
\Phi_2(z) &= \left(-\frac{1}{2} + \frac{3}{2}\,\frac{\Sigma_a \Sigma_{\text{tot}}}{|\varkappa_1|^2}\right)\left[A_1^+ e^{\varkappa_1 z} + A_1^- e^{-\varkappa_1 z}\right] + \\
&\quad + \left(-\frac{1}{2} + \frac{3}{2}\,\frac{\Sigma_a \Sigma_{\text{tot}}}{|\varkappa_2|^2}\right)\left[A_2^+ e^{\varkappa_2 z} + A_2^- e^{-\varkappa_2 z}\right],\\
\Phi_3(z) &= \left(\frac{3}{14}\,\frac{|\varkappa_1|}{\Sigma_{\text{tot}}} - \frac{9}{14}\,\frac{\Sigma_a}{|\varkappa_1|}\right)\left[A_1^+ e^{\varkappa_1 z} - A_1^- e^{-\varkappa_1 z}\right] + \\
&\quad + \left(\frac{3}{14}\,\frac{|\varkappa_2|}{\Sigma_{\text{tot}}} - \frac{9}{14}\,\frac{\Sigma_a}{|\varkappa_2|}\right)\left[A_2^+ e^{\varkappa_2 z} - A_2 e^{-\varkappa_2 z}\right],
\end{aligned}\right\}\qquad (123.7\text{b})$$

where the first subscript $l=0$ of the arbitrary constants has been dropped because it is unnecessary.

Finally, the angular number flux is given by the expression

$$\Phi(z,\mu) = \tfrac{1}{2}\,\Phi_0(z) + \tfrac{3}{2}\,\Phi_1(z)\,P_1(\mu) + \tfrac{5}{2}\,\Phi_2(z)\,P_2(\mu) + \tfrac{7}{2}\,\Phi_3(z)\,P_3(\mu) \qquad (123.8)$$

which contains the $L+1$ arbitrary constants A_j^+, A_j^- appearing in (123.7a).

The terms containing $A_1^+ e^{\varkappa_1 z}$ and $A_1^- e^{-\varkappa_1 z}$ correspond to the asymptotic solution of the integral equation considered in Sect. 118; all other terms, which vary as $\sim e^{-z/l_{\text{tot}}}$, represent the non-asymptotic part.

For $h=1$, the pair of roots $\pm\varkappa_1$ tend to zero and the two corresponding terms in (123.7) become

$$\left.\begin{aligned}
&\text{in } \Phi_0: && A + Bz,\\
&\text{in } \Phi_1: && -\frac{l_{\text{tot}}}{3}\,B,\\
&\text{in } \Phi_l \text{ for } l\geqq 2: && 0,
\end{aligned}\right\}\qquad (123.9)$$

where A and B are two arbitrary constants.

β) Boundary conditions. A few remarks should be added concerning the boundary conditions to be used in connection with the P_L-approximation for odd L.

It should be kept in mind that the exact boundary conditions (112.3) represent an infinite number of conditions which cannot be exactly satisfied in any finite order approximation. Therefore, we should expect that a certain arbitrariness is involved in all the boundary conditions given below.

According to (119.6) at an interface between two media all the harmonic components involved in a P_L-approximation have to be continuous, i.e.

$$\Phi_l^{(A)}(z_s) = \Phi_l^{(B)}(z_s), \quad \text{for } l = 0, 1, 2, \ldots L, \qquad (123.10)$$

where the superscripts A and B refer to the two media and z_s to the (plane) surface separating A from B.

Eqs. (123.10) correspond to $L+1$ equations so that they determine one half of the arbitrary constants involved in the general solution.

The boundary conditions at infinity depend, of course, upon whether or not the neutron current vanishes at infinity.

If there is no current at infinity, then for $z \to \infty$, the general solution must tend to zero and, therefore, all the coefficients A_j^+ of the exponentials $e^{+\varkappa_j z}$ must be zero. Some attention should be paid in the case $h = 1$, to the pair of roots which are zero: $\pm\varkappa_1 = 0$. The corresponding terms in (123.7) are given by (123.9), so that, if the current is zero at infinity, B should be zero while A need not be.

If at infinity there is a source or a sink of neutrons, and $h < 1$, the coefficients A_j^+ with $j \geqq 2$ must be zero, while A_1^+ is determined by the strength of the source at infinity. For $h = 1$ all other coefficients with $j \geqq 2$ are again zero, while the constant B appearing in (123.10) will be given by the strength of the source.

Thus for $h \leqq 1$, one half of the arbitrary constants is determined by the conditions at $+\infty$. Similar remarks hold for $z \to -\infty$.

Finally, we will mention the boundary conditions at a free surface. This is assumed to be the plane $z = 0$ which separates the vacuum on its left from the homogeneous medium on its right. We can stipulate $\frac{1}{2}(L+1)$ conditions, and this is usually done by either one of the two following procedures.

The MARK's boundary conditions require that

$$\Phi(0, \mu_\imath) = 0, \qquad \mu_\imath > 0, \tag{123.11a}$$

where $\mu_\imath$ are the positive roots of

$$P_{L+1}(\mu_\imath) = 0. \tag{123.11b}$$

The MARSHAK's boundary conditions stipulate that

$$\int_0^1 \Phi(0, \mu)\, \mu^{2j-1}\, d\mu = 0 \quad \text{for} \quad j = 1, 2, \ldots, \tfrac{1}{2}(L+1). \tag{123.12a}$$

MARK has proved that the conditions (123.11) are equivalent to replacing the vacuum by a black body ($h = 0$) and to regarding the free surface as an interface where the boundary conditions are given by (123.10)[1] [7].

The strongest argument in favour of MARSHAK's boundary conditions is that they include

$$\int_0^1 \Phi(0, \mu)\, \mu\, d\mu = 0 \tag{123.12b}$$

which expresses that the total number of neutrons entering the system from outside is zero.

The choice between MARK's and MARSHAK's conditions sometimes follows the nature of the problem. As a rule MARSHAK's conditions give better results in the low order approximations, but the accuracy increases faster in the higher approximations when MARK's conditions are used. It is found that usually MARK's conditions become superior in the P_5-, or sometimes in the P_7-approximation. The reader may find a detailed discussion of this point in [7].

γ) An example of the use of the spherical harmonics method. As an example let us discuss very briefly the neutron distribution from a unit plane isotropic δ-source at $z = 0$ in P_3-approximation. The solution may be derived [19] by first solving the problem in which the source is distributed uniformly throughout a slab extending from $z = -b$ to $z = +b$ with a strength equal to S neutrons per $\text{cm}^3 \cdot \text{sec}$, and then passing to the limit in which b tends to zero but $2Sb$ remains finite and equal to 1.

---

[1] See footnote 3, p. 562.

In the central slab, the set of linear differential equations differs from (123.1) only because the first one contains a term $S=\text{const}$ on its right-hand side. Its solution is essentially of the same form as (123.7) except that Φ_0 contains an additive term S/Σ_a. Replacing, for convenience, the exponentials by hyperbolic functions and adding a superscript A to the various quantities, one has

$$\left.\begin{aligned}
\Phi_0^{(A)}(z) &= \frac{S}{\Sigma_a} - A_1\cosh\varkappa_1 z + A_2\cosh\varkappa_2 z,\\
\Phi_1^{(A)}(z) &= A_1\frac{\Sigma_a}{\varkappa_1}\sinh|\varkappa_1 z| + A_2\frac{\Sigma_a}{\varkappa_2}\sinh|\varkappa_2 z|,\\
\Phi_2^{(A)}(z) &= A_1\left(-\frac{3}{2}\frac{\Sigma_{\text{tot}}\Sigma_a}{\varkappa_1^2}+\frac{1}{2}\right)\cosh\varkappa_1 z + A_2\left(-\frac{3}{2}\frac{\Sigma_{\text{tot}}\Sigma_a}{\varkappa_2^2}+\frac{1}{2}\right)\cosh\varkappa_2 z,\\
\Phi_3^{(A)}(z) &= A_1\left(\frac{9}{14}\frac{\Sigma_a}{\varkappa_1}-\frac{3}{14}\frac{\varkappa_1}{\Sigma_{\text{tot}}}\right)\sinh|\varkappa_1 z| + A_2\left(\frac{9}{14}\frac{\Sigma_a}{\varkappa_2}-\frac{3}{14}\frac{\varkappa_2}{\Sigma_{\text{tot}}}\right)\sinh|\varkappa_2 z|.
\end{aligned}\right\}\qquad (123.1$$

In the outer (B) region, the solution (123.7b) holds with the arbitrary constants $A_i^+=0$ because of the boundary conditions at $z=+\infty$:

$$\left.\begin{aligned}
\Phi_0^{(B)}(z) &= A_1^-\,e^{-\varkappa_1(|z|-b)} + A_2^-\,e^{-\varkappa_2(|z|-b)},\\
\Phi_1^{(B)}(z) &= A_1^-\frac{\Sigma_a}{\varkappa_1}\,e^{-\varkappa_1(|z|-b)} + A_2^-\frac{\Sigma_a}{\varkappa_2}\,e^{-\varkappa_2(|z|-b)},\\
\Phi_2^{(B)}(z) &= A_1^-\left(\frac{3}{2}\frac{\Sigma_a\Sigma_{\text{tot}}}{\varkappa_1^2}-\frac{1}{2}\right)e^{-\varkappa_1(|z|-b)} + A_2^-\left(\frac{3}{2}\frac{\Sigma_a\Sigma_{\text{tot}}}{\varkappa_2^2}-\frac{1}{2}\right)e^{-\varkappa_2(|z|-b)},\\
\Phi_3^{(B)}(z) &= A_1^-\left(\frac{9}{14}\frac{\Sigma_a}{\varkappa_1}-\frac{3}{14}\frac{\varkappa_1}{\Sigma_{\text{tot}}}\right)e^{-\varkappa_1(|z|-b)} + A_2^-\left(\frac{9}{14}\frac{\Sigma_a}{\varkappa_2}-\frac{3}{14}\frac{\Sigma_{\text{tot}}}{\varkappa_2}\right)e^{-\varkappa_2(|z|-b)}.
\end{aligned}\right\}\qquad (123.1$$

Thus four arbitrary constants appear in these solutions. They are determined by imposing at the two interfaces $\pm b$ the four boundary conditions (123.10). In passing to the limit $b\to 0$ with $2Sb=\text{const}=1$, one obtains

$$\Phi_0(z) = \Phi_{0\,\text{asy}}(z) + \Phi_{\text{non-asy}}(z) \qquad (123.15\text{a})$$

where

$$\left.\begin{aligned}
\Phi_{0\,\text{asy}}(z) &= \frac{1}{2}\frac{\varkappa_1}{\Sigma_a}\,\frac{1}{1-\left(\frac{3\Sigma_a\Sigma_{\text{tot}}}{\varkappa_1^2}-1\right)\left(\frac{3\Sigma_a\Sigma_{\text{tot}}}{\varkappa_2^2}-1\right)^{-1}}\,e^{-\varkappa_1|z|}\\
&\approx \frac{1}{2}\frac{\varkappa_1}{\Sigma_a}\left(1-\frac{4}{5}\frac{\Sigma_a}{\Sigma_{\text{tot}}}\right)e^{-\varkappa_1 z}
\end{aligned}\right\}\qquad (123.15\text{b})$$

which is correct to terms of the order of $\Sigma_a/\Sigma_{\text{tot}}$, and

$$\Phi_{\text{non-asy}}(z) = \frac{\varkappa_2}{2\Sigma_a}\,\frac{1}{1-\left(\frac{3\Sigma_a\Sigma_{\text{tot}}}{\varkappa_2^2}-1\right)\left(\frac{3\Sigma_a\Sigma_{\text{tot}}}{\varkappa_1^2}1-\right)^{-1}}\,e^{-\varkappa_2|z|} \qquad (123.15\text{c})$$

which corresponds to (118.13b). In particular it should be noticed that $\Phi_{0\,\text{asy}}(z)$ includes the source correction (118.7) valid in the case of weak absorption. This is what we should expect, since according to the considerations given in Sect. 117α, the P_1-approximation is rigorously correct in non-absorbing media; on the other hand both the non-asymptotic and the asymptotic part of the solution tend towards the rigorous expressions valid, even in the case of strong absorption, as the order L of the approximation increases tending to infinity.

Fig. 245 shows a comparison of the neutron distribution produced by an isotropic plane δ-source in the P_1- and P_3-approximations of the spherical harmonics method with the corresponding rigorous solution (Sect. 118) [*19*].

δ) Wick's method. This procedure is sometimes indicated as *method of discrete ordinates*. It was first used by WICK[1] and was later applied by CHANDRASEKHAR to many astrophysical problems[2].

We re-write the transport equation (115.1) for almost isotropic scattering with the coefficients B_0 and B_1 expressed in terms of Σ_s and $\overline{\mu}_0$ by means of (115.8):

$$\mu \frac{\partial \Phi(z,\mu)}{\partial z} + \Sigma_{\text{tot}}\,\Phi(z,\mu) = \Sigma_s \int_{-1}^{+1} \Phi(z,\mu') \frac{1+3\overline{\mu}_0 \mu\mu'}{2}\, d\mu' + \frac{1}{2} S(z,\mu). \tag{123.16}$$

Its general solution is obtained, as usual, by adding to a particular solution of the inhomogeneous equation the general solution of the corresponding homogeneous equation. Therefore, we are led to discussing the general solution of the homogeneous equation. This, according to WICK, may be replaced by a set of linear differential equations obtained by the following procedure: (a) the integral on the right-hand side of (123.16) is approximately expressed as a weighted sum of the values of the integrand corresponding to a conveniently chosen set of values, μ_i, of the variable μ:

$$\left.\begin{aligned} &\int_{-1}^{+1} f(z,\mu)\, d\mu = \sum_{i=1}^{\mathfrak{N}} H_i f(z,\mu_i); \\ &-1 \leqq \mu_1 \leqq \mu_2 \leqq \cdots \leqq \mu_{\mathfrak{N}} \leqq +1. \end{aligned}\right\} \tag{123.17}$$

(b) The angular number flux $\Phi(z,\mu)$ is required to fulfil Eq. (123.16) only for the $\mathfrak{N}$ chosen values μ_i of μ.

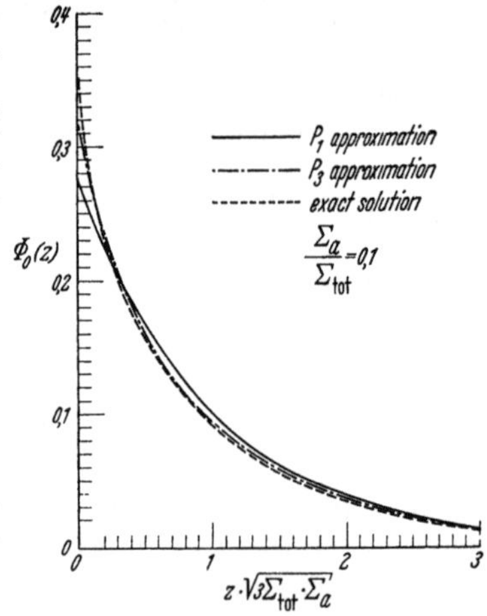

Fig. 245. Neutron distribution from an isotropic plane δ-source calculated according to the spherical harmonics method [*19*].

Although the procedure can be carried on very far without specifying the values of μ_i and H_i which are finally adopted in Eq. (123.17), we prefer to decide from the beginning—for the sake of definiteness—to use the Gauss integration formula[3]. In this case the μ_i are the $L+1$ roots (L must be odd) of the equation

$$P_{L+1}(\mu_i) = 0. \tag{123.18}$$

These roots as well as the corresponding weights are universal constants, the values of which have been collected in Table 91 for L up to 5.

Thus, according to the points (a) and (b) stated above, the homogeneous transport equation can be replaced by the following set of $L+1$ linear differential equations:

$$\left.\begin{aligned} &\mu_i \frac{d}{dz} \Phi(z,\mu_i) + \Sigma_{\text{tot}}\,\Phi(z,\mu_i) = \frac{\Sigma_s}{2} \sum_{r=1}^{L+1} H_r\, \Phi(z,\mu_r)\,(1+3\,\overline{\mu}_0 \mu_i \mu_r); \\ &\qquad i = 1, 2, \ldots L+1 \end{aligned}\right\} \tag{123.19}$$

where the μ_i are the roots of Eq. (123.18).

This procedure bears a very close connection to the spherical harmonics method since the set of Eqs. (115.4) for odd L takes on the form (123.19) if the

[1] See footnote 2, p. 562.

[2] S. CHANDRASEKHAR: Radiative Transfer. Oxford 1950.

[3] C.F. GAUSS: Werke Vol. 3, p. 163; RIEMANN-WEBER, edited by R. v. MISES: Die Differential- und Integralgleichungen der Mechanik und Physik, Vol. I, Chap. 8. Braunschweig: Friedr. Vieweg and Sohn 1925.

linear transformation

$$\Phi(z,\mu_\iota) = \sum_{l=0}^{L} \frac{2l+1}{2} \Phi_l(z)\, P_l(\mu_\iota) \tag{123.20}$$

is applied to it[1].

As an illustration of the corresponding physical meaning, let us write Eq. (123.19) in the P_1-approximation. According to Table 91, one has $\mu_1 = \frac{-1}{\sqrt{3}}$, $\mu_2 = \frac{1}{\sqrt{3}}$ and $H_1 = H_2 = 1$ so that we obtain

$$\left.\begin{aligned} \frac{l_s}{\sqrt{3}} \frac{d\Phi^+}{dz} &= \frac{-1}{N-1} \Phi^+(z) = \Phi^+(z) \frac{1-\bar\mu_0}{2} + \Phi^-(z) \frac{1-\bar\mu_0}{2}, \\ -\frac{l_s}{\sqrt{3}} \frac{d\Phi^-}{dz} &= \frac{-1}{N-1} \Phi^-(z) - \Phi^-(z) \frac{1-\bar\mu_0}{2} + \Phi^+(z) \frac{1-\bar\mu_0}{2}, \end{aligned}\right\} \tag{123.21}$$

where $\Phi^+(z) = \Phi\left(z, \frac{1}{\sqrt{3}}\right)$, $\Phi^-(z) = \Phi\left(z, \frac{-1}{\sqrt{3}}\right)$ and $\frac{\Sigma_s}{\Sigma_{\text{tot}}} = \frac{N-1}{N} = h$. These equation may be interpreted as follows: they represent particles moving in one dimension with a scattering mean free path $l_s/\sqrt{3}$ and a probability $1/N$ of being captured at each collision. When they are not captured, i.e. in $(N-1)/N$ of the cases, they are scattered elastically with a probability $(1+\bar\mu_0)/2$ for continuing their motion in the original direction and a probability $(1-\bar\mu_0)/2$ for having switched to the opposite direction. The quantities $\Phi^+(z)$ and $\Phi^-(z)$ represent the flux of particles moving from left to right and from right to left, respectively. Thus, the P_1-approximation of WICK's method corresponds to FERMI's one-dimensional model considered in Sect. 122.

Table 91. *Weights and roots for Gauss integration formula.*

L	μ_ι	H_ι
1	$\pm \frac{1}{\sqrt{3}} = \pm 0.57735$	1; 1
3	± 0.3400 ± 0.8611	0.6521; 0 6521 0.3479; 0 3479
5	± 0.2386 ± 0.6612 ± 0.9325	0.4679; 0.4679 0.3608; 0.3608 0.1713; 0.1713

We now want a solution of (123.19) of the form[2]

$$\Phi(z,\mu_\iota) = g(\mu_\iota)\, e^{-\varkappa z} \tag{123.22a}$$

where $\varkappa$ is a constant. Inserting (123.22a) into (123.19) a first time, one finds that $g(\mu_\iota)$ should have the form

$$g(\mu_\iota) = \frac{\Sigma_{\text{tot}} + a\mu_\iota}{\Sigma_{\text{tot}} - \varkappa\mu_\iota} \tag{123.22b}$$

[1] We start from the set of differential equations (115.4) of the spherical harmonics method, we multiply the first of these equations by $P_0(\mu_\iota)$ the second by $P_1(\mu_\iota), \ldots,$ the L-th by $P_L(\mu_\iota)$ and we sum with respect to l from 0 to L. With the help of the following relations

$$(2l+1)\cos\vartheta\, P_l(\cos\vartheta) = l\, P_{l-1}(\cos\vartheta) + (l+1)\, P_{l+1}(\cos\vartheta),$$

$$\sum_\iota H_\iota P_l(\mu_\iota) = \int_{-1}^{+1} P_l(\mu)\, d\mu = 2\delta_{l0}; \qquad \sum_\iota H_\iota \mu_\iota P_l(\mu_\iota) = \int_{-1}^{+1} \mu P_l(\mu)\, d\mu = \tfrac{2}{3}\delta_{l1},$$

$$\sum_\iota H_\iota \Phi(z,\mu_\iota) = \int_{-1}^{+1} \Phi(z,\mu)\, d\mu = \Phi_0(z); \qquad \sum_\iota H_\iota \mu_\iota \Phi(z,\mu_\iota) = \int_{-1}^{+1} \mu\,\Phi(z,\mu)\, d\mu = \Phi_1(z)$$

the equations (123.19) are readily found.

[2] M. AGENO: Nuovo Cim. 3, 3 (1946).

where a is a constant. Substituting (123.22b) into (123.19) a second time, we find

$$\Sigma_{\text{tot}} + a\mu_\iota = \frac{\Sigma_s}{2}\sum_{r=1}^{L+1} H_r \left\{\frac{\Sigma_{\text{tot}} + a\mu_r}{\Sigma_{\text{tot}} - \varkappa\mu_r} + 3\bar{\mu}_0\mu_\iota\mu_r \frac{\Sigma_{\text{tot}} + a\mu_r}{\Sigma_{\text{tot}} - \varkappa\mu_r}\right\}$$

where the unknowns are a and $\varkappa$. This relation should hold for all $L+1$ values of μ_ι; thus one has

$$\Sigma_{\text{tot}} = \frac{\Sigma_s}{2}\sum_{r=1}^{L+1} H_r \frac{\Sigma_{\text{tot}} + a\mu_r}{\Sigma_{\text{tot}} - \varkappa\mu_r}, \tag{123.23a}$$

$$a = \frac{3}{2}\Sigma_s\bar{\mu}_0 \sum_{r=1}^{L+1} H_r\mu_r \frac{\Sigma_{\text{tot}} + a\mu_r}{\Sigma_{\text{tot}} - \varkappa\mu_r}, \tag{123.23b}$$

from which it follows that

$$a = 3\bar{\mu}_0\Sigma_{\text{tot}}^2 \frac{1-h}{\varkappa} = 3\bar{\mu}_0 \frac{\Sigma_a\Sigma_{\text{tot}}}{\varkappa}, \tag{123.24a}$$

$$\frac{\Sigma_s}{2}\,\frac{l_{\text{tot}}^2\varkappa^2 + 3\bar{\mu}_0(1-h)}{l_{\text{tot}}^2\varkappa^2 + 3\bar{\mu}_0 h(1-h)} \sum_{r=1}^{L+1} \frac{H_r}{\Sigma_{\text{tot}} - \varkappa\mu_r} = \Sigma_{\text{tot}}. \tag{123.24b}$$

The latter equation has $L+1$ different non-zero roots which are in pairs of opposite sign: $\pm\varkappa_1, \pm\varkappa_2, \ldots, \pm\varkappa_{\frac{1}{2}(L+1)}$. Therefore, the general solution of the set of Eqs. (123.19) is given by

$$\Phi(z,\mu_\iota) = \frac{h}{2}\sum_{r=1}^{L+1} A_r \frac{\Sigma_{\text{tot}} + 3\bar{\mu}_0\Sigma_{\text{tot}}^2 \frac{1-h}{\varkappa_r}\mu_\iota}{\Sigma_{\text{tot}} - \varkappa_r\mu_\iota}\, e^{-\varkappa_r z}, \qquad i = 1, 2, \ldots, L+1 \tag{123.25}$$

where the A_r are arbitrary constants.

The neutron flux

$$\Phi_0(z) = \int_{-1}^{+1} \Phi(z,\mu)\,d\mu = \sum_{\iota=1}^{L+1} H_\iota\,\Phi(z,\mu_\iota)$$

reduces to

$$\Phi_0(z) = \sum_{r=1}^{L+1} A_r e^{-\varkappa_r z} \tag{123.26}$$

by virtue of (123.23a).

Eq. (123.25) defines the angular number flux only for the $L+1$ μ_ι-values which are roots of Eq. (123.18). It is assumed that the actual angular number flux in directions $\mu \neq \frac{\Sigma_{\text{tot}}}{\varkappa_r}$ is given by the analytical continuation of (123.25):

$$\Phi(z,\mu) = \frac{h}{2}\sum_{r=1}^{L+1} A_r \frac{\Sigma_{\text{tot}} + 3\bar{\mu}_0\Sigma_{\text{tot}}^2 \frac{1-h}{\varkappa_r}\mu}{\Sigma_{\text{tot}} - \varkappa_r\mu}\, e^{-\varkappa_r z}. \tag{123.27}$$

This distribution has the inconvenience of being singular for $\mu = \Sigma_{\text{tot}}/\varkappa_r$; such a behaviour is obviously not correct and is due to the approximations adopted.

In the case of isotropic scattering ($\bar{\mu}_0 = 0$), Eq. (123.24b) gives $a = 0$ and Eq. (123.24a) reduces to

$$\frac{\Sigma_s}{2}\sum_{r=1}^{L+1} \frac{H_r}{\Sigma_{\text{tot}} - \varkappa\mu_r} = 1. \tag{123.28}$$

It can be shown that the roots of this equation are identical with the roots given in Table 89 (p. 564), as we should expect because of the equivalence of the two corresponding sets of Eqs. (115.4) and (123.19), which holds, of course, when they are applied to the same angular distribution.

The boundary conditions to be adopted in WICK's method can readily be found. At infinity they are essentially the same as in the spherical harmonics method. At an interface between two media, we should require that

$$\Phi^{(A)}(z_s, \mu_i) = \Phi^{(B)}(z_s, \mu_i), \quad \text{for } i = 1, 2, \ldots, L+1, \tag{123.29}$$

and on a free surface we should stipulate that

$$\Phi(0, \mu_i) = 0 \quad \text{for all positive } \mu_i. \tag{123.30}$$

These are just MARK's boundary conditions (123.11a).

A comparison between the spherical harmonics and the discrete coordinates methods shows that for plane geometries there is, in general, not much difference in their convenience of application. As we shall see below, WICK's method may present some advantages of simplicity when applied to certain problems.

There are, however, some difficulties, which do not exist in the case of the spherical harmonics method, in extending this procedure the other geometries. An adaptation of the method of discrete ordinates for dealing with spherical systems has been suggested by CHANDRASEKHAR, but such an adaptation is entirely equivalent to the spherical harmonics method [7].

ε) A few examples of the use of Wick's method. As an example let us discuss MILNE's problem, the rigorous solution of which has been given in Sect. 118 ε. A homogeneous medium fills the half-space on the right-hand side of the plane $z=0$ which bounds it towards the vacuum. Neutron sources placed at $z=+\infty$ continuously supply neutrons which diffuse towards the surface $z=0$ where they escape into the vacuum.

The $L+1$ arbitrary constants appearing in (123.27) are determined by imposing the boundary conditions at $z=+\infty$ and $z=0$. The first set of conditions are fulfilled by taking the coefficients $A_r = A_j^+$ of the non-asymptotic terms with $\varkappa_j > 0$ equal to zero (i.e. $A_j^+ = 0$ for $j = 2, 3, \ldots \frac{1}{2}(L+1)$). The coefficient of the term containing $\exp[+\varkappa_1 z]$ (A_j^+ for $j=1$) should be different from zero, since it is determined by the strength of the source at infinity.

The boundary conditions at $z=0$ are obtained by introducing (123.27) into the expression (123.30). Thus one obtains $\frac{1}{2}(L+1)$ linear equations which allow the determination of the remaining A^- arbitrary constants ($j=1, 2, \ldots \frac{1}{2}(L+1)$).

For isotropic scattering in the l.s. ($\overline{\mu}_0 = 0$), these $\frac{1}{2}(L+1)$ equations become

$$\frac{A_1^+}{\Sigma_{\text{tot}} - \varkappa_1 \mu_i} + \sum_{j=1}^{\frac{1}{2}(L+1)} \frac{A_j^-}{\Sigma_{\text{tot}} + \varkappa_j \mu_i} = 0 \qquad \text{for all } \mu_i > 0, \tag{123.31}$$

where A_1^+ should be considered as known, since it is determined by the behaviour at infinity.

The numerical values of the constants A_j^- are readily obtained in the case of a non-capturing medium. In this case the two terms containing $\pm\varkappa_1 \to 0$ degenerate, as in the spherical harmonics method, into a linear expression in both variables z and μ. Using the same notation as in (123.9), the angular number flux (123.27) becomes, for $\overline{\mu}_0 = 0$ and $\Sigma_{\text{tot}} = \Sigma_s$,

$$\Phi(z, \mu) = A + B(z - l_s \mu) + \frac{\Sigma_s}{2} \sum_{j=2}^{\frac{1}{2}(L+1)} \frac{A_j^- e^{-\varkappa_j z}}{\Sigma_s - \varkappa_j \mu} \tag{123.32}$$

where A and B as well as A_j^- ($j=2, 3, \ldots \frac{1}{2}(L+1)$) are arbitrary constants. Eq. (123.32) clearly shows the separation between the asymptotic and the non-asymptotic part of the solution (Sect. 118).

We may now proceed with the actual computation of the coefficients. In the P_1-approximation ($L=1$), the sum disappears from Eq. (123.32) and the asymptotic solution should fulfil the $\frac{1}{2}(L+1)=1$ boundary conditions

$$\Phi\left(0, \frac{1}{\sqrt{3}}\right) = A + B\left(0 - \frac{l_s}{\sqrt{3}}\right) = 0 \tag{123.33a}$$

which lead to the following value of the extrapolated end point

$$\frac{d}{l_s} = \frac{1}{\sqrt{3}}.$$

This is just what is obtained from the one-dimensional model of Sect. 122.

In the P_3-approximation, $\frac{1}{2}(L+1)=2$ boundary conditions of type (123.31) should be imposed; according to the data of Table 91 (p. 570) they are

$$\Phi(0, 0.3400) = 0; \qquad \Phi(0, 0.8611) = 0$$

and give the function

$$\left.\begin{aligned}\Phi(z,\mu) &\\ = A\Big[1 + 1.4409\,(\Sigma_s\cdot z - \mu) &- \\ - \frac{0.16808}{1-1.9720\,\mu}\,e^{-1{,}9720\frac{z}{l_s}}\Big]&\end{aligned}\right\} \tag{123.33b}$$

and an extrapolation end point

$$\frac{d}{l_s} = 0.694.$$

In the P_5-approximation, from the three boundary conditions

$$\Phi(0, 0.2386) = 0; \qquad \Phi(0, 0.6612) = 0;$$
$$\Phi(0, 0.9325) = 0$$

it follows that

$$\Phi(z,\mu) = A\left[1 + 1.42059\,(\Sigma_s\cdot z - \mu) - \frac{0.03596}{1-1.2252\mu}\,e^{-1.2252\frac{z}{l_s}} - \frac{0.14387}{1-3.2030\mu}\,e^{-3.2030\frac{z}{l_s}}\right] \tag{123.33c}$$

and

$$\frac{d}{l_s} = 0.7039$$

which is very close to the exact value (118.30a).

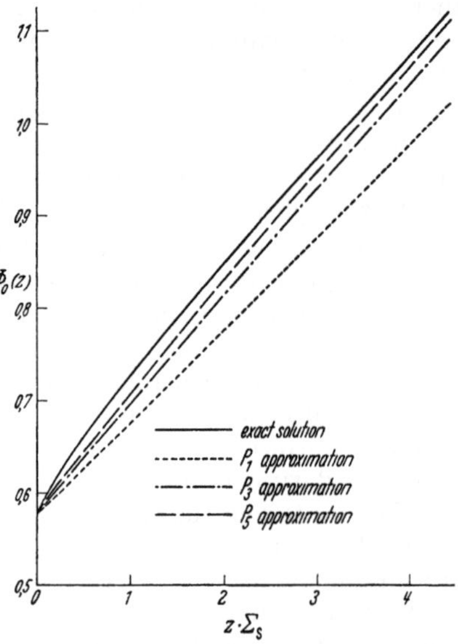

Fig. 246. Flux distribution in MILNE's problem according to WICK's method.

Fig. 246 shows a comparison of the flux distribution computed by means of (123.33) with the corresponding rigorous expression (Sect. 118) [*19*].

The knowledge of the angular number flux at the free surface of a medium deserves a considerable practical interest since it allows the computation of the angular distribution of the neutrons coming out of the surface. This is given by

$$\Phi(0, \cos\vartheta)\cdot\cos\vartheta\, d(\cos\vartheta) \tag{123.34}$$

where the factor $\cos\vartheta$ arises from the fact that, according to the definition (61.13) of the angular number flux,

$$\Phi(z,\mu) = v\,N_0(z,\mu)$$

represents the number of neutrons, at the point z, moving in the considered direction μ. The number of neutrons which cross in one second the free surface at $z=0$ moving in the solid angle $d\omega = 2\pi d \cos\vartheta$ is obtained by multiplying the value of the neutron density $N_0(z, \cos\vartheta)$, at points very close to $z=0$, by the component $v\cos\vartheta$ of the velocity in the direction perpendicular to the plane $z=0$.

Thus, we see that, in general, the Lambert law is violated whenever $\Phi(0,\mu)$ is not a constant with respect to μ.

For isotropic scattering in the l.s. ($\overline{\mu}_0=0$), WICK has given the very simple formula

$$\Phi(0,\cos\vartheta) \approx \frac{\prod\limits_{i=1}^{\frac{1}{2}(L+1)} (\cos\vartheta + |\mu_i|)}{\prod\limits_{i=1}^{\frac{1}{2}(L+1)} (\Sigma_{\text{tot}} + |\varkappa_i| \cos\vartheta)} \tag{123.35}$$

which involves the absolute values of the roots of both, Eq. (123.18) and the determinantal equation.

Fig. 247 shows a comparison of $\Phi(0,\mu)$ computed under various approximations by means of (123.35), with the rigorous expression[1].

In the P_1-approximation, Eq. (123.35) reduces to

$$\Phi(0,\cos\vartheta) \approx \frac{\cos\vartheta + \frac{1}{\sqrt{3}}}{\Sigma_{\text{tot}} + \varkappa_1 \cos\vartheta}$$

which, in the case of very weak absorption ($\varkappa_1 \to 0$), reduces to FERMI'S law (122.20).

The calculation of $\Phi(0,\mu)$ can readily be done in the P_1-approximation even in the case of almost isotropic scattering. Assuming that a source emitting S neutrons/cm² sec is distributed uniformily throughout the medium, one obtains

$$\Phi(0,\cos\vartheta) = 1 + \sqrt{3}\cos\vartheta\left[(1-\overline{\mu}_0)\frac{N-1}{N} + \overline{\mu}_0\right] + \frac{\Phi_0(\infty)}{\Phi_0(0)}\,\frac{1}{N-1} \tag{123.36}$$

where $\Phi_0(\infty)$ and $\Phi_0(0)$ are the values of the flux very far from, and very close to the surface [see Eq. (123.38) below]. In the case of weakly absorbing media $\left(N \gg 1,\ h = \frac{N-1}{N} \to 1\right)$, the third term becomes negligibly small and the quantity in square brackets tends to unity. Thus, the Fermi angular distribution holds even in the case of almost isotropic scattering provided the absorption is small. This result, however, is due to the cancellation of two terms in (123.36), which could not have been predicted a priori.

Finally, we should mention the solution of the following problem which is related to the albedo problem (Sect. 120). In the medium considered in MILNE'S problem, the neutrons are produced uniformly everywhere at a constant rate of S_0 neutrons/cm³ sec. Under these conditions the neutron flux tends to a finite value $\Phi_0(\infty)$ as z tends to ∞, while at the surface it has the value $\Phi_0(0)$. We want the ratio

$$\eta = \frac{\Phi_0(\infty)}{\Phi_0(0)}\,. \tag{123.37}$$

WICK has shown that, for $\overline{\mu}_0 = 0$ one has

$$\eta = \sqrt{N+1} + 1 \tag{123.38a}$$

---

[1] See footnote 2, p. 562.

which is independent of the value of L. Thus, Eq. (123.38a) should be valid even for $L \to \infty$, i.e. it is a rigorous result. If $\bar{\mu}_0 \neq 0$, the problem is much more involved; AGENO[1] has found, for the P_5-approximation, the result

$$\eta = \sqrt{N}\left\{(1-\bar{\mu}_0)^{\frac{1}{2}} + \frac{1+0.219\bar{\mu}_0}{\sqrt{N}} + \frac{0.5-0.204\bar{\mu}_0+0.204\bar{\mu}_0^2}{(1-\bar{\mu}_0)^{\frac{1}{2}}N} + \cdots\right\}. \quad (123.38\text{b})$$

In Fig. 248, the value of N is plotted versus $3\bar{\mu}_0$ for three values of the ratio (123.37); these curves may be used for deriving one of the three quantities $\bar{\mu}_0$, η and N from the experimental values of the other two.

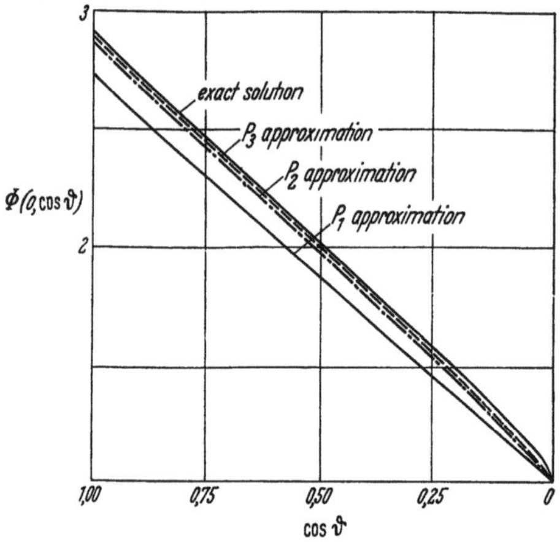

Fig. 247. Flux distribution of the neutrons emitted from a free surface according to WICK's method.

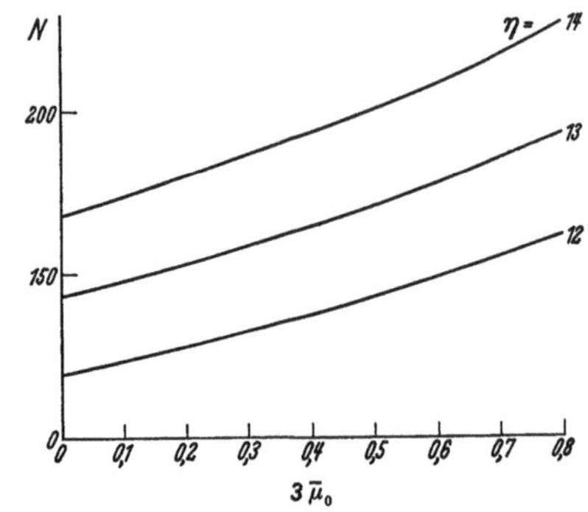

Fig. 248. Average number of collisions N versus 3 times the persistence of velocity $\bar{\mu}_0$ for three values of the ratio (123.37) (AGENO).

124. Time dependent problems in P_1-approximation.

The solution of the time dependent equation (114.6) has already been discussed in Sect. 81 in connection with the problem of the slowing down of neutrons above 1 ev as a function of time. The considerations given there have been extended, in Sect. 111 through the chemical bond region, down to thermal energies. Here we shall treat the time dependent Boltzmann equation for a one-velocity group in the P_1-approximation, and discuss under what conditions it reduces to the time dependent diffusion equation.

For the sake of simplicity, let us refer again to plane symmetry problems in media which scatter the neutrons isotropically in the l.s. The time dependent Boltzmann equation for a one-velocity group is readily derived from (115.1):

$$\mu\frac{\partial\Phi(z,\mu,t)}{\partial z} + \Sigma_{\text{tot}}\,\Phi(z,\mu,t) = \frac{\Sigma_s}{2}\int_{-1}^{+1}\Phi(z,\mu,t)\,d\mu - \frac{1}{v}\frac{\partial\Phi(z,\mu,t)}{\partial t}. \quad (124.1)$$

The P_1-approximation is obtained by introducing the two-harmonics expression

$$\Phi(z,\mu,t) = \tfrac{1}{2}\Phi_0(z,t) + \tfrac{3}{2}\Phi_1(z,t)\cdot\mu \quad (124.2)$$

into (124.1), multiplying the equation thus obtained by either $P_0\,d\mu$ or by $P_1\,d\mu$ and, then, integrating with respect to μ from -1 to $+1$. As a result one obtains the following two differential equations:

$$\left.\begin{aligned}\frac{\partial\Phi_1}{\partial z} + \Sigma_{\text{tot}}\,\Phi_0 &= \Sigma_s\,\Phi_0 - \frac{1}{v}\frac{\partial\Phi_0}{\partial t},\\ \Sigma_{\text{tot}}\,\Phi_1 + \frac{1}{3}\frac{\partial\Phi_0}{\partial z} &= -\frac{1}{v}\frac{\partial\Phi_1}{\partial t}.\end{aligned}\right\} \quad (124.3)$$

[1] See footnote 2, p. 570.

The harmonic component Φ_1 is eliminated by differentiating the first equation with respect to time and the second equation with respect to z and subtracting. The final result has the form of a *telegraph equation* and runs:

$$\frac{l_{\text{tot}}\,v}{3}\,\frac{\partial^2 \Phi_0}{\partial z^2} - \Sigma_a\, v\, \Phi_0 = \left(1 + \frac{\Sigma_a}{\Sigma_{\text{tot}}}\right)\frac{\partial \Phi_0}{\partial t} + \frac{l_{\text{tot}}}{v}\,\frac{\partial^2 \Phi_0}{\partial t^2}\,. \tag{124.4}$$

The structure of this equation is different from that of the time dependent diffusion equation which is only obtained in the limit

$$l_{\text{tot}} \to 0,\ \Sigma_a \to 0 \quad \text{and} \quad v \to \infty \tag{124.5a}$$

so that the diffusion coefficient

$$D_0 = \frac{l_{\text{tot}}\,v}{3} \tag{124.5b}$$

and the neutron mean life for capture

$$T = \frac{1}{\Sigma_a \cdot v} \tag{124.5c}$$

both remain finite. Under these conditions, Eq. (124.4) becomes

$$D_0\,\frac{\partial^2 \Phi_0}{\partial z^2} - \frac{1}{T}\,\Phi_0 = \frac{\partial \Phi_0}{\partial t}\,. \tag{124.6}$$

The telegraph equation shows the phenomenon of *retardation*: as we shall see below, its solution has a well defined front wave propagating at a finite velocity, in addition to a residual perturbation which persists at all points traversed by the wave front [*19*].

According to the time dependent diffusion equation (124.6), on the contrary, an instantaneous neutron source immediately produces a perturbation at all points in space.

The conditions (124.5) under which the telegraph equation (124.4) reduces to the diffusion equation (124.6) are rather well satisfied in the classical cases of heat conduction and molecular diffusion, so that the retardation time can be ignored. In the case of neutrons, the mean free path is so long, that it becomes necessary to investigate whether or not the retardation time should be taken into account.

The difference between the two Eqs. (124.4), (124.6) can be illustrated by comparing their solutions—in a homogeneous non-capturing medium—corresponding to an initial distribution different from zero only inside the interval $z_1 < z < z_2$:

$$\left.\begin{aligned} \Phi_0(z,0) &= \mathfrak{F}(z) && \text{for } z_1 < z < z_2,\\ \Phi_0(z,0) &= 0 && \text{elsewhere},\\ \frac{\partial \Phi_0}{\partial t} &= 0 && \text{everywhere}, \end{aligned}\right\} \quad \text{for } t = 0. \tag{124.7}$$

The solution of the telegraph equation is[1] [*19*]

$$\left.\begin{aligned} \Phi_0(z,t) = \frac{1}{2}\,e^{-\frac{\Sigma_{\text{tot}}v}{2}t}\Bigg\{&\mathfrak{F}\left(z + \frac{vt}{\sqrt{3}}\right) + \mathfrak{F}\left(z - \frac{vt}{\sqrt{3}}\right) + \\ &+ \frac{\sqrt{3}}{v}\int\limits_{z-vt/\sqrt{3}}^{z+vt/\sqrt{3}} \mathfrak{F}(z')\left[\frac{\Sigma_{\text{tot}}\,v}{2} + \frac{\partial}{\partial t}\right] I_0\left(\sqrt{\frac{3}{4}\,\Sigma_{\text{tot}}^2\left[\frac{v^2 t^2}{3} - (z-z')^2\right]}\right)\Bigg\}\,dz' \end{aligned}\right\} \tag{124.8}$$

[1] A. G. WEBSTER: Partial differential equations of Mathematical Physics. New York: G. E. Stechert 1933.

where I_0 is the Bessel function of imaginary argument. The first two terms represent waves which propagate, in opposite directions, at the velocity $v/\sqrt{3}$. The neutron flux is strictly zero at a point z until the time

$$\text{either} \quad t = \frac{\sqrt{3}}{v}(z - z_2) \quad \text{or} \quad t = \frac{\sqrt{3}}{v}(z_1 - z), \tag{124.9}$$

according to whether the considered point is on the right or the left of the interval z_1, z_2. Once the perturbation has reached z, the integrand $\mathfrak{F}(z')$ is different from zero, and the third term in Eq. (124.8) contributes a perturbation which persists indefinitely. This is actually the only term to survive when a time long enough has elapsed that the waves corresponding to the first two terms have passed away.

Since a certain fraction of the neutrons emitted by the source can always reach the point z without suffering any collision at all or only very small deflections (Sects. 77 and 80), we should expect that the velocity of propagation of the neutron waves is equal to the neutron velocity. The fact that the velocity of propagation is given by $v/\sqrt{3}$ is a consequence of the use of the P_1-approximation. As we have pointed out in Sect. 80, the neutrons that move from the source to the point z under consideration without suffering any collision have an angular distribution strongly peaked in the forward direction and, therefore, cannot be described by a P_1-approximation which involves only the first two spherical harmonics.

The problem of the propagation of the perturbation produced by these neutrons at great distances from the source is similar to that of deep penetration, and, therefore, special methods are required for its treatment (Sect. 80) [*19*].

The solution of the time dependent diffusion equation (124.6) corresponding to a plane isotropic δ-source is obtained from (81.8) and (81.10) with $D_0(t) = D_0 = \text{const}$:

$$N_0(z, t) = \frac{1}{v}\,\Phi_0(z, t) = N_0\,\frac{e^{-\frac{z^2}{4D_0 t} - \frac{t}{T}}}{(4\pi D_0 t)^{\frac{1}{2}}}. \tag{124.10}$$

The solution satisfying the initial conditions (124.7) is obtained by applying Eq. (71.1) to (124.10) with $v N_0 = 1$. The result

$$\Phi_0(z, t) = \int_{z_1}^{z_2} \mathfrak{F}(z')\,\frac{e^{-\frac{(z-z')^2}{4D_0 t} - \frac{t}{T}}}{(4\pi D_0 t)^{\frac{1}{2}}}\,dz' \tag{124.11}$$

represents a neutron flux which is different from zero for $t > 0$ at any distance from the source.

In this case, however, a *diffusion time* can be defined, similar to the slowing down time defined in (64.15). According to (124.10), at a given point z, the neutron density starts from zero at $t = 0$, increases, reaching a maximum at

$$t_{\text{diff}} = \frac{\sqrt{\frac{4z^2}{D_0 T} + 1} - 1}{\frac{4}{T}}, \tag{124.12a}$$

and then decreases again. For non-capturing media ($T \to \infty\, L \to \infty$), Eq. (124.12a) becomes

$$t_{\text{diff}} = \frac{z^2}{2D_0}. \tag{124.12b}$$

From the values of diffusion coefficients given in Table 96 (p. 583) and Table 100 (p. 585), we see that the diffusion time at one meter distance from the source is 25 milliseconds in graphite and 140 milliseconds in water.

As we have seen above, the *retardation time* is given by z/v, apart from a numerical factor of the order one. Therefore, the ratio of these two times is of the order of

$$\frac{t_{\mathrm{ret}}}{t_{\mathrm{diff}}} \approx \frac{2D_0}{zv} = \frac{2}{3}\frac{l_s}{z}. \tag{124.13}$$

Thus, at distances from the source large compared to the scattering mean free path l_s, the retardation time is negligible with respect to the diffusion time and, therefore, the diffusion equation (124.6) gives an adequate description of the time dependence of neutron diffusion.

The solution (124.10) of the diffusion equation, referred to a source strength of 1 neutron/cm² emitted at point z' and time t', is sometimes indicated as the *time dependent diffusion kernel*. It allows to calculate—by means of integrations with respect to z' and t'—the solution of Eq. (124.6) corresponding to any arbitrary distribution of time dependent sources with plane symmetry. The *time dependent diffusion kernels* corresponding to other simple geometries (Table 92) may easily be derived from $G_{\mathrm{pl}}(z, t, z', t')$ by the same procedure followed for the case of the time independent diffusion kernels (Table 88, p. 554) and transport kernels (Table 85, p. 510).

An interesting case is that of a neutron source of strength varying sinusoidally with time. The situation is similar to that met with in a classical problem of heat propagation; as we shall see below, the neutron density has the character of a damped wave with a velocity of propagation depending upon the frequency.

Table 92. *Time dependent diffusion kernels* [19].

Geomtry	Notation	Source normalization	Kernel
Plane	$G_{pl}(z, t, z' t')$	1 neutron/cm² at (x', t')	$\frac{e^{-\frac{\lvert z-z'\rvert^2}{4D_0(t-t')}-\frac{t-t'}{T}}}{[4\pi D_0(t-t')]^{\frac{1}{2}}}$
Point	$G_p(r, t, r', t')$	1 neutron at (r', t')	$\frac{e^{-\frac{\lvert \boldsymbol{r}-\boldsymbol{r}'\rvert^2}{4D_0(t-t')}-\frac{t-t'}{T}}}{[4\pi D_0(t-t')]^{\frac{3}{2}}}$
Line	$G(r, \varphi, t, r', \varphi', t')$	1 neutron/cm at (r', φ', t')	$\frac{e^{-\frac{\varrho^2}{4\pi D_0(t-t')}-\frac{t-t'}{T}}}{[4\pi D_0(t-t')]^{\frac{3}{2}}}$ $\varrho^2 = r^2 + r'^2 - 2rr'\cos(\varphi-\varphi')$
Spherical shell .	$G_s(r, t, r', t')$	1 neutron per shell of radius r' at time t'	$\frac{e^{-\frac{t-t'}{T}}}{[4\pi D_0(t-t')]^{\frac{1}{2}}} \times$ $\times\left[e^{-\frac{\lvert r-r'\rvert^2}{4D_0(t-t')}} - e^{-\frac{\lvert r+r'\rvert^{\cdot}}{4D_0(t-t')}}\right]$
Cylindrical shell	$G_l(r, t, r', t')$	1 neutron/cm of shell of radius r' at time t'	$\frac{e^{-\frac{\lvert r-r'\rvert^2}{4D_0(t-t')}-\frac{t-t'}{T}}}{4\pi D_0(t-t')} \cdot I_0\left(\frac{rr'}{2D_0(t-t')}\right)$

I_0 is the modified Bessel function of the first kind.

For the sake of simplicity, we refer to a plane isotropic δ-source embedded in a homogeneous infinite medium, whose intensity oscillates with angular frequency Ω. We can write

$$S(z,t) = \delta(z)\, e^{i\Omega t} \tag{124.14}$$

where it is understood that only the real part of this function—as well as of the corresponding neutron distribution—should be used. By introducing (124.14) into (124.6), the following solution is readily obtained [19]:

$$N_0(z,t) = e^{i\Omega t}\, \frac{L_\Omega}{2D_0}\, e^{-\frac{z}{L_\Omega}} \tag{124.15}$$

which has the same form as the distribution from a stationary source of strength $e^{i\Omega t}$ (Table 88, p. 554) but with the diffusion length $L = \varkappa_0^{-1}$ being replaced by a *complex relaxation distance*

$$L_\Omega = \frac{1}{\sqrt{\varkappa_0^2 + i\,\frac{\Omega}{D_0}}} = \frac{1}{\sqrt{\left(\Sigma_a + i\,\frac{\Omega}{v}\right)\frac{v}{D_0}}}\,. \tag{124.16}$$

With the notations

$$\varrho^2 = \sqrt{\varkappa_0^4 + \left(\frac{\Omega}{D_0}\right)^2}, \tag{124.17}$$

$$L_{\Omega r} = \sqrt{\frac{2}{\varrho^2 + \varkappa_0^2}}, \tag{124.18a}$$

$$L_{\Omega i} = \sqrt{\frac{2}{\varrho^2 - \varkappa_0^2}}, \tag{124.18b}$$

one has

$$\frac{1}{L_\Omega} = \frac{1}{L_{\Omega r}} + \frac{i}{L_{\Omega i}}, \tag{124.19}$$

and the solution (124.15) becomes

$$N_0(z,t) = \frac{L_\Omega}{2D_0}\, e^{i\left(\Omega t - \frac{z}{L_{\Omega i}}\right)}\, e^{-\frac{z}{L_{\Omega r}}}. \tag{124.20}$$

This expression represents a wave of angular frequency Ω, which propagates at the velocity

$$v_w = \Omega L_{\Omega i} = \Omega \sqrt{\frac{2}{\varkappa_0^2 - \varrho^2}}, \tag{124.21a}$$

Table 93. *Propagation velocity, wave length and attenuation length of thermal neutron waves* [19].

	$\frac{\Omega}{D_0} \gg \varkappa_0^2$ i.e. $\Omega \gg \frac{1}{T}$	$\frac{\Omega}{D_0} \ll \varkappa_0^2$ i.e. $\Omega \ll \frac{1}{T}$ [1]
v_w	$\left(\frac{2D_0\,\Omega}{1 - \frac{\varkappa_0^2 D_0}{\Omega}}\right)^{\frac{1}{2}}$	$2\varkappa_0 D_0 \left[1 + \frac{1}{8}\left(\frac{\Omega}{D_0 \varkappa_0^2}\right)^2\right]$
λ_w	$2\pi \left(\frac{2D_0}{\Omega - \varkappa_0^2 D_0}\right)^{\frac{1}{2}}$	$\frac{4\pi \varkappa_0 D_0}{\Omega}\left[1 + \frac{1}{8}\left(\frac{\Omega}{D_0 \varkappa_0^2}\right)^2\right]$
$L_{\Omega r}$	$\left(\frac{2D_0}{\Omega}\,\frac{1}{1 + \frac{\varkappa_0^2 D_0}{\Omega}}\right)^{\frac{1}{2}}$	$\frac{1}{\varkappa_0}\left[1 - \frac{1}{8}\left(\frac{\Omega}{D_0 \varkappa_0^2}\right)^2\right]$

[1] $T = (\Sigma_a \cdot v)^{-1}$.

which has the wavelength

$$\lambda_w = \frac{2\pi}{\Omega} v_w = 2\pi L_{\Omega i} = 2\pi \sqrt{\frac{2}{\varkappa_0^2 - \varrho^2}}, \tag{124.21b}$$

and whose intensity falls off by a factor e over the attenuation length $L_{\Omega r}$ (124.18a).

Table 94. *Numerical examples of parameters characterising thermal neutron waves in graphite* ($D_0 = 1.98 \times 10^5$ cm²/sec, $\varkappa_0 = 1/L = 0.02$ cm⁻¹) [19].

	$\Omega = 10$ sec⁻¹	$\Omega = 10^3$ sec⁻¹
v_w [cm/sec]	7.9×10^3	2.1×10^4
λ_w [cm]	5.0×10^3	132
$L_{\Omega r}$ [cm]	50	19

Table 93 shows the expressions for v_w, λ_w and $L_{\Omega r}$ in two extreme cases, and Table 94 shows two examples of the actual values of these same quantities for thermal neutrons diffusing in graphite.

In Sect. 132 we shall treat a method for measuring the diffusion parameters of thermal neutrons based upon the production and detection of neutron waves.

II. Measurements of the diffusion of thermal neutrons.

125. General remarks and comparison between various procedures. Let us now discuss the measuring procedures used for determining the various parameters entering into the description of the diffusion of thermal neutrons throughout a weakly absorbing medium. In principle only three quantities are necessary for such a description, namely the macroscopic absorption cross section Σ_a, the macroscopic scattering cross section Σ_s and the persistence of velocity μ_0, since all other parameters such as the diffusion length, the diffusion coefficient etc., can be expressed in terms of these three quantities. According to equation (117.21a), the diffusion length L is given by the following expression which is correct to terms of first order in Σ_a/Σ_s

$$\left.\begin{aligned} \frac{1}{L^2} &= \varkappa_0^2 = (3\Sigma_{\text{tot}}\Sigma_a)(1-\bar{\mu}_0)\left[1 - \frac{4}{5}\frac{\Sigma_a}{\Sigma_{\text{tot}}} + \frac{\Sigma_a \bar{\mu}_0}{\Sigma_{\text{tot}}(1-\bar{\mu}_0)} - \cdots\right] \\ &= 3\Sigma_{\text{tr}}\Sigma_a + 3\Sigma_a^2(1-\bar{\mu}_0)(1-k) \qquad \left(k = 0.80\frac{\bar{\mu}_0}{1-\bar{\mu}_0}\right). \end{aligned}\right\} \tag{125.1}$$

The diffusion coefficient D for the neutron flux is expressed in terms of L and Σ_a by Eq. (117.20b):

$$D = \Sigma_a L^2 \text{ cm}. \tag{125.2}$$

This can be transformed into

$$L = \sqrt{\frac{D}{\Sigma_a}} = \sqrt{D_0 T} \tag{125.3}$$

by introducing, according to (114.1c), the diffusion coefficient D_0 for the neutron density

$$D_0 = v\,D \text{ cm}^2/\text{sec}, \tag{125.4}$$

and the mean life for capture of the neutrons in the medium

$$T = \frac{1}{\Sigma_a v}. \tag{125.5}$$

If the capture cross section Σ_a follows the $1/v$ law

$$\Sigma_a(v) = \frac{\gamma}{v} \qquad (\gamma = \text{const}), \tag{125.6}$$

the mean life $T = \gamma^{-1}$ is constant and independent of the molecular structure of the medium (Sect. 96). According to (115.9b) and (117.10), for weakly absorbing media, one has

$$D_0 = \tfrac{1}{3}\, l_{\text{tr}}\, v. \tag{125.7}$$

Finally, we recall that the transport mean free path occurs in the expression of the extrapolated end point, which, for weakly absorbing media, is given by (118.30a), (118.36):

$$d \approx 0.71\, l_{\text{tr}}. \tag{125.8}$$

Thus, for weakly absorbing media, one might think of computing all these quantities from the measured values of Σ_a and Σ_{tr}. Such a procedure, however, presents many inconveniences. First, from transmission experiments in good geometry, one determines Σ_s, not Σ_{tr}, the derivation of which would involve the calculation of $\bar{\mu}_0$ from the angular distribution of the scattered neutrons. This depends in a rather complicated way upon the molecular structure of the medium (Part D, Divisions II and III).

Furthermore, the determination of the capture cross section by transmission experiments in bad geometry becomes practically impossible in the case of weakly absorbing materials such as heavy water, beryllium, graphite and even light water.

Finally, we should recall that the various quantities mentioned above are defined for a one-velocity group, while in the case of thermal neutrons they are intended as averages over the velocity distribution of the neutron flux (Sect. 115). Therefore, the values of the parameters to be used in diffusion problems, depend upon the neutron spectrum at the point where the measurements are actually made. If these extend over a wide spatial region, the constancy of the neutron spectrum with respect to position should be verified, otherwise the values of the parameters would not be well defined, differing from point to point. We shall come back to this remark in Sect. 128.

The kind of average to be used in diffusion problems is indicated by Eq. (115.12) which can readily be derived under the assumption—very rough indeed—that the neutron spectrum is represented with sufficient accuracy by the Maxwellian velocity distribution (101.2). Thus, for example, the macroscopic absorption cross section to be used in diffusion problems, if the medium follows the $1/v$ law, is given by

$$\bar{\Sigma}_a = \frac{\int \frac{\gamma}{v}\, v\, M_n(v)\, dv}{\int v\, M_n(v)\, dv} = \frac{\gamma}{\bar{v}} \tag{125.9a}$$

which, according to (23.8), becomes

$$\bar{\Sigma}_a = \frac{\gamma}{\frac{2}{\sqrt{\pi}}\, v_0} = \frac{1}{1.128}\, \Sigma_a(v_0). \tag{125.9b}$$

We recall that, according to (23.14) or (118.13), the value of the absorption cross section obtained from transmission experiments of a Maxwellian beam (Sect. 136 γ) is

$$\sigma_{\text{th}} = \frac{2}{\sqrt{\pi}}\, \sigma_a(v_0) = 1.128\, \sigma_a(v_0).$$

Thus, the absorption cross section to be used in diffusion problems is $(1.128)^{-1}$ times the absorption cross section at the most probable velocity v_0, and $(1.128)^{-2}$ times the average absorption cross section derived from transmission experiments.

These various remarks show that one can not rely on the values of the diffusion parameters computed from the scattering and absorption cross sections of the atoms of the medium and that special methods are needed for their direct experimental determination.

The various relationships collected above are, nevertheless, extremely usefull since they provide the possibility of checking in detail the more or less accurate fulfilment of the assumptions underlying their derivation; the whole picture of thermal neutron diffusion given in Sect. 28 is thus experimentally tested.

The methods for measuring the various diffusion parameters can be divided into two categories, the first one including all measurements based on the determination of the spatial distribution produced by constant sources of neutrons (steady state assemblies), the second category comprising experiments based on the measurement, as a function of time, of the neutron density produced by neutron sources which vary according to some simple law.

The measurements of the first type are discussed in Sects. 126 to 128, those of the second type in Sects. 130 to 132. Sect. 129 refers to the so-called "diffusion cooling" arising from the fact that thermal neutrons have a wide energy spread and that the diffusion coefficient (125.7) is actually a function of their velocity.

The two categories of methods are hardly comparable since they provide information which overlap only in part. Static methods allow the direct determination of the transport mean free path (Sect. 126), the macroscopic absorption cross section Σ_a (Sect. 127) and the diffusion length L (Sect. 128), while time-dependent methods provide the neutron mean life T (Sects. 130 and 131), the diffusion coefficient D_0 (Sects. 131 and 132), and the diffusion cooling coefficient C (Sect. 131) defined in Sect. 129. From the value of C, the thermalization time constant t^0_{th} can be derived. According to the definition given in Sect. 111 [see Eq. (111.6b)], this parameter enters into an approximate description of the approach of the neutron spectrum to thermal equilibrium with the medium. A comparison of the value of t^0_{th} deduced from the diffusion cooling coefficient C with the result of direct measurements (Sect. 111) provides a further means for testing the correctness of our picture of thermal neutron diffusion.

A comparison of the various procedures, with particular regard to the experimental determination of the diffusion length L by means of the static method (Sect. 128) and the pulsed method in finite media (Sect. 131), has been made by von Dardel and Sjöstrand[1]; their conclusions are in favour of the latter method mainly because it provides a more detailed description of the diffusion of thermal neutrons and it requires a smaller amount of material.

Table 95. *Diffusion parameters of thermal neutrons in low capture materials.* Hughes [*12*].

Material	Density (gm/cm³)	$\mathfrak{N}$ (molecules/cm³)	L (cm)	σ_{tr} (b/molecule)	σ_a (2200 m/sec) (mb/molecule)	l_{tr} (cm)	Σ_a^{-1} (cm)
H_2O	1.00	0.0334×10^{24}	*2.85*	62.9	*660*	0.48	51.2
D_2O	1.10	0.0331	*171*	12.6	0.93	*2.40*	36700
Be	1.85	0.1235	*20.8*	*5.65*	10.1	1.43	906
C	1.60	0.0860	*50.8*	4.24	4.65	*2.74*	2820

[1] G. von Dardel and N. G. Sjöstrand: Progress in nuclear energy. Physics and Mathematics, Vol. 2, p. 183. London and New York: Pergamon Press 1958.

Table 96. *Summary of measurements of diffusion parameters of thermal neutrons in* H_2O *at 22° C*[1].

Authors	T (μsec)	σ_a (2200 m/sec) (barns)	D_0 cm^2/sec	L (cm)	$C\times 10^{-5}$ cm^4/sec	l_{tr} cm	Method	Section
Sisk (1951)		(0.68 ± 0.01)		2.67 ± 0.02		0.48 ± 0.01	extrapolated end point and spacial distribution	126
Hamermesh et al. (1953)		0.329 ± 0.004					$\sigma_B/\sigma_H = 2292 \pm 27$ $\sigma_B = 755 \pm 3$ b	127
Harris et al. (1953)		0.332 ± 0.007					pile oscillation technique	127
Baker et al. (1958)		0.3271 ± 0.0042					$\sigma_B/\sigma_H = 2352 \pm 28$ b $\sigma_B = 769.4 \pm 3.8$ b	127
Berthelot et al. (1947)				2.77 ± 0.04			space distribution produced by negative source	128α
DeJuren et al. (1953)				2.76 ± 0.07			space distribution at great distance from Raγ+Be source	128β
Wilson et al. (1944)				2.77 ± 0.04			space distribution in measuring column	128γ
Von Dardel et al. (1956)	212 ± 3	(0.321 ± 0.005)					pulsed source, infinite medium	130
Meads et al. (1956)	203.3 ± 2.6	(0.335 ± 0.004)					pulsed source, infinite medium	130
Manley et al. (1942)	197 ± 10 [2]	(0.345) [2]					pulsed source, large geometry	130
Scott et al. (1954)	213 ± 4	(0.320 ± 0.008) [2]					pulsed source-finite geometry: $\omega^2 = 0.006 - 0.018\ cm^{-2}$	131
Von Dardel et al. (1954)	204.4 ± 2.0	(0.333 ± 0.003)	36340 ± 750	2.72 ± 0.03	7.3 ± 1.5		pulsed source-small geometry: $\omega^2 = 0.1 - 0.7\ cm^{-2}$	131
Antonov et al. (1956)	207 ± 6	(0.329 ± 0.010)	35000 ± 1000	2.7 ± 0.1	4 ± 1		pulsed source-small geometry: $\omega^2 = 0.09 - 0.93\ cm^{-2}$	131
Bracci et al. (1956)	202 ± 6	0.337 ± 0.010	34850 ± 1100	2.66 ± 0.11	3 ± 1		pulsed source-small geometry: $\omega^2 = 0.09 - 0.96\ cm^{-2}$	131
Campbell et al. (1957)	208		34800		0		pulsed source-small geometry: $\omega^2 = 0.08 - 1.1\ cm^{-2}$	131

[1] Quantities given in brackets are calculated from the various equations given above.
[2] Corrected values (see footnote 1, p. 582).

Table 97. *Summary of measurements of diffusion parameters of thermal neutrons in* D_2O *at 22° C.*

Authors	T (μ sec)	σ_a (2200 m/sec) (barns)	$D_0 \times 10^{-5}$ (cm^2/sec)	L (cm)	$C \times 10^{-5}$ cm^4/sec	l_{tr} (cm)	Method	Section
AUGER et al. (1947)						2.63 ± 0.12	extrapolated end point	126
KASH and WOODS						2.52 ± 0.04	extrapolated end point	126
SARGENT et al. (1947)				171 ± 20			sigma pile	128γ
RAYEWSKI and HOROWITZ (1956)			2.00 ± 0.00			2.45 ± 0.07	modulated source	132

Table 98. *Summary of measurements of diffusion parameters of thermal neutron in* Be [1].

Authors	ϱ (g/cm³)	T (μ sec)	σ_a (2200 m/sec) (barns)	$D_0 \times 10^{-5}$ (cm^2/sec)	L (cm)	$C \times 10^{-5}$ (cm^4/sec)	l_{tr} (cm)	Method	Section
HUGHES et al. (1946)	1.85		$(10.2 \pm 0.5) \times 10^{-3}$		20.8 ± 0.5			sigma pile	128γ
GERASEVA et al. (1956)	1.78		$(8.8 \pm 1.5) \times 10^{-3}$		22.1 ± 1.0				128γ
ANTONOV et al. (1956)	1.78	4000		1.22 ± 0.5		3 ± 1		pulsed source: finite geometry	131
CAMPBELL and STELSON (1957)		≈ 6700		1.25		0		pulsed source: finite geometry	

Table 99. *Summary of measurements of diffusion parameters of thermal neutrons in* BeO [1].

Authors	ϱ (g/cm³)	T (μ sec)	σ_a (2200 m/sec) (barns)	$D_0 \times 10^{-5}$ cm^2/sec	L (cm)	$C \times 10^{-5}$ (cm^4/sec)	l_{tr} (cm)	Method	Section
KOECHLIN et al.	2.92				32.7 ± 0.5			sigma pile	128γ
MANI et al.	2.95	7600 ± 100	7.5 ± 0.1	1.18 ± 0.02	(29.9 ± 0.6)	3.85 ± 0.08	(1.434)	pulsed source: finite geometry	131

[1] Quantities given in brackets are calculated from the various equations given above.

The modulated source method, described in Sect. 132, is similar to the pulsed source method, but it requires—in its present form—a much larger amount of material and a more involved technique.

Table 95 contains the values of the diffusion parameters given by HUGHES [*12*]; the directly measured constants[1] are italicized, the other values are calculated from the various equations given above.

The Tables 96 to 100 summarize the results of the various experiments described in the following sections. They are similar to a table given by VON DARDEL and SJÖSTRAND[2] for the case of light water. Table 103 (p. 596) shows a selection—made by BAKER and WILKINSON—of the most accurate determinations in water, and the corresponding average values.

Table 100. *Summary of measurements of diffusion parameters of thermal neutrons in graphite* *.

Authors	ϱ g/cm³	T μ sec	σ_a (2200 m/sec) barns	$D_0\times10^{-5}$ cm²/sec	L cm	$C\times10^{-5}$ cm⁴/sec	l_{tr} cm	Method	Section
AUGER et al. (1947)							2.40 ± 0.18	extrapolated end point (Dy)	126
HEREWARD et al. (1947)	1.55				51 ± 3			sigma pile	128γ
ERTAUD et al. (1948)	1.71				45.7 ± 3			sigma pile	128γ
CARBLOM (1953)	1.65				45.4 ± 0.2			sigma pile	128γ
BECKURTS (1957)	1.60				40.5 ± 0.7			sigma pile	131
ANTONOV et al. (1956)	1.67	12.200 ± 500		1.98 ± 0.03	(49 ± 2)	12.0 ± 1.9		pulsed source: finite geometry	131
BECKURTS (1957)	1.60			2.13 ± 0.017	(40.9 ± 0.3)	16.3 ± 2.5	(2.58 ± 0.02)	pulsed source: finite geometry	131
RAYEVSKI and HOROWITZ (1956)							2.55 ± 0.09	modulated source	132

* Quantities given in brackets are calculated from the various equations given above.

126. Transport mean free path measurements by means of steady state assemblies. Although many static methods may be devised for measuring the transport mean free path (72.3)

$$l_{tr}=\frac{l_s}{1-\bar{\mu}_0}\qquad(\bar{\mu}_0=\overline{\cos\vartheta}),\qquad(126.1)$$

usually only two are those actually employed[3]: the first is based on the determination of the extrapolated end point (125.8), the second on the change of the diffusion length with the introduction of known amounts of a convenient absorber (for example B_2O_3).

α) The extrapolated end point method. In this case the experimental conditions should approach as much as possible those adopted in the derivation of Eqs. (118.28), (118.30) and (118.36): (a) the medium

[1] P. AUGER, A.M. MUNN and B. PONTECORVO: Canad. J. Res. A **25**, 143 (1947). Cf. Sect. 126. — H.G. HEREWARD, G.C. LAURENCE, H.R. PANETH and B.W. SARGENT: Canad. J. Res. A **25**, 15 (1947). Cf. Sect. 128.— B.W. SARGENT: The Science and Engineering of Nuclear Power, Vol. II, p. 58: Addison-Wesley 1949. Cf. Sect. 128. — HUGHES, WALLACE and NOBLES: Unpublished Argonne work. Cf. Sect. 128. — E. FERMI: Nuclear Physics, p. 193. Revised Edition. Chicago, Ill.: Chicago University Press 1950. Cf. Sect. 128.

[2] See footnote 1, p. 582.

[3] J. HOROWITZ: Suppl. del Nuovo Cim. **4**, 1335 (1956).

should fill an half-space and its boundary towards the vacuum (actually air or a black absorber) should be plane and sharply defined. (b) The neutron sources should be located at very large distances from the boundary so that a rather extended interval of distances is at one's disposal where the non-asymptotic part of the solution is small (Sect. 118γ and δ).

The experiment consists of measuring the thermal neutron flux as a function of the distance z from the medium surface. The experimental points are then fitted to the theoretical flux distribution by adjusting d, or both d and L if the diffusion length is not known with sufficient accuracy.

The value of d represents the distance beyond the medium free-surface where the asymptotic part of the solution of the transport equation crosses the z axis. In non capturing media, Eq. (118.28a) reduces to (118.28b), i.e., $\Phi_{0\,\mathrm{asy}}(z)$ behaves as a straight line.

The fitting of the experimental points to the theoretical flux distribution can be done at various degrees of accuracy. In the simpler experiments, only the asymptotic part of the solution is used. Then the points corresponding to measurements made very close to the surface $z=0$ should not be used in the fitting. An estimate of the distance from the boundary at which the influence of the non-asymptotic part starts to be appreciable can be made by means of the graphs of Figs. 241 whenever the order of magnitude of $h=\frac{\Sigma_s}{\Sigma_s+\Sigma_a}$ is known.

In more refined determinations, the fitting of the experimental points is done by taking into account the non-asymptotic part of the solution; an example of this type of measurement treatment is given later (see Fig. 251).

The measurements should be performed with a density probe (Sect. 134) thin enough to produce a very small perturbation of the neutron distribution (Sect. 135). It is true that here we are only interested in relative measurements; but, even in the case of non-capturing media, close to the surface the neutron density decreases faster than linear because of the non-asymptotic term, and goes suddenly to zero at $z=0$. As a result, a probe thick enough to produce an appreciable neutron depression will tend to give too large a value of d.

This method was used for the first time by AMALDI and FERMI[1] for determining the value of d in paraffin. These authors used the cadmium difference procedure (Sect. 133) employing Rh-detectors of 29 cm² area and 3 mg/cm² or 5.7 mg/cm² thickness corresponding to an absorption of 0.2 and 0.4% respectively for thermal neutrons incident perpendicularly upon their surface. In the approximation in which the angular distribution of the incident neutrons may be represented by Eq. (122.20), the fraction of the neutrons absorbed is obtained by multiplying the figures given above with $\sqrt{3}$. Thus, even taking into account their large area, these detectors were thin enough for neglecting the effect of neutron depression (Sect. 135).

On the other hand, the neutron capture in the medium—which produces a curvature of $\Phi_{0\,\mathrm{asy}}(z)$—and the non-asymptotic part of the solution—which diminishes the value of the flux near the surface—were overlooked by these authors who extrapolated their measurements by a straight line which passed too close to the point at $z=0$. Thus, a value of d was found which was too low by about 30% with respect to the results of successive measurements.

Experiments of the same type in heavy water and graphite were made, sometime later, by AUGER, MUNN and PONTECORVO[2]. Since in the case of heavy

[1] E. AMALDI and E. FERMI: Ric. Sci. 7 (1), 454 (1936). — Phys. Rev. 50, 899 (1936).
[2] P. AUGER, A.M. MUNN and B. PONTECORVO: Canad. J. Res. A 25, 143 (1947).

water only 13 litres were available, it was necessary to design the experimental set-up in such a way that the leakage of neutrons through the sides of the heavy water container was reduced; this was obtained by embedding the apparatus in a graphite pile (Fig. 249).

The aluminium tank containing the heavy water had an internal diameter of 29.9 cm, a wall thickness of 1.5 mm, and an height of 16 cm. The graphite pile was 250 cm high and of square cross section 188 cm on each side; its upper surface was covered with cadmium sheets 1 mm thick. The source of neutrons (2.4 g of Ra mixed with beryllium) was placed on the vertical axis of the pile about one meter below the top of the graphite.

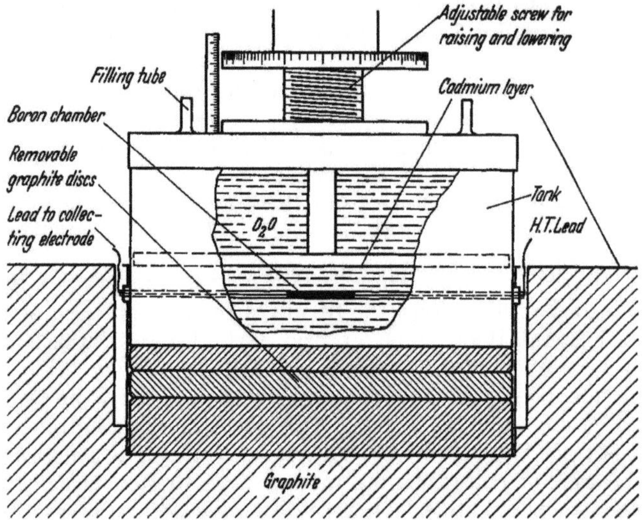

Fig. 249. The experimental set-up used by AUGER *et al.* for measuring the extrapolated end point in heavy water.

The heavy water boundary was defined by the lower surface of a cadmium layer 1 mm thick sprayed on a horizontal steel disc, 8 mm thick, which was fixed on a vertical rod. By means of a micrometric screw, the disk could be moved over a distance of 9 cm.

The neutrons were detected by means of a BF_3 ionization chamber having a 4 mm outside diameter, which was housed at the centre of an aluminium tube having a 6.35 mm outside diameter and wall thickness 0.85 mm, and which crossed the tank along the diameter 4.45 cm above the bottom.

In order to vary the distance between the chamber and the cadmium surface and at the same time to keep the cadmium on level with the graphite surface, the disk was displaced by means of the micrometric screw and the tank raised or lowered to compensate for the former displacements. Graphite disks of graded thicknesses were used below the tank in order to bring about the necessary compensation. In this way the neutron density was explored at a number of distances between the cadmium boundary and the boron chamber.

The following remarks should be added for the design of this experiment: (a) The Raα+Be source was placed at so great distance from the heavy water tank—97 cm and 132 cm in different runs—that the fast neutrons were completely slowed down in graphite and that practically no epi-cadmium neutrons were present in the heavy water. (b) The transverse dimensions of the heavy water tank were large in comparison with the transport mean free path in heavy water.

(c) The distance z from the free surface could be varied over an interval much larger than the transport mean free path. (d) The relaxation length of thermal neutrons in the graphite pile can be computed by means of Eq. (121.19b); for $L=51$ cm (Table 100, p. 585) and $2a=2b=188$ cm, one obtains $L_{00}=33$ cm, i.e., a value smaller than the diffusion length in heavy water. Under these conditions the escape of neutrons through the sides of the tank is essentially determined by the dimensions of the graphite pile, and, therefore, the asymptotic part of the solution is expected to show an appreciable curvature. (e) The ionization chamber was equivalent in absorption to about 1 mg of boron; thus the corresponding flux depression was sufficiently small to neglect its variation with the distance z from the free surface of the heavy water.

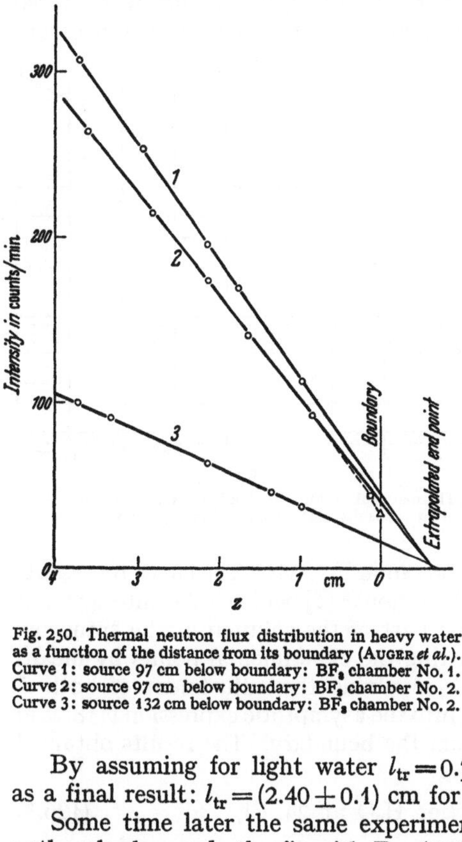

Fig. 250. Thermal neutron flux distribution in heavy water as a function of the distance from its boundary (AUGER *et al.*). Curve 1: source 97 cm below boundary: BF_3 chamber No. 1. Curve 2: source 97 cm below boundary: BF_3 chamber No. 2. Curve 3: source 132 cm below boundary: BF_3 chamber No. 2.

Fig. 250 shows the results of three sets of measurements and the straight lines used by the Canadian group for their fitting. The point at z equal to about 3 mm was measured in order to clearly detect the departure from linearity due to the non-asymptotic part of the solution which should be present near the boundary. The triangle at $z=0$ was computed by means of the theoretical formula of PLACZEK and MARK (Fig. 241).

From the intercept of the straight lines with the z axis, AUGER *et al.* derived $d=(1.64\pm 0.06)$ cm, from which it follows—according to (125.8)—that $l_{tr}=(2.31\pm 0.09)$ cm. This result refers to a mixture of 99.4% D_2O and 0.6% H_2O. If r is the volume fraction of light water contained in heavy water, the measured transport mean free path is given by

$$\frac{1}{(l_{tr})_{\text{measured}}}=\frac{1-r}{(l_{tr})_{D_2O}}+\frac{r}{(l_{tr})_{H_2O}}. \quad (126.2)$$

By assuming for light water $l_{tr}=0.32$ cm and $r=0.6\%$, AUGER *et al.* gave as a final result: $l_{tr}=(2.40\pm 0.1)$ cm for 100% D_2O.

Some time later the same experimental results were re-examined by other authors[1] who made the fit with Eq. (126.2a), took into account the non-asymptotic part of the solution near the boundary and computed the impurity correction (126.2) using the value (126.6) for light water. The result obtained by such a procedure is

$$l_{tr}=(2.63\pm 0.12)\ \text{cm} \quad (\text{in } 100\%\ D_2O \text{ at } 22^\circ\ C) \quad (126.3)$$

which agrees satisfactorily with results of other measurements [cf. Eq. (126.16)].

Similar experiments were performed in graphite by AUGER *et al.* by detecting the neutrons, in different runs, with a BF_3 ionization chamber or with a Dy detector; the latter consisted of 14 mg of dysprosium oxide deposited upon a disk

[1] F. J. SISK: ORNL 933 Series B (1951).

of 1.6 cm diameter. The agreement between the two results is satisfactory:

$$\left.\begin{array}{l} d=(1.72\pm 0.13)\ \text{cm with Dy detector} \\ d=(1.69\pm 0.06)\ \text{cm with BF}_3\ \text{chamber}\end{array}\right\}\ (\text{in graphite: } \varrho=1.68\ \text{g/cm}^3). \quad (126.4)$$

The small difference between these two values is expected because of the shorter mean free path of epi-cadmium neutrons.

The experiments made by SISK[1], although similar to those described above, cover such a large range of values of z as to allow the simultaneous determinations of the extrapolated end point and the diffusion length in light water at 24° C.

The light water was contained in a tank of 183 × 183 × 183 cm³ volume mounted on the thermal column of the Oak Ridge Graphite Reactor. The thermal neutrons coming out of the graphite column passed upward through the tank and reached the boundary of the medium. This consisted of a cadmium absorber 0.5 mm thick carried by an aluminium support of 25 cm diameter. The measurements were made by means of the cadmium difference method using In-foils of 50 mg/cm² thickness and 0.5 cm² area. The exact foil position with respect to the cadmium absorber defining the boundary was carefully measured so that the estimated probable error was less than 0.1 mm.

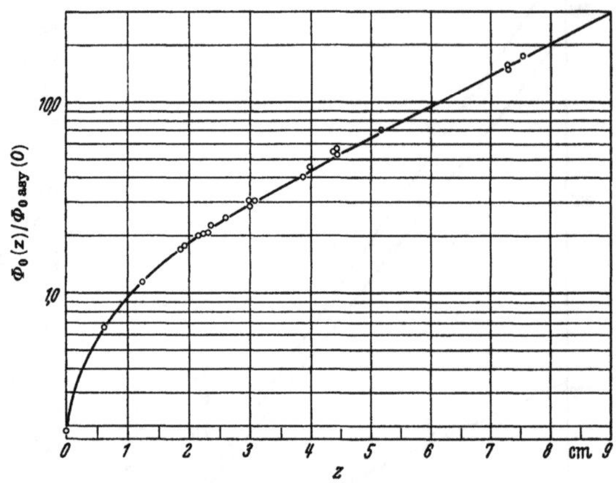

Fig. 251. Thermal neutron flux distribution in light water as a function of the distance from its boundary (SISK).

A semilogarithmic plot of the results is given in Fig. 251. The curve represents the results of LE CAINE's theoretical calculations[2] [6] which take into account the contribution from the non-asymptotic part of the solution of the transport equation. It was obtained by SISK by fitting the experimental points by means of a successive approximation procedure involving the adjustment of both L and d. These are the parameters entering into the asymptotic expression (118.28a) which is valid at very large distances from the boundary. The results obtained by SISK by such a procedure are:

$$L=(2.67\pm 0.02)\ \text{cm} \qquad (\text{for } H_2O \text{ at } 24°\,\text{C}), \tag{126.5}$$

$$l_{tr}=(0.48\pm 0.01)\ \text{cm}; \qquad \sigma_{tr}=(62.4\pm 1.5)\ \text{barns} \tag{126.6}$$

and the ratio

$$\frac{-p(0)}{\Phi_{0\,\text{asy}}(0)}=0.87 \tag{126.7}$$

between the values of the non-asymptotic part and the asymptotic part (hyperbolic) of the solution at the boundary $z=0$.

From the values (126.5), (126.6) of L and l_{tr} SISK also derives the absorption mean free path and absorption cross section in light water:

$$l_a=(44.2\pm 1.3)\ \text{cm}; \qquad \sigma_a=(0.68\pm 0.01)\ \text{barns}. \tag{126.8}$$

[1] See footnote 1, p. 588.
[2] J. LE CAINE: Canad. J. Res. A **28**, 242 (1950).

The value of the capture cross section is definitely larger by about a factor 2, than the results obtained by other methods (Sect. 127). The origin of such a discrepancy is not clear.

β) The boron poisoning method. This method consists of measuring the change in the diffusion length of thermal neutrons after introducing various amounts of boron poison in the medium under investigation. The diffusion length of a weakly absorbing medium is given by (125.1) where the second term always represents a very small correction; for example, in the experiment described below, this term does not surpass 0.1% of the first one. Thus, an experimental determination of the change undergone by L, if Σ_a is changed by a known amount, allows the determination of Σ_{tr}.

This method has the advantage over the extrapolated end point method that the measurements are made far from the medium boundaries where the flux depression produced by the detector is unimportant. In practice, however, it can only be applied to liquid media in which the poisoning material can be dissolved in the form of a convenient salt.

The boron poisoning method was used for the first time by KASH and WOODS [1] who measured the relaxation length of thermal neutrons in a cylindrical column of D_2O with various amounts of B_2O_3 dissolved in it. From the variation of the relaxation length with boron concentration, the effective value (126.2) of l_{tr} and the hydrogen impurity r were determined for the particular D_2O used.

The neutron flux in a cylindrical column with a source distributed on one of its bases is given by Eq. (121.20) which, at sufficiently large distance z from the source, reduces to its fundamental mode

$$\Phi_0(z) = A\, J_0\left(\frac{\lambda_{01}}{R}\, r\right) \sinh \frac{\bar{h} - z}{b_{01}}. \tag{126.9}$$

The relaxation length b_{01} is given by

$$\frac{1}{b_{01}} = \frac{1}{L^2} + \left(\frac{\lambda_{01}}{R}\right)^2 \tag{126.10}$$

where

$$\frac{\lambda_{01}}{R} = \frac{2.4048}{R + 0.71\, l_{tr}} \tag{126.11}$$

is a constant sometimes referred to as the *radial buckling*; it can be determined by fitting the results of radial measurements to a zero-order Bessel function J_0.

Similarly the relaxation length b_{01} can be determined by fitting the results of flux measurements made along the axis of the cylinder with the hyperbolic function $\operatorname{Sin} \frac{\bar{h} - z}{b_{01}}$ (Sect. 128).

Combining (125.1) with (126.10), we obtain

$$3\Sigma_{tr}\Sigma_a(\mathrm{B}) + 3\Sigma_{tr}\Sigma_a(\mathrm{D_2O}) + \left(\frac{\lambda_{01}}{R}\right)^2 = \frac{1}{b_{01}^2} - 3\Sigma_a^2(1 - \bar{\mu}_0)(1 - k). \tag{126.12}$$

Thus Σ_{tr} can be determined from the slope of the straight line obtained by plotting the right-hand side of Eq. (126.12) against $\Sigma_a(\mathrm{B})$. Since the term involving Σ_a^2 is small compared to $1/b_{01}^2$, accuracy in $\bar{\mu}_0$ is not required here. With $\bar{\mu}_0 = 0.15$ [see Eq. (126.18) below] one obtains

$$3(1 - \bar{\mu}_0)(1 - k) = 1.$$

[1] S.W. KASH and D.C. WOODS: Phys. Rev. **90**, 564 (1953).

The measurements were made in a cadmium covered aluminium tank 153 cm in diameter and 183 cm high containing D_2O, which was placed on top of a graphite pedestal, 153 cm in diameter and 183 cm high, with a water boiler reactor at its centre. The reactor was operated at a power level of about $\frac{1}{2}$ watt to provide a flux of about 10^6 thermal neutrons/cm² sec at the base of the D_2O tank.

Flux measurements were made, along the axis of the tank and along a central diameter, with In foils 100 mg/cm² thick using the cadmium difference method (Sect. 133).

Six different concentrations ranging from zero to approximately 150 mg of B_2O_3 per litre of solution were used. For $\Sigma_a(\mathrm{B})$, one should use the average value (115.12b); assuming that the neutrons have a Maxwellian velocity distribution at the temperature T of the D_2O solution, $\Sigma_a(\mathrm{B})$ was computed by multiplying the value

$$\sigma_{\mathrm{B}} = \frac{\sqrt{\pi}}{2}\,(750 \pm 10)\left(\frac{293}{T}\right)^{\frac{1}{2}} \text{barns} \tag{126.13}$$

with the density of boron atoms.

From the results given in Table 101, we see that $\lambda_{01}/\bar{R}$ is independent of the B_2O_3 concentration within the accuracy of the measurements, so that it appears justified to use its average value over all concentrations, $(0.03095 \pm 0.00010)\ \mathrm{cm}^{-1}$.

Table 101. *Results of measurements of relaxation length in boron poisoned* D_2O *cylinder*[1] (KASH and WOODS).

B_2O_3 concentration (mg/l)	Average temperature (°C)	$\Sigma_a(B)\times 10^3$ (cm⁻¹)	Average $\frac{\lambda_{01}}{\bar{R}}\times 10^2$ (cm⁻¹)	Average $b_{01}^{-1}\times 10^2$ (cm⁻¹)	$(b_{01}^{-2}-\Sigma_a^2)\times 10^3$ (cm⁻²)
0	23.6	0	3.107 ± 0.016	3.291 ± 0.018	1.083 ± 0.012
46.9 ± 0.2	22.9	0.536 ± 0.008	3.100 ± 0.012	4.163 ± 0.018	1.733 ± 0 015
72.3 ± 0.1	23.3	0 827 ± 0.011	...	4.561 ± 0.022	2.079 ± 0.020
97.8 ± 0.2	23.3	1.119 ± 0.015	3.101 ± 0.010	4.938 ± 0.029	2.437 ± 0.029
123.5 ± 0.5	23.4	1.413 ± 0.020	3.082 ± 0.020	5.270 ± 0.019	2.775 ± 0.020
146.8 ± 0.5	26.1	1.671 ± 0.023	3.086 ± 0.016	5.573 ± 0.013	3.103 ± 0.014

It should be noticed that the free atom scattering cross sections of boron and oxygen are approximately 4 barns each. For the maximum B_2O_3 concentration used in this experiment, the B_2O_3 macroscopic scattering cross section amounts to approximately $3\cdot 10^{-5}\ \mathrm{cm}^{-1}$. Even with a factor 2 or 3 for molecular effects, this value is very small compared to Σ_{tr} which is approximately $0.4\ \mathrm{cm}^{-1}$. Thus, the change in l_{tr} due to the introduction of B_2O_3 into D_2O is negligible.

The values of $\frac{1}{b_{01}^2} - \Sigma_a^2$ are plotted against $\Sigma_a(\mathrm{B})$ in Fig. 252. From the slope and the intercept of the straight line, KASH and WOODS obtain

$$l_{\mathrm{tr}} = (2.49 \pm 0.04)\ \mathrm{cm} \tag{126.14}$$

$$\Sigma_a(D_2O) = (1.04 \pm 0.10)\cdot 10^{-4}\ \mathrm{cm}^{-1}. \tag{126.15}$$

Since the contribution of other impurities to $\Sigma_a(D_2O)$ is negligibly small, one can use the measured value of the macroscopic absorption cross section for determining the fraction of hydrogen contained in the D_2O used and thereby correct l_{tr} according to (126.2). By converting $\Sigma_a(D_2O)$ into $\Sigma_a(2200\ \mathrm{m/sec})$ according to

[1] The errors listed are standard deviations.

(125.96) and taking[1] $\sigma_a(D_2O) = (1.2 \pm 0.1)$ mb the authors deduced $r = (0.30 \pm 0.05) \cdot 10^{-2}$. From (126.2) with l_{tr} (H_2O) given by (126.6), they finally obtain

$$l_{tr} = (2.52 \pm 0.04)\ \text{cm} \quad (\text{for } 100\%\ D_2O \text{ at } 24^\circ\,C) \tag{126.16}$$

in satisfactory agreement with the result (126.3), of the extrapolated end point method. Finally from (101.1) averaged over a Maxwellian distribution, KASH and WOODS obtain

$$\left.\begin{aligned} \sigma_s(D_2O) &= (14.0 \pm 0.3)\ \text{b} \\ \Sigma_s(D_2O) &= (0.467 \pm 0.010)\ \text{cm}^{-1} \end{aligned}\right\} \tag{126.17}$$

which, combined with (126.16) and (126.1), gives

$$\bar{\mu}_0 = 0.15 \pm 0.03 \quad (\text{for } 100\%\ D_2O \text{ at } 24^\circ\,C). \tag{126.18}$$

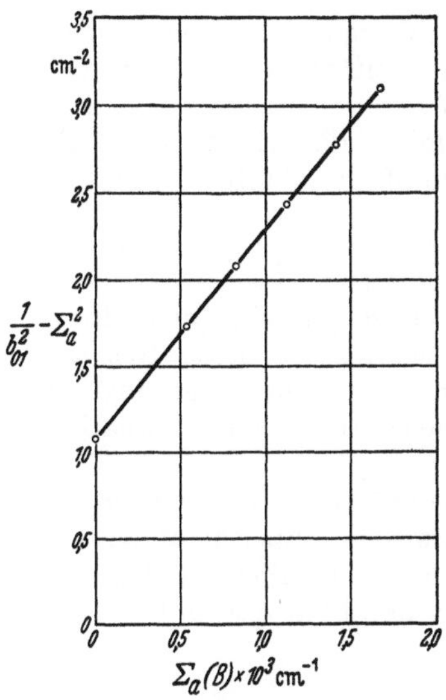

Fig. 252. Relaxation length as a function of boron absorption (KASH and WOODS).

127. Measurements of neutron capture by means of steady state assemblies. In order to express the capture of thermal neutrons during their diffusion through a medium, it is sometimes convenient to use, instead of the macroscopic absorption cross section Σ_a, either the mean life (125.5) or the average number of collisions (120.4) undergone by a thermal neutron before being captured,

$$N = \frac{\Sigma_{\text{tot}}}{\Sigma_a} = \frac{l_a}{l_{\text{tot}}}. \tag{127.1}$$

The mean life has the great advantage over the other two parameters that its value is independent of both the molecular structure and the temperature of the medium whenever the capture cross section Σ_a follows the $1/v$ law (125.6). The average number of collisions N may be thought of as a measure of the extent to which thermal neutrons can be described as a gas diffusing indefinitely throughout the medium.

The macroscopic absorption cross section Σ_a is the quantity more frequently used; it can be measured by various diffusion procedures. The first one is based on the experimental determination of the diffusion length of thermal neutrons in a medium in which the transport mean free path is known. The method is very sensitive to variations of Σ_a, especially if Σ_a is small, since according to (125.3), L tends to infinity as Σ_a goes to zero. Examples of determinations of this type have already been given in Sect. 126: in light water, SISK obtains the value (126.8) for the absorption mean free path $l_a = \Sigma_a^{-1}$, and in heavy water containing 0.3% H_2O, KASH and WOODS find the value (126.15) for Σ_a. A third example of measurements of this type is given in Sect. 128 for the case of beryllium [see Eq. (128.20)].

Another procedure for measuring Σ_a is based on the determination of the total number of thermal neutrons present in an infinite medium as a function of the concentration of a poisoning material such as boron. The main limitation of such a procedure is that in practice it can be applied only to liquid media.

[1] L. KAPLAN, G.R. RINGO and K.E. WILZBACH: Phys. Rev. **87**, 785 (1952); see Table 13, p. 65.

Since T and Σ_a are inversely proportional to one another, and the method described below gives the ratio of the mean life in the pure medium to the mean life in the poisoned medium, the results can be expressed indifferently either in terms of the absorption cross sections or of the corresponding mean lives.

The procedure, first used by FRISCH, VON HALBAN and KOCH[1], can be described as follows. A source of fast neutrons emitting Q_0 neutrons per second is embedded in an homogeneous medium extending to infinity in all directions. We assume that all the nuclei present in the medium have a $1/v$ absorption cross section and that the macroscopic absorption cross section is so small that practically all the neutrons enter into the thermal energy region. This is what actually occurs in media such as water, heavy water, graphite etc., even if they have been slightly poisoned (Sects. 68 and 69). Then we can write

$$Q_0 T = \int n_{\text{th}}(\boldsymbol{r})\, d\boldsymbol{r} \tag{127.2}$$

which expresses that the integral of the density $n_{\text{th}}(\boldsymbol{r})$ of thermal neutrons extended over all space is just equal to the number of neutrons emitted by the source per second, multiplied by the corresponding mean life T.

Thus, in principle, the mean life T could be determined directly from (127.2). This procedure, however, would require a knowledge of the absolute values of both Q_0 and

$$N_{\text{th}} = \int n_{\text{th}}(\boldsymbol{r})\, d\boldsymbol{r}. \tag{127.3}$$

This integral can readily be measured only on a relative scale, i.e., apart from an unknown proportionality constant; various procedures for doing this are described in Sect. 133. It may be recalled here that an absolute determination of N_{th} combined with a known value of T allows, on the contrary, a determination of the source strenght Q_0[2].

The determination of the value of Σ_a (or T) can be reduced to a *relative measurement* of the integrals (127.3) obtained by repeating the experiment with a convenient $1/v$ absorber such as boric acid dissolved in the medium. If $\mathfrak{N}_{\text{B}}$, $\mathfrak{N}'_{\text{H}}$ and $\mathfrak{N}_{\text{H}}$ denote the number of atoms per cm³ of boron and hydrogen in the solution and in pure water, we obtain from (127.3)

$$\frac{N_{\text{th}}(0)}{N_{\text{th}}(\text{B})} = \frac{T(0)}{T(\text{B})} = \frac{\mathfrak{N}'_{\text{H}}}{\mathfrak{N}_{\text{H}}} + \frac{\Sigma_a(\text{B})}{\Sigma_a(\text{H})} \tag{127.4}$$

where $\Sigma_a(\text{H}) = \mathfrak{N}_{\text{H}}\,\sigma_a(\text{H})$, $\Sigma_a(\text{B}) = \mathfrak{N}_{\text{B}}\,\sigma_a(\text{B})$. Thus a plot of the left-hand side of (127.4) versus the macroscopic cross section of the boron introduced into the medium should be represented by a straight line, the slope of which is the absorption mean free path in the pure medium.

Because of the importance of an accurate determination of the capture cross section of hydrogen for thermal neutrons (Sect. 29), a number of authors have made measurements of this type in light water. Their results are collected in Table 102. A correction was made for the detector perturbation of the neutron density only in the four more recent experiments the results of which should be considered as more reliable than the three previous ones.

While the majority of the authors used a small BF_3 counter as a thermal neutron detector, HAMERMESH, RINGO and WEXLER used the activation of I^{128} ($T_{\frac{1}{2}} = 25$ min) produced by radiative capture in iodine dissolved in water as NaI. The method consisted of irradiating two solutions, the first of which was NaI

[1] O.R. FRISCH, H.v. HALBAN and J. KOCH: Kgl. danske Vidensk. Selsk. **15**, 10 (1937).

[2] This procedure is described in Sect. 15 of the article by H.H. BARSCHALL: Detection of neutrons — Vol. XLV of this Encyclopedia.

in H_2O, and the second $NaI+H_3BO_3$ in H_2O. These were prepared by dissolving reagent grade NaI in approximately 700 litres of distilled water and then adding reagent grade H_3BO_3 to half of the iodine solution. The concentrations of B_2O_3 and NaI were accurately determined by chemical analysis to be 2.681 g/litre and 41.48 g/litre respectively.

Sodium iodide rather than manganese salts (Sect. 133) was used because of its chemical compatibility with boric acid in an aequeous solution. The solutions were made slightly basic by adding small volume of NH_4OH in order to prevent air oxydation of iodine. The concentration of the detector was chosen so that its absorption [σ_a(NaI) = 7.4 barns] represented about 5% of that of hydrogen.

Each solution, in turn, was placed in a tank and irradiated for 36 min. The volume of the tank was $92\times 61\times 61$ cm³ and the beam of neutrons from the thermal column of the Argonne heavy water reactor was brought close to its centre through a teflon tube about 5 cm in diameter and 60 cm long.

Table 102. *Ratio of boron to hydrogen capture cross section for thermal neutrons.*

Authors	σ_B/σ_H
FRISCH, VON HALBAN and KOCH[1] . . .	1940 ± 30
SCHULTZ and GOLDHABER[2]	1954 ± 24
FENNING, GRAHAM and SELIGMAN[3] . .	2160 ± 30
KUBITSCHEK[4]	2380 ± 70
WHITEHOUSE and GRAHAM[5]	2270 ± 30
HAMERMESH, RINGO and WEXLER[6] . .	2292 ± 27
BAKER and WILKINSON[7]	2352 ± 28

After irradiation, the solution was stirred for 8 min before two samples, about 20 cm³ each, were placed in two glass jacketed annular Geiger counters.

The results were corrected for: (a) thermal neutron absorption in the teflon tube (0.1%); (b) thermal neutron loss through the entrance hole (0.06%); (c) thermal neutron absorption by NaI. No correction was made for the contribution to the activation of iodine arising from capture of resonance neutrons which, however, is very small.

The capture cross section of the boron used in the experiment was carefully measured by comparison with a boron standard the absorption of which was measured absolutely (Table 11 of Sect. 27). By combining the ratio σ_B/σ_H given in Table 102 with

$$\sigma_B(2200\ \text{m/sec}) = (755\pm 3)\ \text{b}$$

HAMERMESH *et al.* find

$$\sigma_H(2200\ \text{m/sec}) = (0.329\pm 0.004)\ \text{b} \tag{127.5}$$

which agrees very well with the value

$$\sigma_H(2200\ \text{m/sec}) = (0.332\pm 0.007)\ \text{b} \tag{127.6a}$$

obtained by HARRIS *et al.*[8] using a pile oscillator technique (Sect. 91, p. 374).

As slow neutron detectors BAKER and WILKINSON used unmounted discs of boron-loaded nuclear emulsions of 6 mm diameter and 100 μ thickness. The perturbation in this case is small and can be calculated owing to the simple

[1] See footnote 1, p. 593.
[2] LEROY G. SCHULTZ and M. GOLDHABER: Phys. Rev. 67, 202 (1945).
[3] F.W. FENNING, G.A.R. GRAHAM and H. SELIGMAN: BR-135 (1943).
[4] H. KUBITSCHEK: CP-972 (1943).
[5] W.J. WHITEHOUSE and G.A.R. GRAHAM: Canad. J. Res. A 25, 261 (1947).
[6] B. HAMERMESH, G.R. RINGO and S. WEXLER: Phys. Rev. 90, 603 (1953).
[7] A.R. BAKER and D.H. WILKINSON: Phil. Mag. 3, 647 (1958).
[8] S.P. HARRIS, C.O. MUEHLHOUSE, D. ROSE, H.P. SCHROEDER, G.E. THOMAS and S. WEXLER: Phys. Rev. 91, 125 (1953).

form of the detectors. The neutron distribution—due to a Po α+Be source—was determined by counting under the microscope the individual events due to the (n, α) boron-reaction in 54 detecting discs in each of two tanks of height 122 cm and diameter 153 cm, one filled with distilled water and the other filled with an aqueous solution of boric acid. The value of σ_B/σ_H given in Table 102 was deduced by counting a total of about 200000 tracks.

Using for σ_B the value obtained for the standard Harwell boron[1]

$$\sigma_B(2200\,\text{m/sec}) = (769.4 \pm 3.8)\,\text{b}, \tag{127.6b}$$

Baker and Wilkinson deduce

$$\sigma_H(2200\,\text{m/sec}) = (0.3271 \pm 0.0042)\,\text{b}. \tag{127.6c}$$

These authors have then attempted a comparison of the various data from which σ_H can be derived in order to arrive at the best current value for this constant. Table 103 shows the data considered by these authors. The ratio σ_B/σ_H obtained by other authors has a been corrected to correspond to the standard Harwell boron (127.6b) as suggested by the work of Green *et al.*[2]

$$\frac{\sigma_B(\text{U.S.})}{\sigma_B(\text{Harwell})} = 0.986 \pm 0.003\,.$$

The values of L have been corrected for buckling where necessary (Sect. 129), and have been reduced to 22° C by means of (131.3c). Temperature corrections also have been applied to the values of D_0 in accordance with Eq. (131.3b)—and to the values of T—in accordance with water density variation.

The lower part of Table 103 shows the average values of the various parameters and the corresponding values of σ_H obtained by the three independent methods; the value deduced from the ration σ_B/σ_H is based upon the use of the value (127.6b) for σ_B. Since these three results are in excellent mutual agreement and since they are independent of each other, the *final mean value* deduced by Baker and Wilkinson as the best value of σ_H is

$$\sigma_H(2200\,\text{m/sec}) = (0.3315 \pm 0.0017)\,\text{b}. \tag{127.6d}$$

The same procedure was used by other authors for measuring the ratio of the capture cross sections of manganese and hydrogen[3] and of lithium and boron[4].

A systematic investigation of the absorption cross sections of many elements has been made by Lapoint and Rasetti[5] who used essentially the same procedures described above in a simplified form. Instead of an infinite volume, these authors used a limited amount of solution which contained at its centre a BF_3 counter for determining, on an arbitrary scale, the local density of thermal neutrons. The solution was always exposed to the same source of fast neutrons always placed in the same position. When an element of absorption cross section σ_a was dissolved in the water surrounding the counter with a concentration c, a decrease in the number of counts per minute was recorded. The determination of σ_a was made by measuring the concentration c_B of boron which produced the same decrease in the counting rate; then the relation

$$c_B \cdot \sigma_B = c \cdot \sigma_a$$

holds.

[1] P. A. Egelstaff: J. Nucl. Energy 5, 41 (1957).
[2] A. Green, D. J. Littler, E. E. Loackett, V. G. Small, A. H. Spurwy and E. Bowell: J. Nucl. Energy 1, 144 (1954).
[3] R. Ballini, A. Berthelot and C. Smeets: C. R. Acad. Sci., Paris 225, 328 (1947).
[4] F. W. Fenning, G. A. R. Graham and H. Seligman: Canad. J. Res. A 25, 73 (1947).
[5] C. Lapoint and F. Rasetti: Phys. Rev. 58, 554 (1940).

Table 103. *Survey of determinations of the capture cross section of hydrogen for thermal neutrons* (BAKER and WILKINSON).

Sect. 127	σ_B/σ_H	H_2O at 22° C: Sect. 128	H_2O at 22° C: L (cm)	H_2O at 22° C: Sect. 131	H_2O at 22° C: D_0 (cm²/sec)	H_2O at 22° C: Sects. 130 and 131	H_2O at 22° C: T (μsec)
HAMERMESH et al.	2325 ± 28	DE JUREN and ROSENWASSER	2.735 ± 0.015	VON DARDEL and SJÖSTRAND	36340 ± 750	MEADS et al.	$203.4 \pm 2\ 6$
HARRIS et al.	2306 ± 51	BARKOV et al.	2.684 ± 0.020	ANTONOV et al.	34870 ± 1000	VON DARDEL and SJOSTRAND	204.4 ± 2.0
BAKER and WILKINSON	2352 ± 28	WRIGHT and FROST	2.700 ± 0.010	BRACCI and COCEVA	34850 ± 1100	ANTONOV at al.	206.5 ± 4.2
Average $\sigma_B/\sigma_H = 2338 \pm 18$		Average $L = 2.707 \pm 0.012$		CAMPBELL and STELSON	34670 ± 1000	BRACCI and COCEVA	202 ± 6
				DIO and SHOPPER	35440 ± 600	Average $T = 204.3 \pm 1.4$	
				Average $D_0 = 35420 \pm 370$			
		$T = (206.9 \pm 2.8)$ μsec					
$\sigma_H = (0.3291 \pm 0.0030)$ b		$\sigma_H = (0.3293 \pm 0.0045)$ b				$\sigma_H = (0.3335 \pm 0.0023)$ b	

Two precautions should be used in the actual measurements. The thickness of the medium should be larger than a few diffusion lengths in all directions around the counter; the concentrations c and c_B of the various dissolved materials should be small enough to produce a negligible change in the macroscopic scattering cross section: $\Sigma_s(c) \approx \Sigma_s(c_B) \approx \Sigma_s(0)$.

Finally, we should mention that the *average number of collisions* (127.1) can be determined by means of measurements of the ratio (120.22) which is expressed by Eq. (123.38b) in terms of N and $\bar{\mu}_0$ (Fig. 248). The experimens tal procedure for measuring η ha- already been described in Sect. 120 and is illustrated in Fig. 244. From the measured value of η, one can either deduce N, if the persistence of velocity $\bar{\mu}_0$ is already known, or viceversa deduce $\bar{\mu}_0$, if N is known, for example from the measured values of Σ_s and Σ_a. Since η is related to the experimental albedo by Eq. (120.27b) or (120.28), we give in Table 104 the values of β_e for a few low-capture materials[1].

Table 104. *The experimental albedo of low capture materials for thermal neutrons* (STEPHENSON[1]).

Material	Density g/cm³	β_e
Water	1.00	0.82
Heavy water . .	1.10	0.97
Beryllium . . .	1.84	0.85
Beryllium oxide	2.80	0.93
Graphite . . .	1.60	0.93

128. Diffusion length measurements in steady state assemblies. Among the various measurements that can be performed on the diffusion of thermal neutrons, that

[1] R. STEPHENSON: Introduction to nuclear energy. New York: McGraw-Hill 1954.

of the diffusion length L is the most important one. Firstly, L is the only parameter characteristic of the medium which enters into the asymptotic part of the solution of the transport equation. Secondly, from a determination of L and of the transport mean free path, the absorption cross section can be derived even for the case of extremely weak capture. A first example of this type of procedure has been given in Sect. 126β for the case of heavy water, a second one is discussed at the end of this section for the case of beryllium. These two examples —actually rather different from one another in many details—serve as an illustration of the general remark made at the beginning of Sect. 127; very often, the object of a diffusion length measurement is just to obtain the macroscopic capture cross section of the medium in which the neutron diffusion takes place.

A rough classification of the various procedures used for measuring the diffusion length can be made according to the geometry of the medium and the type of source of thermal neutrons adopted. The procedures described in subsection γ) are, generally speaking, the more accurate, especially when the thermal column of a reactor is used as source of thermal neutrons.

α) Experiments with negative sources of thermal neutrons. The idea of a diffusion length of slow neutrons was already implicit in the work of BJERGE and WESTCOTT[1] who found that the number of slow neutrons is reduced to one half after diffusion through 1.6 cm of water. These views were clearly expressed by VON HALBAN and PREISWERK[2] who may be considered as the first to have attempted a measurement of L for thermal neutrons. Their method was qualitatively the same as that adopted, almost at the same time, by AMALDI and FERMI[3] who used Eq. (122.7) for expressing the escape probability $P(z)$ in terms of L.

In order to obtain a source of thermal neutrons with an ideal geometry,—for example, a geometry as close as possible to a plane δ-source—both these groups of authors used *negative sources* consisting of sheets of strong absorbers such as cadmium. The reduction of the activation of a flat thermal neutron detector of very large transversal dimensions is given by Eq. (122.7) as a function of the distance z from an infinitely extending plane sheet of thick cadmium. The results of these measurements were in error by about 25% with respect to later more accurate results.

Essentially the same principle was used by other authors[4,5] who used spherical geometries instead of the plane one. Thus, for example, BERTHELOT *et al.* have measured—with Mn detectors—the flux distribution of thermal neutrons, in a large water tank containing a Ra α+Be source at its centre, a first time without and a second time with a thermal neutron absorber of spherical symmetry surrounding the source. The difference between the two neutron distributions represents the distribution of thermal neutrons captured by the spherical absorber. According to the theorem expressed by Eq. (121.5) (p. 553), at distances larger than the radius of the absorber, the distribution of the absorbed neutrons should be the same—apart from a constant factor— as if all the absorption were concentrated on its centre of symmetry. Thus the neutron distribution due to the spherical negative source should be

$$\Phi_0(r) \propto \frac{e^{-r/L}}{r}. \qquad (128.1)$$

[1] T. BJERGE and C.H. WESTCOTT: Proc. Roy. Soc. Lond., Ser. A **150**, 709 (1935).
[2] H. v. HALBAN and P. PREISWERK: Nature, Lond. **136**, 951 (1935). — J. Phys. Radium **8**, 29 (1937).
[3] E. AMALDI and E. FERMI: Ric. Sci. **7** (1), 56, 454 (1936). — Phys. Rev. **50**, 899 (1936).
[4] G.R. GAMERTSFELDER and M. GOLDHABER: Phys. Rev. **62**, 556 (1942).
[5] A. BERTHELOT, R. COHEN and H. REEL: C. R. Acad. Sci., Paris **225**, 406 (1947).

Two types of spherical absorbers were used in the experiment done by the French group. One consisted of a spherical thin-walled glass container of 16.5 cm diameter filled with a solution of cadmium sulfate (53.9 g/litre of solution); the other was a thin shell of 16.2 cm diameter, of pyrex glass, containing 12% of B_2O_3, filled with water. In both cases the Ra α+Be source was placed at the centre of the absorbing sphere and the measurements were made in the surrounding water starting from a distance r sufficiently large to ensure that the non-asymptotic part of the solution of the transport equation present in the vicinity of the absorber surface was practically reduced to zero. The results obtained by such a procedure are

$$\left.\begin{array}{l} L = (2.77 \pm 0.04)\ \text{cm for neutrons absorbed in Cd} \\ L = (2.88 \pm 0.07)\ \text{cm for neutrons absorbed in B} \end{array}\right\} \left.\begin{array}{c}\text{and diffusing}\\ \text{in water.}\end{array}\right\} \qquad (128.2)$$

The first of these two values may be compared with the value (126.5) obtained by SISK. The larger value of L found for the neutrons absorbed by boron is attributed to the contribution of epi-cadmium neutrons.

β) Experiments with distributed sources of thermal neutrons. Two cases should be distinguished according to whether the medium has infinite or finite dimensions. If a point source of thermal neutrons were available, the asymptotic part of the solution of the transport equation in an infinite medium would be given by Eq. (128.1). Then, by measuring the flux distribution as a function of r and correcting for the non-asymptotic part of the solution, one would obtain

$$\overline{r^2} = \frac{\int\limits_0^\infty \Phi_0(r)\, r^4\, dr}{\int\limits_0^\infty \Phi_0(r)\, r^2\, dr} = 6L^2. \qquad (128.3)$$

In practice, thermal neutrons are always produced by slowing down neutrons of higher energy; therefore, as a source of thermal neutrons, one should use the slowing down density $Q(r, E_m)$ of the neutrons having an energy equal to the migration energy E_m at which they are considered to enter the thermal group (Sects. 110 and 111). Then, according to (71.1) or (118.3), the thermal neutron flux is given by

$$\Phi_0(r) = \int Q(\boldsymbol{r}', E_m)\, \frac{e^{-\frac{|\boldsymbol{r}-\boldsymbol{r}'|}{L}}}{|\boldsymbol{r}-\boldsymbol{r}'|^2}\, d\boldsymbol{r}'. \qquad (128.4)$$

Following the same argument as was used in Sect. 77 for deriving Eq. (77.4), we obtain

$$\overline{r^2_{\rm th}} = \overline{r^2_{E_m}} + 6L^2 \qquad (128.5)$$

which might be used, in principle, for deriving L from measurements of $\overline{r^2_{\rm th}}$ and $\overline{r^2_{E_m}}$ (Sects. 83 and 84). In practice, however, by means of measurements of the types as those in Sects. 83 and 84, the slowing-down density actually measured corresponds to the resonance of the detector used in the experiment; this is usually a Cd-covered In-foil corresponding to an energy of $E_R = 1.46$ ev. Thus Eq. (128.5) is replaced by

$$\frac{1}{6}\left(\overline{r^2_{\rm th}} - \overline{r^2_{\rm In}}\right) = \Delta\tau + L^2 \qquad (128.6)$$

where $\Delta\tau$ is the increase undergone by the age when the neutrons are slowed down from 1.46 ev to $E_m \approx 0.1$ ev. This quantity is a rather small correction which can be computed; therefore, one might think of using Eq. (128.6) for determining L^2. As we have seen in Sects. 83 and 110 it is preferable to use Eq. (128.6) for determining $\Delta\tau$ from direct measurements of $\overline{r^2_{\text{th}}}$, $\overline{r^2_{\text{In}}}$ and L.

Another procedure, also based on Eq. (128.4), has been used by DE JUREN and ROSENWASSER[1]. According to the remark given in Sect. 77, at great distances from the source, the slowing-down density is given by Eq. (77.1). In Sects. 83 and 84 we have seen that such a prediction is in satisfactory agreement with all the measurements of the distribution of resonance neutrons at very great distances from the source. Thus, for r very large, we may adopt the following expression for representing the distribution of the source of thermal neutrons:

$$\frac{S}{D} = \frac{Q}{D} = K\frac{e^{-r/l'}}{r^2} \tag{128.7}$$

where K is an inessential constant related to the strength of the fast neutron source.

By introducing this expression into (128.4), one obtains

$$\Phi_0(r) = K\frac{L}{2D}\frac{e^{-r/L}}{r}\left[C - \mathrm{Ei}(-\alpha_1 r) + \mathrm{Ei}(-\alpha_2 r)\right] \tag{128.8a}$$

where C is a constant,

$$\mathrm{Ei}(-\alpha r) = \int_{\alpha r}^{\infty}\frac{e^{-x}}{x}\,dx$$

is the exponential integral, and

$$\alpha_1 = \frac{1}{l'} - \frac{1}{L}, \qquad \alpha_2 = \frac{1}{l'} + \frac{1}{L}.$$

As $r\to\infty$, the distribution (128.8a) becomes

$$\Phi_0(r) \to C_1\frac{e^{-r/L}}{r} - k\frac{l'^2L^2}{L^2 - l'^2}e^{-r/l'} \tag{128.8b}$$

where C_1 and k are two new constants. If $L > l'$, the first term dominates at sufficiently large radii and the slope of a semilogarithmic plot of $r\,\Phi_0(r)$ is given by $1/L$.

This procedure has been applied by DE JUREN and ROSENWASSER to the measurement of photoneutrons from the Ra γ+Be and $\mathrm{Sb}^{124}\gamma$+Be source (Sect. 49) diffusing in water. The detectors were In foils $2.0\times3.4\ \mathrm{cm}^2$ in area and $76\ \mathrm{mg/cm}^2$ thick which were used, with and without a cadmium filter for determining both, the relaxation length of the In-resonance neutrons, and the thermal flux distribution—the latter by means of the cadmium difference procedure (Sect. 133).—With the Ra γ+Be source, the measurements of resonance neutrons extended from $r = 8$ cm to $r = 13$ cm and the measurements of thermal neutrons from $r = 16$ cm to $r = 22$ cm. The results are

$$l' = 2.45\ \text{cm}, \tag{128.9a}$$

$$L = (2.763 \pm 0.015)\ \text{cm} \quad \text{(in water at } 22^\circ\,\text{C)}. \tag{128.9b}$$

The measurements with the $\mathrm{Sb}^{124}\gamma$+Be source extended over a smaller range of distances for intensity reasons. The results

$$l' = 2.1\ \text{cm}, \tag{128.10a}$$

$$L = (2.76 \pm 0.07)\ \text{cm} \quad \text{(in water at } 22^\circ\,\text{C)} \tag{128.10b}$$

[1] J.A. DE JUREN and H. ROSENWASSER: J. Res. Nat. But. Stand. **51**, 203 (1953).

are in satisfactory agreement with (128.9). The value (128.9b) compares satisfactorily with the results of other methods (126.5), (128.2).

The $Sb^{124}\gamma+Be$ source was also used by BARKOV *et al.*[1] who measured the thermal neutron distribution in water by means of the usual Cd-difference procedure (0.6 mm Cd) applied to In foils 100 mg/cm² thick. The linear dimensions of the sources employed in this experiment are given in Table 54 (p. 322). The water tank was a cylinder of 80 cm diameter and 100 cm height. For r varying between 15 and 35 cm from the source, the neutron flux was very accurately represented by the first term of Eq. (128.8b). The results obtained both with tap water,

$$L = (2.68 \pm 0.02)\ \text{cm} \quad (\text{in tap } H_2O \text{ at } 19°\,\text{C}), \tag{128.10c}$$

and with distilled water,

$$L = (2.69 \pm 0.02)\ \text{cm} \quad (\text{in distilled } H_2O \text{ at } 19°\,\text{C}), \tag{128.10d}$$

are slightly below the value (128.9b); according to BAKER and WILKINSON (Table 103, p. 596) the difference decreases when taking into account small temperature and buckling corrections (Sect. 129).

The procedures described above involve infinite media, i.e. media with linear dimensions much larger than L in all directions. This requirement can easily be fulfilled only if—as in light water—the diffusion length is short and/or the medium is available in any desired amount. In media with very small absorption, such as graphite, beryllium or heavy water, the diffusion length is so large as to make the condition of infinite dimensions in all directions unpractical, especially if one considers the amounts of some of these materials usually at one's disposal.

Thus one is led to consider *finite media* with simple geometries such as cylindres or parallelepipeds. The arrangement is then indicated in the literature as a *sigma pile* because the usual object of the diffusion measurements is to obtain the absorption cross section of the diffusing material.

The Argonne standard graphite pile described in Sect. 84 can be used as a sigma pile; by introducing into the diffusion equation, as source of thermal neutrons, the empirical expression (84.8), the thermal neutron flux along the vertical axis z of the pile is readily obtained [*12*]:

$$\left.\begin{aligned}
&10^6\,\varphi_{\text{th}} = 44.5\,f(r_1) + 82.0\,f(r_2) + 6.8\,f(r_3)\\
&f(r_i) = \sum_{l,\,m\ \text{odd}} L_{lm}\, e^{\left(\frac{r_i}{2L}\right)^2}\left[e^{\frac{z}{L_{lm}}}\left(1-\frac{2}{\sqrt{\pi}}\int_0^{y_1} e^{-y^2}\,dy\right) + e^{-\frac{z}{L_{lm}}}\left(1+\frac{2}{\sqrt{\pi}}\int_0^{y_2} e^{-y^2}\,dy\right)\right]\\
&y_1 = \frac{z}{r_i} + \frac{r_i}{2L_{lm}},\quad y_2 = \frac{z}{r_i} - \frac{r_i}{2L_{lm}};\quad r_1 = 31.4,\ r_2 = 44.4,\ r_3 = 67.0,\\
&\frac{1}{L_{lm}^2} = \frac{\pi^2}{a^2}(l^2+m^2) + \frac{1}{L^2};\quad 2a = 215\ \text{cm}.
\end{aligned}\right\} \tag{128.11}$$

The expressions (84.8), (128.11) may now be used for determining the diffusion length L by comparing them with the measured space distributions of both In-resonance neutrons and thermal neutrons. These equations however, contain many terms so that they do not lend themselves very well to this type of analysis. Only at distances greater than 60 cm from the source, does the thermal flux decrease exponentially with distance, so that the experimental conditions become essentially identical to those described in the following subsection.

[1] L.M. BARKOV, V.K. MAKARIN and K.N. MUKHIN: J. Nucl. Energy 2, 94 (1957).

γ) Experiments with thermal neutrons flowing into a finite medium of simple geometry. The ideal conditions for an experiment of the type described above are attained when the thermal neutrons flow into a measuring column—of parallelepipedal or cylindrical shape—through one of its bases. Then the neutron density distribution throughout the column is given by Eq. (121.15) or Eq. (121.20): apart from minor differences due to the different shapes, both these expressions are sums of an infinite number of modes, the relaxation lengths of which become shorter as the order of the harmonic increases. At sufficiently large distance from the source only the fundamental mode survives and the flux measured along the axis of the assembly decreases exponentially with a relaxation length whose value is bounded to the diffusion length by a very simple relation: Eq. (121.16) in the case of the parallelepiped, and Eq. (121.21) in the case of the cylinder.

In designing an experiment of this type, some attention should be paid to avoiding that an appreciable number of fast neutrons or epi-cadmium neutrons enter the region where the measurements of the thermal neutron flux are actually performed. In the more primitive experimental set-ups a Ra α+Be—or other source — is placed on the axis of the column and the measurements of the thermal neutron flux are started at a distance from the source large enough to make the contribution of fast neutrons or epi-cadmium neutrons negligibly small; these are the experimental conditions mentioned at the end of subsection β.

In more refined experiments, the source of fast neutrons is placed inside a pile in which the neutrons are slowed down to thermal energies and from which they reach the measuring column through a pedestal of graphite of the same cross-sectional area.

Under these conditions, the neutrons diffusing through the graphite pedestal lose the higher harmonics of the distribution before entering the measuring column, with the result that the exponential decrease of the thermal neutron flux—characteristic of the first harmonic—begins close to the base of the column. Since the presence of such a pedestal allows the use of the lower part of the measuring column, this arrangement is particularly recommended when the diffusing material is at one's disposal in limited amounts only.

The adoption of this type of arrangement, however, has the disadvantage of producing a remarkable reduction of the thermal neutron flux. Therefore in practice, it can be adopted only if a neutron source of very high intensity is available. The ideal source for experiments of this type is a thermal column of a reactor, which—because of the filtering effect described in Sect. 102—eliminates all neutrons of energy higher than the corresponding crystalline cut-off (Table 76 on p. 436).

For accurate measurements of the diffusion length with the method described above, a number of precautions should be kept in mind [*19*], [*12*].

(1) A suitable analysis procedure of the measurements should be used for taking into account the contribution of higher harmonics. By a convenient choice of the coordinates of the neutron detectors, it is possible to eliminate a few of the important harmonics; for example, in a square column, the $(1, m)$ and $(l, 1)$ harmonics vanish at $x = a/6$ and $y = a/6$.

(2) Since the column has always a finite length h, an "end correction" of the observed intensities should be introduced; this is given by the term in brackets in Eq. (121.15).

(3) Since the results are sensitive to the adopted transverse dimensions of the column, transverse neutron distributions should be measured. If only one

harmonic is present, the measured transverse flux distribution is strictly the product of two cosines in the case of a rectangular cross section and is strictly a Bessel function of zero-order in the case of a circular cross section of the column.

The transverse dimensions of the column to be used in Eq. (121.15) or (121.20) are those deduced from the transverse flux measurements. They are expected to be equal to the transverse geometrical dimensions of the column plus one extrapolated end point distance d for each free surface. The agreement is often rather poor for various reasons. In particular, in the case of liquid media, the walls of the container act in some way as "reflectors" which produce an increase in the value of d.

A similar effect can arise from neutrons scattered back to the measuring column by the walls or surrounding objects. These spurious effects should be kept as small as possible. The contribution of the thermal neutron component is eliminated by wrapping the assembly with cadmium sheets of about $\frac{1}{2}$ mm thickness.

(4) After corrections (1) and (2) have been applied, a semilogarithmic plot of the flux measurements along the axis is drawn. This should be fitted by a straight line the coefficients of which can be determined by the least square method. Its slope represents the reverse of the relaxation length of the fundamental mode.

(5) The diffusion length is obtained from the relaxation length of the fundamental mode by making use of Eq. (121.16) or (121.21) in accordance with the geometry of the measuring column. A good determination of L requires that the transverse dimensions of the column are very large, otherwise $1/L^2$ is given by a small difference between two large terms.

(6) Finally for accurate determinations one should verify that the neutron spectrum does not change appreciably along the axis of the measuring column. If the capture cross section is appreciable, because of the $1/v$ law, the velocity distribution will tend to "harden" as the neutrons proceed in their diffusion through the medium. On the other hand, if the column is made of microcrystalline material, the neutrons of wavelength longer than the corresponding cut-off will penetrate more easily than those of higher energy. Since the diffusion length obtained by experiments of this type represents an average over the neutron spectrum, any change of the velocity distribution reflects in a change of L, with the result that there is no region of exponential behaviour of the fundamental harmonic.

The constancy of the neutron spectrum at various positions along the axis of the diffusion column can be verified by measurements of the neutron temperature similar to those described in Sect. 108.

In order to reduce the end correction mentioned under (2) and obtain flux measurements over a wide range of distances, the length h of the diffusing column should be made as long as possible. On the other hand, the danger of an appreciable change of the velocity distribution increases with increase of the distance over which the neutron flux is measured, so that a compromise should be adopted.

Since the flux measurements are performed along the axis of the diffusing column where the gradient is small, no special precaution must be adopted for avoiding flux depressions produced by the detector which, therefore, does not need to be very thin.

Water. The diffusion length of thermal neutrons as a function of water temperature has been measured by WILSON, BRAGDON and KANNER[1] and by

[1] V.C. WILSON, E.W. BRAGDON and H. KANNER: CP-2306 (1944).

WRIGHT and FROST[1]. The first group of authors used a cylindrical tank, 47 cm diameter and 56 mm in depth, placed on a 10 cm thick hexagonal pedestal of graphite on top of the thermal column of the CP-2 pile of the Argonne Laboratory. The measurements were made using In foils as detectors at two water temperatures: $t = 27.2^\circ$ C and $t = 93.5^\circ$ C. From the results the authors derive the following linear expression

$$L = (2.644 + 0.0061 \cdot t^\circ \text{C})\ \text{cm} \quad (\text{in } H_2O) \tag{128.12}$$

which is affected by an accidental error of about $\pm 1.5\%$.

The origin of this temperature dependence has been examined by RADKOWSKY[2] who concludes that it is not clear what is the reason for an appreciable change of the scattering properties of water with temperature. Only a small part of the scattering cross section is coherent: at room temperature, about 2 barns which should be compared to a total scattering cross section of about 50 barns. Furthermore, the rotational levels of the water molecules are so closely spaced that even at room temperature the molecules should behave like classical rotators: the first rotational level is only 0.006 ev. On the other hand, the lowest vibrational level is so high (0.2 ev) that it cannot be excited even at the highest temperatures to be reached in reactors. The discrepancy between theory and experiments is reduced 3—4% by taking into account the Doppler effect of the translational degrees of freedom[3]. According to DEUTSCH[4], for a Maxwellian neutron spectrum, a good fit of the experimental expression (128.12) is obtained by assuming that the hydrogen transport cross section is proportional to $1/v$. The author derives the following expression of L in terms of the centigrade temperature t° and the density of water $\varrho(t^\circ)$:

$$L = \frac{2.719}{\varrho(t^\circ)} \left[\frac{t^\circ + 273.1}{295.1}\right]^{\frac{1}{2}} \left[\frac{1.0614}{1 + 0.0614\left[\frac{t^\circ + 273.1}{295.1}\right]^{\frac{1}{2}}}\right]^{\frac{1}{2}} \text{cm}. \tag{128.13}$$

The coefficients are adjusted to give the value of the diffusion length obtained by VON DARDEL and SJÖSTRAND at $t^\circ = 22^\circ$ C (Sect. 131) and represents very satisfactorily the results of WILSON *et al.*[5] considered above, as well as the results of other authors which are reviewed in Sect. 131 (Tables 108 to 110) Eqs. (131.3a and b).

The measurements have been extended to higher temperatures by WRIGHT and FROST[1] who placed a stainless steel vessel—30 cm in diameter and 30 cm in height—in contact with the end of a beryllium thermal column installed through the shielding of the TTR. The vessel was designed for maximum temperature and pressure of 316° C and 170 atmospheres; its wall thickness of 6.5 cm produced an attenuation of the neutron flux by a factor 60. As detectors these authors used foils—2.5 cm in diameter and 5 mm thick—of fired aluminium ceramic containing Dy_2O_3 with 10 mg/cm² Dy.

The results, given in Table 105 (p. 604), show that at high temperature L increases faster than according to the linear expression (128.12). The low value at $t = 25^\circ$ C is attributed by the authors to the perturbation of the neutron distribution produced by the detectors.

[1] W.B. WRIGHT and R.T. FROST: Kapl-WBW-2 (1956).
[2] A. RADKOWSKY: ANL-4476, p. 89 (1944).
[3] V. BENZI: Report of the Com. Naz. Ric. Nucl. 6 (1958).
[4] R.W. DEUTSCH: Nucl. Sci. and Engng. **1**, 252 (1956).
[5] See footnote 1, p. 602.

Heavy water has been investigated by two Canadian groups[1-3] who have performed two sets of measurements: in the first the heavy water was poisoned with lithium carbonate—1.57 g per litre of solution, i.e., $7.70 \cdot 10^{-4}$ atoms of lithium per molecule of water (99.4% D_2O; 0.6% H_2O)—so that the diffusion length was strongly reduced. In the second experiment 99.4% D_2O was used.

In the experiment using the poisoned heavy water[1], 113 litres of solution were contained in a cylindrical tank of 61 cm internal diameter placed on top of a graphite pile $186 \times 186\ \text{cm}^2$ in horizontal section and 274 cm high. A Raα+Be source (2.2 curies) was placed on the axis of the assembly 100 cm below the top of the graphite pile. Such a distance corresponds to about five slowing-down lengths (Sect. 84) and, therefore, the neutrons entering the bottom of the tank containing the solution, were almost entirely thermal (>99% in number). Fig. 253 shows the radial flux distribution obtained with a small BF_3 ionization chamber in three horizontal planes; the first, taken arbitrarily as $z=0$, was very close to the bottom of the tank, the second and the third were at $z=12$ cm and $z=24$ cm. The curves represent averages of measurements taken along eight radii spaced at 45° intervals.

Table 105. *Temperature dependence of the diffusion length in water* (WRIGHT and FROST).

Temperature °F	Temperature °C	L cm
77	25	2.714 ± 0.10
374	190	3.927 ± 0.023
530	277	5.360 ± 0.037

The reader interested in the details of the analysis of the experimental curves is referred to the original paper. Here we limit ourselves to reporting that: (a) five terms of the series (121.20) were necessary for a good fit of the data at $z=0$, three at $z=12$ cm and only two at $z=24$ cm. (b) If the boundary is taken at

$$\bar{R} = (30.50 + 0.05 + 1.65)\ \text{cm} = 32.2\ \text{cm}, \quad (128.14)$$

the relaxation length of the second and third components turns out to be imaginary. In Eq. (128.14) the first term is the geometrical radius of the tank, the third represents the extrapolated end point (Sect. 126α) and the second one the effect of neutron diffusion by the tank walls (aluminium: 3.3 mm thick). By taking $\bar{R}=32.5$ cm, this difficulty is overcome, giving the final result

$$\left.\begin{aligned} L = 22.7\ \text{cm}\quad &(\text{in lithium carbonate solution in heavy water:}\\ &7.70\times 10^{-4}\ \text{lithium atoms per water molecule}).\end{aligned}\right\} \quad (128.15)$$

Fig. 254 shows the axial distribution; the unfilled circles represent measurements with the boron chamber and the filled circles some measurements with dysprosium detectors. The curve has been computed by means of (121.20) with $A_{10}=808$; $A_{11}=-144$; $A_{12}=47$; $A_{13}=-15$; $A_{14}=4$; $\bar{R}=32.5$ cm, $\bar{h}=37.7$ cm, and L given by (128.15). The height $\bar{h}$ which gave the best fit is appreciably larger than the geometrical one ($h=34.6$ cm) even after correction for the extrapolated end point. This effect is interpreted by the authors as due to back diffusion of the tank cover.

In the experiment on heavy water[2,3], 2600 and 3500 kg of 99.4% D_2O were used in two successive sets of measurements. These were contained in a tank,

[1] H. G. HEREWARD, G. C. LAWRENCE, A. M. MUNN, H. R. PANETH and B. W. SARGENT: Canad. J. Res. A **25**, 26 (1947).

[2] B. W. SARGENT, D. V. BAKER, P. E. CAVANAGH, H. G. HEREWARD and N. J. NIEMI: Canad. J. Res. A **25**, 134 (1947).

[3] B. W. SARGENT: p. 58 of Vol. II: The Science and Engineering of nuclear power, edit. by C. GOODMAN. Cambridge, Mass.: Addison-Wesley Press Inc. 1949.

having a (79.9 ± 0.3) cm internal diameter placed on a graphite pedestal having approximately the same diameter and a height of 28 cm. This pedestal surmounted a rectangular parallelepiped of graphite 279×186 cm² in horizontal section and 228 cm high. The neutron source was placed on the axis of the assembly 86 cm below the base of the tank. It consisted of a block of beryllium at the end of an X-ray tube operated at 2 million volts peak and 1.4 mA. Because of the very low energy of the neutrons emitted by this type of source (Sects. 48 and 51), the neutrons reaching the graphite pedestal were completely thermalized. As explained above, the pedestal allowed the amplitudes of the higher harmonics to fall off relative to the first component.

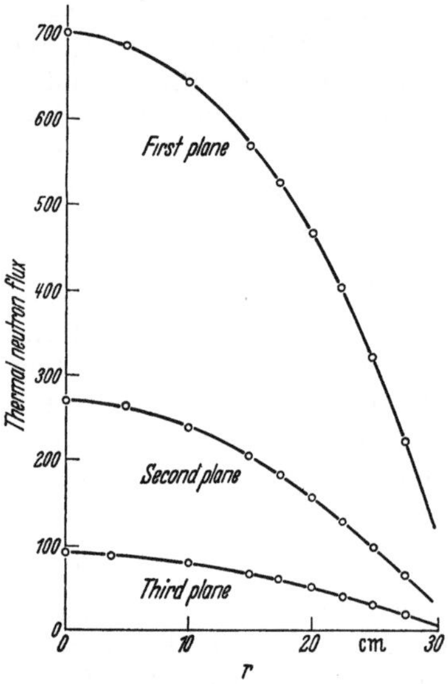

Fig. 253. The radial distribution of thermal neutron flux in a solution of lithium carbonate in D_2O (HEREWARD *et al.*).

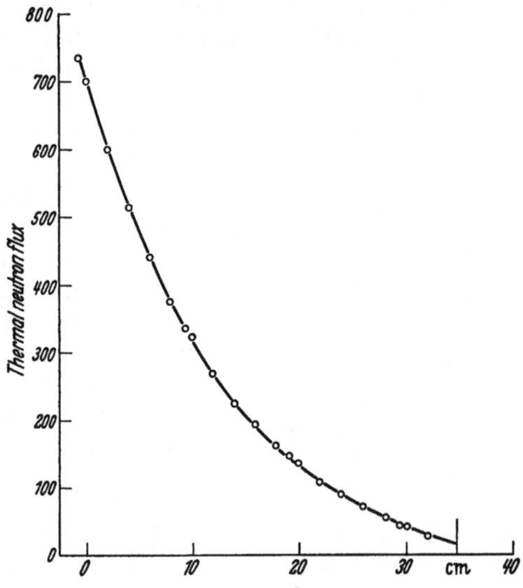

Fig. 254. The axial distribution of thermal neutron flux in a solution of lithium carbonate in D_2O (HEREWARD *et al.*).

By and analysis of the experimental curves, similar to that followed in the case of the lithium carbonate solution in heavy water, and after correction for 0.6% H_2O impurity, the Canadian group finds

$$L = (171 \pm 20) \text{ cm} \quad (\text{in } 100\% \ D_2O \text{ at } 22^\circ \text{C}). \tag{128.16}$$

Beryllium has been investigated by two groups of authors[1,2]. HUGHES *et al.* used as source of neutrons the thermal column of the Argonne pile. A consideration of the various errors involved in a diffusion length measurement showed that with the 1360 kg of beryllium available, a reasonable cross sectional area for the diffusion column was 100×100 cm². Thus, the z dimension of the beryllium column was 65 cm, long enough to extend the measurements to about two relaxation lengths L_{00} of the fundamental mode. Under these conditions, the errors in L arising from the determination of the transverse dimensions of the pile and from the relaxation length L_{00} are expected to be both of the order of 1%, thus yielding a final error of about 3%.

[1] HUGHES, WALLACE and NOBLES (1946): Unpublished Argonne work, reported in Sect. 8-4 of Ref. [*12*].

[2] L.A. GERASEVA, A.V. KAMAYEV, A.K. KRASIN and I.G. MOROSOV: p. 13 of Ref. [*15c*].

The beryllium column was mounted on a graphite pedestal of the same transverse dimensions (100×100 cm²) which was placed on the top surface of the thermal column of the Argonne pile. The whole structure was wrapped in cadmium.

The transverse flux distribution was investigated on a horizontal plane 10 cm above the base of the beryllium column. It was found that a 50 cm graphite pedestal was sufficient to give an accurate pure cosine distribution (Fig. 255), while with smaller pedestals an appreciable contribution from higher harmonics was present.

The possible variation of the neutron spectrum was investigated by measuring the boron absorption by means of In-detectors in the graphite pedestal and at various positions in the beryllium column. The boron absorbers were sheets of pyrex glass which transmitted about 50% of the incident thermal neutrons. Since the same result was found in all investigated positions, the authors concluded that the effects of filtration and hardening mentioned above were negligibly small.

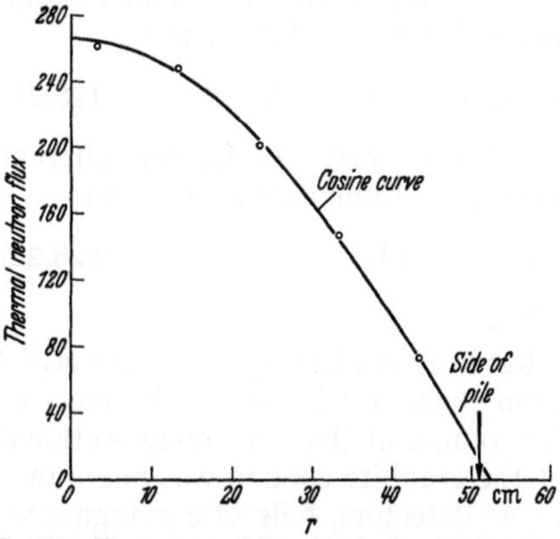

Fig. 255. The transversal flux distribution in a beryllium diffusing column (HUGHES *et al.*).

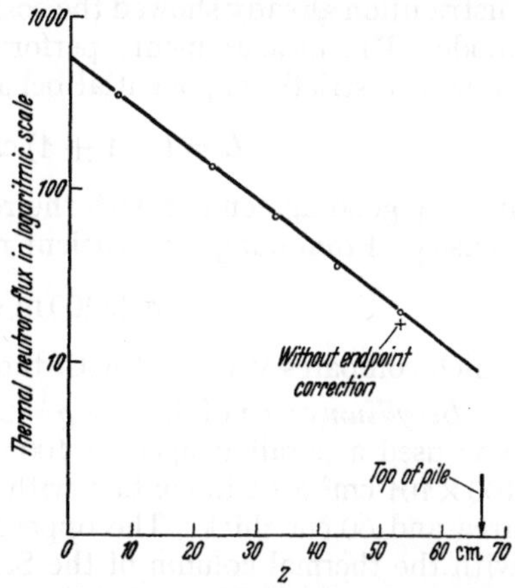

Fig. 256. Flux distribution along the axis in a beryllium diffusing column: z = distance from pedestal (HUGHES *et al.*).

Fig. 256 shows a semilogarithmic plot of the results of thermal flux measurements made with In-foils along the axis of the beryllium column. The relaxation length deduced from the slope of the straight line is $L_{00} = 15.6$ cm which, combined with the value $a = 52.4$ cm, derived from Fig. 255, gives as a final result

$$L = (20.8 \pm 0.5)\text{ cm} \quad (\text{in Be of } \varrho = 1.85\text{ g/cm}^3). \tag{128.17}$$

The transport cross section of beryllium has been calculated by SACHS [*12*] to be

$$\sigma_{\text{tr}} = 5.65\text{ b} \tag{128.18}$$

from the measured scattering cross section of Be, by taking into account the angular distribution of the scattered neutrons expected from Bragg scattering. Introducing (128.17) and (128.18) into (125.1), one derives

$$\overline{\sigma}_a = 9.05\text{ mb} \tag{128.19}$$

from which it follows that

$$\sigma_a\,(2200\text{ m/sec}) = (10.2 \pm 0.5)\text{ mb}, \tag{128.20}$$

a value which represents one of the best determinations of the absorption cross sections in Be [see Table 13, Sect. 29, p. 65 and Eq. (128.22)].

A similar experiment has been made by GERASEVA *et al.*[1] who used 1500 kg of beryllium of very high purity arranged in the form of a parallelepiped of $100 \times 100 \times 84$ cm³ volume and $\varrho = 1.78$ g/cm³ mean density. The measuring column was placed with its z axis horizontal, on a metal support 60 cm high from the floor so that one of its faces of 100×100 cm² area was in contact with a graphite column of the same cross sectional area and 30 cm thick. The opposite face of this layer was in contact with a graphite pile of 220×220 cm² cross sectional area containing a Ra γ+Be source emitting 3×10^7 neutrons per second.

The transverse distribution was measured—by means of Dy detectors 100 mg/cm² thick — in three vertical planes at distances $z = 14$ cm, 34 cm and 54 cm from the surface of the graphite connecting column. At $z = 14$ cm, the transverse distribution already showed the cosine behaviour characteristic of the fundamental mode. The measurements performed along the axis of the beryllium column showed a strictly exponential behaviour from which the authors derived

$$L = (22.1 \pm 1)\ \text{cm} \quad (\text{in Be of } \varrho = 1.78\ \text{g/cm}^3) \tag{128.21}$$

in very good agreement with the result (128.17) after correction for the different density. From danger coefficient measurements, the same authors obtained

$$\sigma_a(2200\ \text{m/sec}) = (8.8 \pm 1.5)\ \text{mb} \tag{128.22}$$

which compares very well with the value (128.20).

Beryllium oxide of density $\varrho = 2.92$ g/cm³ has been studied by KOECHLIN *et al.*[2] who used a parallelepiped of $100 \times 101 \times 90$ cm³ volume with one of its faces of 100×101 cm² area in contact with a graphite column of the same cross sectional area and 60 cm thick. The opposite face of the graphite column was in contact with the thermal column of the Saclay pile. As detectors, foils of a manganese-nickel alloy (90% Mn, 10% Ni) of 8 cm² area and 0.1 mm thick were used. The result

$$L = (32.7 \pm 0.5)\ \text{cm} \quad (\text{in BeO of } \varrho = 2.92\ \text{g/cm}^3) \tag{128.23}$$

agrees satisfactorily with the value quoted by STEPHENSON[3].

Graphite has been investigated by many authors[4-8]. HEREWARD *et al.*[5] used a rectangular parallelepiped 185.8 cm square and 153.6 cm high for a total weight of 9 tons of commercial graphite of average pile density 1.55 g/cm³. A Ra α+Be source (1.2 curie of Ra) was placed on the axis of the pile at 18 cm from one of its faces of 186×154 cm² area, which was covered with a layer of paraffin 12 cm thick. The measurements were performed with dyprosium detectors at three distances from the source: $z = 79$ cm, $z = 116$ cm and $z = 153$ cm.

The amplitudes A_{lm} were found from the amplitudes of the trigonometric functions required for fitting the experimental data along each axis of the transverse planes. The following harmonics had to be considered in the first and second planes: (0, 0), (1, 0), (0, 1), (2, 0), (0, 2). In the third plane the harmonics (2, 0) and (0, 2) completely disappeared and the harmonics (1, 0) (0, 1) were very small.

[1] See footnote 2, p. 605.

[2] J.C. KOECHLIN, J. MARTELLY and V.P. DUGGAL: p. 20 of Ref. [*15c*].

[3] R. STEPHENSON: Introduction to Nuclear Engineering. New York: McGraw-Hill 1954.

[4] W. BOTHE and P. JENSEN: Z. Physik **122**, 749 (1944).

[5] H.G. HEREWARD, G.C. LAWRENCE, H.R. PANETH and B.W. SARGENT: Canad. J. Res. A **25**, 15 (1947).

[6] A. ERTAUD, R. BEAUGE, H. FAUQUEZ, H. DE LABOULAY, C. MERCIER and L. VAUTHRY: Report C.E.A. No 3 (1948).

[7] E.O. KLEMA, R.H. RITCHIE and T. ARNETTE: ORNL-1340 (1952).

[8] L. CARBLOM: Ark. Fysik **6**, 335 (1953).

The value of the relaxation length was derived from the variation undergone by A_{00} in passing from one of the three transverse planes to the other. The diffusion length was finally derived from L_{00} by means of (121.16).

In this experiment an appreciable contribution to the thermal flux measurements was due to fast and epi-cadmium neutrons reaching the transverse planes, where the measurements were performed. Therefore, a correction was computed by means of Eq. (128.4) in two successive steps; the first step consisted of adopting the result of the age theory (Sects. 75 and 84) for the slowing-down density to be introduced into (128.4). The second step consisted of taking into account the deviations of transport theory from age theory. The final result was

$$L = (51 \pm 3)\ \text{cm} \quad (\text{in graphite: } \varrho = 1.55\ \text{g/cm}^3). \tag{128.24}$$

The Saclay group[1] measured, with thin Mn detectors, the thermal neutron flux in a graphite pile of $210.7 \times 210.7 \times 233.5\ \text{cm}^3$ and average pile density $\varrho = 1.71\ \text{g/cm}^3$ containing a Ra α + Be source of 3.8 curies; their result is

$$L = (45.7 \pm 0.3)\ \text{cm} \quad (\text{in graphite: } \varrho = 1.71\ \text{g/cm}^3). \tag{128.25}$$

The experiments made by CARBLOM[2] differ from the two previous ones mainly because he used the neutrons of the D+D reaction produced by bombarding a heavy ice target with deuterons of 200 kev. The graphite pile had a volume of $201.6 \times 194.4 \times 194.4\ \text{cm}^3$ and an average density $\varrho = 1.65\ \text{g/cm}^3$. Using as detectors 90% Mn-10% Ni foils of $2 \times 2\ \text{cm}^2$ area and 0.2 mm thickness, this author obtains

$$L = (45.4 \pm 0.2)\ \text{cm} \quad (\text{in graphite of } \varrho = 1.65\ \text{g/cm}^3). \tag{128.26}$$

Referred to a graphite density $\varrho = 1.60\ \text{g/cm}^3$, the results (128.24) to (128.26) become $L = (49.4 \pm 3)$ cm, $L = (48.8 \pm 0.3)$ cm and $L = (46.8 \pm 0.2)$ cm, respectively. The residual differences may be due in part to differences in purity and/or structure of the materials used by the various authors.

129. Deviations from the one-velocity picture. Diffusion cooling and diffusion heating. The transport undergone by neutrons from a high-energy pulsed source during their slowing down to 1 ev has been considered in Sect. 81 as a function of time in the diffusion theory approximation. The theoretical predictions there derived, have been compared with the results of a few experiments in Sect. 85 and have been extended, in Sect. 111, down to thermal energies through the intermediate region where the molecular structure of the medium begins to be important.

In Sects. 130 and 131, a few experiments done with thermal neutrons from pulsed sources will be discussed in detail and their results will be compared with the theoretical predictions derived from time-dependent diffusion theory.

Before doing that, however, we should point out the existence of deviations from the predicted behaviour which do not originate from the use of the diffusion approximation, but from the treatment of the transport phenomena undergone by thermal neutrons by means of one-velocity theories.

These deviations may be described—very roughly indeed—in the frame work of the diffusion theory, as a physical effect presented by the gas of thermal neutrons diffusing throughout the medium.

The diffusion coefficient entering the diffusion equation (81.5) is defined by Eq. (81.3) as an average over the neutron velocity spectrum; its value is indepen-

[1] See footnote 6, p. 608.
[2] See footnote 8, p. 608.

dent of the space coordinates whenever the probable number of neutrons per unit volume and per unit time interval splits into two factors according to Eq. (81.6).

The description of thermal neutrons as a monoenergetic group diffusing without exchanging energy with the medium is now dropped and replaced by the picture used in Sect. 117γ: the neutron spectrum is extended over a wide energy interval, but the scattering takes place without energy exchanges between neutrons and medium. Then, from the simple fact that the diffusion coefficient (125.7) is a function of the velocity, it follows that at those points inside the medium where there is a net outflux, the spectrum will be depleted of the neutrons for which $D_0(v)$ is very large while at those points where there is a net influx, the spectrum will be deplete of the neutrons corresponding to small values of $D_0(v)$.

A more detailed qualitative description of this effect requires, of course, a specification of the dependence of $D_0(v)$ upon v. Since in almost all cases of practical interest $l_{tr}(v)$ is a constant or a monotonically increasing function of v, we can restrict our considerations to cases in which $D_0(v)$ is an increasing function of v.

Then the spectrum distortion mentioned above will follow very simple rules: at points with a net outflux the spectrum will abound of neutrons in the low velocity tail and at points with a net influx it will abound of neutrons in its high velocity part.

If we drop the assumption that the neutrons keep their velocity during diffusion, the energy exchanges with the medium will tend to restore the thermal equilibrium distribution. The result will be a different value for the average kinetic energy of the neutrons at different points: this will be larger than $\frac{3}{2}kT_n$ at points with a net influx and smaller than $\frac{3}{2}kT_n$ where there is a net outflux. Following VON DARDEL[1,2] who was the first to point out the existençe of such an effect, we shall talk of *diffusion cooling* or *diffusion heating* according to whether the average kinetic energy is higher or lower than $\frac{3}{2}kT_n$. Here $T_n(\approx T_0)$ is the neutron temperature which would be reached in a weakly absorbing infinite medium, where the neutrons are slowed down *uniformly* throughout its entire volume.

In order to clarify the mechanism of this effect, ANTONOV *et al.*[3] have considered a two-group model of a behaviour described by the following diffusion equations:

$$\left.\begin{aligned}\left(D_1\nabla^2-\frac{1}{T}-\beta_1\right)N_1+\beta_2N_2&=\frac{\partial N_1}{\partial t},\\\left(D_2\nabla^2-\frac{1}{T}-\beta_2\right)N_2+\beta_1N_1&=\frac{\partial N_2}{\partial t}.\end{aligned}\right\}\tag{129.1}$$

Here $N_1(\boldsymbol{r},t)$ and $N_2(\boldsymbol{r},t)$ are the probable numbers of neutrons per unit volume and unit time interval belonging to the two groups, D_1 and D_2 the two corresponding diffusion coefficients and β_1 (and β_2) the probabilities per unit time that a neutron of group 1 (or group 2) is transformed into a neutron of group 2 (or group 1) as a result of its interaction with the medium.

If the assumption is made that both groups fulfil the same boundary conditions, the eigenfunctions $R_{lmn}(\boldsymbol{r})$ and the eigenvalues ω_{lmn}^2, corresponding to a

[1] G. F. VON DARDEL: Trans. Roy. Inst. Technol. No, **75** (1954). — Phys. Rev. **94**, 1272 (1954).

[2] G. F. VON DARDEL and N. G. SJÖSTRAND· Phys. Rev. **96**, 1245 (1954).

[3] A. V. ANTONOV, A. I. ISAKOFF, I. D. MURIN, B. A. NEUPOCOYEV, I. M. FRANK, E. L. SHAPIRO and I. V. SHTRANICH. p. 3 of Ref. [*15c*].

certain geometry of the medium and source, will also be the *same* for both groups and the solution of Eqs. (129.1) is readily obtained.

The total number of neutrons

$$N(\boldsymbol{r}, t) = N_1(\boldsymbol{r}, t) + N_2(\boldsymbol{r}, t)$$

splits into a sum of infinite terms corresponding to increasing values of ω_{lmn}^2, and after a time sufficiently long it reduces to its fundamental mode (Sect. 81, p. 306). If the assumption is made that under these conditions the equilibrium is established between the two groups, so that

$$N_1\beta_1 = N_2\beta_2, \tag{129.2}$$

then $N(\boldsymbol{r}, t)$ decreases exponentially with time and we can write

$$N(\boldsymbol{r}, t) \approx \mathrm{e}^{-\left(\frac{1}{T}+F\right)t}$$

where F is a function of $D_1, D_2, \beta_1, \beta_2$ and ω^2. If the neutron leakage of both groups is small compared to the rapidity of mixing of the two groups, i.e., if the condition

$$\frac{(D_1 - D_2)\,\omega^2}{\beta_1 + \beta_2} < 1 \tag{129.3}$$

is fulfilled, then F can be developed into a power series of ω^2 with the result

$$\bar{\gamma} = -\frac{1}{N(\boldsymbol{r}, t)}\frac{\partial N(\boldsymbol{r}, t)}{\partial t} = \frac{1}{T} + \bar{D}\omega^2 - C\omega^4 + G\omega^6 \tag{129.4}$$

where

$$\bar{D} = \frac{\beta_2 D_1 + \beta_1 D_2}{\beta_1 + \beta_2} \tag{129.5}$$

represents, according to (129.2), the average diffusion coefficient:

$$\bar{D} = \frac{N_1 D_1 + N_2 D_2}{N_1 + N_2}, \tag{129.6}$$

and

$$C = \frac{\beta_1\beta_2 (D_1 - D_2)^2}{(\beta_1 + \beta_2)^3} \tag{129.7}$$

depends on the spread $|D_1 - D_2|$ of the diffusion coefficients of the two groups. The third term of the expansion (129.4)

$$G\omega^6 = \frac{\beta_1\beta_2(\beta_1 - \beta_2)(D_1 - D_2)^3}{(\beta_1 + \beta_2)^5}\,\omega^6$$

may be assumed to be negligibly small, since it contains, besides the cube of the left-hand side of Eq. (129.3), a factor $(\beta_1 - \beta_2)$ which also should be small whenever the two neutron groups have properties very close to one another.

Eq. (129.4) differs from Eq. (81.14) only because the diffusion coefficient D_0 is replaced by

$$D_0 = \bar{D} - C\omega^2. \tag{129.8}$$

The same type of reasoning although more complicated, may be applied to the case of thermal neutrons the continuous spectrum of which is represented by a Maxwellian—or almost Maxwellian—velocity distribution. The second term appearing in (129.8) corresponds to diffusion cooling or diffusion heating according to its sign. Its order of magnitude can be evaluated by means of the following arguments which refine and extend the considerations given on p. 501 and 502 of Sect. 111 in order to derive Eq. (111.9).

Since problems in which the neutron density also depends on the space coordinates are considered here, we may start by defining the average kinetic energy of the neutrons at the point $\boldsymbol{r}$ and time t,

$$\bar{E}(\boldsymbol{r}, t) = \frac{\int\limits_0^\infty E \cdot N_0(\boldsymbol{r}, v, t)\, dE}{\int\limits_0^\infty N_0(\boldsymbol{r}, v, t)\, dE,}. \tag{129.9}$$

Under the assumption that the variables $\boldsymbol{r}$ and v can be separated according to (81.6) and that $n_0(E, t)$ is normalized to 1 neutron over the whole spectrum, $\bar{E}(\boldsymbol{r}, t)$ turns out to be independent of $\boldsymbol{r}$ so that

$$\bar{E}(t) = \int\limits_0^\infty E\, n_0(E, t)\, dE \tag{129.10}$$

is identical to (111.5).

The equation describing the rate of total energy loss at the point $\boldsymbol{r}$ and time t is given by

$$\left.\begin{aligned} -\frac{d}{dt}[N_0(\boldsymbol{r}, t)\, \bar{E}(t)] &= N_0(\boldsymbol{r}, t) \int\limits_0^\infty \Sigma_s(E)\, \overline{(E - E')}\, v\, n_0(E, t)\, dE + \\ &+ N_0(\boldsymbol{r}, t) \int\limits_0^\infty \Sigma_a(E)\, E\, v\, n_0(E, t)\, dE + \omega^2 N_0(r, t) \int\limits_0^\infty D_0(E)\, E\, n_0(E, t)\, dE \end{aligned}\right\} \tag{129.11}$$

where the three successive terms arise respectively from scattering, absorption and neutron diffusing off point $\boldsymbol{r}$. The latter term contains the geometrical buckling ω^2 defined by the relation

$$\nabla^2 N_0(\boldsymbol{r}, t) = \omega^2 N_0(\boldsymbol{r}, t). \tag{129.12}$$

We notice that ω^2 is a constant only in the case of finite media in which a single harmonic of the expansion (121.15) or (121.20) is excited. The third term in (129.11) is readily obtained by transforming the rate of energy loss due to diffusion

$$\int E\, n_0(E, t)\, \frac{\partial N_0(\boldsymbol{r}, t)}{\partial t}\, dE$$

by means of (81.5) and (129.12).

Similarly, the decay rate of the total number of neutrons at the position $\boldsymbol{r}$ is given by

$$\left.\begin{aligned} -\frac{d}{dt} N_0(\boldsymbol{r}, t) &= N_0(\boldsymbol{r}, t) \int \Sigma_a(E) \cdot v\, n_0(E, t)\, dE + \\ &+ \omega^2 N_0(\boldsymbol{r}, t) \int D_0(E)\, n_0(E, t)\, dE \end{aligned}\right\} \tag{129.13}$$

where the two terms arise from capture and outflow of neutrons respectively.

The two Eqs. (129.11) and (129.13) can be used to eliminate $N_0(\boldsymbol{r}, t)$; under the assumption that $\Sigma_a(v)$ follows the $1/v$ law (125.6) one obtains[1-3]

$$\left.\begin{aligned} -\frac{d\bar{E}(t)}{dt} &= \int\limits_0^\infty \Sigma_s(E)\, \overline{(E - E')}\, v\, n_0(E, t)\, dE + \\ &+ \omega^2 \int\limits_0^\infty D_0(E)\, [E - \bar{E}(t)]\, n_0(E, t)\, dE. \end{aligned}\right\} \tag{129.14}$$

[1] F. L. Shapiro: p. 82 of Ref. [15c].

[2] K. Beckurts: Nuclear Sci. and Engng. 2, 516 (1957).

[3] G. S. Mani, N. Umakanth, S. B. D. Iyengar and R. Ramanna: Proc. Ind. Acad. Sci. 45, 215 (1957).

It should be noticed that Eq. (129.14) reduces to Eq. (111.4) for $\omega^2=0$, i.e., in the case of an infinite medium where the neutrons are distributed uniformly throughout its entire volume, or also in the case of a non-uniform distribution, whenever one refers to the total number of neutrons present in the whole infinite medium.

Thus, the first term appearing on the right-hand side of Eq. (129.14) can be replaced by the right-hand side of Eq. (111.6b):

$$\int_0^\infty \Sigma_s(E)\,\overline{(E-E')}\,v\,n_0(E,t)\,dE = \frac{1}{t_{\text{th}}^0}\left(\overline{E}(t)-E_0\right) = \frac{1}{t_{\text{th}}^0}\,\frac{3}{2}\,\mathrm{k}\left(T(t)-T_0\right). \tag{129.15a}$$

The second term in (129.14) can readily be expressed in terms of the mean energy carried off by each outflowing neutron:

$$\langle E\rangle_D = \frac{\int_0^\infty D_0(E)\,E\,n_0(E,t)\,dE}{\int_0^\infty D_0(E)\,n_0(E,t)\,dE}. \tag{129.15b}$$

Choosing for $n_0(E,t)$ a Maxwellian distribution of velocities at the instantaneous temperatures $T(t)$,

$$n_0(v,t)\,dv = \frac{4}{\sqrt{\pi}}\,\frac{1}{v_0^3}\,e^{-\frac{v^2}{v_0^2}}\,v^2\,dv, \qquad v_0=\left[\frac{2\mathrm{k}T(t)}{m}\right]^{\frac{1}{2}}, \tag{129.16}$$

and expressing the numerator of (129.15b) in terms of the derivative of the denominator with respect to $v_0(T)$, BECKURTS[1] obtains

$$\langle E\rangle_D = \frac{3}{2}\,\mathrm{k}T + (\mathrm{k}T)^2\,\frac{d}{d(\mathrm{k}T)}\ln D_0(T). \tag{129.17}$$

Eq. (129.14) finally becomes

$$-\frac{d\overline{E}(t)}{dt} = \frac{3}{2}\,\mathrm{k}\,\frac{1}{t_{\text{th}}^0}\left[T(t)-T_0\right] + \omega^2(\mathrm{k}T)^2\,\frac{dD_0(T)}{d(\mathrm{k}T)}. \tag{129.18}$$

The equilibrium temperature T_{eq} is now derived by introducing into (129.18) the condition $d\overline{E}/dt=0$; for T very close to the medium temperature T_0 we obtain, to a first approximation[2],

$$T_{\text{eq}}-T_0 = -\frac{2}{3}\,t_{\text{th}}^0\,T_0^2\left(\frac{dD_0}{dT}\right)_{T=T_0}\omega^2. \tag{129.19}$$

This equation quantitatively expresses the effect of diffusion cooling—or diffusion heating—in terms of $(dD_0/dT)_{T=T_0}\omega^2$; for $\omega^2=0$ one has $T_{\text{eq}}=T_0$ (Sect. 111), otherwise $T_{\text{eq}}\neq T_0$. Since usually $dD_0/dT>0$, we should expect to have diffusion cooling where the buckling ω^2 is positive and diffusion heating where ω^2 is negative[2].

The latter condition occurs with stationary neutron sources whenever the buckling is given by $-1/L^2$. Therefore, the methods to measure, for example, the diffusion length by means of steady state assemblies (Sect. 128) refer to thermal neutrons which are slightly "hotter" than the moderator. In these cases, however, the effect is usually very small and may be neglected. As we shall see in Sect. 131, diffusion cooling, on the contrary, may be appreciable when the buckling ω^2 is positive and large; these conditions occur in a small portion of the medium.

[1] See footnote 2, p. 611.

[2] G. F. VON DARDEL and N. G. SJÖSTRAND: Progress in Nuclear energy. Physics and Mathematics, Vol. 2, p. 183. London and New York: Pergamon Press 1958.

The diffusion coefficient of the neutrons at equilibrium may be expressed in terms of $D_0(T_0)$ by means of the trunkated expansion in powers of $T_{eq} - T_0$:

$$D_0(T_{eq}) = D_0(T_0) + \left(\frac{dD_0}{dT}\right)_{T=T_0} (T_{eq} - T_0) + \cdots. \tag{129.20a}$$

By introducing (129.19) into this equation, we see that, to a first approximation, we can write

$$D_0(T_{eq}) = D_0(T_0) - C\omega^2 \tag{129.20b}$$

where

$$C = \frac{2}{3} t_{th}^0 T_0^2 \left(\frac{dD_0}{dT}\right)^2_{T=T_0}. \tag{129.21}$$

This equation can be used for deriving the value of the thermalization time constant t_{th}^0 from measurements of C and $(dD_0/dT)_{T_0}$. In Sect. 131, we shall discuss the experimental determination of C and $D_0(T_0)$; by repeating the measurements for various medium temperatures, dD/dT_0 may be derived. One might object that the quantity $(dD_0/dT)_{T_0}$ appearing in (129.21) refers to a change of the neutron temperature by keeping the medium temperature constant. It may, however, be expected that the two temperature coefficients $(dD_0/dT)_{T_0}$ and dD_0/dT_0 are not very different from each other—apart from the effect of thermal expansion—since the thermal velocities of the neutrons are much larger than those of the nuclei which are bound and often heavier than the neutron[1].

A particular case of certain interest[2,3] not only because of its simplicity, but also because it is actually met in a few practical cases (Sect. 131), is that of a medium with a transport mean free path independent of neutron velocity and, therefore, of temperature

$$l_{tr}(v) = l_{tr}(T) = \text{const}. \tag{129.22}$$

Then, according to (125.7), the energy dependence of the diffusion coefficient is only due to the factor v, and since

$$\int_0^\infty (E - \overline{E}) \cdot v \frac{4}{\sqrt{\pi}} \frac{1}{v_0^3} e^{-\frac{v^2}{v_0^2}} v^2 \, dv = \frac{1}{3} \overline{v} \overline{E}, \quad \overline{v} = \frac{2}{\sqrt{\pi}} v_0(T), \quad \overline{E} = \frac{3}{2} kT, \tag{129.23}$$

we obtain the following expression for the second term on the right hand side of Eq. (129.14):

$$\omega^2 \int_0^\infty D_0(E) \left[(E - \overline{E}(t)\right] n_0(E, t) \, dE = \frac{\omega^2}{3} \overline{D}_0(T) \overline{E}(t) \tag{129.24}$$

where

$$\overline{D}_0(T) = \frac{l_{tr} \overline{v}}{3} = \frac{1}{3} l_{tr} \frac{2}{\sqrt{\pi}} v_0(T) = \frac{2}{3} l_{tr} \left(\frac{2k}{\pi m}\right)^{\frac{1}{2}} T^{\frac{1}{2}}. \tag{129.25}$$

Thus we may write, instead of (129.18),

$$-\frac{dT}{dt} = \frac{\overline{D}_0(T)\,\omega^2}{3} T + \frac{1}{t_{th}^0} \left(T(t) - T_0\right), \tag{129.26}$$

which expression is rigorous under the assumptions (129.16) and (129.22). In the present case the equilibrium temperature is that root of the equation

$$\frac{\omega^2}{3} \overline{D}_0(T_{eq}) T_{eq} + \frac{1}{t_{th}^0} (T_{eq} - T_0) = 0 \tag{129.27}$$

[1] See footnote 2, p. 612.
[2] See footnote 1, p. 611.
[3] See footnote 3, p. 611.

which turns out to be close to T_0. The approach of $T(t)$ to T_0 is described by the expression

$$T(t) - T_{eq} \propto e^{-t/t_{th}} \tag{129.28}$$

with

$$\frac{1}{t_{th}} = \frac{\overline{D}_0(T_{eq})}{3}\,\omega^2 + \frac{1}{t_{th}^0}. \tag{129.29}$$

This equation shows that, in a finite medium, the time constant for approach to equilibrium is larger than in an infinite medium; this conclusion could be reached, in general, by a closer study of Eq. (129.18). When condition (129.22) is fulfilled, we obtain from (129.20b), instead of (129.21), the relation[1]

$$\frac{C\,\omega^2}{\overline{D}_0(T_0)} = 1 - \left(\frac{T_{eq}}{T_0}\right)^{\frac{1}{2}} \tag{129.30}$$

where use has been made of the fact that $\overline{D}_0(T) \propto v_0(T) \propto T^{\frac{1}{2}}$. This equation can be used for determining T_{eq} from measurements of C and $\overline{D}_0(T_0)$. Once T_{eq} is known, one can use Eq. (129.27) for deriving t_{th}^0. Since this quantity is independent of the geometrical buckling, we can compute it for the simple case of very large sizes of the medium ($\omega^2 \approx 0$, $T_{eq} \approx T_0$). From Eq. (129.27) and (129.30), we obtain

$$C = \frac{1}{6}\overline{D}_0^2(T)\,t_{th}^0 \quad (\text{for } l_{tr} = \text{const}) \tag{129.31}$$

a relation which could have been obtained directly by introducing (129.25) into (129.21) under condition (129.22). If the latter restriction is abandoned, we can write in general

$$C = \chi\, C_{l_{tr}=\text{const}} \tag{129.32}$$

where χ is a factor taking into account the variation of l_{tr} with neutron temperature. From (129.21) and (129.31), it follows that

$$\chi = \left[1 + 2\frac{d\ln D_0(T)}{d\ln T}\right]^2. \tag{129.33}$$

Finally we notice that, for $\Sigma_s(E) = \Sigma_s = \text{const}$, Eq. (129.15a) can be transformed into

$$\langle\overline{(E-E')}\rangle = \int_0^\infty \overline{(E-E')}\, v\, n_0(E,t)\, dE = \frac{\overline{E}(t) - E_0}{t_{th}^0\,\Sigma_s} \tag{129.34}$$

which expresses the mean loss of energy per collision in terms of quantities readily measurable by means of the methods described in Sect. 131.

130. Experiments with pulsed sources embedded in an infinite medium. In this section we shall consider the case of a pulsed neutron source embedded in a homogeneous medium, the dimensions of which are so large in all directions that the loss of neutrons through its surface can be neglected. The experiment usually consists of measuring, as a function of time, the neutron density $N_0(\boldsymbol{r}, t)$ at many positions $\boldsymbol{r}$ distributed throughout the medium. The more relevant features of the techniques used in experiments of this type have been mentioned very briefly in Sect. 85; more details may be found in the review article by VON DARDEL and SJÖSTRAND[2].

[1] See footnote 3, p. 611.

[2] G. VON DARDEL and M. G. SJÖSTRAND: Progress in nuclear energy. Physics and Mathematics, Vol. 2, p. 183. New York and London: Pergamon Press 1958.

Once $N_0(\boldsymbol{r}, t)$ has been measured, some of its integral properties are usually computed, as, for example, its spatial moments (Sect. 71). The moment of zero order, i.e., the integral of $N_0(\boldsymbol{r}, t)$ extended over all space, represents the decay curve of the total number of neutrons present in the medium, which is only determined by the mean life of the neutrons:

$$N_{00}(t) = \int N_0(\boldsymbol{r}, t)\, d\boldsymbol{r} \propto e^{-t/T}. \tag{130.1}$$

If, for the sake of definiteness, we assume that the source can be treated—at least approximately—as a point source, the average of the square of the distance from the source

$$\overline{r^2}(t) = \frac{\int\limits_0^\infty N_0(\boldsymbol{r}, t)\, r^2\, d\boldsymbol{r}}{\int\limits_0^\infty N_0(\boldsymbol{r}, t)\, d\boldsymbol{r}} \tag{130.2}$$

is coupled to the generalized age (111.1) by Eq. (76.17). Therefore, the time dependence of $\overline{r^2}(t)$ should be similar to the curve of Fig. 225, (p. 500), and allows the derivation of the age from energies larger than 1 ev where steady state measurements are easily performed (Sects. 83 and 84) through the chemical region, down to thermal energies.

Two remarks should be made: First both Eqs. (130.1) and (130.2) are rigorous; therefore, the determination of T and τ based on their use does not involve the elementary diffusion theory as it occurs in the procedures described in Sect. 131. Secondly, since the parts of the medium at great distance from the source give a larger contribution to (130.2) than to (130.1), until now procedures as those described above have been employed only for the determination of the mean life but not of the generalized age $\tau(t)$.

Instead of integral properties, as those mentioned above, one could think of using the function $N_0(\boldsymbol{r}, t)$ itself for deriving the values of the three quantities T, $D_0(T_0)$ and C. This procedure, however, meets some difficulty because these quantities are expected to have different values at different points as a consequence of the fact that the buckling—defined by means of Eq. (129.12)—depends upon the distance from the source. If the $\nabla^2 N_0(\boldsymbol{r}, t)$ is positive—as usually occurs near the source—the neutron temperature will fall below the medium temperature, while the reverse situation will be found at great distances from the source. An example of such a behaviour is considered in detail below (Tables 106 and 107). As we shall see in Sect. 131, these difficulties are avoided in the case of finite media where—after a sufficiently long time—only one mode of the neutron distribution survives. Therefore, the most convenient use of measurements made in infinite media consists of deriving some integral properties of $N_0(\boldsymbol{r}, t)$, such as (130.1) or (130.2).

The condition of infinite dimensions in all directions can be easily fufilled in the case of light water or of a solution of absorbing materials in light water. With other media, such as heavy water, beryllium or graphite, the amount of material needed is so large as to make this type of experiments unpractical. In these cases, the use of finite media (Sect. 131) is a much more practical approach.

Thus it appears justified that the measurements reported below all refer to the determination of the neutron mean life in light water or in solutions of borax in light water.

The mean life of thermal neutrons in light water was measured by von Dardel and Watner[1] who used the neutrons emitted by a D+D source ($E_d = 150$ kev)

[1] G. von Dardel and A.W. Watner: Phys. Rev. 91, 1284 (1953).

placed at the centre of a water tank of $97\times 97\times 107$ cm³ volume. The detectors were cylindrical brass counters, 1 cm in diameter, filled with BF_3, which extended through the entire depth of the tank and thus automatically performed an integration of the neutron density in the vertical direction; the integration in the radial direction was made numerically. The effect of the neutron absorption in the counters was investigated by comparing the results obtained with the brass counters with those obtained with aluminium counters filled with BF_3 at a pressure such that they had the same total absorption per unit volume as water. The effect of the absence of scattering in the detector should disappear in the integration over the whole volume of the medium. This may be seen by noticing that: (a) the total number of neutrons $N_{00}(t)$ has essentially the same properties of the neutron density observed in an infinite medium where the neutrons are produced uniformly throughout its entire volume and (b) that in the case of a uniform source distribution the theorem given on p. 544 of Sect. 119δ will hold.

The results, $T=(212\pm 3)$ μsec in water at 22.6° C, is higher than the value obtained by other authors [see Eq. (130.7) and Table 96] probably because of an inadequate background correction[1].

A similar experiment has been made by a group of the Clarendon Laboratory[2] who bombarded, with deuterons of 450 kev a heavy ice target placed at the centre of a 120 cm × 120 cm × 180 cm light water tank. The neutron decay was followed by detecting either the neutrons with a BF_3 counter, or the 2.2 Mev photons emitted in capture of thermal neutrons by hydrogen (Sect. 29) by means of a liquid scintillation counter mounted on the end of a long light guide.

The main advantages of a detector of this type are its great sensitivity, and the fact that both absorption and scattering in the scintillator and in the light guide are so close to those of water that the disturbances produced by the detector are very small. Moreover, because of the large range of the γ-rays in water (≈ 25 cm), the detector integrates over a large volume, so that the effect of any small disturbance in the neutron density due to the detector itself is further reduced. The main disadvantage is that the scintillator does not give the same pulse height for each γ-ray detected and so its sensitivity depends critically on the stability of the electronic equipment.

The neutron decay was followed by counting the γ-rays—or the neutrons—produced in an initial 100 μsec pulse of fast neutrons in successive 50 μsec intervals, the whole cycle being repeated every 1600 μsec. The neutrons were monitored with a BF_3 counter placed in the tank 25 cm from the target and diametrically opposite to the scintillator.

The perturbation produced by the tube through which the deuterons reached the target was estimated experimentally by introducing a dummy vertically below it. The thickness of the walls of the dummy was chosen so as to give the same capture cross section for slow neutrons as that of the actual target tube.

The measurements were performed using both the γ-ray detector and the BF_3 counter placed along the horizontal and the vertical axes through the target. Fig. 257 shows the neutron distribution obtained with the BF_3 counter; the outward diffusion of neutrons with increasing time is clearly shown by these curves.

The decay curves made with the γ-ray detector were strictly exponential at all position. Slight deviations from strict exponentials were observed in the measurements with BF_3 counters; the effect, however, did not prevent the assignement of a lifetime to each position.

[1] See footnote 2, p. 614.

[2] R. E. MEADS, C. J. ENGLAND, C. H. COLLIE and G. C. WEEKS: Proc. Phys. Soc. Lond. A, **69** 469 (1956).

Table 106 shows the values of the effective mean life as a function of the distance r from the source. Its variation is due to the effect of diffusion as explained above. The following type of argument[1] can be used for justifying that the decay curve is exponential—or almost exponential—at each point. The counting rate $C(r, t)$ of a small scintillator, when placed at the point $\boldsymbol{r}$ inside the medium, is given by

$$C(\boldsymbol{r}, t) = A \int \frac{1}{T} N_0(\boldsymbol{r}, t) \frac{e^{-\frac{|\boldsymbol{r}-\boldsymbol{r}'|}{l}}}{|\boldsymbol{r}-\boldsymbol{r}'|^2} d\boldsymbol{r} \tag{130.3}$$

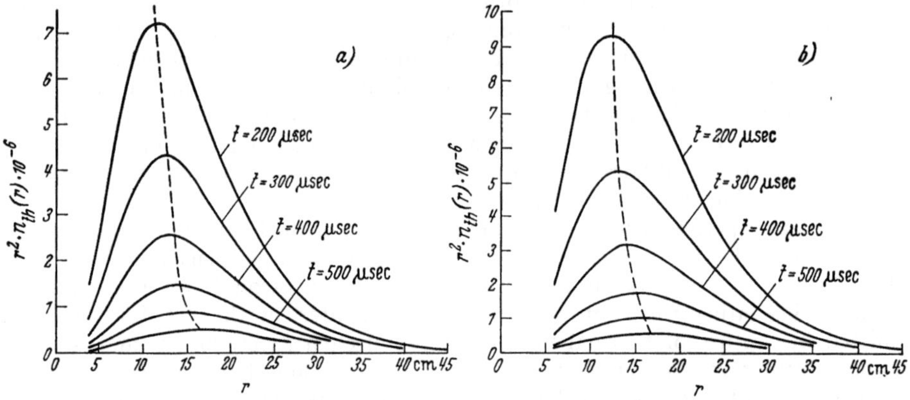

Fig 257 a and b. Distribution of thermal neutrons due to a D + D source (E_d = 450 kev) in water as a function of time: (a) in the horizontal plane, (b) vertically below the target (MEADS *et al.*).

Table 106. *Effective neutron mean life in water (in* µsec*) as a function of distance from the source* (MEADS *et al.*).

(1)	(2)	(3)	(4)	(5)	(6)
10	174.0 ± 0.6	171.7 ± 0.7	178.1	171.5	164.6
15	194.6 ± 0.4	192.6 ± 0.7	—	202.5	187.4
20	204.0 ± 0.2	201.7 ± 0 4	203.5	232.3	212.9
25	208.0 ± 0.84	—	—	261.0	236.0
30	211.1 ± 0.7	206.2 ± 0.85	211.8	278 0	249.3
35	207.4 ± 1.5	—	—	287.7	258.2
40	206.9 ± 0.8	205.3 ± 1.4	210.2	289.5	258.5
45	204.2 ± 0.7	202.2 ± 1.9	—	—	—
50	204.7 ± 0.5	201 9 ± 0.5	207.1		
55	203.1 ± 1.5	202.3 ± 1.8			
60	200.5 ± 0.85	197.6 ± 0.4			
65	201.5 ± 1.4	—			
70	203.7 ± 0.5	197.0 ± 0.8			
75	201.1 ± 1				
80	199.0 ± 1.4				

(1) Distance in cm of detector from target; (2) average life in µsec measured with γ counter in horizontal plane through target; (3) average life measured with γ-ray counter in horizontal plane through target with dummy target in position below target; (4) average life measured with γ counter vertically below target; (5) approximate average life measured with neutron counter in horizontal plane; (6) approximate average life measured with neutron counter vertically below the target.

The errors quoted are the probable errors derived from the statistics. The neutron measurements have a negligible statistical error.

[1] C. H. COLLIE, R. E. MEADS and E. E. LOCKETT: Proc. Phys. Soc. Lond. A **69**, 464 (1956).

where A is a constant depending upon the efficiency of the counter, and l is the mean free path of the 2.2 Mev γ-rays in water. According to (81.4) and (81.5) we can write, to a first approximation,

$$N_0(\boldsymbol{r}, t) \approx N_0(\boldsymbol{r}, 0) + \left(\frac{\partial N_0}{\partial t}\right)_{t=0} \cdot t = N_0(\boldsymbol{r}, 0) + D_0 [\nabla^2 N_0(\boldsymbol{r}, t)]_{t=0} \cdot t$$

from which it follows that

$$N_{0a}(\boldsymbol{r}, t) \approx N_0(\boldsymbol{r}, 0)\, e^{-\frac{t}{T}} (1 + \eta t) \tag{130.4}$$

where

$$\eta(\boldsymbol{r}) = D_0 \frac{\int \nabla^2 N_0(\boldsymbol{r}', 0) \frac{e^{-\frac{|\boldsymbol{r}-\boldsymbol{r}'|}{l}}}{|\boldsymbol{r}-\boldsymbol{r}'|^2} d\boldsymbol{r}'}{\int N_0(\boldsymbol{r}', 0) \frac{e^{-\frac{|\boldsymbol{r}-\boldsymbol{r}'|}{l}}}{|\boldsymbol{r}-\boldsymbol{r}'|^2} d\boldsymbol{r}'}. \tag{130.5}$$

Eq. (130.4) can be written

$$N_0(\boldsymbol{r}, t) \approx N_0(\boldsymbol{r}, 0)\, e^{-\left(\frac{1}{T} - \eta\right) t} \tag{130.6}$$

and shows that the effect of diffusion, in a very rough first approximation, reduces to a change of the decay constant: instead of being $1/T$, the latter turns out to be $(1/T) - \eta$ with $\eta(\boldsymbol{r})$ given by (130.5).

From the data of Table 106, the Claredon Laboratory group[1] derives the value of T by three different methods:

1. By integration of the γ-ray scintillator data throughout the tank. Beyond 35 cm from the target, the measurements at a given time were represented by means of Eq. (82.4) which was used for the extrapolation to infinity. A correction was necessary for the perturbation produced by the target tube. The final result is

$$T = (203.3 \pm 2.6)\ \mu\text{sec}. \tag{130.7}$$

2. By integration of the BF_3 counter data throughout the tank; an extra correction was necessary for the neutron absorption in the counter. The result is

$$T = (201.9 \pm 0.13)\ \mu\text{sec} \tag{130.8}$$

where the error assigned is the statistical error of the measurements.

3. By correcting the effective mean lives observed at a distance r for the effect of neutron diffusion by means of the approximate expressions (130.5) and (130.6). The corrected values are given in Table 107 from which we see that—as we should expect—they are almost independent of the distance from the source.

Of these three data the first one—i.e. Eq. (130.7) —is considered by the authors as the more reliable one.

Similar experiments have been made by a group of the Atomic Energy Establishment of India[2].

Essentially the same procedure can be used for measuring the capture cross section of any material soluble in water. Until now it has been applied only to

---

[1] See footnote 2, p. 616.

[2] R. RAMANNA, G. S. MANI, P. K. IYENGAR, S. B. IYENGAR and B. V. JOSHI: J. Nucl. Energy 2, 145 (1956).

boron[1,2] whose cross section for thermal neutron capture is particularly important (Sect. 27). COLLIE *et al.*[2] start by writing the decay constant of thermal neutrons in an infinite boron solution in the form

$$-\frac{1}{N_0(\boldsymbol{r},t)}\frac{\partial N_0(\boldsymbol{r},t)}{\partial t}=(\mathfrak{N}_{\mathrm{H}}'\sigma_{\mathrm{H}}+\mathfrak{N}_{\mathrm{B}}\sigma_{\mathrm{B}})\bar{v}+\eta(\boldsymbol{r}) \tag{130.9}$$

where $\mathfrak{N}_{\mathrm{H}}'$ and $\mathfrak{N}_{\mathrm{B}}$ are the number of protons and boron atoms per cm³ of solution, $\bar{v}=2200$ m/sec, and $\eta(\boldsymbol{r})$ represents the effect of diffusion which is assumed to be the same in the solution and in water. Thus Eq. (130.9) can be transformed into

$$-\frac{1}{N_0(\boldsymbol{r},t)}\frac{\partial N_0(\boldsymbol{r},t)}{\partial t}=A+\mathfrak{N}_{\mathrm{B}}'\sigma_{\mathrm{B}}\bar{v} \tag{130.10}$$

where

$$A=\mathfrak{N}_{\mathrm{H}}\sigma_{\mathrm{H}}\bar{v}+\eta(\boldsymbol{r})$$

is independent of the boron concentration, and

$$\mathfrak{N}_{\mathrm{B}}'=\mathfrak{N}_{\mathrm{B}}\left[1+\frac{\sigma_{\mathrm{H}}}{\sigma_{\mathrm{B}}}\,\frac{\mathfrak{N}_{\mathrm{H}}'-\mathfrak{N}_{\mathrm{H}}}{\mathfrak{N}_{\mathrm{B}}}\right]. \tag{130.11}$$

Table 107. *Correction of effective mean life of thermal neutrons due to diffusion in water* (MEADS *et al.*).

Distance from target cm	Calc. correction to life time μsec	Corrected average life time μsec
12	+15.97	201.5
15	+ 2.87	197.8
20	− 8.27	198.5
25	−11.31	202.3
30	−12.74	201.7
35	−12.41	201.1
40	− 8.50	202.6
45	− 8.15	201.5
		Mean 200.7 μsec

The quantity in brackets differs from unity by about 0.1% for reasonable concentrations of borax ($\leqq 4\times10^{-6}$ gram of borax per gram of solution); thus the ratio $\sigma_{\mathrm{H}}/\sigma_{\mathrm{B}}$ need not be known very accurately to use (130.11). According to (130.9) and (130.10), the effective decay constant of the neutrons observed at a given distance r from the source should be a linear function of $\mathfrak{N}_{\mathrm{B}}'$ with a slope determined by σ_{B}. By repeating the measurements at various borax concentrations, the Oxford group finds

$$\text{at } r=20\text{ cm}\qquad \sigma_{\mathrm{B}}=(758.1\pm1.6)\text{ barns},$$

$$\text{at } r=30\text{ cm}\qquad \sigma_{\mathrm{B}}=(761.6\pm2.5)\text{ barns},$$

$$\text{at } r=40\text{ cm}\qquad \sigma_{\mathrm{B}}=(762.9\pm1.7)\text{ barns}.$$

The weighted average of these three values is

$$\sigma_B=(761\pm2)\text{ barns}, \tag{130.12}$$

in good agreement with the results of other measurements (Table 11 on p. 61).

131. Experiments with pulsed sources and finite media: determination of *T*, *D*$_0$ and *C*. The neutron density produced by a pulsed source of fast neutrons in a finite medium is given by Eq. (81.12) as a function of $\boldsymbol{r}$ and t. As we have shown in Sect. 111, this expression can be used even after such a long time since the emission from the source, that the neutrons have passed the chemical bond region and reached thermal energies.

This is true on the diffusion theory approximation when $\tau(t)$ is interpreted as the generalized age (111.1). This quantity is represented qualitatively in

[1] See footnote 1, p. 615.
[2] See footnote 1, p. 617.

Fig. 225 from which we see that, for t large enough to be in the thermal region, we may write

$$\tau(t) = \tau_m + D_{\text{asy}}\, t. \tag{131.1}$$

Using this expression for $\tau(t)$, Eq. (81.12) can be rewritten in the form

$$N_0(\boldsymbol{r}, t) = \sum_{l,m,n} A_{lmn} R_{lmn}(\boldsymbol{r})\, e^{-\omega_{lmn}^2 \tau_m}\, e^{-\left(\frac{1}{T} + D_{\text{asy}}\, \omega_{lmn}^2\right) t}. \tag{131.2a}$$

Since the geometrical buckling ω_{lmn}^2, increases as the order (l, m, n) of the successive harmonics increase, Eq. (131.2a), for t sufficiently large, reduces to its fundamental mode, and we can write (Sect. 81, p. 306)

$$N_0(\boldsymbol{r}, t) \to A\, R(\boldsymbol{r})\, e^{-\omega^2 \tau_m}\, e^{-\left(\frac{1}{T} + D_{\text{asy}}\, \omega^2\right) t} \quad \text{(for large } t\text{)}. \tag{131.2b}$$

A slight improvement of this expression is obtained by replacing D_{asy} by the expression (129.8) which takes into account—to a first approximation—the modification of the neutron spectrum arising from the velocity dependence of the diffusion coefficient (Sect. 129). Thus the decay constant of the thermal neutron distribution at a given position inside a finite piece of moderator is given by the quadratic form in ω^2,

$$\bar{\gamma} = -\frac{1}{N_0(\boldsymbol{r}, t)} \frac{\partial N_0(\boldsymbol{r}, t)}{\partial t} = \frac{1}{T} + D_0 \omega^2 - C\omega^4. \tag{131.3a}$$

Thus, four parameters enter into Eq. (131.2b); three of them, T, D_0 and C, depend only on the intrinsic properties of the medium, the fourth, τ_m, depends also upon the initial energy of the neutrons.

The principle of the experiment is the following one: The medium is given an elementary shape for which the geometrical buckling is readily computed; usually cylindrical or parallelepipedal forms are used. The fast neutron source and the detector are place outside the medium in order not to influence the geometry. The decay curve is measured of the neutrons escaping from the medium after a certain time t has elapsed since the time of an instantaneous pulse of fast neutrons emitted by the source.

By repeating the measurements for various values of the geometrical buckling, a curve of $\bar{\gamma}$ versus ω^2 is obtained, which, according to (131.3a) can approximately be represented by a quadratic form. The values of T, D_0 and C are determined which give the best fit of the experimental points.

The value of the age to thermal energies τ_m can also be determined, provided the factor $A\,R(\boldsymbol{r})$ appearing in Eq. (131.2b) is maintained constant when the buckling is changed; under these conditions a semilogarithmic plot of the counting rate at a given time t multiplied by $\exp\left[\left(\frac{1}{T} + D_{\text{asy}}\, \omega^2\right) t\right]$ versus ω^2 is a straight line of angular coefficient $-\tau_m$. If the source and detector positions are fixed, and the source strength is normalized by a fast neutron monitor, A and $R(\boldsymbol{r})$ remain constant provided that the geometry is changed symmetrically with respect to the source and detector[1].

While many experiments have been made for determining T, D_0, and C, only in the case of BeO has a determination of τ_m been actually performed [see Eq. (131.4) below].

[1] G. RAMANNA, G.S. MANI, P.K. IYENGAR, S.B.D. IYENGAR and B.V. JOSHI: p. 24 of Ref. [*15c*]. — J. Nucl. energy, **2**, 145 (1956).

It should be noticed that the procedure described above is based on the diffusion theory approximation.

For accurate results, two points have to be considered with some care. First, neutrons scattered by the walls of the laboratory or surrounding objects should be avoided. Therefore, the medium and the detector should always be wrapped with cadmium sheets and surrounded by boron carbide layers to eliminate the slow neutron background. When fast neutrons are also present in an appreciable amount in the background, a wall of water or paraffin is placed, outside the absorbers mentioned above, all around the experimental set-up.

The second point to be considered in designing experiments of this type, refers to the interpretation of the results which is based on the assumption that only the fundamental mode (131.2b) is present. The intensity of the higher harmonics is strongly reduced with respect to the intensity of the fundamental mode by starting the measurements with a sufficiently long delay. A residue of higher harmonics is usually visible in the decay curve as a deviation from a strictly exponential law observed at the initial part of the measurements (see, for example, Fig. 259).

The influence of the higher harmonics, however, can be further reduced by a judicious choice of the experimental conditions. For the sake of definiteness, let us consider the case of a cylindrical medium. The target of the accelerator should be situated on the axis of the arrangement so that, for symmetry reasons, no harmonic mode having radial nodal planes through the axis will be excited. By a suitable arrangement of the detector, it is possible to eliminate the influence of one further set of harmonic modes, and this possibility should be used to suppress the harmonic mode with the smallest decay constant. VON DARDEL and SJÖSTRAND, for example, have adopted the following criteria. If the height of the cylinder is small, the dominant harmonic modes have cylindrical nodal surfaces. These modes are eliminated by putting the detector below the bottom of the cylinder; a cadmium mask with an opening cut out according to the function $r J_0\left(\frac{2.40 r}{\bar{R}}\right)$ was inserted in front of the BF_3 counter in contact with the base of the cylinder.

If the height of the cylinder is large, the mode with the smallest decay constant is that with one nodal plane perpendicular to the axis of the cylinder. This mode can be suppressed by placing the detectors (for example two or more BF_3 counters) symmetrically with respect to the nodal plane, on the side of the cylinder.

BRACCI and COCEVA[1] obtain the suppression of the third harmonic and the cancellation of the second one by a different procedure. The neutrons were detected by means of a scintillation counter, having a mixture of ZnS and BN as phosphor, placed very close to the lateral surface of the medium cylinder. The neutrons escaping from the medium reached the phosphor through a small hole opened in the absorbing screen (0.2 mm Cd+10 mm boron carbide) coating the cylinder. Two readings were taken at two detector positions corresponding to the two nodal planes of the third longitudinal harmonic. Since at these positions the second harmonic mode has equal values but opposite sign, the corresponding effect was cancelled by averaging the two readings.

The necessity of rejecting the first part of the decay curve sets a practical upper limit to the range of values of ω^2 which can be adopted in the actual measurements, since the smaller the geometrical dimensions of the medium are,

[1] A. BRACCI and C. COCEVA: Nuovo Cim. **4**, 59 (1856).

the smaller the neutron intensity is. This practical lower limit of the medium's dimensions is usually larger than that derived from the use of the diffusion theory approximation.

This procedure, first used by MANLEY, HAWORTH and LUEBKE[1]; has been developed to its full extend by VON DARDEL and SJÖSTRAND[2,3], SCOTT, THOMSON and WRIGHT[4] and by ANTONOV *et al.*[5] and has been applied by many other authors. A review of all these papers has been made by VON DARDEL and SJÖSTRAND[6].

Table 108. *Summary of neutron sources and detectors used by various authors in the investigation of finite media with pulsed sources.*

Authors	Neutron source	Detector	Material
MANLEY *et al.*	D+D 300 kev	BF_3 counters	H_2O
VON DARDEL and SJÖSTRAND	D+D 150 kev	BF_3 counters	H_2O 10° C H_2O 35° C
SCOTT *et al.*	(γ, n) (γ, f) in uranium with the γ-rays of betatron	B^{10} lined proportional counter	H_2O H_2O + boric acid H_2O+Ag NO_3
ANTONOV *et al.*	D+T 400 kev	BF_3 counters	C Be H_2O 23° C H_2O 80° C
RAMANNA *et al.*	D+Be 900 kev	LiI scintillation counter	BeO
BRACCI and COCEVA	D+D 300 kev	ZnS+BN phosphor	H_2O
CAMPBELL and STELSON	Be (p, n) 5.5 Mev	LiI or antracene crystal detecting γ-ray from capture in Cd	H_2O
BECKURTS	D+T 250 Mev	BF_3 counters	C
DIO and SCHOPPER			H_2O 19° C H_2O 49° C H_2O 75° C

Table 108 summarizes the investigated media and the type of neutron sources and detectors used by the various authors. Figs. 258 and 259 show the decay curves obtained for light water at 23° C and graphite by ANTONOV *et al.* The departure from a strict exponential law due to higher harmonics is clearly visible in the curves corresponding to the smallest value of the geometrical buckling.

Figs. 260 and 261 present plots of the decay constant $\bar{\gamma}$ versus ω^2 in water and graphite; they both show the quadratic behaviour to be expected in accordance with (131.3a). The values of the diffusion parameters obtained by these authors for the various investigated media are summarized in Table 109, which also indicates the amounts of materials involved in experiments of this type.

Light water is the material most thoroughly investigated by this method. From the experimental point of view, it has two advantages: small amounts

[1] J.H. MANLEY, L.J. HAWORTH and E.A. LUEBKE: Phys. Rev. **61**, 152 (1942).
[2] G. VON DARDEL: Trans. Roy. Inst. Technol. No. **75** (1954).
[3] G. VON DARDEL and N.G. SJÖSTRAND: Phys. Rev. **96**, 1245 (1954).
[4] F.R. SCOTT, D.B. THOMSON and W. WRIGHT: Phys. Rev. **95**, 582 (1954).
[5] A.V. ANTONOV, A.I. ISAKOFF, I.D. MURIN, B.A. NEUPOCOYEV, I.M. FRANK, F.L. SHAPIRO and I.V. SHTRANICH: p. 3 of Ref. [*15c*].
[6] G. VON DARDEL and N.G. SJÖSTRAND: Progress in nuclear energy. Physics and Mathematics, Vol. 2, p. 183. New York and London: Pergamon Press 1958.

of material are needed and the geometry can be easily changed by varying the water level in a container.

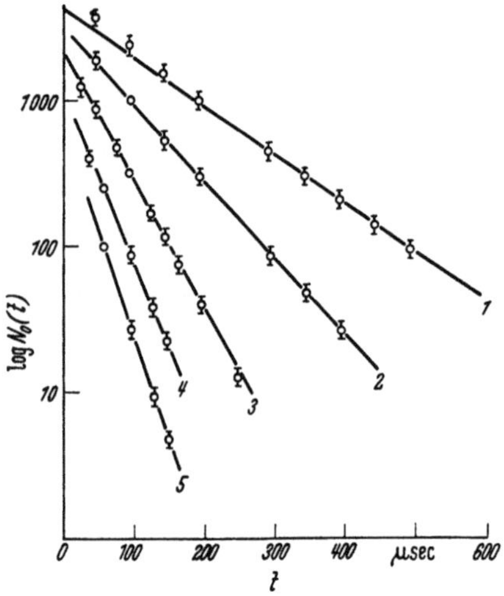

Fig. 258. Decay curves of thermal neutrons in H_2O cylinders of $2R = 186$ mm at 23° C: *1*, $h = 20$ cm $\bar{\gamma}^{-1} = 127$ μsec; *2*, $h = 7.32$ cm $\bar{\gamma}^{-1} = 81$ μsec; *3* $h = 4.25$ cm $\bar{\gamma}^{-1} = 48.9$ μsec; *4*, $h = 3.35$ cm $\bar{\gamma}^{-1} = 38$ μsec; *5*, $h = 2.75$ cm $\bar{\gamma}^{-1} = 29$ μsec. (ANTONOV *et al.*).

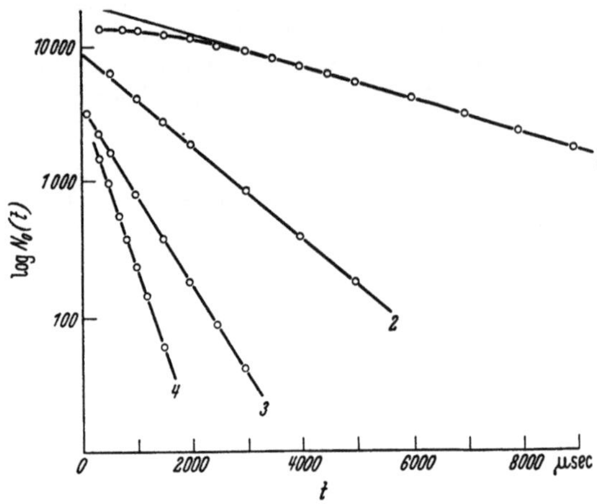

Fig. 259. Decay curves of thermal neutrons in graphite prisms of various sizes: *1*, 200 × 180 × 140 cm³ $\bar{\gamma}^{-1} = 3570$ μsec; *2*, 120 × 80 × 80 cm³ $\bar{\gamma}^{-1} = 1310$ μsec; *3*, 60 × 60 × 60 cm³ $\bar{\gamma}^{-1} = 680$ μsec; *4*, 40 × 40 × 40 cm³ $\bar{\gamma}^{-1} = 350$ μsec (ANTONOV *et al.*).

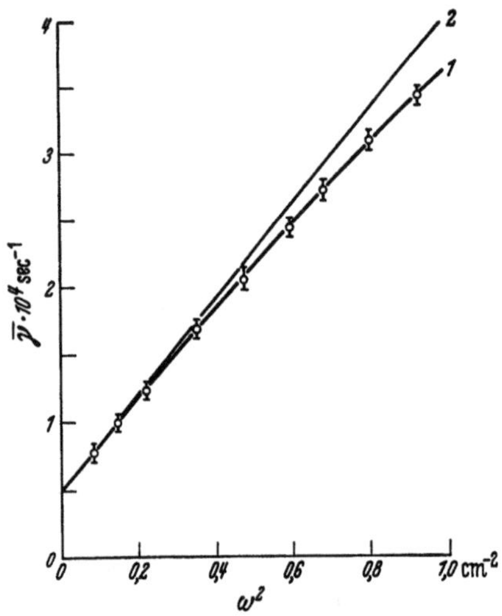

Fig. 260. The decay constant $\bar{\gamma}$ versus the geometrical buckling in H_2O at 23° C; curve 1: $\bar{\gamma} = \frac{1}{T} + D_0\,\omega^2 - C\,\omega^4$, curve 2: tangent to 1 at point $\omega^2 = 0$. (ANTONOV *et al.*).

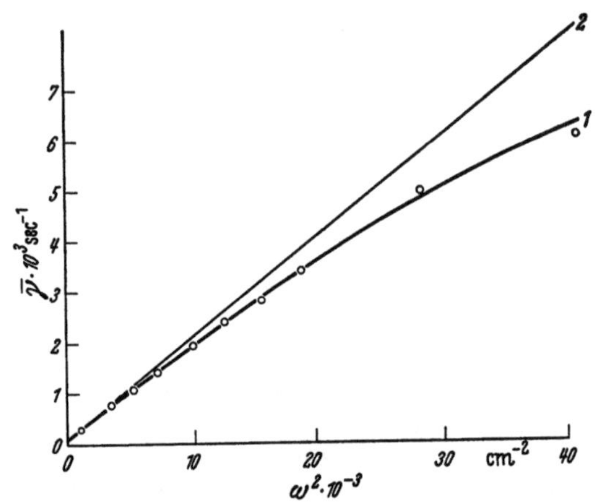

Fig. 261. The decay constant $\bar{\gamma}$ versus the geometrical buckling in graphite; curve 1: $\bar{\gamma} = \frac{1}{T} + D_0\,\omega^2 - C\,\omega^4$, curve 2. tangent to 1 at point $\omega^2 = 0$. (ANTONOV *et al.*).

The values of the diffusion parameters of water at 22° C obtained by the various authors indicated in Table 108 are summarized in Table 96 (p. 583) together with the results obtained by other procedures. The best current values of D_0, L and T suggested by BAKER and WILKINSON, are given in Table 103 (p. 596).

While the agreement between the results of various authors is very satisfactory for the diffusion coefficient D_0 and the mean life T, the values of the diffusion cooling coefficient C are widely spread. In particular, no evidence for a term in ω^4 is found by CAMPBELL and STELSON[1] whose plots of $\bar{\gamma}$ versus ω^2 can very satisfactorily be fitted by a linear expression in ω^2 for both H_2O and Be (Fig. 262).

Table 109. *Values of diffusion parameters of thermal neutrons in various materials* (ANTONOV *et al.*).

Substance and density	Dimensions of systems	$D_0 \times 10^{-5}$ cm²/sec	$C \times 10^{-5}$ cm⁴/sec	T µsec	L cm
Beryllium 1.78 g/cm³	from $72 \times 60 \times 56$ cm³ $\omega^2 = 0.730 \times 10^{-2}$ cm^{-2} to $20 \times 20 \times 20$ cm³ $\omega^2 = 6.118 \times 10^{-2}$ cm^{-2}	1.22 ± 0.05	3 ± 1	4000[2]	22.1[2]
Graphite 1.67 g/cm³	from $200 \times 180 \times 140$ cm³ $\omega^2 = 1.012 \times 10^{-3}$ cm^{-2} to $20 \times 20 \times 40$ cm³ $\omega^2 = 41.290 \times 10^{-3}$ cm^{-2}	1.98 ± 0.03	12.0 ± 1.9	12200 ± 500	49 ± 2.0
Water 23° C	from 0.75×10^3 ml to 5.6×10^3 ml	0.35 ± 0.01	0.04 ± 0.01	207 ± 6	2.7 ± 0.1
Water 80° C	$\omega^2 = 0.086 - 0.94$ cm^{-2}	$0.45^{+0.04}_{-0.02}$	$0.13^{+0.05}_{-0.07}$	212 ± 6	$3.1^{+0.2}_{-0.1}$

VON DARDEL and SJÖSTRAND[3] notice that their rather large value of the diffusion cooling coefficient C agrees—even better than one would expect—with the theoretical value derived by introducing into (129.21) $t_{\text{th}}^0 = 6.3$ µsec (Table 61, p. 339) and $\frac{1}{D_0} \frac{dD_0}{dT_0} = 0.42\%$. We recall that the expression (111.8) for the thermalization time t_{th}^0 has been derived by treating water as a monoatomic gas of mass

Table 110. *Temperature coefficient of the diffusion coefficient D_0 in water.*

Authors	Temperature range (°C)	$\frac{1}{D_0} \frac{dD_0}{dT} \times 10^2$
VON DARDEL and SJÖSTRAND[4,3]	10–35	0.42 ± 0.04
ANTONOV *et al.*[5]	23–80	0.50 ± 0.2
DIO and SCHOPPER[6]	19–75	0.37 ± 0.09

$A = 18$ and that the value obtained by such a procedure is in rough agreement with the results of measurements at room temperature (see Sect. 111, p. 503).

The experimental values obtained for the temperature coefficient of the diffusion coefficient, obtained by changing the medium temperature, are given in Table 110. We should recall here the remark made at the end of Sect. 129 that this quantity corresponds only approximately to that appearing in Eq. (129.21).

As pointed out by ANTONOV *et al.*[5], the interpretation of the values of $\frac{1}{D_0} \frac{dD_0}{dT_0}$ reported in Table 110 requires an appreciable variation of $\bar{\mu}_0$ with temperature.

[1] E.C. CAMPBELL and P.H. STELSON: ORNL-2076, p. 32 (1956).
[2] This quantity was computed from $T = L^2/D_0$ where $L = 22.1$ cm is the value (128.21) obtained by GERASYEVA *et al.*, footnote 2, p. 605.
[3] See footnote 6, p. 622.
[4] See footnote 3, p. 622.
[5] See footnote 5, p. 622.
[6] W.H. DIO and E. SCHOPPER: Nuclear Phys. 6, 175 (1958).

For example, by referring to the measurements of these authors (Table 109), when water is heated from $T = 296°$ K to $T = 353°$ K, the neutron velocity—corresponding to thermal equilibrium with the medium—increases by a factor 1.1. Therefore, of the 29% increase undergone by D_0 through this temperature interval, only 10% is due to the factor v; the remaining 18% (+14% and −9%) should be due to an increase of l_{tr}. Now, heating increases the scattering means free path by 2.6% because of the decrease of the water density and by about 6% because of the decrease of the scattering cross section (see Sect. 101, Fig. 189, p. 428). Thus only one half of the observed increase in l_s can be explained by these two causes. The poor accuracy of the measurements does not permit the Russian authors to consider this statement as definitively proved. Similar conclusions, however, are reached also by other authors (Sect. 128, p. 603).

The temperature dependence of the diffusion coefficient has been investigated by the same method also by DIO and SCHOPPER[1] whose results are collected in Table 111. The values of the diffusion length are computed by means of Eq. (125.3). The following expressions satisfactorily represent the data of Table 111:

$$D_0 = 35050 + (t° - 19) \times (130 \pm 30)\ \mathrm{cm^2/sec}, \tag{131.3b}$$

$$L = 2.70 + (t° - 19) \times (0.0054 \pm 0.0022)\ \mathrm{cm}, \tag{131.3c}$$

Table 111. *Temperature dependence of the diffusion coefficient in water* (DIO and SHOPPER).

Temperature °C	D_0 (cm²/sec)	L (cm)
19	35050 ± 600	2.70 ± 0.05
49	38550 ± 1000	2.85 ± 0.07
75	42250 ± 1000	3.00 ± 0.07

which are in good agreement with the results of the other authors [Table 110 and Eq. (128.12)]. The values of L given in Table 111 also agree very well with Eq. (128.13).

Beryllium has been investigated by ANTONOV *et al.*[2] (Table 109) and by CAMPBELL and STELSON[3]. Their results are summarized in Table 98 together with the results obtained by other methods. The two groups agree satisfactorily for the values of the diffusion coefficient D_0 but not for the diffusion cooling coefficient C; as in the case of H_2O, the Oak Ridge group finds no evidence for a diffusion cooling term (Fig. 262). The reason for such a discrepancy is unclear, especially if one considers that the measurements were performed over the same interval of values of the geometrical buckling.

The determination of the mean life T is very inaccurate in both experiments because the intercept of the curve of $\bar{\gamma}$ versus ω^2 with the ordinate axis is so close to zero that a small uncertainty in the slope of the curve produces an appreciable error in the mean life. The value given by ANTONOV *et al.* is derived by introducing into Eq. (125.3) their value of D_0 and the value (128.21) of the diffusion length obtained by GERASEVA *et al.* The value attributed to CAMPBELL and STELSON in Table 98 is the reserve of the constant term in the linear expression:

$$\bar{\gamma} = (0.15 \cdot 10^3 + 12.5 \cdot 10^5 \omega^2)\ \mathrm{sec^{-1}}$$

which, according to these authors, provides the best fit to the experimental points of Fig. 262.

Fig. 263 shows the time-space distribution of thermal neutrons leaking out of the top of a beryllium block 23 cm × 28.6 cm × 58 cm volume following a 1000 μsec burst of fast neutrons. As neutron detector these authors used an anthracene

[1] See footnote 6, p. 624.
[2] See footnote 5, p. 622.
[3] See footnote 1, p. 624.

scintillator in which the capture γ-rays emitted by a small disk of cadmium where detected. The curves show the shape of the distribution at successive 100 µsec time intervals. They should be of the form

$$\Phi(x,t) \propto \sum_n A_n \sin\left(n \frac{\pi}{2} \frac{x}{x_0}\right) e^{-\omega_n^2 t}$$

and serve to illustrate the presence of higher harmonics which decay more rapidly then the fundamental mode, as it can be seen from the gradual shift to the symmetrical fundamental mode.

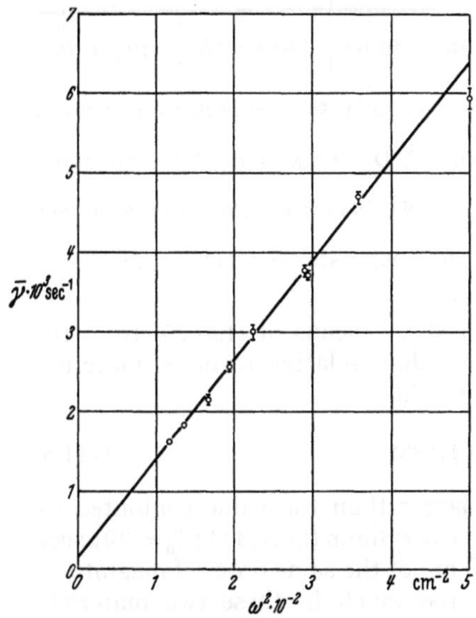

Fig. 262. Decay constant of thermal neutrons in beryllium (CAMPBELL and STELSON).

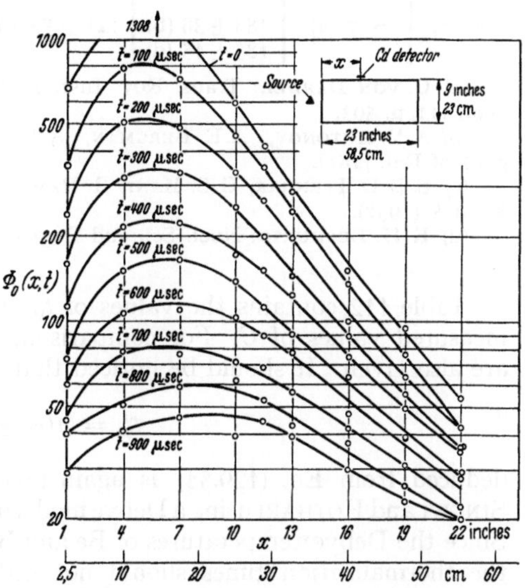

Fig. 263. Time-space distribution of thermal neutrons leaking from top of beryllium block 23 cm × 28.6 cm × 58.5 cm (CAMPBELL and STELSON).

An attempt has been made by VON DARDEL and SJÖSTRAND[1] to derive some information on the thermalization time t_{th}^0 from the value of C found by ANTONOV *et al.* By assuming that the transport mean free path in beryllium is velocity independent, the diffusion coefficient D_0 turns out to be proportional to $T^{\frac{1}{2}}$ and dD_0/dT can, therefore, be computed. From this and $C = 3 \cdot 10^5$ cm^4/sec, VON DARDEL and SJÖSTRAND derive $t_{\text{th}}^0 = 125$ µsec, i.e., a value much larger that that derived by treating beryllium either as a monatomic gas [Table 61, p. 339; Eq. (111.8)] or as a Debye crystal (Sect. 111). This means that the bond between atoms makes an energy transfer considerably more difficult than in the two models considered in Sect. 111.

Beryllium oxide has been investigated by various groups of the Atomic Energy Establishment of India[2-4] whose results are summarized in Table 98, 112 and 113. The transport mean free path is derived by means of Eq. (81.7), and the diffusion length by means of (125.3).

[1] See footnote 6, p. 622.
[2] See footnote 1, p. 620.
[3] S. B. D. IYENGAR, G. S. MANI, R. RAMANNA and N. UMAKANTH: Proc. Ind. Acad. Sci. **45**, 215 (1957).
[4] S. B. D. IYENGAR, G. S. MANI, R. RAMANNA and N. UMAKANTH: Proc. Ind. Acad. Sci. **45**, 224 (1957).

The value

$$\tau_m = (73 \pm 3)\ \text{cm}^2 \tag{131.4}$$

was obtained by using a stilbene scintillator as a fast neutron monitor in order to normalize the source strength of the neutron bursts when the buckling was changed.

Table 112. *Relaxation time constant t_{th} in various materials.*

Size	Water	Graphite	Beryllium	Beryllium oxide		
	∞	∞	∞	∞	40×40×40 cm³	20×20×20 cm³
t_{th} (μsec)	~7 (a)	183 ± 38 (b) 185 ± 45 (d)	121 ± 50 (b)	165 ± 10 (c)	150 ± 8 (c)	119 ± 6 (c)

(a) G. VON DARDEL: Trans. Roy. Inst. Technol. No. **75** (1954); see Sect. 85, p. 339 and Sect. 111, p. 503.

(b) A.V. ANTONOV, A.E. BERGMAN, A.I. ISAKOV, I.D. MURIN and B.A. NENPOKOV: p. 87 of Ref. [*15c*].

(c) S.B.D. IYENGAR, G.S. MANI, R. RAMANNA and N. UNAKANTH: Proc. Ind. Acad. Sci. **45**, 215 (1957).

(d) K.H. BECKURTS: Nucl. Sci. and Engng. **2**, 516 (1957); see Sect. 111, p. 504.

Table 112 contains the values of t_{th} derived by means of (129.29) from the measured values of C. For comparison, the values relative to other materials are also given. It should be noticed that the value

$$t_{th}^0 = (165 \pm 10)\ \mu\text{sec} \tag{131.5}$$

deduced from Eq. (129.31) is again much larger than the value computed by SINGWI and KOTHARI using a Debye model of the beryllium (Sect. 111: $t_{th}^0 = 29\ \mu$sec). Since the Debye temperatures of Be and BeO are of the same order of magnitude, the thermalization times should not differ too much in these two materials. Actually this is found to be true for the experimental values of t_{th}^0 which, however, differ both by a factor of the order of five from the theoretical result. The origin of such a large discrepancy remains to be understood.

Table 113. *Values of various parameters in* BeO *as a function of temperature* (MANI *et al.*).

T_0		297° K	353° K	413° K
D_0 (cm²/μsec)		0.118 ± 0.002	0.136 ± 0.003	0.160 ± 0.003
C		0.385 ± 0.008	0.75 ± 0.02	1.24 ± 0.03
	Size			
t_{th} (μsec)	∞	165 ± 10	244 ± 15	291 ± 17
	40 × 40 × 40 cm³	150 ± 8	206 ± 14	231 ± 16
	20 × 20 × 20 cm³	119 ± 6	146 ± 10	149 ± 10

In a successive paper, the measurements were repeated at higher temperature. The results, given in the upper part of Table 113, show that $D_0(T_0)$ is roughly proportional to $T_0^{\frac{1}{2}}$, so that one can conclude that the maximum variation of l_{tr} in the explored range of temperatures is of the order of (15 ± 10)%. Thus it appears justified to use for the interpretation of the experimental data the various expressions derived in Sect. 129 under the assumption (129.22).

From Table 113, it is seen that the diffusion cooling coefficient C increases very rapidly and about linearly with medium temperature.

The values of the relaxation time given in Table 112 are computed from the values of C by means of Eqs. (129.29) to (129.31); therefore, they also increase by increasing the medium temperature, and give the somewhat paradoxical result that for large values of the geometrical buckling the equilibrium temperature T_{eq} falls off with increasing medium temperature because the increased diffusion cooling effect outweighs the direct heating of the neutrons by the medium.

The whole procedure of computing t_{th} was checked by the same authors who measured it, at room temperature and for a size of 40 cm × 40 cm × 20 cm, by means of a transmission experiment as a function of time through a $1/v$ absorber (Sect. 111). Using a silver foil (0.1178 ± 0.0004) cm thick and Table 117 (p. 658) for the absorption curve of a Maxwellian beam, they found $t_{th} = 140$ μsec which is in very good agreement with the value $t_{th} = 132$ μsec computed from the value of C at room temperature for the size of the assembly used.

Graphite has been investigated by ANTONOV *et al.*[1,2] and BECKURTS[3] whose results are collected in Table 100. By transforming the results of the Russian group from the actual density $\varrho = 1.67$ g/cm³ to $\varrho = 1.60$ g/cm³, one obtains $D_0 = (2.07 \pm 0.03) \cdot 10^5$ cm²/sec and $C = (13.6 \pm 2.1) \cdot 10^5$ cm⁴/sec; the deviations from the values obtained by BECKURTS, although small, are outside the indicated errors and may in part be due to differences in the structure of the graphite.

Table 114. *Equilibrium temperature and thermalization time constant in graphite* (ANTONOV *et al.*).

Sizes	180×140×120 cm³		40×40×20 cm³	
	measured	computed	measured	computed
T_{eq} K°	320 ± 30	295 ± 10	255 ± 50	225 ± 15
t_{th} (μsec)	209 ± 30	191 ± 35	154 ± 65	145 ± 25
Mean transfer of energy in one collision	(4.7 ± 0.7) %	(5.1 ± 1.0) %	(5.2 ± 1.7) %	(5.9 ± 1.5) %

Table 114 shows the values of the equilibrium temperature T_{eq}, the relaxation time obtained from direct measurements of the absorption in a silver foil as a function of time, compared with the values computed from the diffusion cooling coefficient C. The table also contains the values of the relative mean energy transfer per collision (Sect. 129). The authors have also made an estimate of the factor χ, Eq. (129.33), by comparing the experimental value of C with the value of this constant computed from Eq. (129.31). They find

$$\chi \approx 0.94.$$

The value of t_{th}^0 obtained by BECKURTS by a measurement of the absorption in a silver foil as a function of time has already been reported in Sect. 111 [Eq. (111.12)]; it agrees satisfactorily with the value computed from Eq. (129.31) for $\varrho = 1.6$ g/cm³: $t_{th}^0 = (210 \pm 30)$ sec.

An absolute determination of the absorption cross section of various materials can also be made by means of the procedures described above: the material under investigation is uniformly mixed with a weakly absorbing medium and the mean life of thermal neutrons is measured in the poisoned medium thus obtained. For obvious practical reasons, this method has been applied until

[1] See footnote 6, p. 622.

[2] A. V. ANTONOV, A. A. BERGMAN, A. I. ISAKOV, I. D. MURIN and B. A. NEUPOCOYEV: p. 82 of Ref. [*15c*].

[3] K. H. BECKURTS: Nucl. Sci. and Engng. 2, 516 (1957).

now only to materials soluble in water. Typical experiments of this kind have been made by SCOTT *et al.*[1] for the case of boron and silver, and by VON DARDEL and SJÖSTRAND[2] for the case of boron. The second group of authors obtains, as weighted average of various measurements,

$$\sigma_B(2200\ \mathrm{m/sec}) = (764 \pm 3)\ \mathrm{barns} \tag{131.6}$$

which is compared with other precision measurements in Table 11 Sect. 27.

132. Experiments with harmonically modulated sources. According to WEINBERG and NOEDERED [*19*], WIGNER was the first to point out that the diffusion coefficient D_0 and the diffusion length L can be determined by measuring the wavelength and the attenuation constant of neutron waves propagating in a weakly absorbing medium (Sect. 124). Such a procedure, analogous to the Angström method for measuring thermal conductivity, received very little attention until, in very recent times, RAYEVSKI and HOROWITZ[3] determined the diffusion coefficient D_0 in heavy water[4] and graphite[5].

Although the method can be applied to modulated sources combined with media of any geometry, we shall refer, for the sake of simplicity, to the case of an isotropic point source of thermal neutrons embedded in an homogeneous infinite medium with an intensity given by

$$S(\boldsymbol{r},t) = [S_0 + \delta S_0 \cdot e^{i\Omega t}]\,\delta(\boldsymbol{r}). \tag{132.1}$$

Then, according to the considerations given at the end of Sect. 124, the solution of the corresponding diffusion equation (124.6) can be written in the form

$$N_0(\boldsymbol{r},t) = N_0(\boldsymbol{r}) + \delta N_0(\boldsymbol{r}) \cdot e^{i\Omega t} \tag{132.2a}$$

where

$$N_0(\boldsymbol{r}) = \frac{S_0}{4\pi D_0}\,\frac{e^{-\frac{r}{L}}}{r} \tag{132.2b}$$

and

$$\delta N_0(\boldsymbol{r}) = \frac{\delta S_0}{4\pi D_0}\,\frac{e^{-\frac{r}{L_\Omega}}}{r}. \tag{132.2c}$$

We recall that the telegraph equation (124.4) is derived from the time-dependent Boltzmann equation in the P_1-approximation and reduces to the time-dependent diffusion equation (124.6) whenever the conditions (124.5) are fulfilled. These may be re-stated, in the present case of the harmonically modulated source (132.1) in the form

$$\frac{\Sigma_a}{\Sigma_s} \ll 1 \quad \text{and} \quad \frac{l_{\mathrm{tot}}}{v} \ll \frac{1}{\Omega}. \tag{132.3}$$

The first one of these relations limits the absorption cross section of the medium, while the second establishes an upper limit to the frequency of modulation

$$\Omega \ll \frac{v}{l_{\mathrm{tot}}}. \tag{132.4}$$

[1] See footnote 4, p. 622.
[2] G. VON DARDEL and N. G. SJÖSTRAND: Phys. Rev. **96**, 1566 (1954).
[3] J. HOROWITZ: Suppl. del Nuovo Cim. **4**, 1335 (1956).
[4] V. RAYEVSKI and J. HOROWITZ: C. R. Acad. Sci., Paris **238**, 1993 (1954).
[5] V. RAYEVSKI and J. HOROWITZ: p. 42 of Ref. [*15c*].

The variable part (132.2c) of the solution (132.2a) depends on the complex diffusion length (124.16)

$$\frac{1}{L_\Omega^2} = \frac{v}{D_0}\left(\Sigma_a + i\frac{\Omega}{v}\right) = \frac{\Omega}{D_0}\left(\frac{1}{\Omega T} + i\right) \tag{132.5}$$

which, for

$$\frac{v\Sigma_a}{\Omega} = \frac{1}{\Omega T} \ll 1$$

i.e., if

$$\Omega \gg \frac{1}{T}, \tag{132.6}$$

can be expanded in powers of $(\Omega T)^{-1}$ giving the result

$$\frac{1}{L_\Omega} = \sqrt{\frac{\Omega}{2D_0}}\left\{\left(1 + \frac{1}{2}\frac{1}{\Omega T} + \frac{1}{8}\frac{1}{(\Omega T)^2}\right) + i\left(1 - \frac{1}{2}\frac{1}{\Omega T} + \frac{1}{8}\frac{1}{(\Omega T)^2} - \cdots\right)\right\}. \tag{132.7}$$

The modulation rate of the solution

$$M(r) = \frac{\delta N_0(r)}{N_0(r)} = F(r)\, e^{-r/L_\Omega} \tag{132.8}$$

is given by the product of a real function $F(r)$ of r and the complex function $\exp[-r/L_\Omega]$: this means the $M(r)$ is not in phase with the source. The modulus and argument of $M(r)$ are given by

$$|M(r)| = F(r)\exp\left[-r\sqrt{\frac{\Omega}{2D_0}}\left(1 + \frac{1}{2}\frac{1}{\Omega T}\right)\right], \tag{132.9a}$$

$$\varphi = \arg M(r) = -r\sqrt{\frac{\Omega}{2D_0}}\left(1 - \frac{1}{2}\frac{1}{\Omega T}\right). \tag{132.9b}$$

Thus, for a given value of r, $\ln M(r)$ and $\arg M(r)$ are linear functions, respectively, of the variables

$$X_m = \sqrt{\Omega}\left(1 + \frac{1}{2}\frac{1}{\Omega T}\right) \tag{132.10}$$

and

$$X_\varphi = \sqrt{\Omega}\left(1 - \frac{1}{2}\frac{1}{\Omega T}\right) \tag{132.11}$$

and have the same slope:

$$\frac{d\ln|M(r)|}{dX_m} = -\frac{r}{\sqrt{2D_0}}, \tag{132.12a}$$

$$\frac{d\varphi}{dX_\varphi} = -\frac{r}{\sqrt{2D_0}}. \tag{132.12b}$$

The method consists of measuring, for a fixed distance r from the source, $M(r)$ and then plotting $\ln|M(r)|$ versus X_m and/or $\varphi(r)$ versus X_φ: according to (132.12a, b), the corresponding slopes allow for the determination of D_0.

α) Heavy water. In the case of heavy water, a neutron source of the $Sb^{124}\gamma + Be$ type was embedded in the tank of the Saclay reactor with the uranium bars removed. The antimony consisted of a rod 5 mm in diameter placed along the axis of two beryllium quadrants, diametrically opposite one another, of 32 mm internal diameter and 47 mm external diameter. The modulation of the source was obtained by rotating between the antimony rod and the beryllium quadrants two γ-ray absorbers consisting of 12 mm thick uranium quadrants.

The neutron source thus obtained emitted $5 \cdot 10^5$ neutrons/sec and had an amplitude of peak-to-peak modulation of 70%. Its external dimensions were 100 mm in height and 48 mm in diameter.

The following remarks should be added. The source emits fast neutrons which, however, reach thermal energies in a time very short compared to both T and $2\pi/\Omega$. Therefore, one can take as a source of thermal neutrons the distribution of the neutrons of about 1 ev given by age theory as if it were established instantaneously at the time $t=0$.

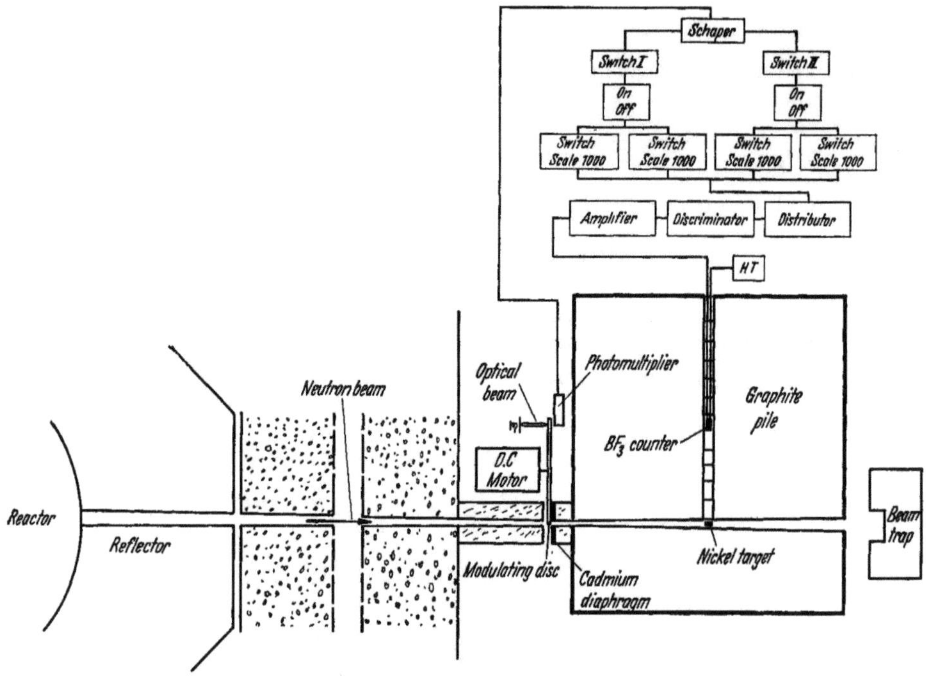

Fig. 264. Horizontal cross section of the experimental device used by RAYEVSKI and HOROWITZ for neutron wave propagation measurements in graphite.

The situation is in some way similar to that encountered in the static diffusion length measurements (Sect. 128β). Also in the present case the measurements should be started at a distance r larger than $\sqrt{\tau_{\rm th}}$ from the source.

The correction due to the finite dimensions of the medium were negligibly small.

The neutrons were detected by means of a BF_3 counter placed at various distances varying between 30 and 90 cm from the source. The modulation frequencies were varied between 25 and 81 Hz.

From the measurements of $|M(r)|$, RAYEVSKY and HOROWITZ obtain

$$D_0 = (2.00 \pm 0.05) \times 10^5 \text{ cm}^2/\text{sec} \tag{132.13}$$

from which it follows, using (125.7), that

$$l_{\rm tr} = (2.45 \pm 0.07) \text{ cm} \quad (\text{in } D_2O \text{ at } 13^\circ \text{C}). \tag{132.14}$$

β) Graphite. The experimental device used for the measurements in graphite is shown in Fig. 264. A beam of neutrons issuing from a radial channel of the

Saclay reactor and modulated by a rotating cadmium disk impinges on a nickel target placed on the axis of a graphite stack of $160 \times 160 \times 180$ cm³ volume and $\varrho = (1.625 \pm 0.003)$ g/cm³ density. The neutron density was measured by means of a BF_3 counter placed on the graphite stack axis which formed an angle of 90° with the incident neutron beam.

NICKEL was chosen as a scatterer because of its large scattering cross section ($\Sigma_s = 1.6$ cm^{-1}); its capture cross section ($\Sigma_a = 0.36$ cm^{-1}) is small enough so as not to attenuate prohibitively the intensity of the scattered neutrons. The nickel target was a cylinder, 15 mm high and 18.5 mm in diameter, which scattered about 90% of the incident thermal neutrons. It represented the modulated point source of thermal neutrons.

The disk used for modulating the beam is made up of cadmium sandwiched between two sheets of duraluminium. Its speed could be adjusted—with a stability of 0.3%—between 600 and 6000 rpm. A five-lobe disk was used for obtaining frequencies of modulation between 50 and 500 Hz; for other frequencies, other disks were used, so that a maximum frequency of 10000 Hz was reached.

The output of the BF_3 counter was connected to four scalers which recorded respectively $C_1 + C_2$, $C_3 + C_4$, $C_2 + C_3$ and $C_4 + C_1$, where C_1, C_2, C_3 and C_4 are the counting rates in the four successive quarters of a period measured starting from the instant at which the neutron source passes through its maximum value. Two electronic switches, triggered by pulses transmitted from a photomultiplier operated by light flashes coming from slits in the disk, provided the distribution of the pulses of the counter to the four scalers.

From the counting rates one readily derives the amplitude of the two components of the modulated part of the solution (132.2), one in phase and the other at 90° with respect to the neutron source,

$$m_{0^\circ} = \frac{(C_4 + C_1) - (C_2 + C_3)}{C}; \qquad m_{90^\circ} = \frac{(C_3 + C_4) - (C_1 + C_2)}{C},$$

$$C = C_1 + C_2 + C_3 + C_4.$$

A number of corrections should be applied to the results of such an experiment:

(a) Higher harmonics are present in the periodic variation of the modulated source. In practice only the third harmonic should be considered.

(b) The finite dimensions of the graphite pile give rise to reflected waves for which a small correction is needed. This can be easily computed by means of a judicious superposition of images.

(c) The time of flight of the neutrons traveling from the rotating disk to the nickel scatterer gives rise to a distorsion of the neutron spectrum reaching the nickel cylinder, with respect to the Maxwellian velocity distribution of thermal neutrons emerging from the reactor.

(d) The anisotropy of the thermal neutron source and the counter dimensions should also be taken into account.

(e) The method of modulating the intensity of the neutron source produces an oscillation in the position of the source having the same frequency as the fundamental mode.

Fig. 265 shows a plot of $|M|$ versus X_m for various distances r of the BF_3 counter from the source. The corresponding slopes are plotted versus r in Fig. 266, from which one can deduce the value of the diffusion coefficient using of Eq. (132.12a). Since the mean temperature of the graphite was 20° C, one can take $v = 2198$ m/sec, so that one obtains

$$l_{tr} = (2.55 \pm 0.09)\ \text{cm} \quad (\text{in graphite: } \varrho = 1.60\ \text{g/cm}^3). \tag{132.15}$$

The results of the measurements of the phase shift φ are less accurate than those referring to $|M(r)|$; they are plotted in Figs. 267 and 268 from which one derives

$$l_{tr} = (2.70 \pm 0.18)\ \text{cm}, \quad (\text{in graphite: } \varrho = 1.60\ \text{g/cm}^3). \tag{132.16}$$

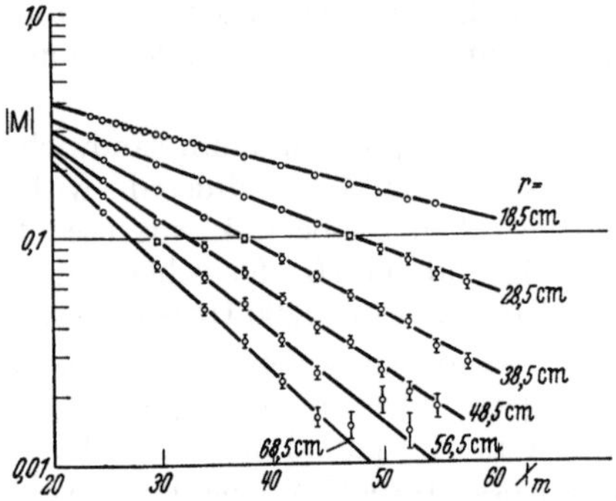

Fig. 265. Semilogarithmic plot of the modulus of the modulation rate versus $X_m = \sqrt{\Omega}\left(1 + \frac{1}{2\Omega T}\right)$; T is taken equal to 1.3×10^{-2} sec (Rayevski and Horowitz).

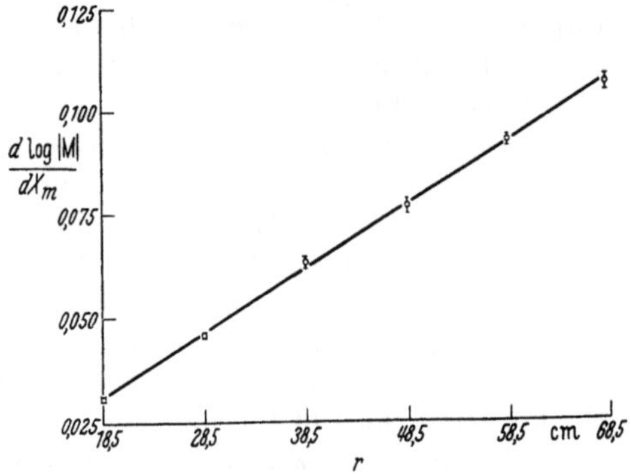

Fig. 266. The slope of the straight lines of Fig. 265 versus the counter distance r from the source (Rayevski and Horowitz).

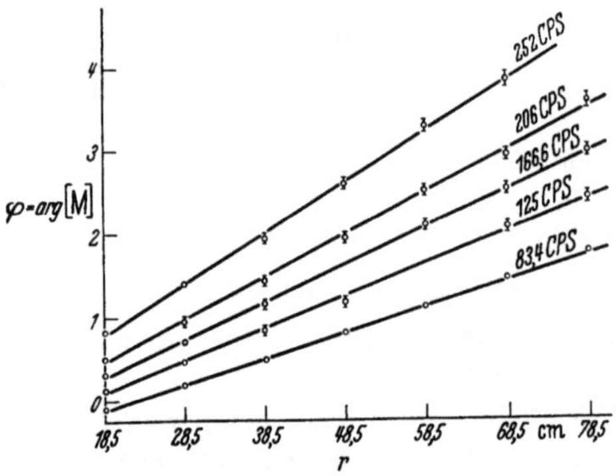

Fig. 267. The argument of the modulation ratio versus the counter distance r from the source for various values of Ω (Rayevski and Horowitz).

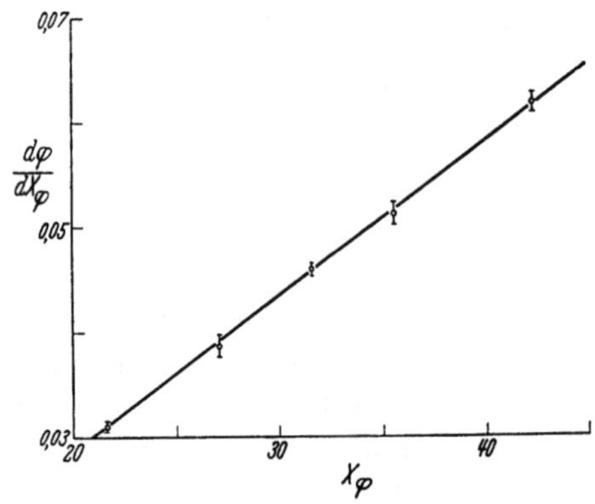

Fig. 268. The slope of the straight lines of Fig. 267 versus X_φ (Rayevski and Horowitz).

III. Detection and absorption of thermal neutrons.

133. Thermal neutron detectors. Thermal neutrons can be detected by means of any nuclear process produced by slow neutrons[1]: the disintegration produced in boron and other light elements (Sects. 17 and 27), the fission of U^{235} and possibly of Pu^{239} and Th^{232} [2], the instantaneous emission of γ-rays following the neutron capture in medium and heavy nuclei (Sect. 16 and 21)[3], the instantaneous emission ($< 10^{-3}$ sec) of low energy electrons (70 kev) originating in slow neutron

[1] H. H. Barschall: Detection of neutrons, Vol. XLV of this Encyclopedia.

[2] J. A. Wheeler: Fission of heavy elements, Vol. XLI, Part 2 of this Encyclopedia.

[3] R. F. Fleischmann: Z. Physik **97**, 242, 265 (1935); **111**, 281 (1938). — Phys. Z. **36**, 806 (1935); **39**, 879 (1938).

Table 115. *Absorption and activation cross sections of a few elements frequently used in experiments on thermal neutrons* [14].

Element	Isotope (%)	σ_a (2200 m/sec) (barns)	Half life	σ_{act} (2200 m/sec) (barns)	f
$_{25}$Mn	Mn^{55} (100)	13.2 ± 0.4	2.58 h	13.4 ± 0.3	1
$_{45}$Rh	Rh^{103} (100)	150 ± 7	4.5 min	12 ± 2	1
			44 sec	140 ± 30	1
$_{47}$Ag		62 ± 2			
	Ag^{107} (51.35)	30 ± 2	2.3 min	44.9	1
	Ag^{109} (48.65)	84 ± 7	270 d	2.8 ± 0.5	1
			24.2 sec	110 ± 20	1
$_{48}$Cd		2550 ± 100			1.3
	Cd^{113} (12.26)	20800 ± 800			1.3
$_{49}$In	In^{113} (4.23)		49 d	56 ± 12	1
			72 sec	2.0 ± 0.6	1
	In^{115} (95.77)		54.1 min	145 ± 15	1
			13 sec	52 ± 6	1
$_{64}$Gd		46000 ± 2000			0.85
	Gd^{157} (15.68)	160000 ± 60000			0.85
$_{66}$Dy		1100 ± 150			1
	Dy^{164} (28.18)		1.3 min	2600 ± 300	1
			139 min	<1000	1
$_{79}$Au	Au^{197} (100)	98.0 ± 1.0	2.7 d	96 ± 10	1

capture by gadolinium[1] and the β- (or γ-)rays emitted by radioactive bodies produced through (n, γ) processes in many elements such as Mn, Rh, Ag, In, Dy, Au etc. (Table 115). Details on the construction and use of these detectors may be found elsewhere [2-4] [*12*], [*17a*]. Here we recall that, according to the definitions given in Sect. 89, a denotes the *total activation* of a detector, i.e., the total number of processes produced per second in the *neutron-sensitive* material of the detector, when exposed to slow neutrons under well defined conditions and A denotes the corresponding total measured activity. In the case of instantaneous detectors A represents the recorded counting rate, while for the case of activated foils, A is the initial counting rate—after infinite exposure to neutrons—recorded by the system used to detect the β- (or γ-)rays emitted by the detector.

Since the total measured activity is always in part due to neutrons of energy of the order of 1 ev or more, the activity due to thermal neutrons A_{th} is obtained by means of the *cadmium difference method* (Sects. 24 and 26): the foil is exposed to neutrons at a given position, first uncovered and a second time covered with a cadmium layer 0.2 to 0.6 mm thick and the corresponding activities A and A_{Cd} are measured. The thermal neutron activity is given by

$$A_{th} = A - F_{Cd} A_{Cd} \tag{133.1}$$

where F_{Cd} is the factor, introduced in Sect. 89, for correcting the absorption undergone by epi-cadmium neutrons in the cadmium filter (Figs. 164 and 165). In the approximation in which the cadmium absorption can be described in terms of the cadmium cut-off (26.20) or (86.1), the thermal neutron activity (133.1) can be attributed to that part of the neutron spectrum which extends below $E_{Cd} = 0.4$ ev. As we have seen in Sects. 105 to 107, as long as the absorption in the moderator used for slowing down the neutrons, is very small, the neutron spectrum

[1] E. AMALDI and F. RASETTI: Ric. Sci. **10**, 115 (1939).
[2] See footnote 1, p. 633.
[3] B. B. ROSSI and H. H. STAUB: National Nuclear Science Series—Experimental techniques. New York: McGraw-Hill 1949.
[4] A. C. GRAVES and D. K. FROMAN: National Nuclear Science Series—Experimental techniques. New York: McGraw-Hill 1952.

in this energy region is rather well represented by a Maxwellian distribution corresponding, eventually, to a temperature T_n different from that of the moderator (Sect. 108).

The various detectors mentioned above are usually separated into *1/v detectors* and *non-1/v detectors* according to the behaviour of the corresponding activation cross section for $E \leqq E_{\text{Cd}}$. The great majority of the detectors follow the $1/v$ law in this narrow energy region: not only boron and the other light elements collected in Table 12 (p. 62), but also Mn, Rh (Table 7, p. 52 and Fig. 11), Ag, In (Table 9 and Fig. 15, p. 57), Au (Table 10 and Fig. 16, p. 59) etc. follow this law with sufficient accuracy fot $E < E_{\text{Cd}}$.

Strong deviations from the $1/v$ behaviour originate from resonances around and below E_{Cd}; typical examples of non-$1/v$ detectors are provided by the γ-rays emitted in neutron capture by cadmium the first resonance of which is at 0.18 ev (Table 8 and Fig. 12, p. 54) and by the conversion electrons emitted by gadolinium whose first resonance is at 0.030 ev (p. 343).

The relative importance of the epi-cadmium term $F_{\text{Cd}} A_{\text{Cd}}$ in (133.1) depends on the conditions of irradiation as well as on the behaviour of the activation cross section below and above E_{Cd}. Its relative value can be estimated from the cadmium ratio discussed in Sect. 90; Table 65 (p. 364) shows $R_{\text{Cd}}(X)$ for a number of nuclei irradiated at a given position in the Argonne heavy water pile, and Figs. 166 and 167 show the variation of $R_{\text{Cd}}(1/v)$ with detector position in two typical cases: inside the reflector of the Brookhaven graphite pile and inside the thermal column of the Argonne heavy water pile.

As we have already mentioned in Sect. 23, the property of $1/v$ detectors which makes them preferable to non-$1/v$ detectors for a great majority of experiments, is that their activation is proportional to the neutron density if the condition of no self-protection (89.3) is fulfilled. In fact, according to (89.4), the activation due to thermal neutrons of a thin detector, is given by

$$a_{\text{th}} = x\, S\, \mathfrak{N} \int_0^{E_{\text{Cd}}} \sigma_a(E)\, \varphi_0(E)\, dE = x\, S\, \mathfrak{N} \int_0^{E_{\text{Cd}}} \sigma_a(v)\, v\, n_0(E)\, dE \tag{133.2}$$

which for

$$\sigma_a(v) \cdot v = k = \text{const}, \tag{133.3}$$

gives

$$a_{\text{th}} = x\, S\, \mathfrak{N}\, k\, n_{\text{th}} \tag{133.4}$$

where

$$n_{\text{th}} = \int_0^{E_{\text{Cd}}} n_0(E)\, dE \tag{133.5}$$

is the neutron density integrated over the energy. According to (89.6), one has

$$A_{\text{th}} = \varepsilon\, x\, S\, \mathfrak{N}\, k\, n_{\text{th}} = \varepsilon\, x\, S\, \mathfrak{N} \cdot \sigma_a(v) \cdot v\, n_{\text{th}} \tag{133.6}$$

i.e. the measured activity due to thermal neutrons (133.1) of a thin $1/v$ detector is proportional to the thermal neutron flux defined by

$$\varphi_{\text{th}}(v) = v \cdot n_{\text{th}}. \tag{133.7a}$$

One has to notice, however, that because of (133.3), Eq. (133.6) establishes a proportionality between A_{th} and the neutron density n_{th}; it may be read as a proportionality between A_{th} and $\varphi_{\text{th}}(v)$ defined by (133.7a) with v chosen *arbitrarily*, provided it is the *same* velocity as that which appears as the argument of $\sigma_a(v)$. The convention is usually made of taking

$$v = v_0 = 2200 \text{ m/sec} \tag{133.7b}$$

and therefore the values of $\sigma_a(v)$ and $\varphi_{\rm th}$ reported in the literature always refer to this velocity [see Eq. (23.15)] [*12*], [*14*].

For intensity reasons, the thickness of the detectors used in many experiments is not small enough to fulfill the condition (89.3). Then the behaviour of a $1/v$ detector may deviate appreciably from that described above and a few corrections must be introduced in order to derive from its measured activity the activity which one would observe with an infinitely thin detector exposed to neutrons under the same conditions.

The correction for *self-protection* takes into account the fact that the atoms belonging to internal layers of a thick detector are exposed to a neutron flux smaller than that impinging upon the atoms of the external layers (Sect. 134). The correction for *hardening* arises from the fact that the neutrons of lower energy are preferentially absorbed in the first layers so that the neutron spectrum reaching the internal layers of the detector is harder than that impinging on its surface (Sect. 136). These effects are enhanced by the *obliqueness* of the neutrons which produces an increase of the effective thickness of the detector. Finally, in *internal experiments*, the finite absorption of thermal neutrons taking place in a thick detector gives rise in the surrounding medium to a *flux depression* with respect to the neutron distribution present at the same position in absence of any detector (Sect. 135).

The *non-$1/v$ detectors* may show, at least in certain cases, a few advantages with respect to $1/v$ detectors because of the very large values of both their thermal neutron cross section $\sigma_{a\,\rm th} = \sigma_a(2200\ \text{m/sec})$ and cadmium ratio. When the thickness of a non-$1/v$ detector is small according to Eq. (88.3), its thermal neutron activation is given by [*12*]

$$a_{\rm th} = \varkappa\, S\, \mathfrak{N}\, f\, \sigma_a(v_0)\, \varphi_{\rm th}(v_0) \tag{133.8}$$

where

$$f = \frac{\int n_0(v)\cdot v\,\sigma_a(v)\,dv}{\int n_0(v)\, v\,\frac{k_0}{v}\,dv} = \frac{1}{n_{\rm th}}\int n_0(v)\,\frac{\sigma_a(v)}{\frac{k_0}{v}}\,dv \tag{133.9}$$

and

$$k_0 = v_0\,\sigma_a(v_0)\,, \qquad v_0 = 2200\ \text{m/sec} \tag{133.10}$$

is a constant different for each non-$1/v$ material. The values of f computed by replacing $n_0(v)$ in (133.9) by the Maxwell distribution of velocities at room temperature [*14*] are given for cadmium and gadolinium as examples in the last column of Table 115.

Thus, for relative measurements the experimental determination of the thermal activation $A_{\rm th}$ of an infinitely thin detector—even if it does not follow the $1/v$ law—can be considered as a determination of $n_{\rm th}$ or $\varphi_{\rm th}$ on relative scales. In some cases, however, an absolute determination of $a_{\rm th}$ or $n_{\rm th}$ is required. This involves a knowledge of the efficiency of detection ε which can be obtained by the procedures described in the articles quoted on p. 356, footnote 3.

In certain internal experiments the quantities of interest are not the activities $A(\boldsymbol{r})$ or $A_{\rm th}(\boldsymbol{r})$ resulting from the exposure of a detector at a given position $\boldsymbol{r}$, but rather their volume integrals extended over the whole moderator. Examples of a situation of this type have been discussed at length in Sect. 83 and 84 in connection with the problem of the determination of the moments of the spatial distribution of neutrons of a given energy.

The volume integral of the activity of a detector plays also a role in certain methods used for the absolute calibration of a neutron source, which have been mentioned in Sect. 127 and are treated in detail in references[1,2]. Here we limit ourselves to recalling that the volume integral may be determined by various procedures. The first one is that described in Sect. 127: the activity of a foil—or other neutron detectors such as a small BF_3 counter—is measured at a number of positions distributed throughout the moderator. Then the neutron distribution, obtained by graphical interpolation, is integrated to give the corresponding volume integral.

A second procedure that can be used in a water moderator consists of dissolving a convenient salt of the element which becomes radioactive due to neutron capture[3,4] (Sect. 127). The more frequently used material is manganese sulfate ($MnSO_4$) to which we shall make reference below. By neutron capture, Mn^{55} (Table 115) is transformed into Mn^{56}, which is β-active and has a half-life $T_{\frac{1}{2}} = 2.58$ h. The manganese activity is produced with a spatial distribution identical to that of the thermal neutron density, apart from a small contribution due to epi-cadmium neutrons. If after irradiation the solution is violently stirred until the activity is uniformly spread, the activity of a given portion of the solution is proportional to the volume integral of the neutrons captured by the manganese.

In a third procedure[5] the integration of the activity of a thin detector over space coordinates is transformed into an integration with respect to time by moving the detector, during irradiation, through the moderator according to a convenient law which depends upon the half-life of the detector. Finally a fourth procedure consists of using a thin foil extending radially from the immediate vicinity of an isotropic point neutron source to great distances from it and whose transverse width is proportional to the square of the distance from the source[5]. If the counting system is arranged so that the β-rays emitted by any point of the foil surface have the same probability to be recorded, the total measured activity will be proportional to the volume integral of the neutron density.

134. Neutron probes. The expression *neutron probe* („Neutronensonde") was introduced by BOTHE[6] for indicating a detector used in internal experiments, such as, for example, the exploration of the neutron distribution inside a moderator; following this author we shall call *density probe* a detector which provides information on the number of neutrons per cm^3 present at a certain position, and *current probe* a detector which allows to determine the number of neutrons crossing in 1 sec 1 cm^2 surface placed at a give position with a certain orientation.

Although, in principle, any neutron detector can be used, the activated foils constitute by far the most convenient type of probes, mainly for geometrical reasons. For example, we have seen in Sects. 83 and 84, that BF_3 counters of very small dimensions have been used for exploring the distribution of thermal neutrons in various media. But, usually, counter dimensions are so large that the neutron density varies appreciably from point to point on their surface. Under these conditions the counting rate does not correspond to the value of the neutron flux at a given point, but rather the some average value over the superficial layer of the moderator surrounding the counter. The thickness of a foil, on the contrary, even when very large with respect to absorption, is always so small that the geometrical effect mentioned above is negligible, especially if the foil is oriented with its surface perpendicular to the gradient of the neutron

---

[1] See footnote 1, p. 633.
[2] See footnote 4, p. 634.
[3] H. L. ANDERSON, E. FERMI and L. SZILARD: Phys. Rev. **56**, 284 (1939).
[4] R. O. O'NEIL and G. SCHARF-GOLDHABER: Phys. Rev. **69**, 368 (1946).
[5] F. G. HOUTERMANS: Phys. Z. **43**, 496 (1942).
[6] W. BOTHE: Z. Physik **120**, 437 (1943).

density during exposure. Therefore, in this and in the following sections, when talking of probes, we shall always refer to activated foils.

α) A density probe would be ideal if its measured activity A were proportional to the neutron density at the position of irradiation; therefore A must be independent of the angular distribution and of the velocity spectrum of the neutrons, and of the orientation of the foil. According to (133.6), an infinitely thin $1/v$ detector of thermal neutrons behaves as an ideal density probe; the finite thickness of the foil used in actual experiments gives rise, however, to deviations from such an ideal behaviour as we have already mentioned in Sect. 133.

In this section we shall discuss in detail the effects arising from neutron self-protection and obliqueness and from absorption of electrons in the probe. The hardening of the neutron spectrum will be considered in Sect. 136 and the effect of the perturbation of the neutron distribution produced by the probe absorption in Sect. 135.

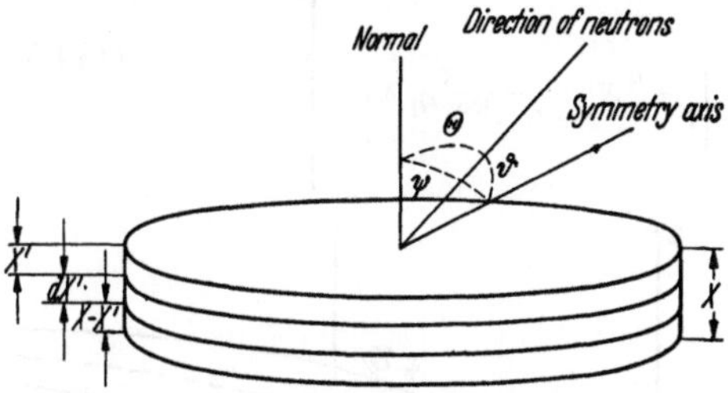

Fig. 269. The normal to the surface of a thick foil forms an angle ψ with the axis of symmetry of the neutron distribution.

A considerable simplification, without appreciable reduction in generality, is obtained by introducing the following two assumptions: the neutron distribution is symmetric about an axis of rotation and the dimensions of the foil are so small that the variation of the flux over its surface is negligible. Thus, at the location of the foil, the neutron flux can be written in the form

$$\Phi(v, \boldsymbol{\Omega}) = \frac{1}{4\pi} \sum_l (2l+1)\, \Phi_l(v)\, P_l(\cos\vartheta) \tag{134.1}$$

where ϑ is the angle between the direction of neutron motion considered and the symmetry axis of the corresponding distribution. From Fig. 269 we see that

$$\cos\vartheta = \cos\psi\cos\Theta - \sin\psi\sin\Theta\cos\varphi \tag{134.2}$$

where ψ is the angle between the normal to the foil surface and the axis of symmetry, and Θ is the angle of incidence of the neutrons considered with respect to the foil normal.

The activation of a layer between X' and $X'+dX'$ inside the foil, resulting from neutrons of velocity v moving in the solid angle $d\boldsymbol{\Omega}$ is given by

$$da = S\,\Phi(v,\vartheta)\, \mathrm{e}^{-\frac{KX'}{\cos\Theta}} K_a\, dX'\, dv\, d\boldsymbol{\Omega} \tag{134.3}$$

where K_a is the activation coefficient which, in general, may differ from the neutron absorption coefficient K. In order to compute the measured activity, we should take into account the absorption of the β-rays in the foil by multiplying (134.3) with the exponential factor (89.8) and then integrating over the foil thickness. In doing this, however, one has to pay attention to the fact that, because of β-ray absorption, the activity of the foil measured on its side 1 is generally different from that measured on its side 2 and that each one of these activities originates in part from neutrons which have entered the foil from the same side, and in part from neutrons which have entered the foil from the opposite side. Thus we can write

$$dA_1 = \omega w\, S\{\Phi(v,\vartheta)\, s_s(\Theta) + \Phi(v,\pi-\vartheta)\, s_o(\Theta)\}\, dv\, d\boldsymbol{\Omega}, \qquad (134.4\text{a})$$

$$dA_2 = \omega w\, S\{\Phi(v,\pi-\vartheta)\, s_s(\Theta) + \Phi(v,\vartheta)\, s_o(\Theta)\}\, dv\, d\boldsymbol{\Omega} \qquad (134.4\text{b})$$

where the two opposite directions have been associated together. The factors ω and w have been defined in Sect. 89: the first factor arises from the counter geometry with respect to the foil, the second from the absorption of the electrons emitted by the foil in the counter window. The two functions

$$\left.\begin{aligned} s_s(\Theta) &= \frac{K\,|\cos\Theta|}{K+\frac{\mu}{\varrho}|\cos\Theta|}\left[1-\mathrm{e}^{-\left(\frac{\mu}{\varrho}+\frac{K}{|\cos\Theta|}\right)X}\right], \\ s_o(\Theta) &= \frac{K\,|\cos\Theta|}{K-\frac{\mu}{\varrho}|\cos\Theta|}\left[\mathrm{e}^{-\frac{\mu}{\varrho}X}-\mathrm{e}^{-\frac{K}{|\cos\Theta|}X}\right] \end{aligned}\right\} \qquad (134.5)$$

give a measure, as a function of the angle of incidence Θ, of the efficiency of detection of the electrons emitted from the same (s_s) and the opposite (s_o) side as that from which the neutrons entered the foil. Fig. 270 shows a plot of $s_s(\Theta)$ and $s_o(\Theta)$ versus $\cos\Theta$ for $KX=0.26$ and $\mu/\varrho=10K$, as would occur for a Rh probe of thermal neutrons, $X\approx 0.3$ g/cm² thick. The dotted line represents

$$s(\Theta) = \tfrac{1}{2}\left[s_s(\Theta)+s_o(\Theta)\right] \qquad (134.6\text{a})$$

which is an even function of Θ and is of interest in the following. It may be noticed that for $\mu/\varrho=0$, the asymmetry due to the β-ray, absorption disappears:

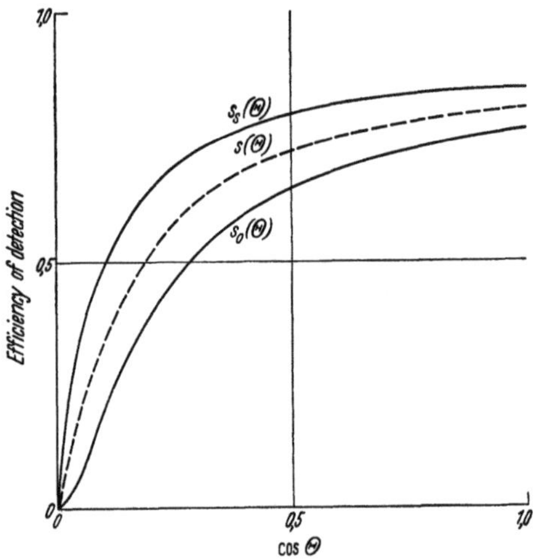

Fig. 270. Efficiency of detection as a function of angle of incidence of the neutrons.

$$s(\Theta) = s_s(\Theta) = s_o(\Theta) = |\cos\Theta|\left[1-\mathrm{e}^{-\frac{KX}{|\cos\Theta|}}\right] \qquad (134.6\text{b})$$

and the two expressions (134.4a, b) are equal to one another and represent the activation of the probe, save of the constant factor ωw. For KX and $\frac{\mu}{\varrho}X$ both so small with respect to 1 that the power series of the exponentials in (134.5) can be truncated at the corresponding linear terms, both $s_s(\Theta)$ and $s_o(\Theta)$ reduce to KX.

The activity actually measured on the two sides of the probe A_1 and A_2 are obtained, in general, by integration of (134.4) over the solid angle and the velocity spectrum; the result may be expressed as series derived by introducing (134.1) into (134.4). Disregarding, for the moment, the velocity dependence, and noticing that, according to (134.4), dA_2 is obtained from dA_1 by exchanging Θ with $\pi-\Theta$, we conclude that the series representing A_2 differs from that representing A_1 only in the sign of the odd terms.

This remark is very useful since it allows for the derivation of the following rule, even without performing the integrations. Let us take the arithmetic average of (134.4a) and (134.4b):

$$\overline{dA} = \tfrac{1}{2}(dA_1 + dA_2). \tag{134.7}$$

The result

$$\overline{dA} = \omega w\, S s(\Theta) \left\{\frac{\Phi(v,\vartheta) + \Phi(v,\pi-\vartheta)}{2}\right\} dv\, d\Omega \tag{134.8}$$

shows that, whenever the neutron flux can be expressed, at the location of the probe, as the sum

$$\Phi(v,\vartheta) = \Phi_0(v) + \Phi_{\mathrm{odd}}(v,\vartheta) \tag{134.9}$$

of one part independent of direction and another part which is an odd function of ϑ, the quantity in curled brackets in (134.8) reduces to $\Phi_0(v)$. Therefore[1], if condition (134.9) is fulfilled, the arithmetic average of the activities of a *thick detector* measured on both sides,

$$\bar{A} = \tfrac{1}{2}(A_1 + A_2), \tag{134.10}$$

is proportional to $\Phi_0(v)$ and therefore to the density of the neutrons of velocity v.

According to (117.8) and (117.22) the diffusion approximation corresponds to beaking off the series (134.1) after the linear term in $\cos\vartheta$,

$$\Phi(v,\vartheta) = \tfrac{1}{2}\Phi_0(v) + \tfrac{3}{2}\Phi_1(v_0)\cos\vartheta. \tag{134.11}$$

Therefore we may conclude that the average activity (134.10) is proportional to $\Phi_0(v)$ whenever the diffusion approximation is adequate to describe the flux distribution. In this case, the integration is easily carried through with the result[2]

$$\left.\begin{matrix} A_1 \\ A_2 \end{matrix}\right\} = \omega w \left\{\tfrac{1}{2}\Phi_0 \alpha_0 \pm \tfrac{3}{2}\Phi_1 \alpha_1 \cos\psi\right\} \tag{134.12}$$

where

$$\left.\begin{aligned} \alpha_0 = \frac{\lambda\nu}{\beta}&\Big\{(1-e^{-\beta})(1+e^{-\nu}) - \frac{\nu}{\beta}\Big(\ln\frac{\nu+\beta}{\gamma} + e^{-\beta}\ln\frac{|\nu-\beta|}{\gamma}\Big) + \\ &+ \Big[\nu(1-e^{-\beta}) - \frac{\nu}{\beta}(1+e^{-\beta})\Big]\mathrm{Ei}(-\nu) + \frac{\nu}{\beta}\times \\ &\times\big[\mathrm{Ei}(-(\nu+\beta)) + e^{-\beta}\mathrm{Ei}(-(\nu-\beta))\big]\Big\}, \\ \alpha_1 = \frac{1}{2}\frac{\lambda\nu}{\beta}&(1+e^{-\beta})\left[1-(1-\nu)e^{-\nu} + \nu^2\,\mathrm{Ei}(-\nu)\right] - \frac{\nu}{\beta}\alpha_0, \end{aligned}\right\} \tag{134.13}$$

with

$$\nu = KX, \qquad \lambda = \frac{K_a}{K}, \qquad \beta = \frac{\mu}{\varrho}X$$

and

$$-\mathrm{Ei}(-y) = \int_y^\infty e^{-t}\frac{dt}{t} > 0 \tag{134.14}$$

is the exponential integral function[3].

Instead of measuring A_1 and A_2 and taking their average according to (134.10), some authors[4,5] prefer to measure, for example, A_1 and compute a correction

[1] E. AMALDI and E. FERMI: Ric. Sci. 7 (1), 454 (1936). — Phys. Rev. 50, 899 (1936).
[2] M.A. VIGON: Z. Naturforsch. 8a, 727 (1953).
[3] E. JAHNKE and F. EMDE: Tables of functions, p. 1 and 6. New York: Dover Publications, 1945.
[4] J.E. HILL, L.D. ROBERTS and G. MCCAMMON: AECD-3390.
[5] A.G. GRAVES and D.K. FROMAN: National nuclear science series—Experimental techniques. New York: McGraw-Hill 1952.

in order to obtain A_2 and A. This can be done in the diffusion approximation by computing the coefficient of the term in $\cos\vartheta$ by means of (115.9a) which show that that Φ_1 is proportional to $d\Phi_0/dz$. The constant of proportionality is obtained by a successive approximation procedure in which one starts by taking, as the zero approximation, the measured values of $A_1(z)$ proportional to $\Phi_0(z)$. The more direct procedure of measuring both A_1 and A_2 and taking their average (134.10) is, however, so simple that one cannot see any reason to use these more indirect methods.

The expressions (134.13) become very simple in a few special cases[1]:

1. The electron absorption is negligibly small ($\beta \ll 1$). Then, according to (134.6b), the asymmetry arising from β-ray absorption disappears and the activity (134.12) reduces to the corresponding activation, save of the constant factor ωw. The odd terms in the series representing A_1 and A_2 all go to zero; so one obtains

$$\alpha_0 = \lambda[1 - e^{-\nu}(1-\nu) + \nu^2 \operatorname{Ei}(-\nu)], \quad (134.15\,a)$$

$$\alpha_1 = 0. \quad (134.15\,b)$$

This result was first derived by BOTHE[2] who disregarded the β-ray absorption. Fig. 271 shows a plot of α_0/λ versus $\nu = KX$; it represents the average probability of absorption of a neutron by a layer of thickness X when exposed to an isotropic distribution of neutrons with constant absorption coefficient (one-velocity group approximation). If even the neutron absorption is very small ($\nu \ll 1$), we find

$$\alpha_0 = 2\lambda\nu.$$

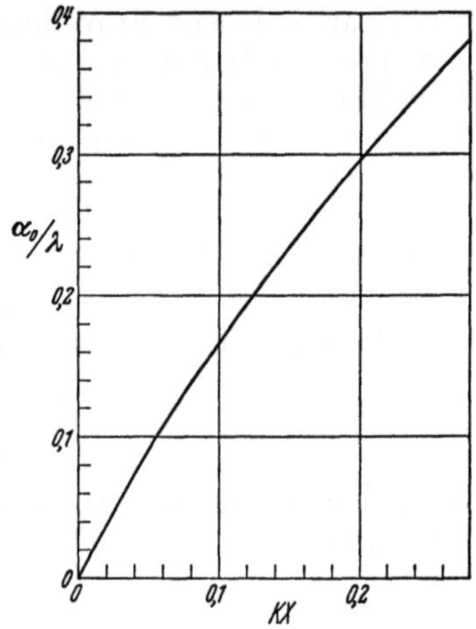

Fig. 271. The function $\alpha_0(KX)/\lambda$ versus $\nu = KX$: Eq. (134.15a).

2. The electron absorption is very large ($\beta \gg 1$) and the neutron absorption very small ($\nu \ll 1$). Then α_1 is of second order in ν and $\alpha_0 = 2\lambda\,\frac{\nu}{\beta}$.

3. Both neutrons and electrons are strongly absorbed ($\beta \gg 1$, $\nu \gg 1$). Then if $\beta \gg \nu$

$$\alpha_0 = \frac{\lambda\nu}{\beta}, \qquad \alpha_1 = \frac{1}{2}\,\frac{\lambda\nu}{\beta};$$

if $\nu \gg \beta$

$$\alpha_0 = \frac{1}{2}\lambda, \qquad \alpha_1 = \frac{1}{3}\lambda.$$

We shall now add a few remarks on the velocity dependence of $\varphi_0(v)$. In the case of resonance detectors with a single sharp resonance at $E = E_R$ (Sect. 89) the computation of the integral with respect to v can be performed by noticing that $K(E)$ is a rapidly varying function of E, while $\varphi_0(E)$ varies only very slowly [see Eq. (86.2)]. Therefore $\varphi_0(E)$ can be replaced by $\varphi_0(E_R)$ with the result that A is proportional to $\varphi_0(E_R)$:

$$A_{Cd} \propto \varphi_0(E_R). \quad (134.16)$$

In the case of a thick $1/v$ probe of thermal neutrons the computation of the integral with respect to v is rather complicated since both the absorption coefficient $K(v)$ and the neutron flux vary appreciably in the region below 0.4 ev; this case will be considered in Sect. 136.

[1] See footnote 2, p. 640.
[2] See footnote 6, p. 637.

β) A current probe can be obtained by covering one or the other of the two sides of the foil with a filter which eliminates the neutrons responsible for its activation. Under these conditions the measured activity is due only to the neutrons impinging on the free probe side. The main inconvenience of such a procedure arises from the neutron absorption in the filter which produces an appreciable perturbation of the neutron distribution, unless the dimensions of both the detector and the filter are very small with respect to the neutron mean free path in the surrounding medium (see Sect. 135).

A variant of the prevoius procedure which can be used when $\frac{\mu}{\varrho} \ll K$ consists in combining the detector and the filter in a single foil thick enough, with respect to $(\mu/\varrho)^{-1}$, that the activity A_i measured on its side i ($i=1$ or 2) arises only from neutrons entering the foil from the same side.

When such an extreme condition is not attained, one can express the activities A_1 and A_2 by means of the following approximate relations[1]

$$\left.\begin{aligned} A_1 &= a\,\Phi^{(1)} + b\,\Phi^{(2)}, \\ A_2 &= b\,\Phi^{(1)} + a\,\Phi^{(2)}, \end{aligned}\right\} \tag{134.17}$$

where $\Phi^{(i)}$ represents the neutron flux incident on side i of the probe. The coefficients a and b can be computed by integrating (134.4) with respect to $d\Omega$ over the half-space; their values depend not only on KX and $\frac{\mu}{\varrho}X$ but also on the angular distribution of the neutrons and the orientation of the foil. Therefore they can be considered as constants, typical of the probe used, only in a rather rough approximation. The ratio a/b can be determined experimentally by measuring A_1 and A_2 when, for example, by means of a convenient filter, one has $\Phi^{(2)}=0$. Once a and b are known, the values of $\Phi^{(1)}$ and $\Phi^{(2)}$ may be derived from the measured values of A_1 and A_2, since one has

$$\left.\begin{aligned} A_1 + A_2 &= (a+b)\,(\Phi^{(1)} + \Phi^{(2)}), \\ A_1 - A_2 &= (a-b)\,(\Phi^{(1)} - \Phi^{(2)}). \end{aligned}\right\} \tag{134.18}$$

The first of these two relations is identical to (134.10). This procedure has been used in some of the experiments on the extrapolated length described in Sect. 126.

135. The flux depression produced by a neutron probe. In this section we shall discuss the depression of the neutron flux produced by an absorbing foil in the surrounding medium paying special regard to its influence on the neutron measurements performed by means of probes. Since the "one velocity group" picture is adopted, the conclusion reached in the following may be approximately correct only for the case of thermal neutrons. The same problem is met for epi-cadmium neutrons; in that case, however, the treatment of the problem requires that the slowing down of the neutrons accompanying their diffusion is taken into account (Sect. 89, p. 356).

The two following effects arising from flux depression may be of considerable practical importance: (a) If the foil is used as a probe, its activation turns out to be smaller than that which one would observe if it were irradiated in the unperturbed flux. (b) If two or more foils are activated simultaneously, the flux depression produced by one of them will influence the activation of the other ones whenever their mutual distance is not large enough.

[1] See footnote 1, p. 640.

While the second of these two effects can be avoided by placing the absorbing foils at sufficiently large distances one from the other, the effect of a foil on its own activation always needs a correction which is expected to depend on the properties of the surrounding medium and on the total number of neutrons absorbed in the probe.

Furthermore it may be noticed that, since the perturbation of the neutron distribution extends over a certain distance, the actually measured activity of the probe does not depend on the value of $n_0(\boldsymbol{r}, v)$ at the foil position but rather on some average value of the neutron density over a rather large region. This kind of averaging is not important if we are only interested in relative measurements made in the same medium and the unperturbed neutron density is constant or almost constant. But it affects appreciably the comparison of the neutron densities in two different media, and even in the same medium when they are measured at positions where their first and second space derivatives are markedly different.

In the following we shall consider the very simple problem of the perturbation produced by an absorbing foil placed in a homogeneous medium in which the neutron flux $n_0(\boldsymbol{r}, v) = n_0(v)$ is *homogeneous* and *isotropic*. Thus the correction computed for such a case will be adequate only if the unperturbed neutron distribution varies slowly over the space region covered by the flux depression.

Let us use the relative perturbation of the neutron density at the position $\boldsymbol{r}$,

$$\varkappa(\boldsymbol{r}) = \frac{n_0 - n(\boldsymbol{r})}{n_s(0)} \tag{135.1}$$

in order to express quantitatively the flux depression in the vicinity of the absorbing foil. Here n_0 is the unperturbed neutron density, $n(\boldsymbol{r})$ the perturbed neutron density at $\boldsymbol{r}$ with respect to the centre of the foil, and n_s the value of $n(\boldsymbol{r})$ at the foil centre ($\boldsymbol{r} = 0$). The relative density perturbation at the foil surface

$$\varkappa_s(\boldsymbol{r}) = \frac{n_0 - n_s(\boldsymbol{r})}{n_s(0)} \tag{135.2}$$

should be averaged over the probe surface in order to derive the relative perturbation of the probe activation

$$\varkappa_a = \frac{a_0 - a}{a}. \tag{135.3}$$

Here a_0 is the activation which one would obtain if the probe where exposed to the unperturbed flux, and a is the activation actually observed. Once $\varkappa_a$ is known, the activation corrected for flux depression is given by

$$a_0 = (1 + \varkappa_a)\, a. \tag{135.4}$$

BOTHE[1] was the first to tackle this problem for a spherical probe of radius R; he used elementary diffusion theory for treating the case $R \gg l_{\mathrm{tr}}$ and an approximate procedure, based on the consideration of the successive collisions in the surrounding medium (see below) in the opposite extreme case $R \ll l_{\mathrm{tr}}$. When R is much smaller than l_{tr} the perturbation of the neutron distribution in the vicinity of the probe varies rapidly over distances of the order of l_{tr} and therefore the use of the elementary diffusion theory is not permissible. BOTHE's

[1] W. BOTHE: Z. Physik **120**, 437 (1943).

results, with l_s replaced by l_{tr} as suggested by TITTLE[1], are

$$\left.\begin{aligned} \varkappa_a &= \left[\frac{3}{4}\frac{R}{l_{tr}}\frac{L}{R+L} - \frac{1}{2}\right]\alpha, & R &\gg l_{tr}, \\ \varkappa_a &= 0.34\frac{R}{l_{tr}}\alpha, & R &\ll l_{tr} \end{aligned}\right\} \tag{135.5}$$

where L is the diffusion length of the neutrons in the medium and α the average probability for a neutron to be absorbed in a single detector traversing. BOTHE had suggested that Eqs. (135.5) could be used even in the case of a plane foil by taking for α the expression (134.15a) [see Eq. (135.23) below] and for R the value $\frac{2}{3}$ of the foil radius. The Bothe-Tittle approximation[1] consists of using the second equation (135.5) with R equal to the radius of the foil; we shall come back to this point in the discussion of Eq. (135.41).

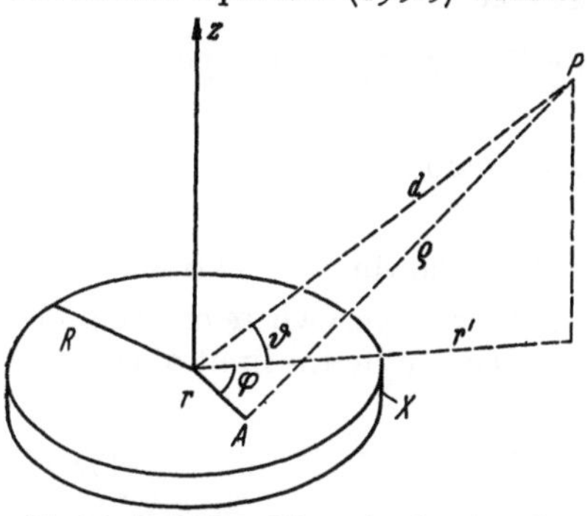

Fig. 272. Definition of the various lengths and angles appearing in Eq. (135.8).

The flux depression produced by a plane foil was first discussed by CORINALDESI[2] who also used the diffusion theory approximation and gave the results in the form of graphs computed for a few values ($\frac{1}{2}$, 1, $\frac{3}{2}$, 2 and ∞) of the ratio of the radius R of the foil to the diffusion length L. The obliqueness of the neutrons crossing the foil was taken into account by multiplying the foil thickness by $\sqrt{3}$ as suggested by the approximate procedure mentioned in Sect. 122 (see also footnote 1, p. 551).

The problem was reconsidered later by VIGON and WIRTZ[3] who by means of the elementary diffusion theory treated both spherical and plane detectors with $R \gg l_{tr}$ for small as well as for large absorption. As an example, the case of plane foils will be discussed here in detail.

For very weak absorption ($KX \ll 1$), the perturbation will be small and the neutron density can be taken as equal to n_s over all the thickness of the probe. The numbers of neutrons absorbed per second by the surface element dS is given by

$$d Q_0 = n_s v K X d S \tag{135.6}$$

and can be treated as a *negative point source* (Sect. 118β). Thus, using Eq. (118.3) and the expression of the point diffusion kernel given in Table 88 (p. 554), we can write

$$\Delta n = n_0 - n(z) = T \int_S \frac{d Q_0}{4\pi L^2} \frac{1}{\varrho'} \mathrm{e}^{-\frac{\varrho'}{L}} \tag{135.7}$$

where $T = (\Sigma_a v)^{-1}$ is the neutron mean life in the surrounding medium and (Fig. 272)

$$\varrho' = \sqrt{d^2 + r^2 - 2\,d\cdot r\cos\vartheta\cos\varphi}. \tag{135.8}$$

[1] C.W. TITTLE: Nucleonics 8 (6), 5 (1951); 9 (1), 60 (1951). — Phys. Rev. 80, 756 (1950).
[2] E. CORINALDESI: Nuovo Cim. 3, 131 (1946).
[3] M.A. VIGON and K. WIRTZ: Z. Naturforsch. 9a, 286 (1954).

When the point P is on the z axis ($\varrho=\sqrt{r^2+z^2}$), the integral can immediately be evaluated so that, according to (135.1) and (135.2), one obtains

$$\varkappa(z)=\frac{3}{2}\frac{L}{l_{tr}}\left(e^{-\frac{z}{L}}-e^{-\frac{\sqrt{z^2+R^2}}{L}}\right)KX, \qquad (135.9a)$$

$$\varkappa_s(0)=\frac{3}{2}\frac{L}{l_{tr}}\left(1-e^{-\frac{R}{L}}\right)KX. \qquad (135.9b)$$

These results are based on the assumption that the neutron density has the value $n_s=n(0)$ over the whole surface of the probe; this is generally not true. The variation of n over the probe surface can be taken into account by considering Eq. (135.7) for $z=0$. Then, instead of (135.8), one has

$$\varrho'=\sqrt{r^2+r'^2-2rr'\cos\varphi},$$

which can be used in order to directly compute the average of the perturbation

$$\overline{(\Delta n)}_{z=0}=\frac{1}{\pi R^2}\int_0^{2\pi}\int_0^R \Delta n\, r'\,dr'\,d\varphi. \qquad (135.10)$$

The integration can be easily carried out in the following two extreme cases.

$R\gg L$. For $R/L\to\infty$ the neutron density is the same everywhere on the probe surface with the result that $\bar{\varkappa}_s$ is equal to $\varkappa_s$ and is given by the limit of (135.9b) for $R\to\infty$

$$\bar{\varkappa}_s=\varkappa_s=\frac{3}{2}\frac{L}{l_{tr}}KX. \qquad (135.11)$$

$R\ll L$. The exponential in (135.7) can be replaced by 1, so that the integral can easily carried out with the result that

$$\bar{\varkappa}_s=\frac{3}{2}\frac{R}{l_{tr}}KX\cdot W \qquad (135.12)$$

where

$$W=\frac{8}{3\pi}=0.85. \qquad (135.13)$$

It may be noticed that, apart from the factor W, Eq. (135.12) is identical to (135.9b) when the exponential is replaced by $1-R/L$. Since the average has been computed assuming the factor n_s of (135.6) as constant over the probe surface, Eq. (135.12) constitutes only a first approximation and the correct value of W is expected to be closer to 1 than (135.13).

For intermediate values of R/L the following approximate interpolation formula has been suggested by the same authors[1]:

$$\bar{\varkappa}_s=\frac{3}{2}\frac{L}{l_{tr}}\left(1-e^{-0.85\frac{R}{L}}\right)KX. \qquad (135.14)$$

When the absorption in the probe is finite, the neutron density can no longer be taken as constant and equal to n_s over the probe thickness. The perturbed neutron flux may be expanded in a series of Legendre polynomials of the angle Θ between flux direction and probe surface normal:

$$\Phi(\Theta)=\frac{1}{4\pi}\sum_l(2l+1)\,\Phi_l\cdot P_l(\cos\Theta). \qquad (135.15)$$

[1] See footnote 3, p. 644.

In the case of elementary diffusion theory (Sect. 117) one has

$$\Phi_0 = n\,v, \qquad \Phi_1 = -\frac{l_{\rm tr}}{3}\,\mathrm{grad}\,\Phi_0, \qquad \Phi_i = 0 \text{ for } i = 2, 3, \ldots. \tag{135.16}$$

The average probability α for a neutron to be absorbed in a single traversing of the probe is connected to the current densities J^+ and J^-, of the neutrons impinging on and emerging from the probe surface, by the relation[1]

$$\frac{J^+ - J^-}{J^+} = \alpha \quad \text{or} \quad 1 - \alpha = \frac{J^-}{J^+}. \tag{135.17}$$

Here

$$\left.\begin{aligned} J^+ &= 2\pi \int_0^1 \Phi(\Theta)\,|\cos\Theta|\,d(\cos\Theta) = \sum_l \frac{2l+1}{2}\,\Phi_l\,J_l^+,\\ J^- &= 2\pi \int_{-1}^0 \Phi(\Theta)\,|\cos\Theta|\,d(\cos\Theta) = \sum_l \frac{2l+1}{2}\,\Phi_l\,J_l^-,\end{aligned}\right\} \tag{135.18}$$

with

$$\left.\begin{aligned} J_l^+ &= \int_0^1 P_l(\cos\Theta)\,|\cos\Theta|\,d(\cos\Theta),\\ J_l^- &= \int_{-1}^0 P_l(\cos\Theta)\,|\cos\Theta)\,d(\cos\Theta).\end{aligned}\right\} \tag{135.19a}$$

For the first two terms one obtains

$$J_0^+ = J_0^- = \tfrac{1}{2}; \qquad J_1^+ = -J_1^- = \tfrac{1}{3}. \tag{135.19b}$$

The quantity α may be expressed in terms of the probe activation

$$\alpha = \frac{a}{S J^+} \tag{135.20}$$

with

$$a = 2\pi S \int_0^{\pi/2} \Phi(\Theta)\,|\cos\Theta|\left(1 - \mathrm{e}^{-\frac{KX}{\cos\Theta}}\right) d(\cos\Theta) = S \sum_l \frac{2l+1}{2}\,\Phi_l\,a^l, \tag{135.21}$$

$$a^l = \int_0^{\pi/2} P_l(\cos\Theta)\,|\cos\Theta|\,d(\cos\Theta).$$

The first two terms can immediately be evaluated:

$$\left.\begin{aligned} a^0 &= \frac{1}{2}\{1 - (1 - KX)\,\mathrm{e}^{-KX} + (KX)^2\,\mathrm{Ei}(-KX)\},\\ a^1 &= \frac{1}{3}\left\{1 - \left(1 - \frac{KX}{2} - \frac{(KX)^2}{2}\right)\mathrm{e}^{-KX} - \frac{(KX)^3}{2}\,\mathrm{Ei}(-KX)\right\}.\end{aligned}\right\} \tag{135.22}$$

It may be noticed that for $\mu/\varrho = 0$ and $\lambda = 1$, Eq. (134.15a) reduces to a_0 apart from a factor $\frac{1}{2}$ arising from the different limits of integration adopted in the two cases. According to (134.15b) $\alpha_1 = 0$, while according to the second equation (135.22), a^1 is different from zero; in the present case the gradient of the neutron density arises from the probe absorption and therefore has different signs on the two sides of the probe. From (135.22) one sees that, for $KX \ll 1$,

$$a^0 \approx KX, \qquad a^1 \approx \tfrac{1}{2} KX.$$

If the expansion of J^+ and a are broken off, each after its first term, one has

$$\alpha = \frac{\frac{1}{2} a^0 \Phi_0}{\frac{1}{4} \Phi_0} = 2 a^0 = \alpha_0. \tag{135.23}$$

---

[1] See footnote 3, p. 644.

We may now use again (135.7) with the negative source (135.6) replaced by

$$d Q_0 = C \cdot v \, d S \tag{135.24}$$

where C is a constant which should be adjusted to satisfy the condition (135.17) at the probe surface. The integration over dS gives, for points on the z axis,

$$\Delta n = n_0 - n(z) = \frac{3}{2} \frac{L}{l_{tr}} C \left[e^{-\frac{z}{L}} - e^{-\frac{\sqrt{z^2+R^2}}{L}} \right] \tag{135.25}$$

which for $z=0$ reduces to

$$n_s = n_0 - \frac{3}{2} \frac{L}{l_{tr}} C \left(1 - e^{-\frac{R}{L}}\right); \qquad (\text{grad}\, n)_s = -\frac{3}{2} \frac{C}{l_{tr}}. \tag{135.26}$$

Combining these relations with (135.16) and (135.18), (135.19) we obtain

$$J^+ = \tfrac{1}{4} (n_s + C)\, v, \qquad J^- = \tfrac{1}{4} (n_s - C)\, v$$

which, introduced into (135.17), yields

$$C = \frac{n_0 \alpha}{2 - \alpha + \frac{3}{2} \frac{L}{l_{tr}} \left(1 - e^{-\frac{R}{L}}\right) \alpha}. \tag{135.27}$$

Introducing this expression in (135.25) one obtains

$$\varkappa(z) = \frac{3}{2} \frac{\alpha}{2-\alpha} \frac{L}{l_{tr}} \left(e^{-\frac{z}{L}} - e^{-\frac{\sqrt{z^2+R^2}}{L}} \right), \tag{135.28a}$$

$$\varkappa_s(0) = \frac{3}{2} \frac{\alpha}{2-\alpha} \frac{L}{l_{tr}} \left(1 - e^{-\frac{R}{L}}\right). \tag{135.28b}$$

In order to find $\varkappa_a$ we notice that

$$\frac{1}{1+\varkappa_a} = \frac{a}{a_0} = \frac{J^+ \alpha}{J_0 \alpha} = \frac{\frac{1}{4} n v + \frac{1}{6} l_{tr} (\text{grand}\, n v)_s}{\frac{1}{4} n v}$$

from which it follows that

$$\varkappa_a = \left\{ \frac{3}{4} \frac{L}{l_{tr}} \left(1 - e^{-\frac{R}{L}}\right) - \frac{1}{2} \right\} \alpha. \tag{135.29}$$

The correction for the variation of n over the probe surface can also be computed in this case by means of (135.10). The result, similar to (135.14), now becomes

$$\bar{\varkappa}_s = \frac{3}{2} \frac{\alpha}{2-\alpha} \frac{L}{l_{tr}} \left(1 - e^{-0.85 \frac{R}{L}}\right). \tag{135.30}$$

According to the definition (135.20), α depends upon the constant C; the formulas (135.28) to (135.30) are usually employed by replacing α by its first approximation (135.23).

The same problem can readily be treated using the transport theory[1] [*19*] provided the foil is very thin and the medium scatters isotropically in the l.s. The integral equation (113.3), in the absence of a foil and with a neutron source of strength-density s_0 distributed uniformly through out the medium, can readily be solved, giving the result

$$\Phi_0(\boldsymbol{r}) = \overline{\Phi}_0 = \frac{s_0}{\Sigma_a} = \text{const}. \tag{135.31}$$

[1] T.H.R. SKYRME: Canad. deel Rep. MS 91.

The thin foil, which is assumed to have a scattering cross section equal to zero, is now introduced into the medium; it absorbs $\Sigma_{a\,\text{foil}}\,\Phi_0(\boldsymbol{r})$ neutrons/cm³ sec and, therefore, acts as a negative source of this strength distributed throughout its volume V. The integral equation (113.3) can now be written

$$\Phi_0(\boldsymbol{r}) = \int\limits_{\substack{\text{all}\\ \text{space}}} [\Sigma_s\,\Phi_0(\boldsymbol{r}') + s_0]\,\frac{e^{-\frac{|\boldsymbol{r}-\boldsymbol{r}'|}{l_{\text{tot}}}}}{4\pi\,|\boldsymbol{r}-\boldsymbol{r}'|^2}\,d\boldsymbol{r}' - \Sigma_{a\,\text{foil}} \int\limits_V \Phi_0(\boldsymbol{r}')\,\frac{e^{-\frac{|\boldsymbol{r}-\boldsymbol{r}'|}{l_{\text{tot}}}}}{4\pi\,|\boldsymbol{r}-\boldsymbol{r}'|^2}\,d\boldsymbol{r}'. \quad (135.32)$$

If the foil is so small that it hardly absorbs any neutrons, one can try to solve Eq. (135.32) by successive approximations. The first approximation is obtained by introducing the zero approximation (135.31) for $\Phi_0(\boldsymbol{r}')$ into the right-hand side of (135.32). The result is

$$\Delta n = \frac{1}{v}\{\overline{\Phi}_0 - \Phi_0(\boldsymbol{r})\} = \frac{\Sigma_{a\,\text{foil}}}{4\pi v}\int\limits_V \frac{e^{-\frac{|\boldsymbol{r}-\boldsymbol{r}'|}{l_{\text{tot}}}}}{|\boldsymbol{r}-\boldsymbol{r}'|^2}\,d\boldsymbol{r}' \quad (135.33)$$

which should be compared to (135.7) valid in the diffusion approximation.

If the foil is a small sphere of radius R with its centre at $r=0$, the integrals can readily be carried out. The depression of the neutron density at the centre of the sphere with respect to the density $n_0 = s_0/T$ at great distance, is given by the following expressions

$$\Delta n = \frac{s_0}{T}\,\frac{\Sigma_{a\,\text{foil}}}{\Sigma_{\text{tot}}}\left(1 - e^{-\frac{R}{l_{\text{tot}}}}\right) \quad \text{by transport theory,} \quad (135.34\text{a})$$

$$\Delta n = \frac{s_0}{T}\,\frac{\Sigma_{a\,\text{foil}}}{\Sigma_a}\left[1 - e^{-\frac{R}{L}}\left(1 + \frac{R}{L}\right)\right] \quad \text{by diffusion theory.} \quad (135.34\text{b})$$

One should keep in mind that these expressions do not give the correct depression in the interior of the foil since they have been derived by using the kernels valid for a single medium [*19*]. If $\frac{R}{l_{\text{tot}}} \ll 1$, then Eqs. (135.34a, b) become, respectively,

$$\Delta n = \frac{s_0}{T}\,(1 - \Sigma_{a\,\text{foil}} R) \quad \text{by transport theory,} \quad (135.35\text{a})$$

$$\Delta n = \frac{s_0}{T}\left(1 - \frac{3}{2}\,\Sigma_{a\,\text{foil}}\,\Sigma_{\text{tot}}\,R^2\right) \quad \text{by diffusion theory,} \quad (135.35\text{b})$$

two expressions which considerably differ from one another, as we should expect, since diffusion theory cannot be applied at distances of the order of l_{tot} (Sect. 118) from perturbations such as sources or absorbers.

Later the problem was reexamined by MEISTER[1] who treated the case of plane detectors by essentially the same procedure as that followed by BOTHE to derive the second equation (135.5). Instead of using the integral equation of transport theory, he considers the effect of the successive scattering processes under the following simplifying assumptions. The surrounding medium is only composed of non-absorbing heavy nuclei ($\sigma_a \ll \sigma_s$), so that the scattering of the neutrons can be taken as isotropic in the l.s. ($l_{\text{tr}} = l_s$). Furthermore, since $R \ll l_s$, the perturbation $n_0 v - n v$ of the neutron flux is small with respect to $n_0 v$ and therefore not only $n_0 v$ but even $n v$ can be treated as isotropic and constant over the probe surface.

[1] H. MEISTER: Z. Naturforsch. **11a**, 347 (1956).

Firstly, the perturbation produced at the surface element dS_2 by the absorption in the surface element dS_1 is computed, and then it is integrated with respect to dS_1 and dS_2 over the probe surface. If the neutrons absorbed in dS_1 were not absorbed, they would have a finite probability of crossing dS_2 after one, two, three, ... collisions in the surrounding medium. Therefore, the perturbation of dS_2 can be written as a series, the first term of which corresponds to neutrons which, if they had emerged from dS_1, would cross dS_2 after a single collision, the second term corresponds to neutrons that would cross dS_2 after two collisions, etc. One can recognize however that if l is the distance between dS_1 and dS_2, the first term of the series is proportional to l/l_s, the second to $(l/l_s)^2$, and so on. Considering that

$$\frac{l}{2} \leqq R \ll l_s,$$

one can conclude that the first approximation, based on considering only the first term of the series, i.e. only upon the effect of the first collision, is not too bad.

The result obtained by such a procedure is[1]

$$\varkappa(z) = \frac{1}{2-\psi(KX)} \left\{\psi(KX) - \zeta\,\psi\left(\frac{KX}{\zeta}\right) + \frac{R}{l_s}\,\varphi_0(KX)\right\} \tag{135.36}$$

where

$$\left.\begin{aligned} \zeta &= \frac{z}{\sqrt{z^2+R^2}}, \\ \psi(KX) &= 1 - e^{-KX} - KX\,\mathrm{Ei}(-KX), \\ \varphi_0(KX) &= \int_0^1 K(1-y^2)\,\frac{2y(1-y^2)}{(1+y^2)^2}\left(1 - e^{-KX\frac{1+y^2}{2y}}\right) dy; \\ K(y^2) &= \int_0^1 \frac{dx}{\sqrt{(1-x^2)(1-y^2x^2)}}. \end{aligned}\right\} \tag{135.37}$$

The functions $\psi(KX)$ and $\varphi_0(KX)$ are plotted in Fig. 273 and 274 versus KX[1].

The perturbation along the probe surface was also computed by means of this procedure, with the result

$$\varkappa(r) = \frac{1}{2-\psi(KX)} \left\{\psi(KX) + \frac{2}{\pi}\,\frac{R}{l_s}\,\mathrm{E}\left(\frac{r^2}{R^2}\right)\varphi_0(KX)\right\} \tag{135.38}$$

where r is the distance from the probe centre and

$$\mathrm{E}(y^2) = \int_0^{\pi/2} \sqrt{1 - y^2\sin^2\alpha}\,d\alpha. \tag{135.39}$$

Finally one has

$$\varkappa_a(0) = \frac{1}{2}\,\frac{R}{l_s}\,\varphi_0(KX), \tag{135.40}$$

$$\bar{\varkappa}_a = \frac{4}{3\pi}\,\frac{R}{l_s}\,\varphi_0(KX). \tag{135.41}$$

The latter expression can be compared with BOTHE's formula (135.5), second equation; even if R is taken to be the radius of the plane foil (and not $\frac{2}{3}$ of it)[2] the value of $\bar{\varkappa}_a$ given by the second equation (135.5) is about 25% lower than that given by (135.41).

---

[1] See footnote 1, p. 648.

[2] See footnote 1, p. 644.

All the considerations given above refer to the case of monoenergetic neutrons of velocity v; therfore averages should be taken of the quantities α, ψ and φ_0 over the neutron spectrum before comparing with experiments. If the probe follows the $1/v$ law and the neutron spectrum is Maxwellian, one can use the following approximate relations, where K_0 indicates the value of K for $v = v_0 = \sqrt{2kT/m}$:

$$\bar{\alpha}_0(KX) = \frac{\sqrt{\pi}}{2}\,\alpha_0(K_0X) = 0.886\,\alpha_0(K_0X) \qquad (135.42)$$

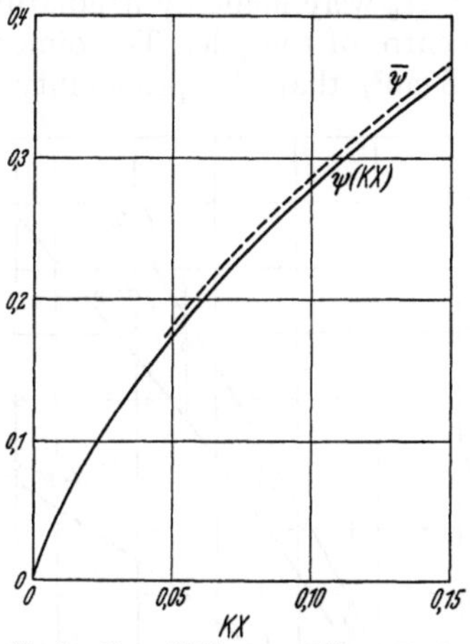

Fig. 273. The function $\psi(KX)$ versus KX; see Eq. (135.37).

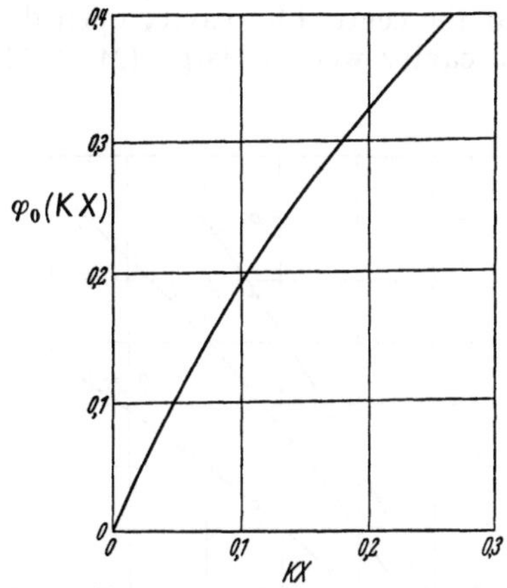

Fig. 274. The function $\varphi_0(KX)$ versus KX; see Eq. (135.37).

which is accurate within 0.5% fov $K_0X < 0.5$[1];

$$\left.\begin{aligned}
\bar{\psi}(KX) &\approx \frac{2}{\sqrt{\pi}}\,\psi(K_0X) = 1.128\,\psi(K_0X) && \text{for } K_0X \ll 1,\\
\bar{\psi}(KX) &\approx 1.04\,\psi(K_0X) && \text{for } K_0X = 0.05,\\
\bar{\psi}(KX) &\approx 1.02\,\psi(K_0X) && \text{for } K_0X = 0.15,
\end{aligned}\right\} \qquad (135.43\,\text{a})$$

$$\left.\begin{aligned}
\bar{\psi}(KX) &\approx \psi_0(K_0X)\\
\bar{\varphi}_0(KX) &\approx \varphi_0(K_0X)
\end{aligned}\right\} \quad \text{for larger values of } K_0X. \qquad (135.43\,\text{b})$$

The approximate expressions (135.43 b) are accurate within 12% and are sufficient for most practical uses[2].

The experimental investigation of the various effects described above was startet by BOTHE[3] who studied the dependence of the activation of flat Dy_2O_3 probes in paraffin and aluminium for two values of R (0.7 and 1.4 cm) as a function of their thicknesses X. Similar experiments were made in graphite by KLEMA and RITCHIE[4] who used Au and In foils of radius $R = 1.9$ cm and thickness x ranging from 0.025 to 0.127 mm. These authors detected, by means of a scintillation counter, the γ-rays emitted by the probe after irradiation, so that the correction due to absorption of the measured radiation in the foil itself was negligible.

The problem was studied rather carefully by MEISTER[1,5] who used In foils of 0.6, 0.9 and 1.2 cm radii and thicknesses ranging from 0.02 to 0.3 g/cm² placed

[1] H. MEISTER: Z. Naturforsch. **10**a, 669 (1955).
[2] See footnote 1, p. 648
[3] See footnote 1, p. 643.
[4] E.D. KLEMA and R.H. RITCHIE: Phys. Rev. **87**, 167 (1952).
[5] H. MEISTER: Z. Naturforsch. **11**a, 356 (1956).

in paraffin and graphite. This author obtained a thermal neutron distribution uniform—and isotropic—with good accuracy over a sufficiently wide region, by using two Ra α+Be sources (500 millicurie each) and irradiating the foils in the saddle point of the neutron distribution, i.e. in the position, along the line joining the two sources, where the neutron density had its minimum.

The measurements in graphite were performed in a pile $99\times108\times132$ cm³ with the two sources placed on the vertical axis at a mutual distance of 48 cm. The measurements of the unperturbed activity A_0 were made by irradiating the foils at the centre of a cavity opened in the centre of the pile. The dimensions of the cavity were so large ($21.8\times21.8\times35.8$ cm³) that the probability for a

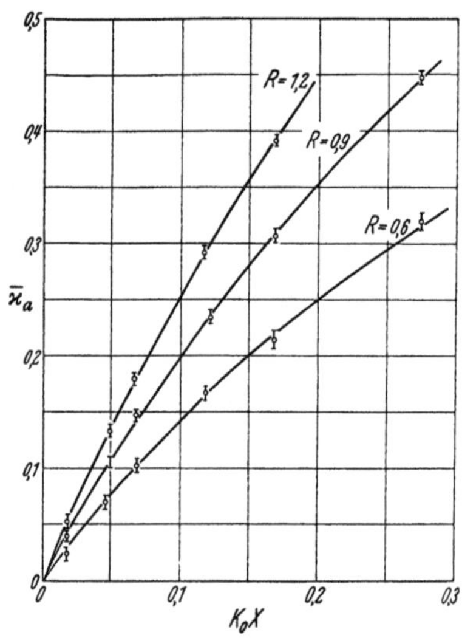

Fig. 275. Relative perturbation of the activation of In foils in paraffin (MEISTER).

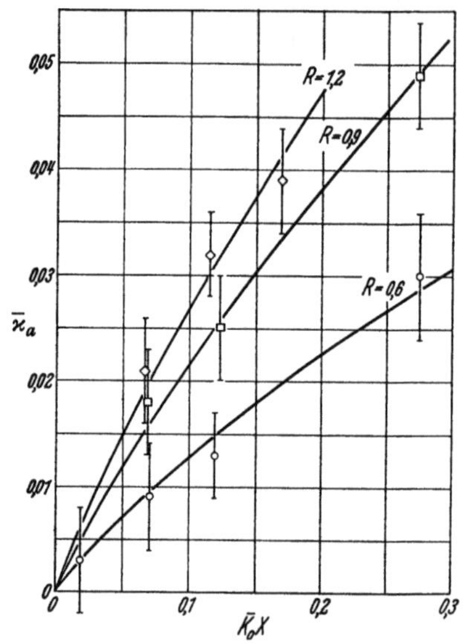

Fig. 276. Relative perturbation of the activation of In foils in graphite (MEISTER).

neutron that had crossed the probe to reach it again was negligibly small. The isotropy of the neutron flux at the centre of the cavity was checked by measuring A_0 as a function of the orientation of the probe.

In the case of paraffin a good uniformity of the neutron distribution was obtained by placing two sources at a distance of 20 cm from the two opposite faces of 40×50 cm² area of a paraffin block 17 cm long, and surrounding the whole set-up with a paraffin reflector 5 cm thick.

According to the cadmium difference procedure, the activities due to thermal neutrons were computed by means of (133.1) for $x_{\mathrm{Cd}}=0.6$ mm.

Figs. 275 and 276 show the measured values of $\bar{\varkappa}_a$ versus K_0X obtained in paraffin and graphite. For K_0 MEISTER uses (0.94 ± 0.6) cm²/g corresponding to $v_0=(2300\pm700)$ m/sec; this value of the most probable velocity was obtained from absorption measurements in Au made in the graphite pile described above in accordance with the procedure given in Sect. 108. The curves given in Fig. 275 and 276 have been obtained by fitting, by a least square method, the expression

$$\bar{\varkappa}_a=\mathrm{const}\cdot\alpha_0(K_0X)$$

Table 116. *Values of the constant* $\bar{\varkappa}_a/\alpha_0$ (MEISTER).

Paraffin R (cm)	0.6	0.9	1.2
Measured value	0.958 ± 0.06	1.346 ± 0.08	1.711 ± 0.10
BOTHE and TITTLE Eq. (135.5)	0.400	0.728	1.000
CORINALDESI	—	—	1.60
SKYRME	0.815	1.120	1.350
VIGON and WIRTZ Eq. (135.30)			
$W=1$	0.495	0.938	1.271
$W=0.849$	0.345	0.721	1.007
Semiempirical formula (135.44)	0.932	1.375	1.708
Graphite R (cm)	**0.6**	**0.9**	**1.2**
Measured value	0.087 ± 0.015	0.148 ± 0.015	0.179 ± 0.020
BOTHE and TITTLE Eq. (135.5)	0.078	0.118	0.157

to the experimental points. The values obtained for the constant by means of such a procedure are compared in Table 116 with the values computed by introducing the following parameter values in the various formulas given above

$$\text{for paraffin}\quad \left.\begin{aligned} l_{\rm tr} &= (0.40 \pm 0.02)\ \text{cm} \\ L &= (2.40 \pm 0.10)\ \text{cm} \end{aligned}\right\} \varrho = (0.88 \pm 0.01)\ \text{g/cm}^3,$$

$$\text{for graphite}\quad l_{\rm tr} = (2.60 \pm 0.10)\ \text{cm}.$$

Since the theoretical values of $\varkappa_a/\alpha_0(K_0 X)$ agree very poorly with the corresponding experimental results, MEISTER suggested to use for

$$1.5 \leqq R/l_{\rm tr} \leqq 3$$

the following semiempirical expression

$$\bar{\varkappa}_a = \left\{\frac{3}{4}\,\frac{L}{l_{\rm tr}}\left(1 - e^{-\frac{R}{L}}\right) - 0.06 \pm 0.06\right\}\bar{\alpha}; \qquad \bar{\alpha} = 0.886\,\alpha_0(K_0 X) \tag{135.44}$$

which differs from (135.29) only for the value of the constant.

The graphite data for $R=0.6$ and $R=0.9$ mm appear to be in agreement, within the experimental errors[1], with the curves computed by means of (135.41) and (135.43b).

The perturbation of the neutron density along the axis of the probe was also measured by MEISTER[2] in graphite. The foils producing the perturbation were

$$\begin{array}{llll} \text{In} & R = 1.5\ \text{cm}; & X = 127.6\ \text{mg/cm}^2; & K_0 X = 0.120 \\ \text{In} & R = 0.9\ \text{cm}; & X = 130.2\ \text{mg/cm}^2; & K_0 X = 0.122 \\ Dy_2O_3 & R = 1.5\ \text{cm}; & X = 30.2\ \text{mg/cm}^2; & K_0 X = 0.075. \end{array}$$

Secondary probes of Dy_2O_3 were used for exploring the perturbation; they were sufficiently thin and small ($X=32$ mg/cm², $K_0X=0.067$, $R=0.5$ cm) to give an activity proportional to the neutron density at the position of irradiation.

The results obtained with the two In probes are plotted as a function of z in Fig. 277; the continuous curves for small values of z are computed by means of (135.36), (135.43b) and the dotted curves, for large values of z, are computed by means of (135.28a), (135.23), (135.42). Fig. 278 shows the variation of the density perturbation along the probe surface[2]. The curves have been computed by means of (135.38), (135.43b).

[1] See footnote 1, p. 648.
[2] See footnote 5, p. 650.

136. The absorption curves of thermal neutrons. The absorption curves of thermal neutrons are obtained by measuring the activities $A_{\text{th}}(0)$ and $A_{\text{th}}(X)$ (133.1) of a detector—usually a plane foil—of thickness X_d (in g/cm²), successively exposed to the same beam of thermal neutrons, a first time uncovered and the second time covered with an absorber layer of thickness X (g/cm²). The absorption curve is given by the transmission of the absorber

$$T(X) = \frac{A_{\text{th}}(X)}{A_{\text{th}}(0)} \tag{136.1}$$

as a function of X. The shape of the curve $T(X)$ depends on the absorption (and possibly scattering) cross section of the absorber, on the activation cross section and thickness of the detector, on the spectrum and angular distribution of the incident neutrons etc.

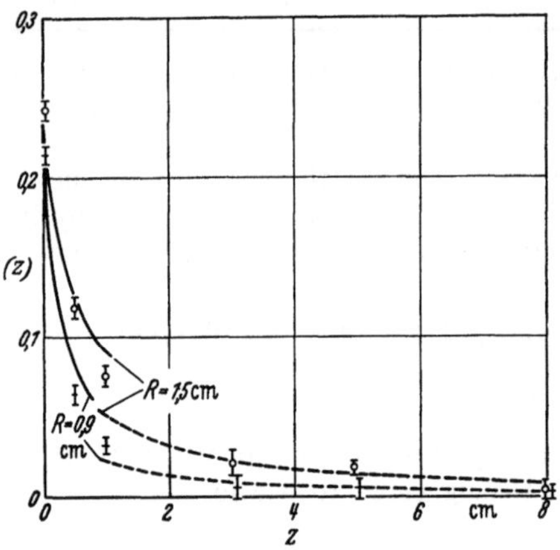

Fig. 277.

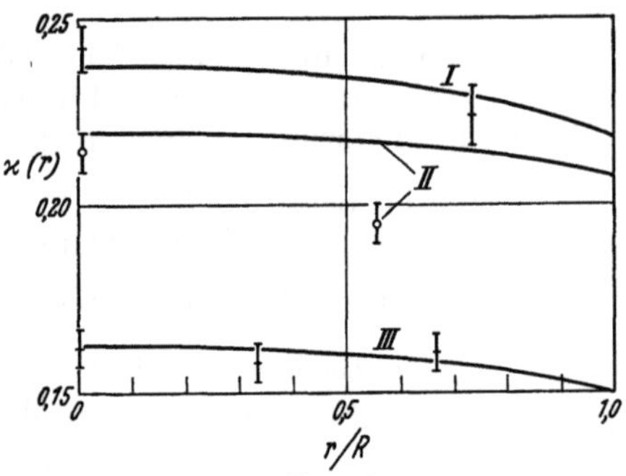

Fig. 278.

Fig. 277. Relative density perturbation produced by In probes along their axis: — Eq. (135.36); - - - Eq. (135.28a).

Fig. 278. Relative density perturbation produced by probes along their surface: — Eq. (135.38). I. In probe, $R = 1.5$ cm, $K_0 X = 0.120$. II. In probe, $R = 0.9$ cm, $K_0 X = 0.122$. III. Dy_2O_3 probe $R = 1.5$ cm, $K_0 X = 0.075$.

Let us first consider the case of monoenergetic neutrons (one-velocity-group approximation) and discuss the influence of both, the angular spread of the incident neutrons and the thickness of the detector, in a few very simple extreme cases. Then we shall examine, still for monoenergetic neutrons, the effect of the finite thickness of the detector and the corresponding correction due to the absorption of β-rays. We shall obtain, in the case of monoenergetic neutrons, results of general validity which can be used, with only minor changes, also in the case of epi-cadmium neutrons (Sect. 94) or of thermal neutrons with a Maxwellian velocity distribution. Finally the more realistic case of a Maxwellian velocity distribution in connection with $1/v$ absorbers and detectors will be examined.

α) Absorption curves of monoenergetic neutrons. If the monoenergetic neutrons constitute a parallel beam which falls perpendicularly upon the surface of the absorber (and detector) the absorption curve is exponential and independent of the detector thickness:

$$T(X) = e^{-KX}.$$

The absorption coefficient is given by

$$K = \frac{\Sigma}{\varrho} = \frac{\mathfrak{N}\sigma}{\varrho}$$

where σ is the absorption cross section of the absorber if the transmission experiment is made in "bad geometry", i.e., with the detector foil placed immediately behind the absorber; σ on the other hand represents the total cross section $\sigma_a + \sigma_s$ if the experiment is made in "good geometry" i.e. under such geometrical conditions that all the neutrons scattered in the absorber, even under a very small angle, miss the detector.

In the following, for the sake of simplicity, we shall neglect the contribution to K arising from scattering, i.e. we shall assume that the experiments are performed in bad geometry and that $\sigma_s \ll \sigma_a$ so that the scattering does not produce any appreciable change in the thicknesses actually crossed by the neutrons in the absorber and detector.

If the beam has an angular spread, the absorption curve is no longer exponential. The two following simple cases will be discussed in some detail:

1. The absorber and the detector are exposed to an isotropic neutron flux Φ_0 as occurs inside a moderator in a region of constant neutron density. Then the neutrons incident per *unit time* and *unit area* of the absorber in the solid angle $d\Omega = 2\pi\, d(\cos\Theta)$ forming the angle Θ with the normal to the absorber surface, is given by the *Lambert law*

$$d\Phi_L = 2\Phi_0 \cos\Theta\, d(\cos\Theta). \tag{136.2}$$

2. The angular distribution of the incident neutrons is slightly anisotropic in the direction of the normal to the absorber surface. In this case we may use the angular distribution

$$d\Phi_G = \frac{2a}{2+a}\Phi_0(1 + a\cos\Theta)\cos\Theta\, d(\cos\Theta). \tag{136.3}$$

This is a generalization of the angular distribution (122.20) derived by FERMI [*8*] by means of the approximate procedure described in Sect. 122 for the thermal neutrons diffusing out of a moderator surface; Eq. (136.3) reduces to (122.20) for $a = \sqrt{3}$.

The shape of the absorption curve also depends on the detector thickness. We shall first consider two opposite extreme cases: the detector is *infinitely thick*, i.e., it is a *black detector* which absorbs all the incident neutrons; or the detector is *infinitely thin* so that it absorbs an infinitely small fraction of the incident neutrons.

The absorption curves obtained with a black detector are given by

$$\mathfrak{C}_i(KX) = \int_0^{\pi/2} d\Phi_i\, e^{-\frac{KX}{\cos\Theta}} \tag{136.4}$$

where $i = L$ or G according to the choice made between the two angular distributions (136.2), (136.3).

The absorption curve obtained with a thin detector is given by

$$C_i(KX) = \int_0^{\pi/2} \frac{d\Phi_i}{\cos\Theta}\, e^{-\frac{KX}{\cos\Theta}} \Bigg/ \int_0^{\pi/2} \frac{d\Phi_i}{\cos\Theta} \tag{136.5}$$

where the factor $\frac{1}{\cos\Theta}$ arises from the probability $\frac{K_d\, dX_d}{\cos\Theta}$ of activation of the detector of thickness dX_d. From (136.4) and (136.5) one obtains

$$C_i(KX) = \frac{\mathfrak{C}_i'(KX)}{\mathfrak{C}_i(KX)} \tag{136.6}$$

where $\mathfrak{C}_i'$ means the derivative of $\mathfrak{C}_i$ with respect to its argument. Introducing (136.2) into (136.4) we obtain

$$\mathfrak{C}_L(KX) = e^{-KX}(1 - KX) - (KX)^2 \operatorname{Ei}(-KX) \tag{136.7}$$

where

$$-\operatorname{Ei}(-y) = \int_y^\infty e^{-t} \frac{dt}{t} > 0 \tag{136.8}$$

is the exponential integral function[1]. From (136.6), (136.7) we derive

$$C_L(KX) = e^{-KX} + KX \operatorname{Ei}(-KX). \tag{136.9}$$

It may be useful to notice that

$$1 - \mathfrak{C}_L(KX)$$

represents the total activation of the layer of thickness X so that it is identical to twice the quantity a^0 defined in (135.22) and identical to (134.15) when one introduces in this equation $\lambda = 1$.

Similar expression for $\mathfrak{C}_G(KX)$ and $C_G(KX)$ can be obtained by using the generalized Fermi law (136.3) instead of the Lambert law. Fig. 279 shows a comparison of C_L and C_F with the exponential function.

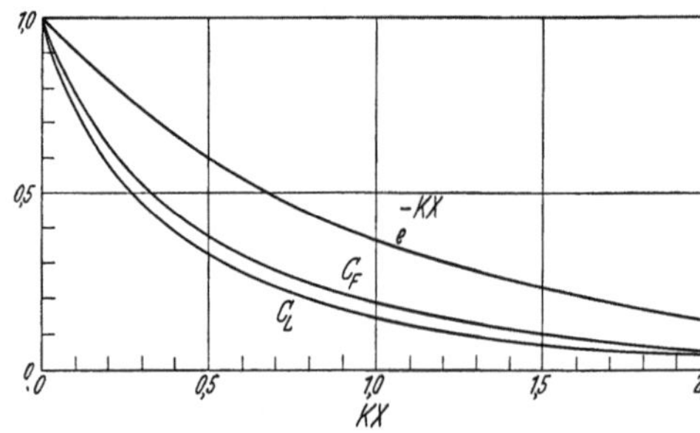

Fig. 279. Absorption curves of monoenergetic neutrons obtained with an infinitely thin detector for various angular distributions of the incident neutrons.

β) Corrections arising from the finite thickness of the detector and the β-ray absorption. Eq. (136.6) is an example of the relation existing in general between differential and integral absorption curves. It has already been used in Sect. 94 for the discussion of the absorption curves of resonance neutrons, and it will be used again, below in this section, in the discussion of the absorption curves of Maxwellian beams.

Another remark of general interest is that the integral curves allow the derivation of the absorption curve for detectors of finite thickness X_d. For the sake of definiteness we shall refer again to the present case of monoenergetic neutrons. The total numbers of neutrons transmitted by the absorber and by the combination of absorber and detector are given by $\mathfrak{C}_i(KX)$ and $\mathfrak{C}_i(KX + K_d X_d)$, respectively. Therefore the activation of the detector when placed behind an absorber of thickness X, is given by the difference

$$\mathfrak{C}_i(KX) - \mathfrak{C}_i(KX + K_d X_d)$$

and the absorption curve, normalized as usual to 1 for $X = 0$, is given as a function of X by the relation

$$\frac{\mathfrak{C}_i(KX) - \mathfrak{C}_i(KX + K_d X_d)}{1 - \mathfrak{C}_i(K_d X_d)}. \tag{136.10}$$

All these relations do not take into account the absorption of the β-rays in the detector. Therefore they should only be used if the thickness of the detector is so small that its activation is constant throughout its thickness. Then, according to (89.6), the measured activity A differs from the activation a by an unessential constant (89.11).

[1] E. Jahnke and F. Emde: Tables of functions. New York· Dover Publications 1945.

If the detector does not satisfy the condition of noself-protection (89.3), i.e., if the detector is so thick that the activation varies appreciable throughout its thickness, the activity A can be computed by means of Eq. (89.9) where $\alpha(x)\,dx$ represents the activation of the layer located between x and $x+dx$ inside the detector. In the present case, $\alpha(x)$ is given by the differential absorption curve (136.6) so that we can write

$$A(X) = \omega w \int_0^{X_d} C_\iota(KX + K_d Y)\, e^{-\frac{\mu}{\varrho} Y}\, dY \qquad (136.11)$$

an expression which can easily be evaluated if μ/ϱ and K are known. If the integrand in (136.11) is too complicated for an analytical integration, the integral can either be computed numerically or by representing $C_\iota(KX)$ by means of the empirical formula (89.10). If a single term of this equation constitutes a sufficient approximation inside the integration interval, the correction factor arising from self-absorption of β-rays is given by

$$\frac{\int_0^{X_d} e^{-\left(\overline{K}+\frac{\mu}{\varrho}\right)Y}\, dY}{\int_0^{X_d} e^{-\overline{K}Y}\, dY} = \frac{\overline{K}}{\overline{K}+\frac{\mu}{\varrho}} \, \frac{1-e^{-\left(\overline{K}+\frac{\mu}{\varrho}\right)X_d}}{1-e^{-\overline{K}X_d}}. \qquad (136.12)$$

γ) Absorption curves of Maxwellian beams. We shall now consider the absorption curves obtained when the neutron spectrum is Maxwellian

$$d\varphi_0(v) \propto v^3\, e^{-\frac{v^2}{v_0^2}}\, dv$$

and both the capture cross section of the absorber and the activation cross section of the detector follow the $1/v$ law. This is the only case of considerable practical interest for which a few general rule can be derived; if the absorber or the detector has a resonance in the thermal energy region or very close to it (as e.g. happens in Cd and Gd) the absorption curve should be computed numerically.

We start again from the case of a parallel beam; the absorption curves for an infinitely thick and an infinitely thin detector are respectively given by

$$\mathcal{M}(K^0X) = \int_0^\infty Y\, e^{-Y-K^0XY^{-\frac{1}{2}}}\, dY, \qquad (136.13)$$

$$M(K^0X) = \frac{2}{\sqrt{\pi}} \int_0^\infty Y^{\frac{1}{2}}\, e^{-Y-K^0XY^{-\frac{1}{2}}}\, dY, \qquad (136.14)$$

where

$$Y = \frac{E}{kT} = \frac{v^2}{v_0^2}, \qquad (136.15)$$

and K^0 is a constant defined by

$$K(v) = \frac{K^0}{Y^{\frac{1}{2}}}. \qquad (136.16\text{a})$$

It represents the value of the absorption coefficient of the absorber for $Y=1$, i.e. for $v=v_0$,

$$K^0 = \frac{\mathfrak{N}}{\varrho}\, \sigma_a(v_0). \qquad (136.16\text{b})$$

Between (136.13) and (136.14) there holds, of course, a relation similar to (136.6). The corresponding initial slopes are

$$\mathcal{M}'(0) = -\frac{\sqrt{\pi}}{2}; \quad M'(0) = -\frac{2}{\sqrt{\pi}}. \tag{136.17}$$

It may be noticed that the initial slope of the differential curve corresponds to the absorption coefficient

$$\frac{2}{\sqrt{\pi}} K(v_0) = \frac{\mathfrak{N}}{\varrho} \frac{2}{\sqrt{\pi}} \sigma_a(v_0) \tag{136.18}$$

which is in agreement with (23.14) and (118.13).

The slopes of the two curves (136.13), (136.14) decrease (in absolute value) with increasing absorber thickness, an effect usually indicated as *hardening of the spectrum* [5], [12]. It obviously arises from the preferential absorption of the more absorbable neutrons; because of the $1/v$ law these are the neutrons of lower energy present in the spectrum. This effect is illustrated in Fig. 280, which shows a comparison of the neutron flux of a Maxwellian beam detected with a thin $1/v$ detector with that of the same beam filtered through a $1/v$ absorber for a 50% total transmission. Thus, in the case of the differential curve (136.14), the absorption coefficient starts from the initial value (136.17) and becomes gradually smaller; it reaches and maintains, with rather good accuracy, the value (136.16b) in the interval[1]

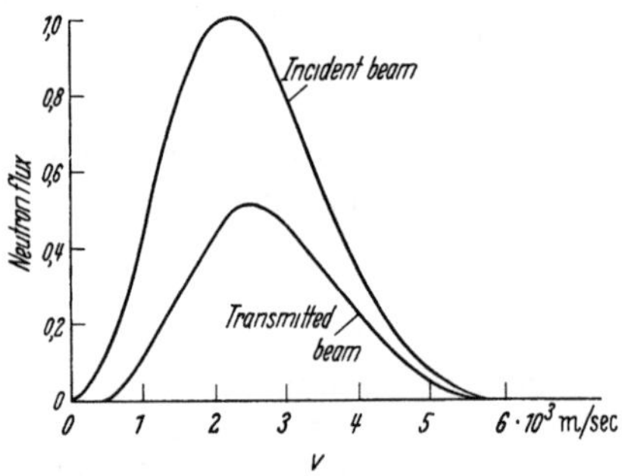

Fig. 280.

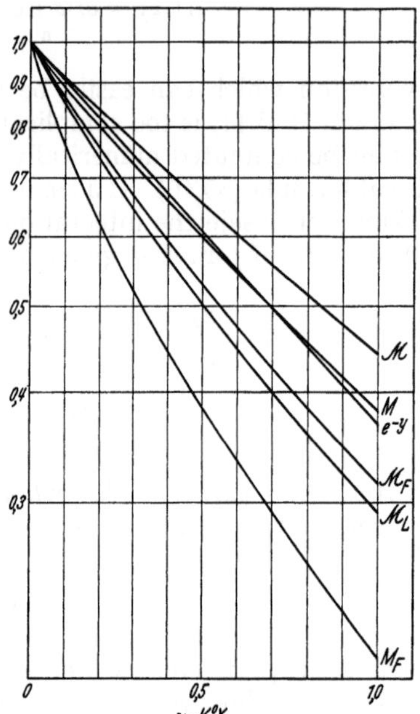

Fig. 281.

Fig. 280. The hardening of the Maxwell distribution for a transmission of 50% calculated for a $1/v$ absorber and $1/v$ detector (HUGHES).

Fig. 281. Absorption curves of a Maxwellian beam. $M \triangleq$ parallel beam, thin detector, $\mathcal{M} \triangleq$ parallel beam, thick detector; $M_F \triangleq$ angular Fermi law, thin detector; $\mathcal{M}_F \triangleq$ angular Fermi law, thick detector, $\mathcal{M}_L \triangleq$ Lambert law, thick detector.

$$0.6 \geqq M(K^0 X) \geqq 0.4. \tag{136.19}$$

This property is very useful because, by choosing the thickness of the absorber so that the condition (136.19) is fulfilled, one directly obtains the absorption coefficient, and, from it, the absorption cross section corresponding to the most probable velocity v_0.

[1] H. H. GOLDSMITH and F. RASETTI: Phys. Rev. **50**, 328 (1936).

Table 117. *Absorption curves of a Maxwellian beam in a 1/v absorber with a 1/v detector.*

y	$\mathcal{M}_L(y)$	$M(y)$	$\mathcal{M}(y)$	$\mathcal{M}_F(y)$	$M_F(y)$
0.00	1.0000	1.0000	1.0000	1.0000	1.0000
0.01	0.9829	0.9888	0.9912	0.9850	0.9612
0.02	0.9664	0.9778	0.9825	0.9705	0.9314
0.03	0.9506	0.9670	0.9739	0.9564	0.9050
0.04	0.9353	0.9564	0.9653	0.9427	0.8810
0.05	0.9205	0.9459	0.9569	0.9293	0.8587
0.1	0.8527	0.8959	0.9161	0.8671	0.7652
0.2	0.7395	0.8063	0.8408	0.7608	0.6282
0.3	0.6476	0.7282	0.7729	0.6724	0.5283
0.4	0.5711	0.6596	0.7114	0.5975	0.4507
0.5	0.5063	0.5989	0.6557	0.5333	0.3883
0.6	0.4510	0.5450	0.6051	0.4778	0.3372
0.7	0.4031	0.4968	0.5589	0.4295	0.2947
0.8	0.3615	0.4538	0.5169	0.3874	0.2588
0.9	0.3252	0.4153	0.4784	0.3508	0.2281
1.0	0.2932	0.3802	0.4431	0.3191	0.2018

Table 117 shows the numerical results obtained by ZAHN[1]. For $y = K^0 X \gg 1$ the following asymptotic series may be used[2];

$$M(y) \to \frac{4}{\sqrt{3}}\left(\frac{1}{2}y\right)^{\frac{2}{3}} e^{-3(y/2)^{\frac{2}{3}}}\left[1 + \frac{17}{36}\left(\frac{2}{y}\right)^{\frac{2}{3}} - \cdots\right]. \qquad (136.20)$$

If the angular distribution of the Maxwellian neutrons is given by Eq. (136.2), the corresponding integral absorption curve turns out to be

$$\mathcal{M}_L(y) = \int_0^\infty z\, e^{-z}\, dz \int_0^{\pi/2} 2\cos\Theta \sin\Theta\, d\Theta\, e^{-y z^{\frac{1}{2}}/\cos\Theta}. \qquad (136.21)$$

With the new variable $t = z\cos^2\Theta$, Eq. (136.21) is transformed into the following expression which is very useful for numerical computations:

$$\left.\begin{aligned} \mathcal{M}_L(y) &= \int_0^\infty z\, e^{-z}\, dz \int_0^z z^{-1}\, dt\, e^{-y t^{-\frac{1}{2}}} \\ &= \int_0^\infty dt\, e^{-y t^{-\frac{1}{2}}} \int_t^\infty e^{-z}\, dz = \int_0^\infty e^{-t - y t^{-\frac{1}{2}}}\, dt = f_0(y), \end{aligned}\right\} \qquad (136.22)$$

where f_0 is defined in footnote[1].

The corresponding differential absorption curve is obtained by applying (136.6). Similar procedures can be used for computing $\mathcal{M}_F$ and M_F the values of which are given in Table 117 and Fig. 281.

[1] C.T. ZAHN: Phys. Rev. 52, 67 (1937). This author has investigated the properties of integrals of the form

$$f_n(x) = \int_0^\infty y^n e^{-y - x y^{-\frac{1}{2}}}\, dy.$$

If $2n$ is an integer, $f_n(x)$ can be put in the form $F(x) + G(x)\ln x$, where $F(x)$ and $G(x)$ are two power series in x. Once, e.g., $f_0, f_{\frac{1}{2}}, f_1$ have been computed by means of such series, other integrals of the same type may be derived using the relation

$$f_n(x) = n f_{n-1}(x) + \tfrac{1}{2} x f_{n-\frac{3}{2}}(x).$$

Table 117 has been taken from ZAHN's paper, but seven values have been corrected by G.C. WICK in the third and fourth digit.

[2] O. LAPORTE: Phys. Rev. 52, 72 (1937).

Sometimes the deviations of the absorption curves for Maxwellian beams from the exponential behaviour are expressed by defining the *effective neutron energy*; this, according to BETHE [*5*], represents the energy of a monoenergetic beam of neutrons which would give the same absorption in the considered absorber layer that is actually observed with the Maxwellian beam. Its value depends not only on the thicknesses of the absorber and the detector, but also on the angular distribution of the incident neutrons.

The corrections for the finite thickness of the detector and the self-absorption of the β-rays may be applied using the same procedure as that described for monoenergetic neutrons.

Bibliography.

[*1*] AJZENBERG, F., and T. LAURITSEN: Energy levels of light nuclei IV. Rev. Mod. Phys. **24**, 321 (1952).

[*2*] AJZENBERG, F., and T. LAURITSEN: Energy levels of light nuclei V. Rev. Mod. Phys. **27**, 77 (1955).

[*3*] BLATT, J.M., and V.F. WEISSKOPF: Theoretical nuclear physics. New York: John Wiley & Sons; London: Chapman & Hall 1952.

[*4*] BETHE, H.A., and R.F. BACHER: Nuclear Physics. Part A. Stationary states of nuclei. Rev. Mod. Phys. **8**, 82 (1936).

[*5*] BETHE, H.A.: Nuclear Physics. Part B. Nuclear Dynamics, Theoretical. Rev. Mod. Phys. **9**, 69 (1937).

[*6*] CASE, K.M., F. DE HOFFMANN and G. PLACZEK: Introduction to the Theory of Neutron Diffusion; Vol. I. Los Alamos: Los Alamos Scientific Laboratory 1953.

[*7*] DAVISON, B., and I.B. SYKES: Neutron Transport Theory. Oxford: Clarendon Press 1957.

[*8*] FERMI, E.: Sul moto dei neutroni nelle sostanze idrogenate. Ric. Sci. **7** (2), 13 (1936), and AECD-NP-2385 (1951).

[*9*] FERMI, E.: Nuclear Physics. Notes compiled by JAY OREAR, A.H. ROSENFELD and R.A. SCHLUTER. Chicago 1949.

[*10*] HANSON, A.O., R.F. TASCHEK and J.H. WILLIAMS: Monoenergetic neutrons from charged particles reactions. Rev. Mod. Phys. **21**, 635 (1949).

[*11*] HOLLANDER, J.M., I. PERLMAN and G.T. SEABORG: Table of isotopes. Rev. Mod. Phys. **25**, 469 (1953).

[*12*] HUGHES, D. J.: Pile neutron research. Cambridge Mass.: Addison Wesley 1953.

[*13*] HUGHES, D. J.: Neutron optics, Interscience Publishers. New York 1954.

[*14*] HUGHES, D. J., and J.A. HARVEY: Neutron cross sections. Washington: U.S. Government Printing Office 1955.

[*15*] International Conference on peaceful Uses of Atomic Energy, Geneva 1955: (a) Vol. II—Physics; Research Reactors. — (b) Vol. IV—Cross sections important to reactor design.—(c) Vol. V—Physics of reactor design.

[*16*] MARSHAK, R. E.: Theory of the slowing down of neutrons by elastic collision with atomic nuclei. Rev. Mod. Phys. **19**, 185 (1947).

[*17*] SEGRÈ, E.: Experimental nuclear physics, Vol. I. New York: John Wiley & Sons; London: Chapman & Hall 1953: (a) Detection methods by H. STAUB. — (b) Passage of radiation through matter by H. BETHE and J. ASHKIN. — (c) Nuclear moments and statistics by N.F. RAMSEY. — (d) Nuclear two body problems and elements of nuclear structure by N.F. RAMSEY. — (e) Charged particle dynamics and optics, relative isotopic abundances of elements, atomic masses by K.T. BAMBRIDGE.

[*18*] SEGRÈ, E.: Experimental nuclear physics, Vol. II. New York: John Wiley & Sons; London: Chapman & Hall 1953: (a) A survey of nuclear reactions by P. MORRISON. — (b) The neutron by B.T. FELD.

[*19*] WEINBERG, A.M., and L.C. NODERER: Theory of neutron chain reactions. AECD-3471, and CF-51-5-98, Vol. I, Chap. I—IV: Diffusion and slowing down of neutrons. Oak Ridge National Laboratory 1951.

Penetration and Diffusion of X Rays.

By

U. FANO, L. V. SPENCER and M. J. BERGER*.

With 57 Figures.

1. Introduction. In the early period of X-ray research, the radiation sources were discharge tubes with applied potentials well below 100 kev. X rays from these sources interact with matter primarily through the photoelectric effect. Their penetration is then limited by photoelectric absorption and is adequately described by an exponential law.

X rays from sources with higher applied potentials and energetic γ rays from radioactive substances[1] interact with matter through additional processes, primarily Compton scattering. Compton scattering gives rise to secondary X rays which travel further and may themselves undergo Compton scattering. There also occur pair production and annihilation, and bremsstrahlung by pair electrons and by electrons recoiling from Compton scattering. Altogether there arises a complex of penetrating secondary radiations whose study becomes part of the study of X-ray penetration.

This complex phenomenon can be analyzed into a succession of statistically independent elementary interactions between X-ray photons and matter, or between secondary fast particles and matter. Each elementary process subdivides the energy of the incident photon or particle. The radiation is thus progressively degraded in energy and deflected in various directions, and penetration becomes associated with degradation and diffusion.

The present article deals particularly with X-ray penetration under conditions where photons are likely to experience repeated Compton scattering before their eventual absorption. This problem has been studied systematically in recent years with the goal of determining how the X rays from a given source become distributed throughout a material. The results of this study have applications in radiology and nuclear engineering, particularly in reactor shielding design [*72*].

This article treats matter as an amorphous assembly of atoms. Coherent scattering by crystal lattices[2] is thus specifically disregarded, even though it could be taken into account, at least for microcrystalline matter, without substantial change of the theory. The atoms of matter will be assumed to be in their normal neutral state. The penetration in ionized matter at high tempera-

* The editing and the preparation of figures and tables have been contributed by Mrs. IDA H. REINGOLD. The preparation of this article and the development of much of the material reported in it have been supported by the U.S. Navy, Office of Naval Research and by the U.S. Atomic Energy Commission, Reactor Division.

[1] The term "X rays" is understood here to indicate any electromagnetic radiation of frequency higher than that of ultraviolet light. The term "γ rays" is understood to indicate X rays emitted by nuclei. The distinction is not relevant for our purpose and the two terms will be used interchangeably with preference for "X rays".

[2] Cf. Vol. XXXII of this Encyclopedia.

tures, such as occurs in stellar atmospheres and in thermonuclear processes, will not be considered[1].

Because photoelectric absorption predominates completely at low photon energies, and penetration theory is elementary under these conditions, attention is directed in this article to photon energies above ~ 20 kev. On the high energy side, this article does not cover the cascade shower processes which are dealt with in cosmic ray studies[2]. In this connection it is recalled that the cascade shower process, which involves electrons and positrons besides X rays, becomes predominant above 10 Mev in heavy elements, and above 100 Mev in light ones. Theories developed for the study of cascade showers in cosmic rays rely on assumptions about the probability of interactions with matter which are adequate only at energies of the order of 1 Gev or more. Below this energy there is a gap in which penetration phenomena are qualitatively known and understood but have not yet been calculated in detail. A few detailed experimental studies which have been made at energies up to 300 Mev will be reviewed in this article.

The treatment of X-ray penetration is thus limited in this article mostly to conditions where the penetration of corpuscular secondaries and of their bremsstrahlung has a minor influence and does not have to be studied in detail. There is, however, an important connection with the penetration and diffusion of corpuscular radiation from the standpoint of theory. On the one hand, early results on the slowing-down and deep penetration of neutrons have been incorporated in the theory of X-ray penetration. Conversely, the results of X-ray studies have much influence on studies of neutron[3] and electron[4] penetration. The presentation in this article has been designed so that many sections can be readily adapted for application to corpuscular radiations.

In the field thus delimited, the theory of the combined effect of multiple interaction processes has been developed to a great extent deductively, from a knowledge of elementary processes. Basic experimentation on multiple process phenomena has served largely as a subsidiary to verify the results of calculations. Accordingly, this article deals mostly with theoretical methods and their application. The theory encounters a serious difficulty because the probabilities of elementary processes depend on the variables in a manner not reducible to simple yet realistic analytic form. This limitation rules out a completely analytic solution; but a description of events which is at least partially analytic is indispensable for a qualitative understanding and a reasonably effective method of calculation. For this reason analytic and numerical procedures become closely integrated in the theory of X-ray penetration.

Part A surveys the background of information on the elementary processes with emphasis on the aspects most relevant to the study of multiple processes. Part B describes how the theory of multiple processes is constructed from data on the elementary processes. Part C reports the procedures and the results of numerous calculations of X-ray distributions, including exploratory studies of a wide range of problems and systematic tabulations for certain types of problems.

---

[1] The penetration and diffusion of radiation, mostly of optical and near-optical frequencies, in stellar atmospheres has been studied extensively in astrophysics, see e.g. Ref. [*22*]. In these processes the energy degradation of photons may be treated as a corrective factor.

[2] Cf. the article by G. Molière in Vol. XLVI of this Encyclopedia.

[3] Cf. the preceding article of E. Amaldi. — See also Ref. [*24*] and in particular J.E. Wilkins, R.L. Hellens and P.F. Zweifel: Neutron Slowing Down in Hydrogenous Media, Proc. Inter. Conf. Peaceful Uses of Atomic Energy, Geneva, 1955, and S.S. Holland and P.I. Richards: J. Appl. Phys. **27**, 1042 (1956).

[4] Cf. the article of R.D. Birkhoff in Vol. XXXIV of this Encyclopedia. See in particular L.V. Spencer: Phys. Rev. **98**, 1597 (1955).

The main theoretical developments have succeeded only in describing X-ray distributions in infinite homogeneous media. The effects of inhomogeneities and boundaries have been investigated primarily by a series of Monte Carlo (random sampling) calculations which provide by now a fairly substantial body of knowledge. These studies are reported in part D.

Experimental information is reported together with the theory where appropriate. Experiments that do not relate directly to existing calculations are reported separately in Part E. This last part also discusses briefly those aspects of experimental procedures and techniques which are particularly relevant to the observation of X-ray distributions.

A. Elementary processes.

2. Classification of X-ray induced processes. The elementary processes experienced by X rays as they travel through matter result from the electric and magnetic forces which the radiation exerts on the electric charges and currents within matter. To classify the processes we distinguish in the first place the different carriers of electricity on which the X rays can act, namely: (I) the *atomic electrons*, ordinary constituents of matter; (II) the *electrons and positrons* ("real" or "virtual") that can be induced in empty space where there is an electric field, particularly in the proximity of atomic nuclei, through the related processes of pair production and vacuum polarization; (III) the *nucleons*, ordinary constituents of nuclear matter, that is protons and also neutrons which, like protons, carry internal currents giving rise to a small magnetic moment; (IV) the *mesons* ("real" or "virtual"), a variety of unstable charge and current carriers which are induced in the space surrounding nucleons.

As a complementary basis for classifying elementary processes we consider their end effects on the incident X rays. The X-ray action brings about forced current oscillations within matter, with the driving frequency of the X rays. To the extent that the proper frequencies of the atomic systems match the X-ray frequency, the forced oscillations are in resonance and give rise to energy transfer from the radiation to the medium. This constitutes effect (A), outright *absorption* of X rays (one photon in each elementary process). Irrespective of resonance, an induced oscillating current radiates X rays in all directions, thus producing effect (B), *scattering*, which normally involves one photon scattered out per photon removed from the incident beam. Since scattering results from forced oscillation combined with re-radiation it constitutes a second order process. But it does not depend strongly on resonance so that in certain energy regions it is more important than absorption. When no part of the incident X-ray energy is retained by internal motions of a portion of matter, the scattering is *elastic* (B'); the X rays scattered by different current elements within that portion of matter have then a fixed phase relationship and interfere, which means that the scattering is *coherent*. On the other hand, when some energy is left to the internal motions and the scattered photon's energy is correspondingly reduced, the scattering is *inelastic* (B''); the intensities scattered by different current carriers, set in motion with respect to one another, are then distinguishable and must add up without interference, which means that the scattering is *incoherent*[1]. The induced oscillating currents may also radiate two or more photons

[1] The distinction between elastic and inelastic depends on what one considers as an isolated portion of matter. For example, Compton scattering is usually regarded as inelastic because it contributes energy to the motion of an individual electron with respect to its surrounding matter. However, the electron might also be regarded as an isolated particle unrelated to any surrounding system; from this point of view Compton scattering is an elastic process between a photon and the electron.

per photon removed from the incident beam; this constitutes effect (*C*), a *multi-photon process*. The additional photons are mostly of very low energy. These higher order processes have an overall effect which is usually much smaller, in ratio $\sim e^2/\hbar c$, than the effect of ordinary absorption and scattering; accordingly,

Table 1. *Classification of elementary X-ray effects.*

Type of effect / Current carriers	Absorption A	Scattering: Elastic B'	Scattering: Inelastic B''	Multi-photon processes C
I Atomic electrons	Photo-electric effect $\sigma \sim Z^5/137^4$	Rayleigh scattering $\sigma \sim Z^2$ (low energy limit)	Compton scattering $\sigma \sim Z/137$ Low energy limit: [1] Thomson scattering $\frac{d\sigma}{d\Omega} = Z\left(\frac{e^2}{mc^2}\right)^2 \times \times \frac{1}{2}(1+\cos^2\vartheta)$	Two-photon Compton scattering
II Electron positron field	Pair production $\sigma \sim Z^2/137^2$	Delbrück scattering $\sigma \sim (Z/137)^4$		
III Nucleons	Photonuclear $(\gamma, n), (\gamma, p)$, etc. $\sigma \sim Z$	(γ, γ) $\sigma \sim Z^2$ Low energy limit: [1] Thomson scattering $\frac{d\sigma}{d\Omega} = \left(\frac{Z^4}{A^2}\right)\left(\frac{1}{1822^2}\right) \times \times \left(\frac{e^2}{mc^2}\right)^2 \frac{1}{2}(1+\cos^2\vartheta)$	$(\gamma, \gamma)'$	
IV Mesons	Photomeson production	Modified (γ, γ)		

Heavy margins indicate the effects of major importance for this article.

[1] Thomson scattering by electrons is scattering of photons with $h\nu \ll mc^2$ and with electrons acting as free. It is inelastic from the standpoint of the atom, but elastic from the standpoint of the electron as a free particle. In fact, the condition $h\nu \ll mc^2$ and the condition that electrons are effectively free from atomic binding are not quite compatible.

Thomson scattering by nuclei is scattering of photons with $h\nu \ll 1822\,Amc^2$ with the nucleus acting as a single, rigid, free particle. It is elastic from the standpoint of the nucleus, as well as from the standpoint of the atom if combined coherently with Rayleigh and Delbrück scattering.

multi-photon effects are looked upon as radiative corrections and will be disregarded in this article.

Combining each of the four classes of current carriers (I to IV) with each of three types of end effects (A, B' and B'') gives the different types of elementary processes indicated in Table 1. Three of these processes, namely the photoelectric effect, Compton scattering and pair production, predominate in probability over all others, the first at lower energies (i.e. for photons below 0.05 to 0.5 Mev), the second at intermediate, the third at high energies (i.e. above 10 to 100 Mev) (Fig. 1). The treatment in this article will be limited to these three processes. Rayleigh scattering may also be of some importance at low energies [*22*]. Nuclear processes are much less probable than the electronic processes, except within narrow spectral bands. Delbrück scattering tends to be obscured by Rayleigh and nuclear scattering processes and has been difficult to observe. Process II B'' could be regarded as a radiative correction to pair production and has not been studied. Photomeson production occurs only for very high energy $\gtrsim 150$ Mev) X rays; it is about 1000 times less frequent than the electronic processes and will be disregarded in the present treatment[1].

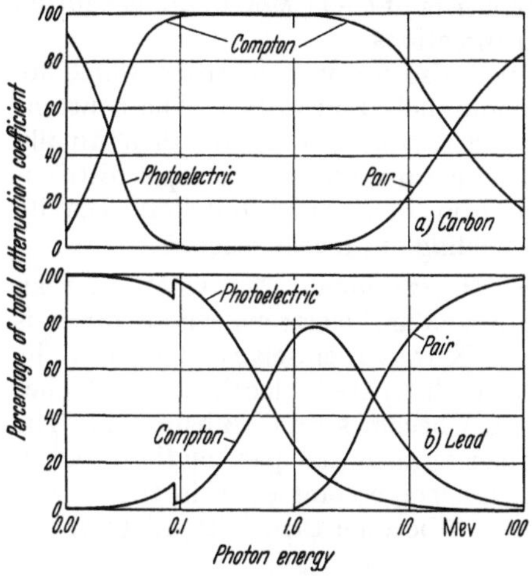

Fig. 1. Relative contributions of various photon interactions to the total attenuation coefficient for carbon and lead (WHITE, private communication).

The coherent meson process IV B would appear as a modification of the nuclear (γ, γ) scattering.

The probability of processes induced by X rays may depend on the polarization of the X rays. This dependence is usually weak and therefore polarization is of very minor importance in practical applications. Nevertheless numerous polarization effects have been studied experimentally and theoretically.

3. Secondary processes. This article deals with penetration phenomena which are controlled by the primary X rays and by secondary Compton-scattered X rays. Other secondary radiations have a subordinate role in most of the problems to be treated but may nevertheless take on importance in a specific context. One must therefore have in mind a variety of elementary processes associated with secondary radiations. Processes involving the emission of secondary X rays are of most direct importance in this connection, because this emission, if sufficiently intense, has to be treated as an additional X-ray source.

α) Emission of secondary X rays. The photoelectric absorption of X rays in an atom is often followed by *fluorescence*, that is, by emission of an X-ray photon of the characteristic line spectrum of that atom. The energy of these photons ranges up to about 100 kev for the K series of heavy elements. The probability that in a photoelectric process a photon is emitted rather than an Auger electron is called "fluorescence yield"; it increases with increasing energy of the photon to be emitted and ranges from very low to nearly one. For the theory of fluorescence see volume XXXI of this Encyclopedia; a collection of data has been given by BROYLES et al.[2]. The effect of fluorescence on the overall process of energy dissipation by X rays will be important only if the incident photon energy is not too much larger than the energy of the fluorescent photons.

[1] However, photoproduction of neutral pions, followed by their disintegration into two γ-rays, constitutes an appreciable source of high energy photons ($\gtrsim 70$ Mev) which may travel in opposite direction to the incident X rays as though they resulted from backscattering.

[2] C. D. BROYLES, D. A. THOMAS and S. K. HAYNES: Phys. Rev. **89**, 715 (1953).

Because most of the energy of incident X rays is given to secondary electrons or positrons, the emission of further X rays as *bremsstrahlung* or also as characteristic X rays may be very important. Indeed bremsstrahlung becomes predominant when the energy of the secondary electrons approaches 10 Mev in heavy materials or 100 Mev in light materials; we have here the cascade-shower mechanism. For a treatment of X-ray emission by electrons see Ref. [*45*] and this Encyclopedia, Vol. XXXV, p. 409ff.; no collection of critical quantitative data over all energy ranges is available.

Positrons, generated by X rays through pair production, always end by producing more X rays as *annihilation radiation*. Except for occasional annihilation in flight or near nuclei, this radiation consists of 0.511 Mev photons emitted in pairs in opposite but otherwise random directions.

In the case of high energy X rays one must also consider neutrons, which are released in appreciable amounts through photonuclear processes. The neutrons are eventually captured by other nuclei, and *capture γ rays* are then usually released, one or more photons per process, with substantially isotropic distribution. For example, photoneutrons released from deuterium contribute heavily to the propagation of X rays in the water shielding of nuclear reactors.

β) Energy transport by secondary corpuscular radiations. X rays of less than 1 Mev transfer nearly all their energy to secondary electrons whose range is much smaller than the mean free path of the X rays themselves. Therefore the energy given to electrons may be regarded as dissipated "locally"[1]. At higher energies the range of electrons (and positrons) becomes quite substantial. The problem of electron penetration becomes then an essential part of the problem of X-ray penetration; indeed the two problems merge completely in the case of cascade showers. All electron-induced processes become then relevant to X-ray penetration[2].

Similarly, secondary neutrons are of importance because they travel long distances; but the solution of neutron penetration problems involving degradation has so far only been achieved for selected problems[3], and is handicapped by the lack of exact knowledge of elementary cross sections. On the other hand, protons and other charged particles ejected by X rays through photonuclear processes take, on the whole, a minor fraction of the incident energy and do not travel far.

4. Probability of X-ray processes. The cross sections for elementary processes induced by X rays are usually measured in "good" geometry, which is called "narrow-beam" geometry in the X-ray literature. However, the knowledge of multiple processes has reached the point where "poor"—or "broad beam" geometry experiments may provide the best data on elementary cross sections[4].

[1] The analysis of the distribution of energy dissipated by a radiation concerns usually the density of energy dissipation averaged over a volume, e.g. over the volume of an instrument, which need not be very small. The type of analysis which is required depends on the size of this volume, and this size depends, in turn, on the purpose at hand. The term "locally" may be used to mean "within the volume of interest".

[2] The theory of cascade showers has been treated by G. Molière in Vol. XLVI of this Encyclopedia.

[3] Cf. the preceding article by E. Amaldi, Part C.

[4] Measurements of the residual X-ray intensity after very deep penetration provide a sensitive index of the total probability of interactions; a 1% difference in this probability changes by 10% the probability of traveling 10 mean free paths without interaction. For very deep penetration, it is difficult to obtain perfect narrow beam geometry. Other geometries, specifically those with spherical symmetry, lend themselves better to theoretical analysis. The data discussed in Sect. 19 on the deep penetration of γ rays in lead with spherical geometry have yielded indirectly a very accurate determination of the "narrow beam" attenuation coefficient [*36*].

The single most important datum is the total probability of all processes which may be produced by photons of a given energy. This probability is measured by observing the attenuation curve $I(x)$ of a monochromatic narrow beam as a function of the thickness x of an absorber. Because the elementary processes in different layers of an absorber occur independently, successive layers of equal thickness must produce equal attenuation. This is stated by the well known law

$$dI = -\mu\, dx\, I, \quad I(x) = e^{-\mu x} I(0). \tag{4.1}$$

The quantity μ is called the narrow-beam attenuation coefficient[1], to be expressed in units reciprocal to the thickness (cm^{-1} or, if the thickness is expressed in mass per unit area, cm^2/g). When μ is expressed in cm^2/g it may be called mass attenuation coefficient. The thickness $(\log 2)/\mu$ which reduces $I(x)/I(0)$ to $\frac{1}{2}$ is called "half-value layer", the thickness $1/\mu$ which reduces it to $e^{-1}=0.368$ is the "mean free path" of the photons, also called "relaxation length".

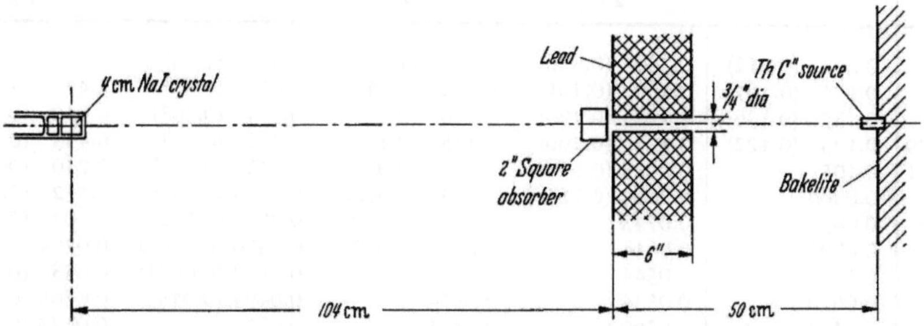

Fig. 2. Experimental arrangement of COLGATE for measuring attenuation coefficients [23].

Among the few modern, accurate measurements of μ is the work of COLGATE [23], which extends over six photon energies from 0.411 to 17.5 Mev and eight elements from C to U. A typical experimental arrangement is shown in Fig. 2. The X rays were detected by a scintillation spectrometer with a differential pulse analyzer to eliminate the detection of stray radiation. Despite the good geometry, narrow beam conditions were difficult to attain because, in this energy range, coherently scattered X rays remain close in direction to the incident beam. To evaluate this effect, apparent values of μ were measured under progressively better geometric conditions corresponding to increasing source-to-detector distance, and the actual value was derived by extrapolation to infinite distance. For a review of other important measurements see Ref. [96].

The probabilities that a photon suffers an interaction process in different atoms of an absorber generally add up without mutual disturbance[2]. Therefore the attenuation coefficient μ is determined by the sum of the cross sections σ for all atoms in a layer of absorber of unit area and unit thickness. If there are N identical atoms in this unit volume, then

$$\mu = N\sigma. \tag{4.2}$$

If the absorber is a chemical compound or mixture, its attenuation coefficient μ is an average of the coefficients for the constituent elements weighted in proportion to the abundance of each element by volume. For the mass attenuation coefficient, the weighting is in proportion to the abundance by weight.

[1] The term "attenuation coefficient" is used here instead of "absorption coefficient" to emphasize that μ includes the effect of X-ray removal by scattering in addition to outright absorption by processes of class A.

[2] Exception must be made for the effect of chemical binding upon the contribution of valence electrons to Rayleigh scattering and for the coherent combination of Rayleigh scattering by different atoms, which matter only under special circumstances, particularly in crystals.

At present, practical applications require cross section data with an accuracy of about 1%. Existing data, theoretical and experimental, attain or approach this accuracy over most of the range of interest. However, there are limited ranges where the accuracy is only of the order of 10%.

For details on the cross sections for individual processes, see Ref. [45] and the pertinent chapters of this Encyclopedia[1]. The total cross section for all elementary processes between 10 kev and 100 Mev, and the relevant aspects of the partial cross sections, are the objects of a critical continuing survey and tabulation of data [96]. The following paragraphs summarize the results of that survey, to which the reader is referred for further information. Table 2 gives

Table 2. *Total attenuation coefficient μ (cm^2/g), according to* WHITE [96].

Energy (Mev)	H_2O	Al	Fe	Sn	Pb
	Density (g cm^{-3})				
	1.0	2 7	7.9	7 3	11.3
0.05	0.222 (0.212)	0.360 (0.326)	1.93 (1.83)	10.5 (10.2)	
0.1	0.171 (0.167)	0.169 (0.161)	0.372 (0.344)	1.65 (1.58)	5.47 (5.29)
0.2	0.137 (0.136)	0.122 (0.120)	0.146 (0.138)	0.324 (0.303)	0.942 (0.896)
0.279	0.123 (0.122)	0.107 (0.106)	0.115 (0.111)	0.182 (0.171)	0.445 (0.416)
0.411	0.105	0.0915 (0.0906)	0.0925 (0.0903)	0.112 (0.107)	0.210 (0.196)
0.5	0.0966	0.0844 (0.0840)	0.0840 (0.0828)	0.0924 (0.0886)	0.152 (0.145)
0.66	0.0855	0.0743	0.0734 (0.0725)	0.0744 (0.0721)	0.107 (0.102)
1.0	0.0706	0.0614	0.0599 (0.0595)	0.0576 (0.0568)	0.0704 (0.0684)
1.28	0.0625	0.0544	0.0525	0.0502 (0.0491)	0.0580 (0.0568)
1.33	0.0610	0.0534	0.0514	0.0494 (0.0485)	0.0565 (0.0555)
2.76	0.0413	0.0369	0.0371	0.0373	0.0429 (0.0425)
3	0.0396	0.0353	0.0360	0.0367	0.0423 (0.0421)
5	0.0301	0.0282	0.0313	0.0355	0.0426
10	0.0219	0.0229	0.0294	0.0383	0.0489

Entries in parentheses have been obtained by subtracting the contribution of coherent scattering. This contribution is not significant at high energies where a single entry is given. An application of the reduced attenuation coefficients is indicated in note 1 on p. 736.

selected values from Ref. [96] which are required for the interpretation of the numerical results presented in this article.

α) Photoelectric absorption. X rays with photon energy below 5 to 10 kev are of little practical importance because their photoelectric absorption is exceedingly strong. Information on this process is scarce, not only experimentally, but also theoretically because the usual hydrogen-like approximation of the wave functions has low accuracy in this range. The situation improves above 10 kev but the accuracy remains of the order of 10% only, up to about 50 kev. The photoelectric absorption is a rapidly increasing function of the atomic number. This is because high-Z atoms have a strong nuclear attraction, so that inner-shell electrons are subject to strong forces capable of resisting the X-ray disturbance, a condition essential for X-ray absorption. The absorption decreases rapidly as the photon energy E increases, except that it jumps up at the "edges" where the energy becomes sufficient to eject inner-shell electrons. Well above the K-edge, $\sigma_{\text{photo}}(E)$ is roughly proportional to $E^{-3.5}$, but then the down-trend decreases to about E^{-1} in the relativistic range, for $E \gtrsim 500$ kev [2]. From about

[1] Cf. the contribution of R.D. EVANS to Vol. XXXIV, and several articles in Vol. XXXI.

[2] For $E \gtrsim 500$ kev the absorption takes place very near to the nucleus where the electrons move with relativistic velocity. Here the electron density is much larger than predicted by non-relativistic mechanics. Indeed it diverges near the nucleus like $r^{\sqrt{1-Z^2/137^2}-1}$ instead of being constant; from this follows the slower decrease of $\sigma_{\text{photo}}(E)$.

50 to 500 kev σ_{photo} is known to about 1%. At still higher energies the information becomes quite poor again, but here the photoelectric absorption is overshadowed by other processes.

β) Compton effect. The ordinary theory of this effect deals with X-ray scattering by electrons assumed to be free from bonds with other particles and initially at rest. These assumptions are reasonably well fulfilled under most conditions where the Compton effect is important (see Fig. 1). The elementary cross section depends on the energy and direction of the scattered photon, but these variables are related by the Compton relationship. The photon energy can be indicated by the corresponding wavelength in units of the Compton wavelength $\left(\frac{h}{mc} = 2.43\times 10^{-10}\,\text{cm}\right)$,

$$\lambda = \frac{m c^2}{E} \tag{4.3}$$

where $mc^2 = 0.511$ Mev. Calling λ and λ' the wavelengths of the incident and scattered photons respectively, and ϑ the deflection from the direction of incidence, the differential cross section for unpolarized photons is given by the Klein-Nishina formula[1]

$$\frac{d\sigma}{d\lambda'\,d\Omega} = \frac{1}{2}\left(\frac{e^2}{m c^2}\right)^2\left(\frac{\lambda}{\lambda'}\right)^2\left[\frac{\lambda'}{\lambda} + \frac{\lambda}{\lambda'} + (\lambda'-\lambda)(\lambda'-\lambda-2)\right]\delta(1-\cos\vartheta+\lambda-\lambda'), \tag{4.4}$$

where $\delta(x)$ is the Dirac function and represents the Compton law of wavelength shift, $d\Omega$ is an element of solid angle, and,

$$\left(\frac{e^2}{m c^2}\right)^2 = 7.94\times 10^{-26}\,\text{cm}^2. \tag{4.5}$$

When the momentum transfer from the photon to the electron is $\ll mc$, that is, either for low photon energy ($\lambda \gg 1$) and any deflection, or for small deflection ($1-\cos\vartheta \ll \lambda$) and any energy, (4.4) reduces to the Thomson cross section for scattering with negligible wavelength shift

$$\frac{d\sigma_{\text{Th}}}{d\lambda'\,d\Omega} = \frac{1}{2}\left(\frac{e^2}{m c^2}\right)^2 (1+\cos^2\vartheta)\,\delta(1-\cos\vartheta+\lambda-\lambda'). \tag{4.6}$$

The integral cross section σ_{Th} obtained from (4.6) is

$$\sigma_{\text{Th}} = \frac{8\pi}{3}\left(\frac{e^2}{m c^2}\right)^2 = 6.653\times 10^{-25}\,\text{cm}^2, \quad \text{for } \lambda \gg 1. \tag{4.7}$$

At higher photon energies the integral of (4.4) decreases steadily but not very rapidly according to the formula

$$\sigma = \frac{3}{8}\sigma_{\text{Th}}\,\lambda\left[(1-2\lambda-2\lambda^2)\log\left(1+\frac{2}{\lambda}\right) + 2\,\frac{1+9\lambda+8\lambda^2+9\lambda^3}{(\lambda+2)^2}\right], \tag{4.8}$$

which reduces to

$$\sigma \sim \frac{3}{8}\sigma_{\text{Th}}\,\lambda\log\frac{3.28}{\lambda}, \quad \text{for } \lambda \ll 1. \tag{4.9}$$

The radiative corrections to (4.4) are of the order of 1% or less. Various, though not all, aspects of the expression have been experimentally verified over a wide range of λ and ϑ at least to this accuracy[2]. Further details, references

[1] O. Klein and Y. Nishina: Z. Physik **52**, 852 (1929).

[2] R. Hofstadter and J.A. McIntyre: Phys. Rev. **76**, 1269 (1949); **78**, 24; **80**, 631 (1950). — W.C. Parkinson: Phys. Rev. **76**, 1348 (1949). — W.G. Cross and N.F. Ramsey: Phys. Rev. **80**, 929 (1950).

and very extensive graphical data on the Compton law and on the Klein-Nishina formula are given in Ref. [*64*].

The assumption of free electrons initially at rest that underlies the Klein-Nishina formula is approximately valid only if the momentum transferred to the electron greatly exceeds the initial momentum of the electron's motion within an atom. In terms of the wavelengths of the incident radiation (λ) and of the atomic electron (λ_{el})[1], this condition is

$$\frac{\lambda}{2\sin\frac{\vartheta}{2}} \ll \lambda_{el}. \tag{4.10}$$

When this condition is not satisfied, Compton scattering is complicated by the binding of the electrons and by their motion within the atom. As a result, Compton scattering becomes less probable than predicted by the Klein-Nishina law, and the sharp Compton relationship between λ, λ' and ϑ is relaxed.

The corrections for the binding effects can be analyzed in the impulse approximation, in which one first considers the probability of scattering by a certain angle as given by the usual cross section. The momentum transferred from the photon to the electron equals approximately $2\frac{mc}{\lambda}\sin\frac{\vartheta}{2}$, provided $\frac{1}{\lambda}(1-\cos\vartheta) \ll 1$. One may then ask whether upon receiving this momentum the bound electron will actually absorb energy—thus completing an inelastic process—and, if so, how much energy it will absorb. The probability that the scattering is inelastic is represented for each material by a function of the momentum transfer called *incoherent scattering function*. This function is poorly known and is discussed in an appendix to Ref. [*96*]. If the scattering is inelastic the Compton law is relaxed to the extent that the wavelength shift may be anywhere within the approximate range

$$\lambda' - \lambda \sim (1-\cos\vartheta)\left[1 \pm \frac{\lambda}{\lambda_{el}\sin\frac{\vartheta}{2}}\right] \tag{4.11}$$

instead of being rigorously equal to $1-\cos\vartheta$. For further information and references see Ref. [*64*].

γ) Rayleigh (coherent) scattering. The scattering of X rays by atomic electrons becomes predominantly elastic (i.e. coherent) just under the conditions of low momentum transfer where the inelastic Compton scattering becomes unlikely, that is, when (4.10) is violated and

$$\frac{\lambda}{2\sin\frac{\vartheta}{2}} \gtrsim \lambda_{el}. \tag{4.12}$$

If each atomic electron contributed to elastic scattering independently of the others, the onset of this process for low momentum transfer would compensate exactly for the reduction of Compton scattering. This follows from the circumstance that (4.12) insures the applicability of the Thomson formula (4.6) which represents the elastic scattering by free electrons when the wavelength shift is disregarded.

However, the various atomic electrons contribute *coherently* to elastic scattering, whose probability is thereby boosted greatly by constructive interference. If the constructive interference extends only to the Z electrons of each atom, the probability increase amounts to a factor f^2 (where f is called the *form factor*); f^2 rises to Z^2 for $\frac{\lambda}{\left|2\sin\frac{\vartheta}{2}\right|}$ much larger than the atomic diameter. For recent calculations and a discussion of the form factor see Ref. [*65*]. To the extent that the atoms of a material are packed regularly, the constructive interference

[1] The wavelength λ_{el} of an atomic electron is, of course, of the order of magnitude of its orbit.

extends to the electrons of different atoms. The total intensity of coherent scattering is thereby increased still further and becomes concentrated in a narrow beam in the forward direction plus other beams in directions dictated by the packing of atoms, as in the extreme case of Bragg reflection by crystal lattices.

Even for high energy photons there is a narrow cone of forward directions ($\vartheta \sim 0$) where coherent scattering predominates, but the total cross section integrated over this cone is negligible for most purposes. For low energy X rays the coherent scattering constitutes much or most of the scattering, but is overshadowed by photoelectric absorption. Thus Rayleigh scattering predominates among the X-ray processes only under special circumstances as in the case of Bragg reflection.

Thus far we have visualized Rayleigh scattering simply as the coherent combination of the Thomson scattering by all electrons, each of them regarded as free. This model disregards the anomalous dispersion effects due to the forces that bind the electrons and cause the photoelectric absorption. These effects are far from negligible, but have not been evaluated extensively or in detail[1]. A program of detailed calculations for Rayleigh scattering by Pb above 100 kev is in progress[2] and some accurate experiments are available in this range[3].

δ) Pair production. This process is energetically possible only for photons with energy above 1.02 Mev, sufficient to provide the rest-mass energies of the electron and positron that are created. The cross section for pair production in matter starts from zero at the threshold energy $E' = 2mc^2$ and rises as the photon energy E increases, at first slowly and then rapidly, i.e. faster than $E - E'$. Eventually the rise flattens out and the cross section approaches a maximum value for very high energy photons. In the rising portion, the cross section for pair production in materials of different atomic number Z is roughly proportional to Z^2, but the rise flattens out earlier in high-Z than in low-Z materials, so that the maximum values increase with Z a little more slowly than in proportion to Z^2. The maximum values are of the order of $Z^2\, 10^{-26}$ cm². Owing to the Z^2 factor, the energy at which pair production becomes the predominant process is lower in heavy materials than in light materials (e.g. at ~ 5 Mev in Pb and ~ 50 Mev in C, see Fig. 1).

Most of the pair production processes occur in the inner part of the atom where the electric field of the nucleus predominates, and the nucleus takes up a recoil momentum with the atomic electrons remaining rigidly attached to it[4]. The process can also be induced by the field surrounding an individual electron, which then recoils out of the atom. The frequency of pair production near an electron is comparable to that near a nucleus with $Z = 1$. Accordingly, all electrons together contribute to pair production, as compared to the nuclei, in a ratio of the order of (but *not* equal to) $Z/Z^2 = 1/Z$.

---

[1] See e.g., L. G. Parratt and C. F. Hempstead: Phys. Rev. **94**, 1593 (1954).

[2] G. E. Brown *et al.*: Proc. Roy. Soc. Lond., Ser. A **227**, 51, 59 (1954); Ser. A **234**, 387 (1956).

[3] A. K. Mann: Phys. Rev. **101**, 4 (1956).

[4] Pair production processes are classified operationally according to the effect they have on an atom, rather than according to the kind of particle which takes part in them. The treatment of "pair production in the screened field of a nucleus" implies that the process has no direct effect on the internal motion of the atom. That is, the process is elastic from the standpoint of the atom and the atom behaves in it as a rigid body. The "screened field" includes the average (static) field of the electrons. On the other hand, when pair production has a direct effect on the intra-atomic motion, the process is inelastic from the standpoint of the atom. The disturbance of intra-atomic motion manifests itself as an electron recoil, because electrons are light, and one speaks then of pair production "in the field of an electron".

The most detailed and complete calculations of pair production rely on the Born approximation and are accurate only for high energy photons and low-Z materials ([45], p. 256ff.). Corrections valid for high-Z materials have been worked out[1]. At lower energies a few isolated values have been computed. Experiments have verified the theory in the main and provide isolated additional data. Judicious interpolation utilizing all the available data provides cross sections for all materials and energies, which are dependable to within a very few percent [96].

ε) Nuclear processes. Information on photonuclear absorption and scattering is still fragmentary and of low accuracy. Nuclear theory provides a basis for estimates rather than accurate calculations (except, to some extent, for the deuteron). Experimentation has been limited severely by the lack of adequate sources of mono-energetic X rays. Most experiments have been conducted with continuous spectra of maximum energies up to 25 Mev.

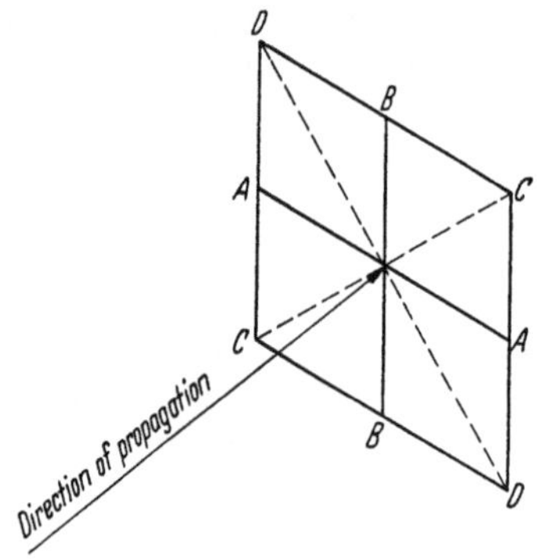

Fig. 3. Diagram illustrating the definition of the Stokes parameters.

Photonuclear absorption by most nuclei has a maximum probability for 15 to 20 Mev photons. The maximum cross section varies roughly in proportion to the atomic number; thus it increases more slowly than the pair production cross section and is comparatively more important for light than for heavy elements. For carbon this maximum amounts to about 10% of the total cross section for electronic processes. Data on photonuclear absorption in this energy range are collected in recent papers[2].

At lower photon energies, below the nuclear disintegration threshold, absorption and scattering are concentrated in sharp resonance lines. Within the lines the nuclear cross section can be very large, well in excess of the cross section for electronic processes[3]. However, the aggregate importance of nuclear processes averaged over any appreciable spectral range is rather small.

ζ) Polarization effects. The intensity and the degree and kind of polarization of an X-ray beam can be identified together by a minimum of 4 parameters. The Stokes parameters $I^{[i]}$ ($i=0, 1, 2, 3$) are convenient for this purpose[4]. Of these, $I^{[0]}$ indicates the intensity of the flux and is identical with the quantity I previously introduced; $I^{[1]}$ indicates the difference between the intensity transmitted by an ideal filter accepting only X rays polarized in a zero-azimuth plane (A in Fig. 3) and the intensity transmitted by the same filter at 90° azimuth (B), $I^{[2]}$ indicates the corresponding difference for the azimuths 45° (C) and 135° (D); $I^{[3]}$ indicates the intensity difference for filters transmitting right-circular and left-circular polarization. The Stokes parameters $I^{[1]}$, $I^{[2]}$ and $I^{[3]}$ vanish for unpolarized radiation. The set of parameters $I^{[i]}$ may be replaced by any orthogonal substitution. In particular a rotation of the 0°-azimuth plane by an angle φ induces, for the parameters $I^{[1]}$ and $I^{[2]}$, the same substitution as the components of plane vectors experience when the coordinates are rotated by 2φ,

$$I^{[1]\prime} = I^{[1]} \cos 2\varphi + I^{[2]} \sin 2\varphi, \qquad I^{[2]\prime} = -I^{[1]} \sin 2\varphi + I^{[2]} \cos 2\varphi. \tag{4.13}$$

---

[1] H. Davies, H. A. Bethe and L. C. Maximon: Phys. Rev. **93**, 788 (1954). — Cf. also Vol. XXXI, this Encyclopedia.

[2] R. Nathans and J. Halpern: Phys. Rev. **93**, 437 (1954). — R. Montalbetti, L. Katz and J. Goldemberg: Phys. Rev. **91**, 659 (1953).

[3] E. Fuller and E. V. Hayward: Phys. Rev. **101**, 64 (1956); **98**, 1537 (1955), and private communication. This reference provides, on the basis of scattering experiments, much evidence on all photonuclear processes up to 40 Mev.

[4] See e.g., U. Fano: J. Opt. Soc. Amer. **39**, 859 (1949). — Phys. Rev. **93**, 121 (1954).

The total cross section of any X-ray induced process depending on polarization is represented correspondingly by a set of four parameters $\sigma^{[i]}$, of which $\sigma^{[0]}$ is the cross section for unpolarized X rays. The total probability of the process is given, as a 4-vector product, by

$$\sum_{i=0}^{3} \sigma^{[i]} I^{[i]} = \boldsymbol{\sigma} \cdot \boldsymbol{I}. \tag{4.14}$$

A process which scatters or in any way modifies the radiation is represented by a 4×4 differential cross section matrix $\|\sigma^{[ij]}\|$. Thus we write the Stokes parameters of scattered radiation, at a distance r from the scatterer, in the form

$$I^{[i]}_{\text{scatt}} = r^{-2} \sum_{j=0}^{3} \sigma^{[ij]} I^{[j]}_{\text{incoming}} \tag{4.15}$$

or, in tensor notation, as the dyadic product

$$\boldsymbol{I}_{\text{scatt}} = r^{-2} \|\sigma\| \cdot \boldsymbol{I}_{\text{incoming}}. \tag{4.16}$$

The total cross section for each elementary process in an isotropic material is independent of the polarization of the incident X rays. That is, of the 4 components of the cross section $\sigma^{[i]}$ all but $\sigma^{[0]}$ vanish. Similarly, the attenuation coefficient μ of (4.1), when regarded as a polarization matrix like $\|\sigma\|$ in (4.16), is simply a multiple of the unit matrix.

On the other hand, the cross sections for X-ray induced processes in which a particle or photon is ejected in a specified direction or with a specified polarization are, in general, dependent on the X-ray polarization. Thus, the cross section for photoelectric absorption with ejection of an electron in a specified direction has two non-vanishing components, $\sigma^{[0]}$ and $\sigma^{[1]}$, if the zero-azimuth plane is chosen to contain the electron direction[1]. Similarly the cross section for production of an electron-positron pair in specified directions should depend on the polarization, but this effect has not been observed.

The polarization dependence of Compton scattering has been calculated in various degrees of detail and expressed in various forms[2]. The matrix form of the complete Klein-Nishina cross section is given here for the case where the electron spin orientation is random before the scattering and is not observed afterwards,

$$\left.\begin{aligned}
&\left\|\frac{d\sigma}{d\lambda'\,d\Omega}\right\| = \frac{1}{2}\left(\frac{e^2}{m c^2}\right)\delta(1-\cos\vartheta-\lambda'+\lambda)\left(\frac{\lambda}{\lambda'}\right)^2 \times \\
&\times \left\|\begin{matrix}
\frac{\lambda'}{\lambda}+\frac{\lambda}{\lambda'}+(\lambda'-\lambda)(\lambda'-\lambda-2) & (\lambda'-\lambda)(\lambda'-\lambda-2) & 0 & 0 \\
(\lambda'-\lambda)(\lambda'-\lambda-2) & 2+(\lambda'-\lambda)(\lambda'-\lambda-2) & 0 & 0 \\
0 & 0 & 2(1-\lambda'-\lambda) & 0 \\
0 & 0 & 0 & \left(\frac{\lambda'}{\lambda}+\frac{\lambda}{\lambda'}\right)(1-\lambda'+\lambda)
\end{matrix}\right\|.
\end{aligned}\right\} \tag{4.17}$$

Here the plane of scattering is taken as zero-azimuth for the polarizations of both the incident and scattered photons. A different choice of zero-azimuth for the incident (or scattered) photon causes a linear substitution in the two center columns (or rows) of this matrix.

When X rays interact with an anisotropic material, even the total cross sections for elementary processes may depend on the polarization, and are properly represented as 4-component quantities $\sigma^{[i]}$. The narrow-beam attenuation coefficient is then a 4×4 matrix $\|\mu\|$, like $\|\sigma\|$ in (4.15) and (4.16); it represents not only the attenuation but also the changes of polarization, if any, of the radiation Eq. (4.1) remains valid with the understanding that I is a 4-component quantity and $\|\mu\|$ an operator applied to it.

The matrix $\|\mu\|$ is conveniently resolved into a symmetric and an antisymmetric part. The antisymmetric part represents the optical activity of the material and is quite negligible for X rays. The symmetric part represents differences in absorption of X rays with different polarizations. This part can be expressed in simple manner in terms of the total cross section

[1] Theory F. SAUTER: Ann. Physik **11**, 454 (1931); experiment: W.H. MCMASTER and F.L. HEREFORD: Phys. Rev. **95**, 723 (1954).

[2] Y. NISHINA: Z. Physik **52**, 869 (1929). — W. FRANZ: Ann. Physik **33**, 689 (1938). — U. FANO: J. Opt. Soc. Amer. **39**, 859 (1949). — F.W. LIPPS and H.A. TOLHOEK Physica, Haag **20**, 395 (1954).

components $\sigma^{[i]}$,

$$\|\mu_{\text{sym}}\| = N \begin{vmatrix} \sigma^{[0]} & \sigma^{[1]} & \sigma^{[2]} & \sigma^{[3]} \\ \sigma^{[1]} & \sigma^{[0]} & 0 & 0 \\ \sigma^{[2]} & 0 & \sigma^{[0]} & 0 \\ \sigma^{[3]} & 0 & 0 & \sigma^{[0]} \end{vmatrix}. \tag{4.18}$$

The matrix can be diagonalized by an orthogonal substitution of the Stokes parameters and has the eigenvalues $\sigma^{[0]}$, $\sigma^{[0]}$, and $\sigma^{[0]} \pm \sqrt{\sigma^{[1]^2} + \sigma^{[2]^2} + \sigma^{[3]^2}}$. Thus there will be one type of polarization (e.g. right circular) which has the maximum absorption, another one opposite to it (left circular) which is absorbed least, whereas Stokes parameters pertaining to polarizations "intermediate" between these two (i.e. linear polarizations) decay with average absorption. The non-vanishing off-diagonal elements in the first column of (4.18) indicate that some polarization is generated by differential filtration even though the incident X rays are unpolarized.

Magnetized materials are anisotropic with respect to X rays because the electron spins within them are oriented prevalently in the direction of magnetization and the total cross section for Compton scattering depends on the spin orientation. Differential filtration of 2.62 Mev γ rays by an Armco-iron magnet core has been observed[1]. From the detailed calculation of the Klein-Nishina cross section it follows that $\sigma^{[1]}$ and $\sigma^{[2]}$ vanish after integration over the azimuth of the scattering plane, that

$$\sigma^{[3]} = \frac{3}{8}\,\sigma_{\text{Th}}\,\lambda\left[(1+\lambda)\log\left(1+\frac{2}{\lambda}\right) - 2\left(\frac{5+4\lambda+\lambda^2}{(\lambda+2)^2}\right)\right] S_x, \tag{4.19}$$

where S_x represents the degree of spin polarization in the direction of incidence, and that $\sigma^{[0]}$ is given by the usual formula (4.8). The Gunst-Page experiment has verified the value of $\sigma^{[3]}$ to within $\sim 10\%$.

η) Energy absorption. As has been pointed out in Sect. 3β, the absorption of X-ray energy in a medium is a two stage process, involving the transfer of energy from the X-ray flux to corpuscular radiation followed by further dissipation of this energy by corpuscular radiation. Under the conditions considered in this article, the travel of the corpuscular radiation is usually negligible, so that as a rule we will consider only the first stage of the absorption process[2]. The usual attenuation coefficient μ is not the appropriate quantity for calculating the energy dissipation because it represents only the removal of photons from the beam, without reference to the subsequent disposal of their energy. To evaluate the rate of energy dissipation one should weight the probability of each photon-induced process by the average fraction of the photon's energy which is dissipated locally.

To this end one should resolve the attenuation coefficient μ into the contributions of different processes and reduce each contribution to account for the estimated escape of energy. Thus for Compton scattering one must discount the energy of the scattered photon, for the photoelectric effect the energy of fluorescence, etc. The total μ corrected in this manner has been called in the past the "true" absorption coefficient and more recently the *energy absorption coefficient*. We denote it by μ_D, or μ_d in the case of air. This quantity is of great importance for X-ray dosimetry. Compilations of μ_D can be found in Refs. [*30*] and [*63e*]. An extract from these data is given in Table 3.

The product of the X-ray flux and of the energy absorption coefficient, integrated over all spectral energies, represents the energy transferred from X rays to secondary electrons and will be denoted as "energy dissipation" per unit mass or per unit volume, depending on the units in which μ_D is expressed.

[1] S.B. Gunst and L.A. Page: Phys. Rev. **92**, 970 (1953).

[2] The motion of corpuscular radiation has been taken into account in the treatment of the penetration of 40-Mev bremsstrahlung in Sect. 19α, and is also important in the experiments under cascade shower conditions discussed in Sect. 29.

Instead of the amount of energy transferred from X rays to electrons, radiological physicists usually consider two related quantities: (a) The amount of energy locally absorbed by the medium from secondary electrons (absorbed dose), or (b) the amount of ionization which the secondary electrons produce in a suitable probe material, usually air (exposure dose[1]). In contrast to (a), (b) does not take into account the displacement of the secondary electrons.

Table 3. *Energy absorption coefficient* $\mu_D (cm^2/g)$. The values for water and air are from Ref. [*63e*], and those for the other materials from Ref. [*30*].

E (Mev)	H_2O	Air (μ_d)	Al	Fe	Pb
0.02	0.523	0.516			
0.05	0.0394	0.0384			
0.1	0.0252	0.0231	0.0371	0.219	2.16
0.2	0.0300	0.0268	0.0275	0.0485	0.586
0.5	0.0330	0.0297	0.0287	0.0293	0.0901
1	0.0311	0.0280	0.0269	0.0261	0.0384
2	0.0260	0.0234	0.0277	0.0219	0.0248
5	0.0190	0.0173	0.0180	0.0198	0.0272
10	0.0155	0.0144	0.0167	0.0213	0.0328

If the travel of secondary electrons is negligible, the energy dissipation computed with the energy absorption coefficient μ_D is equal to the absorbed dose. The energy dissipation computed with the energy absorption coefficient μ_d for air and divided by the mean energy required for ionization, $\varepsilon \approx 34$ ev/ion pair, is equal to the exposure dose.

The comparison between theory and experiment will most frequently be made in terms of dimensionless ratios such as build-up factors. For these the difference between energy dissipation, absorbed dose, and exposure dose tends to vanish.

B. General theory.

5. Qualitative aspects. Given a distribution of X-ray sources within a material, the problem of penetration consists of determining the intensity and spectrum of the X rays which flow in each direction at each point of the material. It is sufficient to solve the problem separately for sources of monoenergetic photons of different energies because radiations from different sources propagate independently, and to solve the problem for *point sources* of collimated photons because any source can be regarded as an aggregate of point collimated elements. For example, an aggregate of identical point sources distributed uniformly over a plane constitutes a *plane source*. Conversely any point source can be represented mathematically as a superposition of plane sources (Sect. 6α). This superposition simplifies the penetration problem in infinite uniform materials, because the

[1] The following definitions have been adopted by the International Committee on Radiation Units [*63e*]:

"*Absorbed dose* of any ionizing radiation is the energy imparted to matter by ionizing particles per unit mass of irradiated material at the place of interest.

The unit of absorbed dose is the *rad*. One rad is 100 ergs per gram."

"*Exposure dose* of X or γ radiation at a certain place is a measure of the radiation that is based on its ability to produce ionization.

The unit of exposure dose is the roentgen (r). One roentgen is an exposure dose of X or γ radiation such that the associated corpuscular emission per 0.001293 gram of air produces, in air, ions carrying 1 electrostatic unit of quantity of electricity of either sign."

X-ray flux generated by a plane source in such a material is uniform over any plane parallel to the source. Most studies have dealt with X rays from high-symmetry sources: plane perpendicular, plane isotropic or point isotropic.

X-ray photons arriving at a point directly from a source are called *primaries*. Secondary X rays of different energies arise from the processes reviewed in Part A. Therefore, X-ray penetration involves photons of various energies even when the primaries are monoenergetic.

α) Penetration of primaries. Differential filtration. At any point of a material the flux of primary X rays can be easily calculated, since the primaries are removed, in their travel from their source to the point, according to the simple, exponential, narrow-beam law (4.1). Despite the complications introduced by secondary radiation, the exponential attenuation of the primaries usually constitutes the single most important influence on X-ray penetration.

Table 4. *Energy E_m at which the total attenuation coefficient attains its minimum value μ_m.*

Material	μ_m cm² g⁻¹	E_m Mev	Material	μ_m cm² g⁻¹	E_m Mev
H_2O	0.0161	51.0	W	0.0402	3.65
Al	0.0210	21.0	Pb	0.0419	3.62
Fe	0.0294	9.95	U	0.0439	3.75
Sn	0.0354	4.56			

Primary photons of different energies are attenuated at rates corresponding to their respective values of the coefficient μ. The X rays with higher μ, which are usually of lower energy, are removed first, i.e. filtered out, and are said to be "softer". Fig. 4 shows an example of differential filtration; curve (b) represents the attenuation of a narrow beam of X rays with continuous spectrum and its slope, which indicates an average value of μ over the spectrum at each point, decreases as the attenuation progresses. Secondary X rays are usually softer than the primaries and contribute less to the overall penetration.

However, at high energies, where pair production predominates, μ decreases with decreasing energy. Therefore, when the source photons have very high energy, there will arise secondaries with the energy indicated in Table 4 which are harder than the primaries and contribute most to the deep penetration.

β) Accumulation of secondaries. Broad-beam attenuation. The production and penetration of secondary X rays constitutes the main non-trivial feature in the analysis of X-ray penetration. Compton scattering is the main source of secondaries, especially in the range from 50 kev to 10 Mev and in low-Z materials; e.g., for 1 Mev photons in water the average number of Compton processes prior to photoelectric absorption is ~ 14. The photon energy is reduced (degraded) in each process. This reduction is much smaller, on the average, for low than for high energy photons, according to (4.4). Thus photons accumulate in the lower portion of the spectrum, but this accumulation is limited on the low energy side by the onset of photoelectric absorption.

The accumulation of secondaries is apparent in the experiments of broad-beam attenuation, where an uncollimated or poorly collimated X-ray beam penetrates a thick absorber. Here the narrow-beam attenuation of the primaries is partly compensated by the accumulation of secondaries, especially in the initial stages of penetration, so that the semi-logarithmic plot of intensity *vs.* thickness, curve (c) of Fig. 4, has increasing slope.

As an index of the accumulation of secondaries one considers the ratio of the total amount of X rays at any one point to the amount of primaries only. This ratio is called the *build-up factor*. For this purpose one may take the "amount" of X rays to mean flux of photons, flux of energy or some other quantity. Thus an appropriate build-up factor should be defined for each device used to measure the X rays. Fig. 5 shows the results of an exposure dose build-up study in simple spherical geometry with a point source of Co^{60} γ rays in a large water tank.

γ) *Quasi-equilibrium.* The progressive accumulation of secondary X rays, in the course of penetration away from the source, constitutes a sort of transient process and one might expect it to be followed at greater depth by some steady state[1]. Indeed, steady states where softer radiations continually arise from harder ones, as photo-electrons are generated by X rays, are familiar and are called states of radiative equilibrium. They are analogous to the equilibrium state of a radioactive substance which is steadily generated by a "parent" activity with longer half-life.

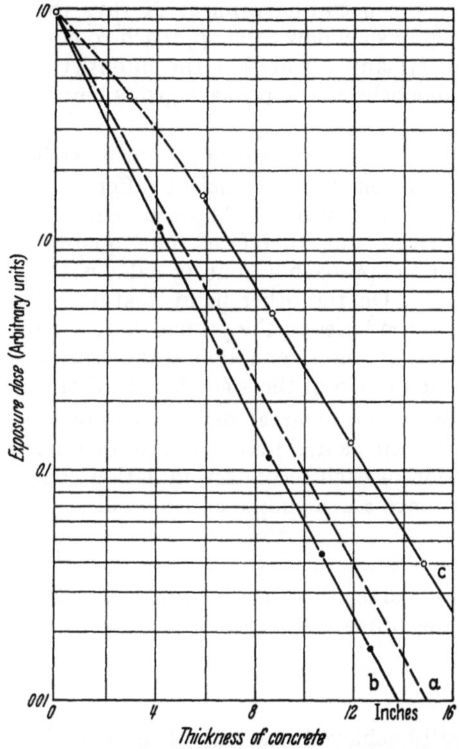

Fig. 4. Differential filtration curves in concrete (WYCKOFF, private communication): (a) Monoenergetic narrow beam. (b) Narrow beam with a continuous spectrum. (c) Broad beam with a continuous spectrum.

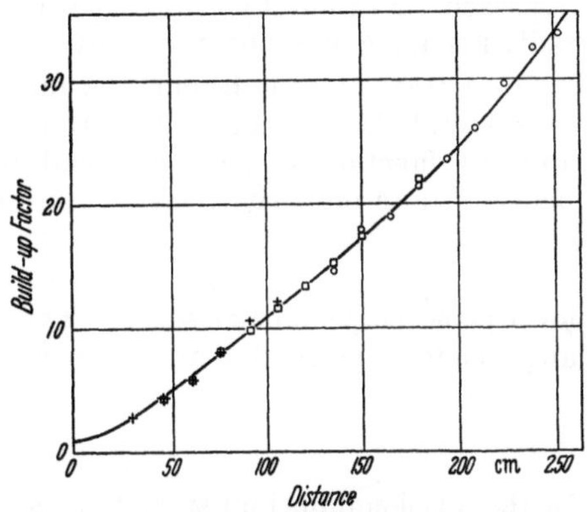

Fig. 5. Exposure-dose build-up factors; point-isotropic Co^{60} source in water; +, □ ionization data for 0.33 and 4.75 curie sources, ○ Geiger counter data for 4.75 curie source; WHITE [95].

X-ray penetration does *not* lead to a typical state of equilibrium. To point up its pecularities, we indicate first the main properties of equilibrium states, namely: (1) The intensity ratio of secondary to primary radiation approaches a maximum as the depth of penetration increases. (2) The quality of the secondary radiation becomes independent of the depth of penetration, because the intensities of its various spectral components bear, in the limit, constant ratios to the intensity of the primaries. (3) Therefore, the intensity of the whole complex radiation in equilibrium is controlled by the progressive attenuation of the primaries, which is exponential in the case of X rays. The intensity ratio between any two spectral components approaches its equilibrium value at a rate inversely proportional to

[1] Note that the concepts of "transients" and "steady state" are applied with respect to *spatial* variations.

the difference of their attenuation coefficients. The eventual intensity ratio is similarly related to this difference[1].

The quasi-equilibrium characterizing the penetration of X rays differs from a true equilibrium in the following respects. In the Compton effect, some of the X rays have very nearly the same energy, the same direction, and the same attenuation coefficient after as before the scattering. In fact, the attenuation coefficient of some secondary X-ray components differs only infinitesimally from the attenuation coefficient of the most penetrating primary X rays that are present. This circumstance does not merely slow down the approach to equilibrium; it actually suppresses its features (1) and (3) in that the intensity of secondary X-ray components relative to a monochromatic primary grows beyond any limit.

Still, within a finite thickness of absorber, some sort of equilibrium must be attained among the secondary X rays whose attenuation coefficient differs from that of the most penetrating component by a sufficiently large amount. In fact, the relative intensity of most secondaries and, therefore, the "quality" of most of the secondary radiation turns out to approach a limiting value within a finite thickness as indicated in feature (2). On the other hand, features (1) and (3) do not hold. The intensity ratio of secondaries to the primaries does not approach a maximum value, but keeps increasing; the attenuation of the primary radiation alone does not fully control the attenuation of the secondary radiation.

In a state of equilibrium, the flow of radiation at great distances from the source would be described as the product of an exponential function of the distance and of a function of the energy and direction of each secondary radiation component. More properly, the function of distance z is a simple exponential,

$$e^{-\mu z} \tag{5.1}$$

for a plane perpendicular source. For point sources there is an additional solid-angle factor $1/r^2$, so that the intensity of the primaries varies like

$$\frac{e^{-\mu r}}{r^2}. \tag{5.2}$$

In the quasi-equilibrium state that is actually achieved, the flux of most of the secondary radiation at great distances is described as the product of a modified exponential function of the distance and of a function of the energy and direction of each component. If ω is a unit vector in the direction of photon travel, z is the photon distance from a plane source, and E_0, E, refer to the photon energy initially and as measured at z, the flux of photons $I(z, \omega, E)$ takes the form

$$I(z, \omega, E) \sim f(z)\, g(\omega, E) \tag{5.3}$$

when z is sufficiently large and E sufficiently smaller than E_0.

At large distances most of the energy is carried by the softer secondary components. One can, therefore, use the form (5.3) of the flux to evaluate approximately the quantities of practical interest. These quantities vary, as functions

[1] In the model of radioactive decay of two substances with decay constants λ_1 and λ_2, $dI_1 = -\lambda_1 I_1\, dt$, $dI_2 = (\lambda_1 I_1 - \lambda_2 I_2)\, dt$, we have: $I_1(t) = I_1(0)\, e^{-\lambda_1 t}$, $I_2(t) = I_1(0)\,\{e^{-\lambda_1 t} - e^{-\lambda_2 t}\} \times \frac{\lambda_1}{\lambda_2 - \lambda_1}$ and $\frac{I_2(t)}{I_1(t)} = \{1 - e^{-(\lambda_2 - \lambda_1)t}\}\frac{\lambda_1}{\lambda_2 - \lambda_1}$. Both the limiting ratio $\frac{\lambda_1}{\lambda_2 - \lambda_1}$, and the time $\frac{1}{(\lambda_2 - \lambda_1)}$ required for $\frac{I_2(t)}{I_1(t)}$ to reach 63% (i.e. $1 - e^{-1}$) of the limiting ratio, are inversely proportional to $\lambda_2 - \lambda_1$.

of distance, in proportion to $f(z)$. Any X-ray measurement, performed at z, gives a result approximately proportional to $f(z)$. Therefore the ratio of $f(z)$ to the exponential function (5.1), or of an analogous $f(r)$ to the exponential function of (5.2), represents the variation of the build-up factor as a function of the distance from the source.

The qualitative considerations leading to (5.3) have been verified to a considerable extent by detailed calculations reported in later sections, and also by direct observation. Fig. 6 shows the results of an experiment by E. HAYWARD [42] in which the spectrum of secondary electrons in water surrounding a γ-ray source was measured at various distances from the source. This is compared with a plot of the number of electrons of various energies ε which should be represented according to Eq. (5.3) by $f(r)\int dE \int_{4\pi} d\omega \times g(\omega, E)\, k_{el}(E, \varepsilon)$ where $k_{el}(E, \varepsilon)$ indicates the probability of production of an electron of energy ε by a photon of energy E. Qualitatively, the shape of the plot becomes independent of the penetration depth at great depths as predicted by (5.3).

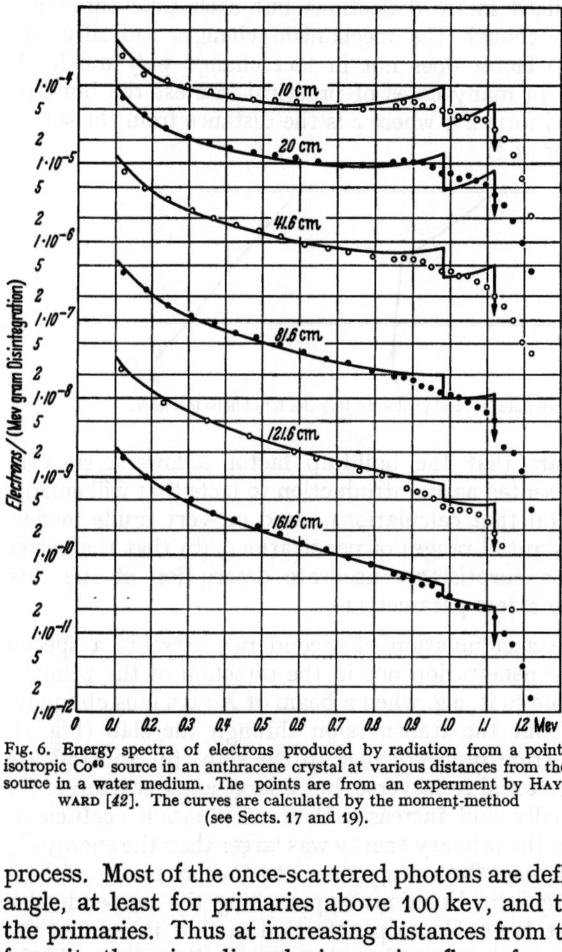

Fig. 6. Energy spectra of electrons produced by radiation from a point-isotropic Co^{60} source in an anthracene crystal at various distances from the source in a water medium. The points are from an experiment by HAYWARD [42]. The curves are calculated by the moment-method (see Sects. 17 and 19).

δ) General trend of the build-up factor. As the primary photons travel away from their source, more and more of them experience a first Compton scattering process. Most of the once-scattered photons are deflected only through a moderate angle, at least for primaries above 100 kev, and thus keep traveling along with the primaries. Thus at increasing distances from the source, but still not too far from it, there is a linearly increasing flux of once-scattered secondaries, accumulated along the path between the source and the point under consideration. In this same region, not too far from the source, most of the photons that have undergone multiple scattering have returned toward the source from greater distances and should be distributed with fairly uniform density. Therefore, the variation of the build-up factor near the source is essentially linear, with the rate of increase determined by the accumulation of once-scattered photons.

As the distance from the source increases further, qualitative arguments indicate that the build-up factor does not depart rapidly from linearity. Most of the multiply scattered photons are greatly degraded in energy, and therefore much less penetrating than the primaries or the once-scattered photons; they are also fairly randomly distributed in direction. Therefore their presence contributes, in the main, to raise the total flux of X rays at each point, roughly in proportion to the combined local flux of primaries and once-scattered photons, but it does not contribute much to the variations of the build-up factor from point to point.

In the eventual state of quasi-equilibrium far from the source, the penetration mechanism is quite different, being determined by the accumulation of secondaries that have experienced many deflections but each time through a small angle only. Here, even though the mechanism changes radically, the resulting trend of the build-up factor does not in fact change very much. It will be shown in Sect. 11 that in many cases of practical interest the build-up factor approaches the analytical form $z^{\bar{\varkappa}_0}$, where z is the distance from the source and $\bar{\varkappa}_0$ is a constant of the order of 1.

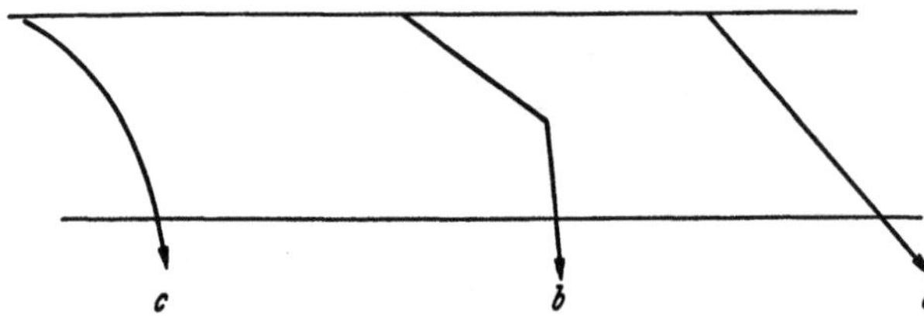

Fig. 7. Transmission through slabs; typical photon trajectories for oblique incidence.

These considerations indicate that the build-up factor follows a smooth, predictable trend. They are presented as an introduction to facts that will emerge in following sections, namely: (a) that calculations based on very crude models achieve fair accuracy over substantial ranges of penetrations, (b) that the better types of calculations achieve a surprisingly accurate description of the flux distribution by calculating only a few parameters.

ε) Oblique penetration. The accumulation of secondaries presents a special aspect when one deals with the penetration not in the direction of the primary photons but in a direction oblique to it, e.g. when a beam of X rays falls obliquey) on a slab and one inquires about the transmission through the slab (Fig. 7). Secondaries scattered in a direction perpendicular to the slab [(b) and (c)] are favored in that they have less material to traverse, but a large angle scattering (b) reduces their energy substantially and increases their attenuation coefficient. (This argument does not apply if the primary energy was larger than the energy E_m of Table 4 and the scattering brings the energy down to the proximity of E_m.) Repeated small-angle scattering into directions approaching the perpendicular to the slab [(c), Fig. 7] is most favorable to penetration in that it reduces the pathlength in the slab without much energy loss, but it constitutes an unlikely sequence of events. Accumulation of these secondaries proceeds quite slowly and becomes the dominant mechanism of penetration only for extremely thick slabs (see Sect. 11 δ).

ζ) Polarization effects. Compton-scattered X rays are polarized, to some extent, in a direction perpendicular to the plane of scattering, as shown by (4.17), even if the incident X rays are unpolarized. This partial polarization causes a subsequent scattering to be not quite random in azimuth, in the sense that any three successive free paths of a photon have a tendency to be coplanar. This effect tends to increase the overall penetration. Monte Carlo calculations indicate that the polarization achieved as a result of multiple scattering becomes appreciable only for low energy radiation (< 0.5 Mev). Since this low energy radiation is not very penetrating the overall effect is small. See Sect. 19β for results of a detailed calculation.

6. Transport equation. The flow of X rays, secondaries as well as primaries, is completely determined as the solution of a transport equation. This equation describes the balance between the numbers of photons of given energy and direction entering and leaving an infinitesimal cylinder, shown in Fig. 8.

The balance may be formulated for conditions where the X-ray source is constant in time and, therefore, the flow of photons throughout the medium is also constant. Transient effects due to source variations in the course of time have not been studied much. Alternatively, one may consider only the time-integral of the source output, without any stipulation about its constancy, and the resulting time-integral of the flux at each point. Time-constant and time-integral treatments are equivalent. The following treatment refers to a time-constant situation.

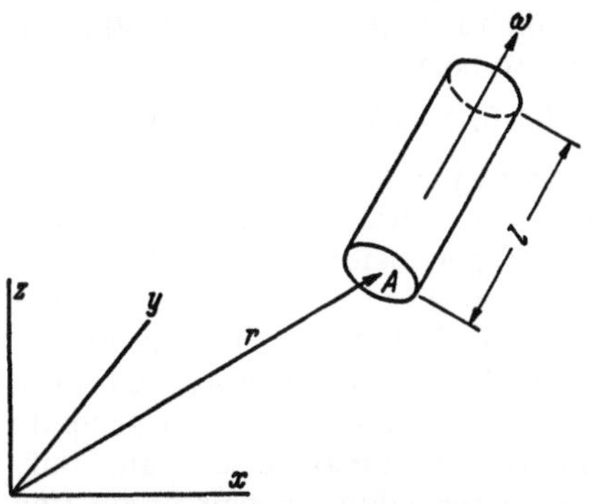

Fig. 8. Infinitesimal cylinder in the derivation of the transport equation. The difference between the currents into and out of the cylinder must equal the difference between absorption and production of photons within the cylinder.

Consider a point $\boldsymbol{r}$ and an infinitesimal right cylinder with a base area A centered at $\boldsymbol{r}$ and with a height l, whose lateral surface is parallel to a direction $\boldsymbol{\omega}$. We define the flux $I(\boldsymbol{r}, \boldsymbol{\omega}, \lambda)\, d\lambda\, d\boldsymbol{\omega}$ as the number of photons with wavelengths between λ and $\lambda + d\lambda$, and with directions between $\boldsymbol{\omega}$ and $\boldsymbol{\omega} + d\boldsymbol{\omega}$, which cross a unit area of the base of the infinitesimal cylinder per unit time. (The wavelength in Compton units, $\lambda = mc^2/E$, is used to indicate the energy because this choice simplifies the treatment of the Compton effect.)

The net number of photons with specified direction and energy leaving the cylinder per unit time is represented by

$$A\, I(\boldsymbol{r} + l\,\boldsymbol{\omega}, \boldsymbol{\omega}, \lambda) - A\, I(\boldsymbol{r}, \boldsymbol{\omega}, \lambda) \tag{6.1a}$$

which has the differential form

$$A\, l\, \boldsymbol{\omega} \cdot \operatorname{grad} I(\boldsymbol{r}, \boldsymbol{\omega}, \lambda). \tag{6.1b}$$

Three factors contribute to this netout flow. One is the narrow-beam attenuation in the whole volume of the cylinder, which amounts, according to (4.1), to $-\mu(\lambda)\, l\, I(\boldsymbol{r}, \boldsymbol{\omega}, \lambda) A$. The second is scattering of photons of direction $\boldsymbol{\omega}'$ and energy mc^2/λ' into the given direction and energy, $\boldsymbol{\omega}$ and mc^2/λ. This scattering, if it happens anywhere within the cylinder volume Al, contributes a positive outflow from the cylinder. It depends on the product of the flux $I(\boldsymbol{r}, \boldsymbol{\omega}', \lambda')$ and of the probability $k(\boldsymbol{\omega}, \lambda, \boldsymbol{\omega}', \lambda')$ of photon scattering into $\boldsymbol{\omega}$ and λ from $\boldsymbol{\omega}'$ and λ', per unit path through the medium and per unit $d\boldsymbol{\omega}$ and $d\lambda$. This product must be integrated over all incident directions $\boldsymbol{\omega}'$ and wavelengths λ'. The third contribution arises from the source, if any, of photons of the given energy and direction within the cylinder. If the photons are produced with a source density $S(\boldsymbol{r}, \boldsymbol{\omega}, \lambda)$, per unit volume, per unit time, per steradian and per unit λ, the corresponding excess flow out of the cylinder will be $A\, l\, S(\boldsymbol{r}, \boldsymbol{\omega}, \lambda)$.

The transport equation is obtained by equating the difference (6.1) to the sum of the three factors contributing to it. Each of these quantities is proportional to the volume $A\,l$ of the cylinder. Therefore this volume is irrelevant to the problem and can be factored out. The equation is then

$$\left.\begin{aligned} \boldsymbol{\omega}\cdot\operatorname{grad} I(\boldsymbol{r},\boldsymbol{\omega},\lambda) &= -\mu(\lambda)\, I(\boldsymbol{r},\boldsymbol{\omega},\lambda) + \\ &+ \int_0^{\lambda} d\lambda' \int_{4\pi} d\boldsymbol{\omega}'\, k(\boldsymbol{\omega},\lambda,\boldsymbol{\omega}',\lambda')\, I(\boldsymbol{r},\boldsymbol{\omega}',\lambda') + S(\boldsymbol{r},\boldsymbol{\omega},\lambda). \end{aligned}\right\} \tag{6.2}$$

The upper limit of the integral in (6.2) has been set at $\lambda'=\lambda$ because inelastic processes can only lower the photon energy, i.e. because $k(\boldsymbol{\omega},\lambda,\boldsymbol{\omega}',\lambda')$ vanishes for $\lambda'>\lambda$; this turns out to simplify the mathematical treatment.

The Eq. (6.2) is quite general; the actual functions μ, k and S to be entered in it depend on the processes considered, on the range of variables which is of interest, and on the degree of approximation which is required. In most applications only Compton scattering without reference to polarization has been included in k. This function is then simply equal to the Klein-Nishina cross section $\frac{d\sigma}{d\lambda'\,d\Omega'}$ of (4.4) multiplied by the number NZ of electrons per unit volume. [To apply (4.4) in this connection one must set $\cos\vartheta=\boldsymbol{\omega}\cdot\boldsymbol{\omega}'$ and interchange the symbols λ and λ', which indicate here the wavelength after and before scattering, respectively.] The integration over λ' extends only from $\lambda-2$ or zero, whichever is larger, to λ. Coherent scattering may, of course, be included in k. Other processes, like fluorescence, multiple photon emission by higher order scattering, and even annihilation radiation may also be included in k as processes which generate new photons from incident ones[1]. (Annihilation radiation may be so treated only if the distance traveled by the positron, between the points of pair production and annihilation, is regarded as negligible.)

The source function must also include the production of photons by all secondary processes which cannot be treated in effect as scattering, i.e., which cannot be included in k. In particular, the bremsstrahlung generated by secondary electrons that have traveled a substantial distance must be included in S, and so must the annihilation radiation if the travel of positrons is not negligible. S will then include terms of the form

$$S_{\text{sec}} = \int_0^{\infty} dE \int_{4\pi} d\boldsymbol{\omega}'\, k_{\text{rad}}(\boldsymbol{\omega},\lambda,\boldsymbol{\omega}',E)\, I_{\text{sec}}(\boldsymbol{r},\boldsymbol{\omega}',E), \tag{6.3}$$

where I_{sec} indicates the flux of radiating particles of energy E traveling in the direction $\boldsymbol{\omega}'$ and k_{rad} is the probable number of photons radiated with wavelength λ and direction $\boldsymbol{\omega}$ per unit path of a particle of given energy and direction.

The flux I_{sec} of secondary particles is, in turn, determined as the solution of a transport equation having the same structure as (6.2), with I_{sec} in the place of I, and with a source term like (6.3) where, instead, the X-ray flux I enters to determine the production of secondary particles. Thus the complete transport equation for a system of radiations consists of a system of equations like (6.2), one for each type of radiation, with the fluxes of other radiations entering in their source terms. The cascade-shower process is described by a system of three equations of this kind, one each for photons, electrons and positrons.

α) Simple source functions. The special role of plane sources. A source of monochromatic X rays of wavelength λ_0 is represented by a source function $S(\boldsymbol{r},\boldsymbol{\omega},\lambda)$ that depends on λ through a Dirac function $\delta(\lambda-\lambda_0)$. In the following we shall always consider monochromatic sources unless otherwise stated. An X-ray source collimated in the direction $\boldsymbol{\omega}_0$ is represented by a function with a factor $\delta(\boldsymbol{\omega}-\boldsymbol{\omega}_0)$. For a point source concentrated at $\boldsymbol{r}_0$, S depends on $\boldsymbol{r}$ through $\delta(\boldsymbol{r}-\boldsymbol{r}_0)$. A *point collimated* source function of one photon $\sec^{-1}$ is, therefore,

$$S(\boldsymbol{r},\boldsymbol{\omega},\lambda) = \delta(\boldsymbol{r}-\boldsymbol{r}_0)\,\delta(\boldsymbol{\omega}-\boldsymbol{\omega}_0)\,\delta(\lambda-\lambda_0). \tag{6.4}$$

[1] For the purpose of this extended application, $k(\boldsymbol{\omega},\lambda,\boldsymbol{\omega}',\lambda')$ should be defined more properly as the average number of photons, with given $\boldsymbol{\omega}$ and λ, released per unit volume where there is a unit flux of photons with $\boldsymbol{\omega}'$ and λ'.

A plane perpendicular to a unit vector $\boldsymbol{u}$ is represented by the equation $\boldsymbol{u}\cdot\boldsymbol{r}=\text{const}$; the plane through the point $\boldsymbol{r}_0$ and perpendicular to $\boldsymbol{u}$ has the equation $\boldsymbol{u}\cdot\boldsymbol{r}=\boldsymbol{u}\cdot\boldsymbol{r}_0$. An X-ray source of one photon $\text{cm}^{-2}\,\text{sec}^{-1}$ distributed uniformly on this plane is represented by the *plane collimated* source function

$$S(\boldsymbol{r},\omega,\lambda)=\delta(\boldsymbol{r}\cdot\boldsymbol{u}-\boldsymbol{r}_0\cdot\boldsymbol{u})\,\delta(\omega-\omega_0)\,\delta(\lambda-\lambda_0). \tag{6.5}$$

The point source (6.4) can be represented as a superposition of plane sources (6.5) utilizing the plane wave (Fourier) expansion of the Dirac function

$$\delta(\boldsymbol{r}-\boldsymbol{r}_0)=\frac{1}{(2\pi)^3}\int d\boldsymbol{q}\,e^{i\boldsymbol{q}\cdot(\boldsymbol{r}-\boldsymbol{r}_0)} \tag{6.6}$$

and the trivial identity

$$e^{i\boldsymbol{q}\cdot(\boldsymbol{r}-\boldsymbol{r}_0)}=\int_{-\infty}^{\infty} ds\,e^{iqs}\,\delta(\boldsymbol{\Omega}\cdot\boldsymbol{r}-\boldsymbol{\Omega}\cdot\boldsymbol{r}_0-s), \tag{6.7}$$

where $\boldsymbol{\Omega}$ indicates the direction of the vector $\boldsymbol{q}=q\,\boldsymbol{\Omega}$. We have then

$$\delta(\boldsymbol{r}-\boldsymbol{r}_0)=\frac{1}{(2\pi)^3}\int d\boldsymbol{q}\int_{-\infty}^{\infty} ds\,e^{iqs}\,\delta(\boldsymbol{\Omega}\cdot\boldsymbol{r}-\boldsymbol{\Omega}\cdot\boldsymbol{r}_0-s). \tag{6.8}$$

Notice that the equation $\boldsymbol{\Omega}\cdot\boldsymbol{r}=\boldsymbol{\Omega}\cdot\boldsymbol{r}_0+s$ represents a plane perpendicular to $\boldsymbol{q}$ and at a distance s from the point $\boldsymbol{r}_0$. The right side of (6.8) represents therefore a double linear combination: plane sources perpendicular to $\boldsymbol{q}$ at all distances s from $\boldsymbol{r}_0$ are first combined with coefficients e^{iqs}, and next the combinations so obtained for all vectors $\boldsymbol{q}$ are combined with equal coefficients.

A combination of sources with complex, or real but non-positive, coefficients is physically unrealistic but constitutes a permissible mathematical device. This device is useful in the X-ray penetration problem because the transport equation (6.2) is linear, so that the X-ray flux $I(\boldsymbol{r},\omega,\lambda)$ produced by a given combination of sources $c_1S_1+c_2S_2+\cdots$ equals the combination $c_1I_1+c_2I_2+\cdots$ of the fluxes from the separate sources. The calculation of the flux from each separate plane source is particularly simple when the medium is infinite and homogeneous, that is, when the coefficients μ and k in the transport equation have the same values at all points of space. Under this condition, if the source function depends only on the component of $\boldsymbol{r}$ along $\boldsymbol{\Omega}$, the flux distribution has the same plane symmetry, i.e., depends on the single space coordinate $\boldsymbol{r}\cdot\boldsymbol{\Omega}$. The number of independent variables of the transport equation is thereby reduced by two.

The mathematical form of the transport equation (6.2) is conveniently specialized, for a general plane source (6.5) and an infinite, uniform, medium, by taking the point $\boldsymbol{r}_0$ as the origin of coordinates and the direction $\boldsymbol{u}$ as the z axis. The source function is then $\delta(z)\,\delta(\omega-\omega_0)\,\delta(\lambda-\lambda_0)$, the flux distribution has the form $I(z,\omega,\lambda)$ and $\omega\cdot\operatorname{grad}I$ reduces to $\omega_z\,\partial I/\partial z$. Wavelengths below λ_0 do not occur and (6.2) becomes

$$\left.\begin{aligned}\omega_z\frac{\partial I(z,\omega,\lambda)}{\partial z}&=-\mu(\lambda)\,I(z,\omega,\lambda)+\\&+\int_{\lambda_0}^{\lambda}d\lambda'\int_{4\pi}d\omega'\,k(\omega,\lambda,\omega',\lambda')\,I(z,\omega',\lambda')+\delta(z)\,\delta(\omega-\omega_0)\,\delta(\lambda-\lambda_0).\end{aligned}\right\} \tag{6.9}$$

In the following we shall deal mostly with this simplified equation.

Plane sources are often considered from which the X rays emerge with a directional distribution symmetric about the normal to the plane, instead of being collimated in a general direction $\boldsymbol{\omega}_0$. For example one considers the *plane conical* source

$$S(\boldsymbol{r},\boldsymbol{\omega},\lambda)=\delta(z)\,\frac{1}{2\pi}\,\delta(\omega_z-\cos\vartheta_0)\,\delta(\lambda-\lambda_0), \tag{6.10}$$

with radiation directed along the surface of a cone of aperture ϑ_0, the *plane perpendicular* source

$$S(\boldsymbol{r},\boldsymbol{\omega},\lambda)=\delta(z)\,\frac{1}{2\pi}\,\delta(\omega_z-1)\,\delta(\lambda-\lambda_0), \tag{6.11}$$

with the collimation direction $\boldsymbol{\omega}_0$ perpendicular to the source plane, and the *plane isotropic* source

$$S(\boldsymbol{r},\boldsymbol{\omega},\lambda)=\delta(z)\,\frac{1}{4\pi}\,\delta(\lambda-\lambda_0). \tag{6.12}$$

When any of these sources replaces the collimated source in (6.9), the flux distribution $I(z,\boldsymbol{\omega},\lambda)$ depends only on the z component of $\boldsymbol{\omega}$ owing to the symmetry about the z axis.

If one considers the point collimated source (6.4) directly, rather than as a combination of plane sources, cylindrical coordinates (z,ϱ,β) are convenient, with the origin at $\boldsymbol{r}_0$ and the z axis in the direction $\boldsymbol{\omega}_0$. Eq. (6.4) becomes then

$$S(\boldsymbol{r},\boldsymbol{\omega},\lambda)=\delta(z)\,\frac{1}{2\pi\varrho}\,\delta(\varrho)\,\frac{1}{2\pi}\,\delta(\omega_z-1)\,\delta(\lambda-\lambda_0). \tag{6.13}$$

If the medium is infinite and uniform, the X-ray flux distribution is then symmetric about z, i.e. a function of $\varphi-\beta$ represented by $I(z,\varrho,\vartheta,\varphi-\beta,\lambda)$, where $\vartheta=\arccos\omega_z$ and φ is the azimuth of $\boldsymbol{\omega}$. We have, then,

$$\left.\begin{aligned}\boldsymbol{\omega}\cdot\operatorname{grad} I(z,\varrho,\vartheta,\varphi-\beta,\lambda)&=\cos\vartheta\,\frac{\partial I}{\partial z}+\sin\vartheta\cos(\varphi-\beta)\,\frac{\partial I}{\partial\varrho}-\\&\quad-\sin\vartheta\sin(\varphi-\beta)\,\frac{1}{\varrho}\,\frac{\partial I}{\partial(\varphi-\beta)}.\end{aligned}\right\} \tag{6.14}$$

Finally, when one considers the simple *point isotropic* source, polar coordinates are convenient with the origin at the source position. The source is then represented by

$$\delta(\boldsymbol{r},\boldsymbol{\omega},\lambda)=\frac{1}{4\pi r^2}\,\delta(r)\,\frac{1}{4\pi}\,\delta(\lambda-\lambda_0). \tag{6.15}$$

If the medium is infinite and uniform, the flux distribution depends only on the distance r from the source, on the component of $\boldsymbol{\omega}$ in the direction of $\boldsymbol{r}$, indicated by $\omega_r=\cos\vartheta_r$, and on λ, and we have

$$\boldsymbol{\omega}\cdot\operatorname{grad} I(r,\omega_r,\lambda)=\cos\vartheta_r\,\frac{\partial I}{\partial r}+\sin\vartheta_r\,\frac{1}{r}\,\frac{\partial I}{\partial\vartheta_r}. \tag{6.16}$$

β) Boundary conditions. The transport equation (6.2) may be applied over a limited region of space or over the whole space. In either case it must be complemented by a boundary condition which describes the flux entering the region of interest. When the whole space is considered, the condition is simply that the incoming flux vanishes because there are no "external" sources of radiation, i.e. that any radiation at infinite distance must be flowing outwards ("Ausstrahlungsbedingung").

The boundary condition, which is usually implied tacitly, is essential to establish the uniqueness of the solution of a transport equation. A linear inhomogeneous integro-differential equation, like (6.2), can be regarded as a limit of a linear system of infinitely many algebraic equations. Accordingly, its solution is uniquely defined, to within the addition of solutions —if any—of the corresponding homogeneous equation. Such solutions are excluded by the boundary condition.

γ) Fluctuations. The flux of photons $I(\boldsymbol{r},\boldsymbol{\omega},\lambda)$ represents an average over a conceivable series of experiments in which different numbers of photons may be detected at one point even though the source emission was identical. These fluctuations reflect the different possible outcomes of each among the elementary processes. Fluctuations are particularly striking in the event of cascade shower formation, or in any other chain of secondary processes where several photons or particles arise from a single one. One may then be interested in the correlated detection of two or more photons or particles, besides the average rate of single detections. Studies of fluctuations have been made in connection with showers[1], but are not covered in this article.

δ) Polarization effects. To take into account the influence of polarization on X-ray penetration one may consider at each point of space, instead of the flux $I(\boldsymbol{r}, \boldsymbol{\omega}, \lambda)$, a set of four Stokes parameters $I^{[i]}(\boldsymbol{r}, \boldsymbol{\omega}, \lambda)$, with $i = 0, 1, 2, 3$ (Sect. 4ζ). The parameter $I^{[0]}$ coincides with the flux I considered thus far. Similarly, one may consider a set of four source parameters $S^{[i]}(\boldsymbol{r}, \boldsymbol{\omega}, \lambda)$ instead of the single function S, and represent attenuation and scattering by means of 4×4 matrices $\mu^{[ij]}$ and $k^{[ij]}$ which operate on the parameters $I^{[i]}$ as $\sigma^{[ij]}$ does in (4.15). The analysis of the transport process, carried out as at the beginning of this section, yields a system of four equations obeyed by the parameters $I^{[i]}$, namely

$$\left.\begin{aligned} \boldsymbol{\omega} \cdot \operatorname{grad} I^{[i]}(\boldsymbol{r}, \boldsymbol{\omega}, \lambda) = &- \sum_j \mu^{[ij]}(\lambda)\, I^{[j]}(\boldsymbol{r}, \boldsymbol{\omega}, \lambda) + \\ &+ \int_0^{\lambda} d\lambda' \int_{4\pi} d\boldsymbol{\omega}' \sum_j k^{[ij]}(\boldsymbol{\omega}, \lambda, \boldsymbol{\omega}', \lambda')\, I^{[j]}(\boldsymbol{r}, \boldsymbol{\omega}', \lambda') + S^{[i]}(\boldsymbol{r}, \boldsymbol{\omega}, \lambda)\,. \end{aligned}\right\} \tag{6.17}$$

Notice that the whole system is represented in condensed form by the single equation (6.2) with the understanding that I and S are four-component vectors, and that μ and k indicate matrices operating on the vector I.

Strictly speaking, the initial formulation and interpretation of (6.2) must be derived from the more accurate Eq. (6.17) by an approximation in which the polarization effects are disregarded. Specifically, if the off-diagonal elements of $\mu^{[ij]}$ and $k^{[ij]}$ are negligible, the system (6.17) breaks up into four separate equations, of which the one with $i = 0$ coincides with (6.2) as originally defined.

7. Survey of theoretical procedures. In the transport equation (6.2) or (6.9), the coefficients $\mu(\lambda)$ and $k(\boldsymbol{\omega}, \lambda, \boldsymbol{\omega}', \lambda')$ are complicated functions of the wavelength and of the scattering angle. Simplified analytical schematizations of these functions can be utilized within limited ranges of values of the variables, but any detailed and extensive solution of the transport problem is likely to require some numerical calculation, even for analytically specified μ and k. Numerical solutions have been further emphasized by the requirements of practical applications.

Direct numerical solution of the transport equation is straight-forward in principle but very laborious. Numerical solution by the Monte Carlo (random sampling) procedures is also laborious but has proven valuable. Most progress has been achieved with combinations of numerical methods and of preliminary analytical procedures which reduce the transport problem to manageable proportions.

α) Direct numerical integration. The direct method of solving the transport equation consists of replacing the integral and the derivative in the equation with sums and differences of the values of the flux I at a sufficiently fine mesh of values of the independent variables. Thereby the equation reduces to an algebraic system which can be solved numerically by standard techniques. This approach encounters practical limitations, due to the labor involved, which may exceed the capabilities of automatic computers. Even for simple geometries there are three independent variables (one component each for $\boldsymbol{r}$, $\boldsymbol{\omega}$, and λ), and one must compute a 3-dimensional mesh of values of I. The main question is, then, how fine a mesh is necessary for a given accuracy. In certain ranges of the independent variables I varies rapidly, e.g. as a function of $\boldsymbol{\omega}$ at high energies and great depths of penetration; here, at least, a very fine mesh is required. Some direct numerical calculations have been done for neutrons[2].

[1] Cf. Vol. XLVI, this Encyclopedia.

[2] J. HELLER and H. KELLER: Phys. Rev. **93**, 932 (1954). — D. SCHIFF: Phys. Rev. **99**, 660 (1955). — B.G. CARLSON: U.S. AEC Report LA 1891, 1955.

β) Monte Carlo method. Numerical calculation of the X-ray flux can also be accomplished by constructing model trajectories of photons according to the Monte Carlo random sampling method. A photon trajectory is identified by the location and result of each successive elementary process experienced by the photon up to its eventual absorption. These identifying marks cannot be predicted exactly; only their probability distribution is predictable, owing to the randomness of quantum phenomena. The Monte Carlo method consists of imitating the actual physical process by an analog numerical random process[1]. Each random number of the analog specifies, through a suitable code, a particular feature of one photon process, for example the change of wavelength in the m-th Compton scattering in the n-th sampled trajectory. Thereby a sequence of random numbers is translated into the record of model trajectories.

The flux of photons at any given point within a material under consideration is then evaluated by scoring the model trajectories which pass in the proximity of that point. The data thus obtained are akin to experimental data, in that they are sample data subject to statistical fluctuations, rather than average values determined by analytical calculation. On the other hand, replacing actual experimentation with a model study affords greatly increased flexibility. Thereby the Monte Carlo method constitutes a valuable adjunct to theory and experiment.

The accuracy of the random sampling of trajectories is a slowly increasing function, proportional to the square root, of the sample size. Therefore accurate Monte Carlo calculations become very laborious. This limitation is particularly severe for the purpose of evaluating the flux at positions where it is extremely low, e.g. at large distances from the sources, since only a minute fraction of the trajectories contributes to the flux at such positions. This deficiency of the method can be remedied, to some extent, by devices such as "importance sampling[2]", which incorporate into the calculation information provided by other methods. The Monte Carlo method is treated in Part D.

γ) Semi-analytical procedures. Neither the Monte Carlo method nor the direct numerical solution of the transport equation appear fully adequate for the practical solution of the penetration problem. Thus there is practical, as well as fundamental, interest in pushing the analytical theory of penetration beyond the formulation of the transport equation. The goal is to obtain a workable approximation method or at least some guidance in performing economically direct calculations. Various approaches toward this goal are described in the following sections.

Some of these approaches rely on expansions appropriate to special conditions, for example, to very large distances from all sources (Sect. 11) or to photons that have experienced only a few scattering processes (Sect. 8). Other approaches consist of treating accurately problems that are circumscribed and therefore simpler but nevertheless informative (Sect. 10). These problems relate not so much to the propagation of X rays from one point to another but rather to the energy degradation and to the deflection of X rays irrespective of their propagation in space. In particular the photon distributions in energy only, and in energy and direction, can be studied *per se* and obey equations with fewer variables than (6.2) or (6.9). Simplification is obtained by separation of variables, through expansion in plane waves and spherical harmonics, even when the total number of independent variables is not reduced. The separation is possible in

[1] The name "Monte Carlo" is derived from the fact that the analog process is equivalent to determining the fate of the photon by playing a game of chance.

[2] See Sect. 25.

infinite, homogeneous, isotropic media because the transport process is invariant under translations and rotations (Sect. 9).

Whatever numerical or analytical procedure is followed, it may be convenient to minimize the variations of the coefficients μ and k as functions of their variables. The structure of the transport equation permits a simple transformation which is useful for this purpose. The kernel of the equation may be expressed in the form

$$k(\omega, \lambda, \omega', \lambda') = \frac{f(\lambda')}{f(\lambda)} \bar{k}(\omega, \lambda, \omega', \lambda'), \tag{7.1}$$

where $f(\lambda')$ is an arbitrary function to be chosen so as to minimize the dependence of $\bar{k}$ on λ' and λ. Substitution of (7.1) into (6.2) or (6.9) and multiplication by $f(\lambda)$ yields a new equation with the kernel $\bar{k}$ in place of k and the new dependent variable

$$\bar{I}(\boldsymbol{r}, \omega, \lambda) = f(\lambda)\, I(\boldsymbol{r}, \omega, \lambda). \tag{7.2}$$

This transformation is particularly helpful for X rays above 1 Mev, i.e. for $\lambda \ll 1$. It has been applied with $f(\lambda) = \lambda$ in Ref. [*34c*] and in the extensive calculations described in Part C, and in Ref. [*66*] with $f(\lambda) = \lambda^{1.8}$ which makes $\bar{k}(\lambda, \lambda')$ very nearly constant when $\lambda'/\lambda \gtrsim \frac{1}{2}$ and $\lambda \ll 1$.

Among the approximation methods, no single one has proved adequate over a satisfactorily wide range of penetration distances, photon energies and directions, but each of them solves, in effect, a particular aspect of the penetration problem. Analysis has made it possible to combine different approximation procedures very effectively into a coherent theory. The analytical framework is provided by the theory for simple or limiting cases. The flux distribution for a realistic situation is then mapped out by an interpolation procedure based on the use of a few suitable parameters of the distribution which are calculated numerically.

Such developments have been particularly helpful in solving the transport equation for *infinite homogeneous* media. Within this limitation, and for energies where the cascade process is not important, the possibility of detailed calculations has been demonstrated and extensive applications have been carried out (see Part C). The effects of boundaries and inhomogeneities have been less successfully analyzed. Here the work has been more exploratory and has utilized heavily the Monte Carlo method, often combined with other, analytical or semianalytical, approaches (see Part D).

8. Iteration (expansion in orders of scattering). An approach to the transport equation, which is straightforward in principle, consists of determining separately the flux of the primaries and of photons scattered once, twice, etc. This approach leads to a Neumann-type series in powers of the kernel k of the transport equation.

If we set

$$I = I^{(0)} + I^{(1)} + I^{(2)} \ldots, \tag{8.1}$$

the equation for the primary flux $I^{(0)}$ is obtained by deleting the integral in (6.2). The equation for the once-scattered flux $I^{(1)}$ is obtained by replacing I with $I^{(0)}$ in the integral of (6.2) and with $I^{(1)}$ elsewhere in the equation. In general we write

$$\left.\begin{aligned} &\omega \cdot \operatorname{grad} I^{(n)}(\boldsymbol{r}, \omega, \lambda) = -\mu(\lambda)\, I^{(n)}(\boldsymbol{r}, \omega, \lambda) + \\ &\quad + \int_0^\lambda d\lambda' \int_{4\pi} d\omega'\, k(\omega, \lambda, \omega', \lambda')\, I^{(n-1)}(\boldsymbol{r}, \omega', \lambda')\,[1 - \delta_{n0}] + S(\boldsymbol{r}, \omega, \lambda)\,\delta_{n0}. \end{aligned}\right\} \tag{8.2}$$

The sum of these equations for $n = 0, 1, \ldots, \infty$ is identical with (6.2). If these equations are considered in succession of increasing n, each of them is equivalent to the simple narrow-beam equation (4.1), because $I^{(n-1)}$ is assumed to be known from the solution of the preceding equation so that the integral term represents the source distribution of the n-times scattered photons. The solution of (8.2) is

$$I^{(n)}(\boldsymbol{r}, \boldsymbol{\omega}, \lambda) = \int_0^\infty d\zeta\, e^{-\mu\zeta} S^{(n)}(\boldsymbol{r} - \boldsymbol{\omega}\zeta, \boldsymbol{\omega}, \lambda), \tag{8.3}$$

with

$$S^{(0)} = S(\boldsymbol{r}, \boldsymbol{\omega}, \lambda), \qquad S^{(n)} = \int_0^\lambda d\lambda' \int_{4\pi} d\boldsymbol{\omega}'\, k(\boldsymbol{\omega}, \lambda, \boldsymbol{\omega}', \lambda')\, I^{(n-1)}(\boldsymbol{r}, \boldsymbol{\omega}', \lambda'). \tag{8.4}$$

These formulas reduce the solution of the transport equation to the evaluation of a sequence of multiple integrals. The multiplicity of integration increases rapidly as the multiplicity of scattering increases, so that a detailed calculation for photons having experienced, say, ten scatterings is out of the question. The main and rather successful effort of calculation by order-of-scattering has been carried out by PEEBLES and is reviewed in Sect. 23.

The iteration procedure constitutes an important subsidiary to other approaches. The first few scattering processes have the effect of smoothing out the distribution of X-ray flux as a function of its variables, but the distribution of once- or twice-scattered photons often shows singularities which are best calculated directly from (8.3). These singularities are sharpest when the primaries are monoenergetic and collimated. In this event once-scattered photons with wavelength λ have directions $\boldsymbol{\omega}$ concentrated on the surface of a cone defined by $\boldsymbol{\omega} \cdot \boldsymbol{\omega}_0 = 1 - \lambda + \lambda_0$. The distribution of photons shifted to λ in two successive Compton processes has a discontinuity, described below. After three or more scatterings the singularities are confined to the derivatives of the angular distribution. Related discontinuities appear in the spectral distribution at constant $\boldsymbol{\omega}$.

According to (4.4), and thus disregarding polarization effects, the Compton scattering kernel can be written in the form

$$k(\boldsymbol{\omega}, \lambda, \boldsymbol{\omega}', \lambda') = k(\lambda, \lambda') \frac{\delta(1 - \boldsymbol{\omega}' \cdot \boldsymbol{\omega} + \lambda' - \lambda)}{2\pi}. \tag{8.5}$$

The distribution of twice-scattered photons is characterized by the iterated kernel

$$\left.\begin{aligned} &k^{(2)}(\lambda, \lambda_0, \boldsymbol{\omega} \cdot \boldsymbol{\omega}_0) \\ &\quad = \int_{\lambda_0}^{\lambda} d\lambda' \int_{4\pi} d\boldsymbol{\omega}'\, k(\lambda, \lambda')\, k(\lambda', \lambda_0)\, \delta(1 - \boldsymbol{\omega} \cdot \boldsymbol{\omega}' + \lambda - \lambda') \frac{\delta(1 - \boldsymbol{\omega}' \cdot \boldsymbol{\omega}_0 + \lambda' - \lambda_0)}{(2\pi)^2}. \end{aligned}\right\} \tag{8.6}$$

The Compton relationships represented by the δ-functions in (8.6) require that the directions $\boldsymbol{\omega}_0$, $\boldsymbol{\omega}'$ and $\boldsymbol{\omega}$ of a photon after 0, 1, and 2 scatterings, and the corresponding wavelengths λ_0, λ', and λ be related as shown in Fig. 9 [*93*]. For a given value of $\lambda - \lambda_0$, the angle ϑ between $\boldsymbol{\omega}_0$ and $\boldsymbol{\omega}$ fulfils the condition

$$\arccos \boldsymbol{\omega}_0 \cdot \boldsymbol{\omega} = \vartheta \leqq \vartheta_L = 2 \arccos\left[1 - \tfrac{1}{2}(\lambda - \lambda_0)\right]. \tag{8.7}$$

That is, $\boldsymbol{\omega}$ must lie within the cone of axis $\boldsymbol{\omega}_0$ and aperture ϑ_L, otherwise the iterated kernel $k^{(2)}$ vanishes. For a given direction of $\boldsymbol{\omega}$ within this cone, the direction of $\boldsymbol{\omega}'$ must lie on a cone whose axis $\boldsymbol{u}$ bisects the angle between $\boldsymbol{\omega}_0$ and $\boldsymbol{\omega}$ and whose aperture α is fixed by the condition

$$\cos\alpha = \frac{\cos\frac{1}{2}\vartheta_L}{\cos\frac{1}{2}\vartheta}. \tag{8.8}$$

The azimuth ψ of $\boldsymbol{\omega}'$ on this cone is random and the wavelength λ' is related to it by

$$\lambda' = \tfrac{1}{2}(\lambda_0 + \lambda) + \sin\tfrac{1}{2}\vartheta \sin\alpha \cos\psi. \tag{8.9}$$

For $\vartheta = \vartheta_L$, the aperture α vanishes, λ' equals $\frac{1}{2}(\lambda_0 + \lambda)$ and (8.6) reduces to

$$[k^{(2)}(\omega, \lambda, \omega_0, \lambda_0)]_{\vartheta=\vartheta_L} = \frac{1}{2\pi(\lambda - \lambda_0 - 2)} k\left(\lambda, \frac{1}{2}[\lambda_0 + \lambda]\right) k\left(\frac{1}{2}[\lambda_0 + \lambda], \lambda_0\right). \tag{8.10}$$

This quantity diverges for $\lambda = \lambda_0 + 2$, at which point the cone includes all directions and its edge condenses at $\omega = -\omega_0$. For $\lambda > \lambda_0 + 2$ the cone contracts again and reduces to $\omega = \omega_0$ at $\lambda = \lambda_0 + 4$. For $\vartheta < \vartheta_L$, the iterated kernel can be expressed as an integral over ψ,

$$k^{(2)}(\omega, \lambda, \omega_0, \lambda_0) = \frac{1}{(2\pi)^2 \, 2\cos\frac{1}{2}\vartheta} \int_0^{2\pi} d\psi \, k(\lambda, \lambda') \, k(\lambda', \lambda_0), \tag{8.11}$$

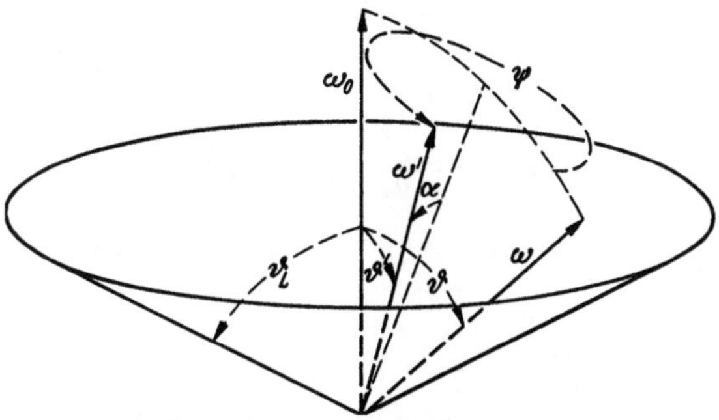

Fig. 9. Geometry of double Compton scattering.

with λ' given by (8.9); this integral can be carried out numerically[1]. Alternatively the kernel can be expressed as an integral over λ',

$$k^{(2)}(\omega, \lambda, \omega_0, \lambda_0) = \frac{1}{(2\pi)^2 \cos\frac{1}{2}\vartheta} \int_{\lambda_m}^{\lambda_M} d\lambda' \frac{k(\lambda, \lambda') \, k(\lambda', \lambda_0)}{\sqrt{\sin^2\frac{1}{2}\vartheta \sin^2\alpha - [\lambda' - \frac{1}{2}(\lambda_0 + \lambda)]^2}}, \tag{8.12}$$

where λ_m and λ_M are the extreme values of λ' according to (8.9). When λ' approaches these limits, the integrand in (8.12) diverges but remains integrable. For ϑ close to its maximum value ϑ_L the radical in the integrand of (8.12) varies much more rapidly than the other factors, so that the other factors may be treated as approximately constant. One finds then the approximate expression

$$k^{(2)}(\omega, \lambda, \omega_0, \lambda_0) \sim k^{(2)}(\lambda, \lambda_0, \omega \cdot \omega_0) = \frac{1}{4\pi\cos\frac{1}{2}\vartheta} k\left(\lambda, \frac{1}{2}[\lambda_0 + \lambda]\right) k\left(\frac{1}{2}[\lambda_0 + \lambda], \lambda_0\right). \tag{8.13}$$

which reduces to (8.10) in the limit $\vartheta = \vartheta_L$; for $\vartheta > \vartheta_L$ both $k^{(2)}$ and $k^{(2)}$ are understood to vanish.

The space distribution of twice-scattered photons is given by expressions analogous to (8.6) or (8.12) whose integrand contains an additional factor that represents the attenuation of once-scattered photons. The evaluation of such expressions offers no additional difficulty.

We list a number of analytic expressions for the first few orders of scattering and for the source geometries considered in Sect. 6, indicating the photon direction ω by polar coordinates (ϑ, φ), using the abbreviation

$$\vartheta_\lambda = \arccos(1 - \lambda + \lambda_0) \tag{8.14}$$

for the deflection in the first scattering, the step function symbol

$$\mathrm{St}(x) = \begin{cases} 0 & \text{for } x < 0 \\ 1 & \text{for } x > 0, \end{cases} \tag{8.15}$$

[1] The Bronwin quadrature formula has been utilized in Ref. [93].

and a function $H(a, z)$ defined by

$$H(a,z)=\begin{cases} e^{-az}, & \text{for } a>0, \quad z>0, \\ -e^{-az}, & \text{for } a<0, \quad z<0, \\ 0, & \text{for } az<0. \end{cases} \tag{8.16}$$

Assuming $\cos\vartheta_0 > 0$, we have then for the source types defined in Sect. 6:

Plane source (6.5) *collimated* in the direction $\boldsymbol{\omega}_0$

$$\left.\begin{aligned} I^{(0)} &= \frac{1}{\cos\vartheta_0}\,\delta(\boldsymbol{\omega}-\boldsymbol{\omega}_0)\,H\left(\frac{\mu_0}{\cos\vartheta_0},\,z\right)\delta(\lambda-\lambda_0), \\ I^{(1)} &= \frac{1}{2\pi\cos\vartheta\cos\vartheta_0}\,\delta(\boldsymbol{\omega}\cdot\boldsymbol{\omega}_0-\cos\vartheta_\lambda)\,k(\lambda,\lambda_0)\,\frac{H\left(\frac{\mu_0}{\cos\vartheta_0},\,z\right)-H\left(\frac{\mu}{\cos\vartheta},\,z\right)}{\frac{\mu}{\cos\vartheta}-\frac{\mu_0}{\cos\vartheta_0}}; \end{aligned}\right\} \tag{8.17}$$

Plane conical source (6.10)

$$\left.\begin{aligned} I^{(0)} &= \frac{1}{2\pi\cos\vartheta_0}\,\delta(\cos\vartheta-\cos\vartheta_0)\,H\left(\frac{\mu_0}{\cos\vartheta_0},\,z\right)\delta(\lambda-\lambda_0), \\ I^{(1)} &= \frac{1}{2\pi\cos\vartheta\cos\vartheta_0}\,[\cos\vartheta-\cos(\vartheta_0+\vartheta_\lambda)]^{-\frac{1}{2}}\,[\cos(\vartheta_0-\vartheta_\lambda)-\cos\vartheta]^{-\frac{1}{2}}\,k(\lambda,\lambda_0)\times \\ &\quad\times\frac{H\left(\frac{\mu_0}{\cos\vartheta_0},\,z\right)-H\left(\frac{\mu}{\cos\vartheta},\,z\right)}{\frac{\mu}{\cos\vartheta}-\frac{\mu_0}{\cos\vartheta_0}}; \end{aligned}\right\} \tag{8.18}$$

Plane perpendicular source (6.11)

$$\left.\begin{aligned} I^{(0)} &= \frac{1}{2\pi}\,\delta(\cos\vartheta-1)\,\delta(\lambda-\lambda_0)\,H(\mu_0,z), \\ I^{(1)} &= \frac{1}{2\pi\cos\vartheta}\,\delta(\cos\vartheta-\cos\vartheta_\lambda)\,k(\lambda,\lambda_0)\,\frac{H(\mu_0,z)-H\left(\frac{\mu}{\cos\vartheta},\,z\right)}{\frac{\mu}{\cos\vartheta}-\mu_0}, \\ I^{(\bar{2})} &= \frac{k^{(\bar{2})}(\lambda,\lambda_0,\cos\vartheta)}{\cos\frac{1}{2}\vartheta_L\cos\vartheta}\,\frac{1}{\frac{\bar{\mu}}{\cos\frac{1}{2}\vartheta_L}-\mu_0}\times \\ &\quad\times\left\{\frac{H(\mu_0,z)-H\left(\frac{\mu}{\cos\vartheta},\,z\right)}{\frac{\mu}{\cos\vartheta}-\mu_0}-\frac{H\left(\frac{\bar{\mu}}{\cos\frac{1}{2}\vartheta_L},\,z\right)-H\left(\frac{\mu}{\cos\vartheta},\,z\right)}{\frac{\mu}{\cos\vartheta}-\frac{\bar{\mu}}{\cos\frac{1}{2}\vartheta_L}}\right\}, \end{aligned}\right\} \tag{8.19}$$

where $\bar{\mu}=\mu\left(\frac{\lambda+\lambda_0}{2}\right)$, $k^{(\bar{2})}$ is given by (8.13), and the expression for $I^{(\bar{2})}$ approximates $I^{(2)}$ when $\lambda\sim\lambda_L$;

Plane isotropic source (6.12)

$$I^{(0)} = \frac{1}{4\pi\cos\vartheta}\,\delta(\lambda-\lambda_0)\,H\left(\frac{\mu_0}{\cos\vartheta},\,z\right); \tag{8.20}$$

Point collimated source (6.13)

$$\left.\begin{aligned} I^{(0)} &= \delta(\cos\vartheta-1)\,\frac{1}{4\pi^2\varrho}\,\delta(\varrho)\,\delta(\lambda-\lambda_0)\,H(\mu_0,z), \\ I^{(1)} &= \delta(\cos\vartheta-\cos\vartheta_\lambda)\,\delta(\varphi-\beta)\,\frac{1}{2\pi\varrho\sin\vartheta}\,\mathrm{St}(z-\varrho\cot\vartheta_\lambda)\,e^{-\mu_0(z-\varrho\cot\vartheta_\lambda)-\mu\varrho/\sin\vartheta_\lambda}\,k(\lambda,\lambda_0); \end{aligned}\right\} \tag{8.21}$$

Point isotropic source (6.15)

$$\left.\begin{aligned} I^{(0)} &= \frac{1}{2\pi}\,\delta(\cos\vartheta_r - 1)\,\frac{1}{4\pi r^2}\,e^{-\mu_0 r}\,\delta(\lambda-\lambda_0),\\ I^{(1)} &= \frac{1}{2\pi\sin\vartheta_\lambda\sin\vartheta_r}\,\mathrm{St}(\cos\vartheta_r - \cos\vartheta_\lambda)\,\frac{1}{4\pi r}\times\\ &\qquad\times e^{-(\mu_0 r\sin\vartheta_r/\sin\vartheta_\lambda)-(\mu r\sin(\vartheta_\lambda-\vartheta_r)/\sin\vartheta_\lambda)}\,k(\lambda,\lambda_0).\end{aligned}\right\} \tag{8.22}$$

9. Fourier-Legendre transformation. The treatment in this section is general and brings out the basic structure of the theory of penetration and diffusion. It provides the framework within which one can derive and inter-relate particular methods and limiting cases, such as the treatment of special simplified problems (Sect. 10), the theory of deep penetration (Sect. 11), the moment-method (Sect. 12) and various approximate procedures (Sects. 13 and 14). In addition, the relationships between the solutions for different source geometries are examined in this section from the standpoint of the general theory, and the treatment of polarization effects is outlined.

The transport process in an *infinite homogeneous isotropic medium* is invariant under translations and rotations. Harmonic analysis of the X-ray flux $I(\boldsymbol{r}, \boldsymbol{\omega}, \lambda)$ takes advantage of this invariance[1].

We consider, without loss of generality, the X-ray flux from a plane monochromatic source, which obeys the transport equation (6.9). The dependence of this flux upon the distance from the source can be represented as a Fourier integral by setting

$$I(z, \boldsymbol{\omega}, \lambda) = \frac{1}{2\pi i}\int_{-i\infty}^{i\infty} dp\, e^{-pz}\, I(p, \boldsymbol{\omega}, \lambda), \tag{9.1}$$

$$I(p, \boldsymbol{\omega}, \lambda) = \int_{-\infty}^{\infty} dz\, e^{pz}\, I(z, \boldsymbol{\omega}, \lambda). \tag{9.2}$$

The transform $I(p, \boldsymbol{\omega}, \lambda)$ is a function of the complex variable p, and is defined by analytical continuation provided the integral in (9.2) exists for a limited range of values of p. The transformation (9.2) utilizes a factor $\exp(pz)$ instead of the more usual $\exp(-iqz)$ because we shall consider primarily real values of p. Multiplication of the transport equation (6.9) by $\exp(pz)$, followed by integration over z—in which the term with the gradient is integrated by parts and the values of I at $z = \pm\infty$ are taken as zero (Sect. 6β)—yields the transformed equation

$$\left.\begin{aligned} -p\,\omega_z\, I(p, \boldsymbol{\omega}, \lambda) &= -\mu(\lambda)\, I(p, \boldsymbol{\omega}, \lambda) + \\ &+ \int_{\lambda_0}^{\lambda} d\lambda' \int_{4\pi} d\boldsymbol{\omega}'\, k(\boldsymbol{\omega}, \lambda, \boldsymbol{\omega}', \lambda')\, I(p, \boldsymbol{\omega}', \lambda') + \delta(\boldsymbol{\omega} - \boldsymbol{\omega}_0)\,\delta(\lambda - \lambda_0).\end{aligned}\right\} \tag{9.3}$$

This transformation has achieved a separation of the variable p, in the sense that (9.3) can be solved *separately for each value of* p. The separation fails when μ and k depend on the position $\boldsymbol{r}$, i.e., when the medium is not infinite and homogeneous. A further separation of variables will be achieved below, assuming that the medium is isotropic, because the transport process represented by (9.3) is then independent of the component of $\boldsymbol{\omega}$ perpendicular to the z axis. Only two independent variables remain inherently related by the transport process, namely the obliquity of X-rays with respect to the z axis, represented by ω_z, and their wavelength.

[1] This point has been emphasized and formulated in group-theoretical language by E. WIGNER: Phys. Rev. **94**, 17 (1954).

Fourier transformation of the general transport equation (6.2), by multiplication with $\exp(-i\,\boldsymbol{q}\cdot\boldsymbol{r})$ followed by integration over $\boldsymbol{r}$, yields directly an equation equivalent to (9.3), namely

$$\left.\begin{aligned} i\,\boldsymbol{q}\cdot\boldsymbol{\omega}\, I(\boldsymbol{q},\boldsymbol{\omega},\lambda) &= -\mu(\lambda)\, I(\boldsymbol{q},\boldsymbol{\omega},\lambda) + \\ &+ \int_0^\lambda d\lambda' \int_{4\pi} d\boldsymbol{\omega}'\, k(\boldsymbol{\omega},\lambda,\boldsymbol{\omega}',\lambda')\, I(\boldsymbol{q},\boldsymbol{\omega}',\lambda') + S(\boldsymbol{q},\boldsymbol{\omega},\lambda), \end{aligned}\right\} \tag{9.4}$$

where

$$I(\boldsymbol{q},\boldsymbol{\omega},\lambda) = \int d\boldsymbol{r}\, e^{-i\boldsymbol{q}\cdot\boldsymbol{r}}\, I(\boldsymbol{r},\boldsymbol{\omega},\lambda), \tag{9.5}$$

$$S(\boldsymbol{q},\boldsymbol{\omega},\lambda) = \int d\boldsymbol{r}\, e^{-i\boldsymbol{q}\cdot\boldsymbol{r}}\, S(\boldsymbol{r},\boldsymbol{\omega},\lambda). \tag{9.6}$$

Eq. (9.4) reduces to (9.3), apart from the greater generality of the source term, if one takes the z axis in the direction of $\boldsymbol{q}$ and sets $q = i p$. The 3-dimensional Fourier transformation underlies the reduction to plane geometry here as in (6.8).

The flux, as a function of the direction $\boldsymbol{\omega}$, can be expanded in a series of orthonormal spherical harmonics $Y_l^m(\vartheta,\varphi)$ [1], indicated concisely by $Y_l^m(\boldsymbol{\omega})$, setting

$$I(p,\boldsymbol{\omega},\lambda) = \sum_{lm} \sqrt{\frac{2l+1}{4\pi}}\, I_l^m(p,\lambda)\, Y_l^m(\boldsymbol{\omega}), \tag{9.7}$$

$$I_l^m(p,\lambda) = \sqrt{\frac{4\pi}{2l+1}} \int_{4\pi} d\boldsymbol{\omega}\, Y_l^{m*}(\boldsymbol{\omega})\, I(p,\boldsymbol{\omega},\lambda). \tag{9.8}$$

When polarization is disregarded, the kernel k of (9.3) depends on $\boldsymbol{\omega}$ and $\boldsymbol{\omega}'$ only through $\boldsymbol{\omega}\cdot\boldsymbol{\omega}'$ and may be expanded, with application of the addition theorem[1] in the form

$$k(\boldsymbol{\omega},\lambda,\boldsymbol{\omega}',\lambda') = \sum_l \frac{2l+1}{4\pi}\, k_l(\lambda,\lambda')\, P_l(\boldsymbol{\omega}\cdot\boldsymbol{\omega}') = \sum_{lm} k_l(\lambda,\lambda')\, Y_l^m(\boldsymbol{\omega})\, Y_l^{m*}(\boldsymbol{\omega}') \tag{9.9}$$

with

$$k_l(\lambda,\lambda') = \int_{4\pi} d\boldsymbol{\omega}\, k(\boldsymbol{\omega},\lambda,\boldsymbol{\omega}',\lambda')\, P_l(\boldsymbol{\omega}\cdot\boldsymbol{\omega}'). \tag{9.10}$$

For Compton scattering, when the kernel is given by (8.5), we have

$$k_l(\lambda,\lambda') = k(\lambda,\lambda')\, P_l(1-\lambda'+\lambda). \tag{9.11}$$

The transport equation (9.3) will be transformed by multiplication with $\sqrt{\frac{4\pi}{2l+1}}\times Y_l^{m*}(\boldsymbol{\omega})$ and integration over $\boldsymbol{\omega}$. Utilizing the expansion (9.9) and the orthonormality of the Y_l^m, and expressing $\cos\vartheta\, Y_l^m$ as a combination of Y_{l+1}^m and Y_{l-1}^m, one finds

$$\left.\begin{aligned} -\frac{p}{2l+1}\left[\sqrt{(l+1)^2-m^2}\, I_{l+1}^m(p,\lambda) + \sqrt{l^2-m^2}\, I_{l-1}^m(p,\lambda)\right] &= -\mu(\lambda)\, I_l^m(p,\lambda) + \\ + \int_{\lambda_0}^{\lambda} d\lambda'\, k_l(\lambda,\lambda')\, I_l^m(p,\lambda') + \sqrt{\frac{4\pi}{2l+1}}\, Y_l^{m*}(\boldsymbol{\omega}_0)\, \delta(\lambda-\lambda_0). \end{aligned}\right\} \tag{9.12}$$

The Eqs. (9.12) with different values of l are interlinked by the terms on the left side. However, equations with different values of m are independent, so that the system of Eqs. (9.12) can be solved *separately for each value of m*. Only the variable λ and the index l, which is a transform variable corresponding to ω_z, remain interlinked, as anticipated above.

[1] Cf. Vol. I of this Encyclopedia, particularly p. 197 and p. 205 Eq. (19.5). Complex conjugation will be indicated by an asterisk, as e.g., in $Y_l^{m*}(\vartheta,\varphi)$.

An alternative form of (9.12) is obtained by adding and subtracting $p I_l^m$ in (9.12) and grouping various terms to yield

$$\left.\begin{aligned}[\mu(\lambda) - p]\, I_l^m(p,\lambda) - \frac{1}{2} p\, \Delta I_l^m(p,\lambda) = \int_{\lambda_0}^{\lambda} d\lambda'\, k_l(\lambda,\lambda')\, I_l^m(p,\lambda) + \\ + \sqrt{\frac{4\pi}{2l+1}}\, Y_l^{m*}(\omega_0)\, \delta(\lambda-\lambda_0),\end{aligned}\right\} \tag{9.13}$$

where Δ indicates a difference operator analogous to the Laplacian in cylindrical coordinates

$$\Delta I_l^m = \frac{\sqrt{(l+1)^2 - m^2}}{l+\frac{1}{2}}\, I_{l+1}^m + \frac{\sqrt{l^2-m^2}}{l+\frac{1}{2}}\, I_{l-1}^m - 2 I_l^m. \tag{9.14}$$

For $l \gg m$ and $l \gg 1$, differences like $I_{l+1}^m - I_l^m$ may be replaced with derivatives and Δ becomes a Bessel operator

$$\Delta = \frac{d^2}{dl^2} + \frac{1}{l}\frac{d}{dl} - \frac{m^2}{l^2}. \tag{9.15}$$

The transformed Eq. (9.12) or (9.13) affords a convenient point of departure for the various approximation treatments of the penetration problem. Deep penetration corresponds to values of p near μ_s [the smallest value of $\mu(\lambda)$ in the range of integration], and this limitation permits a further approximate separation of the variables l and λ (Sect. 11). Expansion of (9.12) into powers of p, which corresponds to a calculation of successive spatial moments of the flux $I(\boldsymbol{r}, \omega, \lambda)$, treats the linkage term as a perturbation and thereby unravels the interlinkage between the equations for different I_l^m's (Sect. 12). The limited problems mentioned in Sect. 7, which deal with simple X-ray degradation, or with degradation and deflection, independently of penetration, correspond to a study of (9.12) in the special case $p=0$ (Sect. 10). Whenever the angular distribution of photons approaches isotropy under the influence of scattering, I_l^m is a rapidly decreasing function of l; the interlinkage term which contains I_{l+1}^m in (9.12) may then be treated as a weak perturbation and (9.12) reduces, in an extreme approximation, to an equation of diffusion theory (Sect. 13).

The general solution of the transport problem in an infinite, uniform, and isotropic medium can be expressed in terms of the solution corresponding to a standard type of plane source. As a standard source we choose $\delta(z)\sqrt{(2l_0+1)/4\pi} \times Y_{l_0}^m(\omega)\,\delta(\lambda-\lambda_0)$, i.e., a plane monochromatic source whose directional distribution is represented by a single spherical harmonic and whose Fourier-Legendre transform is simply $\delta_{l l_0}\,\delta(\lambda-\lambda_0)$. The solution of the transform equation (9.12) or (9.13), modified by replacing its source term with $\delta_{l l_0}\,\delta(\lambda-\lambda_0)$, serves as a Green's function to represent the solution of the transport problem for a general source. This function will be called $G^{(p,m)}(l\lambda, l_0\lambda_0)$, where m is a parameter which cannot be larger than l or l_0. The solution of the transform equation (9.3) in terms of the Green's function is

$$I(p,\omega,\lambda) = \sum_{l l_0 m} \sqrt{\frac{2l+1}{4\pi}}\, Y_l^m(\omega)\, G^{(p,m)}(l\lambda, l_0\lambda_0) \sqrt{\frac{4\pi}{2l_0+1}}\, Y_{l_0}^{m*}(\omega_0). \tag{9.16}$$

The inverse transform of (9.16), which is the solution of (6.9), is

$$I(z,\omega,\lambda) = \sum_{l l_0 m} \sqrt{\frac{2l+1}{2l_0+1}}\, Y_l^m(\omega)\, \frac{1}{2\pi i} \int_{-i\infty}^{i\infty} dp\, e^{-pz}\, G^{(p,m)}(l\lambda, l_0\lambda_0)\, Y_{l_0}^{m*}(\omega_0). \tag{9.17}$$

The more general transform equation (9.4) has a solution analogous to (9.16), namely

$$\left.\begin{aligned} I(\boldsymbol{q},\boldsymbol{\omega},\lambda) &= \int_0^\lambda d\lambda_0 \sum_{l l_0 \hat{m}} \sqrt{\frac{2l+1}{4\pi}}\, Y_l^{\hat{m}}(\hat{\vartheta},\hat{\varphi})\, G^{(-iq,\hat{m})}(l\lambda, l_0\lambda_0) \times \\ &\times \int_{4\pi} d\omega_0 \sqrt{\frac{4\pi}{2l_0+1}}\, Y_{l_0}^{\hat{m}*}(\hat{\vartheta}_0,\hat{\varphi}_0)\, S(\boldsymbol{q},\boldsymbol{\omega}_0,\lambda_0), \end{aligned}\right\} \tag{9.18}$$

where $\hat{\vartheta}$, $\hat{\varphi}$, and $\hat{\vartheta}_0$, $\hat{\varphi}_0$ indicate the polar coordinates of $\boldsymbol{\omega}$ and $\boldsymbol{\omega}_0$ in a system with axis $\boldsymbol{q}$. From this solution one obtains the solution $I(\boldsymbol{r},\boldsymbol{\omega},\lambda)$ of the general transport equation (6.2) by inverse Fourier transformation, but at this point one must consider that the spherical functions in (9.18) pertain to a coordinate system with polar axis parallel to $\boldsymbol{q}$. The integration over the direction $\boldsymbol{q}$ must therefore be preceded by a transformation which relates the photon directions $\boldsymbol{\omega}$ and $\boldsymbol{\omega}_0$ to a new coordinate system independent of the orientation of $\boldsymbol{q}$. This transformation and the subsequent integration over $\boldsymbol{q}$ are described below for the case of a point source.

α) *Connection between the fluxes from plane and point sources.* The flux from a point source is given in terms of the flux from a plane source by a well known and simple formula (Ref. [*24*], p. 64) in the special case that: (1) the sources are isotropic, and (2) one considers only the flux integrated over all photon directions

$$I(\boldsymbol{r},\lambda) = \int_{4\pi} d\omega\, I(\boldsymbol{r},\boldsymbol{\omega},\lambda). \tag{9.19}$$

In this case, to be called case A, one first expresses the integrated flux from a plane source as the flux from a plane distribution of point sources, $I(z,\lambda) = \int_0^\infty 2\pi\varrho\, d\varrho\, I(r,\lambda)|_{r=\sqrt{z^2+\varrho^2}}$. The derivative of this equation provides the desired inverse relationship

$$I(r,\lambda) = \frac{1}{2\pi r}\left[-\frac{\partial}{\partial z} I(z,\lambda)\right]_{z=r}. \tag{9.20}$$

The plane-to-point connection is more complicated when only one of the conditions 1 and 2 obtains (case B), and still more complicated when neither condition obtains (case C). The general plane-to-point connection is provided by the Fourier-transform procedure starting from (9.18). This procedure will be applied in succession to the various cases of increasing complication, beginning with an alternative derivation of the simple result (9.20).

Case A. If the source is monochromatic the integration over λ_0 in (9.18) drops out. If the source is concentrated at the point $\boldsymbol{r}_0 = 0$, S is independent of $\boldsymbol{q}$. Further, if the *source* is also *isotropic* the whole integral over $\boldsymbol{\omega}_0$ reduces to $\delta_{l_0 0}\delta_{m0}$. Finally, *integration of the flux* over $\boldsymbol{\omega}$, as in (9.19), deletes all terms with $l \neq 0$ so that (9.18) reduces to

$$I(\boldsymbol{q},\lambda) = G^{(-iq,0)}(0\,\lambda, 0\,\lambda_0). \tag{9.21}$$

The inverse transform of this equation is

$$\left.\begin{aligned} I(\boldsymbol{r},\lambda) &= \frac{1}{(2\pi)^3}\int d\boldsymbol{q}\, e^{i\boldsymbol{q}\cdot\boldsymbol{r}}\, G^{(-iq,0)}(0\,\lambda, 0\,\lambda_0) \\ &= \frac{1}{(2\pi)^3}\int_0^\infty q^2\, dq \int_{4\pi} d\boldsymbol{\Omega}\, e^{iq\boldsymbol{\Omega}\cdot\boldsymbol{r}}\, G^{(-iq,0)}(0\,\lambda, 0\,\lambda_0). \end{aligned}\right\} \tag{9.22}$$

The integration over $\boldsymbol{\Omega}=\boldsymbol{q}/q$ is elementary and yields

$$\left.\begin{aligned} I(r,\lambda) &= \frac{1}{(2\pi)^3}\int_0^\infty q^2\,dq\,\frac{2\pi}{i\,q\,r}\left[e^{iqr}-e^{-iqr}\right]G^{(-iq,0)}(0\,\lambda,0\,\lambda_0)\\ &= \frac{1}{2\pi r}\,\frac{1}{2\pi i}\int_{-\infty}^{\infty} q\,dq\,e^{iqr}G^{(-iq,0)}(0\,\lambda,0\,\lambda_0)\\ &= \frac{1}{2\pi r}\,\frac{1}{2\pi i}\int_{-i\infty}^{i\infty} p\,dp\,e^{-pr}G^{(p,0)}(0\,\lambda,0\,\lambda_0), \end{aligned}\right\} \tag{9.23}$$

utilizing the symmetry property of the Green's function,

$$G^{(-p,m)}(l\,\lambda,l_0\,\lambda_0) = (-1)^{l-l_0}G^{(p,m)}(l\,\lambda,l_0\,\lambda_0), \tag{9.24}$$

which follows by inspection of (9.12). Comparison of (9.23) with the expression (9.17) of the flux from a plane source leads to (9.20). The derivative in (9.20) is the inverse transform of the factor p in the integral on the right of (9.23).

Case B. Consider now a *point source* at $\boldsymbol{r}_0=0$ *collimated* in the direction $\boldsymbol{\omega}_0$, and the resulting *flux integrated* over $\boldsymbol{\omega}$. The integration of (9.18) over $\boldsymbol{\omega}$ still deletes all terms with l and $\hat{m}$ other than zero. Since $\sqrt{4\pi/(2l_0+1)}\,Y_{l_0}^{0*}(\hat{\vartheta}_0\,\hat{\varphi}_0)$ equals the Legendre polynomial $P_{l_0}(\boldsymbol{\Omega}\cdot\boldsymbol{\omega}_0)$, we have now, instead of (9.22),

$$I(\boldsymbol{r},\lambda) = \frac{1}{(2\pi)^3}\int_0^\infty q^2\,dq\int_{4\pi} d\boldsymbol{\Omega}\,e^{iq\boldsymbol{\Omega}\cdot\boldsymbol{r}}\sum_{l_0} G^{(-iq,0)}(0\,\lambda,l_0\,\lambda_0)P_{l_0}(\boldsymbol{\Omega}\cdot\boldsymbol{\omega}_0). \tag{9.25}$$

The integration over $\boldsymbol{\Omega}$ utilizes the harmonic analysis of the dependence of $\exp(iq\,\boldsymbol{\Omega}\cdot\boldsymbol{r})$ upon the angle between $\boldsymbol{\Omega}$ and $\boldsymbol{u}=\boldsymbol{r}/r$, i.e. the expansion[1]

$$\left.\begin{aligned} e^{iq\boldsymbol{\Omega}\cdot\boldsymbol{r}} &= \sum_L f_L(q\,r)\,P_L(\boldsymbol{\Omega}\cdot\boldsymbol{u})\\ &= \sum_L \frac{2L+1}{2i\,q\,r}\left[e^{iqr}S_{L+\frac12}(-2iqr)+(-1)^{L+1}e^{-iqr}S_{L+\frac12}(2iqr)\right]P_L(\boldsymbol{\Omega}\cdot\boldsymbol{u}), \end{aligned}\right\} \tag{9.26}$$

where f_L is a spherical Bessel function of half-integral order and can therefore be expressed in terms of exponentials and of a standard polynomial of degree L,

$$S_{L+\frac12}(x) = \sum_{\nu=0}^{L}\frac{(L+\nu)!}{\nu!(L-\nu)!}\left(\frac{1}{x}\right)^{\nu}. \tag{9.27}$$

The factor $P_{l_0}(\boldsymbol{\Omega}\cdot\boldsymbol{\omega}_0)$ in (9.25) can also be expressed in terms of harmonics of the orientation of $\boldsymbol{\Omega}$ with respect to $\boldsymbol{u}$, by means of the addition theorem utilized in (9.9). The integration over $\boldsymbol{\Omega}$ reduces then to an application of the orthonormality of spherical functions and yields, in analogy to (9.23),

$$\left.\begin{aligned} I(\boldsymbol{r},\lambda) &= \frac{1}{(2\pi)^3}\sum_{l_0}\int_0^\infty q^2\,dq\,f_{l_0}(q\,r)\,G^{(-iq,0)}(0\,\lambda,l_0\,\lambda_0)\,\frac{4\pi}{2l_0+1}\,P_{l_0}(\boldsymbol{u}\cdot\boldsymbol{\omega}_0)\\ &= \frac{1}{2\pi r}\sum_{l_0}\frac{1}{2\pi i}\int_{-\infty}^{\infty} q\,dq\,e^{iqr}\,S_{l_0+\frac12}(-2i\,q\,r)\,G^{(-iq,0)}(0\,\lambda,l_0\,\lambda_0)\,P_{l_0}(\boldsymbol{u}\cdot\boldsymbol{\omega}_0)\\ &= \frac{1}{2\pi r}\sum_{l_0}\frac{1}{2\pi i}\int_{-i\infty}^{i\infty} p\,dp\,e^{-pr}\,S_{l_0+\frac12}(2p\,r)\,G^{(p,0)}(0\,\lambda,l_0\,\lambda_0)\,P_{l_0}(\boldsymbol{u}\cdot\boldsymbol{\omega}_0). \end{aligned}\right\} \tag{9.28}$$

[1] G.N. WATSON: Bessel Functions, p. 368 and 52. Cambridge: Cambridge University Press 1945. — E. JAHNKE and F. EMDE: Tables of Functions, p. 136. Leipzig: Teubner 1938.

The last expression differs from the corresponding expression for a plane source, i.e., from the integral of (9.17) over ω, by: (a) the factor $1/2\pi r$, which also appears in (9.20), (b) the factor p in the integrand, which also appears in (9.23) and is equivalent to the differential operator in (9.20), and (c) the factor $S_{l_0+\frac{1}{2}}(2pr)$ in the integrand which is of degree l_0 in $1/p$ and may therefore be represented as an l_0-fold integral operator.

At great depths of penetration, i.e., for large values of r, the factor $\exp(-pr)$ in the integrand of (9.28) is a steep function of p, and the integral is given approximately by a steepest descent path in the proximity of a singularity of the Green's function at $p=\mu_s$ (see Sect. 11). The factor $S_{l_0+\frac{1}{2}}(pr)\,p$ is a slowly varying function of p along this path and may be taken as a constant; indeed this constant reduces approximately to $p=\mu_s$ for large values of r, as seen from the analytical form of $S_{l_0+\frac{1}{2}}$ given by (9.27). That is, at large distances from a point source the integrated flux differs from the corresponding flux from a plane source by the simple factor $\mu_s/2\pi r$.

If the source is isotropic but one inquires about the differential flux in a direction ω, one obtains a result analogous to (9.28), namely

$$I(r,\omega,\lambda)=\frac{1}{2\pi r}\sum_l\frac{2l+1}{4\pi}P_l(\boldsymbol{u}\cdot\omega)\frac{1}{2\pi i}\int_{-i\infty}^{i\infty}p\,dp\,e^{-pr}S_{l+\frac{1}{2}}(2pr)\,G^{(p,0)}(l\lambda,0\lambda_0)\,. \tag{9.28'}$$

Case C. If the *point source* is *collimated* and one inquires about the *differential flux*, the starting formula has the full generality

$$\left.\begin{aligned}I(\boldsymbol{r},\omega,\lambda)=\frac{1}{(2\pi)^3}\int_0^\infty q^2\,dq\int_{4\pi}d\boldsymbol{\Omega}\,e^{iq\boldsymbol{\Omega}\cdot\boldsymbol{r}}\sum_{l l_0\hat{m}}\sqrt{\frac{2l+1}{2l_0+1}}\times\\ \times Y_l^{\hat{m}}(\hat{\vartheta},\hat{\varphi})\,G^{(-iq,\hat{m})}(l\lambda,l_0\lambda_0)\,Y_{l_0}^{\hat{m}*}(\hat{\vartheta}_0,\hat{\varphi}_0)\,.\end{aligned}\right\} \tag{9.29}$$

The exponential factor can be expanded as in (9.27) and the integral over q can be transformed as in (9.23) and (9.28)[1],

$$\left.\begin{aligned}I(\boldsymbol{r},\omega,\lambda)=\sum_{Ll l_0\hat{m}}\sqrt{\frac{2L+1}{2l_0+1}}\,\frac{2L+1}{4\pi}\int_{4\pi}d\boldsymbol{\Omega}\,Y_l^{\hat{m}}(\hat{\vartheta},\hat{\varphi})\,Y_{l_0}^{\hat{m}*}(\hat{\vartheta}_0,\hat{\varphi}_0)\,P_L(\boldsymbol{\Omega}\cdot\boldsymbol{u})\times\\ \times\frac{1}{2\pi r}\,\frac{1}{2\pi i}\int_{-i\infty}^{i\infty}p\,dp\,e^{-pr}S_{L+\frac{1}{2}}(2pr)\,G^{(p,\hat{m})}(l\lambda,l_0\lambda_0)\,,\end{aligned}\right\} \tag{9.30}$$

but the integration over $\boldsymbol{\Omega}$ requires more elaborate preliminaries. The spherical harmonics $\hat{Y}_l^{m}(\hat{\vartheta},\hat{\varphi})\,Y_{l_0}^{\hat{m}*}(\hat{\vartheta}_0,\hat{\varphi}_0)$ in (9.18) and in the following formulas depend on the polar coördinates of the directions ω and ω_0 with respect to the axis $\boldsymbol{\Omega}$. To carry out the integration over $\boldsymbol{\Omega}$, these spherical harmonics have to be replaced with functions of new coordinates of ω, ω_0, with respect to a suitable fixed coordinate axis, such as $\boldsymbol{u}=\boldsymbol{r}/r$.[2] Calling $\widetilde{\Theta}$ and $\widetilde{\Phi}$ the coordinates of $\boldsymbol{\Omega}$, $\tilde{\vartheta}$ and $\tilde{\varphi}$ the coordinates of ω and $\tilde{\vartheta}_0$ and $\tilde{\varphi}_0$ the coordinates of ω_0 in a system with axis $\boldsymbol{u}$, one finds

$$\left.\begin{aligned}Y_l^{\hat{m}}(\hat{\vartheta},\hat{\varphi})\,Y_{l_0}^{\hat{m}*}(\hat{\vartheta}_0,\hat{\varphi}_0)=\sum_{JM\tilde{m}_0}(-1)^{\hat{m}-\tilde{m}_0}Y_l^{M+\tilde{m}_0}(\tilde{\vartheta},\tilde{\varphi})\,Y_{l_0}^{\tilde{m}_0*}(\tilde{\vartheta}_0,\tilde{\varphi}_0)\times\\ \times(l\,M+\tilde{m}_0,l_0-\tilde{m}_0|l\,l_0\,J\,M)\sqrt{\frac{4\pi}{2J+1}}\,Y_J^{M*}(\widetilde{\Theta},\widetilde{\Phi})\,(l\,l_0\,J\,0|l\,\hat{m},l_0-\hat{m}),\end{aligned}\right\} \tag{9.31}$$

---

[1] The derivation of (9.30) again utilizes (9.24) and also assumes that $L+l-l_0$ is even Terms with $L+l_0-l$ odd actually drop out, as shown further below.

[2] The function $P_{l_0}(\boldsymbol{\Omega}\cdot\omega_0)$ in (9.25) was similarly replaced by means of the addition theorem with functions of the coordinates of $\boldsymbol{\Omega}$ and ω_0 with respect to $\boldsymbol{u}$.

where the symbols $(\ldots|\ldots)$ indicate the Wigner transformation coefficients of the quantum-mechanical "addition of angular momenta"[1].

When (9.31) is entered in (9.30), the integration over $\boldsymbol{\Omega}$ reduces once more to an application of the orthonormality of spherical functions, all terms with $J \neq L$ and $M \neq 0$ vanish, and (9.30) reduces to

$$\left.\begin{aligned} I(\boldsymbol{r}, \boldsymbol{\omega}, \lambda) = \sum_{l l_0 \tilde{m}_0} \sqrt{\frac{2l+1}{2l_0+1}}\, Y_l^{\tilde{m}_0}(\tilde{\vartheta}, \tilde{\varphi})\, Y_{l_0}^{\tilde{m}_0 *}(\tilde{\vartheta}_0, \tilde{\varphi}_0)\, (-1)^{\tilde{m}_0} \sum_L (l\, \tilde{m}_0, l_0 - \tilde{m}_0 | l\, l_0\, L\, 0) \times \\ \times \frac{1}{2\pi r} \frac{1}{2\pi i} \int_{-i\infty}^{i\infty} p\, dp\, e^{-pr} S_{L+\frac{1}{2}}(2pr) \sum_{\hat{m}} (l\, l_0\, L\, 0 | l\, \hat{m}, l_0 - \hat{m})\, (-1)^{\hat{m}}\, G^{(p, \hat{m})}(l\, \lambda, l_0\, \lambda_0), \end{aligned}\right\} \quad (9.32)$$

in terms of the new coordinates of $\boldsymbol{\omega}$ and $\boldsymbol{\omega}_0$ with polar axis along $\boldsymbol{r} = r\boldsymbol{u}$. All terms of the summation over L vanish for which L lies above $l + l_0$, below $|l - l_0|$, or differs from these limits by an odd number. If L is outside these limits, the Wigner coefficients vanish. If $L - l - l_0$ is odd, the terms with $\hat{m}$ of equal magnitude and opposite sign cancel out, because $G^{(p, -\hat{m})} = G^{(p, \hat{m})}$ but $(l l_0 L 0 | l - \hat{m}, l_0 \hat{m}) = (-1)^{L-l-l_0} (l l_0 L 0 | l \hat{m}, l_0 - \hat{m})$. For the same reason the coefficient of the spherical functions in (9.32) depends only on the magnitude and not on the sign of $\tilde{m}_0$, so that the flux distribution in directions $\boldsymbol{\omega}$ is symmetric with respect to the plane of $\boldsymbol{r}$ and $\boldsymbol{\omega}_0$.

Eq. (9.32) gives the angular distribution of the X-ray flux with respect to the radius vector $\boldsymbol{r}$ from the source to the point of observation; this distribution depends on the direction $\boldsymbol{\omega}_0$ of the source X rays with respect to $\boldsymbol{r}$. An analogous equation gives the distribution of the X-ray flux with respect to the direction of source collimation $\boldsymbol{\omega}_0$; the distribution depends then on the orientation of the point of observation $\boldsymbol{r}$ with respect to $\boldsymbol{\omega}_0$. The new equation is expressed in terms of the polar coordinates ϑ, φ, of $\boldsymbol{\omega}$ and (r, α, β) of $\boldsymbol{r}$ in a system with axis $\boldsymbol{\omega}_0$. Transformation of the product $Y_l^{\tilde{m}_0}(\tilde{\vartheta}, \tilde{\varphi})\, Y_{l_0}^{\tilde{m}_0 *}(\tilde{\vartheta}_0, \tilde{\varphi}_0)$ in (9.32) by a formula equivalent to (9.31) yields

$$\left.\begin{aligned} I(\boldsymbol{r}, \boldsymbol{\omega}, \lambda) = \sum_{l_0 l m} \sqrt{\frac{2l+1}{2L+1}}\, Y_l^{m}(\vartheta, \varphi) \sum_L (l\, m, l_0\, 0 | l\, l_0\, L\, m)\, Y_L^{m *}(\alpha, \beta) \times \\ \times \frac{1}{2\pi r} \frac{1}{2\pi i} \int_{-i\infty}^{i\infty} p\, dp\, e^{-pr} S_{L+\frac{1}{2}}(2pr) \sum_{\hat{m}} (l\, l_0\, L\, 0 | l\, \hat{m}, l_0 - \hat{m})\, (-1)^{\hat{m}}\, G^{(p, \hat{m})}(l\, \lambda, l_0\, \lambda_0). \end{aligned}\right\} \quad (9.33)$$

β) Polarization effects. When polarization is considered, the Fourier transformation of the transport equation (6.17) proceeds like the transformation of (6.2) and yields an equation analogous to (9.4). Here again it is sufficiently general to consider the flux from a plane source, and thus to deal with a transform equation analogous to (9.3). However, the subsequent transformation of direction variables can no longer be carried out by expansion into spherical

---

[1] See e.g. E. U. CONDON and G. SHORTLEY, Theory of Atomic Spectra, London and New York 1935, p. 73, or U. FANO and G. RACAH, Irreducible Tensorial Sets, New York 1959, Sects. 7 and 5. Eq. (9.31) is derived by expressing the initial direct product of spherical harmonics as a sum of irreducible products, of degree J, and then transforming each of these products in correspondence to the change of coordinate axes. The explicit expression of the Wigner coefficients $(j_1 m_1, j_2 m_2 | j_1 j_2 j\, m)$ is

$$\delta_{m, m_1+m_2} \sqrt{\frac{(j_1+j_2-j)!\,(j+j_1-j_2)!\,(j+j_2-j_1)!\,(2j+1)}{(j+j_1+j_2+1)!}} \times$$
$$\times \sum_K \frac{(-1)^K \sqrt{(j_1+m_1)!\,(j_1-m_1)!\,(j_2+m_2)!\,(j_2-m_2)!\,(j+m)!\,(j-m)!}}{K!\,(j_1+j_2-j-K)!\,(j_1-m_1-K)!\,(j_2+m_2-K)!\,(j-j_2+m_1+K)!\,(j-j_1-m_2+K)!}.$$

In this summation K takes on all integral values for which the arguments of the factorial functions involving K are non-negative. Note that $(j_1 j_2 j\, m | j_1 m_1, j_2 m_2) = (j_1 m_1, j_2 m_2 | j_1 j_2 j\, m)$. A numerical tabulation of the Wigner coefficients with $j \leq \frac{9}{2}$ has been made by A. SIMON (ORNL Report 1718, 21 June 1954).

harmonics because the directions of polarization must be considered as well as the photon direction. The expansions (9.7) and (9.9) of the flux and of the kernel must be accordingly generalized.

We take as 0^0-azimuth of X-ray polarizations the plane through a photon direction $\boldsymbol{\omega}$ and the z axis. A scattering process depends not only on the angle $\Theta = \arccos \boldsymbol{\omega}' \cdot \boldsymbol{\omega}$ between the initial and final photon directions, but also on the azimuth angles ψ' and ψ shown in Fig. 10. The three angles ψ', Θ, and ψ constitute a set of Euler angles which relate a coordinate system with axis $\boldsymbol{\omega}'$ to a system with axis $\boldsymbol{\omega}$. The Stokes parameters of the X-ray flux and the kernel of the transport equation are, therefore, to be expanded into functions of the Euler angles which relate different complete systems of coordinates. Such are the transformation functions[1]

$$D^{(l)}_{v'v}(\psi', \Theta, \psi) = e^{i(v'\psi' + v\psi)}\, d^{(l)}_{v'v}(\Theta) \tag{9.34}$$

with

$$d^{(l)}_{v'v}(\Theta) = \sum_r \frac{(-1)^r \sqrt{(l+v')!\,(l-v')!\,(l+v)!\,(l-v)!}}{r!\,(l-v'-r)!\,(l+v-r)!\,(r+v'-v)!} \left(\cos\frac{\Theta}{2}\right)^{2l-v'+v-2r} \left(\sin\frac{\Theta}{2}\right)^{2r+v'-v}. \tag{9.35}$$

These functions are generalizations of the spherical harmonics and reduce to $\sqrt{\dfrac{4\pi}{2l+1}}\, Y_l^v(\Theta, \psi)$ when $v' = 0$.

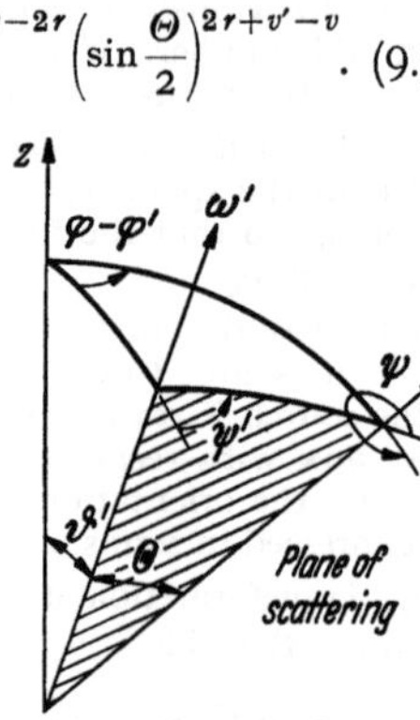

Fig. 10. Geometrical parameters of scattering for polarized radiation.

The main property of the functions D to be utilized is the extension of the addition theorem of $P_l(\boldsymbol{\omega}\cdot\boldsymbol{\omega}')$ which was utilized in (9.9), namely

$$\left.\begin{aligned} D^{(l)}_{v'v}(\psi', \Theta, \psi) &= \sum_m D^{(l)}_{v'm}(\pi, \vartheta', \pi - \varphi')\, D^{(l)}_{mv}(\varphi, \vartheta, 0) \\ &= \sum_m D^{(l)*}_{mv'}(\varphi', \vartheta', 0)\, D^{(l)}_{mv}(\varphi, \vartheta, 0), \end{aligned}\right\} \tag{9.36}$$

where $(\vartheta'\ \varphi')$ are the polar coordinates of $\boldsymbol{\omega}'$ as shown in Fig. 10. This theorem expresses the group property of the transformations between coordinate systems ("frames") with different polar axes, namely that the transformation from the frame $\boldsymbol{\omega}'$ to $\boldsymbol{\omega}$ can be decomposed into two steps: from $\boldsymbol{\omega}'$ to a frame with axis z and from the axis z to $\boldsymbol{\omega}$. One variable of the D's on the right of (9.36) vanishes because the 0^0-azimuth of the frames $\boldsymbol{\omega}'$ and $\boldsymbol{\omega}$ has been chosen to coincide, respectively, with the planes $(z, \boldsymbol{\omega}')$ and $(z, \boldsymbol{\omega})$.

The Stokes parameters, as defined in Sect. 4ζ, are real functions of the azimuth angles of the type $\cos v\psi$ with $v = 0$ or 2 [see Eq. (4.12)], whereas the functions D are complex exponential functions $\exp(iv\psi)$. The simplicity of the addition theorem (9.36) hinges on the use of complex exponentials. To take advantage of (9.36) one should therefore take complex linear combinations of the $I^{[i]}$ and of the matrix elements of the scattering kernel $k^{[ij]}$, such that they depend exponentially on the azimuth variables, and then expand these linear combinations into functions $D^{(l)}_{v'v}$. At the end of the calculation it is convenient to transform back to the ordinary Stokes parameters and matrix elements of the kernel, which are real. We omit the details of the calculation and give its result.

For the Compton effect, with the 0^0-azimuth in the plane of scattering, the kernel $k^{[ij]}(\boldsymbol{\omega}, \lambda, \boldsymbol{\omega}', \lambda')$ is the product of the cross section matrix (4.17) and of the number NZ of electrons per unit volume. This kernel can be factored as in (8.5) where $k(\lambda, \lambda')$ is now a 4×4 matrix whose off-diagonal matrix elements vanish except $k^{[01]}$ and $k^{[10]}$ as in (4.17). The dependence of the kernel on the scattering angle is to be expanded analogously to (9.9) to (9.11), in the form

$$\delta(1 - \boldsymbol{\omega}\cdot\boldsymbol{\omega}' - \lambda + \lambda') = \sum_l (l + \tfrac{1}{2})\, d^{(l)}_{v'v}(\arccos \boldsymbol{\omega}\cdot\boldsymbol{\omega}')\, d^{(l)}_{v'v}(\arccos[1 - \lambda + \lambda']) \tag{9.37}$$

with v' and v equal to 0 or ± 2 for different matrix elements.

The equation analogous to (9.13), which one obtains through the generalized Fourier-Legendre transformation, is

$$\left.\begin{aligned} &[\mu(\lambda) - p]\, I_l^{[i]m}(p, \lambda) - \tfrac{1}{2} p \Delta I_l^{[i]m}(p, \lambda) \\ &\quad = \int_{\lambda_0}^{\lambda} d\lambda' \sum_j k_l^{[ij]}(\lambda, \lambda')\, I_l^{[j]m}(p, \lambda') + S_l^{[i]m}(\omega_0)\, \delta(\lambda - \lambda_0) \end{aligned}\right\} \tag{9.38}$$

[1] See e.g. U. Fano and G. Racah: Irreducible Tensorial Sets. New York 1959.

with the notations defined as follows: The absorption coefficient $\mu(\lambda)$ and the parameter p remain the same as before, since μ would depend on the polarization only if the medium were anisotropic (Sect. 4ζ). The Stokes parameters $I_l^{[i]m}(p,\lambda)$ are defined by

$$\left.\begin{aligned}
I_l^{[0]m}(p,\lambda) &= \int_{4\pi} d\omega\, e^{-im\varphi} d_{m0}^{(l)}(\vartheta)\, I^{[0]}(p,\omega,\lambda),\\
I_l^{[1]m}(p,\lambda) &= \int_{4\pi} d\omega\, e^{-im\varphi} \tfrac{1}{2}\{[d_{m2}^{(l)}(\vartheta) + d_{m,-2}^{(l)}(\vartheta)]\, I^{[1]}(p,\omega,\lambda) -\\
&\qquad - i\,[d_{m2}^{(l)}(\vartheta) - d_{m,-2}^{(l)}(\vartheta)]\, I^{[2]}(p,\omega,\lambda)\},\\
I_l^{[2]m}(p,\lambda) &= \int_{4\pi} d\omega\, e^{-im\varphi} \tfrac{1}{2}\{i\,[d_{m2}^{(l)}(\vartheta) - d_{m,-2}^{(l)}(\vartheta)]\, I^{[1]}(p,\omega,\lambda) +\\
&\qquad + [d_{m2}^{(l)}(\vartheta) + d_{m,-2}^{(l)}(\vartheta)]\, I^{[2]}(p,\omega,\lambda)\},\\
I_l^{[3]m}(p,\lambda) &= \int_{4\pi} d\omega\, e^{-im\varphi} d_{m0}^{(l)}(\vartheta)\, I^{[3]}(p,\omega,\lambda),
\end{aligned}\right\} \tag{9.39}$$

where ϑ and φ are the coordinates of ω. The operator Δ is now defined by

$$\Delta I_l^{[i]m} = \frac{\sqrt{(l+1)^2-v^2}\,\sqrt{(l+1)^2-m^2}}{(l+1)\,(l+\frac{1}{2})}\, I_{l+1}^{[i]m} + \frac{\sqrt{l^2-v^2}\,\sqrt{l^2-m^2}}{l\,(l+\frac{1}{2})}\, I_{l-1}^{[i]m} - 2 I_l^{[i]m}, \tag{9.40}$$

where $v=0$ for $i=0, 3$ and $v=2$ for $i=1, 2$. Note that (9.40) reduces to (9.14) for $v=0$. The matrix $k_l^{[ij]}$ results from the expansion of $k^{[ij]}(\omega,\lambda,\omega',\lambda')$ and is

$$\|k_l^{[ij]}\| = \left\|\begin{matrix}
k^{[00]} d_{00}^{(l)} & k^{[01]} d_{02}^{(l)} & 0 & 0\\
k^{[10]} d_{20}^{(l)} & \begin{matrix}\frac{1}{4}[(k^{[11]}+k^{[22]})\, d_{22}^{(l)} +\\ + (k^{[11]}-k^{[22]})\, d_{2,-2}^{(l)}]\end{matrix} & 0 & 0\\
0 & 0 & \begin{matrix}\frac{1}{4}[(k^{[11]}+k^{[22]})\, d_{22}^{(l)} -\\ - (k^{[11]}-k^{[22]})\, d_{2,-2}^{(l)}]\end{matrix} & 0\\
0 & 0 & 0 & k^{[33]} d_{00}^{(l)}
\end{matrix}\right\| \tag{9.41}$$

where the $k^{[ij]} = NZ\sigma^{[ij]}$, $\sigma^{[ij]}$ is given by (4.17), and the functions $d_{v'v}^{(l)}(\Theta)$ are evaluated at $\Theta = \arccos(1-\lambda+\lambda')$. Finally, the source term is

$$\left.\begin{aligned}
S_l^{[0]m}(\omega_0) &= e^{-im\varphi_0} d_{m0}^{(l)}(\vartheta_0),\\
S_l^{[1]m}(\omega_0) &= e^{-im\varphi_0} \tfrac{1}{2}\{[d_{m2}^{(l)}(\vartheta_0) + d_{m,-2}^{(l)}(\vartheta_0)]\, s^{[1]} - i\,[d_{m2}^{(l)}(\vartheta_0) - d_{m,-2}^{(l)}(\vartheta_0)]\, s^{[2]}\},\\
S_l^{[2]m}(\omega_0) &= e^{-im\varphi_0} \tfrac{1}{2}\{i\,[d_{m2}^{(l)}(\vartheta_0) - d_{m,-2}^{(l)}(\vartheta_0)]\, s^{[1]} + [d_{m2}^{(l)}(\vartheta_0) + d_{m,-2}^{(l)}(\vartheta_0)]\, s^{[2]}\},\\
S_l^{[3]m}(\omega_0) &= e^{-im\varphi_0} d_{m0}^{(l)}(\vartheta_0)\, s^{[3]},
\end{aligned}\right\} \tag{9.42}$$

where the coefficients $s^{[i]}$ indicate the kind of polarization of the source radiation and its degree $\sqrt{s^{[1]2} + s^{[2]2} + s^{[3]2}} \leq 1$. Notice that the source term of (9.38) with $i=0$ coincides with that of (9.13).

The kernel matrix (9.41) has only two off-diagonal elements ,which couple $I_l^{[0]m}(p,\lambda)$ and $I_l^{[1]m}(p,\lambda)$ in the transport equation (9.38). When the source X rays are unpolarized $s^{[i]}$ vanishes for $i \neq 0$; the flux parameters $I_l^{[2]m}$ and $I_l^{[3]m}$ vanish, but the equations for $I_l^{[0]m}$ and $I_l^{[1]m}$ have still to be solved simultaneously. Notice that the parameter $I_l^{[3]m}$ represents the degree of circular polarization and is therefore always uncoupled from the others in isotropic media. The uncoupling of $I_l^{[2]m}$ from the other parameters is made possible by the symmetry of the scattering process with respect to the plane of ω' and ω and by the suitable choice of the definitions (9.39). These definitions simplify under various conditions of symmetry, because $d_{mv}^{(l)}(\vartheta) = (-1)^{m-v} d_{-m,-v}^{(l)}(\vartheta) = (-1)^{l-m} d_{m,-v}^{(l)}(\pi-\vartheta)$. In particular for

$m = 0$ many D functions reduce to spherical harmonics so that (9.39) and (9.42) reduce to

$$I_l^{[i]0}(p, \lambda) = \sqrt{\frac{4\pi}{2l+1}} \int_{4\pi} d\omega\, Y_l^{v*}(\omega)\, I^{[i]}(p, \omega, \lambda), \qquad (9.43)$$

$$S_l^{[i]0}(\omega_0) = \sqrt{\frac{4\pi}{2l+1}}\, Y_l^{v*}(\omega_0)\, s^{[i]}, \qquad (9.44)$$

where $v = 0$ for $i = 0, 3$ and $v = 2$ for $i = 1, 2$. The Eq. (9.38) with $m = 0$ is the only equation that has to be solved under conditions of cylindrical symmetry about the z axis, such as arise when the source is plane conical or plane perpendicular.

10. Problems with the variables separated. This section deals with aspects of X-ray penetration whose treatment presents basically simple problems either because there occurs only one independent variable or because one can eliminate the interlinkages between variables without resort to approximations. These particular aspects of X-ray penetration relate to situations which are physically realizable though not always of practical importance. Their treatment is of interest primarily because it provides building blocks for the solution of more general problems.

α) Energy degradation. We consider here the spectral distribution, irrespective of position and direction, of all X-ray photons in a medium, which results from a succession of scattering processes. This distribution is represented by the integral $I(\lambda)$ of the flux $I(\boldsymbol{r}, \omega, \lambda)$ over the whole medium and over all directions, and coincides with $I_0^0(0, \lambda)$ as given by (9.8).

The quantity $I(\lambda)\, d\lambda$ has dimensions $L T^{-1}$ and represents the average length of path traveled per unit time by photons whose wavelength lies between λ and $\lambda + d\lambda$. This pathlength obeys the Eq. (9.12) with $q = l = 0$, namely

$$\mu(\lambda)\, I(\lambda) = \int_{\lambda_0}^{\lambda} d\lambda'\, k(\lambda, \lambda')\, I(\lambda') + \delta(\lambda - \lambda_0), \qquad (10.1)$$

where the source strength is one photon per unit time[1]. This equation states that the rate of loss $\mu(\lambda)\, I(\lambda)$ of photons of wavelength λ is compensated by the production of such photons by inelastic scattering and by the source.

One arrives at an identical equation by considering a medium where a photon source is distributed uniformly, for example with a radioactive source dissolved in it. In this case $\delta(\lambda - \lambda_0)$ represents a source density of one photon per unit volume and unit time, and $I(\lambda)$ the flux of photons at any one point. An experiment with a distributed source has been made by Faust and Johnson [33] but only the quantity $\int_0^\infty I(\lambda)\, f(\lambda)\, d\lambda$ was measured, $f(\lambda)$ being the spectral response of a counter.

Eq. (10.1) can be solved numerically either by iteration, as in Sect. 8, or more efficiently by direct numerical integration. At $\lambda = \lambda_0$, we have simply $I(\lambda_0) = \delta(\lambda - \lambda_0)/\mu(\lambda_0)$. At somewhat larger values of λ first order iteration yields $I(\lambda) = k(\lambda, \lambda_0)/\mu(\lambda_0)\, \mu(\lambda)$. This value is then used, together with $I(\lambda_0)$, to evaluate $\int_{\lambda_0}^{\lambda} d\lambda'$, and so one proceeds step by step, utilizing the fact that the $\int_{\lambda_0}^{\lambda} d\lambda'$ depends only on values of $I(\lambda')$ for $\lambda' < \lambda$, which are determined earlier in the calculation.

[1] In a time-integral formulation (see p. 680), the source strength is just one photon and $I(\lambda)\, d\lambda$, with the dimension L, represents the average pathlength traveled by that photon within the stated wavelength range.

Details of the numerical procedure are explained in Sect. 16. A large number of calculations have been carried out in this manner, especially as parts of other calculations discussed in Part C. Sample results are shown in Fig. 11.

Fig. 11a shows the spectra resulting from monoenergetic primary photons of different energies in the same medium. At energies sufficiently lower than the source energies the spectra become indistinguishable from one another. This trend arises from the randomness of the energy loss in individual collisions, which erases the "memory" of the initial energy of each photon after a few scatterings. It relates to the quasi-equilibrium discussed in Sect. 5γ. The process of energy

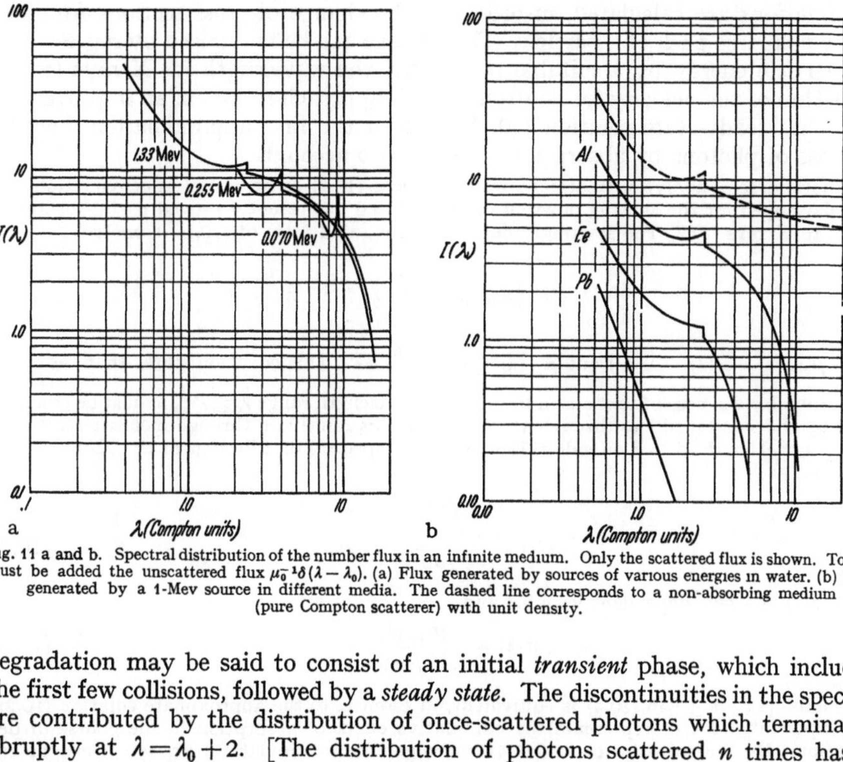

a b

Fig. 11 a and b. Spectral distribution of the number flux in an infinite medium. Only the scattered flux is shown. To this must be added the unscattered flux $\mu_0^{-1}\delta(\lambda-\lambda_0)$. (a) Flux generated by sources of various energies in water. (b) Flux generated by a 1-Mev source in different media. The dashed line corresponds to a non-absorbing medium (pure Compton scatterer) with unit density.

degradation may be said to consist of an initial *transient* phase, which includes the first few collisions, followed by a *steady state*. The discontinuities in the spectra are contributed by the distribution of once-scattered photons which terminates abruptly at $\lambda=\lambda_0+2$. [The distribution of photons scattered n times has a discontinuity in the $(n-1)$st derivative.] The drop at the low-energy end is due to the onset of photoelectric absorption which terminates the sequence of scatterings.

Fig. 11b compares the spectra from sources of equal energy in different media. It shows that, insofar as different media behave in a certain energy range ($\sim$500 kev) as pure Compton scatterers, the corresponding spectral distributions are similar in shape in that range. The drop due to photoelectric absorption occurs at an energy that increases with the atomic number.

Many features of energy degradation processes have been pointed out by PLACZEK[1] in connection with slowing-down of neutrons. If successive steps of degradation, characterized by wavelength shifts $\xi=\lambda-\lambda'$, have approximately equal statistical distributions with a mean $\langle\xi\rangle$, the probability for each photon

[1] G. PLACZEK: Phys. Rev. 69, 423 (1946); see also R. MARSHAK: Rev. Mod. Phys. 19, 185 (1947), the work of ADLER and WICK quoted there, and I. WALLER: Ark. Mat. Astronom. Fys., Ser. A 34, No. 3 (1948).

to land between λ and $\lambda+\delta\lambda$ at the end of one of its steps of degradation is $\delta\lambda/\langle\xi\rangle$.[1] If the photon does land in this interval, its average path length there will be $1/\mu(\lambda)$, so that, for a source of unit photon strength, one obtains the approximate steady state formula

$$I(\lambda) \sim \frac{1}{\mu(\lambda)\langle\xi\rangle}. \tag{10.2}$$

An analytic treatment of the transport equation (10.1), which takes advantage of this line of reasoning, is developed in Ref. [*29*]. "Transient" and "steady state" contributions to $I(\lambda)$ are represented by different terms. Successive approximations are calculated, in orders of the change of cross sections from one to the next free path of a particle. This change can be minimized by representing the particle energy by a suitable parameter, the wavelength for X rays (or the logarithm of the velocity for neutrons). This approach results in an improvement over the steady-state solution (10.2), even in the first approximation, because the loss of photons by absorption is taken into account.

Another, semi-analytic, treatment which utilizes (10.2) as an initial approximation has been developed for the slowing down of electrons[2] and has recently proved useful for application to X rays [*31*]. The starting point is an integrated form of (10.1), namely

$$\int_{\lambda_0}^{\lambda} d\lambda' \, I(\lambda') \left[\mu_{\text{abs}}(\lambda') + \int_{\lambda}^{\infty} d\lambda'' \, k(\lambda'', \lambda')\right] = 1 \tag{10.3}$$

where $\mu_{\text{abs}}(\lambda) = \mu(\lambda) - \mu_{\text{sc}}(\lambda)$ and $\mu_{\text{sc}}(\lambda) = \int_{\lambda}^{\infty} k(\lambda', \lambda)\, d\lambda'$. This equation states that the input of photons by a source of unit strength into the wavelength range $\lambda_0 \leqq \lambda' \leqq \lambda$ must be compensated by the combined loss through absorption in this range and through inelastic scattering to any wavelength $\lambda'' > \lambda$ outside this range. The spectral distribution of the X-ray flux is expressed as

$$I(\lambda) = \frac{e^{-\int_{\lambda_0}^{\lambda} \beta(\lambda')\,d\lambda'}}{\langle\xi\rangle \mu_{\text{sc}}(\lambda)}\, w(\lambda), \tag{10.4}$$

where $w(\lambda)$ is a new dependent variable and

$$\beta(\lambda) = \frac{\mu_{\text{abs}}(\lambda)}{\langle\xi\rangle\,\mu_{\text{sc}}(\lambda)}. \tag{10.5}$$

The factor $[\langle\xi\rangle\mu_{\text{sc}}]^{-1}$ in (10.4) is equivalent, in essence, to the approximate solution (10.2). The exponential factor represents the cumulative effect of absorption in the course of degradation. By the substitution of (10.4) and algebraic manipulation, Eq. (10.3) can be reduced to a form $w(\lambda) + \cdots = 1$, where the dots indicate terms that are small in the expected range of validity of an approximate steady state solution. The small terms have been evaluated for $\lambda > \lambda_0 + 2$ to first order in the parameters $\frac{d}{d\lambda} \frac{\langle\xi^{l+1}\rangle}{\langle\xi^l\rangle}$, which indicate the rate of change of a mean wavelength shift, and $\beta(\lambda)\frac{\langle\xi^{l+1}\rangle}{\langle\xi^l\rangle}$, which indicate the probable absorption per step of degradation, with

$$\langle\xi^l\rangle = \frac{1}{\mu_{\text{sc}}(\lambda)} \int_{\lambda}^{\infty} d\lambda' (\lambda' - \lambda)^l \, k(\lambda', \lambda). \tag{10.6}$$

The result is

$$w(\lambda) = 1 + \frac{1}{2} \frac{d}{d\lambda} \left\| \frac{\langle\xi^2\rangle}{\langle\xi\rangle} - \frac{1}{2} \beta(\lambda) \frac{\langle\xi^2\rangle}{\langle\xi\rangle} + \cdots. \right. \tag{10.7}$$

---

[1] This problem is equivalent to the determination of the probability that a pedestrian step on a line drawn across a sidewalk. The probability equals the length of the pedestrian's foot divided by his average step-length.

[2] L. V. Spencer and U. Fano: Phys. Rev. **93**, 1172 (1954).

The first order corrective terms are of the order of 1% over a substantial range of λ for low-Z materials. Fig. 12 compares spectra obtained by numerical integration of (10.1) with spectra calculated from (10.4) and (10.7).

β) Degradation and deflection. We consider here the photon distribution in energy and direction, irrespective of position in the medium. This distribution $I(\boldsymbol{\omega}, \lambda)$ is represented by the integral of the flux $I(\boldsymbol{r}, \boldsymbol{\omega}, \lambda)$ over the whole medium, and coincides with the transform $I(0, \boldsymbol{\omega}, \lambda)$ as given by (9.5). Accordingly it obeys the Eq. (9.4) with $q=0$, i.e. with its left side set equal to zero.

Expansion of $I(\boldsymbol{\omega}, \lambda)$ into spherical harmonics, according to (9.7) and (9.8), is particularly convenient because the expansion coefficients $I_l^m(\lambda)$ obey the Eq. (9.12) with $p=0$, that is, without interlinkages in l. The equation simplifies further if one takes the z axis of coordinates parallel to $\boldsymbol{\omega}_0$[1], since the source term reduces then to $\delta_{m0}\,\delta(\lambda-\lambda_0)$. Therefore the flux has only expansion coefficients with $m=0$ and the index m may be omitted. The equation becomes now

$$\left.\begin{aligned}\mu(\lambda)\, I_l(\lambda) = \int_{\lambda_0}^{\lambda} d\lambda'\, k_l(\lambda, \lambda')\, I_l(\lambda') + \\ + \delta(\lambda-\lambda_0)\,.\end{aligned}\right\} \tag{10.8}$$

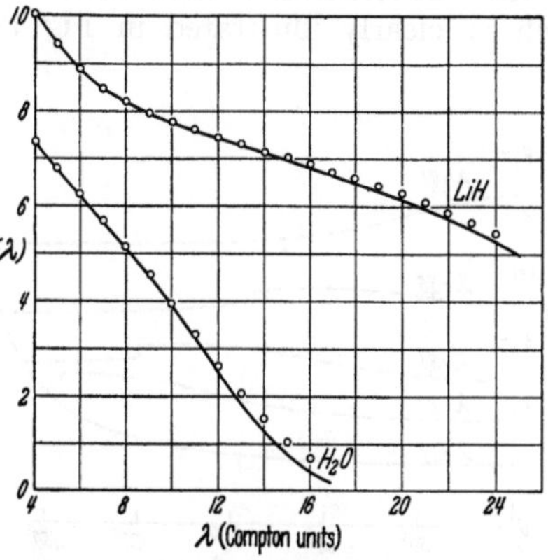

Fig. 12. Comparison of different methods for calculating the spectral distribution of the flux in media of low atomic number. Curves calculated semi-analytically according to (10.4) and (10.7). Points calculated by direct numerical integration. $\lambda_0 = 1$ ($E_0 = 0.511$ Mev).

The calculation of $I_l(\lambda)$ proceeds like that of $I(\lambda) = I_0(\lambda)$ in (α) above, except that $k_l(\lambda, \lambda')$ is an oscillatory function for $l \neq 0$, because it is proportional to $P_l(1-\lambda'+\lambda)$ as shown in (9.11) for Compton scattering. This oscillation produces cancellations in $\int_{\lambda_0}^{\lambda} k_l(\lambda, \lambda')\, I_l(\lambda')\, d\lambda'$ which increase in magnitude with l, and $I_l(\lambda)$ is correspondingly reduced. The decrease of $I_l(\lambda)$ as compared to $I_0(\lambda)$, as λ increases, represents the smearing out of the directional distribution of $I(\boldsymbol{\omega}, \lambda)$ which results from multiple scattering.

Many calculations of coefficients $I_l(\lambda)$ have been made, incidental to other calculations described in part C, but the distributions $I(\boldsymbol{\omega}, \lambda)$ have been reconstructed only in a few exploratory studies. Results are shown in Fig. 13 for a monoenergetic and monodirectional source in Al with $\lambda_0 = 0.1$ [*84*].

In principle it is possible to determine the angular distributions directly by summing a Legendre series; but in practice this is not always feasible and was in fact not done, because of slow convergence of the series. This slow convergence results in part from the discontinuous nature of the first few orders of scattering. If the unscattered radiation is collimated, the singly scattered component is concentrated at a particular obliquity with respect to the axis of collimation, and the twice-scattered component is discontinuous at the largest allowed obliquity [see Eq. (8.7) and Fig. 13][2]. Such discontinuities are particularly prominent for small $\lambda-\lambda_0$ and cannot be accurately represented by a few polynomials. These irregular features were included in Fig. 13 by calculating $I^{(1)}(\vartheta, \lambda)$, $I^{(2)}(\vartheta, \lambda)$,

[1] The coordinates can be chosen freely here, whereas in Sect. 9 they were adapted to the treatment of the space distribution of source and flux.

[2] Higher orders of scattering are discontinuous in successively higher derivatives, but this affects the convergence less.

$I_l^{(1)}(\lambda)$, and $I_l^{(2)}(\lambda)$ with the iteration procedure of Sect. 8, and writing

$$I(\vartheta,\lambda)=I^{(1)}(\vartheta,\lambda)+I^{(2)}(\vartheta,\lambda)+\sum_{l=0}^{\infty}\frac{2l+1}{4\pi}\{I_l(\lambda)-I_l^{(1)}(\lambda)-I_l^{(2)}(\lambda)\}\,P_l(\cos\vartheta). \quad (10.9)$$

The remaining Legendre series in (10.9) converged so that it was sufficient to terminate it at $l_{\max}=3$ so long as $\lambda-\lambda_0$ was of the order of unity or greater. For higher energies ($\lambda-\lambda_0\leqq 1$), $l_{\max}$ was increased, and for $\lambda-\lambda_0=0.1$, it was sufficient to include ten terms. Slow convergence for small $\lambda-\lambda_0$ results from the flux concentration near $\cos\vartheta=1$, which is clearly illustrated in Fig. 13.

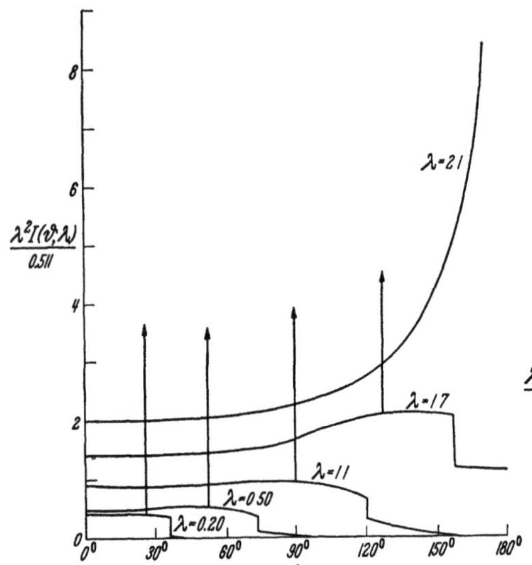

a

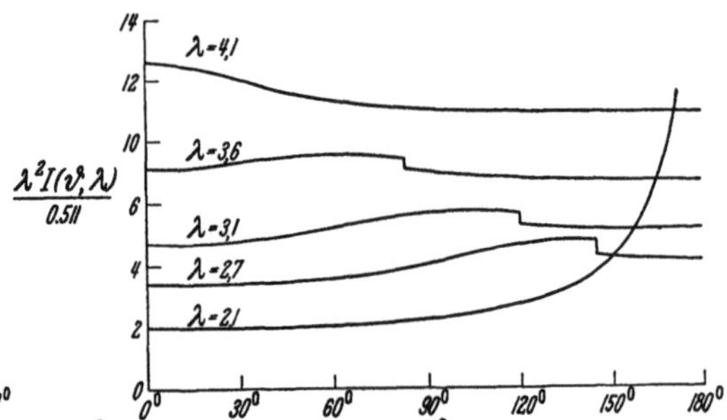

b

Fig. 13 a and b. Directional distribution of the flux in aluminum for various spectral components. $\lambda_0=0.1$ ($E_0=5.11$ Mev). The arrows indicate the once-scattered flux.

A simpler but equally effective procedure consists of separating out not the entire twice-scattered component, but only the discontinuous part which can be evaluated analytically:

$$I(\vartheta,\lambda)=I^{(1)}(\vartheta,\lambda)+I^{(\bar{2})}(\vartheta,\lambda)+\sum_{l=0}^{\infty}\frac{2l+1}{4\pi}\{I_l(\lambda)-I_l^{(1)}(\lambda)-I_l^{(\bar{2})}(\lambda)\}\,P_l(\cos\vartheta), \quad (10.10)$$

where

$$I^{(\bar{2})}(\vartheta,\lambda)=\frac{k\left(\lambda,\tfrac{1}{2}[\lambda_0+\lambda]\right)k\left(\tfrac{1}{2}[\lambda_0+\lambda],\lambda_0\right)}{\mu(\lambda_0)\,\mu\left(\tfrac{1}{2}[\lambda_0+\lambda]\right)\mu(\lambda)\,4\pi\cos\frac{\vartheta}{2}} \quad (10.11)$$

for $\vartheta\leqq\vartheta_L=2\arccos\left[1-\tfrac{1}{2}(\lambda-\lambda_0)\right]$ and vanishes for $\vartheta>\vartheta_L$. The Legendre coefficients $I_l^{(\bar{2})}(\lambda)$ are

$$I_l^{(\bar{2})}(\lambda)=I^{(\bar{2})}(0,\lambda)\,2\pi\int_0^{\vartheta_L}\frac{P_l(\cos\vartheta)}{\cos\frac{\vartheta}{2}}\sin\vartheta\,d\vartheta. \quad (10.12)$$

A typical set of coefficients in (10.10) is given in Table 5.

A small loss of energy is accompanied in general by a small total deflection, even in the case of multiple small angle scattering. The randomness of azimuths in successive scatterings yields a sort of random walk of the photon direction ω, with a resultant total mean square deflection $\langle\vartheta^2\rangle=\sum_{n=1}^{N}\Theta_n^2$, where Θ_n is the deflection in the n-th scattering. For small deflections we have $\Theta_n^2\sim 2(1-\cos\Theta_n)=2(\lambda_n-\lambda_{n-1})$. Therefore $\langle\vartheta^2\rangle=2(\lambda_n-\lambda_0)$ depends only on the total wavelength shift [*34a*].

The angular distribution of photons with a small wavelength shift has nevertheless a large-angle tail, which results from the occasional occurrence of a large number of small deflections in the same direction, such that the total deflection is $\vartheta \sim \sum_{n=1}^{N} \Theta_n = \sum_n \sqrt{2(\lambda_n - \lambda_{n-1})}$. The maximum value of ϑ, for a given N, results from identical partial shifts $\lambda_n - \lambda_{n-1} = (\lambda_N - \lambda_0)/N$. This value,

$$\vartheta_N = N\sqrt{2(\lambda_N - \lambda_0)/N} = \sqrt{2N(\lambda_N - \lambda_0)},$$

increases with increasing N even though $\lambda_N - \lambda_0$ remains constant. The probability of deflections in the large-angle tail can be calculated by solving (10.8) analytically with approximations appropriate to multiple small deflections[1].

Table 5. *Legendre expansion coefficients of the flux from a* 1.28-Mev *source* ($\lambda_0 = 0.4$) *in water, for the spectral component with wavelength* $\lambda = 0.616$.

l	$\mu_0 I_l$	$\mu_0 I_l^{(1)}$	$\mu_0 I_l^{(2)}$	$\mu_0(I_l - I_l^{(1)} - I_l^{(2)})$
0	1.346	0.819	0.394	0.133
1	1.060	0.642	0.312	0.105
2	0.590	0.345	0.182	0.063
3	0.110	0.023	0.054	0.034
4	−0.260	−0.227	−0.031	−0.002
5	−0.394	−0.339	−0.055	0.000

The first and main approximation consists of regarding the probabilities of interaction $\mu(\lambda)$ and $k_0(\lambda, \lambda')$ as effectively constant in the limited wavelength range which one considers. The scattering probability may be made more nearly constant by the transformation (7.1). Specifically for Compton scattering we take

$$k_l(\lambda, \lambda') = k_0(\lambda, \lambda')\, P_l(1 - \lambda + \lambda') \sim C\, P_l(1 - \lambda + \lambda'), \tag{10.13}$$

where

$$C = k_0(\lambda, \lambda) = \frac{3}{4} N Z \sigma_{Th} = 0.300 \frac{Z}{A} \text{cm}^2/\text{g} = 0.300 \frac{Z}{A} \varrho \text{ cm}^{-1}. \tag{10.14}$$

Eq. (10.8) becomes now

$$\mu I_l(\lambda) = C \int_{\lambda_0}^{\lambda} d\lambda'\, P_l(1 - \lambda + \lambda')\, I_l(\lambda') + \delta(\lambda - \lambda_0). \tag{10.15}$$

Since the coefficients of this equation depend only on the wavelength difference $\lambda - \lambda'$ and not on λ itself, its solution is represented appropriately as a combination of steady state solutions $\exp[\eta(\lambda - \lambda_0)]$ by a Laplace transformation of the variable λ [*29*]. Multiplication of (10.15) by $\exp[-\eta(\lambda - \lambda_0)]$ and integration over λ yields

$$\mu I_l(\eta) = C I_l(\eta) \int_0^{\infty} d(\lambda - \lambda')\, e^{-\eta(\lambda - \lambda')} P_l(1 - \lambda + \lambda') + 1, \tag{10.16}$$

with

$$I_l(\eta) = \int_0^{\infty} d(\lambda - \lambda_0)\, e^{-\eta(\lambda - \lambda_0)} I_l(\lambda). \tag{10.17}$$

It is understood that $P_l(1 - \lambda + \lambda')$ vanishes for $\lambda - \lambda' > 2$, in the integral of (10.16), but this limitation is unimportant since small wavelength shifts $\lambda - \lambda'$ contribute most to the integral[2].

A further approximation is made in the evaluation of the integral of (10.16) based on the circumstance that the angular distribution $I(\vartheta, \lambda)$ is sharply peaked at $\vartheta \sim 0$. This circumstance implies that large values of the transform parameter l are of interest and that

[1] T. WELTON: Oak Ridge Nat. Lab. Memo 56-1-9 (unpublished).

[2] The fact that the distribution $I(\vartheta, \lambda)$ is to be discussed for small values of $\lambda - \lambda_0$ implies that large values of the transform variable η are of interest, and therefore that $\exp[-\eta(\lambda - \lambda')]$ is very small except for $\lambda - \lambda' \ll 1$.

$I_l(\eta)$ is a smooth function of l. Therefore l can be treated as a *continuous variable* and the Legendre polynomial can be replaced with its Bessel function approximation

$$P_l(1-\lambda+\lambda') \sim J_0\left(\bar{l}\sqrt{2(\lambda-\lambda')}\right), \tag{10.18}$$

with

$$\bar{l} = l + \tfrac{1}{2}. \tag{10.19}$$

With this further approximation we have

$$\int_0^\infty d(\lambda-\lambda')\, e^{-\eta(\lambda-\lambda')} P_l(1-\lambda+\lambda') \sim \int_0^\infty d\lambda\, e^{-\eta\lambda} J_0\left(\bar{l}\sqrt{2\lambda}\right) = \frac{1}{\eta} e^{-\bar{l}^2/2\eta}. \tag{10.20}$$

The solution of (10.16) is then

$$I_l(\eta) = \frac{1}{\mu - \frac{C}{\eta} e^{-\bar{l}^2/2\eta}}, \tag{10.21}$$

and its inverse transform yields the approximate large-angle distribution

$$I(\vartheta,\lambda) = \sum_l \frac{2l+1}{4\pi} P_l(\cos\vartheta) \frac{1}{2\pi i} \int_{-i\infty}^{i\infty} d\eta \frac{e^{\eta(\lambda-\lambda_0)}}{\mu - \frac{C}{\eta} e^{-(l+\frac{1}{2})^2/2\eta}}. \tag{10.22}$$

The distribution $I(\vartheta,\lambda)$ is a peaked function of ϑ so that the summation over l is very slowly converging. A much more rapidly converging expression can be obtained by the transformation[1]

$$\sum_l \frac{l+\frac{1}{2}}{2\pi} f(\bar{l}) P_l(\cos\vartheta) = \frac{1}{\sqrt{2}\,\pi} \int_\vartheta^{2\pi-\vartheta} \frac{d\chi}{\sqrt{\cos\vartheta - \cos\chi}} \sum_{m=0}^\infty (-1)^m R(\chi + 2\pi m) \tag{10.23}$$

with

$$R(\chi) = \frac{1}{2\pi} \int_{-\infty}^\infty d\bar{l}\, \bar{l}\, f(\bar{l}) \sin(\bar{l}\chi); \tag{10.24}$$

in our problem

$$f(\bar{l}) = I_l(\eta) = \frac{1}{\mu - \frac{C}{\eta} e^{-\bar{l}^2/2\eta}}. \tag{10.25}$$

The integral in (10.24) reduces to the contributions of the residues of poles of $f(\bar{l})$ lying at

$$\bar{l} = \pm i K = \pm i\sqrt{2\eta \log(\eta\mu/C)}, \tag{10.26}$$

and yields

$$R(\chi) = \frac{\eta}{\mu} e^{-K\chi}. \tag{10.27}$$

Because η and, hence, K are understood to be large, the Σ_m in (10.23) may be approximated by the term with $m=0$ and the $\int d\chi$ by integration over a small interval near $\chi=\vartheta$. Thereby (10.23) reduces to

$$\sum_l \frac{l+\frac{1}{2}}{2\pi} I_l(\eta) P_l(\cos\vartheta) = \frac{1}{\sqrt{2\pi K \sin\vartheta}} \frac{\eta}{\mu} e^{-K\vartheta}. \tag{10.28}$$

(When ϑ approaches π, (10.23) does not actually diverge but approaches $2\pi \frac{\eta}{\mu} e^{-K\chi}$.) Entering (10.28) into (10.23), we have

$$\left.\begin{aligned} I(\vartheta,\lambda) &= \frac{1}{\sqrt{2\pi\sin\vartheta}} \frac{1}{2\pi i} \int d\eta \frac{\eta}{\mu} \frac{e^{\eta(\lambda-\lambda_0) - \vartheta\sqrt{2\eta\log(\eta\mu/C)}}}{\left[2\eta\log\left(\frac{\eta\mu}{C}\right)\right]^{\frac{1}{4}}} \\ &= \frac{1}{\sqrt{2\pi\sin\vartheta}} \frac{1}{\mu}\left(\frac{C}{\mu}\right)^{\frac{7}{4}} \frac{1}{2\pi i} \int du\, u \frac{e^{ux - 2\sqrt{\nu x}\sqrt{u\log u}}}{(2u\log u)^{\frac{1}{4}}} \end{aligned}\right\} \tag{10.29}$$

[1] L.V. Spencer: Phys. Rev. **90**, 146 (1953).

where $u=\eta\mu/C$, $x=C(\lambda-\lambda_0)/\mu$, and $\nu=\frac{1}{2}\vartheta^2/(\lambda-\lambda_0)$ represents the minimum number of scatterings required to produce a deflection ϑ with a fixed total wavelength shift $\lambda-\lambda_0$.

The integral in (10.29) may be evaluated by the method of steepest descent, even though in the region of interest ($\lambda-\lambda_0 \ll 1$ and $\vartheta \sim 1$) the saddle point is broad. We take a path through the saddle point $u=u_0$ of the exponential factor in the integrand, thus disregarding the second derivative of the other factors in addition to making the usual steepest descent approximations. The result is conveniently expressed using $\log u_0$ as a parameter which is defined in terms of λ and ϑ by the saddle point equation

$$\frac{u_0 \log u_0}{(\log u_0+1)^2}=\frac{\nu}{x} \tag{10.30}$$

and is given in Table 6 for the applicable range of values,

$$I(\vartheta,\lambda)=\frac{1}{2\pi\mu}\sqrt{\frac{C}{2\mu\sin\vartheta}}\,\frac{\vartheta^{\frac{3}{2}}}{(\lambda-\lambda_0)^2}\,\mathrm{e}^{-\frac{\vartheta^2}{2(\lambda-\lambda_0)}\log u_0\left[1-\frac{1}{(\log u_0)^2}\right]}\left(1+\frac{2}{\log u_0}+\cdots\right). \tag{10.31}$$

For values of $\frac{\vartheta^2}{2(\lambda-\lambda_0)}$ that are not extremely large it may be convenient to utilize an expansion of $I(\vartheta,\lambda)$ in orders of scattering. The integrand of (10.22) may be expanded into powers of C, after which one finds in place of (10.27), (10.28) and (10.31),

$$R(\chi)=\frac{1}{\sqrt{2\pi}\,\mu}\sum_n\left(\frac{C}{\mu\eta}\right)^n\left(\frac{\eta}{n}\right)^{\frac{3}{2}}\chi\,\mathrm{e}^{-\frac{\eta}{2n}\chi^2}, \tag{10.32}$$

$$\sum_l\frac{l+\frac{1}{2}}{2\pi}I_l(\eta)\,P_l(\cos\vartheta)=\frac{1}{2\pi\mu}\,\frac{\vartheta}{\sin\vartheta}\sum_n\left(\frac{C}{\mu\eta}\right)^n\frac{\eta}{n}\,\mathrm{e}^{-\frac{\eta}{2n}\vartheta^2}, \tag{10.33}$$

$$I(\vartheta,\lambda)=\frac{1}{2\pi\mu}\sqrt{\frac{\vartheta}{\sin\vartheta}}\left(\frac{C}{\mu}\right)^2\sum_{n>\nu}\frac{1}{n(n-2)!}\left[\frac{C}{\mu}\left(\lambda-\lambda_0-\frac{\vartheta^2}{2n}\right)\right]^{n-2}. \tag{10.34}$$

Table 6. *Solution of the saddle-point equation (10.30).*

$\frac{\nu}{x}$	u_0	$\frac{\nu}{x}$	u_0
1.0	4.12	20	144
2.0	9.37	40	317
4.0	21.6	70	599
7.0	42.0	100	895
10	64.0		

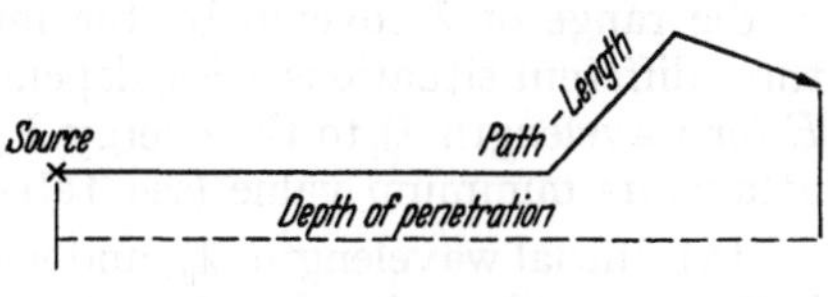

Fig. 14. Connection between pathlength and depth of penetration.

γ) Distribution in pathlength and energy. Straight-ahead approximation. We deal here with a variable not considered thus far, namely the length of path t traveled by a photon, regardless of the direction of the path at various stages of the travel and of the position attained with respect to the source. The pathlength equals the "age" of a photon multiplied by the light velocity. The study of the distribution in pathlength, especially for long pathlengths, is an important preliminary to the study of penetration in depth. As long as no large-angle scattering occurs, the depth of penetration departs but little from the pathlength (see Fig. 14), so that the pathlength distribution constitutes a "straight-ahead approximation" to the distribution-in-depth.

The approximation disregards the interlinkage which exists between the depth of penetration and the direction of the X-ray flow but does not exist between pathlength and direction. In the complete problem of deep penetration (Sect. 11), the interlinkage can be so treated that the final solution is formally the same as in the pathlength problem, with a smaller effective cross section for scattering.

The distribution of photons in pathlength, direction and energy, $I(t, \omega, \lambda)$, obeys a transport equation which is established much in the same manner as (6.9) and differs from it by the replacement of the z coordinate with the pathlength t and of the term $\omega_z \frac{\partial I}{\partial z}$ with $\frac{\partial I}{\partial t}$,

$$\left.\begin{aligned}\frac{\partial I(t,\omega,\lambda)}{\partial t} &= -\mu(\lambda)\, I(t,\omega,\lambda) + \int_{\lambda_0}^{\lambda} d\lambda' \int_{4\pi} d\omega'\, k(\omega,\lambda,\omega',\lambda')\, I(t,\omega',\lambda') + \\ &\quad + \delta(t)\,\delta(\omega-\omega_0)\,\delta(\lambda-\lambda_0).\end{aligned}\right\} \tag{10.35}$$

Integration of this equation over all photon directions gives

$$\frac{\partial I(t,\lambda)}{\partial t} = -\mu(\lambda)\, I(t,\lambda) + \int_{\lambda_0}^{\lambda} d\lambda'\, k(\lambda,\lambda')\, I(t,\lambda') + \delta(t)\,\delta(\lambda-\lambda_0). \tag{10.36}$$

The variables t and λ are separated by a Laplace transformation in t. Multiplication of (10.36) by $\exp(pt)$ [1] and integration over t from 0 to ∞ yields

$$[\mu(\lambda) - p]\, I(p,\lambda) = \int_{\lambda_0}^{\lambda} d\lambda'\, k(\lambda,\lambda')\, I(p,\lambda') + \delta(\lambda-\lambda_0). \tag{10.37}$$

This equation coincides with the Eq. (9.13) for $I_0^0(p, \lambda)$ except that the interlinkage term ΔI_0^0 is absent here. It differs from the simple energy degradation equation (10.1) only by the replacement of $\mu(\lambda)$ with $\mu(\lambda)-p$, and it can be solved in principle like (10.1).

The parameter p, which is subtracted from μ, represents a "negative absorption", i.e. it acts as a regeneration coefficient. If p exceeds $\mu(\lambda)$ we have a net negative absorption and the equation becomes singular. Thus we must exclude all real values of $p \geqq \mu_s = \mu(\lambda_s)$ where μ_s indicates the smallest value of $\mu(\lambda)$ in the range of λ covered by the integration. In this connection the following three different situations arise, depending on the relationship of the photon energy E (or wavelength λ) to the energy E_m (wavelength λ_m) at which the coefficient μ attains its minimum value (see Table 2):

(A) Initial wavelength, λ_0, and wavelength of photons under considerations, λ, both larger than λ_m, i.e. $\lambda_m < \lambda_0 < \lambda$; we have, then, $\lambda_s = \lambda_0$, $\mu_s = \mu(\lambda_0) = \mu_0$;

(B) $\lambda_0 \leqq \lambda_m \leqq \lambda$ so that $\lambda_s = \lambda_m$ and $\mu_s = \mu(\lambda_m)$;

(C) $\lambda_0 < \lambda < \lambda_m$, so that $\lambda_s = \lambda$, $\mu_s = \mu(\lambda)$.

Eq. (10.37) shows that, as p approaches μ_s though remaining smaller than it, $I(p, \lambda_s)$ becomes very large, but remains finite. All values of $I(p, \lambda)$ for $\lambda \geqq \lambda_s$ will also be large because, as $\lambda' = \lambda_s$ is included in the integral, $I(p, \lambda_s)$ acts as a source of scattered photons. The solution of (10.37) for $p \sim \mu_s$ is important because it determines, through the inverse Laplace transformation, how likely photons are to travel a very long path without absorption. General properties of the solutions of (10.37) and of their inverse transforms $I(t, \lambda)$ are discussed in Ref. [*29*]. The following discussion deals with the situation for $p \sim \mu_s$.

When $\mu_s - p$ is small, an iteration approach would converge poorly for $\lambda \gtrsim \lambda_s$, because each step of iteration introduces a factor $\mu(\lambda) - p$ in the denominator. Direct numerical integration also becomes laborious because it must proceed through a large number of steps each over a small range of λ. A different approach becomes possible, however, because the distribution of photons with $\lambda \sim \lambda_s$ consists primarily of photons that have experienced a large number of

[1] Notice that this transform variable has a sign opposite to that most frequently used.

scatterings at $\lambda' \sim \lambda_s$, each with a small shift. The solution of (10.37) will have, then, the steady-state characteristics which have been discussed and exploited in Ref. [29] and outlined in Sect. 10α. It can be worked out either by the method of successive approximations of Ref. [29], or, more directly, by setting $k(\lambda, \lambda') = k(\lambda, \lambda) = C$, as in (10.13), on the ground that the shift $\lambda - \lambda'$ is very small and $k(\lambda, \lambda')$ does not vary rapidly with λ'.

With this approximation (10.37) becomes

$$[\mu(\lambda) - p]\, I(p, \lambda) = C \int_{\lambda_0}^{\lambda} d\lambda'\, I(p, \lambda') + \delta(\lambda - \lambda_0) \tag{10.38}$$

and can be reduced to the differential form

$$\frac{d}{d\lambda}[\mu(\lambda) - p]\, I(p, \lambda) = C\, I(p, \lambda) + \frac{d}{d\lambda}\delta(\lambda - \lambda_0), \tag{10.39}$$

whose solution is

$$I(p, \lambda) = \frac{1}{\mu_0(\lambda) - p}\left[\frac{C}{\mu(\lambda) - p}\, e^{C\int_{\lambda_0}^{\lambda} \frac{d\lambda'}{\mu(\lambda') - p}} + \delta(\lambda - \lambda_0)\right]. \tag{10.40}$$

Notice that the dependence of $\mu(\lambda)$ upon λ is of decisive importance here, whereas it was disregarded in (10.15), because $\mu(\lambda) - p$ is small in the range of interest so that a small relative change of $\mu(\lambda)$ constitutes a major change of $\mu(\lambda) - p$.

Eq. (10.39) and its solution (10.40) apply only when the whole interval of integration, from λ_0 to λ, is confined to $\sim \lambda_s$[1]. If λ_0 lies within such an interval but $\lambda \gg \lambda_s$, the solution (10.40) may be utilized up to a value $\lambda = \lambda_1$ such that $\mu(\lambda_1) - p$ is sufficiently larger than $\mu_s - p$, and then the exact Eq. (10.37) may be integrated numerically for $\lambda > \lambda_1$. If also $\lambda_0 \ll \lambda_s$, one considers the approximate Eq. (10.39) only for $\lambda \sim \lambda_s$, disregarding its source term. The equation becomes then homogeneous, with the solution

$$I(p, \lambda) = \frac{K}{\mu(\lambda) - p}\, e^{C\int^{\lambda} \frac{d\lambda'}{\mu(\lambda) - p}}. \tag{10.41}$$

The constant of integration K must be determined by integrating the exact Eq. (10.37) numerically from λ_0 to the proximity of λ_s. This integration must utilize realistic values of $k(\lambda, \lambda')$ but may replace $\mu(\lambda) - p$ with $\mu(\lambda) - \mu_s$, because the difference $\mu_s - p$ is negligible as compared to $\mu(\lambda) - \mu_s$ until λ approaches λ_s and the approximation $k(\lambda, \lambda') = C$ becomes valid. The numerical integration can then be continued analytically according to (10.41) as soon as the numerical $I(\mu_s, \lambda)$ is seen to obey the approximate (10.39) with $k(\lambda, \lambda') = C$ and no source term. This procedure shows that K is independent of p provided λ_0 is sufficiently smaller than λ_s.

For $\lambda > \lambda_1$ numerical integration of (10.37) may resume, again with $\mu(\lambda) - \mu_s$ in the place of $\mu(\lambda) - p$ because $\mu(\lambda) - p$ is much larger than $\mu_s - p$. Therefore the resulting values of $I(\lambda, p)$ for $\lambda > \lambda_1$ depend on p only through the values of (10.40) or (10.41) at $\lambda \sim \lambda_1$, which serve as initial conditions for the integration beyond λ_1. We have here the state of *quasi-equilibrium* outlined in Sect. 5γ. In conclusion, the probability of finding a photon of given wavelength with an unusually long pathlength is determined by the integral in the exponent of (10.40) or (10.41) over an interval adjacent to λ_s which becomes increasingly narrow as p approaches μ_s.

[1] The range of validity of $k(\lambda', \lambda) \sim C$ can be increased by means of the transformation indicated by (7.1) and (7.2).

Characteristic types of behavior of $I(p, \lambda)$ for $p \sim \mu_s$ correspond to various typical trends of $\mu(\lambda)$ for $\lambda \sim \lambda_s$:

(I) For $\mu(\lambda)$ constant and equal to $\mu_0 = \mu_s$, (10.40) gives for $\lambda > \lambda_0$

$$I(p, \lambda) = \frac{C}{(\mu_0 - p)^2} \, e^{\frac{C(\lambda - \lambda_0)}{\mu_0 - p}}. \tag{10.42}$$

However, the assumption $\mu(\lambda) = \text{const}$ is hardly ever realistic[1].

(II) For $\mu(\lambda)$ rising linearly, we have $\lambda_s = \lambda_0$, $\mu_s = \mu(\lambda_0) = \mu_0$, $\mu(\lambda') = \mu_0 + \dot{\mu}_0(\lambda' - \lambda_0)$, and (10.40) gives for $\lambda > \lambda_0$

$$I(p, \lambda) = C \frac{[\mu_0 - p + \dot{\mu}_0(\lambda - \lambda_0)]^{\frac{C}{\dot{\mu}_0} - 1}}{(\mu_0 - p)^{\frac{C}{\dot{\mu}_0} + 1}}. \tag{10.43}$$

This formula gives a realistic dependence on p whenever $\lambda_0 > \lambda_m$.

(III) For $\mu(\lambda)$ falling linearly, as it does for $\lambda' \sim \lambda < \lambda_m$, we have $\lambda_s = \lambda$, $\mu(\lambda') = \mu_s - \dot{\mu}_s(\lambda - \lambda')$ and (10.41) gives

$$I(p, \lambda) = \frac{K}{(\mu_s - p)^{\frac{C}{|\dot{\mu}_s|} - 1}}. \tag{10.44}$$

(IV) For $\mu(\lambda)$ varying quadratically, near to its minimum at $\lambda = \lambda_m$, we have $\lambda_s = \lambda_m$, $\mu_s = \mu(\lambda_m) = \mu_m$, $\mu(\lambda') = \mu_m + \frac{1}{2}\ddot{\mu}_m(\lambda' - \lambda_m)^2$, and (10.41) gives

$$I(p, \lambda) = \frac{K}{\mu(\lambda) - p} \exp\left\{\frac{C}{\sqrt{\frac{1}{2}\ddot{\mu}_m(\mu_m - p)}}\left[\pi - \arctan\sqrt{\frac{\mu_m - p}{\frac{1}{2}\ddot{\mu}_m(\lambda - \lambda_m)^2}}\right]\right\}. \tag{10.45}$$

This formula gives a realistic dependence on p when $\lambda_0 \ll \lambda_m < \lambda$.

Consider now the inversion of the Laplace transformation, which gives the distribution of photons in pathlength and energy,

$$I(t, \lambda) = \frac{1}{2\pi i} \int_{-i\infty}^{i\infty} dp \, e^{-pt} I(p, \lambda). \tag{10.46}$$

Fig. 15 shows a diagram of the plane of the complex variable p on which a heavy line indicates the section of the real axis, $p \geqq \mu_s$, where I is singular. At the left of μ_s, on the real axis, I rises rapidly on the approach to μ_s as shown by (10.42) to (10.45), whereas the other factor in (10.46), $\exp(-pt)$, falls rapidly toward the right. As a result, the product $\exp(-pt)\, I(p, \lambda)$ has a minimum on the real axis of p, on the left of μ_s and the closer to μ_s the larger t is. This minimum is a saddle point of the integrand of (10.46). Because $I(p, \lambda)$ has no other singularity in the plane of p, there is a path of steepest descent for the integration following the qualitative trend indicated in the diagram.

[1] Eq. (10.42) is of interest for values of the variables such that the mean number of scattering in the range λ_0 to λ, represented by $C(\lambda - \lambda_0)/(\mu_0 - p)$, is much larger than 1. On the other hand, the actual variation of $\mu(\lambda)$ may be disregarded only if $[\mu(\lambda) - \mu_0]/(\mu_0 - p)$ is in fact much smaller than 1. The number $C(\lambda - \lambda_0)/(\mu_0 - p)$ may be expressed in the form

$$\frac{C}{\left\langle \frac{d\mu}{d\lambda} \right\rangle} \frac{\mu(\lambda) - \mu_0}{\mu_0 - p} = \frac{C}{\langle \mu \rangle} \frac{1}{\frac{d \log \mu}{d\lambda}} \frac{\mu(\lambda) - \mu_0}{\mu_0 - p},$$

where the symbols $\langle \ldots \rangle$ indicate suitable averages over the range from λ_0 to λ. Inspection of the data in Tables 4 and 7 shows that the conditions $C(\lambda - \lambda_0)/(\mu_0 - p) \gg 1$ and $[\mu(\lambda) - \mu_0]/(\mu_0 - p) \ll 1$ are hardly ever compatible.

Evaluation of (10.46) by the method of steepest descent[1] yields the asymptotic trend of $I(t, \lambda)$ for large t and for $\lambda > \lambda_0$, in each of the cases listed above,

$$\text{(I)} \qquad I(t, \lambda) \sim e^{-\mu_0 t + 2\sqrt{C(\lambda-\lambda_0)t}} \frac{[C(\lambda-\lambda_0)\,t]^{\frac{1}{4}}}{2\sqrt{\pi}\,(\lambda-\lambda_0)}, \tag{10.47}$$

$$\text{(II)} \qquad I(t, \lambda) \sim e^{-\mu_0 t} \frac{t^{\varkappa_0}}{\Gamma(\varkappa_0+1)}\, C\,[\dot{\mu}_0(\lambda-\lambda_0)]^{\varkappa_0-1}, \tag{10.48}$$

with

$$\varkappa_0 = \frac{C}{\dot{\mu}_0}, \tag{10.49}$$

$$\text{(III)} \quad I(t, \lambda) \sim K\, e^{-\mu(\lambda)t} \frac{t^{\varkappa}}{\Gamma(\varkappa+1)}, \tag{10.50}$$

with

$$\varkappa = \frac{C}{\left|\dfrac{d\mu}{d\lambda}\right|}, \tag{10.51}$$

$$\left.\begin{aligned} \text{(IV)}\; I(t, \lambda) &\sim \frac{K}{\sqrt{3\pi}}\, e^{-\mu_m t + 3(\tau\mu_m t)^{\frac{1}{3}}} \times \\ &\times \frac{\tau\mu_m}{(\tau\mu_m t)^{\frac{5}{6}}} \frac{e^{-\frac{C}{\frac{1}{2}\ddot{\mu}_m(\lambda-\lambda_m)}}}{\mu(\lambda)-\mu_m}, \end{aligned}\right\} \tag{10.52}$$

with

$$\tau = \frac{\pi^2 C^2}{2\mu_m \ddot{\mu}_m}. \tag{10.53}$$

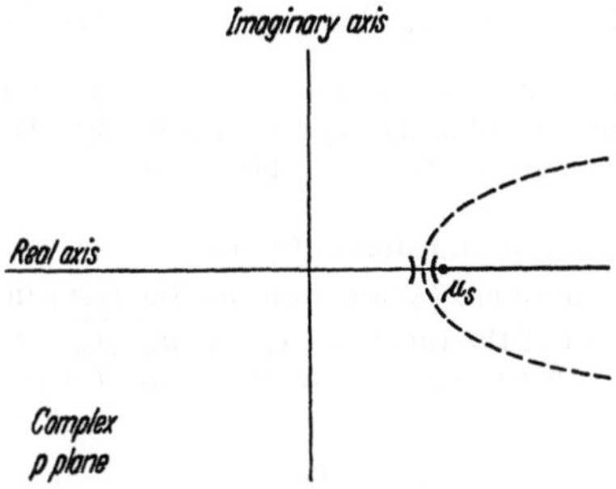

Fig. 15. Topography of the function $e^{-pt} I(p, \lambda)$ in the complex p-plane. The dashed curve indicates a path of steepest descent.

In the calculation of (10.52) the parameter p has been set equal to μ_m in the factors of (10.46) that are singular at $p = \mu(\lambda)$ instead of at $p = \mu_m$; this procedure is valid provided λ is sufficiently larger than λ_m.

Table 7. *Values of* $\frac{1}{\mu}\frac{d\mu}{d\lambda} = \frac{d\log\mu}{d\lambda}$.

Source energy E_0 (Mev)	H_2O	Al	Fe	Sn	Pb	U
10.0	7.5	4.1				
8.0	6.6	4.0	0.48			
6.0	5.5	3.8	1.2			
4.0	4.0	3.3	2.0	0.40		
3.0	3.1	2.7	2.1	0.86	0.33	0.22
2.0	2.1	2.0	1.8	1.4	1.7	2.0
1.0	0.96	0.96	0.89	1.1	1.7	1.9
0.8	0.73	0.73	0.72	0.92	1.6	1.8
0.6	0.52	0.52	0.54	0.83	1.5	1.6
0.4	0.32	0.32	0.37	0.77	1.35	1.4
0.3	0.22	0.22	0.32	0.74	1.2	1.2
0.2	0.13	0.155	0.31	0.72	0.95	0.97
0.15	0.086	0.145	0.33	0.70	0.77	0.82

Values of $d\mu/d\lambda$ to be entered in (10.49) and (10.51) can be obtained from Tables 7 and 2. Notice that (10.48) and (10.52) have the form $f(t)\,g(\lambda)$ to be expected for a state of quasi-equilibrium according to (5.2). These asymptotic trends for large t constitute the significant information obtained by solving the transform equation (10.37) through an approximation valid only for $p \sim \mu_s$.

It may be worth giving the inverse transforms of the expressions (10.42) to (10.45) in greater detail than is provided by their asymptotic forms.

[1] This Encyclopedia, Vol. II, p. 236.

(I) With $\mu_0 - p = u$ and $C(\lambda - \lambda_0) = -a$, (10.42) becomes $u^{-2}\exp(-a/u)$, the transform of the Bessel function J_1, and we have

$$I(t, \lambda) = i\sqrt{\frac{C t}{\lambda - \lambda_0}}\, e^{-\mu_0 t} J_1\left(2i\sqrt{C(\lambda - \lambda)\, t}\right). \tag{10.54}$$

(II) With $\mu_0 - p = u$ and $\dot{\mu}_0(\lambda - \lambda_0) = a$, (10.43) becomes $C u^{-\varkappa_0 - 1}(u + a)^{\varkappa_0 - 1}$, the transform of a Laguerre, or confluent hypergeometric, function and we have

$$I(t, \lambda) = C\, t\, e^{-\mu_0 t} F_1\left(1 - \varkappa_0, 2, -\dot{\mu}_0(\lambda - \lambda_0)\, t\right). \tag{10.55}$$

The expansion of (10.55) or (10.54) in powers of t constitutes an expansion in orders of scattering. The asymptotic forms (10.47) and (10.48) correspond to highly multiple scattering. The complete functions (10.54) and (10.55) describe the transition from one special condition to the other one.

(III) Eq. (10.50) already represents the complete inverse transform of (10.44).

(IV) Setting $\mu(\lambda) - p = \mu(\lambda) - \mu_m$ in the denominator of (10.45) and entering the resulting expression in (10.46) one obtains the inverse transform of the function $\exp\sqrt{4\tau\mu_m/(\mu_m - p)}$, where τ is given by (10.53). The resulting function can be expressed in the form of a real integral suited for numerical evaluation,

$$\left.\begin{aligned} I(t, \lambda) &= K \frac{e^{-\mu_m t}}{\mu(\lambda) - \mu_m} \frac{1}{2\pi i} \int_{-i\infty}^{i\infty} d(\mu_m - p)\, e^{(\mu_m - p)t + \sqrt{\frac{4\tau\mu_m}{\mu_m - p}}} \\ &= K \frac{e^{-\mu_m t}}{\mu(\lambda) - \mu_m} \frac{\tau\mu_m}{(\tau\mu_m t)^{\frac{1}{3}}} \frac{1}{2\pi i} \int_{-i\infty}^{i\infty} du\, e^{\left(u + \frac{1}{\sqrt{u}}\right)(4\tau\mu_m t)^{\frac{1}{3}}} \\ &= K \frac{e^{-\mu_m t}}{\mu(\lambda) - \mu_m} \frac{d}{dt} \frac{2}{\pi} \int_0^\infty \frac{dv}{1 + v^2} \exp\left\{\frac{3 - v^2}{(1 + v^2)^{\frac{2}{3}}} (\tau\mu_m t)^{\frac{1}{3}}\right\}, \end{aligned}\right\} \tag{10.56}$$

where $u = |u|\, e^{i\varphi} = (\mu_m - p)\, t^{\frac{2}{3}} (4\tau\mu_m)^{-\frac{1}{3}}$ and $v = \operatorname{tg}\frac{1}{2}\varphi$; the last integral represents integration of the variable u along the path $|u| = (2\cos\frac{1}{2}\varphi)^{-\frac{2}{3}}$ on which $u + 1/\sqrt{u}$ is real. In the more detailed calculation described in Sect. 21, integrals analogous to (10.56) have been evaluated successfully by expansion in powers of t up to values of $\mu_m t = 180$.

δ) Distribution in pathlength, energy and direction. The distribution $I(t, \omega, \lambda)$ can be studied by expanding the transport equation (10.35) into spherical harmonics. This expansion, combined with a Laplace transformation and with a choice of z axis parallel to ω_0 as in Sect. 10β, leads to

$$[\mu(\lambda) - p]\, I_l(p, \lambda) = \int_{\lambda_0}^{\lambda} d\lambda'\, k_l(\lambda, \lambda')\, I_l(p, \lambda') + \delta(\lambda - \lambda_0), \tag{10.57}$$

which differs from (10.37) only by the replacement of the kernel $k(\lambda, \lambda')$ with the kernel $k_l(\lambda, \lambda')$ from (9.10). The equations for different l can, of course, be solved separately by the procedures of Sect. 10β since (10.57) differs from (10.8) only by having $\mu(\lambda) - p$ on the left side instead of $\mu(\lambda)$.

The distribution in pathlength, energy, and direction has been studied in some detail for high energy X rays by Foldy and Osborn [*34*] and especially by Ogievetskii [*66*]. The problem was treated as a straight-ahead approximation to the penetration problem for a plane perpendicular source, i.e., the pathlength variable of (10.35) was interpreted as a depth of penetration (see p. 706 and 713, 714). Most of the calculations were performed analytically utilizing the approximation (10.13) combined with the scattering cross-section transformation (7.1).

For small and moderate pathlengths a solution is given in Ref. [66] by the order of scattering procedure of Sect. 8. Conversely, long pathlengths are attained primarily by photons that have experienced multiple scattering with small wavelength shifts in the range $\lambda \sim \lambda_s$, as in the problem of Sect. 10γ where deflections were not considered. Multiple small wavelength shifts imply multiple small deflections, and therefore yield a smooth but peaked angular distribution.

Consider first, for the purpose of orientation, the approximation in which the dependence of μ and k on λ is disregarded, even though the variations of μ are nearly always essential for the distribution of photons with long pathlengths, as pointed out on p. 708. The Laplace transform $I(p, \vartheta, \lambda)$ can then be obtained from (10.22), (10.29), or (10.34) by replacing μ with $\mu - p$. If one utilizes the expansion in orders of scattering, obtained from (10.34), the inversion of the Laplace transform is straightforward and yields

$$I(t, \vartheta, \lambda) = \frac{1}{2\pi} \sqrt{\frac{\vartheta}{\sin\vartheta}}\, e^{-\mu t} \sum_{n > \frac{\vartheta^2}{2(\lambda-\lambda_0)}} \frac{(C t)^n}{n!} \frac{\left(\lambda - \lambda_0 - \frac{\vartheta^2}{2n}\right)^{n-2}}{n(n-2)!}. \tag{10.58}$$

An expression appropriate to highly multiple scattering is obtained by evaluating the integral in (10.22), through the residue at its pole, before summing over l. The pole lies at the root of the equation $\eta = [C/(\mu - p)] \exp(-\bar{l}^2/2\eta)$, which is approximately

$$\eta \sim \frac{C}{\mu - p}\, e^{-\frac{1}{2}\bar{l}^2 \frac{\mu-p}{C}} \sim \frac{C}{\mu - p} - \frac{1}{2}\bar{l}^2 \tag{10.59}$$

for $C/(\mu - p) \gg \bar{l}^2$. It follows then from (10.22) that

$$\left.\begin{aligned} I(p, \vartheta, \lambda) &\sim \sum_l \frac{2l+1}{4\pi} P_l(\cos\vartheta) \frac{C}{(\mu-p)^2}\, e^{\frac{C(\lambda-\lambda_0)}{\mu-p} - \frac{1}{2}\left(l+\frac{1}{2}\right)^2(\lambda-\lambda_0)} \sim \\ &\sim \frac{C}{(\mu-p)^2}\, e^{\frac{C(\lambda-\lambda_0)}{\mu-p}} \frac{e^{-\frac{\vartheta^2}{2(\lambda-\lambda_0)}}}{2\pi}. \end{aligned}\right\} \tag{10.60}$$

Thus one obtains a Gaussian distribution of photon directions, resulting from a random walk with total mean square deflection $\langle\vartheta^2\rangle = 2(\lambda - \lambda_0)$ (see p. 703), but only when multiple scattering predominates and the absorption, being constant, fails to modify the angular distribution. These conditions are hardly realistic since they imply that $\mu(\lambda) - \mu(\lambda_0) \gg \mu(\lambda_0) - p$ for all λ of interest, and that, in terms of the pathlength variable, $[\mu(\lambda) - \mu(\lambda_0)]\, t \gg \sqrt{C(\lambda - \lambda_0)\, t} \gg 1$.

To consider a linear variation of $\mu(\lambda)$, as in case II of Sect. 10γ, one treats (10.57) again by Laplace transformation as in (10.17) ff., to obtain

$$(\mu_0 - p)\, I_l(p, \eta) - \dot{\mu}_0 \frac{\partial I_l}{\partial \eta} = \frac{C}{\eta}\, e^{-\bar{l}^2/2\eta} + 1, \tag{10.61}$$

where the term $\partial I_l/\partial\eta$ stems from the variation of $\mu(\lambda)$. Eq. (10.61) has the solution

$$I_l(p, \eta) = \frac{1}{\dot{\mu}_0} \int_0^\infty du\, e^{-\frac{\mu_0 - p}{\dot{\mu}_0} u + \frac{C}{\dot{\mu}_0} \int_0^u du' \frac{e^{-\bar{l}^2/2(\eta+u')}}{\eta+u'}}, \tag{10.62}$$

whose inverse transform is

$$I(t, \vartheta, \lambda) = \sum_l \frac{2l+1}{4\pi} P_l(\cos\vartheta)\, e^{-\mu_0 t} \frac{1}{2\pi i} \int d\eta\, e^{\eta(\lambda-\lambda_0) + \frac{C}{\dot{\mu}_0} \int_0^u du \frac{e^{-(l+\frac{1}{2})^2/2(\eta+u)}}{\eta+u}}. \tag{10.63}$$

The integrations in (10.62) or (10.63) cannot be performed analytically, but one can obtain the derivatives $[\partial^n I_l/\partial(\bar{l}^2)^n]_{\bar{l}=0}$, which represent moments of the angular distribution, i.e. mean values of $(\frac{1}{2}\vartheta^2)^n$. These moments have an analytic form similar to (10.55) and serve to evaluate the departure of the angular distribution (10.63) from a Gaussian by the moment fitting method of Part C [66].

11. Deep penetration. This section deals with the law of variation of X-ray intensity at great distances from a source. This law is represented by the asymptotic trend of the function $f(z)$ of (5.2) for large values of the depth of penetration z. According to Sect. 5γ the deep penetration is controlled by a continuing process of production and removal of the most penetrating secondaries. Therefore our study centers on solving the transport equation in the range of variables corresponding to the most penetrating secondaries, and we can utilize approximation methods appropriate to that range.

Consider the penetration of X rays from a plane source in an infinite isotropic medium, starting from the Fourier transform of the relevant transport equation

$$\left.\begin{aligned} -p\,\omega_z\, I(p,\omega,\lambda) &= -\mu(\lambda)\, I(p,\omega,\lambda) + \\ &+ \int_{\lambda_0}^{\lambda} d\lambda' \int_{4\pi} d\omega'\, k(\omega,\lambda,\omega',\lambda')\, I(p,\omega',\lambda) + \delta(\omega-\omega_0)\,\delta(\lambda-\lambda_0). \end{aligned}\right\} \quad (9.3)$$

The most penetrating secondaries have wavelengths close to the value λ_s defined by the condition that $\mu(\lambda_s)=\mu_s$ is the smallest value of $\mu(\lambda)$ in the range of integration. These secondaries travel in directions ω pointing away from the source, such that $\vartheta\sim 0$ and $\omega_z\sim 1$. The values of the parameter p corresponding to deep penetration are determined by an inspection of the singularities of the transport equation (9.3) analogous to that performed in Sect. 10γ for the pathlength problem. Eq. (9.3) is singular for $p\omega_z \geqq \mu_s$, for the same reason that (10.37) is singular for $p \geqq \mu_s$, and deep penetration corresponds, like long pathlength, to values of p only a little smaller than μ_s. The *critical range of variables* corresponding to the most penetrating secondaries is, therefore, $\lambda\sim\lambda_s$, $\omega_z\sim 1$, $p\sim\mu_s$. More specifically it will be seen that, for any given value of p, those values of λ and ω_z are of interest for which $\mu(\lambda)-p\omega_z$ remains comparable to $\mu(\lambda_s)-p$. The essential information developed in this section corresponds to the limiting case $\mu_s - p \to 0$.

The study thus delimited proves inadequate when the source X rays have wavelength $\lambda_0=\lambda_s$ but a direction ω_0 very oblique with respect to the z axis. Hardly any of the secondaries belong then in the critical range $\lambda\sim\lambda_s$, $\omega_z\sim 1$. We have here a problem which is essentially unsolved and is discussed briefly at the end of this section.

It has often been surmised that a small obliquity of the X-ray flow with respect to the direction $\vartheta=0$ would have a negligible effect on the deep penetration, so that (9.3) could be adequately solved by setting $\omega_z=1$ in the coefficient on its left side ("straight-ahead approximation"). This approximation, utilized in numerous studies (see e.g. [*14*], [*27*], [*34a*], [*66*][1]), identifies the depth of penetration with the pathlength, as pointed out on p. 706 and therefore leads to the results described in Sect. 10γ and δ. In fact, the straight-ahead approximation is hardly ever realistic, as shown by the following argument.

Compare the variations of the coefficient $p\omega_z$ of (9.3), which result from small angle scattering, with the simultaneous variations of the other coefficients of the equation, namely $\mu(\lambda)$ and $k(\omega,\lambda,\omega',\lambda')$. If $\delta\lambda$ indicates the wavelength shift in a small-angle scattering, the variation of $p\omega_z$ according to the Compton law is of the order of $\mu_s\,\delta\lambda$, when $p\sim\mu_s$ and $\omega_z\sim 1$. The accompanying variation of the kernel k is of the order $(\delta\lambda)^2$, and therefore negligible, and the variation $\mu(\lambda)$ may be indicated as $(d\mu/d\lambda)_{\lambda=\lambda_s}\,\delta\lambda$. The relative influence upon penetration

[1] Many results were originally obtained by E. Greuling and others in unpublished Manhattan Project reports.

of changes of obliquity and changes of absorption is therefore indicated by the ratio

$$\frac{\delta(p\,\omega_z)}{\delta\mu} \sim \frac{\mu_s}{(d\mu/d\lambda)_{\lambda=\lambda_s}} = \left(\frac{d\log\mu}{d\lambda}\right)^{-1}_{\lambda=\lambda_s}. \tag{11.1}$$

The data in Table 7 show that this ratio is of the order of 1 for most materials and energies of interest, and becomes infinite at the energy of minimum absorption, i.e. at $\lambda_s = \lambda_m$. Therefore small obliquities of X-ray flow can hardly ever be disregarded. Notice also that the variations of $p\omega_z$, being of first order in $\delta\lambda$, have greater influence than any variation of $\mu(\lambda)$ represented by its second, or higher, derivatives. That is, *only the first derivative* $d\mu/d\lambda$ need be considered for the purpose of calculating the influence of obliquity upon deep penetration.

Even though the straight-ahead approximation is inadequate, the problem of deep penetration is solved by reduction to an "equivalent straight-ahead problem". The reduction results from showing that obliquity has the same effect as a decrease of the scattering cross section. Consider, for example, the Fourier-Legendre transform

$$\left.\begin{aligned}&[\mu(\lambda) - p]\, I_l^m(p,\lambda) - \frac{1}{2} p\Delta I_l^m(p,\lambda)\\ &\qquad = \int_{\lambda_0}^{\lambda} d\lambda'\, k_l(\lambda,\lambda')\, I_l^m(p,\lambda) + \sqrt{\frac{4\pi}{2l+1}}\, Y_l^{m*}(\omega_0)\,\delta(\lambda-\lambda_0)\end{aligned}\right\} \tag{9.13}$$

of the transport equation. The influence of obliquity is represented in this equation by the term $-\frac{1}{2}p\Delta I_l^m$, which is absent in the corresponding "straight-ahead equation" (10.37). The angular distribution of the most penetrating secondaries is sharply peaked at $\vartheta \sim 0$, so that its coefficients I_l^m vary slowly with l, and ΔI_l^m is small. However, the other terms of (9.13) are also small. What matters is the ratio among different terms of the equation, for example the ratio $p\Delta I_l^m/(\mu-p)\, I_l^m$ which is of the order of -1. Once an estimate of this ratio is obtained, the term $-\frac{1}{2}\Delta p I_l^m$ in (9.13) may be replaced with a term proportional to I_l^m. The proportionality constant can be treated formally as a modification of either the attenuation coefficient or of the scattering probability.

The evaluation of the ratio $p\,\Delta I_l^m/(\mu-p)\, I_l^m$ constitutes an eigenvalue problem, which was identified and solved, in the main, by WICK in his work on neutron penetration [98][1]. Two kinds of procedure are possible: one involves extensive numerical work and is suited for detailed calculation; the other is analytic and is suited to determine the characteristics of the singularity of the transform $I(p,\omega,\lambda)$ at $p=\mu_s$. Data on the singularity complement numerical results when the inverse transform $I(z,\omega,\lambda)$ is eventually calculated by the formula (9.1) (see Sect. 21).

A numerical procedure starts with an assumed trial estimate of the mean obliquity $\langle\omega_z\rangle = I_1^0/I_0^0$ for each value of λ and p in the critical range. Since $\frac{1}{2}\Delta I_0^0 = I_1^0 - I_0^0 = -(1-\langle\omega_z\rangle)\, I_0^0$, the initial estimate reduces the Eq. (9.13) for $l=m=0$ to the straight-ahead form (10.37) and thus permits its solution. [Alternatively one may consider the Eq. (9.12) with $l=m=0$ and eliminate from it I_1^0 by the relation $I_1^0 = \langle\omega_z\rangle\, I_0^0$.] If I_0^0 and $I_1^0 = \langle\omega_z\rangle\, I_0^0$ are regarded as known, the Eq. (9.12) with $l=1$ and $m=0$ gives I_2^0. All successive equations, with $m=0$ and increasing l, become unraveled, thereby allowing sequential determination of the I_l^0's. (For $m\neq 0$ the same procedure applies starting with $l=|m|$.) However, the set of I_l^0's obtained in this manner diverges for large l,

[1] See also I. WALLER: Ark. Math. Astronom. Fys., Ser. A **34**, No. 3, 4 and 5 (1948).

unless the correct value for I_1^0/I_0^0 had been chosen at the start. Thus the solution of the system of the Eq. (9.12) with all values of l becomes an eigenvalue problem, which can be solved by trial and error or by iteration. A rapid procedure of iteration is described in Sect. 20.

The analytical procedure, which is developed in the remainder of this section, relies on separation of the energy and obliquity variables. The separation requires that the variables be suitably transformed; it holds only in the critical range of variables $\lambda \sim \lambda_s$, $\vartheta \sim 0$, $p \sim \mu_s$, but with increasingly good approximation when this range becomes progressively narrow in the limit $p \to \mu_s$. The angular distribution of the X rays, and therefore the estimate of $\Delta I_l^m/(\mu - p)\, I_l^m$ is determined by an eigenvalue problem having the form of a quantum mechanical single-variable Schroedinger equation.

α) Reduction to the equivalent straight-ahead problem. Consider the Fourier transform equation for a plane source, in the form (9.13), for a narrow range of values of $\lambda \sim \lambda_s$ and for $\mu_s - p \ll \mu_s$. In this range the scattering kernel may be taken as constant, as in (10.15), and only linear variations of $\mu(\lambda)$ matter as pointed out in the introductory discussion; the linear variations are characterized by $\dot{\mu}_s = (d\mu/d\lambda)_{\lambda=\lambda_s}$. The transform parameter p can be taken as equal to μ_s unless it appears in the combination $\mu_s - p$. The problem must be considered separately for each value of the azimuthal parameter m, but only the solution for $m = 0$ is shown to be important, i.e. the flux distribution at deep penetration acquires cylindrical symmetry; accordingly we assume for simplicity $m = 0$ and drop the index m unless otherwise noted. With these conventions (9.13) takes the form

$$\left.\begin{aligned}(\mu_s - p)\, I_l(p,\lambda) + \dot{\mu}_s(\lambda - \lambda_s)\, I_l(p,\lambda) - \tfrac{1}{2}\mu_s\, \Delta I_l(p,\lambda)\\ = C\int_{\lambda_0}^{\lambda} d\lambda'\, P_l(1-\lambda+\lambda')\, I_l(p,\lambda') + P_l(\cos\vartheta_0)\,\delta(\lambda-\lambda_0).\end{aligned}\right\} \tag{11.2}$$

The source term has been entered in this equation as it was entered in (10.38), with the understanding that the source wavelength λ_0 lies within the critical range of approximate integration $\lambda \sim \lambda_s$. If λ_0 is much smaller than λ_s one should disregard the source term in (11.2) and consider solutions of the resulting homogeneous equation, with a constant factor to be adjusted in analogy to the factor of (10.40) so as to fit a separate solution at $\lambda \ll \lambda_s$ valid below the critical range of integration.

Since the kernel of this integral equation depends only on the wavelength difference $\lambda - \lambda'$ and not on λ or $\lambda - \lambda_s$, Laplace transformation of the variable λ is appropriate, as in (10.17). However (11.2) contains one term with the coefficient $\dot{\mu}_s(\lambda - \lambda_s)$, and this term causes the transform of (11.2) to be a differential equation like (10.61), instead of being algebraic like (10.16). Since $I_l(p,\eta)$ is understood to be a smooth, slowly varying, function of the index l, this index may be treated as a continuous variable and replaced, where desirable, with $\bar{l} = l + \frac{1}{2}$, as in (10.19). Utilizing also the approximate Eq. (10.20) we obtain the transform of (11.2) in the form

$$(\mu_s - p)\, I_l(p,\eta) - \dot{\mu}_s \frac{\partial I_l}{\partial \eta} - \frac{1}{2}\mu_s\, \Delta I_l(p,\eta) = \frac{C}{\eta}\, e^{-\bar{l}^2/2\eta}\, I_l(p,\eta) + P_l(\cos\vartheta_0). \tag{11.3}$$

This equation differs from (10.61) by the interlinkage term with ΔI_l which prevents separate integration in η for each value of l.

The transform of the scattering kernel depends on the combination of variables $\bar{l}^2/2\eta$. This combination stems from the circumstance that a large value of the transform parameter η implies scattering with small wavelength shift, and therefore small deflection angle, which influences only the large-$\bar{l}$ components of the angular distribution. The effect of scattering depends accordingly on the ratio of the transform variables η and $\bar{l}^2$, or of the corresponding primitive variables $\lambda-\lambda_s$ and ϑ^2, rather than on their separate values. For this reason it is convenient to replace the index $\bar{l}$ with the combination

$$\sigma = \frac{\bar{l}}{\sqrt{2\eta}} = \frac{l+\frac{1}{2}}{\sqrt{2\eta}}, \tag{11.4}$$

regarded as a continuous variable, to express $I_l(p,\eta)$ in the form $I(p,\sigma,\eta)$, and to approximate $P_l(\cos\vartheta_0)$ by $J_0(\sqrt{2\eta}\sigma\vartheta_0)$. Eq. (11.3) becomes accordingly

$$\left.\begin{aligned}(\mu_s-p)\,I(p,\sigma,\eta)-\dot{\mu}_s\left[\frac{\partial I(p,\sigma,\eta)}{\partial\eta}-\frac{1}{2\eta}\sigma\frac{\partial I(p,\sigma,\eta)}{\partial\sigma}\right]-\frac{1}{4}\frac{\mu_s}{\eta}\Delta I(p,\sigma,\eta)\\ =\frac{C}{\eta}\,e^{-\sigma^2}I(p,\sigma,\eta)+J_0(\sqrt{2\eta}\,\sigma\vartheta_0),\end{aligned}\right\} \tag{11.5}$$

where

$$\Delta I(p,\sigma,\eta)=\left(\frac{\partial^2}{\partial\sigma^2}+\frac{1}{\sigma}\frac{\partial}{\partial\sigma}-\frac{m^2}{\sigma^2}\right)I(p,\sigma,\eta) \tag{11.6}$$

in accordance with (9.15)[1].

Eq. (11.5) can be solved by separation of variables because its terms, except the source term, can be separated into two groups, one of which depends only on η and the other only on σ. Multiplication of (11.5) by η and regrouping of its terms yields

$$\left.\begin{aligned}&\eta\left(\mu_s-p-\dot{\mu}_s\frac{\partial}{\partial\eta}\right)I(p,\sigma,\eta)\\ &\quad=\left(\frac{1}{4}\mu_s\Delta-\frac{1}{2}\dot{\mu}_s\sigma\frac{\partial}{\partial\sigma}+C\,e^{-\sigma^2}\right)I(p,\sigma,\eta)+\eta\,J_0(\sqrt{2\eta}\,\sigma\,\vartheta_0).\end{aligned}\right\} \tag{11.7}$$

The homogeneous portion of this equation has solutions each of which is the product of a function of η and of a function of σ. The solution of the complete inhomogeneous equation is conveniently expressed as a superposition of such products, namely in the form

$$I(p,\sigma,\eta)=\sum_r I^{\langle r\rangle}(p,\sigma,\eta)=e^{\frac{1}{2}\frac{\dot{\mu}_s}{\mu_s}\sigma^2}\sum_r a_r(p,\eta)\,u_r(\sigma). \tag{11.8}$$

In this equation $u_r(\sigma)$ indicates an eigenfunction of

$$\left(\frac{1}{4}\mu_s\Delta+\frac{1}{2}\dot{\mu}_s-\frac{1}{4}\frac{\dot{\mu}_s^2}{\mu_s}\sigma^2+C\,e^{-\sigma^2}\right)u_r(\sigma)=\bar{C}_r u_r(\sigma) \tag{11.9}$$

with eigenvalue $\bar{C}_r$, $a_r(p,\eta)$ has to be determined by substituting (11.8) in (11.7), and the factor $\exp\left(\frac{1}{2}\frac{\dot{\mu}_s}{\mu_s}\sigma^2\right)$ has been split off from $u_r(\sigma)$ to make (11.9) self-adjoint. Eq. (11.9) is the Wick eigenvalue equation which determines the characteristics of angular distributions at great depths; it has the form of the Schroedinger equation for the motion of a particle in a two-dimensional potential well with cylindrical symmetry and with the profile $\frac{1}{2}\frac{\dot{\mu}_s^2}{\mu_s}\sigma^2-C\,e^{-\sigma^2}$.

---

[1] The dependence of ΔI on m is indicated here explicitly, but m is understood to vanish unless otherwise stated.

Upon substitution of (11.8) and multiplication by $\exp\left(-\frac{1}{2}\frac{\dot{\mu}_s}{\mu_s}\sigma^2\right)$, (11.7) becomes

$$\left.\begin{aligned}&\sum_r u_r(\sigma)\,\eta\left(\mu_s - p - \dot{\mu}_s\frac{\partial}{\partial\eta}\right)a_r(p,\eta)\\&=\sum_r a_r(p,\eta)\left(\frac{1}{4}\mu_s\Delta+\frac{1}{2}\dot{\mu}_s-\frac{1}{4}\frac{\dot{\mu}_s^2}{\mu_s}\sigma^2+C\,\mathrm{e}^{-\sigma^2}\right)u_r(\sigma)+\mathrm{e}^{-\frac{1}{2}\frac{\dot{\mu}_s}{\mu_s}\sigma^2}\eta\,J_0\left(\sqrt{2\eta}\,\sigma\vartheta_0\right)\\&=\sum_r a_r(p,\eta)\,\overline{C}_r\,u_r(\sigma)+\mathrm{e}^{-\frac{1}{2}\frac{\dot{\mu}_s}{\mu_s}\sigma^2}\eta\,J_0\left(\sqrt{2\eta}\,\sigma\vartheta_0\right).\end{aligned}\right\}\quad(11.10)$$

Eq. (11.9) has been utilized here; the orthogonality of its eigenfunctions is exploited by multiplying (11.10) with $u_r(\sigma)$ and integrating over the differential element $\sigma\,d\sigma$ of the cylindrical variable σ, which yields

$$\left(\mu_s - p - \dot{\mu}_s\frac{\partial}{\partial\eta}\right)a_r(p,\eta) = \frac{\overline{C}_r}{\eta}a_r(p,\eta) + \int_0^\infty \sigma\,d\sigma\,\mathrm{e}^{-\frac{1}{2}\frac{\dot{\mu}_s}{\mu_s}\sigma^2}u_r(\sigma)\,J_0\left(\sqrt{2\eta}\,\sigma\vartheta_0\right). \quad (11.11)$$

This equation coincides with (10.61) for $l=0$ except for the replacement of the cross section C with the eigenvalue $\overline{C}_r$ and for the difference of the source term. It may also be regarded as the Laplace transform of the pathlength (or straight-ahead) equation (10.38), with $\overline{C}_r$ in the place of C, with a different source term, and with a linear approximation to $\mu(\lambda)$. The eigenvalues $\overline{C}_r$ are all smaller than C, and accordingly the replacement of C with $\overline{C}_r$ represents the influence of obliquity upon penetration by the adoption of a smaller effective cross section for the production of lower energy X rays.

Therefore the most important term in (11.8), in fact the only important term for deep penetration, will be the one corresponding to largest eigenvalue of (11.9), to be called $\overline{C}_0$. In a Schroedinger equation with the form (11.9), the energy eigenvalue corresponds to $-\overline{C}_r$, so that the ground state of the particle corresponds to the largest eigenvalue $\overline{C}_0$. For $m\neq 0$, the last term of (11.6) corresponds to a "centrifugal potential" of the Schroedinger problem, so that the highest eigenvalue is much lower than the eigenvalue $\overline{C}_0$ for $m=0$. For this reason, components of the X-ray distribution with $m\neq 0$ contribute negligibly to very deep penetration.

The highest eigenvalue $\overline{C}_0$ is readily evaluated by approximation procedures. The variational method with the trial eigenfunction

$$u_0(\sigma) = 2\sqrt{\alpha}\,\mathrm{e}^{-\alpha\sigma^2} \quad (11.12)$$

requires α to be a root of the equation

$$4\frac{C}{\mu_s}\frac{1}{(2\alpha+1)^2} + \left(\frac{\dot{\mu}_s}{\mu_s}\right)^2\frac{1}{4\alpha^2} = 1 \quad (11.13)$$

and yields the eigenvalue as a function of α

$$\overline{C}_0 = C\frac{\alpha}{\alpha+\frac{1}{2}} - \frac{1}{8}\frac{\dot{\mu}_s^2}{\mu_s\alpha} + \frac{1}{2}\dot{\mu}_s - \frac{1}{2}\mu_s\alpha. \quad (11.14)$$

In the special case $\dot{\mu}_s=0$, these results reduce to

$$\alpha = \sqrt{\frac{C}{\mu_s}} - \frac{1}{2}, \quad (11.13')$$

$$\overline{C}_0 = \left(\sqrt{C} - \tfrac{1}{2}\sqrt{\mu_s}\right)^2. \quad (11.14')$$

Calculations of $\overline{C}_0$ have also been carried out by iteration procedures of the types discussed in Sect. 20 and in App. C of Ref. [28]. Results are given in Table 8 for $\dot{\mu}_s \neq 0$ and in Table 9 for $\dot{\mu}_s = 0$. Further eigenvalues of (11.9) have not been calculated. At most one further eigenvalue $\overline{C}_1$, for $m=0$, and the highest eigenvalue $\overline{C}_0$ for $m=1$, may be positive for realistic values of C, μ_s, and $\dot{\mu}_s$. All other eigenvalues are negative, and therefore represent effective probabilities of dissipation, rather than of production, of lower energy X rays.

Let us return now to the expansion (11.8) of the transform $I(p, \sigma, \eta)$. The coefficients $a_r(p, \eta)$ can be obtained by straightforward solution of (11.11), but it may be preferable to consider directly the inverse transform in λ and ϑ of each

Table 8. *Values of the parameter* $\bar{\varkappa}_0$ [Eq. (11.22)].

$\dot{\mu}_0/\mu_0$	C/μ_0									
	0.02	0.05	0.1	0.25	0.5	1.0	2.0	4.0	6.0	8.0
0.02	0.0200	0.0516	0.1095	0.374	2.63	13.00	42.3	113.0	189.5	272.0
0.05	0.01939	0.0494	0.1052	0.332	1.328	5.49	17.21	45.5	76.5	108.9
0.1	0.01846	0.0473	0.0988	0.293	0.871	2.98	8.84	23.0	38.5	54.7
0.25	0.01617	0.0411	0.0844	0.231	0.547	1.439	3.81	9.48	15.69	22.2
0.5	0.01341	0.0338	0.0678	0.1798	0.388	0.887	2.11	4.96	8.16	11.32
1.0	0.01003	0.0252	0.0506	0.1289	0.265	0.559	1.211	2.67	4.24	5.87
2.0			0.0335	0.0841	0.1696	0.345	0.709	1.477	2.28	3.11
4.0								0.821	1.173	1.674
6.0								0.577	0.870	1.165
8.0								0.447	0.671	0.897

To find $\bar{\varkappa}_0$ for a given material and λ_0, determine $\mu_0 = \mu(\lambda_0)$ from Table 2 or Ref. [96], calculate $\dfrac{C}{\mu_0} = \dfrac{0.300}{\mu_0}\dfrac{Z}{A}$ and take the value of $\dfrac{\dot{\mu}_0}{\mu_0}$ from Table 7.

expansion term $I^{\langle r\rangle}(p, \sigma, \eta)$ and the equation obeyed by such inverse transform. Thereby one keeps in evidence the correspondence with the equation (10.38) of the straight-ahead approximation. In particular, the following quantities can be worked out in some detail:

a) the X-ray flux components integrated over photon directions,

$$I^{\langle r\rangle}(p, \lambda) = [I^{\langle r\rangle}(p, \sigma, \eta)]_{\sigma=0} = \frac{1}{2\pi i}\int_{-i\infty}^{i\infty} e^{\eta(\lambda-\lambda_0)}\, a_r(p, \eta)\, u_r(0)\, d\eta, \tag{11.15}$$

b) the X-ray flux components in the z-direction (perpendicular to the source plane),

$$\left.\begin{aligned} &[I^{\langle r\rangle}(p, \vartheta, \lambda)]_{\vartheta=0} \\ &= \frac{1}{2\pi i}\int_{-i\infty}^{i\infty} e^{\eta(\lambda-\lambda_0)}\, a_r(p,\eta)\, d\eta \int_0^{\infty} \frac{2\eta\,\sigma\, d\sigma}{2\pi}\, e^{\frac{1}{2}\frac{\dot{\mu}_s}{\mu_s}\sigma^2}\, u_r(\sigma)\, [J_0(\sqrt{2\eta}\,\sigma\,\vartheta)]_{\vartheta=0} \\ &= \frac{1}{2\pi i}\int_{-i\infty}^{i\infty} e^{\eta(\lambda-\lambda_0)}\, a_r(p,\eta)\, \frac{\eta}{\pi}\, A_r\, u_r(0)\, d\eta = \frac{1}{\pi}\, A_r\, \frac{\partial}{\partial\lambda} I^{\langle r\rangle}(p, \lambda), \end{aligned}\right\} \tag{11.16}$$

where

$$A_r = \int_0^{\infty} \sigma\, d\sigma\, e^{\frac{1}{2}\frac{\dot{\mu}_s}{\mu_s}\sigma^2}\, \frac{u_r(\sigma)}{u_r(0)} \tag{11.17}$$

is independent of η,

c) the X-ray flux component $I^{\langle 0\rangle}(p, \vartheta, \lambda)$, calculated for general conditions but utilizing the crudely approximate gaussian eigenfunction $u_0(\sigma)$ given by (11.12)

$$\left.\begin{aligned} I^{\langle 0\rangle}(p, \vartheta, \lambda) &= \frac{1}{2\pi i}\int_{-i\infty}^{i\infty} e^{\eta(\lambda-\lambda_0)}\, a_0(p, \eta)\int_0^{\infty}\frac{2\eta\,\sigma\, d\sigma}{2\pi}\, 2\sqrt{\alpha}\, e^{\left(\frac{1}{2}\frac{\dot{\mu}_s}{\mu_s}-\alpha\right)\sigma^2}\times \\ &\quad\times J_0\left(\sqrt{2\eta}\,\sigma\vartheta\right) = \frac{1}{2\pi i}\int_{-i\infty}^{i\infty} e^{\eta(\lambda-\lambda_0)}\, a_0(p, \eta)\,\frac{\eta}{\pi}\,\frac{e^{-\eta\frac{\vartheta^2}{2\alpha-\dot{\mu}_s/\mu_s}}}{2\alpha-\dot{\mu}_s/\mu_s} \\ &= \frac{1}{\pi}\,\frac{1}{2\alpha-\dot{\mu}_s/\mu_s}\,\frac{\partial}{\partial\lambda}\, I^{\langle 0\rangle}\left(p, \lambda-\frac{\vartheta^2}{2\alpha-\dot{\mu}_s/\mu_s}\right), \end{aligned}\right\} \tag{11.18}$$

where α is the root of (11.13). The quantities (11.16) and (11.18) can be obtained as by-products of a calculation of $I^{\langle r\rangle}(p, \lambda)$, which is itself given by (11.15).

The integrated flux components $I^{\langle r\rangle}(p, \lambda)$ obey an equation which is the inverse Laplace transform of (11.11) multiplied by the constant $u_r(0)$. The inversion of the source term of (11.11) is difficult because the transform parameter η appears in the Bessel function; therefore the Laplace inversion will be worked out under the following simplifying alternative assumptions:

(a′) The source is *isotropic*; one must then replace $P_l(\cos\vartheta_0)$ in the initial equation (11.2) with δ_{l0}, and $J_0\left(\sqrt{2\eta}\,\sigma\vartheta_0\right)$ in (11.11) with $\delta(2\eta\sigma^2)=\delta(\sigma)/4\eta\sigma$, so that the source term of (11.11) reduces to $u_r(0)/4\eta$.

(b′) The source is *perpendicular*, so that $\vartheta_0=0$ and the source term of (11.11) remains independent of η and equal to

$$A_r^+\, u_r(0) = \int_0^{\infty}\sigma\, d\sigma\, e^{-\frac{1}{2}\frac{\dot{\mu}_s}{\mu_s}\sigma^2}\, u_r(\sigma)\,. \tag{11.19}$$

(c′) The source has an arbitrary direction coordinate ϑ_0 but one utilizes, for $r=0$, the approximate $u_0(\sigma)$ given by (11.12) so that the source term becomes $u_0(0)\, e^{-\eta\vartheta_0^2/(2\alpha+\dot{\mu}_s/\mu_s)}/(2\alpha+\dot{\mu}_s/\mu_s)$. Under these alternative assumptions the equation obeyed by $I^{\langle r\rangle}(p, \lambda)$ is

$$\left.\begin{aligned} &[\mu_s - p + \dot{\mu}_s(\lambda-\lambda_s)]\, I^{\langle r\rangle}(p, \lambda) = \overline{C}_r\int_{\lambda_0}^{\lambda} d\lambda'\, I^{\langle r\rangle}(p, \lambda') + \\ &+\begin{cases} \frac{1}{4}[u_r(0)]^2\int_{\lambda_0}^{\lambda}\delta(\lambda-\lambda')\, d\lambda' & \text{(a}'\text{)} \\ [u_r(0)]^2 A_r^+\,\delta(\lambda-\lambda_0) & \text{(b}'\text{)} \\ \frac{[u_0(0)]^2}{2\alpha+\dot{\mu}_s/\mu_s}\,\delta\left(\lambda-\lambda_0-\frac{\vartheta_0^2}{2\alpha+\dot{\mu}_s/\mu_s}\right), \quad \text{for } r=0 \text{ only.} & \text{(c}'\text{)} \end{cases} \end{aligned}\right\} \tag{11.20}$$

This equation coincides with the pathlength problem Eq. (10.38), as anticipated, except for the replacement of C with the eigenvalue $\overline{C}_r$, for the linear approximation to $\mu(\lambda)$, and for the source term. The source term of (11.20b′) differs from that of (10.38) only by a constant factor. The source term of (11.20c′) indicates that source obliquity has the same effect as a shift of source energy, within the limits of the approximation which holds only for very small values of ϑ_0 and will be discussed in Sect. 11 δ. Comparison of the source terms of (11.20a′) and (11.20c′) indicates that components of an isotropic source emitted in directions further and further from the perpendicular contribute to deep penetration beginning at lower and lower energies.

β) Integrated flux from a perpendicular source. Eq. (11.20b′), which pertains to the total flux of X rays traveling in all directions and originating from a plane perpendicular source (6.11), differs from the straight-ahead equation (10.38) only by the numerical value of its parameters and by the linear approximation to $\mu(\lambda)$. Its solutions and their inverse Fourier transforms are therefore obtained from the results of Sect. 10γ. Sects. 11γ and 11δ will review the additional information on the angular distribution of X rays, to be obtained from (11.16) and (11.18), and on the influence of the direction of emission from the source, to be obtained by solving (11.20a′) or (11.20c′) instead of (11.20b′).

The solutions of (10.38) have singularities at $p=\mu_s$ of the types $(\mu_s-p)^{-aC}$ or $\exp\left(bC/\sqrt{\mu_s-p}\right)$. The analogous solutions of (11.20b′) have the same singularities with the eigenvalue $\overline{C}_r$ instead of the scattering probability C in the exponent. Therefore, in the expansion (11.8) of the flux transform, $I(p,\sigma,\eta)=\sum_r I^{\langle r\rangle}(p,\sigma,\eta)$, the term with $r=0$ predominates greatly in the limit $p\to\mu_s$ because its exponent factor $\overline{C}_0$ is largest. Since the limiting case is of primary interest in this section, terms with $r\neq 0$ in this expansion will be disregarded in the following.

As in Sect. 10γ, characteristic types of behavior of $I(p,\lambda)$ for $p\sim\mu_s$ correspond to various trends of $\mu(\lambda)$ for $\lambda\sim\lambda_s$. Case I, namely constant $\mu(\lambda)$, need not be considered because it is unrealistic (see note on p. 709). Case II, namely $\mu(\lambda)$ an increasing function of λ, is realistic and permits a rather detailed analytical study because in this case the source wavelength λ_0 coincides with λ_s and thus lies in the critical range. The solution of (11.20b′) is then given by (10.43), with $\lambda_s=\lambda_0$, $\overline{C}_0$ in the place of C and with an additional constant factor $[u_0(0)]^2 A_0^+$ which is a number of the order of one. The inverse transform of this solution, calculated from (9.1) for large depths of penetration z, is analogous to (10.48), namely

$$I(z,\lambda)\sim e^{-\mu_0 z}\frac{z^{\overline{\varkappa}_0}}{\Gamma(\overline{\varkappa}_0+1)}\overline{C}_0\,[\dot{\mu}_0(\lambda-\lambda_0)]^{\overline{\varkappa}_0-1}\,[u_0(0)]^2 A_0^+, \tag{11.21}$$

where

$$\overline{\varkappa}_0=\frac{\overline{C}_0}{\dot{\mu}_0}. \tag{11.22}$$

Table 8 gives $\overline{\varkappa}_0$ as a function of the parameters C, μ_0 and $\dot{\mu}_0$. Eq. (11.21) gives the flux of X rays of wavelength λ integrated over all directions, at a distance z from a plane perpendicular source.

The approximation of great depth and of a linear variation of $\mu(\lambda)$, utilized in (11.21), is valid provided $\lambda-\lambda_0$ is so small and z is so large that

$$[\mu(\lambda)-\mu(\lambda_0)-\dot{\mu}_0(\lambda-\lambda_0)]\,z\ll 1 \tag{11.23}$$

and

$$[\mu(\lambda)-\mu(\lambda_0)]\,z\sim\dot{\mu}_0(\lambda-\lambda_0)\,z\gg 1. \tag{11.24}$$

As λ increases, the condition (11.23) breaks down and (11.21) fails to give the correct spectral distribution at great depths. However, quasi-equilibrium is attained, provided (11.24) holds, and the correct $I(z,\lambda)$ differs from (11.21) by a factor that depends only on λ, so that (11.21) gives the correct variation with depth of all spectral components in quasi-equilibrium, namely $e^{-\mu_0 z} z^{\overline{\varkappa}_0}$. To find the dependence on λ at higher λ, the solution of (9.3) has to be extended using values of $\mu(\lambda)$ and $k(\omega,\lambda,\omega',\lambda')$ that are accurate outside the critical range but setting $p=\mu_0$ since small departures of p from μ_0 are of no significance after quasi-equilibrium is attained. The extension can be carried out by the numerical procedure of Sect. 20. The factor

$$z^{\overline{\varkappa}_0} \tag{11.25}$$

of the variation of $I(z, \lambda)$ with depth represents the asymptotic trend of the build-up factor discussed in Sect. 5 γ; this result is verified by the experimental data shown in Fig. 5, since no allowance is required for the different source geometry according to Sect. 11 δ.

Case III occurs when $\mu(\lambda)$ decreases steadily as λ increases from the source wavelength λ_0 to the wavelength one considers, i.e. when one considers only photon energies above the energy E_m of minimum absorption. In this case we have $\lambda_s = \lambda$; the source wavelength would belong in the critical range only under the condition $[\mu(\lambda_0) - \mu_s - \dot{\mu}_s(\lambda_0 - \lambda_s)]\, z \ll 1$, analogous to (11.23), which is certainly violated at sufficiently great depth. Therefore one solves analytically only the homogeneous portion of (11.20), as in the analogous cases of Sect. 10 γ, and one finds, in analogy to (10.50),

$$I(z, \lambda) = K\, \mathrm{e}^{-\mu(\lambda) z} \frac{z^{\bar{\varkappa}}}{\Gamma(\bar{\varkappa} + 1)}, \tag{11.26}$$

with

$$\bar{\varkappa} = \frac{\bar{C}_0}{\left|\frac{d\mu}{d\lambda}\right|}. \tag{11.27}$$

Determination of the constant of integration K in (11.26) requires an integration procedure valid in the interval between λ_0 and the beginning of the critical range. Such an integration, e.g. by the method of Sect. 20, would utilize realistic values of μ and k outside the critical range but a value of p equal to μ_s. It would be continued analytically as soon as $I(\mu_s, \lambda)$ is seen to obey the Eq. (11.20) with the approximate μ and k and with no source term.

Table 9. *Values of the eigenvalues $\bar{C}_0$ of Eq. (11.9) for $\mu_s = \mu_m$, $\dot{\mu}_s = 0$, and of* $\bar{\tau} = \dfrac{\pi^2 \bar{C}_0^2}{2\mu_m \left(\frac{d^2\mu}{d\lambda^2}\right)_{\lambda_m}}$.

Material	$\bar{C}_0$ cm²/g	$\bar{\tau}$	Material	$\bar{C}_0$ cm²/g	$\bar{\tau}$
H_2O	0.118	0.32	W	0.0619	0.60
Al	0.0942	0.35	Pb	0.0594	0.45
Fe	0.0826	0.85	U	0.0565	0.35
Sn	0.0689	0.62			

Case IV occurs when $\mu(\lambda)$ traverses its minimum in the interval of integration between the source wavelength λ_0 and the wavelength λ one considers. Here again the source wavelength is excluded from the critical range at sufficiently great depths of penetration, unless it happens to coincide exactly with the wavelength λ_m of minimum μ. The eventual asymptotic trend of penetration sets in only when $[\mu(\lambda_0) - \mu_m]\, z \gg 1$, a condition which holds at very great depth even though λ_0 is not much lower than λ. Therefore also in this case one solves analytically only the homogeneous portion of (11.20).

In the Eq. (11.20) with $r = 0$, made homogeneous, one enters then the eigenvalue $\bar{C}_0$ of (11.9), calculated for $\dot{\mu}_s = 0$ and $\mu_s = \mu_m$ and given in Table 9, and one sets $\mu_s = \mu_m$, $\dot{\mu}_s = 0$. The equation thus reduced corresponds then to constant μ, i.e. to Case I, rather than to $\mu(\lambda)$ passing through a minimum as in the Case IV which we are considering. The variation of $\mu(\lambda)$ near its minimum is represented by adding to $\mu_s = \mu_m$ a term $\frac{1}{2}\ddot{\mu}_m(\lambda - \lambda_m)^2$. This term was disregarded in Sect. 11 α, because it does not influence the effect of obliquity on penetration as represented by the eigenvalue C_0 [1], but must be included in the integration along λ. One finds

[1] A more detailed analysis of the effect of obliquity when μ is near its minimum has been given in Ref. [*28*].

then a result analogous to (10.52), namely

$$I(z,\lambda) \sim \frac{K}{\sqrt{3}\,\pi}\, e^{-\mu_m z + 3(\bar{\tau}\mu_m z)^{\frac{1}{3}}}\, \frac{\bar{\tau}\,\mu_m}{(\bar{\tau}\,\mu_m z)^{\frac{5}{6}}}\, \frac{e^{-\frac{\bar{C}_0}{\frac{1}{2}\ddot{\mu}_m(\lambda-\lambda_m)}}}{\mu(\lambda)-\mu_m}, \tag{11.28}$$

where

$$\bar{\tau} = \frac{\pi^2\,\bar{C}_0^2}{2\,\mu_m\left(\frac{d^2\mu}{d\lambda^2}\right)_{\lambda=\lambda_m}} \tag{11.29}$$

is a number of the order of 1 given in Table 9.

As in case III, the value of the constant K in (11.28) has to be determined by separate integration from λ_0 to the beginning of the critical region. Separate integration is also required for $\lambda > \lambda_m$ after breakdown of the condition

$$[\mu(\lambda) - \mu_m - \tfrac{1}{2}\ddot{\mu}_m(\lambda-\lambda_0)^2]\, z \ll 1 \tag{11.30}$$

analogous to (11.23). The validity of (11.28) is also restricted by the conditions analogous to (11.24)

$$[\mu(\lambda_0) - \mu_m]\, z \gg 1, \tag{11.31}$$

$$[\mu(\lambda) - \mu_m]\, z \gg 1. \tag{11.32}$$

Eq. (11.28) shows that the asymptotic trend of the build-up factor, analogous to (11.25), is

$$\frac{e^{3(\bar{\tau}\mu_m z)^{\frac{1}{3}}}}{z^{\frac{5}{6}}}. \tag{11.33}$$

γ) Angular distribution. Eqs. (11.16) and (11.18) yield qualitative information on the angular distribution of X rays with wavelengths in and near the critical range. Additional characteristics of this distribution, like the mean square deflection $\langle\vartheta^2\rangle$ have also simple analytical expressions. More detailed data would require calculation of the eigenfunctions $u_0(\sigma)$, whereas the parameters $u_0(0)$ and A_0 are readily estimated to be numbers of the order of 1.

The derivatives with respect to λ in (11.16) and (11.18) do not influence the inversion of the Fourier transform when λ does not coincide with λ_s, i.e. in Cases II and IV, where $\lambda > \lambda_m$. In these cases one sees, from (10.40) and (10.41) or from (11.21) and (11.28), that taking the derivative multiplies $I(z,\lambda)$ by the factor $(\bar{C}_0 - \dot{\mu})/[\mu(\lambda) - \mu_s]$ which equals $(\bar{\varkappa}_0 - 1)/(\lambda - \lambda_0)$ in Case II and $\bar{C}_0/\frac{1}{2}\ddot{\mu}_m(\lambda-\lambda_m)^2 + 2/(\lambda-\lambda_m)$ in Case IV. The flux in the direction $\vartheta = 0$, which is clearly a maximum, equals the integrated flux multiplied by this factor $(\bar{C}_0 - \dot{\mu})/[\mu(\lambda) - \mu_s]$ and by the numerical factor A_0/π. Therefore the ratio $I(z,\vartheta,\lambda)/I(z,\lambda)$ decreases as λ increases, indicating that the *angular distribution broadens*, in solid angle, in proportion to $\mu(\lambda) - \mu_s$ and independently of z. Notice that this result holds only under the condition of quasi-equilibrium, which is respectively (11.24) and (11.32) for the $\dot{\mu}_s > 0$ and $\dot{\mu}_s = 0$. At lower depths (11.21) and (11.28) must be replaced, respectively, with formulas analogous to (10.55) and (10.56), and the resulting angular distribution does depend on the depth z.

In Case III one considers the angular distribution at the most critical wavelength, namely $\lambda = \lambda_s$, and this distribution clearly must depend on the depth. Taking the derivative with respect to λ of the Fourier transform $I(p,\lambda)$ multiplies it by $(\bar{C}_0 - \dot{\mu}_s)/(\mu_s - p) = |d\mu/d\lambda|(\bar{\varkappa}+1)/(\mu_s - p)$. Upon inverse Fourier transformation, the flux in the direction $\vartheta = 0$ turns out to equal the integrated flux multiplied by $|d\mu/d\lambda|\,z$. Therefore the ratio $I(z,0,\lambda)/I(z,\lambda)$ increases with increasing z, indicating that the angular distribution narrows in proportion to $|d\mu/d\lambda|\,z$ for each given $\lambda < \lambda_m$.

δ) Dependence on source characteristics. When the X rays are emitted in directions oblique to the z axis, their initial obliquity influences deep penetration differently depending on whether or not the source wavelength lies in the critical range. The situation is simpler for high energy X-ray sources ($\lambda_0 < \lambda_s \leqq \lambda_m$, Cases III and IV) than for low energy ones ($\lambda_0 > \lambda_m$, Case II), because high energy X rays are scattered before they become most penetrating and may thereby be deflected into directions close to the z axis. In the deep penetration formulas (11.26) and (11.28), only the constant of integration K depends on the direction of the source X rays but the variation of flux with depth is independent of that direction. This result stems from the fact that very deep penetration hinges on repeated scattering in the critical range after the "memory" of the initial direction has been wiped out. Unless the direction approaches the z axis in the early stages of penetration, roughly while $(\mu_0 - \mu_s)\, z \lesssim 1$, the chance of attaining deep penetration is negligible. The chance of this initial deflection is to be evaluated by calculating the factor K as indicated in the discussion of (11.26) and (11.28), i.e. by integrating the transform equation (9.3) for $p = \mu_s$ from the source wavelength to the critical range. Only one study of the degradation to the initial range has been carried out, comparing perpendicular with isotropic rather than with collimated oblique sources (see Sect. 20 and Ref. [80]).

Lower energy X rays emitted with great obliquity have no significant chance of being deflected close to the z axis without becoming much less penetrating, as pointed out in Sect. 5 ε. The effect of initial obliquity can be estimated readily only when this obliquity is small. One may then consider the solution of (11.20c′) which yields, instead of (11.21),

$$\left.\begin{aligned} I(z,\lambda) \sim \mathrm{e}^{-\left(\mu_0 + \frac{\dot\mu_0 \vartheta_0^2}{2\alpha + \dot\mu_0/\mu_0}\right) z} \frac{z^{\bar\varkappa_0}}{\Gamma(\bar\varkappa_0 + 1)} \times \\ \times \bar C_0 \left[\dot\mu_0 \left(\lambda - \lambda_0 - \frac{\vartheta_0^2}{2\alpha + \dot\mu_0/\mu_0}\right)\right]^{\bar\varkappa_0 - 1} \frac{[u_0(0)]^2}{2\alpha + \dot\mu_0/\mu_0}. \end{aligned}\right\} \tag{11.34}$$

This formula should be accurate when the obliquity ϑ_0 causes no very large additional attenuation over the depth z, i.e. only when $\dot\mu_0 \vartheta_0^2 z/(2\alpha + \dot\mu_0/\mu_0) \lesssim 1$, as discussed below.

If the source is isotropic, only a narrow cone of its X rays contributes appreciably to deep penetration, in fact, according to (11.34) only a cone of aperture $\vartheta_0 \sim \sqrt{(2\alpha + \dot\mu_0/\mu_0)/\dot\mu_0 z}$. Solution of the pertinent equation (11.20a′), shows that an isotropic source generates the same integrated flux at deep penetration as though a fraction

$$\frac{1}{4 A_0 \dot\mu_0 z} \tag{11.35}$$

of its strength were directed perpendicularly to the source plane and the rest did not contribute at all. The same effect applies to the primary radiation, whose flux is $\mathrm{e}^{-\mu_0 z}/2\mu_0 z$ at great depth, so that the build-up factor remains proportional to $z^{\bar\varkappa_0}$ for an isotropic source. These results indicate that a source may be regarded as perpendicular if it is collimated within a cone of aperture $\ll 1/\dot\mu_0 z$, since $2\alpha + \dot\mu_0/\mu_0$ is of the order of 1, and is effectively isotropic if it is uniform over a cone of aperture $\gg 1/\dot\mu_0 z$.

The procedure of Sect. 11 α could be applied, in principle, even when the source obliquity is substantial. In the source term of (11.10) the factor $J_0(\sqrt{2\eta\sigma}\vartheta_0)$ is then a rapidly oscillating function of σ and the integral vanishes unless $u_r(\sigma)$ also oscillates rapidly, which occurs only for large r. Thus the term $r = 0$ of the expansion (11.8), which would contribute effectively to deep penetration, is "excited" by a source of negligible strength and could become dominant only at unrealistic depths. The terms with large r have negative eigenvalues $\bar C_r$ and therefore are ineffective in generating lower energy X rays. It is still unknown how the penetration to great but realistic depths results from a combination of different components. It can be safely assumed only that the term $r=0$ predominates when the obliquity has a small effect.

The results obtained in this section on the deep penetration of X rays from plane sources are readily applied to X rays from other sources utilizing: (a) the relationships between the fluxes from plane and point sources described in Sect. 9α, and (b) the fact that the flux from any source is simply the sum of the fluxes from the various points of the source. As pointed out on p. 695, the integrals over the transform variable p in (9.28) and (9.32) simplify greatly in the limit for large values of the distance r from a point source. They reduce to the integral which one obtains as a function of the distance from a plane source multiplied by $\mu_s/2\pi r$. That is, if the direction of emission from a point and from a plane source makes the same angle with respect to the direction of the point of observation, the fluxes from these sources are related by

$$I(r, \omega, \lambda) \sim \frac{\mu_s}{2\pi r} [I(z, \omega, \lambda)]_{z=r} \quad \text{for } \mu_s r \gg 1. \tag{11.36}$$

The primary X-ray fluxes from point isotropic and plane isotropic sources are respectively $e^{-\mu_0 r}/4\pi r^2$ and $e^{-\mu_0 z}/2\mu_0 z$, in a ratio $\mu_0/2\pi r$. Therefore, for $\lambda_0=\lambda_s$, the build-up factor is exactly equal for point and plane isotropic sources at very deep penetration (see also Sect. 19γ).

12. Moments of the flux distribution. The spectral distribution of the X-ray flux throughout a material can be characterized by its moments. The moments are integrals over all space of the flux at each point $\mathbf{r}=(x, y, z)$ multiplied by powers of the coordinates x, y, z. In the basic case of the flux from a plane source (6.5), only moments of the distribution-in-z need be considered, namely

$$I_n(\omega, \lambda) = \int_{-\infty}^{\infty} d z\, z^n I(z, \omega, \lambda). \tag{12.1}$$

These moments can be expressed as derivatives of the Fourier transform (9.2) evaluated at $p=0$, i.e. they are coefficients of the expansion of $I(p, \omega, \lambda)$ in powers of p,

$$I(p, \omega, \lambda) = \sum_{n=0}^{\infty} I_n(\omega, \lambda) \frac{p^n}{n!}. \tag{12.2}$$

They obey equations which are obtained either by multiplying the transport equation (6.9) by z^n and integrating over z or by expanding its transform (9.3) into powers of p,

$$\left.\begin{aligned} -n\,\omega_z I_{n-1}(\omega, \lambda) &= -\mu(\lambda) I_n(\omega, \lambda) + \\ &+ \int_{\lambda_0}^{\lambda} d\lambda' \int_{4\pi} k(\omega, \lambda, \omega', \lambda') I_n(\omega', \lambda') + \delta_{n0}\,\delta(\omega-\omega_0)\,\delta(\lambda-\lambda_0). \end{aligned}\right\} \tag{12.3}$$

These equations can be treated in succession starting with $n=0$. Each step of the succession is then equivalent to the degradation-deflection problem of Sect. 10β because the interlinkage term $-n\,\omega_z I_{n-1}$ contains the solution of the preceding equation and can be treated as a source term. Legendre expansion of (12.3), or expansion of (9.12) into powers of p, yields the basic moment equation

$$\left.\begin{aligned} \mu(\lambda) I_{nl}^m(\lambda) &= \int_{\lambda_0}^{\lambda} d\lambda'\, k_l(\lambda, \lambda')\, I_{nl}^m(\lambda') + \delta_{n0} \sqrt{\frac{4\pi}{2l+1}}\, Y_l^{m*}(\omega_0)\,\delta(\lambda-\lambda_0) + \\ &+ \frac{n}{2l+1} \left[\sqrt{(l+1)^2-m^2}\, I_{n-1,l+1}^m(\lambda) + \sqrt{l^2-m^2}\, I_{n-1,l-1}^m(\lambda)\right]. \end{aligned}\right\} \tag{12.4}$$

For a given set of values of n, l, and m, (12.4) coincides with (10.8), except for its source term, and can therefore be solved numerically once the source term is known. The source term is completely specified for $n=0$; for $n\neq 0$ it may be regarded as known if the equations are solved in an appropriate succession of increasing n. This succession must take into account that the source term of (12.4) contains Legendre coefficients of the $(n-1)$-st moment which are not of index l but of index $l\pm 1$.

In particular we consider this succession for the determination of the moments of the Green's function $G^{(p,m)}(l\lambda, l_0\lambda_0)$ defined in Sect. 9. These moments are defined as

$$G_n^{(m)}(l\lambda, l_0\lambda_0) = \left[\frac{\partial^n}{\partial p^n} G^{(p,m)}(l\lambda, l_0\lambda_0)\right]_{p=0} \tag{12.5}$$

and obey the Eqs. (12.4) with the source term $\delta_{n0}\sqrt{4\pi/(2l+1)}\,Y_l^{m*}(\omega_0)$ replaced with $\delta_{n0}\,\delta_{ll_0}\,\delta(\lambda-\lambda_0)$. Fig. 16 demonstrates that only a limited number of equations has to be solved to obtain $G_n^{(m)}(l\lambda, l_0\lambda_0)$ for a given set of values of n, l, l_0. The symmetry property (9.24) of the Green's function leads to

$$G_n^{(m)}(l\lambda, l_0\lambda_0) = 0 \quad \text{for} \quad l-l_0+n \text{ odd}. \tag{12.6}$$

The scheme of interlinkages in (12.4) shows further that

$$G_n^{(m)}(l\lambda, l_0\lambda_0) = 0 \quad \text{for} \quad |l-l_0|>n. \tag{12.7}$$

These results are also illustrated by Fig. 16.

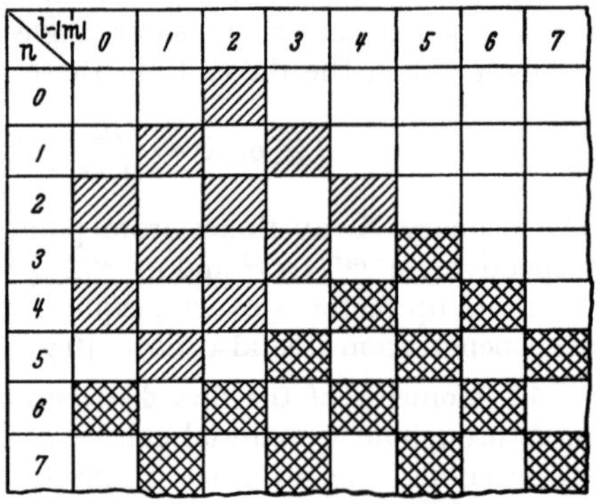

Fig. 16. Moment interlinkage diagram for the case $l_0 = |m|+2$. All moments with n and l corresponding to the singly-lined squares must be calculated to determine the moment with $n=5$ and $l=|m|+1$.

Eq. (12.7) represents an important property of the correlation between the depth of penetration of radiation and the obliquity with respect to the direction of penetration. This correlation is reflected in the checkerboard pattern of Fig. 16, showing that the interlinkage-in-l spreads at a rate of one unit per each unit of n. Particular consequences of (12.7) are: (1) The flux integrated over all directions is represented by the Legendre coefficient with $l=0$; therefore its n-th moment depends only on the first n Legendre coefficients of the source angular distribution. (2) The flux from an isotropic source ($l_0=0$) is made anisotropic by selection in the course of penetration, but its n-th moment has an angular distribution with only n non-vanishing Legendre coefficients, because $G_n^{(0)}(l\lambda, 0\lambda_0)$ vanishes for $l>n$.

The system of Eq. (12.4) makes it possible to calculate a number of Legendre coefficients of several moments of an X-ray distribution with a moderate amount of numerical labor. Details about the calculations are given in Sect. 16. A knowledge of the entire, infinite, sequence of moments would be necessary, in principle, to describe a distribution of X-ray flux. In practice, however, a calculation of high-order moments can be replaced with information drawn from other methods. In particular, the values of the high-order moments depend heavily on the extent of deep penetration, whose trend is known from the analysis of Sect. 11. Many details of the X-ray distribution near the source, which are not reflected clearly in the values of the moments, can be obtained from one or two steps of the iteration procedure of Sect. 8. Part C will show how results contributed by different methods can be effectively utilized to complement a calculation of moments.

The utilization of different lines of evidence may be regarded as an interpolation procedure applied to the Green's function $G^{(p,m)}(l\lambda, l_0\lambda_0)$ as a function of the Fourier transform parameter p. The values of p which are important for the inversion of the transform lie on the real axis between 0 and μ_s. The first few moments of the space distribution determine the trend of G near $p=0$. The deep penetration treatment of Sect. 11 determines the existence and the character of a singularity of G at $p=\mu_s$. Smoothness of G as a function of p, between 0 and μ_s, can be assumed with regard to multiply scattered X-rays. Interpolation is, therefore, possible provided the non-smooth features of the distribution of photons scattered only once or twice are calculated separately by iteration.

α) *Particular source geometries.* The angular distribution of a plane source of X rays collimated in a direction ω_0 is represented by the factor $\delta(\omega-\omega_0)$ in the source term of (12.3) and by $\sqrt{4\pi/(2l+1)}\,Y_l^{m*}(\omega_0)$ in the source term of (12.4). The moments $G_n^{(m)}(l\lambda, l_0\lambda_0)$ of the Green's function pertain to the flux distribution of X rays from a plane source whose angular distribution would be represented by the single spherical harmonic $\sqrt{(2l_0+1)/4\pi}\,Y_{l_0}^{m_0}(\omega)$, in (12.3) and by δ_{ll_0} in (12.4). Therefore the moments I_{nl}^m calculated by solving (12.4) for a *plane collimated* source are the following linear combinations of moments of the Green's function

$$I_{nl}^m(\lambda) = \sum_{l_0} G_n^{(m)}(l\lambda, l_0\lambda_0)\sqrt{\frac{4\pi}{2l_0+1}}\,Y_{l_0}^{m*}(\omega_0), \tag{12.8}$$

where the summation is limited by (12.6) and (12.7). For the other types of plane sources described in Sect. 6, the moments of the X-ray distribution are as follows:

a) *plane conical source* (6.10),

$$I_{nl}^m(\lambda) = \sum_{l_0} G_n^{(0)}(l\lambda, l_0\lambda_0)\,P_{l_0}(\cos\vartheta_0)\,\delta_{m0}, \tag{12.9}$$

(b) *plane perpendicular* source (6.11),

$$I_{nl}^m(\lambda) = \sum_{l_0} G_n^{(0)}(l\lambda, l_0\lambda_0)\,\delta_{m0}, \tag{12.10}$$

(c) *plane isotropic source* (6.12),

$$I_{nl}^m(\lambda) = G_n^{(0)}(l\lambda, 0\lambda_0)\,\delta_{m0}. \tag{12.11}$$

For point sources, the moments can also be expressed in terms of the moments of the Green's function, utilizing the connection between the fluxes from plane and point sources developed in Sect. 9α[1]. Cases of increasing complication will be considered separately but their classification will not be quite the same here as in Sect. 9α. In each case the moments are identified with coefficients of the expansion of the Fourier transform of the flux into powers of the transform parameters, in analogy to (12.2).

For the *integrated flux* $I(r,\lambda)$ from a *point isotropic* source, the Fourier transform is given by (9.21) and has the expansion

$$I(\boldsymbol{q},\lambda) = G^{(-iq,0)}(0\lambda, 0\lambda_0) = \sum_n{}' G_n^{(0)}(0\lambda, 0\lambda_0)\frac{(-iq)^n}{n!}, \tag{12.12}$$

[1] Ref. [*12*] develops an alternative equivalent procedure, in which one first expresses the moments of the X-ray distribution generated by a plane source in terms of those corresponding to a point source, in analogy to the treatment leading to (9.20), and then one niverts the relationship thus obtained.

where $\sum'$ indicates summation over even, non-negative, values in accordance with (12.6). On the other hand the transform is defined by

$$\left.\begin{aligned} I(\boldsymbol{q},\lambda) &= \int d\boldsymbol{r}\, e^{-i\boldsymbol{q}\cdot\boldsymbol{r}} I(r,\lambda) = 4\pi\int_0^\infty r^2\,dr\,\frac{\sin q r}{q r}\, I(r,\lambda) \\ &= \sum_n{}' \frac{(-i q)^n}{(n+1)!}\, 4\pi \int_0^\infty r^2\,dr\,r^n\, I(r,\lambda). \end{aligned}\right\} \tag{12.13}$$

Comparison of the expansions (12.12) and (12.13) yields

$$I_n(\lambda) = 4\pi\int_0^\infty r^2\,dr\,r^n\, I(r,\lambda) = (n+1)\, G_n^{(0)}(0\,\lambda, 0\,\lambda_0), \quad \text{for } n \text{ even.} \tag{12.14}$$

For the *differential flux* $I(r,\omega,\lambda)$ from a *point isotropic* monoenergetic source, the Fourier transform (9.18) can be expanded in the form

$$\left.\begin{aligned} I(\boldsymbol{q},\omega,\lambda) &= \sum_l \frac{2l+1}{4\pi} P_l(\cos\hat{\vartheta})\, G^{(-iq,0)}(l\,\lambda, 0\,\lambda_0) \\ &= \sum_l \frac{2l+1}{4\pi} P_l(\cos\hat{\vartheta}) \sum_{n-l}{}' G_n^{(0)}(l\,\lambda, 0\,\lambda_0)\, \frac{(-i q)^n}{n!}. \end{aligned}\right\} \tag{12.15}$$

On the other hand we have, utilizing the notations of Sect. 9 and in particular the expansion (9.26),

$$\left.\begin{aligned} I(\boldsymbol{q},\omega,\lambda) &= \int d\boldsymbol{r}\, e^{-i\boldsymbol{q}\cdot\boldsymbol{r}} \sum_l \frac{2l+1}{4\pi} P_l(\omega\cdot\boldsymbol{u})\, I_l(r,\lambda) \\ &= \int_0^\infty r^2\,dr \int_0^\pi \sin\hat{\alpha}\,d\hat{\alpha} \int_0^{2\pi} d\hat{\beta} \sum_L f_L(-q r) P_L(\cos\hat{\alpha}) \sum_{l m} Y_l^m(\hat{\vartheta},\hat{\varphi})\, Y_l^{m*}(\hat{\alpha},\hat{\beta})\, I_l(r,\lambda) \\ &= \sum_l P_l(\cos\hat{\vartheta}) \int_0^\infty r^2\,dr\, f_l(-q r)\, I_l(r,\lambda). \end{aligned}\right\} \tag{12.16}$$

The spherical Bessel function $f_l(-q r)$ has the expansion

$$f_l(-q r) = (2l+1) \sum_{n-l}{}' \frac{\tfrac{1}{2}!}{\frac{n-l}{2}!\,\frac{n+l+1}{2}!\,2^n} (-i q r)^n, \tag{12.17}$$

so that (12.16) becomes

$$I(\boldsymbol{q},\omega,\lambda) = \sum_l (2l+1)\, P_l(\cos\hat{\vartheta}) \sum_{n-l}{}' \frac{\tfrac{1}{2}!\,(-i q)^n}{\frac{n-l}{2}!\,\frac{n+l+1}{2}!\,2^n} \int_0^\infty r^2\,dr\,r^n\, I_l(r\,\lambda). \tag{12.18}$$

Comparison of (12.15) and (12.18) yields

$$\left.\begin{aligned} I_{n l}(\lambda) = 4\pi \int_0^\infty r^2\,dr\,r^n\, I_l(r,\lambda) &= \frac{\frac{n-l}{2}!\,\frac{n+l+1}{2}!\,2^n}{n!\,\tfrac{1}{2}!}\, G_n^{(0)}(l\,\lambda, 0\,\lambda_0), \\ &\text{for } n-l \geq 0 \text{ and even.} \end{aligned}\right\} \tag{12.19}$$

For the *integrated flux* $I(z,\varrho,\lambda)$ from the *point collimated* monoenergetic source (6.13), the transform (9.18) reduces to

$$\left.\begin{aligned} I(\boldsymbol{q},\lambda) &= \sum_{l_0} G^{(-iq,0)}(0\,\lambda, l_0\,\lambda_0)\, P_{l_0}(\cos\hat{\vartheta}_0) \\ &= \sum_{l_0} \sum_{N-l_0}{}' G_N^{(0)}(0\,\lambda, l_0\,\lambda_0)\, \frac{(-i q)^N}{N!}\, P_{l_0}(\cos\hat{\vartheta}). \end{aligned}\right\} \tag{12.20}$$

On the other hand, utilizing cylindrical coordinates with axis ω_0 and the integral representation and power series of the Bessel function J_0, we have

$$\left.\begin{aligned} I(\boldsymbol{q},\lambda) &= \int d\boldsymbol{r}\, e^{-i\boldsymbol{q}\cdot\boldsymbol{r}} I(z,\varrho,\lambda) \\ &= \int_{-\infty}^{\infty} dz \int_0^{\infty} \varrho\, d\varrho \int_0^{2\pi} d\beta\, e^{-i[q_z z + q_\varrho \varrho\cos(\varphi-\beta)]} I(z,\varrho,\lambda) \\ &= 2\pi \int_{-\infty}^{\infty} dz\, e^{-iq_z z} \int_0^{\infty} \varrho\, d\varrho\, J_0(-q_\varrho \varrho)\, I(z,\varrho,\lambda) \\ &= \sum_n {\sum_\nu}' \frac{(-i q_z)^n}{n!} \frac{(-i q_\varrho)^\nu}{\frac{\nu}{2}!^2\, 2^\nu}\, 2\pi \int_{-\infty}^{\infty} dz \int_0^{\infty} \varrho\, d\varrho\, z^n \varrho^\nu I(z,\varrho,\lambda). \end{aligned}\right\} \tag{12.21}$$

To compare the expansions (12.20) and (12.21), we express the product $q^N P_{l_0}(\cos\hat{\vartheta}_0)$ in (12.20) as a homogeneous polynomial of degree N in $q_z = q\cos\hat{\vartheta}_0$ and in $q_\varrho^2 = q^2 - q_z^2$,

$$q^N P_{l_0}(\cos\hat{\vartheta}_0) = {\sum_\nu}' B^\nu_{N-\nu,l_0}\, q_z^{N-\nu} q_\varrho^\nu. \tag{12.22}$$

Since $N - l_0$ is non-negative and even, according to (12.20), it is convenient to factor out q^N on the left side of (12.22) in the form $(q_z^2 + q_\varrho^2)\, q^{N-2}$ and thus to establish the recursion formula

$$B^\nu_{n l_0} = B^\nu_{n-2,l_0} + B^{\nu-2}_{n l_0} \quad \text{for } n+\nu > l_0. \tag{12.23}$$

This formula, together with the special case

$$B^\nu_{n,n+\nu} = \frac{(n+\nu)!}{n!\,\frac{\nu}{2}!^2\, 2^\nu} (-1)^{\frac{\nu}{2}}, \tag{12.23'}$$

determines all the coefficients $B^\nu_{n l_0}$. Comparison of (12.20) and (12.21) yields now

$$\left.\begin{aligned} I^\nu_n(\lambda) &= 2\pi \int_{-\infty}^{\infty} dz \int_0^{\infty} \varrho\, d\varrho\, z^n \varrho^\nu I(z,\varrho,\lambda) \\ &= \sum_{l_0} G^{(0)}_{n+\nu}(0\,\lambda, l_0\,\lambda_0) \frac{B^\nu_{n\,l_0}}{B^\nu_{n,n+\nu}} (-1)^{\frac{\nu}{2}} \quad \text{for } \nu \text{ even}. \end{aligned}\right\} \tag{12.24}$$

To deal with the *differential flux* $I(z,\varrho,\vartheta,\varphi-\beta,\lambda)$ from the *point collimated* monoenergetic source (6.13), we first express its transform, given by (9.18), in terms of coordinates with the direction of collimation ω_0 as polar axis. This is done by a formula analogous to (9.31), in terms of Wigner coefficients, which leads to the combination of Green's functions

$$\sum_{\hat{m}} (l\,l_0\,L\,0\,|\,l\,\hat{m}, l_0-\hat{m})(-1)^{\hat{m}}\, G^{(-iq,\hat{m})}(l\,\lambda, l_0\,\lambda_0) = \overline{G}^{(-iq,L)}(l\,\lambda, l_0\,\lambda_0) \tag{12.25}$$

encountered in (9.32) and (9.33). This modified Green's function $\overline{G}$ obeys a system of equations analogous to (9.12), and its moments $\overline{G}^{(L)}_N = \sum_{\hat{m}} (l l_0\, l\, 0\,|\,l\hat{m}, l_0-\hat{m})\,(-1)^{\hat{m}}\, G^{(\hat{m})}_N$ obey a system analogous to (12.4). It is essential for our calculation that each equation of these systems interlinks a $\overline{G}$ with given l and L, and $\overline{G}$'s with $l\pm 1$ and $L\pm 1$; the source term has $L=0$. It follows, in analogy to (12.6) and (12.7), that the moments $\overline{G}^{(L)}_N(l\lambda, l_0\lambda_0)$ vanish unless $N-L$ is even and non-negative. The transform of the differential flux is then found to be

$$\left.\begin{aligned} I(\boldsymbol{q},\boldsymbol{\omega},\lambda) &= \sum_{l l_0 \hat{m}} \sqrt{\frac{2l+1}{2l_0+1}}\, Y_l^{\hat{m}}(\hat{\vartheta},\hat{\varphi})\, G^{(-iq,\hat{m})}(l\lambda, l_0\lambda_0)\, Y_{l_0}^{\hat{m}*}(\hat{\vartheta}_0,\hat{\varphi}_0) \\ &= \sum_{l m l_0 L} \sqrt{\frac{2l+1}{2L+1}}\, Y_l^m(\vartheta,\varphi)\,(l m, l_0 0 | l l_0 L m)\, Y_{l_0}^{m*}(\Theta,\Phi) \sum_{N-L}{}' G_N^{(L)}(l\lambda, l_0\lambda_0)\, \frac{(-iq)^N}{N!}, \end{aligned}\right\} \quad (12.26)$$

where Θ, Φ, are the coordinates of the direction of $\boldsymbol{q}$.

The flux $I(z,\varrho,\vartheta,\varphi-\beta,\lambda)$ is a function of $\varphi-\beta$, rather than of φ and β separately, owing to its cylindrical symmetry and can therefore be expanded into harmonics $Y_l^m(\vartheta,\varphi-\beta)$. Proceeding as in (12.21) we have[1]

$$\left.\begin{aligned} I(\boldsymbol{q},\boldsymbol{\omega},\lambda) &= \int d\boldsymbol{r}\, e^{-i\boldsymbol{q}\cdot\boldsymbol{r}}\, I(z,\varrho,\vartheta,\varphi-\beta,\lambda) \\ &= \int_{-\infty}^{\infty} dz \int_0^{\infty} \varrho\, d\varrho \int_0^{2\pi} d\beta\, e^{-i[q_z z + q_\varrho \varrho \cos(\Phi-\beta)]} \sum_{lm} \sqrt{\frac{2l+1}{4\pi}}\, Y_l^m(\vartheta,\varphi-\beta)\, I_l^m(z,\varrho,\lambda) \\ &= \sum_{lm} \sqrt{\frac{2l+1}{4\pi}}\, Y_l^m(\vartheta,\varphi-\Phi)\, 2\pi \int_{-\infty}^{\infty} dz\, e^{-iq_z z} \int_0^{\infty} \varrho\, d\varrho\, (-i)^m J_m(-q_\varrho \varrho)\, I_l^m(z,\varrho,\lambda) \\ &= \sum_{lm} \sqrt{\frac{2l+1}{4\pi}}\, Y_l^m(\vartheta,\varphi-\Phi) \sum_n \sum_{\nu-|m|}{}' \frac{(-iq_z)^n}{n!}\, \frac{(-iq_\varrho)^\nu}{\frac{\nu-m}{2}!\,\frac{\nu+m}{2}!\,2^\nu} \times \\ &\quad \times 2\pi \int_{-\infty}^{\infty} dz \int_0^{\infty} \varrho\, d\varrho\, z^n \varrho^\nu I_l^m(z,\varrho,\lambda). \end{aligned}\right\} \quad (12.27)$$

To make (12.26) and (12.27) comparable we first give the same form to the harmonic expansions of the angular distribution by expressing $Y_l^m(\vartheta,\varphi)\, Y_L^{m*}(\Theta,\Phi)$ in the form $Y_l^m(\vartheta,\varphi-\Phi)\times Y_L^{m*}(\Theta,0)$. Next we represent the product $q^N Y_L^{m*}(\Theta,0)$ as a homogeneous polynomial in q_z and q_ϱ, in analogy to (12.22)[2]

$$q^N \sqrt{\frac{4\pi}{2L+1}}\, Y_L^{m*}(\Theta,0) = \sum_{\nu-|m|}{}' B_{N-\nu,L}^{\nu m}\, q_z^{N-\nu} q_\varrho^\nu. \quad (12.28)$$

The generalized coefficients $B_{nL}^{\nu m}$ obey the recursion formula (12.23) and have the special value

$$B_{n\,n+\nu}^{\nu m} = \frac{\sqrt{(n+\nu-m)!\,(n+\nu+m)!}}{n!\,\frac{\nu-m}{2}!\,\frac{\nu+m}{2}!\,2^\nu}\,(-1)^{\frac{\nu-m}{2}}. \quad (12.29)$$

The comparison of (12.26) and (12 27) yields then

$$\left.\begin{aligned} I_{nl}^{\nu m}(\lambda) &= 2\pi \int_{-\infty}^{\infty} dz \int_0^{\infty} \varrho\, d\varrho\, z^n \varrho^\nu I_l^m(z,p,\lambda) \\ &= \sum_{l_0 L} (l m, l_0 0 | l l_0 L m)\, (-1)^{\frac{\nu-m}{2}}\, \frac{B_{nL}^{\nu m}}{B_{n\,n+\nu}^{\nu m}}\, \frac{\sqrt{(n+\nu-m)!\,(n+\nu+m)!}}{(n+\nu)!}\, \overline{G}_{n+\nu}^{(L)}(l\lambda, l_0\lambda_0), \\ &\qquad \text{for } \nu-|m| \text{ even and non-negative.} \end{aligned}\right\} \quad (12.30)$$

The moments (12.30) and (12.19) of the fluxes from point collimated and point isotropic sources obey interlinked systems of equations analogous to (12.4). These equations are obtained by integrating over the space coordinates of the transport equation (6.2) with the appropriate expression of the source terms, namely (6.13) or (6.15), and of $\boldsymbol{\omega}\cdot\operatorname{grad} I$, namely

[1] The expansion of J_m in this formula pertains to $m>0$, but this restriction has no effect because $J_{-m}=(-1)^m J_m$ and $J_{-m} Y_l^{-m} = J_m Y_l^m$.

[2] Notice that $\Theta=\hat{\vartheta}_0$ since Θ represents the obliquity of $\boldsymbol{q}$ with respect to $\boldsymbol{\omega}_0$ and $\hat{\vartheta}_0$ the obliquity of $\boldsymbol{\omega}_0$ with respect to $\boldsymbol{q}$.

(6.14) or (6.16), multiplied respectively by $z^n \varrho^\nu$ or r^n. The equation for the point collimated source is

$$\left.\begin{aligned}\mu(\lambda)\, I_{nl}^{\nu m}(\lambda) = &\int_{\lambda_0}^{\lambda} d\lambda'\, k_l(\lambda,\lambda')\, I_{nl}^{\nu m}(\lambda') + \delta_{n0}\,\delta_{\nu 0}\,\delta_{m0}\,\delta(\lambda-\lambda_0) + \\ &+ \frac{1}{2l+1}\Big\{n\left[\sqrt{(l+1)^2-m^2}\, I_{n-1\,l+1}^{\nu m}(\lambda) + \sqrt{l^2-m^2}\, I_{n-1\,l-1}^{\nu m}(\lambda)\right] - \\ &- \frac{\nu+m}{2}\left[\sqrt{(l+2-m)\,(l+1-m)}\, I_{n,l+1}^{\nu-1\,m-1}(\lambda) - \sqrt{(l-1+m)\,(l+m)}\, I_{n,l-1}^{\nu-1\,m-1}(\lambda)\right] + \\ &+ \frac{\nu-m}{2}\left[\sqrt{(l+2+m)\,(l+1+m)}\, I_{n,l+1}^{\nu-1\,m+1}(\lambda) - \sqrt{(l-1-m)\,(l-m)}\, I_{n,l-1}^{\nu-1\,m+1}(\lambda)\right]\Big\}\end{aligned}\right\} \tag{12.31}$$

and that for the point isotropic source is

$$\left.\begin{aligned}\mu(\lambda)\, I_{nl}(\lambda) = &\int_{\lambda_0}^{\lambda} d\lambda'\, k_l(\lambda,\lambda')\, I_{nl}(\lambda') + \delta_{n0}\,\delta_{l0}\,\delta(\lambda-\lambda_0) + \\ &+ (n-l)\,\frac{l+1}{2l+1}\, I_{n-1,\,l+1}(\lambda) + (n+l+1)\,\frac{l}{2l+1}\, I_{n-1,\,l-1}(\lambda).\end{aligned}\right\} \tag{12.32}$$

The interlinkage of (12.32) is analogous to that of (12.4) but the interlinkage of (12.31) is fourfold, in correspondence to the four indices n, ν, l, and m. Notice that the interlinkage is such that one can calculate only moments fulfilling the limitations indicated in (12.19) and (12.25), namely: $n-l \geqq 0$ and even for an isotropic source and $\nu-|m| \geqq 0$ and even for a collimated source.

13. Connection with diffusion theory. One speaks of diffusion when photons or particles migrate by traveling successively in many different, randomly distributed, directions. The directions of any two consecutive free flights are wholly uncorrelated when the scattering is isotropic. When the scattering is anisotropic the directional correlation is lost progressively through multiple scattering, and a nearly random distribution of directions may be achieved just the same. Diffusion implies that any directional correlation is short-lived as compared to the duration of the migration process.

When conditions of diffusion prevail over most of the migration distance from a source to a point $\boldsymbol{r}$, the flux distribution around $\boldsymbol{r}$ can be derived by a suitably simplified theory of the transport process. The extent of the simplification depends on how closely the process resembles an ideal diffusion, exemplified by a random walk with infinitely many free flights of equal length. Accordingly, we may consider a variety of diffusion-type theories of different complexity. In the simplest of these theories the transport equation reduces to the differential form

$$\Delta I(\boldsymbol{r}, t) = \frac{\partial I(\boldsymbol{r}, t)}{\partial t}, \tag{13.1}$$

where I is the flux integrated over all directions, Δ is the Laplace operator, and t is an "age" parameter which increases in the course of the travel.

Diffusion hardly ever predominates in the process of X-ray penetration. Directional correlation persists unless or until a photon's energy is of the order of 100 kev or lower. Below this energy the sequence of scatterings soon terminates in photoelectric absorption, except in materials of extremely low atomic number. In a diffusion process an average migration distance is $\sqrt{n_T}\, L$, where L is the *transport mean free path*, i.e. the average pathlength over which directional correlation persists[1], and n_T is the number of such free paths prior to absorption.

[1] The transport mean free path may be defined as the mean path between successive scatterings divided by the mean value of $1-\cos\Theta$, averaged over the probability of single scattering by an angle Θ.

For photons of energy $\lesssim 100$ kev we have $L \sim 1/NZ\sigma_{Th}$, where NZ is the density of electrons and σ_{Th} is given by (4.7). The wavelength λ increases approximately by 1 for each photon mean free path. Thus, for a photon that escapes absorption until its energy has dropped to 20 kev, corresponding to $\lambda = 25$, we have $n_T \sim 25$ and $\sqrt{n_T} \sim 5$. The distance $5/NZ\sigma_{Th}$ is still no larger than one single mean free path would be for primary photons of energy ~ 2 Mev[1]. Deep penetration can never be described by a diffusion theory because it is due to photons that maintain a strong directional correlation.

For these reasons, diffusion treatments have received little application to X-ray problems. Nevertheless they might prove valuable adjoints, expecially for a theory of boundary effects where the contribution of low energy photons is important.

Under diffusion conditions, the flux distribution $I(\boldsymbol{r}, \boldsymbol{\omega}, \lambda)$ is nearly isotropic in $\boldsymbol{\omega}$, that is, the sequence of its Legendre coefficients $I_l^m(\boldsymbol{r}, \lambda)$ converges rapidly as l increases. The values I_l^m are small for large l, when k_l in (9.12) oscillates so rapidly that the $\int d\lambda'$ is small and does not compensate the absorption represented by the term $\mu(\lambda)\, I_l^m$. For radiations other than X rays, k_l is often small or even zero for $l \neq 0$. A diffusion-type theory is obtained, for example, by clipping off the sequence of equations (9.12) by assuming $I_{l_c+1}^0 = 0$ in the equation for $I_{l_c}^0$, where l_c is a suitable small value of l. The finite sequence of equations with $0 \leq l \leq l_c$ can then be solved more easily than the complete infinite sequence, for example by eliminating all I_l^0's with $l > 0$ and reducing the problem to a single, more complicated, equation for I_0^0.

Another method equivalent to clipping off the sequence of I_l^m's consists of evaluating the $\int d\boldsymbol{\omega}'$ in (9.3) by Gaussian quadrature. This method replaces the integral with a linear combination of the values $I(p, \boldsymbol{\omega}_i, \lambda)$ of the flux in directions $\boldsymbol{\omega}_i$ for which $(\omega_z)_i$ is a zero of the Legendre polynomial $P_{l_c}(\omega_z)$ (Ref. [22], Chap. III). It involves no error if the coefficients $I_l^m(p, \lambda)$ actually vanish for $l > l_c$. One considers then the transport equation (6.9) or (9.3) only for the flux in the special directions $\boldsymbol{\omega}_i$, intending to estimate the flux in a general direction $\boldsymbol{\omega}$ by interpolation. Eq. (9.3) with the variables $\boldsymbol{\omega}$ and λ is thus reduced to a system of l_c equations, one for each value $(\omega_z)_i$, with the single variable λ. This procedure can be regarded as a step to a direct solution of the transport equation by converting it, as in Sect. 7, into a system of algebraic equations. Its equivalence to clipping off the sequence of I_l^m's has been demonstrated in detail by WANG and GUTH[2]. The equivalence holds exactly only in an infinite medium; the formulation of boundary conditions is different in the two approximations. The Gaussian quadrature method is preferable, in this respect, to the clipped-off Legendre expansion because it leads to a better formulation of boundary conditions. Alternate approximate representations of the flux, $I(\boldsymbol{r}, \boldsymbol{\omega}_i, \lambda)$, are obtained by choosing the $\boldsymbol{\omega}_i$'s according to some quadrature formula other than that of GAUSS[3], which implies a different assessment of the relevant features of the physical situation.

In a typical diffusion process, the flux distribution has an isotropic component and a component proportional to $\boldsymbol{\omega} \cdot \boldsymbol{\Omega}$ where $\boldsymbol{\Omega}$ is the direction of the density gradient. Thus there are only two non-vanishing harmonics, I_0^0 and I_1^0, so that the sequence of I_l^0's is clipped off after $l = 1$. One can see from (9.12) that, even though k_l contributes very little for all $l \neq 0$, I_1^0 is never quite negligible because the Eq. (9.12) with $l = 1$ and $m = 0$ contains $p I_0^0$ as an additional source, except when $p = 0$ (uniform flux distribution). That is, the presence of a non-vanishing I_1^0 is essential to a non-uniform flux distribution. The moment equation (12.4) for I_{n0}^0, with $n \neq 0$, has its only source term proportional to $I_{n-1,1}^0$, so that I_{n0}^0 would vanish for $I_{n-1,1}^0 = 0$. With reference to the diagram of Fig. 16 for $m = 0$, only equations corresponding to squares in the first two columns, with $l = 0$ or 1, need to be solved in this diffusion approximation; the solution of equations pertaining to other columns and to $m \neq 0$ yields only corrective effects.

---

[1] Diffusion is much more important for neutrons, whose n_T value often exceeds 100 and whose L is not much smaller at low than at high energy. Electrons are similar to X rays in this respect; their energy loss over one transport mean free path is no less than $\sim 10\%$, so that their n_T value is $\lesssim 10$; also, their L decreases rapidly in the course of degradation.

[2] M.C. WANG and E. GUTH: Phys. Rev. **84**, 1092 (1951), Sect. V.

[3] R.C. O'ROURKE: Ref. [68] has utilized for this purpose the Radau formula in which the set of ω_{z_i}'s includes $\omega_z = \pm 1$.

In this same approximation we consider now the result of eliminating I_1^0 from the first two Eqs. (9.12). The equation for $l=1$, with I_2^0 set equal to zero, is

$$\mu(\lambda)\, I_1^0(\lambda) - \tfrac{1}{3} p\, I_0^0(\lambda) = \int_{\lambda_0}^{\lambda} d\lambda'\, k_1(\lambda, \lambda')\, I_1^0(\lambda') + \cos\vartheta_0\, \delta(\lambda - \lambda_0) \tag{13.2}$$

and its solution can be indicated formally as[1]

$$\left.\begin{aligned} I_1^0(\lambda) &= \int_{\lambda_0}^{\lambda} d\lambda'\, G^{(0,0)}(1\lambda, 1\lambda')\, [\tfrac{1}{3} p\, I_0^0(\lambda') + \cos\vartheta_0\, \delta(\lambda'-\lambda_0)] \\ &= \tfrac{1}{3} p\, \mathcal{G}\, I_0^0 + \cos\vartheta_0\, G^{(0,0)}(1\lambda, 1\lambda_0)\,. \end{aligned}\right\} \tag{13.3}$$

Entering this expression of I_1^0 in the equation for $l=0$ yields

$$[\mu(\lambda) - \tfrac{1}{3} p^2 \mathcal{G}]\, I_0^0(\lambda) = \int_{\lambda_0}^{\lambda} d\lambda'\, k_0(\lambda, \lambda')\, I_0^0(\lambda') + \delta(\lambda - \lambda_0) + p\cos\vartheta_0\, G^{(0,0)}(1\lambda, 1\lambda_0)\,. \tag{13.4}$$

This equation has now the same form as (10.1), except that $\mu(\lambda) - \frac{1}{3}p^2\mathcal{G}$ replaces $\mu(\lambda)$, $\mathcal{G}$ being an integral operator and not merely a function.

This simplified form of the transport equation is not yet equivalent to the differential diffusion equation (13.1). The Laplace transform of a solution of (13.1) depends on p through a characteristic factor $e^{p^2 t}$. Let us consider how the solution of (13.4) may be reduced to this form.

As indicated above, the concept of diffusion implies that the kernel k_1 of (13.2) is, in effect, small so that the GREEN's function $G^{(0,0)}(1\lambda, 1\lambda_0)$ in (13.3) decreases rapidly as $\lambda - \lambda_0$ increases and the directional correlation of the flux dies off rapidly in the course of degradation. Accordingly, the expression $\mathcal{G} I_0^0$ in (13.4), representing an integral over a narrow range of $I_0^0(\lambda')$, will be set approximately equal to a product $g(\lambda)\, I_0^0(\lambda)$, with the understanding that $\frac{1}{3}p^2 g(\lambda) \ll \mu(\lambda)$. Eq. (13.4) has now quite the same structure as (10.1), and its solution can be expressed by the approximate formula (10.4). To the parameter $\beta(\lambda')$ in the exponent of (10.4) must be added now a term $-\frac{1}{3}p^2 g(\lambda')/\langle\xi\rangle \mu_{sc}(\lambda')$. It follows that the solution of (13.4) depends on p, to the zero-th order in the expansion (10.7), according to

$$I_0^0(p, \lambda) \propto \exp\left\{\frac{1}{3} p^2 \int_{\lambda_1}^{\lambda} d\lambda' \frac{g(\lambda')}{\langle\xi\rangle\, \mu_{sc}(\lambda')}\right\} [1 + p\cos\vartheta_0\, g(\lambda_0)]\,. \tag{13.5}$$

The exponential factor in (13.5) depends on p like the transform of the solution of (13.1), the "age" being represented by

$$t = \frac{1}{3} \int_{\lambda_0}^{\lambda} d\lambda' \frac{g(\lambda')}{\langle\xi\rangle \mu_{sc}(\lambda')}\,. \tag{13.6}$$

The linear dependence on p of the source term has also a simple interpretation, because the inverse transform of (13.5), calculated according to (9.1), may be expressed in the form

$$I_0^0(z, \lambda) \propto \left[1 - \cos\vartheta_0\, g(\lambda_0) \frac{\partial}{\partial z}\right] \frac{1}{\sqrt{t}}\, e^{-\frac{z^2}{4t}}\,. \tag{13.7}$$

The operator in brackets represents a small shift of the source point from $z=0$ to $z=\cos\vartheta_0\, g(\lambda_0)$, corresponding to the fact that the diffusion starts only after the collimation of the source photons has been lost. Indeed $\cos\vartheta_0$ represents the degree of collimation of the source in the direction z, and $g(\lambda_0)$ should accordingly be interpreted as the transport mean free path.

The preceding outline of the solution of the transport equation (9.12) according to the diffusion approximation indicates how one might start from this approximation to complement the other approaches to the same equation which have been treated in Sects. 10, 11 and 12. Thus far, the only diffusion treatments of X-ray penetration taking into account energy degradation have been given by

---

[1] Notice that the Green's function $G^{(p,m)}(l\lambda, l_0\lambda_0)$ is diagonal in l when $p=0$, i.e. that $G^{(0,m)}$ contains a factor $\delta_{l l_0}$.

CHANDRASEKHAR[1] [22] and O'ROURKE [68]. They used Gaussian quadrature, starting directly from the transport equation (6.2), with the scattering kernel as given by the low energy approximation (4.6) to the Compton cross section and with the absorption coefficient μ as given by the integral (4.7) of that cross section. Photoelectric absorption was thereby disregarded, which is unrealistic for ordinary matter and for X-ray energies at which the diffusion approximation is reasonable. It would be realistic, however, for hydrogen or for a gas of free electrons.

14. Other approximation methods. Various authors have schematized the Compton process, replacing the broad statistical distributions of wavelength shifts and deflections with fixed shifts and deflections of suitably assigned magnitude. Thus successive Compton scatterings of a photon are assumed to degrade its energy through a definite sequence of steps identified by wavelengths λ_0, $\lambda_1 = \lambda_0 + \xi_1, \ldots, \lambda_n = \lambda_0 + \sum_{i=1}^{n} \xi_i, \ldots$.

In particular, HIRSCHFELDER *et al.* [48], [57] made this assumption and also assumed that photons scattered n times have directions $\boldsymbol{\omega}_n$ lying on the surface of a cone identified by

$$\boldsymbol{\omega}_n \cdot \boldsymbol{\omega}_0 = (1 - \xi_n)\, \boldsymbol{\omega}_{n-1} \cdot \boldsymbol{\omega}_0 = \prod_{i=0}^{n} (1 - \xi_i)\,. \tag{14.1}$$

In fact, a wavelength shift ξ_n requires that $\boldsymbol{\omega}_n \cdot \boldsymbol{\omega}_{n-1} = 1 - \xi_n$; the assumption of random azimuth of the scattering plane then causes the *mean* value $\langle \boldsymbol{\omega}_n \cdot \boldsymbol{\omega}_0 \rangle$ — but not $\boldsymbol{\omega}_n \cdot \boldsymbol{\omega}_0$ itself — to be equal to $\boldsymbol{\omega}_n \cdot \boldsymbol{\omega}_{n-1} \langle \boldsymbol{\omega}_{n-1} \cdot \boldsymbol{\omega}_0 \rangle = (1 - \xi_n) \langle \boldsymbol{\omega}_{n-1} \cdot \boldsymbol{\omega}_0 \rangle$.

From these assumptions, one can proceed readily by the iteration method of Sect. 8, entering in (8.3) and (8.4)

$$I^{(n)}(\boldsymbol{r}, \boldsymbol{\omega}, \lambda) = f^{(n)}(\boldsymbol{r})\, \delta(\lambda - \lambda_n)\, \delta(\boldsymbol{\omega} \cdot \boldsymbol{\omega}_0 - \boldsymbol{\omega}_n \cdot \boldsymbol{\omega}_0)\, \frac{1}{2\pi} \tag{14.2}$$

and

$$S^{(n)}(\boldsymbol{r}, \boldsymbol{\omega}, \lambda) = \delta(\lambda - \lambda_n)\, \delta(\boldsymbol{\omega} \cdot \boldsymbol{\omega}_0 - \boldsymbol{\omega}_n \cdot \boldsymbol{\omega}_0)\, \frac{f^{(n-1)}(\boldsymbol{r})}{2\pi}\, \mu_{sc}(\lambda_{n-1}) \tag{14.3}$$

which permits an elementary integration of the transport problem.

The delicate point in this procedure is the choice of the wavelength shifts ξ_n. One possibility is to take the value which corresponds to the mean energy shift of photons with initial wavelength λ_{n-1} (see [48a]). Thereby one obtains a reasonable estimate for the average course of degradation of all photons (see Sect. 10α). On the other hand, the transmission of photons through a thick absorber depends heavily on photons that experience less-than-average degradation. HIRSCHFELDER *et al.* took into account this bias by the following procedure. Given photons of wavelength λ_0 incident on a slab of thickness D expressed in mean free paths for Compton scattering of the incident photons, one calculates the energy transmitted by once-scattered photons: (a) exactly, and (b) as a function of an assumed standard shift ξ_1. The value of ξ_1 for which these two transmissions are equal, is a function $F(\lambda_0, D)$. The successive shifts are then taken from the equation

$$\xi_n = F(\lambda_{n-1}, D)\,. \tag{14.4}$$

Spot checks indicate that the numerical results obtained by this method lie within 10 to 20% of the results of later more realistic calculations. The approxi-

[1] The main results of CHANDRASEKHAR are for energy independent cross sections but with absorption included. With this restriction he succeeded in generalizing the diffusion approach into an exact solution.

mation is better for heavy than for light elements, probably because of the smaller number of degradation steps.

FAUST [*33*] followed a different procedure. He picked the wavelength shifts ξ_n on the basis of average energy shift, as in Ref. [*48a*], and schematized the directional distribution quite drastically, assigning all photons after each scattering either to the "forward" direction (ω_0) or to the "backward" one ($-\omega_0$). Faust also made measurements with Co^{60} γ rays in a water tank with a counter shielded from one side; his experimental and theoretical results agreed to within $\sim 10\%$.

CORNER and coworkers [*20*] tried a variety of approaches to develop a practical flexible method of estimating X-ray penetration. By representing the X-ray flux at a distance z from a source in the form $I(z) = I(0)\,e^{-\alpha(z)z}$, they focused the attention on the function $\alpha(z)$ which varies slowly with z. This function would be equal to the attenuation coefficient $\mu(\lambda_0)$ of the primaries if there were no scattered photons. In practice it is smaller than $\mu(\lambda_0)$ because of the build-up, and approaches $\mu(\lambda_0)$ at first rapidly and then slowly as z increases. Attempts were made to bracket $\alpha(z)$ between upper and lower limits. In the end a "modified first-order scattering method" proved most practical. In this method the penetration of once-scattered X rays is calculated in detail and the higher order scattering is not included explicitly. Allowance is made for the contribution of multiply scattered photons by adopting a reduced value for the attenuation of once-scattered X rays and an increased value for their rate of energy deposition in a dose-measuring device.

These various approaches have given remarkably serviceable results, considering the extreme crudeness of the assumptions involved. Some arguments tending to explain this fact have been indicated in Sect. 5δ. On the other hand, the accuracy of many of the methods would be difficult to assess if more dependable procedures were not available for intercomparison.

C. Calculation of flux distributions in infinite homogeneous media.

15. Introductory survey. As discussed in Sect. 7, the theory of gamma ray penetration has been developed by first studying simple situations. The pieces of information gained in this manner and described in the preceding sections constitute the "building blocks" of a general theory of gamma ray penetration. We face now the problem of how to put these "building blocks" together to construct the flux distributions which represent the ultimate objective of our study. There exists no unique procedure or prescription which can be said to provide "the" solution to the gamma ray penetration problem. Instead a number of techniques have been developed for incorporating various types of information into mathematical representations of the desired flux distributions.

To determine a flux distribution, we follow a sequence of three operations: (a) The transport equation is suitably transformed so that one can make an exact calculation of a set of parameters characterizing the flux distribution. The calculation of the characteristic parameters is essentially analytical in the spatial and angular variables, but numerical in the energy variable. (b) An initial estimate of the flux distribution is made, based on all available *a priori* information including that which is obtained from the study of simple situations. This estimate serves as a "weight function" in the further development. (c) The weight function is modified in a manner designed to incorporate into it all the information contained in the set of calculated characteristic parameters. The modified weight function then constitutes a representation of the flux distribution. The accuracy of the representation increases with the amount of information, qualitative or numerical, that is utilized.

The quantitative information, as represented by the characteristic parameters, may take various forms. It may be in the form of spatial moments, as discussed in Sect. 16 to 18, or consist of values of a Fourier-transformed distribution as in Sect. 20 and 21. The parameters characterizing the directional distribution are usually spherical-harmonic expansion coefficients. Note that moments, and values of the transform, are readily calculated only for infinite homogeneous media, or for equivalent situations (see Sect. 27β). Part C is therefore limited to this case. Part D contains a discussion of penetration in finite or inhomogeneous media.

The "weight function" represents an initial estimate which embodies *qualitative* information about the flux distribution. Because spatial flux distributions are known to be smooth, positive, unimodal, and predominantly exponential (see Sect. 5), the corresponding weight functions also have these characteristics. In addition, more refined information about deep penetration trends (Sect. 11) and trends near the source (Sect. 8) is sometimes taken into account.

Different techniques which utilize the same qualitative and quantitative information give equally valid representations of a flux distribution[1]. One technique is to be considered superior to another if it allows use of more information or is simpler to apply.

In Sect. 17 orthogonal polynomial expansions are described. With these, the weight function is modified by a multiplicative factor in the form of a polynomial. The most extensive applications have been made with polynomial methods.

Sect. 18 contains a description of the "function-fitting" method, according to which the representation consists of several additive terms, each with the analytic form of the weight function[2]. This technique can be adapted to a wider variety of situations than polynomial methods because it permits a more drastic revision of the initial estimate than does multiplication by polynomials. Function-fitting is particularly useful for constructing distributions from transform parameters (see Sect. 21).

In order to avoid unnecessary detail, the explanation of various methods of calculation, in Sects. 16, 18, 20 and 21, is made with reference to one standard problem: the diffusion and penetration of radiation from a *mono-energetic, plane perpendicular* source. This restriction does not limit the generality of the discussion, because the spirit of approach, and even the detailed methods of fitting are essentially the same for other source types. Only minor modifications are required to take into account geometric characteristics.

Sect. 19 contains results of calculations by the moment-method and comparisons with corresponding experiments. Five elementary source types are considered: the *plane perpendicular*, *point isotropic*, *plane isotropic*, *plane oblique*, and the *point collimated* ("gun") source. Sect. 21 contains results obtained by the Fourier-transform method for the penetration of radiation to great depths.

16. Numerical solution of the moment equations. This section deals with the calculation of the moments of the flux distribution for a *plane perpendicular*

[1] For a comparison of different moment-fitting techniques in a typical example, see Table 12.

[2] In contrast to polynomial methods, "function-fitting" can be viewed as a simple example of approximations in which the corrective function is convoluted with, rather than multiplied by, the weight function. See J. S. LANGER and R. S. VARGA: U.S. AEC Report WAPD-TN-520, November 1955.

source. It is convenient to rewrite Eq. (12.4), for $m=0$, in the form

$$\mu(\lambda)\, I_{nl}(\lambda) = \int_{\lambda_0}^{\lambda} d\lambda\, k_l(\lambda,\lambda')\, I_{nl}(\lambda') + Q\, I_{nl}(\lambda) + \delta_{n0}\,\delta(\lambda-\lambda_0) \qquad (16.1)$$

where Q is an interlinkage operator defined as follows:

$$Q\, I_{nl}(\lambda) = \frac{n}{2l+1}\{(l+1)\, I_{n-1,l+1}(\lambda) + l\, I_{n-1,l-1}(\lambda)\}. \qquad (16.2)$$

The kernel, according to (4.4) and (9.11) is[1]

$$k_l(\lambda,\lambda') = \frac{3}{8} N_A \frac{Z}{A}\varrho\, \sigma_{Th}^2 \left(\frac{\lambda'}{\lambda}\right)^2 \left\{\frac{\lambda'}{\lambda} + \frac{\lambda}{\lambda'} - 2(\lambda-\lambda') + (\lambda-\lambda')^2\right\} P_l(1-\lambda+\lambda'), \qquad (16.3)$$

where N_A is AVOGADRO's number, $\sigma_{Th} = \frac{8\pi}{3}\left(\frac{e^2}{m c^2}\right)^2$ is the Thomson cross section, and Z, A and ϱ are the atomic number, atomic weight and density of the medium.

Before proceeding with the solution of (16.1) we shall introduce modifications designed to facilitate the numerical work. In the first place, the source function can be made non-singular through the separation of the unscattered flux, by setting

$$I_{nl} = I_{nl}^{(0)}\ \text{(unscattered)} + I_{nl}^{(s)}\ \text{(scattered)}. \qquad (16.4)$$

$I_{nl}^{(0)}$ obeys an equation similar to (16.1) but with the scattering integral omitted, i.e. an algebraic equation, whose solution is

$$I_{nl}^{(0)} = \frac{n!\,\delta(\lambda-\lambda_0)}{\mu_0^{n+1}}. \qquad (16.5)$$

Proceeding iteratively (as in Sect. 8) one obtains through the substitution of (16.4) and (16.5) into (16.1) the equation

$$\mu(\lambda)\, I_{nl}^{(s)}(\lambda) = \int_{\lambda_0}^{\lambda} d\lambda'\, k_l(\lambda,\lambda')\, I_{nl}^{(s)}(\lambda') + Q\, I_{nl}^{(s)}(\lambda) + \frac{n!\, k_l(\lambda,\lambda_0)}{\mu_0^{n+1}}. \qquad (16.6)$$

It is advantageous to make the further transformation

$$I_{nl}^{(s)}(\lambda) = \frac{n!}{\lambda\,\mu_0^{n+1}}\, \bar{I}_{nl}(\lambda) \qquad (16.7)$$

which is of the type (7.2) and leads to an integral equation with fewer numerical constants, and with a kernel $\frac{\lambda}{\lambda'} k_l(\lambda,\lambda')$[2]. This kernel is a smoother function than the original kernel $k_l(\lambda,\lambda')$ of (16.6) and therefore more suitable for numerical integration. The new equation is

$$\mu(\lambda)\, \bar{I}_{nl}(\lambda) = \int_{\lambda_0}^{\lambda} d\lambda' \frac{\lambda}{\lambda'} k_l(\lambda,\lambda')\, \bar{I}_{nl}(\lambda') + \mu_0\, Q\, \bar{I}_{nl}(\lambda) + \lambda\, k_l(\lambda,\lambda_0). \qquad (16.8)$$

Usually one requires a linear combination of the moments, rather than the moments themselves, for the construction of flux distribution function. For

[1] The use of the Klein-Nishina cross-section is an approximation because there are additional scattering processes, as discussed in Part A. The approximation is valid because these processes have small cross-sections and do not affect the penetration and diffusion of X rays significantly. For the sake of consistency, scattering processes omitted from $k_l(\lambda,\lambda')$ should also be omitted from $\mu(\lambda)$.

[2] Note that $\bar{I}_{nl}$ is a moment of the scattered energy flux.

example, in the Laguerre polynomial expansion described in Sect. 17β, the expansion coefficients are $a_{nl} = \sum_{m=0}^{n} (-1)^m \binom{n}{m} \bar{I}_{ml}$. It is possible to replace (16.8) by an equivalent system of integral equations (with different interlinkages) that yields directly the desired moment combinations. This reduces round-off error, which is an important consideration when the numerical work is done by slide-rule or desk-computer.

Going on to the numerical solution of the system (16.8), we recall from Sect. 12 that the integral equations for various (n, l) must be solved in a specified sequence. To simplify the notation we suppress indices and write generically

$$\mu(\lambda)\, I(\lambda) = \int_{\lambda_0}^{\lambda} d\lambda'\, K(\lambda, \lambda')\, I(\lambda') + S(\lambda). \tag{16.9}$$

The lower limit λ_0 of the integral is an explicit expression of the fact that $I(\lambda)$ vanishes for $\lambda < \lambda_0$. The function $S(\lambda)$ includes the interlinkage term and the source term in (16.8). The source term $\lambda k(\lambda, \lambda_0)$ included in $S(\lambda)$ vanishes for $\lambda > \lambda_0 + 2$, which makes $I(\lambda)$ discontinuous at $\lambda_0 + 2$.

The integral equation (16.9) is expeditiously solved through approximation by a finite system of algebraic equations. Because (16.9) is of the Volterra type, with λ as the upper limit of the integral, the corresponding algebraic system is triangular and its solution elementary. We evaluate $I(\lambda)$ at a set of discrete points $\lambda_0, \lambda_1, \ldots, \lambda_j, \ldots, \lambda_n$ and replace the integral in (16.9) by a sum:

$$\int_{\lambda_0}^{\lambda_j} d\lambda'\, K(\lambda_j, \lambda')\, I(\lambda') \cong \sum_{i=0}^{j} c_{ij}\, K(\lambda_j, \lambda_i)\, I(\lambda_i) \qquad (0 \leqq j \leqq n) \tag{16.10}$$

where the c_{ij} are the weights required by the appropriate numerical quadrature formula[1]. The result is the system of equations

$$\left.\begin{aligned} \mu(\lambda_0)\, I(\lambda_0) &= S(\lambda_0), \\ \mu(\lambda_j)\, I(\lambda_j) &= \sum_{i=0}^{j} c_{ij}\, K(\lambda_j, \lambda_i)\, I(\lambda_i) + S(\lambda_j) \qquad (1 \leqq j \leqq n) \end{aligned}\right\} \tag{16.11}$$

whose solution is

$$\left.\begin{aligned} I(\lambda_0) &= \frac{S(\lambda_0)}{\mu(\lambda_0)}, \\ I(\lambda_j) &= \frac{\sum_{i=0}^{j-1} c_{ij}\, K(\lambda_j, \lambda_i)\, I(\lambda_i) + S(\lambda_j)}{\mu(\lambda_j) - c_{jj}\, K(\lambda_j, \lambda_j)}. \end{aligned}\right\} \tag{16.12}$$

The number of points on which the numerical integration (16.10) is based should be kept reasonably small, but must provide a mesh fine enough so that the variations of the integrand are described accurately. The choice of the integration intervals $\Delta\lambda_j = \lambda_{j+1} - \lambda_j$ depends on λ_0 and l. For small λ and λ', $K(\lambda, \lambda') = \frac{\lambda}{\lambda'} k_l(\lambda, \lambda') \sim 1 + \left(\frac{\lambda'}{\lambda}\right)^2$ so that $\frac{\delta K}{K-1} \sim \frac{2\,\delta\lambda'}{\lambda'}$. It is therefore required that near the source wavelength, $\Delta\lambda_i$ be small compared to λ_0. The scattering coefficient $k_l(\lambda, \lambda')$ is proportional to the Legendre polynomial $P_l(1 - \lambda + \lambda')$, so that it has l zeros in the interval $\lambda - 2 \leqq \lambda' \leqq \lambda$. The integration interval $\Delta\lambda_j$ should therefore be small compared to the period of oscillation. Note that the average

[1] It has been found sufficient to use quadrature formulas based on parabolic approximation, such as Simpson's rule.

period is $2/l$ and that the interval between zeros is not less than $1-(l-1)\sqrt{\frac{l+2}{l(l^2+2)}}$ [1]. An argument in favor of further shortening the integration interval is the fact that $\bar{I}_{nl}$ itself is also an oscillatory function.

Economy of computing effort can be achieved by not keeping $\Delta\lambda_j$ constant at a permissible maximum value, but making it a function of l and λ. Often it suffices to divide the region of integration into two or three subregions in each of which $\Delta\lambda_j$ is kept constant. For very small values of λ_0, it is advantageous to use logarithmically spaced integration intervals [*19*].

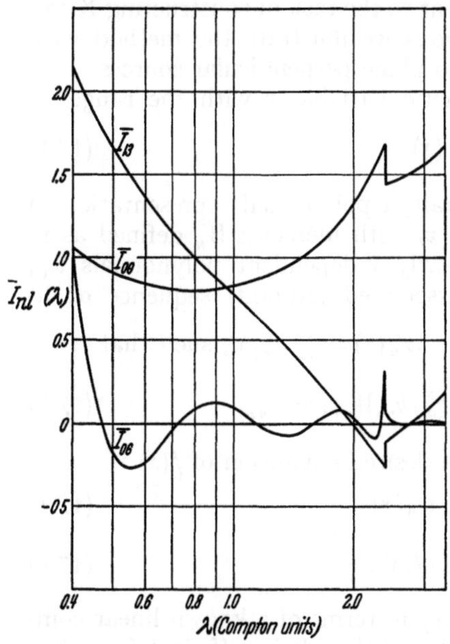

Fig. 17. Energy flux moments $\bar{I}_{nl}(\lambda)$ defined by (16.7), for a plane perpendicular 1.28-Mev source in water.

The computation of moments is feasible with desk computers, and several pilot calculations have been done in this way (Refs. [*83*] and [*85*]). More work has been done with automatic digital computers [*41*] [2]. Fig. 17 shows several $\bar{I}_{nl}$'s for a 1.28 Mev source in water. In these calculations a spacing $\Delta\lambda_j=\frac{2}{37}$ was used in a Simpson-type integration [3]. Note the discontinuities at λ_0+2, where singly-scattered radiation ceases to contribute to the flux, and the oscillations of the moments for $l>0$.

The amount of numerical work required for the computation of the moments is roughly proportional to the square of the magnitude of the highest wavelength considered. If high accuracy is required at very low energies (this will only be the case for low-Z media) a saving of effort will result from supplementing the method of this section with the analytical procedure indicated in Sect. 10α. Another method of economizing the computing effort, which is sufficiently accurate for most purposes, particularly when one is interested in an integral quantity such as the energy dissipation, consists of limiting the exact calculation to wavelengths $\lambda\lesssim\lambda_0+3$ and using extrapolations beyond this point. Such extrapolations are made possible by the establishment of a quasi-equilibrium of the spectrum at low energies (see Sect. 10α). Thus all moments of the same Legendre expansion coefficient (i.e. with the same index l) tend to be proportional to each other in the low energy limit, so that only one of them need be computed exactly. Moreover, the shape of the equilibrium spectrum is quite insensitive to the value of the source energy (see Fig. 11a), so that one universal equilibrium spectrum will suffice for many situations. Finally, the extrapolation of I_{00} can be guided

[1] G. SZEGÖ: Orthogonal Polynomials, p. 116. New York: American Mathematical Society 1939.

[2] An extensive set of moments, for different materials and source energies, is available on microfilm from Nuclear Development Corporation of America, Ref. [*71*].

[3] The calculation was done on the automatic computer of the National Bureau of Standards (SEAC). The mesh was finer than needed for small l. It was convenient in this case to do a little more computing than necessary in order to simplify the programming of the problem for the computer.

by the knowledge that the energy injected by the source into the medium must be equal to the energy dissipated, so that

$$\frac{1}{\lambda_0} = \int_{\lambda_0}^{\infty} \frac{d\lambda'}{\lambda'} \mu_D(\lambda') I_{00}(\lambda'). \tag{16.13}$$

This integral equation may serve to verify the validity of an extrapolation, or it could be used in a direct calculation of I_{00}.

17. Construction of distributions from moments: Polynomial method. *α) General discussion.* We sketch here the general method of orthogonal polynomial expansions, emphasizing those features most pertinent to the task of constructing X-ray flux distributions. In the following subsections we illustrate the method with specific applications, limiting our attention to plane perpendicular sources.

Consider a function $f(x)$ of a single generic variable x with the range R. Moments

$$f_n = \int_R dx\, x^n f(x) \tag{17.1}$$

are assumed to exist and be known. To obtain a polynomial representation of $f(x)$, one first chooses a weight function $W(x)$ with moments W_n defined as in (17.1). Next, one needs a sequence of linearly independent polynomials w_0, $w_1(x), \ldots, w_n(x) = \sum_i b_{ni} x^i$, which has an associated (adjoint) sequence of linearly independent polynomials $w_0^\dagger, w_1^\dagger(x), \ldots, w_m^\dagger(x) = \sum b_{mj}^\dagger x^j$, such that

$$\int_R dx\, W(x)\, w_m^\dagger(x)\, w_n(x) = \sum_{ji} b_{mj}^\dagger b_{ni} W_{j+i} = \delta_{mn}. \tag{17.2}$$

In terms of $W(x)$ and its polynomials, the desired expansion of $f(x)$ is

$$f(x) = W(x) \sum_n a_n w_n(x) \tag{17.3}$$

with

$$a_n = \int_R dx\, w_n^\dagger(x)\, f(x). \tag{17.4}$$

Since $w_n^\dagger(x)$ is a polynomial, (17.4) expresses a_n in terms of a known linear combination of the moments f_n. Thus, a knowledge of the f_n's is sufficient for determining the representation (17.3).

Given $W(x)$, there is still some freedom in the choice of the functions w_n and $w_n^\dagger$ [1]. There is no reason to restrict the generality of the w_n's, which will therefore be assumed to contain all powers of x up to x^n. In regard to the $w_n^\dagger$'s, two cases commonly arise: (1) Successive moments $f_0, f_1, f_2, \ldots$ are to be utilized, in which case $w_n^\dagger = w_n$ and the two polynomial systems are said to be self-adjoint. (2) Because of the symmetry properties of $f(x)$ only even (or odd) moments f_n are to be utilized, in which case the $w_n^\dagger$'s must similarly be even (or odd), so that the two polynomial sets are different and form a bi-orthogonal system [2].

[1] Note that the w_n's can be functions other than polynomials, in which case (17.3) is a more general type of functional expansion, similar to the function-fitting representation in Sect. 18. The $w_n^\dagger$'s must be polynomials if we want to determine the expansion coefficients a_n from moments. We shall in the further discussion make the assumption that the w_n's are also polynomials.

[2] In case (1), (17.4) for $n \leq N$, supplies $\frac{(N+1)(N+2)}{2}$ conditions which exactly specify $w_0, w_1, \ldots, w_N$. In case (2), on the other hand, (17.4) yields $(N+1)^2$ conditions, while $(N+1)(N+2)$ conditions are required to specify $w_0, w_1, \ldots, w_N$ and $w_0^\dagger, w_1^\dagger, \ldots, w_N^\dagger$ completely. The additional $N+1$ conditions represent an arbitrariness in the normalization. Each w_n can be multiplied by a constant, provided that the corresponding $w_n^\dagger$ is divided by the same constant.

In the practical situations of interest here, only a few moments $f_0 \dots f_n$ are available. With these it is possible to obtain the first few terms of the expansion (17.3), i.e. terms for which $w_n(x)$ is of degree less than or equal to N. These suffice if the a_n decrease rapidly with increasing n[1].

The significance of the expansion (17.3) is illustrated by the following argument: The first term of the series contains a polynomial of degree zero, i.e. the constant $a_0 w_0$. Thus the weight function $W(x)$ itself represents the initial approximation of $f(x)$ to within the factor $a_0 w_0$. If $W(x)$ were chosen just equal to the unknown function $f(x)$, the expansion would reduce to a single term. The successive terms in the expansion may be described as corrective terms that are added to the initial approximation $W(x) \approx f(x)$ so that successive moments of the approximation function

$$W(x)\,[a_0\,w_0 + a_1\,w_1(x) + \cdots] \tag{17.5}$$

will coincide with successive moments of the actual function $f(x)$.

The weight function $W(x)$ should be carefully adapted to the *a priori* information regarding $f(x)$, such as trends for large and small x. In fact, for convergence of the sequence in (17.5) it is *mandatory* that $W(x)$ resemble $f(x)$ so closely that the correction factor is representable by a polynomial of low degree.

According to the qualitative description given in Sect. 5 δ, spatial γ-ray flux distributions have a basic exponential trend which is modified by a smooth and slowly varying "build-up" factor. We identify the exponential trend with the weight function and the corrective polynomial with the build-up factor.

The choice of an exponential weight function has the following interpretation in transform space: As discussed in Sect. 11, the transformed flux distribution always has a singularity at $p = \mu_s$. For values of p approaching this singularity the transformed distribution rises sharply. The polynomial approximation, when transformed, also has poles at $p = \mu_s$ in the neighborhood of which there is a similar rise.

Although we are dealing with an infinite medium, it is possible to restrict the range of the spatial variable to $0 \leqq z \leqq \infty$ provided that backscattering is unimportant[2]. This may be the case for spectral components with λ close to λ_0 or for penetration in high-Z materials which strongly absorb backscattered radiation. The case $0 \leqq z \leqq \infty$ is treated in subsection β) and the general case $-\infty \leqq z \leqq \infty$ in γ). Finally, δ) describes polynomial expansions of angular distributions.

β) *Spatial distributions: Negligible backscatter.* If the range R may be limited to $0 \leqq z \leqq \infty$ and if the predominant trend $e^{-\mu_s z}$ is taken as the weight function[3], the appropriate self-adjoint polynomials are Laguerre polynomials

$$L_n(\mu_s z) = \sum_{j=0}^{n} (-1)^j \binom{n}{j} \frac{(\mu_s z)^j}{j!}. \tag{17.6}$$

[1] The question of convergence in the limit of indefinitely large n is difficult to answer; we shall not attempt a discussion of this point, since in practice we are always limited to a small number of moments. In general, enough is known about the distributions to be considered so that one can find weight functions yielding at least asymptotic convergence. This has been investigated through trial expansions of Bessel and Laguerre functions, which are similar to the desired unknown functions. See also H. CRAMER, Mathematical Methods of Statistics, Princeton, 1946, p. 227, for the case of a Gaussian weight function.

[2] This schematization is particularly applicable to the source type discussed in this section, namely the plane perpendicular source which emits photons in the direction of positive z.

[3] It is not absolutely necessary to have the coefficient in the weight function exactly equal to μ_s. In fact, some computational advantage is gained by using a coefficient a little larger than μ_s in which case the successive terms in the expansion are more likely to have the same sign.

Thus if spatial moments $I_{nl}(\lambda)$, $(n=0, 1, \ldots, N)$ are available to describe the l-th Legendre coefficient $I_l(z, \lambda)$, one makes the expansion

$$I_l(z, \lambda) = e^{-\mu_s z} \sum_{n=0}^{N} a_{nl}(\lambda)\, L_n(\mu_s z) \tag{17.7}$$

where

$$a_{nl}(\lambda) = \mu_s \int_0^\infty dz\, L_n(\mu_s z)\, I_l(z, \lambda) = \sum_{j=0}^{n} (-1)^j \binom{n}{j} \frac{\mu_s^{j+1} I_{jl}}{j!} = \lambda^{-1} \sum_{j=0}^{n} (-1)^j \binom{n}{j} \bar{I}_{jl}(\lambda).$$

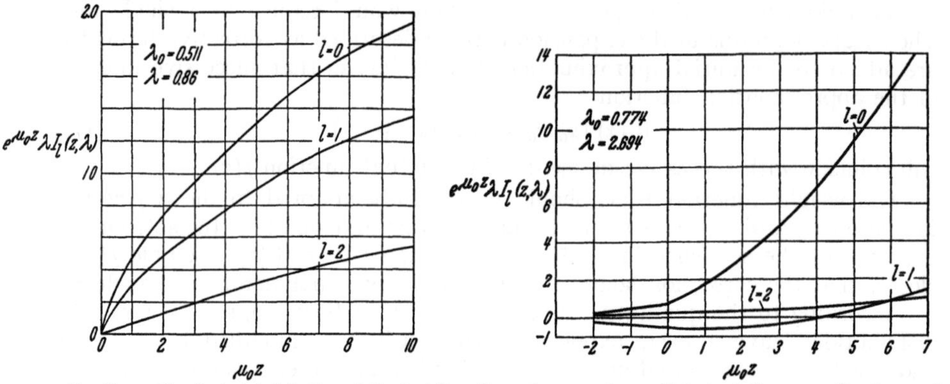

Fig. 18a and b. Spatial distribution of the first three Legendre expansion coefficients of the energy flux from a plane perpendicular source: (a) Case of negligible back-scatter: 1-Mev source in lead. (b) General case: 0.66-Mev source in water.

Fig. 18a presents sample distributions calculated in this way, for the penetration of 1-Mev X rays in lead. Table 10 presents corresponding moments $\bar{I}_{nl}$, defined as in (16.7), together with Laguerre expansion coefficients $a_{nl}(\lambda)$ to indicate the typical convergence characteristics of this type of polynomial expansion.

Table 10. *Moments $\bar{I}_{nl}(\lambda)$ and Laguerre expansion coefficients $a_{nl}(\lambda)$ for the spectral component with wavelength $\lambda = 0.86$ from a 1-Mev plane perpendicular source in lead.*

n	λa_{n0}	$\bar{I}_{n0}$	λa_{n1}	$\bar{I}_{n1}$	λa_{n2}	$\bar{I}_{n2}$
0	0.412076	0.412076	0.270641	0.270641	0.068826	0.068826
1	−0.283088	0.695164	−0.191283	0.461924	−0.062640	0.131466
2	−0.060956	0.917296	−0.036844	0.616363	−0.002406	0.191700
3	−0.029788	1.108260	−0.017860	0.751818	−0.000394	0.249922
4	−0.016532	1.281312	−0.010133	0.876016	−0.000548	0.305978
5	−0.009492	1.442668	−0.005994	0.992545	−0.000586	0.359752

It is instructive to consider the transform of the Laguerre expansion (17.7),

$$I_l(p, \lambda) = \frac{1}{(\mu_s - p)} \sum_{n=0}^{N} a_{nl}(\lambda) \left(\frac{p}{\mu_s - p}\right)^n. \tag{17.8}$$

The predominant feature of $I(p, \lambda)$, namely the singularity at $p = \mu_s$ is properly taken into account by this expansion because each of the terms has a pole at $p = \mu_s$. Therefore expansion (17.8) will converge better than a Taylor expansion about $p=0$, which has moments as coefficients [as in Eq. (12.2)].

There are circumstances in which other weight functions may be useful. For example, near $z=0$ the flux of photons scattered n times is proportional to z^n. Accordingly one could calculate the flux of low orders of scattering by iteration (see Sect. 8) and use a weight function $(\mu_s z)^\xi e^{-\mu_s z}$, $\xi = 1, 2, \ldots$, (together with self-adjoint "associated Laguerre" polynomials) in a moment calculation of the flux contained in the remaining orders of scattering. It is also possible to incorporate into the weight function more information about the deep penetration trend of the distributions. For example, one could adjust the approximation so that $z^N W(z)$ has the correct asymptotic trend, where z^N is the highest power of z occurring in the expansion. This procedure has not found practical use.

γ) Spatial distributions: General case. We now allow for the possibility of backscattering, so that a range $-\infty \leqq z \leqq \infty$ is required. There is little *a priori* information available about the spatial distribution of the backscattered flux ($z<0$) so that the choice of a suitable weight-function in this region is not obvious. This difficulty may be avoided by generalizing the procedure in 17β in such a way that no initial specification of the backscattered flux is required. We split the flux distribution into a symmetric and an antisymmetric component, such that the sum of the two components describes the flux in the forward direction, and their difference the backscattered flux. This leads to an accurate determination of the flux everywhere except far behind the source, where the difference between the two components becomes very small.

We consider the *plane perpendicular* source $\frac{1}{2\pi}\,\delta(z)\,\delta(\lambda-\lambda_0)\delta(\omega_z-1)$ as the sum of two sources: (1) a symmetrical source component $\frac{1}{4\pi}\,\delta(z)\,\delta(\lambda-\lambda_0)\times\{\delta(\omega_z-1)+\delta(\omega_z+1)\}$ which injects equal photon fluxes in opposite directions perpendicular to the plane $z=0$, and which gives rise to a photon distribution $I^S(z,\omega_z,\lambda)$; (2) an anti-symmetric source component $\frac{1}{4\pi}\,\delta(z)\,\delta(\lambda-\lambda_0)\,\{\delta(\omega_z-1)-\delta(\omega_z+1)\}$ which injects a photon flux equal to that of the symmetric source on the positive side of $z=0$, and a "negative" photon flux of equal magnitude on the negative side, and which gives rise to a photon distribution $I^A(z,\omega_z,\lambda)$. The two flux components I^S and I^A, and their Legendre expansion coefficients, I_l^S and I_l^A, have the symmetry properties

$$\left.\begin{aligned} I^S(z,\omega_z,\lambda) &= I^S(-z,-\omega_z,\lambda)\\ I_l^S(z,\lambda) &= (-1)^l\, I_l^S(-z,\lambda)\end{aligned}\right\} \tag{17.9}$$

$$\left.\begin{aligned} I^A(z,\omega_z,\lambda) &= -I^A(-z,-\omega_z,\lambda)\\ I_l^A(z,\lambda) &= (-1)^{l+1}\, I_l^A(-z,\lambda).\end{aligned}\right\} \tag{17.10}$$

It will thus be sufficient to determine these functions for $z \geqq 0$, and we shall in the remainder of this sub-section take z to mean $|z|$. The Legendre expansion coefficients of the total flux can therefore be expressed in the form

$$\left.\begin{aligned} I_l(z,\lambda) &= I_l^S(z,\lambda) + I_l^A(z,\lambda)\\ I_l(-z,\lambda) &= (-1)^l\,\{I_l^S(z,\lambda) - I_l^A(z,\lambda)\}.\end{aligned}\right\} \tag{17.11}$$

It follows from (17.9) and (17.10) that

$$\left.\begin{aligned} I_{nl}^S(\lambda) &= \int_0^\infty z^n\, I_l^S(z,\lambda)\,dz = \tfrac{1}{2}\,I_{nl}(\lambda) &&\text{if } n+l \text{ is even}\\ &= 0 &&\text{if } n+l \text{ is odd}\\ I_{nl}^A(\lambda) &= \int_0^\infty z^n\, I_l^A(z,\lambda)\,dz = 0 &&\text{if } n+l \text{ is even}\\ &= \tfrac{1}{2}\,I_{nl}(\lambda) &&\text{if } n+l \text{ is odd.}\end{aligned}\right\} \tag{17.12}$$

We note that, according to the discussion in Sect. 12, the I_{nl}^S's and the I_{nl}^A's each form a closed system, in the sense that according to the moment interlinkage equations only I^S_{nl}'s contribute as source terms to the computation of other I^S_{nl}'s, and only I^A_{nl}'s contribute to other I^A_{nl}'s.

The moment-fitting problem has now been separated into two parts, namely the construction of $I^S_l(z, \lambda)$ from the "even" moments and of $I^A_l(z, \lambda)$ from the "odd" moments[1].

In order to determine appropriate polynomial expansions for $I^S_l(z, \lambda)$ and $I^A_l(z, \lambda)$, we next examine our information concerning these functions. The predominant spatial trend must be $e^{-\mu_s z}$ regardless of l. Furthermore, *all* Legendre coefficients I^S_l and I^A_l tend to a finite limit as $z \to 0$. Finally, alternate moments rather than consecutive moments must be used in determining the spatial expansion coefficients. For example, the expansion for $I^S_0(z, \lambda)$ must utilize the moments $\frac{1}{2} I_{00}(\lambda)$, $\frac{1}{2} I_{20}(\lambda)$, $\frac{1}{2} I_{40}(\lambda)$, ..., and the expansion for $I^A_0(z, \lambda)$ the moments $\frac{1}{2} I_{10}(\lambda)$, $\frac{1}{2} I_{30}(\lambda)$, $\frac{1}{2} I_{50}(\lambda)$, ...

It is not possible to use Laguerre polynomials, which require consecutive moments; therefore we are led to consider analogous systems which are bi-orthogonal rather than self-adjoint. In particular, we consider polynomial sets $U^k_n(\mu_s z)$ and $U^{k\dagger}_n(\mu_s z)$ which are bi-orthogonal with respect to $(\mu_s z)^k e^{-\mu_s z}$. The $U^{k\dagger}_n$'s should contain every other power of z in correspondence with the use of alternate moments. The lowest moment used in the construction determines the value of k. Writing $x = \mu_s z$ for simplicity, the orthogonality relations become

$$\int_0^\infty dx\, x^k e^{-x} U^{k\dagger}_m(x)\, U^k_n(x) = \delta_{mn}. \tag{17.13}$$

With these polynomials we make the following expansions:

$$\left.\begin{aligned} I^S_l(z, \lambda) &= e^{-\mu_s z} \sum_n s^k_{nl}(\lambda)\, U^k_n(\mu_s z), \qquad & k &= \begin{cases} 0, & l \text{ even} \\ 1, & l \text{ odd}, \end{cases} \\ I^A_l(z, \lambda) &= e^{-\mu_s z} \sum_n a^k_{nl}(\lambda)\, U^k_n(\mu_s z), \qquad & k &= \begin{cases} 0, & l \text{ odd} \\ 1, & l \text{ even} \end{cases} \end{aligned}\right\} \tag{17.14}$$

with

$$\left.\begin{aligned} s^k_{nl} &= \mu_s \int_0^\infty dz\, (\mu_s z)^k\, U^{k\dagger}_n(\mu_s z)\, I^S_l(z, \lambda), \qquad & k &= \begin{cases} 0, & l \text{ even} \\ 1, & l \text{ odd}, \end{cases} \\ a^k_{nl} &= \mu_s \int_0^\infty dz\, (\mu z)^k\, U^{k\dagger}_n(\mu_s z)\, I^A_l(z, \lambda), \qquad & k &= \begin{cases} 0, & l \text{ odd} \\ 1, & l \text{ even}. \end{cases} \end{aligned}\right\} \tag{17.15}$$

The coefficients $s^k_{nl}(\lambda)$, $a^k_{nl}(\lambda)$ are the desired linear combinations of alternate moments.

---

[1] It is possible to reverse the order of procedure, by first summing over l and then carrying out the moment-fitting in the space variable. In this case the moment-fitting would have to be done with moments of the form

$$I^S_{2n}(\omega_z, \lambda) = \int_0^\infty dz\, z^{2n} I^S(z, \omega_z, \lambda) = \frac{1}{2} \sum_{l=0}^\infty \frac{4l+1}{4\pi} I_{2n,2l}(\lambda)\, P_{2l}(\omega_z)$$

$$I^S_{2n+1}(\omega_z, \lambda) = \int_0^\infty dz\, z^{2n+1} I^S(z, \omega_z, \lambda) = \frac{1}{2} \sum_{l=0}^\infty \frac{4l+3}{4\pi} I_{2n+1,2l+1} P_{2l+1}(\omega_z)$$

and similar expressions for the I^A_n's.

The polynomials have the form

$$\left.\begin{aligned} U_n^{k\dagger}(x) &= \sum_{j=0}^{n} (-1)^j \binom{n}{j} \frac{x^{2j}}{(2j+k)!}, \\ U_n^k(x) &= \frac{(-1)^{n+k}}{2^n n!}\, e^x \left(\frac{d}{dx}\right)^{2n+k} e^{-x} x^n \sum_{j=0}^{n} \frac{(n+j)!}{j!(n-j)!} \frac{1}{(2x)^j} \end{aligned}\right\} \tag{17.16}$$

and have the recursion property

$$U_{n+1}^k(x) = \frac{1}{2(n+1)}\left[(2n+1+k) - x + x\frac{d}{dx}\right] U_n^k(x). \tag{17.16'}$$

Table 11. *First few polynomials $U_n^{k\dagger}$, U_n^k of the bi-orthogonal system associated with the weight function $x^k e^{-x}$.*

$U_0^{0\dagger} = 1$	$U_0^0 = 1$
$U_1^{0\dagger} = 1 - \frac{x^2}{2!}$	$U_1^0 = \frac{1}{2}(1-x)$
$U_2^{0\dagger} = 1 - 2\frac{x^2}{2!} + \frac{x^4}{4!}$	$U_2^0 = \frac{1}{8}(3 - 5x + x^2)$
$U_3^{0\dagger} = 1 - 3\frac{x^2}{2!} + 3\frac{x^4}{4!} - \frac{x^6}{6!}$	$U_3^0 = \frac{1}{48}(15 - 33x + 12x^2 - x^3)$
$U_4^{0\dagger} = 1 - 4\frac{x^2}{2!} + 6\frac{x^4}{4!} - 4\frac{x^6}{6!} + \frac{x^8}{8!}$	$U_4^0 = \frac{1}{384}(105 - 279x + 141x^2 - 22x^3 + x^4)$
$U_0^{1\dagger} = 1$	$U_0^1 = 1$
$U_1^{1\dagger} = 1 - \frac{x^2}{3!}$	$U_1^1 = \frac{1}{2}(2-x)$
$U_2^{1\dagger} = 1 - 2\frac{x^2}{3!} + \frac{x^4}{5!}$	$U_2^1 = \frac{1}{8}(8 - 7x + x^2)$
$U_3^{1\dagger} = 1 - 3\frac{x^2}{3!} + 3\frac{x^4}{5!} - \frac{x^6}{7!}$	$U_3^1 = \frac{1}{48}(48 - 57x + 15x^2 - 3)$
$U_4^{1\dagger} = 1 - 4\frac{x^2}{3!} + 6\frac{x^4}{5!} - 4\frac{x^6}{7!} + \frac{x^8}{9!}$	$U_4^1 = \frac{1}{384}(384 - 561x + 207x^2 - 26x^3 + x^4)$

The sum in the expression for $U_n^k(x)$ is the function $S_{n+\frac{1}{2}}(2x)$ defined by (9.27), and the expression $e^{-x} S_{n+\frac{1}{2}}$ equals $\sqrt{2x/\pi}\, K_{n+\frac{1}{2}}(x)$, where K is a Bessel function with imaginary argument[1]. Table 11 lists the first few U_n^0, $U_n^{0\dagger}$ and U_n^1, $U_n^{1\dagger}$ polynomials.

The analogy between this type of expansion and the Laguerre expansion discussed in the preceding sub-section is particularly striking in transform space, where application of the symmetry properties (17.9) and (17.10) leads to the

[1] G.M. WATSON: Bessel Functions, p. 78–80. Cambridge: Cambridge University Press 1945. Note that the U_n and V_n polynomials defined by Eq. 17 in Ref. [*83*] correspond to the polynomials U_n^0, $x\,U_n^2$, respectively, defined here. Note also that the U_n^m polynomials utilized in Ref. [*85*] represent only *every other* associated polynomial system defined by (17.13) and (17.16). The polynomials $U_n^k(x)$ defined here are solutions of the *third* order differential equation

$$\left\{x\frac{d^3}{dx^3} + (k+1-3x)\frac{d^2}{dx^2} + 2(x-k-1)\frac{d}{dx} - 2n\right\} U_n^k = 0.$$

following expressions analogous to (17.8):

$$\left.\begin{aligned}\int_0^\infty dz\,[e^{pz}+(-1)^k e^{-pz}]\,I_l^S(z,\lambda) &= 2\frac{p^k\mu_s^{1-k}}{\mu_s^2-p^2}\sum_n s_{nl}^k(\lambda)\left[\frac{p^2}{\mu_s^2-p^2}\right]^n, \quad k=\begin{cases}0, l \text{ even}\\1, l \text{ odd}\end{cases}\\ \int_0^\infty dz\,[e^{pz}+(-1)^k e^{-pz}]\,I_l^A(z,\lambda) &= 2\frac{p^k\mu_s^{1-k}}{\mu_s^2-p^2}\sum_n a_{nl}^k\left[\frac{p^2}{\mu_s^2-p^2}\right], \quad k=\begin{cases}0, l \text{ odd}\\1, l \text{ even}.\end{cases}\end{aligned}\right\} \tag{17.17}$$

These differ from (17.8) mainly in that p^2 has replaced p, so that they are either even or odd in p corresponding to symmetry or antisymmetry in the space variable.

Fig. 18b presents several functions $I_l(z,\lambda)$ as calculated with expressions (17.14) and (17.15) for a 0.66 Mev γ-ray source in water. $I_0(z,\lambda)$ is of special interest, being the total flux integrated over all directions. Note that all I_l's tend to zero for negative z.

δ) Angular distributions. In the preceding two sub-sections, polynomial expansions have been used to determine the coefficients $I_l(z,\lambda)$. To obtain directional distributions $I(z,\vartheta,\lambda)$ it is necessary to combine the I_l's, i.e. to evaluate the sum

$$I(z,\vartheta,\lambda)=\frac{1}{2\pi}\sum_l\left(l+\frac{1}{2}\right)I_l(z,\lambda)\,P_l(\vartheta). \tag{17.18}$$

For a *plane perpendicular* source this problem can be treated in a similar manner to the degradation and deflection problem of Sect. 10β. Thus we separate the low orders of scattering and write

$$\left.\begin{aligned}I(z,\vartheta,\lambda) &= I^{(0)}(z,\vartheta,\lambda)+I^{(1)}(z,\vartheta,\lambda)+I^{(\bar 2)}(z,\vartheta,\lambda)+\\ &+\frac{1}{2\pi}\sum_{l=0}^{l_{\max}}\left(l+\frac{1}{2}\right)I_l^{(\bar 2+)}(z,\lambda)\,P_l(\vartheta)\end{aligned}\right\} \tag{17.19}$$

where $I^{(0)}(z,\vartheta,\lambda)$, $I^{(1)}(z,\vartheta,\lambda)$, and $I^{(\bar 2)}(z,\vartheta,\lambda)$ are given by (8.19). Thereby we express analytically the discontinuities of the first few orders of scattering. The Legendre coefficients $I_l^{(\bar 2+)}(z,\lambda)$ of the residual angular distribution are defined as

$$I_l^{(\bar 2+)}(z,\lambda)=I_l(z,\lambda)-I_l^{(0)}(z,\lambda)-I_l^{(1)}(z,\lambda)-I_l^{(\bar 2)}(z,\lambda), \tag{17.20}$$

where $I_l^{(0)}$, $I_l^{(1)}$, and $I_l^{(\bar 2)}$ are Legendre coefficients of the corresponding distributions (8.19). Instead of calculating these coefficients it is preferable to obtain $I_l^{(\bar 2+)}(z,\lambda)$ directly by the polynomial method from its moments

$$I_{nl}^{(\bar 2+)}(\lambda)=I_{nl}(\lambda)-I_{nl}^{(0)}(\lambda)-I_{nl}^{(1)}(\lambda)-I_{nl}^{(\bar 2)}(\lambda). \tag{17.21}$$

The difference $I_{nl}(\lambda)-I_{nl}^{(0)}(\lambda)$ on the right of this equation coincides with the moment $I_{nl}^{(s)}$ defined in (16.4) and calculated by the method of Sect. 16. The moments $I_{nl}^{(1)}(\lambda)$ and $I_{nl}^{(\bar 2)}(\lambda)$ are obtained from the interlinked equations

$$\mu(\lambda)\,I_{nl}^{(1)}(\lambda)=\frac{n}{2l+1}\left[(l+1)\,I_{n-1\,l+1}^{(1)}(\lambda)+l\,I_{n-1\,l-1}^{(1)}(\lambda)\right]+\frac{k_l(\lambda,\lambda_0)}{\mu(\lambda_0)^{n+1}}\,n!, \tag{17.22}$$

$$\left.\begin{aligned}\mu(\lambda)\,I_{nl}^{(\bar 2)}(\lambda) &= \frac{n}{2l+1}\left[(l+1)\,I_{n-1\,l+1}^{(\bar 2)}(\lambda)+l\,I_{n-1\,l-1}^{(\bar 2)}(\lambda)\right]+\\ &+\frac{1}{2\pi}I_l^{(\bar 2)}(\lambda)\,\mu(\lambda)\,\frac{\left[\dfrac{1}{\mu(\lambda_0)}\right]^{n+1}-\left[\dfrac{1-\frac{1}{2}(\lambda-\lambda_0)}{\mu(\frac{1}{2}[\lambda+\lambda_0])}\right]^{n+1}}{\dfrac{1}{\mu(\lambda_0)}-\dfrac{1-\frac{1}{2}(\lambda-\lambda_0)}{\mu(\frac{1}{2}[\lambda+\lambda_0])}}\,n!,\end{aligned}\right\} \tag{17.23}$$

where $I_l^{(2)}(\lambda)$ is given by (10.12). Sample angular distributions calculated by these methods are presented in Fig. 19, for a 0.66 Mev source in water. Note the irregular features of the distribution due to singly and doubly scattered flux which are superimposed on the smooth distribution of multiply scattered flux.

18. Construction of distributions from moments: Function-fitting method. α) *Introduction.* The range of problems to which the method of polynomial expansion can be applied is limited by the rather stringent requirements for specifying the weight-function. In the problem treated in Sect. 17 (plane perpendicular source) the choice of a suitable weight-function was obvious; but if the source were plane oblique, for example, rather than plane perpendicular, the choice would be less obvious; in particular, it would be difficult to select the proper coefficient (taking the place of μ_s) in the exponential weight-function (see Sects. 5 ε and 11 δ). Another method has therefore been developed which is applicable in a wider variety of circumstances because it requires a less completely specified weight-function. Only the functional form of the weight-function, but not the numerical values of the parameters occuring in it, need be specified. The method has been called function-fitting because it leads to the representation of the flux distribution function as a sum of functions. For example, a spatial distribution may be represented as a sum of exponentials with different coefficients.

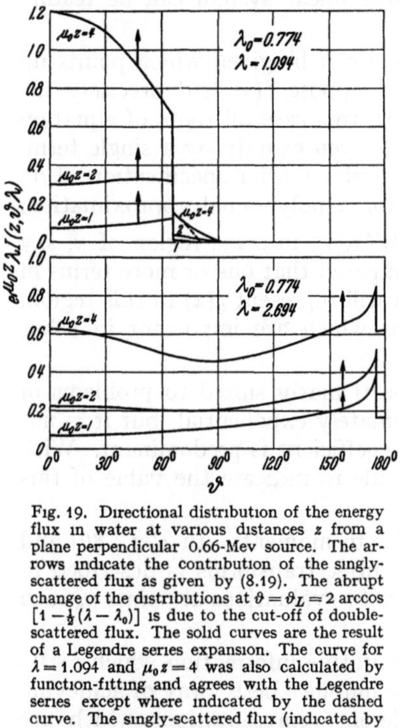

Fig. 19. Directional distribution of the energy flux in water at various distances z from a plane perpendicular 0.66-Mev source. The arrows indicate the contribution of the singly-scattered flux as given by (8.19). The abrupt change of the distributions at $\vartheta=\vartheta_L=2\arccos[1-\frac{1}{2}(\lambda-\lambda_0)]$ is due to the cut-off of double-scattered flux. The solid curves are the result of a Legendre series expansion. The curve for $\lambda=1.094$ and $\mu_0 z=4$ was also calculated by function-fitting and agrees with the Legendre series except where indicated by the dashed curve. The singly-scattered flux (indicated by arrows) and the discontinuous part of the double-scattered flux were calculated analytically [see Eq. (17.19)].

As the function-fitting method can be used when the characteristic numerical parameters are not moments, we shall present it in a rather general form. Consider a function $f(x)$ to be determined from a set of known numerical parameters f_n which result from a sequence of operations O_n applied to $f(x)$. To obtain a representation of $f(x)$ we choose a weight function $W(x,\beta)$, dependent on an adjustable parameter β. We then write

$$f(x)=\sum_i \alpha_i W(x,\beta_i) \tag{18.1}$$

with the intention of determining the β_i as well as the α_i from the equations

$$f_n=\sum_i \alpha_i O_n W(x,\beta_i), \quad n=0,1,2,\ldots. \tag{18.2}$$

In general, Eqs. (18.2) are non-linear and difficult to solve; this problem can be circumvented, however, by choosing W such that $O_n W$ factors in a particular way, namely $O_n W=V_n V_i' \beta_i^n$. With such a choice, (18.2) becomes

$$\bar{f}_n=\sum_i \bar{\alpha}_i \beta_i^n \tag{18.3}$$

where $\bar{\alpha}_i=\alpha_i V_i'$, $\bar{f}_n=\frac{f_n}{V_n}$. These equations, while non-linear, are nevertheless easily solved (see [22]). If n runs from zero to $(2N-1)$, the β_i are roots of the

N-th degree algebraic equation

$$\begin{vmatrix} \beta^N & \bar{f}_N & \bar{f}_{N+1} & \cdots & \bar{f}_{2N-1} \\ \vdots & & & & \\ \beta^2 & \bar{f}_2 & \bar{f}_2 & \cdots & \bar{f}_{N+1} \\ \beta & \bar{f}_1 & \bar{f}_2 & \cdots & \bar{f}_N \\ 1 & \bar{f}_0 & \bar{f}_1 & \cdots & \bar{f}_{N-1} \end{vmatrix} = 0. \tag{18.4}$$

Once the β_i have been evaluated, the remaining linear system can be readily solved to obtain the $\bar{\alpha}_i$ and from them the α_i.

An interpretation of the representation (18.1) can be given which points up the contrast with polynomial representations: Suppose $f(x)$ had precisely the analytic form $W(x, \beta)$, for some value $\beta = \beta'$. In this case all roots of equation (18.4) would be identical, so that $f(x)$ would be given exactly by a single term. Note that this exact result would be obtained without prior specification of β', while such a specification would be required for a polynomial approximation.

More commonly, $f(x)$ may have the form $W(x, \beta')$ in some region of x, say for x asymptotically large. It can then be anticipated that one or more terms in the sum (18.1) will have parameters $\beta_i \sim \beta'$ and will represent $f(x)$ in this region, while the other terms constitute modifications which are important in other regions of x.

One can now see why function fitting is particularly suited to problems of X-ray penetration. The attenuation is approximately exponential, but it is not always apparent in advance which attenuation coefficient is predominant. With function fitting the moments themselves are made to indicate the value of this important parameter.

Furthermore, the operators O_n need not refer to moments. In Ref. [*80*] and in Sect. 21 an application of the method to Fourier transforms is described in which $W(x, \beta)$ has the form $e^{\{\beta y(x)\}}$ and O_n refers to evaluations at points x_n such that $y(x_n) - y(x_{n-1})$ is a constant[1].

It is necessary to take into account the qualitative information which exists regarding the unknown function; but there is no unique "best" way of achieving this. Different weight functions lead to different representations which are equally valid if the same amount of information is used. (For an example see Table 12.)

There are restrictions on β which depend on the specific form of $W(x, \beta)$. For example, in Eqs. (18.5) below, only real positive values have been regarded as acceptable. Often the solution of Eq. (18.4) will yield values which are unacceptable, being perhaps negative, imaginary, or too widely separated. The corresponding representations must almost always be rejected as not consistent with known distribution characteristics. Such a situation is evidence that there is something wrong with the weight-function, and further study will often corroborate this. Considerable searching may be necessary to obtain an acceptable weight function, but when there are several such functions they lead to representations that are in close agreement. Thus the function-fitting technique

[1] Another example is given in Ref. [*81*], in which the O_n are combinations of derivatives evaluated at a particular value of x. This is a logical extension of the application of moments, since the moments of a function may be considered derivatives of its transform evaluated for $p = iq = 0$.

appears to work in an all-or-nothing fashion and is usually accurate when applicable[1].

β) Spatial distributions. As in Sect. 17γ, we consider the flux distribution $I(z, \omega_z, \lambda)$ to be separated into symmetric and antisymmetric components: $I^S(z, \omega_z, \lambda)$ and $I^A(z, \omega_z, \lambda)$. These are in turn described by their Legendre coefficients $I_l^S(z, \lambda)$, $I_l^A(z, \lambda)$, which are functions of positive z only and are characterized by the moment systems (17.12). Using an exponential weight function we write

$$\left.\begin{matrix} I_l^S(z, \lambda) \\ I_l^A(z, \lambda) \end{matrix}\right\} = \sum_i \alpha_{i\,l}\, e^{-\frac{\mu_0 z}{\beta_{i\,l}}}. \tag{18.5}$$

By equating corresponding moments on both sides of Eqs. (18.5), sets of equations are obtained having the form (18.3), whose solution yields the α's and β's.

Table 12. *Results of moment-fitting by different methods.*

The tabulated quantity is $\exp(\mu_0 z)\,\lambda I_0(z, \lambda)$ at the wavelength $\lambda = 0.181$ and various distances z from a plane perpendicular 10-Mev source in lead. In all cases the symmetric and antisymmetric components of $\lambda I_0(z, \lambda)$ have been fitted separately with the use of four moments each. The behavior of the first two function-fitting representations is governed at large penetration distances by scale parameters $\beta = 1.21$ and $\beta = 1.19$, respectively, which correspond closely to the ratio $\mu_0/\mu_m = 1.20$. The Gaussian weight function was used to indicate the consequences of choosing an inappropriate weight function.

	Polynomial expansion	Function-fitting		
		Weight function		
$\mu_0 z$	$e^{-\mu_0 z}$	$e^{-\mu_0 z/\beta}$	$\left(1+\frac{\mu_0 z}{\beta}\right) e^{-\mu_0 z/\beta}$	$e^{-(\mu_0 z/\beta)^2}$
−10	0.009	0.001	−0.005	3.59
−7	0.025	−0.0003	−0.005	−0.091
−4	−0.007	−0.0003	−0.002	−0.21
−2	−0.008	−0.0001	−0.001	−0.015
0	−0.013	−0.0004	0.008	−0.014
2	0.205	0.194	0.191	0.176
4	0.515	0.523	0.528	0.352
7	1.42	1.43	1.42	1.41
10	3.29	3.20	3.18	5.28

For some spectral components the exponential weight function may not be acceptable, but greater flexibility in fitting can be conveniently achieved by multiplying the exponential by a suitable low order polynomial. A convenient type of weight function is

$$e^{-x} x^n \sum_{j=0}^{n} \frac{(n+j)!}{n!\,(n-j)!}\,\frac{1}{(2x)^j} = e^{-x} x^n S_{n+\frac{1}{2}}(2x) = x^{n+\frac{1}{2}} K_{n+\frac{1}{2}}(x) \tag{18.6}$$

where $K_{n+\frac{1}{2}}$ is the Bessel function mentioned on p. 744. The first n odd powers of x in the expansion of this function vanish.

Table 12 compares values of $\lambda I_0(z, \lambda)$ obtained using (18.5) and (18.6) with the corresponding polynomial approximations for a 10-Mev source in lead. The agreement is good up to very deep penetrations, where the function-fitting results are somewhat more reliable.

[1] By contrast, a representation by polynomial expansions is always possible but may be unsuspectedly poor if some characteristic such as a sudden rise or discontinuity of the unknown function has not been included in the weight function. The apparent convergence of the first few terms of an expansion is not necessarily a reliable indicator of the quality of the representation.

γ) Angular distributions. In the calculation of the directional distributions, function fitting has particular advantages for photons which have not lost much energy ($\lambda - \lambda_0 \ll 1$). The Legendre polynomial representation of these distributions (with weight function unity) converges slowly even after the separation and exact treatment of the first few orders of scattering (see Sect. 10β).

We base our calculations on the Legendre coefficients $I_l^{(2+)}(z, \lambda)$ [as defined by Eq. (17.20)] which may be obtained either by polynomial expansion or by function fitting in the space variable. Linear combinations of $I_l^{(2+)}(z, \lambda)$, called angular moments, are most useful for function fitting. They are defined by

$$M_m^{(2+)}(z, \lambda) = 2\pi \int_0^2 d(1-\omega_z)\,(1-\omega_z)^m\, I^{(2+)}(z, \omega_z, \lambda).$$

The first few are

$$\left.\begin{aligned} M_0^{(2+)} &= I_0^{(2+)}\\ M_1^{(2+)} &= I_0^{(2+)} - I_1^{(2+)}\\ M_2^{(2+)} &= \tfrac{4}{3} I_0^{(2+)} - 2 I_1^{(2+)} + \tfrac{2}{3} I_2^{(2+)}\\ M_3^{(2+)} &= 2 I_0^{(2+)} - \tfrac{18}{5} I_1^{(2+)} + 2 I_2^{(2+)} - \tfrac{2}{5} I_3^{(2+)} \text{ etc.}\end{aligned}\right\} \tag{18.7}$$

Perhaps the most obvious representation of $I^{(2+)}$ involves the functions $\exp\left(-\frac{1-\omega_z}{\beta_i}\right)$; however, the known slope discontinuity of thrice-scattered radiation at $\omega_z = \cos\left[3 \arccos \frac{\lambda - \lambda_0}{3}\right]$ suggests a trapezoidal representation

$$\left.\begin{aligned} I^{(2+)} &= \sum_i \frac{\alpha_i}{\beta_i}\left[1 - \frac{(1-\omega_z)}{\beta_i}\right] && \text{if } \omega_z \geqq 1 - \beta_i,\\ &= 0 && \text{if } \omega_z < 1 - \beta_i \end{aligned}\right\} \tag{18.8}$$

with $\beta_0 = 1 - \cos\left[3 \arccos \frac{\lambda - \lambda_0}{3}\right]$. The other parameters are determined by function-fitting from the angular moments $M_n^{(2+)}$. If one starts with N equations of the type (18.3) one can by pairwise subtraction obtain $N-1$ new equations from which α_0 has been eliminated, and which can again be put into the form (18.3).

As illustrated in Fig. 19, there is close agreement between the results of function-fitting and of polynomial representations. Only at large angles is there a noticeable discrepancy. Function fitting according to (18.8) describes the slope change of the angular distribution more realistically, but is incorrect in artifically cutting off the distribution.

19. Applications of the moment method and comparisons with experiments. Extensive calculations have been done for *plane perpendicular, plane isotropic* and *point isotropic* sources, but only pilot calculations for *plane oblique* and *point collimated* (gun) sources[1]. Spatial distributions and energy spectra have received much attention, whereas angular distributions have been computed only in isolated instances.

The experimental coverage reflects to some extent the state of the calculations. A summary description of various experiments is given in Table 13. The most

[1] The development of the theory for the point collimated source is recent and not yet quite complete. Only energy dissipation distributions but no spectra have so far been calculated. The discussion in subsection 19ε is therefore preliminary in nature. *Note added in proof.* Systematic calculations for plane oblique sources have recently been made by L. V. Spencer and J. Lamkin, U.S. Nat. Bu. of Standards Report 5944, July 17, 1958 (unpublished), and additional ones are in progress.

accurate work has been done with point isotropic radioactive sources surrounded by attenuating material. Both spectra and integral fluxes have been carefully measured as functions of the distance from the source. Experimental data for other source types, and for angular distributions, are more limited. Thus the

Table 13. *Survey of experiments.*

Ref. No.	Author	Source	Source energy E_0 (Mev)	Geometry	Medium	Type of information	Measuring device
[*1*]	BEACH *et al.*	Cs^{137}	0.661	PLP PTI	H_2O	$I(r, \lambda)$	NaI(Tl) crystal spectrometer
[*2*]	BEACH *et al.*	Na^{24}	2.76	PTI	H_2O Hg	$I_D(r)$	Anthracene crystal spectrometer
[*26*]	ELLIOT *et al.*	Co^{60}	1.33 1.17	PTI	Pb	$I_D(r)$	X-ray film
[*32*]	FAUST	Co^{60}	1.33 1.17	PTI PLI	H_2O	$I(r, \lambda)$ $I(z, \lambda)$	Geiger counter; estimates of spectra by Pb filtration
[*33*]	FAUST and JOHNSON	Co^{60}	1.33, 1.17		H_2O	$I(\lambda)$	Geiger counter
[*36*]	GARRETT and WHYTE	Co^{60}	1.33, 1.17	PTI	Pb Fe	$I_d(r)$	Ionization chamber
[*42*]	HAYWARD	Co^{60}	1.33, 1.17	PTI	H_2O	$I_D(r)$	Anthracene crystal spectrometer
[*77*]	ROYS *et al.*	N^{16}	6.2	PTI	H_2O	$I_D(r)$	NaI(Tl) crystal spectrometer
[*94*]	WEISS and BERNSTEIN	Co^{60} Hg^{203}	1.33, 1.17 0.279	PTI	H_2O	$I(r, \lambda)$ $I(z, \lambda)$	Ionization chamber and Geiger counter
[*95*]	WHITE	Co^{60}	1.33, 1.17	PTI	H_2O	$I_d(r)$	Ionization chamber and Geiger counter
[*97*]	WHYTE	Co^{60}	1.33, 1.17	PTI	concrete	$I(r, \omega_r, \lambda)$	NaI(Tl) crystal spectrometer
[*101*]	ZENDLE *et al.*	Bremsstr. spectrum	Top energy 37 Mev	PLP	H_2O	$I_d(z)$	Anthracene crystal spectrometer and ionization chamber
[*54*]	KIRN *et al.*	Co^{60} Cs^{137} Au^{198}	1.33, 1.17 0.661 0.411	PLP PLO	Pb, concrete and polyethylene	$I_d(z)$	Ionization chamber
[*50*]	HUBBELL *et al.*	Bremsstr. spectrum	Top energies 8 and 10 Mev	PLP	Pb	$I(z, \vartheta, \lambda)$	NaI (Tl) crystal spectrometer

Abbreviations: PLP = Plane perpendicular; PLO = Plane oblique; PLI = Plane isotropic; PTI = Point isotropic.

comparisons between theory and experiment cover so far only part of the range of basic situations. Satisfactory agreement has been the outcome of these comparisons, however, and no outstanding discrepancy has been found.

In this section theory and experiment will be discussed jointly and compared for each problem where this is possible. The normalization of the figures and tables is consistent with the conventions of the Appendix.

α) Plane perpendicular source. Typical spectra generated by 0.5-Mev and 3-Mev sources in a low-Z material (water) are shown in Figs. 20a and b. Figs. 20c

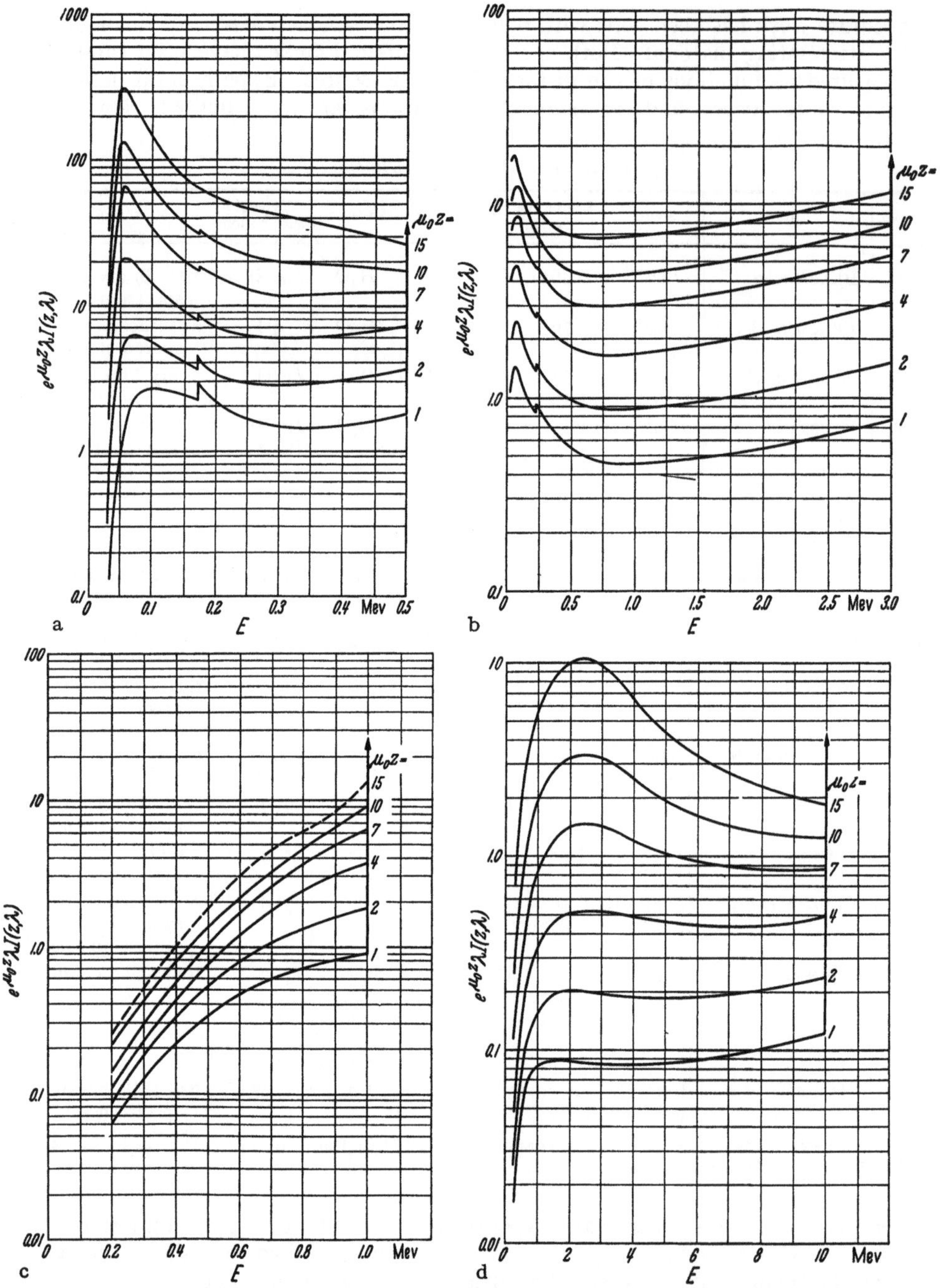

Fig. 20 a—d. Spectral distributions of the energy flux at various distances z from a plane perpendicular source. The arrows indicate the unscattered component $e^{\mu_0 z}\lambda I^{(0)}(\lambda, z) = \lambda_0 \delta(\lambda - \lambda_0)$, (Goldstein and Wilkins, Ref. [41]): (a) 0.5-Mev source in water. (b) 3-Mev source in water. (c) 1-Mev source in lead. (d) 10-Mev source in lead.

and d contain spectra in a high-Z material (lead) generated by 1-Mev and 10-Mev sources. (In lead, a source energy of 1 Mev is well below, and a source energy of 10 Mev well above the energy E_m of minimum attenuation.)

The low-energy components of the flux in lead are greatly reduced in comparison with the corresponding components in water, due to increased photo-electric absorption. The spectra in water have a peak at about 0.05 Mev which is caused by the interplay of photo-electric absorption and the accumulation of soft scattered photons. The spectral shape *at low energies* changes little with increasing penetration; this is an illustration of the quasi-equilibrium discussed in Sect. 5γ. But *near the source energy* there is a continuing change of the spectra with increasing penetration depth such that the relative importance of the high-energy components is reduced.

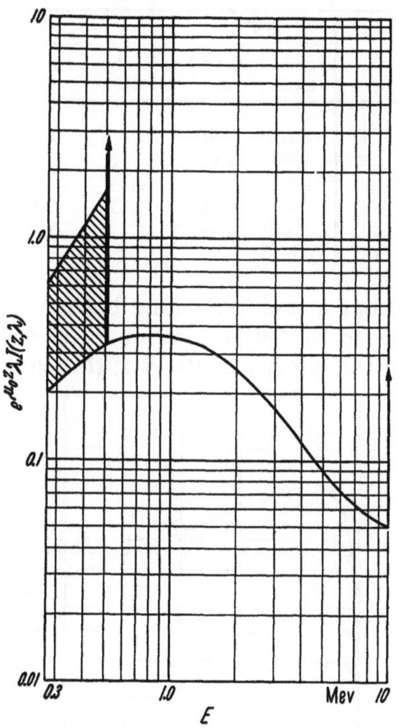

Fig. 21. Contribution of scattered annihilation radiation (shaded area) to the flux at a distance $\mu_0 z = 4$ from a 10-Mev plane perpendicular source in lead. (Ref. [9].)

The spectra in Fig. 20c exhibit a peak in the neighborhood of 2 Mev which is correlated with the minimum of the attenuation coefficient near 3 Mev. Such a peak is characteristic of spectra which include the energy E_m of minimum attenuation and is further discussed in Sect. 21.

In the calculations whose results are shown in Figs. 20a — d, secondary radiation in the form of bremsstrahlung, annihilation quanta and fluorescence radiation has been neglected. Little is known about the bremsstrahlung contribution; a rough estimate indicates that in lead approximately 6% of the energy of a 5-Mev source and 15% of the energy of a 10-Mev source go into bremsstrahlung[1].

Table 14. *Percentage increase of the energy dissipation I_d due to annihilation radiation, at various distances from plane perpendicular sources* (GOLDSTEIN [39]).

Material	E_0 (Mev)	$\mu_0 z=1$ %	$\mu_0 z=7$ %	$\mu_0 z=15$ %
H_2O	10	(3.9)	(3.4) (estimated)	(3.3)
	8	3.5	3.0	2.8
	6	3.3	2.6	2.4
Fe	10	7.5	6.2	5.4
	8	7.2	5.8	4.9
	6	6.6	4.8	4.1
Pb	10	6.1	5.3	4.2
	8	6.1	5.0	3.9
	6	5.9	4.6	3.8

The spectrum of annihilation radiation has been investigated in some detail [9], [39]. It has an upper limit and sharp peak at 0.511 Mev. Fig. 21 shows how it contributes to the spectrum from a 10-Mev source in lead. Table 14 shows the percentage contributions of the annihilation radiation to the exposure dose in various media. The size of the contribution does not depend strongly on the atomic number of the medium, and is typically of the order of 5 to 6% for source energies up to 10 Mev. Fluorescence radiation in lead has been estimated to increase the exposure dose by approximately 2% for a 0.05-Mev source [39].

Table 15 contains values of the energy build-up factor B_E. The source energies and media correspond to those of the spectra in Figs. 20a — d. In all cases the

[1] S.T. COHEN and E.H. PLESSET: Rand Report, RM-39, May 20, 1948.

build-up factor is a monotonically increasing function of the penetration depth. The build-up for a 0.5-Mev source in water is particularly rapid for two reasons: scattered components of the flux are not absorbed except at very low energies, and scattering causes only a small fractional reduction of the photon energy. For a 10-Mev source in lead, the build-up is slow until an appreciable amount of scattered 3-Mev components accumulate and dominate the further penetration. The characteristic features of the build-up are further illustrated by Fig. 22 which indicates by means of a contour diagram the dependence of the energy dissipation build-up factor on the source energy, atomic number of the medium and penetration distance.

Table 15. *Energy build-up factor B_E for plane perpendicular sources* (GOLDSTEIN and WILKINS [41]).

$\mu_0 z$	H_2O $E_0=0.5$ Mev	H_2O $E_0=3.0$ Mev	Pb $E_0=1.0$ Mev	Pb $E_0=10.0$ Mev
1	2.75	1.58	1.89	1.17
2	4.49	2.09	2.68	1.36
4	9.11	3.16	4.45	1.77
7	20.2	4.79	7.57	2.51
10	34.5	6.44	11.2	3.44
15	70.6	9.23	18.2	5.48

Fig. 23 compares calculated and experimental values of the energy absorbed in water from secondary electrons resulting from a bremsstrahlung beam with an upper limit of 40 Mev. The calculation, due to BRYSK [19], involves the following steps: (a) calculation of the moments of the X-ray distribution; (b) deter-

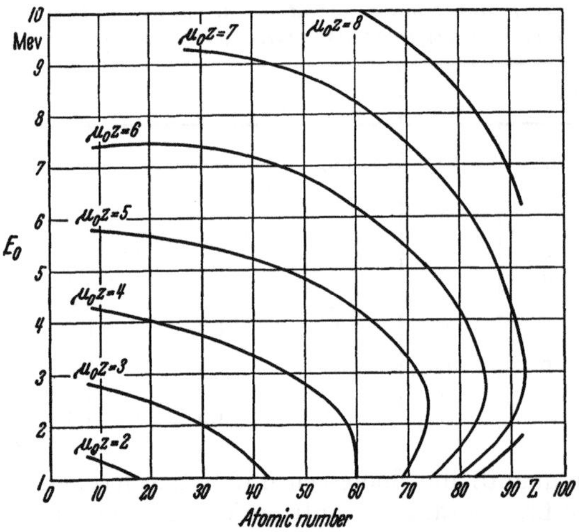

Fig. 22.

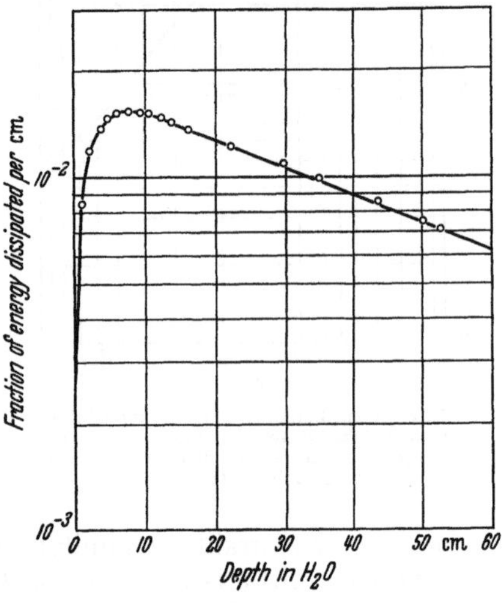

Fig. 23.

Fig. 22. Dependence of penetration on source energy and material The contour lines for given penetration distances z indicate the source energy E_0 and the atomic number Z for which the exposure dose build-up factor $B_d(z)$ is 3 for plane perpendicular sources. (Based on [41].)

Fig. 23. Distribution-in-depth of energy absorption from a betatron X-ray beam. Solid curve: calculation for 40 Mev peak energy, BRYSK [19]. Points· experiments with 37 Mev peak energy, ZENDLE *et al.* [101].

mination, with the use of these moments, of the source distribution of fast electrons; (c) solution of the electron transport equation to obtain the moments of the electron spatial distribution[1]; (d) integration of these moments over the electron energy spectrum weighted by the electron stopping power of water, to obtain moments of the distribution of absorbed energy; and (e) construction, with the

[1] Cf. H.W. LEWIS: Phys. Rev. **78**, 529 (1950). The moments were obtained numerically by an integration procedure similar to that used for the X-ray flux.

use of these moments, of a polynomial representation of the distribution. The weight function $e^{-\mu_0 z} - e^{-Az}$ was used, the parameter A being determined from the moments[1].

The experimental results in Fig. 23, due to ZENDLE *et al.* [101] were obtained by the combination of two series of measurements: (1) the main experiment consisted of the deter-

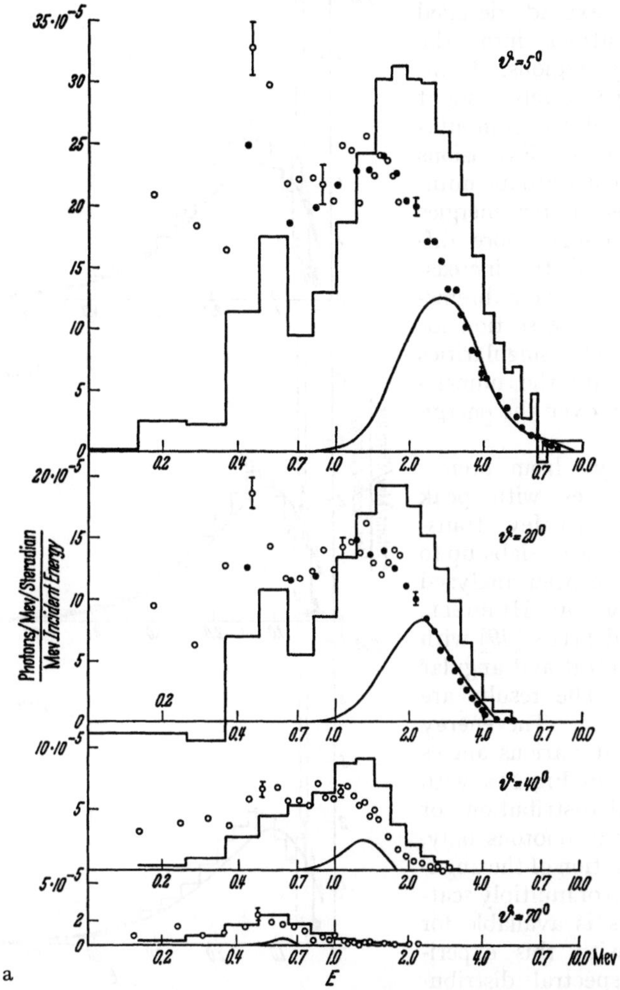

Fig. 24 a and b. Spectra of X rays transmitted by Pb barriers; plane perpendicular bremsstrahlung sources, HUBBELL *et al.* [50]. (a) Spectra observed in various directions; 6" barrier, 10 Mev peak source energy. Solid and open circles: NaI(Tl) pulse-height distributions in different runs. Histograms: Photon spectra obtained after correction for detector response. Solid curves Calculated spectra of singly-scattered photons.

mination of the energy absorption along the axis of a broad but finite beam of bremsstrahlung; (2) auxiliary measurements, also along the beam axis, were made with annular beams. On the basis of these, the results of the main experiment could be extrapolated to the case of

[1] The spatial distribution of the electrons is dominated by the distribution of the X-ray flux, having the same shape but a small displacement away from the source. Only close to the source a transition effect occurs that is associated with the build-up of the secondary electrons. In cases where energy transport by electrons is not a secondary process but constitutes a major effect, the moment-fitting problem is more difficult and different procedures must be followed. Cf. L.V. SPENCER: Phys. Rev. **98**, 1597 (1955).

an infinitely broad beam such as is assumed in BRYSK's calculation. The extrapolated ionization chamber measurements were carefully related to energy absorption through calibrations with a calorimeter, so that *absolute* values were obtained of the energy absorption in a gram of water per unit source energy. The agreement with the theory is excellent, in the transition region (caused by the build-up of secondary electrons) as well as at great penetration depths.

BRYSK's work is the first attempt to extend detailed X-ray calculations into the cascade energy regions. It involves comparatively slight modifications of the computational techniques. Extensions to media of higher atomic number and higher source energies are feasible though more difficult, because of the increasing role of secondary bremsstrahlung; the cross section for this process has singularities which complicate the numerical integration over the energy variable [67].

The X rays from bremsstrahlung sources with peak energies up to 10 Mev, transmitted through lead slabs up to six inches, have been analyzed experimentally by HUBBELL, HAYWARD and TITUS [49] with regard to spectral and angular distributions. The results are shown in Fig. 24. The energy distributions at various angles are compared in Fig. 24a with the calculated distribution for singly scattered photons only, since no calculation of the angular distribution of multiply scattered X rays is available for comparison with this experiment. The spectral distributions, obtained by numerical integration over all directions of emergence, are compared in Fig. 24b with the calculated distributions from Ref. [41], modified to include the annihilation radiation according to Ref. [9]. Both experimental and theoretical spectra peak at about 1 Mev, i.e. substantially below the energy of minimum attenuation. The excess of the experimental results over the calculated data in the upper energy range may be attributed to the neglect, in the theory, of secondary bremsstrahlung generated in the lead absorber. The deficiency of photons observed in the lowest energy range may be attributed to the presence of the exit surface, which was also neglected in the calculation.

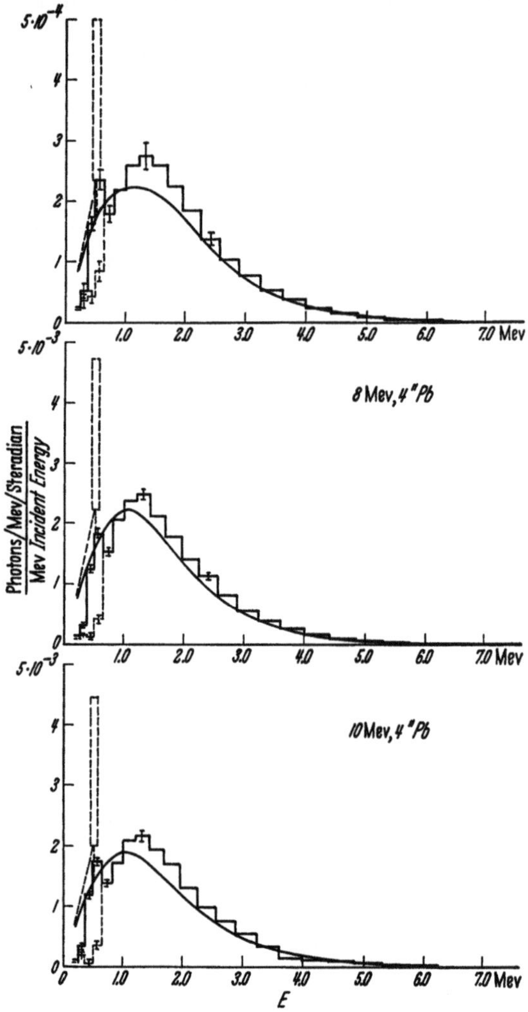

Fig. 24 b. Spectra integrated over angular distribution. Solid histograms: experimental data. Solid curves: calculation for infinite medium Dotted rectangle and curve: calculated spectra of unscattered and scattered annihilation X rays emerging from finite barrier. Dotted histogram: Experimental data minus calculated contribution of annihilation radiation.

β) Point isotropic source. Fig. 25 presents point-isotropic build-up factors for Co^{60} radiation (1.17 and 1.33 Mev) in water, iron and lead. The corresponding experimental points represent ionization chamber measurements or equivalent information obtained by G. R. WHITE [*95*] and C. GARRETT and G. N. WHYTE [*36*][1]. In order to obtain the experimental build-up factors, one must take the ratio of the total ionization to the ionization produced by unscattered flux. The latter quantity is difficult to measure directly, and must be estimated indirectly through a calibration of the source strength. This calibration constitutes one of the chief difficulties of the experiments.

From a different point of view, the experimental data in Fig. 25 may be considered as a very accurate determination of the attenuation coefficients at 1.33 Mev because the experimental build-up factor at deep penetrations is extremely sensitive to small changes in the assumed value of μ_0. According to GARRETT and WHYTE, the discrepancies between the theory and their results, which are of the order of 5% and perhaps twice as large as the experimental accuracy, could be made to vanish if the attenuation coefficient at 1.33 Mev were increased by 0.4% for iron, and decreased by 0.2% for lead, compared to the values in the standard tabulation in Ref. [*96*].

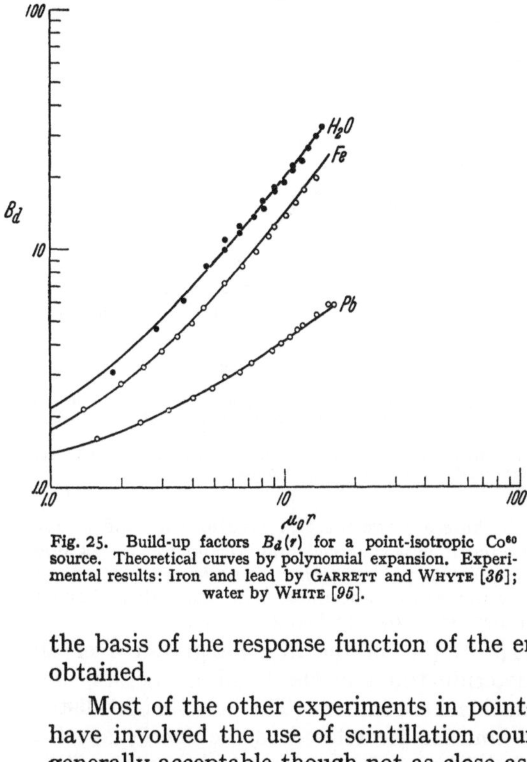

Fig. 25. Build-up factors $B_d(r)$ for a point-isotropic Co^{60} source. Theoretical curves by polynomial expansion. Experimental results: Iron and lead by GARRETT and WHYTE [*36*]; water by WHITE [*95*].

Another experiment in point-isotropic geometry with Co^{60} radiation in lead has been carried out by J. O. ELLIOT *et al.* [*26*] who used photographic films as detectors. Measured build-up factors were compared with those calculated on the basis of the response function of the emulsion, and excellent agreement was obtained.

Most of the other experiments in point-isotropic geometry listed in Table 14 have involved the use of scintillation counters. The agreement with theory is generally acceptable though not as close as in Fig. 25. The occasional discrepancies appear to be correlated with difficulties in matching experimental and theoretical conditions.

In order to trace the influence of the source geometry, a plot is presented in Fig. 26 of the energy flux build-up factor ratio

$$\frac{B_E^{PTI} \quad \text{(point isotropic)}}{B_E^{PLP} \quad \text{(plane perpendicular)}}.$$

[1] G. R. WHITE used Geiger counters to extend her measurements from 11 to 16 mean free paths. The Geiger counter measurements were normalized at 11 mean free paths to cavity chamber results. The composite set of experimental values thus obtained is representative of cavity chambers because the spectrum is mostly in quasi-equilibrium beyond 11 mean free paths. GARRETT and WHYTE used sources of different strengths and chambers of different sizes to cover the whole range of penetration.

The ratio is higher than one for great penetrations and smaller very close to the source. This general phenomenon follows from geometric considerations[1]. For extremely great penetrations the build-up factor ratio tends to a limiting value which is represented by $\frac{1}{2} A_0 (\dot{\mu}_0/\mu_0)$ according to Sect. 11 δ.

In the course of multiple scattering, photons become polarized even when the source radiation is unpolarized. As indicated in Sect. 5ζ, this results in a shift of the center of gravity of the flux distribution away from the source. An accurate calculation has been made to verify this effect and determine its magnitude in point-isotropic geometry [86]. Of the four Stokes parameters describing the polarized radiation, the total intensity $I^{[0]}(r, \lambda)$ was of prime interest. In an isotropic medium (with random orientation of the electron spins) the circular polarization parameter $I^{[3]}$ remains identically zero, and $I^{[2]}$ vanishes because of symmetry. The moment equations derived from (6.17) for a point isotropic source are similar

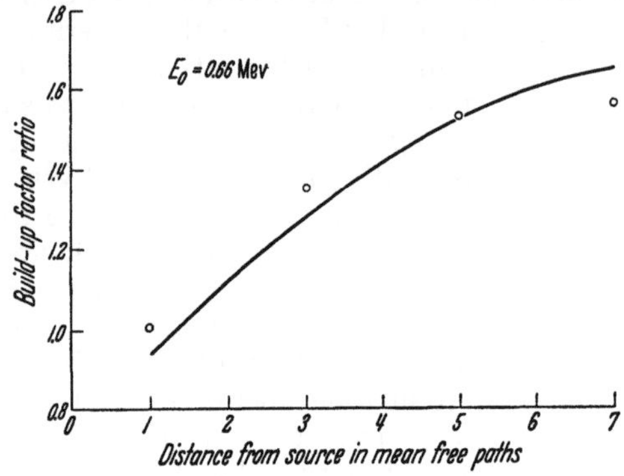

Fig. 26.

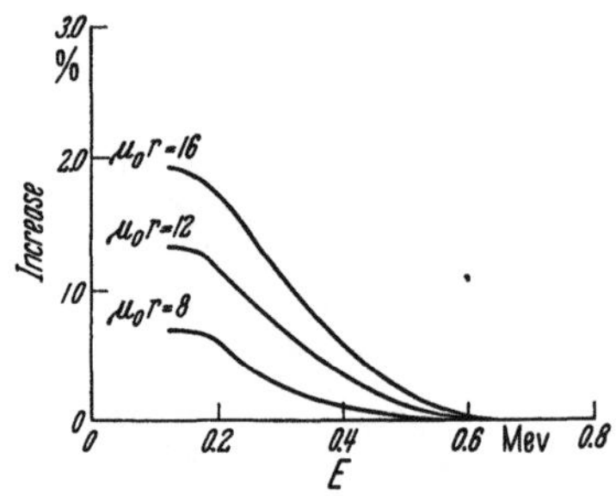

Fig. 27.

Fig. 26. Ratio of the energy build-up factors in point-isotropic and plane-perpendicular geometry; 0.66-Mev source in water. Theoretical curve by interpolation from [41]; experimental points from BEACH *et al.* [1].

Fig. 27. Increase, due to polarization, of the flux $I^{[0]}(r, \lambda)$ from a point-isotropic 1.28-Mev source in water. The curves level off when the flux becomes nearly isotropic at low energies [86].

to (12.32), but for each pair of indices (n, l) there are two integral equations instead of one which couple $I_{nl}^{[0]}$ and $I_{nl}^{[1]}$. This system of equations was solved numerically for a 1.28 Mev source in water. A comparison calculation neglecting polarization effects was also carried out according to (12.32), with exactly the same numerical procedure. The small flux changes due to polarization, determined in this manner, are shown in Fig. 27.

Measurements of X-ray spectra $I(r, \lambda)$ (i.e. of the flux integrated over all directions) are more difficult than determinations of the build-up factors, and have usually been carried out with scintillation counters. The pulse-height distributions obtained with a scintillation counter are related to the energy spectra according to the integral equation (28.1). As an alternative to solving this equation, the theoretical results can be weighted in proportion to the detector efficiency for a direct comparison with the "raw" data. This has been done to obtain a comparison with the spectral measurements of HAYWARD [42] for Co^{60} radiation in water (Fig. 6). Her results, made absolute through calibration of the source strength, consist of spectra of photon-ejected electrons in an anthracene crystal. WEISS and BERNSTEIN [94] made spectral measurements on Co^{60} radiation in water with a NaI crystal. The response function of their spectrometer was sufficiently peaked so that at low energies the experimental pulse height distribution was practically identical with the X-ray spectrum. They were able to demonstrate the following characteristics of the radiation predicted by the theory: (1) There is a peak at about 0.05 Mev (corresponding to the theoretical

[1] For a discussion of this effect, see R. ARONSON: Nuclear Development Associates Memo 15C-37 (1954).

curves in Fig. 20). (2) The shape of the low-energy part of the spectrum is almost independent of the source-detector distance and of the source energy.

A few angular distributions have been calculated for the low-energy components of the flux generated by Co^{60} radiation in water [*85*] and are compared in Fig. 28 with experimental data. The experimental conditions are different enough so that the comparison is rough, but the shapes of the experimental distributions are consistent with the theory.

γ) Plane isotropic source. The fluxes from plane and point isotropic sources are closely related as discussed in Sect. 9, p. 693 ff. and elsewhere. In particular, the ratio of build-up factors for these two sources approaches unity at large distances from the source, as shown at the end of Sect. 11. This ratio must also equal one at the source because each of the build-up factors equals one, owing to the

Table 16. *Ratio of energy dissipation build-up factors in plane isotropic and point isotropic geometry, in water.*

$$\frac{B_d^{\mathrm{PLI}}(z)}{B_d^{\mathrm{PTI}}(r)_{r=z}}$$

$\mu_0 z$	E_0 (Mev)			$\mu_0 z$	E_0 (Mev)		
	0.5	1	2		0.5	1	2
0	1.00	1.00	1.00	0.4	1.82	1.53	1.37
0.001	1.24	1.18	1.13	1.0	1.80	1.50	1.35
0.004	1.30	1.22	1.16	4	1.42	1.29	1.21
0.01	1.36	1.27	1.19	10	1.21	1.14	1.10
0.04	1.50	1.36	1.26	15	1.15	1.10	1.05
0.1	1.62	1.43	1.31	∞	1.00	1.00	1.00

singularity of the unscattered radiation at the source position. However, the singularity is only logarithmic for a plane isotropic source, so that the build-up factor differs from one appreciably even at very small distances from the source. Values of the build-up factor ratio $\frac{B_d^{PLI}(z)}{B_d^{PTI}(r)\big|_{r=z}}$ for sources of various energies in a water medium are given in Table 16.

Spectra in plane-isotropic geometry do not differ qualitatively from spectra for other source types and there has been little effort to study them experimentally. Directional distributions of various spectral components have been calculated for a 1-Mev plane isotropic source in water [*3*]. Illustrative results are given in Fig. 29 for a high-energy flux component. The sequence of curves demonstrates the progressive concentration of the flux in the forward direction as the penetration increases.

δ) Plane oblique source. Spectral distributions generated by 0.66-Mev sources in water have been calculated by function-fitting methods [*5*]. Fig. 30 compares spectra at various distances from a source with obliquity $\vartheta_0 = 60°$ with corresponding spectra from a plane perpendicular source ($\vartheta_0 = 0°$). Note that with increasing source obliquity the high-energy flux components become weaker for large positive z, because they have suffered little deflection, maintain much of the original source obliquity, and therefore have to pass through more material. For small positive z and negative z, however, they become stronger because their obliquity keeps them near the source and facilitates scattering toward the rear of the source.

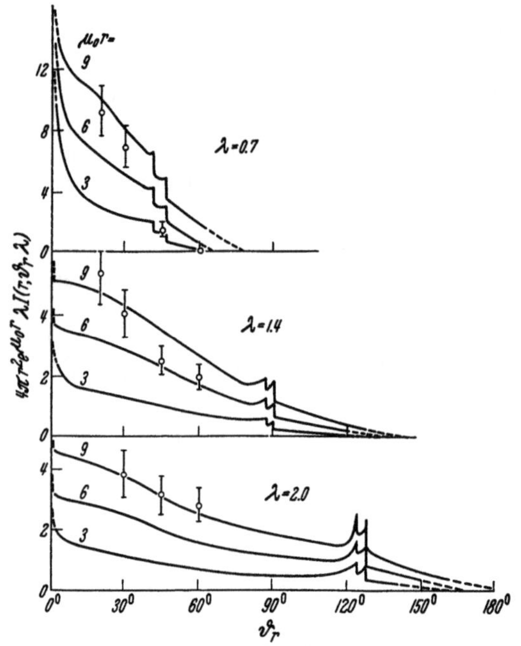

Fig. 28.

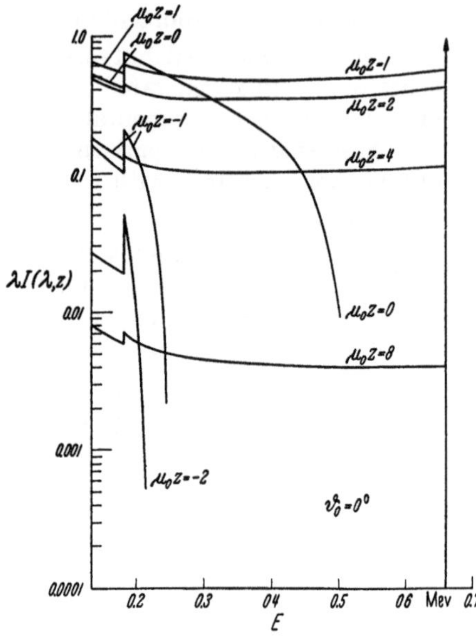

Fig. 30 a.

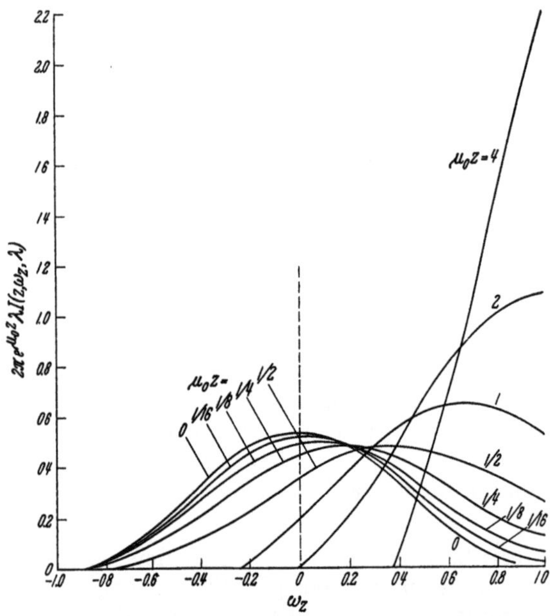

Fig. 29.

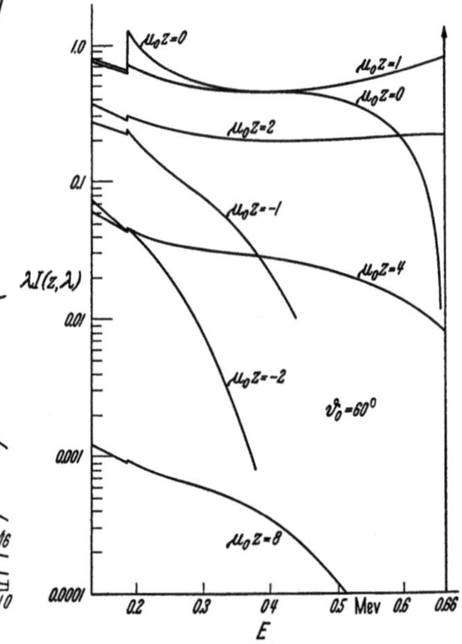

Fig. 30 b.

Fig. 28. Angular distribution of the energy flux at various distances from a point-isotropic Co^{60} source. Solid curves: Calculation for water [66]; the irregularities of the distributions reflect the behavior of the singly-scattered flux. Experimental values (Ī) from WHYTE [97] for penetration in concrete, normalized so that the integral over all angles agrees with the infinite-medium theory.

Fig. 29. Calculated angular distribution of the energy flux (spectral component with $E = 0.864$ Mev) at various distances from a plane isotropic 1-Mev source in water [6].

Fig. 30 a and b. Calculated spectral distribution of the energy flux at various distances from a plane oblique 0.66-Mev source in water [5]. (a) Source obliquity $\vartheta_0 = 0°$. (b) Source obliquity $\vartheta_0 = 60°$.

Fig. 31 gives the integrated energy flux as a function of the initial obliquity ϑ_0 for different penetrations. The curves for positive z bend down as the increasing source obliquity prevents the deep penetration of more and more of the high-energy components. The curves for negative z have the opposite behavior, bending upwards as increasing obliquity favors backscattering.

There are as yet no experiments with plane oblique sources comparable with the calculations for an infinite medium, but the penetration through *slabs* has been measured by Kirn, Kennedy and Wyckoff [*54*][1]. Their results for a concrete medium afford some basis of comparison with the theory. Fig. 32 contains semilogarithmic plots of the experimental and theoretical build-up factors *vs.* the oblique penetration distance for obliquity angles $\vartheta_0 = 0°$ and 60°.

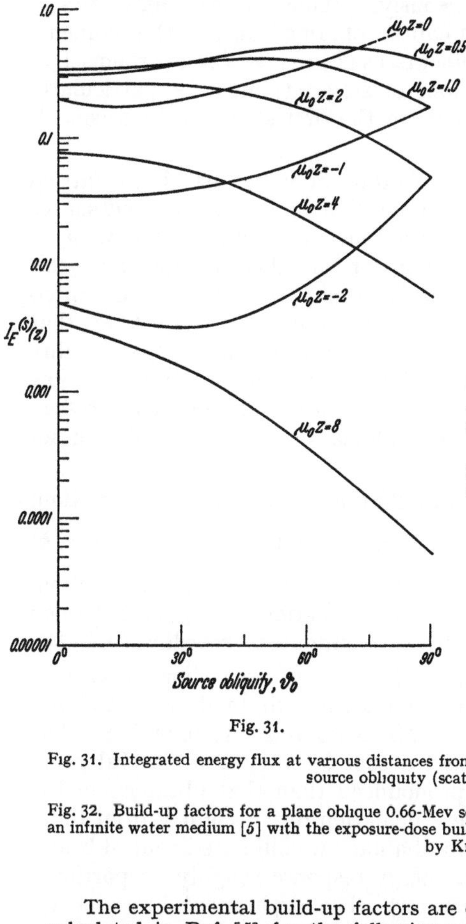

Fig. 31.

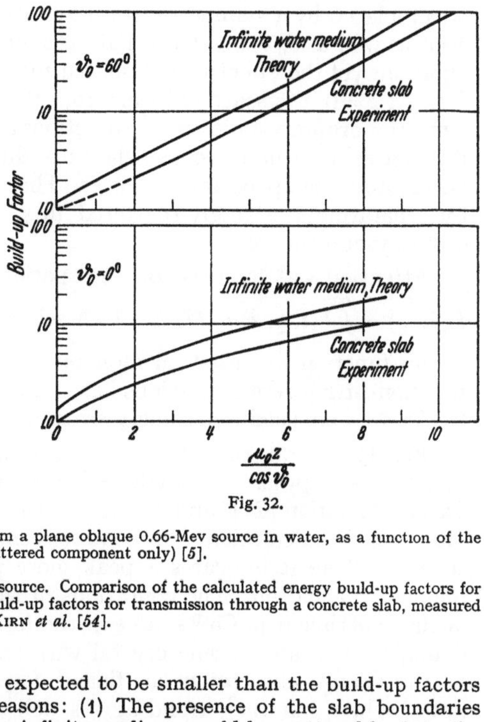

Fig. 32.

Fig. 31. Integrated energy flux at various distances from a plane oblique 0.66-Mev source in water, as a function of the source obliquity (scattered component only) [*5*].

Fig. 32. Build-up factors for a plane oblique 0.66-Mev source. Comparison of the calculated energy build-up factors for an infinite water medium [*5*] with the exposure-dose build-up factors for transmission through a concrete slab, measured by Kirn *et al.* [*54*].

The experimental build-up factors are expected to be smaller than the build-up factors calculated in Ref. [*5*] for the following reasons: (1) The presence of the slab boundaries leads to the removal of photons which in an infinite medium would be scattered back to the detector (for a discussion of this effect, see Sect. 26β). (2) Concrete absorbs low energy photons somewhat more strongly than water. (3) The experimental source contained components much softer than the nominal energy (Co^{60}) owing to scattering in the source; for deep penetrations, this leads to an overestimate of the unscattered flux that should be used in the determination of the build-up factor. All three effects tend to reduce the build-up factor by a fractional amount that is rather independent of the distance from the source. Finally, the calculations pertain to the integrated energy flux, and the measurements to the exposure dose. This difference is not serious because the slab boundary in effect removes those low-energy photons for which the difference would be most significant.

---

[1] *Note added in proof:* See also N.E. Scofield, R.L. Lynn and W.E. Kreger, U.S. Naval Radiological Defense Laboratory Report TR 272, Oct., 31, 1958.

ε) Point collimated (gun) source. We consider now the flux $I(z, \varrho, \lambda)$ at depth z and radial distance ϱ from the initial "line of fire". The calculations to be described are based on the use of a two-dimensional lattice of moments[1] $I_n^\nu(\lambda)$ computed numerically according to (12.31).

The construction of the flux distribution from the moments has been achieved in two stages: (1) Moment fitting in the variable z to obtain "partial moments" $I^\nu(z, \lambda)$; (2) moment-fitting in the variable ϱ to obtain from the $I^\nu(z, \lambda)$'s the flux distribution $I(z, \varrho, \lambda)$. Such a two-stage procedure is considerably easier than moment-fitting in both variables simultaneously, although it has difficulties of its own. Good functional representations can be obtained only if the moments used are quite precise. As the partial moments used in stage (2) are themselves obtained by moment-fitting, the calculations of stage (1) must be particularly accurate so that the amplification of error from the first stage to the second is not serious[2].

A gun source calculation for 1.28 Mev photons in water has been carried out [*82*] with the following simplification. The ordinary point collimated source was replaced by a symmetric source emitting photons with initial directions $\vartheta_0 = 180°$ as well as $\vartheta_0 = 0°$. Only half as many moments are then required for a calculation, yet the effect of the source modification was approximately accounted for. At $z = 0$, the scattered flux from the symmetric source is exactly twice that from the ordinary point-monodirectional source. In the region $\mu_0 z > 1$ the flux differences are small because they are due to back-scattered radiation with low energy and small penetrating power. The accuracy of the calculation was increased by separating the singly-scattered flux, and calculating it analytically for an ordinary gun source.

Attention was focussed on the spatial distribution of the energy dissipation, $I_D(z, \varrho) = \int_{\lambda_0}^{\infty} I(z, \varrho, \lambda)\, \mu_D(\lambda)\, E\, d\lambda$. A set of sixteen moments I_{Dn}^ν $(\nu, n = 0, 2, 4, 6)$ of this function was used for moment-fitting. In both stages of the construction, function-fitting representations based on the weight function $e^{-x} S_{n+\frac{1}{2}}$, as defined by (9.27), were used. The singly-scattered flux was computed according to (8.21).

Fig. 33a shows the radial distribution of the energy dissipation at various penetration depths. For small penetrations, comparisons are made with a Monte Carlo calculation [*11*], and an experiment by Marasco and Thurlow [*58*]. The Monte Carlo calculation is in good agreement except for very small lateral penetrations where it indicates a peak more pronounced than that obtained in the moment calculation. The conditions of the calculation differ slightly from those of the experiment (a Co^{60} source in the form of a narrow collimated conical beam, detection with anthracene crystal with an energy response roughly proportional to the energy absorption coefficient for water). The experimental results are not absolute, but their trend agrees well with that of the calculations.

Singly-scattered radiation determines some of the characteristic features of the energy dissipation distributions in Fig. 33a, including the common intercept of the curves at $\varrho = 0$, and the discontinuity of the curve for $z = 0$ at $\varrho = 0$. Fig. 33b gives the energy dissipation by singly-scattered radiation for small values of $\mu_0 \varrho$ and $\mu_0 z$. As z approaches zero, the slope of the distributions becomes steeper near $\varrho = 0$ until a limiting discontinuous curve is reached.

---

[1] The angular distribution of the flux will not be considered so that the indices l and m have the value zero and will therefore be suppressed.

[2] There is the possibility of an alternative procedure in which one begins by constructing partial moments $I_n(\varrho, \lambda)$, but experience has shown this to be more difficult [*82*].

20. Numerical solution of the Fourier transform of the transport equation. This section outlines a procedure, developed in Ref. [*80*], for the numerical solution of the transport equation. The procedure is particularly suited to the treatment of deep penetration. Its point of departure is the transformed Eq. (9.3) which pertains specifically to flux distributions generated by plane sources but is in fact adequate to treat any flux distribution in an infinite, homogeneous, and isotropic medium. This section is further restricted to flux distributions from conical sources of the type (6.10). More general types of plane source yield flux distributions that are initially not symmetric about the z axis, but become symmetric nevertheless at deep penetration as pointed out in Sect. 11. For a conical source, Eq. (9.3) becomes

$$\left.\begin{aligned} &\mu[(\lambda) - p\,\omega_z]\, I(p, \omega_z, \lambda) \\ &= \int_{\lambda_0}^{\lambda} d\lambda' \int_{4\pi} d\omega'\, k(\omega, \lambda, \omega', \lambda')\, I(p, \omega'_z, \lambda') + \\ &+ \frac{1}{2\pi}\, \delta(\omega_z - \cos\vartheta_0)\, \delta(\lambda - \lambda_0)\,. \end{aligned}\right\} \tag{20.1}$$

Deep penetration corresponds to values of p approaching μ_s, as discussed in Sect. 11. We seek a procedure to solve (20.1) numerically, which is valid for $p \sim \mu_s$ but does not require the limiting assumption involved in the asymptotic analytical solution of Sect. 11.

For a given value of p the equation is to be solved progressively for increasing values of λ, beginning at the source wavelength λ_0,

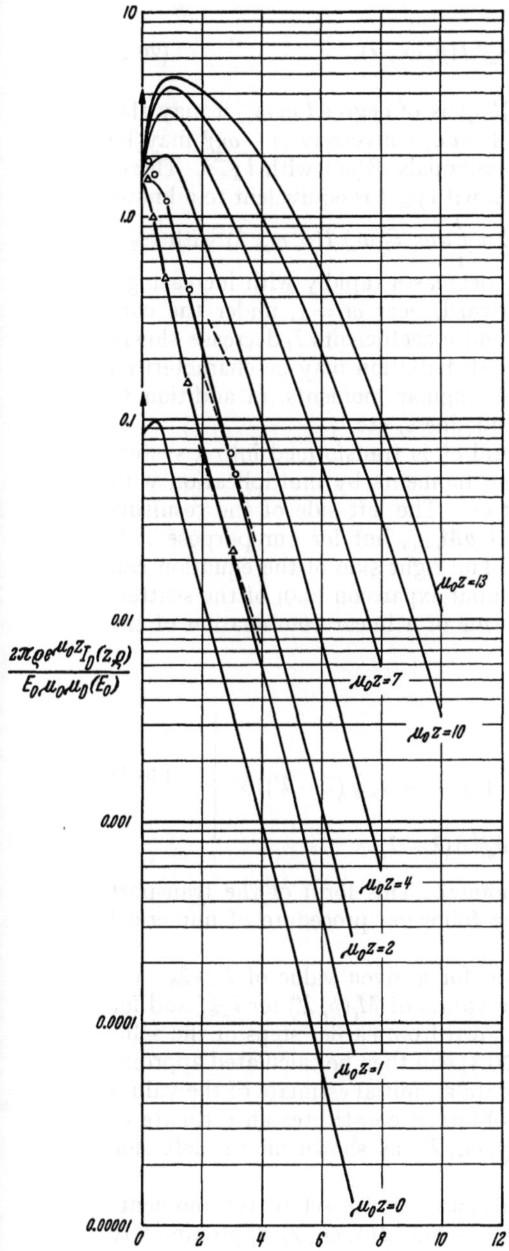

a

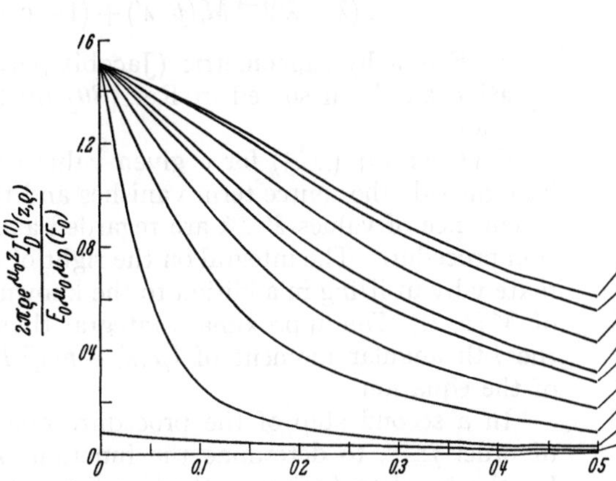

b

Fig. 33a and b. Distribution of energy dissipation about the axis of a point collimated 1.28 Mev source in water, at various penetration depths z. The ordinate is dimensionless [with $E_0 \mu_0 \mu_D(E_0) = 0.00236$ Mev cm^{-2}]. (a) Solid curves: Calculated by moment method [*82*]. △ and ○: calculated by Monte Carlo [*11*]. Dashed curves: from experiments by MARASCO and THURLOW [*58*] with Co^{60} source, at $\mu_0 z = 1$, 2, and 3, normalized to agree with theory for $\mu_0 z = 2$ and $\mu_0 \varrho = 3$ (●). (b) Contribution of singly-scattered X rays, calculated from (8.21).

as in the moment-method calculations. At each new value of λ one must determine the angular distribution of flux, i.e., the dependence of I on ω_z.

The angular distribution may be characterized by the mean values of the successive powers of $1-\omega_z=1-\cos\vartheta=2\sin^2\frac{1}{2}\vartheta\sim\frac{1}{2}\vartheta^2$, which are called angular moments,

$$M_j(p,\lambda)=\int_{-1}^{1}d\omega_z\,(1-\omega_z)^j\,I(p,\omega_z,\lambda). \tag{20.2}$$

Notice that, since the Legendre polynomial $P_l(\omega_z)$ is of degree l in ω_z, it may also be expressed as a polynomial of degree l in $1-\omega_z$; conversely $(1-\omega_z)^j$ may be expressed as a linear combination of the polynomials $P_l(\omega_z)$ with $l\leqq j$. Therefore knowledge of all angular moments $M_j(p,\lambda)$ with $j\leqq s$ is equivalent to a knowledge of the Legendre coefficients $I_l(p,\lambda)=2\pi\int_{-1}^{1}d\omega_z\,P_l(\omega_z)\,I(p,\omega_z,\lambda)$ with $l\leqq s$. However, the magnitude of the moments M_j decreases rapidly with increasing j when the angular distribution is peaked forward, near $\omega_z=1$, under the conditions of deep penetration, whereas the Legendre coefficients I_l decrease slowly with increasing l. It follows that the angular distribution may be characterized adequately by the knowledge of its first few angular moments, in addition to qualitative features such as smoothness, boundedness, etc.

The Eq. (20.1) with the two variables λ and ω_z is transformed into a system of interlinked single-variable equations among moments by multiplication with $(1-\omega_z)^j$ $(j=0,1,2,\ldots)$ and integration over ω_z. The left side of the resulting equation may be expressed as $[\mu(\lambda)-p]\,M_j+p\,M_{j+1}$, but for our purpose it is conveniently left in the form of an integral. The right side of the equation can be transformed utilizing the Legendre polynomial expansion (9.9) of the scattering probability; it is thereby expressed in terms of a linear combination of the moments M_i with $i\leqq j$. One finds

$$\left.\begin{aligned}&\int_{-1}^{1}d\omega_z(1-\omega_z)^j[\mu(\lambda)-p\,\omega_z]\,I(p,\omega_z,\lambda)\\&\quad=\int_{\lambda_0}^{\lambda}d\lambda'\,k_0(\lambda,\lambda')\sum_{i=0}^{j}\binom{j}{i}^2F\left(-i,j+1,j-i+1;\tfrac{1}{2}(\lambda-\lambda')\right)\times\\&\quad\quad\times(\lambda-\lambda')^{j-i}M_i(p,\lambda')+(1-\cos\vartheta_0)^j\,\delta(\lambda-\lambda_0),\end{aligned}\right\} \tag{20.3}$$

where F is a hypergeometric (Jacobi) polynomial. This form of the transport equation has been solved in Ref. [*80*] by the following procedure of numerical iteration.

Consider Eq. (20.3) for a given value j and for a given value of $\lambda>\lambda_0$. On its right side the source term vanishes and the values of $M_i(p,\lambda')$ for $i\leqq j$ and for a sequence of values $\lambda'<\lambda$ are regarded as known from earlier steps of the solution procedure. The integral on the right of (20.3) can then be calculated approximately by utilizing in addition to the known data an initial estimate of the values of $M_i(p,\lambda)$. The approximate integral thus obtained constitutes an estimate of the j-th angular moment of $[\mu(\lambda)-p\omega_z]\,I(p,\omega_z,\lambda)$, as shown on the left side of the equation.

In a second step of the procedure one utilizes a finite set of the moments of order $j\leqq s$, to determine the function $[\mu(\lambda)-p\omega_z]\,I(p,\omega_z,\lambda)$ approximately by the function-fitting method of Sect. 18. The number of moments required for this purpose, namely $s+1$, may be as low as four or even two. The function thus obtained yields, upon division by $\mu(\lambda)-p\omega_z$, an estimate of the desired function $I(p,\omega_z,\lambda)$.

From this estimate, a new set of moments $M_j(p, \lambda)$ with $j \leqq s$ is calculated. (This step is facilitated when the earlier function-fitting procedure for determining $[\mu - p\omega_z]\, I$ has utilized such a weight function that the integral (20.2) can be performed analytically.) The new set of estimated M_j is then entered on the right side of (20.3) to serve as the point of departure for a repetition of the procedure outlined thus far.

This procedure is to be repeated until the difference between one set of estimated M_j and the preceding one is regarded as sufficiently small. As few as two cycles of iteration have been found sufficient. Once an acceptable estimate of the moments $M_j(p, \lambda)$ and of the function $I(p, \omega_z, \lambda)$ has been obtained for one value of λ, the procedure is started again for a somewhat larger wavelength.

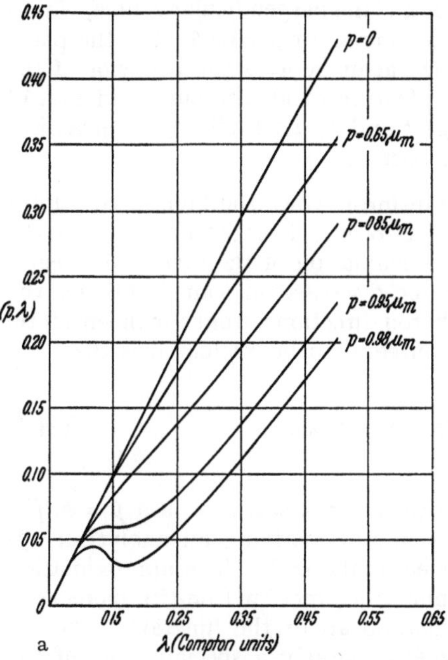

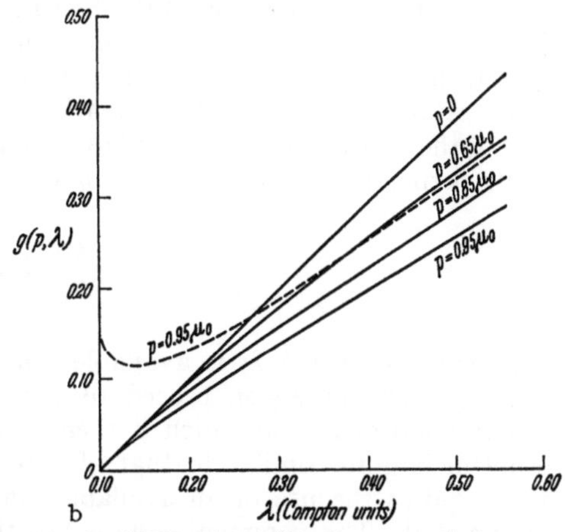

Fig. 34 a and b. Mean deflection parameter $g(p, \lambda)$ defined by (20.4) [80]: (a) 10.2-Mev plane perpendicular source in lead. (b) 5.1-Mev plane perpendicular source (solid curves) and plane isotropic source (dashed curve) in iron.

A characteristic parameter of the distribution function $I(p, \omega_z, \lambda)$ is the "mean deflection" $g(p, \lambda)$ defined by

$$g(p, \lambda) = \frac{M_1(p, \lambda)}{M_0(p, \lambda)} = 1 - \frac{I_1(p, \lambda)}{I_0(p, \lambda)} = \langle 1 - \omega_z \rangle \sim \frac{1}{2} \langle \vartheta^2 \rangle. \tag{20.4}$$

As discussed in Sect. 11, a correct estimation of the ratio $g(p, \lambda)$, or of the equivalent "mean obliquity" $\langle \omega_z \rangle = 1 - g(p, \lambda)$, leads without further difficulty to a calculation of the flux integrated over the angular distribution $M_0(p, \lambda) = I_0(p, \lambda)$. Therefore, the determination of $g(p, \lambda)$ by the procedure indicated in this section is equivalent to the solution of the eigenvalue problem in Wick's asymptotic procedure of Sect. 11. Actually the procedure of this section can be utilized to solve the Wick problem itself (see App. C of [28]).

In Fig. 34a curves of $g(p, \lambda)$ vs. λ are given for selected values of p between zero and $0.98\mu_s$ [1]. They pertain to a plane perpendicular 10.22 Mev source in lead. Note that g goes linearly to zero as $\lambda \to \lambda_0$, conforming to the behavior of the singly-scattered flux component for which $g = \lambda - \lambda_0$. The curves cor-

[1] We recall that according to the discussion of Sect. 10 (see Fig. 15) only the region $p \leqq \mu_s$ is of interest.

responding to large values of p dip to a minimum at $\lambda=0.16$. At this energy the attenuation coefficient reaches its minimum value, the penetrating power of the photons is almost entirely dependent upon their directions, and the concentration of the angular distribution around $\omega_z=1$ is therefore strongest. It can be seen from Sect. 11 that the minimum value of g approaches zero at a rate proportional to $\sqrt{\mu_s-p}$ as $p\to\mu_s$.

In Fig. 34b, curves of $g(p, \lambda)$ vs. λ are shown for 5.11 Mev radiation in iron (i.e. with the source energy *below* the energy at which the attenuation coefficient is a minimum). Various curves are given for a plane perpendicular source, with p ranging up to 0.95, and for a plane isotropic source for $p=0.95$. For the plane isotropic source g remains finite at λ_0, which again reflects the behavior of the singly-scattered component. A noteworthy feature of all the curves in Fig. 34 is their tendency to become parallel for large p and λ. This reflects the establishment of the quasi-equilibrium described in Sect. 5.

21. Numerical inversion of the Fourier transform. Once the Fourier transform $I(p, \omega, \lambda)$ of the flux from a plane source has been calculated numerically, the spatial distribution is to be obtained by the evaluation of the inversion integral (9.1). A procedure for this evaluation will be discussed in some detail for the case of the flux $I(z, \lambda)$, integrated over all photon directions, from a monoenergetic plane perpendicular source. The inversion integral to be evaluated is then

$$I(z, \lambda)=\frac{1}{2\pi i}\int_{-i\infty}^{i\infty} dp\, e^{-pz} I(p, \lambda). \tag{21.1}$$

Assume that $I(p, \lambda)$ is available numerically for a set of values $p=p_j (j=1, 2, \ldots)$. The inversion procedure consists of first deriving an analytical approximation of $I(p, \lambda)$ which is then inverted analytically. The approximation problem is quite similar to that of determining a function from its moments[1], and requires the use of all available information about the function to be approximated. The comments below are therefore largely a specialization of the arguments in Sect. 15. The quality of the analytical approximation depends on the choice of the points p_j and of the weight function used in the representation.

It is advantageous to locate the points p_j on the real axis, for the following reasons[2]: (1) For real p, the function $I(p, \lambda)$ is real and can be obtained from a single real set of Eqs. (9.12); for complex p it would be necessary to solve a dual system of equations for both real and imaginary parts of $I(p, \lambda)$. (2) The singularities of $I(p, \lambda)$ as well as the saddle-points of the integrand of (21.1) are located on the real axis. (3) Except for the points of singularity, $I(p, \lambda)$ is a smoother and better-behaved function on the real axis than elsewhere in the complex plane, and is therefore more easily approximated.

If the integral (21.1) were evaluated along a path of steepest descent, the major contribution would arise from the interval of integration near the saddle point of the integrand. This point lies at the value of p specified, for given z,

[1] The two problems are formally equivalent for equidistant p_j, i.e. for $p_j=(j+1)\,\Delta p$, in which case

$$\int dz\, e^{p_j z} I(z, \lambda)=\frac{1}{\Delta p}\int d(e^{\Delta p z})\,(e^{\Delta p z})^j\, I(z, \lambda).$$

[2] This reasoning also applies to transport phenomena for other types of radiation. However, in the similar problem of neutron penetration, H. HURWITZ and P.F. ZWEIFEL, AEC Report KAPL-1269, have successfully used imaginary p_j.

by the relation

$$z = -\frac{\partial}{\partial p} \ln I(p, \lambda). \tag{21.2}$$

Therefore it is important to know $I(p, \lambda)$ accurately in the vicinity of the saddle point, no matter what method of integration is actually used. The spacing of the points p_j on the real axis should reflect this "concentration of information" in the saddle point region, at least to the extent that the saddle point is straddled by a pair of points p_j. The location of the saddle point need only be known with limited accuracy for this purpose.

The analytical representation of $I(p, \lambda)$ can be conveniently obtained by the function-fitting technique described in Sect. 18. The weight function should not only have the correct singularity at $p = \mu_s$ but should also have a known inverse. Three cases must be distinguished depending on the relation of λ_0 and λ to the wavelength λ_m at which the attenuation coefficient is a minimum. The following representations have been found useful, all of which employ a weight function of the form $\exp\left(\beta\gamma(p)\right)$ [1]:

1. $\lambda_m < \lambda_0$. The controlling singularity has the form $(\mu_0 - p)^{-\beta}$. The transform can be represented by

$$I(p, \lambda) = \sum_i \alpha_i (\mu_0 - p)^{-\beta_i} \tag{21.3}$$

which has the inverse

$$I(z, \lambda) = e^{-\mu_0 z} \sum_i \frac{\alpha_i}{\Gamma(\beta_i)} z^{\beta_i - 1}. \tag{21.4}$$

2. $\lambda_0 < \lambda < \lambda_m$. The controlling singularity has the form $(\mu - \beta)^{-\beta(\lambda)}$, with a λ-dependent exponent in contrast with case 1. The transform can be represented by

$$I(p, \lambda) = \sum_i \alpha_i [\mu(\lambda) - p]^{-\beta_i} \tag{21.5}$$

which has also an inverse of the form (21.4).

3. $\lambda_0 < \lambda_m \leqq \lambda$. The controlling singularity has the form $\exp[\beta/\sqrt{\mu_m - p}]$, and we take

$$I(p, \lambda) = \sum_i \alpha_i e^{\frac{\beta_i}{\sqrt{\mu_m - p}}}. \tag{21.6}$$

The inverse can be calculated by expanding the exponentials in (21.6):

$$I(z, \lambda) = e^{-\mu_m z} \sum_i \alpha_i \left\{ \sum_{n=0}^{\infty} \frac{n}{2} \frac{\beta_i^n (z)^{\frac{n}{2} - 1}}{n! (n/2)!} \right\}. \tag{21.7}$$

(See also Sect. 10γ for additional discussion of these inverse transforms.)

Energy spectra and spatial distributions obtained by the Fourier transform method are shown in Fig. 35a for a 10.22 Mev source in lead, and Fig. 35b for a 5.11 Mev source in iron. The main features of the spectra in lead are the following: (a) Spectral components with energies in the range $E_m < E < E_0$, i.e. with wavelengths $\lambda < \lambda_m$ behave asymptotically as $\{(\mu_m z)^{\bar{\varkappa}(\lambda)} e^{-(\mu - \mu_m) z}\} e^{-\mu_m z}$ where $\bar{\varkappa}(\lambda)$ is an eigenvalue parameter given by (11.27). [One of the parameters β_i

[1] See Sect. 11 for the derivation of the singularities here used. Note also that the representations given here are modified in Ref. [80] by additional factors which take into account to some extent the trend of the distribution for small p and which add some flexibility to the procedure for interpolating between the $I(p_j, \lambda)$.

of (21.5) will have approximately the value $\bar{\varkappa}+1$.] As E approaches E_m from above, the exponential factor in the brackets tends to unity, and $\bar{\varkappa}(\lambda)$ increases. (2) At $E=E_m$ the asymptotic trend is $\{z^{-\frac{5}{6}}\, e^{3(\frac{1}{4}\bar{\tau}\mu_m z)^{\frac{1}{3}}}\}\, e^{-\mu_m z}$, which is similar to (11.33) but has an additional factor $\frac{1}{4}$ in the exponent in correspondence to the value $\frac{1}{2}\pi$, rather than π, of the bracket in (10.45). The value of $\bar{\tau}$ is given in Table 9 and relates to one of the β_i in (21.6). (3) At energies $E<E_m$ a quasi-equilibrium is established, all spectral components in this region having the common trend $\{z^{-\frac{5}{6}}\, e^{3(\bar{\tau}\mu_m z)^{\frac{1}{3}}}\}\, e^{-\mu_m z}$ given by (11.33), so that the spectral shape does not change with the depth of penetration.

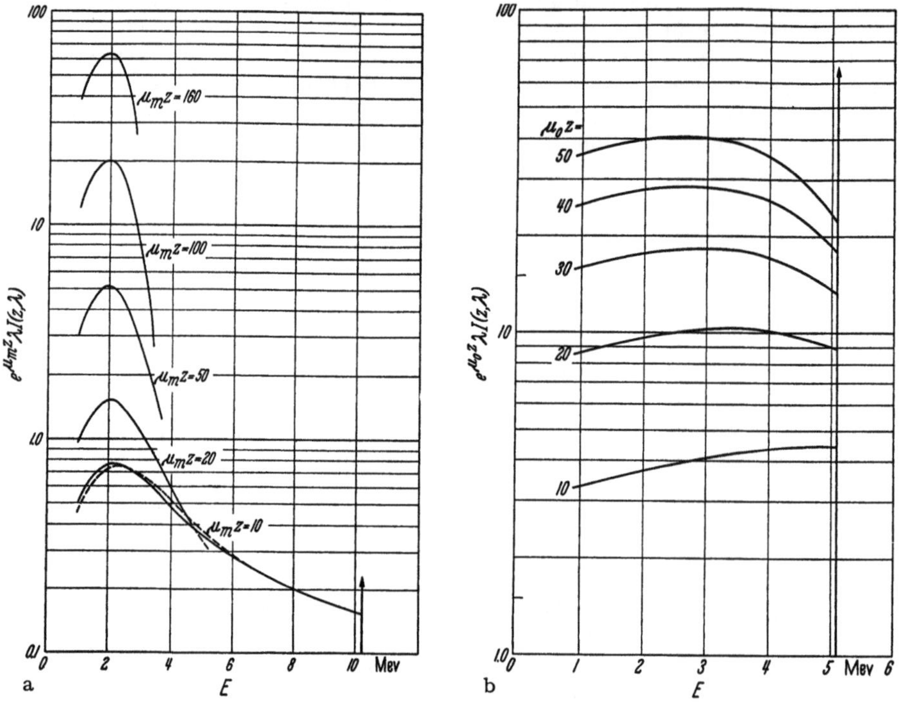

Fig. 35 a and b. Energy flux at various distances from a plane perpendicular source, calculated by the Fourier transform method [80]. (a) 10 2 Mev source in lead. Dashed curve from moment calculation for $\mu_m z=10$. (b) 5.1 Mev source in iron.

These results exemplify the trend toward quasi-equilibrium. Because the low-energy components "build up" as $z^{-\frac{5}{6}}\, e^{3(\bar{\tau}\mu_m z)^{\frac{1}{3}}}$ but the λ_m component as $z^{-\frac{5}{6}}\, e^{3(\frac{1}{4}\bar{\tau}\mu_m z)^{\frac{1}{3}}}$, there must be an energy region in which a transition occurs from one asymptotic penetration law to the other. This transition region becomes narrower with increasing penetration, so that the spectrum becomes very steep at E_m. At lower energies, in the region of quasi-equilibrium, there appears the most striking feature of the spectra, namely the maximum at about $E\sim 2$ Mev.

While the main virtue of the Fourier transform method is to describe very deep penetration, it is also serviceable for intermediate penetrations, where it yields results in close agreement with those of the polynomial method, as seen in Fig. 35a.

The expansion of exponentials in (21.6) into powers of β_i, followed by term-by-term inversion, corresponds roughly to an "orders of scattering" evaluation, because β_i represents essentially the effective scattering constant $\overline{C}_0$ of Sect. 11. Thus the number of terms required for convergence, which turned out to be about 15 for $\mu_m z = 150$, gives the number of scatterings important at that depth of penetration.

The curves for the flux in iron, shown in Fig. 35b, are similar to those for lead, and exhibit the same characteristic features, including the transition region near E_0 and the low energy region of quasi-equilibrium. Again the peak of the spectra is at ~ 2 Mev, i.e. well below E_0.

D. Effects of boundaries and inhomogeneities.

22. Survey of methods. α) *Lack of analytical theory.* The facility and accuracy with which the transport equation can be solved depends on the extent to which the angular, spatial and energy variables can be separated. This separation can be achieved by exploiting invariance or symmetry properties inherent in a given problem. In Sect. 9 it has been shown that the translational and rotational invariance of an infinite homogeneous isotropic scattering medium allows one to separate the spatial and angular variables through Fourier and Legendre transformations. The presence of boundaries or inhomogeneities makes a separation more difficult because the amount of inherent symmetry is reduced. This is one of the reasons why the analytical theory of boundary effects is still in a rudimentary state.

The astrophysical problem of reflection of radiation by a stellar atmosphere has been solved exactly by AMBARZUMIAN and by CHANDRASEKHAR who utilized the invariance of the law of reflection with respect to the addition of a layer of arbitrary thickness to a semi-infinite atmosphere[1]. These authors also applied related invariance arguments to obtain exact solutions for the reflection and transmission of radiation by plane-parallel atmospheres of finite thickness. These developments have been carried out with the use of energy-independent cross-sections, so that they apply to very soft radiation only and are outside the scope of this article. The extension to the case of energy-dependent cross-sections is more difficult and has not yet been explored.

In an infinite medium with uniform composition and a density variation in one dimension (e.g. a plane-parallel atmosphere) the flow across an infinite plane perpendicular to the direction of stratification depends only on the amount of mass between the source and the plane but not on the density distribution. By virtue of this invariance property, one can—in effect—replace the stratified medium by an equivalent "re-scaled" medium with constant density to which the standard theory for infinite homogeneous media applies. This transformation is trivial for problems with plane sources, but is shown in Sect. 27β to be possible also for other source geometries.

β) *Iterative solutions.* One general approach to the solution of boundary problems consists of abandoning macroscopic transport theory (in which the effects of all photon interactions are considered jointly) and of studying instead the effects of individual interactions by an expansion into "orders of scattering", as in Sect. 8. The integro-differential transport equation is thereby reduced to a chain of differential equations for the components of radiation that have been scattered once, twice, ..., n times. Boundary conditions can be applied relatively

[1] S. CHANDRASEKHAR [*22*]. This treatise also contains a bibliography and description of AMBARZUMIAN's work. See also V. KOURGANOFF: Basic Methods in Transfer Problems. Oxford 1952.

easily to the differential equations for the various orders of scattering. The solutions thus obtained are formally exact but difficult to evaluate, because the determination of the flux of radiation scattered n times requires the evaluation of an n-fold integral for each of the variables. Analytical or numerical evaluation of these integrals has been found impractical for $n > 3$, while as many as six orders of scattering in high-Z materials, and 10 to 15 orders in low-Z materials, may contribute substantially to the total flux. Therefore the higher orders of scattering must be estimated by some kind of extrapolation. This is feasible only if the number of contributing orders of scattering is small, either because of strong photo-electric absorption (in high-Z materials), or because of the small size of the scatterer.

γ) Monte Carlo method. Most of the available solutions of boundary problems are results of the Monte Carlo method, i.e. the random sampling of photon trajectories[1]. The general features of this method have already been indicated in Sect. 7. It may be considered as a kind of analog experiment, carried out with pencil and paper or with the aid of an electronic computing machine. Its salient features are great flexibility and adaptability to complex boundary conditions on the one hand, laboriousness of the required calculations on the other. It is an inductive method, in that the general principles governing the effects of boundaries and inhomogeneities must be established through the analysis of a number of specific schematized problems. For limited purposes, or for exploratory calculations, hand computations are feasible, but extensive applications of the method appear to be tied to the use of automatic computing machines. The programming of Monte Carlo calculations for an automatic computer requires only a modest amount of analysis because they are largely repetitive.

Compared to a physical experiment, a "Monte Carlo experiment" has various advantages. Unwanted effects in the form of background radiation are absent, radiation sources with any desired energy and geometry can be assumed, and the interpretation of results is not obscured by the complexity of the detector response. On the other hand, the problem of statistical fluctuations is more serious, because the intensity of "Monte Carlo sources" is generally much lower than that of physical sources. In a hand computation it is not practicable to sample more than about 100 photon trajectories, and even with an advanced automatic computer a sample of the order of 10^6 for a given problem may reach the limit of feasibility, while a radioactive source with a strength of 100 millicuries, for example, will emit 2.22×10^{11} photons per minute. Refined statistical techniques are therefore desirable to extract a maximum amount of information from a limited sample of photon trajectories.

δ) Qualitative discussion. By boundary effect we mean—roughly speaking—the difference between the flux in a bounded medium and the flux at the same location in a corresponding infinite homogeneous medium[2]. Boundary effects are often due to the absence of photons which in an infinite medium would have crossed the same surface repeatedly but in a bounded medium are unable to do so. These photons are low in energy and diffuse in direction so that boundary

[1] Many physicists and mathematicians have contributed to the development of the Monte Carlo method. Much relevant material is collected in Refs. [*60*], [*61*]. J. VON NEUMANN [*60c*], S. ULAM and E. FERMI should be mentioned among the originators of the method, and H. KAHN [*60b*], [*61d*], [*51*], [*52*] has made important contributions to its further development.

[2] Problems involving inhomogeneous media can in general be resolved into sequences of problems involving homogeneous media. See e.g. p. 772.

effects may be expected to have the following characteristics: (1) They are confined mainly to the low-energy part of the photon spectrum, and the change of the integrated energy flux is small. (2) Because of the small penetrating power of the low-energy spectral components, boundary effects are local, i.e. confined to regions not farther than one to two mean free paths from the boundary, and are furthermore rather independent of the radius of curvature of the boundary.

The analysis of a number of Monte Carlo solutions in Sect. 26 confirms this general picture[1]. It appears that in many circumstances modifications of the integrated energy flux or energy dissipation by boundaries amount to 10 to 20% in low-Z materials, but are much smaller in high-Z materials or for higher source energies.

The indicated features suggest the possibility of separating boundary problems into two parts: (1) The penetration of the radiation from the source to the vicinity of a boundary, and (2) the effect of the boundary. One can often solve problem (1) for an infinite homogeneous medium by the various methods described in parts B and C of this article, and then superimpose a solution of problem (2) obtained by random sampling. Thereby one takes advantage both of the efficiency of the analytical method and of the flexibility of the Monte Carlo method. The technique of "correlated random-sampling", described in Sect. 25 ε, is particularly useful for implementing such an approach.

In summary, a well-developed theory of boundary effects is still lacking. Only some of the principal physical features have been explored. The following presentation has therefore the nature of a progress report, and in a number of instances possible lines of attack rather than fully worked-out theories will be discussed. Many experimental results on boundary effects are contained in the radiological and engineering literature (see e.g., Refs. [*46*] and [*72*]), but usually under conditions that are complex and not specified with enough precision to allow detailed comparisons with calculations. There is the further complication that the boundary effects are often of the same order of magnitude as the experimental errors. Some experiments are discussed and compared with calculations in Sect. 26, and a description of back-scattering experiments will be found in Sect. 30.

23. Iterative methods. The iterative solution of the transport equation for an infinite homogeneous medium has been given in Sect. 8. For bounded media, the nature of the solution remains the same, but instead of integrating with respect to the distance ζ between collisions [as in Eq. (8.3)], one must integrate with respect to the space variables to which the boundary conditions apply. For example, given a plane-parallel barrier extending between $z=0$ and $z=a$, the flux inside the barrier is given by the recursion formula

$$\left.\begin{aligned} \omega_z I^{(n)}(z,\omega_z,\lambda) &= \int_0^z dz'\, e^{-\frac{\mu}{\omega_z}(z-z')}\, S^{(n)}(z',\omega_z,\lambda) \quad \text{if} \quad \omega_z>0,\\ &= -\int_z^a dz'\, e^{-\frac{\mu}{\omega_z}(z-z')}\, S^{(n)}(z',\omega_z,\lambda) \quad \text{if} \quad \omega_z<0, \end{aligned}\right\} \tag{23.1}$$

where $S^{(0)}=S(z,\omega_z,\lambda)$ is the source function, and

$$S^{(n)}(z,\omega_z,\lambda)=\int_0^\infty d\lambda' \int_{4\pi} d\boldsymbol{\omega}'\, k(\boldsymbol{\omega},\lambda,\boldsymbol{\omega}',\lambda')\, I^{(n-1)}(z,\omega_z',\lambda'). \tag{23.2}$$

---

[1] A situation which does not fit into this classification is described in Sect. 26 δ.

Using equations equivalent to (23.1) and (23.2), PEEBLES and PLESSET [*69a, b*] have made extensive calculations of the number of photons and the energy transmitted through iron and lead barriers. The contributions of the first three orders of scattering were calculated analytically, but those of the higher orders of scattering were estimated. The estimation was based on an extrapolation to higher orders of scattering of the ratio of two probabilities. The numerator of the ratio was the probability that a photon would be transmitted after exactly n scatterings, and the denominator was the probability that a photon would reach and survive the n-th collision in an infinite medium. The denominator was evaluated numerically up to the 12th order of scattering. The ratio is a more smoothly varying function than either the denominator or the numerator, and tends toward a constant value as the order of scattering increases. The result of an extrapolation pertaining to the transmission of 2.56-Mev photons through lead is shown in Fig. 36.

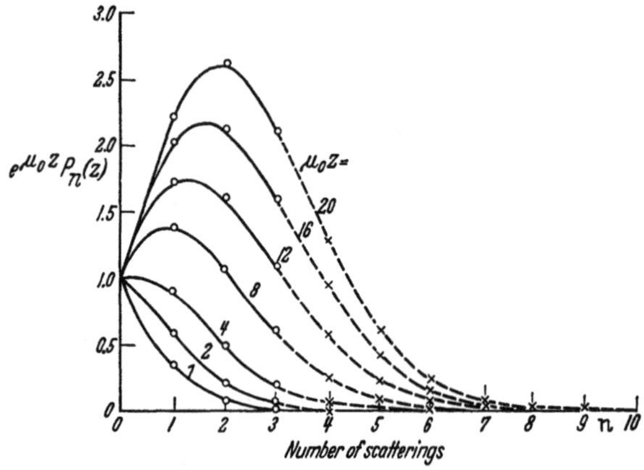

Fig. 36. Extrapolation of an "orders of scattering" calculation of the transmission of 2.55-Mev photons incident perpendicularly on lead slabs. Ordinate proportional to the probability $P_n(z)$ that a photon will be transmitted through a slab of thickness z after exactly n scatterings. Circles indicate calculated values for $n=0$, 1, 2 and 3. Crosses and dashed portions of the curves are extrapolated. (PEEBLES [*69b*].)

PEEBLES and PLESSET also used an alternative procedure in which penetration through a barrier was analyzed into successive penetrations through different layers of the barrier, each layer being thin enough so that the probability of three or more scatterings in it was negligible. The essence of their method is summarized below. For additional details see Ref. [*69*].

The transmission and reflection by a very *thin* uniform barrier can be expressed in terms of a transmission operator T and a reflection operator R which transform the incident flux I into the transmitted flux TI and the reflected flux RI. The solution of the corresponding problem for a *thick* barrier, be it uniform or made up of layers of different materials, can easily be obtained in terms of the elementary transmission and reflection operators. Let the thick barrier consist of N adjoining elementary barriers characterized by operators T_j and $R_j (j = 1, 2, \dots N)$. Let I_j^+ and I_j^- be the fluxes in the forward and backward directions, respectively, at the interface of the j-th and $(j+1)$-st elementary barrier. They are related by the following difference equations:

$$I_j^+ = T_j I_{j-1}^+ + R_j I_j^-, \qquad 1 \leqq j \leqq N; \tag{23.3}$$

$$I_j^- = T_{j+1} I_{j+1}^- + R_{j+1} I_j^+, \quad 0 \leqq j \leqq N-1. \tag{23.4}$$

Without loss of generality, the incident flux I may be assumed to enter the barrier on one side only, so that

$$I_0^+ = I \quad \text{and} \quad I_N^- = 0. \tag{23.5}$$

The solution of Eqs. (23.3) to (23.5) is particularly simple when $N=2$. In this case

$$I_2^+ = T_2 (1 - R_1 R_2)^{-1} T_1 I \tag{23.6}$$

and

$$I_0^- = T_1 R_2 (1 - R_1 R_2)^{-1} T_1 I + R_1 I. \tag{23.7}$$

Thus the composite barrier consisting of two elementary barriers is characterized by a transmission operator

$$T^* = T_2(1 - R_1 R_2)^{-1} T_1, \tag{23.8}$$

and a reflection operator

$$R^* = T_1 R_2 (1 - R_1 R_2)^{-1} T_1 + R_1. \tag{23.9}$$

By repeated pairwise combination, solutions for composite barriers of any thickness can be built up. Such an approach would appear to be particularly useful for dealing with inhomogeneous barriers consisting of layers of different materials.

In numerical applications, the flux distributions I_j^+ and I_j^- can be represented, in the range of interest, by a set of $i \times k$ discrete values corresponding to i energies and k directions, in which case the operators T_j and R_j are matrices of order $(i \times k)^2$. Accurate calculations require the use of a fine grid so that the necessary matrix operations are laborious. The fraction of radiation reflected by a barrier is generally small, so that it is permissible to evaluate the inverse $(1 - R_1 R_2)^{-1}$ needed in Eqs. (23.8) and (23.9) by an expansion in powers of R_1 and R_2 in which only a few terms are kept.

A typical grid used in the calculations of Peebles and Plesset consisted of 49 points (7 energies $\times$ 7 directions, non-equidistantly spaced). The thickness of each elementary barrier was kept small enough so that the third and higher orders of scattering could be disregarded in computing the elementary transmission operator. Backscattering from one elementary barrier to another was neglected, so that the term $(1 - R_1 R_2)^{-1}$ in (23.8) could be replaced by the unit matrix. Thus the composite transmission operator for a barrier consisting of N elementary barriers was approximated by the product $T_N T_{N-1} \ldots T_1$ of elementary transmission operators. The results obtained with this approximation are in good agreement with the results of the extrapolated "orders of scattering" calculations.

Table 17. *Comparison of different methods for computing energy build-up factors for a plane perpendicular 5.11 Mev source.*

a) Infinite medium, moment method: GOLDSTEIN and WILKINS [*41*]; b) Slab, infinite medium results adjusted for boundary effect according to Table 21. c) Slab, "orders of scattering" method: PEEBLES, as quoted in [*41*].

$\mu_0 z$	Iron			Lead		
	a)	b)	c)	a)	b)	c)
1	1.31	1.28	1.32	1.20	1.19	1.20
2	1.54	1.60	1.65	1.37	1.36	1.41
4	2.34	2.27	2.35	1.83	1.82	1.91
7	3.50	3.38	3.55	2.77	2.76	2.9
10	4.90	4.72	5.05	4.05	4.02	4.4
15	7.40	7.11	7.90	8.0	7.9	8.7

GOLDSTEIN and WILKINS [*41*] have extracted from the data of PEEBLES and PLESSET the values of the energy build-up factor, and have compared them with the values obtained by the moment-method. Typical results for iron and lead slabs are shown in Table 17. The agreement between the two methods is on the whole good, but the "orders of scattering" build-up factors for a barrier of finite thickness are consistently somewhat higher than the moment-method results when the latter have been adjusted downwards to take into account the boundary effect. The discrepancy appears to be due to slight differences in the source geometries and elementary cross-sections assumed in the two calculations.

24. Basic Monte Carlo procedures. The application of the Monte Carlo method is equivalent to playing a game of chance whose outcome provides an estimate of the X-ray transport process under consideration. The rules of the game can be formulated so that the game closely resembles the physical process. Such a procedure is called straight-forward sampling and forms the subject of the present section. In Sect. 25 we shall discuss more elaborate techniques by which the statistical fluctuations of Monte Carlo estimates can be reduced. These techniques include the "biasing" of the game of chance to make events of interest happen more often, and the combination of analytical with Monte Carlo calculations.

α) *Sampling of photon histories.* The history (trajectory) of a photon undergoing multiple scattering can be described by an array of coordinates

$$\left.\begin{array}{l} \boldsymbol{r}_1,\ \boldsymbol{r}_2,\ldots,\ \boldsymbol{r}_n,\ldots,\ \boldsymbol{r}_N \\ \boldsymbol{\omega}_1,\ \boldsymbol{\omega}_2,\ldots,\ \boldsymbol{\omega}_n,\ldots,\ \boldsymbol{\omega}_N \\ \lambda_1,\ \lambda_2,\ldots,\ \lambda_n,\ldots,\ \lambda_N \end{array}\right\} \tag{24.1}$$

where $\boldsymbol{r}_n$ indicates the position, $\boldsymbol{\omega}_n$ the direction and λ_n the wavelength of the photon *immediately before the n-th collision*[1]. All collisions up to and including the $(N-1)$st are scatterings whereas the N-th is an absorption which ends the history. For the formal description of Monte Carlo procedures it is convenient to associate with the variables $\boldsymbol{r}_n$, $\boldsymbol{\omega}_n$ and λ_n an additional variable w_n which indicates photon survival. If all collisions up to and including the $(n-1)$st in a given history are scatterings, $w_n=1$; otherwise $w_n=0$; by definition $w_1=1$.

The set of coordinates $(\boldsymbol{r}_n, \boldsymbol{\omega}_n, \lambda_n, w_n)$ representing the state of a photon before the n-th collision will be denoted as $\boldsymbol{\alpha}_n$. A photon history consisting of a sequence of states $\boldsymbol{\alpha}_1, \boldsymbol{\alpha}_2, \ldots, \boldsymbol{\alpha}_n, \ldots, \boldsymbol{\alpha}_N$ can be regarded as generated by a Markov process. That is to say, the process is determined by a transition probability density $\psi(\boldsymbol{\alpha}_{n+1}|\boldsymbol{\alpha}_n)$ which governs transitions from one state $\boldsymbol{\alpha}_n$ to the next state $\boldsymbol{\alpha}_{n+1}$. The random sampling of a photon history involves the following sequence of choices. First one selects the state $\boldsymbol{\alpha}_1$ before the first collision from a specified distribution $F_1(\boldsymbol{\alpha})$ representing the initial conditions. Then one selects $\boldsymbol{\alpha}_2$ from the distribution $\psi(\boldsymbol{\alpha}_2|\boldsymbol{\alpha}_1)$, $\boldsymbol{\alpha}_3$ from $\psi(\boldsymbol{\alpha}_3|\boldsymbol{\alpha}_2)$, and so on, until the photon is eventually absorbed and the survival coordinate assumes the value zero[2].

In relating the transition probability density $\psi(\boldsymbol{\alpha}_{n+1}|\boldsymbol{\alpha}_n)$ to the X-ray scattering and absorption cross sections, we shall for the time being assume the medium to be infinite and homogeneous. Generalizations to inhomogeneous or bounded media are simple and will be described later. The transition $\boldsymbol{\alpha}_n \to \boldsymbol{\alpha}_{n+1}$ can be broken up into a number of separate events:

1. Survival. A collision results either in absorption or in scattering[3]. With probability $1-\frac{\mu_{sc}(\lambda_n)}{\mu(\lambda_n)}$ the n-th collision is allowed to result in absorption, the

[1] In the Monte Carlo literature an alternative method of description is often used in which a history is characterized by the states of the photon *immediately after* successive collisions.

[2] The selection of a state from a distribution involves the selection of various random variates from different probability distributions. The expression "selecting a random variate x from a probability distribution $f(x)$" is used as a shorthand description of the following procedure. A set of numbers is prepared such that the frequency of numbers with magnitudes between x and $x+dx$ is proportional to $f(x)\,dx$. An arrangement is made for picking an element of this set such that each element has the same chance of being chosen. This is called random sampling or selection. The element actually selected is called a random variate x. In practice it is sufficient in Monte Carlo calculations to prepare only one set of numbers corresponding to the distribution $f(x)=1$. Such numbers are called random numbers. It will be shown in Sect. 24β that one can obtain from them, by suitable transformations, random variates from any distribution $f(x)$.

[3] The absorption of a photon gives rise to secondary X rays and corpuscular radiations (see Sect. 4). When the further propagation of these radiations is important, the Monte Carlo method can be extended to include them. In regard to secondary X rays such an extension is straightforward. The Monte Carlo treatment of electrons is complicated by the fact that an electron history usually consists of many hundreds or even thousands of collisions so that a direct imitation of the physical process is out of the question. It becomes necessary to construct an artificial random walk such that each step combines the effect of many collisions of electrons with atoms [T. Sidei, T. Higasimura and K. Kinosita: Monte Carlo Calculations of the Multiple Scattering of the Electron. Mem. Fac. Engng., Kyoto Univ. **19**, 220 (1957)]. For Monte Carlo calculations of photon-electron cascades in the cosmic radiation see R. Wilson (Refs. [*60a*] and [*99*]), and for similar calculation applicable to the response function of NaI scintillation counters see J. G. Campbell and A. J. F. Boyle [*21*].

new survival coordinate is chosen as $w_{n+1}=0$, and the history is terminated. With probability $\frac{\mu_{sc}(\lambda_n)}{\mu(\lambda_n)}$ the n-th collision is allowed to be a scattering, $w_{n+1}=1$, and the history is continued.

2. *Wavelength change.* The wavelength λ_{n+1} after the n-th collision is selected from the normalized Klein-Nishina distribution $\frac{k(\lambda_{n+1},\lambda_n)}{\mu_{sc}(\lambda_n)}$.

3. *Change of direction.* The wavelength change determines the scattering deflection[1],

$$\Theta = \arc\cos(1-\lambda_{n+1}+\lambda_n). \tag{24.2}$$

All azimuthal deflections resulting from a scattering being equally likely (provided the effects of polarization are disregarded), the azimuthal scattering angle χ_n is determined by selecting at random an angle between zero and 2π. In the coordinate representation $\boldsymbol{\omega}=(\sin\vartheta\cos\varphi, \sin\vartheta\sin\varphi, \cos\vartheta)$ the dependence of the photon direction after the n-th collision on the deflection angles Θ_n and χ_n is given by the following spherical-trigonometric relations:

$$\cos\vartheta_{n+1} = \cos\vartheta_n\cos\Theta_n + \sin\vartheta_n\sin\Theta_n\cos\chi_n, \tag{24.3}$$

$$\sin(\varphi_{n+1}-\varphi_n) = \frac{\sin\Theta_n\sin\chi_n}{\sin\vartheta_{n+1}}, \tag{24.4}$$

$$\cos(\varphi_{n+1}-\varphi_n) = \frac{\cos\Theta_n-\cos\vartheta_n\cos\vartheta_{n+1}}{\sin\vartheta_n\sin\vartheta_{n+1}}. \tag{24.5}$$

4. *Change of position.* The distance (pathlength) ζ_n between the n-th and $(n+1)$st collision has the distribution function

$$f(\zeta_n) = \mu(\lambda_{n+1})\, e^{-\mu(\lambda_{n+1})\zeta_n}. \tag{24.6}$$

The $(n+1)$st collision takes place at

$$\boldsymbol{r}_{n+1} = \boldsymbol{r}_n + \zeta_n\boldsymbol{\omega}_{n+1}. \tag{24.7}$$

Steps (1) to (4) can be summarized by the formula

$$\left.\begin{aligned}
\psi(\boldsymbol{\alpha}_{n+1}|\boldsymbol{\alpha}_n) &= \frac{k(\boldsymbol{\omega}_{n+1},\lambda_{n+1},\boldsymbol{\omega}_n,\lambda_n)}{\mu(\lambda_n)}\,\mu(\lambda_{n+1})\, e^{-\mu(\lambda_{n+1})(\boldsymbol{r}_{n+1}-\boldsymbol{r}_n)\cdot\boldsymbol{\omega}_{n+1}}; \\
&\qquad\qquad w_n = w_{n+1} = 1, \\
&= 1-\frac{\mu_{sc}(\lambda_n)}{\mu(\lambda_n)} \qquad ; \quad w_n=1,\ w_{n+1}=0, \\
&= 0 \qquad ; \quad w_n = 0.
\end{aligned}\right\} \tag{24.8}$$

The distribution $F_1(\boldsymbol{\alpha})$ of states immediately before the first collision is obtained as an integral over the source distribution $S(\boldsymbol{r},\boldsymbol{\omega},\lambda)$:

$$F_1(\boldsymbol{\alpha}) = \int_0^\infty d\zeta\,\mu(\lambda)\, e^{-\mu(\lambda)\zeta}\, S(\boldsymbol{r}-\zeta\boldsymbol{\omega},\boldsymbol{\omega},\lambda). \tag{24.9}$$

In particular, for a point-collimated source,

$$S(\boldsymbol{r},\boldsymbol{\omega},\lambda) = \delta(\boldsymbol{r}-\boldsymbol{r}_0)\,\delta(\boldsymbol{\omega}-\boldsymbol{\omega}_0)\,\delta(\lambda-\lambda_0), \tag{24.10}$$

the initial distribution takes on the simplified form

$$F_1(\boldsymbol{\alpha}) = \mu(\lambda)\, e^{-\mu(\lambda)(\boldsymbol{r}-\boldsymbol{r}_0)\cdot\boldsymbol{\omega}_0}\,\delta(\boldsymbol{\omega}-\boldsymbol{\omega}_0)\,\delta(\lambda-\lambda_0). \tag{24.11}$$

---

[1] One may proceed in the reverse order, first selecting the scattering deflection Θ_n by random sampling from the Klein-Nishina angular distribution whereupon the wavelength λ_{n+1} is determined by (24.2).

From a photon history in an infinite homogeneous medium one can easily derive a corresponding history in an infinite inhomogeneous medium. If the scattering collisions are regarded as due to Compton scattering by free electrons, the wavelength coordinates and direction coordinates of a history, i.e. the second and third rows of the array (24.1), do not depend on the nature of the medium; nor do the distances between successive collisions provided they are expressed in units of mean free paths. The transformation of a history required to account for a change of medium therefore consists of two steps only: new survival probabilities must be computed with the use of the appropriate space-dependent scattering and attenuation coefficients, and the distances between collisions (in absolute units) must be appropriately shortened or stretched. Let $\hat{\zeta}_n$ denote the distance between collisions, $\hat{\boldsymbol{r}}_n$ a spatial coordinate and $\hat{\mu}(\lambda, \hat{\boldsymbol{r}})$ the attenuation coefficient in the inhomogeneous medium. The distance $\hat{\zeta}_n$ can be obtained as the solution of the equation

$$\mu(\lambda_{n+1})\,\zeta_n = \int_0^{\hat{\zeta}_n} \hat{\mu}(\lambda_{n+1}, \hat{\boldsymbol{r}}_n + \boldsymbol{\omega}_{n+1}\zeta)\, d\zeta\,, \tag{24.12}$$

and the position coordinates in the inhomogeneous medium are

$$\hat{\boldsymbol{r}}_{n+1} = \hat{\boldsymbol{r}}_n + \hat{\zeta}_n\,\boldsymbol{\omega}_{n+1}. \tag{24.13}$$

By a bounded medium, as distinct from an infinite medium, we mean one with an "absorbing" boundary that can be crossed by a photon only once in its history. In other words, the photon, upon escaping through an absorbing boundary, cannot return to the medium, so that in effect it must be counted as lost or absorbed. Two conditions must be satisfied to make a boundary absorbing. The region outside the boundary must be empty of matter (or filled with a perfectly absorbing material) so that there is no chance of a scattering event which could deflect an escaped photon back into the medium. The boundary surface must be convex so that the re-entry of an escaped photon into the medium without scattering is impossible. Whether or not the second condition is fulfilled may depend on the direction of the escaping photon. In random sampling, the presence of an absorbing boundary can be taken into account by selecting successive states as if the medium were infinite, but terminating histories as soon as the trajectory crosses an absorbing boundary. Thus histories in a bounded medium can be obtained without further computation as shortened versions of histories in an infinite medium.

β) Random sampling techniques [*52*], [*60c*], [*61a*], [*61e*]. The sampling of photon histories requires the selection of random variates from a number of different distributions. This is accomplished most easily with the use of *random numbers* which are uncorrelated and distributed uniformly between zero and one. K-digit decimal random numbers can be obtained by dividing into sections of length K a sequence of random digits in which each digit occurs with frequency 1/10, and where consecutive digits are independent of each other in all possible combinations. Such randomness is of course an ideal which can only be approached. Tables of random numbers are available which have passed suitable statistical tests[1].

[1] Rand Corporation: A Million Random Digit Numbers with 100000 Normal Deviates. Glencoe, Illinois: Free Press 1955. — M. G. KENDALL and B. B. SMITH: Random Sampling Numbers, Tracts for Computers IV. Cambridge 1954. — For a discussion of the statistical aspects of random sampling see M. G. KENDALL: The Advanced Theory of Statistics, Vol. 1, pp. 196—203. London 1947.

With an automatic computer it is more convenient to generate random numbers as they are needed than to take them from a large stored table. There are many arithmetical methods for generating so-called "pseudo-random" numbers which are sufficiently random although produced systematically [*61a*]. The following is an example. Let $a_0, a_1, a_2, \ldots$ be a sequence of 8-decimal-digit integers, with a_0 an arbitrary non-zero integer, and with a_{n+1} equal to the remainder obtained upon dividing $23a_n$ by (10^8+1). A sequence of numbers generated in this manner is periodic with a large period, 5882352 different numbers being produced before repetition occurs. When these numbers are divided by 10^8 one obtains random numbers $\varrho_n = 10^{-8} a_n$ lying between zero and one which—according to various statistical tests—appear to be random. There are many similar schemes, adapted to the characteristics of particular computers in regard to the number-base and the number of digits contained in a random number.

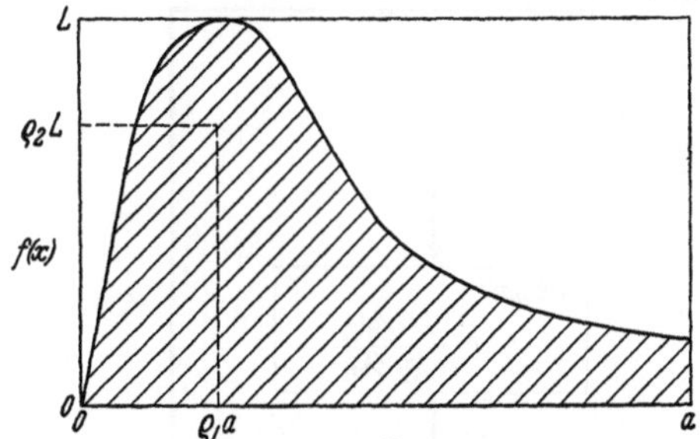

Fig. 37. Rejection method for selecting a random variate from a probability distribution $f(x)$. The choice of a random number pair ϱ_1 and ϱ_2 determines a point $(\varrho_1 a, \varrho_2 L)$. If this point lies in the shaded area under the curve $f(x)$, $x_1 = \varrho_1 a$ is selected as random variate.

One general method for selecting random variates depends on the fact that any random variable, no matter what its distribution, can be expressed as a function of another random variable that is distributed uniformly between zero and one. Let $f(x)$ denote a generic probability distribution, and

$$F(x) = \int_x^\infty f(x')\,dx' \tag{24.14}$$

the corresponding cumulative probability distribution. The cumulative probability F is itself a random variable and is distributed uniformly between zero and one, its distribution function being $f(x)\left|\frac{dx}{dF}\right| = 1$. A random variate x can therefore be obtained indirectly through the selection of a random number ϱ (between zero and one) and calculation of the inverse $x = F^{-1}(\varrho)$. For example, the selection of the distance ζ_n between successive collisions from the distribution (24.6) depends on the cumulative distribution

$$F(\zeta_n) = \mu(\lambda_{n+1}) \int_{\zeta_n}^\infty e^{-\mu(\lambda_{n+1})\zeta}\,d\zeta = e^{-\mu(\lambda_{n+1})\zeta_n}, \tag{24.15}$$

so that the selection rule is

$$\zeta_n = -\frac{1}{\mu(\lambda_{n+1})} \log \varrho. \tag{24.16}$$

The chief drawback of the "cumulative probability" method is that the calculation of the inverse $F^{-1}(\varrho)$ may become rather laborious.

Another general selection method is the rejection technique [*60c*], [*61c*], illustrated by Fig. 37. We consider a probability distribution $f(x)$ limited to a finite range, $0 \leqq x \leqq a$, and denote the maximum of $f(x)$ in that range by L. Two random numbers, say ϱ_1 and ϱ_2, are chosen and the quantity $x_1 = \varrho_1 a$ is computed. If $\varrho_2 L \leqq f(x_1)$, x_1 is accepted as a random variate from the distribution $f(x)$. If $\varrho_2 L > f(x_1)$ the random numbers ϱ_1 and ϱ_2 are rejected and the selection process is repeated with a new pair of random numbers. The probability of accepting x_1 is equal to $\int_0^{f(x_1)/L} d\varrho_2 = f(x_1)/L$ so that the frequency distribution of the sampled variates is $f(x)$ as required. The rejection technique becomes inefficient if too large a fraction of the random number pairs must be rejected. The rejection rate depends on the characteristics of $f(x)$, and is equal to $1-\frac{1}{aL}\int_0^a f(x)\,dx = 1-\frac{1}{aL}$.

It is possible to combine the advantages of the two selection methods described above. To select a random variate from a distribution $f(x)$ one first selects a random variate from a related distribution $g(x)$ by the "cumulative probability" method and then decides by the rejection technique whether or not to accept that number as a random variate from

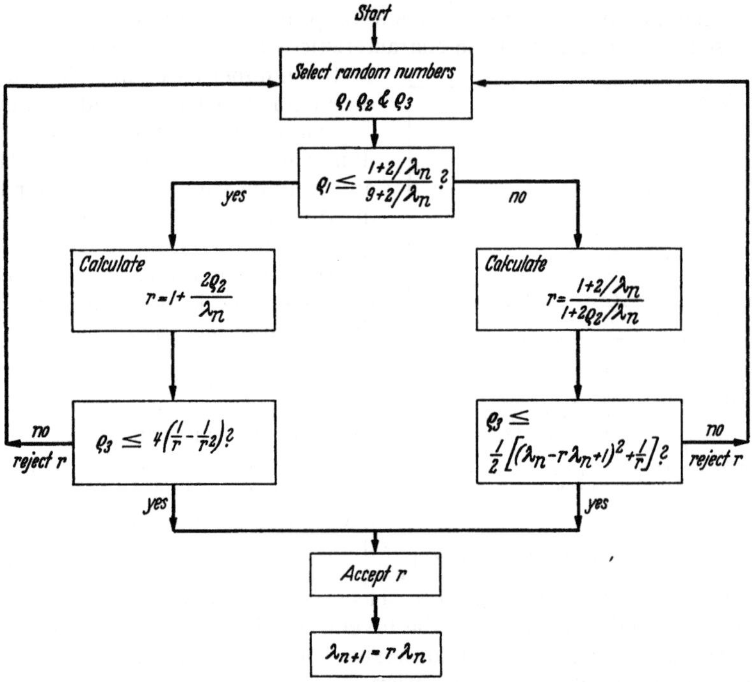

Fig. 38. Flow diagram of the procedure for sampling wavelength changes from the Klein-Nishina distribution (KAHN [52]).

$f(x)$. Thus one first computes $x_1=G^{-1}(\varrho_1)$ where ϱ_1 is a random number and $G(x)$ is the cumulative distribution associated with $g(x)$. Then one picks another random number, say ϱ_2, and tests whether $\varrho_2 L' \leqq f(x_1)/g(x_1) = h(x_1)$, where L' is the maximum of $h(x)$ in the range $0 \leqq x \leqq a$. If this inequality is satisfied, x_1 is accepted as a random variate from $f(x)$. If not, the random numbers ϱ_1 and ϱ_2 are rejected and the procedure is repeated with a new pair of random numbers. The probability of accepting x_1 is $\int_0^{h(x_1)/L'} d\varrho_2 = h(x_1)/L'$, and the probability of obtaining it in the first place is $g(x_1)$, so that x_1 is in fact a random variate from the distribution $f(x)=g(x)\,h(x)$. If $g(x)$ resembles $f(x)$ closely, the quantity $h(x_1)/L'$ is generally not much below one, so that the acceptance rate is high. One should therefore use an auxiliary distribution $g(x)$ which is as similar as possible to $f(x)$ but which allows easy computation of the inverse $G^{-1}(\varrho_1)$.

An application and slight elaboration of this scheme has been suggested for selecting wavelength changes from the Klein-Nishina distribution (H. KAHN [52]). Omitting an irrelevant normalization constant, one can express this distribution in the form

$$k(r,\lambda_n) = \frac{\lambda_n+2}{9\lambda_n+2}\,g_1(r)\,h_1(r) + \frac{8\lambda_n}{9\lambda_n+2}\,g_2(r)\,h_2(r), \tag{24.17}$$

where

$$r = \frac{\lambda_{n+1}}{\lambda_n}, \qquad g_1 = \frac{\lambda_n}{2}, \qquad h_1 = 4\left(\frac{1}{r} - \frac{1}{r^2}\right),$$

$$g_2 = \frac{\lambda_n + 2}{2r^2}, \qquad h_2 = \frac{1}{2}\left[(1 - \lambda_n r + \lambda_n)^2 + \frac{1}{r}\right].$$

With relative frequencies $\frac{\lambda_n+2}{9\lambda_n+2}$ and $\frac{8\lambda_n}{9\lambda_n+2}$ one samples from $f_1(r) = g_1(r)\, h_1(r)$ or $f_2(r) = g_2(r)\, h_2(r)$, respectively. A flow chart of the resulting sampling procedure is given in Fig. 38. It can be shown that the average number of random number triplets required to select a random variate λ_{n+1} is equal to 3.0 for $\lambda_n = 0.05$, decreases to a minimum of 1.6 for $\lambda_n = 0.5$, and rises to a limiting value of 1.7 as $\lambda_n \to \infty$.

γ) Estimation by straightforward sampling. An illustrative problem. By sampling a set of histories one obtains an aggregate of states

$$\{\boldsymbol{\alpha}_n^{(h)}\}, \qquad \begin{array}{l} n = 1, 2, \ldots, n_h \\ h = 1, 2, \ldots, H, \end{array} \tag{24.18}$$

where the index h labels the history. This aggregate forms a statistical estimate of the multi-dimensional probability distribution $F(\boldsymbol{\alpha}_1, \boldsymbol{\alpha}_2, \ldots, \boldsymbol{\alpha}_n \ldots)$ for finding a photon in a sequence of specified states before successive collisions. The Monte Carlo method may thus be regarded as a numerical technique for evaluating X-ray distributions by an "orders of scattering" approach. It is more general, however, than the "orders of scattering" equations formulated in Sects. 8 and 23 in that it provides estimates of the correlation among successive states in a history.

When this correlation is of no interest, it is sufficient to consider the distributions $F_1(\boldsymbol{\alpha}), F_2(\boldsymbol{\alpha}), \ldots, F_n(\boldsymbol{\alpha}), \ldots$ of states before the first, second, ... n-th collision, etc., irrespective of the history to which each state belongs. The distribution function $F_n(\boldsymbol{\alpha})$ is obtained from $F(\boldsymbol{\alpha}_1, \boldsymbol{\alpha}_2, \ldots, \boldsymbol{\alpha}_n, \ldots)$ by integration over all independent variables except $\boldsymbol{\alpha}_n$, and can be estimated from the sub-aggregate of states $\{\boldsymbol{\alpha}_n^{(h)}\}$, $h = 1, 2, \ldots, H$. For states $\boldsymbol{\alpha}$ with survival coordinate $w = 1$, $F_n(\boldsymbol{\alpha})$ is equal to $\mu(\lambda)\, I^{(n)}(\boldsymbol{r}, \boldsymbol{\omega}, \lambda)$ where $I^{(n)}(\boldsymbol{r}, \boldsymbol{\omega}, \lambda)$ is the n-th order flux component defined by (8.1)[1]. The sum of the distribution functions for the different orders of scattering,

$$F(\boldsymbol{\alpha}) = \sum_{n=1}^{\infty} F_n(\boldsymbol{\alpha}), \tag{24.19}$$

correspondingly equals $\mu(\lambda)\, I(\boldsymbol{r}, \boldsymbol{\omega}, \lambda)$ and is called a *collision density*.

Various Monte Carlo estimation procedures will be described in this and the next section with reference to the following problem which does not require knowledge of the correlation among successive states, and which therefore can be formulated in terms of the collision density (24.19). We consider a medium which is completely enclosed by a boundary Σ, and a radiation source distributed in some manner through the medium or on its boundary. The boundary Σ is assumed to be "absorbing" for all photons coming from the interior of the medium, i.e. a return of escaped photons into the medium is impossible. It is desired to determine the amount and composition of the radiation that emerges from the medium through a portion Σ_1 of the boundary which will be designated as the "target surface".

[1] This equality holds, if $I^{(n)}(\boldsymbol{r}, \boldsymbol{\omega}, \lambda)$ is interpreted as the time-integral of the flux corresponding to the total emission of one photon. As stated on p. 680 the transport equation may be interpreted either as a relation between time-integrals of the source emission and of the flux, or as a relation between time-constant emission and flux.

To visualize the illustrative problem more clearly, one may keep in mind the special case of a plane-parallel slab of thickness a confined between the planes $z=0$ and $z=a$ which form the boundary Σ. Given a beam of radiation incident on the slab in the positive z-direction, one may consider the source to be concentrated in the plane $z=0$. The target surface Σ_1 is identified with the plane $z=a$ for the case of transmission, and with the plane $z=0$ for the case of reflection.

The introduction of absorbing boundaries is not only a convenient schematization for direct applications, but is also useful as an intermediate step in the solution of more complicated problems involving non-absorbing boundaries. For example, the "composite-barrier" formalism of Sect. 23 can be used for the calculation of the flux through the interface between two different adjoining slabs, once the laws of transmission and reflection have been established for simple slabs with absorbing boundaries.

Within the context of the present problem it is appropriate to separate the set A of all possible photon states into three categories,

$$A = A_0 + A_1 + A_2, \tag{24.20}$$

where A_0 is the set of all states $\boldsymbol{\alpha}_n$ ($n=1, 2, \ldots, N$) with space coordinates $\boldsymbol{r}_n$ in the interior of the region enclosed by Σ; A_1 is the set of all terminal states $\boldsymbol{\alpha}_N$ such that the transition $\boldsymbol{\alpha}_{N-1}\to\boldsymbol{\alpha}_N$ involves a crossing of the target surface Σ_1; and A_2 is the set of all terminal states $\boldsymbol{\alpha}_N$ such that the transition $\boldsymbol{\alpha}_{N-1}\to\boldsymbol{\alpha}_N$ involves a crossing of a part of the boundary Σ other than Σ_1.

The amount τ of radiation emerging from the medium across the target surface Σ_1 can be expressed as a weighted integral over the collision density:

$$\tau = \int_{A_1} T(\boldsymbol{\alpha})\, F(\boldsymbol{\alpha})\, d\boldsymbol{\alpha}, \tag{24.21}$$

where $T(\boldsymbol{\alpha})$ is a scoring weight (usually understood to be a function of $\boldsymbol{\omega}$ and λ only) which takes into account the type of quantity whose transport across Σ_1 is of interest. For example, in the slab-transmission problem, the following scoring weights could be used:

$T(\boldsymbol{\alpha})$	Interpretation of τ
1	Average number of photons transmitted per incident photon (= probability of transmission).
$\dfrac{0.511}{\lambda}$	Average amount of energy (in Mev) transmitted per incident photon (= probability of transmission $\times$ average energy of the transmitted photons).
$\dfrac{0.511\,\mu_d(\lambda)}{\lambda\ \cos\vartheta}$ (μ_d in cm^2/g)	Exposure dose (in Mev/g) recorded by an air-equivalent isotropic detector located behind a slab, due to an incident beam of 1 photon/cm^2.

A Monte Carlo estimate of τ can be obtained by "scoring" a set of sampled histories. The score t_h associated with history h depends on the manner in which the history ends, and is defined as follows:

$$\left.\begin{aligned} t_h &= 0 && \text{if } \boldsymbol{\alpha}_N^{(h)} \text{ lies in } A_0 \text{ (absorption inside the medium)},\\ &= T(\boldsymbol{\alpha}_N^{(h)}) && \text{if } \boldsymbol{\alpha}_N^{(h)} \text{ lies in } A_1 \text{ (crossing of the target surface } \Sigma_1),\\ &= 0 && \text{if } \boldsymbol{\alpha}_N^{(h)} \text{ lies in } A_2 \text{ (leakage across part of the boundary } \Sigma \text{ other than } \Sigma_1). \end{aligned}\right\} \tag{24.22}$$

A statistical estimate of τ is provided by the average score

$$\langle t\rangle = \frac{1}{H}\sum_{h=1}^{H} t_h. \tag{24.23}$$

The variance of this estimate can be formally expressed as

$$\sigma^2 = \frac{1}{H}\left\{\int\limits_{A_1} T^2(\boldsymbol{\alpha})\, F(\boldsymbol{\alpha})\, d\boldsymbol{\alpha} - \tau^2\right\}. \tag{24.24}$$

Two sources of statistical error contribute to this variance: (a) fluctuations of the number of photons transmitted across the target surface, and (b) fluctuations of the properties (energy, direction, etc.) of the transmitted photons. Analyzing the statistical errors from this point of view, one may regard the score t_h as the product of two factors,

$$t_h = \varepsilon_h\, T(\boldsymbol{\alpha}_N^{(h)}), \tag{24.25}$$

where ε_h is a random variable which takes on the value one if the history h ends with the photon crossing the target surface, and the value zero otherwise. The mean value of ε_h is equal to the probability of emergence through the target surface,

$$p = \int\limits_{A_1} F(\boldsymbol{\alpha})\, d\boldsymbol{\alpha}. \tag{24.26}$$

Let $\langle T\rangle$ denote the average, and $\langle T^2\rangle$ the mean square average of $T(\boldsymbol{\alpha}_N^{(h)})$, the averages being evaluated over all possible terminal states $\boldsymbol{\alpha}_N^{(h)}$ that can occur in case of transmission, i.e. when $\varepsilon_h = 1$. The corresponding averages of t_h are

$$\langle t\rangle = (1-p)\cdot 0 + p\langle T\rangle \tag{24.27}$$

and

$$\langle t^2\rangle = (1-p)\cdot 0 + p\langle T^2\rangle. \tag{24.28}$$

From these equations one can deduce an alternative expression for the variance more informative than (24.24). Of particular interest is the relative standard deviation $\sigma/\langle t\rangle$, which is a measure of the precision of the estimate $\langle t\rangle$. One finds that

$$\frac{\sigma}{\langle t\rangle} = \sqrt{\frac{\langle t^2\rangle - \langle t\rangle^2}{\langle t\rangle^2 H}} = \sqrt{\frac{1-p}{pH} + \frac{\dfrac{\langle T^2\rangle}{\langle T\rangle^2} - 1}{pH}}. \tag{24.29}$$

Both terms under the square-root on the right-hand side of (24.29) are non-negative. The first represents the fluctuations of the number of transmitted photons; the second represents the fluctuations of the properties of the transmitted photons and vanishes when the scoring weight $T(\boldsymbol{\alpha})$ is constant.

25. Refinements of the Monte Carlo method. Estimation by straightforward sampling is inefficient in circumstances where only a small subset of the sampled photon histories contributes significant information toward the solution of the problem at hand. It may happen that only a minute fraction of the photons emitted by the source reach the target surface. In shielding problems, for example, it would not be unusual for the transmission probability p to have a value as low as 10^{-4}. According to (24.29), at least $H = 10^6$ histories would then be required for an estimate $\langle t\rangle$ of the emergent radiation with a relative standard deviation $\sigma/\langle t\rangle = 0.1$.

There are numerous techniques for increasing the efficiency of a Monte Carlo calculation[1] some of which will be described in this section in reference to the illustrative problem of Sect. 24γ. They depend on increasing the amount of information extracted from each history, on modifications of the sampling scheme to increase the fraction of significant histories, or on a re-formulation of the problem to make best use of pre-existing information from other sources such as infinite-medium theory. The various techniques are described separately, but it may be advantageous to use more than one technique on the same problem.

α) Analytical averaging. The amount of random sampling can be reduced by a partially analytical treatment such as replacing one of the random variables by its expected value. For example, instead of allowing chance to decide whether a collision results in an absorption or in a scattering, one can permit the photon to survive all collisions in the sampling process. To compensate for this bias one associates with the state $\boldsymbol{\alpha}_n$ an "expected survival" weight factor $W_s(\boldsymbol{\alpha}_n)$ which lies between zero and one, and which—in effect—takes the place of the random variable w_n. $W_s(\boldsymbol{\alpha}_n)$ is the average fraction of the photons with an energy history corresponding to (24.1) that survive all collisions up to and including the $n-1$-st. Therefore

$$W_s(\boldsymbol{\alpha}_1) = 1, \quad W_s(\boldsymbol{\alpha}_{n+1}) = W_s(\alpha_n) \frac{\mu_{s\iota}(\lambda_n)}{\mu(\lambda_n)}. \tag{25.1}$$

The analytical calculation of absorption makes it necessary to terminate histories by an artificial cut-off, e.g. when a sufficiently large value of λ is reached or when the weight W_s becomes sufficiently small. With the possibility of a zero-score due to premature absorption eliminated, the score t_h assigned to the h-th history is no longer defined by (24.22), but by

$$\left.\begin{aligned} t_h &= T(\boldsymbol{\alpha}_N^{(h)})\, W_s(\boldsymbol{\alpha}_N^{(h)}) && \text{if } \boldsymbol{\alpha}_N^{(h)} \text{ lies in } A_1, \\ &= 0 && \text{if } \boldsymbol{\alpha}_N^{(h)} \text{ lies in } A_2. \end{aligned}\right\} \tag{25.2}$$

One obtains, with a given number of histories, a larger number of non-zero sources than by straightforward sampling estimation such that the average score $\langle t\rangle$ is the same but the relative standard deviation $\sigma/\langle t\rangle$ is reduced.

Instead of treating the crossing of a boundary by a photon as a chance event, one can calculate analytically after each collision the probability that the photon will emerge from the medium through the target surface before making another collision. The radiation transport is then computed as a weighted sum over the entire set of states in the sample; this estimation procedure has been called "collision density method" [4]. The probability that a photon will escape across the target surface after the n-th collision is

$$\left.\begin{aligned} p(\boldsymbol{r}_n, \boldsymbol{\omega}_{n+1}, \lambda_{n+1}) &= e^{-\mu(\lambda_{n+1})s_n} && \begin{array}{l}\text{if the direction vector } \boldsymbol{\omega}_{n+1} \\ \text{points toward the target surface,}\end{array} \\ &= 0 && \text{otherwise,} \end{aligned}\right\} \tag{25.3}$$

where s_n is the pathlength from the point $\boldsymbol{r}_n$ of the n-th collision to the target surface in the direction $\boldsymbol{\omega}_{n+1}$. The score assigned to history h is now defined as

$$T_h = \sum_{n=0}^{N_h-1} T(\boldsymbol{\omega}_{n+1}, \lambda_{n+1})\, p(\boldsymbol{r}_n, \boldsymbol{\omega}_{n+1}, \lambda_{n+1}). \tag{25.4}$$

The first term in this series pertains to the escape of unscattered radiation, s_0 being the distance from the source to the target surface. The summation in

[1] For a comprehensive survey, see H. KAHN [52].

(25.4) extends over all states of a history except the terminal state. The contribution from each collision depends on its closeness to the target surface. Whereas in straightforward sampling only an actual crossing of the target surface counts (which can happen only once in a history), the collision density method gives credit for a "near miss" and allows several contributions to each score T_h so that the sampling fluctuations are reduced.

The collision density method can be extended so that not only the last free flight of the photon before its emergence from the medium is treated analytically, but also the survival probability and changes of wavelength and direction associated with the last collision prior to emergence. This method introduces in the estimate of the collision density $F(\boldsymbol{\alpha})$ one step of the analytical iteration procedure indicated in Sect. 8. In order to formulate the iteration, we need the transport equation for $F(\boldsymbol{\alpha})$ which can be derived most easily by an "orders of scattering" treatment. The distribution function $F_n(\boldsymbol{\alpha})$ of states immediately before the n-th collision obeys the recursion relation

$$F_{n+1}(\boldsymbol{\alpha}) = \int\limits_{A_0} \psi(\boldsymbol{\alpha}|\boldsymbol{\alpha}')\, F_n(\boldsymbol{\alpha}')\, d\boldsymbol{\alpha}', \quad n \geqq 1 \tag{25.5}$$

which is the analog of (8.3) and (8.4) adapted to the boundary conditions of the illustrative problem of Sect. 24γ. By summing (25.5) over all orders of scattering one obtains the transport equation

$$F(\boldsymbol{\alpha}) = \int\limits_{A_0} \psi(\boldsymbol{\alpha}|\boldsymbol{\alpha}')\, F(\boldsymbol{\alpha}')\, d\boldsymbol{\alpha}' + F_1(\boldsymbol{\alpha}), \tag{25.6}$$

where $F_1(\boldsymbol{\alpha})$ is the initial distribution specified by (24.9) or (24.11).

When the right-hand side of (25.6) is substituted for $F(\boldsymbol{\alpha})$ into (24.21), the expression for the radiation transport across the target surface becomes

$$\left.\begin{aligned} \tau &= \int\limits_{A_1} T(\boldsymbol{\alpha}) \Big[\int\limits_{A_0} \psi(\boldsymbol{\alpha}|\boldsymbol{\alpha}')\, F(\boldsymbol{\alpha}')\, d\boldsymbol{\alpha}' + F_1(\boldsymbol{\alpha})\Big] d\boldsymbol{\alpha} \\ &= \int\limits_{A_0} T^*(\boldsymbol{\alpha})\, F(\boldsymbol{\alpha})\, d\boldsymbol{\alpha} + \int\limits_{A_1} T(\boldsymbol{\alpha})\, F_1(\boldsymbol{\alpha})\, d\boldsymbol{\alpha}, \end{aligned}\right\} \tag{25.7}$$

where

$$\left.\begin{aligned} T^*(\boldsymbol{\alpha}) &= \int\limits_{A_1} T(\boldsymbol{\alpha}')\, \psi(\boldsymbol{\alpha}'|\boldsymbol{\alpha})\, d\boldsymbol{\alpha}' \\ &= \int\limits_{\lambda}^{\infty} d\lambda' \int\limits_{4\pi} d\omega'\, T(\omega', \lambda')\, \frac{k(\omega', \lambda', \omega, \lambda)}{\mu(\lambda)}\, p(\boldsymbol{r}, \omega', \lambda'), \end{aligned}\right\} \tag{25.8}$$

with $p(\boldsymbol{r}, \omega', \lambda')$ defined by (25.3).

The second term on the right-hand side of (25.7) can usually be evaluated analytically. The sampling estimate of the first term is obtained by assigning to the h-th history the score

$$t_h = \sum_{n=1}^{N_h - 1} T^*(\boldsymbol{\alpha}_n^{(h)}), \tag{25.9}$$

and an estimate of τ is given by

$$\langle t \rangle = \frac{1}{H} \sum_{h=1}^{H} t_h + \int\limits_{A_1} T(\boldsymbol{\alpha})\, F_1(\boldsymbol{\alpha})\, d\boldsymbol{\alpha}. \tag{25.10}$$

Another extension of the collision density method consists of treating analytically the space variable associated with the initial photon collision. Instead of actually sampling the location of the first collision, one may carry it along as an analytical variable $\boldsymbol{r}_1^{(h)}$, with the space coordinate $\boldsymbol{r}_n^{(h)}$ of each state in history h depending on $\boldsymbol{r}_1^{(h)}$. An improved estimate is obtained by averaging the terms $T^*(\boldsymbol{\alpha}_n^{(h)})$ in (25.9) over all values of $\boldsymbol{r}_1^{(h)}$ for which the space coordinates of all states in the history remain within the bounded medium. In this manner one eliminates fluctuations arising from the variability of the pathlengths traversed by the photons when they are unscattered and most penetrating.

In problems involving only one space coordinate it has been found feasible to treat absorption as well as *all* spatial displacements analytically. This is equivalent to computing the probability that a photon with a given directional and wavelength history will be transmitted across the target surface, and then taking an average over all possible sequences of wavelengths and directions by random sampling. Such an approach has been used to calculate the transmission and reflection of radiation by slabs [*7*].

β) Importance sampling [*38a*], [*38b*], [*52*], [*60b*]. In order to increase the fraction of the sample photon histories which are "important", i.e. contribute most of the information needed for a Monte Carlo estimate, one abandons sampling by a stochastic process which is the direct analog of the physical process; instead, one generates histories by a biased process that has no direct physical counterpart and leads to a biased collision density $\hat{F}(\boldsymbol{\alpha})$. The bias should be chosen so that $\hat{F}(\boldsymbol{\alpha})$ is concentrated as much as possible in the scoring region A_1 and is large for those states for which $T(\boldsymbol{\alpha})$ is large. In order to compensate for the bias, one must associate with each state $\boldsymbol{\alpha}_n$ a weight factor $\hat{W}(\boldsymbol{\alpha}_n)$ which depends on the biased initial distribution $\hat{F}_1(\boldsymbol{\alpha})$ and the biased transition probability $\psi(\boldsymbol{\alpha}_{n+1}|\boldsymbol{\alpha}_n)$ characteristic of the importance sampling scheme:

$$\hat{W}(\boldsymbol{\alpha}_1) = \frac{F_1(\boldsymbol{\alpha}_1)}{\hat{F}_1(\boldsymbol{\alpha}_1)}; \qquad \hat{W}(\boldsymbol{\alpha}_{n+1}) = \hat{W}(\boldsymbol{\alpha}_n)\,\frac{\psi(\boldsymbol{\alpha}_{n+1}|\boldsymbol{\alpha}_n)}{\hat{\psi}(\boldsymbol{\alpha}_{n+1}|\boldsymbol{\alpha}_n)}, \qquad n \geqq 1. \tag{25.11}$$

Eq. (24.21) which defines the transport of radiation across the target surface can be re-written in the form

$$\tau = \int\limits_{A_1} \left[\frac{T(\boldsymbol{\alpha})\,F(\boldsymbol{\alpha})}{\hat{F}(\boldsymbol{\alpha})}\right] \hat{F}(\boldsymbol{\alpha})\,d\boldsymbol{\alpha} \tag{25.12}$$

so that τ can be considered the average of a scoring weight $\dfrac{T(\boldsymbol{\alpha})\,F(\boldsymbol{\alpha})}{\hat{F}(\boldsymbol{\alpha})}$ over the biased collision density $\hat{F}(\boldsymbol{\alpha})$. The corresponding Monte Carlo estimate is obtained by assigning to history h a score

$$\left.\begin{aligned} t_h &= T(\boldsymbol{\alpha}_N^{(h)})\,\hat{W}(\boldsymbol{\alpha}_N^{(h)}) && \text{if } \boldsymbol{\alpha}_N^{(h)} \text{ lies in } A_1 \\ &= 0 && \text{if } \boldsymbol{\alpha}_N^{(h)} \text{ lies in } A_0 \text{ or } A_2. \end{aligned}\right\} \tag{25.13}$$

We shall now show how to define an "ideal bias" such that the Monte Carlo estimate would have zero variance and a single history would yield an exact value of τ. In order to determine the ideal bias one would have to solve a boundary problem as difficult as the original problem. Yet the discussion of zero-variance estimates provides indications how to obtain estimates with non-vanishing but small variance.

The variance of a biased-sampling estimate based on the analysis of H histories can be written formally as

$$\sigma^2 = \frac{1}{H}\left\{\int\limits_{A_1} \left[\frac{T(\boldsymbol{\alpha})\,F(\boldsymbol{\alpha})}{\hat{F}(\boldsymbol{\alpha})}\right]^2 \hat{F}(\boldsymbol{\alpha})\,d\boldsymbol{\alpha} - \tau^2\right\}. \tag{25.14}$$

This expression would vanish if the biased collision density were to satisfy the condition

$$\hat{F}(\boldsymbol{\alpha}) = \frac{T(\boldsymbol{\alpha})\,F(\boldsymbol{\alpha})}{\tau}. \tag{25.15}$$

Suppose the biased process is based on an initial distribution

$$\hat{F}_1(\boldsymbol{\alpha}) = \frac{F_1(\boldsymbol{\alpha})\,Q(\boldsymbol{\alpha})}{\int\limits_A F_1(\boldsymbol{\alpha})\,Q(\boldsymbol{\alpha})\,d\boldsymbol{\alpha}} \tag{25.16}$$

and a transition probability density

$$\hat{\psi}(\boldsymbol{\alpha}_{n+1}|\boldsymbol{\alpha}_n) = \frac{\psi(\boldsymbol{\alpha}_{n+1}|\boldsymbol{\alpha}_n)\,Q(\boldsymbol{\alpha}_{n+1})}{\int\limits_A \psi(\boldsymbol{\alpha}_{n+1}|\boldsymbol{\alpha}_n)\,Q(\boldsymbol{\alpha}_{n+1})\,d\boldsymbol{\alpha}_{n+1}}, \tag{25.17}$$

where $Q(\boldsymbol{\alpha})$ is the so-called *importance function* and satisfies the conditions

$$\left.\begin{aligned} Q(\boldsymbol{\alpha}) &= \int\limits_A Q(\boldsymbol{\alpha}')\,\psi(\boldsymbol{\alpha}'|\boldsymbol{\alpha})\,d\boldsymbol{\alpha}' &&\text{if } \boldsymbol{\alpha} \text{ is in } A_0,\\ &= T(\boldsymbol{\alpha}) &&\text{if } \boldsymbol{\alpha} \text{ is in } A_1,\\ &= 0 &&\text{if } \boldsymbol{\alpha} \text{ is in } A_2. \end{aligned}\right\} \tag{25.18}$$

It follows from (25.5) and (25.16) to (25.18) that

$$\hat{F}_2(\boldsymbol{\alpha}) = \int\limits_{A_0} \hat{\psi}(\boldsymbol{\alpha}|\boldsymbol{\alpha}')\,\hat{F}_1(\boldsymbol{\alpha}')\,d\boldsymbol{\alpha}' = \frac{F_2(\boldsymbol{\alpha})\,Q(\boldsymbol{\alpha})}{\int\limits_A F_1(\boldsymbol{\alpha})\,Q(\boldsymbol{\alpha})\,d\boldsymbol{\alpha}}. \tag{25.19}$$

Similarly,

$$\hat{F}_n(\boldsymbol{\alpha}) = \frac{F_n(\boldsymbol{\alpha})\,Q(\boldsymbol{\alpha})}{\int\limits_A F_1(\boldsymbol{\alpha})\,Q(\boldsymbol{\alpha})\,d\boldsymbol{\alpha}} \tag{25.20}$$

for all n, and therefore

$$\hat{F}(\boldsymbol{\alpha}) = \sum_{n=1}^{\infty} \hat{F}_n(\boldsymbol{\alpha}) = \frac{F(\boldsymbol{\alpha})\,Q(\boldsymbol{\alpha})}{\int\limits_A F_1(\boldsymbol{\alpha})\,Q(\boldsymbol{\alpha})\,d\boldsymbol{\alpha}}. \tag{25.21}$$

The denominator of (25.21) can be evaluated with the use of the relation

$$\int\limits_A [F_n(\boldsymbol{\alpha}) - F_{n+1}(\boldsymbol{\alpha})]\,Q(\boldsymbol{\alpha})\,d\boldsymbol{\alpha} = \int\limits_{A_1} F_n(\boldsymbol{\alpha})\,T(\boldsymbol{\alpha})\,d\boldsymbol{\alpha} \tag{25.22}$$

which is a consequence of (25.5) and (25.18). By summing (25.22) over all orders of scattering, one finds that

$$\int\limits_A F_1(\boldsymbol{\alpha})\,Q(\boldsymbol{\alpha})\,d\boldsymbol{\alpha} = \int\limits_{A_1} F(\boldsymbol{\alpha})\,T(\boldsymbol{\alpha})\,d\boldsymbol{\alpha} = \tau. \tag{25.23}$$

It can be seen by combining (25.21), (25.23) and the second of the Eqs. (25.18) that the biased collision density $\hat{F}(\boldsymbol{\alpha})$, generated with the use of the importance function $Q(\boldsymbol{\alpha})$, satisfies the condition (25.15) for every $\boldsymbol{\alpha}$ in A_1 and therefore leads to a zero-variance estimate of τ.

The three Eqs. (25.18) defining $Q(\boldsymbol{\alpha})$ can be combined to yield the integral equation

$$Q(\boldsymbol{\alpha}) = \int\limits_{A_0} Q(\boldsymbol{\alpha}')\,\psi(\boldsymbol{\alpha}'|\boldsymbol{\alpha})\,d\boldsymbol{\alpha}' + Q_1(\boldsymbol{\alpha}), \tag{25.24}$$

which holds for $\boldsymbol{\alpha}$ in A_0, and whose "source function" is[1]

$$Q_1(\boldsymbol{\alpha}) = \int\limits_{A_1} T(\boldsymbol{\alpha})\,\psi(\boldsymbol{\alpha}'|\boldsymbol{\alpha})\,d\boldsymbol{\alpha}'. \tag{25.25}$$

---

[1] The source function is identical with the scoring weight $T^*(\boldsymbol{\alpha})$ used in (25.9) for evaluating the radiation transport by the collision density method.

The kernel of (25.24) has the same functional form as the kernel of the transport equation (25.6) for the collision density $F(\boldsymbol{\alpha})$, but the arguments of the kernel are interchanged. Whereas the kernel $\psi(\boldsymbol{\alpha}|\boldsymbol{\alpha}')$ in (25.6) represents a transition from $\boldsymbol{\alpha}'$ to $\boldsymbol{\alpha}$, the kernel $\psi(\boldsymbol{\alpha}'|\boldsymbol{\alpha})$ in (25.24) represents the inverse transition from $\boldsymbol{\alpha}$ to $\boldsymbol{\alpha}'$. The two integral equations are called *adjoint* to each other. The solution of either of them can be used, according to (25.23), for the evaluation of τ.

Knowledge of the exact importance function would imply that the problem under investigation is already completely solved. But even with an approximate importance function $\tilde{Q}(\boldsymbol{\alpha})$, the statistical errors of a Monte Carlo estimate can be greatly reduced for a given number of sampled histories. $\tilde{Q}(\boldsymbol{\alpha})$ may be obtained as an approximate solution of (25.24), either analytically or by random sampling[1]. Into the approximate importance function one can incorporate prior information about the solution of the problem obtained by qualitative or semi-analytical considerations. Thus $\tilde{Q}(\boldsymbol{\alpha})$ plays a role analogous to that of the weight function used in Part C for the construction of flux distributions from their moments. To mention a very simple example, histories ending in absorption provide no information about the transport of radiation across the target surface. It is reasonable, therefore, to require that

$$\tilde{Q}(\boldsymbol{\alpha}) = 0 \tag{25.26}$$

whenever the survival coordinate w of the state $\boldsymbol{\alpha}$ is zero. This prohibition of absorption during the sampling process is equivalent to the treatment of absorption by analytical averaging described in Sect. 25α.

γ) Exponential transformation. Let us consider an approximate importance function applicable to the problem of the transmission of radiation through a slab. A photon beam is assumed to be incident in the positive z direction on the face $z=0$ of a slab extending between $z=0$ and $z=a$, and the problem is to calculate the transmission of radiation through the face $z=a$. Taking into account the approximately exponential nature of the attenuation of X rays, one may choose an approximate importance function which, in addition to satisfying condition (25.26), has the form

$$\tilde{Q}(z) = \mathrm{e}^{-c(a-z)}, \tag{25.27}$$

where c is a fixed parameter whose value is discussed below.

As a consequence of (24.7),

$$\tilde{Q}(z_{n+1}) = \tilde{Q}(z_n + \zeta_n \cos\vartheta_{n+1}) \tag{25.28}$$

so that, for given z_n and $\cos\vartheta_{n+1}$, the introduction of the importance function leads to a modification of the distribution of pathlengths ζ_n between the n-th and $(n+1)$-st collisions. The distribution $f(\zeta_n)$, given by (24.6), must be replaced with

$$\left.\begin{aligned} \hat{f}(\zeta_n) &= \frac{f(\zeta_n)\,\tilde{Q}(z_n+\zeta_n\cos\vartheta_{n+1})}{\int\limits_0^\infty f(\zeta_n)\,\tilde{Q}(z_n+\zeta_n\cos\vartheta_{n+1})\,d\zeta_n} \\ &= [\mu(\lambda_{n+1}) - c\cos\vartheta_{n+1}]\,\mathrm{e}^{-[\mu(\lambda_n)-c\cos\vartheta_{n+1}]\zeta_n}. \end{aligned}\right\} \tag{25.29}$$

[1] The requisite sampling techniques are similar to those used in the adjoint problem of evaluating the collision density $F(\boldsymbol{\alpha})$, but the appearance of the kernel $\psi(\boldsymbol{\alpha}'|\boldsymbol{\alpha})$ rather than $\psi(\boldsymbol{\alpha}|\boldsymbol{\alpha}')$ in (25.24) implies that one must follow photon histories in a direction opposite to that of the physical process, starting the histories at the target surface and tracing them back toward the source. (See [*60b*]).

The change from $f(\zeta_n)$ to $\hat{f}(\zeta_n)$ is equivalent to the replacement of the attenuation coefficient $\mu(\lambda)$ with the quantity

$$\hat{\mu}_c(\lambda) = \mu(\lambda) - c\cos\vartheta. \tag{25.30}$$

In order for (25.29) to be a well-behaved distribution function it is necessary that $\hat{\mu}_c(\lambda)$ be non-negative. This is guaranteed if one restricts the parameter c to be smaller than the attenuation coefficient $\mu(\lambda)$ in the wavelength-range covered by the calculation. One can then regard $\hat{\mu}_c(\lambda)$ as a modified attenuation coefficient, which is large if the photon moves backwards ($\cos\vartheta < 0$) and small if it moves forward ($\cos\vartheta > 0$). The use of $\hat{\mu}_c(\lambda)$ thus implies sampling from a process that is biased so as to favor forward motion of the photon and therefore transmission through the slab.

One may employ the modified attenuation coefficient not only for the computation of pathlengths but also for the computation of photon absorption. In this case the biased sampling process has a simple interpretation, being associated with a modified flux distribution

$$\hat{I}_c(z, \cos\vartheta, \lambda) = e^{cz} I(z, \cos\vartheta, \lambda). \tag{25.31}$$

If (25.31) is substituted into the transport equation (6.9), one obtains the transport equation for the modified flux,

$$\left.\begin{aligned} \cos\vartheta\,\frac{\partial \hat{I}_c(z,\cos\vartheta,\lambda)}{\partial z} &= -\left[\mu(\lambda) - c\cos\vartheta\right]\hat{I}_c(z,\cos\vartheta,\lambda) + \\ &+ \int_{\lambda_0}^{\lambda} d\lambda' \int_{4\pi} d\omega'\, k(\omega,\lambda,\omega',\lambda')\,\hat{I}_c(z,\cos\vartheta,\lambda) + \delta(z)\,\delta(\omega-\omega_0)\,\delta(\lambda-\lambda_0), \end{aligned}\right\} \tag{25.32}$$

which differs from (6.9) only through the replacement of $\mu(\lambda)$ by $\mu(\lambda) - c\cos\vartheta = \hat{\mu}_c(\lambda)$.

The probability p of transmission through the slab is proportional to the flux evaluated at the slab boundary $z = a$. If, by using the attenuation coefficient (25.30) one samples a biased collision density proportional to $\hat{I}_c = e^{cz} I$, the resulting biased transmission probability is

$$\hat{p} = e^{ca}\, p, \tag{25.33}$$

where a is the thickness of the slab. The weight factor (25.1) resulting from the analytical treatment of absorption takes the form[1]

$$W_s(\boldsymbol{\alpha}_1) = 1; \quad W_s(\boldsymbol{\alpha}_n) = \prod_{j=1}^{n-1} \frac{\mu_{sc}(\lambda_j)}{\hat{\mu}_c(\lambda_j)}, \quad n \geqq 2. \tag{25.34}$$

To compute the relative standard deviation $\sigma/\langle t\rangle$ of an estimate obtained with the use of the exponential transformation, one must substitute into (24.29) $\hat{p}$ in place of p, and $\langle T^2 W_s^2\rangle/\langle T W_s\rangle^2$ in place of $\langle T^2\rangle/\langle T\rangle^2$. By increasing the value of c, one increases $\hat{p}$ and thereby decreases $\sigma/\langle t\rangle$, but at the same time one increases $\langle T^2 W_s^2\rangle/\langle T W_s\rangle^2$ and thereby increases $\sigma/\langle t\rangle$. The optimal value of c depends on the nature and configuration of the medium and on the source energy.

---

[1] The interpretation of (25.34) is that on the average $1 - \mu_{sc}/\hat{\mu}_c$ photons are "absorbed" and $\mu_{sc}/\hat{\mu}_t$ photons "scattered" per collision. The quantity $1 - \mu_{sc}/\hat{\mu}_c$ can become negative, in which case it represents the generation of additional photons rather than absorption. In this sense the exponential transformation is an analytical analog of the "splitting" method described in the next subsection.

With limited experience, values of the order of $c \sim 0.7$ to $0.9\mu_s$ have been found satisfactory, where μ_s is the minimum value of the attenuation coefficient in the energy region of interest. The exponential transformation could be improved by making the parameter c a function of the wavelength.

δ) Splitting. In the examples discussed above the approximate importance function was simple enough so that new random sampling techniques, beyond those required for straightforward sampling and described in Sect. 24, were not needed. In general this is not the case. Every new problem with its appropriate importance function may require new sampling schemes. For example, for the treatment of deep penetration it is desirable to sample preferentially those photon histories which are characterized by sequences of several collisions with small energy losses. This would mean abandoning the sampling of wavelength changes according to the Klein-Nishina distribution by the scheme outlined in Fig. 38.

It is possible to simplify the sampling procedure greatly, while largely retaining the advantages of importance sampling, through the following device suggested by VON NEUMANN and ULAM. After the importance function $\tilde{Q}(\boldsymbol{\alpha})$ has been decided upon, one samples the initial state $\boldsymbol{\alpha}_1$ of each history from the biased distribution $\hat{F}_1(\boldsymbol{\alpha})$, but all subsequent states according to the unbiased transition probability density $\psi(\boldsymbol{\alpha}_{n+1}|\boldsymbol{\alpha}_n)$ of straightforward sampling. However, when a photon makes a transition from $\boldsymbol{\alpha}_n$ to $\boldsymbol{\alpha}_{n+1}$, it is split into a number of replacements whose individual histories are then followed. The number of replacements is equal to the ratio of importance functions, $\tilde{Q}(\boldsymbol{\alpha}_{n+1})/\tilde{Q}(\boldsymbol{\alpha}_n)$. Each photon carries a weight factor, and the weight of a parent photon is distributed equally among its replacements. By this procedure the average number of photons in state $\boldsymbol{\alpha}$ is made proportional to $F(\boldsymbol{\alpha})\,\tilde{Q}(\boldsymbol{\alpha})$, so that—according to (25.21)—the effect is the same as if one had sampled according to a biased transition probability $\hat{\psi}(\boldsymbol{\alpha}_{n+1}|\boldsymbol{\alpha}_n)$ associated with $\tilde{Q}(\boldsymbol{\alpha})$. The ratio $\tilde{Q}(\boldsymbol{\alpha}_{n+1})/\tilde{Q}(\boldsymbol{\alpha}_n)$ is usually not an integer, but contains an integral part q and a fractional part q'. The appropriate number of replacements can be determined by random sampling. It is chosen as q with probability $1-q'$, and as $q+1$ with probability q', so that the average replacement number is $(1-q')q+q'(q+1)=q+q'=\tilde{Q}(\boldsymbol{\alpha}_{n+1})/Q(\boldsymbol{\alpha}_n)$. If a photon makes a transition to a state of lesser importance, $\tilde{Q}(\boldsymbol{\alpha}_{n+1})$ is smaller than $\tilde{Q}(\boldsymbol{\alpha}_n)$ and the integral part of the replacement number is zero which implies that the history is terminated with probability $1-q'$.

ε) Correlated sampling. It is often advantageous to use the same set of photon histories as the basis of the solution of more than one problem. For example, the influence of the composition of the medium on X-ray penetration can be investigated in this manner. It has been shown in Sect. 24 that one can convert histories in one material into histories in another material by a suitable change of the pathlengths between collisions and of the survival probabilities, all other characteristics of the history remaining unchanged. Let τ_1 and τ_2 denote the radiation transport across the target surfaces of two different media, the configurations of source and medium being the same in the two cases, and let $\langle t_1\rangle$ and $\langle t_2\rangle$ denote the corresponding Monte Carlo estimates. If the histories used for the estimate $\langle t_1\rangle$ are identical with the histories used for the estimate $\langle t_2\rangle$ in regard to wavelengths, directions, and distances between collision in units of mean free paths, the two estimates are highly correlated. Therefore the statistical accuracy of the estimate

$$\langle t_1 - t_2\rangle = \frac{1}{H}\sum_{h=1}^{H}(t_{h1} - t_{h2}) \tag{25.35}$$

of the difference $\tau_1 - \tau_2$ is more accurate than the individual estimates $\langle t_1 \rangle$ and $\langle t_2 \rangle$. The variance σ_{12}^2 of $\langle t_1 - t_2 \rangle$ is given by

$$\sigma_{12}^2 = \sigma_1^2 - 2\varrho\,\sigma_1\,\sigma_2 + \sigma_2^2 \tag{25.36}$$

where σ_1^2 and σ_2^2 are the respective variances and ϱ is the correlation coefficient of $\langle t_1 \rangle$ and $\langle t_2 \rangle$; σ_{12}^2 is small if the correlation coefficient is close to one.

Correlated sampling could similarly be used to determine the effect of the direction of the incident beam on the transmission of X rays through a slab. Instead of defining the slab boundaries by the equations $z=0$ (or a) one uses instead the definition $x \sin\vartheta_1 \cos\varphi_1 + y \sin\vartheta_1 \sin\varphi_1 + z \cos\vartheta_1 = 0$ (or a), where the parameters ϑ_1 and φ_1 are to be identified with the direction of incidence of the beam. In this manner one can run through a whole set of directions of incidence with the same set of histories.

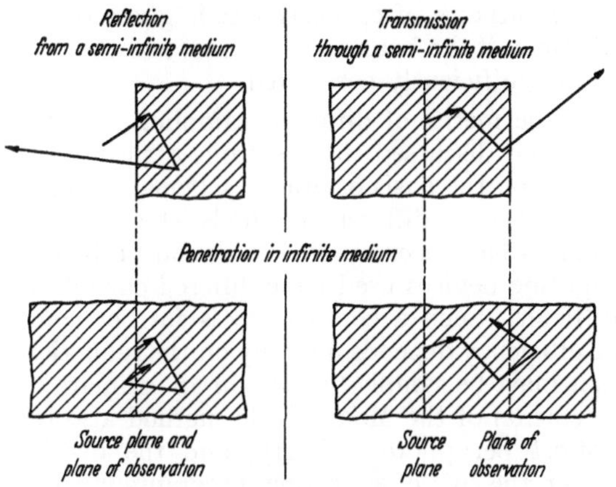

Fig. 39. Comparison of photon trajectories in semi-infinite and infinite media.

Correlated sampling is also a sensitive technique for investigating the effect of changes of the scattering and absorption cross sections. Two correlated calculations are done with different assumed cross sections, all other random elements being kept alike as much as possible. Such a difference technique has been used to establish the effect of the polarization-dependence of the scattering cross section on X ray flux distributions[1].

Perhaps the most important use of correlated sampling consists of making the large body of results for infinite-homogeneous media applicable to the solution of boundary problems. The Monte Carlo method is then required only to supply *corrections* which take into account the presence of boundaries or inhomogeneities. Because these corrections are often small, particularly for the energy flux or dose, the requirements in regard to the accuracy of the Monte Carlo calculation can be relaxed. A further simplification may result from the circumstance that boundary effects often depend little on the distance between the source and the boundary, so that it may be unnecessary to extend the Monte Carlo treatment to deep penetrations. When applied to the illustrative problem of Sect. 24γ, the correlated sampling technique involves the assignment of a double role to the boundary Σ. In one part of the calculation it is regarded as absorbing, and in another part as non-absorbing. The radiation transport through the target surface is expressed in the form

$$\tau = \tau^\infty + (\tau - \tau^\infty), \tag{25.37}$$

where τ^∞ is the radiation transport through the target surface Σ_1 which would occur if the medium were infinite. With τ^∞ known from infinite-medium theory, it is sufficient to estimate the difference $\tau - \tau^\infty$. The initial portion of each sampled

[1] L. V. SPENCER: Unpublished. The small flux changes (of the order of a few percent) thus found are in agreement with the results of a later difference calculation by the moment method (see Fig. 26).

history up to the first crossing of the boundary Σ is analyzed to obtain a score t_h as defined by (24.22). The history is then continued as if the medium were infinite and the photon is allowed to cross Σ repeatedly (see Fig. 39 for an example). The entire history is used to compute a score

$$t_h^\infty = \Sigma\, T(\boldsymbol{\alpha}_N^{(h)}), \tag{25.38}$$

where the sum is taken over all states $\boldsymbol{\alpha}_n^{(h)}$ which are reached by a transition $\boldsymbol{\alpha}_{n-1}^{(h)} \to_n^{(h)}$ involving a crossing of the target surface Σ_1. The quantity

$$\langle t - t^\infty \rangle = \frac{1}{H} \sum_{h=1}^{H} (t_h - t_h^\infty) \tag{25.39}$$

constitutes an estimate of $\tau - \tau^\infty$. The variance of this estimate is given by an expression analogous to (25.35) and is small if the boundary problem and the corresponding infinite-medium problem are sufficiently correlated.

ζ) Efficiency of different schemes. The merits of different sampling schemes and estimation methods have so far been discussed mainly from a theoretical standpoint in the Monte Carlo literature, on the basis of drastic schematizations; but there has been no systematic comparison of different methods when applied in practice. Even when various techniques have been applied to similar problems, the assumed conditions and the computing devices used have differed enough to make a comparison of efficiency difficult. Such a comparison is further complicated by the circumstance that the efficiency depends not only on the reduction of the number of photon histories that must be sampled to obtain an estimate of specified accuracy. Most of the refinements of the Monte Carlo method also require some additional computing effort per photon history, and the overall figure of merit of any scheme depends on the cost of analyzing, programming and running the problem on a computer. In a calculation of the integrated energy flux transmitted through an iron slab 16 mean free paths thick[1] it has been found that the analytical treatment of spatial displacements reduced the computation time by a factor of 2500, and the use of correlated sampling by an additional factor of 80. In problems involving deep penetration, savings of this order of magnitude are not unusual.

26. Monte Carlo solutions of boundary problems. A summary list of problems that have been treated by random sampling, as well as indications of the special techniques used, will be found in Table 18. Most of these problems will be discussed briefly in this section, in some cases together with relevant experimental evidence.

We shall have frequent occasion to relate boundary problems to corresponding infinite-medium problems. By way of example, such a correspondence is indicated in Fig. 39 for the reflection of radiation incident on a semi-infinite medium and for the emergence of radiation from a source in the interior of a semi-infinite medium. By "boundary effect" we mean in this case the ratio of the flux (or dose, build-up factor, etc.) in the semi-infinite medium to the flux at the same location that would prevail if the semi-infinite medium were replaced by an infinite medium, all other conditions remaining unchanged.

α) Reflection. The reflecting power of a bounded medium can be expressed by a variety of parameters. The quantity most commonly used in the *albedo*. The "number albedo" is defined as the ratio of the number of photons emerging from the medium to the number entering. If one is interested in the spectral or

[1] Carried out on the National Bureau of Standards Automatic Computer (SEAC) (Ref. [7]).

Table 18. *Boundary problems treated by the Monte Carlo method.*

Ref. No.	Author(s)	Source energy (Mev)	Geometry	Medium	Type of boundary problem	Method
[*99*]	WILSON	20 to 500		Pb	Electron and photon initiated showers	Srs; random sampling of electron and positron as well as photon trajectories
[*21*]	CAMPBELL and BOYLE	6, 12, 18		NaI	Energy dissipation in NaI spectrometer at very high energies	Srs; random sampling of electron and positron as well as photon trajectories
[*51*]	KAHN	1.28	PLP	H_2O	T	Imp, A, C
[*44*]	HAYWARD and HUBBELL	1.00	PLP, PLO PLI	H_2O, Al Cu, Sn Pb	R	A
[*91*]	THEUS et al.	0.66	PLP	H_2O	R and T	A
		6.00	PLO	Pb	T	A; Imp (biased scatt. cross sec. and exponential transformation)
[*4*]	BERGER	0.66	PLP, PLO	H_2O	R and T	A, C, Corr.
[*70*]	PERKINS	0.66 to 6.0	PLP, PLO	Al	R	A, C
[*7*]	BERGER and DOGGETT	0.66, 1, 4 and 10	PLP	H_2O, Fe Sn, Pb	R and T	A, Analytical calculation of all displacements, Corr.
[*90*]	THEUS and BEACH	6.13	PLP, PLO	Fe	R	Srs; annihilation radiation included
[*8*]	BERGER and DOGGETT	0.66 to 4.45	PTI PCL	NaI	Energy dissipation in NaI spectrometer	Srs
[*6*]	BERGER	1.28	PTI	H_2O	Energy dissipation near interface of two media	A, Corr.
[*11*]	BERGER and SPENCER	1.28	PTI	H_2O	Change of energy dissipation near spherical boundary	A, Corr.
[*11*]	BERGER and SPENCER	1.28	PTC	H_2O	Penetration of collimated beam into semi-infinite medium	A, Corr.
[*102*]	ZERBY	1.3	PLP PLO	Polyethylene Pb	T, composite slab	Srs
Unpublished	KALOS	1 to 6 Mev	PLP	H_2O Pb	T, composite slab	Imp (splitting technique)

R = Reflections by semi-infinite medium
T = Transmission through slab
A = Analytical treatment of absorption
C = Use of collision density method

Imp = Importance sampling
Corr = Correlated sampling
Srs = Straightforward random sampling
PLP = Plane perpendicular

PLO = Plane oblique
PLI = Plane isotropic
PTI = Point isotropic
PTC = Point collimated.

angular distribution of the emerging photons (or on the location of emergence) one can define a corresponding differential albedo whose integral equals the number albedo. The "energy albedo" is the fraction of the total incident X-ray energy which emerges from the medium. It is an integral over the differential

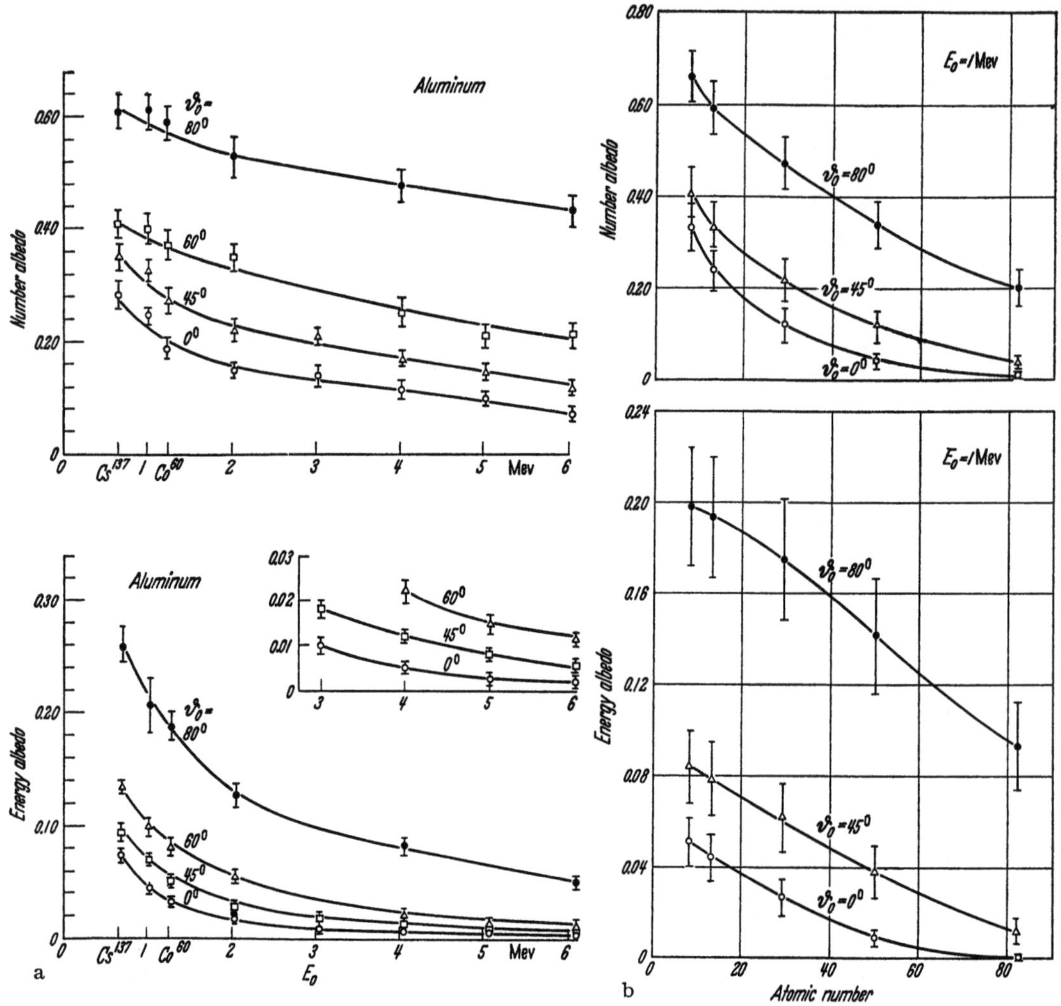

Fig. 40 a and b. Reflection of radiation by a semi-infinite medium, for various directions of incidence ϑ_0 ($\vartheta_0 = 0$ corresponds to perpendicular incidence): (a) Albedo of aluminum as function of the source energy E_0 (Perkins [70]). (b) Albedo as function of the atomic number Z of the medium for 1-Mev radiation (Hayward and Hubbell [44]).

number albedo weighted by the ratios of the energies with which the photons leave and enter the medium. In calculating the albedo, each component of the flux of the incident and emerging X rays has to be multiplied by the cosine of the angle between the X-ray direction and the normal to the surface of the medium; the normal is understood to be directed into the medium for incident X rays and out of the medium for emerging X rays.

Whereas the albedo serves to characterize the surface of a medium as a new source of emerging X rays, different parameters are considered in connection with the response of a detector placed on or near the surface. Detectors often have an equal response to X rays traversing them in any direction. The number of photons traversing the detector is then of interest rather than the number traversing the surface. *Reflection* (or *backscattering*) *coefficients* may be defined in terms of the relative response of a detector, in the presence or absence of an adjacent bounded medium, for a given source-detector configuration. In the following, "energy reflection coefficient" means the fractional increase, induced by the presence of the bounded medium, in the reading of a detector whose response is proportional to the energy of the photons traversing it.

The reflecting power of semi-infinite media is illustrated by Fig. 40. Fig. 40a shows the number-albedo and energy-albedo of aluminum as functions of the energy and direction of the incident radiation, and Fig. 40b the number-albedo

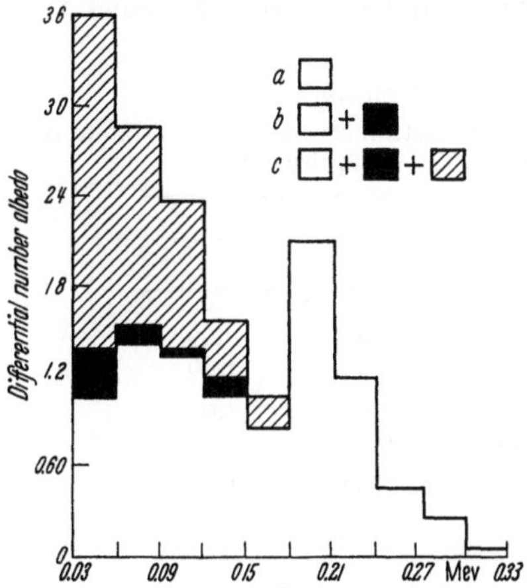

Fig. 41. Spectra of reflected radiation. Plane perpendicular 0.66-Mev source, water medium [4]. (a) Differential number albedo of a slab of thickness $\mu_0 z = 2$. (b) Differential number albedo of a semi-infinite medium. (c) Spectrum of scattered photons traversing the plane $z = 0$ in an infinite medium.

Table 19. *Energy reflection coefficient for semi-infinite media (plane source with energy E_0 and obliquity ϑ_0). The estimated accuracy is* $\pm 10\%$ [7].

Medium	E_0 (Mev)	ϑ_0 0°	ϑ_0 60°
Water	0.4	0.240	0.278
	0.66	0.148	0.189
	1.0	0.081	0.153
	4.0	0.012	0.035
	10	0.002	0.007
Iron	1.0	0.061	0 142
	4.0	0.011	
Tin	1.0	0.022	0.086
	4.0	0.005	
Lead	1.0	0.009	0 042
	4.0	0.001	

and energy-albedo as functions of the atomic number of the medium and the direction of incidence, for 1-Mev radiation. Table 19 contains values of the energy-reflection coefficient for selected energies and media. The exposure-dose reflection coefficients for diverging conical beams from X-ray tubes, as reported in the radiological literature (e.g. [*46*]), are in agreement as to order of magnitude with the results of Table 19. A further analysis of backscattering experiments is given in Sect. 30[1].

The reflecting power of a semi-infinite medium is a decreasing function of the atomic number and of the energy of incidence, and the energy-albedo is much lower than the number-albedo. An explanation of these features can be obtained by an analysis of the spectral and directional distribution of the reflected radiation. Fig. 41 shows the number albedo of water for radiation from a plane perpendicular 0.66-Mev source. Three related energy spectra are contained in Fig. 41, of which (a) pertains to a slab, (b) to a semi-infinite medium and (c) to an infinite medium. The upper end of the spectra is at ~ 0.3 Mev because of the considerable energy loss associated with the reversal of the direction of a photon. There are two peaks, one at ~ 0.2 Mev which is mainly due to singly- and doubly-scattered

[1] *Note added in proof:* For more extensive and systematic calculations see M. J. BERGER and D. J. RASO, U.S. Nat. Bur. of Standards Report 5982, July 25, 1958 (unpublished).

radiation, and another at $\sim$0.05 Mev due to multiply-scattered radiation. For the purpose of comparison, histogram (c) shows the spectrum of the photons that would traverse the same surface if the medium were unbounded. The difference between (c) and (b) represents the contribution of the multiple traversals. By preventing multiple traversals the boundary acts as a sink for low energy photons, and in particular reduces greatly the peak at 0.05 Mev. In media of high atomic number, where soft radiation is present only in very small amounts, the boundary effect is correspondingly small. The difference between histograms (b) and (a) represents photons which have been prevented from being reflected by escaping through the far side of the finite barrier. This difference is small and most apparent at low energies. The reflecting power of a barrier is close to that of a semi-infinite medium, except for very thin barriers. Table 20 shows that

Table 20. *Dependence of reflection on slab thickness.*

The last three columns give the ratio of the energy reflection coefficient for a slab to that of a semi-infinite medium. (Plane source with energy E_0 and obliquity ϑ_0) [7].

Medium	E_0	ϑ_0	Slab thickness, $\mu_0 z$		
			0 5	1 0	2 0
Water	0.66	0°	0.65	0.88	0.99
		60°	0.61	0.96	1.00
Iron	1.00	0°	0.79	0.93	1.00
		60°	0.89	0.98	1.00
Tin	1 00	0°	0 95	0 99	1.00
Lead	1.00	0°	0 97	1.00	1.00

barriers with a thickness of one mean free path (m.f.p.) of high-Z material, or two m.f.p. of low-Z material have practically the same energy reflection coefficient as a semi-infinite medium.

Fig. 42 shows the angular distribution of the reflected X rays resulting from the perpendicular incidence of a beam of Co^{60} radiation on a semi-infinite medium of aluminum. The intensity is roughly proportional to $|\cos\vartheta|$, but the peak of the angular distribution appears to be slightly displaced from the direction normal to the source plane.

At high energies, particularly in high-Z materials for which the albedo would otherwise be insignificant, reflection of radiation in the form of annihilation quanta is important. Such quanta, being emitted isotropically with an energy of 0.51 Mev, have a good chance of emerging from a bounded medium if they are produced near the boundary. It has been found by THEUS and BEACH [*90*] that in the case of 6.13 Mev radiation incident perpendicularly on iron, annihilation quanta contribute 74% of the number-albedo and 81% of the energy-albedo. In lead, annihilation quanta make up almost the entire albedo. It has been calculated [*9*] that $\sim$0.1 annihilation quanta are reflected per 10-Mev photon incident perpendicularly on a semi-infinite lead medium.

β) Transmission. The random sampling calculation of transmission is much more difficult than that of reflection because the fraction of photons emerging from the medium is much smaller, particularly for thick barriers which are important in shielding studies. By means of an interesting example, H. KAHN has pointed up these difficulties and shown how they can be overcome with importance sampling [*51*]. In a calculation of the transmission of 1.28 Mev photons through a water barrier he used two different approaches: (1) the collision density method, and (2) the collision density method reinforced by a biased sampling scheme appropriate for the treatment of deep penetration. Fig. 43 compares the results of the two methods in terms of the build-up factor for energy transmission (fraction of the incident energy that emerges from the opposite side of the slab). For thick slabs the discrepancy between methods (1) and (2) is inconsistent with the indicated statistical standard deviations. For comparison, the results of a build-up factor calculation by the moment-method for an infinite medium are

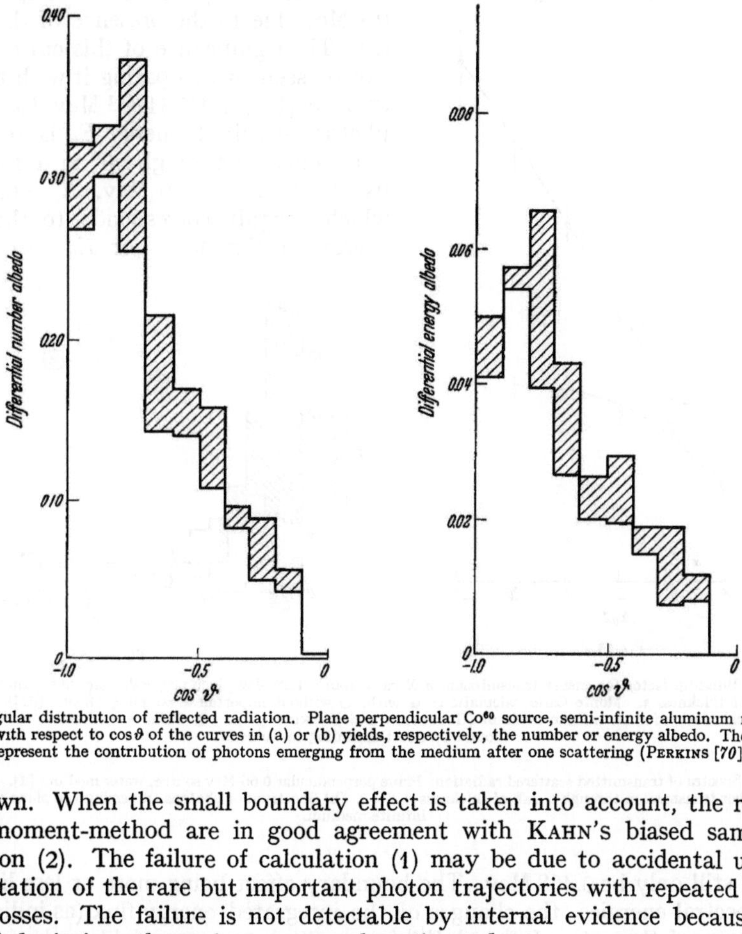

Fig. 42. Angular distribution of reflected radiation. Plane perpendicular Co^{60} source, semi-infinite aluminum medium. Integration with respect to cos ϑ of the curves in (a) or (b) yields, respectively, the number or energy albedo. The shaded areas represent the contribution of photons emerging from the medium after one scattering (PERKINS [70]).

also shown. When the small boundary effect is taken into account, the results of the moment-method are in good agreement with KAHN's biased sampling calculation (2). The failure of calculation (1) may be due to accidental under-representation of the rare but important photon trajectories with repeated small energy losses. The failure is not detectable by internal evidence because the standard deviations themselves were underestimated.

The transmission of beam radiation through slabs has also been treated by the correlated sampling technique discussed in Sect. 25 δ [7]. Let $I_E^{(s)}(z)$ denote the integrated energy flux at a distance z from a plane source in an infinite medium, $I_E^{(s)*}(z)$ the corresponding quantity in a semi-infinite medium extending from $-\infty$ to z, and $I_E^{(s)**}(z)$ the corresponding quantity in a slab extending from 0 to z. (As indicated by the superscript s, all three distributions pertain to *scattered* radiation.) The boundary effect can be summarized by means of the ratios

$$K_1^{(s)}(z) = I_E^{(s)*}(z)/I_E^{(s)}(z) \tag{26.1}$$

and

$$K_2^{(s)}(z) = I_E^{(s)**}(z)/I_E^{(s)}(z). \tag{26.2}$$

The ratio $K_2^{(s)}$ is approximately equal to $K_1^{(s)}$ for penetrations exceeding about two m. f. p. Numerical values of these ratios for plane perpendicular sources and various slab materials will be found in Table 21. Between one and four m.f.p., $K_2^{(s)}$ slowly increases, and from four to sixteen m.f.p. it is almost constant, which indicates the establishment of equilibrium angular and spectral distributions.

The energy spectrum of the X rays transmitted through a slab, together with the corresponding infinite-medium spectrum, is shown in Fig. 44. The difference between the two histograms indicates the loss of photons at energies below 0.3 Mev due to the presence of the boundary. The significance of this cut-off energy can be seen by comparing it with the energy $E' = (1/E_0 + 1/0.511)^{-1}$ Mev to which a photon of initial energy E_0 is reduced if it is deflected through 90° in a single collision. If $E_0 = 0.66$ Mev, $E' = 0.288$ Mev which roughly corresponds to the cut-off energy in Fig. 44. For $E_0 = 10$ Mev, E' would still only be 0.468 Mev. The boundary effect being more or less limited to low spectral energies, the changes of the integrated energy flux (as indicated by departure of the ratios $K_1^{(s)}$ and $K_2^{(s)}$ from unity) are appreciable only for conditions where soft components make a substantial contribution to the spectrum, i.e. in low-Z materials and for low source energies.

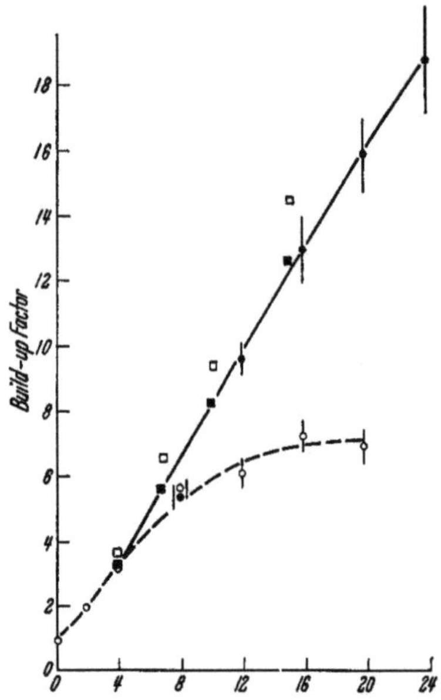

Fig. 43.

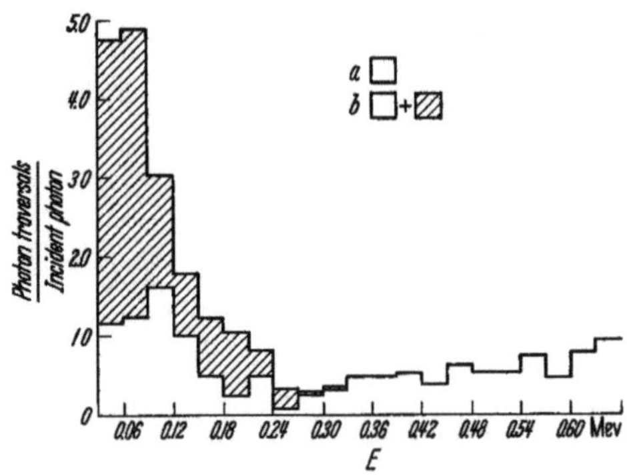

Fig. 44.

Fig. 43. Build-up factor for energy transmission of X rays from a 1.28-Mev plane perpendicular source through a water barrier of thickness z. Monte Carlo calculation: ● with, ○ without importance sampling (KAHN [51]). □ Moment-method calculation for an infinite medium (GOLDSTEIN and WILKINS [39]). ■ Infinite medium values corrected for boundary effect according to Table 21.

Fig. 44. Spectra of transmitted scattered radiation. Plane perpendicular 0.66-Mev source, water medium [4]. (a) Differential number transmission through a slab of thickness $\mu_0 z = 2$. (b) Spectrum of photons traversing the plane $\mu_0 z = 2$ in an infinite medium.

Table 21 is also applicable, in good approximation, to the energy dissipation or the exposure dose. Fig. 45 contains a comparison of theoretical and experimental estimates of the transmission $I_d^{**}(z)/I_d^{**}(0)$ through a slab of thickness z. The experimental transmission curves are lower than the theoretical curves, but have, for $\mu_0 z \gtrsim 2$, the same slope in a semi-logarithmic plot. The downward shift of 10% for Co^{60} radiation in water, and 18% for 0.66-Mev radiation in concrete, is explained by the presence of radiation of less than nominal energy in the source spectra caused by multiple scattering within the sources. This degraded radiation contributes to $I_d^{**}(0)$, but not to $I_d^{**}(z)$ except for small penetrations, because it is quickly absorbed.

For shielding applications it is of interest to know the transmission through composite slabs consisting of layers of different materials. Such calculations have been done by KALOS[1] who used importance sampling via the "splitting" technique. Fig. 46 contains his results on the exposure-dose build-up factor for a 1-Mev beam incident perpendicularly on a layer of three m.f.p. of water followed by a variable thickness of lead. There is an initial drop of the build-up which reflects

Table 21. *Transmission boundary corrections and build-up factors for radiation incident perpendicularly on slabs.* $B_E^* = (B_E - 1)\,K_1^{(s)} + 1$, *and* $B_E^{**} = (B_E - 1)\,K_2^{(s)} + 1$, *where* B_E *is the corresponding infinite medium energy build-up factor (the estimated accuracy of* B_E *is* $\pm 5\%$*)* [7].

Medium	E_0, Mev	$K_2^{(s)}$	$K_1^{(s)}$	$K_1^{(s)} \approx K_2^{(s)}$				B_E^{**}	B_E^*	$B_E^* \approx B_E^{**}$			
		$\mu_0 z$						$\mu_0 z$					
		1	1	2	4	8	16	1	1	2	4	8	16
Water	0.66	0.601	0.663	0 713	0.783	0.785	0.784	1.86	1.95	3.10	5.99	13.3	39.4
	1	0.661	0.720	0.754	0.821	0.828	0.830	1.80	1.87	2.72	5.01	10.5	25.7
	4	0.849	0.885	0.912	0.920	0.926	0.933	1.40	1.42	1.83	2.60	4.21	7.20
Iron	1	0.790	0.798	0.851	0.890	0.895	0.894	1.70	1.71	2.43	4.07	7 80	17.8
	4	0.890	0.910	0.923	0.936	0.932	0.949	1.35	1.35	1.72	2 50	4.17	7.45
	10	0.941	0.959	0.972	0.974	0.978	0.977	1.16	1 16	1.35	1.75	2.80	5.85
Tin	1	0.889	0.911	0.924	0.935	0.938	0.946	1.56	1.56	2 10	3.15	5.31	10.2
	4	0.941	0.926	0.955	0.967	0.974	0.978	1.31	1.31	1 63	2.35	4.12	9.41
	10	0.951	0.960	0.962	0.973	0.971	0.969	1.12	1 12	1.26	1.59	2.75	8.22
Lead	1	0.939	0.951	0.969	0.975	0.979	0.982	1.35	1.35	1.63	2.09	2.87	4.24
	4	0.941	0.977	0.982	0.990	0.992	0.994	1.23	1.23	1.44	1.98	3 28	7.46
	10	0.986	0.990	0.995	0.992	0.994	0.995	1.09	1.09	1.17	1.40	2.17	6.47
Estim. Accuracy (%)		± 5.0	± 2.0	± 1.5	± 1.5	± 2.0	± 2.5						

the quick absorption of the degraded radiation coming through the water as the lead is added. The build-up factor then rises, paralleling the curve of the build-up in pure lead.

According to KALOS, the build-up factor after penetration through a thickness z_1 of *water followed by* a thickness z_2 of *lead* can be represented by the following formula which holds for source energies up to 6 Mev:

$$\left.\begin{aligned} B_d(z_1, z_2) = B_{d,\mathrm{Pb}}(z_2) + \Bigg\{ & \frac{B_{d,\mathrm{H_2O}}(z_1) - 1}{B_{d,\mathrm{Pb}}(z_1) - 1}\, e^{-1.7\mu_0(\mathrm{Pb})\, z_2} + \\ & + \left(\frac{\mu_{sc}}{\mu}\right)_{\mathrm{H_2O}} \left(\frac{\mu}{\mu_{sc}}\right)_{\mathrm{Pb}} \left(1 - e^{-\mu_0(\mathrm{Pb})\, z_2}\right)\Bigg\} \{B_{d,\mathrm{Pb}}(z_1 + z_2) - B_{d,\mathrm{Pb}}(z_2)\}. \end{aligned}\right\} \tag{26.3}$$

When a layer of *lead* of thickness z_2 is *followed by* a layer of *water* of thickness z_1, the resulting build-up factor shows a smooth rise from the value for pure lead toward that for pure water, the rate of approach depending on the source energy. A good fit to the Monte Carlo results is obtained with

$$B_d(z_2, z_1) = B_{d,\mathrm{H_2O}}(z_1) + \frac{B_{d,\mathrm{Pb}}(z_2) - 1}{B_{d,\mathrm{H_2O}}(z_2) - 1} \{B_{d,\mathrm{H_2O}}(z_1 + z_2) - B_{d,\mathrm{H_2O}}(z_1)\}. \tag{26.4}$$

Further results on composite slabs can be found in Refs. [54] and [102].

1 Private communication.

γ) Energy dissipation in bounded media. (i) Spherical boundary. For radiological applications it is of interest to know the spatial distribution of the energy dissipated in a bounded medium. As a simple example, we consider an isotropic point-source at the center of a water sphere of radius R which is in turn surrounded by a perfect absorber (vacuum). The problem is to determine the energy dissipation $I_D^*(r, R)$ at a distance r from the source. Such a schematization might be used to estimate the dose due to a radioactive isotope lodged in a human body.

The solution $I_D^*(r, \infty) = I_D(r)$ for the limiting case of an infinitely large sphere can be found

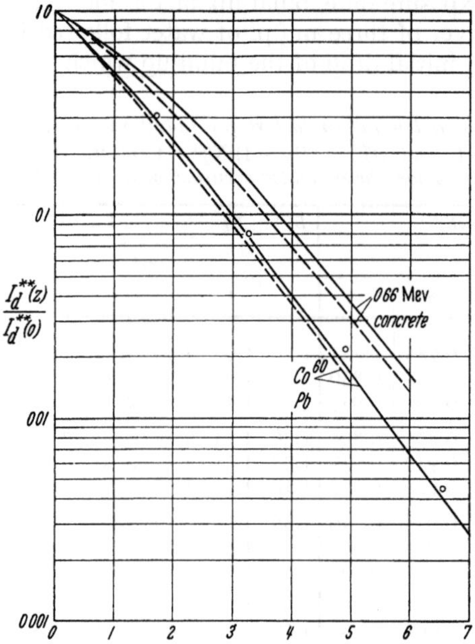

Fig. 45.

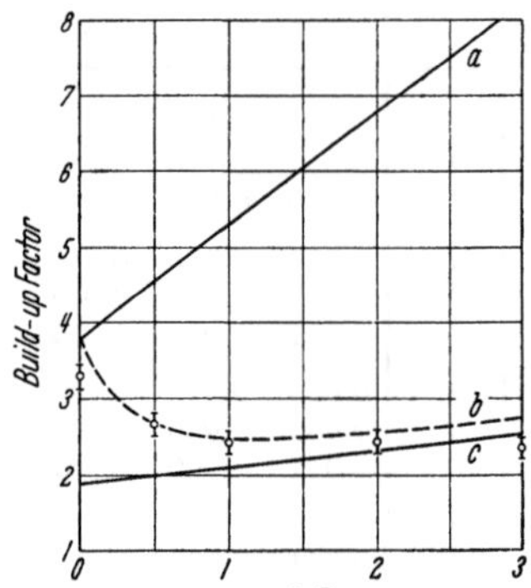

Fig. 46.

Fig. 45. Transmission of radiation from a plane perpendicular source through lead and concrete slabs. Solid curves from moment-method calculation with boundary corrections according to Table 21. Circles from Monte Carlo results of ZERBY [*102*]. Dashed curves from experiments by KIRN *et al.* [*54*].

Fig. 46. Comparison of exposure dose build-up factors for uniform and composite slabs. Plane perpendicular 1-Mev source. (KALOS, private communication). (a) $(3 + \mu_0 z)$ m.f.p. of water. (b) 3 m.f.p of water followed by $\mu_0 z$ m f.p. of lead. (c) $(3 + \mu_0 z)$ m.f.p. of lead. Curves (a) and (c) from a calculation by the moment method corrected for boundary effects according to Table 21. Curve (b) from a Monte Carlo calculation.

by the moment-method. The energy dissipation in the finite sphere is reduced near the boundary, and can be expressed as

$$I_D^*(r\,R) = I_D(r)\left\{[B_D(r) - 1]\,\frac{I_D^{(s)*}(r, R)}{I_D^{(s)}(r)} + 1\right\}. \tag{26.5}$$

Fig. 47 shows the ratio $I_D^*(r, R)/I_D(r)$ for a 1.28 Mev source. The results are based on the use of a build-up factor[1]

$$B_D(r) = 1 + \mu_0 r\,\{1.325\, e^{0.0314\,\mu_0 r} - 0.461\, e^{-0.244\,\mu_0 r}\} \tag{26.6}$$

and a Monte Carlo calculation [*11*] whose results can be summarized by the formula

$$\frac{I_D^{(s)*}(r, R)}{I_D^{(s)}(r)} = 1 - 0.16\, e^{-3.08\,\mu_0 (R-r)}. \tag{26.7}$$

[1] Obtained with the function-fitting method of Sect. 18, see Ref. [*11*].

Notice that the ratio (26.7) depends only on $R-r$, i.e., that the boundary effect is independent of the radius of curvature of the boundary but depends only on the distance from it. It is therefore expected that (26.7) will also be applicable to other types of geometrical boundaries. At the boundary ($r=R$), and for large R, (26.7) should go over into the corresponding ratio $K_1^{(s)}$ for transmission through a slab, as given in Table 21[1]. By interpolation in this table, one obtains a value of $a=0.15$ at a source energy of 1.28 Mev, which is consistent with the value $a=0.16$ in (26.7).

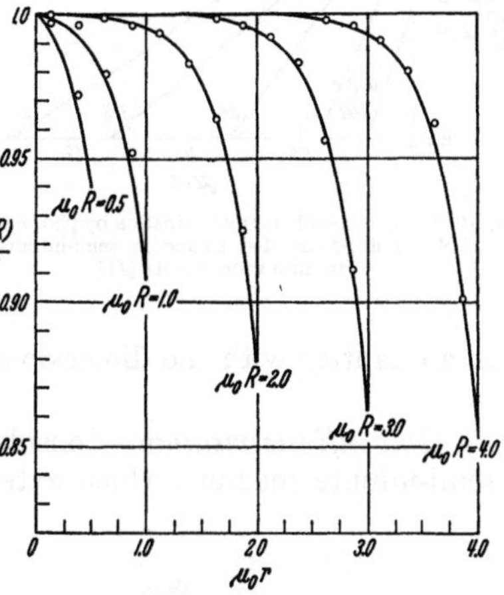

Fig. 47.

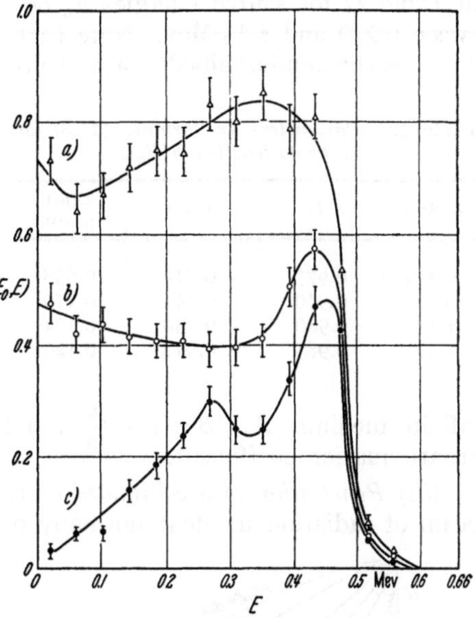

Fig. 48.

Fig. 47. Reduction of energy dissipation near a spherical boundary; 1.28-Mev point isotropic source in water. Points based on direct Monte Carlo results for $I_D^{(s)*}(r, R)/I_D^{(s)}(r)$, solid curves on formula (26.7) [11].

Fig. 48. Response function $P(E_0, E)$ of cylindrical NaI(Tl) crystal for collimated radiation incident axially with energy $E_0 = 0.66$ Mev [8]. $P(E_0, E)\,dE$ is the probability that an incident photon will dissipate an energy E in the crystal. In the notation of Sect. 28β, $P(E_0, E) = Y(E_0)\,F(E_0, E)$.

Crystal dimensions (inches)

	Length	Radius
(a)	2.0	0.75
(b)	0.5	0.25
(c)	8.0	2.5

(ii) Cylindrical boundary. In order to understand the response of a radiation detector, it is frequently necessary to solve a diffusion problem involving a boundary. As an example, consider a NaI scintillation crystal of cylindrical shape, and a narrow collimated beam of photons incident along its axis. The response of such a device is not proportional to the energy of the incident photons, because some photons are not completely absorbed in the crystal but emerge from it with reduced energy after one or more scatterings. Fig. 48 shows calculated spectral distributions of the energy dissipated in a NaI crystal by those photons which eventually escape[2] [8]. The use of such information for the analysis of X-ray spectra is discussed in Sect. 28β.

Monte Carlo results for the fraction p of interacting photons that are completely absorbed in the crystal (photo-fraction) can be summarized with an average

[1] Note that $K_1^{(s)}$ as given in Table 21 applies to the integrated energy flux, while (26.7) applies to the absorbed dose, but the difference is insignificant.

[2] An "orders of scattering" calculation, by MAEDER *et al.* [56], mainly applicable to small and medium-sized crystals, is in good agreement with the Monte Carlo results.

accuracy of 2% by the formula

$$p\left(\sqrt{LR}, E_0\right) = A(E_0) - B(E_0)\, e^{-c(E_0)\sqrt{LR}}, \tag{26.8}$$

which applies to collimated radiation with energies E_0 incident axially on the end-face of cylindrical NaI crystals of radius R and length L. Numerical values of the parameters A, B and C are given in Table 22 for source energies E_0 between 0.279 and 1.33 Mev. Note that $1-A$ is the number albedo for a semi-infinite medium, $A - B = 1 - \frac{\mu_{sc}}{\mu}$, and C is a quantity with the dimensions of an attenuation coefficient.

Table 22. *Parameters in formula* (26.8) *for the photo-fraction* $p(E_0)$.

E_0 (Mev)	$A(E_0)$	$B(E_0)$	$C(E_0)$ (cm^{-1})
0.279	0.973	0.514	0.658
0.661	0.940	0.841	0.219
1.17	0.980	0.934	0.124
1.33	0.985	0.942	0.120

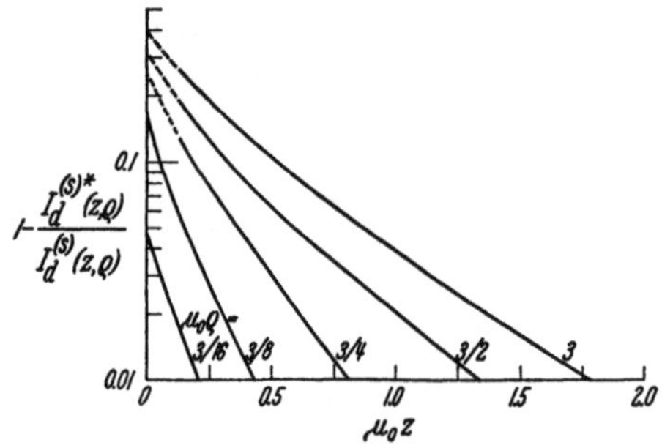

Fig. 49. Comparison of energy dissipation by photons from a 1.28 Mev point-collimated source in semi-infinite and infinite water media [11].

(*iii*) *Penetration of a collimated beam into a semi-infinite medium.* Consider a beam of radiation incident normally on a semi-infinite medium which extends

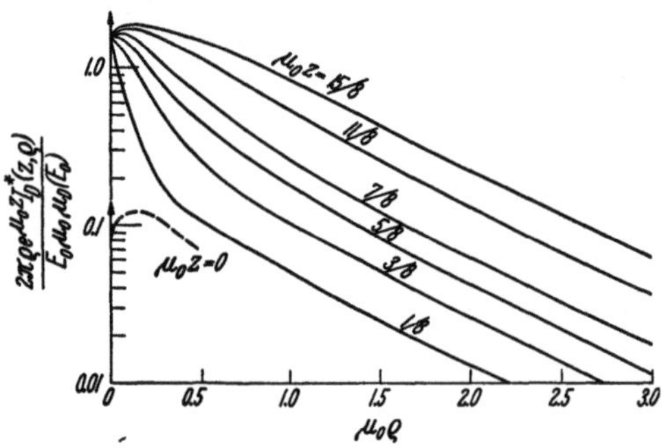

Fig. 50.

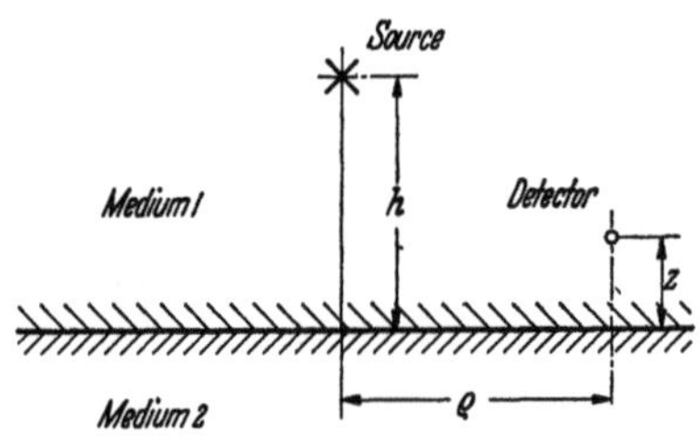

Fig. 51.

Fig. 50. Distribution of energy dissipation about the axis of a 1.28 Mev point-collimated source in a semi-infinite water medium, at various penetration depths z. Normalization as in Fig. 33 [11].

Fig. 51. Geometry of interface problem.

between $z = 0$ and $z = \infty$, with a resulting energy dissipation distribution $I_D^*(z, \varrho)$. The corresponding distribution $I_D(z, \varrho)$ for an infinite medium has already been described in Sect. 19 (Fig. 33a). The boundary effect, i.e., the ratio $I_D^*(z, \varrho/I_D(z, \varrho)$, as calculated by correlated random sampling, is shown in Fig. 49 for 1.28 Mev radiation incident on water. Fig. 50 shows plots of $I_D^*(z, \varrho)$ vs. ϱ for various values of z. These are quite similar in shape to the corresponding infinite-medium curves in Fig. 33a. Fig. 50 was not obtained by a combination of the results in Figs. 33a and 49 (although it is consistent with them) because too much interpolation would have been required. Instead, it was based on an independent correlated random sampling calculation, the correlation utilized being in this case that between the once-scattered and multiply-scattered flux.

δ) Propagation of radiation near the interface of two media. We consider now the situation sketched in Fig. 51. The region $z \geqq 0$ is filled with one kind of medium (No. 1), and the region $z<0$ with a different medium (No. 2). A point-isotropic source is located at height $z=h$ in medium No. 1. We wish to know the energy dissipation in medium No. 1 as a function of z and the lateral distance ϱ between source and detector.

First we take up the case in which medium No. 2 is much denser than medium No. 1, but otherwise has the same scattering and absorption cross-sections[1]. The presence of the dense medium has two opposite effects, which are most

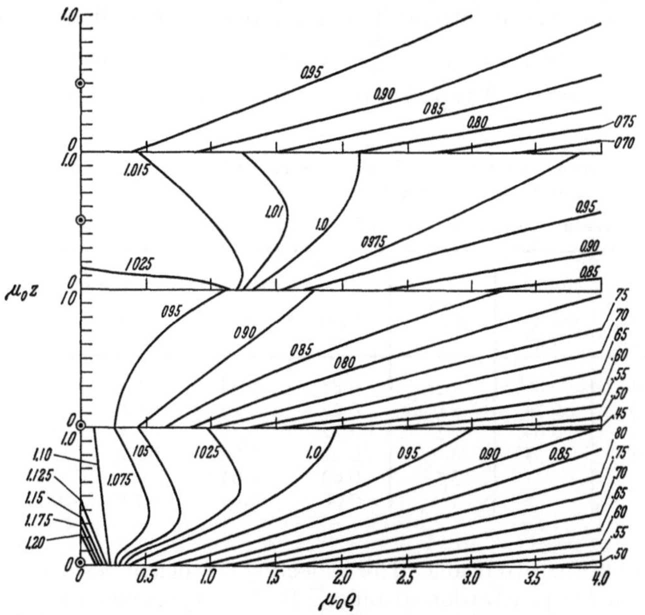

Fig. 52. Energy dissipation near the interface between two media of different density: Contour diagrams of the ratios K and K' (1.28-Mev point isotropic source, water-like medium). Source position indicated by ⊙ [6].

(a) K, (b) K' } $\mu_0 h = 0$ (c) K, (d) K' } $\mu_0 h = 0.5$

pronounced when the detector is at the boundary. Photons have an increased chance of being reflected toward the detector if they enter the dense medium close to the detector, which is likely to occur if the lateral distance ϱ between source and detector is small. On the other hand, photons will be sidetracked and prevented from reaching the detector if the point of entry is far from the detector, which is likely if ϱ is large. Thus there is a region of increase of energy dissipation near the source, and a region of decrease far from the source, compared to the spatial pattern of energy dissipation in a homogeneous medium.

A detailed mapping of the "topography" of the boundary effect has been obtained by correlated sampling for 1.28 Mev radiation in water, the ratio of the densities in the two regions being assumed to be infinite [6]. Results of this calculation are shown in Fig. 52 in the form of a contour map in the (z, ϱ)-plane. When these contours are rotated about a vertical axis through the source, they generate surfaces on which the energy dissipation is changed by a factor $K(z, \varrho)$

[1] This situation could be realized practically by combinations such as air-water, air-concrete or air-aluminum, because the attenuation coefficients (in cm^2/g) are very nearly the same for these materials over a wide range of energies.

compared to the energy dissipation $I_D(r)$ at a distance $r = \sqrt{\varrho^2 + (z-h)^2}$ from the source in an infinite homogeneous medium.

In addition to the limiting case of an abrupt change in density (but no change of chemical composition) another limiting case has been considered in which medium No. 2 is assumed to be a perfect absorber. The correction factor for this situation is denoted by $K'(z, \varrho)$, and its contours are also shown in Fig. 52.

As anticipated in the qualitative discussion, K is greater than one near the source and smaller elsewhere, wheareas K' is smaller than one everywhere because the presence of an absorbing boundary can only lower the flux. K' is smaller than K everywhere, but at intermediate and large distances from the source the difference is small. The quantity $1 - \frac{K'}{K}$ is a kind of reflection coefficient indicating the fraction of the energy dissipation at any one point due to radiation reflected

Table 23. *Comparison of theoretical and experimental results for energy dissipation near the interface separating two media of different density.*

Water: [*6*]. Iron: Titus [*92*]. Air-ground: Soole [*79*].

$\mu_0 \varrho$	K ($\mu_0 h = 0$) Water (calc)	K ($\mu_0 h = 0$) Airground (exp)	$\mu_0 \varrho$	K ($\mu_0 h = 0.16$) Water (calc)	K ($\mu_0 h = 0.16$) Iron (exp)	K ($\mu_0 h = 0.5$) Water (calc)	K ($\mu_0 h = 0.5$) Iron (exp)	$1-K'/K$ ($\mu_0 h = 0$) Water (calc)
0.64	0.91	0.86	1.5	0.87	0.85	0.97	0.97	0.077
1.28	0.75	0.74	2.0	0.78	0.76	0.92	0.93	0.078
1.92	0.64	0.64	2.5	0.72	0.70	0.88	0.89	0.080
2.56	0.057	0.55	3.0	0.67	0.67	0.85	0.86	0.082
			3.5	0.63	0.64	0.82	0.81	0.084
			4.0	0.60	0.61	0.78	0.77	0.086

from the denser of the two media. This reflection coefficient is small and changes only slowly with the penetration distance. It thus represents a boundary effect following the pattern found in the previous examples in this Section.

By contrast, the density-interface boundary effects, as represented by the correction factors K and K', have novel characteristics: (1) The boundary corrections are large, (2) they tend to increase without limit as the penetration distance increases, and (3) they involve predominently high-energy rather than low-energy spectral components. The reason for this is that we are considering the propagation or radiation *along* rather than *across* a boundary, so that the photons are continually affected by the boundary throughout their travel from source to detector.

The theoretical results are compared in Table 23 with experiments by Soole [*79*] and Titus [*92*]. Soole measured the propagation of radiation in the air above the ground. In the experiment of Titus, the dense material was steel, and the light material steel-wool (with a density $\frac{1}{30}$-th that of steel). Because the presence of the boundary affects mainly high-energy spectral components for which Compton-scattering is more important than photo-electric absorption, the experimental ratio K for iron should lie close to the calculated ratio for water. This is in fact the case within the limits of the theoretical and experimental uncertainties.

27. Adaptation of infinite-medium theory to boundary problems. α) *Approximate solution by schematization.* The knowledge of the directional distribution of the flux in an infinite homogeneous medium is in some circumstances sufficient to provide a good approximate solution of boundary problems. As a typical example, we discuss the calculation of the flux transmitted through a plane parallel barrier. The schematization in this case consists of equating the transmitted flux with the flux at the same depth in a corresponding infinite medium integrated over all directions corresponding to emergence from the barrier. The validity of this approximation in a specific case is demonstrated by Fig. 53, which presents a Monte Carlo analysis of the directional distribution of the energy flux at a depth of 4 m.p.f. in an infinite water medium (plane perpendicular 0.66 Mev source). The angular distribution is broken down into three components: (1) the inwardbound flux I_{in} at a depth $z=4$ m.f.p., due to photons passing this depth for the 2-nd, 4-th, ... $(2n)$-th time; (2) the outwardbound flux $I_{\text{out}}^{(1)}$ of photons making a first passage; (3) the outwardbound flux $I_{\text{out}}^{(3+)}$ of photons making the 3-rd, 5-th,... $(2n+1)$-st passage. The flux transmitted through a barrier is equal to $I_{\text{out}}^{(1)}$, whereas according to the schematization, one would set it equal to $I_{\text{out}}^{(1)} + I_{\text{out}}^{(3+)}$, thus incorrectly including photons which have made repeated passages through the boundary plane. In the example of Fig. 53, the resulting overestimate of the emergent energy flux through a barrier is only 4%.

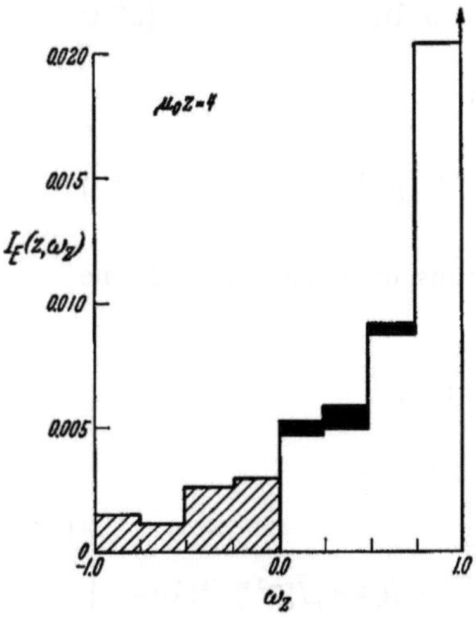

Fig. 53. Analysis of the angular distribution of the integrated energy flux across a surface at 4 m.f.p. from a plane perpendicular 0 66-Mev source in water [7]:
□ $I_{\text{out}}^{(1)}$, first outward traversal;
▨ I_{in}, inward traversals;
■ $I_{\text{out}}^{(3+)}$, multiple outward traversals.

The preceding schematization has been found useful in calculating the flux of annihilation radiation emerging from a semi-infinite medium [9]. Schematizations in the same spirit have also been applied to problems involving partially shielded gamma-ray detectors, in which the effect of the shielding materials was simulated by endowing the detector with a fictitious efficiency depending on the direction from which the radiation reaches it [10].

β) Density scaling method. It is possible to extend the moment-method to the treatment of diffusion in media which vary in density in one space-direction but are infinite and of uniform composition. In this event the attenuation coefficient and the Klein-Nishina differential scattering coefficient depend on the position in space only through a common density factor $D(z)$, where the z-axis is the direction of the density variation. This schematization is applicable to the propagation of radiation in the atmosphere, whose density variation could be represented approximately by an exponential function; similarly, a stratified two-medium configuration such as the air-water or air-concrete combinations discussed on p. 800, could be represented by a step-function $D(z)$, insofar as the differences in composition may be disregarded.

The following treatment applies to a point source collimated in the z-direction. If the variable z is replaced by the "mass thickness"[1]

$$\tau = \int_0^z D(z')\, dz', \tag{27.1}$$

[1] Also called "optical thickness" in the astrophysical literature.

an adaptation of (6.2) and (6.14) yields the transport equation

$$\left.\begin{aligned}\cos\vartheta\frac{\partial I}{\partial\tau}+\frac{1}{D(z)}\left\{\sin\vartheta\cos(\varphi-\beta)\frac{\partial I}{\partial\varrho}-\sin\vartheta\sin(\varphi-\beta)\frac{1}{\varrho}\frac{\partial I}{\partial(\varphi-\beta)}\right\}\\=-\mu(\lambda)\,I(\tau,\varrho,\omega,\lambda)+\int\limits_0^{\lambda}d\lambda'\int\limits_{4\pi}d\omega'\,k(\omega,\lambda,\omega',\lambda')\,I(\tau,\varrho,\omega',\lambda)+\\+\delta(\lambda)\frac{1}{2\pi\varrho}\delta(\varrho)\,\delta(\omega_z-1)\,\delta(\lambda-\lambda_0)\,.\end{aligned}\right\}\tag{27.2}$$

After expanding the flux distribution into spherical harmonics one defines spatial moments of the harmonic expansion coefficients, in analogy to (12.30), as

$$I_{nl}^{\nu m}(\lambda)=2\pi\int\limits_{-\infty}^{\infty}d\tau\int\limits_0^{\infty}\varrho\,d\varrho\,\tau^n\varrho^\nu I_l^m(\tau,\varrho,\lambda)\,.\tag{27.3}$$

It is also convenient to introduce modified moments

$$\tilde I_{nl}^{\nu m}(\lambda)=2\pi\int\limits_{-\infty}^{\infty}d\tau\int\limits_0^{\infty}\varrho\,d\varrho\,\tau^n\varrho^\nu\frac{dz}{d\tau}I_l^m(\tau,\varrho,\lambda)\,.\tag{27.4}$$

By carrying out the requisite integrations over the transport equation (27.2) one obtains the moment-interlinkage equation

$$\left.\begin{aligned}\mu(\lambda)\,I_{nl}^{\nu m}(\lambda)=\int\limits_{\lambda_0}^{\lambda}d\lambda'\,k_l(\lambda,\lambda')\,I_{nl}^{\nu m}(\lambda')+\delta_{n0}\,\delta_{\nu0}\,\delta_{m0}\,\delta(\lambda-\lambda_0)+\\+\frac{n}{2l+1}\left[\sqrt{(l+1)^2-m^2}\,I_{n-1\,l+1}^{\nu m}(\lambda)+\sqrt{l^2-m^2}\,I_{n-1\,l-1}^{\nu m}(\lambda)\right]-\\-\frac{1}{2l+1}\left\{\frac{\nu+m}{2}\left[\sqrt{(l+2-m)(l+1-m)}\,\tilde I_{n\,l+1}^{\nu-1\,m-1}(\lambda)-\right.\right.\\\left.-\sqrt{(l-1+m)(l+m)}\,\tilde I_{n\,l-1}^{\nu-1\,m-1}(\lambda)\right]-\\\left.-\frac{\nu-m}{2}\left[\sqrt{(l+2+m)(l+1+m)}\,\tilde I_{n\,l+1}^{\nu-1\,m+1}(\lambda)-\sqrt{(l-1-m)(l-m)}\,\tilde I_{n\,l-1}^{\nu-1\,m+1}(\lambda)\right]\right\}.\end{aligned}\right\}\tag{27.5}$$

This equation differs from the corresponding Eq. (12.31) for a medium of uniform density only through the appearance of $\tilde I_{n\,l\pm1}^{\nu-1\,m\pm1}(\lambda)$ in place of $I_{n\,l\pm1}^{\nu-1\,m\pm1}$ in the curly brackets. The interlinkage of (27.5) is such that one must first solve the equations for $\nu=0$ and then proceed to successively higher values of ν. When $\nu=0$, one also has $m=0$ (as shown in Sect. 12) so that the terms in curly brackets are accompanied by a factor zero and need not be considered, and the density variation makes no difference. This reflects the circumstance that the set of moments with $\nu=0$ describes the flux from a plane source, in which case density variations are completely taken into account by expressing distances from the source plane in units of the mass thickness.

The moments I_{nl}^{0m} can be utilized to determine by moment-fitting the distribution-in-τ,

$$I_l^{0m}(\tau,\lambda)=2\pi\int\limits_0^{\infty}\varrho\,d\varrho\,I_l^m(\tau,\varrho,\lambda)\,.\tag{27.6}$$

This distribution can then be multiplied by $\tau^n\frac{dz}{d\tau}$ and integrated over τ to yield the modified moments $\tilde I_{nl}^{0m}$ according to (27.4). Knowledge of these moments permits one to solve (27.5) for $\nu=1$. All moments can progressively be determined by this recursion procedure. The practical usefulness of the method may depend on the development of efficient moment-fitting techniques suited to each specific situation.

E. Experimental aspects.

28. Methods and limitations. The experimental study of X-ray penetration and diffusion requires intensity measurements, spectral analysis and measurements of directional distributions, to be performed within an extended medium or on the X rays emerging from a surface. The problem thus formulated can hardly be attacked in its full generality, especially in view of the requirement that the instrumentation itself should not disturb the flow of radiation to any substantial extent. The measurements, particularly of absolute intensity, may be influenced by disturbing factors to an extent that is sometimes amplified by special circumstances. These factors include: (a) The distribution of materials surrounding the source and the detector. For example, it is well known among γ-ray spectroscopists that backscattering even from a distant wall causes a spurious 200 kev component to appear distinctly in γ-ray spectra. (b) Secondary phenomena in the source itself. For example, scattering within a γ-ray source may cause a few percent of the photons to emerge with greatly reduced energy. These low-energy photons may affect the source calibration or the main measurements disproportionately if the detector is particularly sensitive to them or if the experiment involves large angle scattering. (c) Secondary phenomena in the detector. These will be discussed below; they are difficult to evaluate quantitatively and constitute a major obstacle to the absolute calibration of detectors.

X rays are detected and measured by a great variety of instruments which respond to the energy dissipated by secondary electrons traversing their sensitive volume. The detectors may be classified according to whether they measure an integral of the energy delivered by a large number of photons or whether they respond to individual photons, thus acting as counters. In recent years scintillators, which may be used as counters, spectrometers or integrating devices, have largely replaced other detectors. Proportional and Geiger counters have fallen into disuse primarily because of their relatively low efficiency and because their response is a sensitive function of the photon energy. Ionization chambers are convenient because of their simplicity of construction and functioning; their sensitivity has been enhanced by the introduction of vibrating-reed electrometers as current measuring devices.

α) Integrating devices. Ionization chambers and scintillation detectors are both commonly used as integrating devices. Both may be constructed so as to measure the energy dissipated by X rays in a medium under study. A thin-walled ion chamber, which absorbs a very small fraction of the energy of the electrons traversing it, may be used as a probe for this purpose. With proper design, a chamber will disturb the X-ray flow very little, and will yield an ionization current related by theory to the energy dissipated in the medium [*46*], without any need for absolute calibration. A scintillating material can often be selected with a gamma-ray absorption coefficient and stopping power for electrons similar to those of the material of interest. A volume of scintillator, in conjunction with a photomultiplier tube which converts the scintillation light into an electrical current, constitutes a suitable probe for energy dissipation measurements.

β) Spectral analysis. Diffraction spectrometry has not been utilized in the study of penetration and diffusion, despite its high resolution, because of its low efficiency, and because it requires good collimation. Measurements of photon energy spectra with moderate resolution and high efficiency have been made possible by the development of scintillation counters [*78*]. A photon of energy E incident on a scintillation phosphor transfers to secondary electrons within the phosphor an energy $E' \leqq E$. Some fraction of E' goes into a light flash which is then converted by a photomultiplier tube into an electrical pulse to be amplified and measured electronically. The pulse size measurement may be regarded as a measurement of the dissipated energy E'. The pulse size may be expressed on a suitable scale by an energy variable E'' which has a statistical distribution centered at E'. This distribution is approximately Gaussian, with a variance proportional to E' determined by fluctuations in the number of electrons liberated at the cathode of the photomultiplier:

$$G(E', E'') = \frac{1}{\sqrt{2\pi a E'}} \, e^{-(E''-E')^2/2aE'}. \tag{28.1}$$

The parameter a is a characteristic of the spectrometer and must be determined by experimental observations.

The relationship between the observed pulse size E'' and the incident photon energy E is represented by a response function

$$R(E, E'') = \int_0^E dE' \, L(E, E') \, G(E', E''), \tag{28.2}$$

where $L(E, E')$ indicates the probability that an energy dissipation E' results from the incident photon. This probability depends on the mechanism of X-ray interaction and therefore is a function not only of E but of the material and the size of the phosphor. The phosphors used most frequently are thallium-activated sodium iodide and organic crystals such as anthracene. When E is low and the phosphor contains a high-Z element, such as I in NaI(Tl)[1], photoelectric absorption is the predominant process and E' equals E, except for the possible escape of fluorescent X rays, so that $L(E, E') \sim \delta(E - E')$. When the incident photon interacts through a single Compton scattering, which is the rule in thin organic crystals, $L(E, E')$ has a broad distribution which peaks at the largest possible value of E', i.e., $E' = E\Big/\left(1 + \frac{mc^2}{2E}\right)$. If the phosphor is large, many photons may be totally absorbed following multiple Compton scattering, so that $L(E, E')$ again approaches $\delta(E - E')$. For high photon energies and for NaI(Tl), pair production becomes important and $L(E, E')$ would once more equal $\delta(E - E')$ except for the escape of energy by one or two photons of annihilation radiation and, at very high energies, by bremsstrahlung (see Sect. 4). Thus the analysis of the scintillator performance constitutes by itself a complicated problem of penetration and diffusion.

The response function $R(E, E'')$ may be obtained by experimental calibration of a given spectrometer with monochromatic γ-ray sources (see e.g. [*49*])[2]. Fig. 54 shows typical experimental pulse height distributions. Alternatively the factor $L(E, E')$ in (28.2) can also be found by calculation. Under simple conditions, when only the photo-effect matters, or for thin low-Z detectors for which only a single Compton process need be considered, the theoretical calculation of $L(E, E')$ is elementary. Multiple scattering calculations of $L(E, E')$ for NaI crystals are discussed in Sect. 26γ. The function L is conveniently analyzed in the form

$$L(E, E') = Y(E) \, \{p(E) \, \delta(E - E') + F(E, E')\}. \tag{28.3}$$

Here $Y(E)$ indicates the detector efficiency, i.e. the probability that the incident photon experiences at least one interaction; for a simple geometry and a phosphor

---

[1] Thallium-activated NaI crystal.

[2] Concerning the extension of such calibration to high energies with the use of electron beams, see [*55*].

thickness t we have $Y(E)=1-e^{-\mu_0 t}$. The probability that an interaction is followed by total dissipation of the photon energy is called the photofraction and is indicated by $p(E)$, whereas $F(E, E')$ represents the probability distribution of the energy left in the phosphor by escaping photons. Typical values of $p(E)$ are given by (26.9) and sample distributions $Y(E)\,F(E, E')$ in Fig. 48.

In order to extract the spectrum $I(E)$ from the observed pulse height distribution $I_p(E'')$ one must solve the integral equation[1]

$$I_p(E'')=\int_0^\infty R(E, E'')\, I(E)\, dE. \tag{28.4}$$

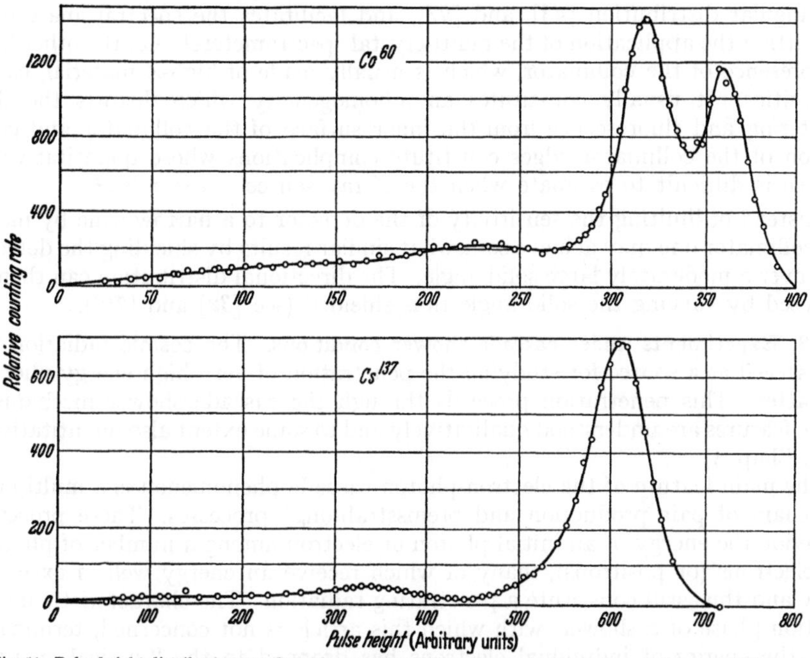

Fig. 54. Pulse-height distributions produced by a collimated X-ray beam incident of a cylindrical NaI(Tl) crystal of 5″ diameter and 4″ length. (KOCH, private communication. See also FOOTE and KOCH [35].)

For numerical solutions of this integral equation see Refs. [2], [25] and [49]. If the spectrum $I(E)$ is known theoretically, one can carry out the integral (28.4) and make a comparison between the theoretical and the experimental pulse height distribution. This procedure was followed in the experiment by E. HAYWARD [42].

For low photon energies, say $E<150$ kev, the spectral analysis with an NaI(Tl) crystal is simplified because in (28.3) we have $F(E, E')\sim 0$. But the analysis of low energy photons in the presence of high energy photons remains complicated because small pulses, with $E'<150$ kev, may result from the interaction of photons with $E>150$ kev. This complication can be circumvented by separating the contributions of low and high energy photons to $I_p(E'')$ through a filter which removes the low energy photons without otherwise changing the spectrum [94].

[1] When the pulse size is classified by an analyzer into a number of channels, 1, 2, ..., n ..., covering the energy intervals $E_{n+1}-E_n$, one replaces $R(E, E'')$ by the distribution-in-channels $R_n(E)=\int_{E_n}^{E_{n+1}} dE''\, R(E, E'')$.

Tin is suitable for this purpose because it has almost the same mass attenuation coefficient as NaI.

Filtration of X rays through various thicknesses of different materials has been considered at times to have promise as a simple independent technique for spectral analysis, but it should now be regarded primarily as a complement to the more powerful scintillation analysis.

γ) Directional detectors. Thus far we have considered only detectors that respond to X rays incident from all directions. The response may be restricted to a narrow cone of directions of incidence by shielding the detector except through the opening of a collimator. Such a detector provides information about the angular distribution [*43*], and [*97*], and facilitates the spectral analysis by permitting the application of the multi-crystal spectrometers[1]. On the other hand the presence of the collimator, which is usually made of high-Z material, causes difficulties. It usually constitutes an inhomogeneity which distorts the flux. Scattering and fluorescence from the inner surface of the collimator and penetration of the collimator edges constitute complications whose quantitative influence is difficult to estimate when the X-ray source is extended[2].

Instead of limiting the sensitivity of the detector to a narrow cone by means of a collimator one may achieve somewhat similar results by shielding the detector only over a moderately large solid angle. The directional distribution can then be analyzed by varying the solid angle thus shielded (see [*32*] and [*79*]).

29. Experiments under cascade shower conditions. The cosmic radiation has long served as a source for studying the penetration of very high energy photons in matter. This penetration proceeds through the cascade shower mechanism, whose features are understood qualitatively and to some extent also quantitatively ([*76*], Chap. 3).

The main feature of the electron-photon cascade phenomenon is a multiplicative chain of pair production and bremsstrahlung[3] processes. These processes subdivide the energy of an initial photon or electron among a number of photons and electrons (or positrons), many of which receive an energy well in excess of 1 Mev and thus still constitute a penetrating radiation. The characteristic multiplication phase of a shower, with which this article is not concerned, terminates when the energy of individual electrons has dropped to the "critical energy" below which bremsstrahlung ceases to be the dominant mode of energy loss ($\sim$100 Mev in water or $\sim$10 Mev in lead) or when photons are no longer absorbed predominantly by pair production (see Fig. 1). Lower energy electrons quickly dissipate their energy in a multitude of excitations and ionizations. For photons, on the other hand, pair production ceases to predominate in the proximity of the energy at which photons are most penetrating (see Table 4). Therefore the penetration of the tail of a cascade shower is a process of the very type that has been discussed at length in this article, especially in Sects. 11 and 21.

Most of the theoretical and experimental studies on cascade shower processes have dealt with the multiplication phase. As indicated in Sect. 1, the theory may be greatly simplified and developed analytically at very high energies, where

[1] A basic two-crystal spectrometer which observes a Compton process has been introduced by R. Hofstadter and J. A. McIntyre: Phys. Rev. **78**, 619 (1950). Multi-crystal spectrometers have been developed which respond only to pair production, see S. A. E. Johannson: Phil. Mag. **43**, 249 (1952) and J. K. Bair and F. M. Maienschein: Rev. Sci. Instrum. **22**, 343 (1951).

[2] Experimental and theoretical estimates have been made by R. Mather (Private communication).

[3] See Sect. 2, and this Encyclopedia, Vols. XXXI and XXXV, p. 409ff.

the integral cross sections for the main processes become energy-independent and those for excitation, ionization and Compton effects may be disregarded. An exploratory calculation by the Monte Carlo method with realistic cross sections and for showers initiated at 300 Mev has given results substantially at variance with theories depending on simplified cross sections [*60a*], [*99*]. Experimentally, the low intensity available from cosmic radiation has not encouraged the study of the deep penetration phase of cascade showers where the multiplication ends[1]. However, a few basic experiments on the deep penetration of cascade showers have been performed with electron accelerators yielding energies above 100 Mev. Thorough investigations have been made of the distribution of energy dissipation in materials ranging from C to Pb exposed to high energy beams over a range of $\sim 10^4$ in the density of energy dissipation. BLOCKER *et al.* [*16*] exposed the material under study to the bremsstrahlung beam from a 330-Mev electron accelerator and inserted a flat, thin, ionization chamber across the beam at various depths within the absorber, so as to sample the energy dissipation. Fig. 55 shows the result of these penetration measurements. KANTZ and HOFSTADTER [*53*] exposed various materials to beams of electrons with fixed energies between 118 and 185 Mev, and sampled the energy dissipation with a small probe consisting of a scintillator with absorption properties similar to those of the absorbing material. They measured not only the energy distribution in depth but also radially around the beam axis.

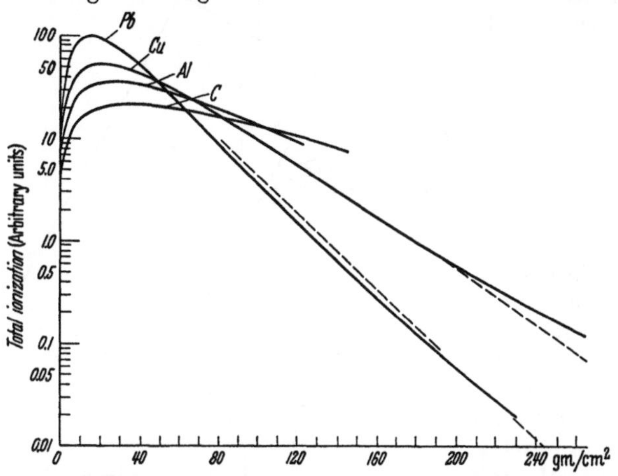

Fig. 55. Ionization vs depth of penetration in various materials, for 330 Mev bremsstrahlung. The dashed curves have slopes corresponding to the minimum attenuation coefficient μ_m (BLOCKER *et al.* [*16*]).

In these experiments the penetration was followed to depths where it is controlled by the secondary X rays with minimum absorption. The slope of semilogarithmic intensity plots was compared with the minimum absorption coefficient of X rays (see Fig. 55). The lesser slope of the intensity plot reflects the build-up of secondaries discussed qualitatively in Sect. 5, but no comparison with more detailed theoretical results has been made. The formula (11.33) for very great depths is probably not sufficiently accurate at the depths attained in these experiments, and no detailed calculations by the semi-asymptotic method of Sects. 20 and 21 are available for the conditions of the experiments.

Three additional experiments have been performed to observe special features of the distribution of radiation in massive absorbers exposed to 330-Mev bremsstrahlung. The distribution-in-depth of the X rays in the spectral band around 15 to 20 Mev, which are active for photonuclear reactions, was sampled using

[1] The role of the most penetrating secondaries in determining the cascade tail of cosmic rays has been emphasized and estimated by K. GREISEN: Phys. Rev. **75**, 1074 (1949). The accumulation of very soft secondaries, in equilibrium or near equilibrium with more penetrating components, has been described by M. AGENO, G. CORTELLESSA and R. QUERZOLI Nuovo Cim. **1**, 453 (1955).

as an indicator the radioactivity induced in C, Cu and Ag foils [88]. The angular distribution of X rays in this energy band emerging from the back face of an absorber was also observed by foil activation [75]. The energy and angular distribution of electrons near the shower maximum in lead has been observed in a cloud chamber[1].

Other detailed experiments on the penetration of high energy X-ray beams include one with 86 and 176 Mev bremsstrahlung beams [59], whose results have not been analyzed in detail, and one with a 40 Mev beam incident of water [101] in which the distribution of energy dissipation was observed both in depth and radially. The depth distribution obtained in this experiment is compared in Sect. 19 with the results of a detailed calculation.

30. Backscattering experiments. The backscattering of X rays from a semi-infinite medium has been investigated experimentally by HAYWARD and HUBBELL [43] and by HINE and McCALL [47], as well as by Monte Carlo calculations (Sect. 26). These two approaches complement one another. Experiments with intense sources can more readily provide a detailed analysis of the radiation emerging from a surface according to spectrum, direction and point of emergence. Monte Carlo calculations, being free from instrumental limitations, can more readily provide absolute data on the total probability of backscattering.

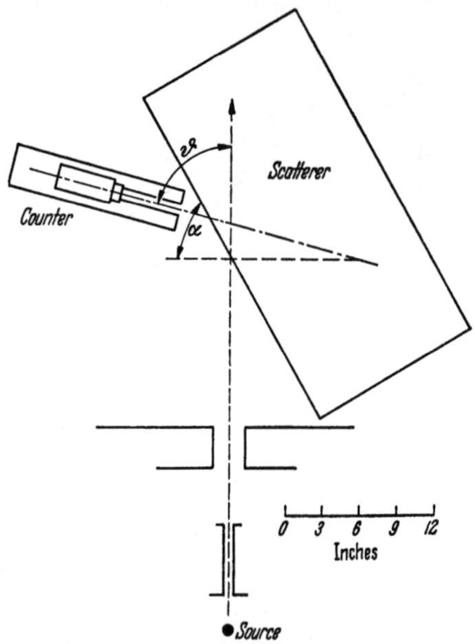

Fig. 56. Experimental arrangement of a backscattering experiment (HAYWARD and HUBBELL [43]).

Both experiments on backscattering analyzed the X rays with NaI(Tl) scintillators and yielded similar qualitative results. In the Hine-McCall experiment a point isotropic source and the detector lay on a line perpendicular to the surface of a material, and the detector observed together the primary and backscattered radiation; the geometry was fixed except for the source-backscatterer distance. In the Hayward-Hubbell experiment both source and detector were collimated; the geometrical variables included not only the angles α and ϑ shown in Fig. 56 but also the positions of the source and detector axes relative to one another and to the scatterer surface. Various features of the backscattered γ rays were revealed by observations performed with different combinations of variables. The results of the Hayward-Hubbell experiment with a Co^{60} γ-ray source are outlined in the following paragraphs.

Whenever the detector was directed so that it could "see" some of the material exposed to the γ-ray beam from the source, the spectral distribution showed two maxima. The origin of the maxima was determined by varying the conditions of observation. The maximum at higher energy was shown to result from singly-scattered X rays, because its energy varied with the angle ϑ according to the Compton law. The second maximum was greatly reduced in intensity when the

[1] K.M. CREWE and E. HAYWARD: Phys. Rev. **80**, 40 (1950).

scatterer was made thin to reduce the opportunity for multiple scattering (Fig 57a). This maximum also disappeared when the collimator axes of source and detector were not coplanar. It follows that the second maximum arises from X rays scattered twice approximately in the same plane. Calculation has actually shown that photons with initial wavelength λ_0 and which have experienced two Compton

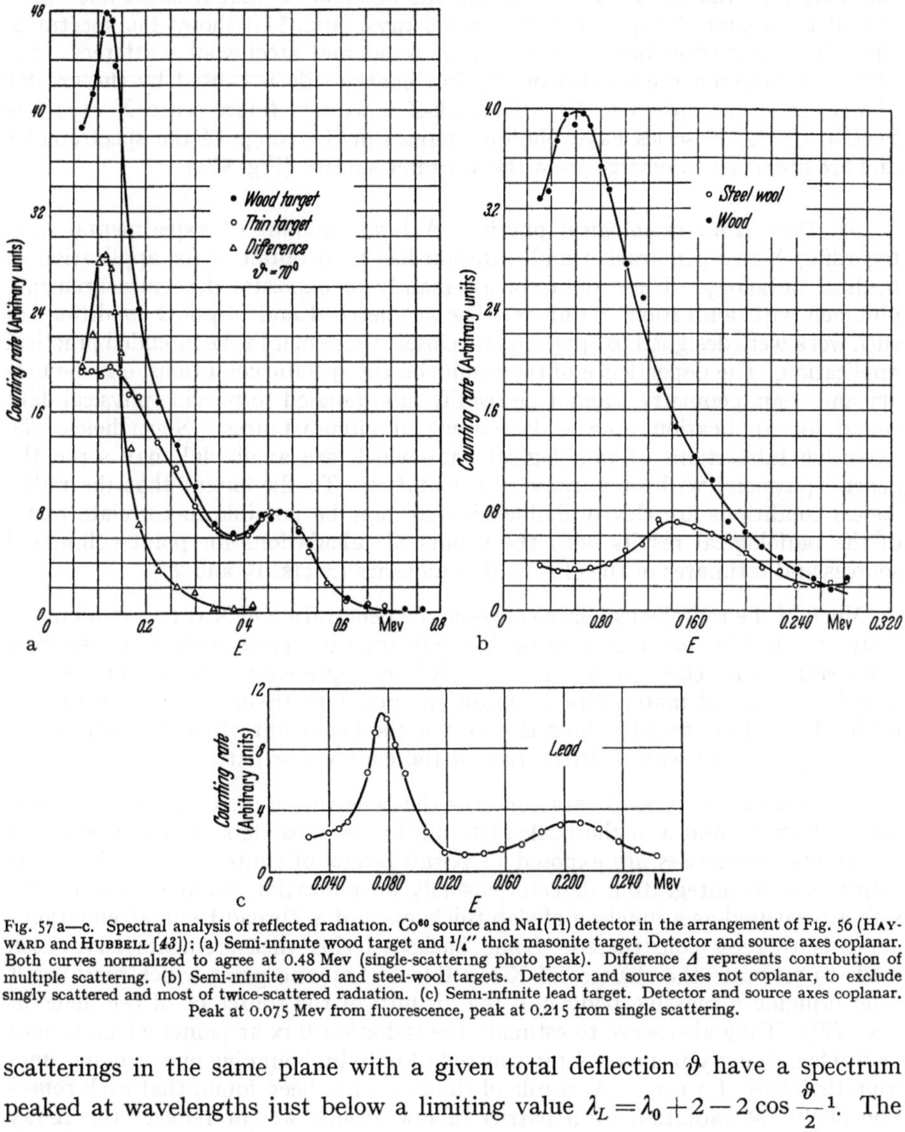

Fig. 57a—c. Spectral analysis of reflected radiation. Co^{60} source and NaI(Tl) detector in the arrangement of Fig. 56 (HAYWARD and HUBBELL [43]): (a) Semi-infinite wood target and $^1/_4$" thick masonite target. Detector and source axes coplanar. Both curves normalized to agree at 0.48 Mev (single-scattering photo peak). Difference Δ represents contribution of multiple scattering. (b) Semi-infinite wood and steel-wool targets. Detector and source axes not coplanar, to exclude singly scattered and most of twice-scattered radiation. (c) Semi-infinite lead target. Detector and source axes coplanar. Peak at 0.075 Mev from fluorescence, peak at 0.215 from single scattering.

scatterings in the same plane with a given total deflection ϑ have a spectrum peaked at wavelengths just below a limiting value $\lambda_L = \lambda_0 + 2 - 2\cos\frac{\vartheta}{2}$[1]. The

[1] This value of λ_L follows from (8.7) for $\vartheta_L = \vartheta$. For λ near λ_L, Fig. 9 and Eqs. (8.7) and (8.8) show that the direction ω' of first scattering cannot depart much from coplanarity with ω_0 and ω, and in the layout of Fig. 56 the photon must emerge from the scatterer traveling along a line nearly coplanar with the axis of the source. For $\lambda \ll \lambda_L$ and fixed ϑ, the lines of flight of the twice scattered photons are not so concentrated near a plane. The same conclusion may be reached by considering that in (8.11) the whole integral over ψ contributes to lines of flight nearly coplanar with the axis of the source when λ is close to λ_L, but only narrow intervals of ψ near 0 and π contribute when $\lambda \ll \lambda_L$.

experiments verify this relationship between the position of the spectral maximum and ϑ. Twice-scattered radiation with $\lambda \ll \lambda_L$ and multiply-scattered radiation constitute a background to the maximum.

The multiply-scattered radiation, which is approximately isotropic within the scatterer, was observed by pointing the detector so that it could not "see" any of the material exposed to primary γ-rays. Fig. 57b shows the spectra of the diffuse radiation backscattered from wood and steel-wool scatterers. The difference between the spectra on the low energy side is caused by the greater photoelectric absorption in Fe. For high-Z scatterers fluorescence X rays (see Sect. 4) of the K series have photon energies in the range of the spectrometer and are observed to emerge from the scatterer surface (Fig. 57c).

31. Data from radiological physics. A large amount of experimental data regarding X-ray penetration and diffusion has been obtained for application in medical radiology. These data consist mainly of exposure dose measurements obtained with ionization chambers. The geometrical and physical conditions of such work were designed to approximate closely those met in the intended practical application. The dependence of the results on the specific conditions, e.g., boundary and beam geometry, cannot be readily disentangled to provide physical data suited for application in a wide variety of circumstances. Nevertheless the extensive tabulations of radiological data which are available[1] may serve the general physicist well for purpose of orientation. To the extent that the radiological conditions are clearly defined it may now be possible to correlate many of the radiological results with the results of calculations for point collimated sources and estimates of the effects of boundaries (Sects. 19 and 26).

Much of the radiological data concerns the penetration of X rays into a human body. Conical beams of X rays of different aperture, cross section and spectral composition are considered. The "body" is represented by a "phantom", namely a block of material of composition similar to tissue. Measurements are made of the dose, mostly along the axis of the beam but often also radially so as to map out the whole distribution of the energy dissipated.

The other main type of data concerns the penetration of X rays through protective barriers and also the backscattering by such barriers. Slabs of material of various thicknesses are exposed to X-ray beams of finite cross section. The response of an integrating detector, usually an ionization chamber, behind the slab is measured as a function of slab thickness and a "broad beam attenuation" curve is thereby constructed. Some backscattering measurements have been made to estimate the changes of X-ray flux at any one point which may result from bringing a massive body into the vicinity of the source or of the detector (see [*17*]). They also serve to estimate the radiation flux at points which cannot be reached directly from an X-ray source but only by bouncing one or more times from the walls of a maze. As a rule of thumb, it has been found that each reflection of diffuse radiation by a barrier of low atomic weight reduces the X-ray flux by an order of magnitude[2]. This is consistent with the results of Monte Carlo calculations (Sect. 26).

---

[1] See Refs. [*37*], [*62*], [*17*], [*63*], [*46*], [*18*]. See also Ref. [*78*], in particular Chap. 12, for related material.

[2] H.O. Wyckoff and C.B. Braestrup: Radiology **51**, 840 (1948).

Appendix. Notation, units and normalization.

α) Variables.

A. Basic.

(a) Position and displacement

$\boldsymbol{r} = (x, y, z)$
$= (z, \varrho, \beta)$
$= (r, \alpha, \beta)$
$t =$ pathlength
$\zeta =$ pathlength between collisions

(b) Direction and deflection

$\boldsymbol{\omega} = (\vartheta, \varphi)$
$\omega_r = \frac{\boldsymbol{\omega} \cdot \boldsymbol{r}}{r} = \cos\vartheta_r$
$\omega_z = \cos\vartheta$

also

$\boldsymbol{\omega} = (\hat{\vartheta}, \hat{\varphi}) = (\tilde{\vartheta}, \tilde{\varphi})$ (p. 693, 695) with reference to special axes.

$\cos\Theta = \boldsymbol{\omega} \cdot \boldsymbol{\omega}'$

(c) Energy

$E =$ photon energy in Mev
$\lambda = \frac{mc^2}{E}$, Compton wavelength
$\xi = \lambda - \lambda'$

(d) The combinations $\mu_0 z$, $\mu_0 \varrho$ etc. indicate distance in mean free paths (m.f.p.).

B. Special values.

(a) Subscript zero indicates that the value of the independent variable pertains to the source X rays.

(b) λ_m, E_m Wavelength, energy of minimum μ for a *material.*

λ_s, E_s Wavelength, energy of smallest μ in the range covered by a given *calculation* or *experiment.*

$\vartheta_\lambda = \arccos(1 - \lambda + \lambda_0)$, deflection angle for once-scattered photons (p. 688).

$\vartheta_L = 2\arccos(1 - \frac{1}{2}[\lambda - \lambda_0])$, limiting deflection for twice-scattered photons (p. 687.)

β) Interaction probabilities per unit path length.

A. Differential probabilities.

$k(\boldsymbol{\omega}, \lambda, \boldsymbol{\omega}', \lambda')\, d\boldsymbol{\omega}\, d\lambda$ scattering from $\boldsymbol{\omega}', \lambda'$ to $(\boldsymbol{\omega}, \boldsymbol{\omega} + d\boldsymbol{\omega})$, $(\lambda, \lambda + d\lambda)$.

$k_l(\lambda, \lambda') = \int_{4\pi} d\boldsymbol{\omega}\, P_l(\boldsymbol{\omega} \cdot \boldsymbol{\omega}')\, k(\boldsymbol{\omega}, \lambda, \boldsymbol{\omega}', \lambda')$ (9.10)

$k(\lambda, \lambda') = k_0(\lambda, \lambda')$

$C = k(\lambda, \lambda) = 0.300 \frac{Z}{A}$ cm²/g (for Compton scattering) (10.14)

$k^{(i)}(\boldsymbol{\omega}, \lambda, \boldsymbol{\omega}_0, \lambda_0)$, $k^{(i)}(\lambda, \lambda_0)$ i-th iterated kernels (p. 687 ff.)

$k^{[ij]}(\boldsymbol{\omega}, \lambda, \boldsymbol{\omega}', \lambda')$, $(i, j = 0, 1, 2, 3)$ scattering with polarization dependence (p. 684).

B. Integral probabilities.

$\mu(\lambda) =$ total attenuation coefficient (p. 666)

$\mu_{sc}(\lambda) = \int_{\lambda}^{\lambda+2} k(\lambda', \lambda)\, d\lambda'$ scattering probability

$\mu_D =$ energy absorption coefficient (p. 673)

$\mu_a =$ energy absorption coefficient in air (p. 673)

$\mu_0 = \mu(\lambda_0)$, $\mu_m = \mu(\lambda_m)$, $\mu_s = \mu(\lambda_s)$, $\mu_1 = \mu\left(\frac{\lambda + \lambda_0}{2}\right)$

$\dot{\mu}_0 = \left.\frac{\partial\mu}{\partial\lambda}\right|_{\lambda=\lambda_0}$ etc.

$\mu^{[ij]}(\lambda)$ $(i, j = 0, 1, 2, 3)$ attenuation coefficient with polarization dependence (p. 672).

γ) Source normalization and notation.

A. Unless otherwise specified, the following source normalizations apply:
Uniformly distributed source, 1 photon cm^{-3} sec^{-1}
Plane source, 1 photon cm^{-2} sec^{-1}
Point source, 1 photon sec^{-1}.

B. $S(\boldsymbol{r}, \boldsymbol{\omega}, \lambda)$, with various geometric specializations (p. 681 ff.).

δ) Distribution functions.

A. *Differential.*
$I(\boldsymbol{r}, \boldsymbol{\omega}, \lambda)\, d\boldsymbol{\omega}\, d\lambda$ = number flux (or flux) (photons cm^{-2} sec^{-1}) (p. 680)
$E\, I(\boldsymbol{r}, \boldsymbol{\omega}, \lambda)\, d\boldsymbol{\omega}\, d\lambda = \lambda\, I(\boldsymbol{r}, \boldsymbol{\omega}, \lambda)\, d\boldsymbol{\omega}\, dE$ energy flux (Mev cm^{-2} sec^{-1}).

B. *Integral.*

(a) Omission of an independent variable implies that it has been integrated out, e.g. $I(z, \lambda) = \int_{4\pi} d\boldsymbol{\omega}\, I(z, \omega_z, \lambda)$.

(b) $\iint \lambda I\, dE\, d\boldsymbol{\omega} = I_E$ integrated energy flux
$\iint \lambda I \mu_D\, dE\, d\boldsymbol{\omega} = I_D$ energy dissipation
$\iint \lambda I \mu_a\, dE\, d\boldsymbol{\omega} = I_a$ energy dissipation in air.

C. *Separation into components.*

$I^{(n)}$	flux of photons scattered n times
$I^{(s)} = I - I^{(0)}$	flux of all scattered photons
$I^{[i]}$ $(i = 0, 1, 2, 3)$	i-th polarization component (Stokes parameter).

D. *Build-up factors.*

$B_E = \frac{I_E}{I_E^{(0)}}$ energy

$B_D = \frac{I_D}{I_D^{(0)}}$ energy dissipation

$B_a = \frac{I_a}{I_a^{(0)}}$ energy dissipation in air.

ε) Transformations.

A. *Correspondence of independent variables.*
Fourier transformation, $\boldsymbol{r} \to \boldsymbol{q}$ (9.5)
$z \to p = i\, q$ (9.2)
Legendre transformation, $\boldsymbol{\omega}(\vartheta, \varphi) \to (l, m)$ (9.8). In small-angle approximation the discrete variable l is replaced by the continuous variable $\bar{l}$ (p. 705).
Laplace transformation, $\lambda \to \eta$ (10.17)
Moment transformation, $(z, \varrho) \to (n, \nu)$, $r \to n$ (Sect. 12).

B. *Notation.*

(a) When a distribution is transformed, its symbol is kept and only the independent variable is changed, $I(\boldsymbol{r}, \boldsymbol{\omega}, \lambda) \to I(\boldsymbol{q}, \boldsymbol{\omega}, \lambda)$.

(b) Legendre and moment transform variables are indicated as subscripts and superscripts $I_{n\,l}^{\nu\, m}$.

(c) When the value of a transform variable is zero, it is frequently omitted, which signifies that the corresponding variable has been integrated out, e.g.,

$$I(\boldsymbol{r}, \lambda) = I_0^0(\boldsymbol{r}, \lambda) = \int d\boldsymbol{\omega}\, I(\boldsymbol{r}, \boldsymbol{\omega}, \lambda).$$

C. Green's function.

(a) $G^{(p,m)}(l\lambda, l_0\lambda_0)$ is the Green's function of the transport equation for a plane source (p. 692).

(b) $G_n^{(m)}(l\lambda, l_0\lambda_0)$ is the corresponding moment transform (12.5).

Bibliography.

[1] BEACH, L.A., R.B. THEUS and W.R. FAUST: Naval Research Lab. Report (NRL-4277), Dec. 1, 1953. Analysis of Scintillation Spectrometer Observations of the Penetration of Cs^{137} Gamma Radiation through Water.

[2] BEACH, L.A., R.B. THEUS and W.R. FAUST: Naval Research Lab. Report (NRL-4581), July 26, 1955. Penetration of Na^{24} Radiation through H_2O and Hg.

[3] BERGER, M.J.: J. Appl. Phys. **26**, 1504 (1955). Angular Distribution of Multiply Scattered Gamma Radiation from a Plane Isotropic Source.

[4] BERGER, M.J.: J. Res. Nat. Bur. Stand. **55**, 343 (1955). Reflection and Transmission of Gamma Radiation by Barriers: Monte Carlo Calculation by a Collision Density Method.

[5] BERGER, M.J.: J. Res. Nat. Bur. Stand. **56**, 111 (1956). Penetration of Gamma Radiation from a Plane Monodirectional Oblique Source.

[6] BERGER, M.J.: J. Appl. Phys. **28**, 1502 (1957). Calculation of Energy Dissipation by Gamma Radiation Near the Interface between Two Media.

[7] BERGER, M.J., and J.A. DOGGETT: J. Res. Nat. Bur. Stand. **56**, 89 (1956). Reflection and Transmission of Gamma Radiation by Barriers: Semianalytic Monte Carlo Calculation.

[8] BERGER, M.J., and J.A. DOGGETT: Rev. Sci. Instrum. **27**, 269 (1956) and J. Res. Nat. Bur. Stand. **56**, 355 (1956). Response Function of NaI(Tl) Scintillation Counters.

[9] BERGER, M.J., J.H. HUBBELL and I.H. REINGOLD: Phys. Rev. **113**, 857 (1959). Contribution of Annihilation Radiation to the Gamma Ray Flux in Lead.

[10] BERGER, M.J., and J.C. LAMKIN: J. Res. Nat. Bur. Stand. **60**, 109 (1958). Sample Calculations of Gamma-Ray Penetration into Shelters: Contributions of Skyshine and Roof Contaminations.

[11] BERGER, M.J., and L.V. SPENCER: Radiation Res. **10**, 552 (1959). Some Radiological Applications of Gamma Ray Transport Theory.

[12] BERGER, M.J., and L.V. SPENCER: Phys. Rev. **113**, 408 (1959). General Relation between Fluxes from Point and Plane Sources.

[13] BETHE, H.A., and J. ASHKIN: Experimental Nuclear Physics, Vol. 1, p. 166. New York· Wiley 1953.

[14] BETHE, H.A., U. FANO and P.R. KARR: Phys. Rev. **76**, 538 (1949). Penetration and Diffusion of Hard X-rays Through Thick Barriers. I. The Approach to Spectral Equilibrium.

[15] BLIZARD, E.P.: Annual Rev. Nucl. Sci. **5**, 73 (1955). Nuclear Radiation Shielding.

[16] BLOCKER, W., R.W. KENNEY and W.K.H. PANOFSKY: Phys. Rev. **79**, 419 (1950). Transition Curves of 330-Mev Bremsstrahlung.

[17] BRAESTRUP, C.B., and H.O. WYCKOFF: Radiation Protection. Springfield, Ill.: Charles Thomas 1958.

[18] British Hospital Physicist Assoc. Data. Brit. J. Radiol. **25**, 88 (1952).

[19] BRYSK, H.: Phys. Rev. **96**, 419 (1954). Spatial Distribution Energy Dissipation by High-Energy X-Rays.

[20] CAVE, L., J. CORNER and R.H.A. LISTON: Proc. Roy. Soc. Lond., Ser. A **204**, 223 (1950). The Scattering of Gamma-Rays in Extended Media. I. Perpendicular Incidence on a Plane Slab.

CORNER, J., and R.H.A. LISTON: Proc. Roy. Soc. Lond., Ser. A **204**, 323 (1950). The Scattering of Gamma-Rays in Extended Media. II. Backscattering of Gamma-Rays from a Thick Slab.

CORNER, J., F.A.G. DAY and R.E. WEIR: Proc. Roy. Soc. Lond., Ser. A **204**, 329 (1950). The Scattering of Gamma-Rays in Extended Media. III. Problems with Spherical Symmetry.

[21] CAMPBELL, J.G., and A.J.F. BOYLE: Austral. J. Phys. **6**, 171 (1953). Energy Resolution of γ-Radiation up to 18 Mev by NaI Scintillation Counters.

[22] CHANDRASEKHAR, S.: Radiative Transfer. Oxford: Oxford University Press 1950.

[23] COLGATE, S.A.: Phys. Rev. **87**, 592 (1952). Gamma-Ray Absorption Measurements.

[24] DAVISON, B., and J.B. SYKES: Neutron Transport Theory. Oxford: Clarendon Press 1957.

[25] EHRLICH, M.: Scintillation Spectrometry of Low-Energy Bremsstrahlung. J. Res. Nat. Bur. Stand. **54**, 107 (1955).
[26] ELLIOT, J.O., R.T. FARRAS, R.D. MYERS and C.F. RAVILIOUS: Phys. Rev. **85**, 1048 (1952). Multiple Gamma Ray Scattering in Lead.
[27] FANO, U.: Phys. Rev. **76**, 739 (1949). Penetration and Diffusion of X-Rays through Thick Barriers. II. The Asymptotic Behavior when Pair Production is Important.
[28] FANO, U.: J. Res. Nat. Bur. Stand. **51**, 95 (1953). Penetration of X- and Gamma Rays to Extremely Great Depths.
[29] FANO, U.: Phys. Rev. **92**, 328 (1953). Degradation and Range Straggling of High Energy Radiations.
[30] FANO, U.: Nucleonics **11**, 8 (1953). Gamma-Ray Attenuation.
[31] FANO, U., and A.T. NELMS: J. Res. Nat. Bur. Stand. **59**, 207 (1957). An Approximate Expression for Gamma-Ray Degradation Spectra.
[32] FAUST, W.R.: Phys. Rev. **77**, 227 (1950). Multiple Compton Scattering II.
[33] FAUST, W.R., and M.H. JOHNSON: Phys. Rev. **75**, 467 (1949). Multiple Compton Scattering.
[34a] FOLDY, L.L.: Phys. Rev. **81**, 395 (1951). Diffusion of High Energy Gamma Rays through Matter. I. Fundamental Equations.
[34b] FOLDY, L.L., and R.K. OSBORN: Phys. Rev. **81**, 400 (1951). Diffusion of High Energy Gamma Rays through Matter. II. Solution of the Diffusion Equation.
[34c] FOLDY, L.L.: Phys. Rev. **82**, 927 (1951). Diffusion of High Energy Gamma-Rays through Matter. III. Refinement of the Solution of the Diffusion Equation.
[35] FOOTE, R.S., and H.W. KOCH: Rev. Sci. Instrum. **25**, 746 (1954). Scintillation Spectrometers for Measuring the Total Energy of X-Ray Beams.
[36] GARRETT, C., and G.N. WHYTE: Phys. Rev. **95**, 889 (1954). Build-up Measurements on Co^{60} Gamma Radiation in Iron and Lead.
[37] GLASSER, O., E.H. QUIMBY, L.S. TAYLOR and J.L. WEATHERWAX: Physical Foundations of Radiology, 2nd edit. New York, N.Y.: Paul B. Hoeber, Inc. 1952.
[38a] GOERTZEL, G.: Oak Ridge National Laboratory Report 434, Dec. 1949. Quota Sampling and Importance Functions in Stochastic Solution of Particle Problems.
[38b] GOERTZEL, G.: Oak Ridge National Laboratory Report 423, Dec. 1949. A Proposed Particle Attenuation Problem.
[39] GOLDSTEIN, H.: Nuclear Development Associates Report NDA 15C-31. Estimates of the Effect of Fluorescence and Annihilation Radiation of Gamma Ray Penetration.
[40] GOLDSTEIN, H.: Fundamental Aspects of Reactor Shielding. Reading, Mass.: Addison-Wesley Publishing Co, Inc. 1958.
[41] GOLDSTEIN, H., and J.E. WILKINS jr.: USAEC Report, NYO-3075 (1954). Calculations of the Penetration of Gamma Rays.
[42] HAYWARD, E.: Phys. Rev. **86**, 493 (1952). The Electron Spectra Produced by a Co^{60} Source in H_2O.
[43] HAYWARD, E., and J.H. HUBBELL: J. Appl. Phys. **25**, 506 (1954). The Backscattering of the Co^{60} Gamma Rays from Infinite Media. See also HAYWARD, E., and J.H. HUBBELL: Nat. Bur. Stand. Report (NBS-2768).
[44] HAYWARD, E., and J.H. HUBBELL: Phys. Rev. **93**, 955 (1955). The Albedo of Various Materials for 1-Mev Photons.
[45] HEITLER, W.: The Quantum Theory of Radiation. Oxford: Oxford University Press 1954.
[46] HINE, G.J., and G.L. BROWNELL: Radiation Dosimetry. New York: Academic Press 1956.
[47] HINE, G.J., and R.C. MCCALL: Nucleonics **12**, No. 4, 27 (1954).
[48a] HIRSCHFELDER, J.O., J.L. MAGEE and M.H. HALL: Phys. Rev. **73**, 853 (1948). The Penetration of Gamma Radiation through Thick Layers. I. Plane Geometry Klein-Nishina Scattering.
[48b] HIRSCHFELDER, J.O., and E.N. ADAMS: Phys. Rev. **73**, 863 (1948). The Penetration of Gamma Radiation through Thick Layers. II. Plane Geometry, Iron and Lead.
[49] HUBBELL, J.H.: Rev. Sci. Instrum. **29**, 65 (1958). Response of a Large Sodium-Iodide Detector to High Energy X-Rays.
[50] HUBBELL, J.H., E. HAYWARD and W.F. TITUS: Phys. Rev. **108**, 1361 (1957). Energy and Angular Distribution of X-Rays Scattered in Lead.
[51] KAHN, H.: Nucleonics **6**, No. 5, 27 and No. 6, 60 (1950). Random Sampling (Monte Carlo) Techniques in Neutron Attenuation Problems.
[52] KAHN, H.: USAEC Report, R-1237 (Apr. 19, 1954). Applications of Monte Carlo.
[53] KANTZ, A., and R. HOFSTADTER: Phys. Rev. **89**, 607 (1953). Electron-Induced Showers in Copper.

[54] KIRN, F. S, R. J. KENNEDY and H. O. WYCKOFF Radiology 63, 94 (1954). The Attenuation of Gamma Rays at Oblique Incidence
[55] KOCH, H. W., and J. M. WYCKOFF: J. Res. Nat. Bur. Stand. 56, 319 (1956). Response of a Sodium Iodide Scintillation Spectrometer to 10- to 20-Million-Electron-Volt Electrons and X-Rays.
[56] MAEDER, D., R. MULLER and V. WINTERSTEIGER: Helv. phys. Acta 27, 3 (1954). Über die Linienform monochromatischer γ-Strahlungen in Szintillationsspektrographen.
[57] MAIGNAN, P.: C. R. Acad. Sci., Paris 230, 2018 (1950). Propagation des Rayons γ.
[58] MARASCO, F. W., and P. E. F. THURLOW: USAEC Report WAPD-RM-219. Application of Gamma Ray Build-Up Data to Shield Design.
[59] MILLER, W., and R. J. KENNEDY: Radiation Res. 4, 360 (1956). Attenuation of 86- and 176-Mev Synchrotron X-Rays in Concrete and Lead.
[60] Monte Carlo Method (Proc. of a Symposium held in 1949), Nat. Bur. Stand. Appl. Math. Ser. No. 12 (1951). Contains, amongst others, the following items ([60a] to [60c]):
[60a] WILSON, R. R.: Showers Produced by Low-energy Electrons and Photons.
[60b] KAHN, H, and T. E. HARRIS: Estimation of Particle Transmission by Random Sampling.
[60c] NEUMANN, J. VON: Various Techniques used in Connection with Random Digits.
[61] Monte Carlo Methods, ed. H. A. MEYER. New York, N.Y.: Wiley 1956. (Proc. of a Symposium held in 1954.) Contains, amongst others, the following items ([61a] to [61e]):
[61a] TAUSSKY, O., and J. TODD: Generation of Pseudo-random Numbers.
[61b] BERGER, M. J.: An Application of the Monte Carlo Method to a Problem in Gamma Ray Diffusion.
[61c] BEACH, L. A., and R. B. THEUS: Stochastic Calculations of Gamma Ray Diffusion.
[61d] KAHN, H.: Use of Different Monte Carlo Sampling Techniques.
[61e] BUTLER, J. W. Machine Sampling from Given Probability Distributions.
[62] MORGAN, R. H. Handbook of Radiology. Chicago Year Book Publishers 1955.
[63a] Nat. Bur. of Standards Handbook No. 50, Supr. of Doc., Washington, 25, D.C. X-Ray Protection Design.
[63b] Nat. Bur. of Standards Handbook No. 54, Supr. of Doc., Washington, 25, D.C. Protection Against Radiations from Radium, Co^{60} and Cs^{137} (Revision of H23).
[63c] Nat. Bur. of Standards Handbook No. 55, Supr. of Doc., Washington, 25, D.C. Protection against Betatron-Synchrotron Radiations up to 100 Million Electron Volts.
[63d] Nat. Bur. of Standards Handbook No. 60, Supr. of Doc., Washington, 25, D.C. X-Ray Protection (Revision of H41).
[63e] Nat. Bur. of Standards Handbook No. 62, Supr. of Doc., Washington, 25, D C. Report of the International Commission of Radiological Units and Measurements (1956).
[64] NELMS, A. T.: Nat. Bur. Stand. Circ. 542 (1953). Graphs of the Compton Energy-Angle Relationship and the Klein-Nishina Formula from 10 Kev to 500 Mev.
[65] NELMS, A. T., and I. OPPENHEIM: J. Res. Nat. Bur. Stand. 55, 53 (1955). Data on the Atomic Form Factor. Computation and Survey.
[66] OGIEVETSKII, I.: J. Exp. Theor. Phys. USSR. 29, 454, 467 (1955). The Theory of the Propagation of Gamma Rays through Matter; and Angular Distribution of Gamma Rays at Great Depths of Penetration in Matter.
[67] OLSON, C A., and L. V. SPENCER. J. Res. Nat. Bur. Stand. 60, 85. Energy Spectra of Cascade Electrons and Photons.
[68] O'ROURKE, R. C.: Phys. Rev. 85, 881 (1952); 89, 999 (1953). Multiple Compton Scattering of Low Energy Gamma Radiation.
[69a] PEEBLES, G. H., and M. S. PLESSET: Phys. Rev. 81, 430 (1951). Transmission of Gamma Rays through Large Thickness of Heavy Materials.
[69b] PEEBLES, G, H: Rand Corporation Report R-240, Dec 1, 1952. Gamma-Ray Transmission through Finite Slabs.
[70] PERKINS, J. F.: J. Appl. Phys. 26, 655 (1955). Monte Carlo Calculations of Gamma-Ray Albedos of Concrete and Aluminum.
[71] PREISER, S., C. JENSEN and H. GOLDSTEIN: Nuclear Development Associates Memo 15C-53, Oct. 20, 1955. Guide to SEAC Printed Output Tapes of Gamma-Ray Penetration Calculations.
[72] PRICE, B. T., C. C. HORTON and K T. SPINNEY: Radiation Shielding. New York, N.Y.: Pergamon Press 1957.
[73] Reactor Handbook, Vol. I-Phys, USAEC (1955), AECD-3645.
[74] ROCKWELL, T.: Reactor Shielding Design Manual. New York: McGraw-Hill 1956.
[75] ROSE, J. W.: Phys. Rev. 82, 747 (1951). Angular Distribution of Photons in Showers in Lead.

[76] Rossi, B.: High Energy Particles. New York, N.Y.: Prentice-Hall 1952.
[77] Roys, P.A., K. Shure and J. J. Taylor: Phys. Rev. **95**, 911 (1954). Penetration of 6-Mev Gamma-Rays in Water.
[78] Siegbahn, K.: Beta- and Gamma-Ray Spectroscopy. New York, N.Y.: Interscience Publishers, Inc. 1955.
[79] Soole, B.W.: Proc. Roy. Soc. Lond., Ser. A **230**, 343 (1955). The Angular Distribution of Multiply Scattered Gamma Radiation.
[80] Spencer, L.V.: Phys. Rev. **88**, 793 (1952). Penetration and Diffusion of X-Rays: Calculation of Spatial Distributions by Semi-Asymptotic Methods.
[81] Spencer, L.V.: Phys. Rev. **90**, 146 (1953). Electron Penetration, Calculation of Peaked Angular Distributions.
[82] Spencer, L.V.: Penetration of Gamma Rays from a Point Collimated Source. (To be published).
[83] Spencer, L.V., and U. Fano: J. Res. Nat. Bur. Stand. **46**, 446 (1951). Penetration and Diffusion of X-Rays. Calculation of Spatial Distributions by Polynomial Expansion.
[84] Spencer, L.V., and F.A. Jenkins: Phys. Rev. **76**, 1885 (1949). Penetration and Diffusion of Hard X-Rays through Thick Barriers. IV. Multiply Scattered γ-Rays: Angular Distribution.
[85] Spencer, L.V., and F. Stinson: Phys. Rev. **85**, 662 (1952). Further Calculations of X-Ray Diffusion in an Infinite Medium.
[86] Spencer, L.V., and C. Wolff: Phys. Rev. **90**, 510 (1953). Penetration and Diffusion of Hard X-Rays: Polarization Effects.
[87] Starfelt, N., and H.W. Koch: Phys. Rev. **102**, 1598 (1956). Differential Cross-Section Measurements of Thin-Target Bremsstrahlung Produced by 2.7- to 9.5-Mev Electrons.
[88] Strauch, K.: Phys. Rev. **81**, 973 (1951). High Energy Photo-Nuclear Reactions.
[89] Tait, J.H., and K.T. Spinney: British Atomic Energy Report, AERE T/R 639 (Feb. 1951). Penetration of Gamma Radiation into Matter.
[90] Theus, R.B., and L.A. Beach: Naval Research Lab. Report (NRL-4701) Feb. 8, 1956. Gamma-Ray Albedo from Iron.
[91] Theus, R.B., L.A. Beach, J.D. Plawchan and W.R. Faust. Naval Research Lab. Report (NRL-4412) Aug. 3, 1954. Stochastic Estimates of γ-Ray Spectral Intensities from Shallow and Deep Penetration.
[92] Titus, W.F.: Nucl. Sci. and Engng. **3**, 609 (1958). Measurement of the Gamma-Ray Dose Near the Interface between Two Media.
[93] Vinti, J.P.: Phys. Rev. **91**, 345 (1953). Theory of the Penetration of γ-Rays through Thin Barriers.
[94] Weiss, M.M., and W. Bernstein: Phys. Rev. **92**, 1264 (1953). Degradation of Gamma-Rays in Water.
[95] White, G.R.: Phys. Rev. **80**, 154 (1950). The Penetration and Diffusion of Co^{60} Gamma-Rays in Water using Spherical Geometry.
[96] White-Grodstein, G.R.: Nat. Bur. Stand. Circ. 583 (1957). X-Ray Attenuation Coefficients 10 kev to 100 Mev.
[97] Whyte, G.N.: Canad. J. Phys. **33**, 96 (1955). Measurements of Spectral and Angular Distributions of Secondary Gamma-Rays in Matter.
[98] Wick, G.C.: Phys. Rev. **75**, 738 (1949). On the Space Distribution of Slow Neutrons.
[99] Wilson, R.R.: Phys. Rev. **86**, 261 (1952). Monte Carlo Study of Shower Production.
[100] Wyckoff, H.O., R.J. Kennedy and W.R. Bradford: Nucleonics **3**, 62 (1948). Broad and Narrow Beam Attenuation in Lead and Concrete.
[101] Zendle, B., H.W. Koch, J. McElhinney and J.W. Boag: Radiation Res. **5**, 107 (1956). Studies of Dose Distributions in Water for Betatron X-Rays up to 37 Mev.
[102] Zerby, C.D.: USAEC Report ORNL-2224 (1956). Transmission of Obliquely Incident Gamma Radiation through Stratified Slab Barriers.

Sachverzeichnis.

(Deutsch-Englisch.)

Bei gleicher Schreibweise in beiden Sprachen sind die Stichwörter nur einmal aufgeführt.

Subject Index.

(English-German.)

Where English and German spelling of a word is identical the German version is omitted.

MIX
Papier aus verantwortungsvollen Quellen
Paper from responsible sources
FSC® C105338

If you have any concerns about our products,
you can contact us on
ProductSafety@springernature.com

In case Publisher is established outside the EU,
the EU authorized representative is:
Springer Nature Customer Service Center GmbH
Europaplatz 3, 69115 Heidelberg, Germany

Printed by Libri Plureos GmbH
in Hamburg, Germany